Programmable Logic Controllers

Principles and Applications

Third Edition

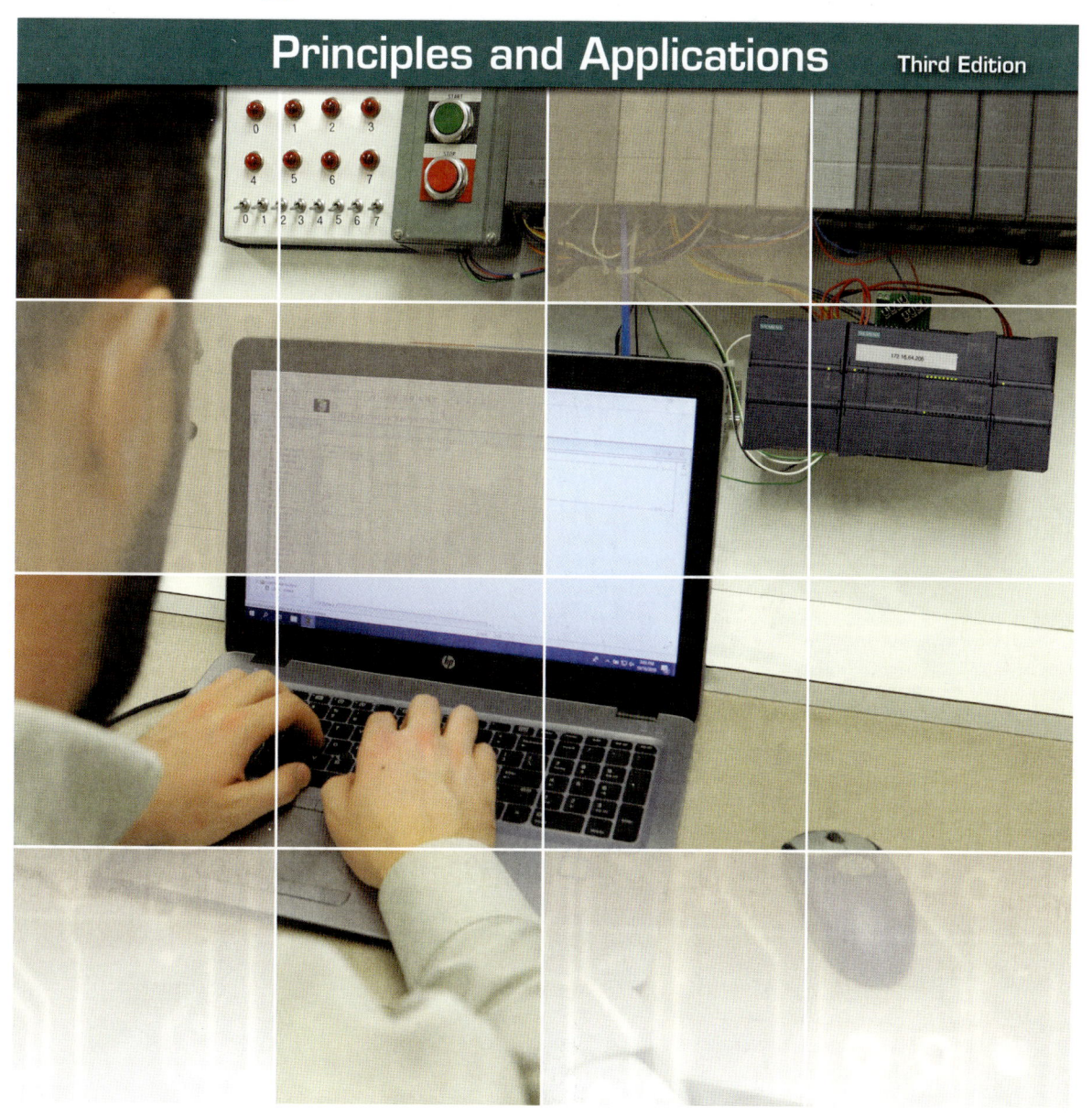

AMERICAN TECHNICAL PUBLISHERS
Orland Park, Illinois

Glen A. Mazur
William J. Weindorf

Programmable Logic Controllers: Principles and Applications contains procedures commonly practiced in industry and the trade. Specific procedures vary with each task and must be performed by a qualified person. For maximum safety, always refer to specific manufacturer recommendations, insurance regulations, specific job-site and plant procedures, applicable federal, state, and local regulations, and any authority having jurisdiction. The material contained herein is intended to be an educational resource for the user. American Technical Publishers assumes no responsibility or liability in connection with this material or its use by any individual or organization.

American Technical Publishers Editorial Staff

Editor in Chief:
    Peter A. Zurlis
Assistant Production Manager:
    Nicole D. Burian
Technical Editor:
    Kyle D. Wathen
Supervising Copy Editor:
    Catherine A. Mini
Copy Editor:
    James R. Hein
Editorial Assistant:
    Lauren D. Bedillion

Art Supervisor:
    Sarah E. Kaducak
Cover Design:
    Steven E. Gibbs
Cover Photo:
    Christopher T. Proctor
Illustration/Layout:
    Nick W. Basham
    Bethany J. Fisher
    Steven E. Gibbs
Digital Media Manager:
    Adam T. Schuldt
Digital Resources:
    Cory S. Butler
    James V. Cashman
    Tim A. Miller

Basofil® is a registered trademark of Basofil Fibers, L.L.C. Excel® is a registered trademark of the Microsoft Corporation. Kevlar® is a registered trademark of DuPont de Nemours, Inc. Micro810™ is a registered trademark of Rockwell Automation, Inc. MicroLogix™ is a registered trademark of Rockwell Automation, Inc. The National Electrical Code® (NEC®) is a registered trademark of the National Fire Protection Association® (NFPA®). The National Fire Protection Association® (NFPA®) is a registered trademark of the National Fire Protection Association®. NFPA 70E®: Standard for Electrical Safety in the Workplace is a registered trademark of the National Fire Protection Association® (NFPA®). Nomex® is a registered trademark of DuPont de Nemours, Inc. PLC-5® is a registered trademark of the Allen-Bradley Company, Inc. PowerPoint® is a registered trademark of Microsoft Corporation. RSLinx™ is a registered trademark of Allen-Bradley Company, Inc. RSLogix™ is a registered trademark of Allen-Bradley Company, Inc. SLC™500 is a registered trademark of Allen-Bradley Company, Inc. Windows® is a registered trademark of Microsoft Corporation. QuickLink, QuickLinks, Quick Quiz, and Quick Quizzes, are either trademarks or registered trademarks of American Technical Publishers.

© 2021 by American Technical Publishers
All rights reserved

3 4 5 6 7 8 9 – 21 – 9 8 7 6 5 4 3

Printed in the United States of America

    ISBN 978-0-8269-1396-8
    eISBN 978-0-8269-9589-6

This book is printed on recycled paper.

Acknowledgments

The authors and publisher are grateful for the technical information and assistance provided by:

Dave McDonald
Automation Systems Estimator
Integrity Integration Resources

ABB Inc.	Lincoln Electric, Inc.
Advanced Assembly Automation Inc.	Lovejoy, Inc.
Atlas Technologies Inc.	PLC Multipoint
Baldor Electric Company	R.C. Flagman Inc.
Carlo Gavazzi Inc.	Rockwell Automation, Inc
Carrier Corporation	Rockwell Automation, Allen-Bradley Company, Inc.
Cleaver Brooks	Saftronics, Inc.
Dreamsquare/Shutterstock.com	Salisbury
Fluke Corporation	SEW-Eurodrive, Inc.
GE Fanuc Automation	Siemens
Ideal Industries, Inc	Worcester Controls Corporation
Jackson Systems LLC	

Table of Contents

CHAPTER 1 — PROGRAMMABLE LOGIC CONTROLLERS AND CRITICAL SAFETY PRACTICES 2

- Chapter Review 49
- PLC Exercises 51

CHAPTER 2 — PLC ELECTRICAL PRINCIPLES, RATINGS, AND CIRCUIT CALCULATIONS 72

- Chapter Review 103
- PLC Exercises 105

CHAPTER 3 — PLC PROGRAMMING SYMBOLS, DIAGRAMS, AND LOGIC FUNCTIONS 118

- Chapter Review 145
- PLC Exercises 147

CHAPTER 4 — PLC HARDWARE, MEMORY, AND OPERATING CYCLE 156

- Chapter Review 175
- PLC Exercises 177

CHAPTER 5 — PLC SYSTEMS, CIRCUITS, AND INTERFACE DEVICES 186

- Chapter Review 227
- PLC Exercises 231

CHAPTER 6 — PLC PROGRAMMING DIAGRAMS, ADDRESSES, AND BIT INSTRUCTIONS — 242

Chapter Review 263

PLC Exercises 265

CHAPTER 7 — PLC PROGRAMMING TIMER AND COUNTER INSTRUCTIONS — 274

Chapter Review 293

PLC Exercises 295

CHAPTER 8 — PLC ANALOG DEVICE INSTALLATION, PROGRAMMING, AND TROUBLESHOOTING — 312

Chapter Review 335

PLC Labs 337

CHAPTER 9 — PLC INSTALLATIONS AND STARTUP — 354

Chapter Review 381

PLC Labs 383

CHAPTER 10 — TROUBLESHOOTING METHODS AND TEST INSTRUMENT OPERATION — 396

Chapter Review 429

PLC Labs 431

Table of Contents

CHAPTER 11 — TESTING AND TROUBLESHOOTING ELECTRICAL DEVICES AND PLC HARDWARE — 438

Chapter Review 471

PLC Labs 473

CHAPTER 12 — TROUBLESHOOTING WITH PLC SOFTWARE — 482

Chapter Review 499

PLC Labs 501

CHAPTER 13 — PLC SYSTEM MAINTENANCE — 510

Chapter Review 527

PLC Labs 529

APPENDIX 545

GLOSSARY 565

INDEX 573

Learner Resources

- Quick Quizzes®
- Illustrated Glossary
- Flash Cards
- Chapter Activities
- PLC Trainer Component Sources
- TECO Resources
- Allen-Bradley® Resources
- SIEMENS Resources
- Programming Software Tutorial Videos
- Internet Resources

Features

Objectives identify specific concepts addressed within each chapter.

QR Codes at the beginning of each chapter enable quick access to Learner Resources.

Industrial application photos facilitate proper safety measures during troubleshooting applications.

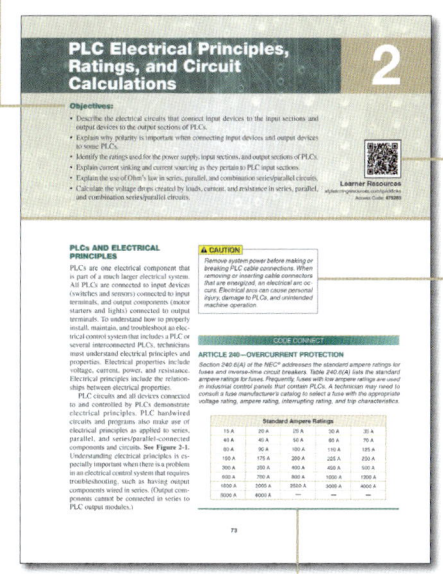

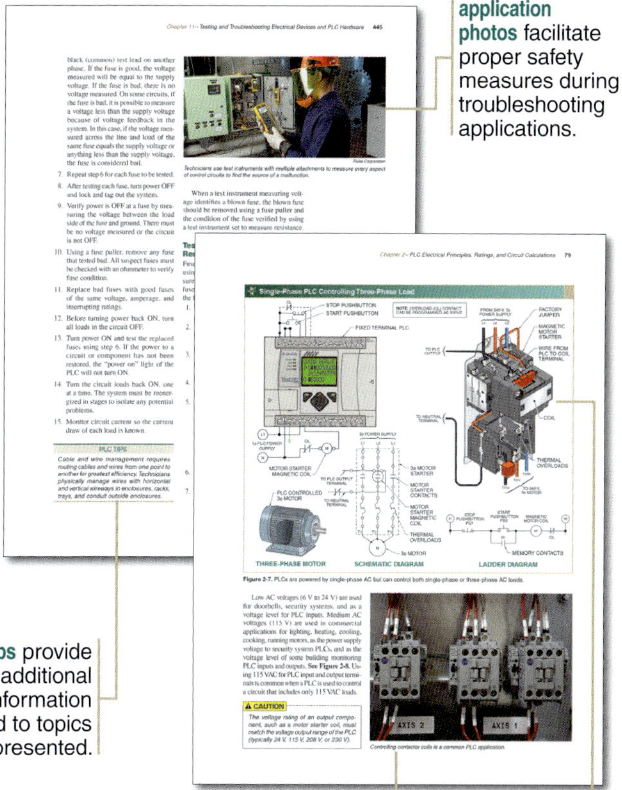

Caution, Warning, and Danger boxes provide specific safety practices or information for certain electrical applications.

Code Connects provide specific code information from the 2020 NEC®.

PLC Tips provide additional information related to topics presented.

Full color photos supplement text and illustrations.

Chapter Reviews reinforce comprehension of chapter content.

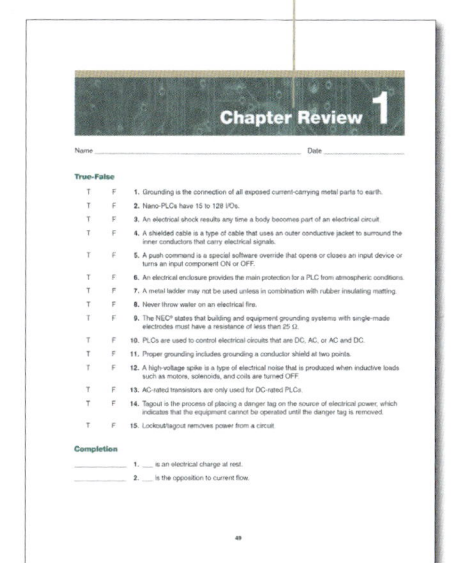

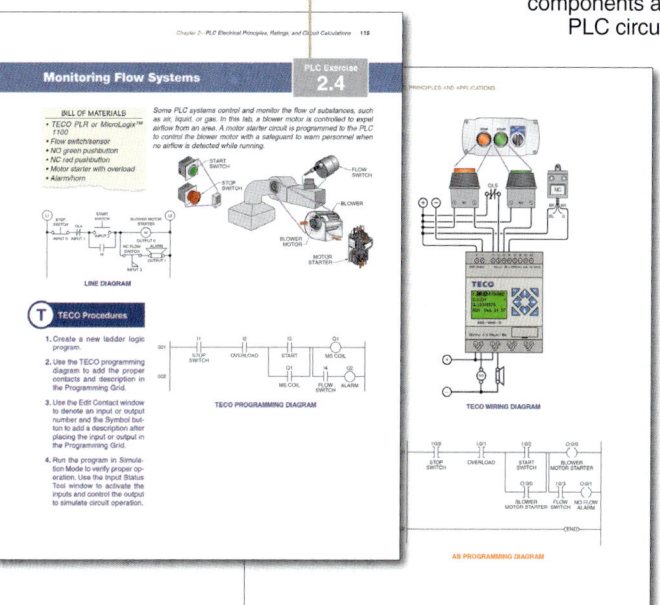

PLC Exercises/Labs reinforce developmental skills to program and wire PLC circuits.

Detailed illustrations show operation of components and PLC circuits.

Learner Resources

The *Programmable Logic Controllers: Principles and Applications* Learner Resources are self-study tools that reinforce the content covered in the book. These online resources can be accessed using either of the following methods:

- Key ATPeResources.com/QuickLinks into a web browser and enter QuickLinks™ access code **475203**.
- Use a Quick Response (QR) reader app to scan the QR Code with a mobile device.

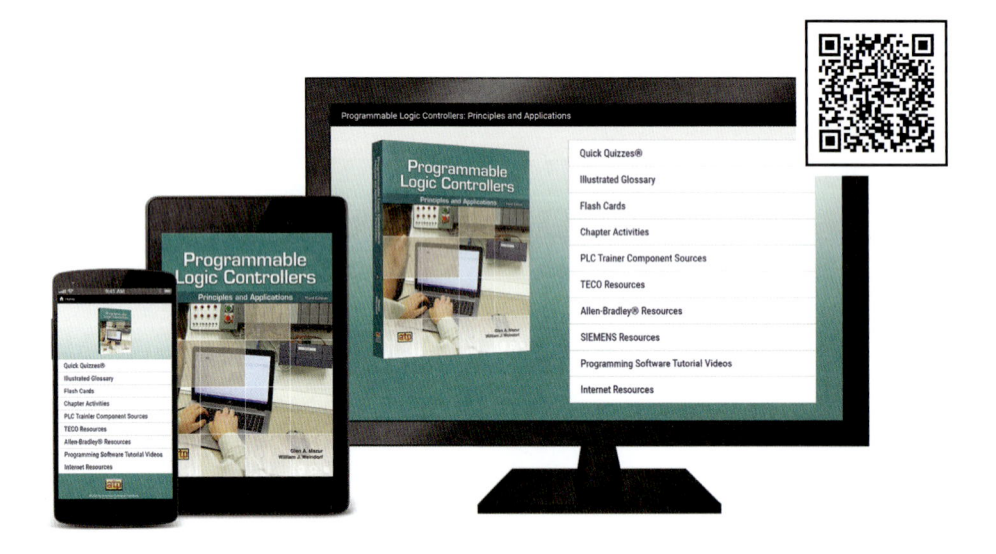

Programmable Logic Controllers: Principles and Applications includes access to online learner resources that reinforce textbook content and enhance learning. The learner resources include the following:

- **Quick Quizzes**® that provide interactive questions for each chapter, with embedded links to highlighted content with the textbook and to the Illustrated Glossary
- **Illustrated Glossary** that serves as a helpful reference to commonly used terms, with selected terms linked to textbook illustrations
- **Flash Cards** that allow a self-study/review of common terms and their definitions
- **Chapter Activities** that provide PDFs of activities based on corresponding chapter content
- **PLC Trainer Component Sources** that list components used to build PLC circuits or trainers
- **TECO Resources** that provide PDFs and weblinks for downloading software and using TECO PLCs
- **Allen-Bradley**® **Resources** that provide PDFs and weblinks for downloading software and using Allen-Bradley® PLCs
- **SIEMENS Resources** that provide weblinks for purchasing SIEMENS software and PLCs, and a PDF document used to assist in programming SIEMENS Step 7 software
- **Programming Software Tutorial Videos** consisting of procedural videos for programming both TECO and Allen-Bradley® software
- **Internet Resources** that provide access to additional resources to support continued learning

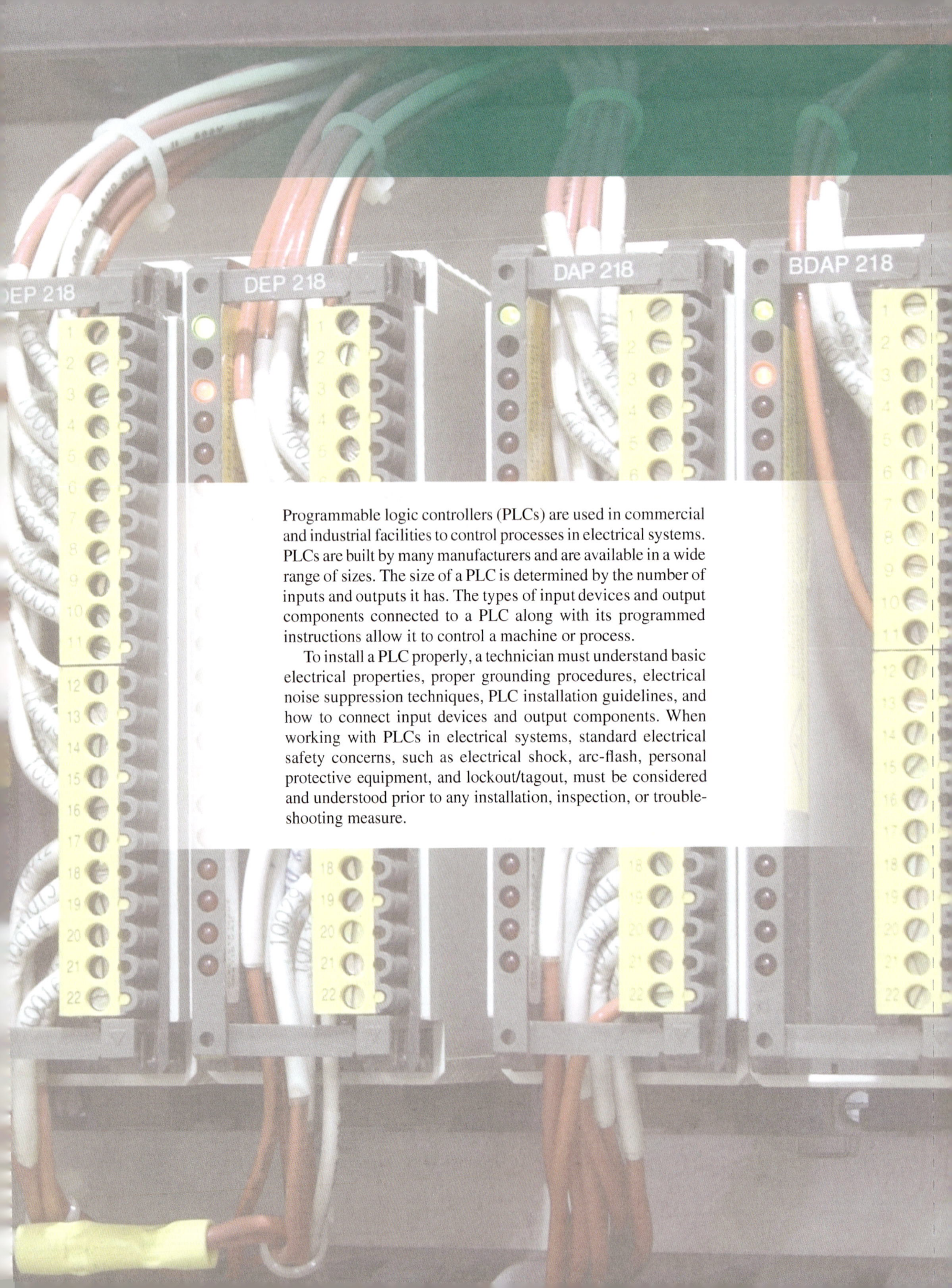

Programmable logic controllers (PLCs) are used in commercial and industrial facilities to control processes in electrical systems. PLCs are built by many manufacturers and are available in a wide range of sizes. The size of a PLC is determined by the number of inputs and outputs it has. The types of input devices and output components connected to a PLC along with its programmed instructions allow it to control a machine or process.

To install a PLC properly, a technician must understand basic electrical properties, proper grounding procedures, electrical noise suppression techniques, PLC installation guidelines, and how to connect input devices and output components. When working with PLCs in electrical systems, standard electrical safety concerns, such as electrical shock, arc-flash, personal protective equipment, and lockout/tagout, must be considered and understood prior to any installation, inspection, or troubleshooting measure.

Programmable Logic Controllers and Critical Safety Practices

Objectives:
- Describe how PLCs are used in everyday environments and industrial operations.
- Identify the sizes of manufactured PLCs and the number of inputs and outputs each size can have.
- List the types of devices and components used in PLC systems.
- List the considerations associated with PLC safety.
- Describe the size designations used by PLC manufacturers.
- Describe the electrical properties of voltage, current, and resistance.
- Explain how to ground PLCs and PLC systems.
- Describe PLC installation guidelines and personal protective equipment (PPE).
- Explain the lockout/tagout process and the need to inspect PLC systems.

Learner Resources
atplearningresources.com/quicklinks
Access Code: 475203

HOW PROGRAMMABLE LOGIC CONTROLLERS WORK

A *programmable logic controller (PLC)* is an industrial computer control system that is preprogrammed to carry out automatic operations. PLCs can control thousands of inputs and outputs; monitor, sequence, time, and count motion- and machine-control applications; collect, distribute, and manage data; and process system diagnostics. PLCs are used in automated industrial processes to increase system performance and minimize human errors.

Depending on the inputs and outputs, a PLC can monitor and record run-time data such as machine productivity or operating temperature, automatically start and stop processes, and generate alarms when machines malfunction. PLCs typically use modular cards, each of which contains a central processing unit (CPU) that stores the memory, such as read-only memory (ROM) and random-access memory (RAM), as well as a type of communication device such as an Ethernet card. Input modules and output modules attach to the PLC, and input modules send signals to the CPU to control the output module devices. Programs are either uploaded to the PLC or downloaded from the PLC to control the industrial application. **See Figure 1-1.** PLCs are capable of handling complicated logic operations and make fault-finding easier. They increase reliability while lowering power consumption and maintenance costs. PLCs also facilitate data collection and documentation capacity by communicating with web browsers and connecting to databases stored in the cloud.

PLCs continuously receive and monitor information from input devices or sensors. Then they process that information and trigger connected output devices based on preprogrammed parameters. PLCs can be adapted to either process electrical signals or monitor and control sensors and actuators in order to carry out preprogrammed commands for various applications. They can be used as stand-alone units that continuously monitor a process or specific machine function. PLCs can also be networked to control an entire production line. **See Figure 1-2.**

Everyday Use of PLCs

PLCs are used to control many items used in everyday environments. For example, PLCs are used to control traffic lights and manage traffic flow 24 hours a day, 365 days a year to allow for timely and direct emergency rescue services. PLCs are also used to control every process in automatic car wash facilities, from the amount of soap-and-water mixture sprayed on a vehicle to the length of time the rotating wipers spin.

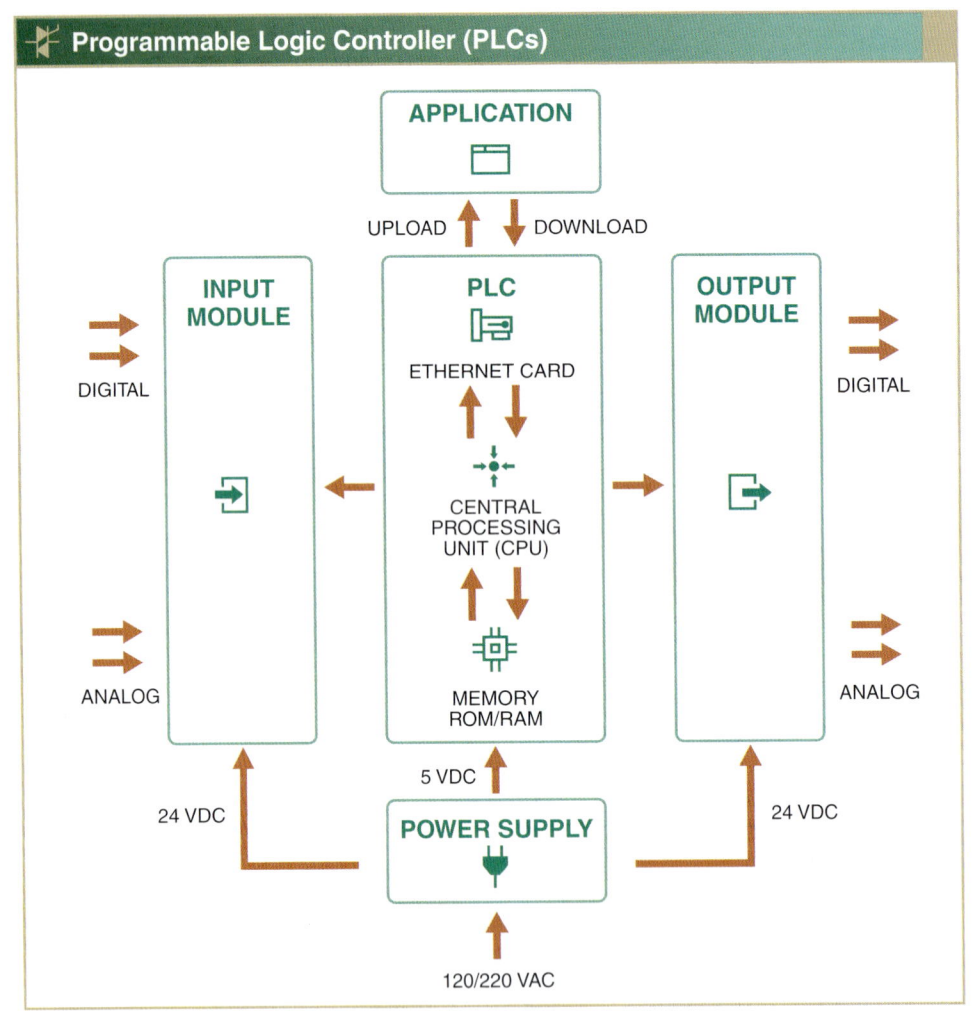

Figure 1-1. PLCs consist of CPUs, input modules, and output modules used to control certain types of industrial applications.

Figure 1-2. PLCs process electrical signals to carry out preprogrammed commands as well as monitor and control sensors and actuators based on preprogrammed application parameters.

PLCs on elevators are loaded with sets of ladder logic instructions that detect the different floors requested and direct the lift to move accordingly. PLCs control conveyor belts on moving walkways and luggage carousels. They also control the belts on large conveyor systems that move raw ingredients through production as they are converted into a variety of food or beverage items.

Industrial Use of PLCs

PLCs are used in hazardous industrial environments to ensure optimal processing and production efficiency. They are used to guide the operation of process valve controls, pump systems, stage air conditioning systems, and electrical substations. PLCs also play a large role in helping manufacturers reduce cycle times and gain batch process savings. They provide rigid control for cycle automation, which minimizes manual intervention. Batch processes are sequential and require time-based or event-based decisions. PLCs automate outputs and enable remote observation of each batch process, which facilitates any needed production process changes.

For example, PLCs control the high-speed production of ointments, fluids, and pills of various shapes and sizes in the pharmaceutical industry. PLCs are also used to monitor and control water-tank quenching systems in the aerospace industry and filling machine control systems in the food-and-beverage industry. **See Figure 1-3.**

In the pulp-and-paper industry, PLCs are used to monitor corrugation machine control systems. They are also used to monitor injection molding control systems in the plastics-and-packaging industry. PLCs are used in the petroleum-and-chemical industry to interface with control systems for pumping stations, pipelines, extraction wells, gas odorizing, flair control, compressor units, and turbine overspeed trips. In the textile industry, PLCs installed on weaving machines receive and record data, including the speed at which raw yarn is processed into a textile. **See Figure 1-4.**

PLC Industrial Processes

Dreamsquare/Shutterstock.com

Figure 1-3. In the food-and-beverage industry, PLCs monitor and control rotary filling machines to fill, cap, and label bottles. Then they transfer those bottles to the next stage in the process.

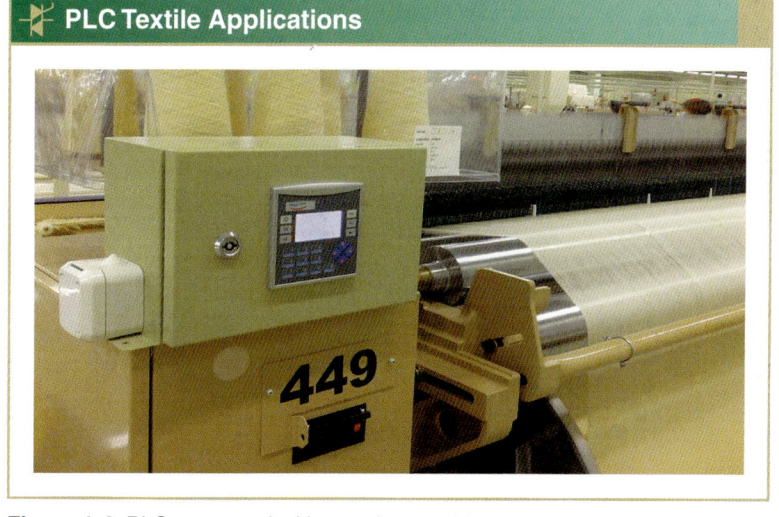

PLC Textile Applications

Figure 1-4. PLCs are used with weaving machines to receive and record data and to control the speed at which yarn is processed into a textile.

In the energy-and-power industry, PLCs and supervisory control and data acquisition (SCADA) systems are used to detect when energy and power consumption are higher or lower than normal. PLCs are the hardware and SCADA systems are the software that gather and analyze real-time data. **See Figure 1-5.** Plant managers use that real-time data to visualize the hourly, daily, weekly, seasonal, and annual consumption trends and identify areas where energy is being wasted.

Supervisory Control and Data Acquisition (SCADA) Systems

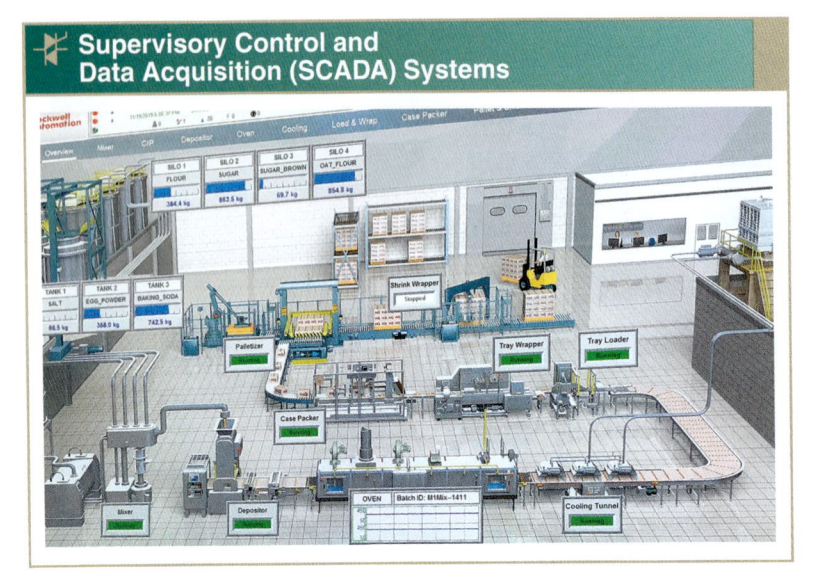

Figure 1-5. SCADA systems integrated with PLCs gather and analyze real-time data to identify trends and enhance process parameters to increase efficiency.

PLCs and SCADA systems are also used to help stabilize the operation of power plants by monitoring load frequency control systems and auxiliary systems and by preventing steam and gas turbines from tripping. They are also used to convert signals from wind speed and direction sensors to control wind turbine operation and to provide critical information, such as predicting sensor failure before it occurs.

In large processing plants, PLCs are used for the automatic startup and shutdown of critical equipment. They ensure that critical equipment cannot be started unless all conditions for a safe start have been met. PLCs also monitor conditions for safe runs and trip equipment when abnormalities are detected. For example, the steam temperature of an industrial boiler is extremely hard to control and requires continuous monitoring and inspection at frequent intervals 24 hours a day, 365 days per year. The use of PLCs and SCADA systems can help prevent catastrophic injuries and destruction of property that may result from unstable boiler systems and human error. Sensors measure the water level, temperature, and pressure in each boiler in a system. If the set values are exceeded, the entire system is shut down, and valves are automatically opened to release steam and pressure. **See Figure 1-6.**

Smart Production Processes

Smart production processes are facilitated by the use of PLCs and SCADA systems. The PLC is used as the central processor for real-time decisions. The SCADA system provides performance data that helps managers better leverage resources, schedule runs, coordinate logistics, and plan for supplier timing to create more efficient processes. Both cement production and glass production use PLCs for smart production.

Cement production involves mixing various raw materials in a kiln. The quality of these raw materials and their proportions significantly impact the quality of the final product. PLCs ensure the use of the right quality and quantity of each raw material within the kiln during the mixing process in order to achieve the best quality product.

Due to the complexity of producing float glass, PLCs are used in tandem with bus technology, which creates a distributed control network. This allows for the gathering of additional data and the control of digital quality and position. PLCs are also used to control the material ratios and the processing of flat glasses.

PLC-Trained Employment Opportunities

A number of jobs are available for technically trained individuals who can operate, program, integrate, and maintain PLCs. Technicians trained to work with PLCs incorporate technical skills from the electrical, mechanical, computer, and manufacturing disciplines. These technicians are employed in machine design, power generation, building automation, plant maintenance, process control, transportation, PLC programming, systems integration, component testing, quality control, technical sales, and related fields. **See Figure 1-7.**

PLC/SCADA Boiler Applications

MOTOR — BOILER — PLC/HMI PANEL

PRESSURE SENSOR

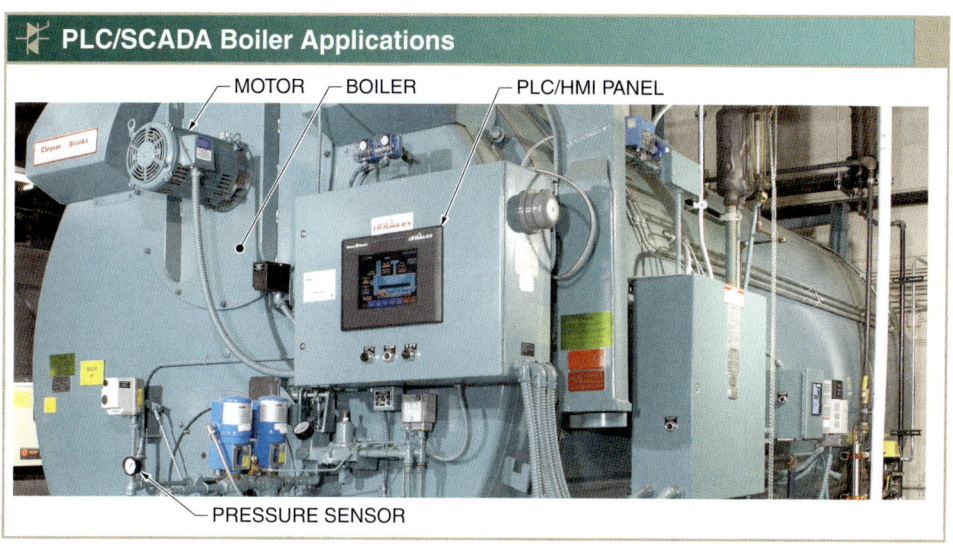

Figure 1-6. PLCs and SCADA systems control boiler operations by measuring water levels, temperatures, and pressures to ensure stable operation.

PLC Technicians

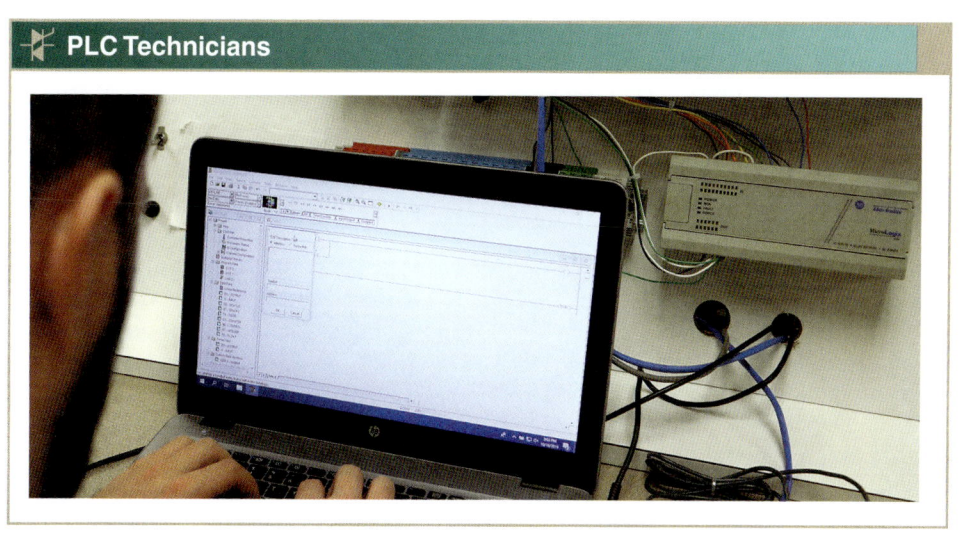

Figure 1-7. Technicians trained to work with PLCs are employed to install and repair I/O networks, data highways, variable-speed drives, and process control equipment. They are also employed to write PLC programs for a variety of automated control systems.

Technicians with advanced PLC training may be offered positions as specialists or troubleshooters. They can install and repair I/O networks, data highways, variable-speed drives, and process control equipment. They can also write PLC programs for a variety of automated control systems ranging from simple on-off controls to robotics. Job descriptions and salaries listed for technicians vary based on the type of industry, the region of the country, and level of experience.

Job titles may include the following:
- Automation Technician
- Aviation Instrument Technician
- Controls Technician
- Electrical Controls Technician
- Field Service Technician
- Instrumentation and Control Technician
- Maintenance Technician
- PLC Programmer
- PLC Technician

PLC SYSTEM SETUP

A variety of PLCs are used in industrial and commercial applications. When setting up a PLC system, an understanding of the similarities and differences between PLC types and the connections between input devices, analog devices, and output components is required to correctly connect the PLC inputs and outputs.

PLCs are manufactured from nano- and micro-sizes that control a few input and output terminals (devices and components) to medium and large sizes that control thousands of input and output terminals (devices and components). PLC system settings can be viewed with integrated programming devices, handheld programming devices, human-machine interfaces, and desktop or laptop computers. Communication between PLCs in a system depends on protocol type, such as Ethernet, ASCII, and Modbus protocols.

A PLC operates as a stand-alone controller for all digital and analog inputs and outputs (I/Os). Input devices operate as either ON or OFF, and control output components that require ON/OFF switching. Analog input devices use varying physical quantities, such as temperature, pressure, and moisture, and convert them to electrical quantities, such as voltage or current. The electrical quantities are then used to control digital or analog outputs, such as DC motors and servo motors. The type of PLC used depends on the number of I/Os needed. For example, nano-PLCs have less than 16 I/Os, micro-PLCs have 16 to 128 I/Os, medium PLCs have 129 to 512 I/Os, and large PLCs have 513 or more I/Os. **See Figure 1-8.**

All PLCs have similar terminal locations and designations. Terminal designations for I/Os can be similar on nano-, micro-, and medium/large PLCs. Medium/large PLCs use rack-mounted I/O modules for system connections. Devices common to all PLC systems include the following:

- power supply modules
- PLCs
- digital input devices
- analog input devices
- output components

POWER SUPPLY MODULES

A *power supply module* is a self-contained, regulated or nonregulated voltage unit that supplies DC or AC voltage to a PLC. A DC-powered PLC can be supplied with a 120/240 V to 24 V power supply module. The internal components of a 120/240 V to 24 V power supply module consist of a transformer, rectifier, and filtering components (capacitor and resistor) to convert AC voltage to DC voltage. **See Figure 1-9.**

The power supply for a small PLC is normally a plug-in module with positive (24 V) and negative (0 V) DC leads or a PLC-mounted power supply with terminal screws. (The required connections are indicated near the screws.) A medium/large PLC typically contains a PLC-mounted power supply that provides power to rack-mounted modules. Power supply modules are connected to terminal strips or terminal blocks that supply power to inputs and outputs.

Terminal Strips

A *terminal strip* is a strip of adjacent terminal screws to which electrical devices are connected. A power supply can be connected to a terminal strip to provide power to a PLC and all inputs and outputs. For example, a 12-circuit, 24-screw terminal strip can be assembled with plate jumpers on the power supply side and spade terminal end connectors on the supplied power side. Ten plate jumpers are used to create two halves on a terminal strip. Five plate jumpers are used to create the positive section, and five plate jumpers are used to create the negative section. Twelve spade terminal end connectors are connected to terminal screws on the supplied power side of the strip to provide both a positive and negative connection point for all circuit inputs and outputs. **See Figure 1-10.**

When using a power supply plug-in wall module, a positive (+) wire marker is attached to the 24 V wire lead and is used to identify the positive section of the terminal strip. The 24 V wire lead (black wire) is connected to the top left terminal screw of the strip. A negative (−) wire marker is attached to the negative wire lead (black wire with white strip) and is used to identify the negative section of the strip. The negative wire lead is connected to the bottom left terminal screw of the strip.

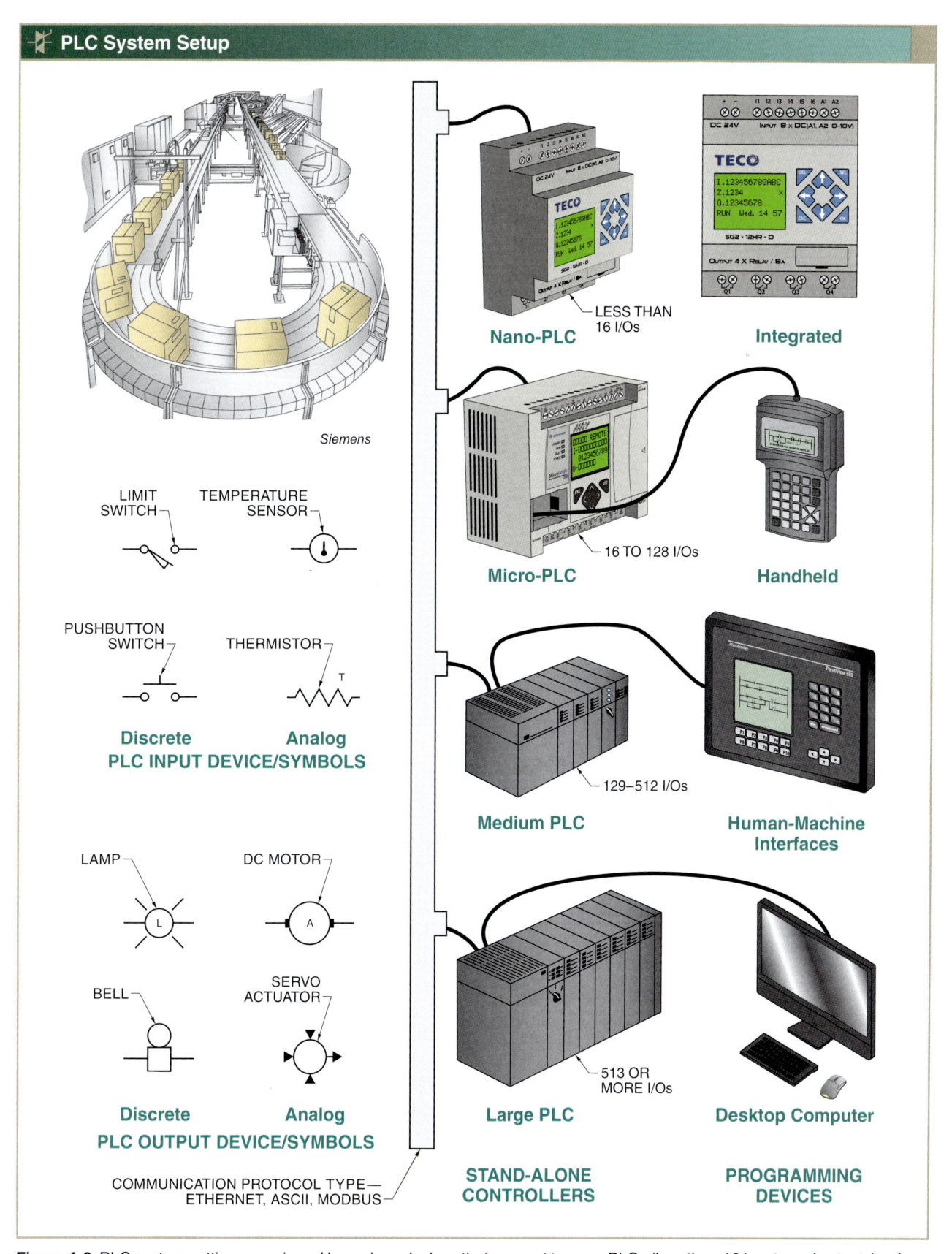

Figure 1-8. PLC system settings are viewed by various devices that connect to nano-PLCs (less than 16 inputs and outputs), micro-PLCs (16 to 128 inputs and outputs), medium PLCs (129 to 512 inputs and outputs), and to large-sized PLCs that control 513 or more inputs and outputs.

PLC Power Supply Modules

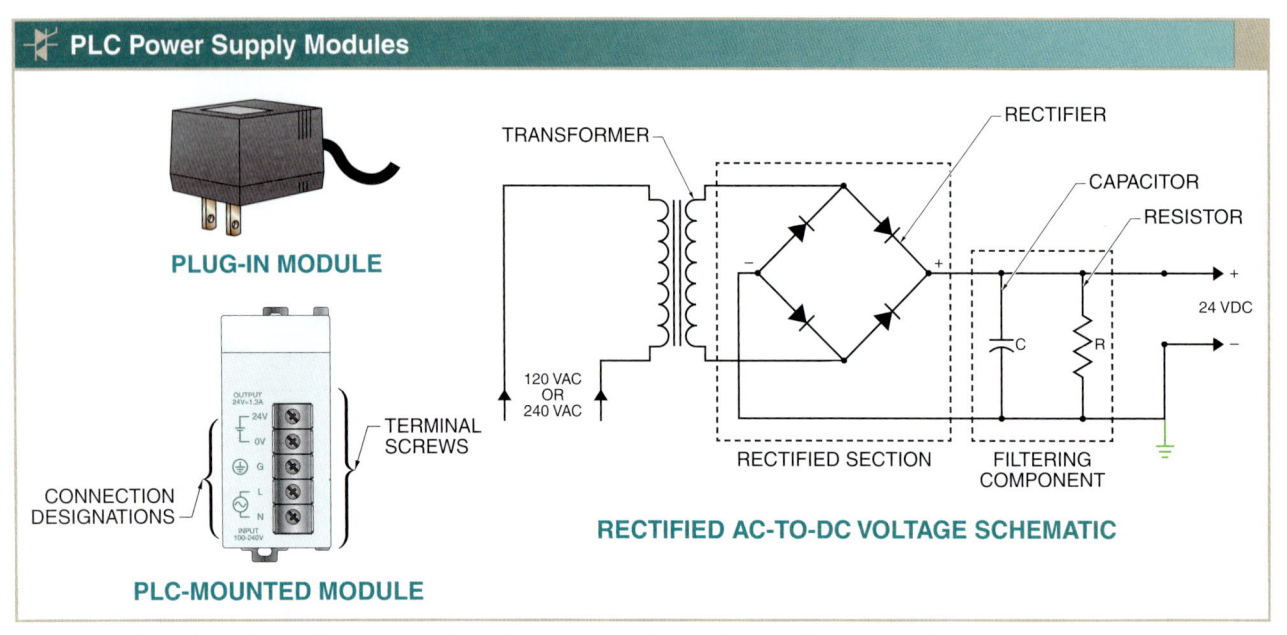

Figure 1-9. A 120/240 V to 24 V power supply module—sourced directly from a 120/240 V main supply—can be a plug-in module with positive (24 V) and negative (0 V) DC wire leads or a PLC-mounted module.

Terminal Blocks

A *terminal block* is a device used to interconnect wires from electrical devices. Terminal blocks are used to manage wire connections between PLC inputs/outputs and electrical devices/components. Most terminal blocks are mounted on DIN-rail systems. Typically, an end anchor locks a row of terminal blocks in place. Each terminal block contains two wire insert terminals, and the wires are held in place with screws. A screw connecting plate enables the two wires to be mechanically connected. **See Figure 1-11.**

PLCs

PLCs used in electrical systems include nano-, micro-, and medium/large PLCs. These can be used with digital and analog devices as well as output components to control various industrial and commercial applications. Common nano-, micro-, and medium/large PLC manufacturers include Allen-Bradley, TECO, and Siemens. Each manufacturer builds a nano-PLC, such as the 24 V TECO SG2, the Allen-Bradley Micro810™, and the Siemens LOGO! 8. Allen-Bradley also builds the MicroLogix™ 1100 micro-PLC.

Nano-PLCs

The 24 V TECO SG2, Allen-Bradley Micro810™, and Siemens LOGO! 8 each contain a supplied power section, an input section, and an output section. Each PLC has power terminals and I/O terminals. The supplied power section is supplied with either 120 VAC or +12/24 VDC from a power supply module. The input section contains four-to-six digital and two-to-four analog input terminals. The output section has four relay contacts. **See Figure 1-12.** *Note:* Other PLC manufacturers have similar terminal designations and I/O arrangements. Check the user manual for I/O designations and power specifications.

Nano-PLC Supplied Power Sections. The supplied power section of a nano-PLC is powered with a VAC or VDC power supply module. For the 24 V TECO SG2, a 24 V supply is connected to the positive (+) 24 DC terminal screw and negative (−) DC terminal screw. Similarly, the Allen-Bradley Micro 810™ has a +DC12 terminal for the positive voltage connection and a −DC12 terminal for the negative voltage connection. The Siemens LOGO! 8 is designed for VAC or VDC and has an L+ terminal for the line voltage connection and an M terminal for the neutral connection.

Nano-PLC Input Sections. The input section of the 24 V TECO SG2 consists of six digital input terminals designated I1, I2, I3, I4, I5, and I6, and two analog input terminals designated A1 and A2. The input section of the Allen-Bradley Micro810™ has one common terminal (COM0) and eight input terminals designated I-00 through I-07. The last four terminals are shared digital and analog input terminals. The input section of the Siemens LOGO! 8 has eight input terminals designated I1 through I8 with terminals I1/I2 and I7/I8 useable for analog input devices.

Nano-PLC Output Sections. The output sections of the 24 V TECO SG2, Allen-Bradley Micro810™, and Siemens LOGO! 8 each have eight output terminals that are grouped into four relay contacts. Each relay contact has a voltage-in terminal (left terminal) and a voltage-out terminal (right terminal). Each in terminal must receive voltage from a power supply module. A terminal jumper can be used to provide voltage for all in terminals. Each holds voltage until a program is run to close the contact and send the voltage through the PLC to the out terminals to power the output components. The 24 V TECO SG2 and Siemens LOGO! 4 relay contacts are labeled Q1, Q2, Q3, and Q4. The Allen-Bradley Micro810™ relay contacts are labeled CM0, CM1, CM2, and CM3 for the in terminals, and O-00, O-01, O-02, and O-03 for the out terminals.

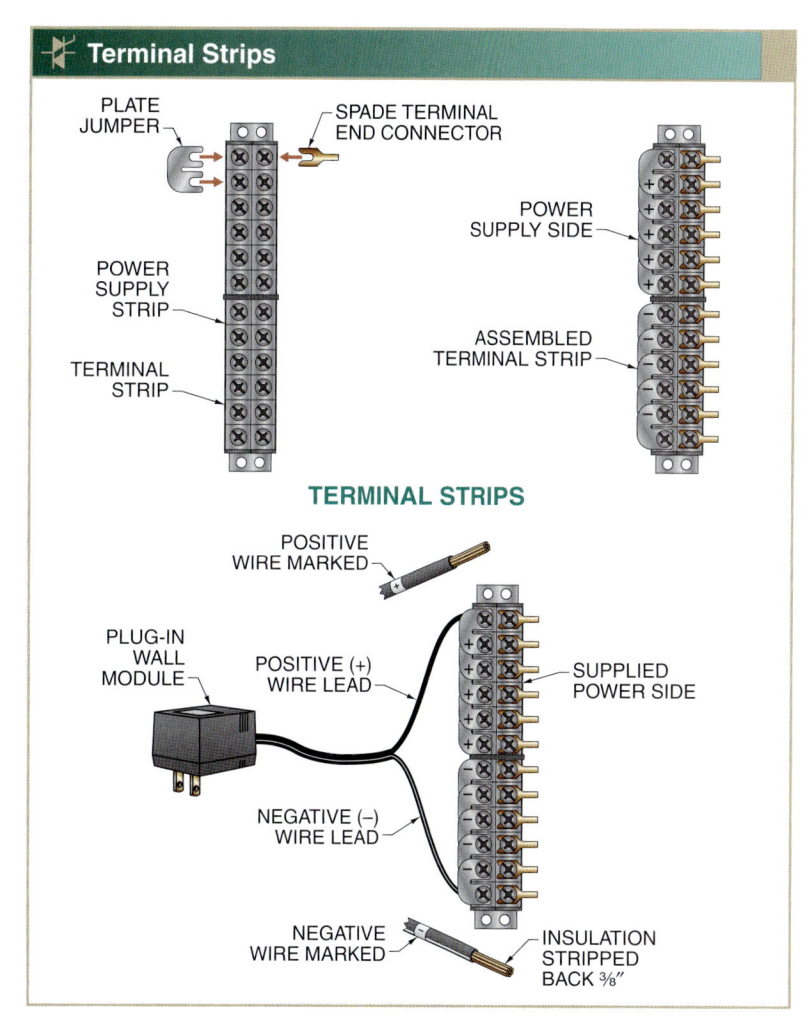

Figure 1-10. A terminal strip can be used with a power supply to provide power for a PLC and all inputs and outputs.

Micro-PLCs

Allen-Bradley's MicroLogix™ 1100 has eighteen I/Os and is powered by +12/24 VDC or 120 VAC from a power supply module. Unlike the nano-PLCs that have both an input and output supplied power section, the MicroLogix™ 1100 has only one. This section supplies power internally to the input side of the PLC, so each input device can be powered from the input power supply. The MicroLogix™ 1100 contains ten digital inputs, two analog inputs, and six relay output contacts. **See Figure 1-13.** *Note:* Other PLC manufacturers have similar terminal designations and I/O arrangements. The user manual should be checked for I/O designations and power specifications.

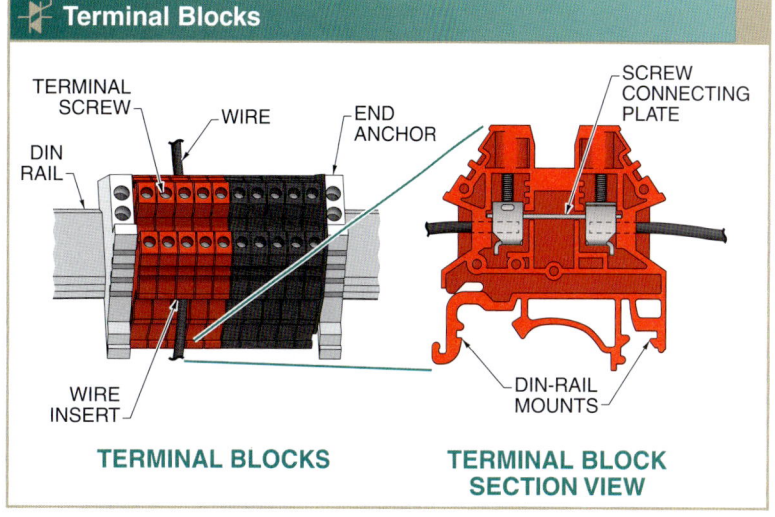

Figure 1-11. Terminal blocks can be used to manage wires between inputs/outputs and electrical devices/components.

Nano-PLC Types

24 V TECO SG2
- SUPPLIED POWER SECTION
  - (+) 24 V TERMINAL SCREW
  - (−) 0 V TERMINAL SCREW
- DIGITAL INPUT TERMINALS
- ANALOG INPUT TERMINALS
- PLC INPUT SECTION
- PLC OUTPUT SECTION
- OUTPUT RELAY CONTACTS
- OUTPUT CONTACTS EACH WITH VOLTAGE "IN" AND VOLTAGE "OUT" TERMINALS

AB MICRO810™
- DIGITAL INPUT TERMINALS
- DC COMMON TERMINALS
- SUPPLIED POWER TERMINALS
- SHARED DIGITAL AND ANALOG INPUT TERMINALS
- TERMINAL JUMPER

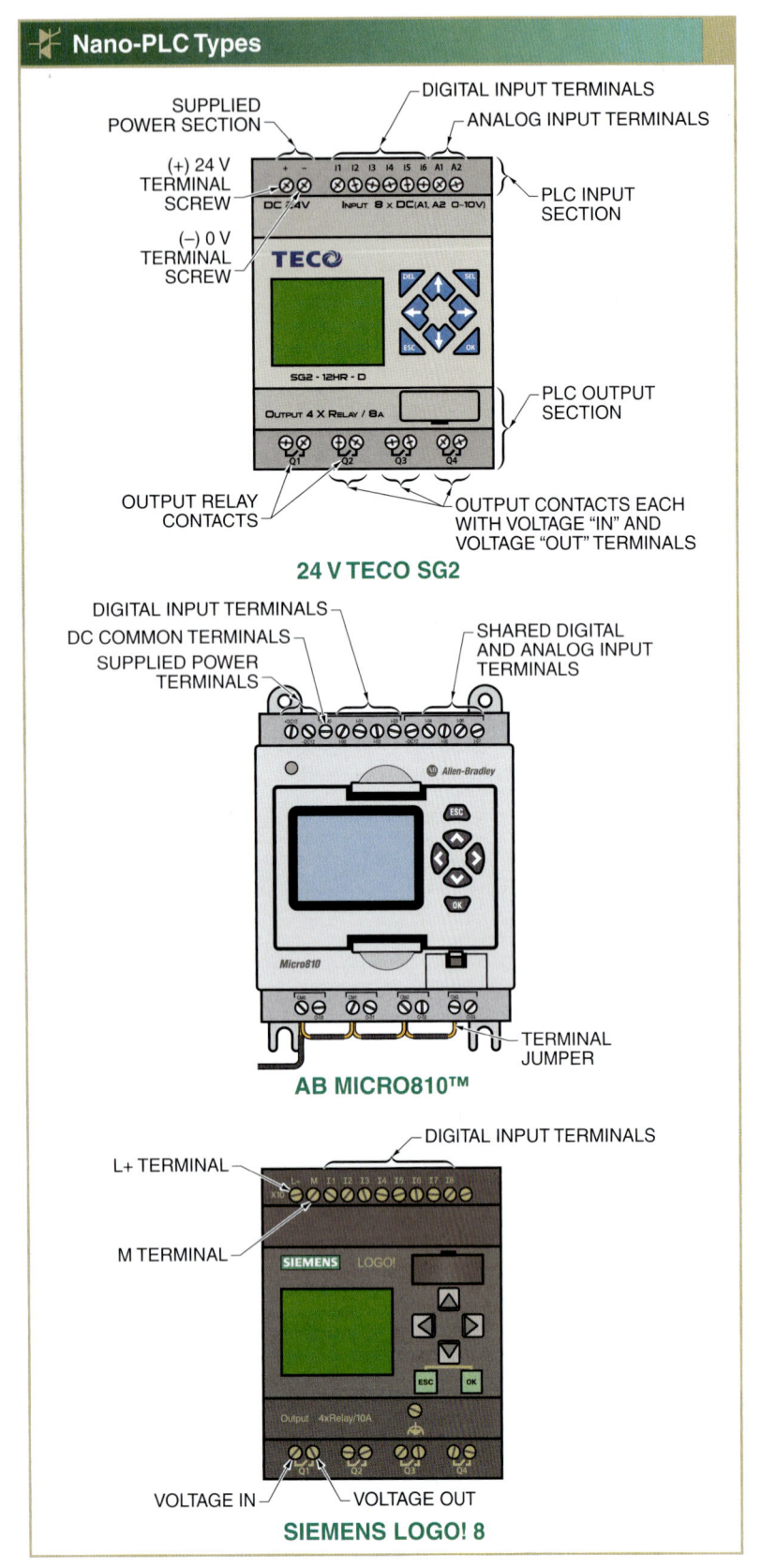

SIEMENS LOGO! 8
- L+ TERMINAL
- M TERMINAL
- DIGITAL INPUT TERMINALS
- VOLTAGE IN
- VOLTAGE OUT

Figure 1-12. Nano-PLCs consist of the 24 V TECO SG2, Allen-Bradley Micro810™, and the Siemens LOGO! 8.

PLC TIPS

I/O module PLC systems allow insertion or removal of terminal blocks while power is applied to the controller.

Micro-PLC Supplied Power Sections. The supplied power section of the MicroLogix™ 1100 micro-PLC has three terminals for VAC- or VDC-supplied power. For VAC-supplied power, a main supply line (L1), a second supply line or neutral (L2/N), and a ground connection are available. For VDC-supplied power, a positive (+) voltage, a negative (−) voltage, and a ground connection are available. To supply power to the MicroLogix™ 1100 PLC, a conductor is connected from a power supply module to the L1/+ terminal; the second line, neutral, or negative voltage conductor is connected to the L2/N/− terminal; and a ground conductor is connected to the ground terminal.

Micro-PLC Input Sections. The input section of the MicroLogix™ 1100 consists of ten digital input terminals, two DC common (DC COM) terminals, two analog input terminals, and one analog common terminal. The digital input terminals designated I/0, I/1, I/2, and I/3 are grouped together to work with the first DC COM terminal. The digital input terminals I/4, I/5, I/6, I/7, I/8, and I/9 are grouped together to work with the second DC COM terminal. Both analog inputs IV1 (+) and IV2 (+) work with an analog common (IA COM) terminal. An input power supply is used to provide power to input devices. The supplied power section provides power internally to the input power supply.

Micro-PLC Output Sections. The output sections of several MicroLogix™ 1100 PLC models have six relay contacts. Each relay contact has a voltage-in terminal and a voltage-out terminal. Each voltage-in is designated as VAC or VDC. The voltage-out terminals are designated as O/0, O/1, O/2, O/3, O/4, and O/5. Each VAC terminal can receive separate voltage from the power supply module to power the output components, or a terminal jumper can be used to send power to each VAC/VDC terminal from the L1 terminal.

Micro-PLC Types

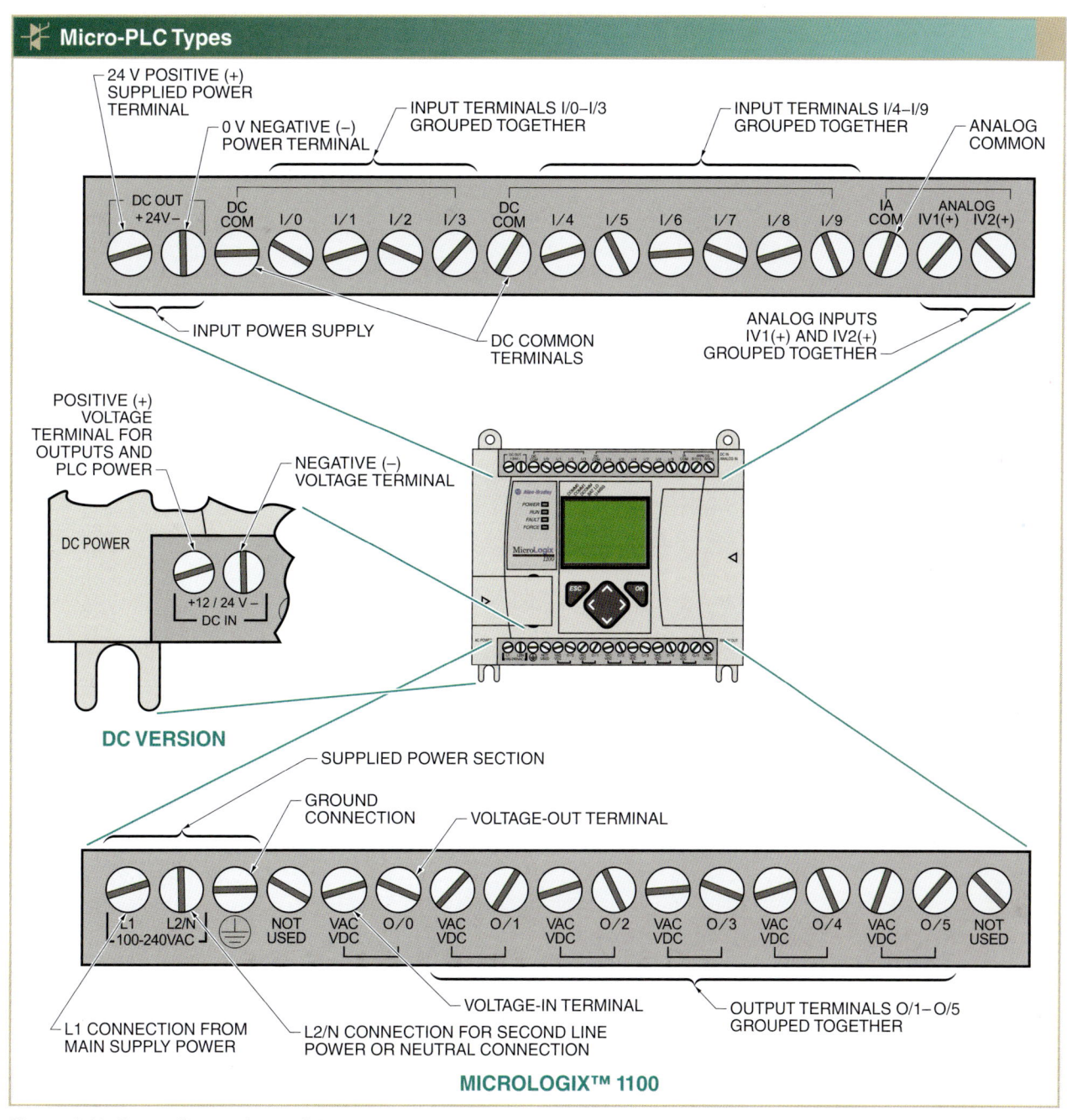

Figure 1-13. Depending on the model number, the MicroLogix™ 1100 may consist of up to ten digital inputs, two analog inputs, and six discrete (relay) output terminals.

Medium/Large PLCs and Rack-Mounted I/O Modules

Medium/large PLCs contain rack-mounted I/O modules that control input devices and output components. The rack-mounted modules have vertical terminal strips for connecting the input devices and output components. During operation, the I/O terminal strips are covered, but a reference chart is provided on the inside of the cover that shows the layout of the input and output terminals.

Rack-mounted modules connect to medium or large PLCs through a rack enclosure. The rack enclosure contains different slots for I/O module circuit boards. Each slot is powered through a backplane inside of the rack enclosure which provides power to each I/O module. The backplane is supplied VAC or VDC power from the PLC power supply module. Power supplies for I/O modules can be rated for 3 A, 4 A, 12 A, or 16 A depending on the type of I/O module and how many are used. I/O modules can consist of input modules, output modules, or combined input and output modules. **See Figure 1-14.**

PLC INPUT AND OUTPUT CONNECTIONS

In order to fully understand how digital input devices, analog input devices, and output components function in a PLC system, it is important to understand how PLC input and output connections are established. All PLCs allow an electrical system's input devices (pushbuttons, pressure switches, and photoelectric switches) and output components (pilot lights, solenoids, motor starters, and VFDs) to be connected to an input section and an output section. How and where the inputs are connected depends on the input device and output component type as well as the specific PLC model used. Generally, inputs are connected in the same manner regardless of the PLC manufacturer or model. Most PLC manufacturers have similar terminal layouts and designations.

PLC input devices are connected to provide voltage to the PLC input section. With a nano-type PLC, the voltage to power the PLC power section and input devices is supplied from a VDC or VAC power supply. The output voltage (VDC or VAC) is supplied from a main power supply at a voltage level of 24 VDC or 24/120/240 VAC based on what the PLC is rated for and the power required by the load.

For example, a simple control application can be programming a PLC to control power for a magnetic motor starter circuit. A *magnetic motor starter* is a starter with an electrically operated mechanical switch (contactor) that includes motor overload protection. A *contactor* is an electrical control device that is designed to control high-power, non-motor loads. Magnetic motor starters include electrical contacts, a coil that opens and closes the contacts, and overload protection. The coil, when energized, closes the contacts inside the magnetic motor starter to supply power to the motor/load. Overload protection can consist of thermal, magnetic, or solid-state device overloads. **See Figure 1-15.**

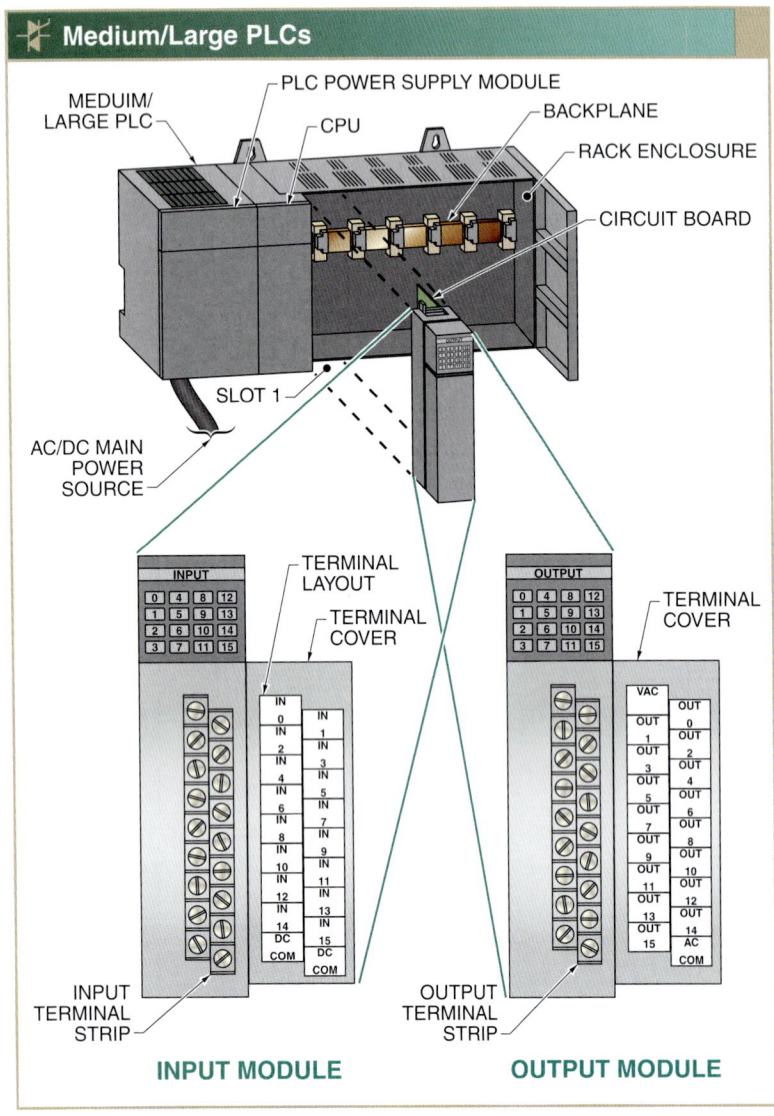

Figure 1-14. I/O modules consist of input modules, output modules, and combined input and output modules.

Nano-PLC Connections

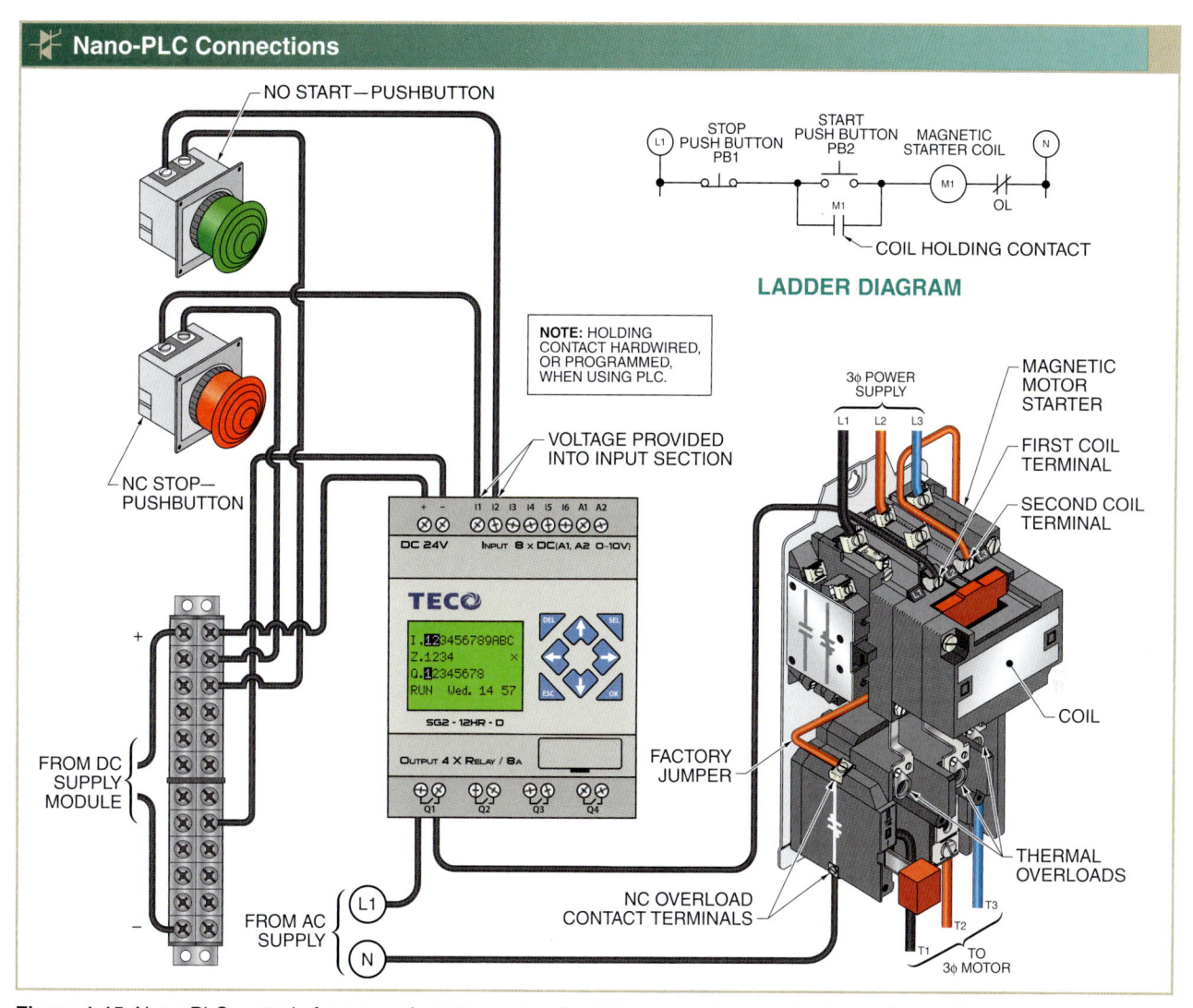

Figure 1-15. Nano-PLC control of a magnetic motor starter circuit involves a start pushbutton, a stop pushbutton, and a magnetic motor starter.

PLC control of a magnetic motor starter circuit contains a start pushbutton, a stop pushbutton, and a magnetic motor starter. The start and stop pushbuttons are the input devices connected to the PLC input section, and the magnetic motor starter coil is controlled by the PLC and powered by the PLC output section. The start and stop pushbuttons can be provided power from a power supply module. The stop pushbutton connects to the first input terminal, and the start pushbutton connects to the second input terminal. The power-in terminal for the output section is provided 120 VAC (L1). The power-out terminal is wired to the first motor starter coil terminal. A factory jumper (wire) connects the second motor starter coil terminal to the first normally closed (NC) overload contact terminal. The neutral wire connects to the other overload contact terminal.

Input devices and output components for a micro-PLC are connected in the same manner. The only differences are that a micro-PLC allows more input devices and output components to be connected, and the input devices are not powered from a separate power supply. The input power supply provides power to the stop and start pushbutton terminals. A jumper is connected between the negative power terminal and the common input terminal. The stop pushbutton is wired to the first input terminal, and the start pushbutton is wired to the second input terminal. An outside power supply provides power to the L1 terminal of the PLC power supply section.

Then a jumper is used to supply power from the L1 terminal to the power-in terminal (VAC) for output 1. The corresponding out terminal (O/1) is wired to the first motor starter coil terminal. A factory jumper connects the second motor starter coil terminal to the first NC overload contact. The neutral wire connects to the other NC overload contact. **See Figure 1-16.**

To program this wiring application, three input instructions and one output instruction are needed. Instructions are programmed on one rung in series, starting with the stop pushbutton, the start pushbutton, and then the magnetic motor starter coil as the output instruction. A rung is programmed in parallel with the start pushbutton and the NO coil holding contact M1.

Micro-PLC Connections

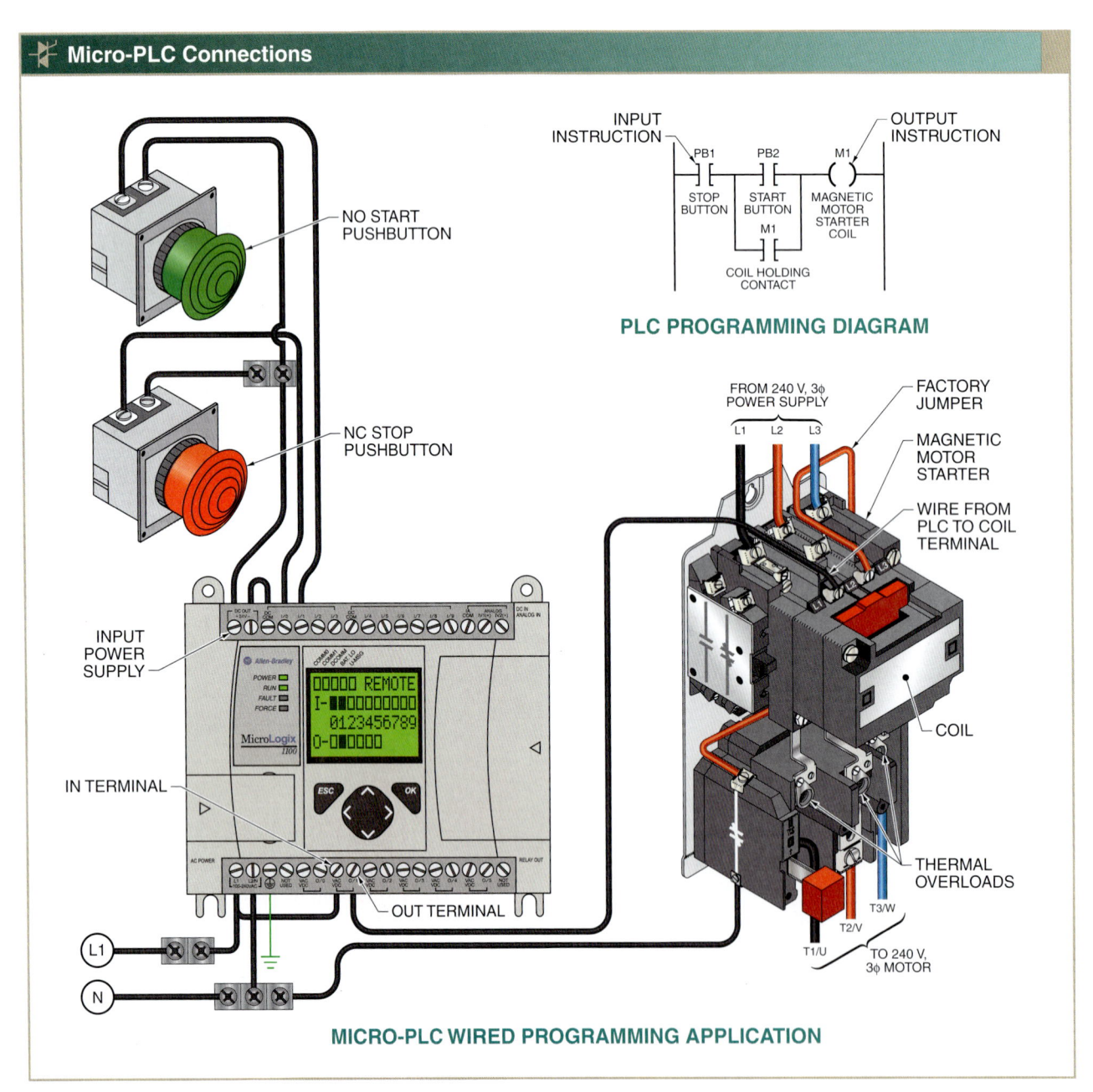

PLC PROGRAMMING DIAGRAM

MICRO-PLC WIRED PROGRAMMING APPLICATION

Figure 1-16. A micro-PLC connects in the same manner as a nano-PLC, but the input devices are not powered from a separate power supply.

Medium and large PLCs can house input, output, or combination input/output rack-mounted modules which enables control of several different types of input devices and output components, such as AC-rated, DC-rated, digital, and analog devices simultaneously. When connecting input devices and output components to a medium/large PLC, a magnetic motor starter circuit can be controlled with one DC-rated input module and one AC-rated output module. The stop and start pushbuttons are powered from a VDC source and are wired to the input module. Supplied power connections can be made through a terminal block or strip.

The stop pushbutton is wired to the input 1 terminal (IN 1), and the start pushbutton is wired to the input 2 terminal (IN 2). The negative is connected to one DC common (DC COM) terminal where a jumper is used to connect both common terminals. The output module is powered with 120 VAC and is connected to the VAC terminal. The output 1 terminal (OUT 1) is wired to the first motor starter coil terminal. A factory jumper connects the second motor starter coil terminal to the first NC overload contact terminal. The output common terminal (AC COM) connects to the other NC overload contact terminal. **See Figure 1-17.**

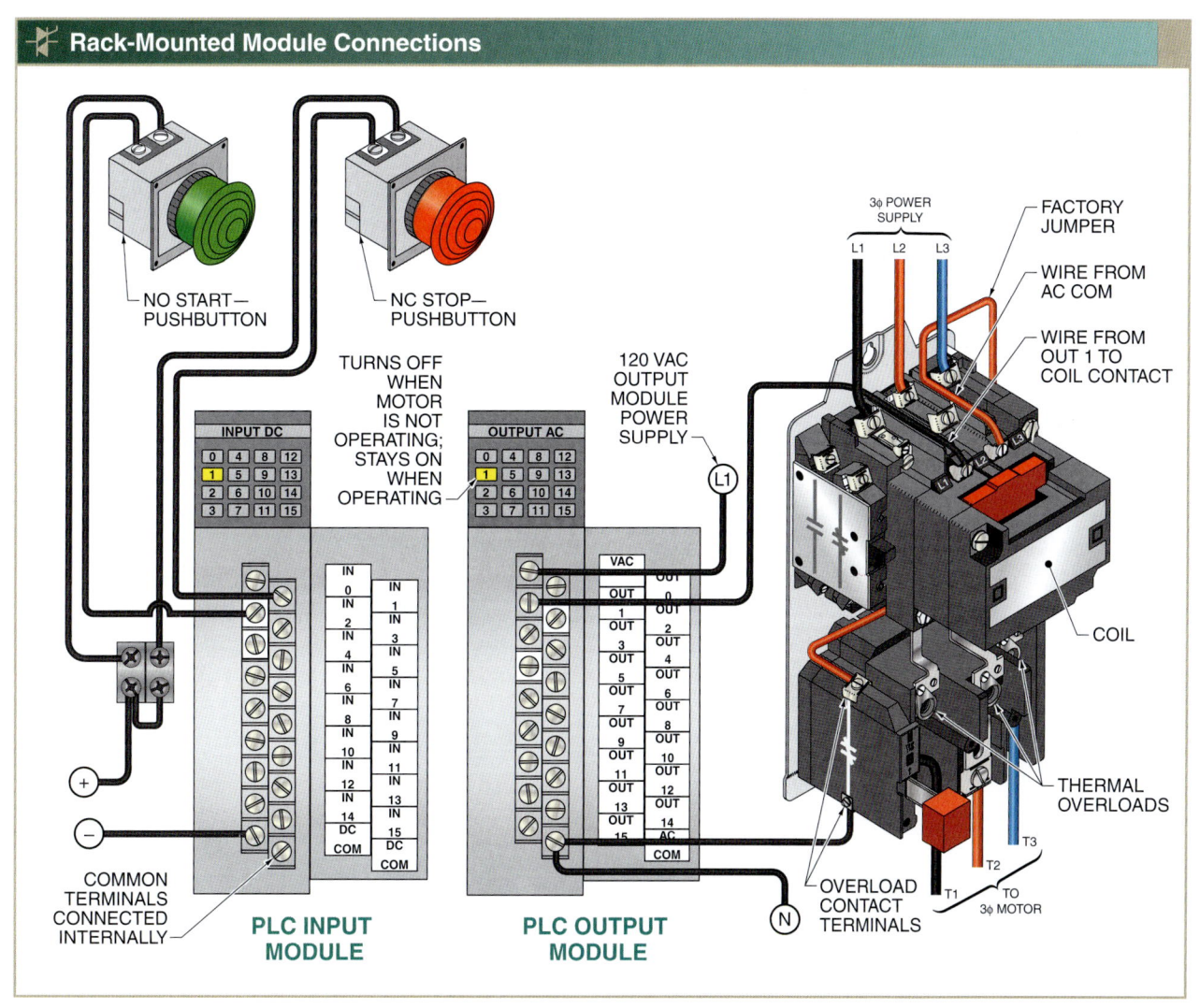

Figure 1-17. A magnetic motor starter circuit can be controlled by using one DC-rated input module and one AC-rated output module.

I/O Switching Devices

Switching devices that enable the operation of PLC applications can be identified as mechanical contacts, such as pushbuttons, limit switches, and selector switches, or as solid-state switches, such as proximity switches, photoelectric switches, and sensors. A mechanical contact uses two wires to connect power and a PLC input terminal, and these contacts are rated to work with AC or DC power. Solid-state switches, such as analog sensors, use three-wire transistors as input switching devices and are either DC-rated, which are only used with DC-rated PLCs, or AC-rated, which are only used with AC-rated PLCs. Transistors consist of current sourcing (PNP/positive) and current sinking (NPN/negative) switching devices.

Similarly, PLC output switching devices consist of mechanical contacts and solid-state switches. Solid-state output switching devices consist of PNP and NPN transistors, silicon-controlled rectifiers (SCRs), and triacs. A load can be controlled by a PLC when the voltage and current ratings of the load are within the PLC switching device ratings. SCRs are used in DC circuits to switch high-level currents. Triacs are used in AC circuits to switch AC devices. **See Figure 1-18.**

PLC Sinking and Sourcing Connections

A PLC that is DC-rated requires an input that is configured for a positive or negative voltage. If a PLC input is designed for a positive voltage (current sourcing), a current sourcing field device is used. If a PLC input is designed for a negative voltage (current sinking), a current sinking field device is used. Regardless of manufacturer, all sourcing and sinking field devices are wired the same. Likewise, some PLC DC outputs can be wired for either a positive or negative voltage output. For example, a PNP current sourcing proximity switch can be used to send a signal to a current sourcing input for control of a sourcing output to energize the load. **See Figure 1-19.**

Nano-PLC inputs are typically only designed for either positive or negative polarity. For a micro-PLC, the inputs can be designed for connection to a positive, to a negative, or to both types of current sinking and sourcing field devices. A micro-PLC can include common terminals (DC COM) for connections between both types. If one DC COM terminal is connected to the positive DC terminal, then its corresponding inputs are the required current sinking devices. Likewise, if the other DC COM terminal is connected to the negative DC terminal, then its corresponding inputs are the required current sourcing devices. **See Figure 1-20.**

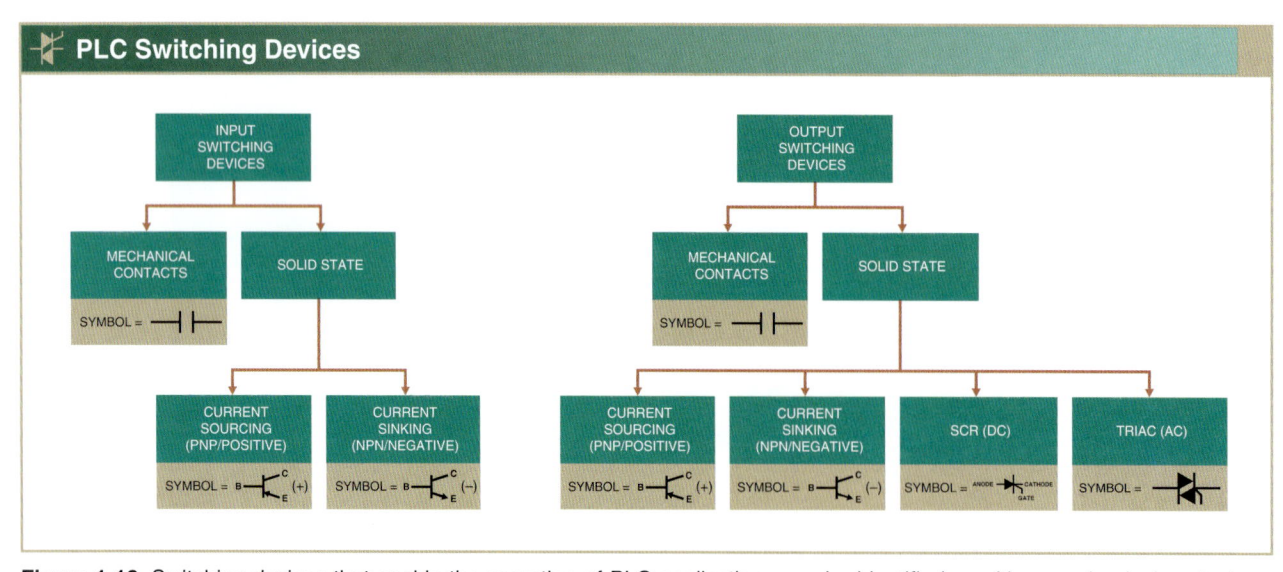

Figure 1-18. Switching devices that enable the operation of PLC applications can be identified as either mechanical contacts or solid-state switches.

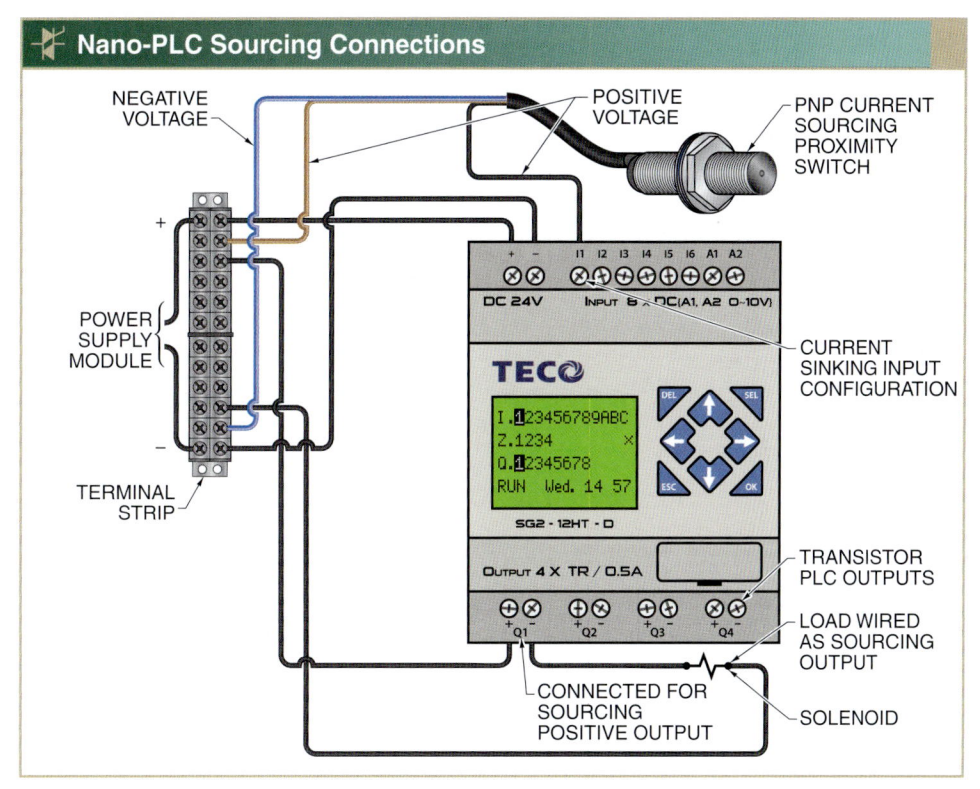

Figure 1-19. A PNP current sourcing proximity switch sends a signal to a current sinking input for control of a sourcing output to energize the load.

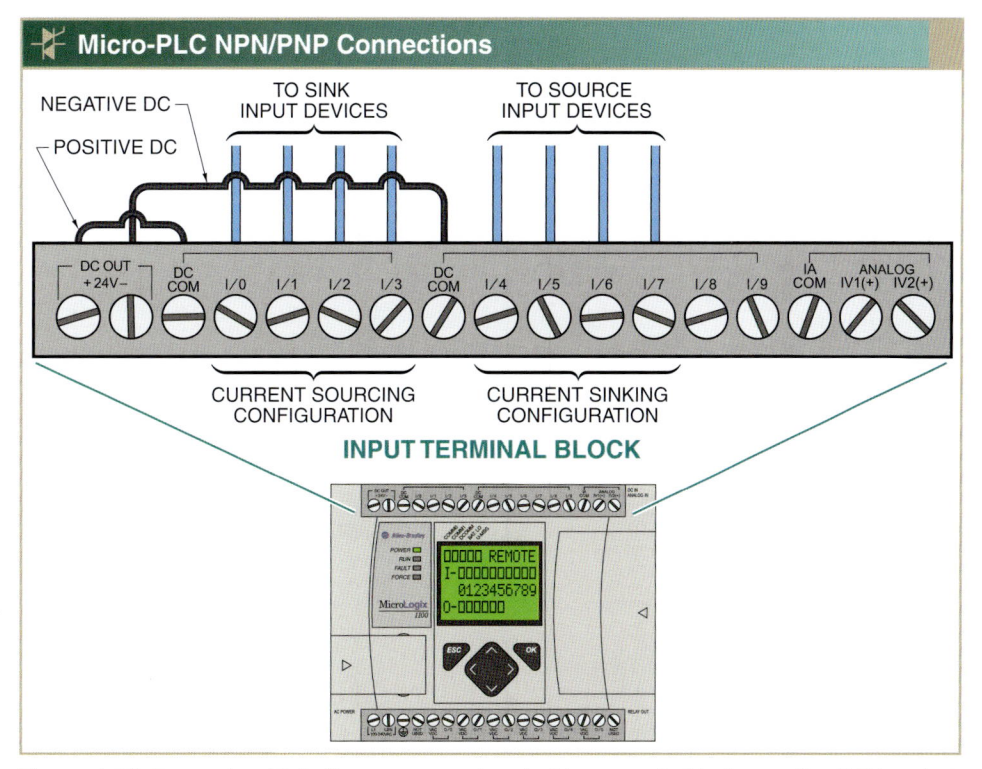

Figure 1-20. On a micro-PLC, if one common terminal is connected to the positive DC terminal, then the corresponding inputs are the required current sinking devices. Likewise, if one common terminal is connected to the negative DC terminal, then the corresponding inputs are the required current sourcing devices.

For medium/large PLCs, the input and output modules are designed for several different input and output types, such as DC positive and negative, AC, analog, digital, source, and sink modules. Input sink/source devices connect to each module type the same as they would on a nano- or micro-PLC based on terminal designations. **See Figure 1-21.**

A sourcing input module connects positive VDC to the PLCs inputs. A sinking input module connects negative VDC to the PLCs inputs. For output components, a sourcing output module supplies positive VDC to power the field output component. Likewise, a sinking output module supplies negative VDC to power the field output component.

Rack-Mounted Sinking and Sourcing Connections

24 VDC SOURCING INPUT MODULE

24 VDC SINKING INPUT MODULE

24 VDC SOURCING OUTPUT MODULE

24 VDC SINKING OUTPUT MODULE

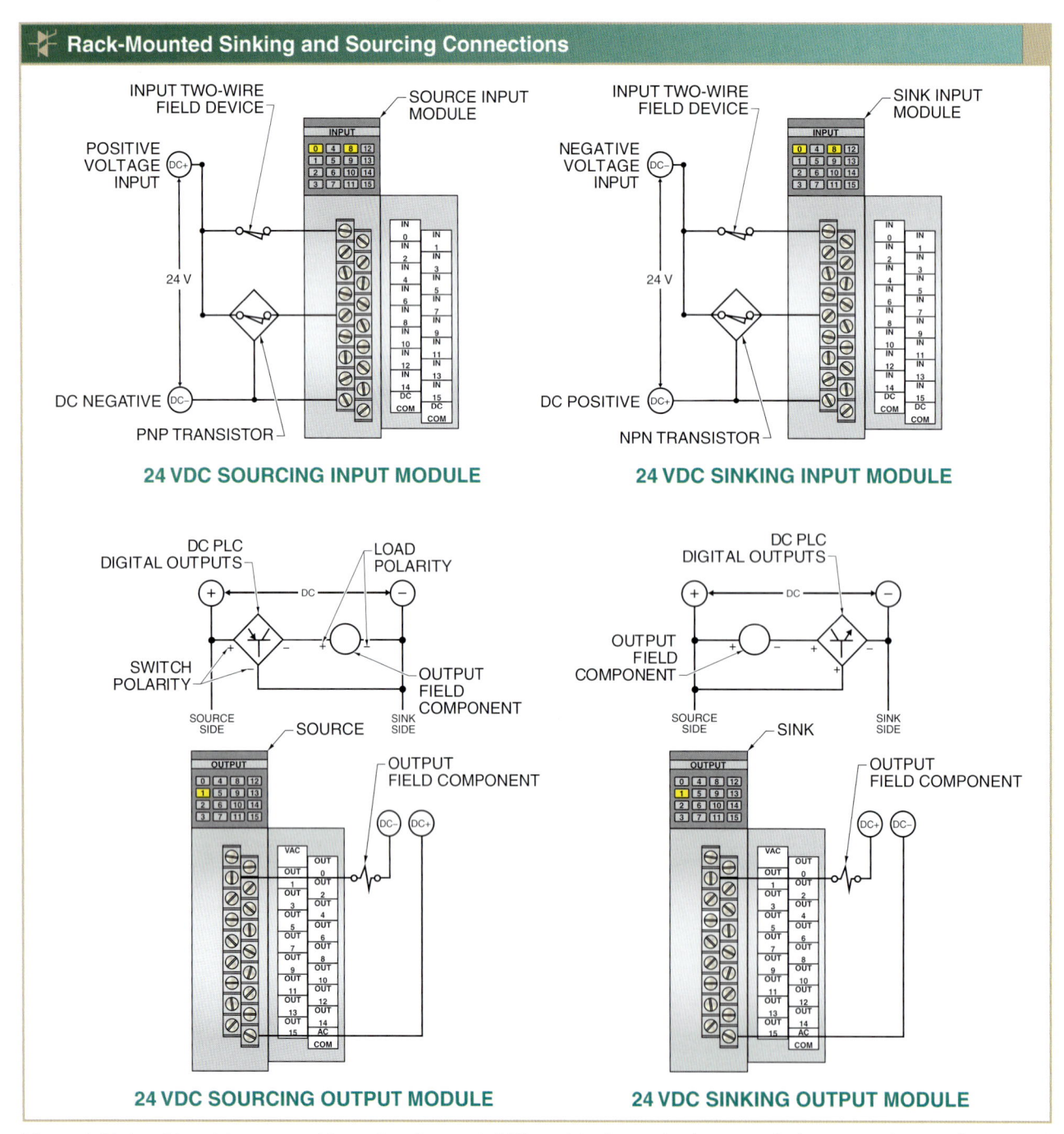

Figure 1-21. Input transistors connect to each rack-mounted I/O module. A sourcing input terminal connects the negative terminal and the collector of a PNP transistor. A sinking input terminal connects the positive terminal and the collector of an NPN transistor.

DIGITAL INPUT DEVICES

A digital input device sends signals to indicate the condition of a system to a PLC input section. Types of digital input devices used in PLC systems include limit switches, photoelectric switches, proximity switches, pushbuttons, and selector switches.

Limit switches are either NO or NC contacts and are activated by a roller operator. A photoelectric switch emits a light beam that is reflected back to the switch. The photoelectric switch is a transistor and is connected as a sourcing or sinking input. A proximity switch uses a magnetic field to sense whether an object is within a specified distance from the switch. A proximity switch can be connected as a sourcing input. Pushbuttons are either NO or NC momentary-actuated switches. Typically, green pushbuttons are NO, whereas red pushbuttons are NC. A selector switch can be used for three-position conditions, such as an OFF position, a normally closed manual (MAN) position, and a normally open automatic (AUTO) position. **See Figure 1-22.**

ANALOG INPUT DEVICES

An analog input device is a type of sensor that sends continuous signals to a PLC input section based on the conditions of the PLC system. Analog inputs can send a continuous signal to a PLC of a specified voltage or current range, such as 0 VDC to 10 VDC or 4 mA to 20 mA. Analog signals can represent environmental conditions such as temperature, humidity, pressure, flow rate, level, distance, and viscosity. PLCs use an internal converter to convert an analog signal into a digital representation of a voltage or current value.

Types of analog input devices included in PLC systems are solar cells and relative humidity/temperature sensors. Two types of solar cells can be a large solar cell, such as a 0 V to 6 V cell, or small solar cell, such as a 0 V to 3 V cell. A relative humidity sensor (or relative humidity transmitter) can send a 0 V to 10 V output signal to a PLC input section to relay the amount of humidity in the air. **See Figure 1-23.**

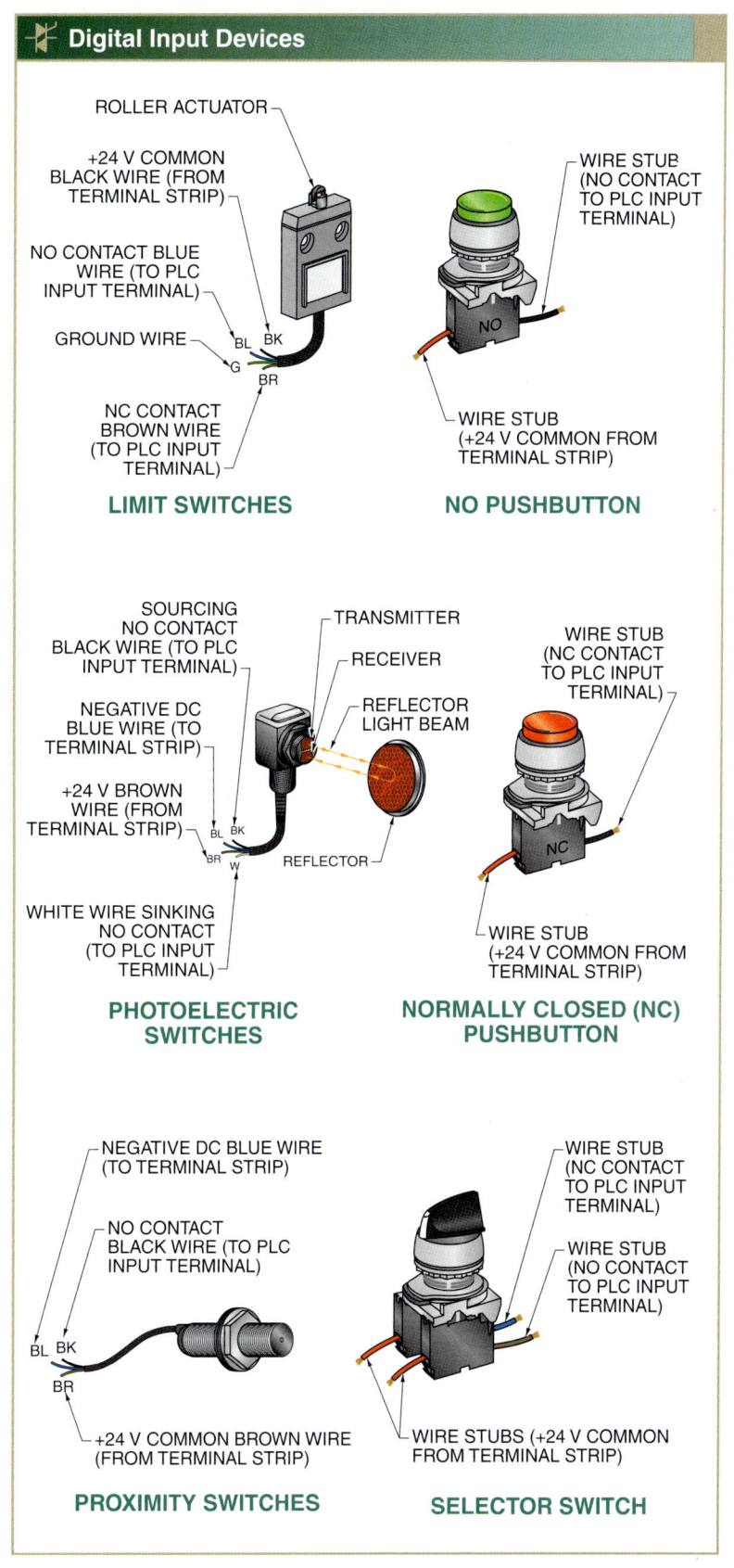

Figure 1-22. Types of digital input devices used in PLC systems include limit switches, photoelectric switches, proximity switches, pushbuttons, and selector switches.

CODE CONNECT

Article 690 and Article 691 of the NEC® cover photovoltaic systems and large-scale photovoltaic electric supply stations. Article 690 applies to photovoltaic systems other than those covered by Article 691. Article 691 covers large photovoltaic systems with a minimum size of 5000 kW (5 MW) that are not controlled by a utility.

Analog Input Devices

SOLAR CELLS

- MAX. 6 V OUTPUT SIGNAL
- BLACK WIRE TO COMMON
- RED WIRE 6 V OUT
- MAX. 3 V OUTPUT SIGNAL
- BLACK WIRE TO COMMON
- RED WIRE 3 V OUT

HUMIDITY/TEMPERATURE SENSORS

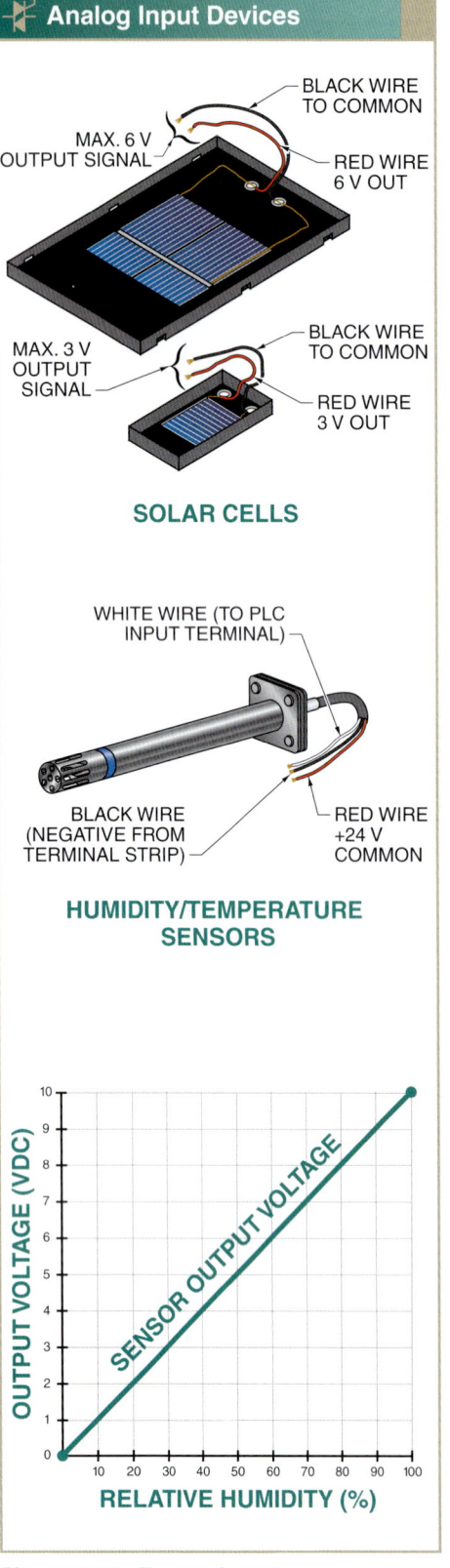

- WHITE WIRE (TO PLC INPUT TERMINAL)
- BLACK WIRE (NEGATIVE FROM TERMINAL STRIP)
- RED WIRE +24 V COMMON

SENSOR OUTPUT VOLTAGE (graph: OUTPUT VOLTAGE (VDC) vs. RELATIVE HUMIDITY (%))

Figure 1-23. Types of analog input devices included in PLC systems are solar cells and relative humidity/temperature sensors.

OUTPUT COMPONENTS

The output components (loads) controlled by a PLC can initiate or adjust system operations. Types of output components include signaling towers, horns/alarms, fans, and contactors.

Signaling towers consist of three or more lights (green, yellow, and red) that are individually controlled through a PLC and operate with 24 VDC. Horns/alarms are audible devices that vary in tones depending on the level of VDC sent by a PLC. Fan blades are capable of producing airflow and rely on a DC motor to function. Contactors are heavy-duty, three-phase industrial power control devices with a fourth set of line connections that are typically used as auxiliary contacts. PLCs connect to contactors by a 24 VDC activation coil, similar to that of a motor starter. **See Figure 1-24.**

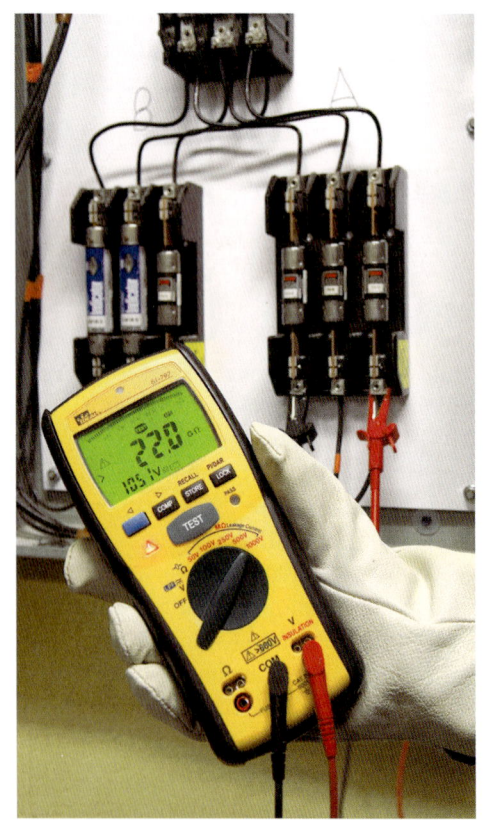

Ideal Industries, Inc

Personal protective equipment (PPE), such as insulating gloves, must be worn while testing power control circuits for proper voltage measurements in order to protect the wearer from electrical shock.

Output Components

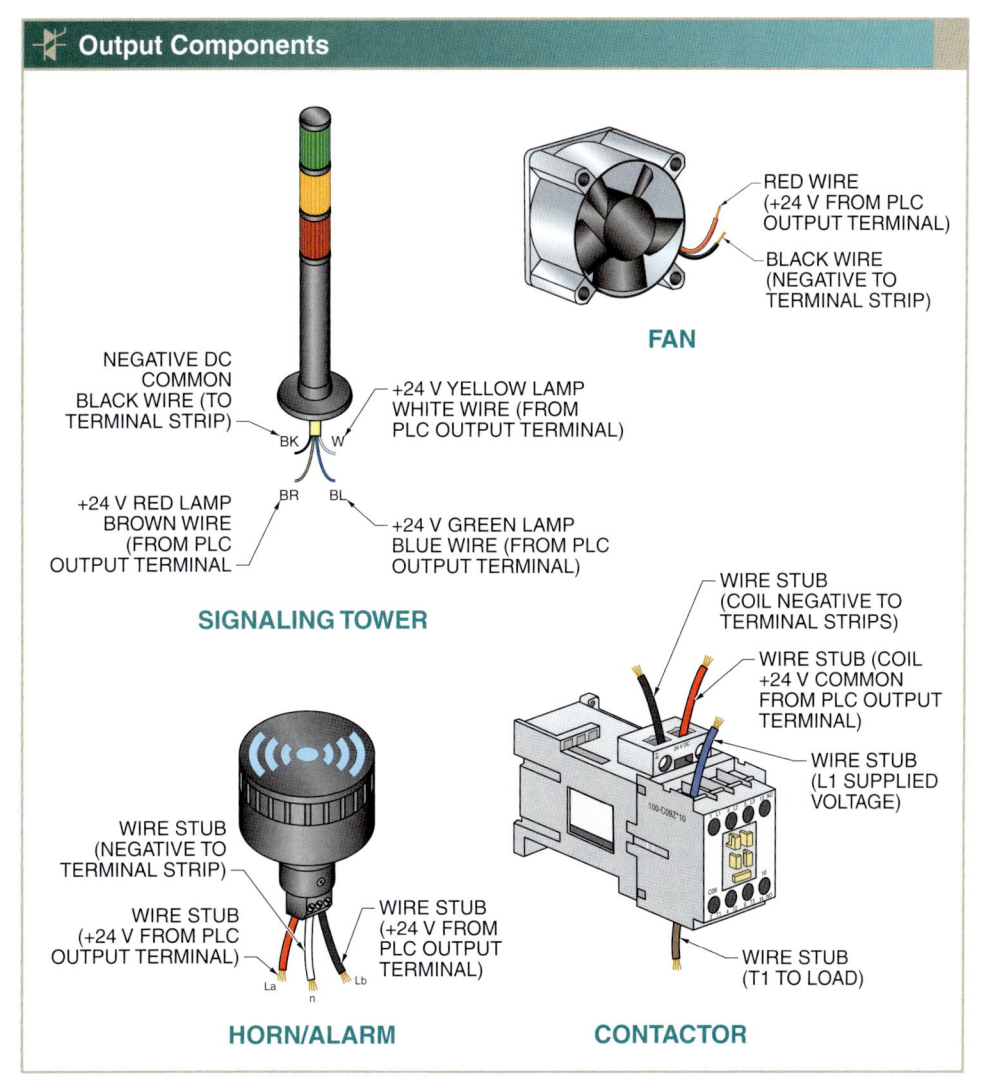

Figure 1-24. Types of output components include signaling towers, horns/alarms, fans, and contactors.

PLC SAFETY CONSIDERATIONS

Any system that includes a PLC can cause an electrical shock. PLC safety considerations include the following:

- All PLCs are considered "open type" electrical equipment and must be mounted in an enclosure.
- Any PLC can overheat and become a fire hazard.
- Any PLC can be programmed with force and disable functions, which if used incorrectly can cause injuries to persons working around a system and/or equipment damage.

> **PLC TIPS**
>
> *PLCs must be installed in an enclosure or cabinet for protection from the outside environment. All cables connected to a PLC must remain in the enclosure.*

Because PLCs are a part of an electrical system, PLCs can be an electrical hazard to anyone installing, working around, or servicing them. In addition to potentially producing an electrical shock to anyone coming in contact with the electrical parts of a PLC system, improperly installed and/or maintained PLCs can

---

CODE CONNECT

Section 90.1 of the NEC® covers the Code's purpose. The purpose of the NEC® is to protect personnel and property from hazards associated with the use of electricity. The NEC® is not intended as a design manual or a set of instructions for untrained persons. Adherence to the NEC® and proper maintenance result in an installation that is basically safe but may not be convenient or adequate. The NEC® is aligned with International Safety Standards.

also become a fire hazard or cause an explosion. **See Figure 1-25.** Technicians must know the proper safety procedures and practices to follow when working on a PLC-controlled (or any) electrical system. Safety procedures and practices include what type of personal protective equipment (PPE) to wear, what types of tools and test instruments to use, and all National Fire Protection Association 70E® safety rules and National Electrical Code® installation rules that must be followed.

When properly installed and programmed, a PLC can control an electrical system to safely and efficiently operate a process. A major advantage of a PLC controlled process is that a PLC can be programmed and reprogrammed as process conditions change. **See Figure 1-26.** Another advantage is that when servicing an electrical control system, most PLCs include force and disable commands that, when properly used, aid in process startups and troubleshooting. When misapplied, force and disable commands can present a serious safety problem in any part of the electrical or mechanical systems controlled by a PLC.

PLC Hazards

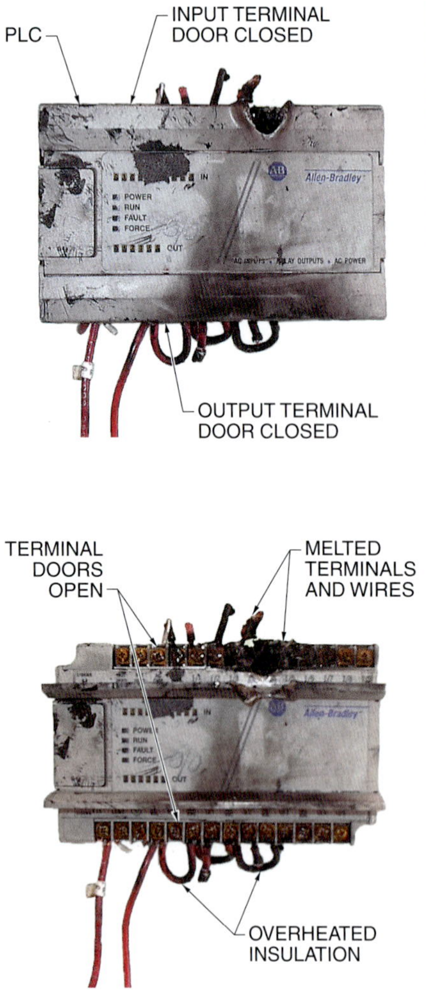

Figure 1-25. Improperly installed and/or maintained PLCs can overheat, leading to fire or explosion.

PLC TIPS

Specialty input and output modules for rack-mounted PLCs can include weight, temperature, and communications modules for specific process conditions in electrical systems.

National Fire Protection Association®

The *National Fire Protection Association® (NFPA®)* is a national organization that provides guidance in assessing the hazards of products of combustion. The NFPA® sponsors the development of *NFPA 70®: National Electrical Code® (NEC®)* and *NFPA 70E®: Standard for Electrical Safety in the Workplace.*

The *National Electrical Code® (NEC®)* is a code book of electrical standards that indicates how electrical systems must be installed and how work must be performed. **See Figure 1-27.** The purpose of the NEC® is to protect electrical workers and equipment from electrical hazards.

The NFPA 70E® is a set of voluntary standards for electrical, safety-related work practices. It covers specific work practices that are required for electrical workers, both qualified and unqualified. A *qualified person* is an individual with the necessary education and training who is familiar with the construction and operation of electrical equipment and devices.

Chapter 1—Programmable Logic Controllers and Critical Safety Practices   25

PLC Advantage

PROCESS TO RECEIVE PRODUCT CHANGEOVER

Siemens

MECHANICAL CONTROL CABINET — HARDWIRED RELAYS MUST BE REPLACED AND CABINET REWIRED FOR PRODUCT CHANGEOVER

PLC CONTROL CABINET — PLC PROGRAMMED FOR PRODUCT CHANGEOVER

⚠ WARNING
Any change to the program of a PLC can dramatically change the operation of a process or machine.

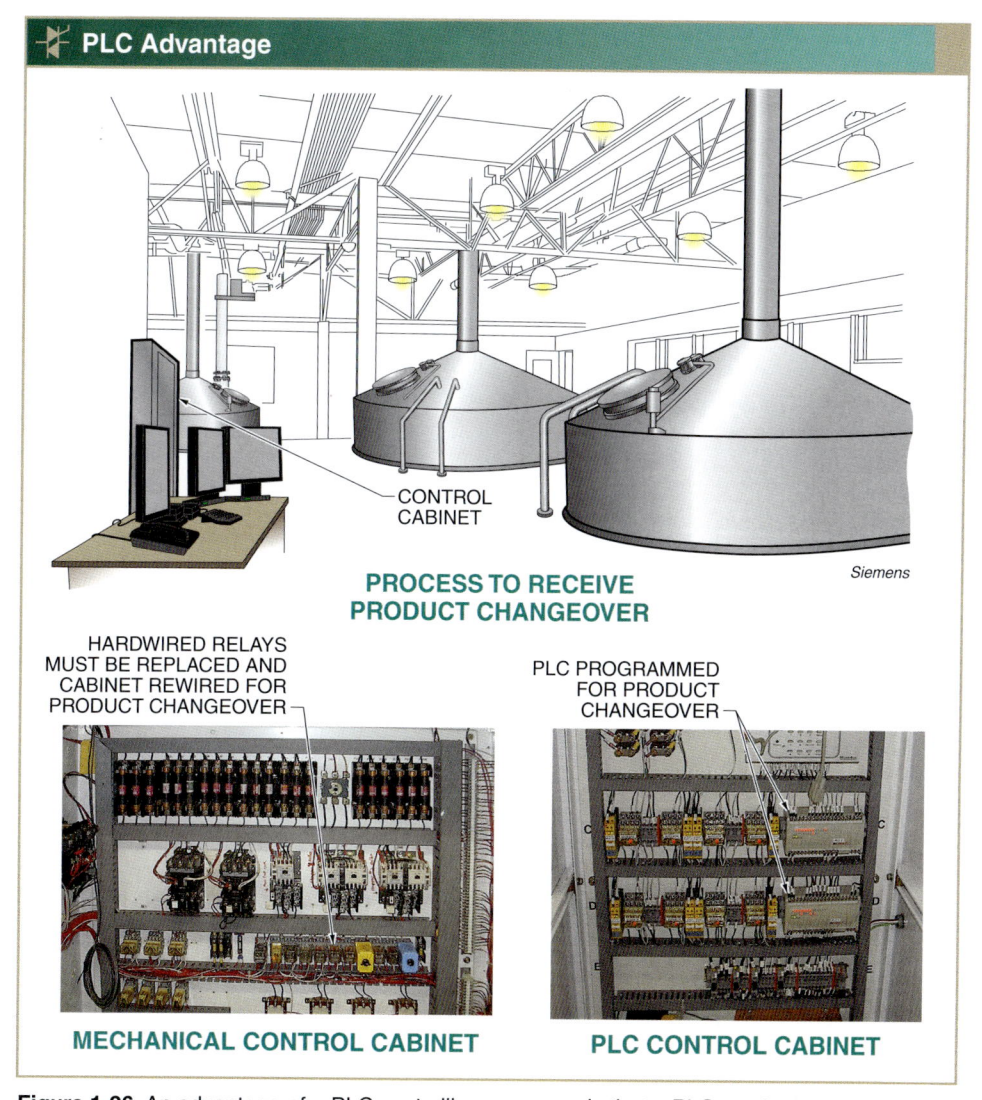

Figure 1-26. An advantage of a PLC controlling a process is that a PLC can be programmed and reprogrammed as process conditions change.

National Electrical Code® (NEC®)

Section	Description
Article 90	Introduction
Chapter 1	General
Chapter 2	Wiring and Protection
Chapter 3	Wiring Methods and Materials
Chapter 4	Equipment for General Use
Chapter 5	Special Occupancies
Chapter 6	Special Equipment
Chapter 7	Special Conditions
Chapter 8	Communication Systems
Chapter 9	Tables

CODE BOOK — NFPA 70 National Electrical Code 2020

NFPA®
1 Batterymarch Park
Quincy, MA 02169

CODE CONNECT
Section 760.1 of the NEC® defines the scope of Article 760, Fire Alarm Systems. The scope includes not only the installation of a fire alarm system but also the circuits controlled and interconnected to a fire alarm system. Circuits controlled and interconnected to a fire alarm system may include elevator recall, fire doors, door release, damper control, smoke control, and fan control.

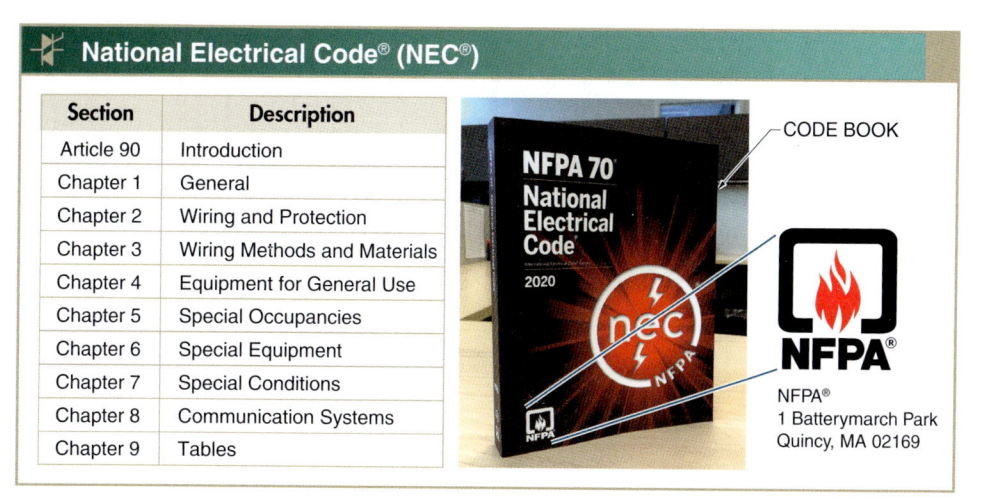

Figure 1-27. The NEC® is a code book of electrical standards that indicates how electrical systems must be installed and how work must be performed.

Force and Disable Safety Considerations

A major safety issue that must be considered when working with a PLC-controlled electrical system is that the PLC program can include a force command. A *force command* is a special software override that simulates opening or closing an input device or turns an output component ON or OFF. Force commands are designed for use when troubleshooting a system. Forcing an input or output device ON or OFF allows for checking an electrical circuit with software assistance. **See Figure 1-28.**

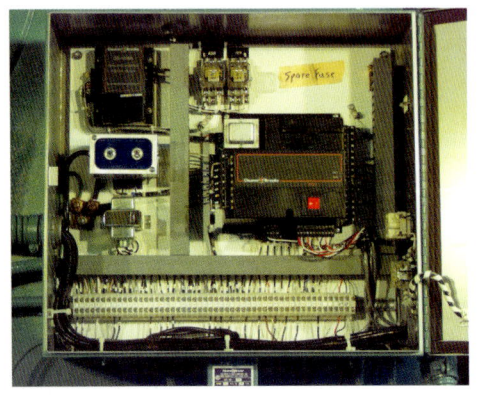

PLCs are mounted in enclosures with terminal blocks, relays, and control and power connections, which allows input and output connections to be tested easily.

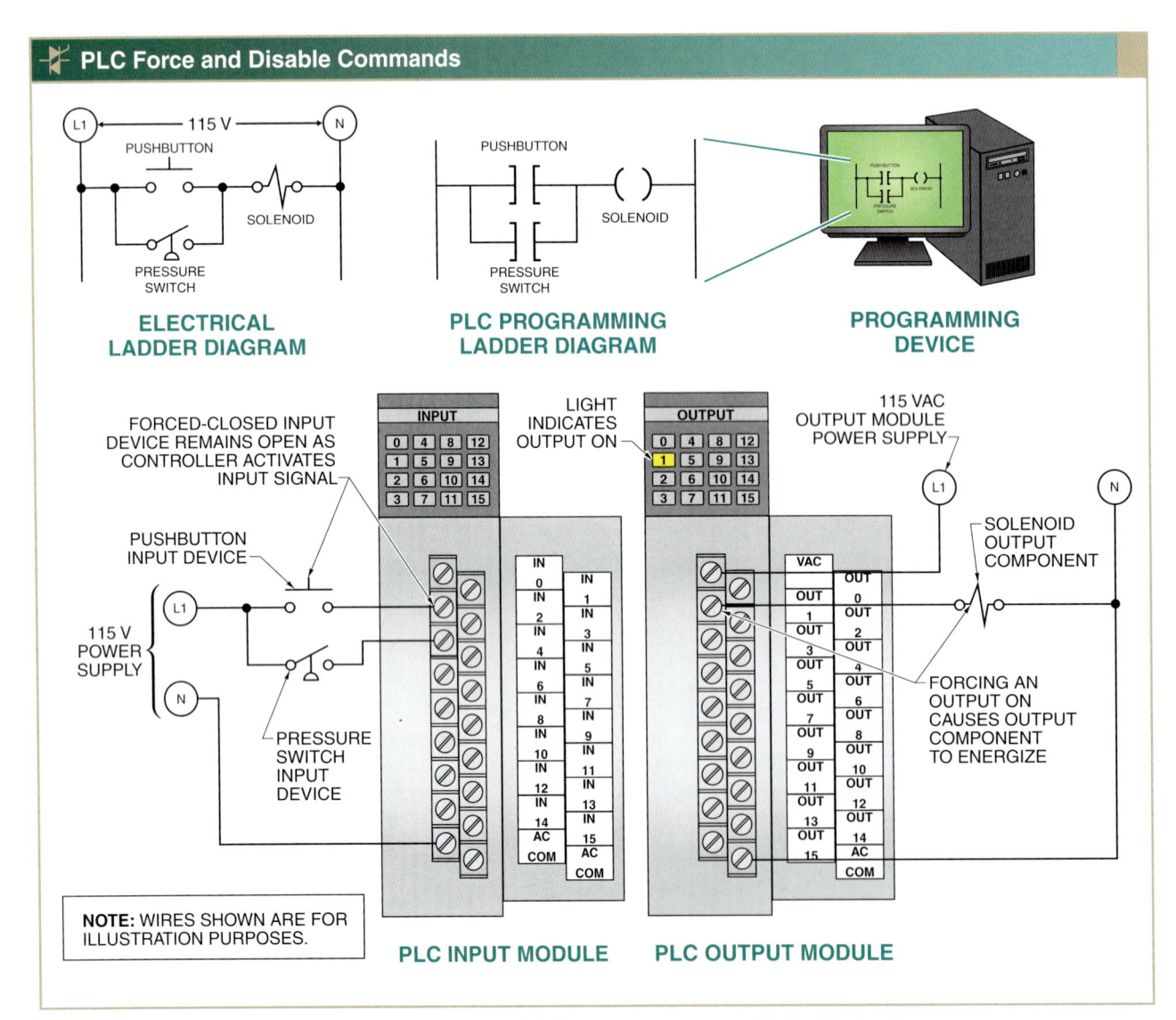

Figure 1-28. PLC force and disable commands are used during system startup and for troubleshooting.

An input device can be forced into an ON condition to test output circuit operation. Forcing an input signal can be used when service is required on a defective input device. If the input device is not critical to the PLC program for facility production, the signal from the defective input device can be forced ON until the device is repaired. The force ON command is removed once the input device is repaired.

An output device turns ON regardless of the programmed logic when the force ON command is used. An output device remains ON until the force command is removed. Extra care must be taken when using force commands because force commands override all safety features and protocols designed into a program.

A *disable command* is a special software override that prevents one or all input or output devices from operating. Disable commands are the opposite of force commands. Care must be taken when using a disable command so that no input device or output component that is critical for a safe system is disabled unless absolutely necessary. Technicians must ensure that all force and disable commands are removed from a PLC before returning a system to normal operation.

Electrical Shock

According to a recent OSHA study, 136 electrical fatalities occurred in the year 2017. Making contact with or exposure to electricity is the sixth most common cause of workplace fatalities. Safe work habits are required when working on or around any electrical devices, components, or PLC systems.

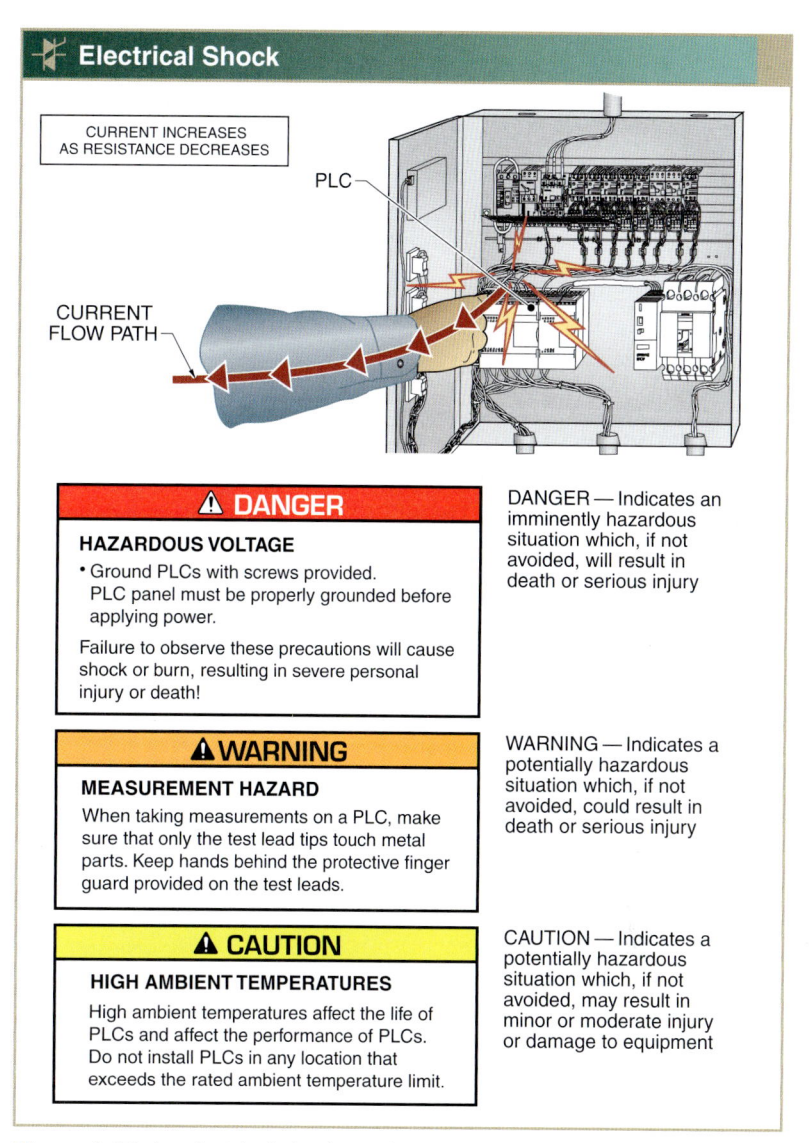

An electrical shock results any time a body becomes part of an electrical circuit. **See Figure 1-29.** Electrical shock varies from a mild shock to fatal current. The severity of an electrical shock depends on the amount of electric current (in mA) that flows through the body, the length of time the body is exposed to the current flow, the path the current takes through the body, the physical size and condition of the body through which the current passes, and the amount of body area exposed to the electric contact. Safety labels are present on electrical cabinets and in electrical systems. They appear as "Danger," "Warning," or "Caution" labels with a description of the hazard involved.

Figure 1-29. An electrical shock results anytime a body becomes part of an electrical circuit.

The amount of current that passes through a body or circuit depends on the voltage and resistance of the completed electrical circuit. During an electrical shock, the body of a person becomes part of an electrical circuit. The resistance a person's body offers to the flow of current varies. Sweaty hands have less resistance than dry hands. A wet floor has less resistance than a dry floor. The lower the resistance, the greater the current flow and greater the severity of shock. **See Figure 1-30.**

Electrical Shock Effects

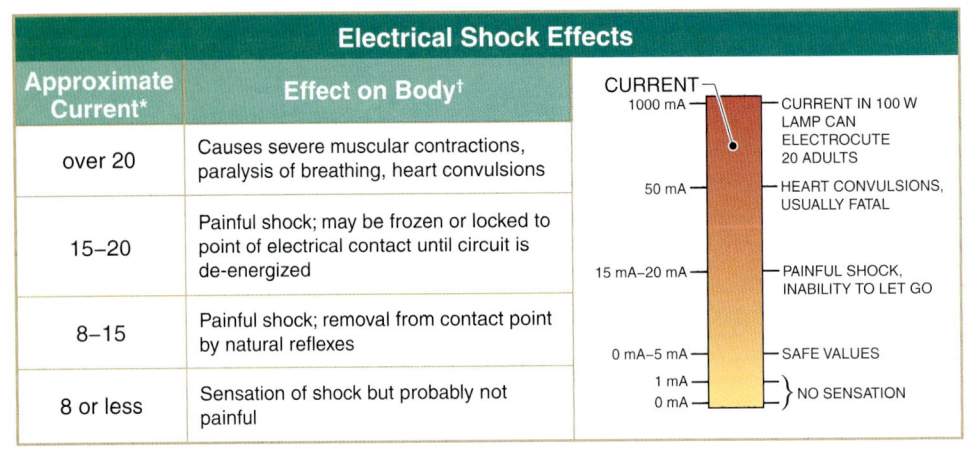

Approximate Current*	Effect on Body†
over 20	Causes severe muscular contractions, paralysis of breathing, heart convulsions
15–20	Painful shock; may be frozen or locked to point of electrical contact until circuit is de-energized
8–15	Painful shock; removal from contact point by natural reflexes
8 or less	Sensation of shock but probably not painful

CURRENT
- 1000 mA — CURRENT IN 100 W LAMP CAN ELECTROCUTE 20 ADULTS
- 50 mA — HEART CONVULSIONS, USUALLY FATAL
- 15 mA–20 mA — PAINFUL SHOCK, INABILITY TO LET GO
- 0 mA–5 mA — SAFE VALUES
- 1 mA
- 0 mA — NO SENSATION

* in mA
† effects vary depending on time, path, amount of exposure, and condition of body

Figure 1-30. Possible effects of electrical shock include the heart and lungs ceasing to function and/or severe burns where the electricity (current) enters and exits the body.

Arc-Flash Boundaries

The NFPA® specifies an arc-flash boundary and two shock protection boundaries. All qualified persons working within the arc-flash boundary are required to use arc-rated protective clothing and equipment. Shock protection boundary distances vary depending on the voltage involved. **See Figure 1-31.**

Electrical work must be performed in compliance with the NEC® and NFPA 70E. The NEC® and NFPA 70E state that technicians must receive safety training that includes the following:

- how to calculate the degree of arc-flash hazard
- how to choose the proper PPE for an application
- which tools to use for a task while being safe
- proper use of arc-flash labels

Approach Boundaries

- ARC-FLASH BOUNDARY
- LIMITED APPROACH BOUNDARY
- RESTRICTED APPROACH BOUNDARY
- RESTRICTED SPACE
- LIMITED SPACE
- DOOR OPEN ON ELECTRICAL CABINET

Voltage*	Limited	Restricted
0 to 50	N/A	N/A
51 to 150	3'-6"	Avoid contact
151 to 750	3'-6"	1'-0"
751 to 15,000	5'-0"	2'-2"

* measured phase to phase

Figure 1-31. The NFPA specifies boundary distances that vary depending on voltage.

ELECTRICAL PROPERTIES

Electricity has three basic properties: current, voltage, and resistance. Ohm's law states that in an electrical circuit voltage is equal to resistance multiplied by current.

Current

Current (I) is the rate at which electrons flow through a conductor. Current is measured in amperes (A). **See Figure 1-32.** An *ampere* is one coulomb passing a given point in an electrical circuit in one second. A *coulomb* is the practical unit of measurement for a specific quantity of electrons.

Current may be direct current or alternating current. *Direct current (DC)* is current that flows in only one direction. *Alternating current (AC)* is current that reverses direction of flow at regular intervals.

PLCs are used to control electrical circuits that have DC, AC, or AC and DC current. Many PLC-controlled systems include DC and AC.

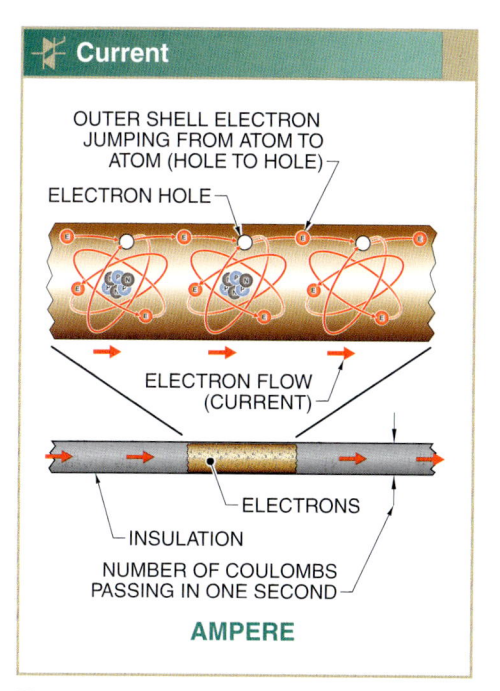

Figure 1-32. Current is the amount of electrons flowing through an electrical circuit and is measured in amperes.

Voltage

Voltage (E) is the electromotive force or pressure in a conductor that allows electrons to flow. Voltage is measured in volts (V). **See Figure 1-33.** Per NFPA 70E®, voltages greater than 50 VAC can cause an electrical shock. To aid in the prevention of electrical shock, the input side of a PLC typically has input terminals that are rated at 24 V or less. The input devices that send low-voltage signals to a PLC include temperature switches, photoelectric/proximity switches, pushbuttons, and pressure switches.

However, unlike the input side of a PLC (low-voltage levels), the output side typically has high-voltage levels (above 50 VAC). The output side of a PLC controls electrical loads such as lamps, solenoids, and motor starters that typically operate at higher voltages. Because power (P) measured in watts is equal to voltage (E) multiplied by current (I), higher voltages allow the amount of current required to be lower for a given power rating. For example, a 120 W lamp rated at 120 V would draw 1 A (120 V × 1 A = 120 W). A 120 W lamp rated at 12 V would draw 10 A (12 V × 10 A = 120 W).

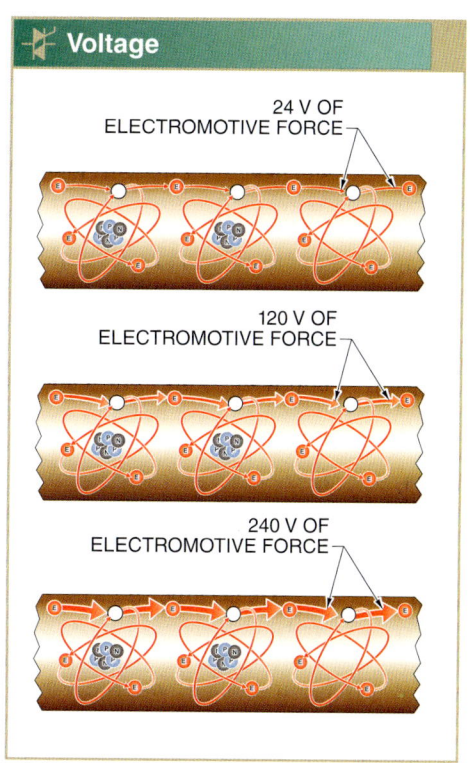

Figure 1-33. Voltage is the amount of electromotive force in a circuit and is measured in volts.

Resistance

Resistance (R) is the opposition to current flow. Resistance is measured in ohms, which is represented by the Greek letter omega (Ω). **See Figure 1-34.** The rule of thumb to remember with resistance is the higher the resistance, the lower the current flow and the lower the resistance, the higher the current flow. Resistance is why technicians must wear insulating electrical gloves and use insulated tools and test instruments when working around a PLC or any energized electrical system. Technicians must remember that the greater the insulation (resistance) between a technician and an electrical circuit, the less the chance of electrical shock.

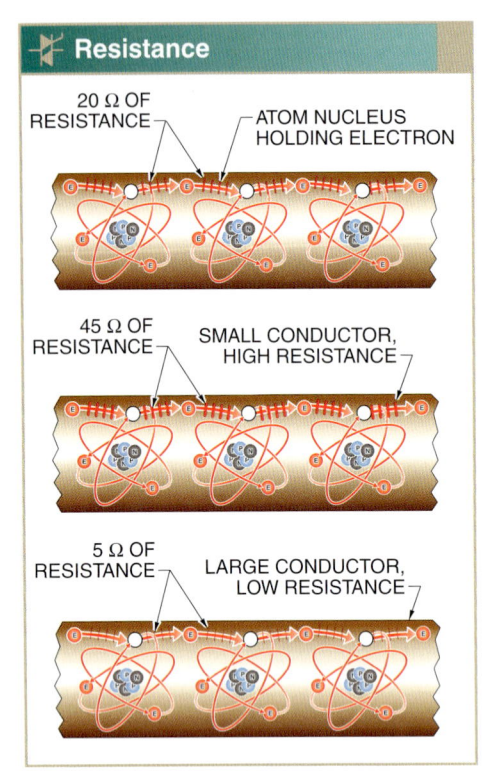

30 PROGRAMMABLE LOGIC CONTROLLERS: PRINCIPLES AND APPLICATIONS

Figure 1-34. Resistance is the opposition to the flow of electrons and is measured in ohms.

GROUNDING SYSTEMS

Electrical circuits are grounded to safeguard equipment and technicians against the hazards of electrical shock. Proper grounding of electrical tools, machines, equipment, and conveying systems is one of the most important factors in preventing hazardous electrical conditions.

Grounding is the connection of all exposed non-current-carrying metal parts to earth. Grounding provides a direct path for unwanted fault current to travel to earth without causing harm to technicians or equipment. Grounding in an electrical system is accomplished by connecting energized systems to grounding elecrode conductors (GECs), such as metal underground pipes, metal frames of buildings, concrete-encased electrodes, or grounding rings/electrodes. **See Figure 1-35.**

⚠ WARNING

Before taking resistance measurements, a circuit must be de-energized or the meter could explode.

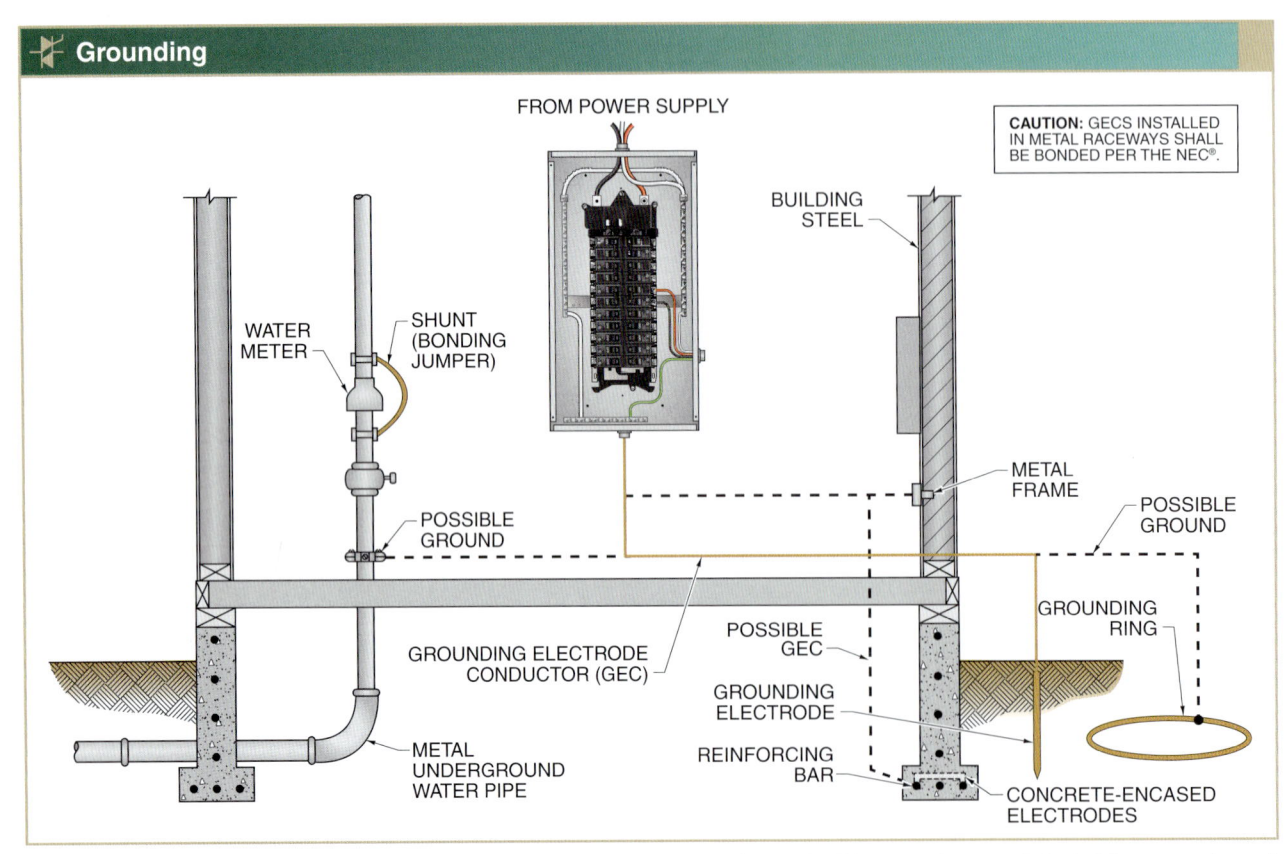

Figure 1-35. Grounding provides a direct path for unwanted fault current to travel to earth without causing harm to technicians or equipment.

The three categories of facility grounding are building grounding, equipment grounding, and electronic equipment grounding. Each grounding category has a different purpose and, when combined, a safe and effective grounding system for technicians and equipment is provided. **See Figure 1-36.**

⚠ WARNING
Neutral-to-ground connections must not be made in any subpanels or equipment. If a neutral-to-ground connection is made, a parallel path for the normal return current from loads is created. This causes current to flow through the ground system and potentially create a dangerous shock situation.

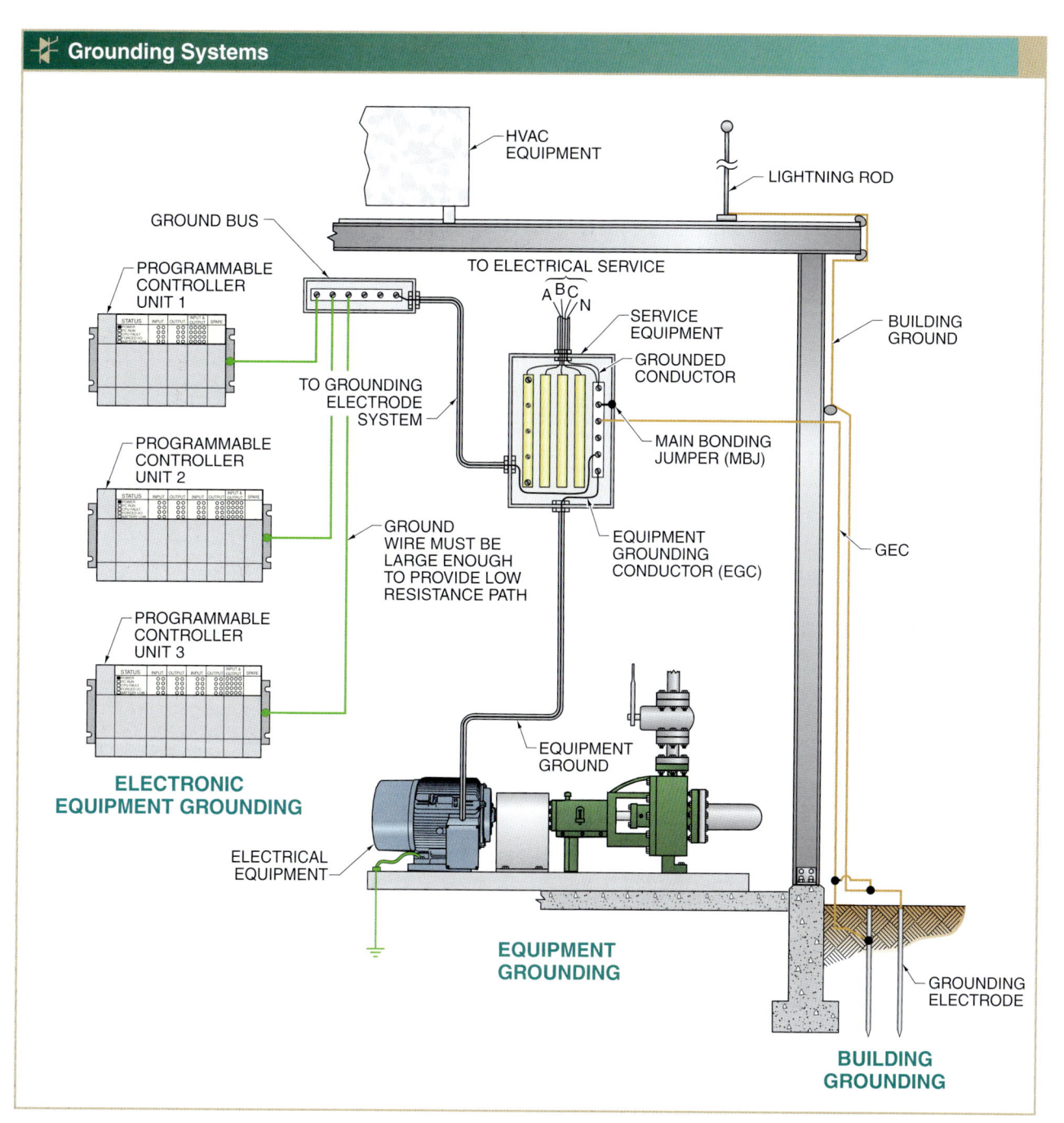

Figure 1-36. Building grounding, equipment grounding, and electronic equipment grounding are used to create a safe working environment for technicians.

CODE CONNECT

Article 708 of the NEC® covers critical operations power systems (COPSs). A COPS is a power system for a designated critical operations area (DCOA) that requires continuous operation, such as public safety, disaster response, or essential business operations. Article 708 covers general information, power sources, overcurrent protection, system performance, and circuit wiring and equipment. COPSs are designed, installed, and maintained to a higher standard to ensure their operation is not disrupted.

Building Grounding

Building grounding is the connection of an electrical system to earth ground through a GEC to grounding electrodes, the metal frame of a building, concrete-encased electrodes, or underground metal water pipes. Building grounding ensures that there is a low impedance (low resistance) grounding path for fault current (short or lightning) to earth ground. A *low impedance ground* is a grounding path that contains very little resistance to the flow of fault current to ground. **See Figure 1-37.**

Building grounding includes protecting the building or outside structures from lightning strikes by providing a path to earth. A lightning rod grounding system must also have low resistance because of the high current requirements created by lightning.

Equipment Grounding

Equipment grounding is the connection of machinery electrical systems to earth ground to reduce the chance of electrical shock by grounding all non-current-carrying exposed metal. The most important reason for equipment grounding is to prevent electrical shock when a person comes in contact with electrical equipment or exposed metal of machinery. **See Figure 1-38.**

Non-current-carrying metal parts that are connected to a grounding system include motors, metal boxes, raceways, enclosures, metal equipment parts (including any metal parts of a PLC), and any metal a technician might touch while working on an electrical circuit. A fault current can exist because of insulation failure or because a current-carrying wire makes direct contact with a non-current-carrying metal part of a system.

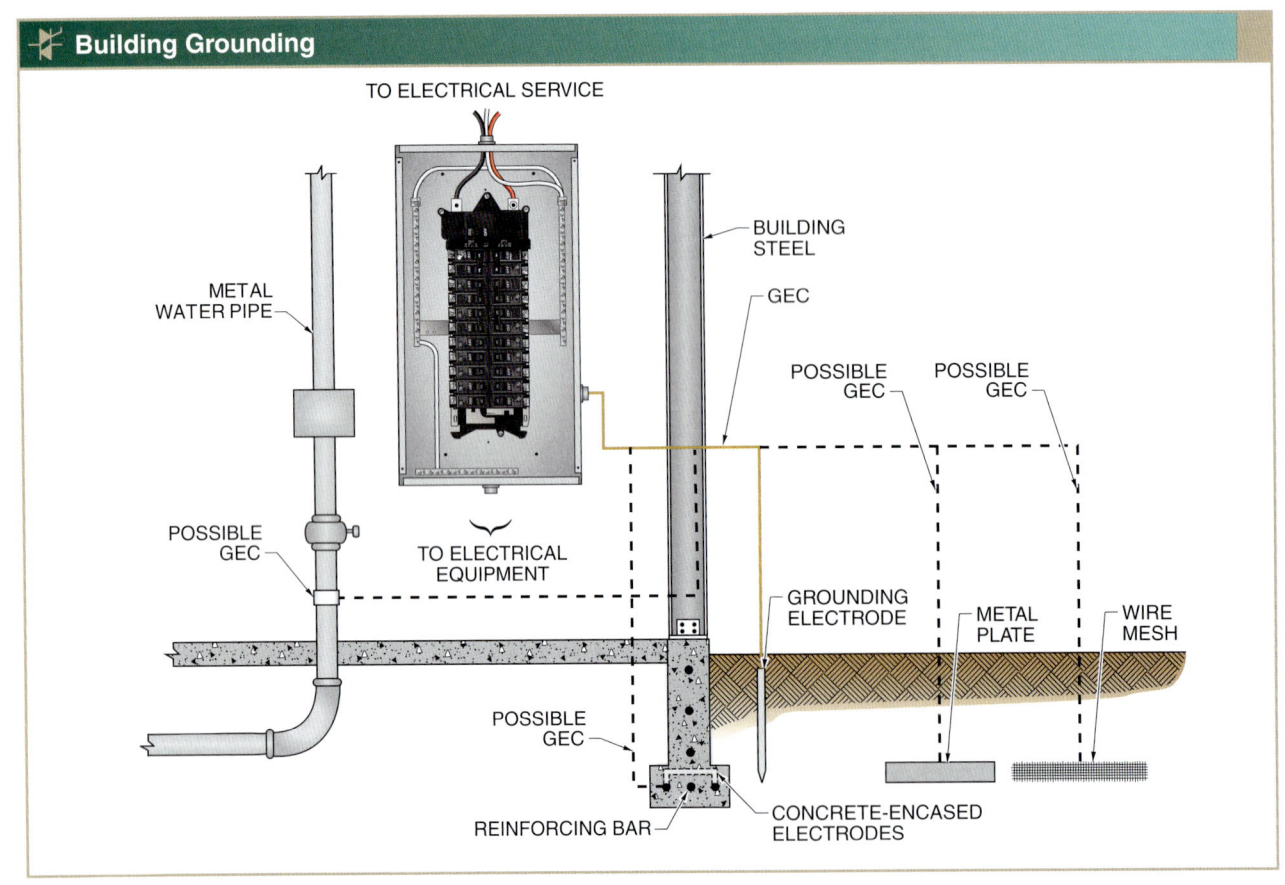

Figure 1-37. Building grounding ensures that there is a low impedance (low resistance) grounding path for fault current (electrical short of lightning) to earth ground.

Equipment grounding also aids in preventing electrical shocks from static electricity and static buildup in equipment. Static electricity can also cause fires and explosions when allowed to accumulate.

Electronic Equipment Grounding

Electronic equipment grounding is the connection of electronic equipment, such as PLCs, to earth ground to reduce the chance of electrical shock through grounding the equipment and all non-current-carrying exposed metal parts. Although electronic equipment grounding is basically the same as equipment grounding, electronic equipment grounding is used to provide a quality ground for electronic systems to reduce electrical noise with PLCs, process control equipment, and other system operations. **See Figure 1-39.**

A quality earth ground used for electronic grounding reduces static electrical charges, which allows signal integrity to be maintained in sensitive electronic equipment such as PLCs, and other electronic control systems. Signal reliability is sometimes difficult to maintain in a PLC system where signals from input devices (as low as 3 V) are routed near power conductors.

Equipment Grounding

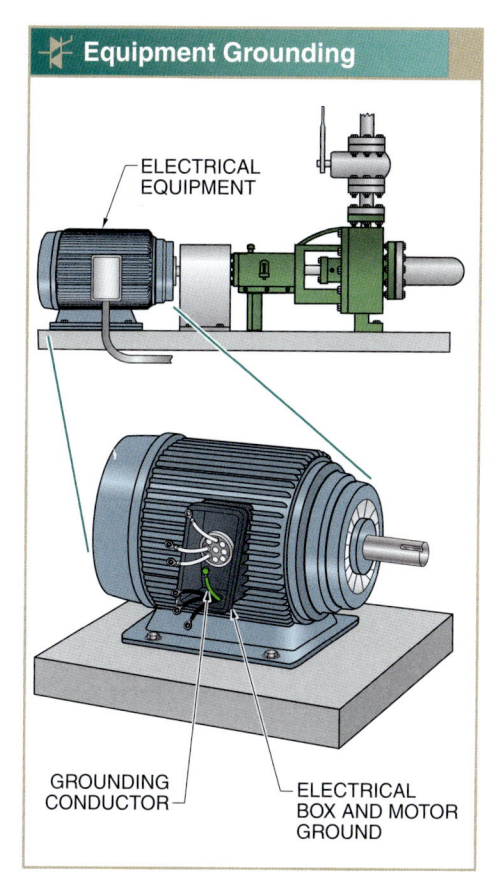

Figure 1-38. Equipment grounding prevents electrical shock when a person comes in contact with electrical equipment or exposed metal of machinery.

Electronic Equipment Grounding

Figure 1-39. Electronic equipment grounding is used to provide a quality ground for electronic systems to enable better communication (less noise) with PLCs, process control equipment, and other facility operations.

> **⚠ CAUTION**
> When chassis tabs of a PLC do not lay flat, use washers as shims before tightening screws so chassis does not warp. Warping the chassis can damage the backplane and interfere with quality grounding connections.

> **CODE CONNECT**
> Section 250.12 of the NEC® addresses clean surfaces on equipment to be grounded and requires the paint or other protective coatings be removed from surfaces or threads to ensure good metal-to-metal contact for electrical continuity. Section 250.12 also permits fittings to be used that are designed to cut through protective coating. When connecting a lug for grounding to a painted back panel in a control cabinet, the paint must be removed/scraped away. When installing a pipe connector on a painted enclosure, a locknut designed to cut through the paint must be used.

> **CODE CONNECT**
>
> *Per article 250 of the NEC®, grounding conductors must be permanently attached to the grounding electrode with bolts or other listed means. Soldering is not permitted.*

The difference between building grounding, equipment grounding, and electronic equipment grounding is that building and equipment grounding are used to prevent electrical shocks and fires. The primary use of building and equipment grounding is to remove fault currents as fast as possible by using fuses or circuit breakers to disconnect power. The NEC® states that building and equipment grounding systems using single-made electrodes must have a resistance of less than 25 Ω. **See Figure 1-40.**

Electronic equipment grounding is primarily used for electronic noise reduction by eliminating noise and other unwanted signal interference. Unwanted current that is removed to ground by electronic grounding systems is typically measured in milliamps (mA) and continues to flow as long as the electronic equipment is connected to a power source. Manufacturers of PLCs and other electronic equipment often specify a grounding system with a resistance of 5 Ω, 3 Ω, 1 Ω, or less. Low-resistance electronic grounding can be accomplished by connecting a PLC to a large grounding conductor that has a low resistance. Large grounding conductors provide quality electronic grounding and good equipment grounding.

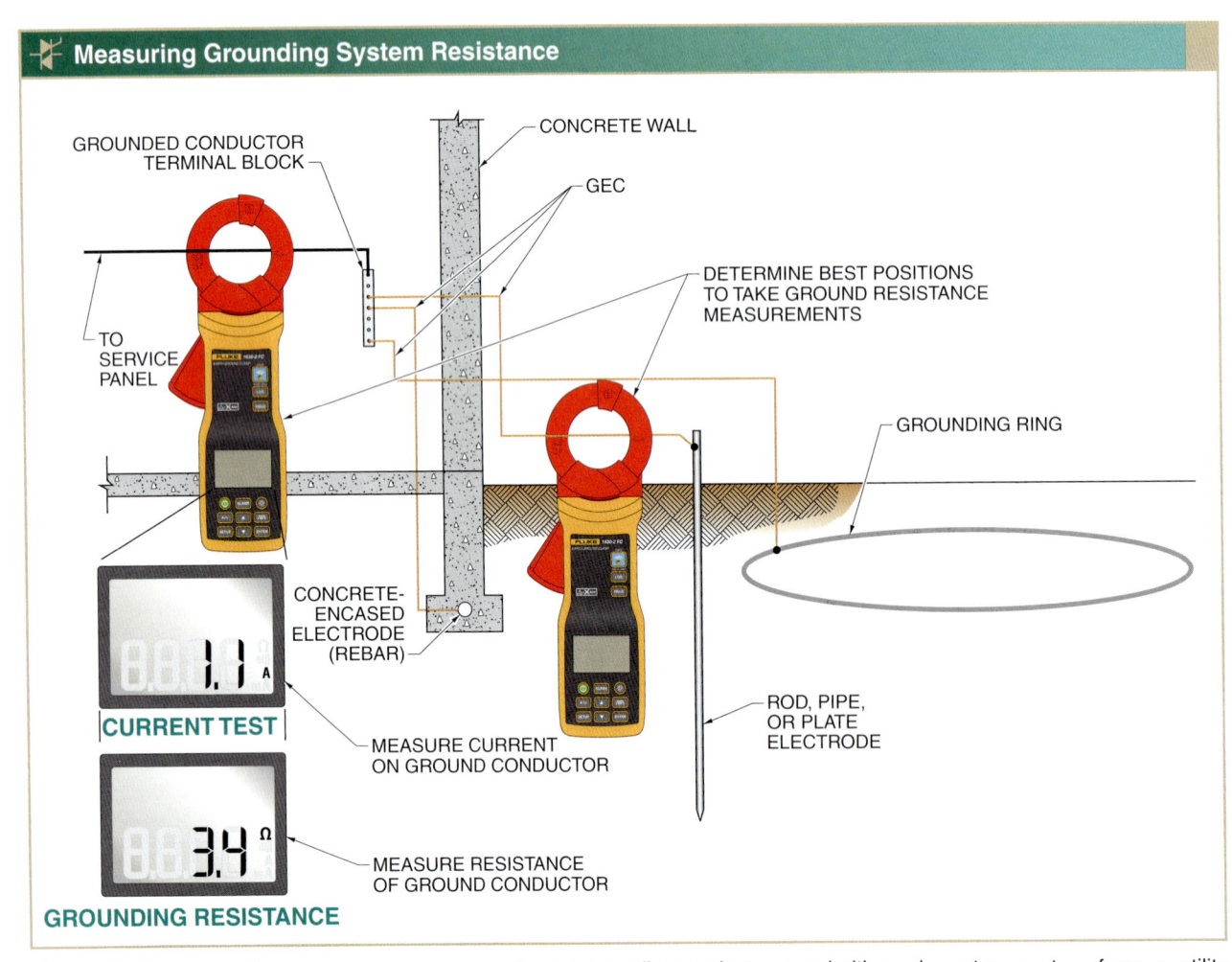

Figure 1-40. Ground resistance measurements are taken on grounding conductors used with service entrances, transformers, utility transmission, and communication (control circuit) grounds.

ELECTRICAL NOISE SUPPRESSION

Electrical noise is any unwanted signal present on power lines. **See Figure 1-41.** Electrical noise enters a PLC system through input devices, output devices, and power supply lines. Placing a PLC away from noise-generating equipment such as motors, motor starters, welding machines, and electric motor drives reduces unwanted noise pickup.

Even when PLCs are placed away from noise-generating devices, noise can reach a PLC through the conductors used to connect the system input devices (switches and sensors) and output components (lights and motors). To prevent noise and false signals from entering PLCs through the routing of power lines and control lines, technicians must not route low-voltage DC signal conductors near high-voltage (120 V) AC signal conductors. When different types of signals must cross, the cables should cross at 90° to minimize noise interference. **See Figure 1-42.**

Although it is impossible to eliminate noise completely from commercial and industrial electrical environments, noise suppression devices suppress noise to an acceptable level and must be included in every PLC installation. Certain sensitive input devices (analog, digital, and thermocouple) require a shielded cable to reduce electrical noise.

A *shielded cable* is a type of cable that uses an outer conductive jacket (shield) to surround the inner conductors that carry the signals. The shield blocks electromagnetic interference. The shield of a cable must be properly grounded to be effective. Proper grounding includes grounding the shield at one point only. A shield grounded at two points tends to conduct current between the two ground locations. **See Figure 1-43.**

PLC TIPS

The drain wire and foil shield of a shielded cable must be grounded at one end of the cable only.

Electrical Noise

AC Sine Wave without Noise   AC Sine Wave with Noise

OSCILLOSCOPE DISPLAYS

SENSOR — POWER DISTRIBUTION CABINET

MOTORS
PLC AND ENCLOSURE POSITIONED AWAY FROM NOISE-GENERATING EQUIPMENT

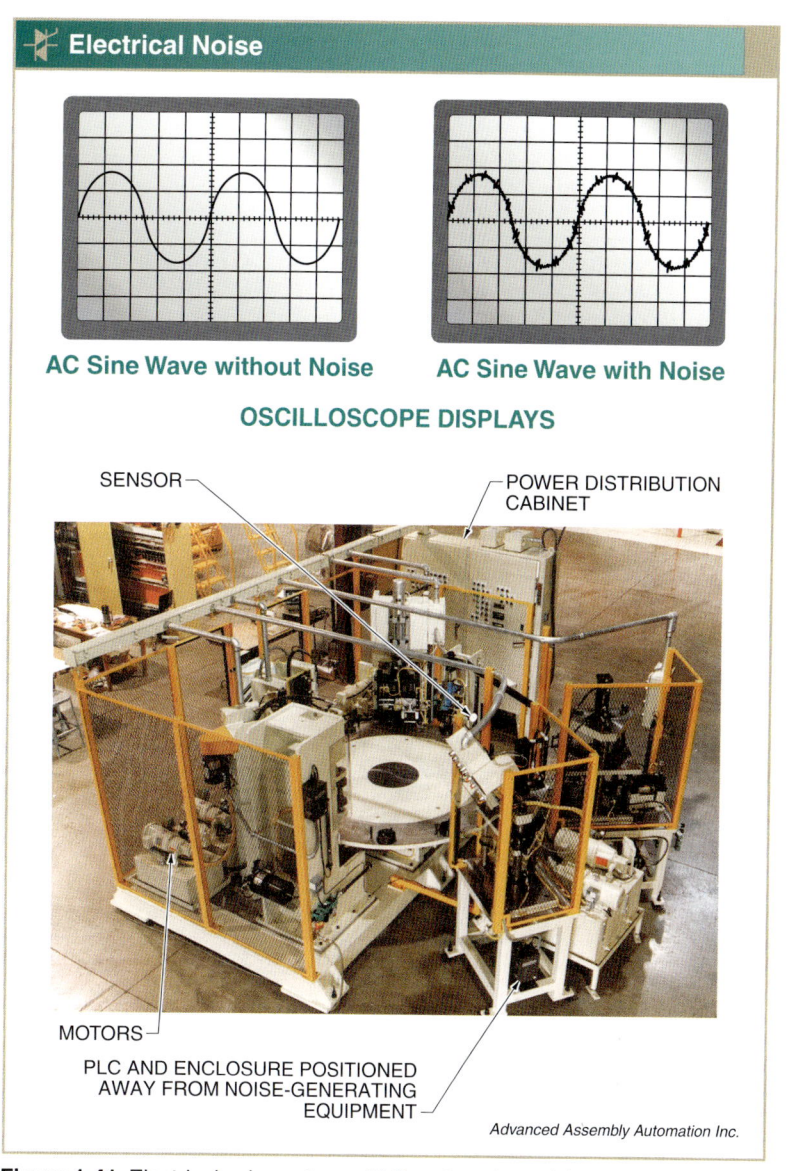

Advanced Assembly Automation Inc.

Figure 1-41. Electrical noise enters a PLC system through input devices, output components, and power supply lines.

Shielded twisted-pair cable must have all insulation and shield trimmed correctly or the cable must have shrink tubing installed properly to avoid accidental grounding or induced noise.

36 PROGRAMMABLE LOGIC CONTROLLERS: PRINCIPLES AND APPLICATIONS

Induced Electrical Noise

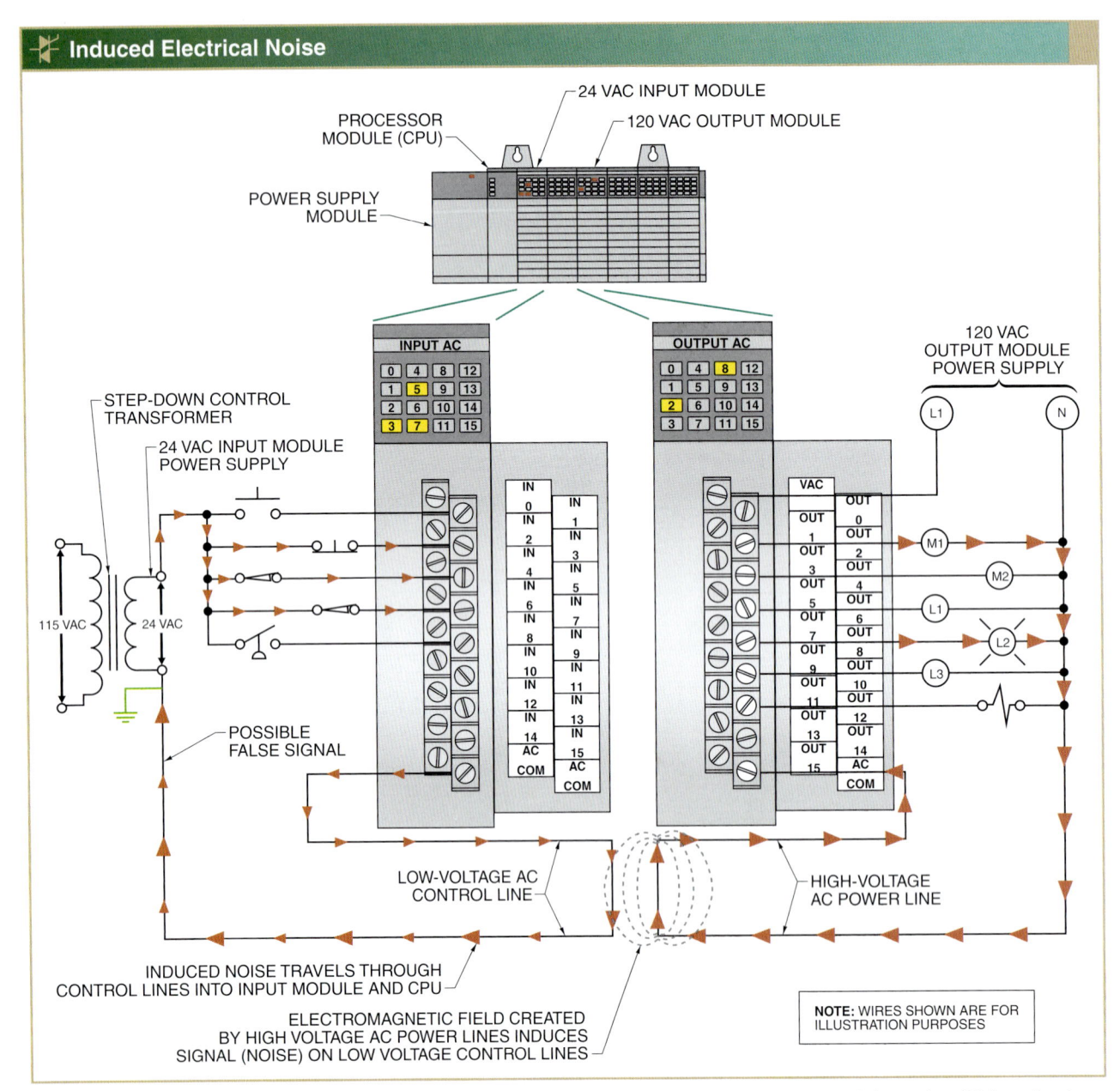

Figure 1-42. To prevent false signals from entering a PLC, input and output lines must cross at right angles (90°) and not run parallel to each other.

Shielded Cables

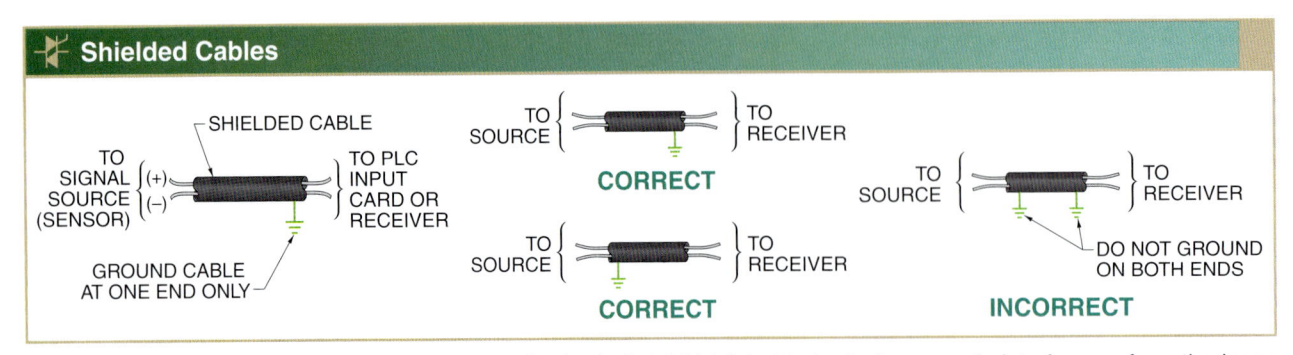

Figure 1-43. A shielded cable uses an outer conductive jacket (shield) to block electromagnetic interference from the inner, signal-carrying conductors.

STATIC ELECTRIC CHARGES

Static electricity is an electrical charge at rest. Proper grounding protects PLCs and related equipment from false operation, failure, or damage from such problems as static electricity and induced noise.

Grounding and shielded conductors reduce unwanted noise. Eliminating or reducing noise in a PLC system is important because low voltages (less than 5 V) can induce low currents into other low voltage control circuits. The signals in low-voltage circuits are seriously affected by electrical noise (magnetic fields) created from static electricity or induced by other electrical circuit wiring.

Induced noise produces intermittent control circuit malfunctions. Twisted wire pairs and coaxial cable help reduce the noise induced into PLC systems. To prevent ground loop problems, only one end of a shielded conductor must be grounded. A grounded shield diverts unwanted induced interference to ground.

PLC TIPS

PLCs are manufactured for Class I, Division 2, Groups A, B, C, D or nonhazardous locations only. National Electrical Manufacturers Association (NEMA) standard 250 and the International Electrotechnical Commission (IEC) publication 60529 explain the degrees of protection provided by various types of electrical enclosures.

A *high-voltage spike* is a type of electrical noise that is produced when inductive loads such as motors, solenoids, and coils are turned OFF. High-voltage spikes typically cause problems for PLCs. High-voltage spikes must be suppressed with snubber circuits to prevent PLC and other problems. A *snubber circuit* is an electrical circuit designed to suppress voltage spikes. **See Figure 1-44.** Typical snubber circuits use metal oxide varistors (MOV), resistors (R) and capacitors (C), or diodes, depending on the specific type of load being protected.

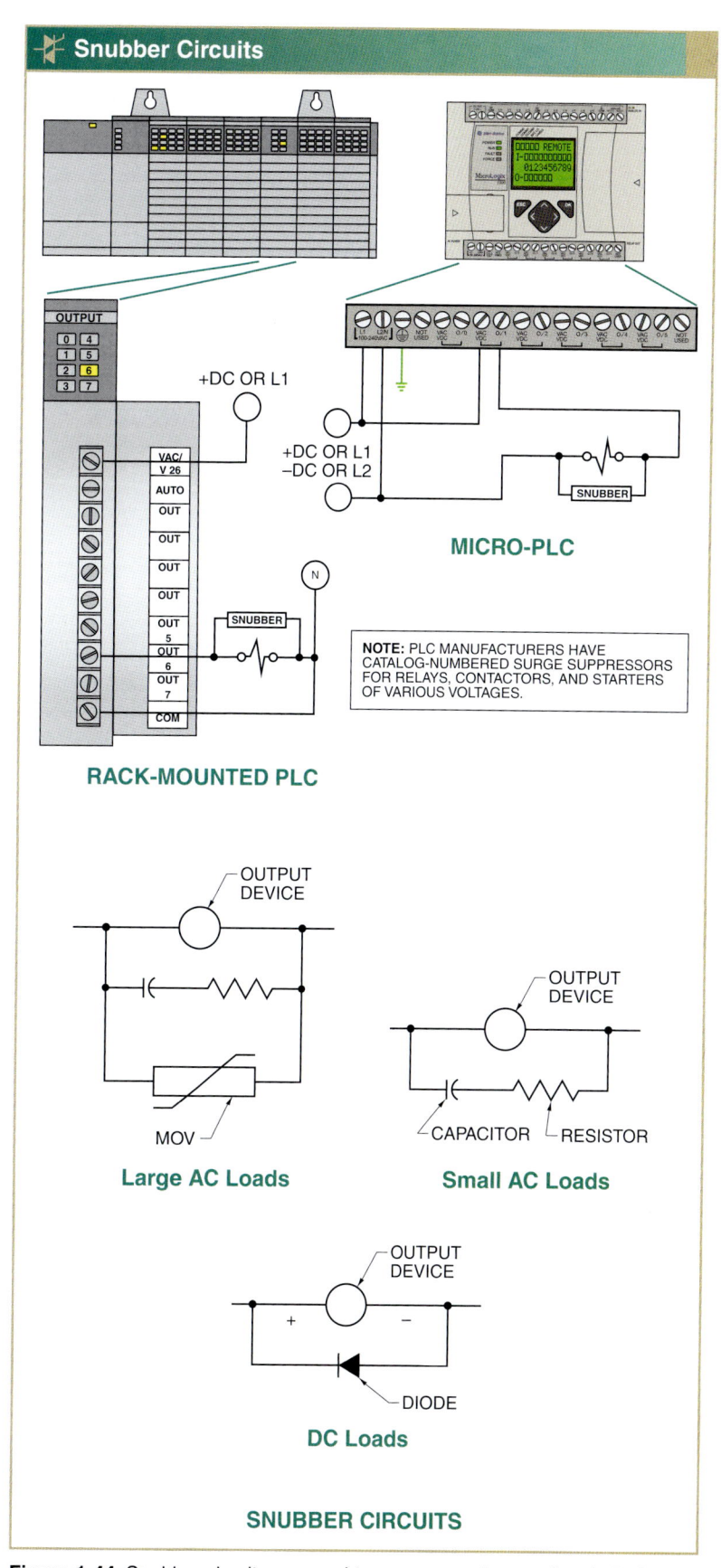

Figure 1-44. Snubber circuits are used to suppress voltage spikes in PLCs.

PLC ENCLOSURES

PLCs are typically placed in enclosures (cabinets) that include additional electrical components such as control transformers, fuses, or circuit breakers. An *electrical enclosure* is a housing that protects wires and equipment and prevents personal injury by accidental contact with energized circuits. An enclosure also provides the main protection for a PLC from atmospheric conditions. Using the proper enclosure prevents problems caused by contamination, moisture, and physical damage. Enclosures are categorized by the protection provided. An enclosure is selected based on the location of the equipment and NEC® requirements. **See Figure 1-45.**

PLC Enclosures

Enclosures

Type	Use	Service	Tests	Comments
1	Indoor	No unusual conditions	Rod entry, rust resistance	
3	Outdoor	Windblown dust, rain, sleet, and ice on enclosure	Rain, external icing, dust, and rust resistance	Does not provide protection against internal condensation or internal icing
3R	Outdoor	Falling rain and ice on enclosure	Rod entry, rain, external icing, and rust resistance	Does not provide protection against dust, internal condensation, or internal icing
4	Indoor/outdoor	Windblown dust and rain, splashing water, hose-directed water, and ice on enclosure	Hosedown, external icing, and rust resistance	Does not provide protection against internal condensation or internal icing
4X	Indoor/outdoor	Corrosion, windblown dust and rain, splashing water, hose-directed water, and ice on enclosure	Hosedown, external icing, and corrosion resistance	Does not provide protection against internal condensation or internal icing
6	Indoor/outdoor	Occasional temporary submersion at a limited depth		
6P	Indoor/outdoor	Prolonged submersion at a limited depth		
7	Indoor locations classified as Class I, Groups A, B, C, or D, as defined in the NEC®	Withstand and contain an internal explosion of specified gases; contain an explosion sufficiently so as an explosive gas-air mixture in the atmosphere is not ignited	Explosion, hydrostatic, and temperature	Enclosed heat-generating devices shall not cause external surfaces to reach temperatures capable of igniting explosive gas-air mixtures in the atmosphere
9	Indoor locations classified as Class II, Groups E or G, as defined in the NEC®	Dust	Dust penetration, temperature, and gasket aging	Surface temperature shall not cause accumulated or airborne dust to ignite
12	Indoor	Dust, falling dirt, and dripping noncorrosive liquids	Drip, dust, and rust resistance	Does not provide protection against internal condensation
13	Indoor	Dust, spraying water, oil, and noncorrosive coolant	Oil explosion and rust resistance	Does not provide protection against internal condensation

ENCLOSURE WITH COOLING UNIT

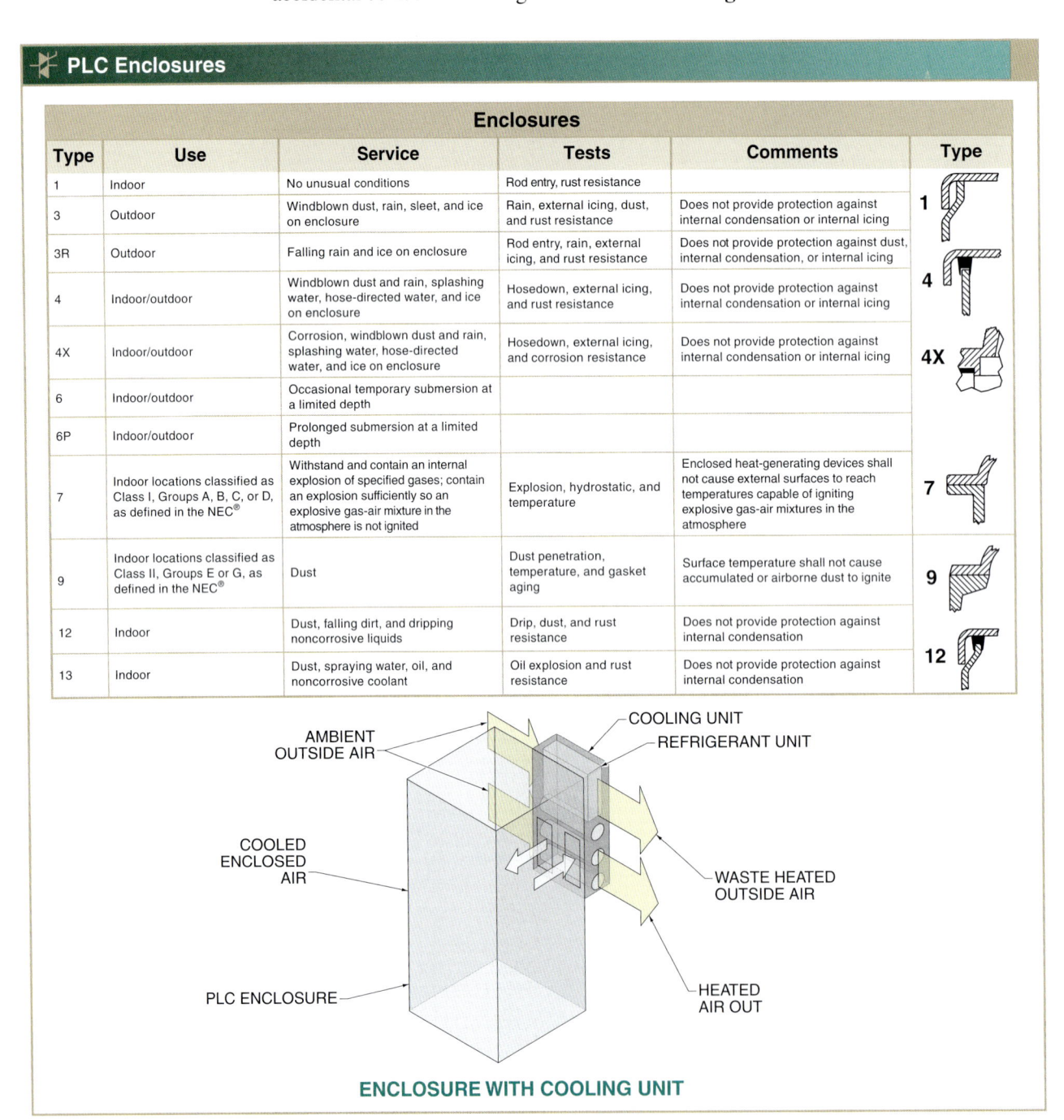

(Labels: AMBIENT OUTSIDE AIR, COOLING UNIT, REFRIGERANT UNIT, COOLED ENCLOSED AIR, WASTE HEATED OUTSIDE AIR, PLC ENCLOSURE, HEATED AIR OUT)

Figure 1-45. Depending on the PLC application, an enclosure with a cooling unit may be required.

ELECTRICAL SAFETY RULES

Technicians must work safely at all times around electrical systems. Electrical safety rules must be followed when working with energized electrical equipment to aid in the prevention of injuries from electrical energy sources. **See Figure 1-46.**

PLC INSTALLATION GUIDELINES

PLC system installation begins with a sufficient number of emergency stops and a master control relay that removes power to the inputs and outputs of a PLC and stops all motion of the machine(s) or process. **See Figure 1-47.** When working with energized PLCs and functioning PLC systems, proper personal protective equipment must be worn.

PLC TIPS

Overtravel, interlocks, and stop pushbutton safety circuits of a machine must be hardwired in series directly to the master control relay, which allows the relay to be de-energized.

PERSONAL PROTECTIVE EQUIPMENT

Personal protective equipment (PPE) is clothing and/or equipment worn by technicians to reduce the possibility of injury in the work area. The use of personal protective equipment is required whenever work occurs on or near energized exposed electrical circuits. *NFPA 70E®: Standard for Electrical Safety in the Workplace* addresses electrical safety requirements for employee workplaces that are necessary for the safeguarding of employees in pursuit of gainful employment. For maximum safety, personal protective equipment must be used as specified in NFPA 70E®, OSHA 29 CFR 1910 Subpart 1—*Personal Protective Equipment* (1910.132 through 1910.138), and other applicable safety mandates.

Electrical Safety Rules

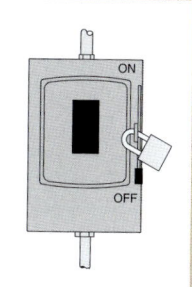

- Always comply with the NEC® and NFPA 70E.
- Use UL® labeled or other certification organization-labeled appliances, components, and equipment.
- Keep electrical grounding circuits in good condition. Ground any conductive component or element that does not have to be energized. The grounding connection must be a low-resistance conductor heavy enough to carry the largest fault current that may occur. 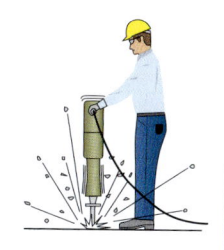
- Turn OFF, lockout, and tagout disconnect switches when working on any electrical circuit or equipment. Test all circuits after they are turned OFF. Insulators may not insulate, grounding circuits may not ground, and switches may not open the circuit.
- Use double-insulated power tools or power tools that include a third conductor grounding terminal, which provides a path for fault current. Never use a power tool that has the third conductor grounding terminal removed.
- Always use protective and safety equipment. 
- Practice emergency procedures and have a plan of action in an emergency situation.
- Check conductors, cords, components, and equipment for signs of wear or damage. Replace any equipment that is not safe.
- Never throw water on an electrical fire. Turn OFF the power and use a Class C or Class ABC fire extinguisher. 
- Work with another individual when working in a dangerous area or with dangerous equipment.
- Learn CPR and first aid.
- Do not work when tired or taking medication that causes drowsiness. 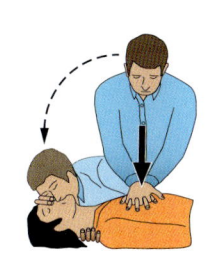
- Do not work in poorly lighted areas.
- Always use nonconductive ladders. Never use a metal ladder when working around electrical equipment.
- Ensure there are no atmospheric hazards such as combustible dust or vapor in the area. A live electrical circuit may emit a spark at any time.
- Use one hand when working on a live circuit to reduce the chance of an electrical shock passing through the heart and lungs.
- Never bypass or disable fuses or circuit breakers.
- Extra care must be taken with an electrical fire because burning insulation produces toxic fumes.

Figure 1-46. Electrical safety rules aid in the prevention of injuries from electrical energy sources.

PLC Installation Guidelines

- Do not apply power to a PLC until the entire operating manual is understood.
- PLC-controlled systems must include a sufficient number of emergency stops that totally stop the operation of a system when a failure occurs that could cause injury.
- Only qualified technicians should have access to PLC parameters. Changing PLC settings affects an entire system.
- Incoming power must be connected to PLC terminals L1 (VAC/VDC) and L2/N.
- Use separate metal conduits for routing input power conductors, input control conductors, and output power conductors.
- PLCs must be housed in a properly rated enclosure with all field input and output cabling entering and exiting the enclosure via an approved fitting for the environmental rating of the enclosure.
- Never touch any internal part of a PLC when power is applied.
- Each wire connected to a PLC must be properly identified and labeled.
- Some PLCs include fault indicator lights that indicate a fault in the system and/or display error codes when a problem exists. Always check the status of fault indicator lights and error codes when servicing PLCs.
- Never use the force or disable commands on a PLC unless the way the commands affect the operation of an entire system is understood.
- All PLCs produce heat that must be dissipated away from the PLC to prevent false operation or equipment damage. Always ensure proper spacing of components within an enclosure and allow proper airflow for cooling. Periodically change or clean filters to ensure proper airflow.

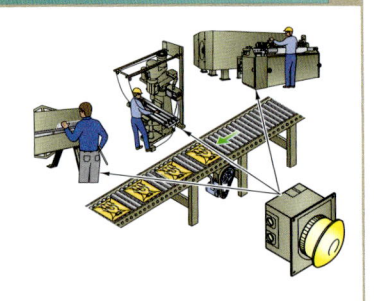

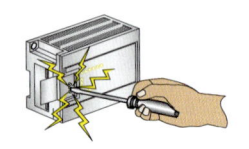

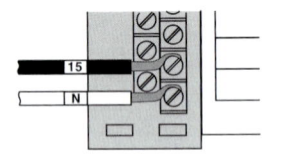

Figure 1-47. PLC system installation begins with a sufficient number of emergency stops and a master control relay that removes power to the inputs and outputs of the PLC and stops all motion of the machine(s) or process.

Per NFPA 70E®, only qualified persons shall perform work on or near energized equipment operating at 50 V or more. All personal protective equipment and tools are selected for at least the operating voltage of the equipment or circuits to be worked on or equipment that is near to the place of work. All PPE, tools, and test equipment must be suited for the work to be performed. Personal protective equipment includes protective clothing, such as arc-rated clothing, and eye protection, such as safety glasses, goggles, face shields, and arc-flash hoods. **See Figure 1-48.** PPE used for electrical shock protection, such as safety glasses and rubber insulating gloves, is worn during PLC system tests and troubleshooting procedures.

Figure 1-48. PPE includes items that protect a technician from electrical and other hazards.

Protective Clothing

When working on a PLC system or any electrical system, the proper protective clothing must be worn. *Protective clothing* is clothing that provides protection from contact with sharp objects, hot equipment, and harmful materials. Protective clothing made of durable material such as denim should be snug yet allow ample movement. Clothing must fit snugly to avoid the danger of becoming entangled in moving machinery. Pockets should allow convenient access but must not snag on tools or equipment. Soiled protective clothing must be washed to reduce the flammability hazard of the clothing.

Arc-rated protective clothing must be worn when working with energized high-voltage electrical circuits. **See Figure 1-49.** Arc-rated protective clothing is made of materials such as Nomex®, Basofil®, and/or Kevlar® fibers. Arc-resistant fibers can also be coated with PVC to increase arc resistance and offer weather resistance. Arc-rated protective clothing must meet the following three requirements:

- Clothing must not ignite and continue to burn.
- Clothing must have an insulating value high enough to allow heat to dissipate through clothing and away from the skin.
- Clothing must provide resistance to the break-open forces generated by the shock wave of an arc.

Eye Protection

Eye protection must be worn to prevent eye or face injuries caused by flying particles, contact arcing, and radiant energy. Eye protection must comply with OSHA 29 CFR 1910.133, *Eye and Face Protection*. Eye protection standards are specified in ANSI Z87.1, *Occupational and Educational Eye and Face Protection*. Eye protection includes safety glasses or goggles, arc-rated face shields, and arc flash hoods. **See Figure 1-50.**

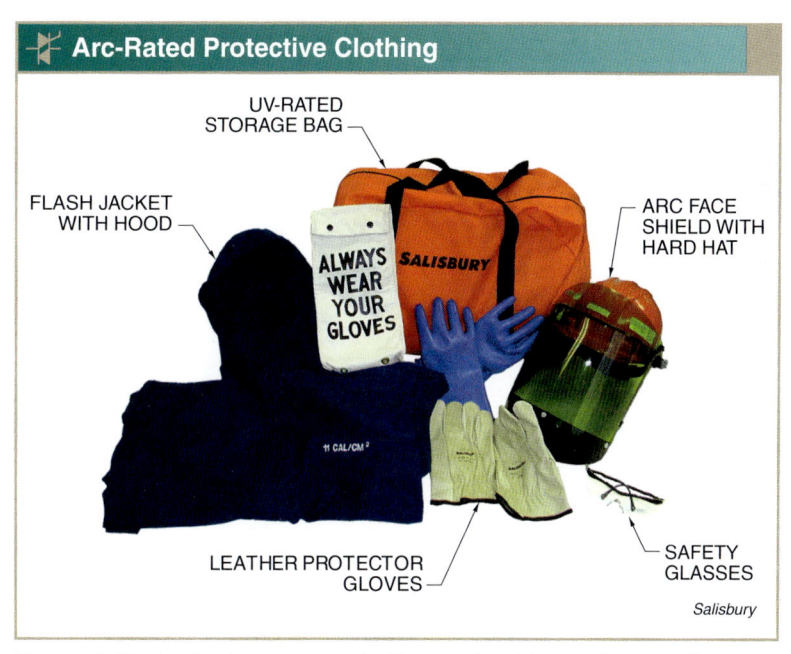

Figure 1-49. Arc-flash protective clothing made of Nomex®, Basofil®, and/or Kevlar® fibers must be used when working with live high-voltage electrical circuits.

⚠ WARNING

Per ASTM F496 specifications, insulating glove cuffs must extend beyond the cuff of the protector gloves by the following:

- *Class 00—0.5″*
- *Class 0—0.5″*
- *Class 1—1″*
- *Class 2—2″*
- *Class 3—3″*
- *Class 4—4″*

Safety glasses are an eye protection device with special impact-resistant glass or plastic lenses, reinforced frames, and side shields. Plastic frames are designed to keep the lenses secured in the frame if an impact occurs and minimize the shock hazard when working with electrical equipment. Side shields provide additional protection from flying objects. Tinted-lens safety glasses protect against low-voltage arc hazards. Non-tinted safety glasses with side shields can be worn when working on PLC systems that are operating in a normally safe environment.

Eye Protection

[Figure showing safety glasses with reinforced frame, impact-resistant lens, and side shields; arc-rated face shield with adjustable headband and plastic shield; arc flash hood with arc blast hood and hood face shield]

SAFETY GLASSES

ARC-RATED FACE SHIELD

ARC FLASH HOOD

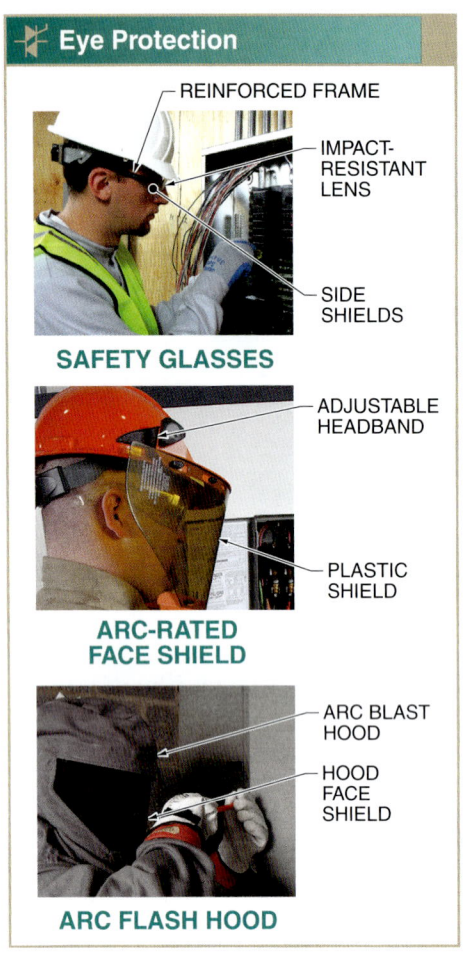

Figure 1-50. Eye protection must be worn to prevent eye or face injuries caused by contact arcing, radiant energy, or flying particles.

Goggles are an eye protection device with a flexible frame that are secured to the face with an elastic headband. Goggles fit snugly against the face to seal the areas around the eyes and can be used over prescription glasses. Goggles with clear lenses protect against small flying particles or splashing liquids. Tinted goggles are used to protect against low-voltage arc hazards.

A *face shield* is an eye and face protection device that covers the entire face with a plastic shield and is used for protection from flying objects. Tinted, arc-rated face shields protect against low-voltage arc hazards. A tinted, arc-rated face shield should be worn when working on any PLC system that has voltages over 50 V or in any environment where arc flashes are possible.

An *arc flash hood* is an eye and face protection device that covers the entire head with plastic and material. Arc blast hoods are used to protect against high-voltage arc blasts. Technicians working with energized equipment must wear arc blast protection per NFPA 70E®.

Safety glasses, arc-rated face shields, and arc flash hood lenses must be properly maintained to provide protection and clear visibility. Lens cleaners are available that clean without risk of lens damage. Pitted or scratched lenses reduce vision and can cause lenses to fail on impact.

PLC TIPS

Arc suppression blankets are available that help protect technicians from the explosive and burning effects of electric arcs and flashes. Arc suppression blankets are placed between the technician and hazardous equipment such as transformer terminals, switchgear, circuit breakers, and high-voltage conductors. Blanket clamp pins are used to hold blankets in place.

LOCKOUT/TAGOUT

Electrical power must be removed when a PLC or any electrical equipment is inspected, serviced, repaired, or replaced. Power is removed and all equipment must be locked out and tagged out to ensure the safety of personnel working on the equipment. **See Figure 1-51.**

Per Occupational Safety and Health Administration (OSHA) standards, equipment is locked out and tagged out before any preventive maintenance or servicing is performed. *Lockout* is the process of removing the source of electrical power and installing a lock, which prevents the power from being turned ON. *Tagout* is the process of placing a danger tag on the source of electrical power, which indicates that the equipment cannot be operated until the danger tag is removed.

Lockout/Tagout Kits

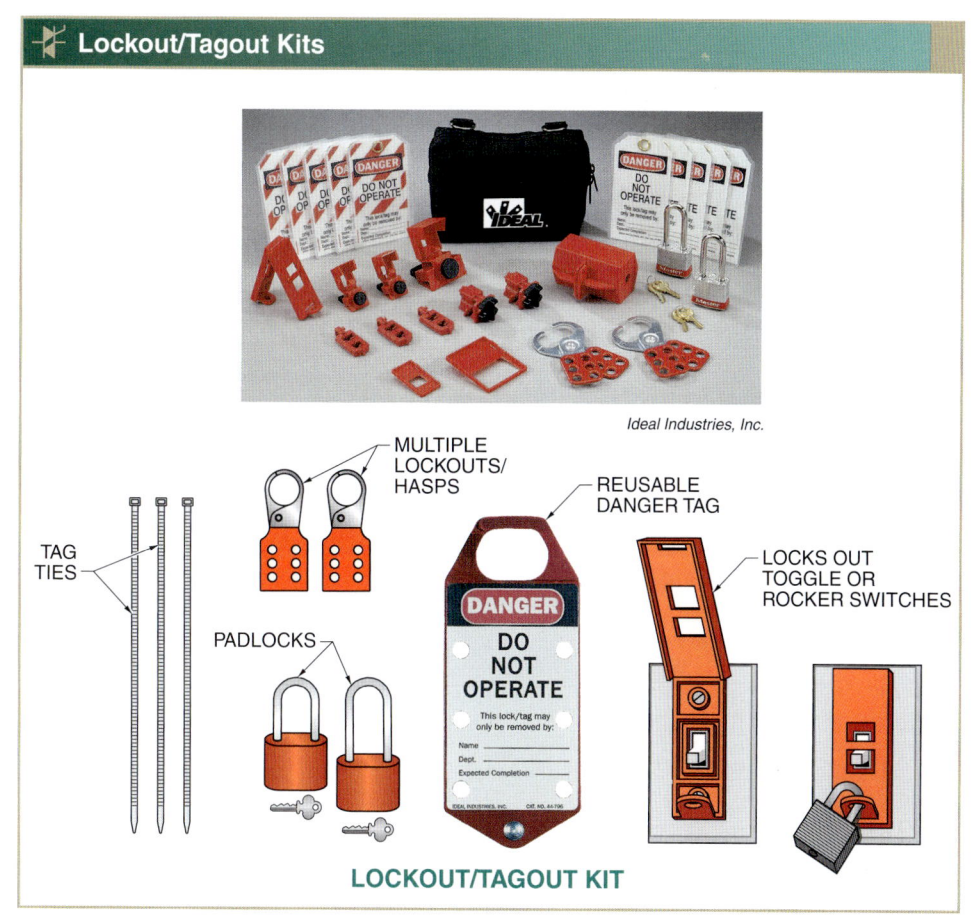

Ideal Industries, Inc.

Figure 1-51. Lockout/tagout kits contain reusable danger tags, tag ties, multiple lockout hasps, magnetic signs, padlocks, and information on lockout/tagout procedures.

CODE CONNECT

Section 110.25 of the NEC® states that if a disconnecting means is required to be locked in the open position, the locking mechanism shall remain in place with or without the lock. Removable lockout/tagout devices for disconnecting means are used to satisfy OSHA requirements for de-energization. Section 110.25 prohibits the use of removable lockout/tagout devices to satisfy NEC® requirements.

A danger tag has the same importance and purpose as a lock and is used alone only when a lock does not fit the disconnect device. A danger tag must be attached at the disconnect device with a tag tie or equivalent and have a space for the name, craft, and other required information. A danger tag must be able to withstand the elements and expected atmosphere for as long as the tag remains in place.

A lockout/tagout is used during the following situations:

- when servicing electrical equipment that does not require power to be ON to perform the service
- when removing or bypassing a machine guard or other safety device
- when the possibility exists of being injured or caught in moving machinery
- when clearing jammed equipment
- when the danger exists of being injured if equipment power is turned ON

Circuit breakers typically require a circuit breaker lock, hasp, lock, and tag before the lockout/tagout is complete.

Lockouts and tagouts do not by themselves remove power from a circuit. An approved company procedure must be followed when applying a lockout/tagout. Lockouts and tagouts are attached only after the equipment is turned OFF and tested to ensure that the power is OFF. A lockout/tagout procedure is required for the safety of technicians due to modern automated equipment hazards. OSHA provides a standard procedure for equipment lockout/tagout. The procedure for lockout/tagout per OSHA is as follows:

1. Prepare for machinery shutdown.
2. Shut down machinery or equipment.
3. Isolate machinery or equipment.
4. Lock out and tag out electrical disconnect and other energy supply devices (valves). **See Figure 1-52.**
5. Release all stored energy (capacitors, pneumatic, and hydraulic).
6. Verify that machinery or equipment is isolated.

A lockout/tagout must not be removed by any other person than the person who installed the lockout/tagout, except in an emergency using company procedures. The authorized person must follow approved procedures. A list of company rules and procedures is given to all technicians who use lockout/tagout.

The following points are important when locking and tagging out a piece of equipment:

- Use both a lockout and tagout when possible.
- Use a tagout when a lockout is impractical. A tagout is used alone only when a lock does not fit the disconnect device.
- Use a lockout hasp when multiple technicians are part of the lockout/tagout.
- Notify all employees affected before using a lockout/tagout.
- Remove all power sources including the primary and secondary.
- Measure for voltage using a voltmeter to ensure that the power is OFF.

Lockout Devices

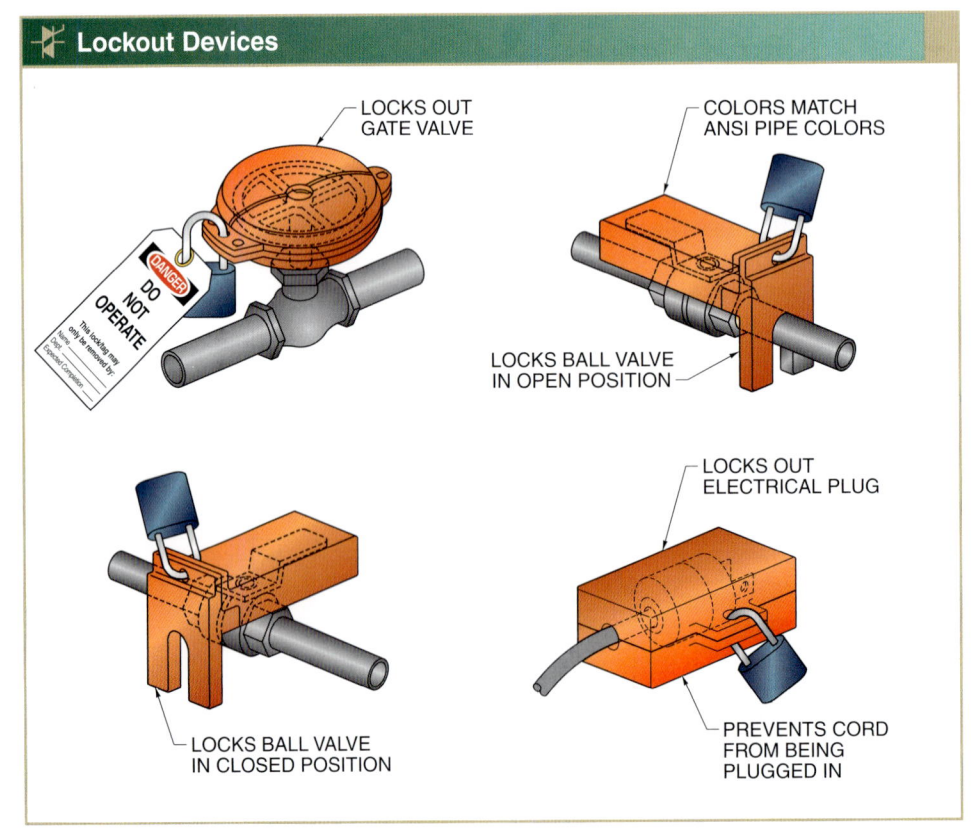

Figure 1-52. Lockout devices resist chemicals, cracking, abrasion, and temperature changes and are available in colors to match American National Standards Institute (ANSI) pipe colors. Lockout devices are sized to fit standard industry equipment.

INSPECTING A PLC SYSTEM

Before applying power to any electrical system, an inspection of the entire system must be performed. A PLC system must be checked for general workmanship (neat wire runs and clean installation) and tight connections. Also, the system must meet or exceed NEC® and local code requirements. A good inspection helps ensure electrical and PLC systems work safely.

An inspection of all parts of a PLC system is required before any power is applied to the system and any time service is performed. Inspection is performed to ensure that each module is in the correct location, securely mounted (with module locks secured where applicable), correctly wired, and properly programmed. **See Figure 1-53.**

To inspect a PLC system, apply the following procedure:

1. Ensure that the PLC and all associated hardware is securely mounted in the proper location. Loose screws and bolts cause vibration, which can cause plug-in circuit boards to loosen, open circuits, and arc. Any arcing circuits are a fire hazard, and care must be taken when an enclosure door is open because the enclosure does not protect the electrical parts from the outside atmosphere when the door is open, no matter what the enclosure rating.

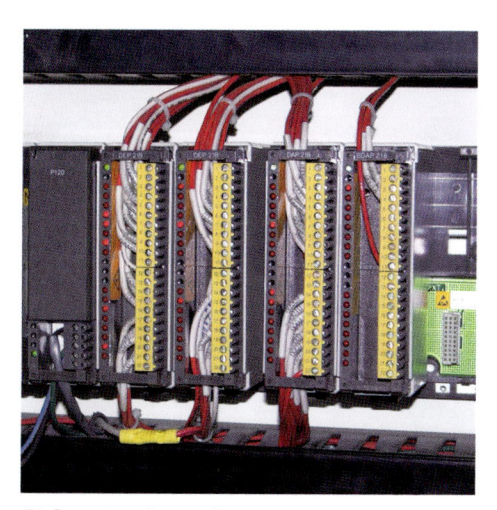

PLC system inspections must occur before any power is turned ON, encompass all hardware attachments, wire connections, and grounding, and ensure that all sections and screws not being used are secured in place.

2. Check all wire connections from the main disconnect to the PLC input devices, output components, expansion and communication cables, and jumper links. A *jumper link* is a device used to select the power rating of a dual-voltage PLC. A jumper link must be in the correct location.

   Check to ensure that all wire connections are tight and that proper wiring procedures are followed. Proper wiring procedures include physically separating the main power lines from the input lines, input lines from the output lines, power lines from communication lines, and AC voltage lines from DC voltage lines. Separating AC and DC voltage lines reduces the possibility of introducing electromagnetic interference onto the lines. Electromagnetic interference can produce false signals and cause program errors.

3. Ensure that the PLC and all exposed non-current-carrying parts are grounded. Proper grounding is an important safety precaution in any electrical installation. **See Figure 1-54.** Proper grounding is especially important in PLC applications because improper grounding can lead to interference being induced into PLCs. Induced interference can cause output devices to be falsely turned ON, which can cause personnel injury and/or equipment damage. Refer to equipment manufacturer grounding recommendations. To prevent problems in a PLC system, the grounding path must be permanent, continuous and uninterrupted, of minimum resistance, and of sufficient size to carry any potential fault current.

> ⚠ **CAUTION**
>
> *When inspecting a de-energized PLC, electrostatic discharge from the human body can damage integrated circuits and semiconductors of the PLC when the backplane connector pins or power supply internal components are touched.*

46 PROGRAMMABLE LOGIC CONTROLLERS: PRINCIPLES AND APPLICATIONS

Inspecting PLCs

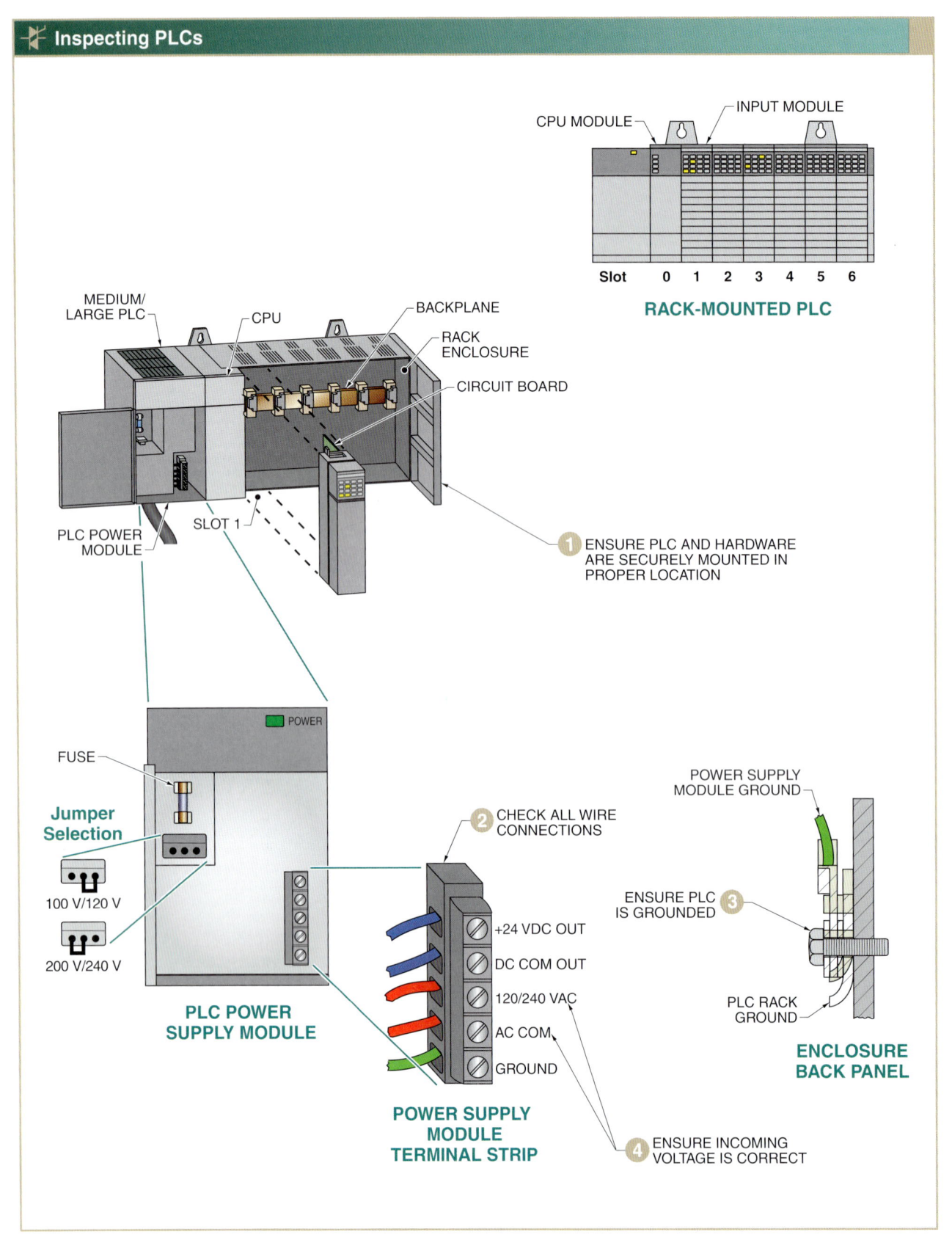

Figure 1-53. A proper inspection of a PLC ensures safe control of an electrical system.

PLC Enclosure Grounding

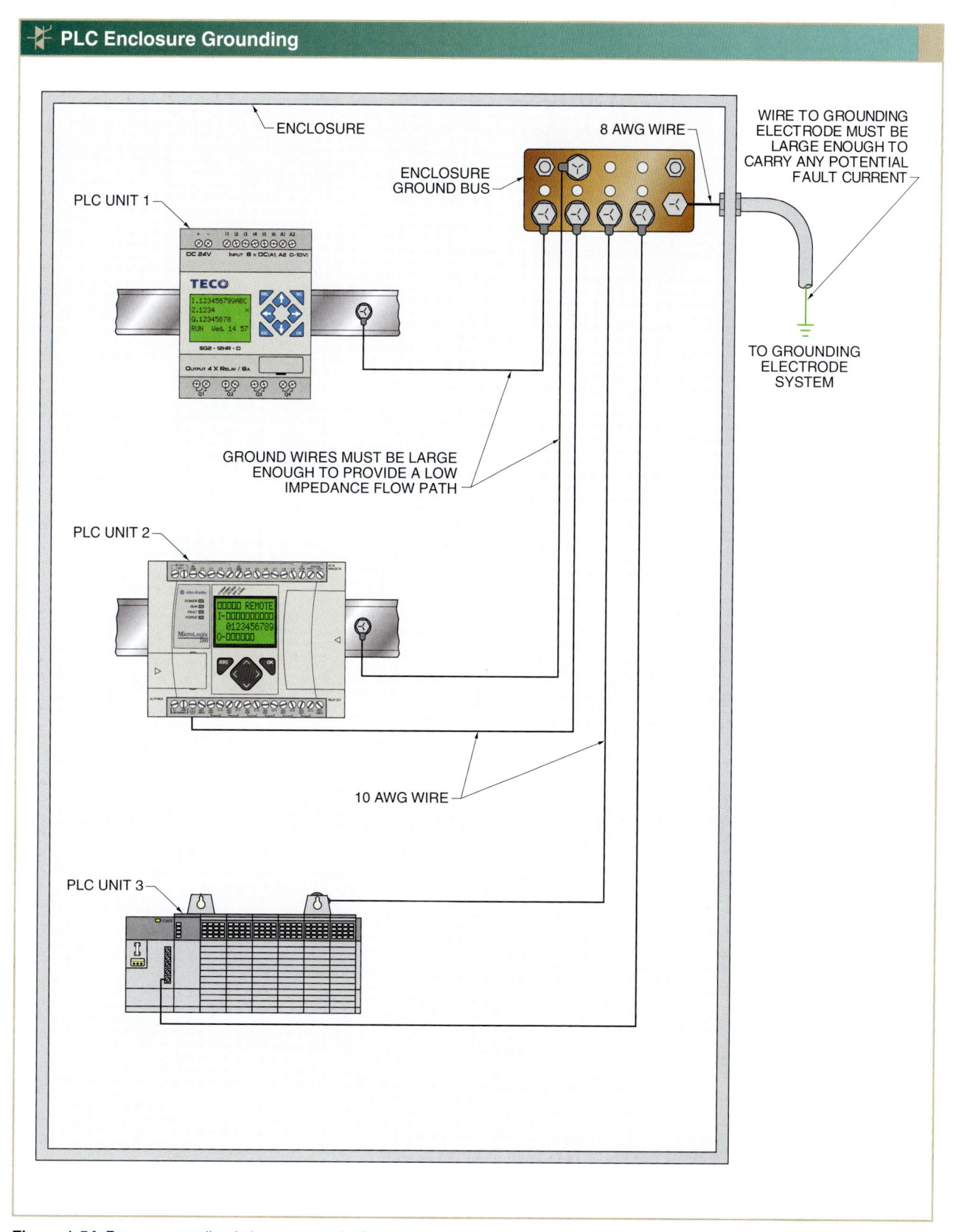

Figure 1-54. Proper grounding is important in PLC applications. Improper grounding can lead to interference (noise) being induced into PLCs, which can cause output devices to be falsely turned ON and put personnel and equipment at risk.

4. Ensure that incoming voltage is correct according to the wiring specifications of the PLC. Excessive voltage damages the circuits in a PLC. Destructive problems occur when 240 V is applied to the terminals of a PLC designed for 120 V. Inadequate voltage causes PLC malfunctions.

Spare parts should be available because many programmable logic controller problems can be solved by the substitution of a board or module. Most PLC manufacturers provide a list of recommended spare parts to stock in-house. The list normally includes a single replacement module for each CPU and power supply and spare I/O modules equal to 10% of the total number used in the system. Output modules fail more often than input modules because of the higher voltages and currents used.

⚠ **WARNING**

When installed in a hazardous (classified) location, all PLC wiring must comply with NEC® Articles 501.10(A) and (B), Wiring Class I Locations.

Chapter Review 1

Name _____ Date _____

True-False

T F **1.** Grounding is the connection of all exposed current-carrying metal parts to earth.

T F **2.** Nano-PLCs have 15 to 128 I/Os.

T F **3.** An electrical shock results any time a body becomes part of an electrical circuit.

T F **4.** A shielded cable is a type of cable that uses an outer conductive jacket to surround the inner conductors that carry electrical signals.

T F **5.** A push command is a special software override that opens or closes an input device or turns an input component ON or OFF.

T F **6.** An electrical enclosure provides the main protection for a PLC from atmospheric conditions.

T F **7.** A metal ladder may not be used unless in combination with rubber insulating matting.

T F **8.** Never throw water on an electrical fire.

T F **9.** The NEC® states that building and equipment grounding systems with single-made electrodes must have a resistance of less than 25 Ω.

T F **10.** PLCs are used to control electrical circuits that are DC, AC, or AC and DC.

T F **11.** Proper grounding includes grounding a conductor shield at two points.

T F **12.** A high-voltage spike is a type of electrical noise that is produced when inductive loads such as motors, solenoids, and coils are turned OFF.

T F **13.** AC-rated transistors are only used for DC-rated PLCs.

T F **14.** Tagout is the process of placing a danger tag on the source of electrical power, which indicates that the equipment cannot be operated until the danger tag is removed.

T F **15.** Lockout/tagout removes power from a circuit.

Completion

_____ **1.** ___ is an electrical charge at rest.

_____ **2.** ___ is the opposition to current flow.

3. ___ is any unwanted signal present on power lines.

4. A(n) ___ is one coulomb passing a given point in an electrical circuit in one second.

5. A(n) ___ is an electrical circuit designed to suppress voltage spikes.

6. A(n) ___ is a special software override that prevents one or all input devices or output components from operating.

7. ___ is clothing and/or equipment worn by technicians to reduce the possibility of injury in the work area.

8. ___ is the electromotive force or pressure in a conductor that allows electrons to flow.

9. ___ is the process of removing the source of electrical power and installing a lock, which prevents the power from being turned ON.

10. A(n) ___ is the practical unit of measurement for a specific quantity of electrons.

Multiple Choice

1. A(n) ___ is a dual-voltage unit used to set the power rating of a PLC.
    A. CPU module
    B. input module
    C. jumper link
    D. snubber circuit

2. ___ is the rate at which electrons flow through a conductor.
    A. Current
    B. Power
    C. Resistance
    D. Voltage

3. A ___ is a self-contained, regulated or nonregulated voltage unit that supplies DC or AC voltage to a PLC.
    A. power supply module
    B. terminal strip
    C. terminal block
    D. motor starter

4. ___ an eye and face protection device that covers the entire face with a plastic shield, providing protection from flying objects.
    A. An arc flash hood is
    B. A face shield is
    C. Goggles are
    D. Safety glasses are

5. ___ is the connection of machinery electrical systems to earth ground to reduce the chance of electrical shock by grounding all non-current-carrying exposed metal.
    A. Building grounding
    B. Electronic equipment grounding
    C. Equipment grounding
    D. System grounding

PLC Familiarization

PLC Exercise 1.1

T TECO Procedures

Answer the following questions based on the nano-PLC shown and the TECO SG2 User Manual found under the TECO Resources tab in the Learner Resources.

1. On which side (input or output) of the PLC is the 24 VDC power supply section?

   _____

2. How many digital input terminals does the PLC have?

   _____

3. How many analog input terminals does the PLC have?

   _____

4. What is the voltage range of the analog inputs?

   _____

5. How many output terminals does the PLC have?

   _____

6. What kind of output contacts does the PLC have?

   _____

7. What is the operating temperature of the PLC?

   _____

8. What is the input voltage rating range?

   _____

9. What is the voltage rating of the outputs?

   _____

10. What is the maximum resistive load of each output?

    _____

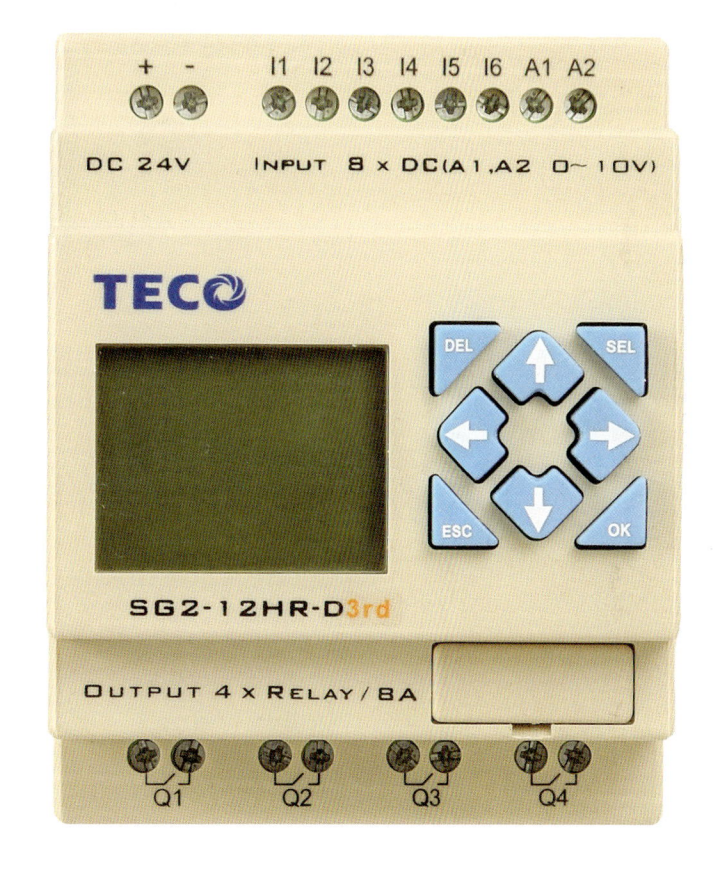

Allen-Bradley® Procedures

Answer the following questions based on the micro-PLC shown and the MicroLogix™ 1100 User Manual found under the Allen-Bradley® Resources tab in the Learner Resources.

11. On which side (input or output) of the PLC does the PLC supply 24 VDC?

    _____

12. On which side (input or output) of the PLC is the 120 VAC power supply section?

    _____

13. How does the PLC provide equipment grounding?

    _____

14. How many total digital input devices does the PLC allow?

    _____

15. Which input terminals are grouped with input terminal I/0 and share a common terminal?

    _____

16. Which input terminals are grouped with input terminal I/4 and share a common terminal?

    _____

17. How many total analog input devices does the PLC allow?

    _____

18. Which analog terminals share a common terminal?

    _____

19. How many total output components does the PLC allow?

    _____

20. Which output terminals are grouped together?

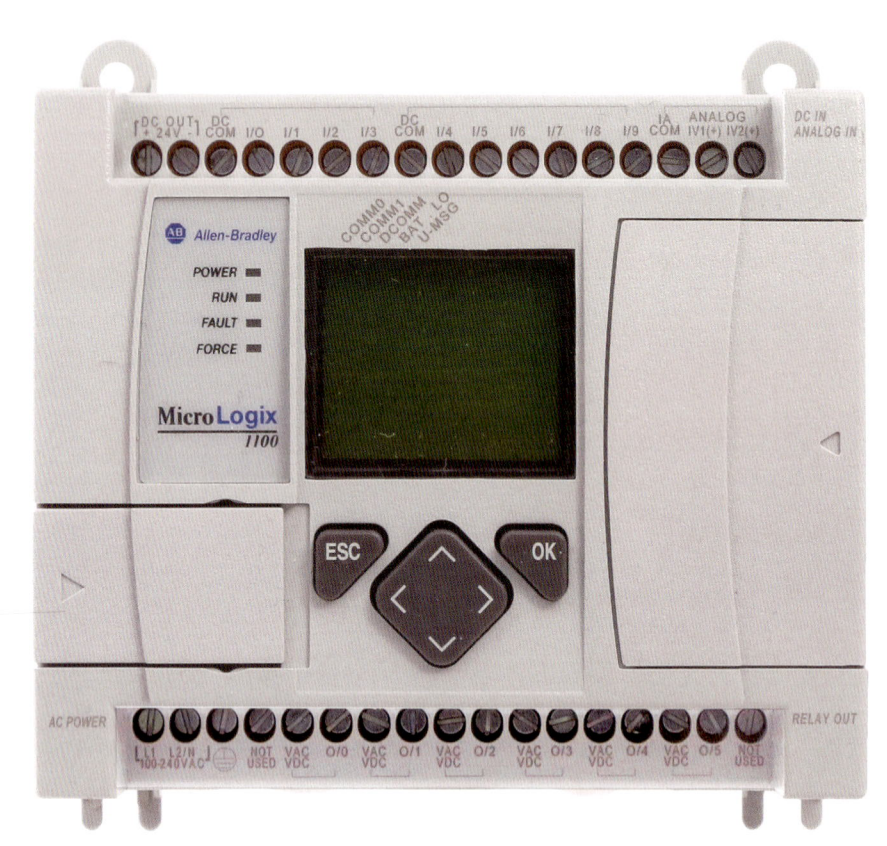

    _____

21. What kind of output contacts does the PLC have?

_____

22. Which PLC status LEDs are present to indicate the PLC's operating state?

_____

23. What is the operating temperature of the PLC?

_____

24. What is the input voltage rating range?

_____

25. What is the maximum controlled load for the PLC outputs in nonhazardous locations?

_____

PLC Model Identification Using TECO Software

PLC Exercise 1.2

BILL OF MATERIALS
- TECO SG2 Client software
- Personal computer

PLC manufacturers offer many types of PLC models based on a machine or on process parameters. A TECO model identification number lists controller type, I/O count, form factor, output type, and input power rating. For this lab, use the TECO SG2 Client software program to select three different PLC model numbers and identify the differences between each model identification number.

Note: To download the free SG2 Client software, go to the Learner Resources and click on the TECO Resources tab.

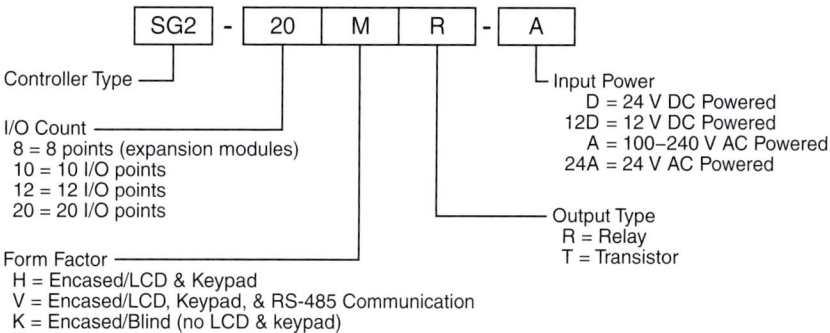

SG2 - 20 M R - A

Controller Type

I/O Count
- 8 = 8 points (expansion modules)
- 10 = 10 I/O points
- 12 = 12 I/O points
- 20 = 20 I/O points

Form Factor
- H = Encased/LCD & Keypad
- V = Encased/LCD, Keypad, & RS-485 Communication
- K = Encased/Blind (no LCD & keypad)
- C = Bareboard
- E = Expansion

Input Power
- D = 24 V DC Powered
- 12D = 12 V DC Powered
- A = 100–240 V AC Powered
- 24A = 24 V AC Powered

Output Type
- R = Relay
- T = Transistor

TECO Procedures

1. Open the TECO SG2 Client program by clicking on the SG2 Client icon.

2. In the program window, select NEW LADDER LOGIC PROGRAM.

3. The screen will display a Select Model Type window. Select model SG2-10HR-A and answer the following:

   a. What is the input power rating?

   _____

   b. How many inputs does this model have?

   _____

   c. How many analog inputs does this model have?

   _____

   d. How many outputs does this model have?

   _____

   e. What type of output contact does this model have?

   _____

   f. What is the current rating for each output?

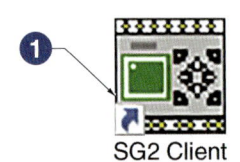

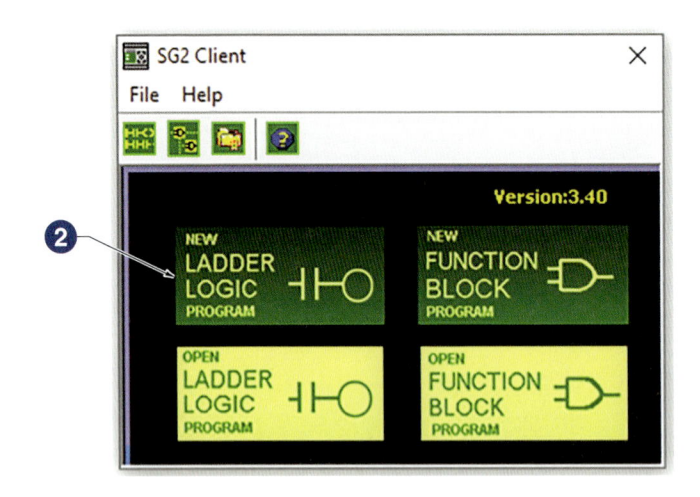

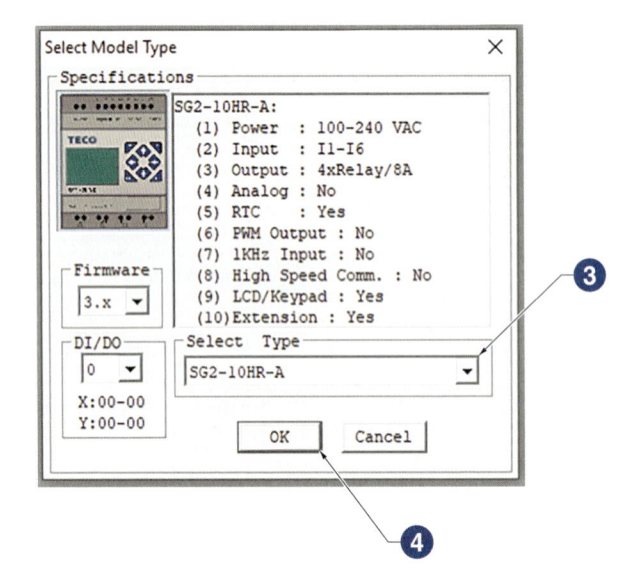

   _____

4. Click on the OK button, and the programming window will appear.

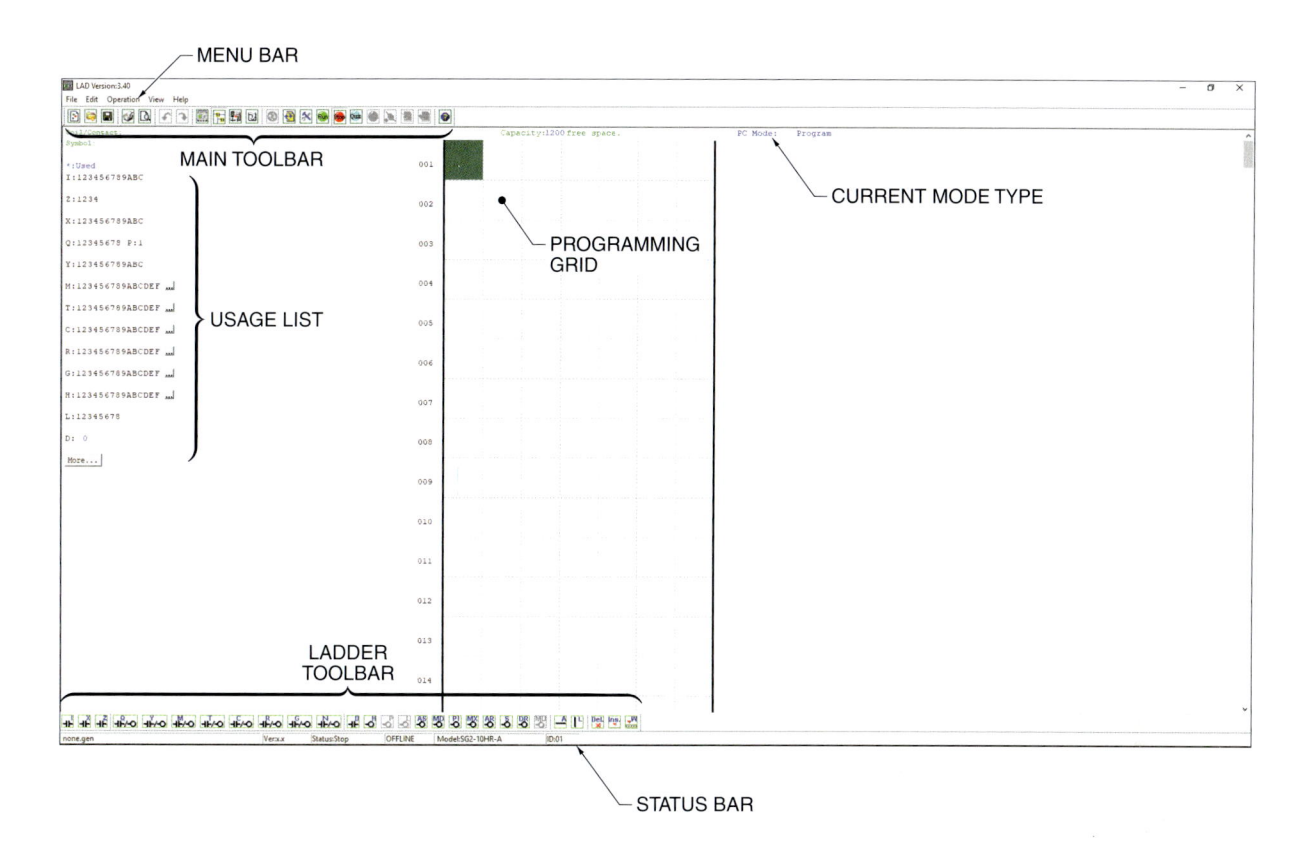

5. At the top of the programming window on the main toolbar, click on the Keypad icon, and then answer the following:

   a. Does the PLC depiction show the input power voltage rating?

   _____

   b. Does the PLC depiction show the output contact type?

   _____

   c. Does the PLC depiction show the current rating of each output?

   _____

   d. Does the PLC depiction show how the inputs and outputs are wired?

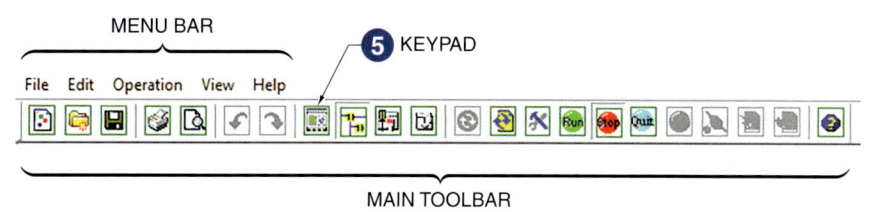

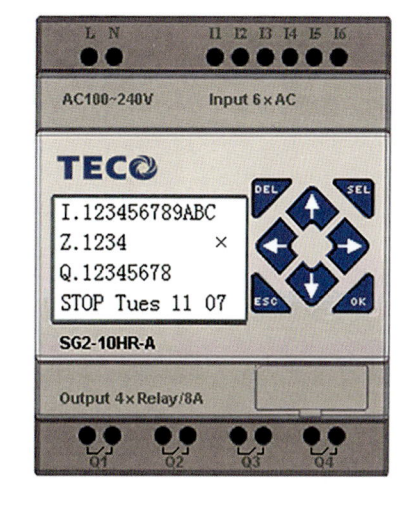

   _____

6. On the main toolbar, click on the Simulator icon to show how the inputs and outputs are wired, and then answer the following:

   a. Is the incoming power wired to power L or N?

   _____

   b. Are the output contacts wired to red or black for power?

   _____

7. Close the SG2 Client program window and then the SG2 Client software window.

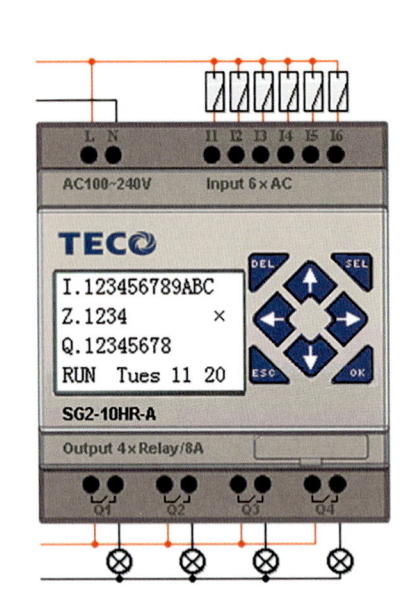

TECO Procedures

1. In the SG2 Client program window, select NEW LADDER LOGIC PROGRAM.

2. The screen will display a Select Model Type window. Select model SG2-12HR-D and answer the following:

   a. What is the input power rating?

   _____

   b. How many inputs does this model have?

   _____

   c. How many analog inputs does this model have?

   _____

   d. How many outputs does this model have?

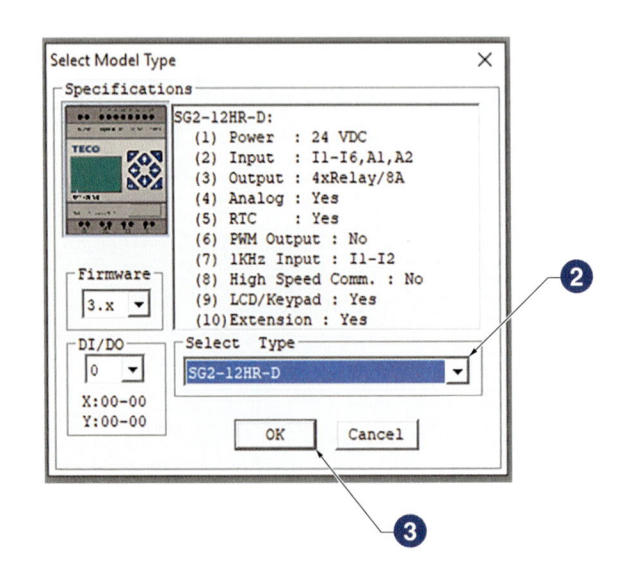

   _____

e. What type of output contact does this model have?

_____

f. What is the current rating for each output?

_____

3. Click on the OK button, and the programming window will appear.

4. At the top of the programming window on the main toolbar, click on the Keypad icon and answer the following:

   a. Does the PLC depiction show the input power voltage rating?

   _____

   b. Does the PLC depiction show the output contact type?

   _____

   c. Does the PLC depiction show the current rating of each output?

   _____

   d. Does the PLC depiction show how the inputs and outputs are wired?

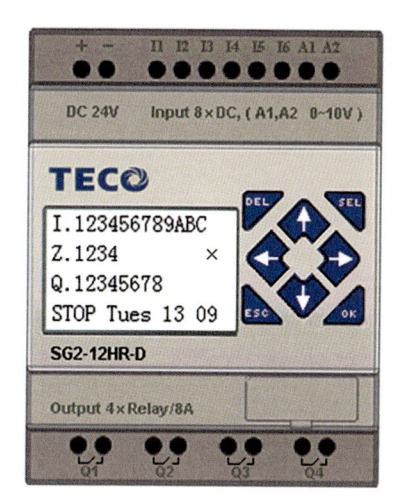

   _____

5. On the main toolbar, click on the Simulator icon to show how the inputs and outputs are wired, and then answer the following:

   a. Are the inputs wired as sink or source?

   _____

   b. Are the output contacts wired to red or black for power?

   _____

6. Close the SG2 Client program window and then the SG2 Client software window.

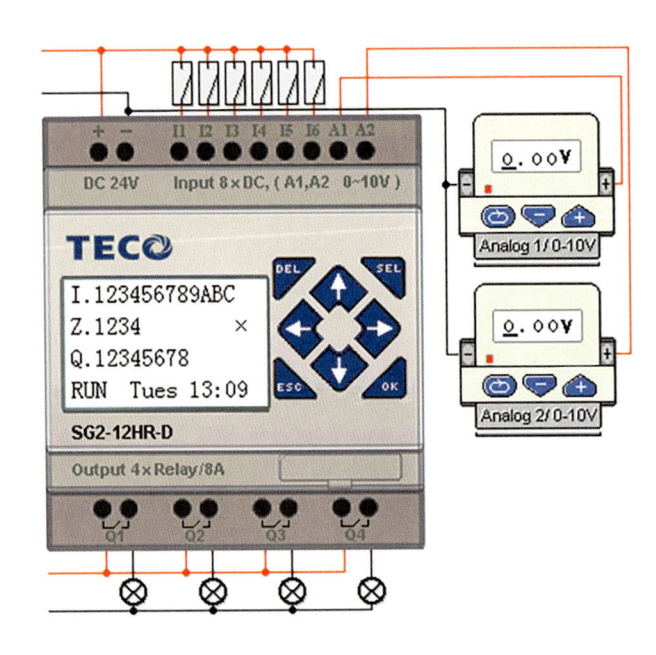

TECO Procedures

1. In the SG2 Client program window, select NEW LADDER LOGIC PROGRAM.

2. The screen will display a Select Model Type window. Select model SG2-12HT-D, and then answer the following:

   a. What is the input power rating?

   _____

   b. How many inputs does this model have?

   _____

   c. How many analog inputs does this model have?

   _____

   d. How many outputs does this model have?

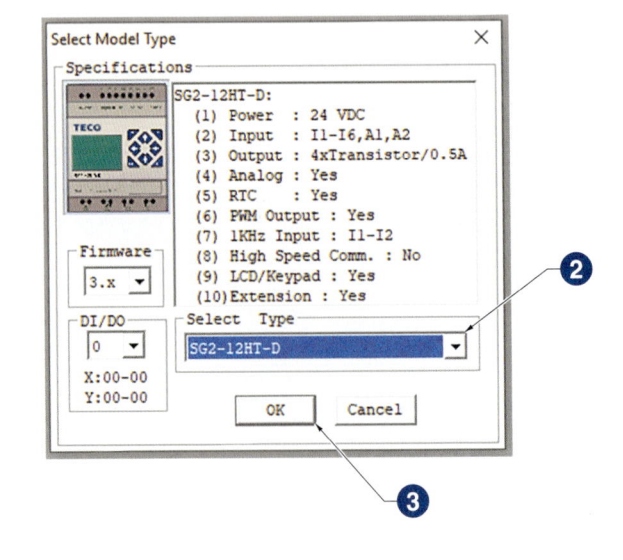

   _____

e. What type of output contact does this model have?

_____

f. What is the current rating for each output?

_____

3. Click on the OK button, and the programming window will appear.

4. At the top of the programming window on the main toolbar, click on the Keypad icon and answer the following:

a. Does the PLC depiction show the input power voltage rating?

_____

b. Does the PLC depiction show the output contact type?

_____

c. Does the PLC depiction show the current rating of each output?

_____

d. Does the PLC depiction show how the inputs and outputs are wired?

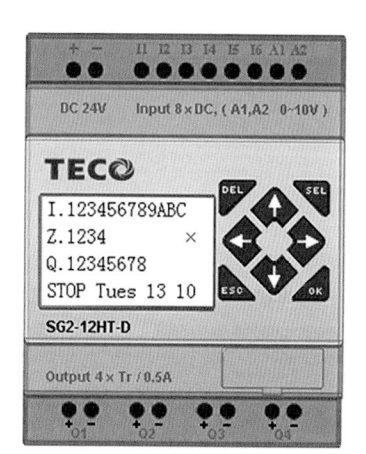

_____

5. On the main toolbar, click on the Simulator icon to show how the inputs and outputs are wired, and then answer the following:

   a. Are the inputs wired as sink or source?

   _____

   b. Are the output contacts wired to red or black for power?

   _____

6. Close the SG2 Client program window and then the SG2 Client software window.

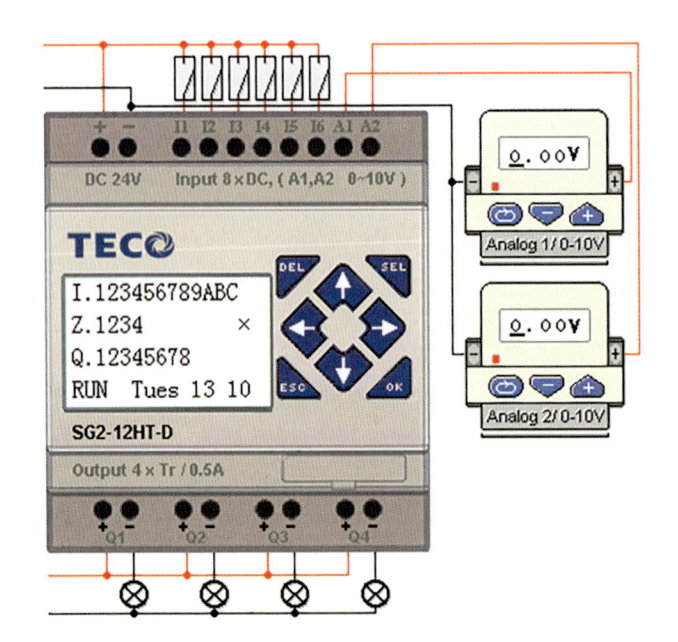

Wiring and Programming PLC Input and Output Connections

PLC Exercise 1.3

BILL OF MATERIALS
- TECO PLR or MicroLogix™ 1000 PLC
- NC red pushbutton
- NO green pushbutton
- Motor starter with overload/contactor
- Load (optional)

For this lab, start and stop pushbuttons will control a motor starter coil to control power to a motor. Use the SG2 Client software and a TECO SG2-12HR-D PLC or the RSLogix Micro Starter Lite software and an AB Micrologix™ 1100 PLC to program and wire a three-wire start/stop motor starter PLC circuit. The output contacts of the PLC may be a mechanical or a solid-state type to control either a DC or AC motor starter.

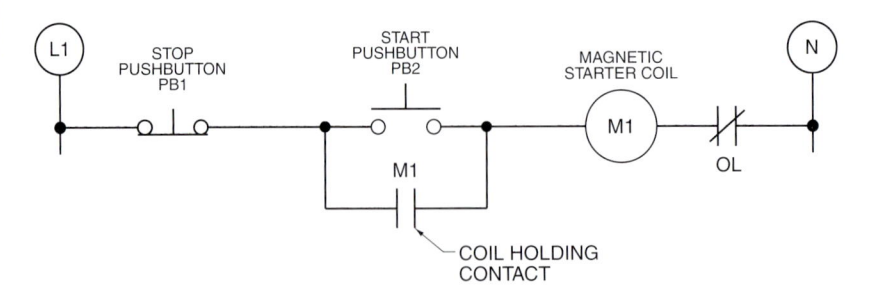

LINE DIAGRAM

TECO Procedures

1. Open the TECO SG2 Client program by clicking on the SG2 Client icon.

2. In the program window, select NEW LADDER LOGIC PROGRAM.

3. Use the TECO programming diagram to add the proper contacts and description in the Programming Grid for the start/stop motor starter circuit. Click and place the input and output instructions from the Ladder Toolbar into the Programming Grid. Inputs are abbreviated I, and outputs are abbreviated Q. The AND Line or OR Line is used to make connections between each I/O.

4. Use the Edit Contact window to denote an Input or Output number and the Symbol button to add a description after placing the input or output in the Programming Grid.

5. Run the program in Simulation Mode to verify proper operation. Click on the Simulator icon on the main toolbar. Use the Input Status Tool window to activate the inputs and control the output to simulate circuit operation.

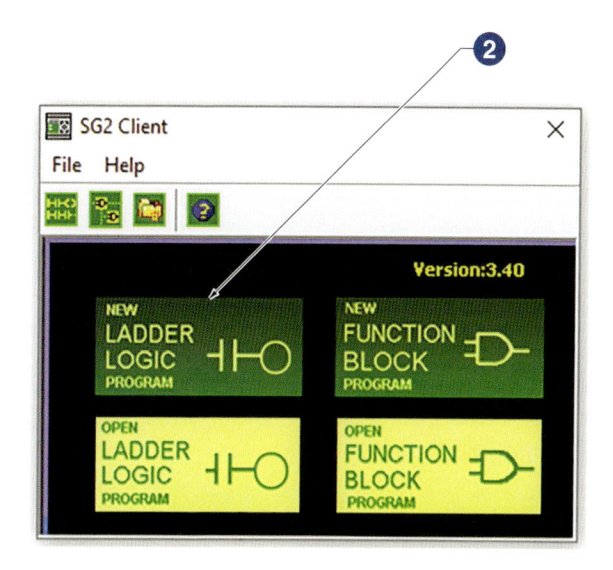

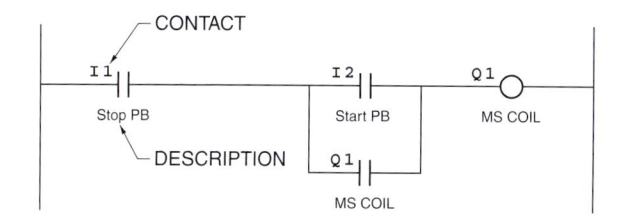

TECO PROGRAMMING DIAGRAM

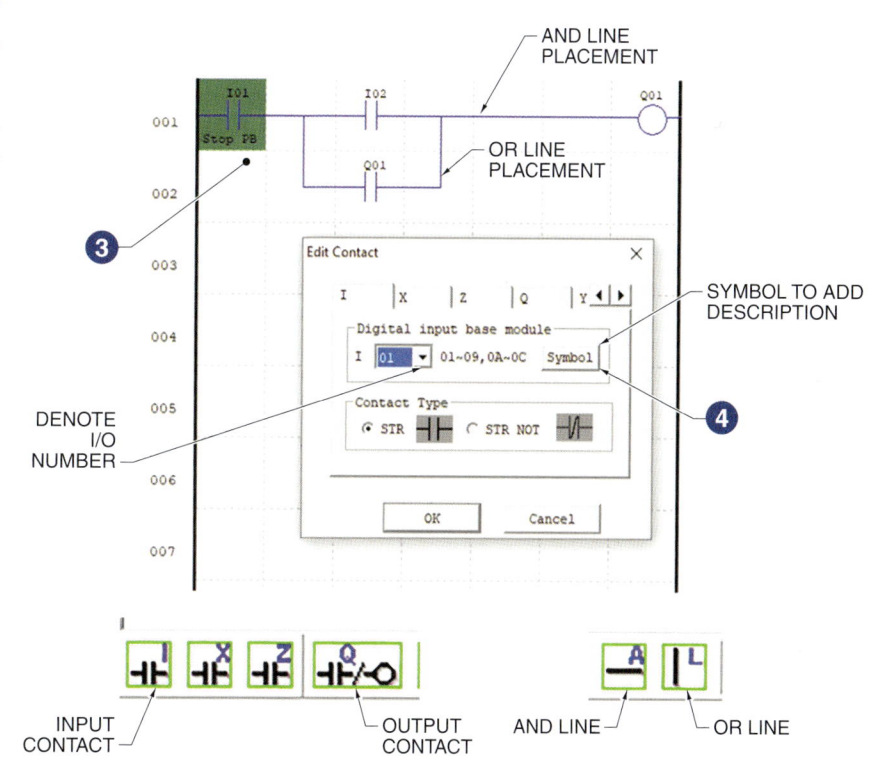

6. Use the wiring diagram to wire the inputs, using an NC red pushbutton as a stop switch and an NO green pushbutton as a start switch.

7. Wire the output components to the PLC, using a motor starter/contactor to control a motor or load.

8. Turn power to the PLC ON and link the com port to establish communication between the PC and the TECO SG2-12HR-D. On the Menu Bar, click Operation and then Link Com Port. Select the port to be used by the PC and click Link.

9. After successfully linking the com port, write the program to the SG2-12HR-D to download the program logic to the PLC. On the Menu Bar, click Operation and then Write.

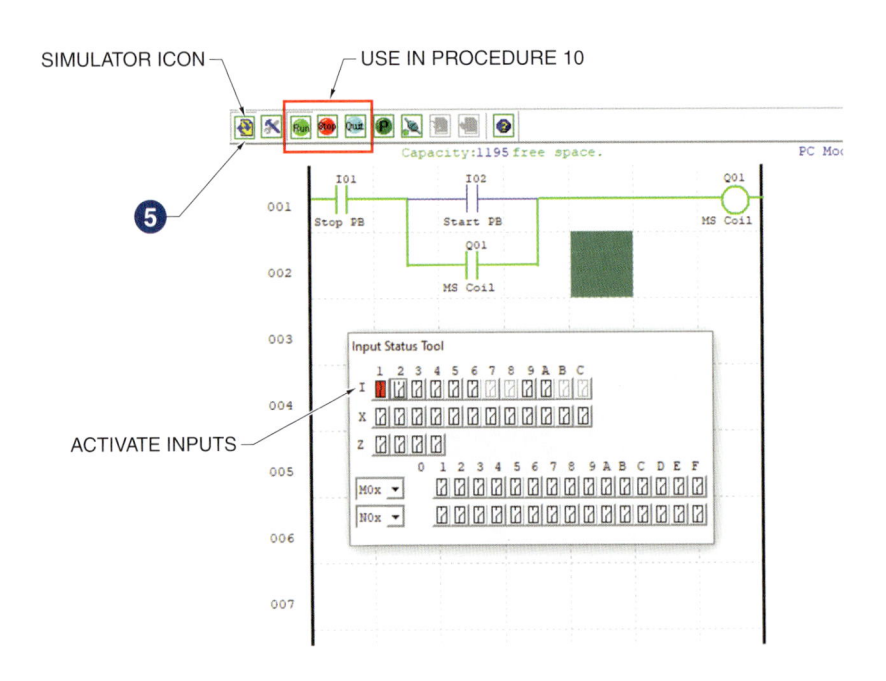

⚠ CAUTION

Ensure all power is OFF to a PLC and circuit before beginning any wiring procedures. Any load connected to the PLC outputs must be within the voltage and current rating of the output contacts.

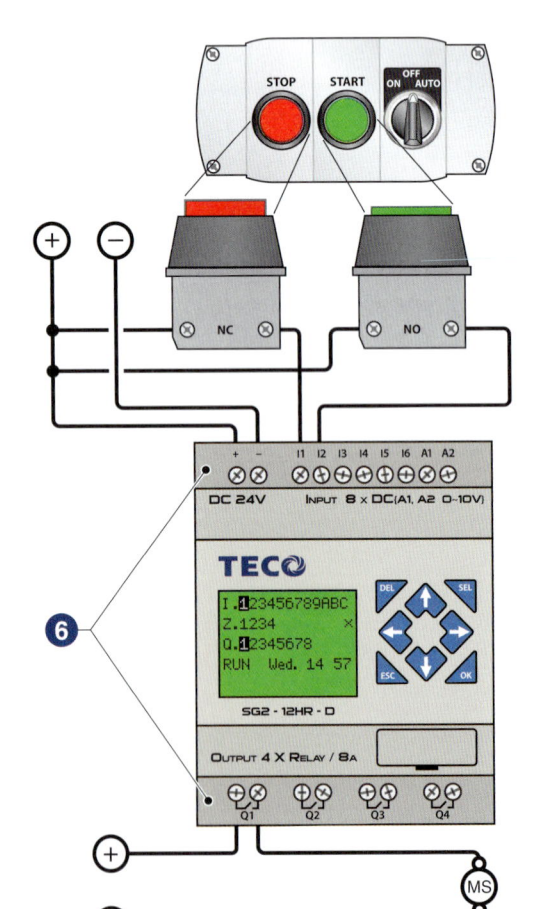

TECO WIRING DIAGRAM

10. Once the program is downloaded to the PLC, place the SG2-12HR-D in Run Mode. Activate the Start pushbutton and monitor the I/Os on the LCD display.

11. Stop the program and use the Save As function to save and place the lab application at the required folder location.

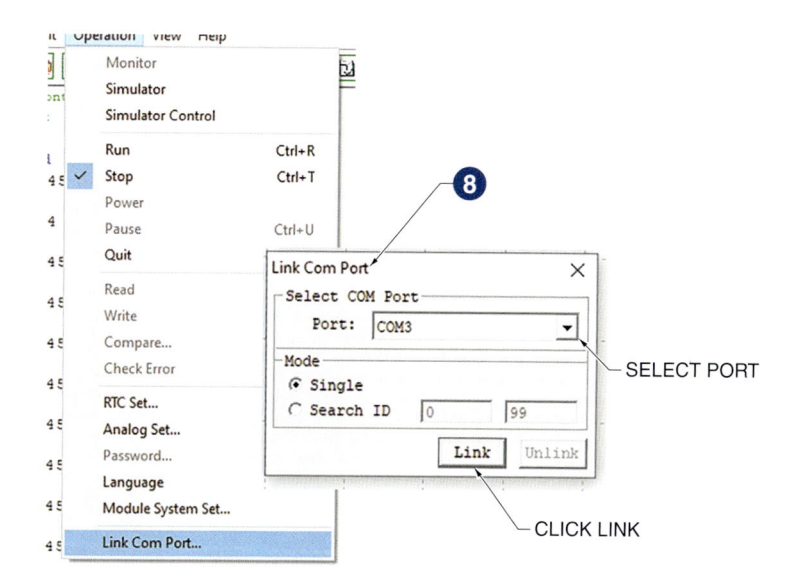

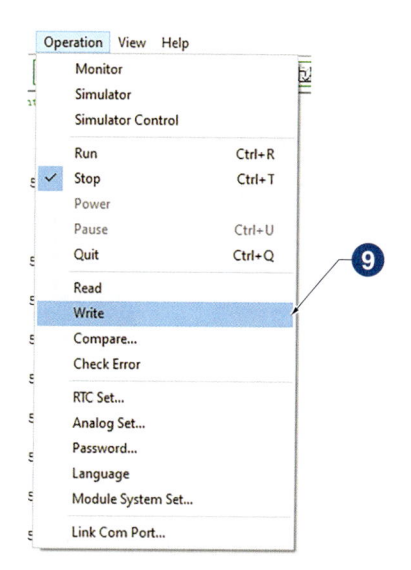

PROGRAMMABLE LOGIC CONTROLLERS: PRINCIPLES AND APPLICATIONS

Allen-Bradley® Procedures

1. Open the RSLogix™ Micro Starter Lite program. In the program window, click File and New... to create a new programming diagram. In the Select Processor Type window, select the type of MicroLogix™ 1100 used in the application and click OK.

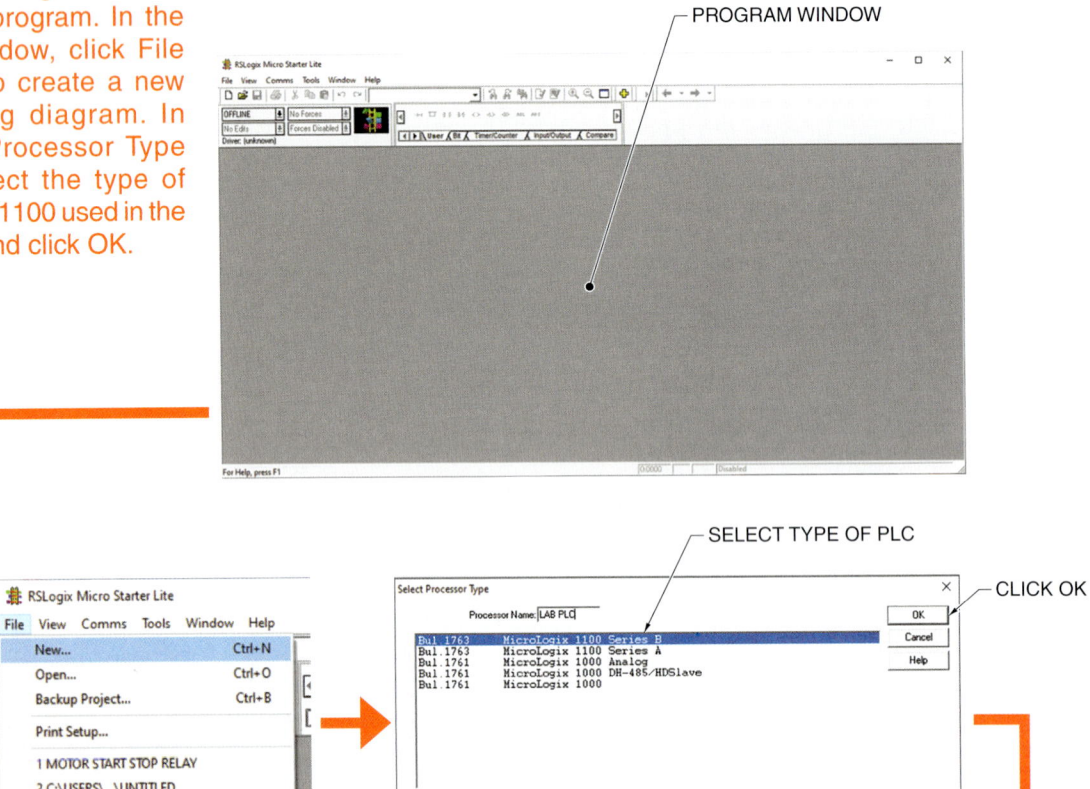

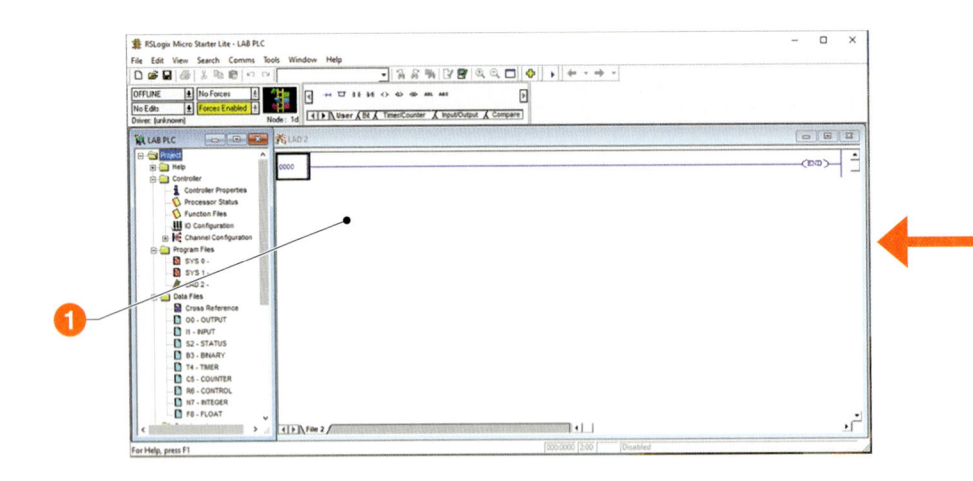

— PROGRAM WINDOW

— SELECT TYPE OF PLC

— CLICK OK

2. Use the AB programming diagram for the start/stop motor starter circuit to add the proper address and description in the LAD 2 programming window. Use the Instruction Toolbar to add a new rung. Drag and drop inputs and the output into the LAD 2 programming window.

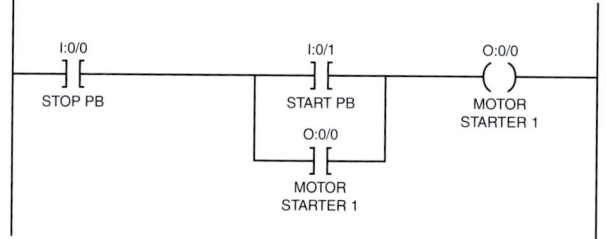

AB PROGRAMMING DIAGRAM

- I:0/0 — STOP PB
- I:0/1 — START PB
- O:0/0 — MOTOR STARTER 1
- O:0/0 — MOTOR STARTER 1

LAD 2 PROGRAMMING WINDOW

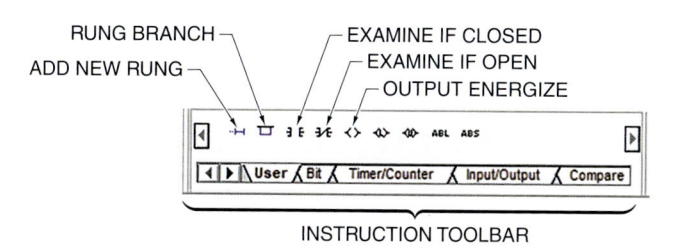

INSTRUCTION TOOLBAR
- RUNG BRANCH
- ADD NEW RUNG
- EXAMINE IF CLOSED
- EXAMINE IF OPEN
- OUTPUT ENERGIZE

3. Verify that the programming diagram has no errors. Use the Edit menu and click Verify Project. Save the program by clicking the File menu and Save as....

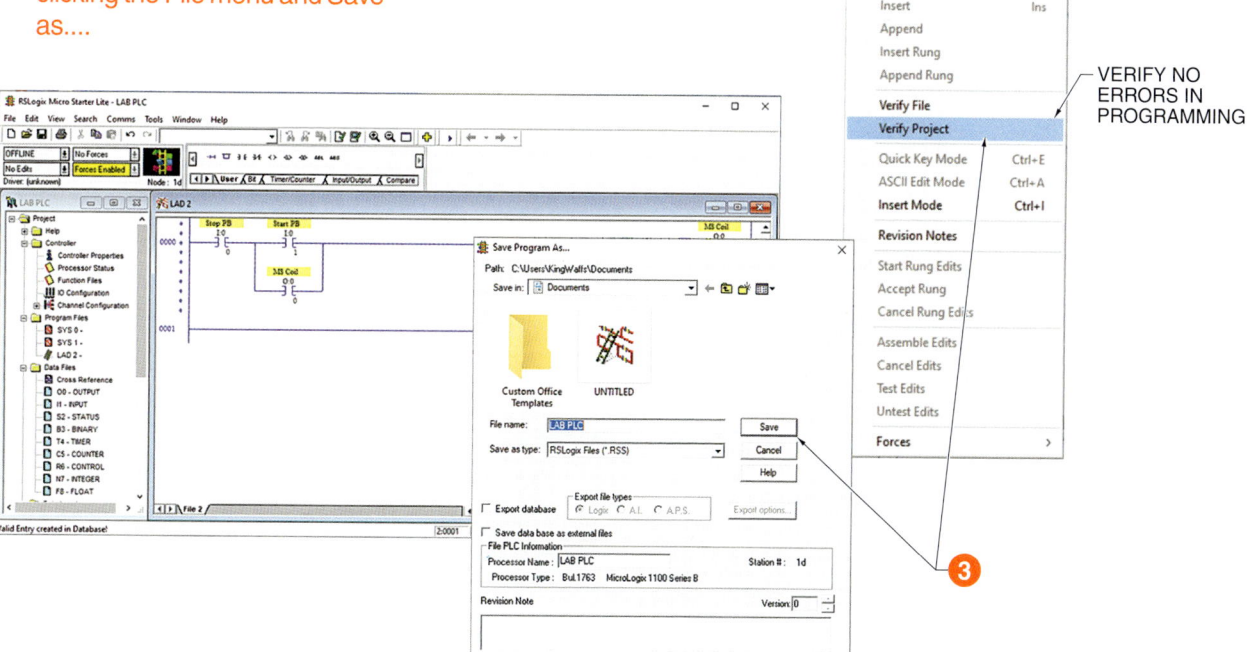

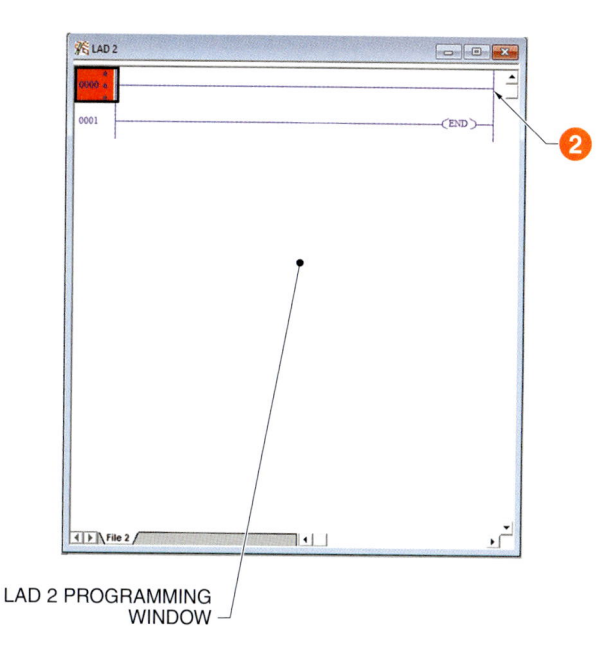

VERIFY NO ERRORS IN PROGRAMMING

4. Use the wiring diagram to wire the inputs. Use an NC red pushbutton as a stop switch and an NO green pushbutton as a start switch. Wire the motor starter coil/contactor to the output side of the MicroLogix™ 1100. Once all connections have been safely made, turn power to the PLC ON.

> **⚠ CAUTION**
>
> Ensure all power is OFF to a PLC and circuit before beginning any wiring procedures. Any load connected to the PLC outputs must be within the voltage and current rating of the output contacts.

5. Use RSLinx™ to establish communication between the PC and the Micrologix™ 1100.

6. In the Online Bar, select Download from the Operational Mode dropdown list. Click Yes to proceed with the download and OK in the Revision Note window.

7. After sucessfully downloading to the PLC, place the PLC in Run Mode. Go online and, in the Online Bar, select Run from the Operational Mode dropdown list.

8. Activate inputs and monitor them to determine correct operation of the output.

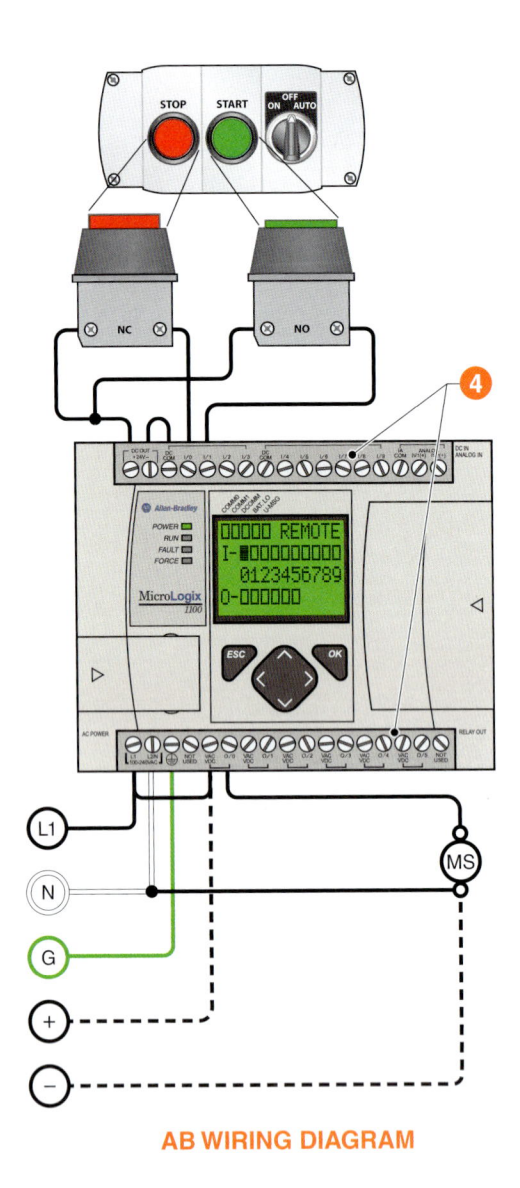

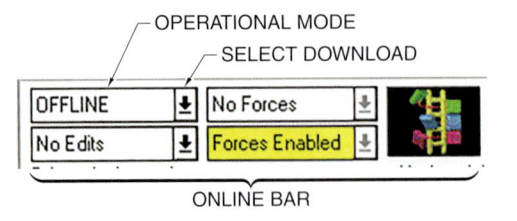

AB WIRING DIAGRAM

ONLINE BAR
- OPERATIONAL MODE
- SELECT DOWNLOAD

Chapter 1—Programmable Logic Controllers and Critical Safety Practices **67**

CLICK

ONLINE BAR

⑥

SELECT RUN

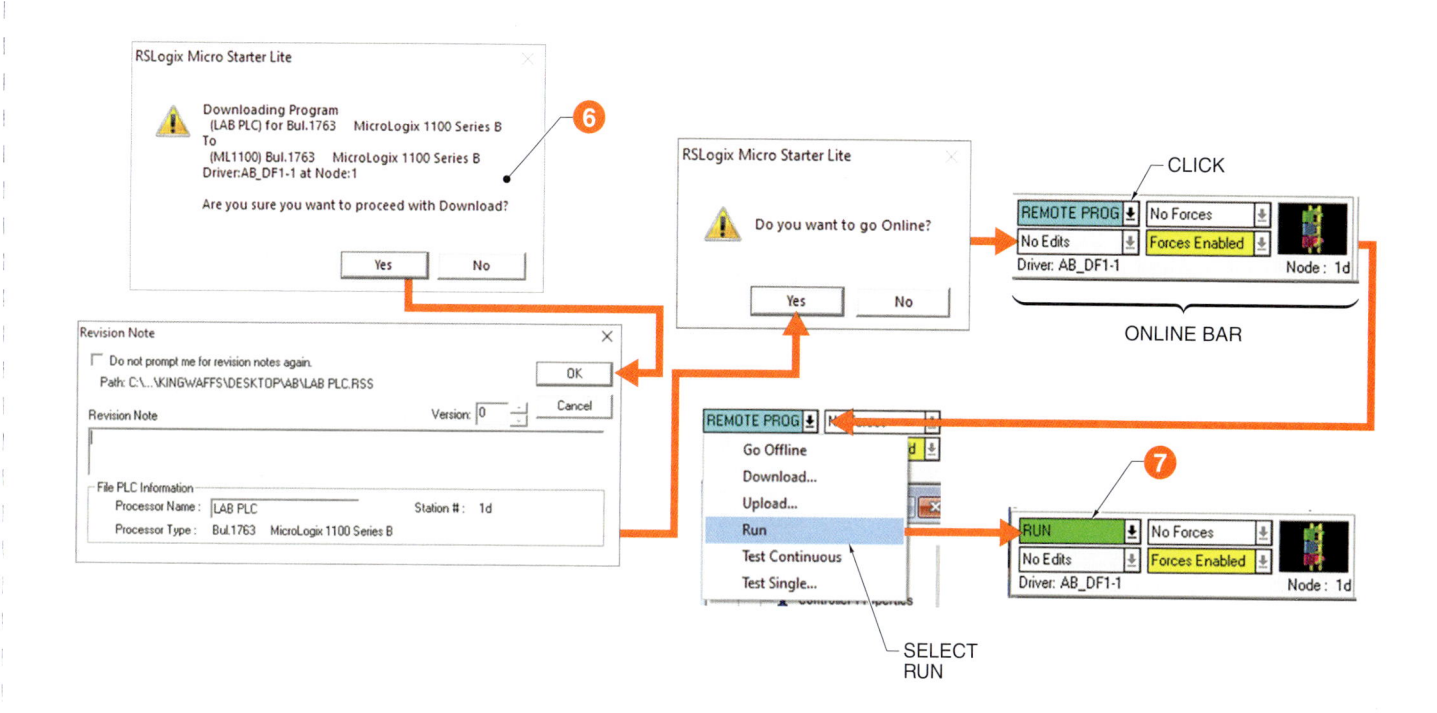

⑦

Light Indicators for Output Operation

PLC Exercise 1.4

BILL OF MATERIALS
- TECO PLC or MicroLogix™ 1100
- One NC stop pushbutton
- One NO start pushbutton
- Three lights (red/green/yellow)
- Motor starter/contactor

Using the PLC circuit from Exercise 1.3, modify the PLC circuit and programming diagram by installing and programming lights to provide a visible indication of a load's operation. A green light indicates the load is ON, and a red light indicates the load is OFF. A yellow light may be added to the circuit to indicate there is power present at the PLC I/Os. The yellow light will not be wired or programmed to the PLC.

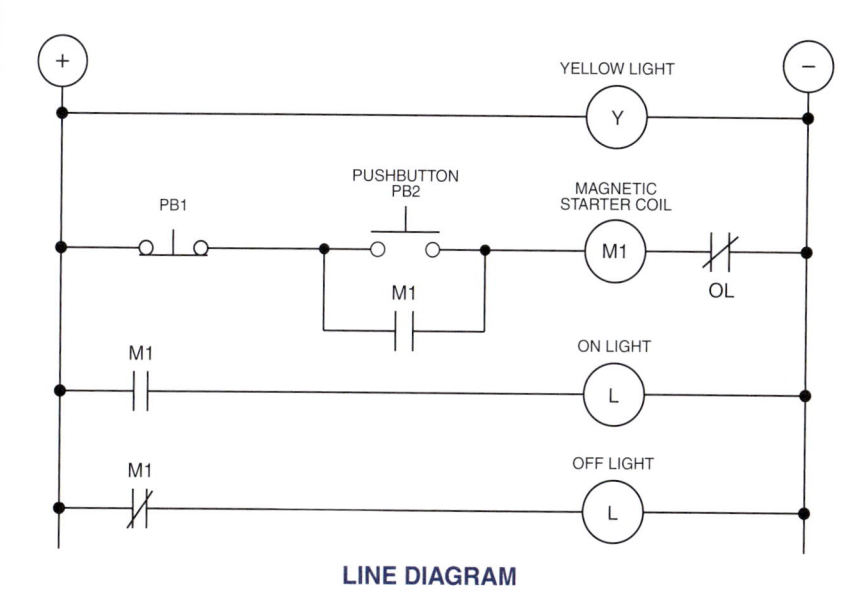

LINE DIAGRAM

TECO Procedures

1. Create a new ladder logic program.

2. Use the TECO programming diagram for the start/stop motor starter circuit with indicator lights to add the proper contacts and description in the programming grid.

3. Edit each contact with a proper number and description.

4. Run the program in Simulation Mode to verify proper operation.

5. Use the wiring diagram to wire the inputs. Be sure to use the NC red pushbutton as the stop switch and the NO green pushbutton as the start switch.

6. Wire the output components to the PLC, using a motor starter/contactor to control a motor or load as output 1, the green light as output 2, and the red light as output 3.

7. Turn power to the PLC ON and link the com port to establish communication between the PC and the TECO SG2-12HR-D.

8. Write the program to the SG2-12HR-D to download the program logic to the PLC.

9. Place the SG2-12HR-D in Run Mode. Activate the start pushbutton and monitor the I/Os on the LCD display.

10. Stop the program and use the Save As function to save and place the lab application to the required folder location.

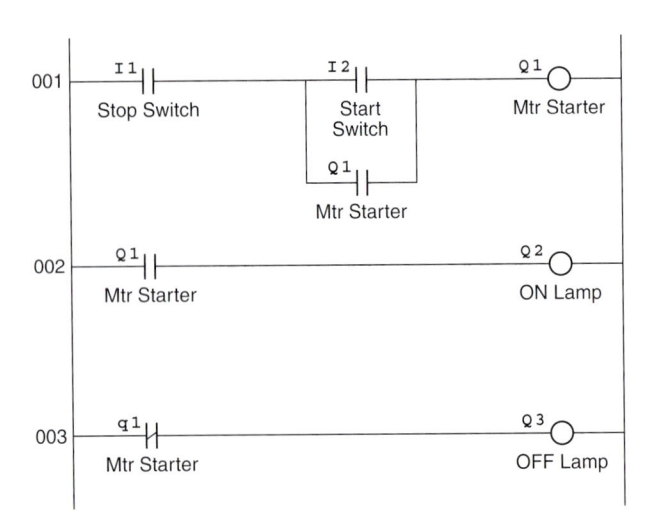

TECO PROGRAMMING DIAGRAM

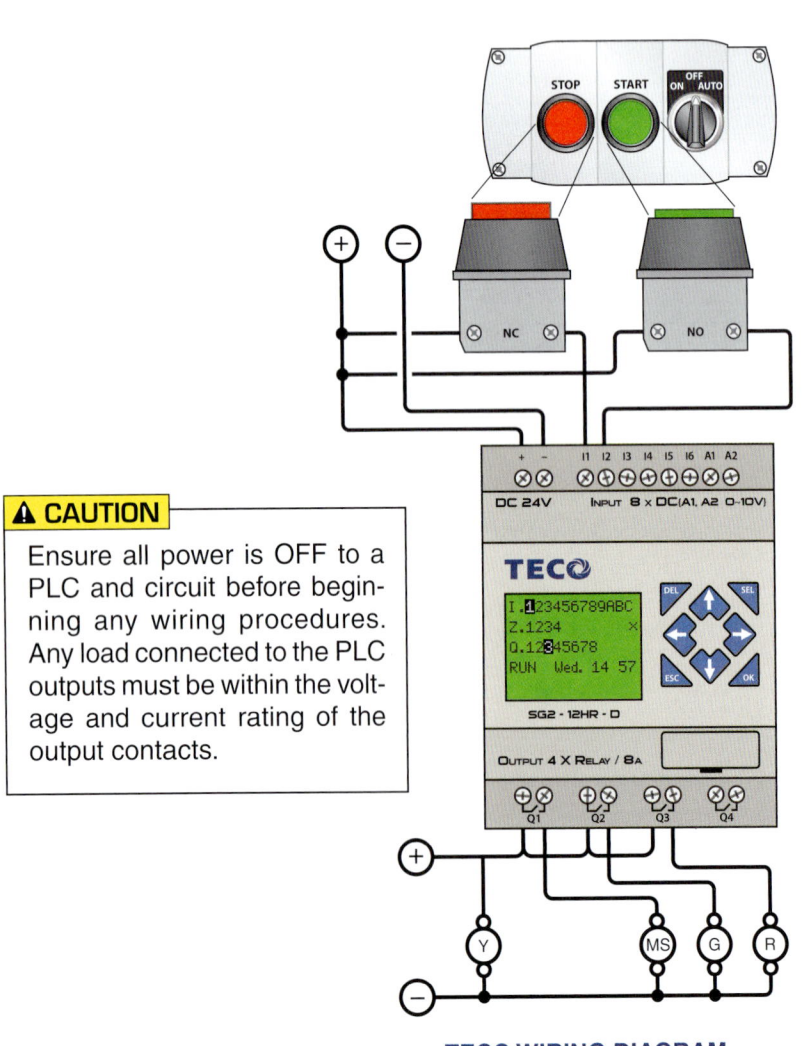

⚠ **CAUTION**
Ensure all power is OFF to a PLC and circuit before beginning any wiring procedures. Any load connected to the PLC outputs must be within the voltage and current rating of the output contacts.

TECO WIRING DIAGRAM

Allen-Bradley® Procedures

1. Start a new programming diagram.

2. Use the AB programming diagram for the start/stop motor starter circuit with indicator lights to add the proper address and description in the programming window.

3. Verify that the programming diagram has no errors by using Verify Project. Save the program.

4. Use the wiring diagram to wire the inputs. Wire an NC red pushbutton as the stop switch to I/0 and an NO green pushbutton as the start switch to I/1. Wire the motor starter coil to the O/0 contact, the green light to the O/1 contact, and the red light to the O/2 contact. Turn power to the PLC ON.

5. Use RSLinx™ to establish communication between the PC and the Micrologix™ 1100.

6. Download the program to the PLC and go online.

7. Place the PLC in Run Mode.

8. Activate inputs and monitor them to determine the correct operation of the circuit.

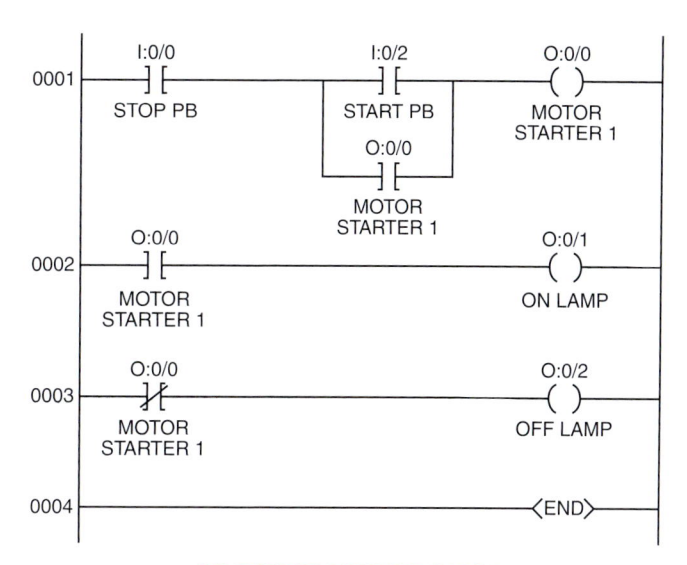

AB PROGRAMMING DIAGRAM

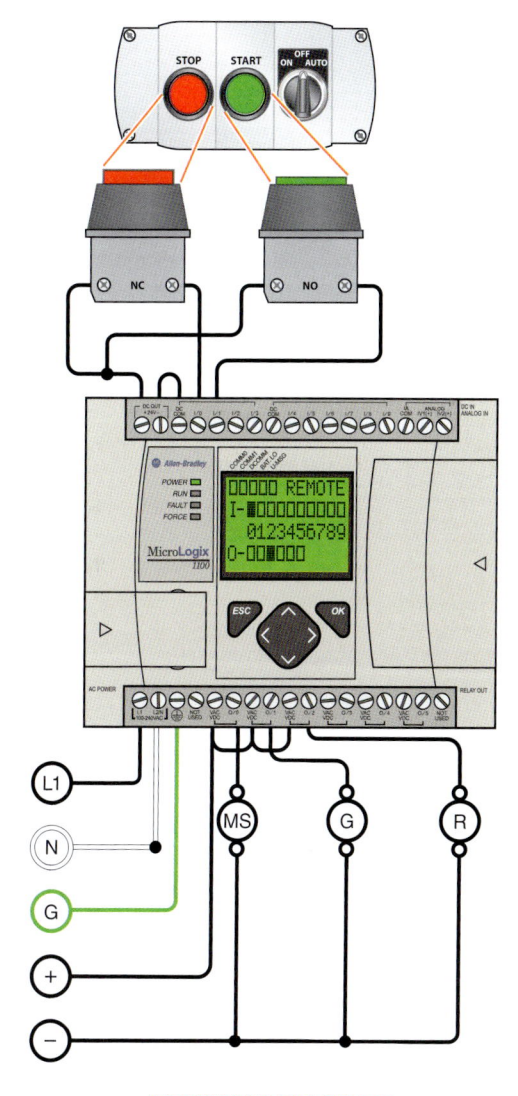

AB WIRING DIAGRAM

SIEMENS Programming and Wiring

Technicians using SIEMENS STEP 7 software and Simatic S7-1200 PLCs can click on the Learner Resources tab labeled "SIEMENS Resources" for the S7-1200 Easy Book for help in programming the equivalent TECO and Allen-Bradley® exercises.

Programming Symbols

Instruction	TECO SG2 Client	AB RSLOGIX	Siemens STEP 7
Input (examine if closed)	─| |─	─] [─	─| |─
Input (examine if open)	─|/|─	─]/[─	─|/|─
Output	─○─	─( )─	─( )─
Output Latch (set)	─(↑)─	─(L)─	─(S)─
Output Unlatch (reset)	─(↓)─	─(U)─	─(R)─
Timer On-Delay	─○─ TO1 *Must select mode when programming	TON Timer On Delay Timer 0 Time Base 1.0 Preset 120 Accum 0 <EN> <DN>	─(TON)─
Timer Off-Delay	─○─ TO1 *Must select mode when programming	TOF Timer Off Delay Timer 1 Time Base 0.01 Preset 215 Accum 0 <EN> <DN>	─(TOF)─
Retentive Timer	─○─ TO1 *Must select mode when programming	RTO Retentive Timer On Timer 2 Time Base 0.001 Preset 58 Accum 0 <EN> <DN>	─(TONR)─
Count Up	─○─ CO1 *Must select mode when programming	CTU Count Up Counter 0 Preset 173 Accum 0 <EN> <DN>	* CTU INT CU Q R CV PV
Count Down	─○─ CO1 *Must select mode when programming	CTD Count Down Counter 1 Preset 250 Accum 0 <EN> <DN>	* CTD INT CD Q LD CV PV

* See Table 6-17/6-18 of S7-1200 Easy Book.

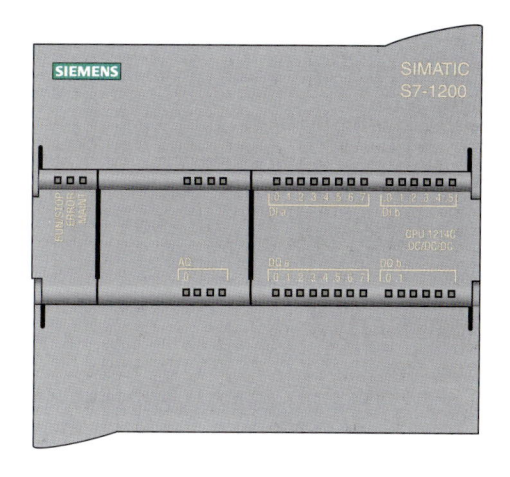

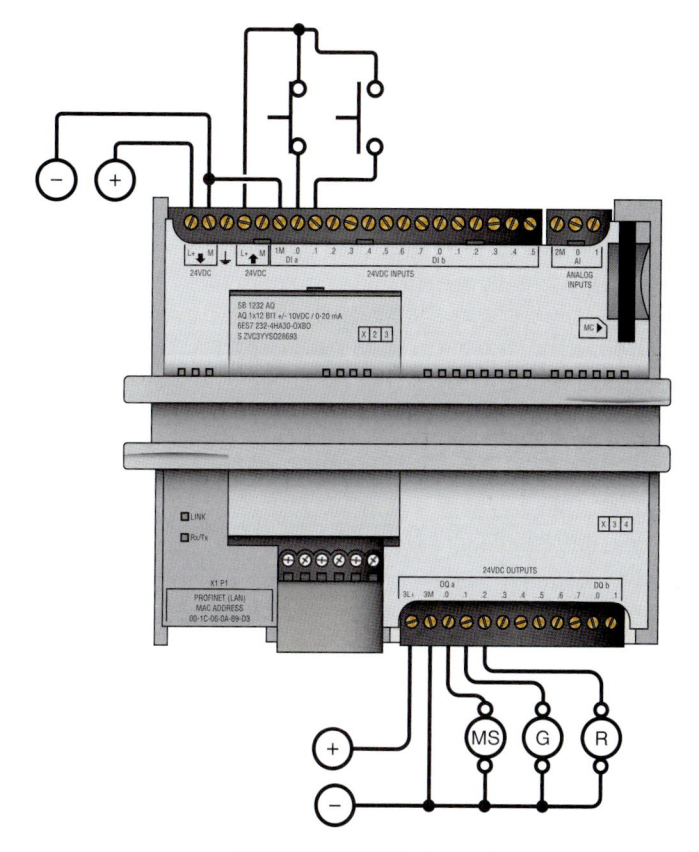

SIEMENS WIRING DIAGRAM

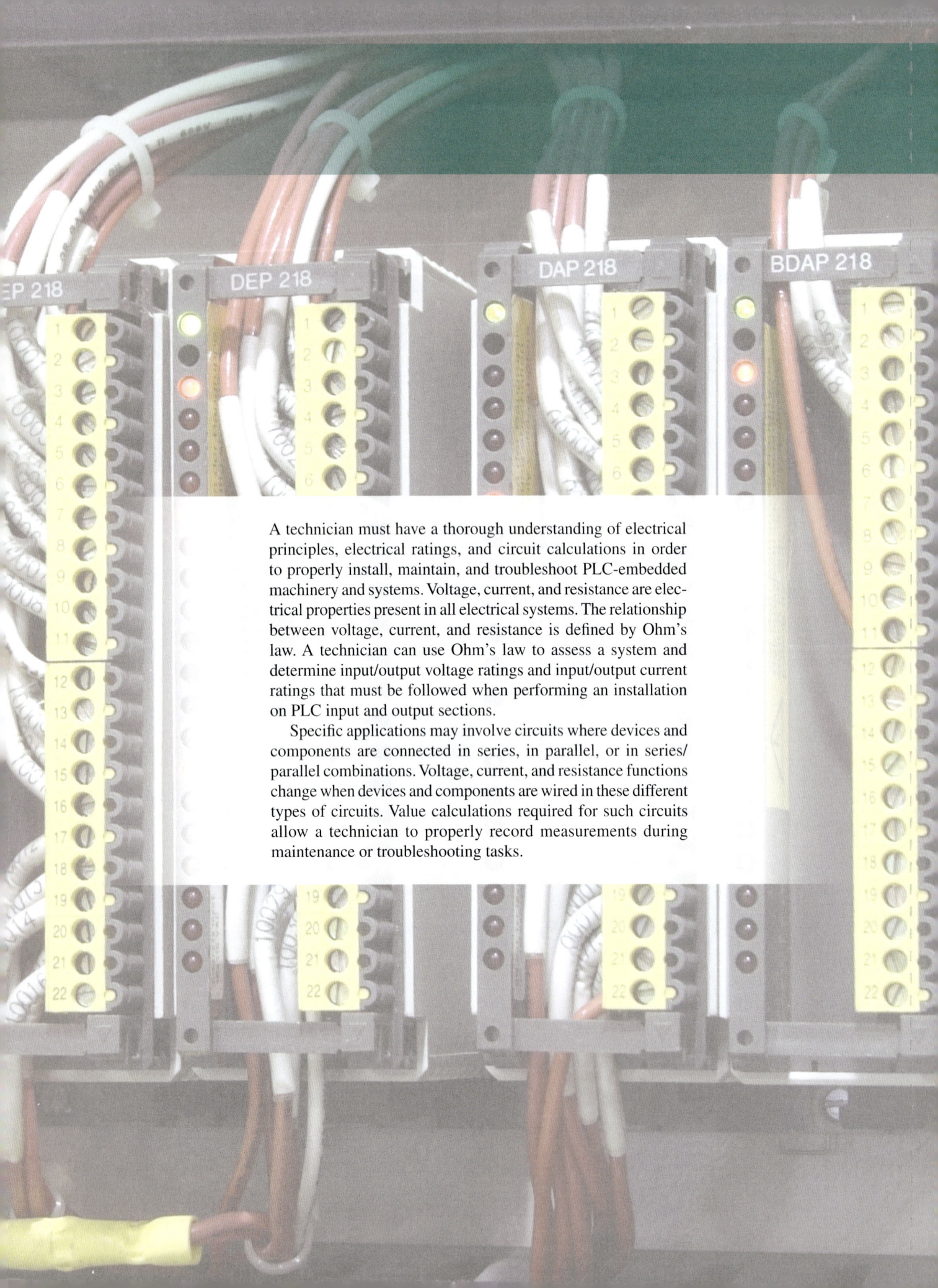

A technician must have a thorough understanding of electrical principles, electrical ratings, and circuit calculations in order to properly install, maintain, and troubleshoot PLC-embedded machinery and systems. Voltage, current, and resistance are electrical properties present in all electrical systems. The relationship between voltage, current, and resistance is defined by Ohm's law. A technician can use Ohm's law to assess a system and determine input/output voltage ratings and input/output current ratings that must be followed when performing an installation on PLC input and output sections.

Specific applications may involve circuits where devices and components are connected in series, in parallel, or in series/parallel combinations. Voltage, current, and resistance functions change when devices and components are wired in these different types of circuits. Value calculations required for such circuits allow a technician to properly record measurements during maintenance or troubleshooting tasks.

PLC Electrical Principles, Ratings, and Circuit Calculations

2

Objectives:

- Describe the electrical circuits that connect input devices to the input sections and output devices to the output sections of PLCs.
- Explain why polarity is important when connecting input devices and output devices to some PLCs.
- Identify the ratings used for the power supply, input sections, and output sections of PLCs.
- Explain current sinking and current sourcing as they pertain to PLC input sections.
- Explain the use of Ohm's law in series, parallel, and combination series/parallel circuits.
- Calculate the voltage drops created by loads, current, and resistance in series, parallel, and combination series/parallel circuits.

Learner Resources
atplearningresources.com/quicklinks
Access Code: 475203

PLCs AND ELECTRICAL PRINCIPLES

PLCs are one electrical component that is part of a much larger electrical system. All PLCs are connected to input devices (switches and sensors) connected to input terminals, and output components (motor starters and lights) connected to output terminals. To understand how to properly install, maintain, and troubleshoot an electrical control system that includes a PLC or several interconnected PLCs, technicians must understand electrical principles and properties. Electrical properties include voltage, current, power, and resistance. Electrical principles include the relationships between electrical properties.

PLC circuits and all devices connected to and controlled by PLCs demonstrate electrical principles. PLC hardwired circuits and programs also make use of electrical principles as applied to series, parallel, and series/parallel-connected components and circuits. **See Figure 2-1.** Understanding electrical principles is especially important when there is a problem in an electrical control system that requires troubleshooting, such as having output components wired in series. (Output components cannot be connected in series to PLC output modules.)

> **⚠ CAUTION**
>
> Remove system power before making or breaking PLC cable connections. When removing or inserting cable connectors that are energized, an electrical arc occurs. Electrical arcs can cause personal injury, damage to PLCs, and unintended machine operation.

CODE CONNECT

ARTICLE 240 — OVERCURRENT PROTECTION

Section 240.6(A) of the NEC® addresses the standard ampere ratings for fuses and inverse-time circuit breakers. Table 240.6(A) lists the standard ampere ratings for fuses. Frequently, fuses with low ampere ratings are used in industrial control panels that contain PLCs. A technician may need to consult a fuse manufacturer's catalog to select a fuse with the appropriate voltage rating, ampere rating, interrupting rating, and trip characteristics.

Standard Ampere Ratings				
15 A	20 A	25 A	30 A	35 A
40 A	45 A	50 A	60 A	70 A
80 A	90 A	100 A	110 A	125 A
150 A	175 A	200 A	225 A	250 A
300 A	350 A	400 A	450 A	500 A
600 A	700 A	800 A	1000 A	1200 A
1600 A	2000 A	2500 A	3000 A	4000 A
5000 A	6000 A	—	—	—

PLC Electrical Circuits

PLC INPUT CIRCUITS

- SWITCHES IN SERIES CONNECTIONS
- SENSORS IN PARALLEL CONNECTIONS
- SWITCHES AND SENSORS IN SERIES/PARALLEL

PLC OUTPUT CIRCUITS

- SOLENOID AND RELAY SERIES CONNECTIONS
- MOTOR AND LIGHT PARALLEL CONNECTIONS
- SOLENOIDS AND LIGHT SERIES/PARALLEL CONNECTIONS

⚠ CAUTION

PLC power supplies can be damaged by voltage surges from other systems that are switching inductive loads such as motors, motor starters, solenoids, and relays ON and OFF. To avoid damaging the power supply of a PLC from systems switching inductive loads, an isolation transformer is used to isolate the PLC power supply from the harmful voltage surges. Also, a surge suppressor can be placed across any operating coil such as a motor starter, relay, solenoid, or contactor that is located in a PLC cabinet or close proximity of a PLC.

SURGE SUPPRESSOR

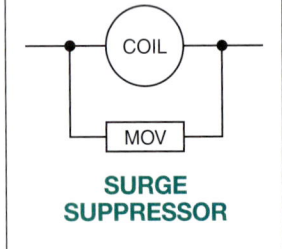

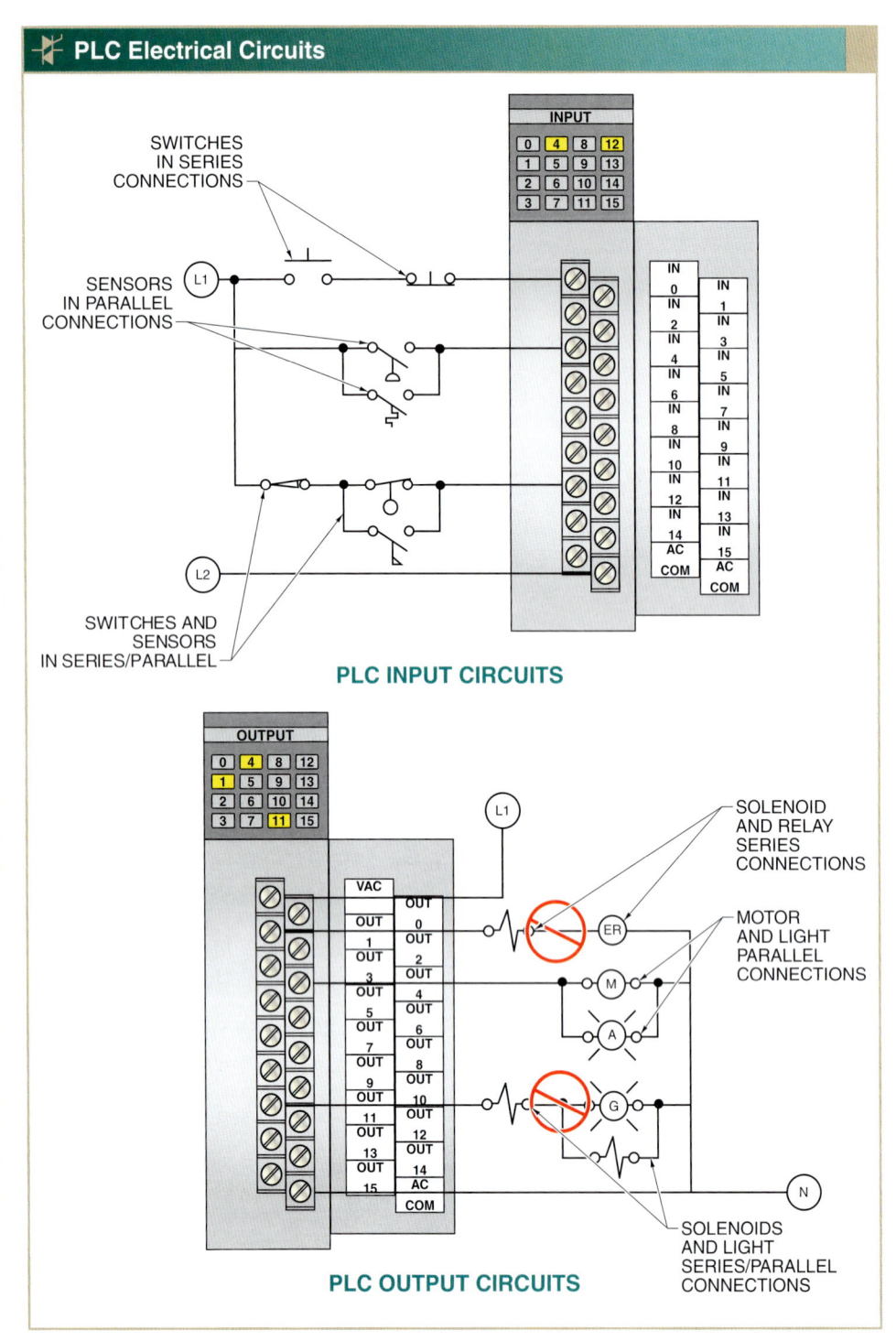

Figure 2-1. Basic rules must be followed when installing output devices to PLCs. Output devices cannot be installed in series from PLC output terminals.

PLC Problems

PLC problems typically fall into one of two categories: software and hardware. A *PLC software problem* is a condition where a PLC is properly installed and all inputs and outputs are working, but there is a problem with the PLC program. The PLC cannot execute the programmed instructions. **See Figure 2-2.**

PLC Software Problems

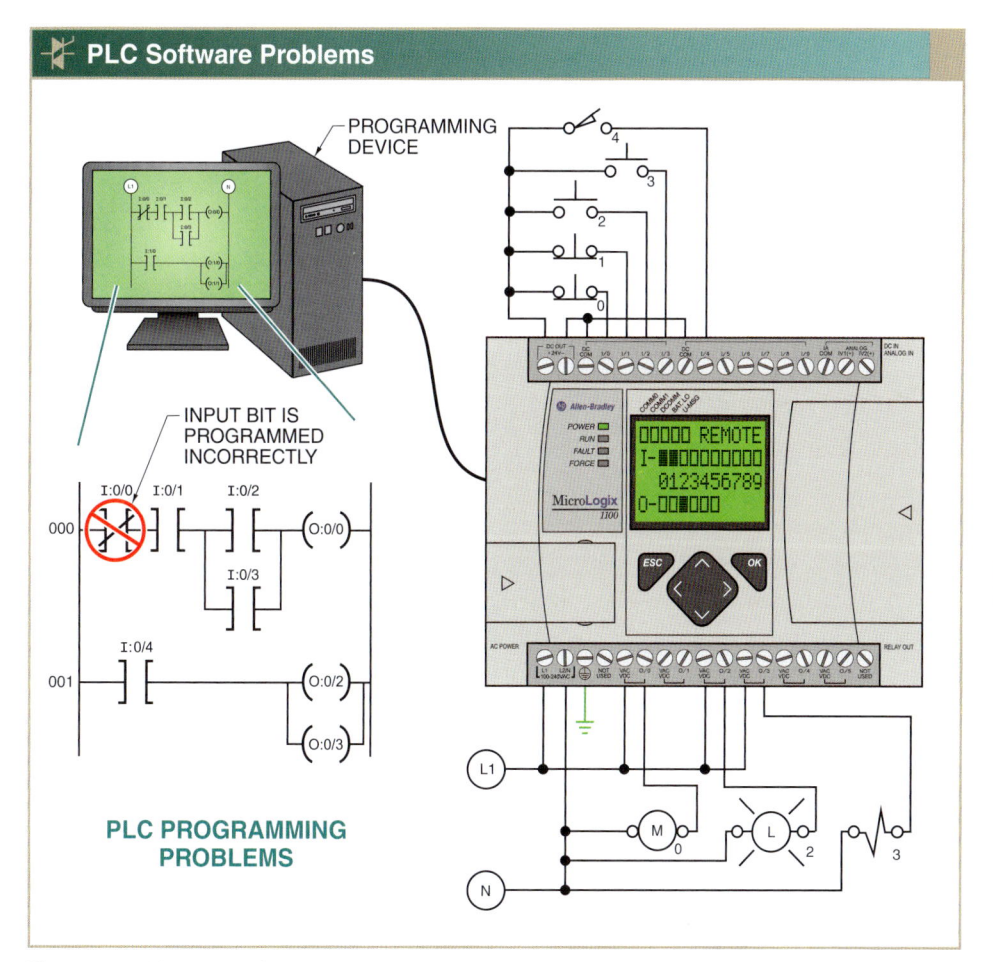

Figure 2-2. When all PLC system hardware is functioning but the system is not operating correctly, the problem may be with the software, such as an incorrect program, incorrect programmed symbols, or the program not being keyed correctly into the PLC.

CODE CONNECT
The NEC® requires ground-fault circuit interrupters (GFCIs) to protect personnel, and ground-fault protection of equipment (GFPE) to protect equipment in specific installations. GFPE is only intended to protect equipment, not personnel. Section 230.95 requires GFPE for solidly grounded wye electric services of more than 150 V to ground but not exceeding 1000 V phase-to-phase for each service disconnect rated 1000 A or more.

Technicians must understand the PLC software (program) for the type and model of PLC used and how the electrical control system is operating. Technicians may encounter input bits inside a new software program that are programmed incorrectly, such as a normally closed (NC) contact programmed as an NC input instruction. PLC programs include symbols that represent series, parallel, and series/parallel-connected input devices and output components. An understanding of series and parallel circuits is a must when working with any PLC or PLC-controlled electrical system. Also, analyzing these circuits can help technicians determine how to construct programming diagrams during installations.

A *PLC hardware problem* is a condition where a PLC program is correct for the application, but the system is not working properly because there is a problem with inputs, outputs, or wiring (loose connections, cross talking, incorrect wiring); a short or open circuit exists somewhere in the system; or another hardware problem exists. **See Figure 2-3.** Understanding electrical principles is important when testing an electrical system or individual components and deciding what corrective action to take if a problem is found. For any troubleshooting to be successful, knowing what the voltage, current, power, and resistance values should be when taking a measurement is essential.

76 PROGRAMMABLE LOGIC CONTROLLERS: PRINCIPLES AND APPLICATIONS

PLC Hardware Problems

⚠ **WARNING**

Chassis, 24 V, and RS-232 grounds are typically connected internally in PLCs. The chassis ground terminal screw must be connected to chassis ground prior to any devices being connected to a PLC. Failure to follow procedures can result in damage to the PLC and personal injury from electrical shock.

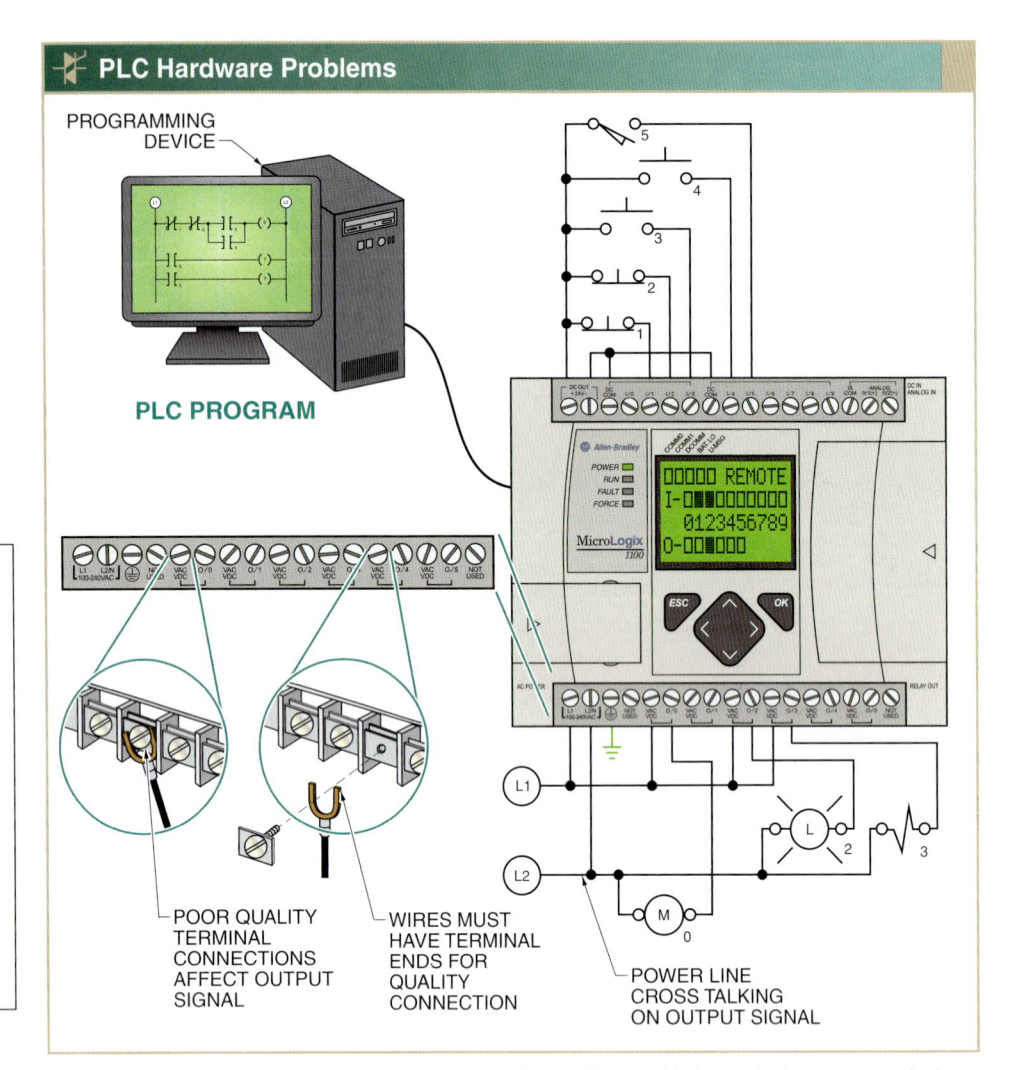

Figure 2-3. A PLC hardware problem is caused by problems with input devices, output devices, wiring, or a short or open somewhere in the system.

VOLTAGE

Voltage (E) is the electromotive force or pressure in a conductor that allows electrons to flow. Voltage is measured in volts (V) and is either direct (DC) or alternating (AC). *DC voltage* is voltage in a circuit that has current that flows in one direction only. *AC voltage* is voltage in a circuit that has current that reverses its direction of flow at regular intervals.

Voltage is produced by electromagnetism (generators), chemicals (batteries), light (photocells), heat (thermocouples), pressure (piezoelectricity), or friction (static electricity). Most electrical systems operate from standard power delivered by a local utility company, but some manufacturers create electrical power using generators. PLCs that control remote or portable equipment can be powered by batteries or a combination of batteries and solar power, such as with portable traffic lights used in a construction zone. **See Figure 2-4.**

Some PLC input devices such as temperature sensors and pressure sensors make use of the varying voltage produced by thermocouples or piezoelectric sensors to provide an analog input signal to a PLC. An *analog signal* is an electronic signal that has continuously changing quantities (values) between defined limits. Analog signals such as a 0 VDC to 10 VDC or 4 mA DC to 20 mA DC are standard analog input signals used in PLC analog input modules.

Portable PLCs

Figure 2-4. The voltage supplied to portable PLCs can come from portable generators, batteries, or solar panels.

PLC TIPS

Connect the black test lead of a test instrument to circuit ground or to the negative side of a DC object and the red test lead to the voltage or positive side of a DC object. Reverse the test leads when a negative sign appears in front of the reading on the display.

DC Voltage Polarity

All DC voltage sources have a positive and negative terminal. The positive and negative terminals establish polarity in a circuit. *Polarity* is the positive (+) or negative (−) state of an object. DC positive polarity is also called the source, or supply. DC negative polarity is also called the common, or sink. **See Figure 2-5.** All points in a DC circuit have polarity. Understanding the positive and negative points of a DC circuit is important when wiring DC devices and components to the input and output terminals of a PLC.

AC Voltage

AC voltage is the most common voltage used to produce work. AC voltage is produced by generators that create an AC sine wave while operating. An *AC sine wave* is a symmetrical waveform that contains 360 electrical degrees and has one positive alternation and one negative alternation per cycle. A *cycle* is one complete positive alternation and one complete negative alternation of a waveform. An *alternation* is half of a cycle. A sine wave reaches a peak positive value at 90°, returns to 0 V at 180°, increases to a peak negative value at 270°, and returns to 0 V at 360°. **See Figure 2-6.**

AC voltage is either single-phase (1ϕ) or three-phase (3ϕ). *Single-phase AC voltage* is a type of electrical supply that contains only one alternating voltage waveform (sine wave). *Three-phase AC voltage* is a type of electrical supply that combines three alternating waveforms (sine waves) each displaced 120 electrical degrees (one-third of a cycle) apart. Three-phase voltage is produced when three coils are simultaneously rotated in a generator.

PLC input terminals are designated numbers starting with 0 and numerically increase (0, 1, 2, 3, and so on) to a set amount. Output terminals are designated numbers in the same manner.

78 PROGRAMMABLE LOGIC CONTROLLERS: PRINCIPLES AND APPLICATIONS

DC Voltage Polarity

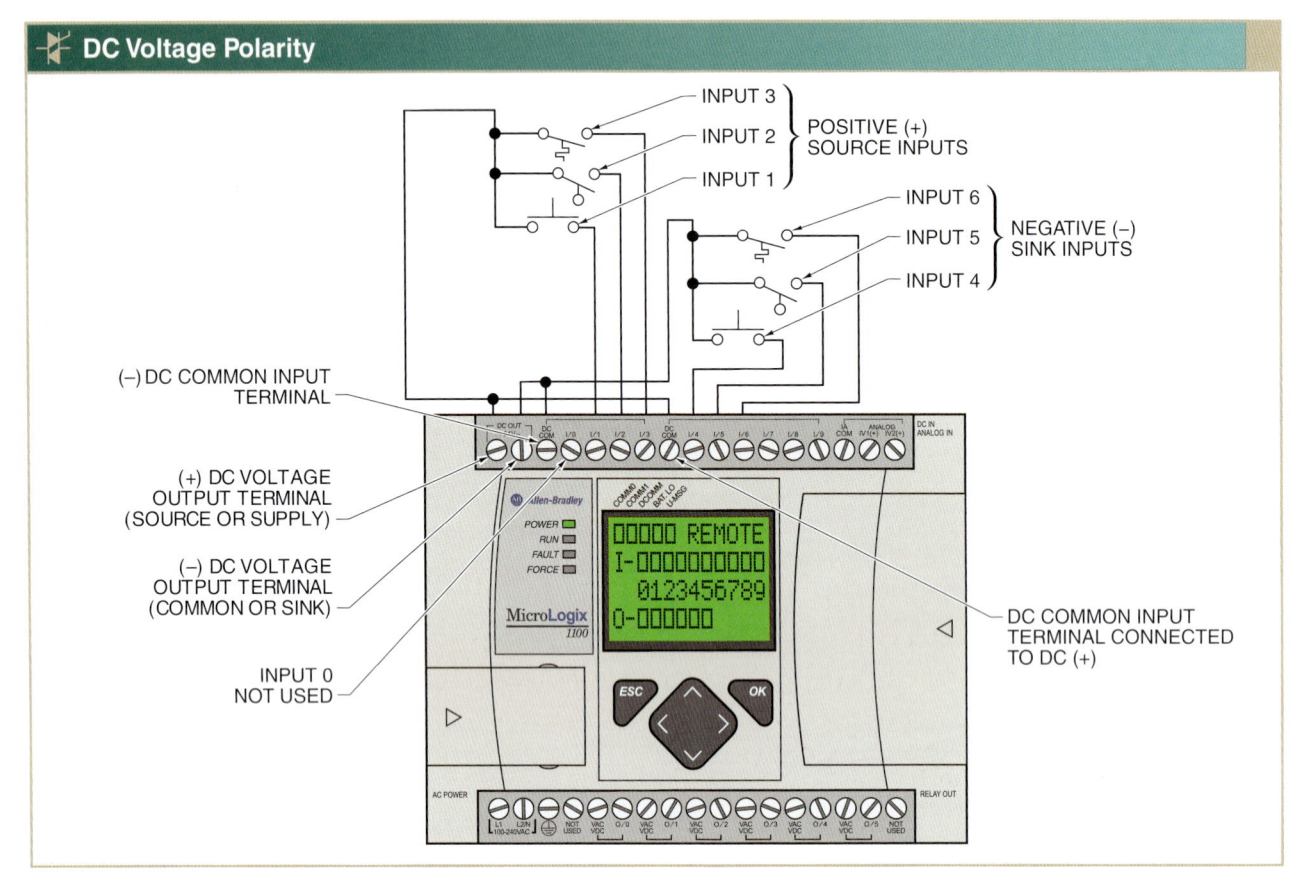

Figure 2-5. All points in a DC voltage circuit have positive (+) or negative (−) polarity.

Single-phase power lines are marked L1 and L2 or N. Three-phase power lines are marked L1 (A or R), L2 (B or S), and L3 (C or T). Single-phase power is used for 115/230 V loads and three-phase power is used for 230/460 V loads. PLCs use single-phase voltage as a power supply and to control any AC-rated inputs or outputs. **See Figure 2-7.** However, single-phase power from the output of a PLC can also be used to control a three-phase motor starter. Likewise, single-phase power from a PLC can be used to control a relay or heating contactor that controls a three-phase heater.

Three-phase AC power is not used directly by PLCs. PLCs send single-phase power to magnetic motor starters, electric motor drives, lighting contactors, or heating contactors that control three-phase power to three-phase loads. In most three-phase power applications, a PLC output is used to control a starter (contactor) coil and the coil is used to open or close the high-power-rated contacts.

AC Voltages

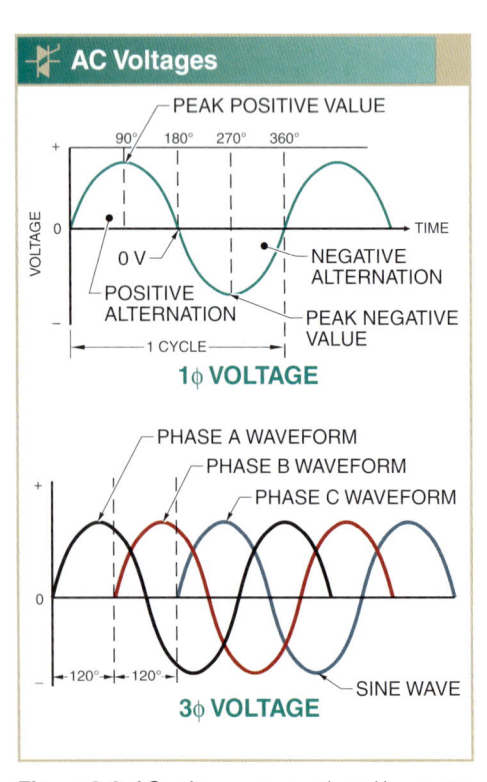

Figure 2-6. AC voltages are produced by generators, which create single-phase or three-phase sine waves while operating.

Chapter 2—PLC Electrical Principles, Ratings, and Circuit Calculations 79

Single-Phase PLC Controlling Three-Phase Load

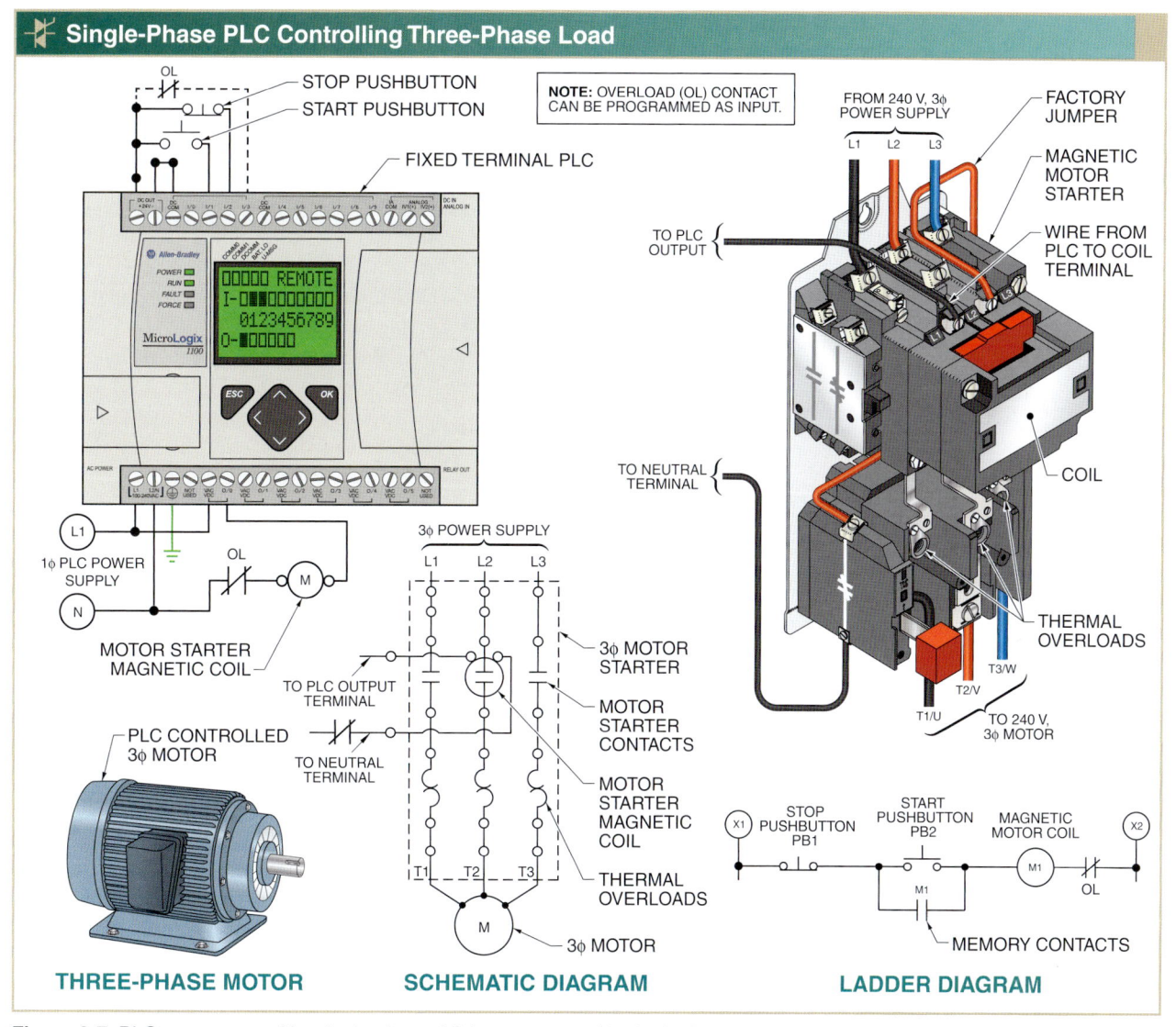

Figure 2-7. PLCs are powered by single-phase AC but can control both single-phase or three-phase AC loads.

Low AC voltages (6 V to 24 V) are used for doorbells, security systems, and as a voltage level for PLC inputs. Medium AC voltages (115 V) are used in commercial applications for lighting, heating, cooling, cooking, running motors, as the power supply voltage to security system PLCs, and as the voltage level of some building monitoring PLC inputs and outputs. **See Figure 2-8.** Using 115 VAC for PLC input and output terminals is common when a PLC is used to control a circuit that includes only 115 VAC loads.

> ⚠ **CAUTION**
>
> The voltage rating of an output component, such as a motor starter coil, must match the voltage output range of the PLC (typically 24 V, 115 V, 208 V, or 230 V).

Controlling contactor coils is a common PLC application.

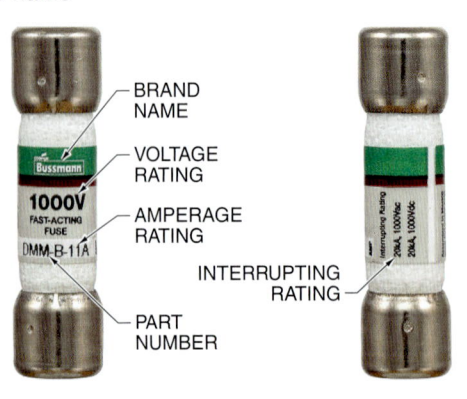

PLC Power Supply Voltage Ratings

All PLCs must receive power to operate. Power supplied to a PLC is used to operate its internal circuitry (CPU) and possibly supply voltage to the output and input terminals of the PLC. PLCs have a supply voltage rating that can be a fixed rating, such as 115 VAC, or a voltage range, such as 85 VAC to 265 VAC. Some PLCs have a dual voltage rating of 85 VAC to 132 VAC or 170 VAC to 265 VAC that is set by the placement of a jumper or switch. **See Figure 2-9.** Even when a PLC has a fixed voltage rating, the voltage applied to the PLC can still vary within the standard +5% to −10% that applies to most fixed-voltage-rated electrical equipment.

PLC Input Voltage Ratings

All PLCs have input devices such as pushbuttons, level switches, and proximity switches connected to input terminals. The input device circuit must be powered by a specified voltage or voltage range. PLC input terminals are available with DC, AC, or DC/AC ratings.

Low-voltage-rated input terminals (12 VDC or 24 VDC) are the most common and preferred type of PLC input terminals. Lower voltages are safer and provide for an easier installation because lower voltages do not require input wires to be run in conduit. Many PLCs provide a 12 VDC or 24 VDC output power terminal that can be used to power input terminals. **See Figure 2-10.**

The AC input terminals of a PLC are typically rated at 115 VAC because 115 VAC is the standard voltage for most single-phase AC circuits and loads. Although 115 VAC-rated PLC input terminals simplify a circuit somewhat because one voltage level (115 VAC) can be used for all PLC wiring (PLC power supply, input and output circuits), the circuit is not as safe as a PLC that uses low-voltage DC for the input terminals.

Standard Voltage Applications

Standard AC Voltages

Application	Level*
Doorbells, security systems	6, 24
Most commercial appliances (refrigerators, large ovens), lighting applications	110, 115, 120
Industrial motors, heating elements	208, 240, 480

Standard DC Voltages

Application	Level*
Flashlights, watches, etc.	1.5, 3
Outdoor power equipment, automobiles, trucks	6, 9, 12, 24, 36
Printing presses, small electric railway systems	125, 250, 600
Large electric railway systems	1200, 1500, 3000

*in volts

Figure 2-8. Low and medium standard AC/DC voltage levels are used in PLC applications.

AC voltages (208 V to 480 V) are used in commercial applications for cooking, heating, and cooling. These AC voltages are also used in industrial applications to convert raw materials into usable products in addition to providing lighting, heating, and cooling for plant/facility personnel.

CODE CONNECT

ARTICLE 240—OVERCURRENT PROTECTION

Section 240.60(C) of the NEC® lists the marking requirements for fuses. All fuses shall be marked with the following:
- *amperage rating*
- *voltage rating*
- *interrupting rating (other than 10,000 A)*
- *current limiting (where applicable)*
- *manufacturer's trademark or name*

An interrupting rating is not required on fuses used for supplementary protection. Fuse manufacturers typically have a specific alphanumerical symbol for each fuse. This alphanumerical symbol can be used to order replacement fuses. A manufacturer typically provides a cross-reference table to facilitate substituting fuses with identical characteristics from different fuse manufacturers.

BRAND NAME
VOLTAGE RATING
AMPERAGE RATING
INTERRUPTING RATING
PART NUMBER

⚠ **WARNING**

Resistance measurements must only be made on de-energized circuits. Test circuits for voltage before taking a resistance measurement.

PLC Power Supply Voltage Ratings

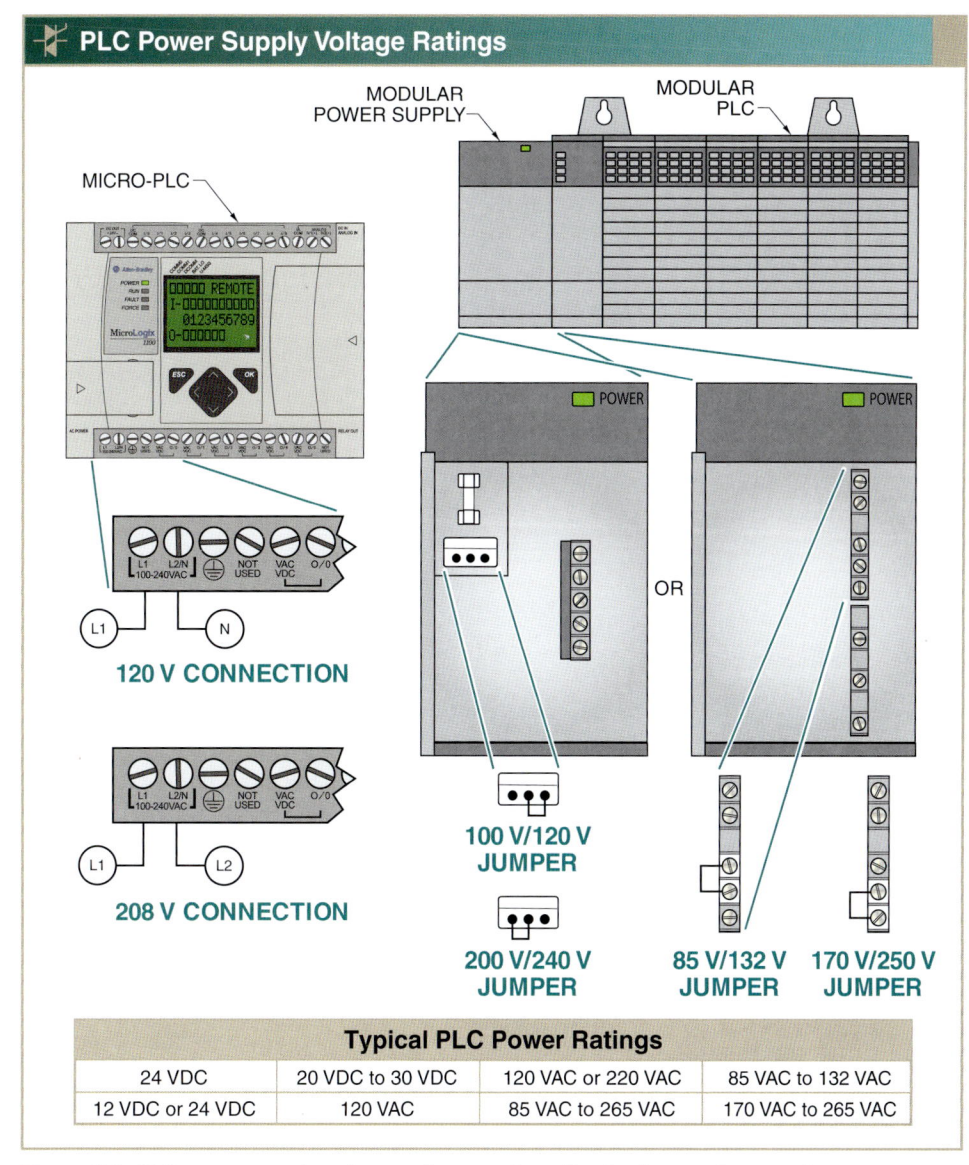

Typical PLC Power Ratings

24 VDC	20 VDC to 30 VDC	120 VAC or 220 VAC	85 VAC to 132 VAC
12 VDC or 24 VDC	120 VAC	85 VAC to 265 VAC	170 VAC to 265 VAC

Figure 2-9. PLC power supply voltage ratings can be a fixed voltage rating, a voltage range, or a dual voltage rating.

CODE CONNECT

Section 430.6(A)(1) of the NEC® covers the use of table values for specific motor related calculations. When calculating conductor ampacity or switch ampere ratings, or when determining the degree of branch circuit short-circuit and ground-fault protection required, table values are used instead of the actual motor nameplate current values. Table values are used because two motors from different manufacturers with identical horsepower and voltage ratings may have different motor nameplate current values. The NEC® determines the most appropriate current values for different motor horsepower and voltage ratings. Technicians should refer to Table 430.247, Table 430.248, Table 430.249, and Table 430.250.

AC-rated input terminals cannot be powered by DC power, even when the voltage level (115 VAC or 115 VDC) is the same. Likewise, DC-rated PLC input terminals cannot be powered by AC power, even when the voltage level (24 VAC or 24 VDC) is the same.

> ⚠ **CAUTION**
> *Always ensure that the function switch position matches the jack connections of the test leads. Test instruments can be damaged when the test lead positions and the function switch setting do not match.*

PLC Output Voltage Ratings

All PLCs have output components such as lamps, solenoids, and motor starters connected to output terminals. The output terminals of a PLC are powered by a specific voltage or voltage range. The type of voltage (AC or DC) and voltage level (12 V, 24 V, and 120 V) used for a PLC are based on the loads controlled by the PLC. Loads that require more power typically have higher voltage ratings because power (P) is equal to voltage (E) multiplied by current (I). The higher the voltage rating is for a load, the lower the current rating is for any given power rating.

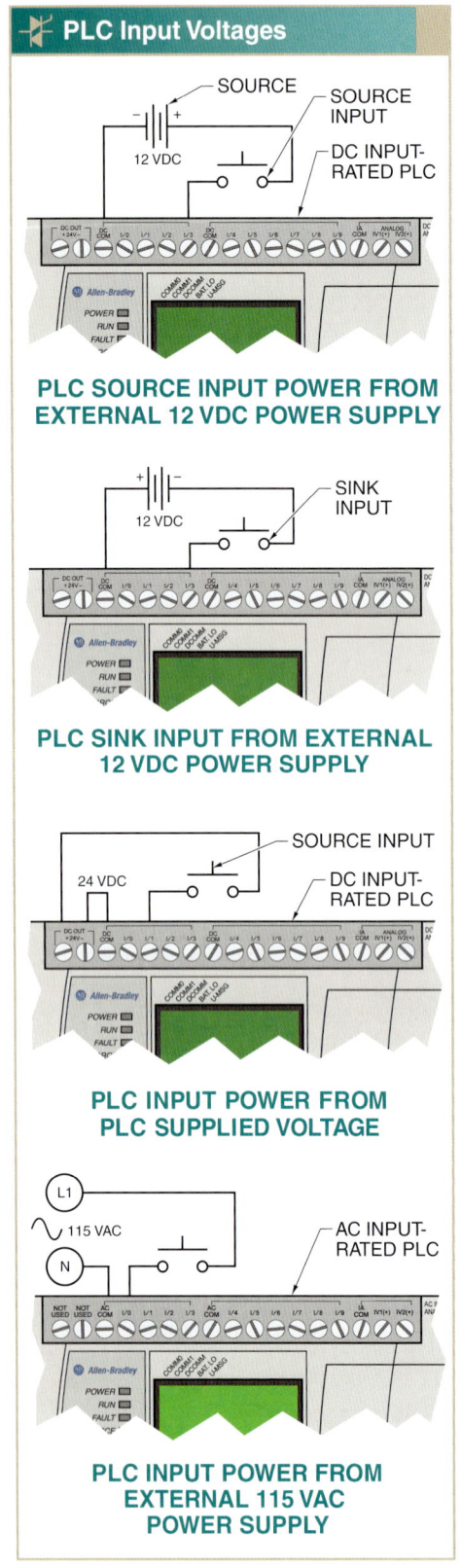

82 PROGRAMMABLE LOGIC CONTROLLERS: PRINCIPLES AND APPLICATIONS

PLC Input Voltages

PLC SOURCE INPUT POWER FROM EXTERNAL 12 VDC POWER SUPPLY

PLC SINK INPUT FROM EXTERNAL 12 VDC POWER SUPPLY

PLC INPUT POWER FROM PLC SUPPLIED VOLTAGE

PLC INPUT POWER FROM EXTERNAL 115 VAC POWER SUPPLY

Figure 2-10. PLC input terminals are available with DC, AC, or DC/AC ratings that typically use low voltages such as 12 VDC or 24 VDC.

PLC TIPS

A PLC input or output (I/O) using a PNP transistor is called a sourcing or positive I/O. A PLC I/O using an NPN transistor is called a sinking or negative I/O.

The output terminal switches of a PLC are either mechanical contact switches or solid-state switches. **See Figure 2-11.** The advantage of a mechanical contact output terminal is that a mechanical contact can switch either AC or DC loads. An AC solid-state output terminal of a PLC can only switch AC loads and a DC solid-state output terminal can only switch DC loads. Solid-state output switches of PLCs include triacs for switching AC circuits and transistors and SCRs for switching DC circuits.

PLC Output Terminal Switching

Output Type	Symbol
Mechanical contact AC or DC switching device; used to switch high- or low-level currents	CONTACT PUSHBUTTON
Transistor NPN or PNP DC switch; used to switch low-level DC	COLLECTOR (C) BASE (B) EMITTER (E) NPN PNP
SCR DC switching device; one SCR used to switch high-level currents	ANODE CATHODE GATE SCR
Triac AC switching device	TRIAC

Figure 2-11. The output terminal switches of a PLC are either mechanical switches or solid-state switches, such as transistors, SCRs, and triacs.

CURRENT

Current (I) is the rate at which electrons flow through a conductor. When a source of power (voltage) is connected to a circuit, current is produced. The more power a load requires, the greater the flow of electrons. For example, a 10 horsepower (HP) dual-voltage-rated motor (230/460 V) draws approximately 28 A when wired for 230 V, but only draws 14 A when wired for 460 V.

Different voltage sources produce different amounts of current. For example, standard AAA, AA, A, C, and D size batteries all produce 1.5 V, but each size delivers a different amount of current. Size AAA batteries deliver the smallest amount of current and size D batteries deliver the highest amount of current. Because of battery current specifications, a load connected to a size D battery operates longer than the same load connected to a size AAA battery.

PLC Input Current Ratings

PLCs are designed for DC or AC input voltages. When using switches with mechanical contacts, such as standard pushbuttons or mechanical limit switches, the switches can be connected to either DC input terminals or AC input terminals.

However, solid-state switches, such as most proximity and photoelectric sensors, must only be used with DC or AC power sources depending on the rating of the solid-state switch. A DC-rated solid-state input switch can only be connected to an input terminal of a DC-rated PLC. An AC-rated solid-state input device can only be connected to the input terminal of an AC-rated PLC.

The amount of current in an input circuit of a PLC is low because the internal circuitry of a PLC input section is the load of the input circuit. The internal circuitry only requires a few mA of current to activate. A typical PLC input section only draws about 5 mA to 20 mA of current. However, low-current inputs of a PLC are capable of controlling high-current outputs. **See Figure 2-12.**

Low current requirements in the input section of a PLC are one major advantage of using a PLC over conventional hardwired circuits. In conventional hardwired circuits, input devices such as limit switches and pressure switches are wired into the same circuit as output devices (lamps and solenoids). An input circuit using hard wiring must switch the same amount of current and voltage as required by the output circuit. For example, a limit switch would have to be wired into a 115 VAC circuit and switch a 1 A (1000 mA) lamp if conventional hard wiring were used instead of a PLC.

PLC TIPS

A motor's power (P) rating will be listed in horsepower (HP) or watts/kilowatts (W/kW). For every 1 HP, there is 746 W of power.

$$P_{HP} = \frac{P_W}{746}$$

$$P_{HP} = \frac{P_{kW}}{0.746}$$

$$P_W = 746 \times P_{HP}$$

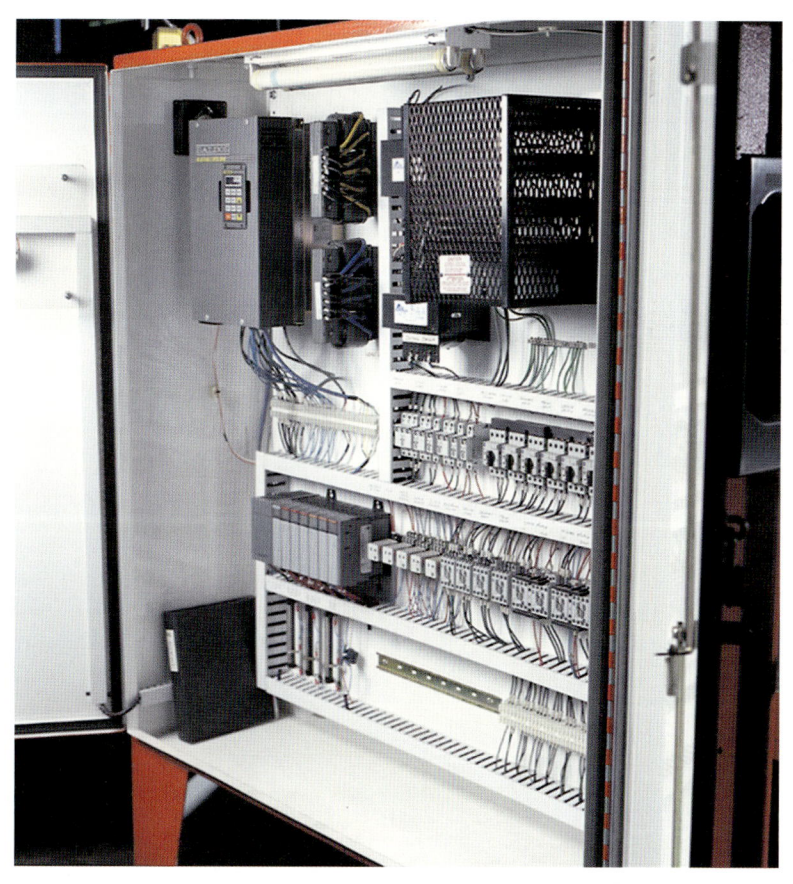

Baldor Electric Company

PLC enclosures can include electric motor drives, motor starters, contactors, transformers, interface devices, solid-state relays, and terminal strips.

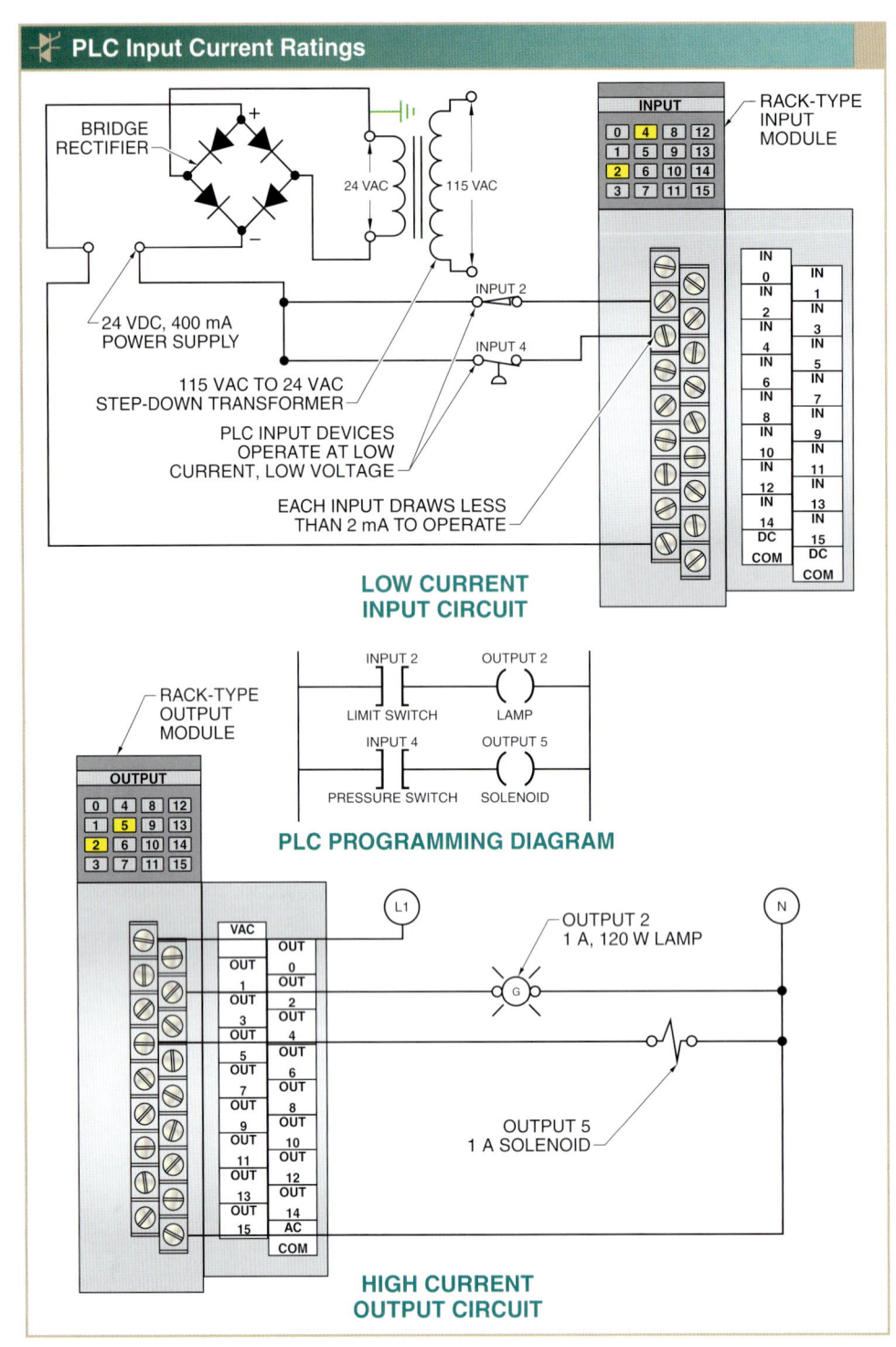

84 PROGRAMMABLE LOGIC CONTROLLERS: PRINCIPLES AND APPLICATIONS

Figure 2-12. Typical PLC input circuitry draws about 5 mA to 20 mA of current.

DC Input Switching

DC power is commonly used with PLC input circuits. The advantage of using DC is that a DC supply is kept to a low level (typically 12 VDC or 24 VDC) for safety. Also, most photoelectric and proximity sensors typically use transistors as the switching element and are rated to operate in a range between 10 VDC to 30 VDC.

The two types of DC input switches are two-wire and three-wire. **See Figure 2-13.** A two-wire input switch circuit uses mechanical switches (limit) or solid-state (photoelectric and proximity) switches. A three-wire input switch circuit uses solid-state switches, such as NPN or PNP transistors.

A three-wire input switch circuit typically uses an NPN (sink) or PNP (source) transistor as the switching device. For most applications, the exact transistor used does not matter as long as the switch is properly connected into the input circuit of the PLC. However, NPN transistor switches are far more commonly used than PNP transistor switches.

Some PLC input terminals are rated as current sink only, some are rated as current source only, and others as current sink/current source. PLC input terminals rated as current sink/current source can be used with NPN or PNP solid-state switching devices.

Two-Wire and Three-Wire Input Device Wiring

TWO-WIRE MECHANICAL LIMIT SWITCH

TWO-WIRE SOLID-STATE PROXIMITY SWITCH

THREE-WIRE SOLID-STATE PROXIMITY SWITCH (PNP/CURRENT SOURCE TYPE)

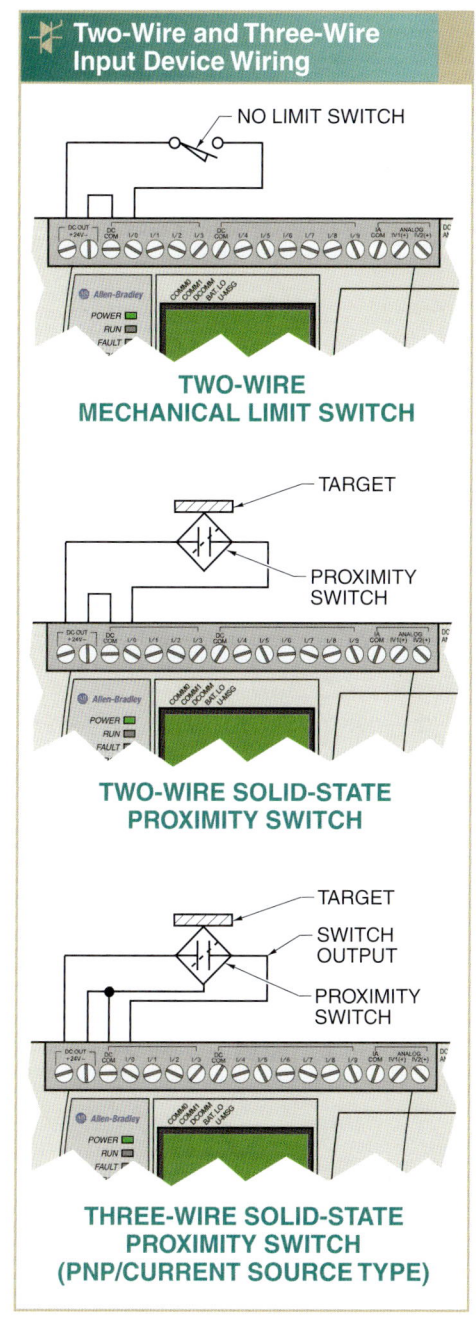

Figure 2-13. A two-wire input switch circuit uses mechanical (pushbutton) or solid-state (photoelectric and proximity) switches, and a three-wire input switch circuit uses solid-state switches.

Some PLC input modules are rated for current sinking and current sourcing switching for simultaneous signaling.

Current Sinking (NPN Transistor Switching). When using an NPN transistor-type input device, the load (PLC input section) is connected between the positive terminal of the supply voltage and the output terminal (collector) of the switch or sensor. When a switch such as a photoelectric or proximity sensor detects a target, electrons flow through the transistor switch, energizing the input terminal. **See Figure 2-14.**

86 PROGRAMMABLE LOGIC CONTROLLERS: PRINCIPLES AND APPLICATIONS

DC Current Sinking Input Circuits

LADDER DIAGRAM

CURRENT SINK INPUT TERMINAL PLC

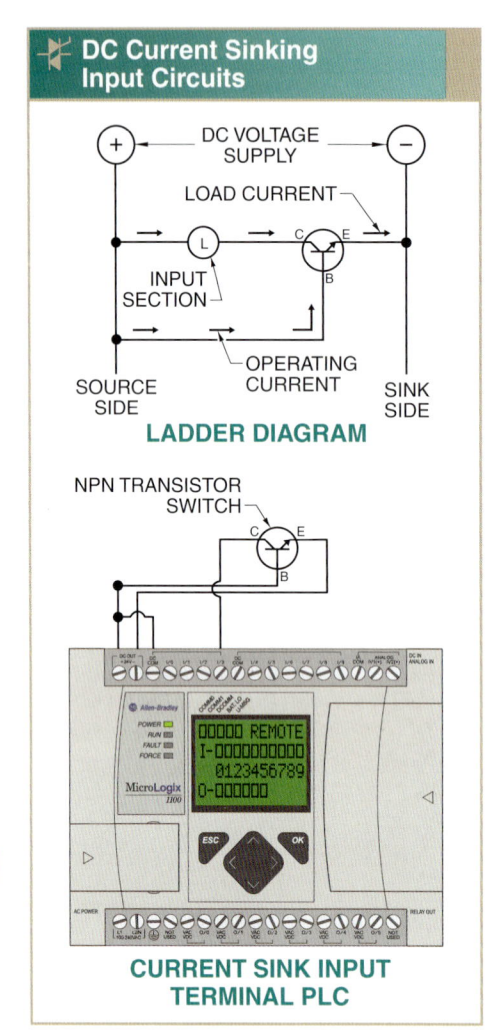

Figure 2-14. Switching devices that use NPN transistors as switching elements are called current sinking, negative switching, or NPN devices.

PLC TIPS

PLC DC input devices can be configured as sinking or sourcing depending on how the DC COM terminal is wired into the input device circuit.

DC Current Sourcing Input Circuits

LADDER DIAGRAM

CURRENT SOURCE INPUT TERMINAL PLC

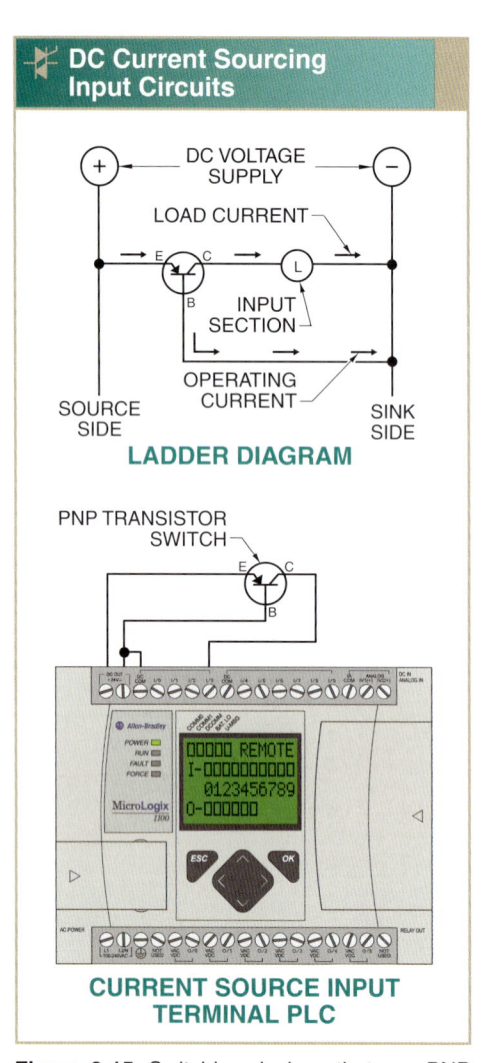

Figure 2-15. Switching devices that use PNP transistors as the switching elements are called current sourcing, positive switching, or PNP devices.

CODE CONNECT

Section 220.87 of the NEC® addresses how to determine, using maximum demand data, if an existing feeder or service has enough ampacity available for additional loads. The maximum demand data for a 1-year period (typically obtained from the power company) can be used, or a 30-day load study using a power logger (set to a 15-minute averaging interval) can be performed to determine the maximum demand. Per Section 220.87, new loads can be added if the maximum demand multiplied by 125% plus the new load does not exceed the ampacity of the feeder or service. Section 220.87 also requires that the existing feeder or service has proper overcurrent protection.

Switching devices that use an NPN transistor as the switching element are called current sink, negative switching, or NPN devices. A *sink* is the negative power supply terminal of a DC-powered PLC. Sinks use conventional flow through an NPN transistor. A current-sinking switch "sinks" the current from the load.

Current Sourcing (PNP Transistor Switching). When using a PNP transistor-type input device, the load (PLC input section) is connected between the negative terminal of the supply voltage and the output terminal (collector) of the switch or sensor. When the switch detects a target, electron flows through the transistor and the PLC input terminal is energized. **See Figure 2-15.**

Switching devices that use PNP transistors as the switching element are called current source, positive switching, or PNP devices. A *source* is the positive terminal of a DC-powered PLC. Sources use electron current flow through a PNP transistor. A current sourcing switch "sources" the current to the load.

PLC Output Current Ratings

With PLC input circuits, the PLC input section is the load. The amount of current that the input switch is required to carry is determined by how much current the PLC input section circuitry needs in order to determine whether an input device is open or closed. The internal circuitry of a PLC requires a small amount of current, typically only a few mA.

However, PLC output components connected to the output section of a PLC determine the amount of current a PLC must carry. The amount of current drawn by an output component can vary from a few mA (small indicating lamps), to several amps (solenoids, higher wattage lamps, small motors, and large-motor motor starters), to hundreds of amps (large three-phase motors and three-phase heating elements). Large current output components must be connected to a PLC through an interface device.

The current rating of PLC output circuitry is listed for the amount of current the internal switches can safely handle when the load is first turned ON ("make" current rating), when the load is operating ("continuous" current rating) and when the load is turned OFF ("break" current rating). **See Figure 2-16.**

When AC loads are running, the break current rating will always be much lower than the make or continuous current ratings because when an AC circuit is turned OFF, a large damaging current will try to arc across any switch that is beginning to open. With AC circuits, the arc dissipates somewhat because AC current is always fluctuating between zero and peak current. However, in a DC circuit there is no fluctuation, making switching DC circuits more damaging. Because of the possible damage, a PLC output relay contact or output switch has a much lower DC rating than AC rating for the same output switch.

⚠ WARNING

The current rating on an electrical switch device is the maximum current rating. It is best to operate a switch at 80% or less than its maximum current rating.

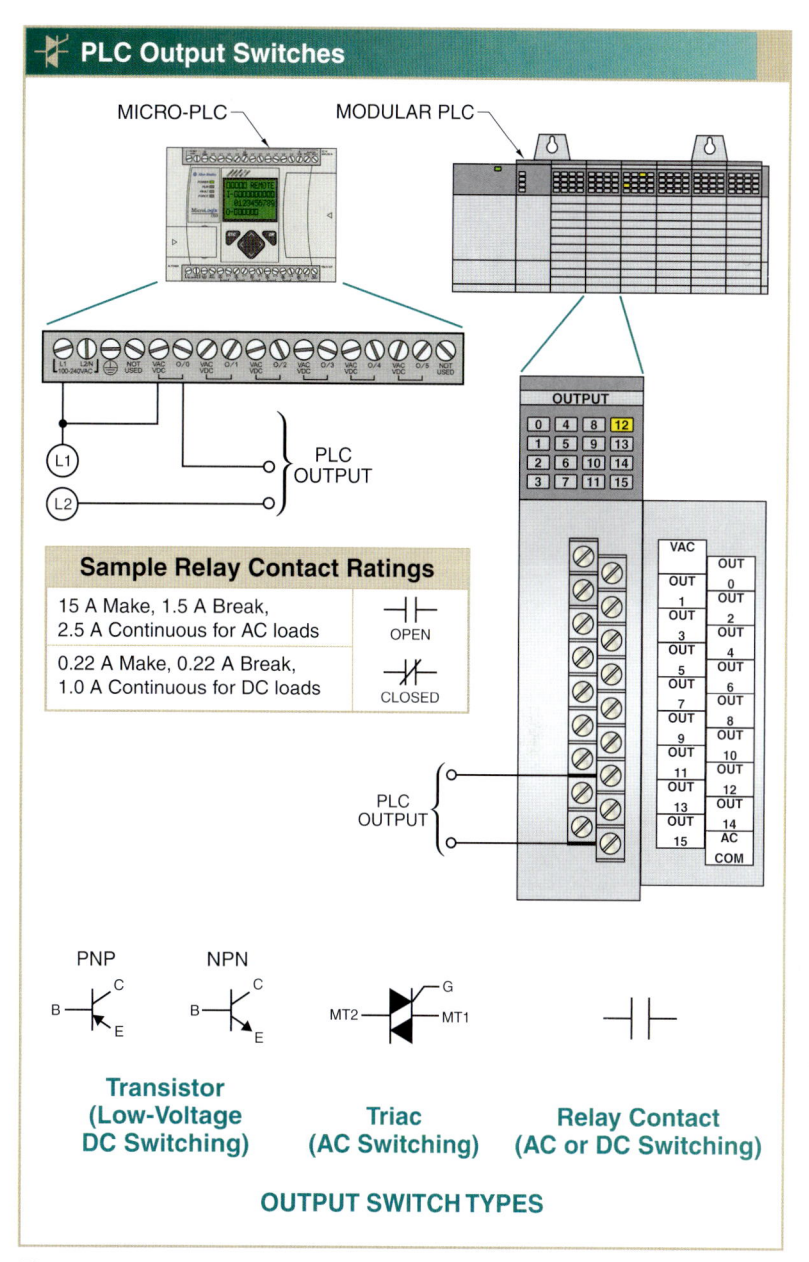

Figure 2-16. The output sections of PLCs use transistors, triacs, and mechanical or solid-state relays for switching.

Inductive Kickback

Any time current flows through a conductor of coils, a magnetic field is produced. Coils are found in motor windings, transformers, and solenoids, which create inductance in an electrical circuit. AC current flow produces an alternating magnetic field around a coil, whereas DC current produces a constant magnetic field around a coil.

When current flow is stopped through a coil, the magnetic field collapses, which induces a current flow into the circuit in the reverse direction. This is referred to as inductive kickback. **See Figure 2-17.**

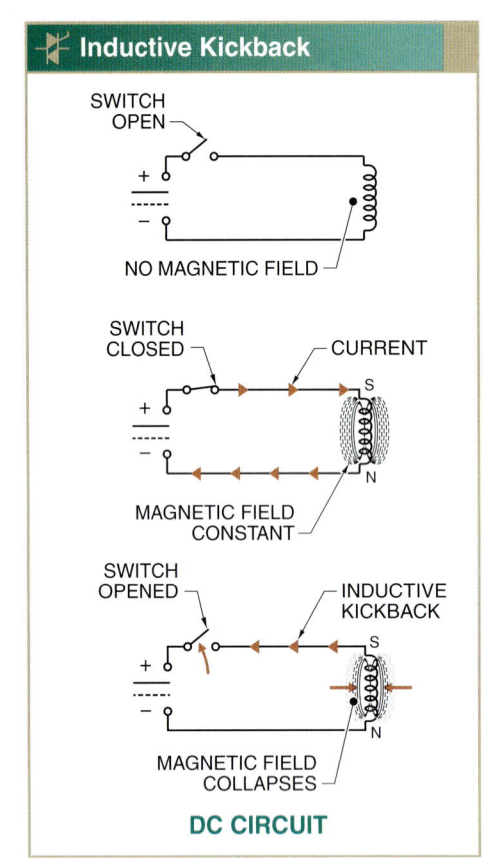

Figure 2-17. Inductive kickback is a current that is induced in the reverse direction in DC series circuits caused by a magnetic field collapse in the coil of a conductor.

RESISTANCE

Resistance (R) is the opposition to current flow. Higher resistance specifications and measurements are expressed using prefixes, as in kilohms (kΩ) and megohms (MΩ); 1 kΩ equals 1000 Ω and 1 MΩ equals 1,000,000 Ω.

Resistance limits the flow of electrons in an electrical circuit and must be understood when installing and/or troubleshooting a circuit that includes a PLC. Technicians must always remember that the higher the resistance, the lower the current flow, and the lower the resistance, the higher the current flow in an electrical circuit.

Components designed to insulate, such as rubber or plastic, have very high resistance. Components designed to conduct, such as wires or switch contacts, have very low resistance. The resistance of insulators decreases when insulators are damaged by moisture and/or overheating. The resistance of conductors (wires) increases when damaged by burning and/or corrosion. Factors that affect the resistance of conductors are size of the wire, length of the wire, material, and operating temperature.

A conductor with a large cross-sectional area has less resistance than a conductor with a small cross-sectional area. **See Figure 2-18.** A large conductor can also carry more current. The longer a conductor is, the greater the resistance of the conductor. Short conductors have less resistance than long conductors of the same american wire gauge (AWG) size. Copper (Cu) is a better conductor than aluminum (Al) and can carry more current for a given AWG size. Temperature also affects resistance. For metals, the higher the temperature, the greater the resistance. The higher the operating temperature of a wire, the lower the ampacity rating of the wire.

Conductor Ratings

AWG	Copper Ampacity	Aluminum Ampacity	Dia*
22	—	—	25.0
20	—	—	32.0
18	—	—	40.0
16	—	—	51.0
14	15	—	64.0
12	20	15	81.0
10	30	25	102.0
8	40	30	128.0
6	55	40	162.0

* in mils (Rounded to nearest whole number.)

Figure 2-18. Copper (Cu) is a better conductor than aluminum (Al) because copper can carry more current for a given size (AWG).

In PLC installations, the resistance of conductors used to wire PLCs, PLC inputs, and PLC outputs must be kept to a minimum. Using copper wire instead of aluminum wire reduces resistance. Making sure all connections are clean and tight also reduces resistance because corrosion and loose connections increase overall circuit resistance.

PLC TIPS
An output load, either mechanical or solid-state, can have a higher current rating than a PLC contact. To increase the operable life of the PLC contact, a control relay (CR) or solid-state relay (SSR) with a high current rating can be controlled to switch the load.

OHM'S LAW

Ohm's law is the direct relationship between voltage (E), current (I), and resistance (R) in an electrical circuit. Ohm's law states that current in a circuit is proportional to voltage and inversely proportional to resistance. Any value of Ohm's law can be found when the other two values are known. The relationship between voltage, current, and resistance can be shown in a pie chart form. If one variable on the pie chart is covered, it reveals the equation that can be used to find the value of the covered variable. **See Figure 2-19.**

The relationship between voltage, current, and resistance is linear. If one variable increases by a certain percentage, then the other two variables increase or decrease by the same percentage. If resistance remains constant in a circuit, then as voltage doubles current doubles or as voltage is halved current is halved. A linear relationship can be described by the following:

If $E = 20$ V
Then $20 \text{ V} \div 20 \text{ }\Omega = 1.0$ A
If $E = 60$ V
Then $60 \text{ V} \div 20 \text{ }\Omega = 3.0$ A
If $E = 100$ V
Then $100 \text{ V} \div 20 \text{ }\Omega = 5.0$ A

Ohm's Law

E = VOLTAGE (IN V)
I = CURRENT (IN A)
R = RESISTANCE (IN Ω)

$$E = I \times R$$
VOLTAGE = CURRENT × RESISTANCE

$$I = \frac{E}{R}$$
CURRENT = $\frac{\text{VOLTAGE}}{\text{RESISTANCE}}$

$$R = \frac{E}{I}$$
RESISTANCE = $\frac{\text{VOLTAGE}}{\text{CURRENT}}$

LINEAR RELATIONSHIP

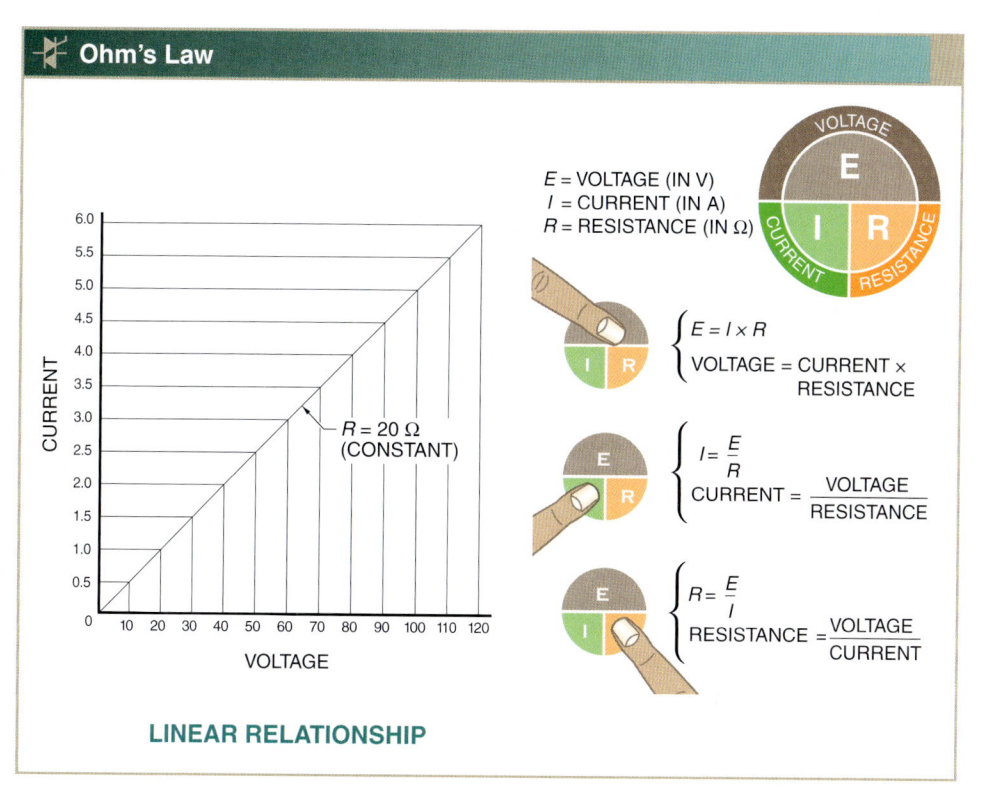

Figure 2-19. Ohm's law states that current (*I*) in a circuit is proportional to voltage (*E*) and inversely proportional to resistance (*R*).

Calculating Voltage Using Ohm's Law

Ohm's law states that voltage (E) in a circuit is equal to resistance (R) times current (I). To calculate voltage using Ohm's law, apply the following formula:

$$E_T = I \times R$$

where
E_T = total voltage (in V)
I = current (in A)
R = resistance (in Ω)

For example, if an electrical circuit has a resistor with a resistance of 3000 Ω, has a light emitting diode (LED) with a resistance of 3000 Ω, and carries a current of 4 mA, what is the total voltage (E_T) of the input circuit? **See Figure 2-20.**

Calculating Voltage

$E_T = I \times R$
$E_R = 0.004\ A \times 3000\ Ω$
$E_R = 12\ V$

$E_{LED} = 0.004\ A \times 3000\ Ω$
$E_{LED} = 12\ V$
$E_T = E_R + E_{LED}$
$E_T = 12\ V + 12\ V = 24\ V$

VALUES ASSIGNED FOR DISCUSSION PURPOSES ONLY

Figure 2-20. Total voltage (E_T) is calculated by multiplying the current (I) by the resistance (R) of a circuit.

Calculating Current Using Ohm's Law

Ohm's law states that current (I) in a circuit is equal to voltage (E) divided by resistance (R). To calculate current using Ohm's law, apply the following formula:

$$I = E \div R_T$$

where
E = voltage (in V)
I = current (in A)
R_T = total resistance (in Ω)

For example, if a PLC output circuit has a load with a resistance of 6000 Ω and a AC voltage of 24 V, what is the current (I) of the output circuit? **See Figure 2-21.**

Calculating Current

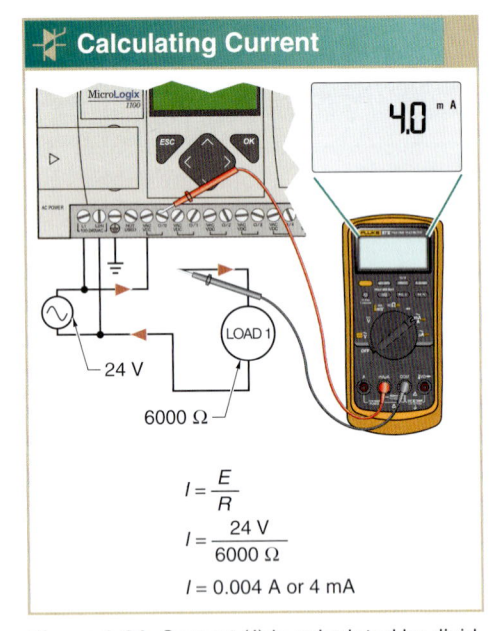

$I = \dfrac{E}{R}$

$I = \dfrac{24\ V}{6000\ Ω}$

$I = 0.004\ A\ or\ 4\ mA$

Figure 2-21. Current (I) is calculated by dividing the voltage (E) by the total resistance (R_T) of a circuit.

Calculating Resistance Using Ohm's Law

Ohm's law states that resistance (R) in a circuit is equal to voltage (E) divided by current (I). To calculate resistance using Ohm's law, apply the following formula:

$$R_T = E \div I$$

where
E = voltage (in V)
I = current (in A)
R_T = total resistance (in Ω)

For example, if a PLC output circuit has an AC voltage of 24 V and a current of 4 mA, what is the total resistance (R_T) of the output load? **See Figure 2-22.**

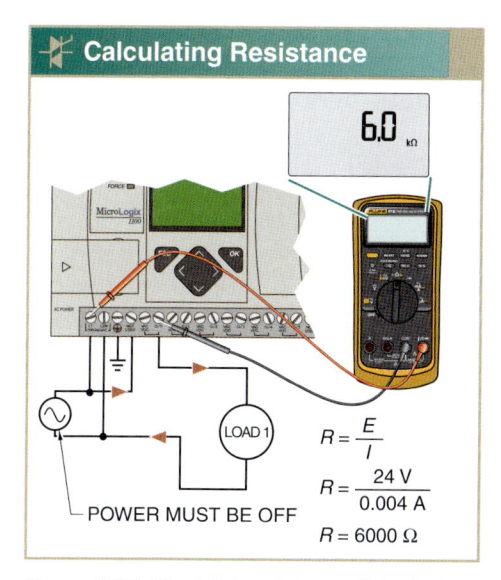

Figure 2-22. The total resistance (R_T) of a load is calculated by dividing the voltage (E) by the current (I) in a circuit.

POWER FORMULA

Power (P) is the rate of energy consumption or conversion in an electrical circuit or system in a given amount of time. The unit of measure for power is the watt (W), which is a rate of energy conversion of 1 joule per second. The power formula is the direct relationship between power (P), voltage (E), and current (I) in an electrical circuit. Similarly, the relationship between power, voltage, and current can be shown in a pie chart form. If one variable on the pie chart is covered, it reveals the equation that can be used to find the value of the covered variable. **See Figure 2-23.**

IMPEDANCE

Impedance (Z) is the total opposition to current flow in an AC series circuit. As in DC circuits, and AC circuits that do not contain a significant amount of inductance and/or capacitance, and are purely resistive, the opposition to current flow is resistance (R). In purely capacitive circuits, that contain capacitors, the opposition to current flow is capacitive reactance (X_C). In AC series circuits, that contain resistance and reactance, the opposition to current flow is impedance. Since this relationship is similar to Ohm's law, impedance can be used in place of resistance in the pie chart. **See Figure 2-24.**

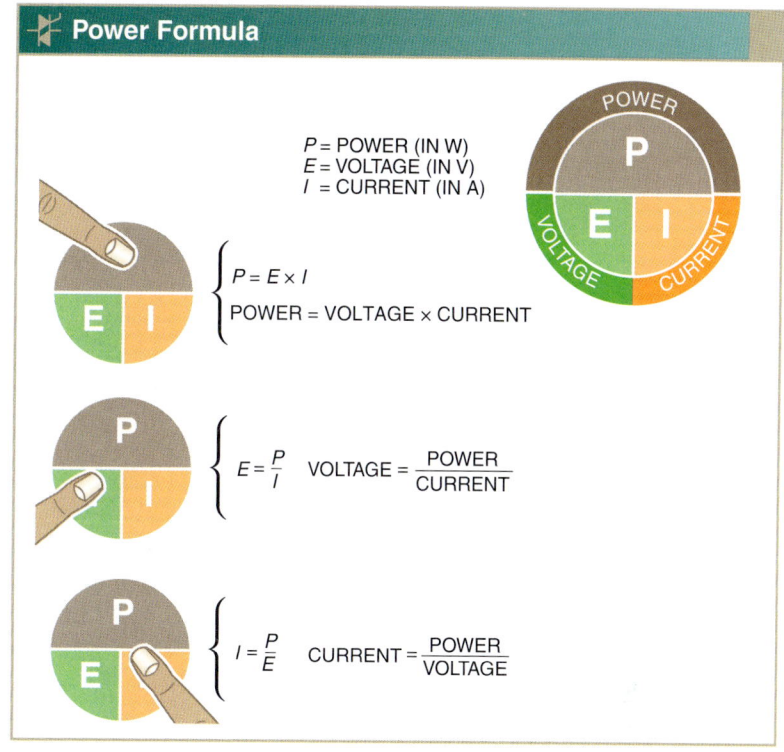

Figure 2-23. The direct relationship between power (P), voltage (E), and current (I) can be shown in a pie chart form.

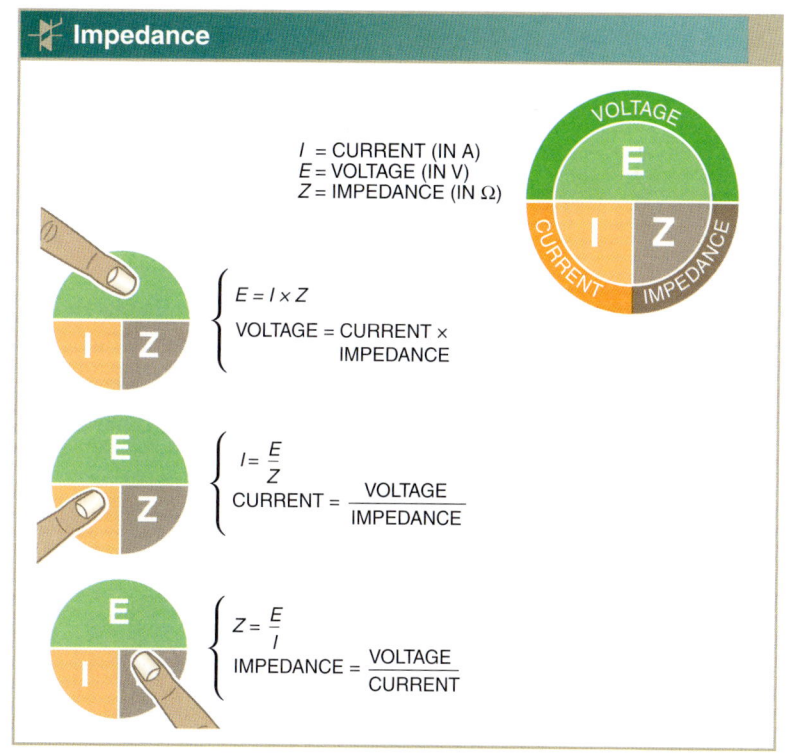

Figure 2-24. The direct relationship between impedance (Z), voltage (E), and current (I) can be shown in a pie chart form.

SERIES CIRCUITS

Fuses, switches, loads, conductors, and other electrical devices and components can be connected in series. *Series-connected devices and components* are two or more devices or components that are connected so that there is only one flow path for current to take. Opening a series circuit at any point stops the flow of current. Current stops flowing any time a fuse blows, a circuit breaker trips, or a switch opens. **See Figure 2-25.**

PLC circuits include series-connected components. All fuses of a PLC are connected in series with the power supply. The power supply feeds power to the input and output sections, with all electrons flowing from the power supply through the fuses. Likewise, some PLC communications cables used to connect separate processors together are connected in series (daisy-chained). **See Figure 2-26.**

Resistance in Series Circuits

The total resistance in a circuit containing series-connected loads equals the sum of the resistances of all the loads. The total resistance in a series circuit increases when loads are added in series and decreases when loads are removed. Loose connections and switching contacts also have resistance, which adds to the total circuit resistance. **See Figure 2-27.**

To calculate the total resistance of a series circuit, apply the following formula:

$$R_T = R_1 + R_2 + R_3 + R_4 + \ldots$$

where
R_T = total resistance (in Ω)
R_1 = resistance 1 (in Ω)
R_2 = resistance 2 (in Ω)
R_3 = resistance 3 (in Ω)
R_4 = resistance 4 (in Ω)

For example, what is the total resistance of a circuit that has 0.5 Ω, 1.0 Ω, 1.0 Ω, and 100 Ω resistances connected in series?

$$R_T = R_1 + R_2 + R_3 + R_4$$
$$R_T = 0.5 + 1 + 1 + 100$$
$$R_T = \mathbf{102.5 \; \Omega}$$

PLC TIPS

Copper and aluminum are both highly conductive materials used in electrical systems. Aluminum is lighter than copper and is primarily used in high power distribution systems. Copper is used for power distribution inside of buildings. The resistance of these conductors is directly proportional to their length. The longer a wire or cable, the greater the resistance.

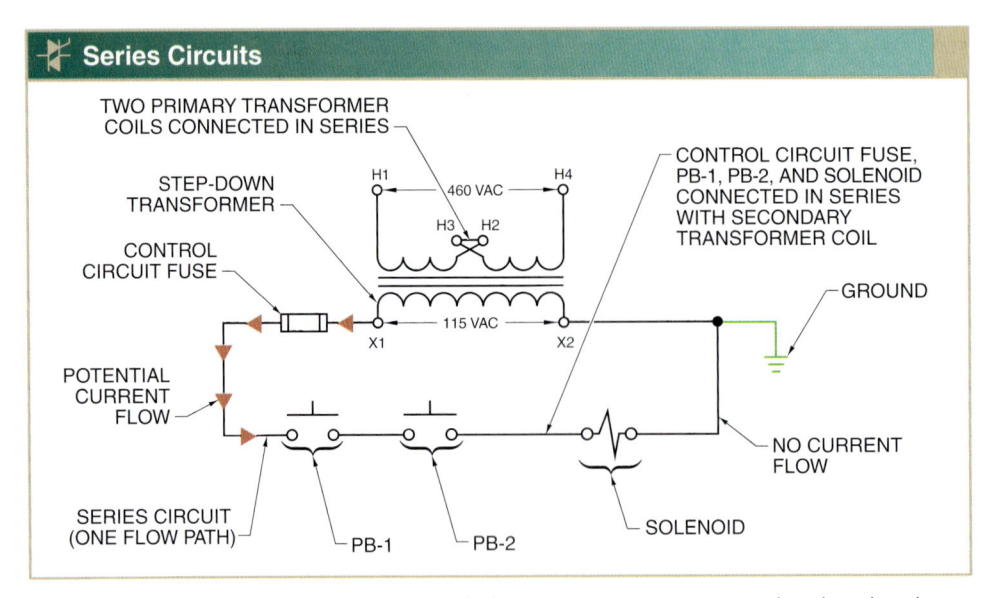

Figure 2-25. Series circuits have two or more devices or components connected so there is only one flow path for current to take.

Series-Connected (Daisy-Chained) Communications Cables

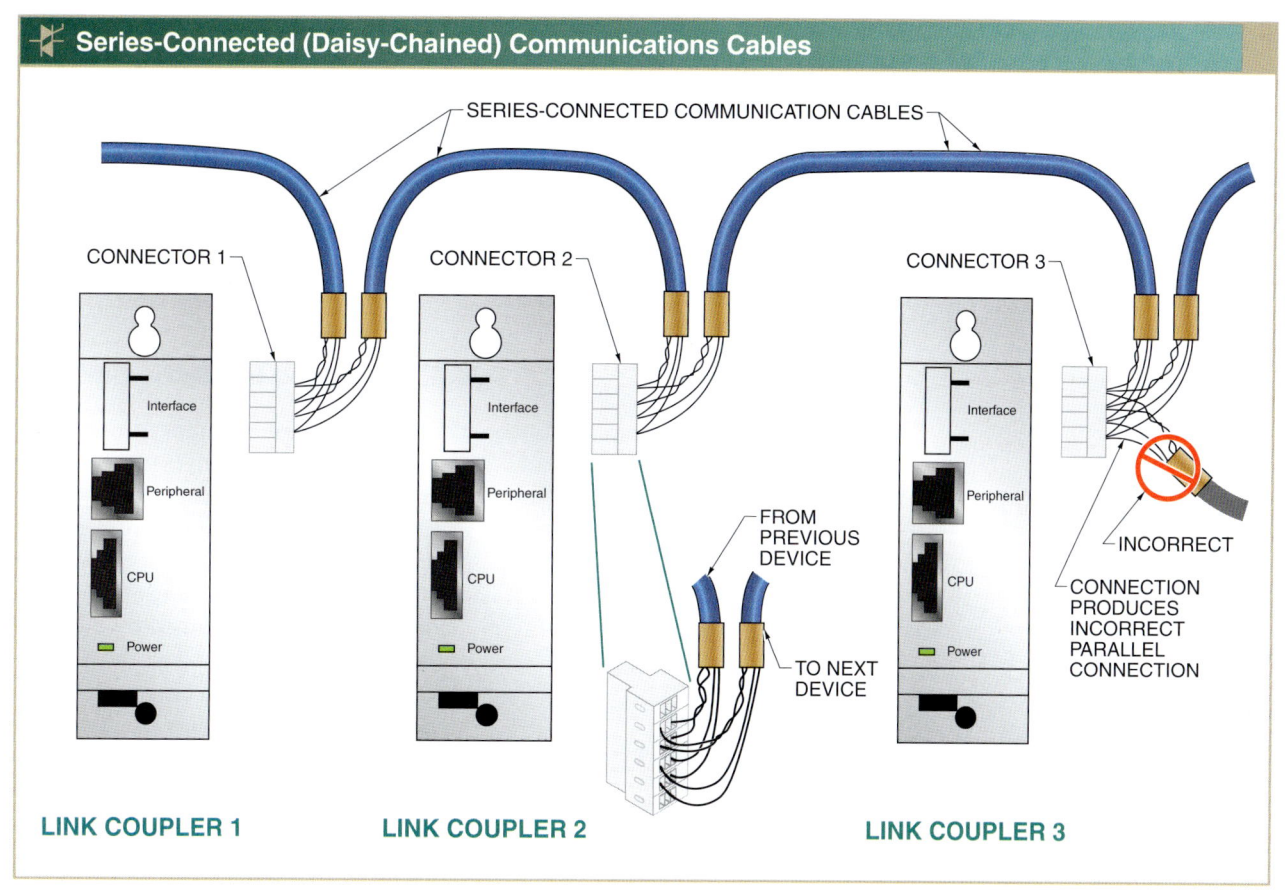

Figure 2-26. PLC communications cables connected in series (daisy-chained) are used to connect individual PLC processors.

Resistance in Series Circuits

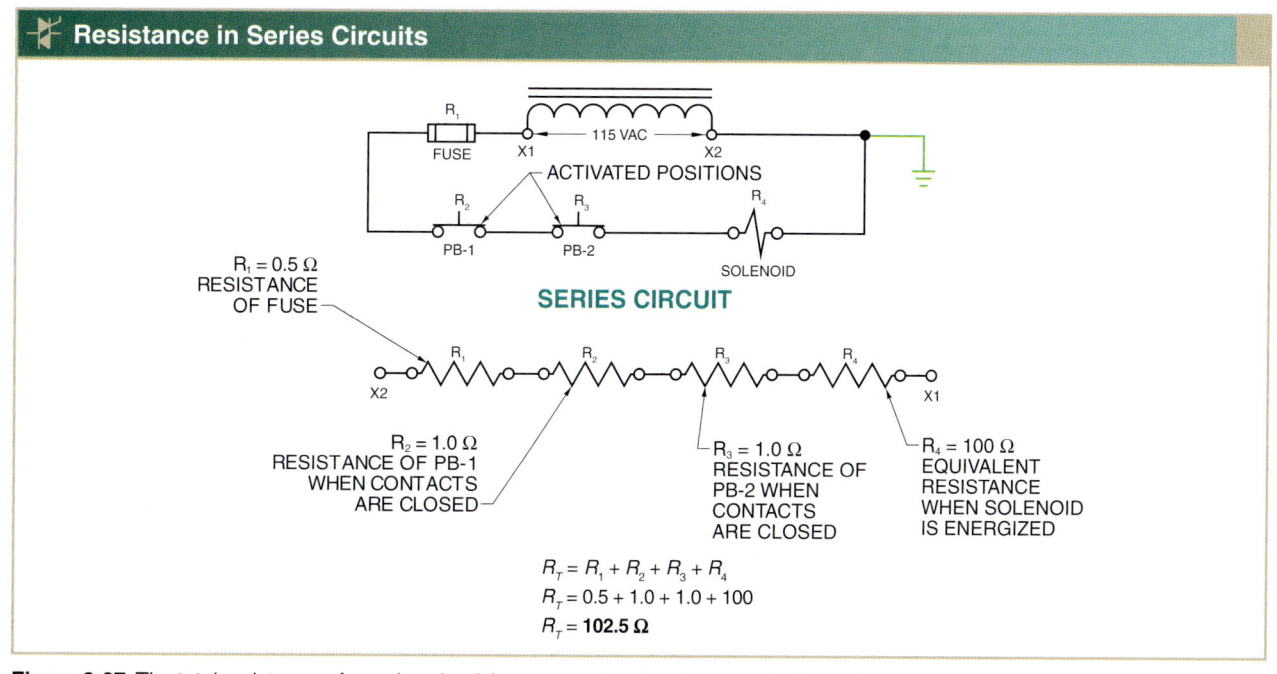

$R_T = R_1 + R_2 + R_3 + R_4$
$R_T = 0.5 + 1.0 + 1.0 + 100$
$R_T = \mathbf{102.5\ \Omega}$

$R_1 = 0.5\ \Omega$ RESISTANCE OF FUSE
$R_2 = 1.0\ \Omega$ RESISTANCE OF PB-1 WHEN CONTACTS ARE CLOSED
$R_3 = 1.0\ \Omega$ RESISTANCE OF PB-2 WHEN CONTACTS ARE CLOSED
$R_4 = 100\ \Omega$ EQUIVALENT RESISTANCE WHEN SOLENOID IS ENERGIZED

Figure 2-27. The total resistance of a series circuit increases when loads are added in series and decreases when loads are removed.

Current in Series Circuits

The current in a circuit containing series-connected loads is the same throughout the circuit. The same amount of current that the load draws is the same amount of current through the switch controlling the load and the fuse protecting the circuit. **See Figure 2-28.** To calculate current in a series circuit, apply the following formula:

$$I = \frac{E}{R}$$

where
I = circuit current (in A)
E = circuit supply voltage (in V)
R = total circuit resistance (in Ω)

For example, what is the current through a series circuit with a supply voltage of 115 V and a total circuit resistance of 102.5 Ω?

$$I = \frac{E}{R}$$
$$I = \frac{115}{102.5}$$
$$I = \mathbf{1.12\ A}$$

PLC TIPS

Resistors are built into circuit designs to limit the amount of current or oppose current flow in specific places. The construction and sizing of resistors allow the resistance value of a circuit to be precisely controlled. They also allow the circuit to dissipate heat in specific sections or generate heat for particular applications.

Voltage Drops in Series Circuits

The total voltage applied across loads connected in series is divided across the individual loads that have resistance, no matter how low the resistance of the component. Each load drops a set percentage of the applied voltage. *Voltage drop* is the amount of voltage consumed by a device or component as current passes through the object. The exact voltage drop across each load depends on the resistance of each specific load. The voltage drops across any two loads are the same if the resistance values of the loads are the same. **See Figure 2-29.**

Current in Series Circuits

[Diagram showing a series circuit with FUSE, PB-1, PB-2, 115 VAC transformer (X1, X2), and SOLENOID, with four ammeters each reading 1.1 A]

$$I = \frac{E}{R}$$
$$I = \frac{115}{102.5}$$
$$I = 1.12\ A$$

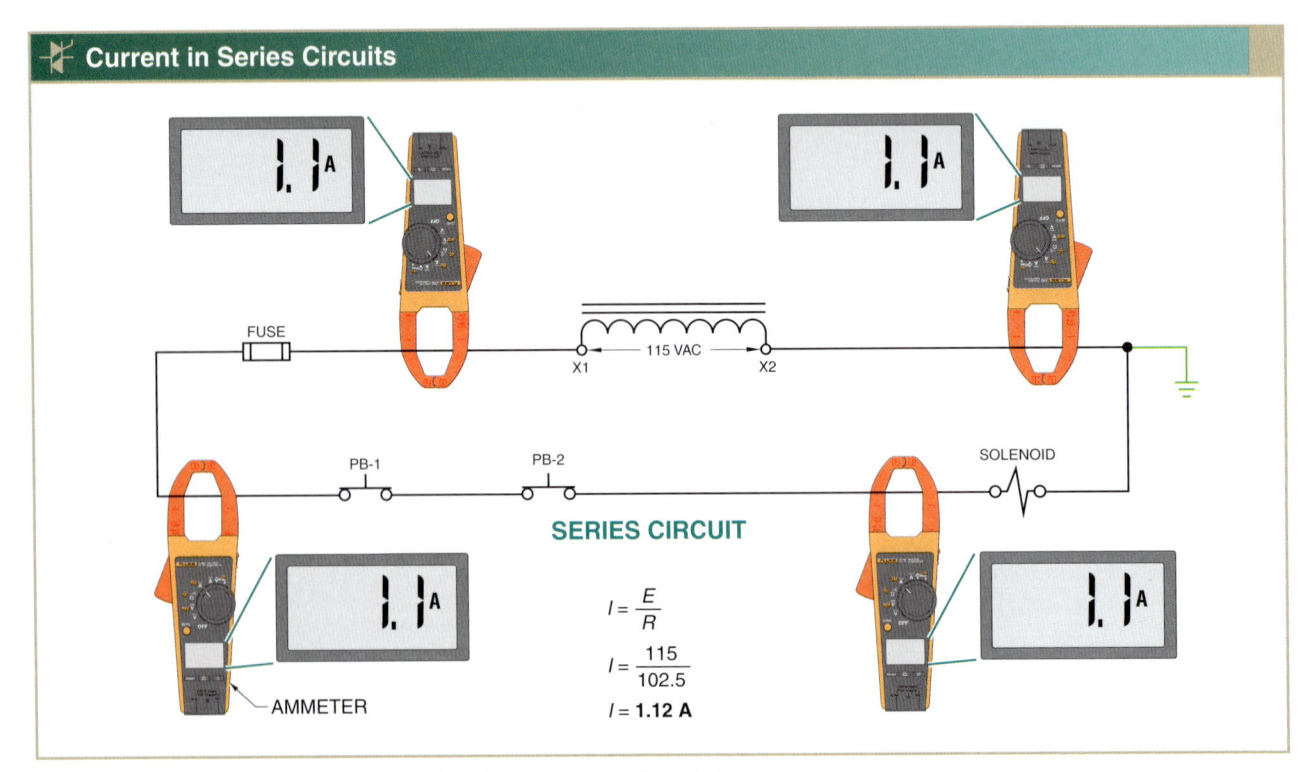

Figure 2-28. Current flow in a series circuit is the same everywhere in the circuit.

Series Circuit Voltage Drops

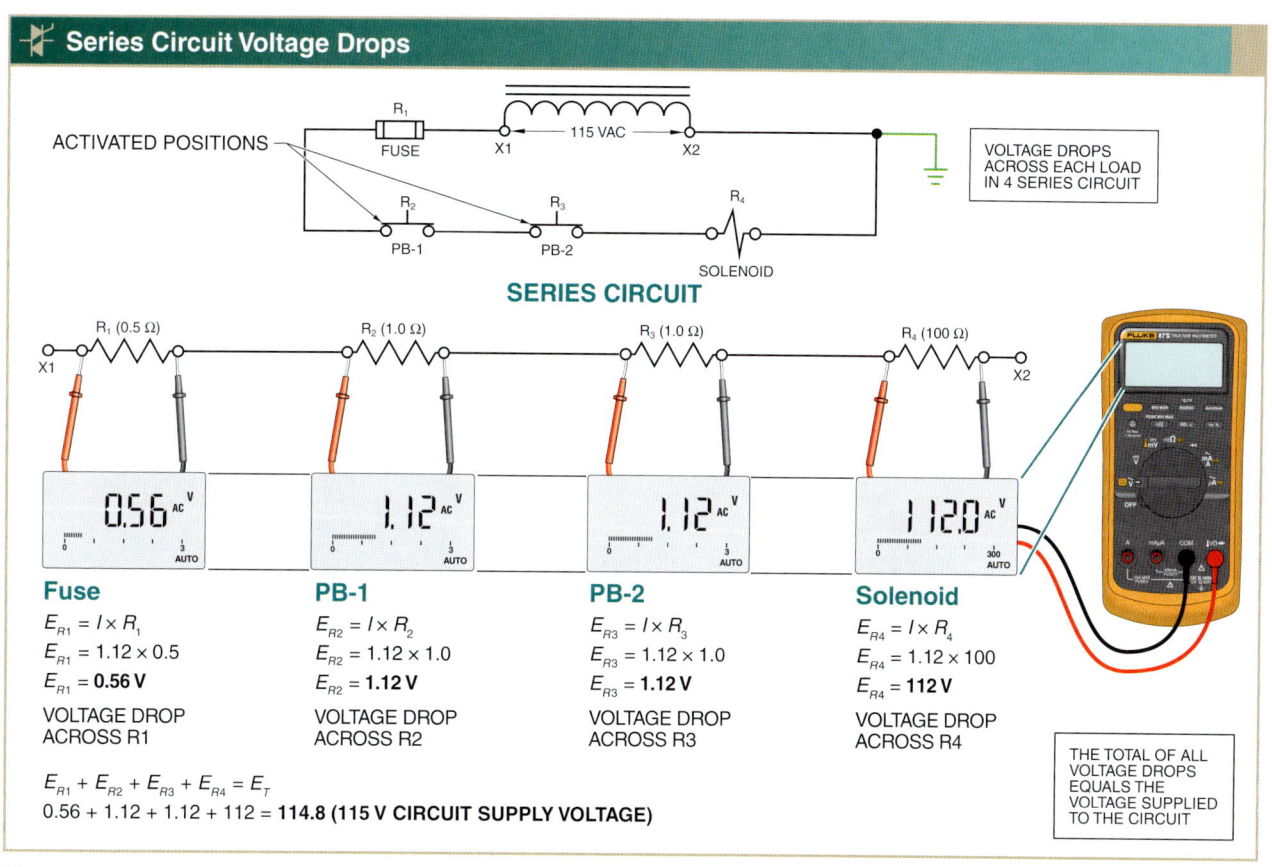

Figure 2-29. Voltage drop is the amount of voltage consumed by a device or component as current passes through it.

To calculate the voltage drop of a load in a series, apply the following formula:

$$E_R = I \times R$$

where
E_R = voltage drop consumed by load (in V)
I = current flow through load (in A)
R = resistance of load (in Ω)

For example, what are the voltage drops created by a fuse (0.5 Ω), PB-1 (1.0 Ω), PB-2 (1.0 Ω), and a solenoid (100 Ω) with a current flow of 1.12 A?

Fuse
$$E_{R1} = I \times R_1$$
$$E_{R1} = 1.12 \times 0.5$$
$$E_{R1} = \mathbf{0.56\ V}$$

PB-1
$$E_{R2} = I \times R_2$$
$$E_{R2} = 1.12 \times 1.0$$
$$E_{R2} = \mathbf{1.12\ V}$$

PB-2
$$E_{R3} = I \times R_3$$
$$E_{R3} = 1.12 \times 1.0$$
$$E_{R3} = \mathbf{1.12\ V}$$

Solenoid
$$E_{R4} = I \times R_4$$
$$E_{R4} = 1.12 \times 100$$
$$E_{R4} = \mathbf{112\ V}$$

The total of all voltage drops equals the voltage supplied to a circuit.

$$E_{R1} + E_{R2} + E_{R3} + E_{R4} = E_T$$
$$0.56 + 1.12 + 1.12 + 112 = \mathbf{114.8\ V}$$

Due to the rounding of the calculated current of 1.12 A, the 114.8 V calculated is the 115 V power supplied to the circuit.

PLC TIPS

Solid-state switches are derated or require heat sinks when used in circuits. Unlike mechanical switches that have close to no voltage drop, solid-state switches have a voltage drop of about 0.5 V to 0.8 V during operation. This distinction is important when calculating voltage drops.

PARALLEL CIRCUITS

Fuses, switches, loads, conductors, and other devices and components can be connected in parallel. *Parallel-connected devices and components* are two or more devices or components that are connected so that there is more than one flow path for current to take. Input devices (switches) and output components (loads) can be connected in parallel. **See Figure 2-30.**

Technicians must take care when working with parallel circuits because current can be flowing in one part of a circuit even though another part of the circuit is off or open. Understanding and recognizing parallel-connected devices and components enables technicians to take proper measurements, make circuit modifications, and troubleshoot circuits in a timely manner.

Resistance in Parallel Circuits

The total resistance in a circuit containing parallel-connected loads is less than the lowest resistance value of any given load. The total resistance decreases if loads are added in parallel and increases if loads are removed. **See Figure 2-31.**

To calculate total resistance of two loads in a parallel circuit or leg of a circuit, apply the following formula:

$$R_T = \frac{R_1 \times R_2}{R_1 + R_2}$$

where
R_T = total resistance (in Ω)
R_1 = resistance 1 (in Ω)
R_2 = resistance 2 (in Ω)

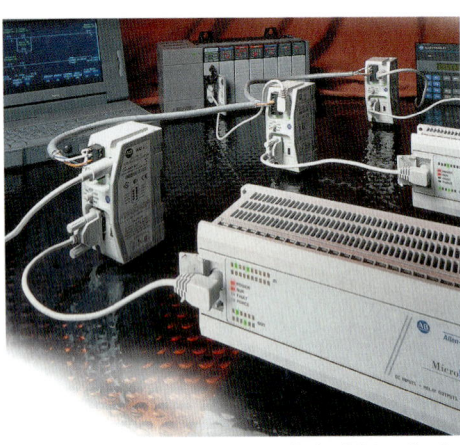

Rockwell Automation, Inc.
PLCs connected to a communication network are connected in parallel for improved communication speed.

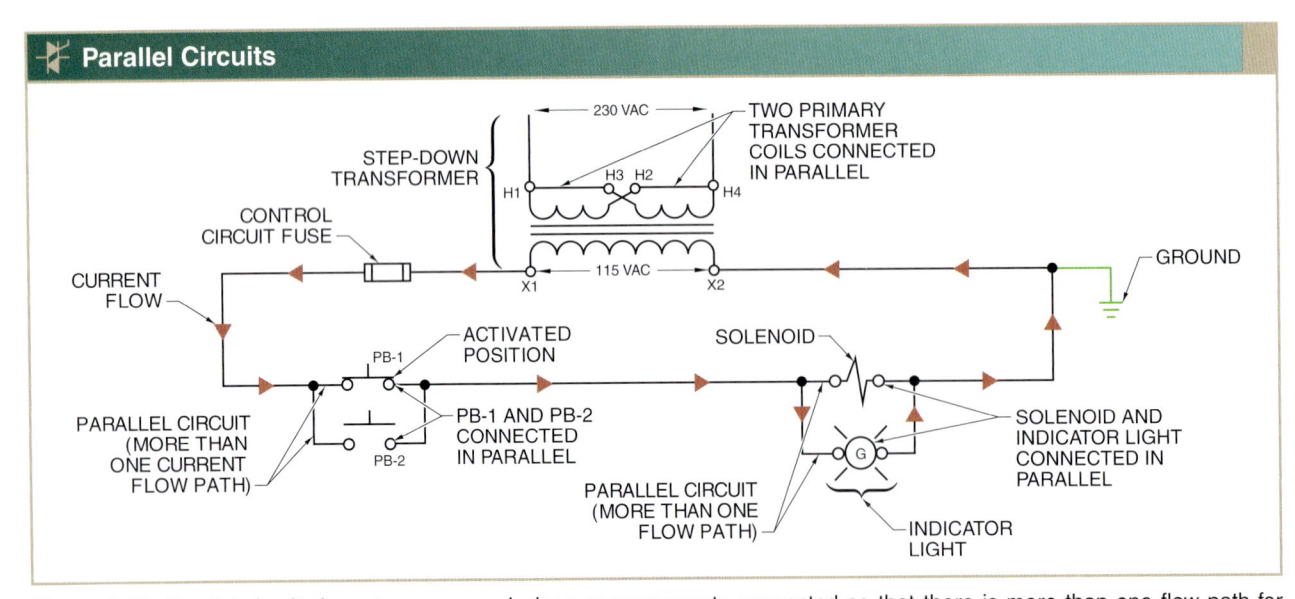

Figure 2-30. Parallel circuits have two or more devices or components connected so that there is more than one flow path for current to take.

Resistance in Parallel Circuits

$$R_T = R_1 + \left(\frac{R_2 \times R_3}{R_2 + R_3}\right) + \left(\frac{R_4 \times R_5}{R_4 + R_5}\right)$$

$$R_T = 0.5 + \left(\frac{1 \times 1}{1 + 1}\right) + \left(\frac{100 \times 200}{100 + 200}\right)$$

$$R_T = 0.5 + \left(\frac{1}{2}\right) + \left(\frac{20{,}000}{300}\right)$$

$$R_T = 0.5 + 0.5 + 66.67$$

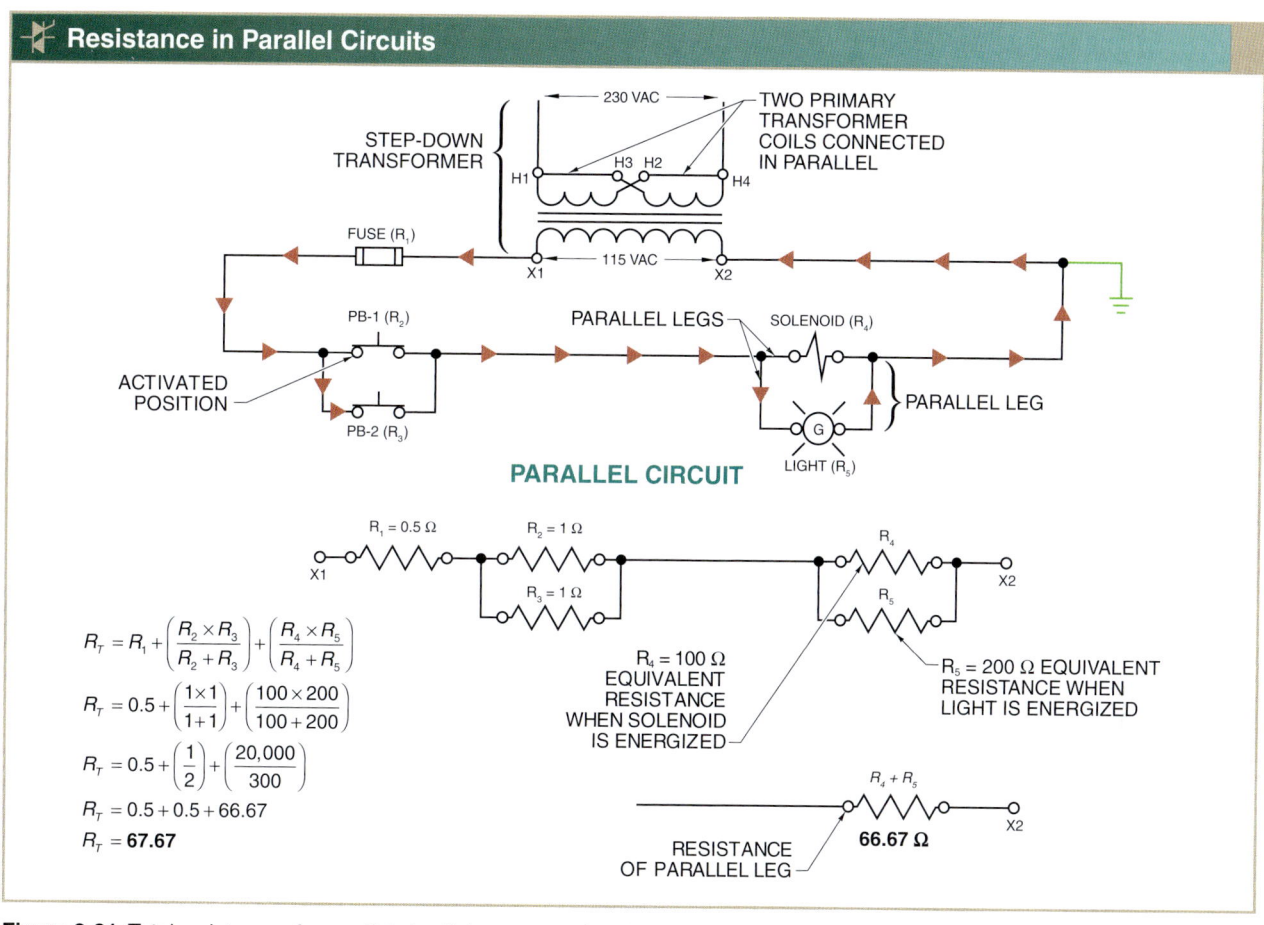

$$R_T = \mathbf{67.67}$$

Figure 2-31. Total resistance of a parallel circuit decreases when loads are added in parallel and increases when loads are removed.

For example, what is the total resistance of a circuit with a 0.5 Ω fuse in series, two 1.0 Ω pushbuttons in parallel, and a parallel leg containing components that have 100 Ω and 200 Ω of resistance?

$$R_T = R_1 + \left(\frac{R_2 \times R_3}{R_2 + R_3}\right) + \left(\frac{R_4 \times R_5}{R_4 + R_5}\right)$$

$$R_T = 0.5 + \left(\frac{1 \times 1}{1 + 1}\right) + \left(\frac{100 \times 200}{100 + 200}\right)$$

$$R_T = 0.5 + \left(\frac{1}{2}\right) + \left(\frac{20{,}000}{300}\right)$$

$$R_T = 0.5 + 0.5 + 66.67$$

$$R_T = \mathbf{67.67}$$

Current in Parallel Circuits

Total current in a circuit containing parallel-connected loads equals the sum of the current through all the loads. Total current increases when loads are added in parallel and decreases when loads are removed. **See Figure 2-32.**

To calculate total current in a parallel circuit or leg, apply the following formula:

$$I_T = \frac{E}{R_T}$$

where

I_T = total circuit current (in A)
E = circuit supply voltage (in V)
R_T = total circuit resistance (in Ω)

For example, what is the total circuit current and the current through two loads of a leg of a circuit that are connected in parallel when one load (solenoid) is 100 Ω and the second load (indicator light) is 200 Ω with circuit voltage of 115 V?

98 PROGRAMMABLE LOGIC CONTROLLERS: PRINCIPLES AND APPLICATIONS

Current in Parallel Circuits

PARALLEL CIRCUIT

PARALLEL CIRCUITS HAVE THE BEGINNING OF COMPONENTS CONNECTED TOGETHER AND THE END OF COMPONENTS CONNECTED TOGETHER

Circuit Current

$$I_T = \frac{E}{R_T}$$

$$I_T = \frac{115}{67.67}$$

$$I_T = 1.7 \text{ A}$$

Solenoid Leg

$$V_{R4+R5} = 1.7 \times 67.67$$

$$V_{R4+R5} = 113.3 \text{ V}$$

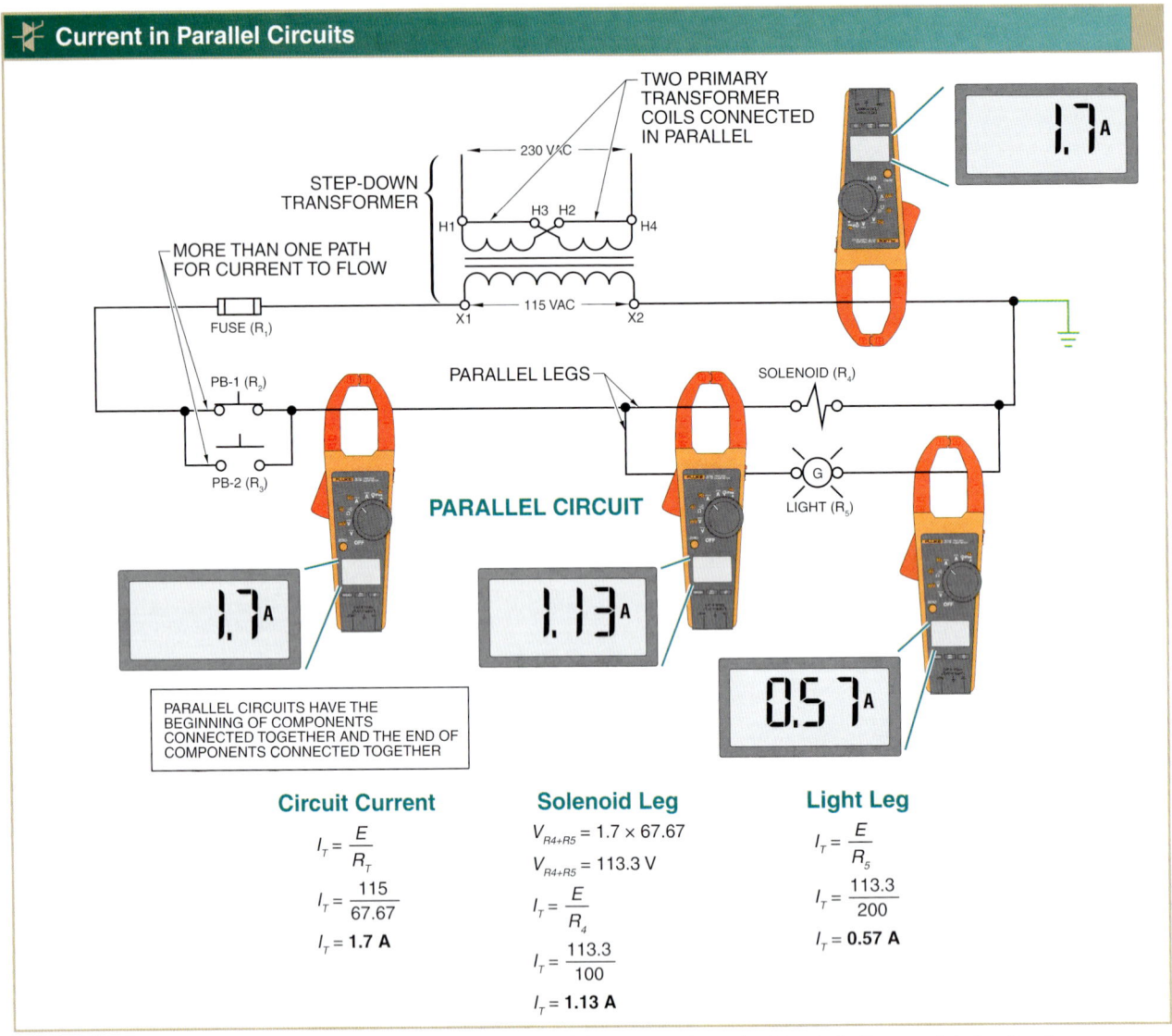

$$I_T = \frac{E}{R_4}$$

$$I_T = \frac{113.3}{100}$$

$$I_T = 1.13 \text{ A}$$

Light Leg

$$I_T = \frac{E}{R_5}$$

$$I_T = \frac{113.3}{200}$$

$$I_T = 0.57 \text{ A}$$

Figure 2-32. Total current in a parallel circuit equals the sum of the current through all loads in the parallel circuit or leg.

Total Circuit Current

$$I_T = \frac{E}{R_T}$$

$$I_T = \frac{115}{67.67}$$

$$I_T = 1.7 \text{ A}$$

Solenoid Leg

$$V_{R4+R5} = 1.7 \times 67.67$$

$$V_{R4+R5} = 113.3$$

$$I_S = \frac{E}{R}$$

$$I_S = \frac{113.3}{100}$$

$$I_S = 1.13 \text{ A}$$

Light Leg

$$I_L = \frac{E}{R}$$

$$I_L = \frac{113.3}{200}$$

$$I_L = 0.57 \text{ A}$$

PLC TIPS

Three types of analog signals include voltage, current, and resistance. Each type is continuous. Variable quantities such as temperatures, pressures, flow rates, and analog proximities can all produce an analog signal. Typically, transduction is the process used by analog sensors to change energy in one form to another.

Voltage Drops in Parallel Circuits

The voltage drop across each load is the same when loads are connected in parallel. The voltage drop across each load remains the same when parallel loads are added or removed. **See Figure 2-33.**

To calculate the voltage drop of connected loads, apply the following formula:

$$E_R = I \times R$$

where
E_R = voltage drop created by load (in V)
I = current flow through load (in A)
R = resistance of load (in Ω)

Parallel Circuit Voltage Drops

PARALLEL CIRCUIT

Fuse
$E_{R1} = I \times R_1$
$E_{R1} = 1.7 \times 0.5$
$E_{R1} = \mathbf{0.85\ V}$
VOLTAGE DROP ACROSS R1

Switches
$E_{R2\&3} = I \times R_{2\&3}$
$E_{R2\&3} = 1.7 \times 0.5$
$E_{R2\&3} = \mathbf{0.85\ V}$
VOLTAGE DROP ACROSS R2 & R3

Solenoid and Indicator Light
$E_{R4\&5} = I_T \times R_{4\&5}$
$E_{R4\&5} = 1.7 \times 66.67$
$E_{R5} = \mathbf{113.3\ V}$
VOLTAGE DROP ACROSS R4 & R5

TOTAL VOLTAGE
$E_T = E_{R1} + E_{R2\&3} + E_{R4\&5}$
$ET = 0.85 + 0.85 + 113.3$
$ET = \mathbf{115\ V}$

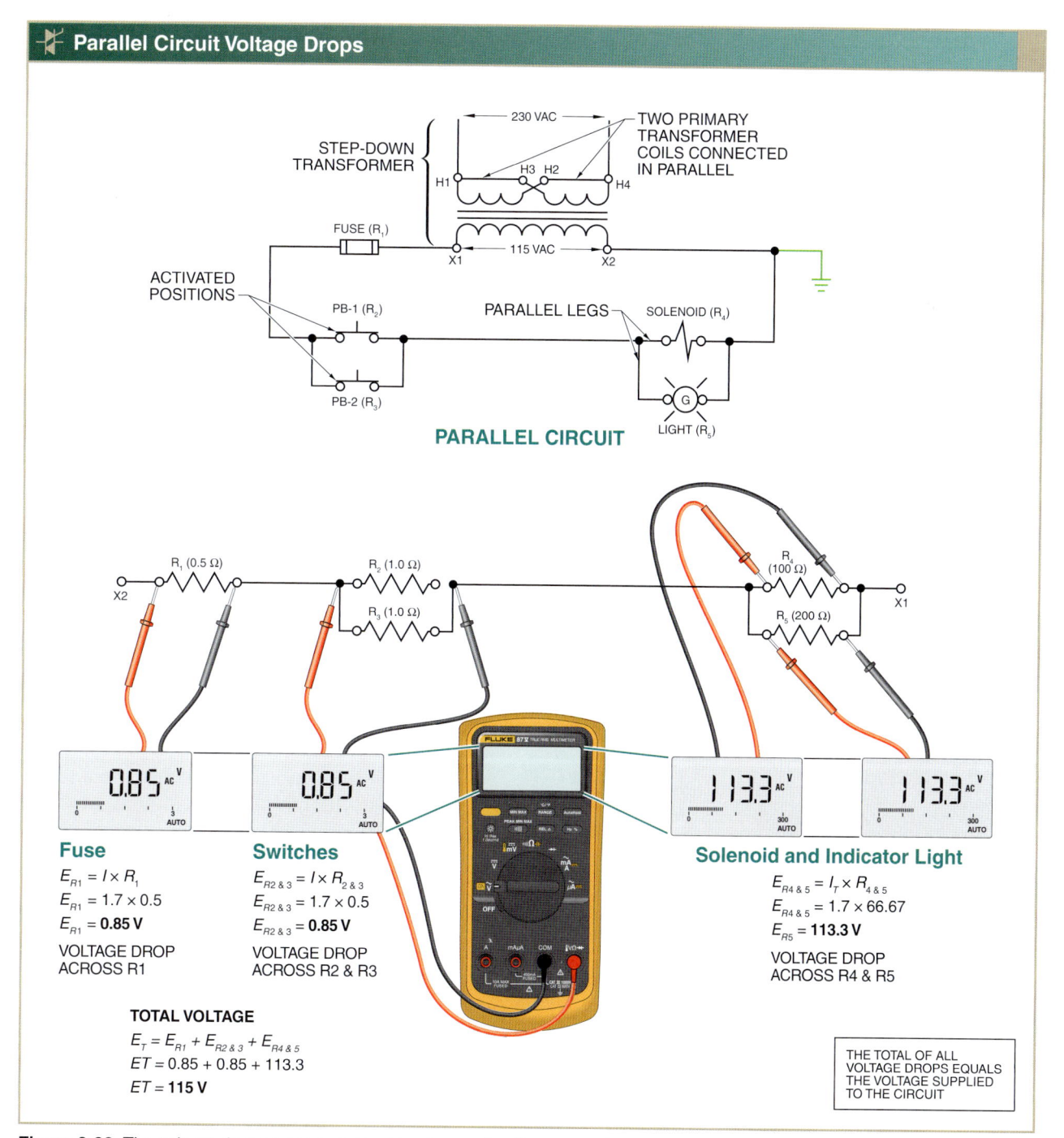

THE TOTAL OF ALL VOLTAGE DROPS EQUALS THE VOLTAGE SUPPLIED TO THE CIRCUIT

Figure 2-33. The voltage drop across individual loads remains the same when parallel loads are added or removed.

For example, what is the voltage drop of the fuse, the two parallel-connected pushbutton switches, and the two parallel-connected loads (solenoid and indicator light) that are all connected to a 115 VAC power supply?

Fuse
$$E_{R1} = I \times R_1$$
$$E_{R1} = 1.7 \times 0.5$$
$$E_{R1} = \mathbf{0.85\ V}$$

Switches
$$E_{R2\ \&\ 3} = I \times R_{2\ \&\ 3}$$
$$E_{R2\ \&\ 3} = 1.7 \times 0.5$$
$$E_{R2\ \&\ 3} = \mathbf{0.85\ V}$$

Solenoid and Indicator Light
$$E_{R4\ \&\ 5} = I_T \times R_{4\ \&\ 5}$$
$$E_{R4\ \&\ 5} = 1.7 \times 66.67$$
$$E_{R4\ \&\ 5} = \mathbf{113.3\ V}$$

Total Voltage
$$E_T = E_{R1} + E_{R2\ \&\ 3} + E_{R4\ \&\ 5}$$
$$E_T = 0.85 + 0.85 + 113.3$$
$$E_T = \mathbf{115\ V}$$

CODE CONNECT

ARTICLE 250 — GROUNDING AND BONDING

Section 250.97 of the NEC® addresses the use of prepunched concentric and eccentric knockouts of electrical enclosures on circuits operating at over 250 volts to ground. If the enclosure containing the prepunched concentric or eccentric knockouts has not been listed as providing a reliable bonding connection, then other devices must be used to provide a reliable connection. The equipment includes bonding-type locknuts, bonding bushings, and grounding locknuts referenced in Section 250.92(B)(4).

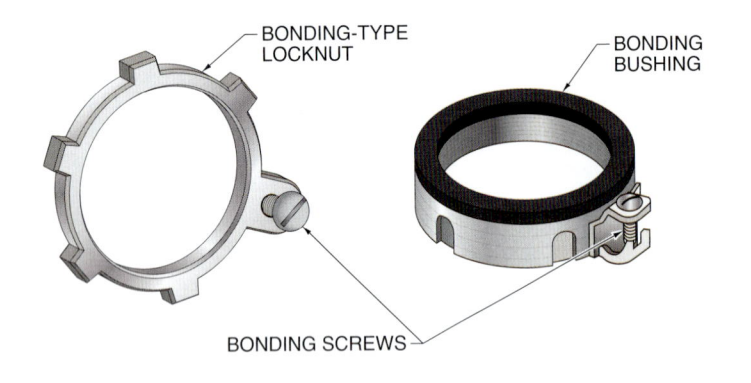

COMBINATION CIRCUITS

Fuses, switches, loads, conductors, and other devices and components can be connected with a combination of series/parallel connections. A *combination circuit* is a combination of series- and parallel-connected devices and components. In most combination circuits, loads such as lights, solenoids, and motors are typically connected in parallel. Fuses, switches, and circuit breakers that control and monitor current flow through loads are connected in series. Understanding combination circuits enables technicians to take proper measurements, make correct connections when troubleshooting, and create circuit modifications. **See Figure 2-34.**

Resistance in Combination Circuits

Resistors and loads, such as solenoids, lights, and motors are often connected in series/parallel combinations. A series/parallel circuit can contain a number of resistors and loads connected in many series/parallel combinations. A series/parallel combination is always equal to one combined total resistance (RT) value. The total resistance equals the sum of the series loads and the equivalent resistance of the parallel combinations. To calculate total resistance in a series/parallel combination that contains three or more resistors, apply the following formula:

$$R_T = \frac{1}{\dfrac{1}{R_1} + \dfrac{1}{R_2} + \dfrac{1}{R_3} + \ldots}$$

where
R_T = total resistance (in Ω)
R_1 = resistance 1 (in Ω)
R_2 = resistance 2 (in Ω)
R_3 = resistance 3 (in Ω)

To calculate total resistance in a series/parallel circuit that contains multiple series/parallel combinations, the circuit must be broken down into the basic series and parallel parts. First, the two or more resistance values are combined into a single resistance value if two or more are connected in series in a parallel leg. Then, two or more resistors in parallel are combined into a single resistance value. Once all parallel values have been combined into one resistance value, the sum of the series resistance values is combined with the sum of the parallel resistance values to acquire the total resistance of the circuit.

Combination Circuits

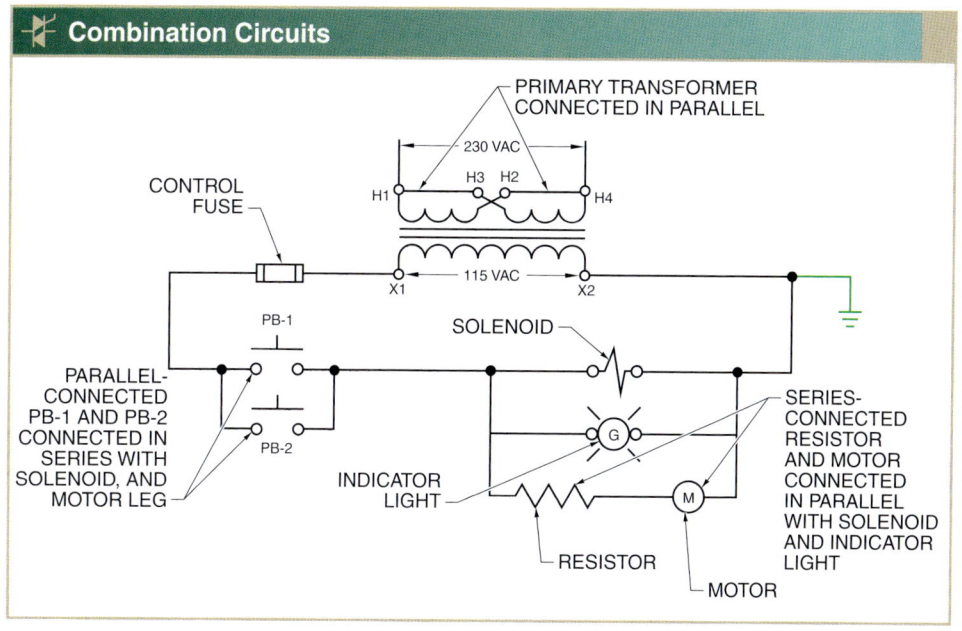

Figure 2-34. Combination circuits can have combinations of series-connected devices, such as fuses, switches, and circuit breakers, connected to combinations of parallel-connected devices, such as lights, solenoids, and motors.

For example, what is the total resistance of a circuit with a 0.5 Ω fuse in series, two 1.0 Ω pushbuttons in parallel, and a parallel leg containing components that have 100 Ω and 200 Ω resistance in parallel with a resistor of 50 Ω and motor of 500 Ω connected in series? **See Figure 2-35.**

Current and Voltage Drop in Combination Circuits

The total current and total voltage in combination circuits follow the same principles in the series portion and parallel portion of an electrical circuit. Current in individual parts of a combination circuit stays the same in series and divides in parallel legs. Voltage when applied across individual parts divides in series and stays the same in parallel legs. **See Figure 2-36.**

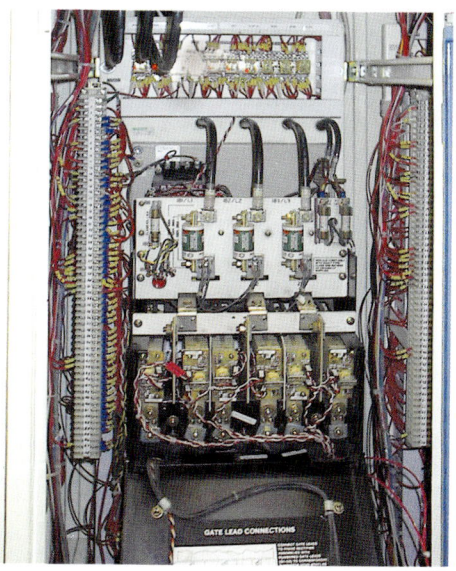

Power control cabinets contain a number of combination circuits that feed power to separate locations at certain voltage levels.

Resistance in Combination Circuits

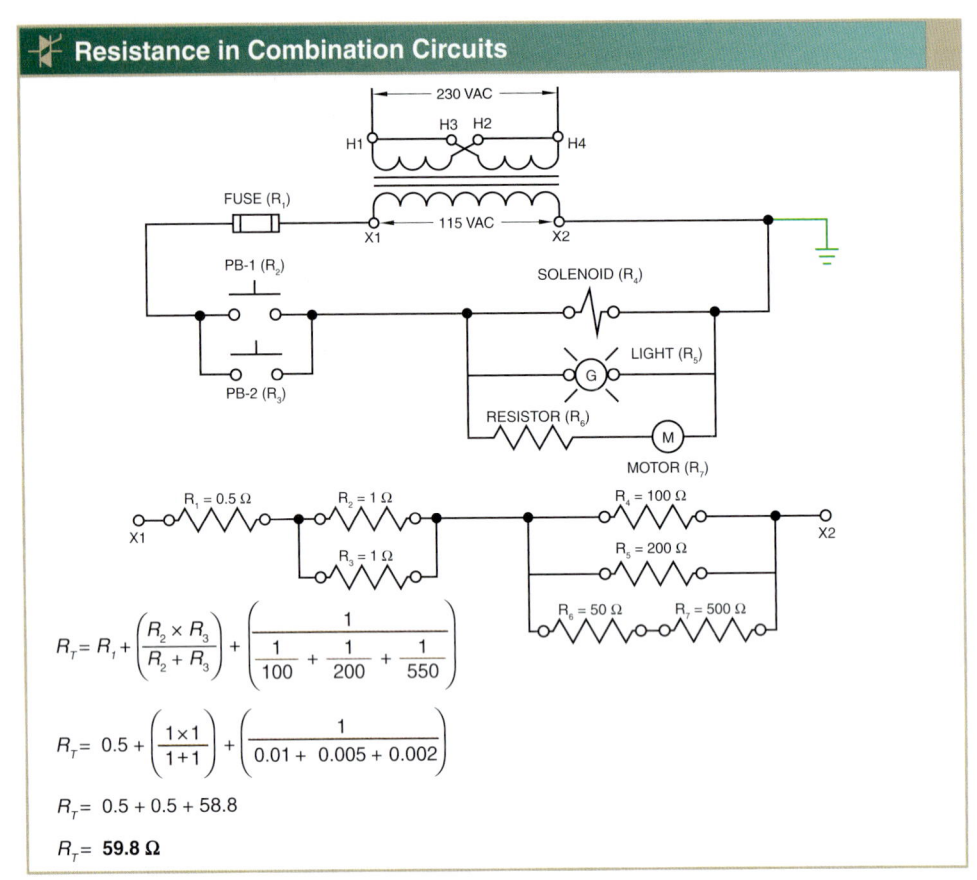

$$R_T = R_1 + \left(\frac{R_2 \times R_3}{R_2 + R_3}\right) + \left(\frac{1}{\frac{1}{100} + \frac{1}{200} + \frac{1}{550}}\right)$$

$$R_T = 0.5 + \left(\frac{1 \times 1}{1+1}\right) + \left(\frac{1}{0.01 + 0.005 + 0.002}\right)$$

$$R_T = 0.5 + 0.5 + 58.8$$

$$R_T = \mathbf{59.8 \ \Omega}$$

Figure 2-35. Total resistance in a combination circuit must be broken down into parts of series and parallel combinations to acquire the total resistance of the circuit.

Current in Combination Circuits

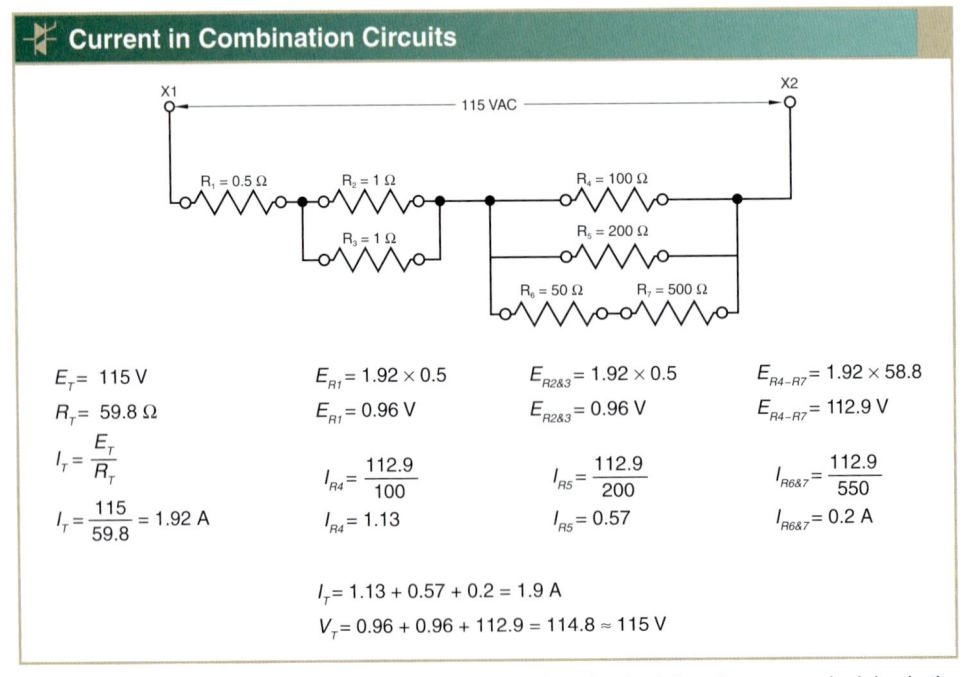

$E_T = 115 \ V$

$R_T = 59.8 \ \Omega$

$I_T = \dfrac{E_T}{R_T}$

$I_T = \dfrac{115}{59.8} = 1.92 \ A$

$E_{R1} = 1.92 \times 0.5$

$E_{R1} = 0.96 \ V$

$I_{R4} = \dfrac{112.9}{100}$

$I_{R4} = 1.13$

$E_{R2\&3} = 1.92 \times 0.5$

$E_{R2\&3} = 0.96 \ V$

$I_{R5} = \dfrac{112.9}{200}$

$I_{R5} = 0.57$

$E_{R4-R7} = 1.92 \times 58.8$

$E_{R4-R7} = 112.9 \ V$

$I_{R6\&7} = \dfrac{112.9}{550}$

$I_{R6\&7} = 0.2 \ A$

$I_T = 1.13 + 0.57 + 0.2 = 1.9 \ A$

$V_T = 0.96 + 0.96 + 112.9 = 114.8 \approx 115 \ V$

Figure 2-36. Total current and total voltage in combination circuits follow the same principles in the series portion and parallel portion of an electrical circuit.

Chapter Review 2

Name _____ Date _____

True-False

T F **1.** An analog signal continuously changes in value between defined limits.

T F **2.** Corrosion decreases resistance in a conductor.

T F **3.** All DC voltage sources have a positive and negative terminal.

T F **4.** Voltage sources with the same voltage level may produce different amounts of current.

T F **5.** For metals, the higher the temperature, the lower the resistance.

T F **6.** When voltage levels are the same, a PLC may operate on either alternating current (AC) or direct current (DC).

T F **7.** A conductor with a large cross-sectional area has less resistance than a conductor with a small cross-sectional area.

T F **8.** Long conductors have less resistance than short conductors of the same size (AWG).

T F **9.** All PLCs must receive power to operate.

T F **10.** A current sourcing input uses the positive terminal of a DC-powered PLC as the source.

T F **11.** Opening a parallel circuit at any point stops the flow of current.

T F **12.** Output components can be connected in series to PLC output modules.

T F **13.** Typically, single-phase power is used for low-voltage loads and three-phase power is used for high voltage loads.

T F **14.** The current in a circuit containing series-connected loads is the same throughout the circuit.

T F **15.** In a parallel-connected circuit, the voltage drop across each load varies when parallel loads are added or removed.

Completion

_____ 1. ___ is electron flow that is unidirectional.

_____ 2. ___ is the amount of voltage consumed by an input device or output component as current passes through the object.

_____ 3. ___ are two or more devices or components that are connected so that there is more than one flow path for current to take.

_____ 4. A(n) ___ is half of a sine wave cycle.

_____ 5. ___ is the opposition to current flow.

_____ 6. A(n) ___ is a condition where a PLC is properly installed and all input devices and output components are working but there is a problem with the PLC program.

_____ 7. ___ is the positive (+) or negative (−) state of an object's connections.

_____ 8. ___ is a type of electrical supply that contains only one alternating voltage waveform.

_____ 9. A(n) ___ is the negative power supply terminal of a DC-powered PLC.

_____ 10. A(n) ___ is a symmetrical waveform that contains 360 electrical degrees and has one positive alternation and one negative alternation per cycle.

Multiple Choice

_____ 1. ___ is produced when three coils are simultaneously rotated in a generator.
   A. Rectified AC current
   B. DC current
   C. Single-phase voltage
   D. Three-phase voltage

_____ 2. Power is equal to ___.
   A. current multiplied by resistance
   B. current multiplied by watts
   C. voltage multiplied by current
   D. voltage multiplied by resistance

_____ 3. Switches that use a(n) ___ as the switching element are called current sink devices.
   A. NPN transistor
   B. PNP transistor
   C. SCR
   D. triac

_____ 4. Current sources use electron flow through a(n) ___.
   A. NPN transistor
   B. PNP transistor
   C. SCR
   D. triac

_____ 5. The ___ rating of PLC output circuitry is the amount of current the internal switches can safely handle when the load is first turned ON, when the load is operating, and when the load is turned OFF.
   A. current
   B. zero
   C. resistance
   D. voltage

Designing PLC Systems

PLC Exercise 2.1

A typical industrial application may need a nano-PLC, a micro-PLC, or I/O modules to control a section of a process used to make a consumable drinking product. A part of the process needs to be designed where the PLC maintains the power to a pump to control the liquid-level in a tank and to enable a safeguard to warn personnel if the tank overflows. For this application, use the detailed drawing to design and install the addition to the PLC system needed for the drinking product.

Developing a Bill-of-Materials

Examine the detailed drawing of the application and complete a bill of materials needed to build the PLC system. In the chart, fill out the required inputs and outputs needed, based on the PLC type and power supply used for the inputs, outputs, and PLC.

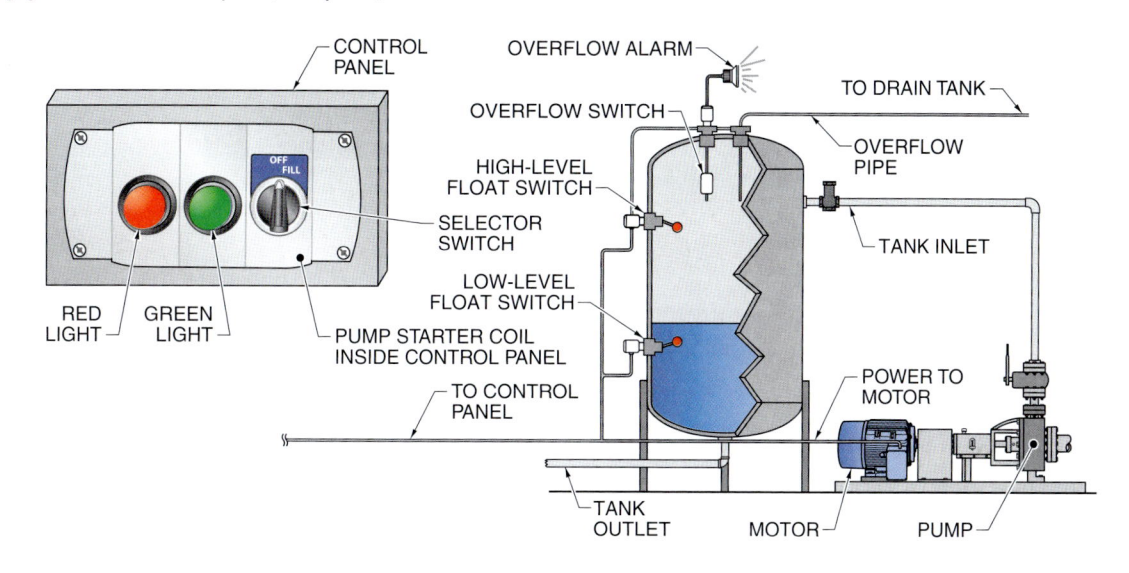

Bill of Materials			
PLC Type	**Inputs**	**Outputs**	**Power Supply (AC/DC)**
TECO			Inputs
MicroLogix™ 1100			Outputs
Medium PLC			PLC

Addressing Inputs and Outputs

After identifying each input and output, examine the line diagram and assign the inputs and outputs with the appropriate address in the chart based on the type of PLC used.

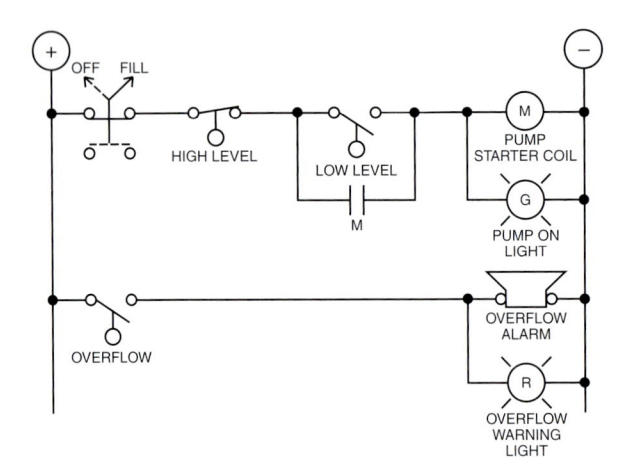

LINE DIAGRAM

Input and Output Addresses

Inputs	TECO Address	MicroLogix™ 1100 Address	I/O Module Address	Outputs	TECO Address	MicroLogix™ 1100 Address	I/O Module Address
Selector switch	I01	I:0/0	I:1/0	Pump starter coil	Q01	O:0/0	O:2/0
High-level switch				Green light			
Low-level switch				Alarm			
Overflow switch				Red light			

Formatting Inputs and Outputs

Refer to the line diagram and the input and output addresses box. Label each input and output with its appropriate address and device label in each programming diagram. When formatting a programming diagram, the actual input device symbol is programmed as a generic OPEN or CLOSED instruction and is labeled with the name of the device. When programming using TECO SG2 client ladder logic, the programming grid can be expanded to 5-contact by going to Edit in the Menu bar and selecting 5-contact.

T TECO Procedures

001 I01 SELECTOR SWITCH — — Q01 PUMP STARTER COIL
002 Q01 PUMP STARTER COIL
003
004

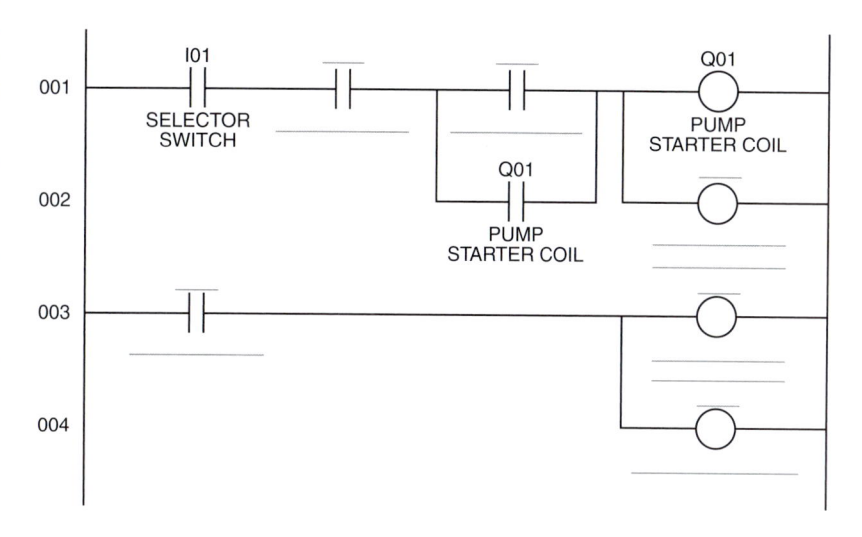

TECO PROGRAMMING DIAGRAM

AB Allen-Bradley® Procedures

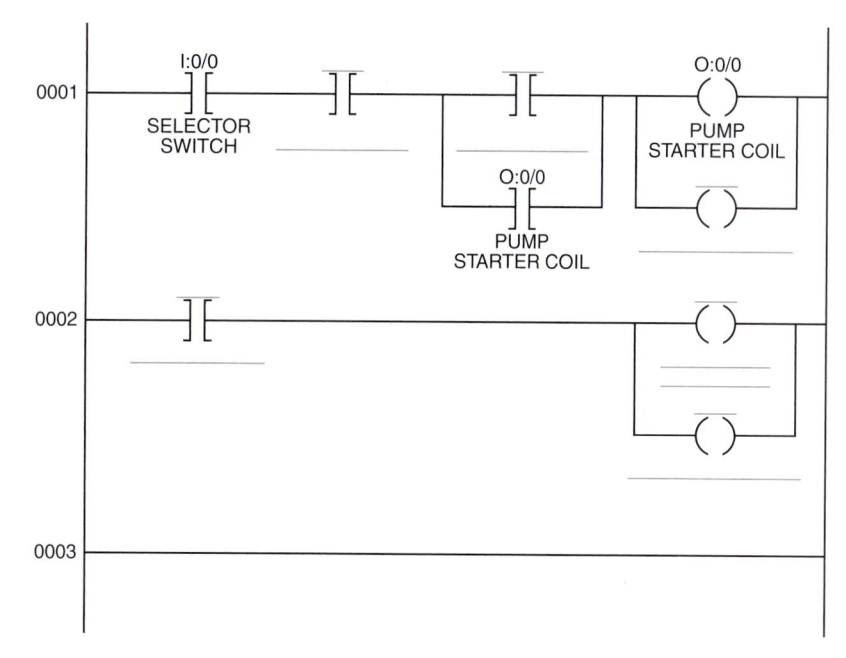

0001 I:0/0 SELECTOR SWITCH — — O:0/0 PUMP STARTER COIL
       O:0/0 PUMP STARTER COIL
0002
0003

AB PROGRAMMING DIAGRAM

Wiring Inputs and Outputs

Input and output terminals appear different depending on the PLC type and manufacturer. On some PLCs, input and output 1 are numbered starting with 1 and increase numerically (1, 2, 3). With other types of PLCs, the first input starts with the number 0 and increases numerically (0, 1, 2). For example, on a TECO SG2-12HR-D PLC, inputs 1 through 4 are labeled I1, I2, I3, and I4. On an AB Micrologix™ 1100 PLC, inputs 1 through 4 are labeled I/0, I/1, I/2, and I/3. On I/O modules, inputs are labeled IN1, IN2, IN3, and IN4, and outputs are labeled OUT1, OUT2, OUT3, and OUT4. Using the wiring diagram for each PLC type, draw the connections from the four inputs and four outputs to the appropriate input/output terminals.

Running PLC Circuits

Wire, program, and run the PLC circuit as designed using the SG2 Client software and a TECO SG2-12HR-D PLC or the AB Micrologix™ 1100 PLC and RSLogix Micro Starter Lite software.

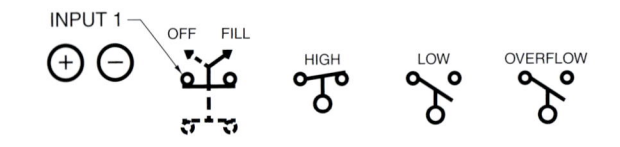

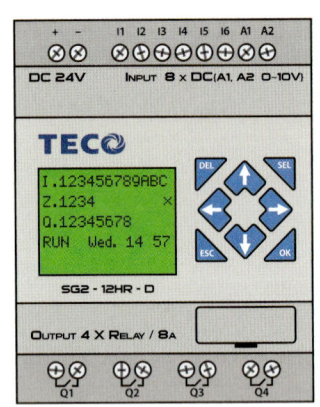

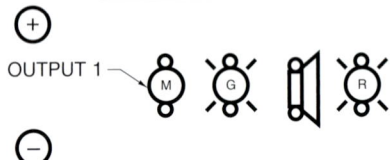

TECO WIRING DIAGRAM

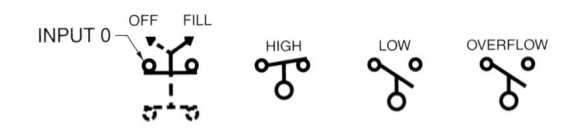

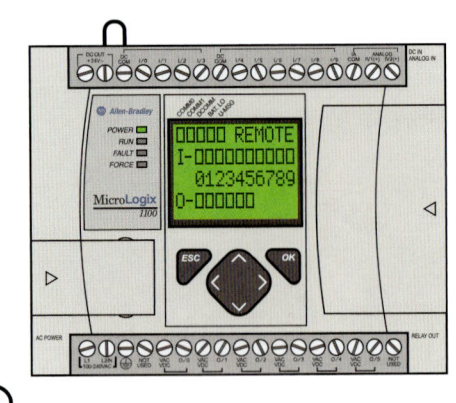

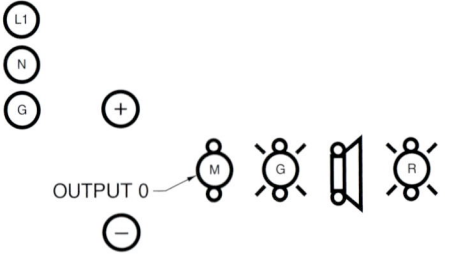

AB WIRING DIAGRAM

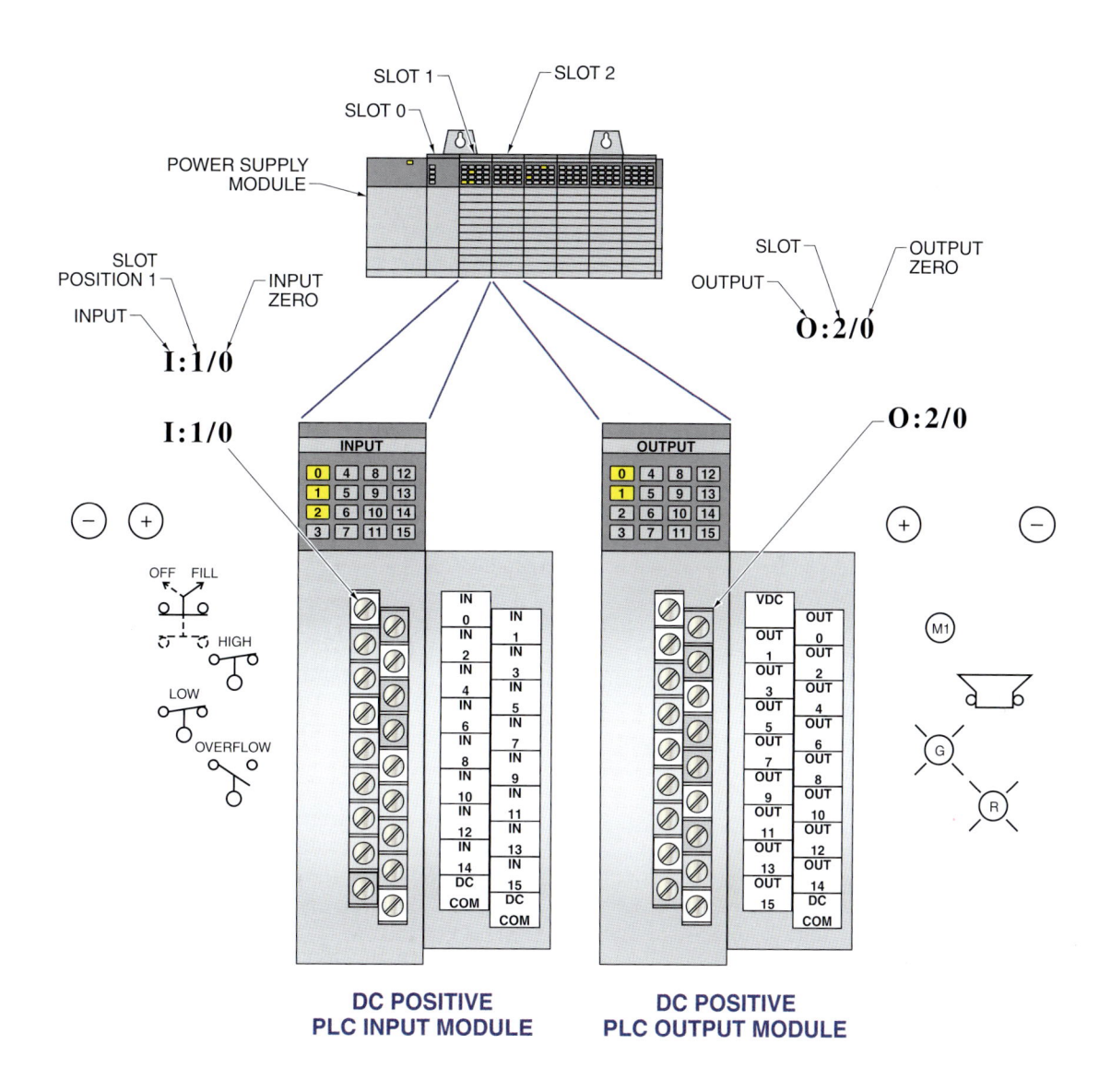

Chapter 2—PLC Electrical Principles, Ratings, and Circuit Calculations 109

Wiring Sinking and Sourcing Input Switching Devices

PLC Exercise 2.2

BILL OF MATERIALS

- Micro-PLC (sink/source rated)
- NC red pushbutton
- NO green pushbutton
- NPN and PNP solid-state switches

PLCs can be rated to receive current sink or current source input switching devices. A three-wire input switch circuit uses solid-state switches, such as NPN or PNP transistors. Photoelectric switches are solid-state NPN or PNP transistors used in many industry applications, such as for detecting materials, counting products, checking and inspecting failures, validating processes, preventing malfunctions, and sorting materials without coming into contact with them. For this lab, use the diagram and bill of materials to properly connect the photoelectric switch and pushbuttons as sinking and sourcing connections.

Wiring Sinking Connections

Using the line diagram, wire the NO pushbutton, NC pushbutton, and photoelectric switch as current sinking connections to the PLC.

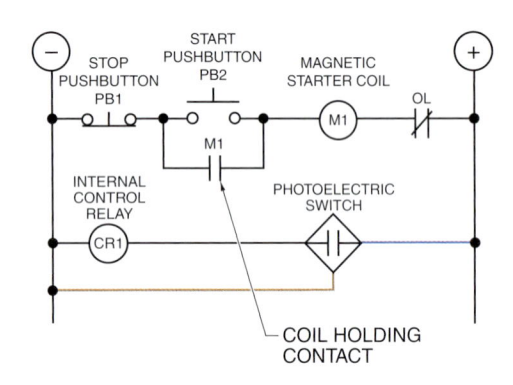

CURRENT SINK LINE DIAGRAM

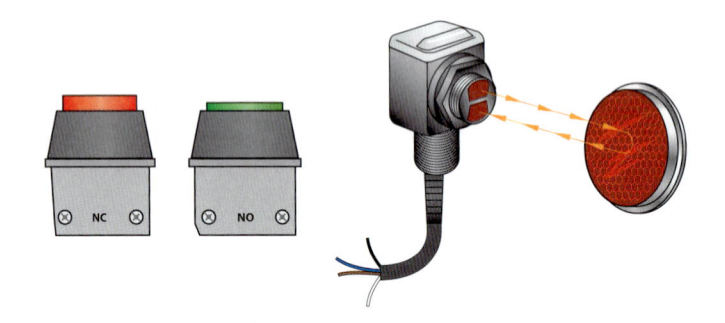

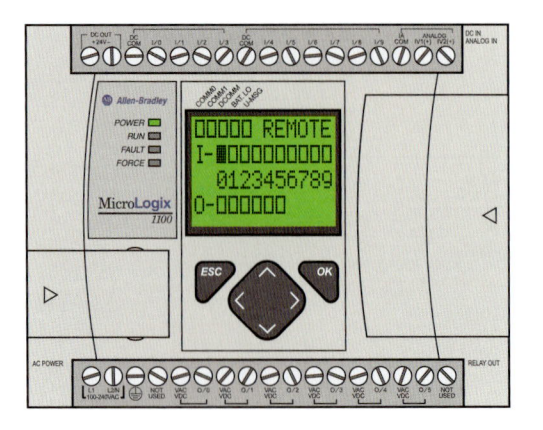

CURRENT SINKING CONNECTIONS

Wiring Sourcing Connections

Using the line diagram, wire the NO pushbutton, NC pushbutton, and photoelectric switch as current sourcing connections to the PLC.

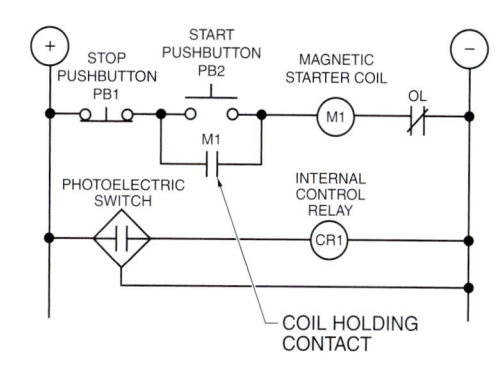

CURRENT SOURCE LINE DIAGRAM

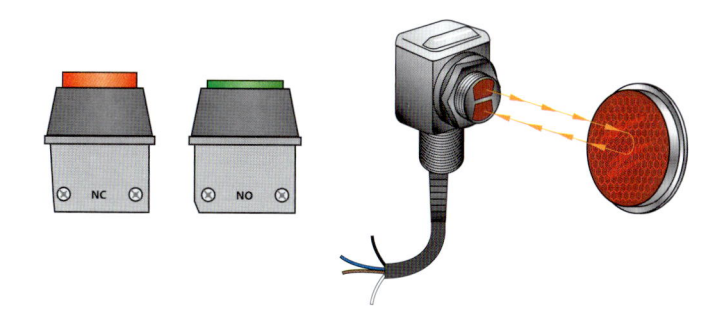

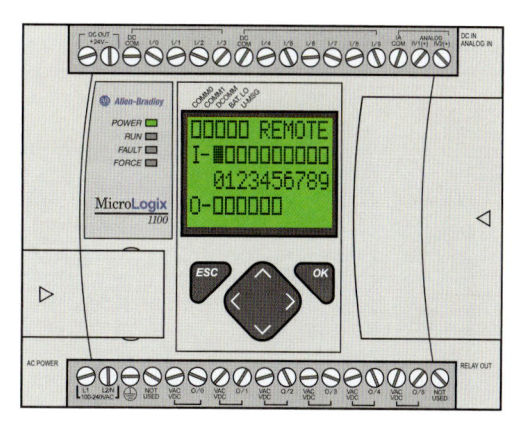

CURRENT SOURCING CONNECTIONS

Pressurized Systems

PLC Exercise 2.3

BILL OF MATERIALS
- TECO PLR or MicroLogix™ 1100 PLC
- Selector switch
- NO pressure/limit switch
- NC pressure/limit switch
- Motor starter with overloads
- Alarm/horn

Types of pressurized systems need to maintain system pressure and have a safety circuit in order to alert personnel of overpressure. In this lab, an ON/OFF selector switch controls a compressor motor starter circuit and alarm circuit. The overloads (OLs) are hardwired normally closed. The compressor motor starter is controlled by a normally closed pressure switch. An alarm circuit is wired to sound an alarm when high pressure is detected.

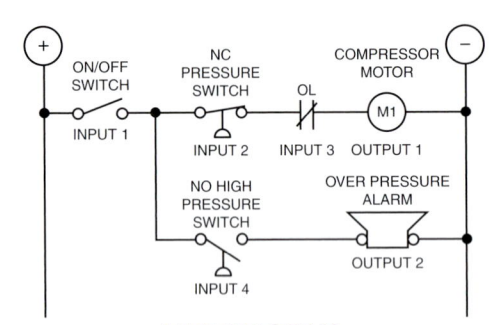

LINE DIAGRAM

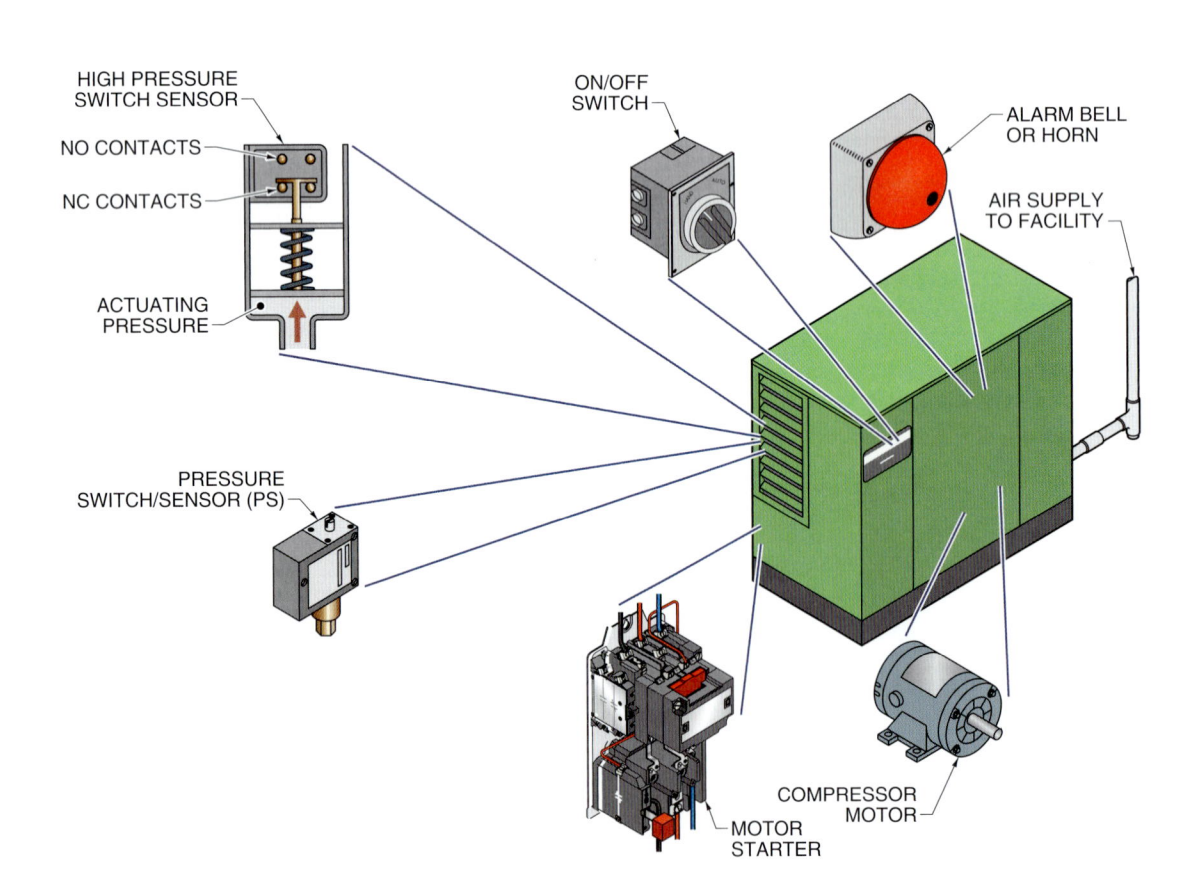

TECO Procedures

1. Create a new ladder logic program.

2. Use the TECO programming diagram to add the proper contacts and description in the Programming Grid.

3. Use the Edit Contact window to denote an input or output number and the Symbol button to add a description after placing the input or output in the Programming Grid.

4. Run the program in Simulation Mode to verify proper operation. Use the Input Status Tool window to activate the inputs and control the output to simulate circuit operation.

5. Use the wiring diagram to wire the input devices and output components to the PLC.

6. Turn power to the PLC ON and link the Com port.

7. Write the program to the SG2-12HR-D to download the program logic to the PLC.

8. Place the SG2-12HR-D in Run Mode. Activate the start pushbutton and monitor the I/Os on the LCD display.

9. Stop the program and use the Save As function to save and place the lab application in the required folder location.

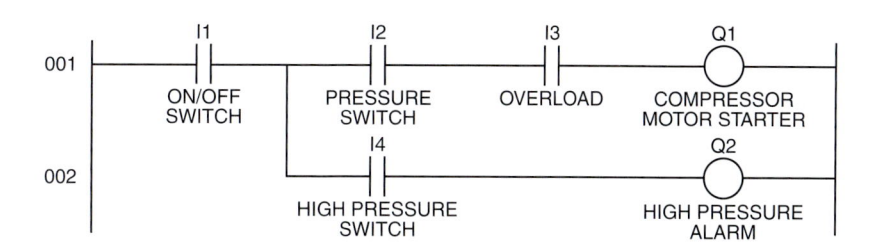

Rung 001: I1 (ON/OFF SWITCH) — I2 (PRESSURE SWITCH) — I3 (OVERLOAD) — Q1 (COMPRESSOR MOTOR STARTER)

Rung 002: I4 (HIGH PRESSURE SWITCH) — Q2 (HIGH PRESSURE ALARM)

TECO PROGRAMMING DIAGRAM

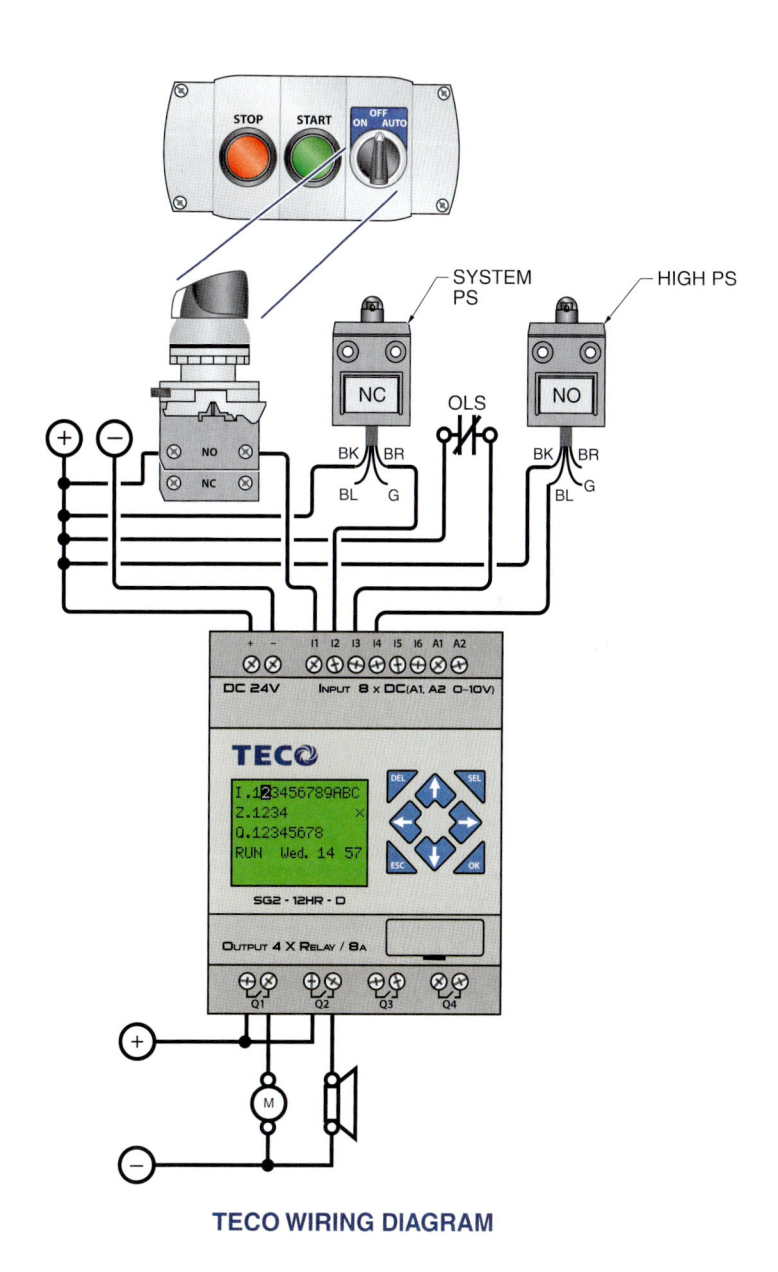

TECO WIRING DIAGRAM

114 PROGRAMMABLE LOGIC CONTROLLERS: PRINCIPLES AND APPLICATIONS

AB Allen-Bradley® Procedures

1. Open the RSLogix™ Micro Starter Lite program and create a new programming diagram.

2. Use the AB programming diagram to add the proper address and description in the LAD 2 programming window.

3. Verify that the programming diagram has no errors and save the project.

4. Use the wiring diagram to wire the input devices and output components to the PLC. Once all connections have been made, turn power to the PLC on.

5. Use RSLinx™ to establish communication between the PC and the Micrologix™ 1100.

6. Download the program logic to the PLC.

7. Place the PLC in Run Mode.

8. Activate input(s) and monitor them to determine if the output(s) are operating correctly.

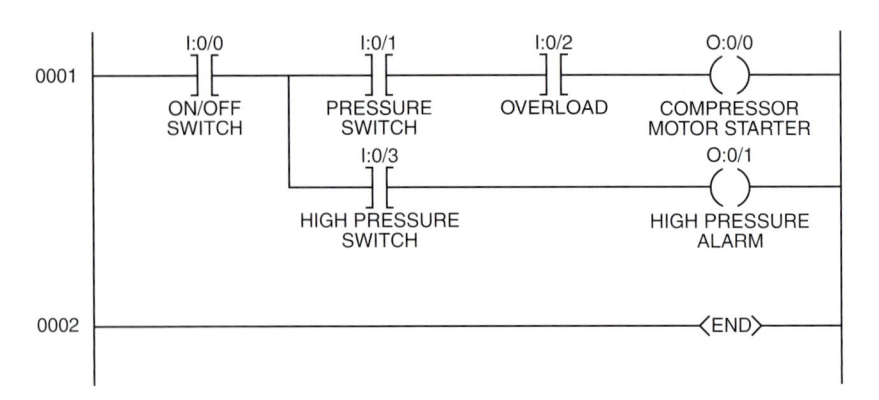

```
      I:0/0        I:0/1        I:0/2         O:0/0
0001 ──┤├──────────┤├───────────┤├────────────( )──
      ON/OFF      PRESSURE     OVERLOAD     COMPRESSOR
      SWITCH      SWITCH                    MOTOR STARTER
       I:0/3                                   O:0/1
      ──┤├──                                 ──( )──
      HIGH PRESSURE                         HIGH PRESSURE
        SWITCH                                 ALARM

0002                                         ⟨END⟩
```

AB PROGRAMMING DIAGRAM

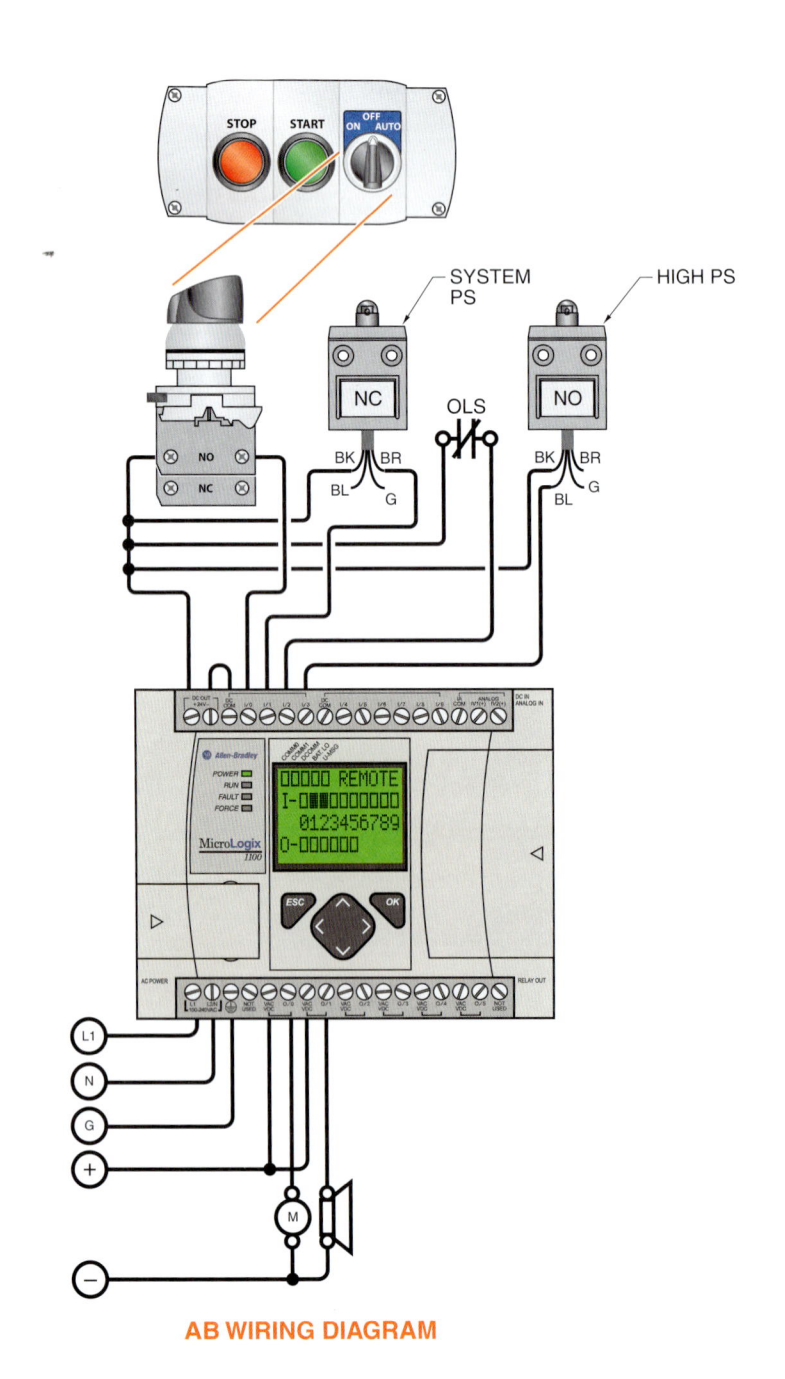

AB WIRING DIAGRAM

Monitoring Flow Systems

PLC Exercise 2.4

BILL OF MATERIALS
- TECO PLR or MicroLogix™ 1100
- Flow switch/sensor
- NO green pushbutton
- NC red pushbutton
- Motor starter with overload
- Alarm/horn

Some PLC systems control and monitor the flow of substances, such as air, liquid, or gas. In this lab, a blower motor is controlled to expel airflow from an area. A motor starter circuit is programmed to the PLC to control the blower motor with a safeguard to warn personnel when no airflow is detected while running.

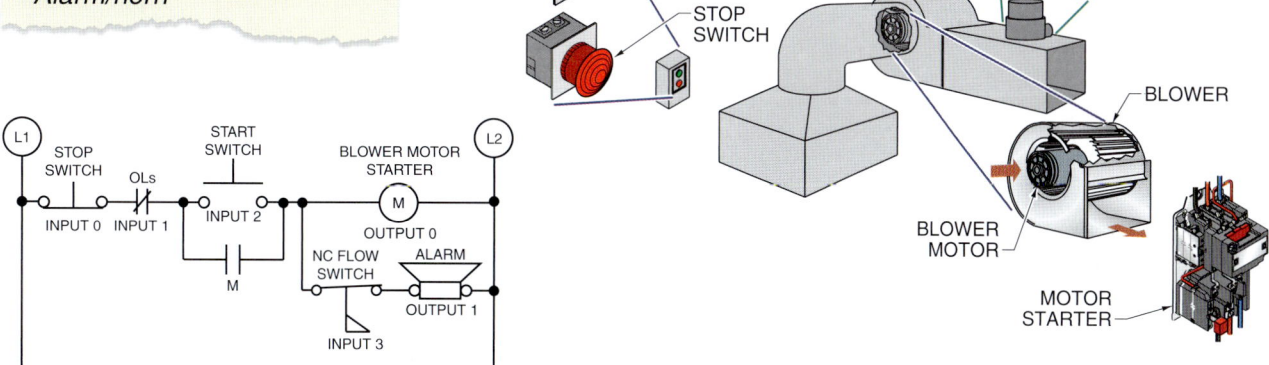

LINE DIAGRAM

TECO PROGRAMMING DIAGRAM

TECO Procedures

1. Create a new ladder logic program.

2. Use the TECO programming diagram to add the proper contacts and description in the Programming Grid.

3. Use the Edit Contact window to denote an input or output number and the Symbol button to add a description after placing the input or output in the Programming Grid.

4. Run the program in Simulation Mode to verify proper operation. Use the Input Status Tool window to activate the inputs and control the output to simulate circuit operation.

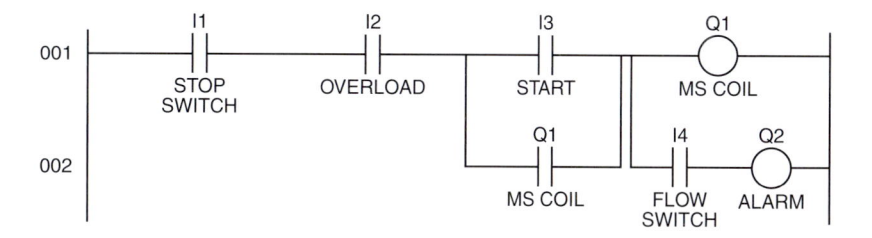

5. Use the wiring diagram to wire the input devices and output components to the PLC.

6. Turn power to the PLC ON and link the com port.

7. Write the program to the SG2-12HR-D to download the program logic to the PLC.

8. Place the SG2-12HR-D in Run Mode. Activate the start pushbutton and monitor the I/Os on the LCD display.

9. Stop the program and use the Save As function to save and place the lab application in the required folder location.

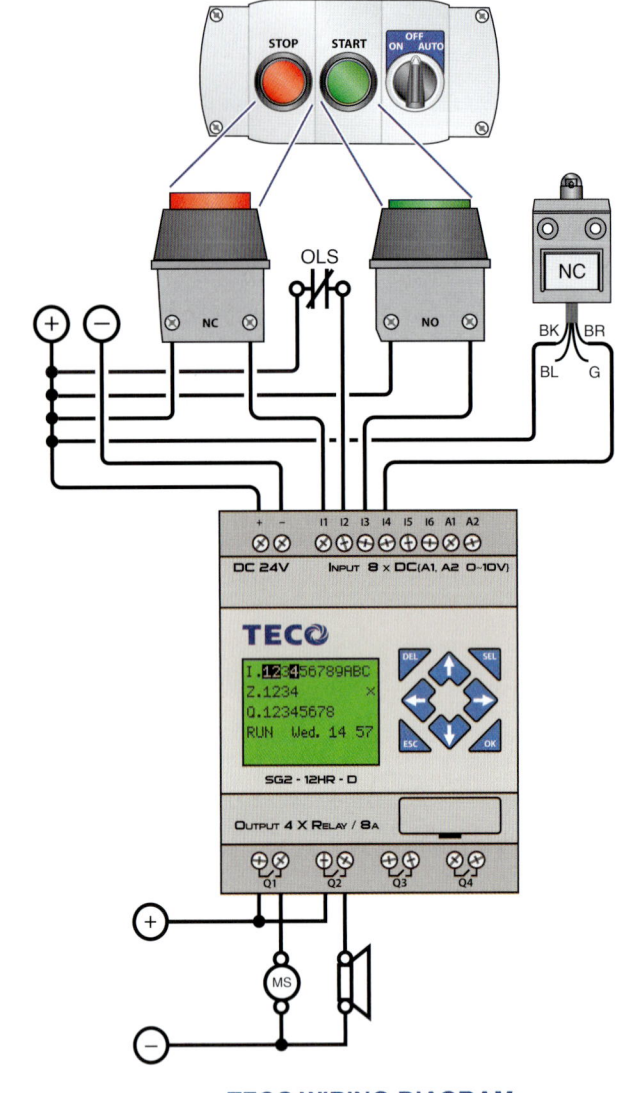

TECO WIRING DIAGRAM

AB Allen-Bradley® Procedures

1. Open the RSLogix™ Micro Starter Lite program and create a new programming diagram.

2. Use the AB programming diagram to add the proper address and description in the LAD 2 programming window.

3. Verify that the programming diagram has no errors and save the project.

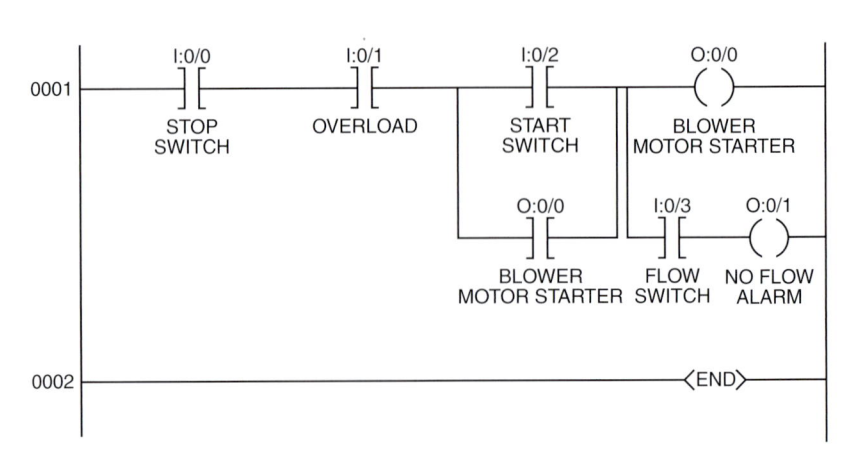

AB PROGRAMMING DIAGRAM

4. Use the wiring diagram to wire the input devices and output components to the PLC. Once all connections have been made, turn power to the PLC on.

5. Use RSLinx™ to establish communication between the PC and the Micrologix™ 1100.

6. Download the program logic to the PLC.

7. Place the PLC in Run Mode.

8. Activate input(s) and monitor them to determine if the output(s) are operating correctly.

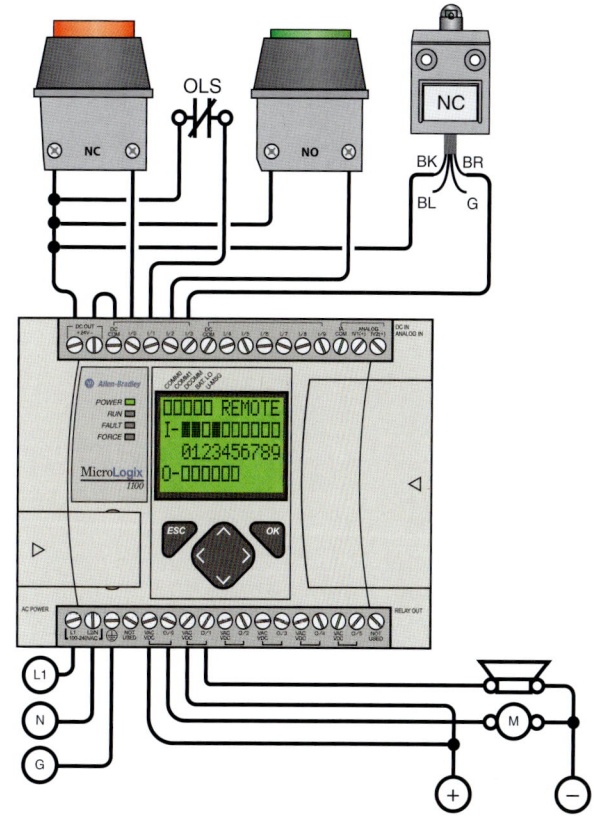

AB WIRING DIAGRAM

S SIEMENS Programming and Wiring

Use the S7-1200 Easy Book to program and wire the equivalent TECO and Allen-Bradley exercises. Reference page 70 for Siemens STEP 7 programming symbols.

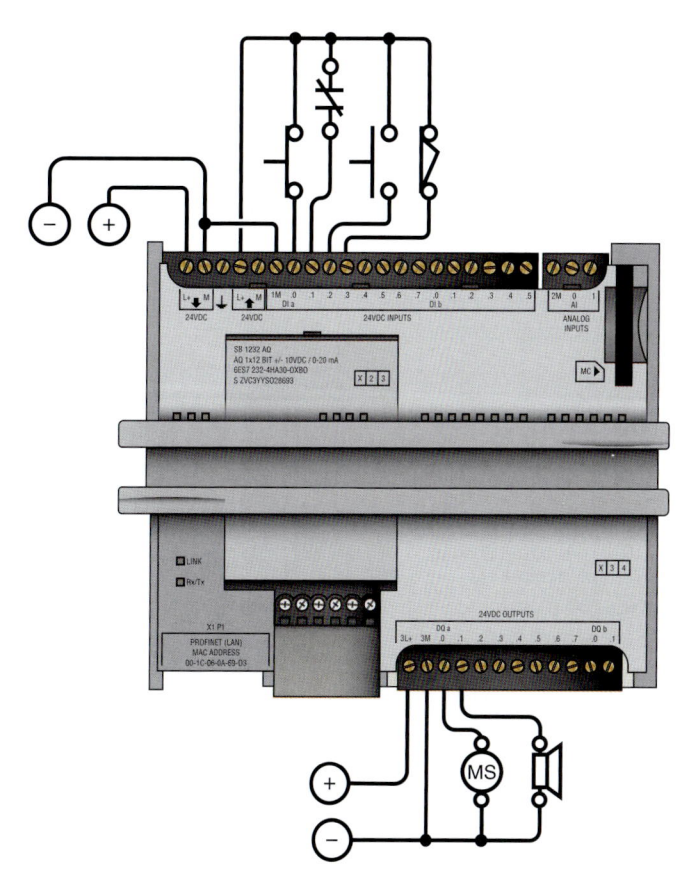

SIEMENS WIRING DIAGRAM

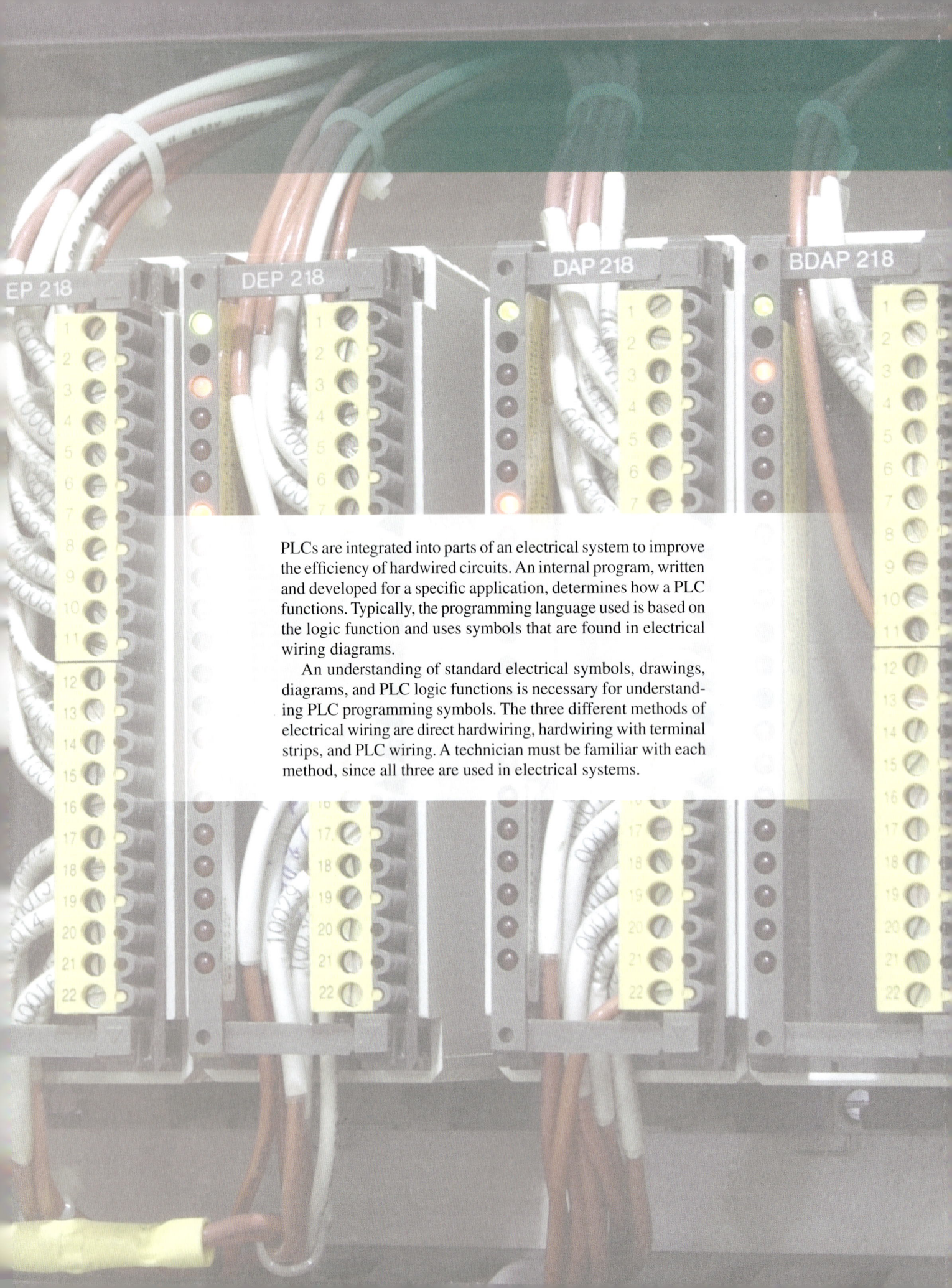

PLCs are integrated into parts of an electrical system to improve the efficiency of hardwired circuits. An internal program, written and developed for a specific application, determines how a PLC functions. Typically, the programming language used is based on the logic function and uses symbols that are found in electrical wiring diagrams.

An understanding of standard electrical symbols, drawings, diagrams, and PLC logic functions is necessary for understanding PLC programming symbols. The three different methods of electrical wiring are direct hardwiring, hardwiring with terminal strips, and PLC wiring. A technician must be familiar with each method, since all three are used in electrical systems.

PLC Programming Symbols, Diagrams, and Logic Functions

3

Objectives:

- Identify the use of electrical and programming symbols in electrical and programming diagrams.
- Explain normally open (NO) and normally closed (NC) for manual and automatic switches.
- Identify the use of PLC programming symbols in software diagrams.
- Explain the differences between pictorial drawings, wiring diagrams, ladder diagrams, and PLC programming diagrams.
- Explain the common logic functions.
- Describe direct hardwiring, hardwiring using terminal strips, and PLC wiring.

Learner Resources
atplearningresources.com/quicklinks
Access Code: 475203

ELECTRICAL SYMBOLS AND DIAGRAMS

All trades have a specific language that must be understood in order for information to be exchanged efficiently. Trade languages include symbols, drawings, diagrams, words, phrases, and/or abbreviations. Because PLCs are a part of larger electrical systems, an understanding of standard electrical symbols and diagrams, as well as an understanding of PLC programming symbols and software diagrams is required in order to program, troubleshoot, and/or work on systems that include PLCs.

Standard Electrical Symbols

A *symbol* is a graphic element that represents a quantity, unit, device, or component. Standard electrical diagrams use specific symbols to represent the electrical devices or components used in electrical circuits. Each symbol is drawn to provide a visual indication of what the device or component is and to show how the device operates. For example, devices such as pushbuttons, limit switches, temperature switches, pressure switches, flow switches, level switches, and foot switches all have specific symbols that are easily recognized. **See Figure 3-1.**

All electrical switches (input devices) are manually, mechanically, or automatically operated. A *manually operated switch* is any switch that requires a person to physically change the state of the switch. Typical manual switches are pushbuttons, toggle switches, selector switches, and foot switches. A *mechanically operated switch* is a switch that detects the physical presence of an object.

The only true mechanical switch is the limit switch. However, because proximity switches (no-touch sensors) were developed to replace mechanical limit switches, proximity switches are typically classified with limit switches. An *automatically operated switch* is a switch that is activated independent of a person or object. Typical automatically operated switches are level, flow, temperature, and pressure instruments.

Understanding whether a switch is manually, mechanically, or automatically operated is important because each switch has a symbol that is used on electrical diagrams. Any normally open (NO) switch that is manually operated will have the operator of the switch (part that moves) drawn above the terminals of the switch. **See Figure 3-2.** The symbol for a normally open switch shows the manual operator moving downward when activated.

Switch Symbols

Device	Part	Abbr.	Symbol	Function/Notes
PUSHBUTTON	Single-circuit, normally open (NO)	PB or PB-NO	MANUAL OPERATOR / TERMINALS	To make (NO) or break (NC) a circuit when manually depressed; one of the simplest and most common forms of control; typical pushbutton consists of one or more contact blocks, an operator, and a legend plate
LIMIT SWITCH (MECHANICAL)	Normally open (NO)	LS	MECHANICAL OPERATOR	To convert mechanical motion into an electrical signal; limit switches accomplish this conversion by using some type of lever to open or close contacts within the limit switch enclosure
TEMPERATURE SWITCH	Normally open (NO)	TEMP SW or TS	TEMPERATURE OPERATOR	To respond to temperature changes; temperature switch may be used to maintain a specified temperature within a process, or to protect against overtemperature conditions
PRESSURE SWITCH	Normally open (NO)	PS	PRESSURE OPERATOR	To open or close contacts in response to pressure changes in media such as air, water, or oil; electrical contacts may be used to start or stop motors of fans, open or close dampers or louvers, or signal a warning light or alarm
FLOW SWITCH	Normally open (NO)	FLS	FLOW OPERATOR	To sense the movement of a fluid; fluid may be air, water, oil, or other gases or liquids; flow switch is a control switch that is usually inserted into a pipe or duct; element will move and activate contacts whenever the fluid flow is sufficient to overcome a spring tension
LEVEL SWITCH	Normally open (NO)	LEVEL SW or LS	LEVEL OPERATOR	To measure and respond to the level of material; material may be water, oil, paint, granules, or other solids; level switches are control devices
FOOT SWITCH	Normally open (NO)	FTS	FOOT OPERATOR	To allow free use of hands while providing for manual control of a machine; many foot switches include a guard to prevent accidental operation

Figure 3-1. Switch symbols are drawn to provide a visual indication of how each switch operates in a circuit.

Normally open switches that are mechanically or automatically operated show the operator drawn below the terminals of the switch. The symbol for a normally open mechanical or automatic switch shows the operator moving upward when activated.

Likewise, any normally closed (NC) switch that is manually operated shows the operator of the switch drawn below the terminals of the switch. The symbol for a normally closed switch shows the manual operator moving downward when activated.

A normally closed switch that is mechanically or automatically operated will have the operator of the switch drawn above the terminals of the switch. The symbol for a normally closed switch shows the mechanical or automatic operator moving upward when activated.

Normally Open (NO) and Normally Closed (NC) Switches						
Device	Part	Abbr.	Symbol/Condition		Function/Notes	
			Normal	Activated	
SWITCH (MANUAL)	Normally open (NO)	S	MANUAL OPERATOR / TERMINALS	MANUAL OPERATOR	Control contacts are contacts that switch low currents; power contacts are contacts that switch high currents	
	Normally closed (NC)	S	TERMINALS / MANUAL OPERATOR	MANUAL OPERATOR		
LIMIT SWITCH (MECHANICAL)	Normally open (NO)	LS	TERMINALS / MECHANICAL OPERATOR	MECHANICAL OPERATOR	To convert mechanical motion into an electrical signal; limit switches accomplish this conversion by using some type of lever to open or close contacts within the limit switch enclosure	
	Normally closed (NC)	LS	MECHANICAL OPERATOR / TERMINALS	MECHANICAL OPERATOR		

Figure 3-2. Any NO switch that is manually operated has the operator of the switch drawn above the terminals, and any NC switch that is manually operated has the operator of the switch drawn below the terminals.

Manually operated switches are drawn on electrical diagrams in the "normal" condition, which is the nonactivated condition (no one touching the switch). Because mechanical and automatic switches can be in an activated or nonactivated condition after the switches are installed in a circuit, the switches can be drawn in either condition on a diagram. For example, a normally open limit switch can be drawn on a diagram as either normally open or normally open, held closed. **See Figure 3-3.**

The reason for drawing a switch on an electrical diagram in the activated position is to better understand circuit operation when troubleshooting. For example, a refrigerator has a lamp inside that is turned ON when the refrigerator door is opened and turned OFF when the door is closed. The opening of the door releases a limit switch so the switch can be returned to normal condition (condition of switch before activation). The closing of the door activates the limit switch so that the normally closed contacts open. **See Figure 3-4.**

For the refrigerator door circuit to operate, a normally closed limit switch is installed in series with the lamp inside the refrigerator. When the door is opened, the limit switch is in normal condition and the normally closed contacts energize the lamp. When the door is closed the limit switch is activated and the normally closed contacts are opened and the lamp is turned OFF. Because the limit switch in a refrigerator door is in an activated position (door closed) for probably 99% of its operating life, the switch is drawn on the diagram as normally closed, held open.

Knowing how switches are drawn (normal condition or activated condition) is important when working on any electrical circuit. Understanding symbols for mechanical and automatic switches simplifies circuit troubleshooting.

In a refrigeration control circuit, two pressure switches are used to provide safety and protect the refrigeration unit and circuits. **See Figure 3-5.** A high-pressure switch/control (HPS or HPC) is used to prevent a high system pressure from damaging the compressor and/or burning out the electric motor by causing an overload condition.

Switch Operator Conditions

Device	Part	Abbr.	Symbol	Notes
LIMIT SWITCH (MECHANICAL)	Normally open (NO)	LS	MECHANICAL OPERATOR	Shown on diagram in normal condition with nothing touching switch operator
	Normally open, held closed (NO)	LS	OPERATOR IN CLOSED CONDITION	Shown on diagram in activated condition in which something (guard in place) is touching switch operator
	Normally closed (NC)	LS	MECHANICAL OPERATOR	Shown on diagram in normal condition with nothing touching switch operator
	Normally closed, held open (NC)	LS	OPERATOR IN CLOSED CONDITION	Shown on diagram in activated condition in which something (guard in place) is touching switch operator
TEMPERATURE SWITCH	Normally open (NO)	TEMP SW	100°F TEMPERATURE OPERATOR / 80°F AIR TEMPERATURE	Shown on diagram in a condition in which the temperature around the switch is less than the actual setting of the temperature switch
	Normally open, held closed (NO)	TEMP SW	100°F OPERATOR IN CLOSED CONDITION / 110°F AIR TEMPERATURE	Shown on diagram in a condition in which the temperature around the switch is greater than the actual setting of the temperature switch
	Normally closed (NC)	TEMP SW	100°F TEMPERATURE OPERATOR / 80°F AIR TEMPERATURE	Shown on diagram in a condition in which the temperature around the switch is less than the actual setting of the temperature switch
	Normally closed, held open (NC)	TEMP SW	100°F OPERATOR IN OPEN CONDITION / 110°F AIR TEMPERATURE	Shown on diagram in a condition in which the temperature around the switch is greater than the actual setting of the temperature switch

Figure 3-3. Electrical switches are drawn in normal condition or activated condition on electrical diagrams.

Refrigerator Light Circuits

L1 — 120 V — N

NORMALLY CLOSED, HELD OPEN LIMIT SWITCH

REFRIGERATOR LIGHT

CLOSED DOOR HOLDS LIGHT LIMIT SWITCH OPEN

NORMAL CONDITION OF SWITCH CONTACTS (NO / NC)

ACTIVATED CONDITION OF SWITCH CONTACTS (NO / NC)

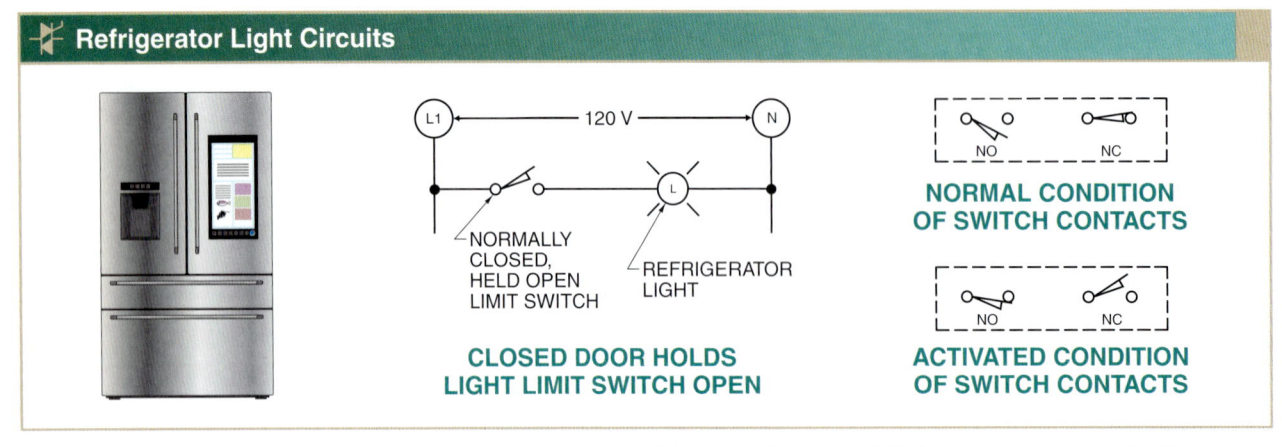

Figure 3-4. Switches are drawn on diagrams in the operating condition the switch is most likely to be in at any point in time.

Refrigeration Controls

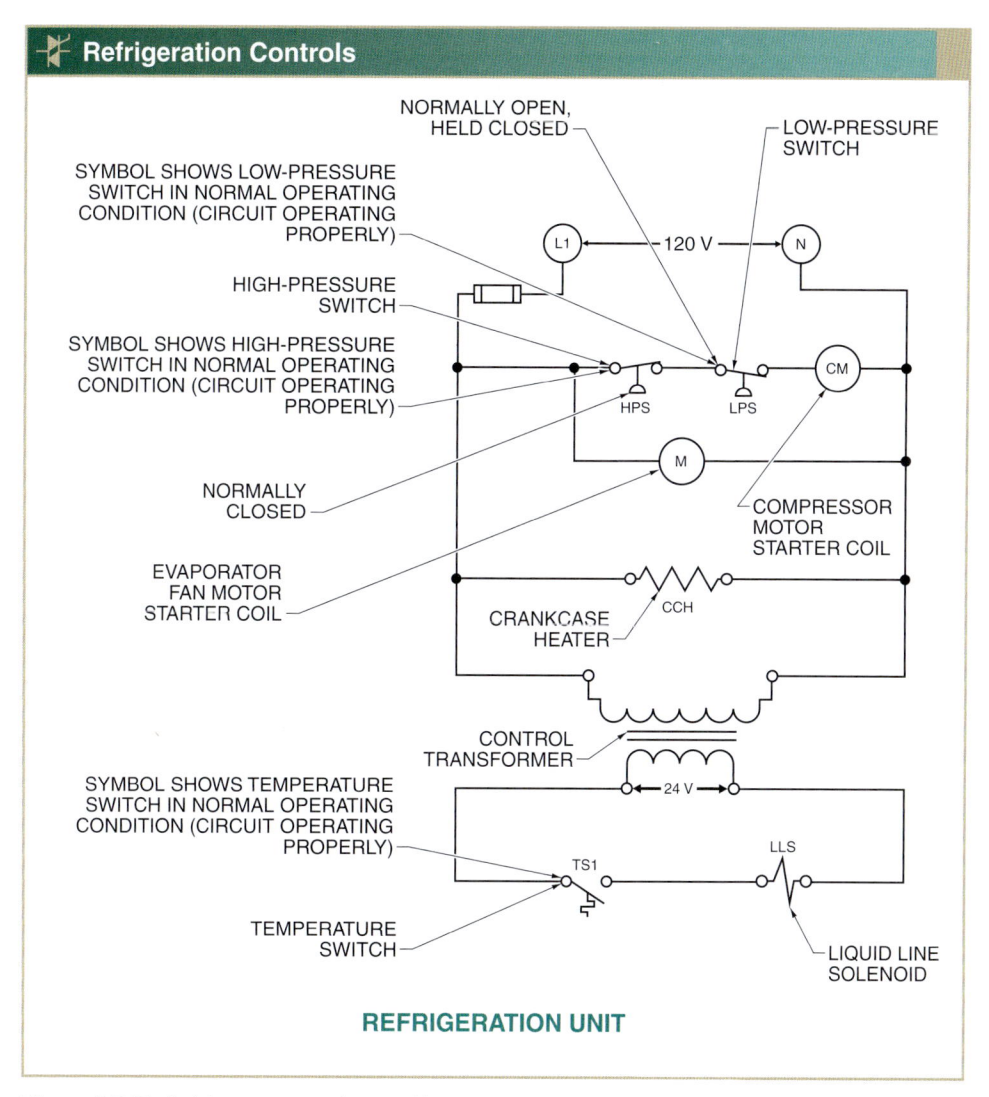

Figure 3-5. Technicians must understand how manual, mechanical, and automatic switch symbols are drawn to troubleshoot circuits effectively.

The normally closed HPS switch contacts open and de-energize the compressor when system pressure becomes too high. The HPS switch is drawn in a normally closed (NC) condition on the diagram because as long as the system is operating properly, the switch will remain in a normally closed condition.

The normally open low-pressure switch/control (LPS or LPC) is used to de-energize a compressor when system pressure drops too low. A low system pressure can damage the compressor and other components. However, the LPS switch is not drawn in a normally open (NO) condition because as long as the system is operating properly, the switch will remain in an activated condition.

Originally all electrical switches used mechanically operated contacts to start and stop the flow of electricity in a circuit. Today, both mechanically operated and solid-state (no moving parts) switches are used to start and stop the flow of electricity in a circuit. The symbol difference between a mechanical switch and a solid-state switch is that a solid-state switch includes a diamond shape around the switch symbol. **See Figure 3-6.**

CODE CONNECT

Section 310.10(G) of the NEC® covers the requirements for parallel conductors. Conductors of this type may be required to supply large ampere loads. The NEC® allows conductors 1/0 AWG and larger to be parallel, but requires that they be the same length, be the same type of material (copper or aluminum), be the same size (area in circular mils), be the same insulation type, and be terminated in the same fashion. These requirements help ensure parallel conductors have the same resistance. Consequently, the total current will be equally divided.

Solid-State Switch Symbols

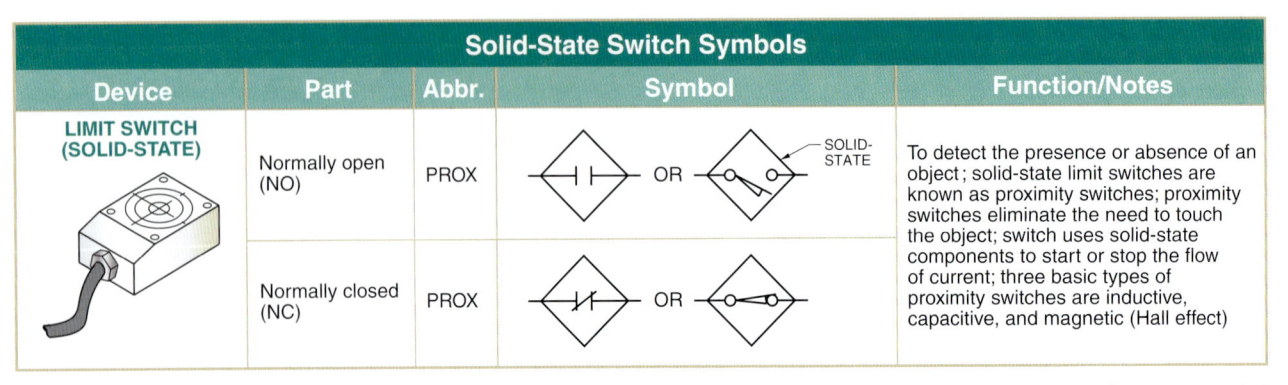

Device	Part	Abbr.	Symbol	Function/Notes
LIMIT SWITCH (SOLID-STATE)	Normally open (NO)	PROX	◇─┤├─◇ OR ◇─○╱○─◇ ─ SOLID-STATE	To detect the presence or absence of an object; solid-state limit switches are known as proximity switches; proximity switches eliminate the need to touch the object; switch uses solid-state components to start or stop the flow of current; three basic types of proximity switches are inductive, capacitive, and magnetic (Hall effect)
	Normally closed (NC)	PROX	◇─┤╱├─◇ OR ◇─○╱○─◇	

Figure 3-6. Solid-state switches start and stop the flow of electricity in a circuit without the use of moving parts. Solid-state switch symbols show the actual switch symbol in a diamond shape.

PLC Programming Symbols

PLC programs are designed using PLC programming software. PLC programming software uses various symbols, letters, and numbers to designate devices and components. Devices such as input relays, timers, and counters all have symbols and identifying addresses (assigned values and numbers).

When designing a program for controlling a circuit using a PLC, device and component symbols are selected and displayed on a computer screen as the program is developed. The symbols are typically selected from a tool palette that is displayed on the screen. Basic symbols used to program a PLC include normally open inputs, normally closed inputs, and standard outputs. **See Figure 3-7.**

Input devices such as pushbutton, limit, and pressure switches are drawn as generic normally open or normally closed contacts. Unlike standard electrical symbols, there is no way to distinguish which type of switch (pressure, temperature, or limit) is being used by looking at a generic, normally open or normally closed PLC input device symbol. However, when programming a PLC contact, a description can be added to the generic symbol. For example, LEVEL SWITCH can be added to the generic symbol for level switch.

Additional information can be included in the description for more clarity. For example, OVERFLOW LEVEL SWITCH 1—TANK 3 can be added to a generic symbol for a level switch that is used for overflow detection in tank 3.

Standard output components include lamps, solenoids, motor starters, and alarms that are used to perform work or give an indication of circuit operation. When using standard electrical symbols, output components have specific symbols that easily identify what the output component is. As with input devices, generic symbols are used to represent output components, and words are added to identify exactly what a component is when programming PLCs.

Most PLC programming software also includes expanded (special) input instructions and output instructions that can be programmed into a control circuit. Expanded input devices include contacts that open or close when a set value (time, count, or pressure) is reached. Expanded output components can latch or unlatch. Expanded output components include timers, counters, and logic functions. Expanded input devices and output components allow for much more circuit flexibility and capability than could be achieved in a standard mechanical relay circuit.

All expanded input devices and output components have a specific symbol that represents the device or component. Like any PLC programming input or output symbol, words are typically added to the symbol to further explain what the symbol represents. Although exact symbols differ slightly from manufacturer to manufacturer, most symbols have similar shapes and designations. **See Figure 3-8.**

PLC Programming Symbols

Basic Programming Symbols	
Component	**Symbol**
Input device (normally open)	─┤ ├─
Input device (normally closed)	─┤/├─
Standard output	─( )─

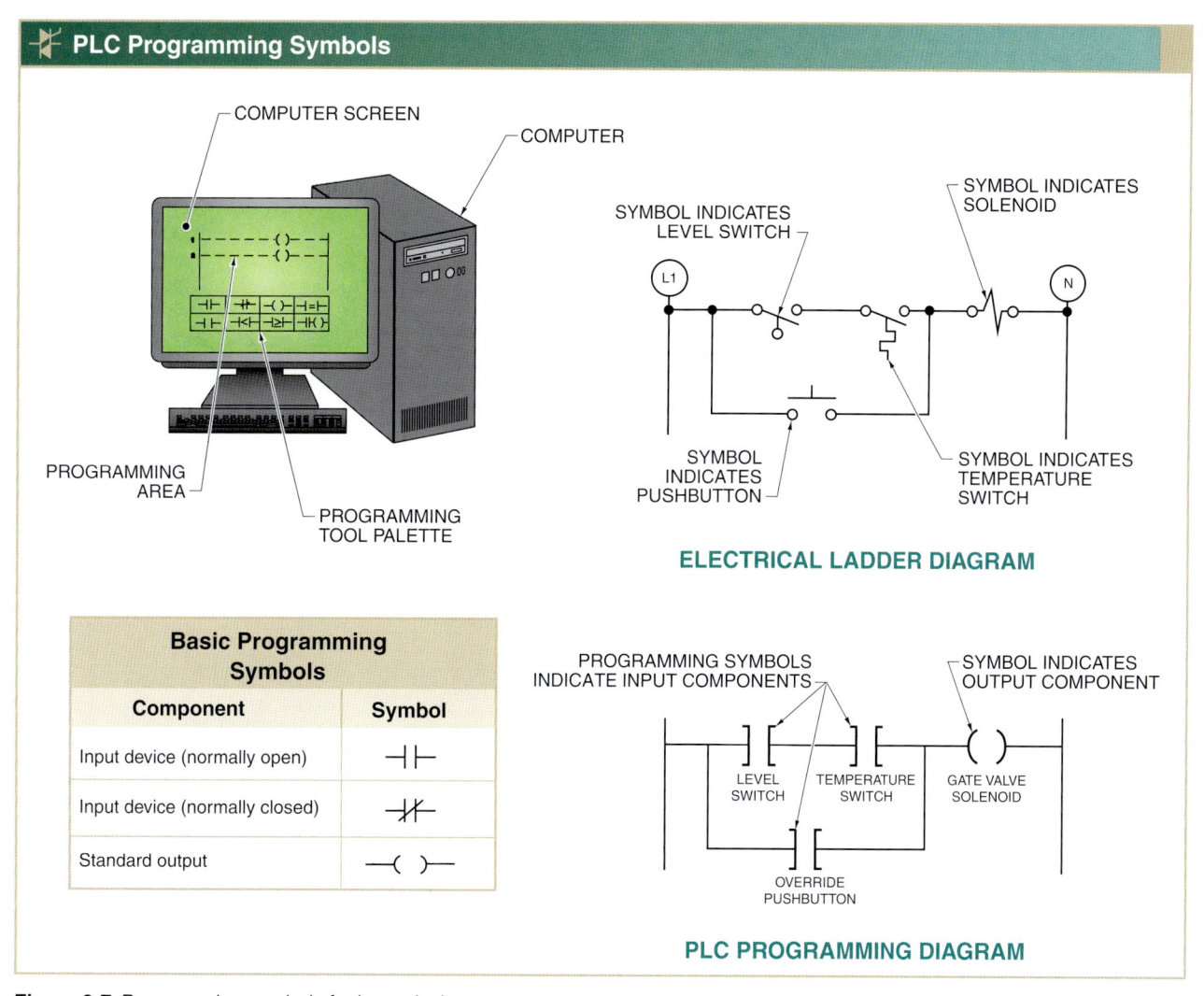

Figure 3-7. Programming symbols for input devices are drawn as generic normally open or normally closed contacts, and all output components are drawn as parentheses.

Bit, Timer, and Counter Instructions. PLC software programs use basic PLC programming symbols, such as bit instructions (relay-type), to represent devices in control circuits. These programs also use basic symbols, such as timer and counter instructions, to control specific device parameters.

Bit instructions are programmed for relay-type devices, such as mechanical switches, and are represented as "examine if closed" (normally open) and "examine if open" (normally closed). Bit instructions programmed for output components include "output energize," "output latch," and "output unlatch." Other versions of bit instructions include "one shot," "one shot rising," and "one shot falling," which are used to control the scan duration of a contact. When a signal is applied to a contact, the contact sends one signal to the output component, regardless of how long the contact (mechanical switch) is pressed.

Timer instructions are used to control durations of signals and are programmed specifically for the PLC application. Timer instructions include "timer on-delay," "timer off-delay," "retentive timer," "reset," and "read high speed clock."

Expanded (Special) Programming Symbols

Component	Symbol
Equal-to contact	—]=[—
Not-equal-to contact	—]≠[—
Greater-than or equal to contact	—]≥[—
Less-than contact	—]<[—
Addition	—(+)—
Subtraction	—(−)—
Multiplication	—(×)—
Division	—(÷)—
End	—(END)—
Set or latch	—(SET)— —(L)—
Reset or unlatch	—(RSET)— —(U)—
Read high speed clock	—(RHC)— or —[RHC]—
Timer	—(TMR)— or —[TMR]—
Counter	—(CNT)— or —[CNT]—

Figure 3-8. Special symbols are used in PLC programming when a PLC has advance control functions.

Control logic is a plan of operation as designed with a PLC program for an electrical circuit or system.

Bit, Timer, and Counter Instructions

BIT

- —] [— EXAMINE IF CLOSED
- —]/[— EXAMINE IF OPEN
- —( )— OUTPUT ENERGIZE
- —(L)— OUTPUT LATCH
- —(U)— OUTPUT UNLATCH
- ONS ONE SHOT
- OSR ONE SHOT RISING
- OSF ONE SHOT FALLING

TIMER AND COUNTER

- TON TIMER ON-DELAY
- TOF TIMER OFF-DELAY
- RTO RETENTIVE TIMER
- RES RESET
- RHC READ HIGH SPEED CLOCK
- CTU COUNT UP
- CTD COUNT DOWN
- HSC HIGH SPEED COUNTER

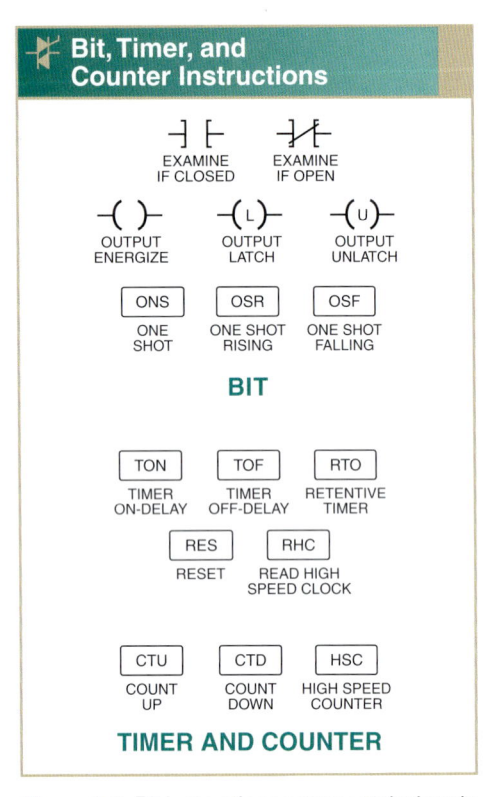

Figure 3-9. Bit instructions represent devices in control circuits. Timer and counter instructions control specific device parameters.

Counter instructions are used with inputs or outputs to count specific parameters in a PLC application. These instructions include "count up," "count down," and "high-speed counter." These counter instructions can be found in various PLC programs from simple to complex and in different forms, but they all operate the same. **See Figure 3-9.**

Math, Compare, Conversion, and Logical Instructions. PLC programming symbols include math instructions, compare instructions, conversion instructions, and logical instructions. These allow numerical data to be used with control logic for accuracy and precision in PLC applications.

Math instructions can compute both basic and advanced forms of math, which include addition, subtraction, multiplication, and division as well as the negate, clear, sine, and tangent functions. Compare instructions can be used in control logic to compare values from different sources. Compare instructions include "equal," "not equal," "greater than," "less than," "greater than or equal to," "less than or equal to," "mask compare for equal," and "limit test."

Both math and compare instructions can be used with timer and counter values in a process to make the calculations more precise. Both math and compare instructions are used in programming diagrams to count numbers of products, weigh products, add temperature changes in food production equipment, and allow any numeric variable into a process. **See Figure 3-10.**

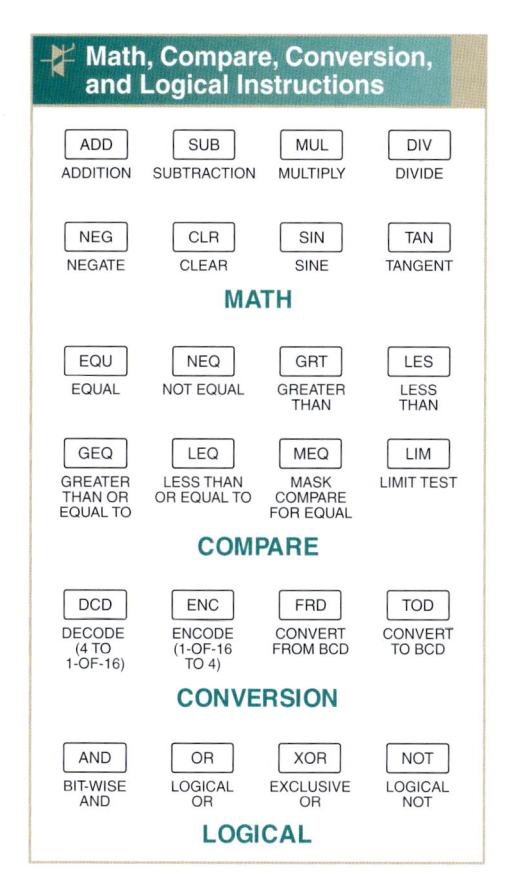

Figure 3-10. Math instructions, compare instructions, conversion instructions, and logical instructions allow numerical data to be used with control logic for accuracy and precision in PLC applications.

"bit shift left," "bit shift right," and "copy function." File instructions can also be programmed for loading words into files, such as "first in first out load/unload" and "last in first out load/unload."

Sequencer instructions are programmed for transferring 16 bits of data to word addresses. A *word address* is a designated name or number in a PLC program for a particular input or output. Sequencer instructions include "sequencer compare," "sequencer output," and "sequencer load."

Move instructions are related to file and sequencer instructions and are represented as "move" and "masked move." Move instructions move program data from a source to a destination inside a PLC program. **See Figure 3-11.**

Conversion and logical instructions are related to math and compare instructions. Conversion instructions change numbers from one form to another. For example, a conversion instruction can convert an integer to a binary coded decimal (BCD). Conversion instructions are represented as "decode (4 to 1-of-16)," "encode (1-of-16 to 4)," "convert from BCD," and "convert to BCD."

Logical instructions perform basic logic operations on words (bits of data) in a PLC program. These are represented as "bit-wise AND," "logical OR," "exclusive OR," and "logical NOT."

File, Sequencer, and Move Instructions. File and sequencer PLC programming symbols include file, sequencer, and move instructions. File instructions shift data one bit at a time to earlier or later locations within a PLC program and are entered as

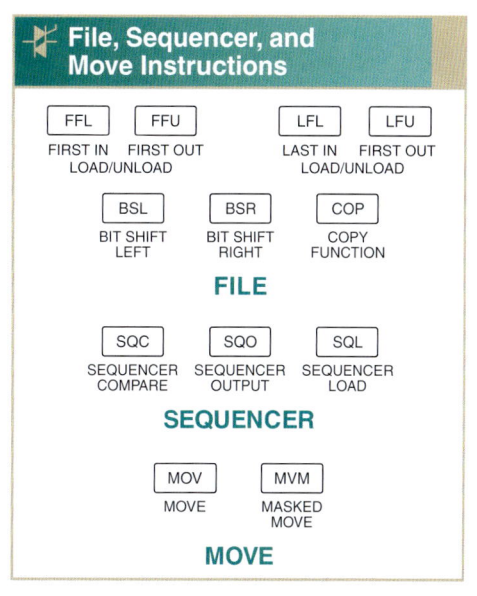

Figure 3-11. File instructions shift data one bit at a time to earlier or later locations within a PLC program, sequencer instructions transfer data to word addresses, and move instructions move data from a source to a destination within a PLC program.

Program Control, Interrupt, and I/O Message Instructions. Program flow PLC programming symbols include program control instructions, interrupt instructions, and I/O message instructions. These instructions allow a technician to alter the sequence of scanning and execution in control logic.

Program control instructions enable a program to jump to and return from a subroutine. Types of program control instructions include "jump to label," "label," "jump to subroutine," "subroutine label," "return from subroutine," "suspend," "temporary end," and "program end."

Interrupt instructions allow a program to automatically interrupt a normal scan and can also scan a subroutine. Interrupt instructions include "interrupt subroutine," "selectable timed start," "user interrupt disable," and "user interrupt enable."

I/O message instructions allow inputs or outputs to be updated immediately, prior to normal scan completion. Types of input and output instructions include "immediate input with mask," "immediate output with mask," and "I/O refresh." **See Figure 3-12.**

Program Control, Interrupt, and I/O Message Instructions

JMP	LBL	JSR	SBR
JUMP TO LABEL	LABEL	JUMP TO SUBROUTINE	SUBROUTINE LABEL

RET	SUS	TND	END
RETURN FROM SUBROUTINE	SUSPEND	TEMPORARY END	PROGRAM END

PROGRAM CONTROL

INT	STS	UID	UIE
INTERRUPT SUBROUTINE	SELECTABLE TIMED START	USER INTERRUPT DISABLE	USER INTERRUPT ENABLE

INTERRUPT

IIM	IOM	REF
IMMEDIATE INPUT WITH MASK	IMMEDIATE OUTPUT WITH MASK	I/O REFRESH

I/O MESSAGE

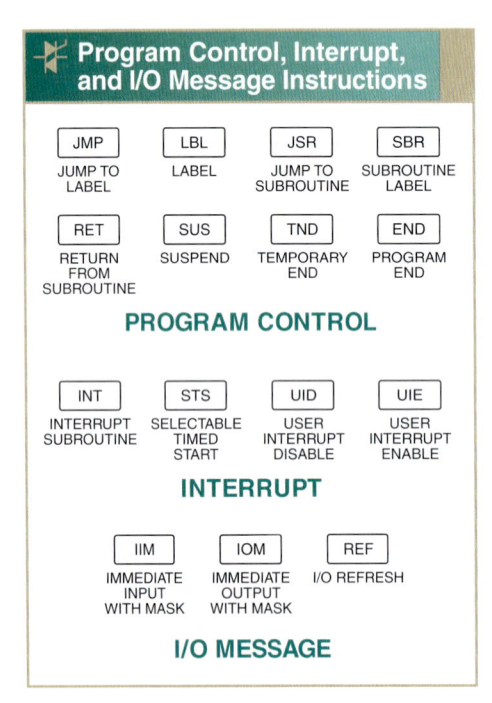

Figure 3-12. Program control instructions, interrupt instructions, and I/O message instructions allow a technician to alter the sequence of scanning and execution in control logic.

"PID" and Communication Instructions. A process control programming symbol is represented by the "PID" instruction. A "PID" instruction is an output instruction used to control the set point of a variable inside control logic. Typical variables include temperature, flow, pressure, and levels. "PID" is an abbreviation for the three elements of an equation used to control these set points. The three elements are proportional control, integral control, and derivative control.

The communication programming symbol includes the "service communication" instruction and the "message" instruction. Common PLC installations in industrial settings are required to have multiple PLCs linked together. "Service communication" and "message" instructions enable all linked PLCs to read data from and transmit data to other PLCs. **See Figure 3-13.**

"PID" and Communication Instructions

PID

PID

SVC	MSG
SERVICE COMMUNICATION	MESSAGE

COMMUNICATION

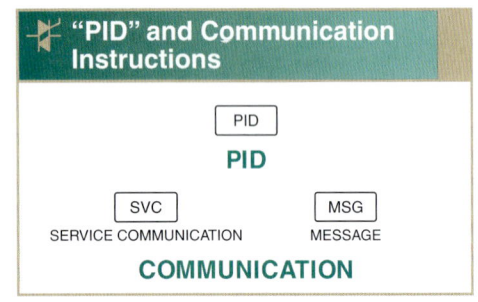

Figure 3-13. A "PID" instruction is an output instruction used to control the set point of a variable inside control logic. Communication instructions enable all linked PLCs to read data from and transmit data to other PLCs.

Pictorial Drawings

A *pictorial drawing* is a drawing that resembles a photograph or three-dimensional picture. A pictorial drawing shows what a device looks like, but does not give accurate dimensions or other specific detail information. PLC pictorial drawings show the layout and position of devices and components used for an application. **See Figure 3-14.**

Device and component symbols are placed as near to true position as possible. Components can be drawn with as much detail as required or in general outline form. Pictorial drawings are helpful when designing a system, when installing system component parts, and when troubleshooting a system. Pictorial drawings are also helpful in showing what additional components can be added to a system and what function a device or component has within the system.

Wiring Diagrams

A *wiring diagram* is a technical diagram that displays the connection of all devices and components to a PLC. Wiring diagrams show the connections as closely as possible. Internal and external connections can be shown in detail to allow tracings. Wiring diagrams are used when installing devices and when troubleshooting. **See Figure 3-15.**

Wiring diagrams are also used to show smaller more specific parts of PLC wiring. For example, thermocouples can be connected to the analog input section of a PLC to provide temperature information to the PLC. A *thermocouple* is a temperature sensor made of two dissimilar metals that are joined at the end where heat is to be measured. The two dissimilar metals produce a voltage output at the nonjoined end that is proportional to the measured temperature. **See Figure 3-16.**

An individual thermocouple can be connected to an analog PLC input, or several thermocouples can be connected in series or parallel combinations. When thermocouples are connected in parallel, a PLC will receive an average of all the thermocouple temperature measurements. When thermocouples are connected in series, a PLC will receive the sum of all the thermocouple temperature measurements.

Rockwell Automation, Inc.

Direct hardwiring of control circuits was very common for many years, but the time spent troubleshooting and making circuit modifications has decreased with the use of terminal strips and input and output module wiring.

Pictorial Drawings

PLC Multipoint

PLC ENCLOSURE PHOTOGRAPH

PICTORIAL DRAWING

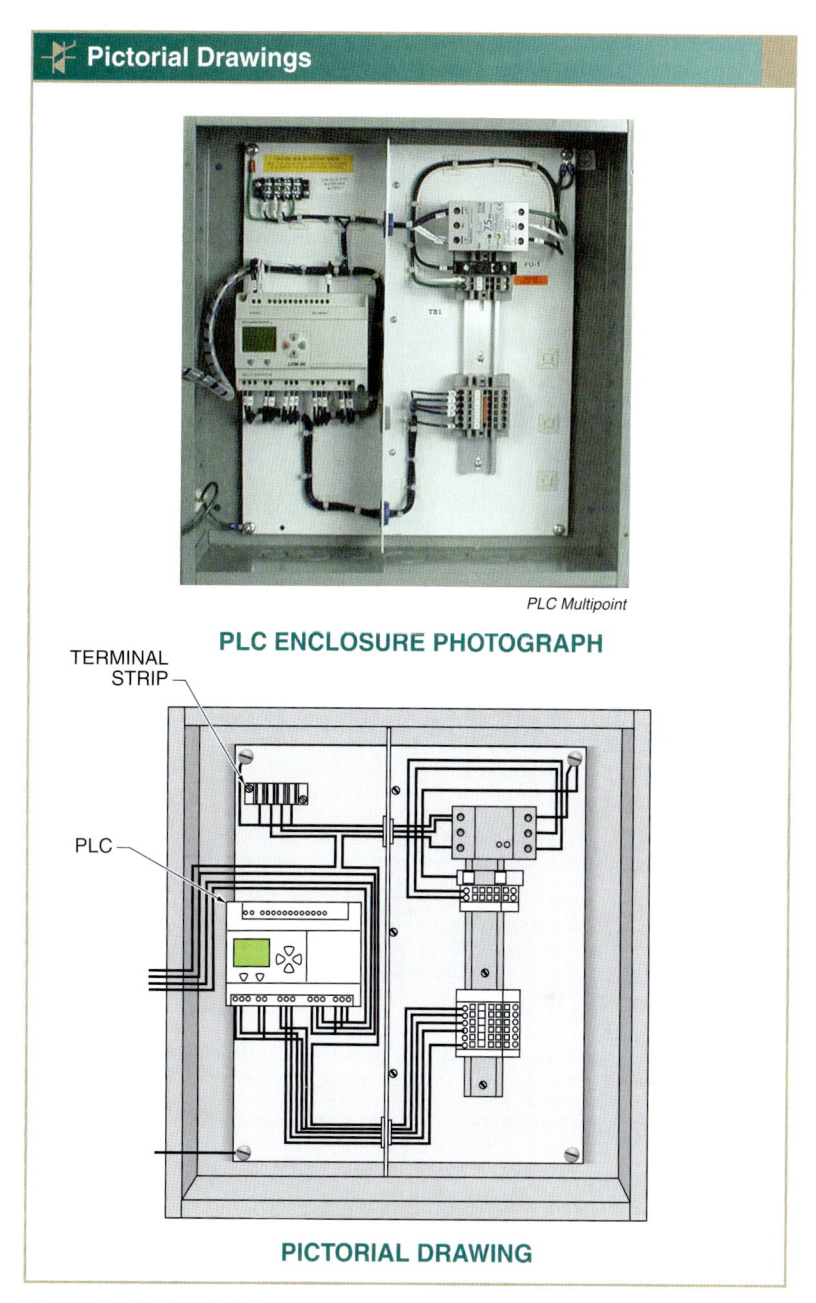

Figure 3-14. Pictorial drawings are used to provide a visual representation of devices, components, and wiring.

Line/Ladder Diagrams

A *line (ladder) diagram,* is a diagram that has a series of single lines (rungs) that indicate the logic of a control circuit and how the control devices are interconnected. Both standard electrical circuits and PLC electrical circuits use line diagrams to display information about control circuits as simply as possible.

130 PROGRAMMABLE LOGIC CONTROLLERS: PRINCIPLES AND APPLICATIONS

PLC Wiring Diagrams

MICRO-PLC WIRING DIAGRAM

RACK-MOUNTED PLC WIRING DIAGRAM

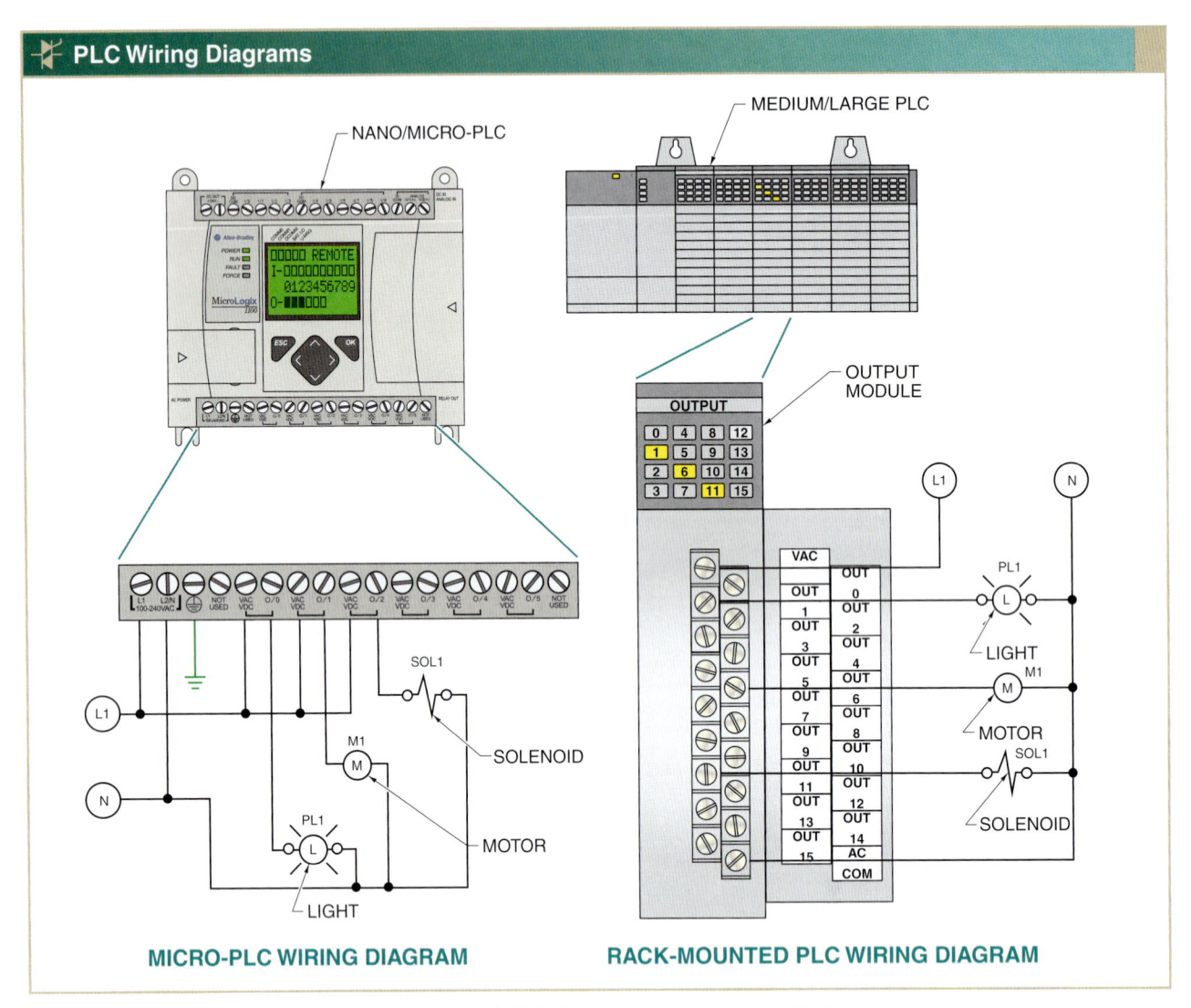

Figure 3-15. Wiring diagrams show the connection of all devices and components to a PLC.

Line diagrams use standard electrical symbols to indicate what types of input devices and output components are being used. PLC programming diagrams use bit instructions that are identified with words to indicate what types of input devices and output components are being used. **See Figure 3-17.**

PLCs use line diagrams as a language for inputting information and for developing equivalent PLC programming diagrams. However, other languages are also used to program PLCs such as Boolean algebra, functional block, and structured text. Line diagrams and Boolean algebra are basic PLC languages. Functional block and structured text are higher-level languages required to execute more powerful operations such as data manipulation, diagnostics, and report generation.

Except for a few differences, PLC programming diagrams are similar to line diagrams. PLC programming diagrams have two vertical power lines that represent L1 and L2 (X1 and X2 when a control transformer is used); however, no voltage potential exists between the two lines. Horizontal lines (rungs) represent the current flow paths between the vertical power lines. Each rung can represent several input devices and an output component. **See Figure 3-18.**

Wiring Thermocouples

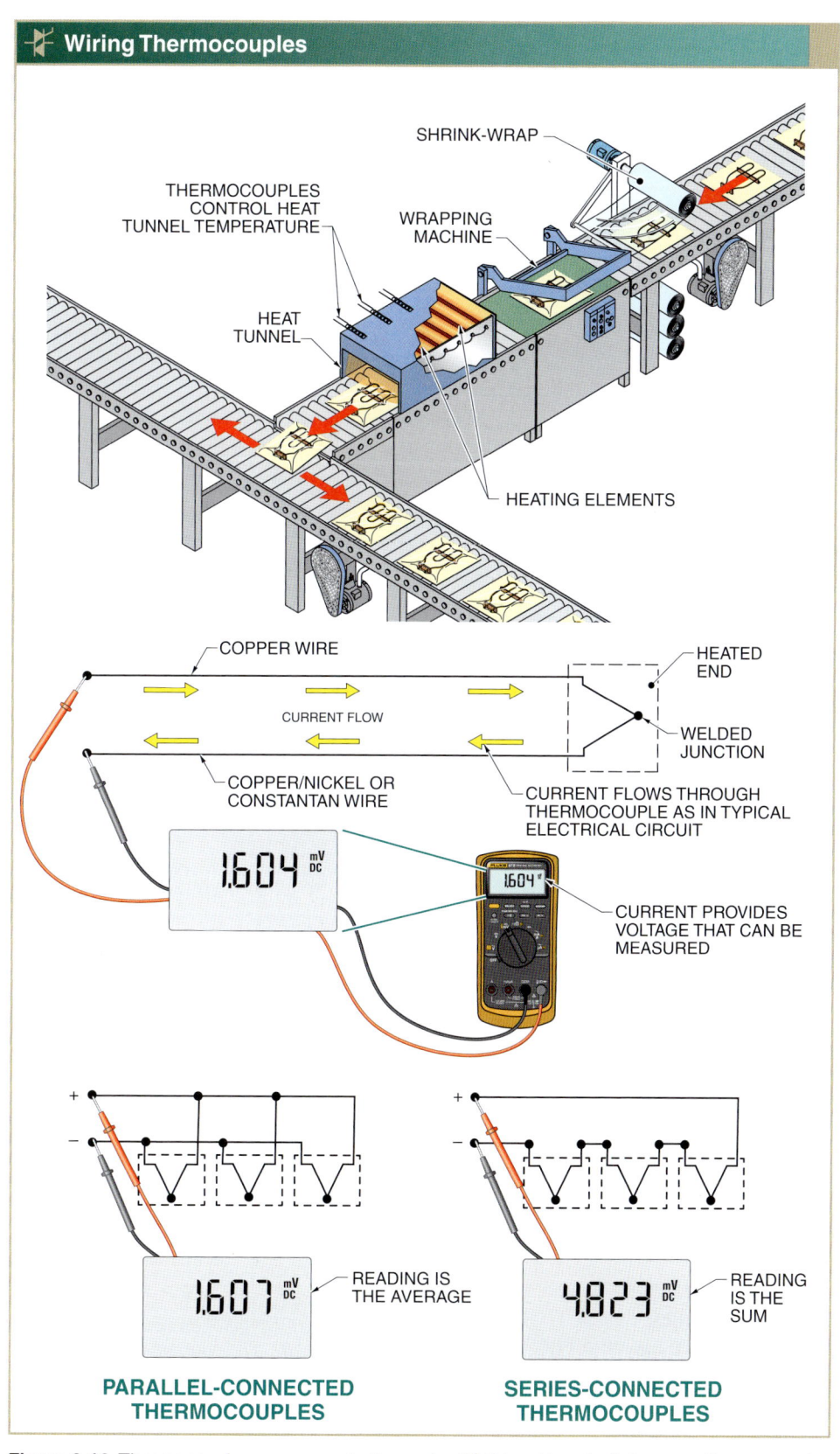

Figure 3-16. Thermocouples are connected to analog PLC input terminals because thermocouples produce a voltage proportional to the measured temperature.

CODE CONNECT

When the bottom of a raceway or cable is run across a roof and exposed to direct sunlight, Section 310.15(B)(2) of the NEC® requires a minimum distance of 7/8″ (23 mm) between the bottom of the raceway or cable and the roof. If the distance is less than 7/8″, 60°F must be added to the ambient temperature for adjustment purposes. The 7/8″ distance is intended to diminish the effects of the heat radiating from the roof. Conduit roof support systems that comply with section 310.15(B)(2) are available from several manufacturers.

132 PROGRAMMABLE LOGIC CONTROLLERS: PRINCIPLES AND APPLICATIONS

Line Diagrams

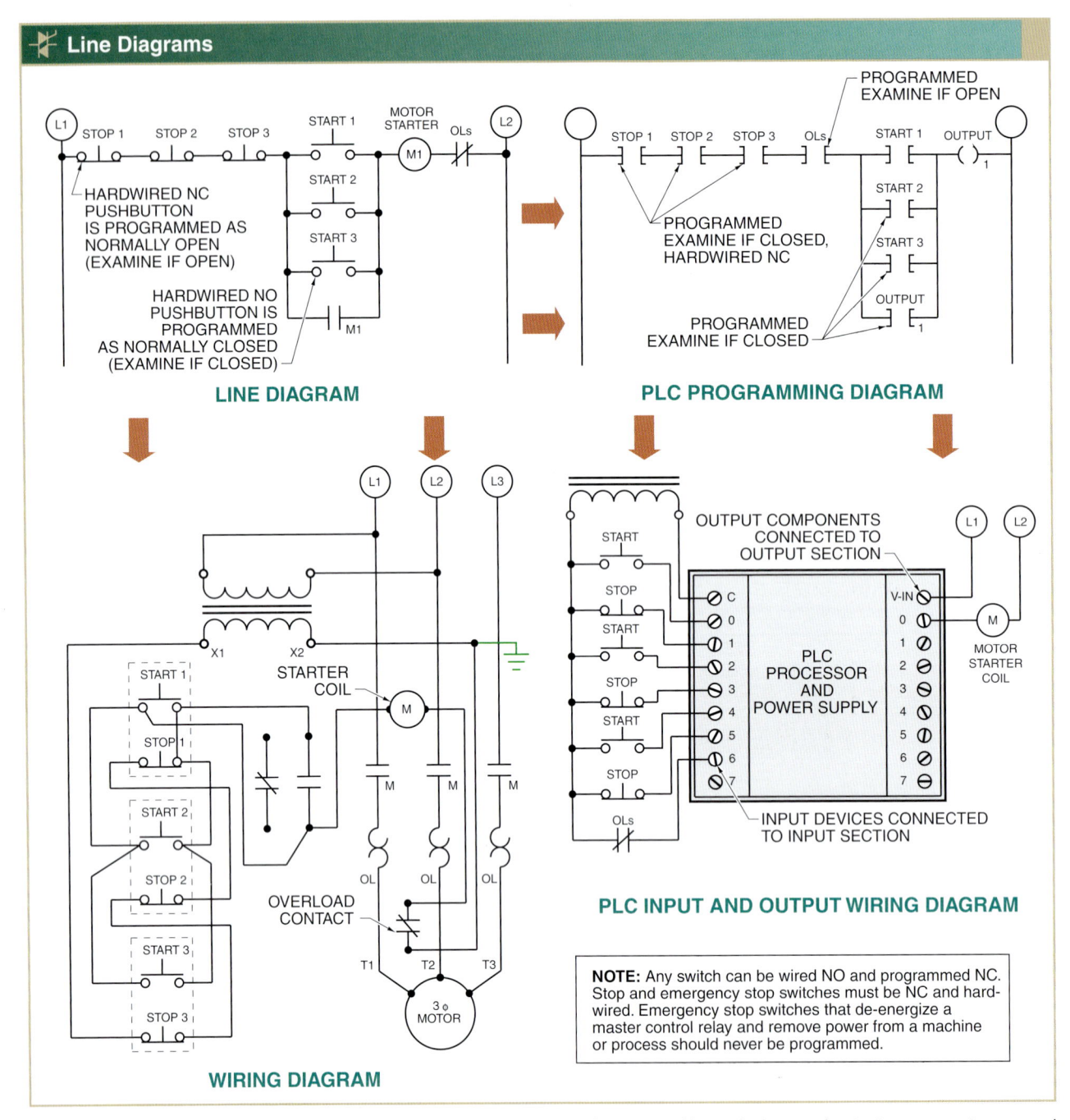

Figure 3-17. Line diagrams use standard electrical symbols to indicate what types of input devices and output components are used in a circuit. Line diagrams are used for inputting information to develop PLC programming diagrams.

NOTE: Any switch can be wired NO and programmed NC. Stop and emergency stop switches must be NC and hardwired. Emergency stop switches that de-energize a master control relay and remove power from a machine or process should never be programmed.

Standard traffic light circuits have only two input switches: CIRCUIT STOP and CIRCUIT START. Likewise, PLC-controlled traffic light circuits have only two input switches: CIRCUIT STOP and CIRCUIT START. Standard traffic light circuits also have three outputs (lamps): RED, YELLOW, and GREEN.

PLC TIPS

Motor starter overloads (heaters) protect the circuits from high temperatures caused by excessive current. The heaters cause the normally closed overload contact to open, thus de-energizing the coil of the motor starter and removing the power that feeds to the motor/load.

Traffic Light Line Diagrams

TRAFFIC LIGHT LINE DIAGRAM

TRAFFIC LIGHT PLC PROGRAMMING DIAGRAM

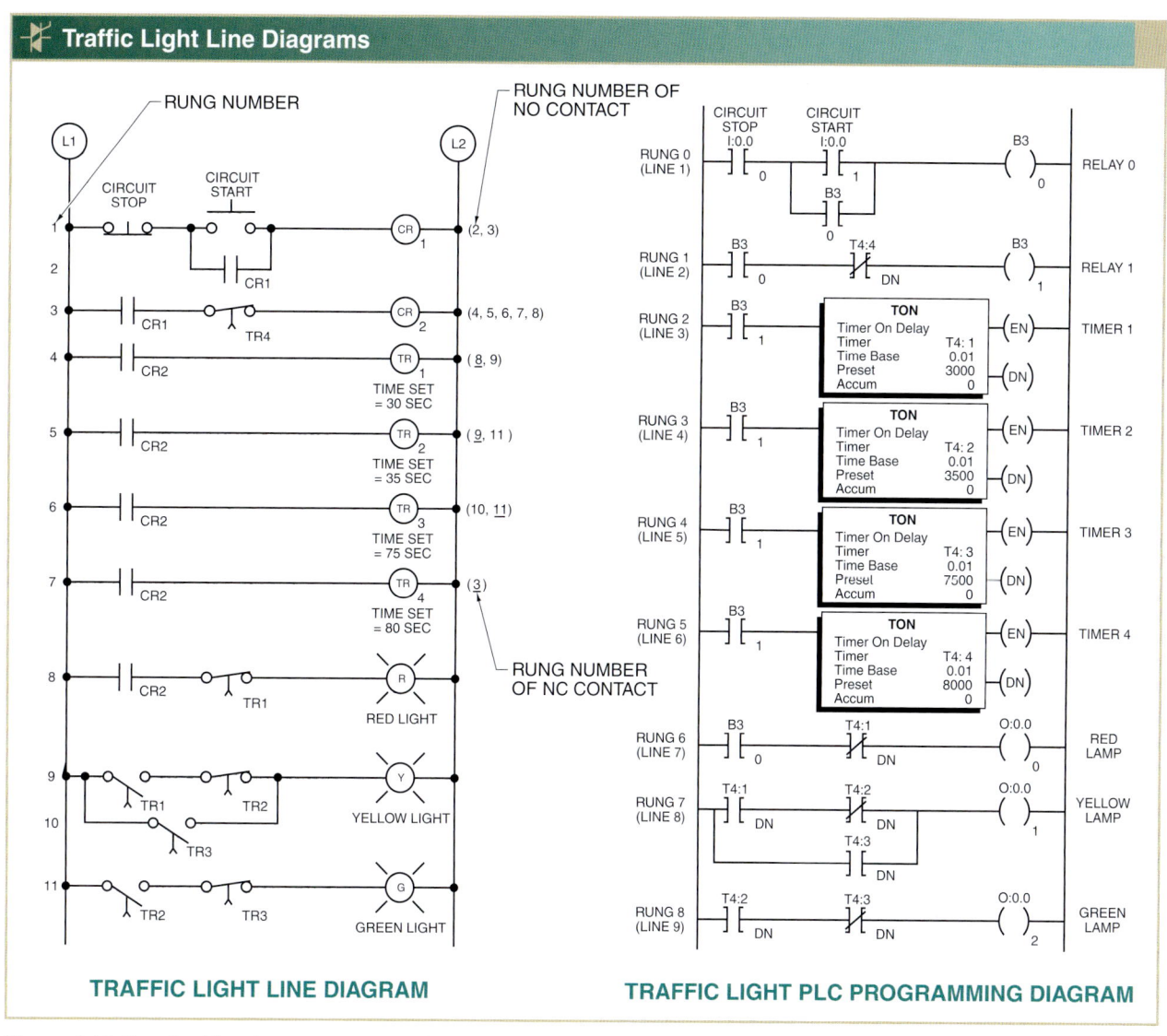

Figure 3-18. Standard line diagrams and PLC programming diagrams are similar, although different symbols are used to represent the same device or component.

One main advantage of using PLCs to control traffic lights compared to hardwired circuits is that PLCs eliminate the need for external control relays (CRs) and timers (TRs). In a standard traffic light circuit, two control relays and four timers need to be wired into the circuit in order for the circuit to operate. However, a PLC-controlled traffic light circuit has no control relays or timers wired into the control circuitry. The control relays and timers are simply programmed into the PLC and exist only as part of the PLC software and processor.

PLC LOGIC FUNCTIONS

All electrical control circuits (hardwired or PLC) that use switches to control loads are comprised of basic switching logic functions. The switching logic functions are used individually or in combination. The five basic switching logic functions are AND, OR, NOT, NOR, and NAND. Normally open (NO) and normally closed (NC) AND and OR switches or contacts connected in series or parallel are used to create the five logic functions. The five basic logic functions are used in ladder diagrams, PLC programming diagrams, and digital logic diagrams. **See Figure 3-19.**

Logic Function	Circuit Description	Ladder Diagrams	PLC Programming Diagrams	Digital Logic/PLC Logic Block Programming Diagrams
AND	**ENERGIZED** Output energized if all inputs are activated **DE-ENERGIZED** Output de-energized if any one input is deactivated			
OR	**ENERGIZED** Output energized if one or more inputs are activated **DE-ENERGIZED** Output de-energized if all inputs are deactivated			
NOT	**ENERGIZED** Output energized if input is not activated **DE-ENERGIZED** Output de-energized if input is activated			
NOR	**ENERGIZED** Output energized if no inputs are activated **DE-ENERGIZED** Output de-energized if one or more inputs are activated			
NAND	**ENERGIZED** Output energized unless all inputs are activated **DE-ENERGIZED** Output de-energized if all inputs are activated			

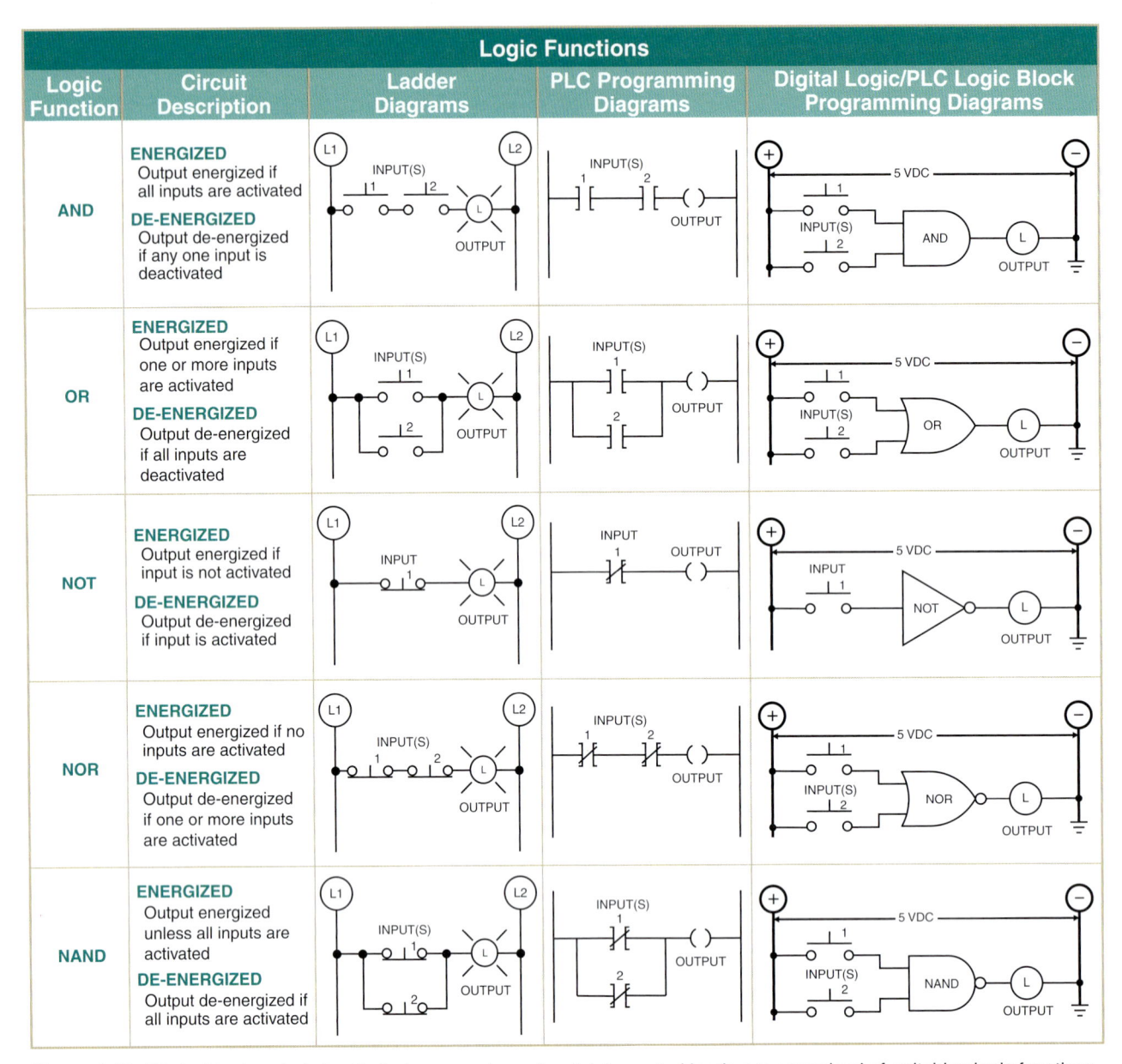

Figure 3-19. All electrical control circuits that use any type of switch to control loads are comprised of switching logic functions.

AND Circuit Logic

AND circuit logic is control logic developed when two or more normally open switches or contacts are connected in series to control a load. A logic circuit is AND logic because input 1 and input 2 must be actuated to energize the output component. The input devices that develop AND logic circuits can be any number of normally open switches (contacts) connected in series. **See Figure 3-20.**

OR Circuit Logic

OR circuit logic is control logic developed when two or more normally open switches or contacts are connected in parallel to control a load. A logic circuit is OR logic because input 1 or input 2 is actuated to energize the output component. The input devices that develop OR logic can be any number of normally open switches (contacts) connected in parallel. **See Figure 3-21.**

AND Circuit Logic

LINE DIAGRAM

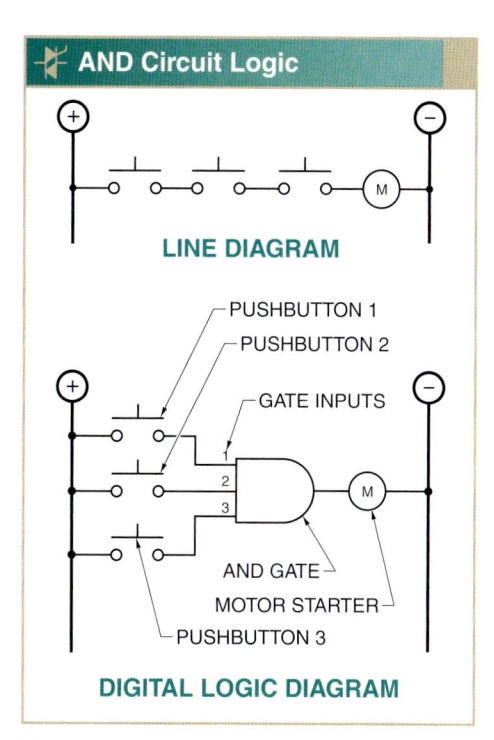

DIGITAL LOGIC DIAGRAM
- PUSHBUTTON 1
- PUSHBUTTON 2
- GATE INPUTS
- AND GATE
- MOTOR STARTER
- PUSHBUTTON 3

Figure 3-20. All input devices of an AND logic circuit must be actuated in order to energize the output component.

OR Circuit Logic

LINE DIAGRAM

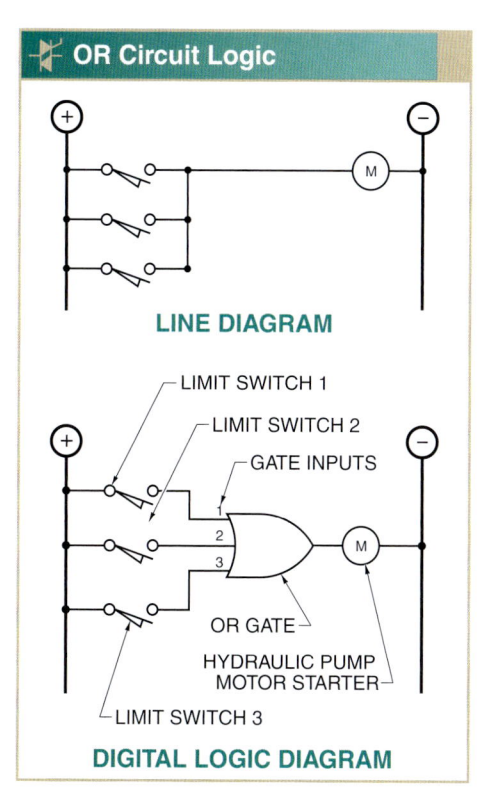

DIGITAL LOGIC DIAGRAM
- LIMIT SWITCH 1
- LIMIT SWITCH 2
- GATE INPUTS
- OR GATE
- HYDRAULIC PUMP MOTOR STARTER
- LIMIT SWITCH 3

Figure 3-21. Only one input device in an OR logic circuit must be actuated in order to energize the output component.

NOT Circuit Logic

NOT circuit logic is control logic developed when a normally closed switch or contact is connected to control a load. A logic circuit is NOT logic because the output component is on when the input device is not actuated (opened). Only one normally closed switch (contact) is used to develop NOT logic. When two or more normally closed contacts are used together, NOR logic or NAND logic is developed. **See Figure 3-22.**

NOT Circuit Logic

LINE DIAGRAM
- LIMIT SWITCH NO HELD CLOSED

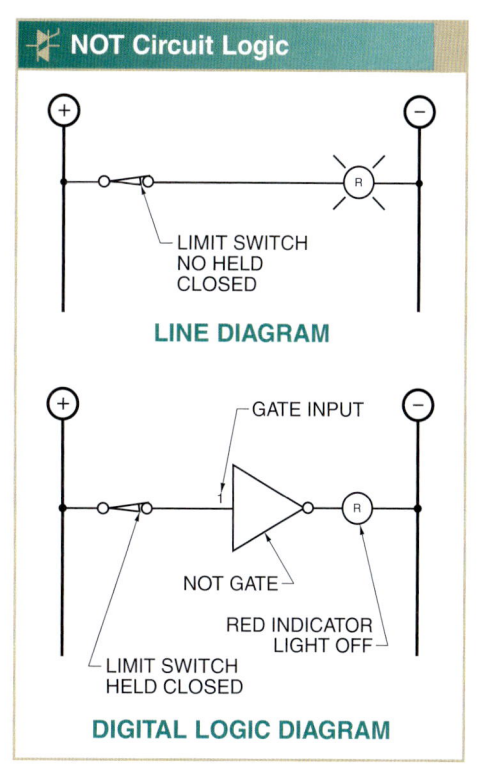

DIGITAL LOGIC DIAGRAM
- GATE INPUT
- NOT GATE
- RED INDICATOR LIGHT OFF
- LIMIT SWITCH HELD CLOSED

Figure 3-22. The input device of a NOT logic circuit must not be actuated in order to energize the output component.

NOR Circuit Logic

NOR circuit logic is control logic developed when two or more normally closed switches or contacts are connected in series to control a load. A circuit logic is NOR logic because the output component is ON when input 1 and input 2 are not actuated. The input devices that develop NOR logic can be any number of normally closed switches (contacts) connected in series. **See Figure 3-23.**

NOR Circuit Logic

[Diagram: Air conditioner unit with labels – AIR CONDITIONER UNIT, ELECTRICAL CONTROLS, CONDENSER COILS, COMPRESSOR AND MOTOR. *Carrier Corporation*]

LINE DIAGRAM

[Diagram: Digital logic diagram with MOTOR TEMPERATURE SWITCH, GATE INPUTS, NOR GATE, COMPRESSOR MOTOR, HIGH-PRESSURE SWITCH]

DIGITAL LOGIC DIAGRAM

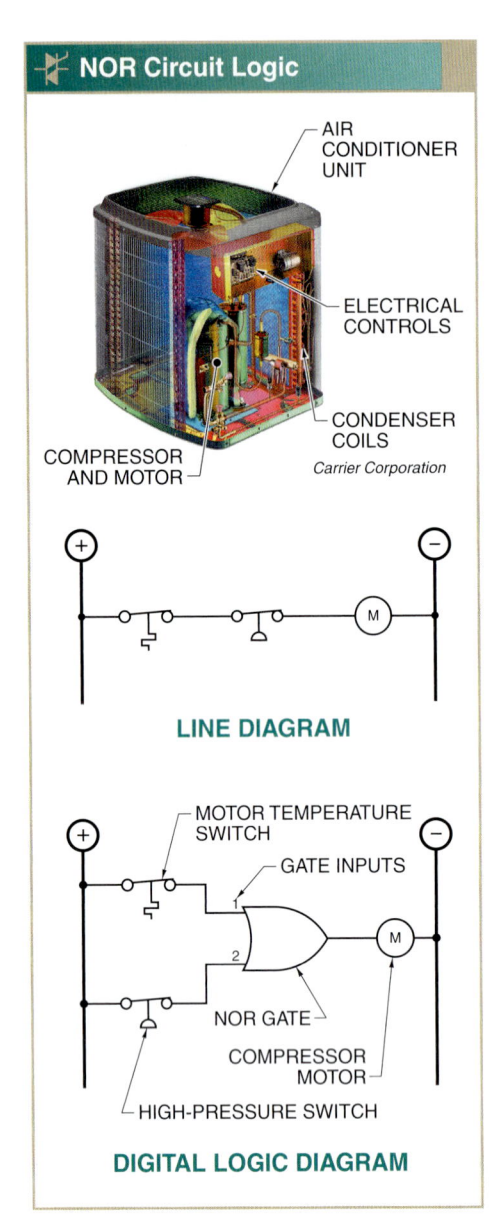

Figure 3-23. All input devices of a NOR logic circuit must not be actuated in order to energize the output component.

NAND Circuit Logic

LINE DIAGRAM

[Diagram: Digital logic diagram with LIMIT SWITCH 1, GATE INPUTS, NAND GATE, SOLENOID, LIMIT SWITCH 2]

DIGITAL LOGIC DIAGRAM

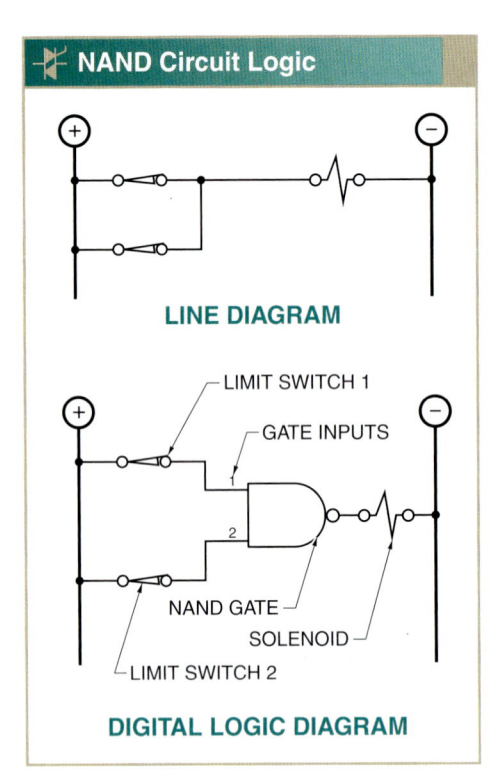

Figure 3-24. Only one input device of a NAND logic circuit must not be actuated in order to energize the output component.

ELECTRICAL WIRING METHODS

PLC-controlled systems have individual electrical devices and components interconnected (wired) to form electrical circuits. There are three basic methods of wiring electrical circuits: direct hardwiring, hard-wiring using terminal strips, and PLC wiring. All three methods have advantages and disadvantages. Because all three methods are used, an understanding of each type is necessary when working on electrical circuits or PLC-controlled systems.

NAND Circuit Logic

NAND circuit logic is control logic developed when two or more normally closed switching contacts are connected in parallel to control a load. A circuit logic is NAND logic because the output component is ON when input 1 or input 2 is not actuated. The input devices that develop NAND logic can be any number of normally closed switches (contacts) connected in parallel. **See Figure 3-24.**

Direct Hardwiring

Direct hardwiring is the oldest and most straightforward wiring method used. *Direct hardwiring* is a wiring method where the power circuit and the control circuit are wired point-to-point. **See Figure 3-25.** *Point-to-point wiring* is a wiring method where each component in a circuit is connected (wired) directly to the next component as specified by wiring or line diagrams.

Chapter 3—PLC Programming Symbols, Diagrams, and Logic Functions **137**

Direct Hardwired Circuits

REVERSING CONTROL CIRCUIT LINE DIAGRAM

REVERSING CIRCUIT WIRING DIAGRAM

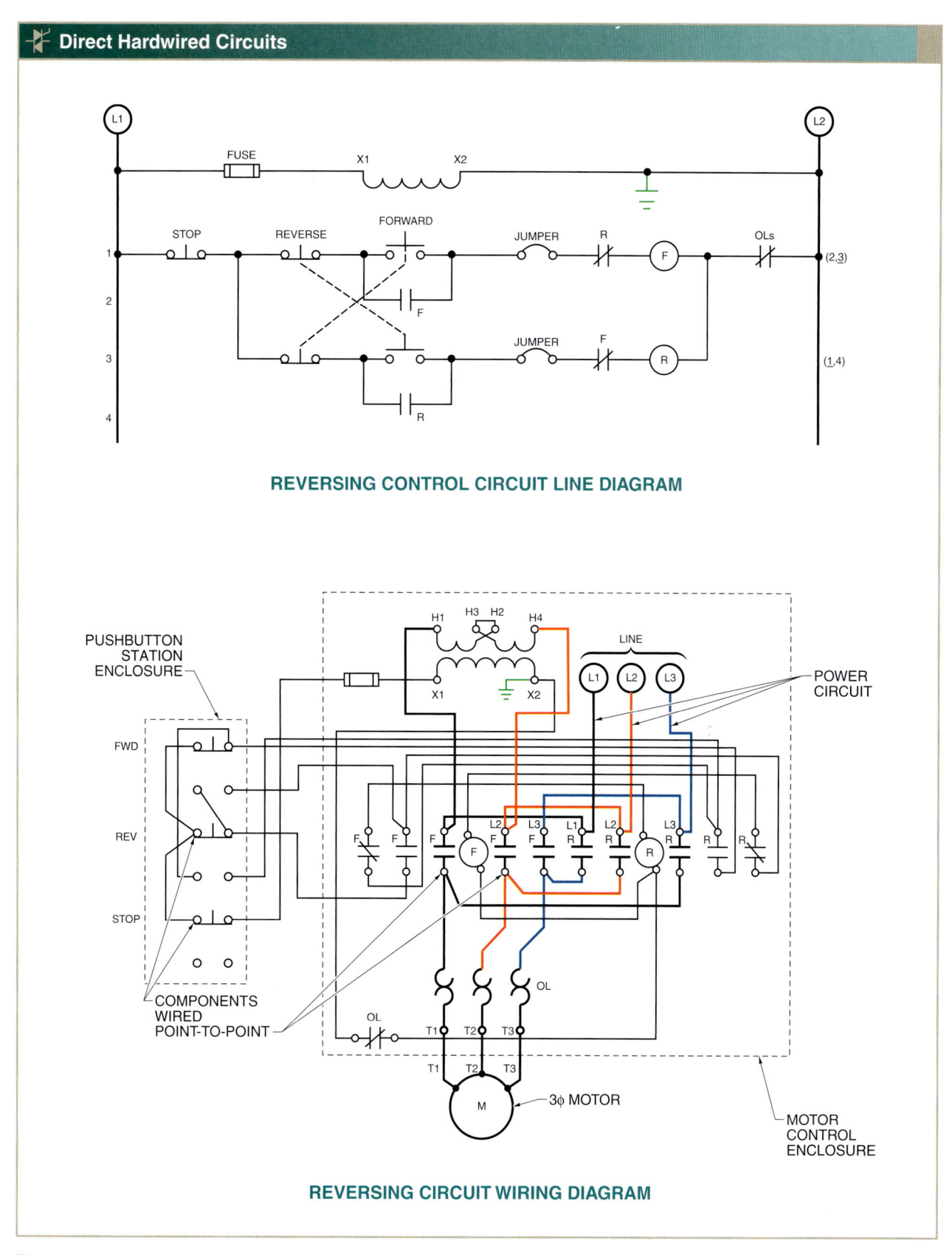

Figure 3-25. In direct hardwired circuits, the power circuit and control circuit are wired point-to-point.

For example, a transformer X1 terminal is connected directly to a fuse, the fuse is connected directly to a stop pushbutton, the stop pushbutton is connected directly to a reverse pushbutton, the reverse pushbutton is connected directly to a forward pushbutton, and so on until the final connection from the overload (OL) contact is made back to transformer X2 terminal.

Direct hardwired circuits can operate for many years. The disadvantage of a direct hardwired circuit is that circuit troubleshooting and circuit modifications are time-consuming. For example, when a problem occurs in a direct hardwired circuit, the wiring diagram of the circuit must be understood. Without a wiring diagram, circuit wiring is determined by tracing each wire throughout the circuit and creating a schematic (sketch). Tracing each wire in a circuit and making a sketch to find the wire with the problem is time-consuming. Time is saved as experience is gained from working on a circuit several times and understanding all device and component operations.

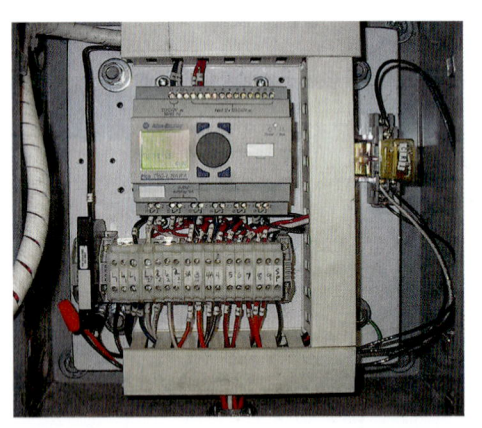

PLC-wired control circuits have all input devices wired to the input sections of PLCs and all output components wired to the output sections of PLCs directly through terminal strips or terminal blocks.

Also, a direct hardwired circuit may be difficult to modify. **See Figure 3-26.** For example, when a forward rotation indicator lamp and a reverse rotation indicator lamp are to be added to a motor control circuit, the exact connection points for the lamps must be located. Once the exact connection points are found, the lamps can be wired into the control enclosure. However, problems can arise when making the actual connection. For example, there may not be enough room under a terminal screw for more wires to be connected.

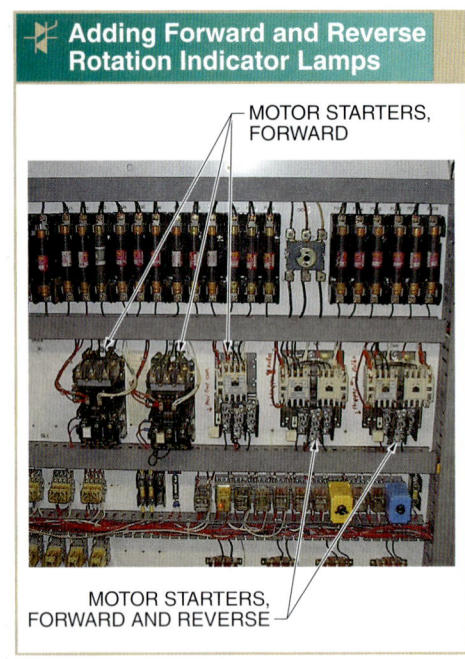

Figure 3-26. To add forward rotation indicator lamps and reverse rotation indicator lamps to a direct hardwired circuit is difficult because the exact connection points for the lamps must be found.

A hardwired circuit modification such as adding forward and reverse indicator lamps may not be a problem if the lamps only require the addition of a couple of new wires. **See Figure 3-27.** When performing an indicator lamp modification, old wires are not required to be moved or removed.

Some circuit modifications, such as adding limit switches, are more difficult. For example, when forward rotation and reverse rotation limit switches are added to a circuit, some wiring must be removed from the circuit. Also, new wiring for the limit switches must be added.

Modifications to Direct Hardwired Circuits

REVERSING CONTROL CIRCUIT LINE DIAGRAM

REVERSING CIRCUIT WIRING DIAGRAM

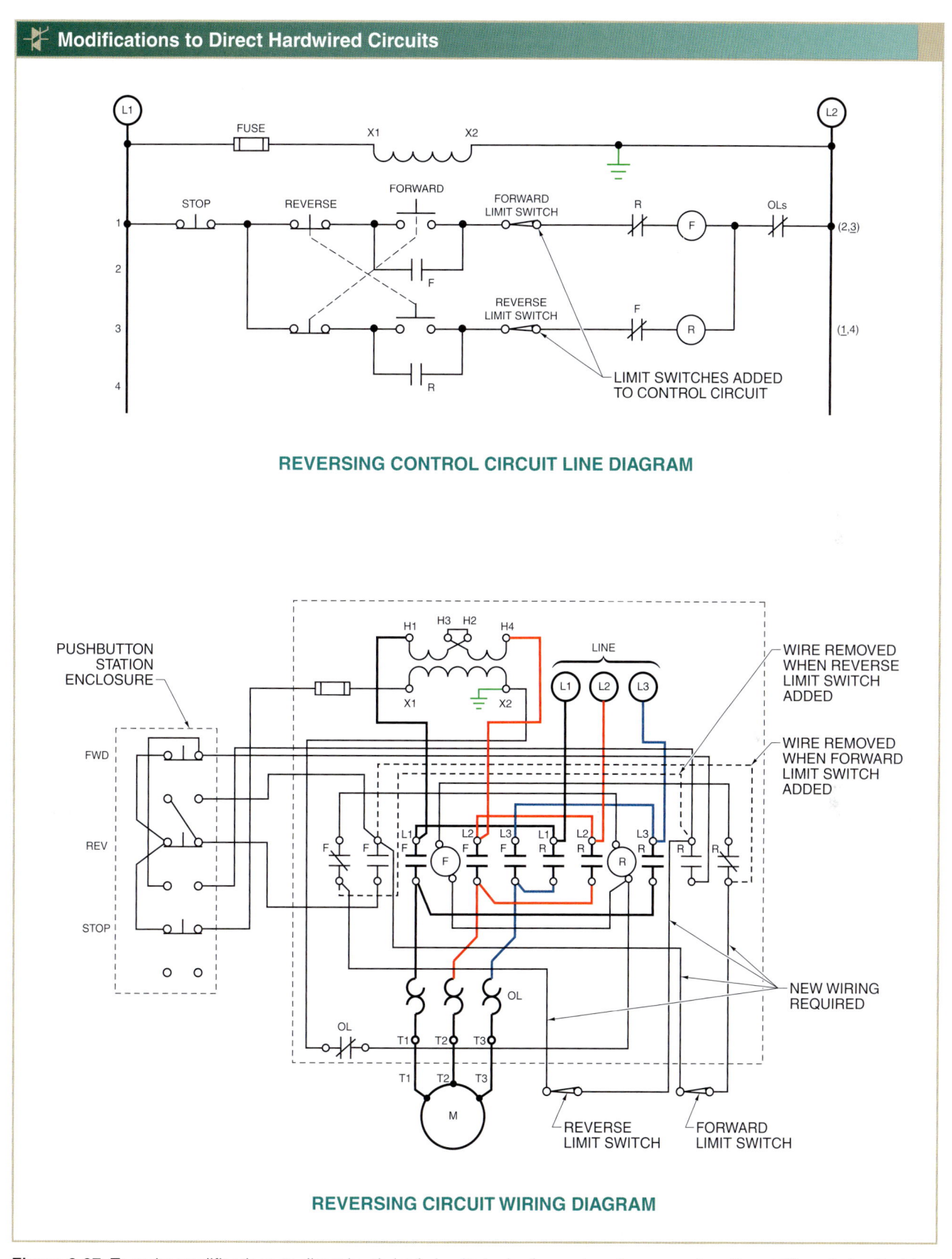

Figure 3-27. To make modifications to direct hardwired circuits typically requires the removal and/or addition of circuit wiring.

Before limit switches can be added, the wires connecting the normally closed interlock contacts of the forward and reversing motor-starter coils to the pushbuttons must be removed (or opened) and the limit switches wired in the created openings. Technicians making circuit modifications must have a wiring diagram of the circuit (or understand the circuit from past experience) in order to know which wires to open and where to locate the new limit switches.

Hardwiring Using Terminal Strips

Hardwiring to a terminal strip allows for easier circuit modifications and simplifies circuit troubleshooting. When using terminal strips for wiring, each wire in the control circuit is assigned a reference number (terminal screw wire number) on the ladder diagram to identify the various wires that connect the components in a circuit. Each reference point is assigned a wire reference number. **See Figure 3-28.**

In the past, wire reference numbers were commonly assigned from the top left to the bottom right. Today however, diagrams typically have the power line on the top left (typically L1 or X1) assigned the number 1, and the power line on the right (typically L2 or X2), the number 2. Using reference numbers 1 and 2 for power supply lines allows control circuit voltage to be measured between terminal 1 and terminal 2 and aids in troubleshooting a circuit. When several connections to a given number are required, jumpers can be added to the terminal strip to provide multiple connection points.

When troubleshooting a circuit that has a terminal strip, a technician can go directly to the terminal strip and take measurements to help isolate the problem. First, a DMM is placed between terminals 1 and 2. When voltage is not correct at terminals 1 and 2, the problem will be found at the transformer or in the ground. When voltage is correct across terminals 1 and 2, the common DMM test lead is left on terminal 2, and the voltage test lead is moved to various terminals until the problem is found.

Terminal strips with wire reference numbers help in troubleshooting. Terminal strips also make circuit modifications easier to perform. Modifications are easier because most, if not all, of the wires required to make a change are disconnected and reconnected at the terminal strip. **See Figure 3-29.**

PLC Wiring

Using a PLC for motor control allows for greater flexibility and monitoring of the motor and control circuit. A PLC can monitor and control all motor control functions, but cannot directly monitor and display motor parameters such as voltage, current, frequency, and power.

When using a PLC to control a circuit, the PLC is part of the control circuit. The power circuit to a component does not change. What does change is that the control circuit input devices (pushbuttons, limit switches, and overload contacts) are wired to a PLC input module, and the control circuit output components (motor starter coils and indicator lamps) are wired to a PLC output module. **See Figure 3-30.**

Circuit operation (logic) is typically programmed using PLC software on a desktop or laptop computer. Once a program is created, the program (circuit) is downloaded to the PLC. PLC software can monitor and display the condition (ON or OFF) of circuit input devices and output components. When changes in the control circuit are required, the changes can be reprogrammed and downloaded without changing the circuit wiring.

When programming a PLC for input devices, the actual input device type (normally open or normally closed) and the way the input device is programmed are critical. The way a PLC is programmed for an input device is critical because when using a PLC, an input device can be wired normally open and programmed either normally open or normally closed. Likewise, an input device can be wired normally closed and programmed either normally closed or normally open.

Hardwiring Using Terminal Strips

REVERSING CONTROL CIRCUIT LINE DIAGRAM

REVERSING CIRCUIT WIRING DIAGRAM

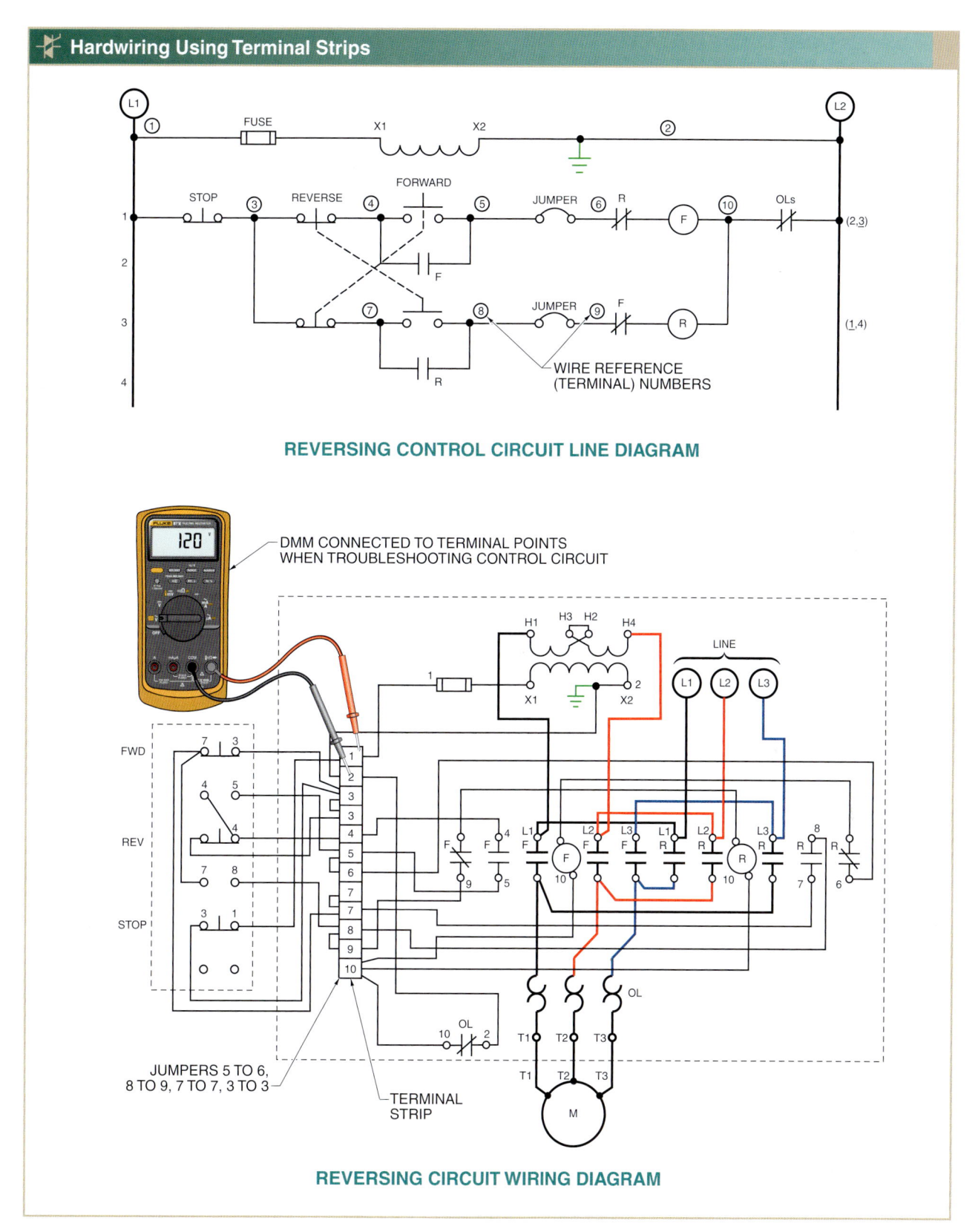

Figure 3-28. Hardwiring using terminal strips simplifies circuit troubleshooting.

142 PROGRAMMABLE LOGIC CONTROLLERS: PRINCIPLES AND APPLICATIONS

Modifications to Hardwired Circuits with Terminal Strips

REVERSING CONTROL CIRCUIT LINE DIAGRAM

REVERSING CIRCUIT WIRING DIAGRAM

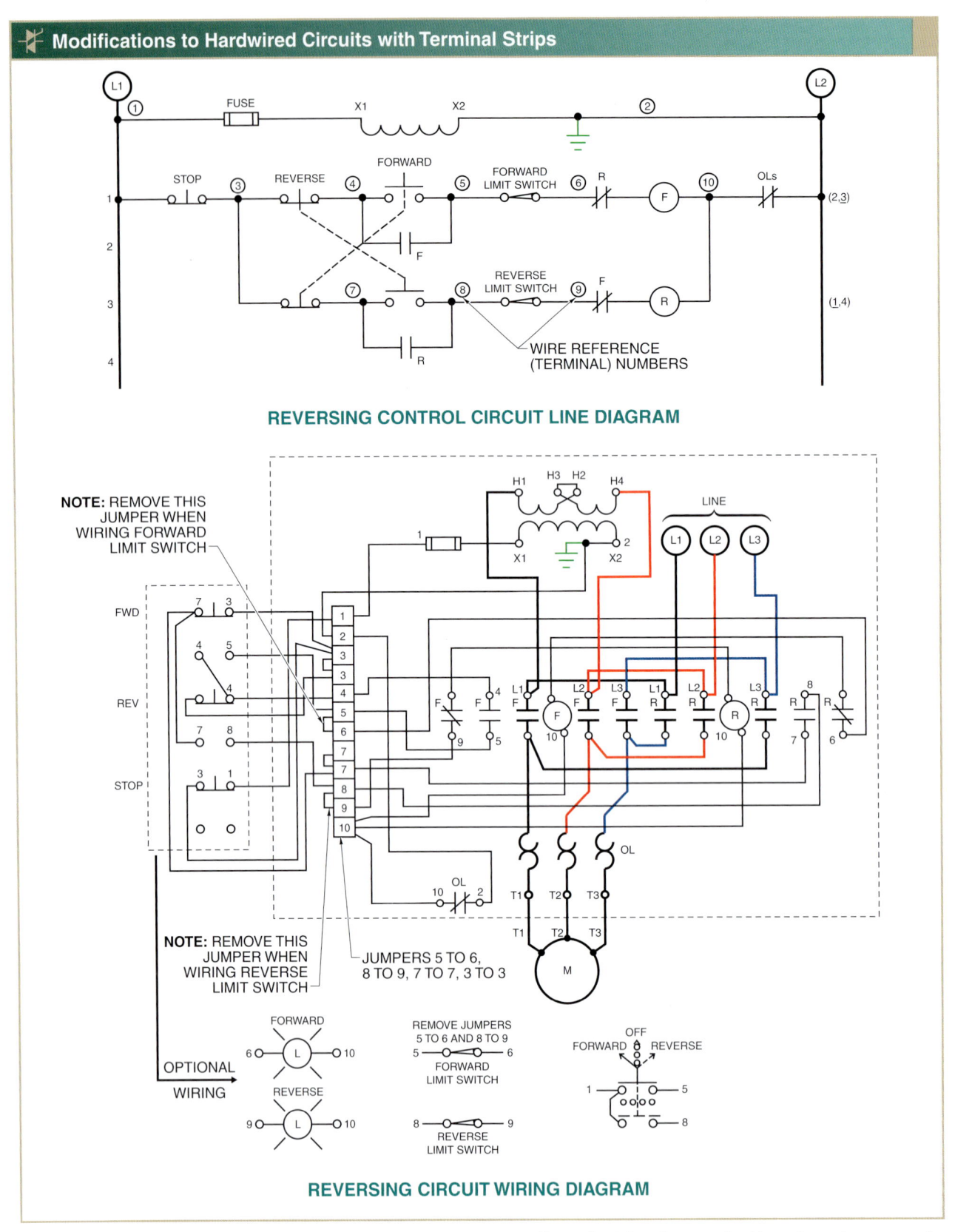

Figure 3-29. Modifications are easier with terminal strips because most, if not all, of the wires required to make a change are disconnected and reconnected at the terminal strip.

PLC-Wired Circuits

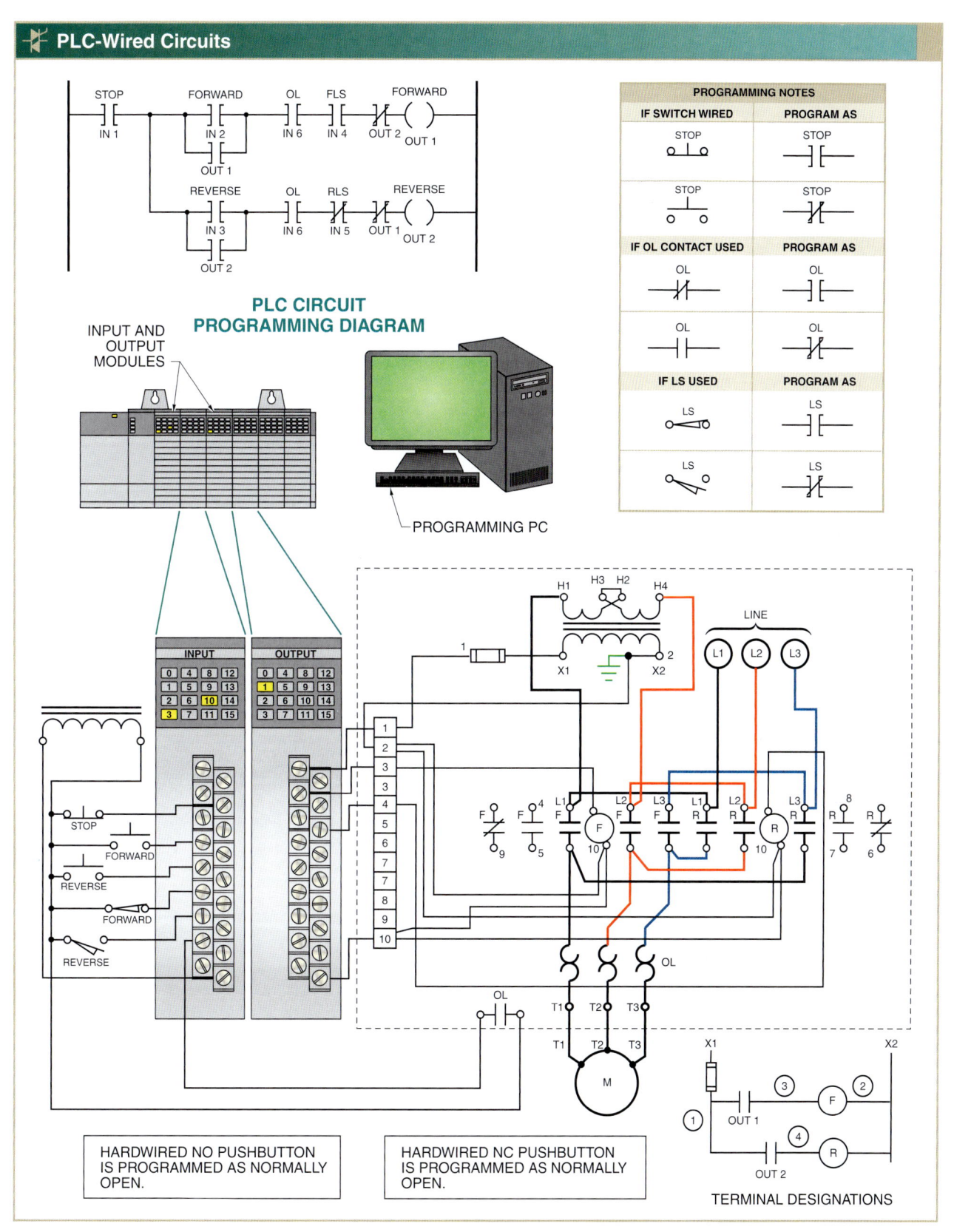

Figure 3-30. When using a PLC to control a circuit, all input devices are wired to the input module of the PLC, and all output components are wired to the output module of the PLC.

Chapter Review 3

Name _____ Date _____

True-False

T　F　**1.** A pictorial drawing shows what a device looks like, but does not give accurate dimensions or other specific detail information.

T　F　**2.** When using a PLC to control a circuit, the PLC is part of the control circuit.

T　F　**3.** An automatically operated switch is a switch that is activated independent of a person or object.

T　F　**4.** Direct hardwiring decreases the difficulty of troubleshooting or modifying a circuit.

T　F　**5.** The symbol for a normally open mechanical or automatic switch shows the operator moving upward when activated.

T　F　**6.** PLC programs are designed using computer-assisted drafting software.

T　F　**7.** There is no way to distinguish which type of switch (pressure, temperature, or limit) is being used by looking at a generic normally open or normally closed electrical symbol.

T　F　**8.** A normally open limit switch can be drawn on a diagram as either normally open or normally open, held closed.

T　F　**9.** The difference between the symbol for a mechanical switch and the symbol for a solid-state switch is that a solid-state switch includes a circle around the switch symbol.

T　F　**10.** Ladder diagrams used standard electrical symbols to indicate what types of input devices and output components are being used.

T　F　**11.** Multiple thermocouples can be connected to the input section of a PLC in parallel or in series.

T　F　**12.** Standard output components include pushbutton, limit, and pressure switches.

T　F　**13.** NAND circuit logic is NAND logic because the output component is ON when input 1 and input 2 are not actuated.

T　F　**14.** Terminal strips increase the ease of circuit modifications because most of the wires required to make a change are disconnected and reconnected at the terminal strip.

T　F　**15.** Manually operated switches are drawn on electrical diagrams in the "normal" condition, which is the activated condition.

Completion

_____　**1.** A(n) ___ is a drawing that resembles a photograph or three-dimensional picture.

_____　**2.** ___ is a wiring method where each component in a circuit is connected directly to the next component as specified by wiring or line diagrams.

_____ 3. A(n) ___ is a graphic element that represents a quantity, unit, device, or component.

_____ 4. A(n) ___ is a temperature sensor made of two dissimilar metals that generates power at the joined ends of the metals where heat is to be measured.

_____ 5. A(n) ___ is a diagram that has a series of single lines that indicate the logic of the control circuit using identifiable symbols and how the control devices are interconnected.

_____ 6. ___ is a wiring method where the power circuit and the control circuit are wired point-to-point.

_____ 7. A(n) ___ is a technical drawing that shows the electrical connection of all devices and components to a PLC.

_____ 8. ___ is control logic developed by activating input 1 or input 2 to energize an output component.

_____ 9. A(n) ___ is any switch that requires a person to physically change the state of the switch.

_____ 10. ___ is control logic developed by activating input 1 and input 2 to energize an output component.

Multiple Choice

_____ 1. With ___ circuit logic, the output component is ON when the input device is not actuated.
   A. NAND
   B. NOT
   C. NOR
   D. OR

_____ 2. With ___ circuit logic, the output component is ON when input 1 and input 2 is not actuated.
   A. AND
   B. NOR
   C. NAND
   D. OR

_____ 3. A(n) ___ switch is a switch that detects the physical presence of an object.
   A. automatically operated
   B. manually operated
   C. mechanically operated
   D. solid-state

_____ 4. When thermocouples are connected in series, a PLC will receive the ___ of all the individual thermocouple temperature measurements.
   A. average
   B. highest
   C. lowest
   D. sum

_____ 5. There are ___ basic methods of wiring electrical circuits.
   A. two
   B. three
   C. four
   D. five

Programming Function Blocks Using TECO Software

PLC Exercise 3.1

BILL OF MATERIALS
- TECO PLR
- Power supply
- Six inputs
- Four outputs

In any PLC system, inputs and outputs must operate in a predetermined manner. To accomplish this, PLCs can be programmed in different ways, such as using ladder logic or function block logic, to meet circuit requirements. In the following lab, use the SG2 Client software to create a new function block program to understand basic logic block functions.

T Programming Function Block Procedures

1. Open the TECO SG2 Client program by clicking on the SG2 Client icon.

2. In the program window, select NEW FUNCTION BLOCK PROGRAM.

3. The screen will display a Select Model Type window. Select model SG2-12HR-D or a different model based on the TECO model used for the application.

4. A Function Block Diagram (FBD) environment programming window will appear. The FBD environment programming window includes a Menu bar, a Tools bar, the programming area, an FBD Tools bar, and a Status bar. The FBD Tools bar includes these icons: Select, Connect, Constants/Connectors, Logic Block, Function Block, All, Cut/Join Connection, Delete, and Comments.

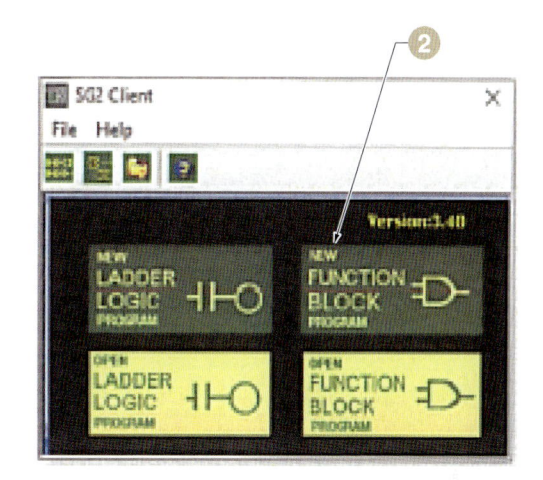

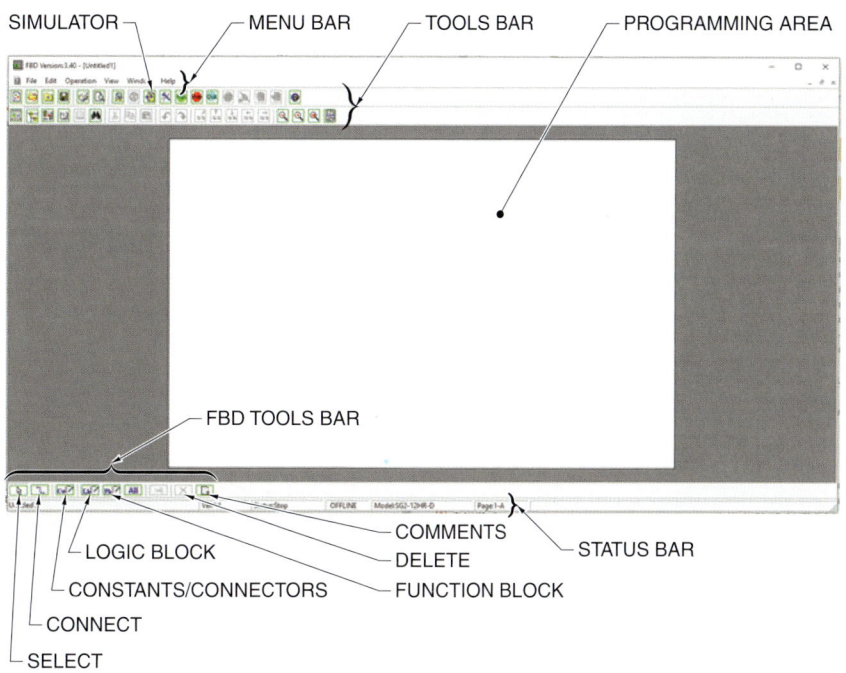

FUNCTION BLOCK DIAGRAM ENVIRONMENT

5. In the Tools bar, click on the Constants/Connectors icon to display the types of inputs and outputs used in the programming environment.

6. Click on the Logic Block icon to display the types of logic blocks that can be used in the programming environment. Logic blocks that can be programmed include AND, OR, NOT, NAND, NOR, and XOR.

7. Click on the Function Block icon to display the types of timers and counters that can be used in the programming area.

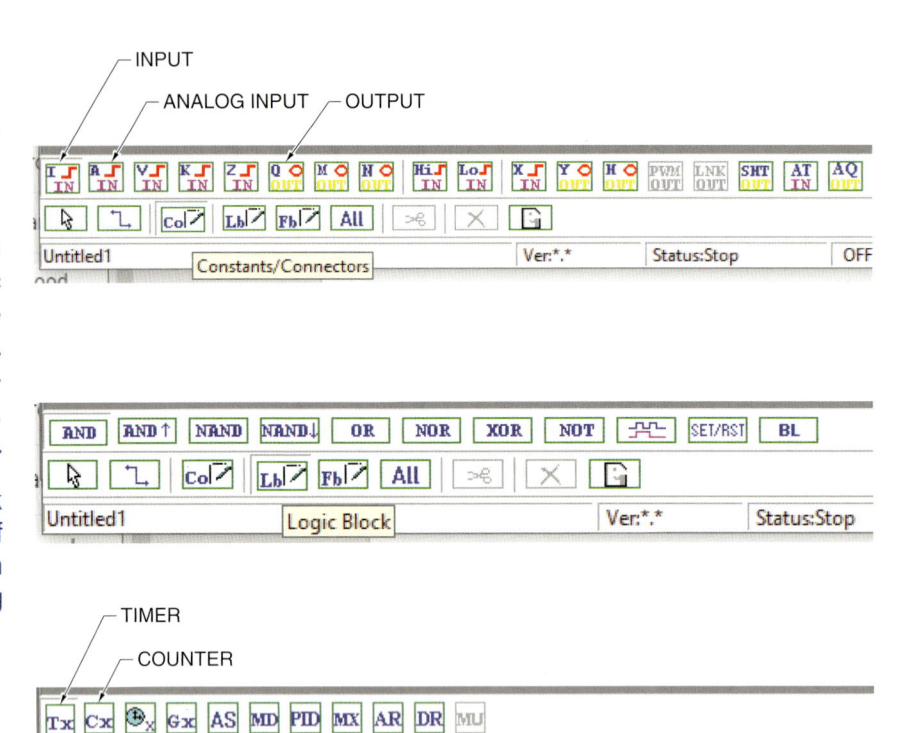

T Programming AND/OR Circuit Logic

Next, use the FBD as a reference to program AND/OR circuit logic. In the FBD Tools bar, select the proper input(s), output(s), and logic blocks to place in the programming area. Click on the Constants/Connectors icon to select the inputs and outputs. Then click on the Logic Block icon to select the AND logic block and the OR logic block. Use the Connect tool to attach each input and output to the logic blocks. After all connections have been made, run the simulator and test each input using the Input Status tool.

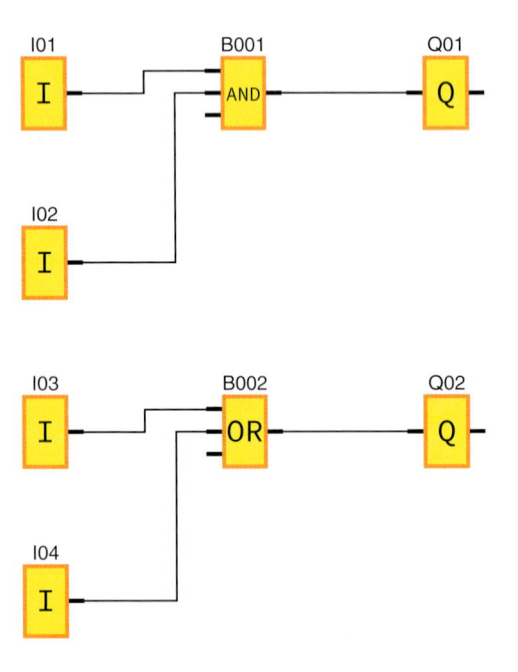

AND/OR FUNCTION BLOCK DIAGRAM

Programming NOT/NOR/NAND Circuit Logic

Next, use the FBD as a reference to program NOT/NOR/NAND circuit logic. In the FBD Tools bar, select the proper input(s), output(s), and logic blocks to place in the programming area. Click on the Constants/Connectors icon to select the inputs and outputs. Then click on the Logic Block icon to select the NOT, NOR, and NAND logic blocks. Use the connect tool to attach each input and output to the logic blocks. After all connections have been made, run the simulator and test each input using the Input Status tool.

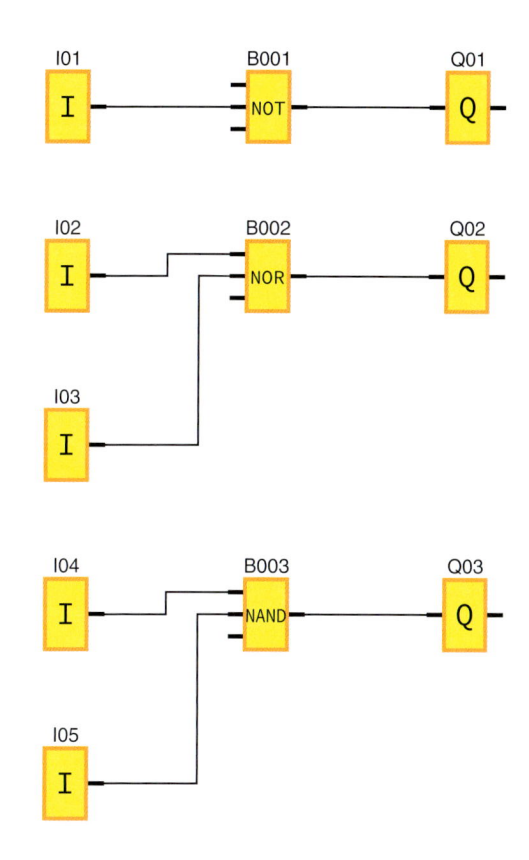

NOT/NOR/NAND FUNCTION BLOCK DIAGRAM

Motor Control with Test Lights

PLC Exercise 3.2

BILL OF MATERIALS

- TECO or nano-PLC
- NC red pushbutton
- NO green pushbutton
- Two test pushbuttons
- Two lights (green/red)
- Motor starter/contactor
- One load (optional)

In this lab, use the SG2 Client software to create a ladder or FBD program to control a motor starter circuit with push-to-test indicator lights. A push-to-test pushbutton is used to test a circuit's motor ON indicator light when the motor is off and to test a circuit's motor OFF indicator light when the motor is on. Use the line diagram as a reference to create the ladder diagram or the FBD for the lab.

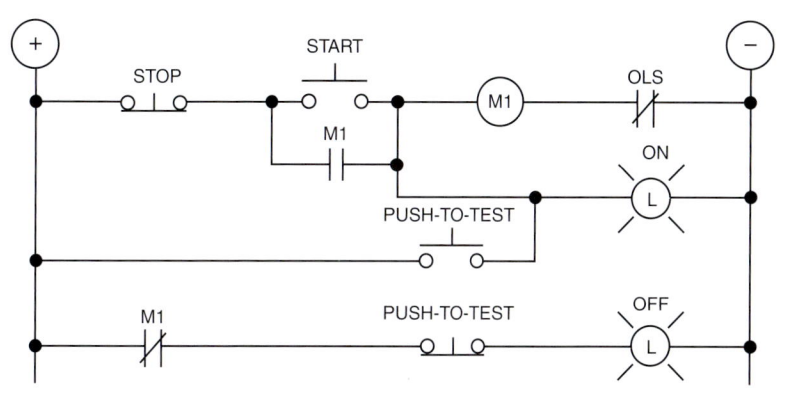

LINE DIAGRAM

TECO Procedures

1. Create a new ladder logic or FBD program.

2. Use the TECO programming diagram or FBD to add the proper contacts and description in the programming grid/area.

3. Use the Edit Contact window to denote an input or output number and the Symbol button to add a description after placing the input or output in the programming grid. For an FBD, use the Comment icon to add the description.

4. Run the program in Simulation Mode to verify proper operation. Use the Input Status tool window to activate the inputs and control the output to simulate circuit operation.

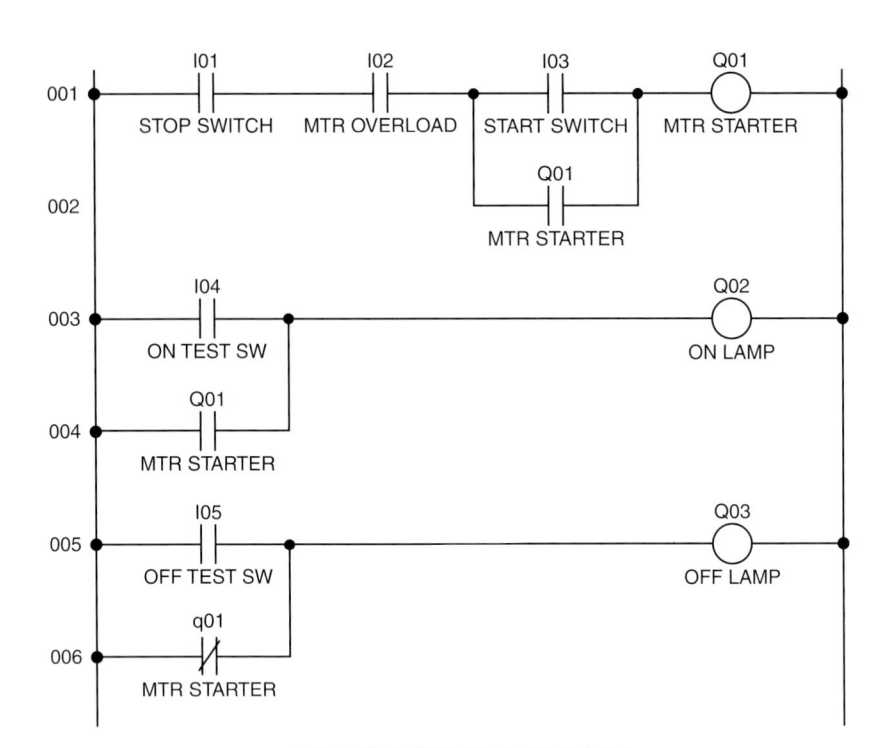

PROGRAMMING GRID/AREA

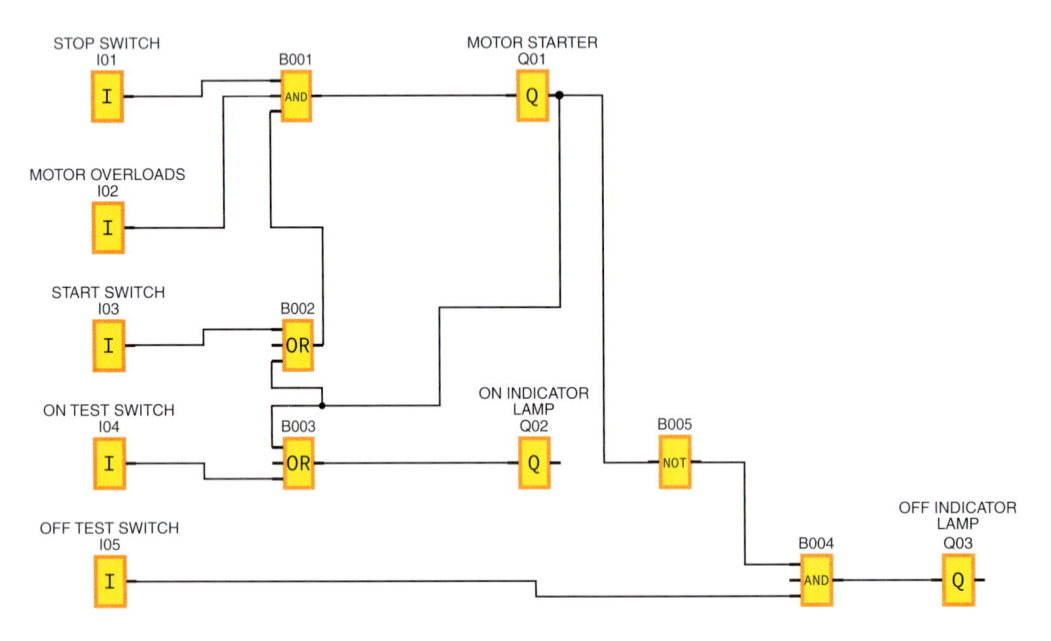

FUNCTION BLOCK DIAGRAM

5. Use the wiring diagram to wire the input devices and output components to the PLC.

6. Turn power to the PLC ON and link the com port.

7. Write the program to the SG2-12HR-D to download the program logic to the PLC.

8. Place the SG2-12HR-D in Run Mode. Activate the start pushbutton and monitor the I/Os on the LCD display.

9. Stop the program and use the Save As function to save and place the lab application in the required folder location.

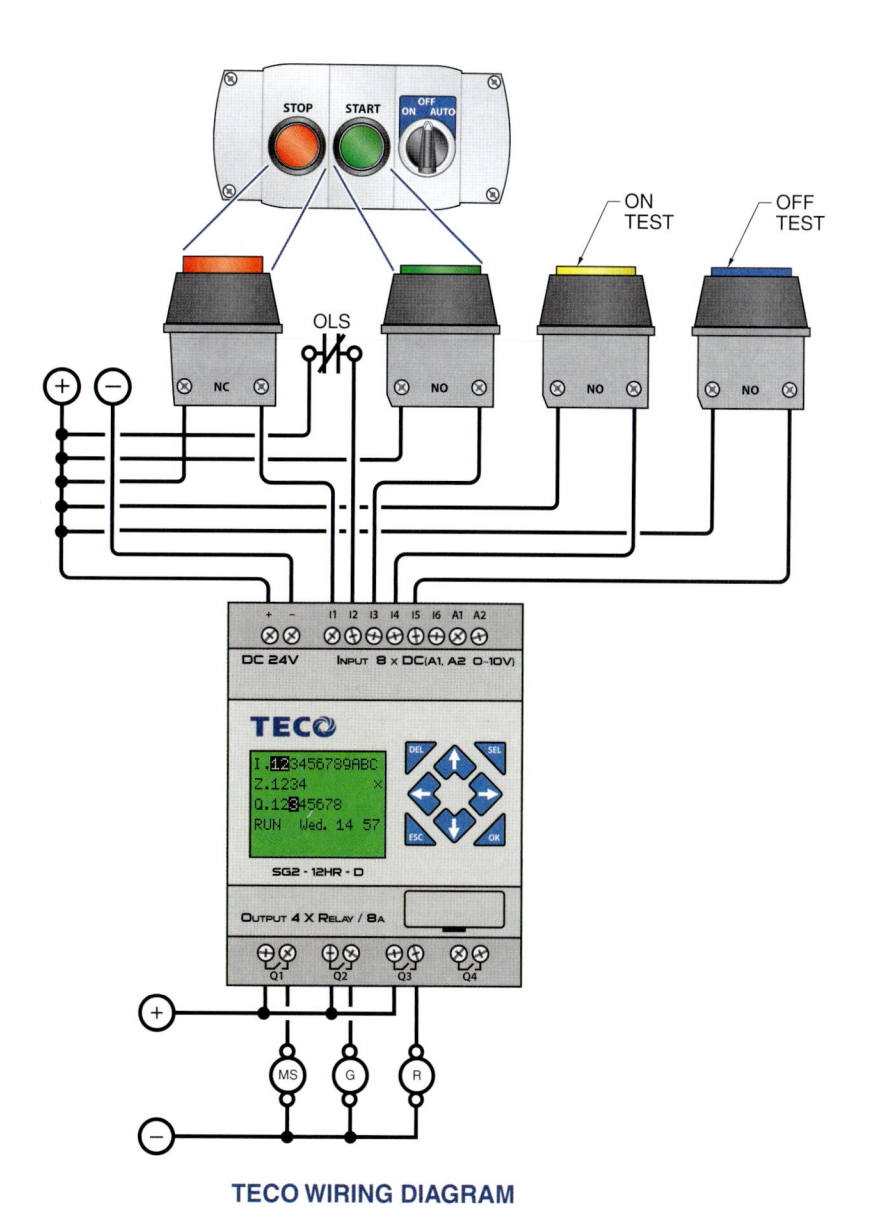

TECO WIRING DIAGRAM

Programming Combination Circuit Logic

PLC Exercise 3.3

BILL OF MATERIALS

- TECO or MicroLogix™ 1100
- Selector switch
- Two limit switches
- One light
- Motor starter/contactor
- One horn/alarm

In building systems, safeguards are in place to keep certain areas safe from hazards. For this lab, a sprinkler system protects an area from fire. When a sprinkler head opens, a solenoid releases a rollup door. The door to the protected area closes due to gravity, and an indication is provided when an alarm bell sounds and an alarm light is ON. The solenoid is de-energized when the door activates the door down limit switch. The alarm bell and alarm light can then be shut OFF.

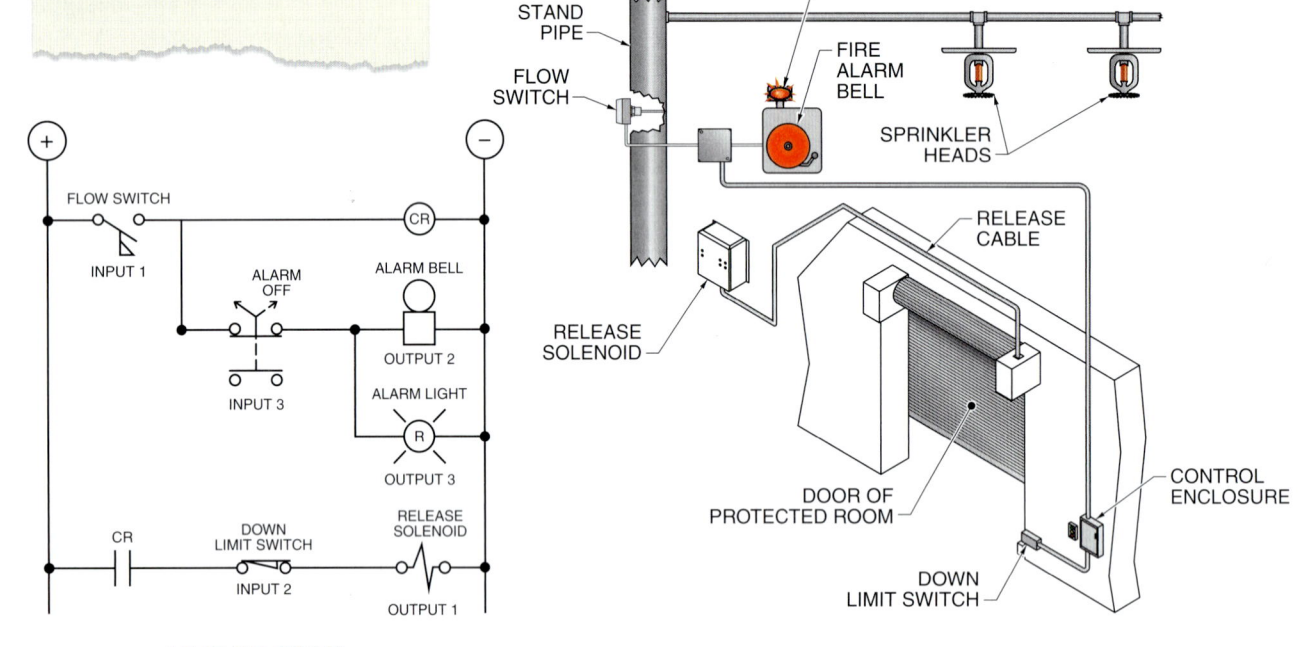

LINE DIAGRAM

TECO Procedures

1. Create a new ladder logic or FBD program.
2. Use the TECO programming diagram or FBD to add the proper contacts and description in the programming grid/area.

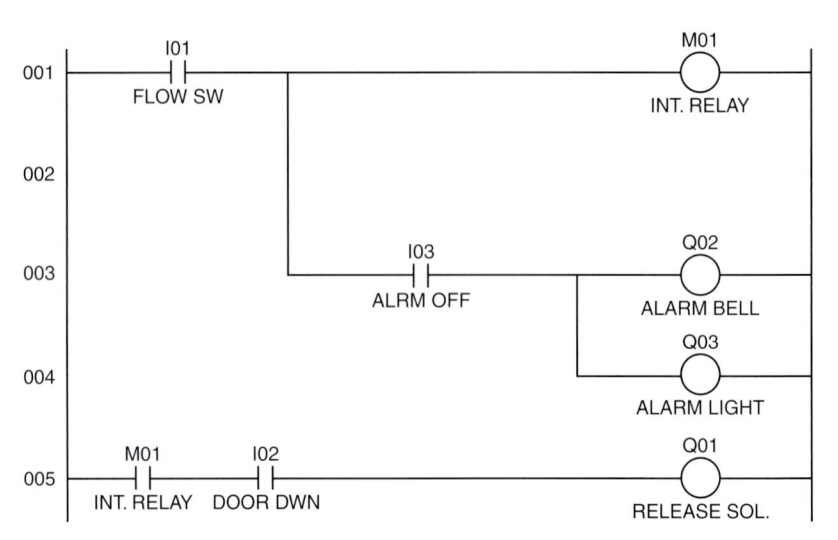

TECO PROGRAMMING DIAGRAM

Rung 001: I01 (FLOW SW) → M01 (INT. RELAY)
Rung 003: I03 (ALRM OFF) → Q02 (ALARM BELL), Q03 (ALARM LIGHT)
Rung 005: M01 (INT. RELAY), I02 (DOOR DWN) → Q01 (RELEASE SOL.)

3. Use the Edit Contact window to denote an input or output number and the Symbol button to add a description after placing the input or output in the programming grid. For an FBD, use the Comment icon to add the description.

4. Run the program in Simulation Mode to verify proper operation. Use the Input Status tool window to activate the inputs and control the output to simulate circuit operation.

5. Use the wiring diagram to wire the input devices and output components to the PLC.

6. Turn power to the PLC ON and link the com port.

7. Write the program to the SG2-12HR-D to download the program logic to the PLC.

8. Place the SG2-12HR-D in Run Mode. Activate the start pushbutton and monitor the I/Os on the LCD display.

9. Stop the program and use the Save As function to save and place the lab application in the required folder location.

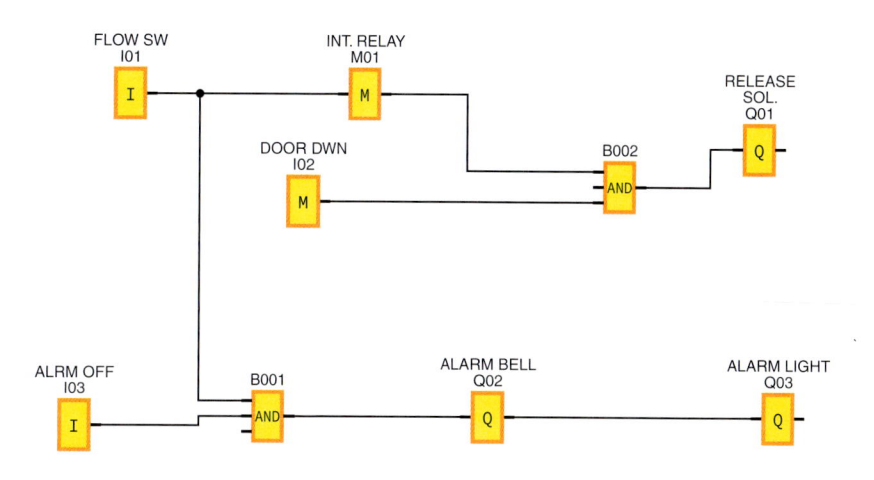

FBD PROGRAMMING AREA

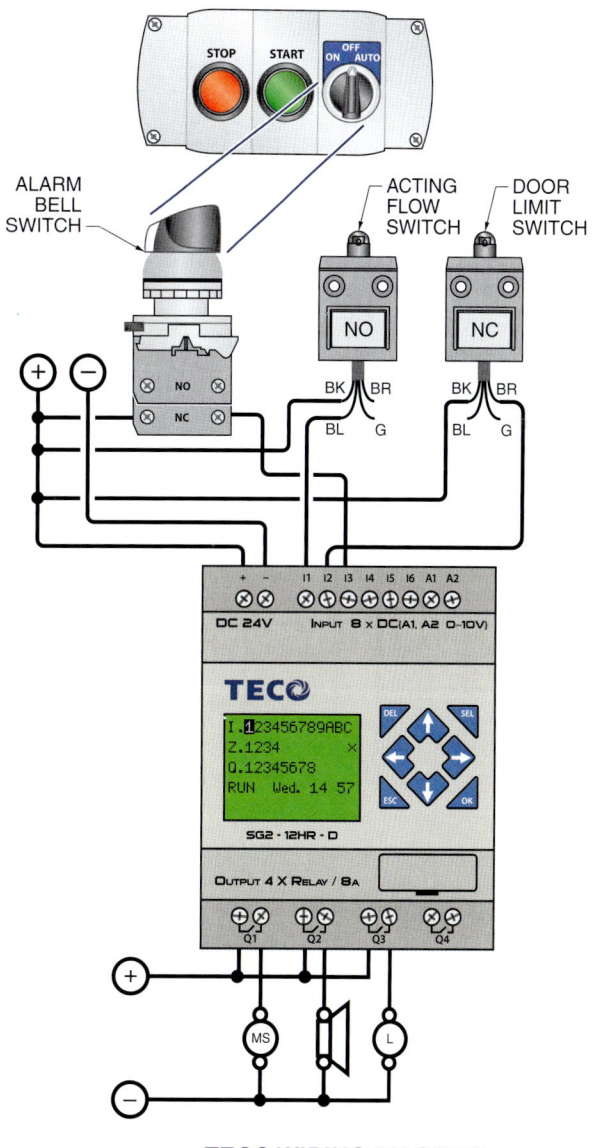

TECO WIRING DIAGRAM

Allen-Bradley® Procedures

1. Open the RSLogix™ Micro Starter Lite program and create a new programming diagram.

2. Use the AB programming diagram to add the proper address and description in the LAD 2 programming window.

3. Verify that the programming diagram has no errors and save the project.

4. Use the wiring diagram to wire the input devices and output components to the PLC. Once all connections have been made, turn power to the PLC on.

5. Use RSLinx™ to establish communication between the PC and the Micrologix™ 1100.

6. Download the program logic to the PLC.

7. Place the PLC in Run Mode.

8. Activate input(s) and monitor them to determine if the output(s) are operating correctly.

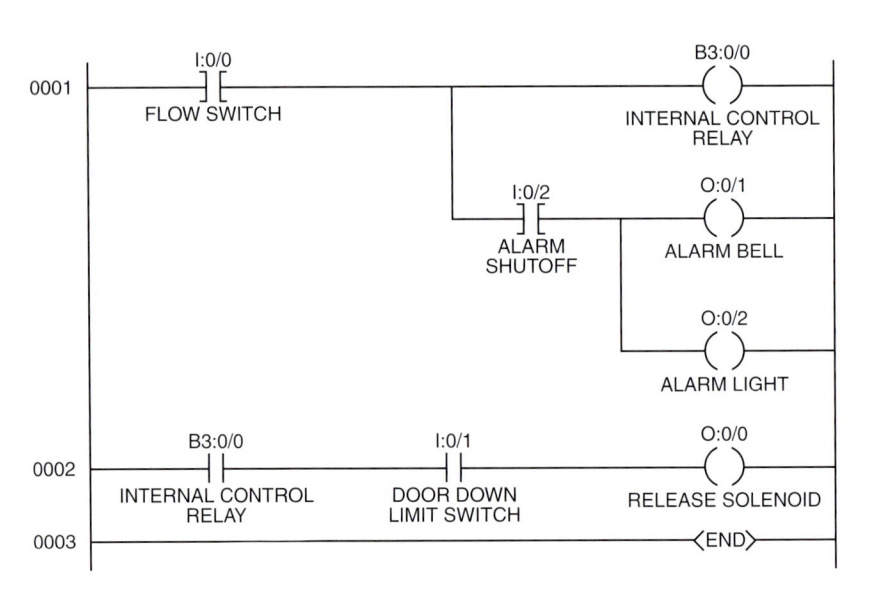

Rung 0001:
- I:0/0 FLOW SWITCH — B3:0/0 INTERNAL CONTROL RELAY
- I:0/2 ALARM SHUTOFF — O:0/1 ALARM BELL
- O:0/2 ALARM LIGHT

Rung 0002:
- B3:0/0 INTERNAL CONTROL RELAY — I:0/1 DOOR DOWN LIMIT SWITCH — O:0/0 RELEASE SOLENOID

Rung 0003: ⟨END⟩

AB PROGRAMMING DIAGRAM

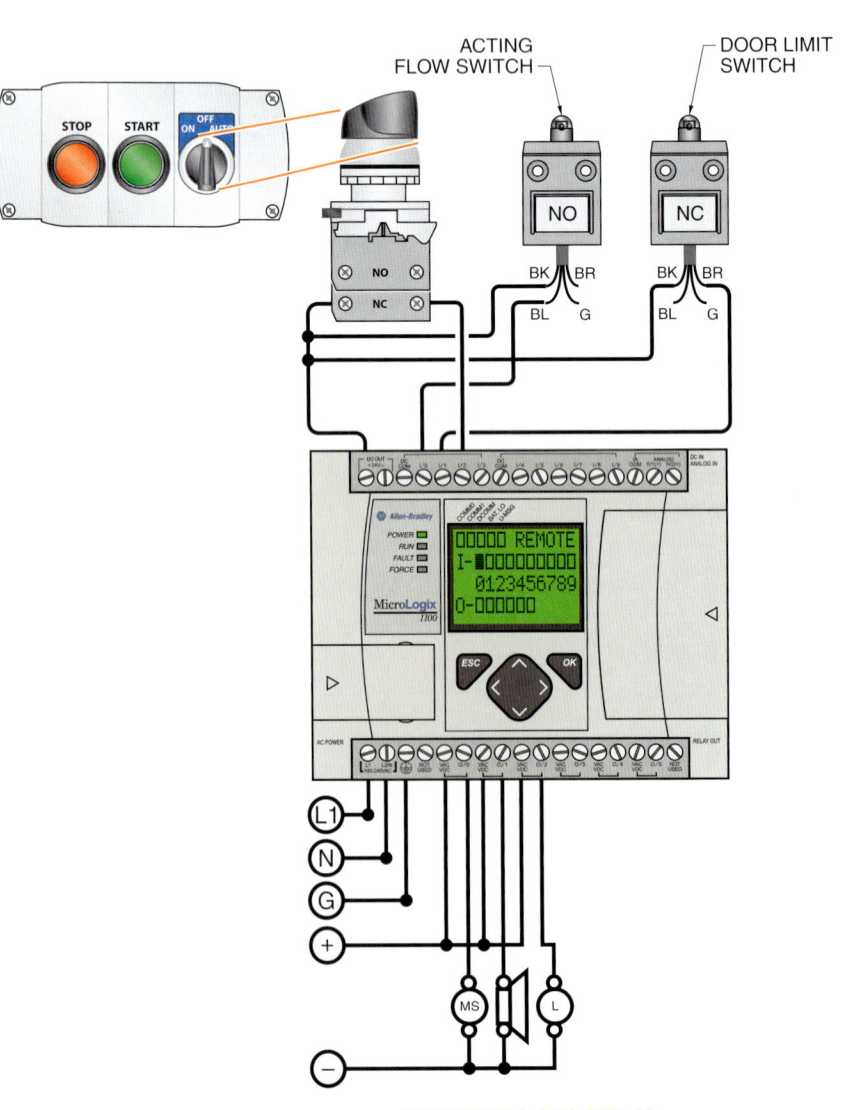

AB WIRING DIAGRAM

SIEMENS Programming and Wiring

Use the S7-1200 Easy Book to program and wire the equivalent TECO and Allen-Bradley exercises. Reference page 70 for Siemens STEP 7 programming symbols.

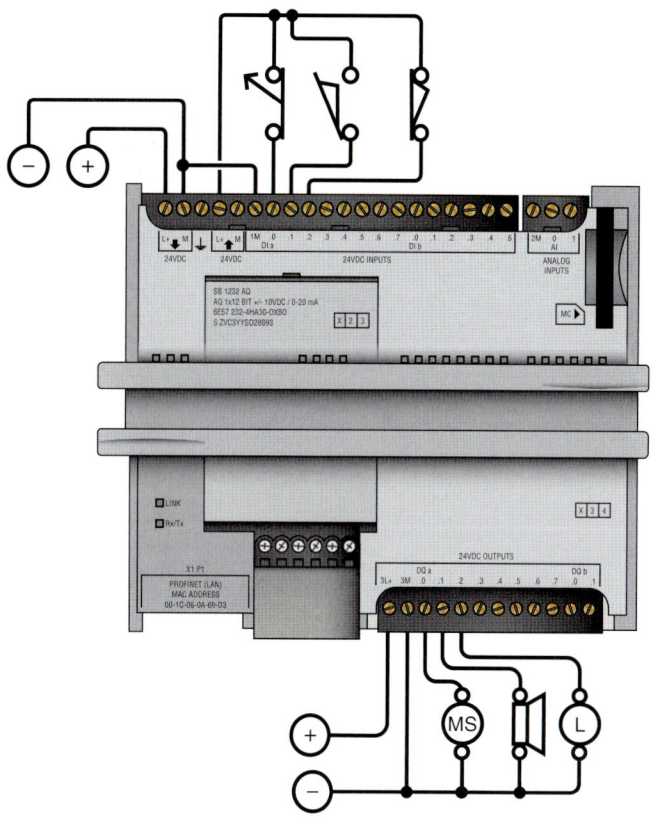

SIEMENS WIRING DIAGRAM

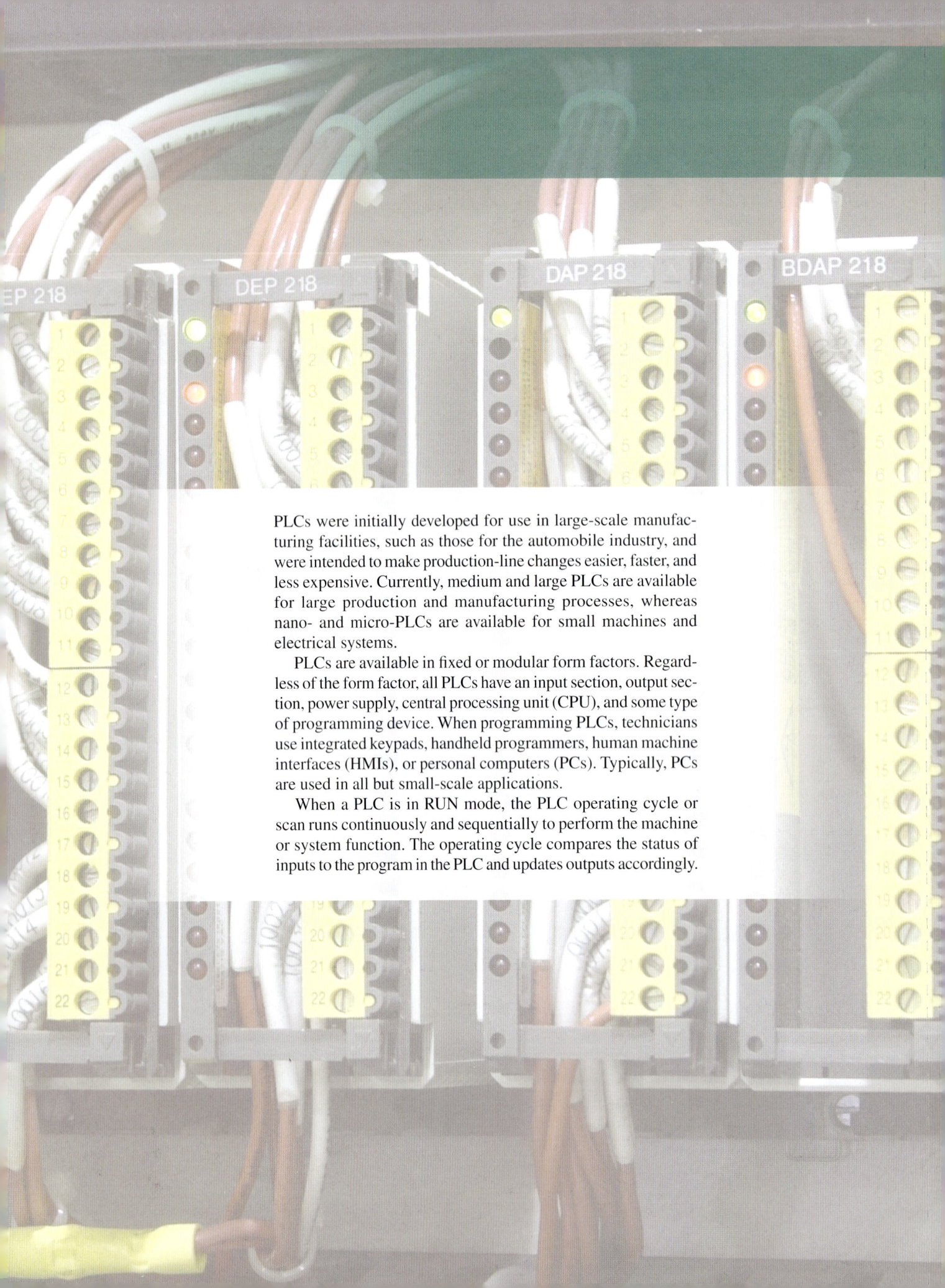

PLCs were initially developed for use in large-scale manufacturing facilities, such as those for the automobile industry, and were intended to make production-line changes easier, faster, and less expensive. Currently, medium and large PLCs are available for large production and manufacturing processes, whereas nano- and micro-PLCs are available for small machines and electrical systems.

PLCs are available in fixed or modular form factors. Regardless of the form factor, all PLCs have an input section, output section, power supply, central processing unit (CPU), and some type of programming device. When programming PLCs, technicians use integrated keypads, handheld programmers, human machine interfaces (HMIs), or personal computers (PCs). Typically, PCs are used in all but small-scale applications.

When a PLC is in RUN mode, the PLC operating cycle or scan runs continuously and sequentially to perform the machine or system function. The operating cycle compares the status of inputs to the program in the PLC and updates outputs accordingly.

PLC Hardware, Memory, and Operating Cycle

4

Objectives:

- Describe the functions and different variations of a PLC's five sections: the input section, the output section, the power supply, the CPU, and programming devices.
- List the typical number of I/Os for the various size classifications of PLCs.
- Explain the different types of PLC memory.
- Describe the operating cycle of a PLC and how it relates to a PLC's memory.

Learner Resources
atplearningresources.com/quicklinks
Access Code: 475203

PLC DEVELOPMENT

Programmable logic controllers (PLCs) were developed in the late 1960s for use in the automobile industry. When automobile model years changed, time-consuming and expensive rewiring of production lines and equipment was required. PLCs were designed to take the place of hardwired relays, timers, counters, and other control devices. Changes that required several days and an entire maintenance department rewiring control cabinets and equipment can now be made in only a few minutes by one technician reprogramming a PLC.

A PLC is similar to a real-time computer and is designed for use in industrial environments that may be hostile. PLCs are designed to withstand electrical noise, vibration, and the wide range of temperatures found in industrial facilities. PLCs require less space and provide more control functionality than the hardwired components (relays and timers) that PLCs replace. **See Figure 4-1.**

Early PLCs had limited functionality and were expensive. However, numerous improvements and innovations have been made, and PLCs are now a cost-effective alternative for an application that requires at least three or four control relays and one or two timers. PLCs are available in a variety of sizes and types from various manufacturers.

PLCs are used in pump applications to control the time of operation by using level sensors that monitor fluid levels.

157

158 PROGRAMMABLE LOGIC CONTROLLERS: PRINCIPLES AND APPLICATIONS

PLC Space Advantage

HARDWIRED ENCLOSURE

PLC-WIRED ENCLOSURE
Cleaver Brooks

Figure 4-1. PLCs require less space and are easier to modify than hardwired control components.

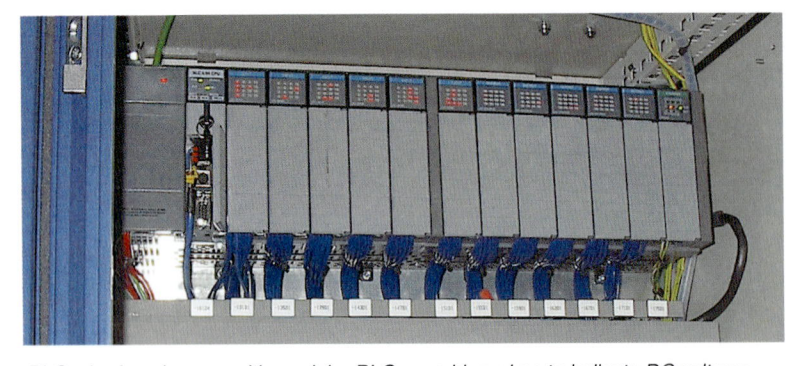

PLC wired enclosures with modular PLCs use blue wires to indicate DC voltage.

PLC SECTIONS

All PLCs have five common sections regardless of the manufacturer, model, or size. The sections common to all PLCs are the input section, output section, power supply, central processing unit (CPU), and programming device. The sections work together to provide a PLC with usable functionality. The sections are found in various shapes, sizes, or form factor (fixed or modular PLC). **See Figure 4-2.** Some PLCs have all five sections integrated into a single package (fixed). Other PLCs are designed with modular components to allow for greater flexibility.

PLC Input Sections

The *input section* of a PLC is a grouping of terminal screws that receives signals from input devices and sends the signals to the CPU. Pushbuttons, selector switches, limit switches, and proximity switches are common input devices. Each input device is connected to a specific screw terminal that corresponds to a specific location (address) in the memory of the CPU. Typically, PLC input sections have light-emitting diodes (LEDs) or status indicators on an LCD display to indicate when an input device is closed (sending signal) and voltage is present at the input terminal. **See Figure 4-3.**

PLC Input Section Internal Circuitry. Input devices typically operate at voltages higher than the voltage used by the CPU (5 V). Optical isolation is used by PLCs to convert high-voltage input signals to a voltage level the CPU can use. Inside AC-powered PLCs, AC input signals are converted to DC by a rectifier, and the DC signal turns on an LED. **See Figure 4-4.** The light from the LED is detected by a phototransistor, and the phototransistor sends a signal to the CPU. In most cases, the phototransistor and the CPU operate at the same voltage level.

PLC TIPS

PLC CPUs have capacitors that store a nonlethal charge for extended periods of time. PLCs must not be mounted in a position where technicians could be injured from a startling reaction.

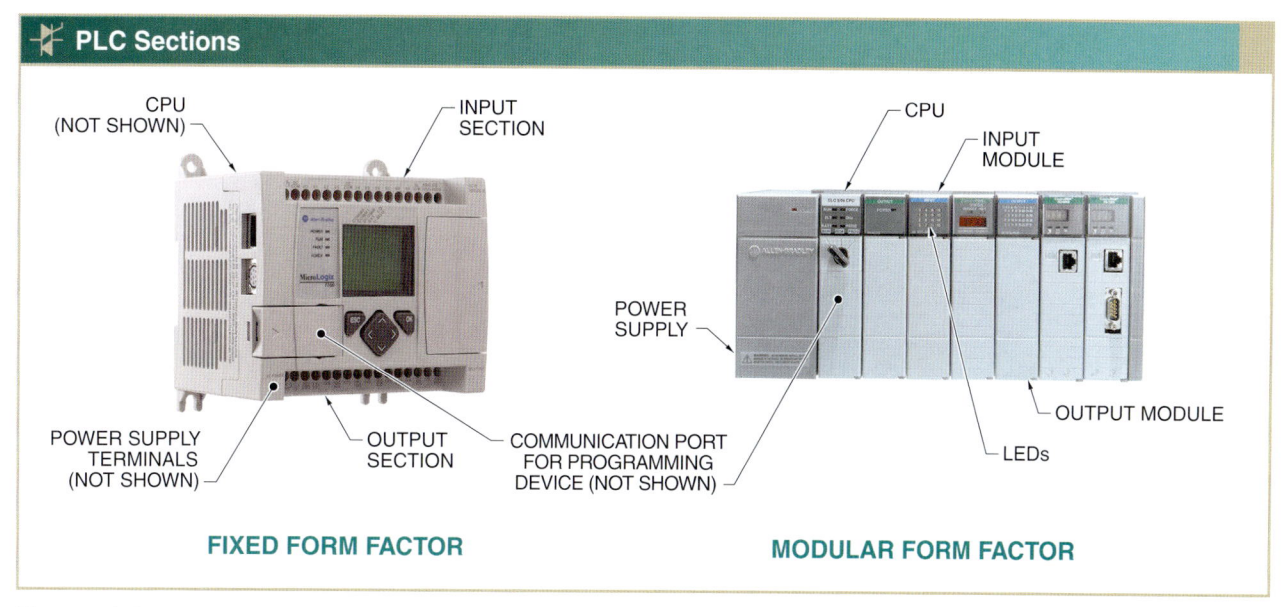

Chapter 4—PLC Hardware, Memory, and Operating Cycle **159**

PLC Sections

FIXED FORM FACTOR
- CPU (NOT SHOWN)
- INPUT SECTION
- POWER SUPPLY TERMINALS (NOT SHOWN)
- OUTPUT SECTION
- COMMUNICATION PORT FOR PROGRAMMING DEVICE (NOT SHOWN)

MODULAR FORM FACTOR
- CPU
- INPUT MODULE
- POWER SUPPLY
- OUTPUT MODULE
- LEDs

Figure 4-2. Regardless of the form factor, all PLCs have five common sections. These sections are the input section, output section, power supply, central processing unit (CPU), and programming device or port.

PLC Input Sections

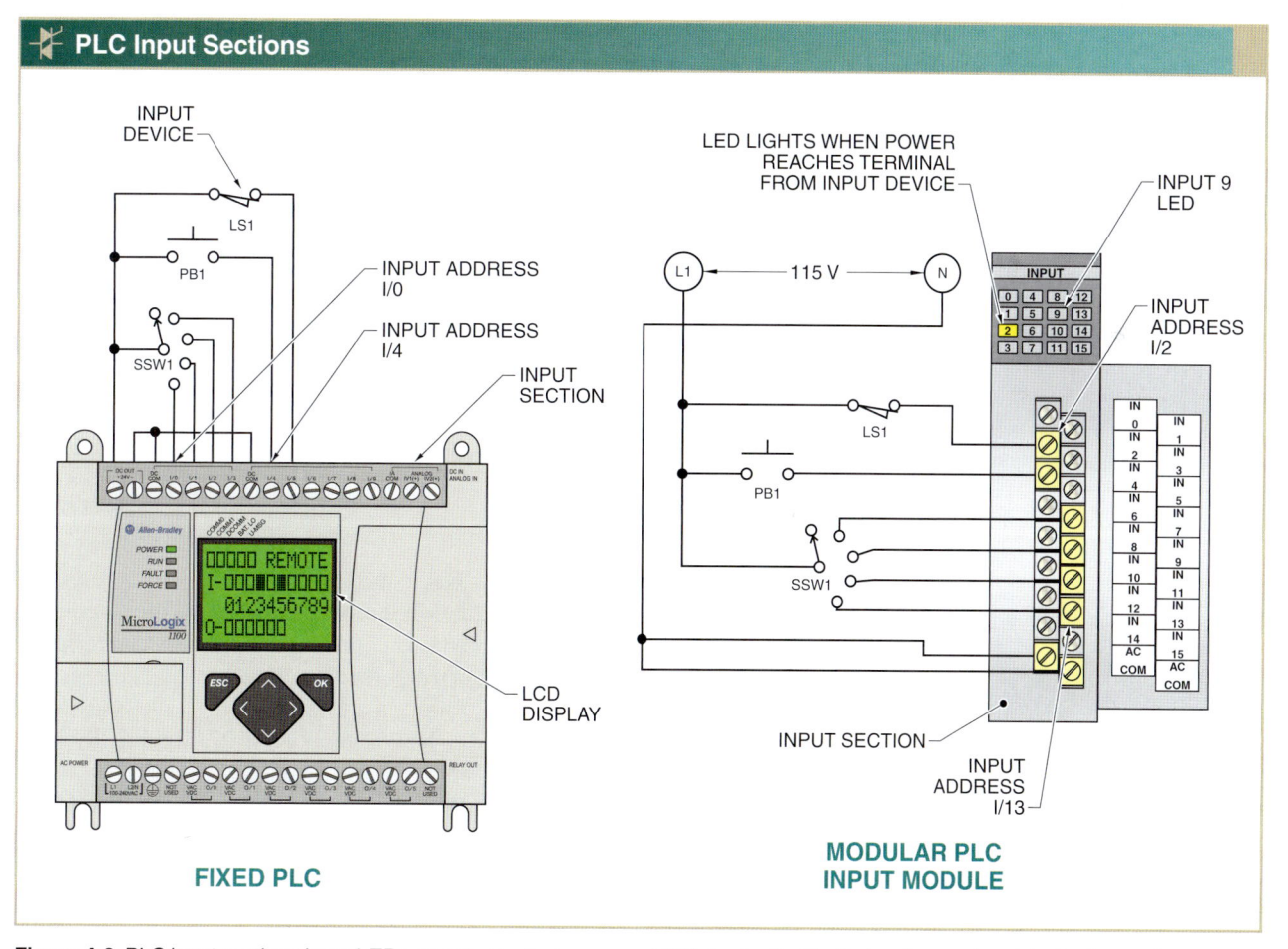

FIXED PLC
- INPUT DEVICE
- LS1
- PB1
- SSW1
- INPUT ADDRESS I/0
- INPUT ADDRESS I/4
- INPUT SECTION
- LCD DISPLAY

MODULAR PLC INPUT MODULE
- LED LIGHTS WHEN POWER REACHES TERMINAL FROM INPUT DEVICE
- INPUT 9 LED
- INPUT ADDRESS I/2
- LS1
- PB1
- SSW1
- INPUT SECTION
- INPUT ADDRESS I/13

Figure 4-3. PLC input sections have LEDs or status indicators on an LCD display that indicate when an input device is sending a signal to a specific address of the input section.

PLC Internal Circuitry

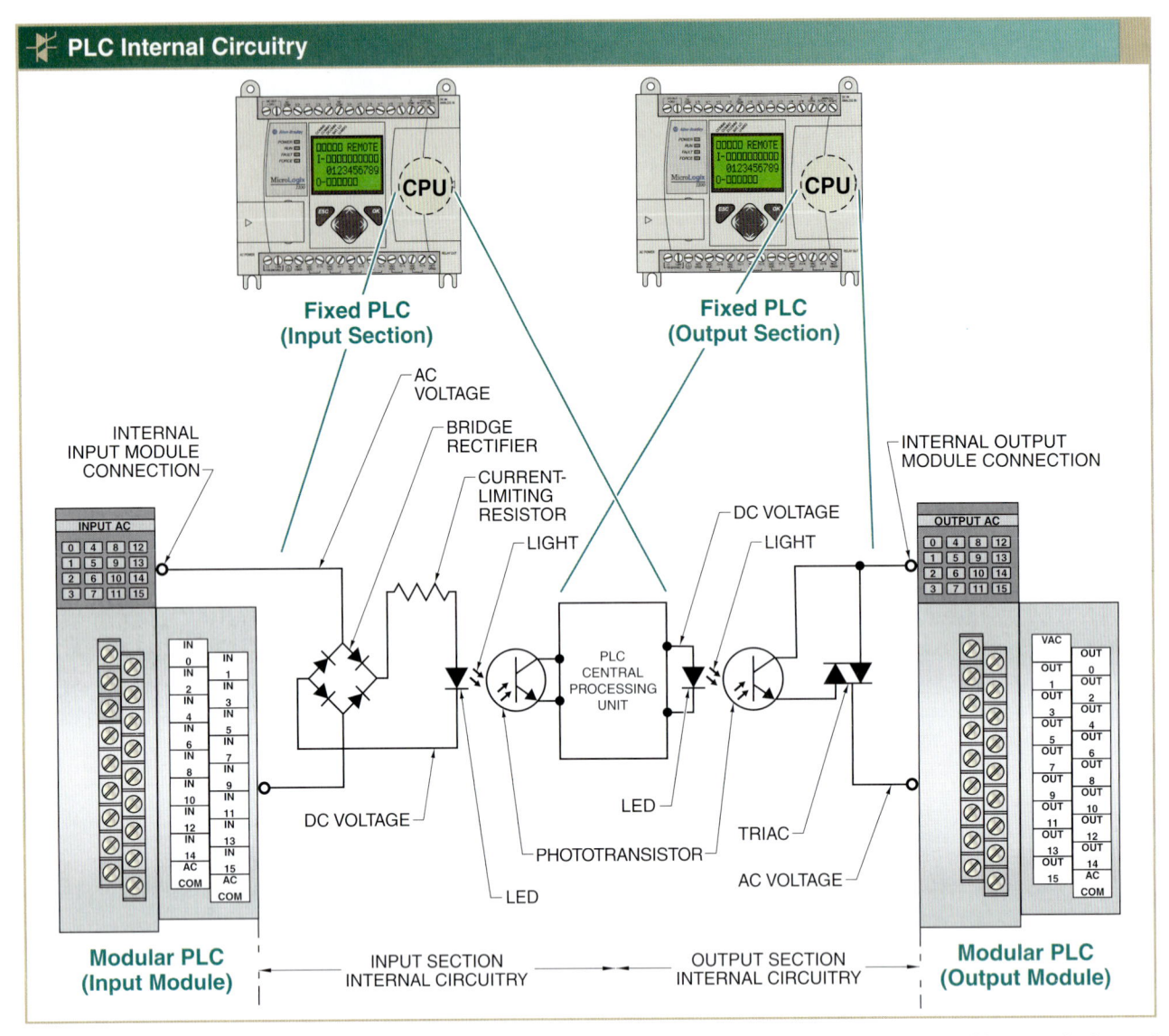

Figure 4-4. The internal circuitry of a PLC input section converts signals to the voltage level required by the CPU, and the internal circuitry of an output section converts signals from the CPU to the voltage level required by output devices.

PLC Input Section Wiring. Conductors are wired from input devices and terminate at PLC input terminals. The input terminals have self-lifting pressure plates or screw-cage clamps to attach the conductors.

Often the insulation color of the conductor indicates the type of signal, either VDC or VAC. A common color scheme in an industrial environment would be blue for DC wiring and red for AC wiring.

Conductors wired from input devices must meet specific requirements. The requirements involve wire types, insulation temperature ratings, the number of wires per terminal, wire size, and terminal wiring torque. These requirements are found in a manufacturer's PLC manual.

Nano- and micro-PLCs typically have integral terminals, which are fixed to the PLC. However, rack-mounted PLCs have removable terminal blocks. These blocks make it easier to terminate conductors and allow a defective PLC to be replaced without disconnecting and reconnecting all the field wiring. **See Figure 4-5.**

Conductor Requirements

Wire Type	Insulation Temperature Rating	Wire Size (2 wire maximum per terminal screw)	
		1 wire per terminal	2 wires per terminal
Solid	Cu-90°C (194°F)	12 to 20 AWG	16 to 20 AWG
Stranded	Cu-90°C (194°F)	14 to 20 AWG	18 to 20 AWG

Wiring torque = 0.56 Nm (5.0 in-lb) rated

FIXED INPUT TERMINALS

REMOVABLE TERMINAL BLOCKS

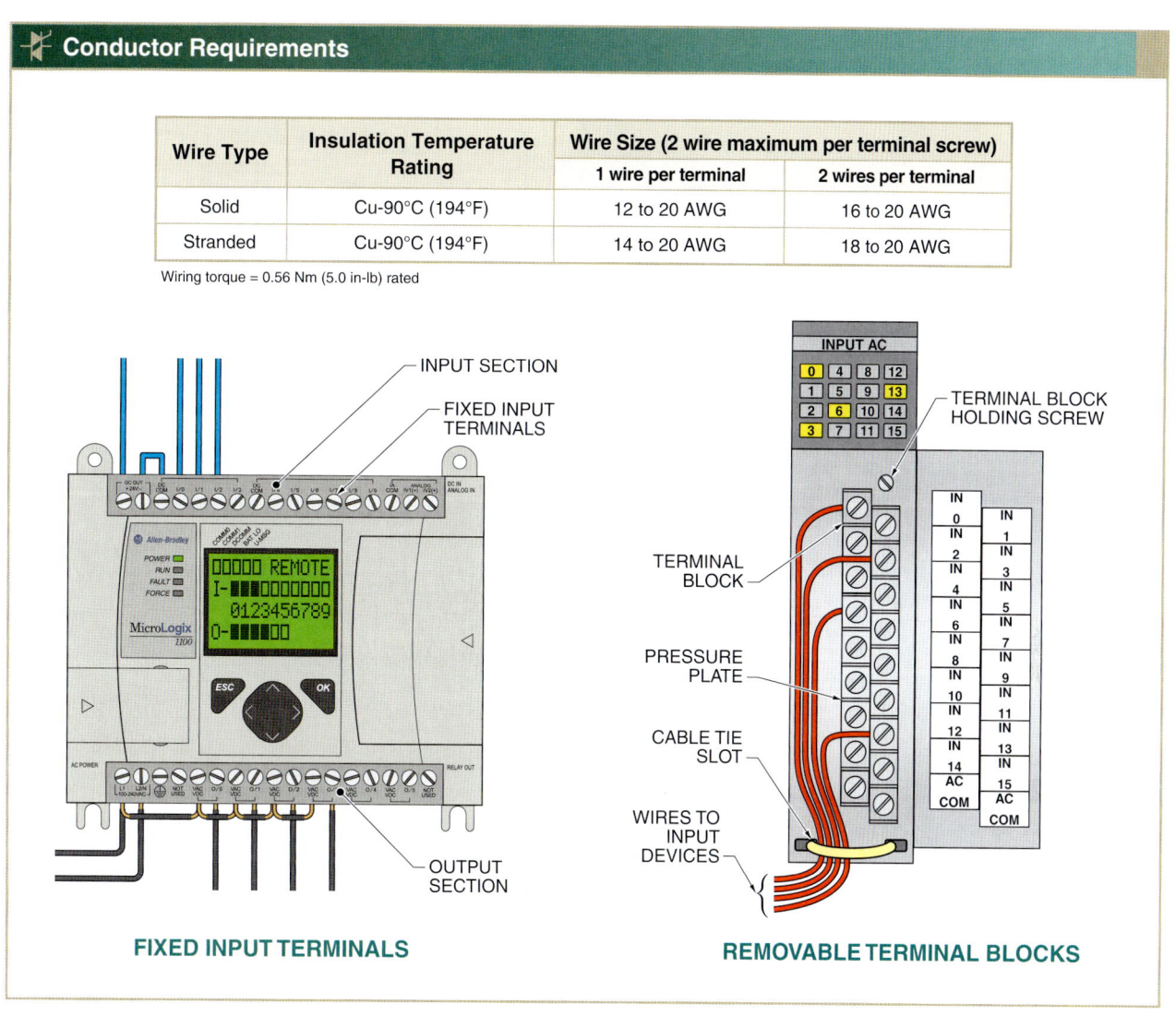

Figure 4-5. When used as wires from input devices to PLC input terminals, conductors are commonly blue for DC and red for AC.

PLC Output Sections

The *output section* of a PLC is the section that sends signals from the CPU to components that perform work. Components such as pilot lights, control relays, motor starters, and solenoids are typical components controlled by PLCs. Each output component is connected to a specific screw terminal of the output section that corresponds to a specific location (address) in the memory of the CPU. Typically, each PLC output screw terminal has an LED or LCD display to indicate when voltage to energize an output component is present at the terminal. **See Figure 4-6.**

PLC Output Section Internal Circuitry

Output components typically operate at a voltage that is higher than the operating voltage of the CPU (5 V). Optical isolation is also used to convert low-level voltage signals from the CPU to a voltage that an output component can use. Inside the PLC, low-level voltage signals are used to turn on an LED. The light from the LED is detected by a phototransistor, which turns on a triac. Triacs are used to actually turn output components ON and OFF. The CPU and the LED of the phototransistor operate at the same voltage level.

CODE CONNECT

Section 300.7(A) of the NEC® covers sealing a conduit when it transitions between areas with different temperatures. When air travels from a warm area to a cold area in a conduit, condensation forms in the cold area. Section 300.7(A) requires the raceway to be sealed to prevent warm air from traveling to the cold area. This is a common issue with large commercial refrigeration units. Explosion-proof seals are not required.

PLC Output Sections

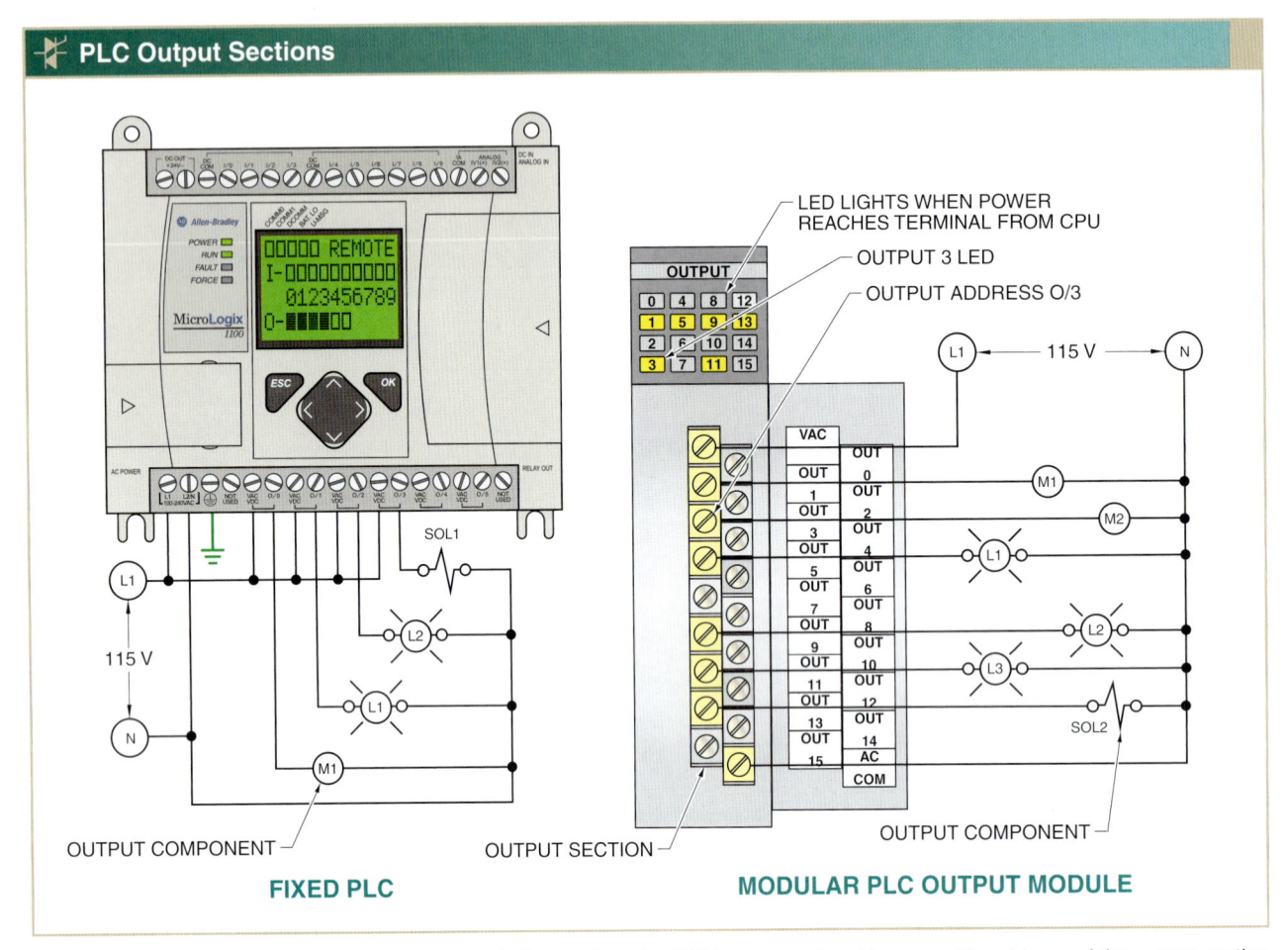

Figure 4-6. PLC output sections have LEDs that indicate when the CPU sends a signal to a specific address of the output section to turn ON an output component.

PLC TIPS

Current requirements for each input device and output component should be listed on a card with a PLC.

PLC Power Supplies

A *power supply* is a section of a PLC that converts the voltage of a power source to the low-level voltage (typically 5 VDC) required by the CPU and the related electronics inside the PLC. **See Figure 4-7.** PLC power supplies are available for any voltage source encountered. Some power supplies can be used with more than one voltage. A jumper or dip switch is used to select the voltage level, 115 VAC or 230 VAC, or the power supply is designed to accept a range of voltages, such as 85 VAC to 132 VAC.

PLC power supply sections (modules) typically use LEDs to indicate when voltage is present. PLC power supplies are also designed to withstand momentary losses of power without affecting the operation of the PLC. *Hold-up time* is the length of time a PLC can tolerate a power loss without affecting operation. Typical hold-up time for PLCs varies from 10 ms (milliseconds) to 3 sec.

⚠ WARNING

Set any dipswitches or input voltage jumpers before applying power to a power supply. Power supply terminal screws are rated for 8.8 in-lb or 7 in-lb of torque depending on the power supply model. Hazardous voltage is present on exposed pins of chassis when power is applied. An electrical arc can occur or anyone making contact with a pin can receive an injury.

PLC Power Supplies

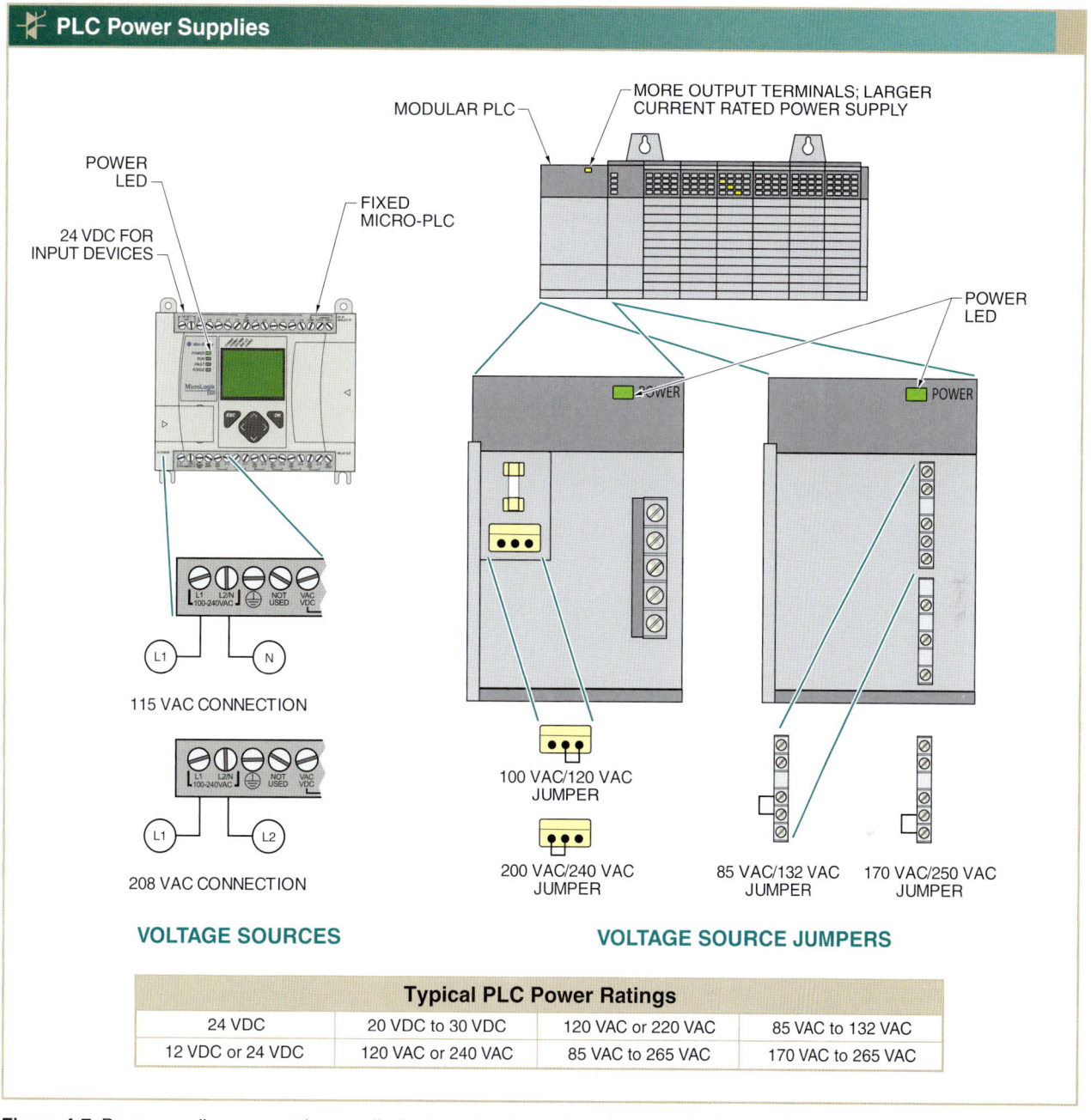

VOLTAGE SOURCES
- 115 VAC CONNECTION
- 208 VAC CONNECTION

VOLTAGE SOURCE JUMPERS
- 100 VAC/120 VAC JUMPER
- 200 VAC/240 VAC JUMPER
- 85 VAC/132 VAC JUMPER
- 170 VAC/250 VAC JUMPER

Typical PLC Power Ratings			
24 VDC	20 VDC to 30 VDC	120 VAC or 220 VAC	85 VAC to 132 VAC
12 VDC or 24 VDC	120 VAC or 240 VAC	85 VAC to 265 VAC	170 VAC to 265 VAC

Figure 4-7. Power supplies convert the supplied voltage to a low voltage level required by the CPU and other electronic circuits.

In addition to various voltage ratings, PLC power supplies have various current ratings. As the number of input devices and output components increase, the current required from the power supply increases. Many PLC power supplies provide 24 VDC for input signals. The voltage and current ratings of the power supply must not be exceeded when installing a PLC.

PLC Central Processing Units

A *central processing unit (CPU)* is a section of a PLC that houses the processor (brain) of the PLC. A user-developed control program (PLC program) resides in the CPU. The CPU receives signals from input devices, such as limit switches or proximity sensors, compares the signals with information in the PLC program, and sends signals through the

PLC TIPS

Siemens S7-1200 PLCs have power supplies with power ratings of 93 VAC to 132 VAC and 187 VAC to 264 VAC.

output section to turn output components ON and OFF. **See Figure 4-8.**

CPUs use a variety of LEDs to indicate processor status. CPUs are rated according to the size of the program memory and scan time. *Program memory* is the maximum number of programming instructions a CPU can hold. Program memory is measured in increments of 1000 (1k) instructions. *Scan time* is the length of time a processor takes to execute a program once. Scan time is measured in milliseconds per 1k (ms/k) of ladder logic programming. Scan time increases as the number of instructions increase.

PLC Programming Devices

A *programming device* is a device with a keypad that is used to write and enter a user-developed control program into a PLC. Programming devices can also be used to monitor and troubleshoot the equipment or process that is being controlled. Programming devices fall into four categories: integrated programming devices, handheld programming devices, human machine interfaces, and personal computers (PCs). Some PLCs can work with multiple programming devices. There are advantages and disadvantages to each type of programming device.

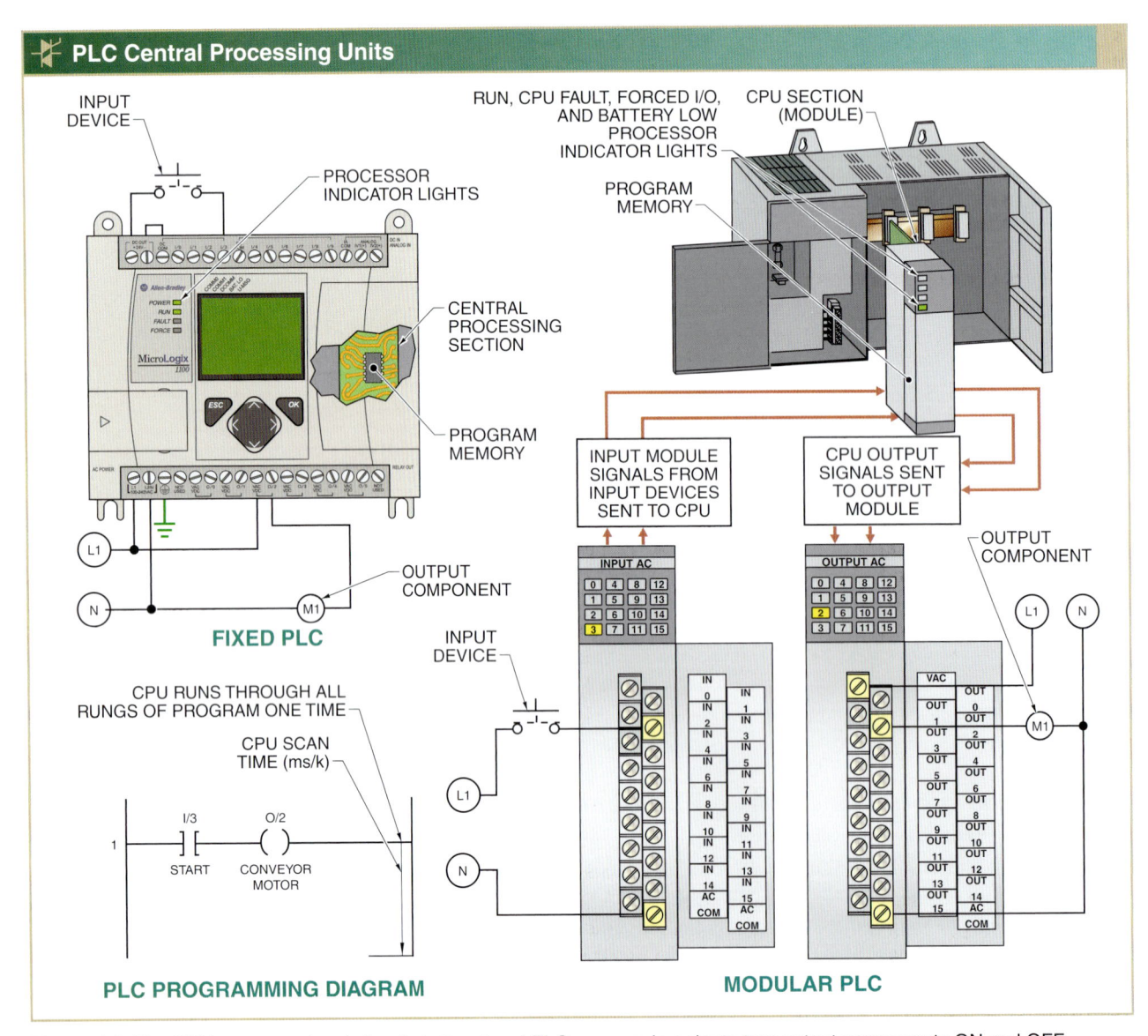

Figure 4-8. The CPU compares input signals to the stored PLC program in order to turn output components ON and OFF.

Integrated Programming Devices. An *integrated programming device* is a device that consists of a small liquid crystal display (LCD) and a small keypad or set of buttons that are part of a PLC. Integrated programming devices are typically found on nano- and pico-sized PLCs. A PLC program is entered into the PLC using the keypad and viewed on the LCD display.

Integrated programming devices are a convenient way to access a PLC program without additional hardware. However, small LCD displays provide only limited viewing, and making any programming changes with an integrated keypad is typically time-consuming. **See Figure 4-9.**

Handheld Programming Devices. A *handheld programming device* is a separate keypad device that is not an integral part of a PLC. A handheld programming device is about the same size as a smart phone but with a larger display and keypad than an integrated programming device. Handheld programming devices connect to a PLC with a cable and connector.

As with integrated programming devices, handheld programming devices are typically used with smaller PLCs. Handheld programming devices are used to enter, copy, display, store, and transfer PLC programs. The ability to copy, store, and transfer a PLC program allows a

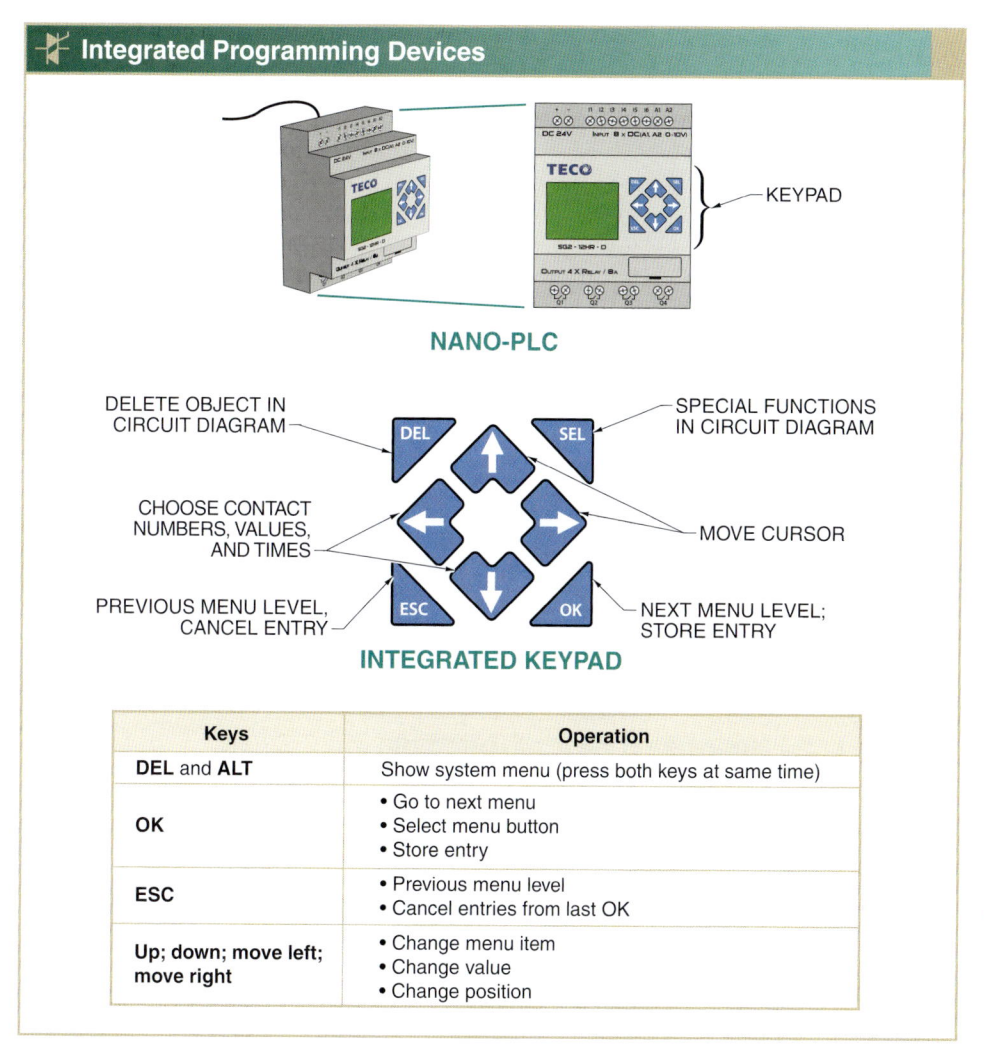

Integrated Programming Devices

NANO-PLC — KEYPAD

INTEGRATED KEYPAD
- DEL — DELETE OBJECT IN CIRCUIT DIAGRAM
- SEL — SPECIAL FUNCTIONS IN CIRCUIT DIAGRAM
- Arrows — CHOOSE CONTACT NUMBERS, VALUES, AND TIMES / MOVE CURSOR
- ESC — PREVIOUS MENU LEVEL; CANCEL ENTRY
- OK — NEXT MENU LEVEL; STORE ENTRY

Keys	Operation
DEL and ALT	Show system menu (press both keys at same time)
OK	• Go to next menu • Select menu button • Store entry
ESC	• Previous menu level • Cancel entries from last OK
Up; down; move left; move right	• Change menu item • Change value • Change position

Figure 4-9. Integrated programming devices provide a convenient way to access a PLC program without additional hardware.

CODE CONNECT

The NEC® defines a nonlinear load as any load where the wave shape of the load current is not proportional to the wave shape of the load voltage. Nonlinear loads contain DC power supplies that draw current in pulses at the peak of the voltage sine wave. Nonlinear loads include compact fluorescent lights, LED lights, electronic equipment, and variable-frequency drives.

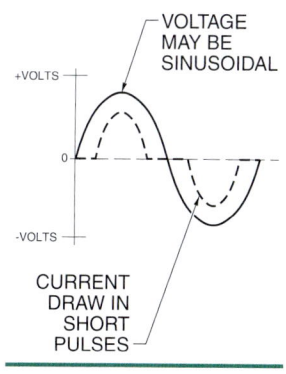

VOLTAGE MAY BE SINUSOIDAL
CURRENT DRAW IN SHORT PULSES

handheld programming device to be used with multiple PLCs having the same type of programming. Although the display and keypad are larger than that of an integrated programming device, viewing and editing are still limited with handheld programming devices. **See Figure 4-10.**

Human Machine Interfaces. A *human machine interface (HMI)* is a color or monochrome display panel with a keypad and buttons that shows the status of a process or application in real time. Communication cables are used to connect HMIs to PLCs. HMIs are also referred to as graphic terminals or operator interfaces.

Typically, HMIs consist of a display panel, which may or may not be a touchscreen, that uses text, graphics, or a combination of the two to represent a process or application. The keypad and buttons provide access to different screens and/or control functions involved with the process shown. All of these features allow an operator to monitor and control the process in real time.

HMIs are designed in a variety of sizes depending on the PLC type and the size of the process or application. They are even available for demanding environments, such as semiconductor clean rooms or hazardous locations described per the NEC®.

Handheld Programming Devices

MICRO-PLC

HANDHELD PROGRAMMING DEVICE

NOTE: The present-day use of handheld programming devices is rare.

DIAGNOSTIC/TROUBLESHOOTING KEYS

EDITING KEYS

INSTRUCTION KEYS

NAVIGATION KEYS

EDITING KEY

EDITING KEY

HANDHELD PROGRAMMING DEVICE KEYPAD

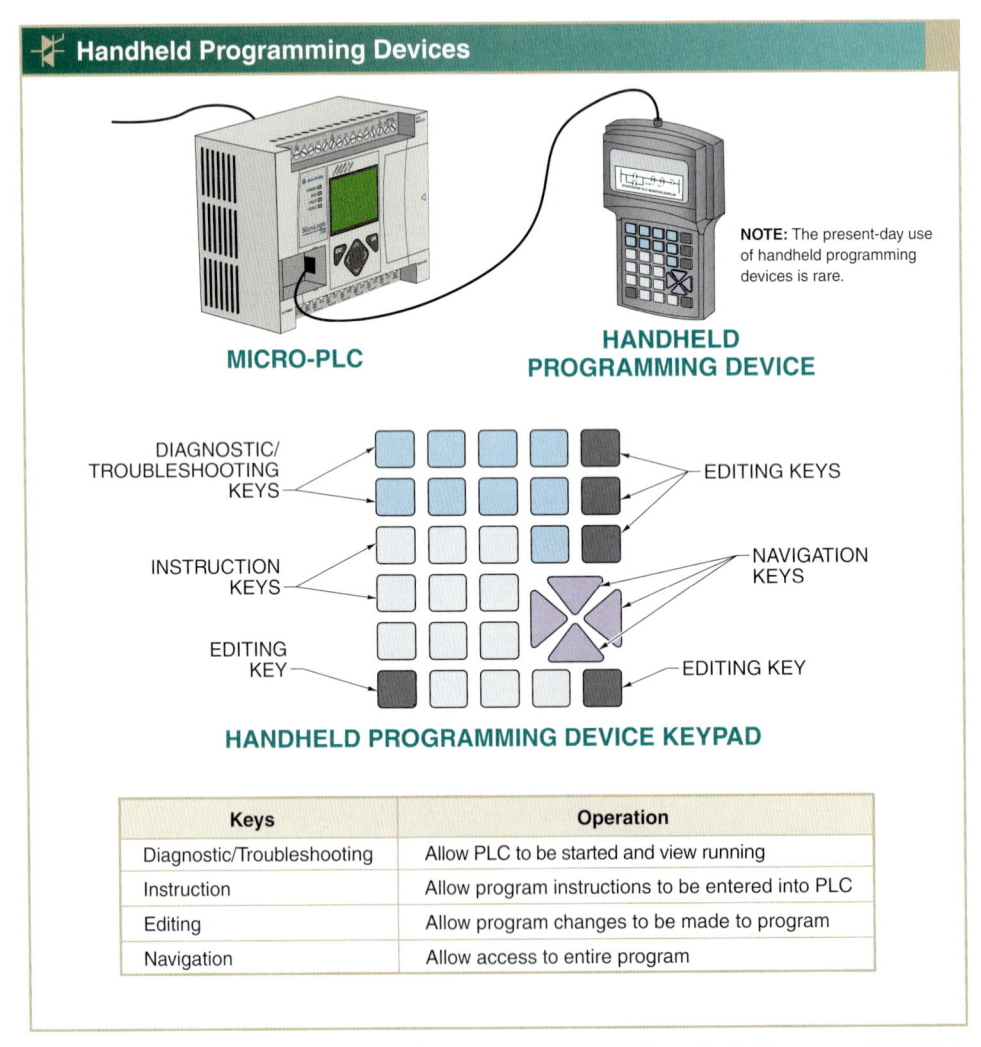

Keys	Operation
Diagnostic/Troubleshooting	Allow PLC to be started and view running
Instruction	Allow program instructions to be entered into PLC
Editing	Allow program changes to be made to program
Navigation	Allow access to entire program

Figure 4-10. Handheld programming devices can copy, store, and transfer PLC programs to multiple PLCs that have the same type of programming.

Although PLCs cannot be directly programmed from HMIs, the parameters of certain instructions can be adjusted from one, such as timer and counter preset values. For this, an HMI may require a separate programming software package. **See Figure 4-11.**

Personal Computers. Personal computers (PCs) are the most common programming devices used with PLCs. Desktop and laptop PCs are used to program PLCs through an interface cable and connectors. All software and interface cables are specific to a PLC family and manufacturer and cannot be used on other PLC families of the same manufacturer or with a PLC of a different manufacturer. PCs are used as programming devices on all sizes of PLCs. **See Figure 4-12.**

In some applications, personal computers (PCs) are used to troubleshoot PLCs that control electric motor drives.

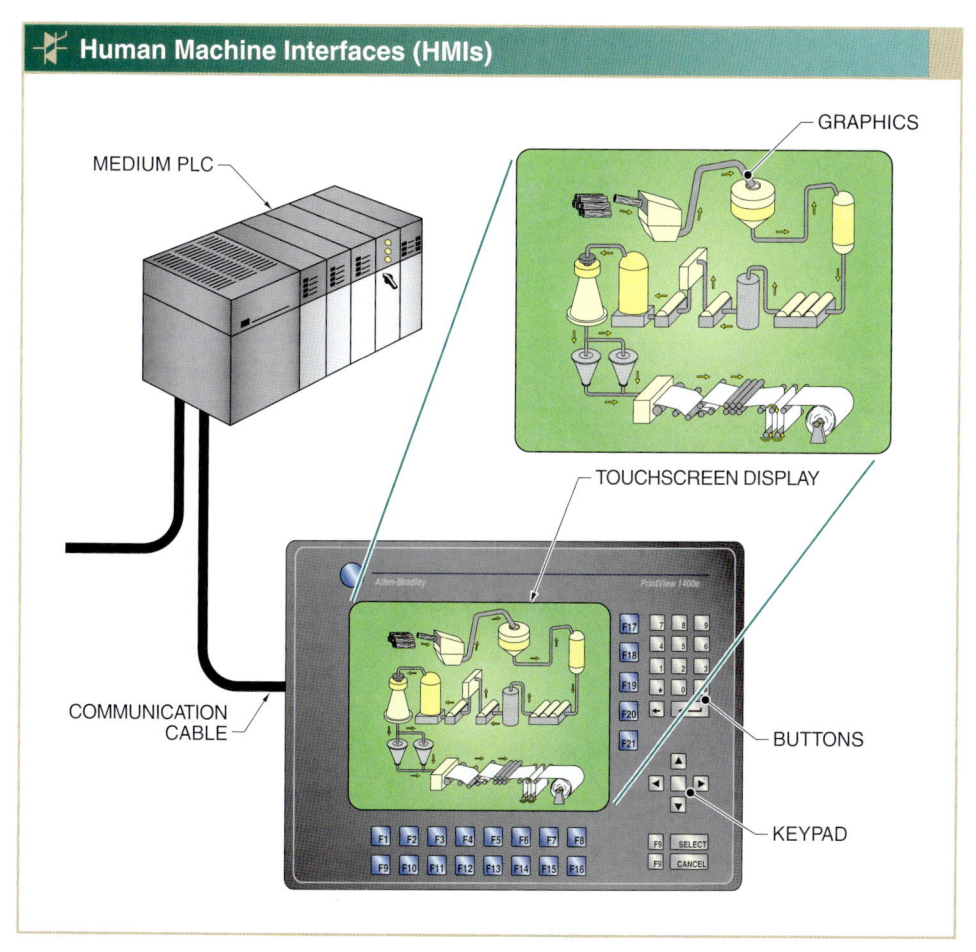

Human Machine Interfaces (HMIs)

Figure 4-11. HMIs allow access to PLC programs and operate similar to PC access but with slightly limited capabilities.

168 PROGRAMMABLE LOGIC CONTROLLERS: PRINCIPLES AND APPLICATIONS

Personal Computer (PC) Programming Devices

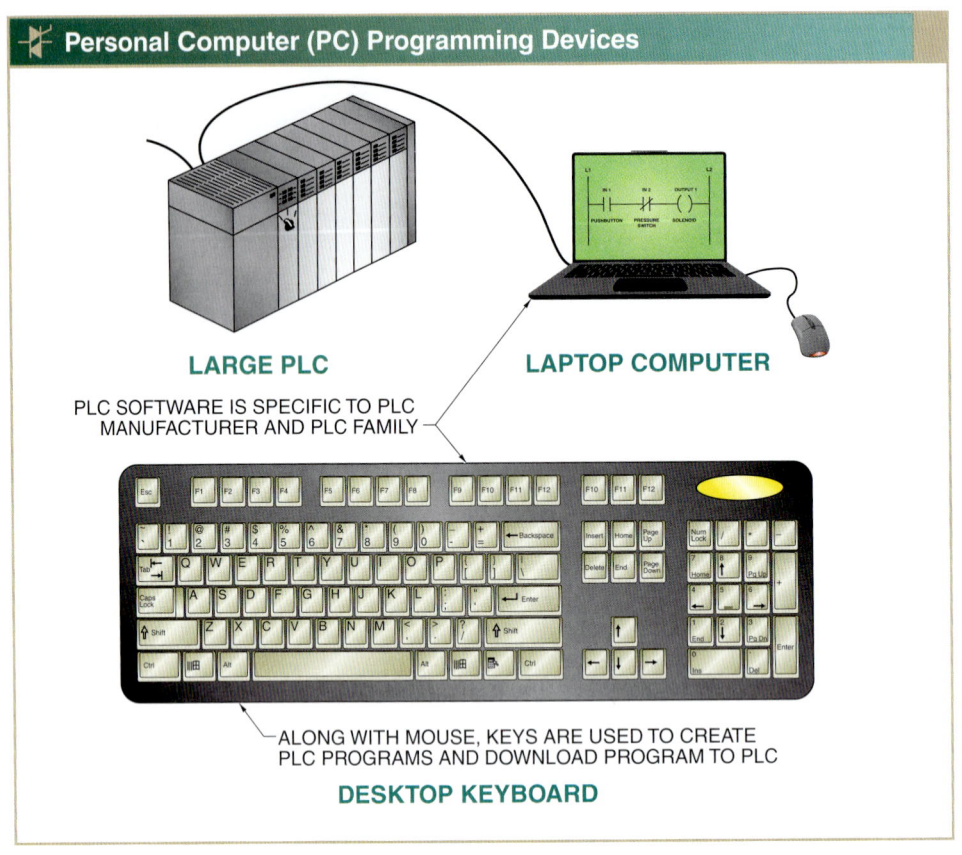

LARGE PLC

LAPTOP COMPUTER

PLC SOFTWARE IS SPECIFIC TO PLC MANUFACTURER AND PLC FAMILY

ALONG WITH MOUSE, KEYS ARE USED TO CREATE PLC PROGRAMS AND DOWNLOAD PROGRAM TO PLC

DESKTOP KEYBOARD

Figure 4-12. PCs are the most common programming devices used with PLCs but require specific interface cables and connectors specific to a PLC family and manufacturer.

In addition to basic programming and monitoring functions, PCs have additional advantages over integrated and handheld programming devices that include the following:

- PCs have large display screens to display multiple lines of PLC programming.
- PCs typically make programming easier by using Windows® software.
- PCs can highlight a PLC program with instruction and rung comments.
- PCs can be used to print out hard copies of a PLC program.
- PCs can copy and store multiple PLC programs at one time.
- PCs can copy PLC programs to various memory formats such as solid-state drives (SSDs), memory sticks, or external hardrives.

The disadvantages of PCs when compared to integrated and handheld programming devices include the following:

- Desktop and laptop PCs are heavier than other programming devices.
- Desktop and laptop PCs and any necessary programming software and interface cables are more expensive than integrated or handheld programming devices.

PLC TIPS

When connecting a PLC to a PC (laptop or desktop), a USB port or Ethernet port is typically used. On older versions of PLCs, the serial port (COM 1) may be used.

PLC CLASSIFICATIONS

PLCs are typically classified by the number of input and output (I/O) terminals. For example, a PLC with 10 input and 8 output terminals would be an 18 I/O PLC. PLCs typically have more input terminals than output terminals. As the number of I/Os increase, the physical size of the PLC must increase.

Typically PLCs are classified into four sizes: nano (pico), micro, medium, and large. **See Figure 4-13.** The maximum number of I/Os (with expansion units) associated with each PLC size is typically less than 16 I/Os for nano-PLCs, 16 I/Os to 128 I/Os for micro-PLCs, 129 I/Os to 512 I/Os for medium PLCs, and more than 512 I/Os for large PLCs.

Most PLC manufacturers have more than one family of PLCs. Individual PLC families share a common shape/footprint, hardware attributes, and programming software. For example, Rockwell Automation has the PLC-5®, SLC™500, and MicroLogix™ families. Siemens Energy and Automation has the S7-1200, and S7-1500 families. Typically, PLC components are not interchangeable between PLC manufacturers or between PLC families of the same manufacturer. Also, manufacturers typically have more than one PLC size within a family, such as with Rockwell Automation's MicroLogix™ family. The MicroLogix™ family of PLCs has the MicroLogix™ 1100 with up to 144 I/Os and the MicroLogix™ 1400 with up to 288 I/Os.

CODE CONNECT

Section 110.3(B) of the NEC® states, "Listed or labeled equipment shall be installed and used in accordance with any instructions included in the listing or labeling." Most PLCs, regardless of the manufacturer or size, are listed and/or labeled by qualified testing organizations. A technician must follow the installation instructions supplied with the PLC in order to be NEC® compliant. If the technician has a question regarding the instructions, the PLC manufacturer's technical support should be contacted. A telephone number or web link is typically noted in the instructions.

PLC Sizes

NANO-PLC — LESS THAN 16 I/Os

MICRO-PLC — 16 I/Os TO 128 I/Os

MEDIUM PLC — 129 I/Os TO 512 I/Os

LARGE PLC — MORE THAN 512 I/Os

Rockwell Automation, Inc.

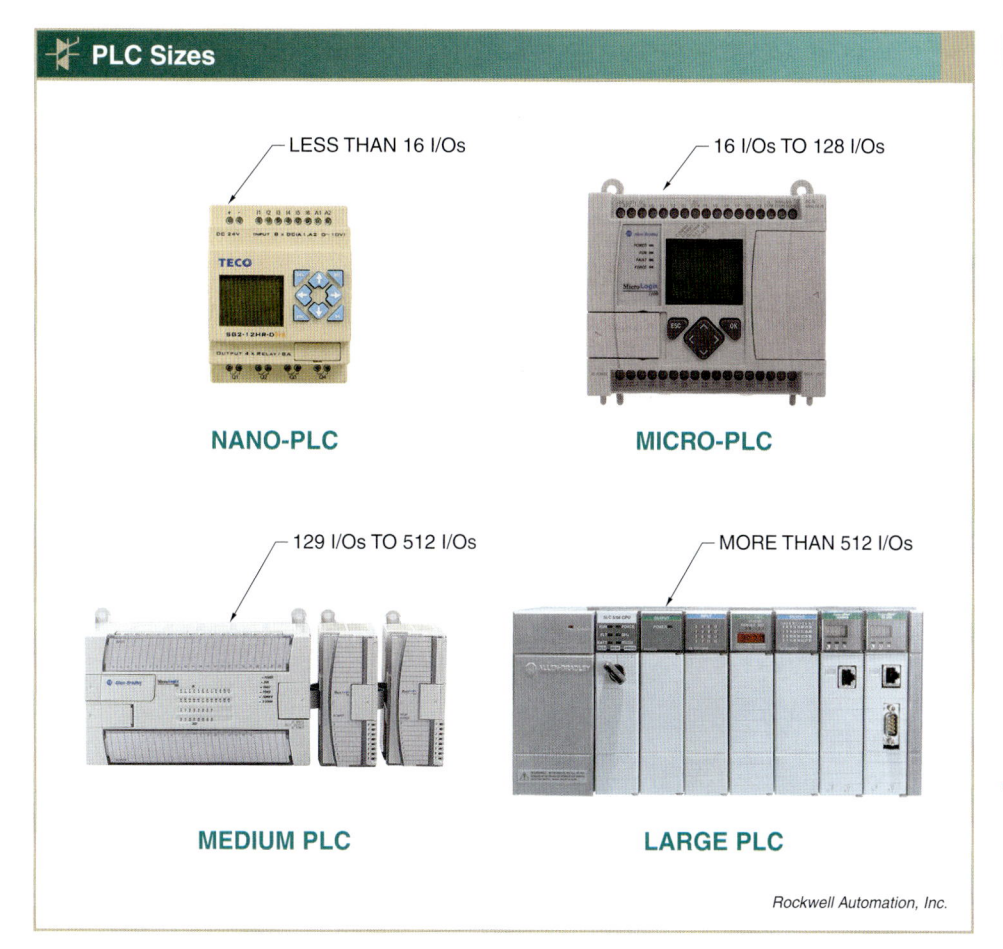

Figure 4-13. The size of a PLC is determined by the number of input and output terminals on the PLC.

Form Factors

Form factor is the physical configuration used to connect the components of a PLC into a housing. The two types of form factors used with PLCs are fixed and modular.

Fixed PLCs

A *fixed PLC* is a PLC that has a set number and type of input and output (I/O) terminals. The input section, output section, power supply, processor, and in some cases the integrated programming device are contained in a common housing. Fixed PLCs are typically used with small machines where physical space is limited.

The number and type of I/Os cannot be changed in a fixed PLC. **See Figure 4-14.** However, the number of I/Os of some fixed PLCs can be increased by incorporating an expansion unit. Expansion units also have a set number and type of I/O terminals. Expansion units are mounted as close to the fixed PLC as possible and are connected via a data bus or data cable.

PLC TIPS
Do not use a cable (data bus) other than the one provided by the PLC manufacturer for connecting expansion units and chassis as data corruption may occur.

Modular PLCs

A *modular PLC* is a PLC that has a variable number of input and output (I/O) terminals based on the number of cards or modules placed into a chassis or rack. A card chassis has slots that accommodate the power supply, processor, and input and output cards. Chassis and racks with various numbers of slots are available. The input terminals, output terminals, power supply, and processor are all separate cards or modules that are assembled in a variety of configurations. **See Figure 4-15.**

Input cards or modules are available with any number of input terminals and input types (analog or digital). Output cards or modules are also available with any number of output terminals and output types (analog, digital, or combination).

Note: Combination digital or analog I/O cards are available that have both inputs and outputs.

Power supplies are available for various voltage sources and with various current capacities. Processors are available with small or large amounts of program memory and with fast scan times. Most modular PLCs can increase I/O capacity by connecting an expansion chassis to the PLC by a data bus. The expansion chassis is typically mounted to the right of the PLC's base chassis but may be mounted at a remote location using a data cable. Modular PLCs are easy to expand and provide more flexibility than fixed PLCs.

Fixed PLCs

FIXED PLC WITH EXPANSION UNIT/MODULE

Labels: CPU STATUS LED, HOUSING, EXPANSION UNITS OR MODULES (16 I/O), FIXED I/O INPUT SECTION, INPUT OR OUTPUT EXPANSION UNIT OR MODULE LED, BASE UNIT, EXPANSION UNIT OR MODULE TERMINAL DOOR, INPUT LED, OUTPUT LED, INPUT OR OUTPUT OR MODULE EXPANSION UNIT LED, COMMUNICATION PORT, FIXED I/O OUTPUT SECTION, DATA BUS, UP TO 6 EXPANSION UNITS OR MODULES CAN BE ADDED

Rockwell Automation, Inc.

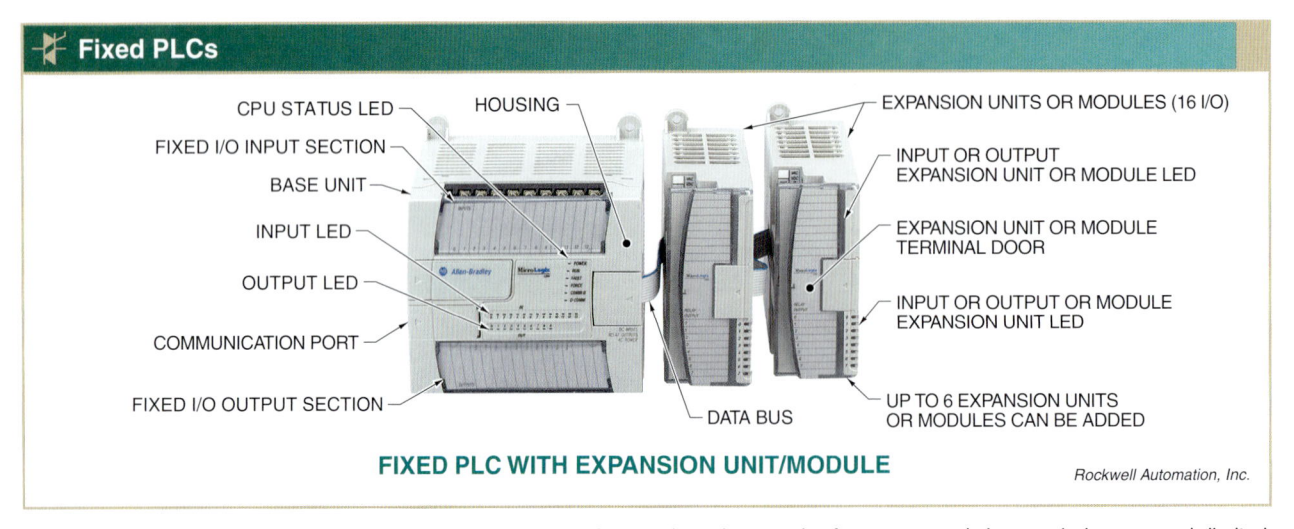

Figure 4-14. Typically, fixed PLCs are used in small applications that require only a couple of sensors or switches, and where space is limited.

Modular PLCs

MODULAR PLC WITH EXPANSION CHASSIS

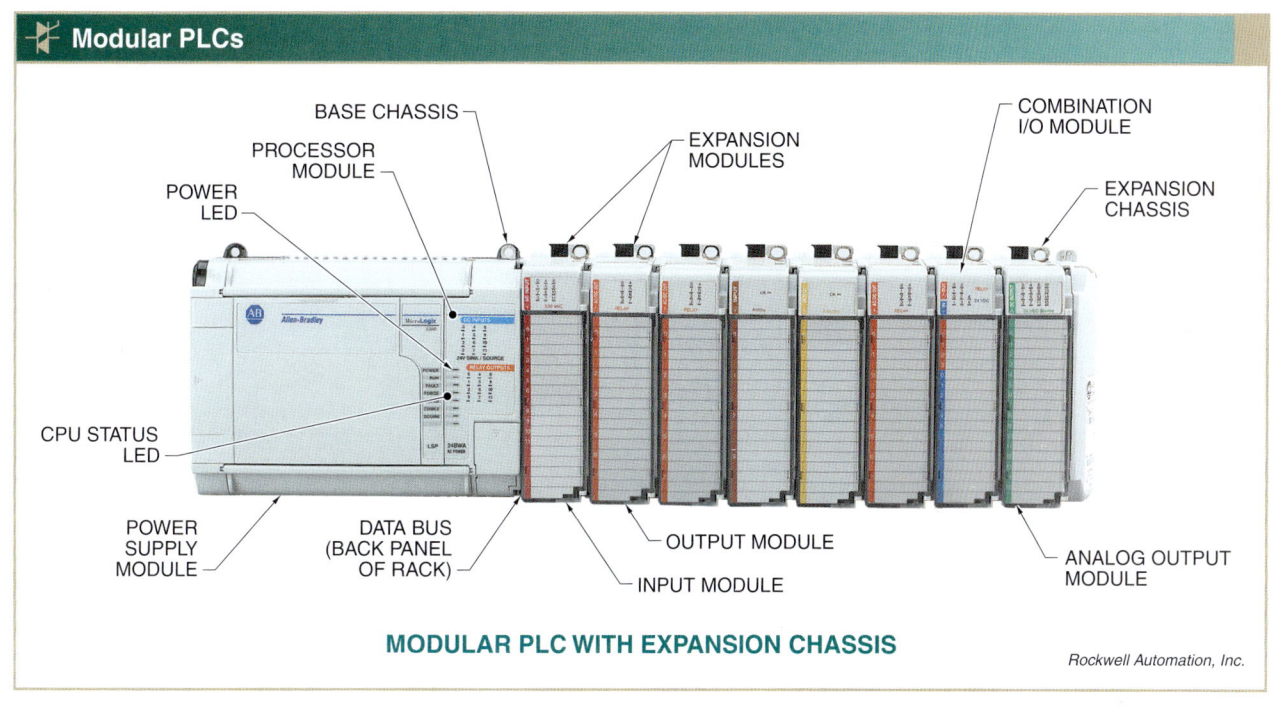

Rockwell Automation, Inc.

Figure 4-15. Modular PLCs are made up of individual sections that can be added to a PLC to expand the PLC's I/O and power capabilities for almost any application.

PLC MEMORY

The processor of a PLC contains memory in which the PLC program and related data are stored. *Memory* is a part of the CPU where program files are loaded for execution and data files are stored for fast access. *Data* is information that is stored in the memory of a CPU. The two areas in the memory of a PLC are the program files and the data table files. Program files are files that contain the user-developed control program (PLC program). A *data table file* is the section of PLC memory that contains the status of the CPU, inputs and outputs, timer and counter preset and accumulated values, and other program instruction values. **See Figure 4-16.**

Memory can be volatile or nonvolatile. In the event of a power failure, volatile memory loses stored data while nonvolatile memory remains intact. Most PLCs contain both types of memory.

Random Access Memory

Random access memory (RAM) is a type of memory that permits accessing the storage medium to store and retrieve program files and data table files. The PLC program is run while stored in RAM memory. A battery or capacitor is used to provide back-up power so data will not be lost when normal power is turned OFF or lost.

Electrically Erasable Programmable Read-Only Memory

Electrically erasable programmable read-only memory (EEPROM) is a type of nonvolatile memory that is retained when power is lost. EEPROM is used in PLCs to provide a backup for the RAM memory. A copy of the PLC RAM program is stored in EEPROM memory. Typically, data is transferred from RAM to EEPROM upon loss of power. At power up, data (including the PLC program) is transferred from EEPROM to RAM. The EEPROM can be an integral part of a PLC or a separate unit that plugs into a PLC socket. Many PLCs have EEPROM chips referred to as memory modules.

CODE CONNECT

Section 110.26(E)(1)(a) of the NEC® requires a dedicated space above an electrical installation that is free of any piping, duct, or foreign equipment. The space must be equal to the width and depth of the equipment, and it must extend from the floor to the structural ceiling or 6 feet above the equipment, whichever is lower.

172 PROGRAMMABLE LOGIC CONTROLLERS: PRINCIPLES AND APPLICATIONS

PLC Memory

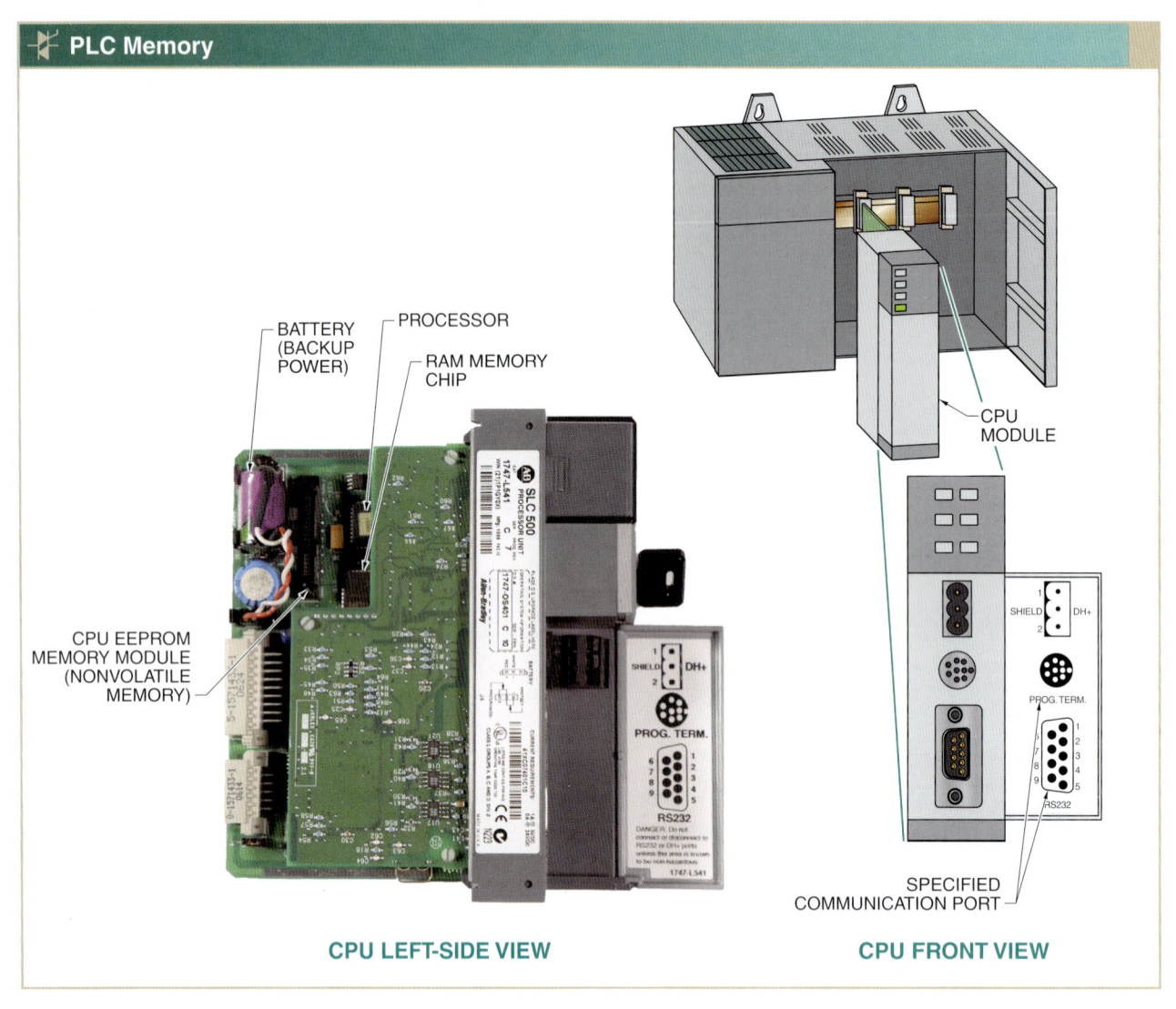

CPU LEFT-SIDE VIEW

CPU FRONT VIEW

Figure 4-16. Memory in a PLC is used for program files that contain the user PLC program and data table files, which contain the status of inputs and outputs, timer preset and accumulated values, counter preset and accumulated values, and other program instruction values.

The standard unit of memory for a PLC is the word. A *word* is a unit of memory that consists of 16 bits. A *bit* is the smallest unit of memory. **See Figure 4-17.** Bits are numbered right to left starting at 0. A *nibble* is a group of 4 bits. A *byte* is a group of 8 bits.

Each bit corresponds to a specific location on the processor chip of a PLC. A "1" denotes the presence of voltage at a bit location. A "0" denotes the absence of voltage at a bit location. Data (information stored in memory) is represented by a series of 1s and 0s. As with computers, PLCs manipulate all data using 1s and 0s.

PLC OPERATING CYCLE

When placed in the RUN mode, a PLC starts an operating cycle. The operating cycle of a PLC continues until the PLC is taken out of RUN mode. The operating cycle of a PLC consists of a series of actions performed sequentially and continuously. **See Figure 4-18.**

The operating cycle is also referred to as the scan or the PLC scan. The five sections of the operating cycle are the input scan, program scan, output scan, service communications, and housekeeping and overhead.

Input Scan

The *input scan* is the section of the operating cycle during which the PLC examines the input devices for the absence or presence of voltage and records a 0 or 1 at the corresponding input data table location.

Program Scan

The *program scan* is the section of the operating cycle during which the PLC examines the PLC program, compares the program to the status of the inputs, uses the comparison to determine what output components will be energized or de-energized, and records a 0 or 1 at the corresponding output data table location.

Output Scan

The *output scan* is the section of the operating cycle during which the PLC energizes or de-energizes output components based on the information in the output data table.

Service Communications

Service communications is the section of the operating cycle during which the PLC communicates with other devices, such as handheld programmers or PCs.

Housekeeping and Overhead

Housekeeping and overhead is the section of the operating cycle during which the PLC performs memory management and updates timers and counters.

PLC operating cycles occur in a matter of milliseconds. The length of an operating cycle is determined by the size of the program, the number and type of instructions, and the amount of service communications, housekeeping, and overhead required.

PLC TIPS

Handheld programming device memory modules must be handled by the ends of the carrier or edges of the plastic housing. Do not expose memory modules to electrostatic charges. Electrostatic charges destroy the data in memory modules.

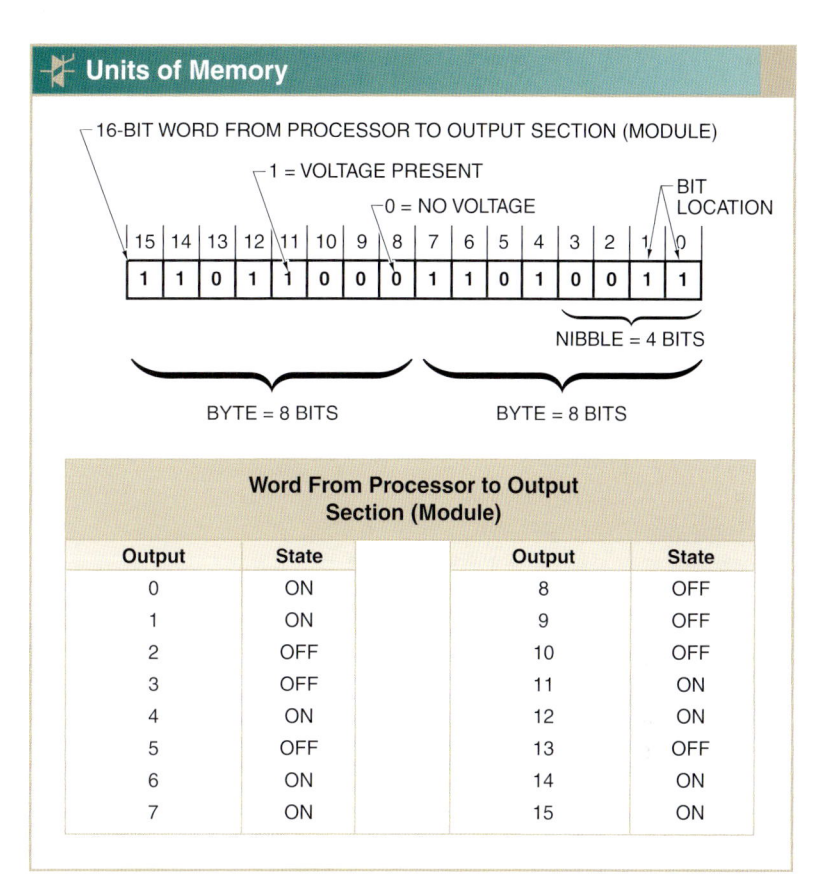

Units of Memory

16-BIT WORD FROM PROCESSOR TO OUTPUT SECTION (MODULE)
1 = VOLTAGE PRESENT
0 = NO VOLTAGE
BIT LOCATION

15	14	13	12	11	10	9	8	7	6	5	4	3	2	1	0
1	1	0	1	1	0	0	0	1	1	0	1	0	0	1	1

NIBBLE = 4 BITS
BYTE = 8 BITS   BYTE = 8 BITS

Word From Processor to Output Section (Module)

Output	State	Output	State
0	ON	8	OFF
1	ON	9	OFF
2	OFF	10	OFF
3	OFF	11	ON
4	ON	12	ON
5	OFF	13	OFF
6	ON	14	ON
7	ON	15	ON

Figure 4-17. A word is a unit of memory in a PLC and consists of 16 bits. A bit is the smallest unit of memory.

Industrial processes utilize large modular PLCs, and often more than one, to run complex electrical systems.

174 PROGRAMMABLE LOGIC CONTROLLERS: PRINCIPLES AND APPLICATIONS

PLC Operating Cycle

① INPUT SCAN

VOLTAGE IS PRESENT AT I/3 BECAUSE LS1 IS CLOSED; A 1 IS PLACED AT BIT #3 CORRESPONDING TO I/3

Input Data Table

15	14	13	12	11	10	9	8	7	6	5	4	3	2	1	0
0	0	0	0	0	0	0	0	0	0	0	0	1	0	0	0

INSTRUCTION I/3 IS TRUE BECAUSE A 1 IS AT BIT #3; INSTRUCTION O/2 IS TRUE BECAUSE INSTRUCTION I/3 IS TRUE

② PROGRAM SCAN

I/3 — LIMIT SWITCH 1
O/2 — INDICATOR LIGHT 3

INSTRUCTION O/2 IS TRUE AND A 1 IS PLACED AT BIT #2 CORRESPONDING TO O/2

Output Data Table

15	14	13	12	11	10	9	8	7	6	5	4	3	2	1	0
0	0	0	0	0	0	0	0	0	0	0	0	0	1	0	0

③ OUTPUT SCAN

OUTPUT O2, INDICATION LIGHT 3, IS TURNED ON BECAUSE A 1 HAS BEEN PLACED AT BIT #2

④ SERVICE COMMS

⑤ HOUSEKEEPING AND OVERHEAD

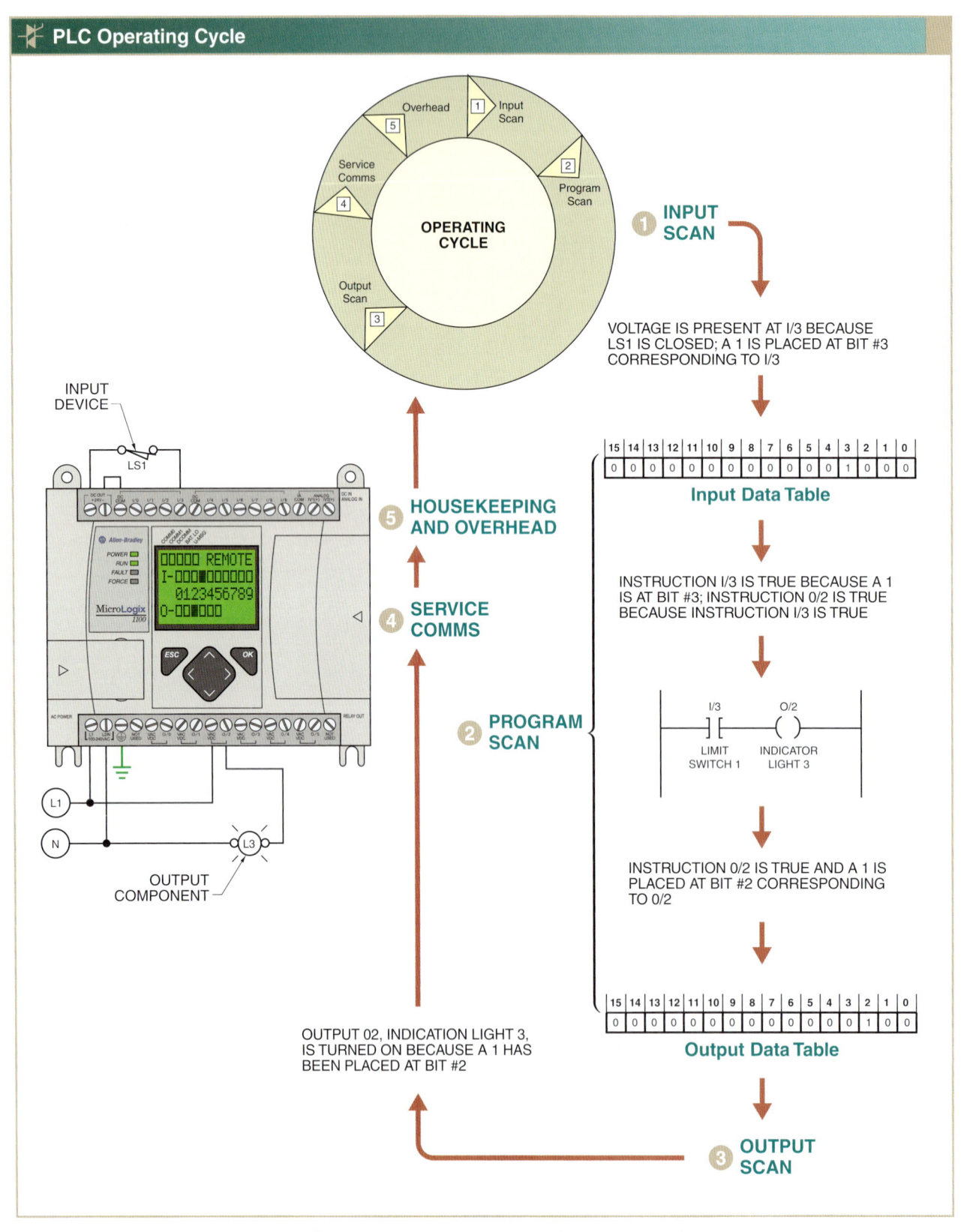

Figure 4-18. The operating cycle of a PLC runs continuously and sequentially while data tables are updated as inputs and outputs change state.

Chapter Review 4

Name _____ Date _____

True-False

T F 1. Integrated programming devices are typically found on nano-PLCs.

T F 2. Personal computers (PCs) are the most common programming devices used with PLCs.

T F 3. Input devices typically operate at lower voltages than those used by the CPU of a PLC.

T F 4. An output component typically operates at a voltage lower than the operating voltage of a PLC CPU.

T F 5. PCs are used as programming devices on all sizes of PLCs.

T F 6. The two types of form factors used with PLCs are fixed and modular.

T F 7. Housekeeping and overhead is the section of the operating cycle during which a PLC performs memory management and updates timers and counters.

T F 8. All PLCs have three common sections regardless of the manufacturer, model, or size.

T F 9. Pushbuttons, selector switches, limit switches, and proximity switches are common input devices.

T F 10. The standard unit of memory for a PLC is the byte.

T F 11. PLC components are not typically interchangeable between manufacturers.

T F 12. EEPROM is used in PLCs to provide power failure backup for RAM memory.

T F 13. CPUs are rated according to program memory size and scan time.

T F 14. The number and type of I/Os cannot be changed in a fixed PLC.

T F 15. PLCs are a cost-effective alternative for an application that requires a control relay and a timer.

Completion

_____ 1. A(n) ___ is a device with a keypad that is used to write and enter a user-developed control program into a PLC.

_____ 2. ___ is the length of time a PLC can tolerate a power loss without affecting operation.

_____ 3. A(n) ___ is a separate keypad device that is not an integral part of a PLC.

_____ 4. ___ is the maximum number of programming instructions a CPU can hold.

_____ 5. ___ is a chip inside a CPU that stores data at specific addresses.

_____ 6. A(n) ___ is a group of 4 bits.

_____ 7. ___ is the section of the operating cycle during which a PLC communicates with other devices, such as handheld programmers or PCs.

_____ 8. The ___ is the section of the PLC that houses the processor (brain) of the PLC.

_____ 9. ___ is the physical configuration used to connect the components of a PLC into a housing.

_____ 10. A(n) ___ is a PLC that has a variable number of I/O terminals based on the number of cards or modules placed into a chassis or rack.

Multiple Choice

_____ 1. Which of the following is not a PLC size classification?
   A. large
   B. micro
   C. medium
   D. small

_____ 2. Which of the following is not a component that a PLC would typically control directly?
   A. control relay
   B. pilot light
   C. motor starter
   D. motor

_____ 3. Which of the following units is the smallest unit of memory?
   A. bit
   B. byte
   C. nibble
   D. word

_____ 4. Which of the following is not a section common to all PLCs?
   A. programming device
   B. expansion chassis
   C. central processing unit
   D. power supply

_____ 5. Which of the following is not used to program PLCs?
   A. human machine interfaces
   B. handheld programming devices
   C. integrated programming devices
   D. digital interface autopanels

Two-Motor Starter Circuits

PLC Exercise 4.1

BILL OF MATERIALS
- TECO PLR or Micrologix™ 1100 PLC
- NC red pushbutton
- NO green pushbutton
- Limit switch
- Motor starter/contactor
- Light

Different types of industries use conveyor equipment to send products from one point to another. Typically, a PLC will control motor starter circuits with multiple motors to run conveyor belts. For this lab, a motor control circuit is wired with two motors. A limit switch, overloads, and stop switch are hardwired NC, and the start switch is hardwired NO. An indicator light is installed to turn on when one or both overloads trip.

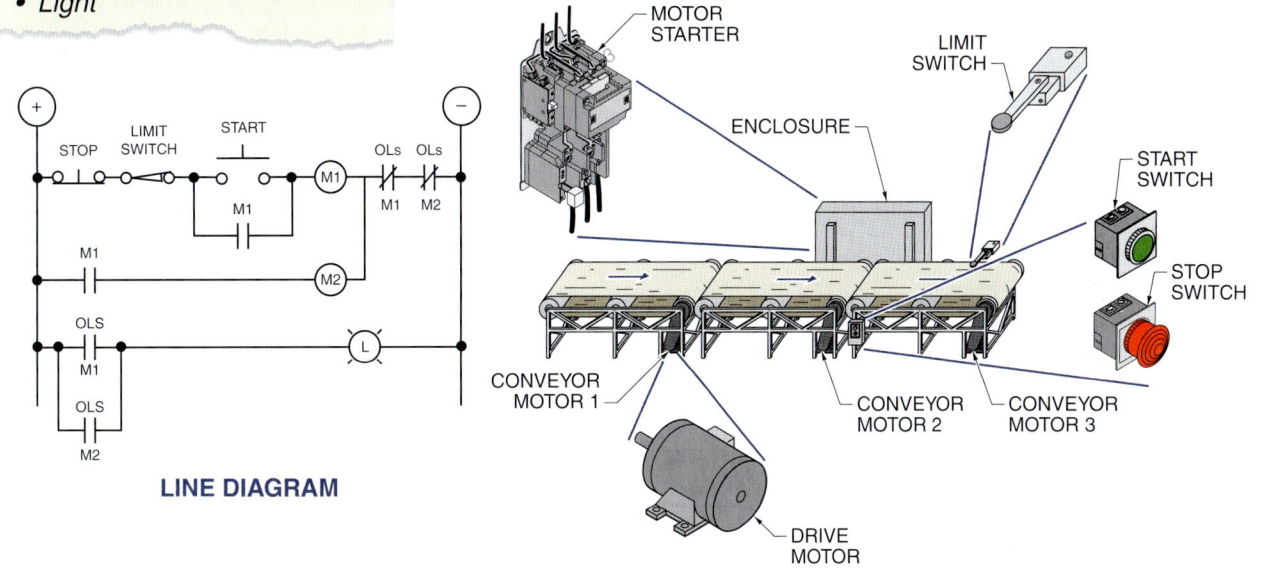

LINE DIAGRAM

TECO Procedures

1. Create a new ladder logic or function block diagram program.

2. Use the TECO programming diagram or FBD to add the proper contacts and description in the programming grid/area. To switch from a 3-contact programming grid to a 5-contact programming grid, select 5-contact from the Edit drop down list in the Menu bar.

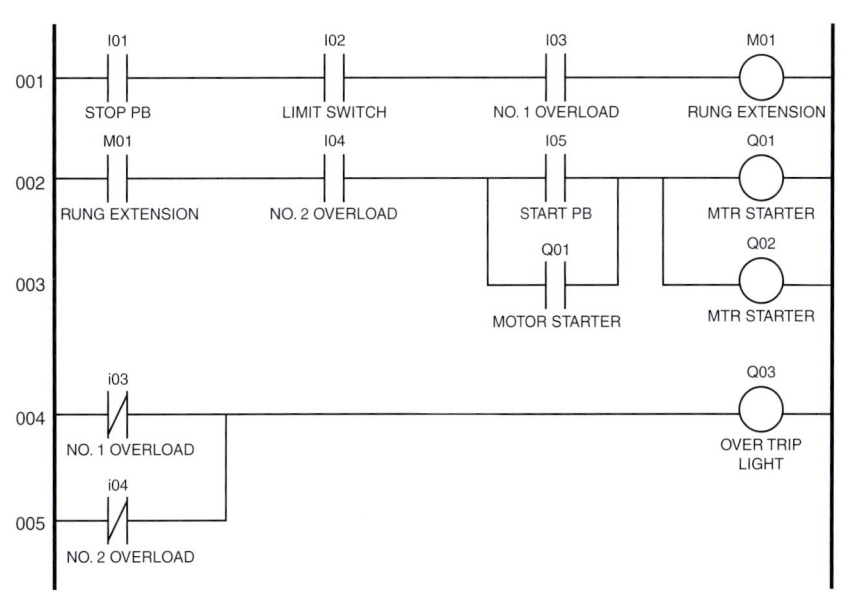

TECO PROGRAMMING DIAGRAM

Rung	Contacts	Output
001	I01 (STOP PB) — I02 (LIMIT SWITCH) — I03 (NO. 1 OVERLOAD)	M01 (RUNG EXTENSION)
002	M01 (RUNG EXTENSION) — I04 (NO. 2 OVERLOAD) — I05 (START PB)	Q01 (MTR STARTER)
003	Q01 (MOTOR STARTER)	Q02 (MTR STARTER)
004	i03 (NO. 1 OVERLOAD) /	Q03 (OVER TRIP LIGHT)
005	i04 (NO. 2 OVERLOAD) /	

177

3. Use the Edit Contact window to denote an input or output number and the Symbol button to add a description after placing the input or output in the programming grid. For an FBD, use the Comment icon to add the description.

4. Run the program in Simulation Mode to verify proper operation. Use the Input Status tool window to activate the inputs and control the output to simulate circuit operation.

5. Use the wiring diagram to wire the input devices and output components to the PLC.

6. Turn power to the PLC ON and link the com port.

7. Write the program to the SG2-12HR-D to download the program logic to the PLC.

8. Place the SG2-12HR-D in Run Mode. Activate the start pushbutton and monitor the I/Os on the LCD display.

9. Stop the program and use the Save As function to save and place the lab application in the required folder location.

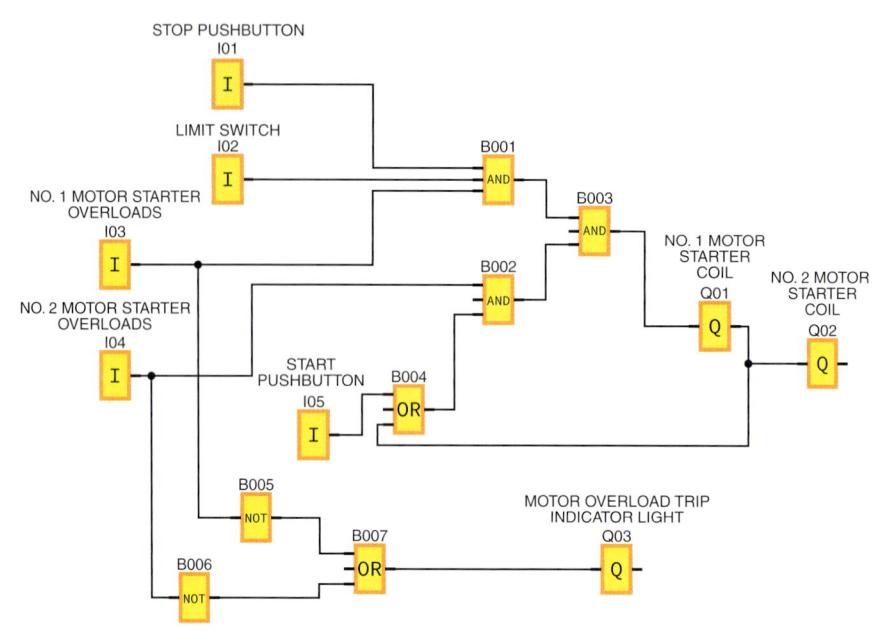

FUNCTION BLOCK DIAGRAM

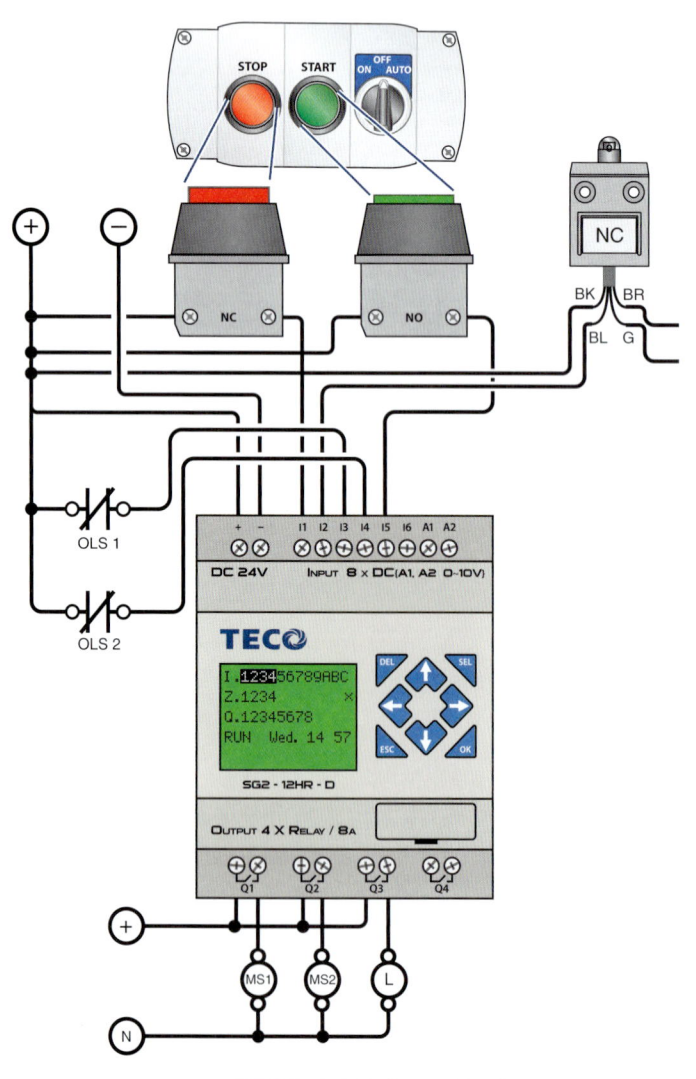

TECO WIRING DIAGRAM

Chapter 4—PLC Hardware, Memory, and Operating Cycle **179**

Allen-Bradley® Procedures

1. Open the RSLogix™ Micro Starter Lite program and create a new programming diagram.

2. Use the AB programming diagram to add the proper address and description in the LAD 2 programming window.

3. Verify that the programming diagram has no errors and save the project.

4. Use the wiring diagram to wire the input devices and output components to the PLC. Once all connections have been made, turn power to the PLC ON.

5. Use RSLinx™ to establish communication between the PC and the Micrologix™ 1100.

6. Download the program logic to the PLC.

7. Place the PLC in Run Mode.

8. Activate input(s) and monitor to determine if the output(s) are operating correctly.

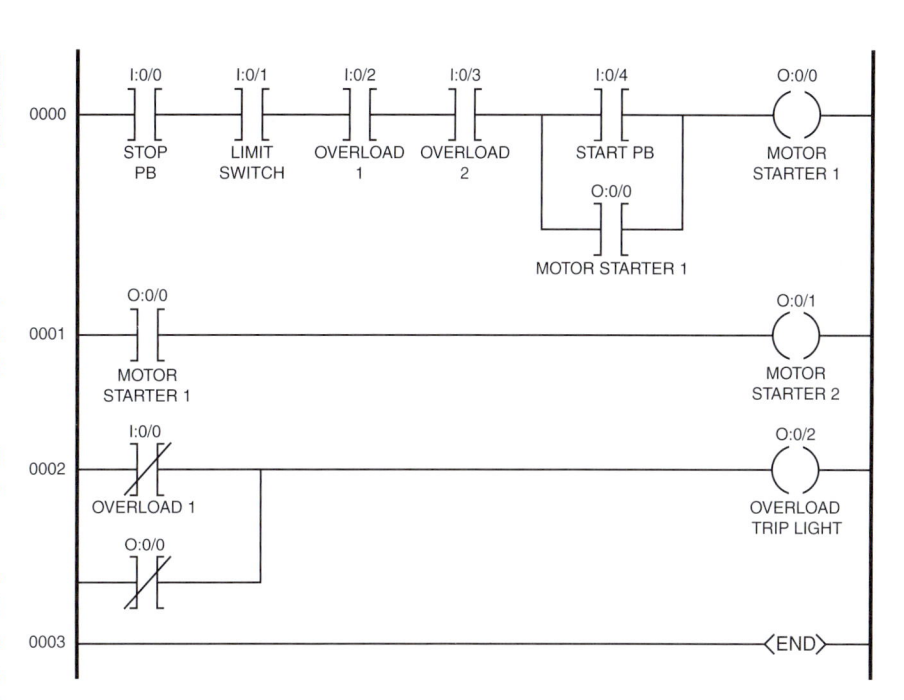

AB PROGRAMMING DIAGRAM

Rung 0000:
- I:0/0 STOP PB
- I:0/1 LIMIT SWITCH
- I:0/2 OVERLOAD 1
- I:0/3 OVERLOAD 2
- I:0/4 START PB (parallel with O:0/0 MOTOR STARTER 1)
- O:0/0 MOTOR STARTER 1

Rung 0001:
- O:0/0 MOTOR STARTER 1
- O:0/1 MOTOR STARTER 2

Rung 0002:
- I:0/0 OVERLOAD 1 (NC), parallel with O:0/0 (NC)
- O:0/2 OVERLOAD TRIP LIGHT

Rung 0003: ⟨END⟩

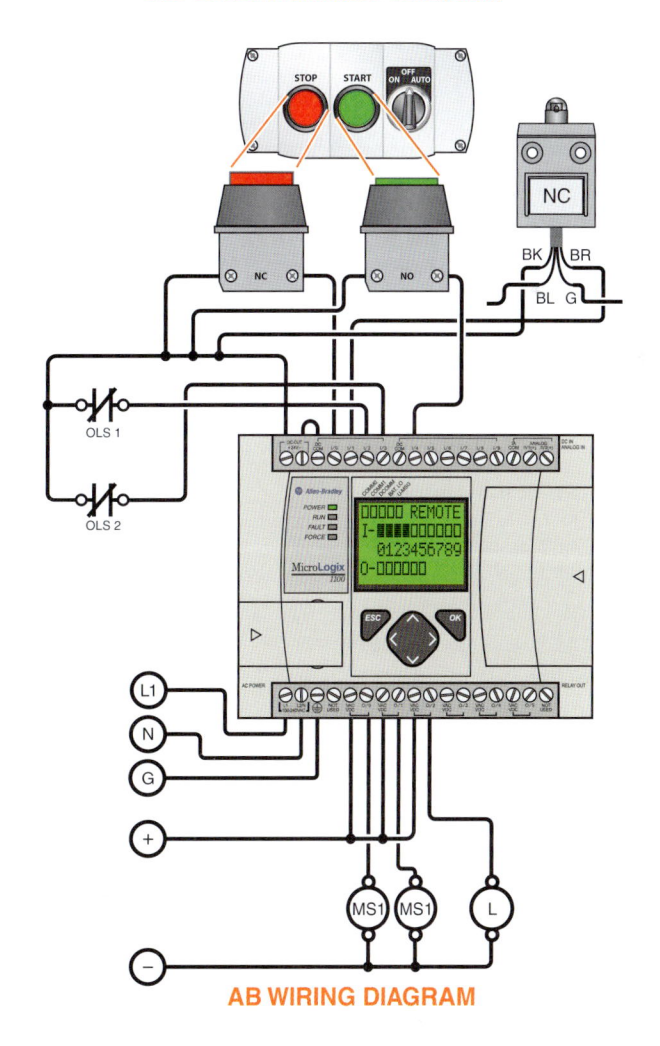

AB WIRING DIAGRAM

Two-Motor Starter Circuits with Control Relays

PLC Exercise 4.2

BILL OF MATERIALS

- TECO PLR or Micrologix™ 1100 PLC
- Two NC red pushbuttons
- Two NO green pushbuttons
- Two motor starters/contactors

In different types of packaging applications, the motor running a roller or conveyor may need to be controlled from separate stations. The use of control relays built into PLC logic helps with these types of tasks. For this lab, an internal control relay is programmed to facilitate a motor circuit that controls two motors from two different stations. Two stop switches are hardwired NC, and two start switches are hardwired NO in parallel. Each motor starter is connected to its own output contact.

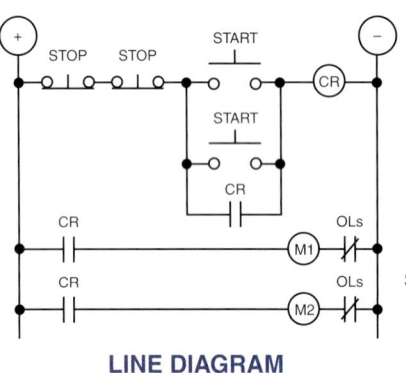

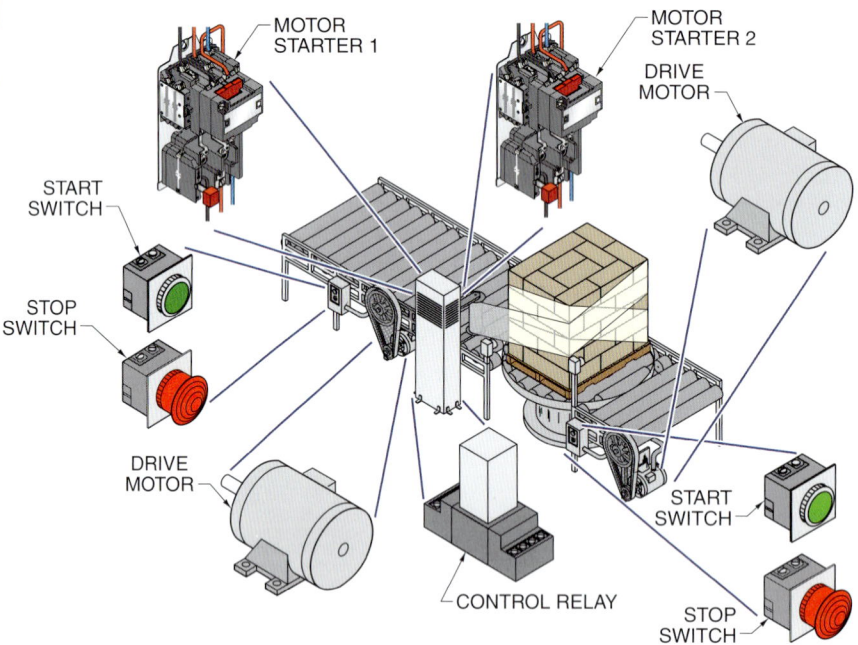

LINE DIAGRAM

TECO Procedures

1. Create a new ladder logic or function block diagram program.

2. Use the TECO programming diagram or FBD to add the proper contacts and description in the programming grid/area.

3. Use the Edit Contact window to denote an input or output number and the Symbol button to add a description after placing the input or output in the programming grid. For an FBD, use the Comment icon to add the description.

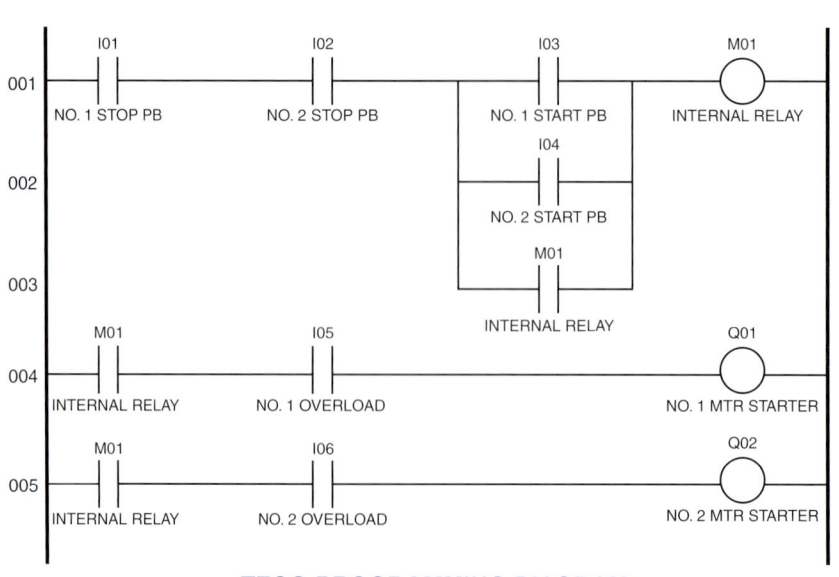

TECO PROGRAMMING DIAGRAM

4. Run the program in Simulation Mode to verify proper operation. Use the Input Status tool window to activate the inputs and control the output to simulate circuit operation.

5. Use the wiring diagram to wire the input devices and output components to the PLC.

6. Turn power to the PLC ON and link the com port.

7. Write the program to the SG2-12HR-D to download the program logic to the PLC.

8. Place the SG2-12HR-D in Run Mode. Activate the start pushbutton and monitor the I/Os on the LCD display.

9. Stop the program and use the Save As function to save and place the lab application in the required folder location.

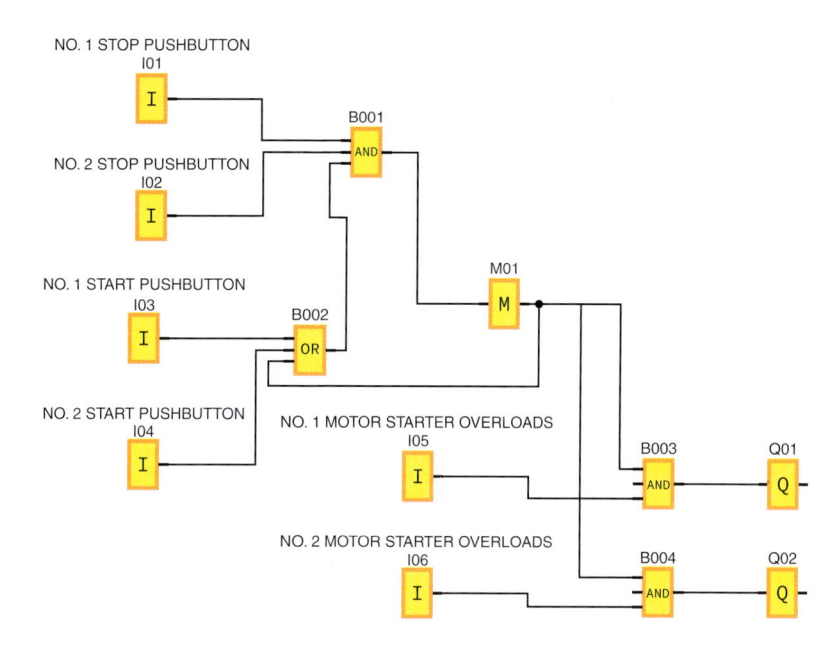

FUNCTION BLOCK DIAGRAM

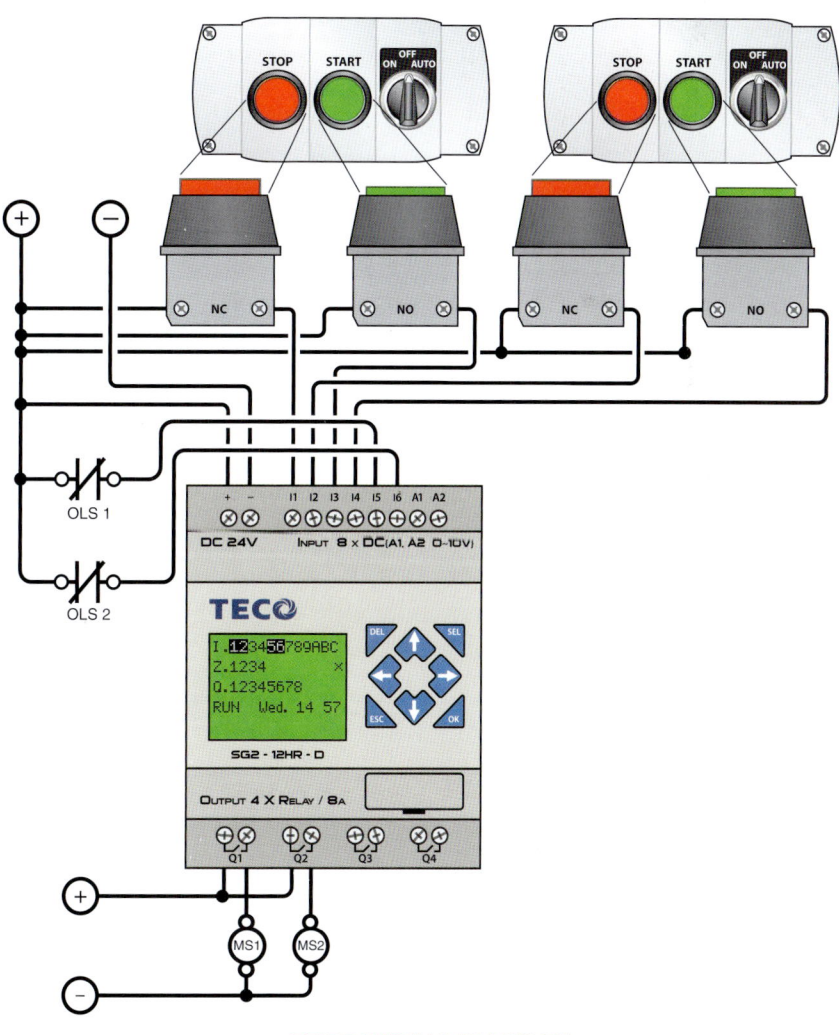

TECO WIRING DIAGRAM

Allen-Bradley® Procedures

1. Open the RSLogix™ Micro Starter Lite program and create a new programming diagram.

2. Use the AB programming diagram to add the proper address and description in the LAD 2 programming window.

3. Verify that the programming diagram has no errors and save the project.

4. Use the wiring diagram to wire the input devices and output components to the PLC. Once all connections have been made, turn power to the PLC ON.

5. Use RSLinx™ to establish communication between the PC and the Micrologix™ 1100.

6. Download the program logic to the PLC.

7. Place the PLC in Run Mode.

8. Activate input(s) and monitor them to determine if the output(s) are operating correctly.

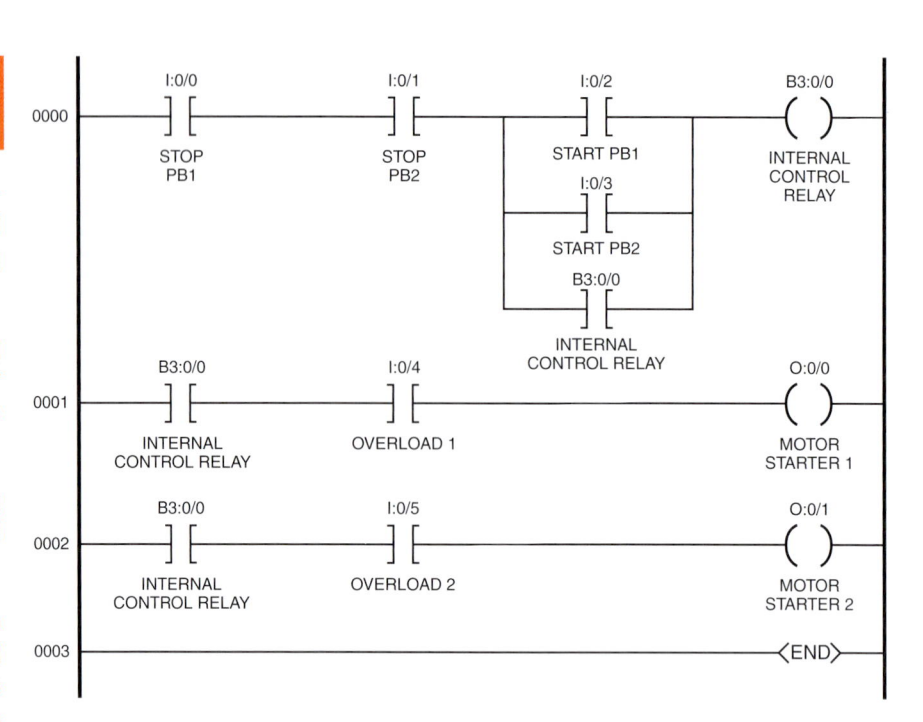

AB PROGRAMMING DIAGRAM

Rung 0000:
- I:0/0 STOP PB1 —| |—
- I:0/1 STOP PB2 —| |—
- I:0/2 START PB1 —| |— (parallel with I:0/3 START PB2, and B3:0/0 INTERNAL CONTROL RELAY)
- B3:0/0 INTERNAL CONTROL RELAY —( )—

Rung 0001:
- B3:0/0 INTERNAL CONTROL RELAY —| |—
- I:0/4 OVERLOAD 1 —| |—
- O:0/0 MOTOR STARTER 1 —( )—

Rung 0002:
- B3:0/0 INTERNAL CONTROL RELAY —| |—
- I:0/5 OVERLOAD 2 —| |—
- O:0/1 MOTOR STARTER 2 —( )—

Rung 0003: ⟨END⟩

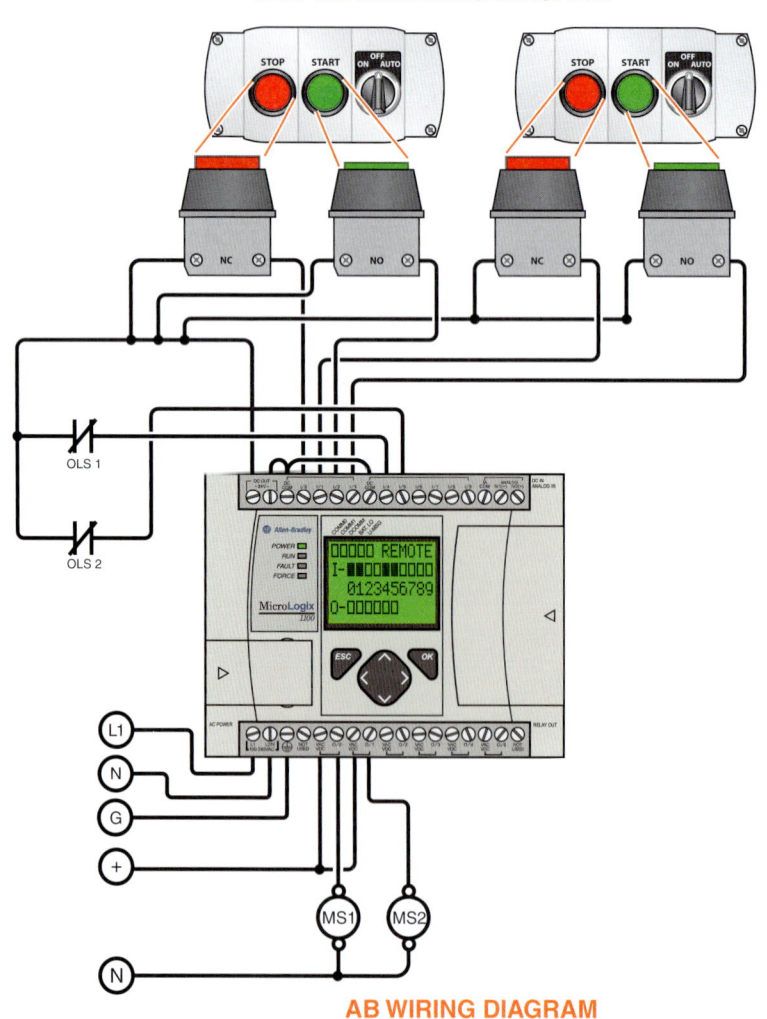

AB WIRING DIAGRAM

Reversing Starter Circuits

PLC Exercise 4.3

BILL OF MATERIALS
- TECO PLR or Micrologix™ 1100 PLC
- NC red pushbutton
- Two NO green pushbuttons
- Forward/reversing motor starter

In conveyor applications, a motor may have to be controlled to propel a conveyor belt forward or backward. In this lab, a reversing motor starter circuit is programmed. An overload and stop pushbutton are wired NC. The forward and reverse start pushbuttons and wired NO.

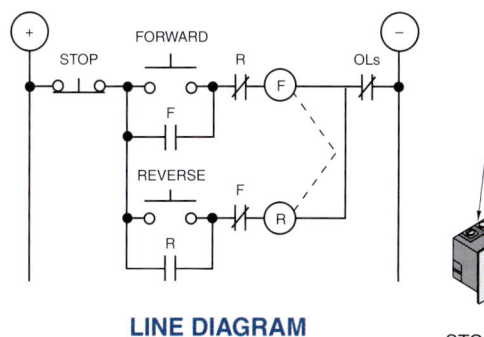

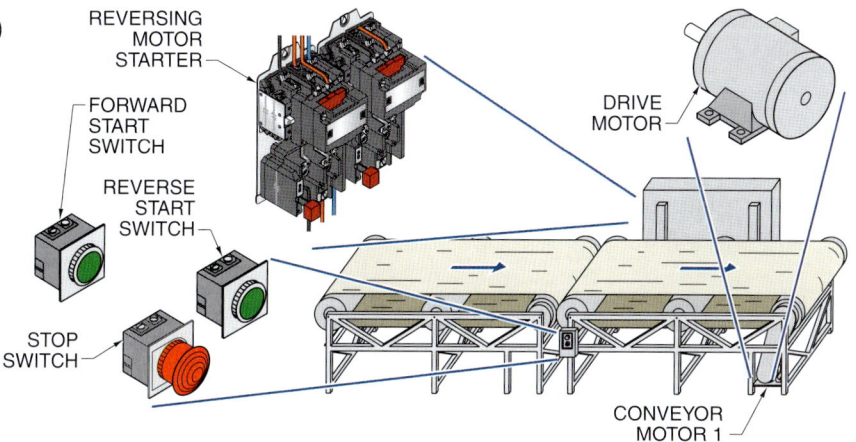

LINE DIAGRAM

T TECO Procedures

1. Create a new ladder logic or function block diagram program.

2. Use the TECO programming diagram or FBD to add the proper contacts and description in the programming grid/area.

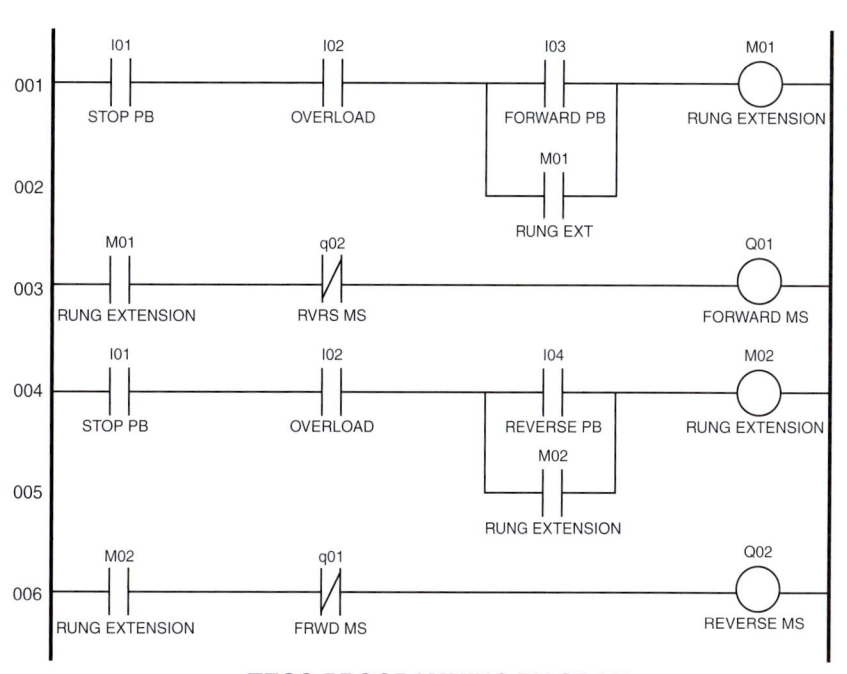

TECO PROGRAMMING DIAGRAM

Rung 001: I01 (STOP PB) — I02 (OVERLOAD) — I03 (FORWARD PB) — M01 (RUNG EXTENSION)
Rung 002: M01 (RUNG EXT)
Rung 003: M01 (RUNG EXTENSION) — q02 (RVRS MS, NC) — Q01 (FORWARD MS)
Rung 004: I01 (STOP PB) — I02 (OVERLOAD) — I04 (REVERSE PB) — M02 (RUNG EXTENSION)
Rung 005: M02 (RUNG EXTENSION)
Rung 006: M02 (RUNG EXTENSION) — q01 (FRWD MS, NC) — Q02 (REVERSE MS)

3. Use the Edit Contact window to denote an input or output number and the Symbol button to add a description after placing the input or output in the programming grid. For an FBD, use the Comment icon to add the description.

4. Run the program in Simulation Mode to verify proper operation. Use the Input Status Tool window to activate the inputs and control the output to simulate circuit operation.

5. Use the wiring diagram to wire the input devices and output components to the PLC.

6. Turn power to the PLC ON and link the com port.

7. Write the program to the SG2-12HR-D to download the program logic to the PLC.

8. Place the SG2-12HR-D in Run Mode. Activate the start pushbutton and monitor the I/Os on the LCD display.

9. Stop the program and use the Save As function to save and place the lab application in the required folder location.

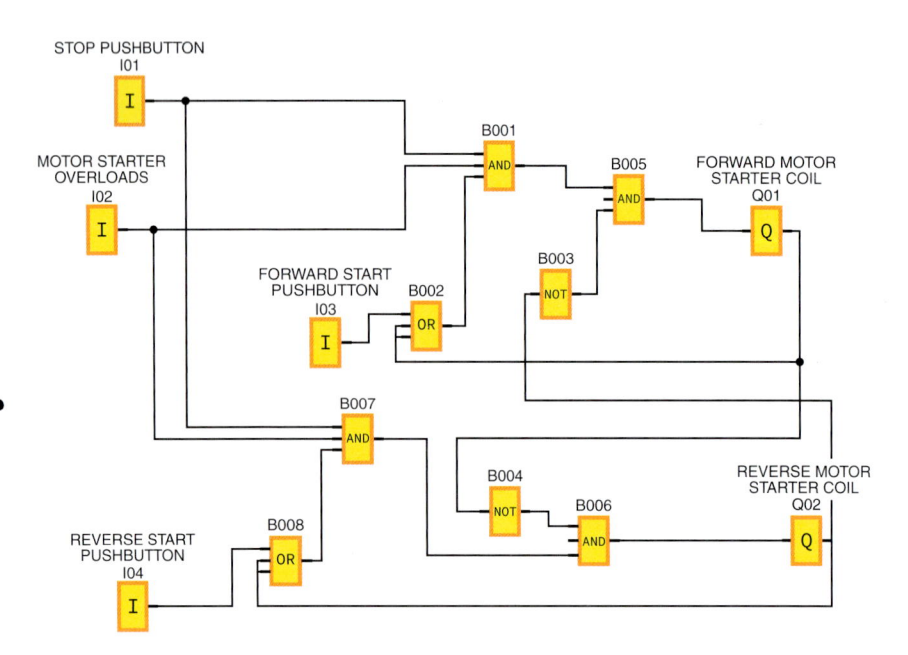

FUNCTION BLOCK DIAGRAM

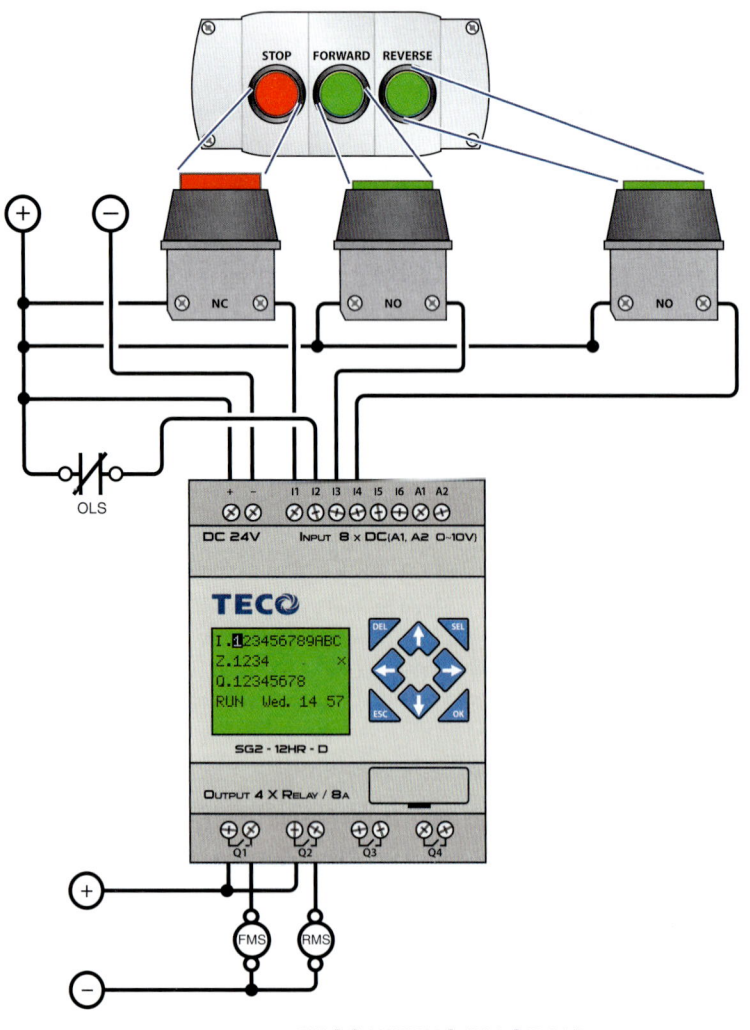

TECO WIRING DIAGRAM

Allen-Bradley® Procedures

1. Open the RSLogix™ Micro Starter Lite program and create a new programming diagram.

2. Use the AB programming diagram to add the proper address and description in the LAD 2 programming window.

3. Verify that the programming diagram has no errors and save the project.

4. Use the wiring diagram to wire the input devices and output components to the PLC. Once all connections have been made, turn power to the PLC ON.

5. Use RSLinx™ to establish communication between the PC and the Micrologix™ 1100.

6. Download the program logic to the PLC.

7. Place the PLC in Run Mode.

8. Activate input(s) and monitor them to determine if the output(s) are operating correctly.

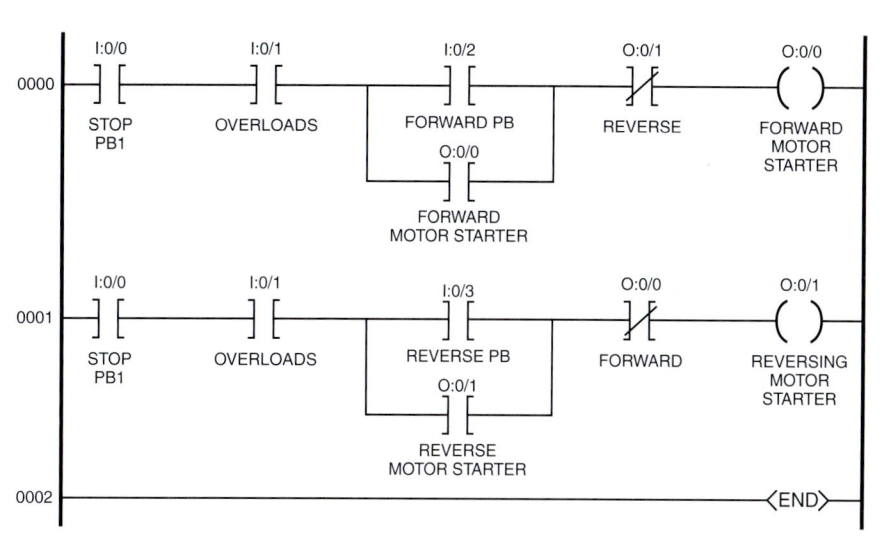

AB PROGRAMMING DIAGRAM

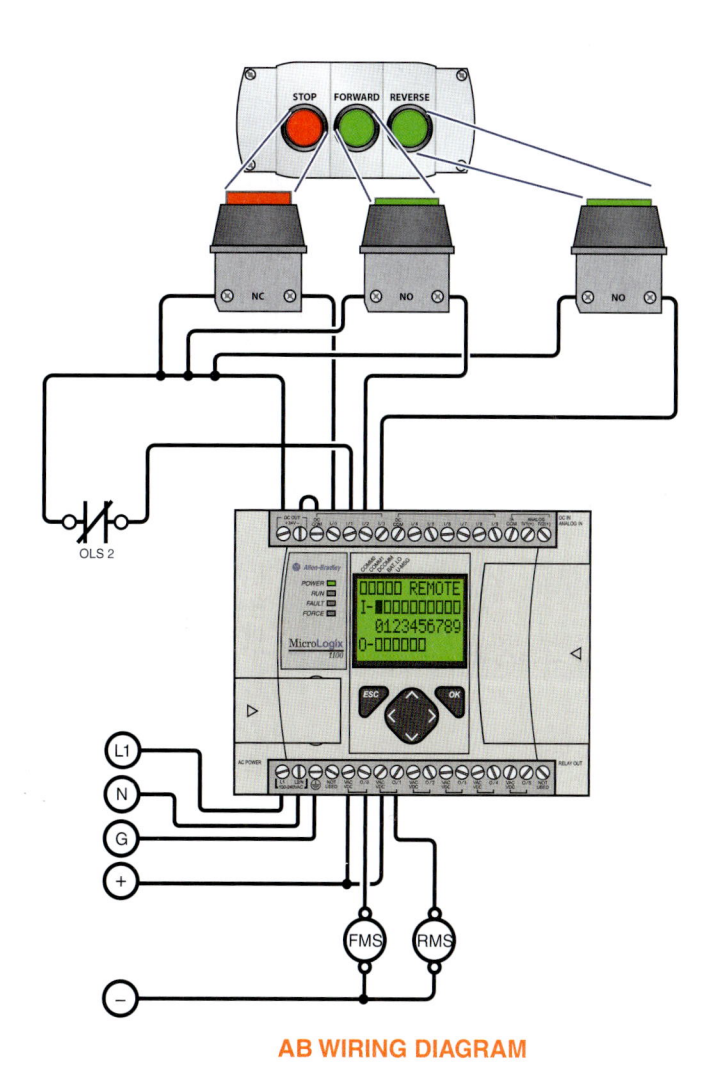

AB WIRING DIAGRAM

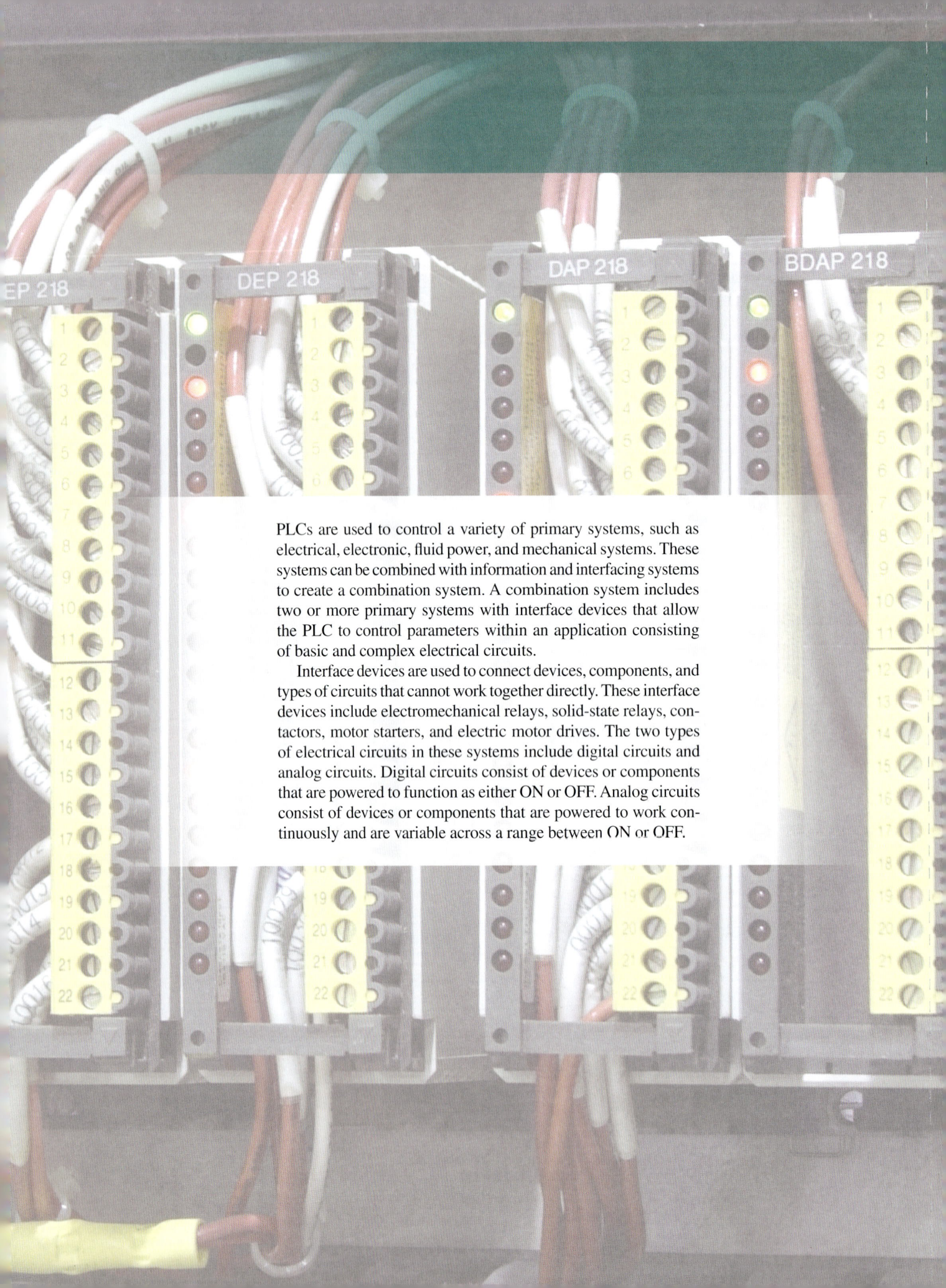

PLCs are used to control a variety of primary systems, such as electrical, electronic, fluid power, and mechanical systems. These systems can be combined with information and interfacing systems to create a combination system. A combination system includes two or more primary systems with interface devices that allow the PLC to control parameters within an application consisting of basic and complex electrical circuits.

Interface devices are used to connect devices, components, and types of circuits that cannot work together directly. These interface devices include electromechanical relays, solid-state relays, contactors, motor starters, and electric motor drives. The two types of electrical circuits in these systems include digital circuits and analog circuits. Digital circuits consist of devices or components that are powered to function as either ON or OFF. Analog circuits consist of devices or components that are powered to work continuously and are variable across a range between ON or OFF.

PLC Systems, Circuits, and Interface Devices

5

Objectives:

- Describe the primary systems used to transmit and control energy.
- Explain what an interface device is and when this device is required in an electrical circuit.
- Identify the differences between electromechanical relays and solid-state relays and why they are used with input and output devices.
- Describe how contactors and motor starters are used as interface devices and how they are properly connected to the output sections of PLCs.
- Explain the differences between digital and analog electrical circuits.
- Describe how PLCs and electric motor drives operate in analog circuits.
- Identify the types of signals used by analog input devices.
- Identify the types of signals used by analog output components.

Learner Resources
atplearningresources.com/quicklinks
Access Code: **475203**

PLC SYSTEMS

Work produces light, sound, motion, heat, and more. *Work* is the movement of a load (in pounds) over a distance (in feet). Work is accomplished in an organized, consistent, and repetitive manner by using a system. PLCs are used to control systems (machines or processes) in a predetermined order based on the PLC program. When a PLC is used to control a system, system operation can be monitored and easily changed by reprogramming the process control circuit.

Primary Systems

There are several types of primary systems used to produce work. A *primary system* is a system that transmits and controls the movement of some form of energy. Primary systems include electrical, electronic, fluid power, mechanical, and informational systems. **See Figure 5-1.** Primary systems are used individually or in combination, or are interfaced to perform the required work.

Electrical Systems. An *electrical system* is a primary system that produces work by transmitting and controlling the flow of electricity through conductors (wires). Either alternating current (AC) or direct current (DC) is used to produce the required power. **See Figure 5-2.**

Primary Systems

ELECTRICAL SYSTEM

ELECTRONIC SYSTEM

Atlas Technologies Inc.
FLUID POWER SYSTEM

SEW-Eurodrive, Inc.
MECHANICAL SYSTEM

Figure 5-1. Primary systems transmit and control the movement of energy.

187

CODE CONNECT

Article 750 of the NEC® covers energy management systems. Energy management systems monitor and control energy usage in a building. Section 750.20 of the NEC® prohibits energy management systems from overriding control of alternate power sources for fire pumps, health care facilities, emergency systems, legally required standby systems, and critical operations power systems.

Electrical Systems

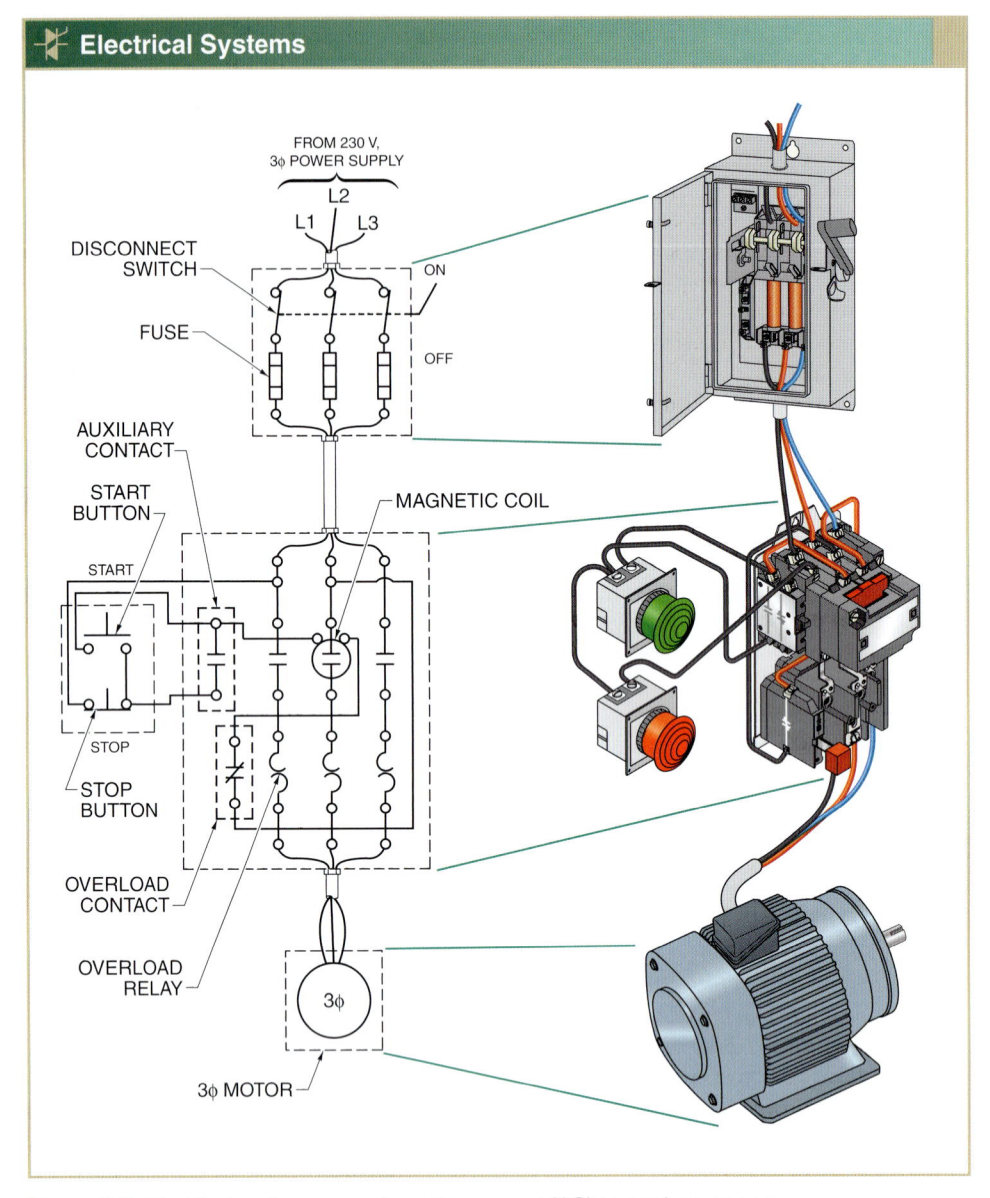

Figure 5-2. Electrical systems use alternating current (AC) to produce power.

Electronic Systems. An *electronic system* is a primary system in which electricity is monitored and controlled to send and/or receive information, produce sound or vision, store data, control circuits, or perform other work. Electronic circuits use either analog or digital electrical signals. An *analog signal* is an electronic signal that has continuously changing quantities (values) between defined limits. A *digital signal* is an electronic signal that has two specific quantities that change in discrete steps (ON or OFF). **See Figure 5-3.**

Fluid Power Systems. A *fluid power system* is a primary system that produces work by transmitting fluid (gas or liquid) under pressure through pipework. A *pneumatic system* is a fluid power system that transmits power using a gas (air). A *hydraulic system* is a fluid power system that transmits power using a liquid (typically oil). **See Figure 5-4.**

Electronic Systems

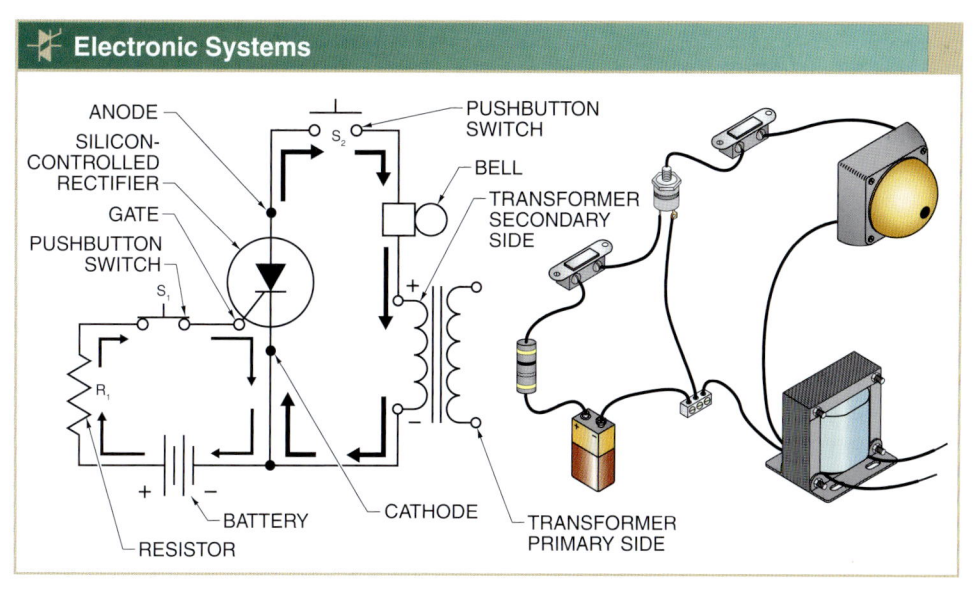

Figure 5-3. Electronic systems monitor and control electricity to send and/or receive information, produce audible or visual signals, store data, control circuits, or perform other work.

Fluid Power Systems

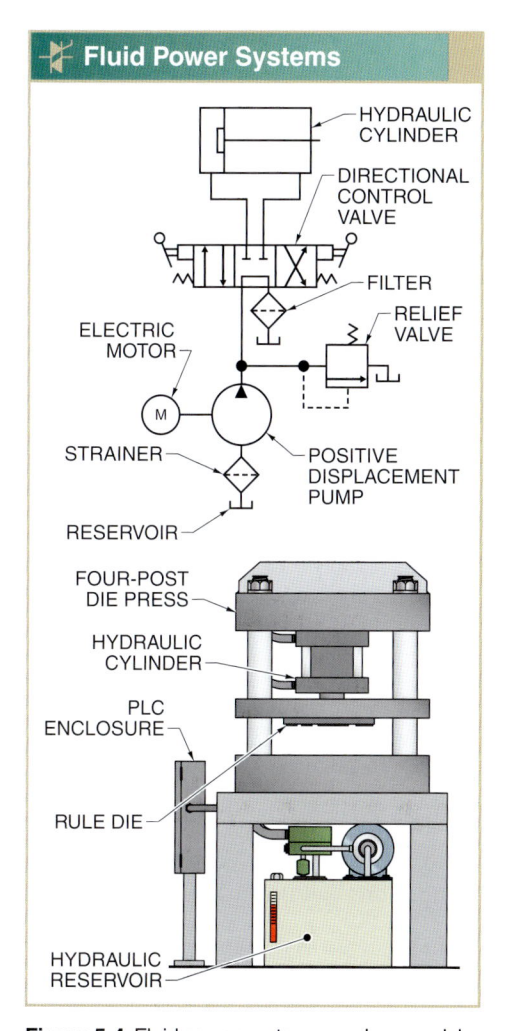

Figure 5-4. Fluid power systems produce work by transmitting fluid under pressure through pipework.

Mechanical Systems. A *mechanical system* is a primary system in which power is transmitted through gears, belts, chains, shafts, couplings, and linkages. **See Figure 5-5.** Mechanical energy can be transmitted through multiple devices, such as from belt to gear to shaft.

PLC TIPS

Over 80% of mechanical system failures are related to vibration. Successful vibration monitoring programs compare vibration readings from permanently installed and remote measuring terminals connected to PLCs to determine when bearing problems are developing.

Informational Systems. An *informational system* is a primary system that monitors operations, displays quantities (values), and indicates the status of machines and processes. Electrical, electronic, fluid power, and mechanical quantities and conditions are displayed, indicating the status of the circuit, process, or application. The informational variables that are displayed and recorded include time, temperature, speed, weight, voltage, current, power, flow rate, level, volume, counts, color, brightness, and pressure. **See Figure 5-6.**

Mechanical Systems

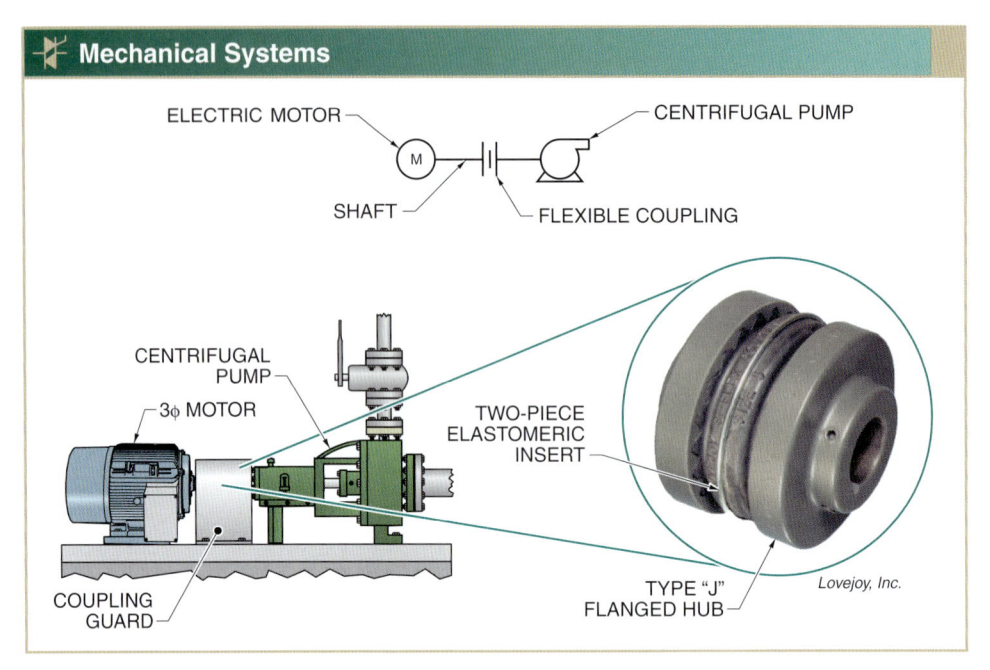

Figure 5-5. Mechanical systems transmit power using a motor that turns a shaft coupled to a centrifugal pump.

Informational Systems

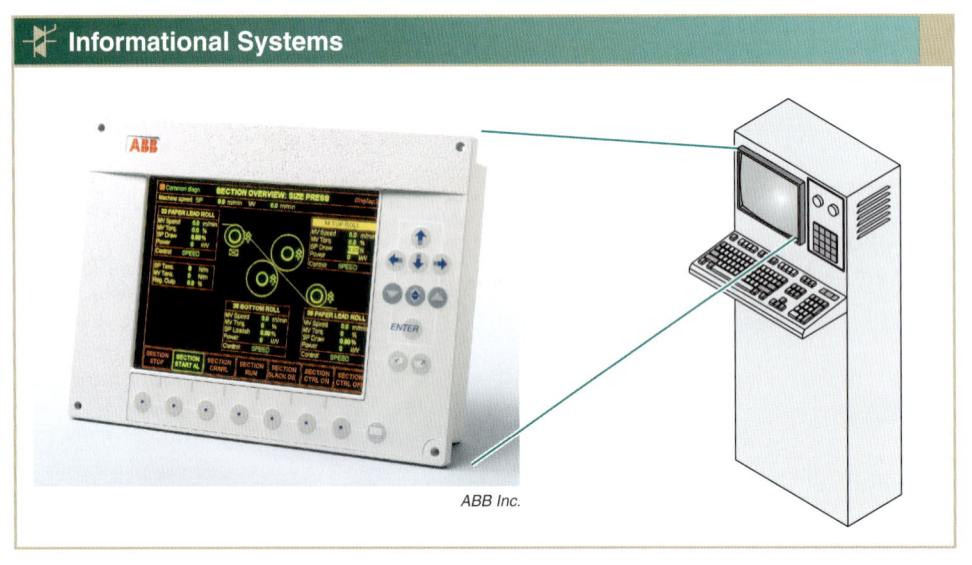

Figure 5-6. Informational systems display electrical, electronic, fluid power, and mechanical quantities and conditions to indicate the status of a circuit, process, or application.

Interfacing Systems. An *interfacing system* is a system or device that allows primary systems to work and/or communicate as one. Primary systems operating at one voltage level (12 V, 115 V) or of one type (electrical or hydraulic) are interconnected so that various parts of the systems can operate together. Interfacing systems allow parts of primary systems that cannot directly work together to be compatible. **See Figure 5-7.**

Combination Systems. A *combination system* is a system that interconnects two or more primary systems to combine the individual advantages of each system to meet the requirements of a given application. If the parts are compatible, the systems can be directly connected. However, combination systems are typically connected using interface devices. **See Figure 5-8.**

System Interfacing

System interfacing permits devices and components of various levels of voltage, current, and power to work together as a system. An *interface device* is an item that allows variously rated components to be used together in the same circuit. Interface devices convert one form or level of energy (electrical, electronic, fluid power, etc.) to another form. With interface devices, almost any machine or process change can be accommodated. **See Figure 5-9.**

Combination Systems

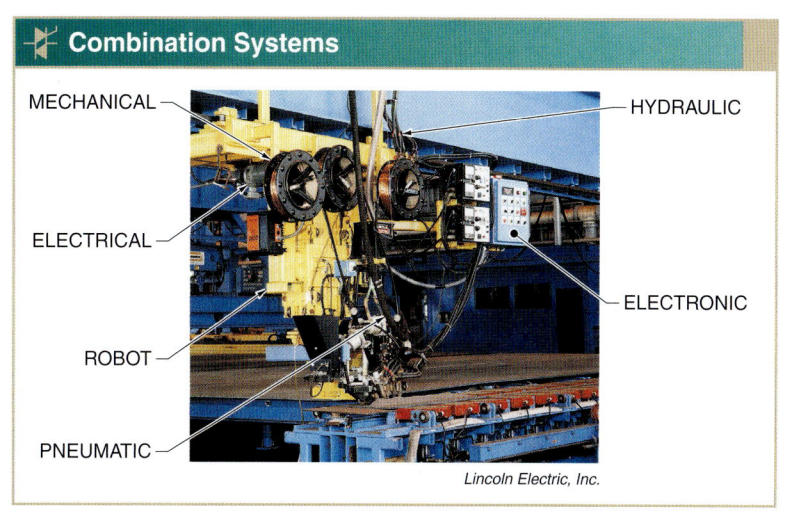

Lincoln Electric, Inc.

Figure 5-8. Combination systems such as industrial robots interconnect two or more primary systems (hydraulic, pneumatic, mechanical, electrical, electronic, digital, or welding), combining the individual advantages of each system to meet the requirements of a given application.

Interfacing Systems

CODE CONNECT

Section 110.22(A) of the NEC® requires that a disconnect be marked with the reason for disconnection. The markings must be legible, durable for the disconnect location/environment, and specific as to the purpose of the disconnect. For example, marking a disconnect with the word "FAN" is not acceptable. Marking the same disconnect "SUPPLY FAN #2" is acceptable. The NEC® does not require a disconnect to be marked if its purpose is obvious due to its location/environment.

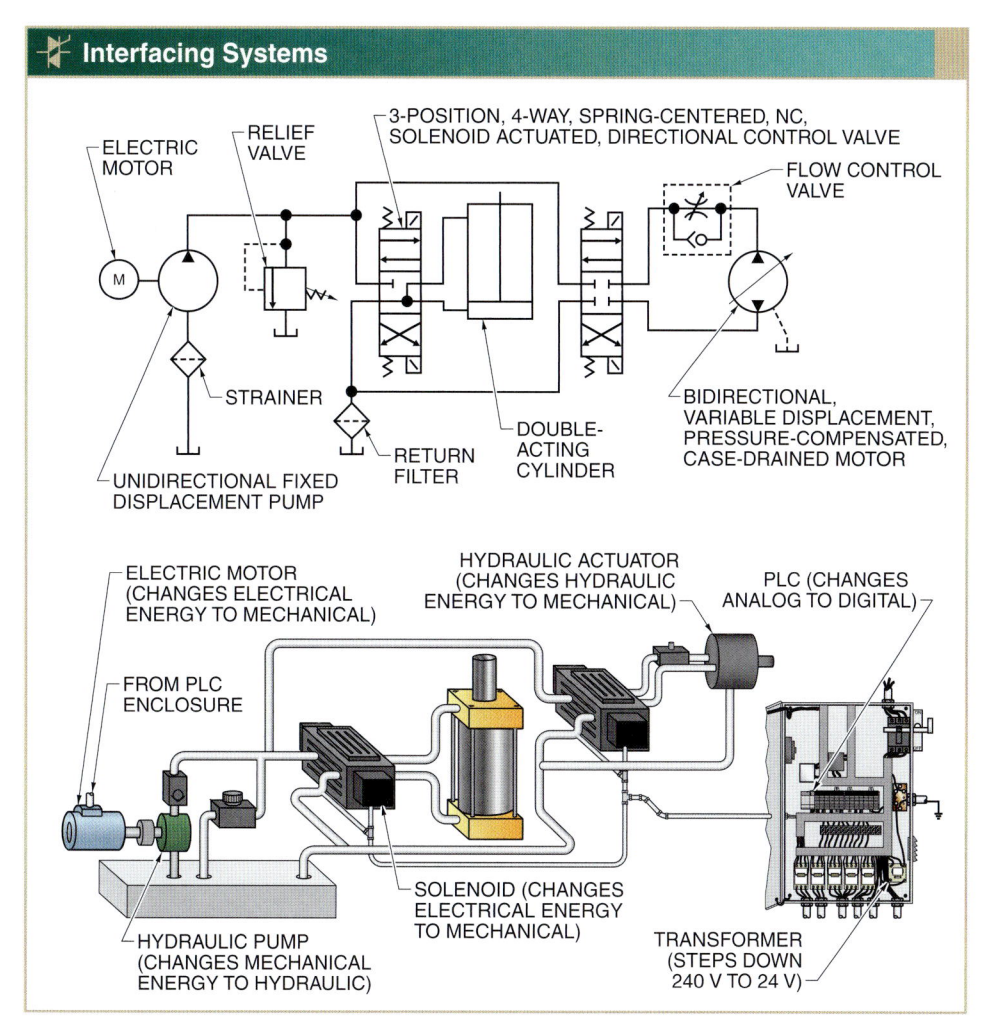

Figure 5-7. Interfacing systems interconnect primary systems.

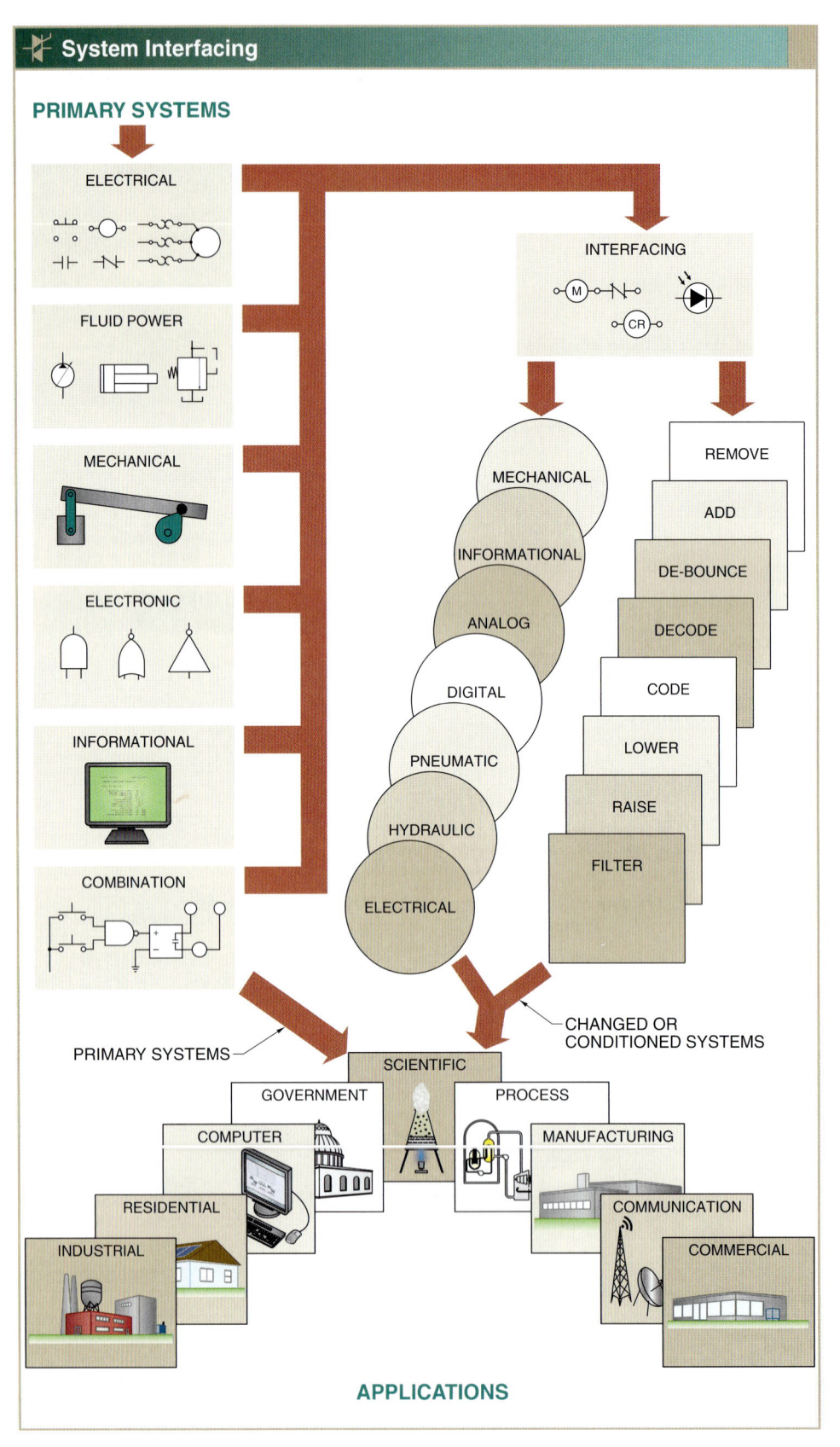

Figure 5-9. System interfacing permits devices and components of various levels of voltage, current, and power to work together as a single system.

Interface devices change or condition a system in several ways. For example, the change can be from a 5 VDC voltage level used by a digital electronic circuit to a 115 VAC voltage level used by solenoid-operated hydraulic valves being controlled by a digital circuit. In the case of a solenoid-operated hydraulic valve, an interface device receives a digital signal and uses the signal to ultimately turn the hydraulic valve ON. A solid-state relay is the interface device typically used for turning a hydraulic valve ON and OFF.

Circuit conditioning is required any time the existing electrical power is not at the proper phase, voltage, or current level for the application. The electrical power may need to be filtered, raised, lowered, or changed to correct the condition. Electronic systems often require voltage or current level adjustments, signal coding, signal de-bouncing, or filtering of the electrical power. Fluid power systems may require that pressure be raised or lowered, that flow rate be raised or lowered, or that temperature be lowered. Typical circuit conditioning in an electrical system requires changing (rectifying) AC to DC. Bridge rectifiers are the most efficient and common rectifier used in single-phase (1ϕ) rectification circuits. **See Figure 5-10.**

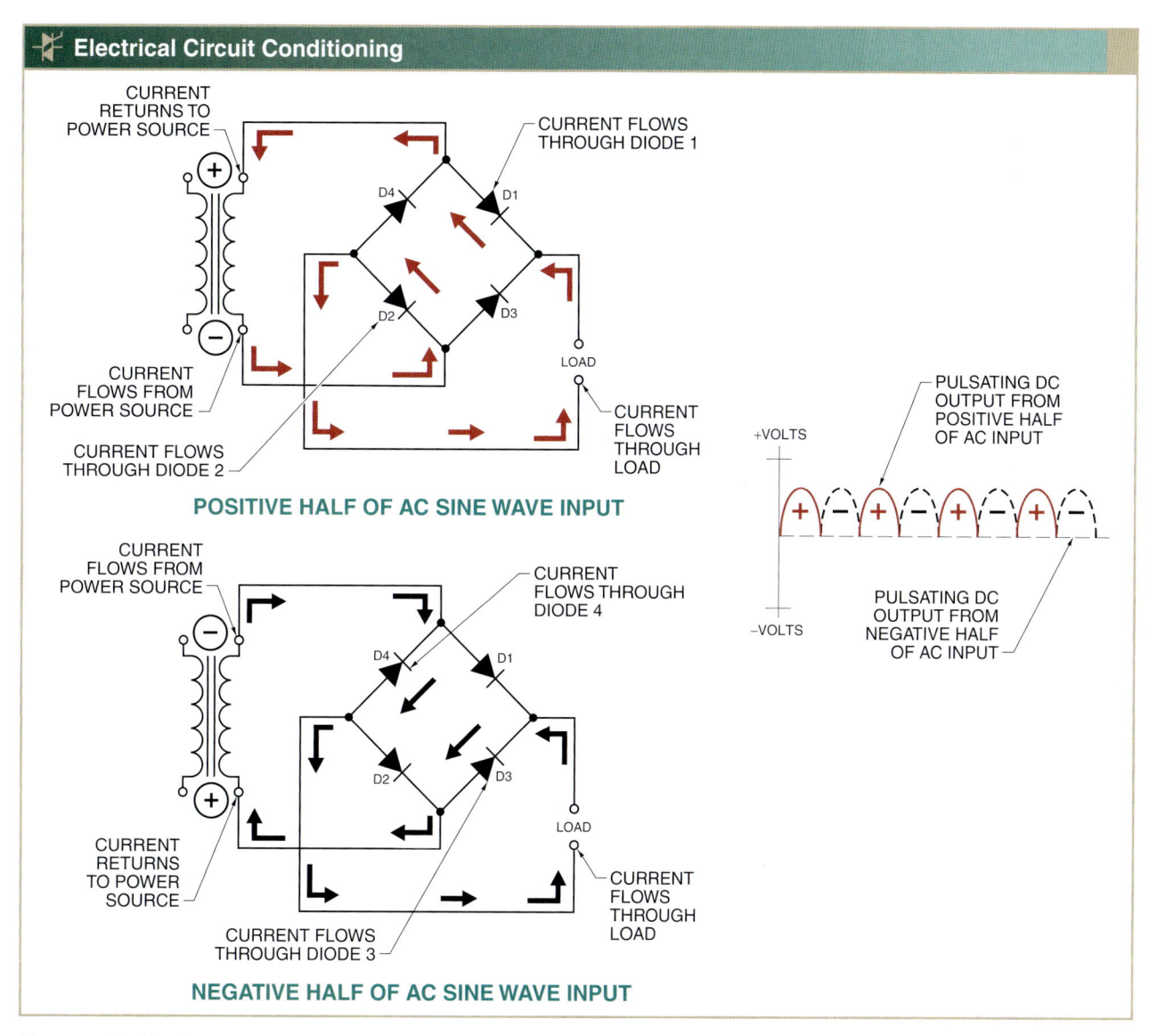

Figure 5-10. Circuit conditioning is required any time the existing electrical power is not at the proper phase, voltage, or current level for the application.

In any system using a PLC, interface devices are required any time an input device is not directly compatible with the input section of the PLC. Likewise, an interface device is required any time an output component is not directly compatible with the output section of the PLC. For example, transformers, magnetic motor starters, and solenoid-operated fluid power valves are required interface devices for a PLC controlling a baking process. **See Figure 5-11.**

In a baking process application, a PLC input module requires a 24 VDC input signal from each of the 16 inputs (I:0 through I:15). A 24 VDC input signal is preferred over a 115 VAC input signal because a 24 VDC signal is safer (lower voltage).

When 115 VAC is the available supply voltage to a baking system, a transformer with a bridge rectifier is required to stepdown a 115 VAC to 24 VDC. A transformer and bridge rectifier are typical interface devices that allow two parts of an electrical circuit that normally could not work together to work together. When a transformer is not used and 24 VDC PLC input devices and input modules are connected directly to a 115 VAC supply, the PLC input module will burn up and the input devices will be damaged or destroyed.

In the baking process, the PLC output module is rated to control (energize and de-energize) 115 VAC loads. The PLC output module can control loads rated less than 115 VAC as long as the current rating of the loads is less than the maximum PLC output module rating. However, the PLC output module cannot be used to control loads rated at a higher voltage than the output module rating because the higher voltage would cause arcing at the PLC output contacts and damage or destroy the output section of the PLC. Also, the current rating of the output loads cannot exceed the PLC output section or module current rating.

During the baking process, when the conveyor and oven motors are 230 VAC/460 VAC 3ϕ motors, the motors cannot be connected directly to a PLC output module. To control high-voltage motors, the PLC must be connected to the motors through an interface device. Typically, high-voltage motor interface devices are magnetic motor starters that have 24 VDC or 115 VAC rated coils and 230/460 VAC rated power contacts. The 115 VAC coil allows the output module of the PLC to interface with the coil, and the 230/460 VAC contacts allow the motor starter to interface with 3ϕ motors.

To develop the linear force required to push the product (bread) into the oven, fluid power (pneumatic) cylinders are used. In order for the PLC electrical system to control the pneumatic system, a three-position, four-way, spring-centered, normally closed, solenoid-actuated, pushbutton-override directional control valve is required as the interface. The solenoids develop a magnetic field that is used to move the spool that is inside the directional valve to direct air to either end of the cylinders.

Interface devices change or condition a system. **See Figure 5-12.** Interface devices can perform the following functions:
- multiply the number of available output contacts
- allow one voltage level to control another voltage level
- allow a small current to control a large current
- allow small DC electronic signals to control AC circuit loads of any size
- change one voltage level to another voltage level
- change the control logic of a circuit

Electrical Power Consumption

Electrical energy is energy created with the flow of an electric charge. All electrical output devices, whether rotating or nonrotating, convert electrical energy into another form of energy, such as mechanical energy, heat, light, or sound. This changed energy form is then used to perform work.

Approximately 62% of all electrical energy is converted into rotary motion by motors. These motors, which are used in large numbers, are available in sizes exceeding 500 HP. Even small motors use large amounts of electrical energy compared to other types of output devices.

System Interface Devices

BREAD-BAKING APPLICATION

PNEUMATIC SCHEMATIC

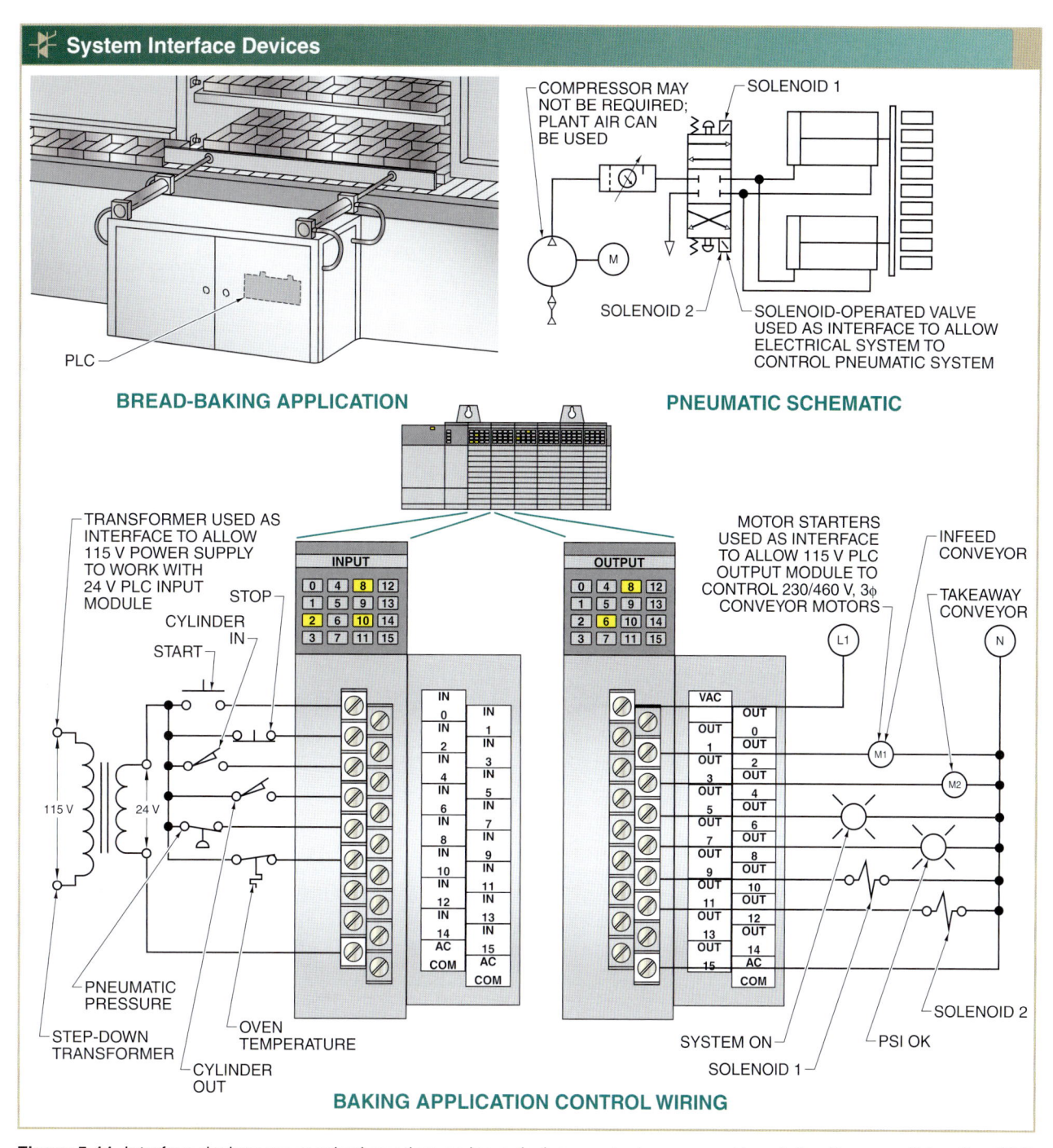

BAKING APPLICATION CONTROL WIRING

Figure 5-11. Interface devices are required any time an input device or output component is not directly compatible with a PLC.

Standard three-phase (3ϕ) motors use the largest amounts of electrical energy in commercial and industrial applications. They use approximately 85% of all energy consumed by motors and approximately 52.7% of the total electrical energy consumed. Most motor manufacturers offer energy-efficient 3ϕ motors in addition to standard 3ϕ motors because of the amount of energy the standard motors consume. Standard motors have an average efficiency of 89%. So it is usually worth paying the approximately 20% higher cost for an energy-efficient 3ϕ motor if it is to be operated long-term.

Interface Device Uses

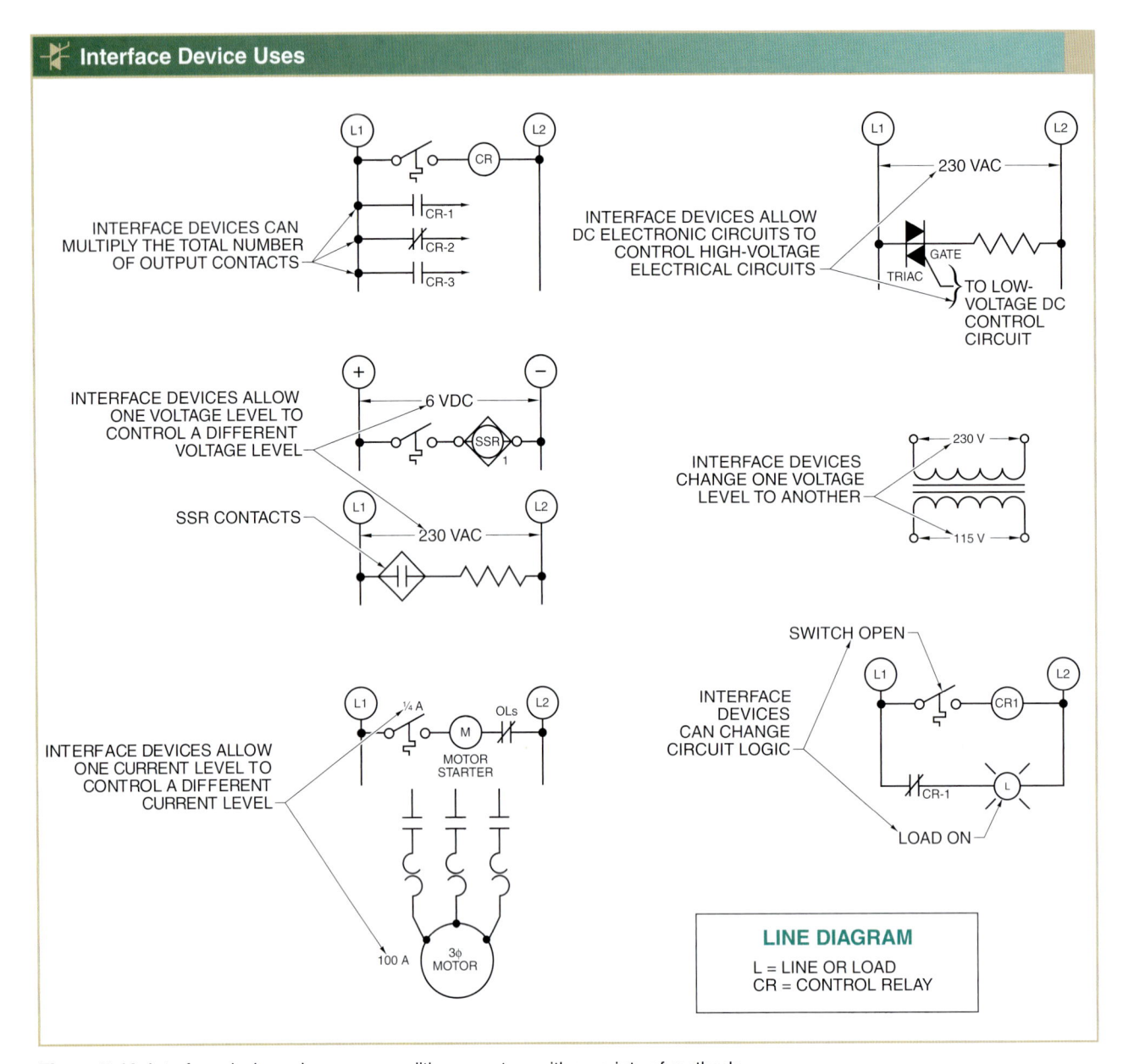

Figure 5-12. Interface devices change or condition a system with a variety of methods.

Lighting accounts for the second-largest use of electrical energy. Approximately 20% of all electrical energy is converted into light. Light is produced by lamps (bulbs). The most common lamps used in commercial and industrial lighting are fluorescent lamps and high-intensity discharge (HID) lamps. Types of HID lamps include low-pressure sodium, mercury vapor, metal halide, and high-pressure sodium. HID lamps are also the most common lamps used for exterior lighting.

Electrical energy is also used to produce heat, linear motion, audible signals, and visual signals. Approximately 18% of all electrical energy is converted into one of these four output types. This group involves a large number of output devices that consume very little energy compared to rotating output devices. Electric heating is the exception because producing heat from electricity is clean, safe, and convenient but requires large amounts of energy. **See Figure 5-13.**

Electrical Energy Consumption

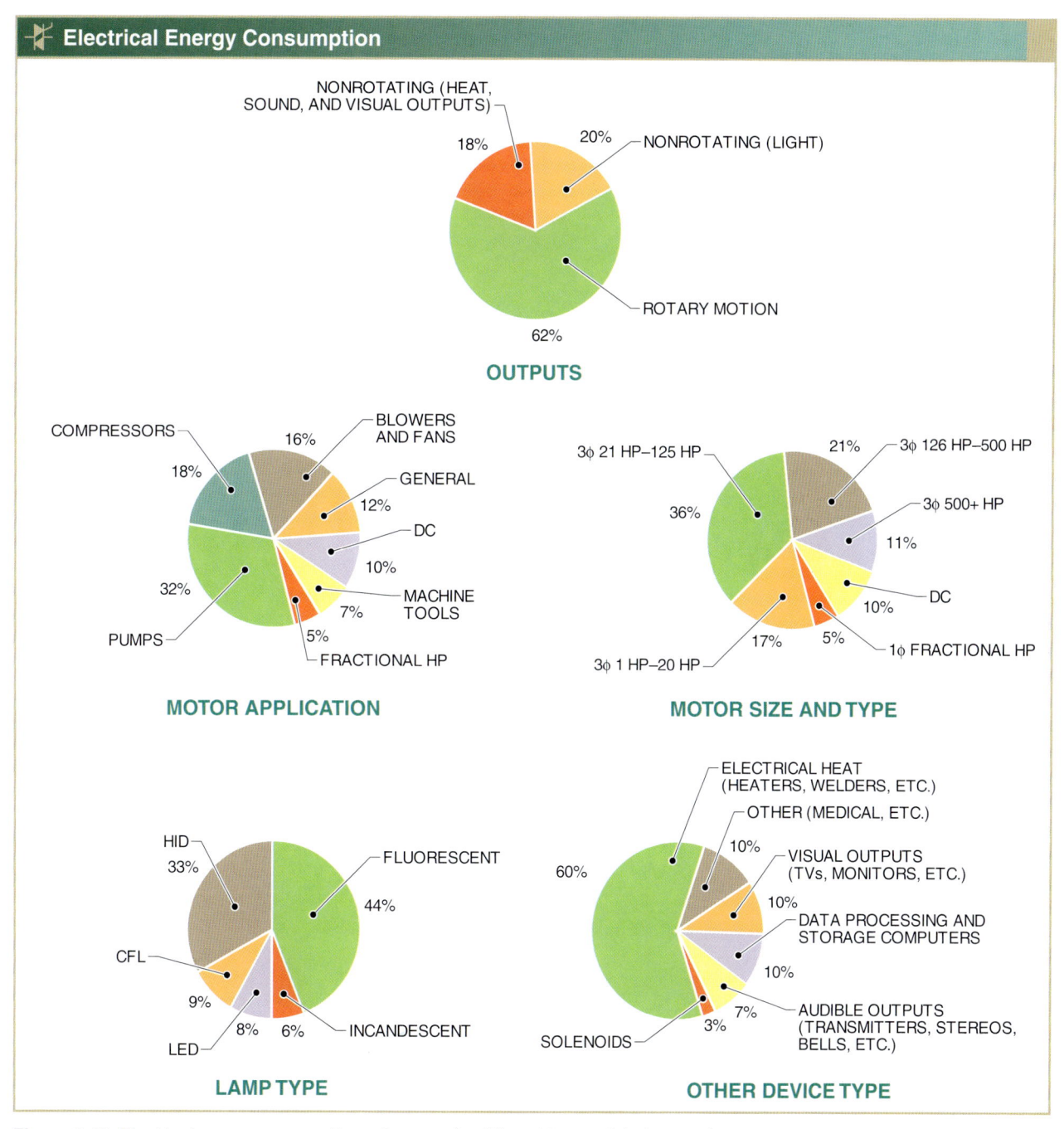

Figure 5-13. Electrical energy consumption rates vary for different types of devices and motors.

ELECTRICAL CIRCUITS

Electrical systems are made up of individual circuits and combinations of basic and complex circuits. Electrical systems that are controlled by PLCs typically include several individual circuits and combination circuits. PLCs are used to reduce cost, simplify wiring, and monitor circuit operation.

Basic Electrical Circuits

Basic electrical circuits must include a source of electricity (battery, generator, or solar cell), a method of controlling the flow of electricity (switch), and a component (load) that converts electrical energy into another form of energy (light, heat, sound, rotating motion, or linear force).

198 PROGRAMMABLE LOGIC CONTROLLERS: PRINCIPLES AND APPLICATIONS

In addition, a basic circuit must also include a protection device (fuse or circuit breaker) to ensure that the circuit operates safely and within designed electrical limits. For example, a basic electrical circuit used to produce heat includes a power supply (voltage rating will vary depending on the design of the unit); a selector switch and temperature switch for controlling the flow of electricity; heating elements as the load; and circuit breakers and/or fuses for protecting the system from overcurrents. **See Figure 5-14.**

Basic Electrical Circuits

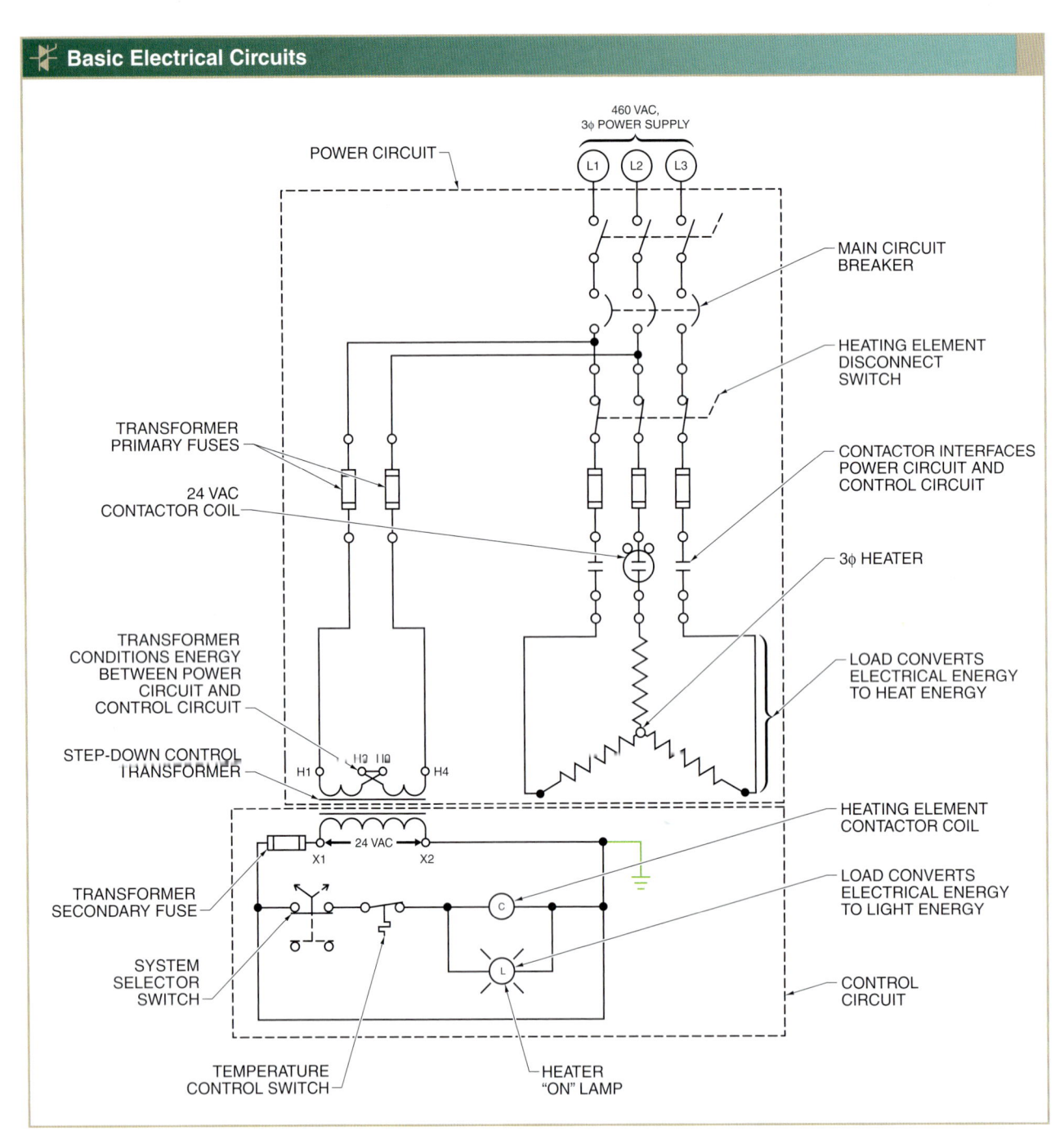

[Diagram showing a basic electrical circuit with 460 VAC, 3φ power supply. Labeled components include: POWER CIRCUIT, MAIN CIRCUIT BREAKER, HEATING ELEMENT DISCONNECT SWITCH, TRANSFORMER PRIMARY FUSES, 24 VAC CONTACTOR COIL, CONTACTOR INTERFACES POWER CIRCUIT AND CONTROL CIRCUIT, 3φ HEATER, TRANSFORMER CONDITIONS ENERGY BETWEEN POWER CIRCUIT AND CONTROL CIRCUIT, LOAD CONVERTS ELECTRICAL ENERGY TO HEAT ENERGY, STEP-DOWN CONTROL TRANSFORMER (H1, H2, H3, H4), HEATING ELEMENT CONTACTOR COIL, TRANSFORMER SECONDARY FUSE, LOAD CONVERTS ELECTRICAL ENERGY TO LIGHT ENERGY, SYSTEM SELECTOR SWITCH, CONTROL CIRCUIT, TEMPERATURE CONTROL SWITCH, HEATER "ON" LAMP.]

Figure 5-14. Basic electrical circuits must include a source of electricity, a method of controlling the flow of electricity, and a component that converts electrical energy into some other usable form of energy. A circuit must also include protection device(s) to ensure that the circuit operates safely and within designed electrical limits.

Interface devices are also required. A transformer is required to step-down 460 VAC to 24 VAC and to allow the higher voltage power circuit to work with the lower voltage control circuit. A heating contactor is required to allow a single-phase low-voltage control circuit to control three-phase high-voltage heating elements.

A heating element control circuit is connected to a PLC so that the PLC controls the heating elements through its programming. The PLC requires two input terminals, one for the selector switch and one for the temperature switch, and two output terminals, one for the heating contactor coil and one for the heater "ON" lamp. **See Figure 5-15.** Even though the hardwired heating element circuit and the PLC-controlled heating element circuit both operate in the same manner, there are differences between the two circuits.

When a heating element is hardwired into a circuit, the heating contactor coil and heater ON lamp are wired in parallel. In a PLC-controlled circuit, the heating contactor coil and heater ON lamp are each wired to separate output terminals, and the output terminals are programmed to be in parallel. By connecting the heating contactor coil and heater ON lamp to different output terminals, the current of each output component passes only through the assigned output terminal. In addition, each output component can be monitored separately and forced ON or OFF using the PLC program.

In a hardwired heating element circuit, the current of the heating contactor coil and heater ON lamp passes through both the selector switch and temperature switch. Both switches must be rated for the current level of both loads. In a PLC-controlled circuit, selector switches and temperature switches do not carry the current of the loads but only the small amount of current drawn by the PLC input module. The selector switch and temperature switch contacts can be rated for a very low current and should operate without failure.

In a hardwired heating element circuit, the circuit cannot easily be changed or modified without rewiring the control circuit.

In a PLC-controlled circuit, the circuit can be reprogrammed without rewiring the input devices or output components and also allows for a centralized location when starting any system troubleshooting. Input and output module status lamps indicate which input devices are closed and which output components are ON.

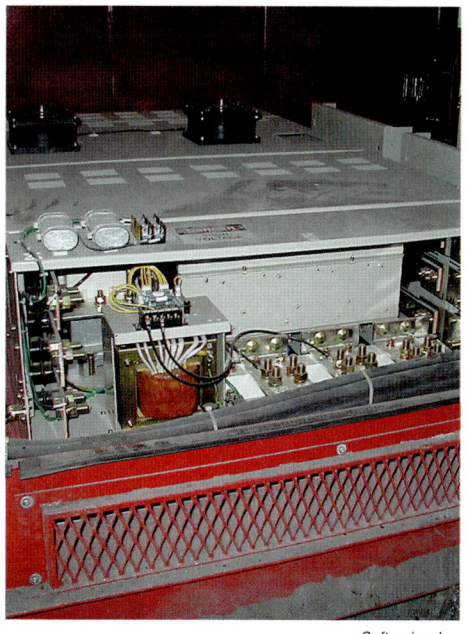

Saftronics, Inc.
Control transformers are interface devices that step-down voltages in an electrical enclosure for PLC, electric motor drive, and other solid-state equipment usage.

CODE CONNECT
Section 430.43 of the NEC® covers the automatic restart of a motor after an overload (heater) trip. Section 430.43 requires that a motor overload device that can automatically restart a motor be approved for the specific motor it protects but forbids installing an overload device that automatically restarts the motor.

Improving Basic Electrical Circuits

All electrical circuits begin as basic circuits. In a basic electrical circuit, electricity is delivered from the power supply, protected from overcurrents and overloads, controlled by switches, and used to produce work through output components. Basic circuits can be improved by adding protection (phase-loss protection); instrumentation for monitoring circuit parameters; and controls for loads (dimming lamps, timing and sequencing load operations); as well as by interconnecting all basic circuits in a system. **See Figure 5-16.**

200 PROGRAMMABLE LOGIC CONTROLLERS: PRINCIPLES AND APPLICATIONS

Basic PLC Electrical Circuits

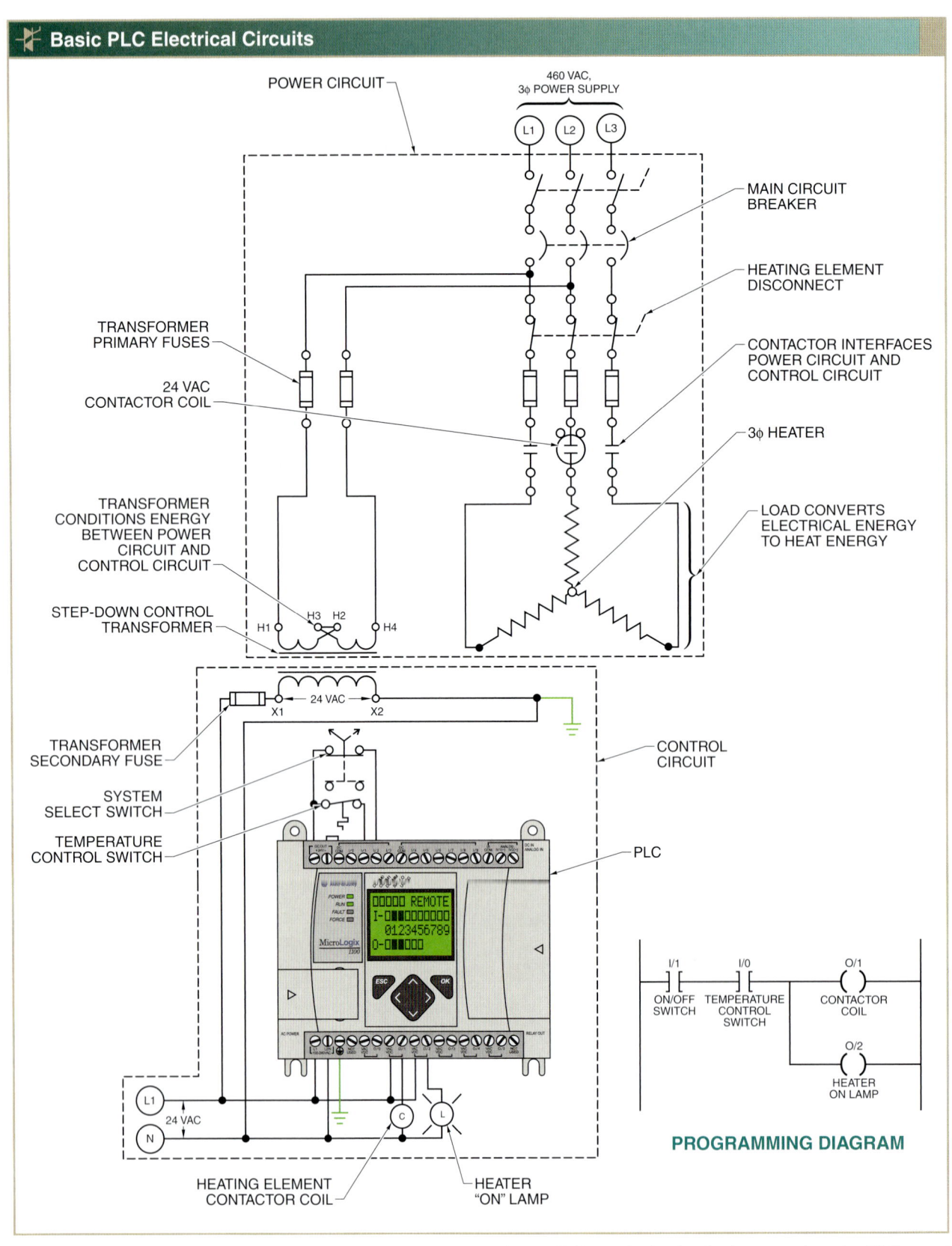

Figure 5-15. Basic PLC-controlled circuits with multiple loads wired in parallel have the loads wired separately to the PLC output terminals and the output terminals programmed in parallel.

Improving Basic Electrical Circuits

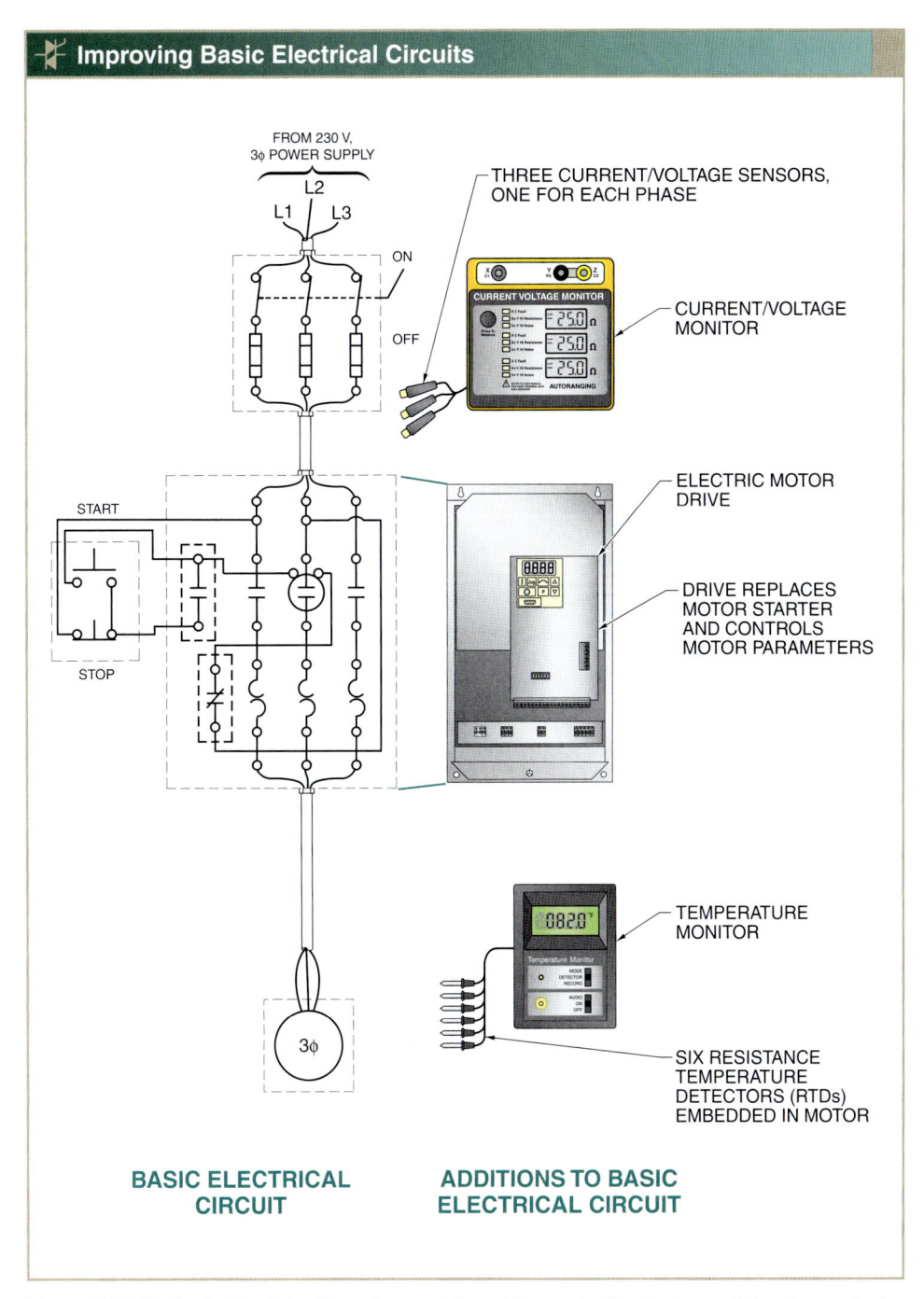

Figure 5-16. Basic electrical circuits are improved by adding protection, instrumentation for monitoring circuit parameters, and finer controls, as well as by interconnecting all basic circuits into a system.

PLC TIPS

Originally, PLCs were designed to replace relays. However, the relays the PLCs replaced were not interface devices but relays that were used to create circuit logic, such as AND, OR, NOT, NOR, and NAND logic.

Complex Electrical Circuits

The more complex an electrical circuit is in a system, the greater the need for consolidated controls, such as with network control systems. A *network control system* is a system of computers, terminals, databases, and PLCs connected by communication lines.

In a network, input devices such as pushbuttons, limit switches, temperature switches, smoke detectors, and pressure switches, and output components such as motors, lamps, solenoids, and alarms are interconnected to PLCs, PCs, and/or other electronic control and monitoring devices. PLCs are used to monitor the input device circuits and control the output component circuits. PCs, along with PLCs, are added to allow the gathering, monitoring, controlling, and displaying of system data. PLCs and PCs are interconnected through communication lines to form various types and levels of networks. **See Figure 5-17.**

On a plant floor, input devices and output components are connected to PLC input and output modules (or sections). Various input module types allow analog and digital input devices to send information to the PLC. An analog input device sends a continuously changing variable signal to a PLC or PC. Temperature, pressure, flow, and level sensors are common analog input devices for a PLC-controlled system. ON/OFF switches such as pushbuttons, limit switches, and toggle switches are common digital input devices. A digital input device is a device that is either ON or OFF (open or closed). **See Figure 5-18.**

An output component is used to produce work, light, heat, or sound or to display information. Output components use the majority of electrical power in a system because output components such as lamps, motors, and heating elements produce the required work. Because output components use far more power than input devices, output components often require an interface device (relay, contactor, or motor starter) between the output component and the PLC output module.

The more complex the machine or process, the higher the level of control required. PLCs allow this higher level of control.

Interfacing Circuits

All electric loads must be controlled. Control circuits are used to control loads that vary from simple to complex. An ON/OFF switch is used for the simple control of low-power loads, but PLCs, computers, and networks are used for complex control applications. The application parameters, load size, cost limitations, and environment in which the load is operated determine the type of control circuit used.

Once all loads (lamps, motors, or heating elements) are identified, the exact type of control devices and output components required to make the circuit operate as designed are selected. Low-power output components are connected directly to the output terminals of the PLC as long as the output component voltage, current, and power ratings are less than the maximum rating of the PLC output module. An interface device is required when the power requirements of the load are higher than the ratings of the output module. A *high-power load* is a load whose voltage and current rating are greater than the voltage and current rating of the PLC output. The most common interface devices used to allow PLCs to control high-power loads are relays, contactors, and motor starters. **See Figure 5-19.**

Relays. Relays typically have low-power-rated contacts and are used for switching low-power loads such as alarms, small solenoids, and very small (milliwatt) motors. Low-power loads can be connected directly to a PLC as long as the load is less than the PLC output section rating. However, by using a relay between the load and the PLC, the circuit and system is better protected. Even loads like solenoids can produce high transient voltages and shorts. When a solenoid is connected directly to a PLC and a fault occurs, the PLC output section can be damaged and will have to be replaced and all output components (and inputs on nonmodular PLCs) rewired.

Contactors. Contactors have high-power-rated contacts and are used for switching high-power loads such as lamps (fluorescent and HID), heating elements, transformers, and small motors (less than ¼ HP) that have built-in overload protection. Contactors are required between the PLC output section and high-power loads.

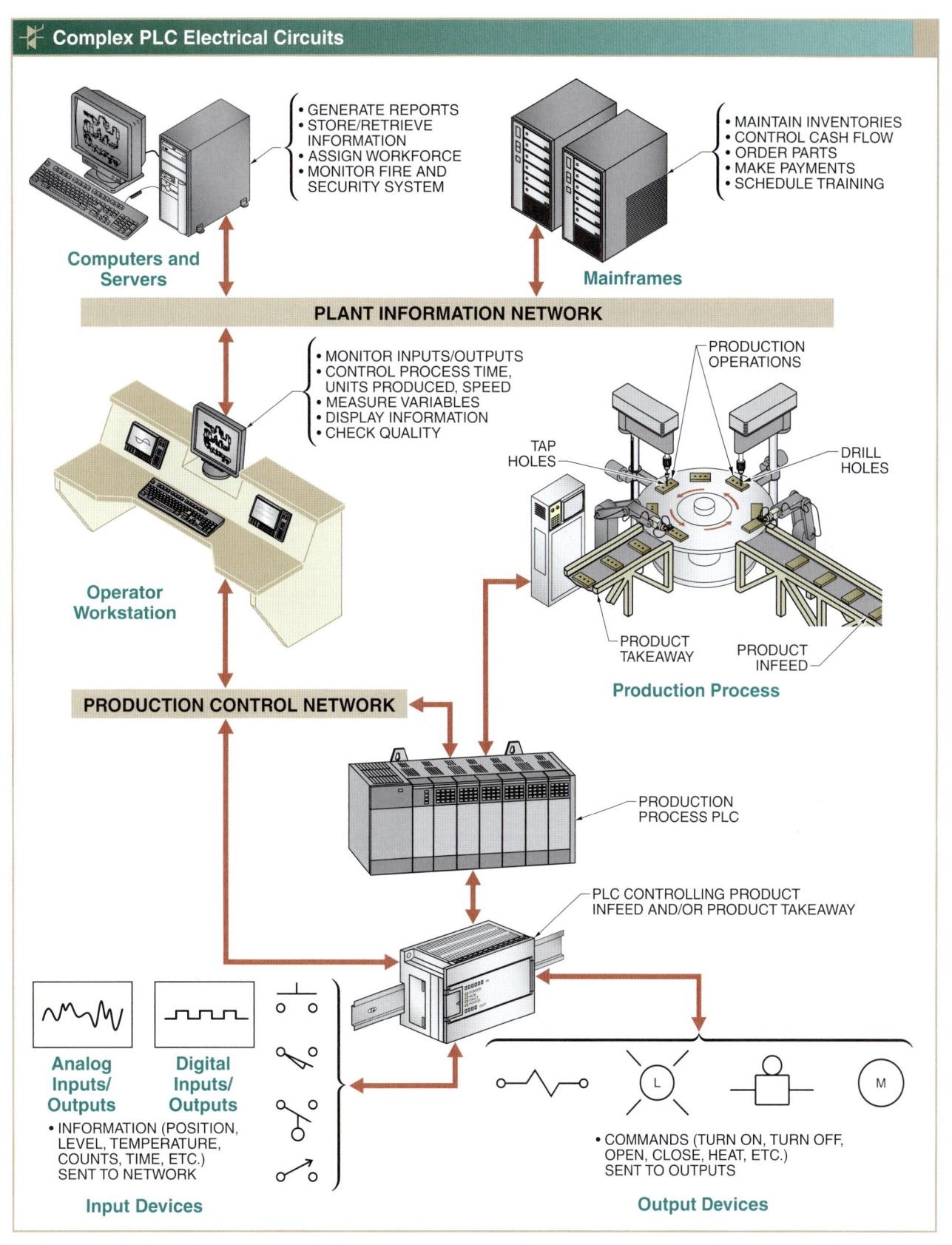

Chapter 5 — PLC Systems, Circuits, and Interface Devices   203

Figure 5-17. PLCs and PCs are interconnected through communication lines to form various types and levels of networks.

Input Devices and Output Components

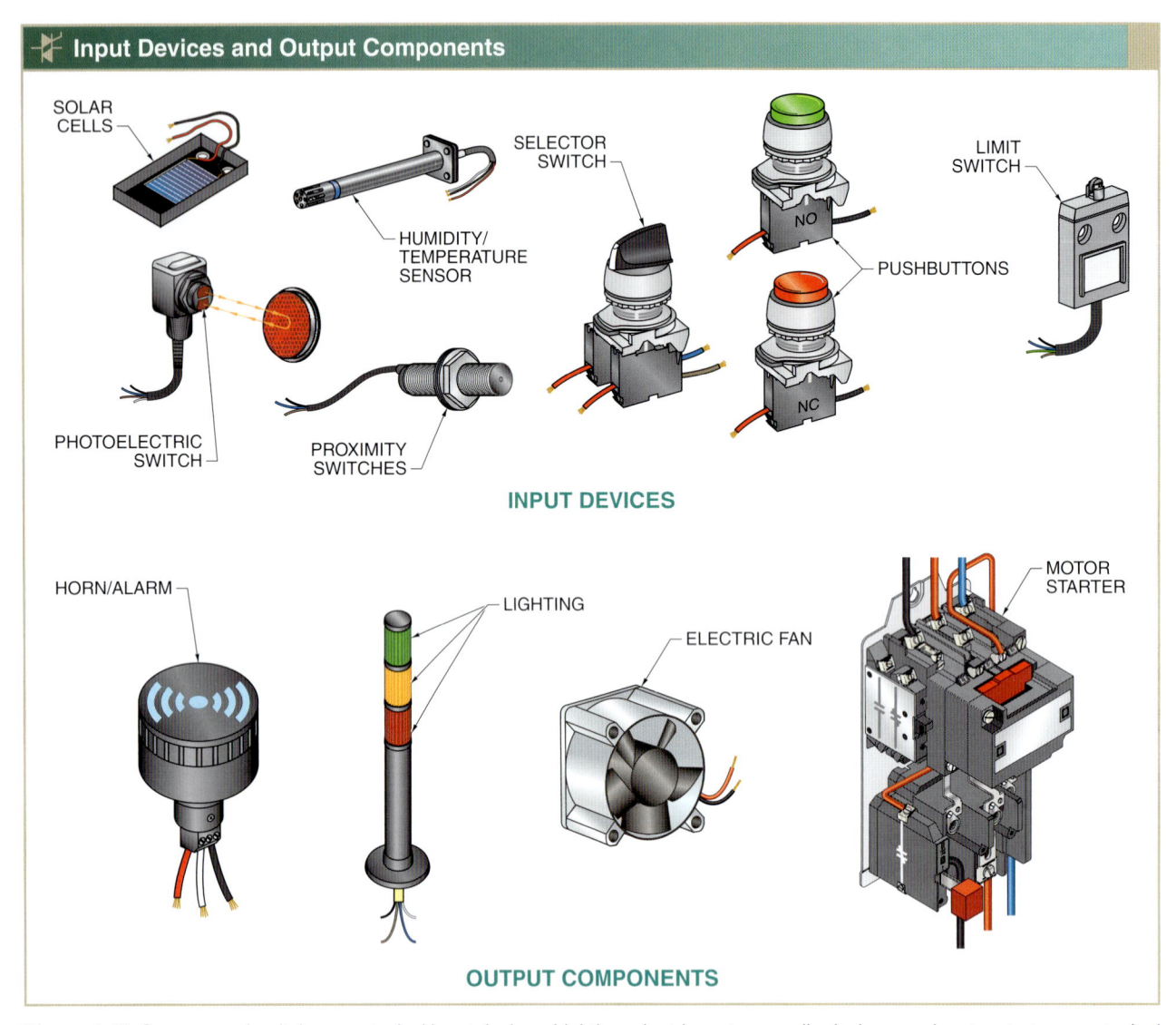

Figure 5-18. Sensors and switches are typical input devices. Lighting, electric motors, audio devices, and motor starters are typical output components.

Motor Starters. Motor starters are used for switching most motors over ¼ HP and all motors over 1 HP. Contactors and motor starters are available in sizes that switch loads ranging from a few amperes to several thousand amperes. Motor starters are required between the PLC output section and high-power loads.

INTERFACE DEVICES

There are several interface devices used in electrical circuits to connect components that cannot be directly connected. Some relays and all contactors and motor starters have high-power contacts that can be used to control high-power-rated loads. The contacts of interface devices can be mechanical or solid-state. Mechanical contacts have been used for years and their design allows them to absorb a certain amount of abuse. Mechanical contacts are used in electromechanical relays, contactors, and motor starters. Solid-state contacts have increased in use and are now commonly used with electric motor drives (variable-frequency drives, AC drives, and DC drives) as well as with solid-state relays.

⚠ **DANGER**
Contactor and motor starter interface devices can be controlled by a PLC using 12 VDC or 24 VDC. However, they can be connected to a 230 VAC, 460 VAC, or 575 VAC output component.

Output Component Interface Devices

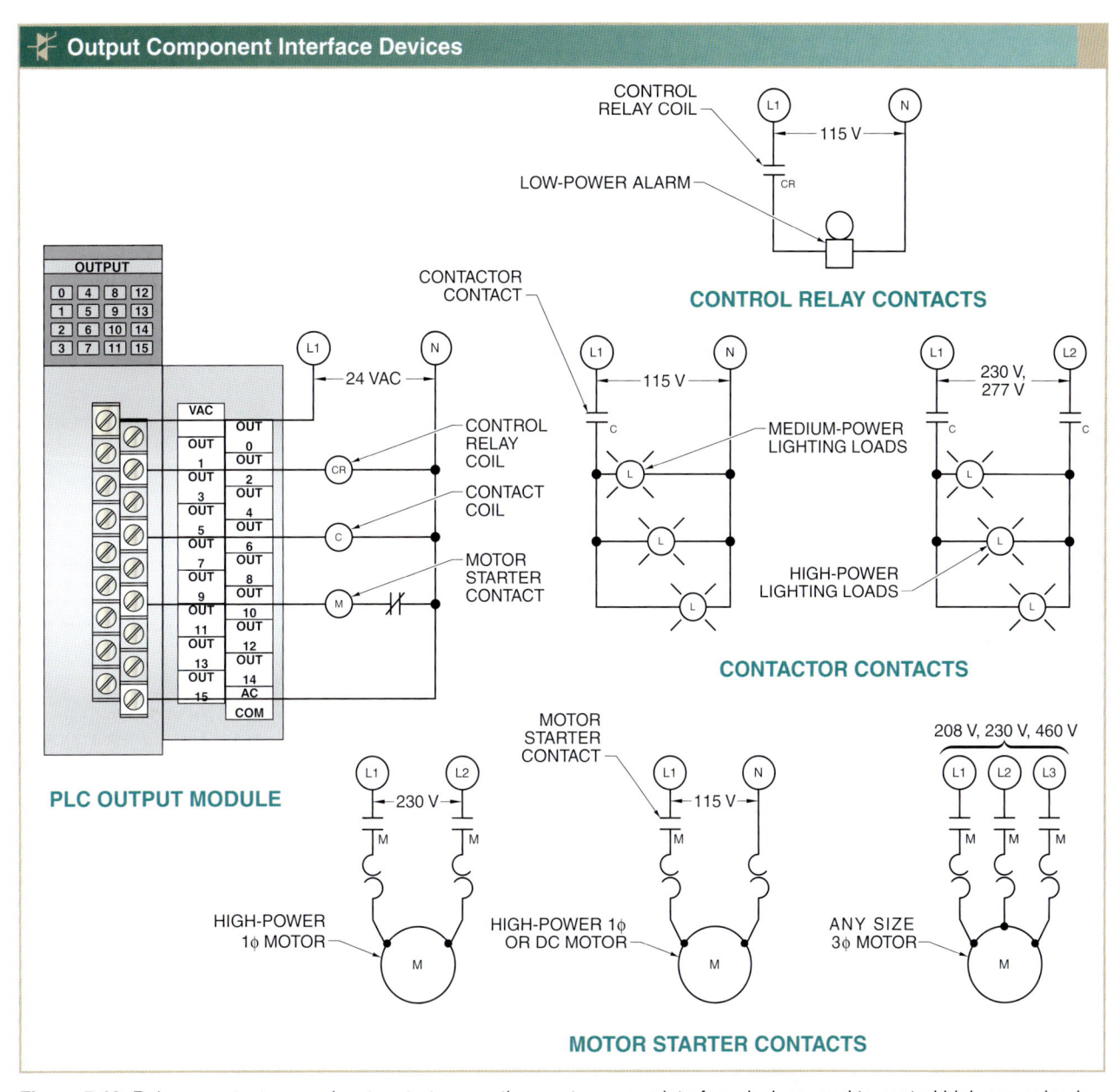

Figure 5-19. Relays, contactors, and motor starters are the most common interface devices used to control high-power loads.

Electromechanical Relays

A *control relay* is a device that controls an electrical circuit by opening and closing contacts in another circuit. Control relays can be electromechanical or solid-state. An *electromechanical relay* is a switching device that has sets of contacts that are closed by magnetic force. In an electromechanical relay circuit, a coil is used to close normally open contacts with a magnetic force that develops when the coil is energized. Control relays are described by the number of poles, throws, and breaks the relay has. **See Figure 5-20. See Appendix.**

A *pole (contact)* is a completely isolated circuit that a relay can switch. A single-pole contact can carry current through only one circuit at a time. A double-pole contact can carry current through two circuits simultaneously. In a double-pole contact, the two circuits are mechanically connected to open or close simultaneously and are electrically insulated from each other.

Electromechanical Relays

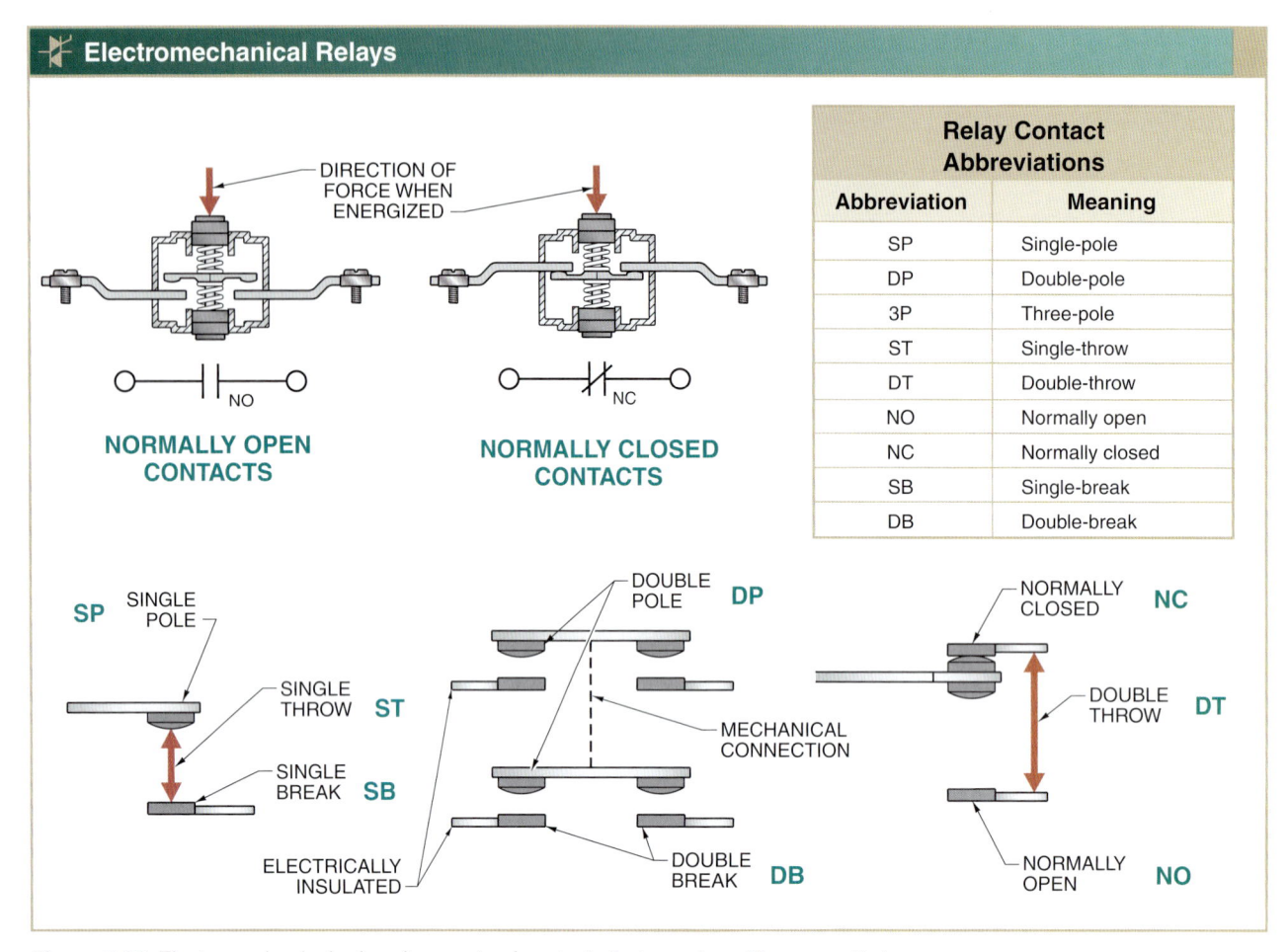

Relay Contact Abbreviations	
Abbreviation	Meaning
SP	Single-pole
DP	Double-pole
3P	Three-pole
ST	Single-throw
DT	Double-throw
NO	Normally open
NC	Normally closed
SB	Single-break
DB	Double-break

Figure 5-20. Electromechanical relays have sets of contacts that are closed by magnetic force.

On electrical prints, mechanical connections are represented by dashed lines connecting the poles. Control relays are available with one pole to twelve poles. A *throw* is the number of closed contact positions per pole. A single-throw contact can only control one circuit. A double-throw contact can control two circuits.

A *break* is a place on a contact that opens or closes an electrical circuit. A *single-break (SB) contact* is a contact that breaks (opens) an electrical circuit in one place. A *double-break (DB) contact* is a contact that breaks (opens) an electrical circuit in two places. All contacts are designed to be single-break or double-break. Single-break contacts are used when switching low-power devices such as horns and lights. Double-break contacts are used when switching devices such as solenoids.

Solid-State Relays

A *solid-state relay* is a relay that uses electronic switching devices in place of mechanical contacts. In a solid-state relay, the coil is replaced by an input circuit and the contacts are replaced with a solid-state switching component (transistor, silicon-controlled rectifier [SCR], or triac). **See Figure 5-21.**

The input circuit of a solid-state relay is activated by applying a voltage to the circuit that is higher than the specified pick-up voltage of the relay. Most solid-state relays have an input voltage range of 3 VDC to 32 VDC. The voltage range allows a single solid-state relay to be used with most electronic circuits and PLC output modules.

The switched circuit (output) of a solid-state relay is controlled by the input signal. Most solid-state relays use a triac (for AC switching) or an SCR (for DC switching) as the output switching component. Triacs and SCRs change from an OFF-state (contacts open) to an ON-state (contacts closed) very quickly when the input is energized. The gate of a solid-state relay is switched ON when the relay receives the proper input signal.

The type of relay used for an application depends on the life expectancy, electrical requirements, and cost requirements of the application. **See Figure 5-22.** Electromechanical relays are typically rated for 250,000 operations, and solid-state relays can be rated for millions of operations. Either type of relay can be used in applications that require a nominal amount of switching operations and where switching time is not important. However, in applications that require thousands of switching operations a day, such as with flashing lights and high-speed production lines, solid-state relays are used.

Solid-State Relays

AC OUTPUT SOLID-STATE RELAY

∼ = AC
⎓ = DC

DC OUTPUT SOLID-STATE RELAY

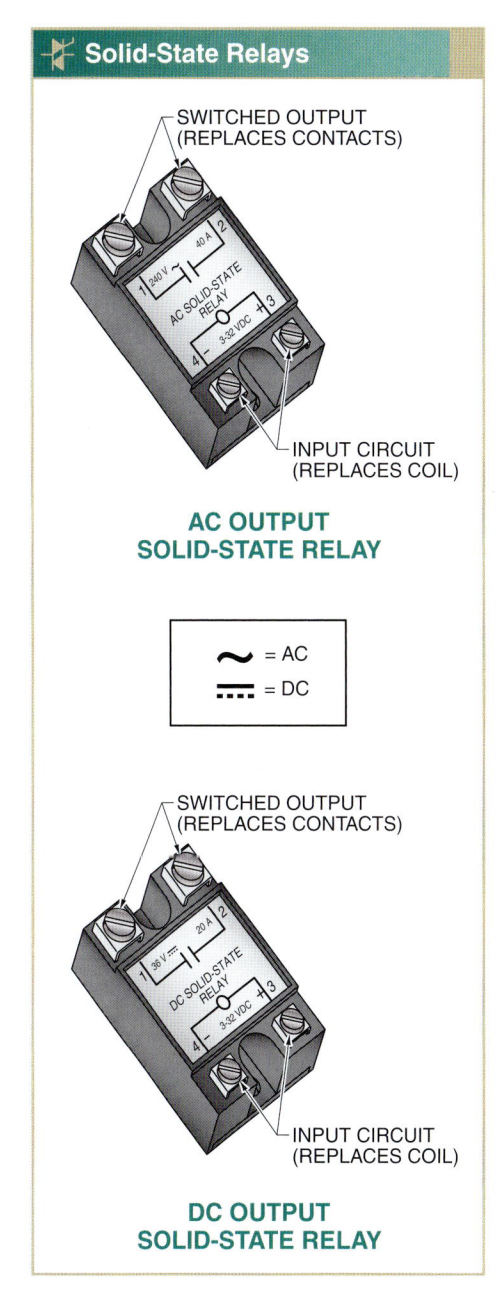

Figure 5-21. Solid-state relays have an input voltage range, such as 3 VDC to 32 VDC, that allows a single solid-state relay to be used with most electronic circuits and PLC output modules.

Advantages and Limitations of Relays

Electromechanical	
Advantages	**Limitations**
Multipole, multithrow contact arrangements	Contact wear; limited life
AC and DC contacts	Rapid switching or high currents shorten life
Low cost	Electromagnetic noise generation
No contact voltage drop; no heat sink required	Poor performance when switching high-inrush circuits
Resistant to transient voltages	
No OFF-state leakage current	

Solid-State	
Advantages	**Limitations**
Long life	Typically one set of contacts per relay
No contacts to wear	Voltage drop across relay requires heat sink
No electromagnetic interference	Only AC or DC compatibility
Resistant to shock and vibration	OFF-state leakage current
Fast switching capabilities	Narrow frequency range (40 Hz to 70 Hz)
Zero switching and instant-on switching modes	
Logic compatible to PLCs	

Figure 5-22. The type of relay used depends on the life expectancy, electrical requirements, and cost requirements of the application.

208 PROGRAMMABLE LOGIC CONTROLLERS: PRINCIPLES AND APPLICATIONS

Although there are applications using low-power devices in which the output load can be directly connected to a PLC output section or module safely, any load that can cause a problem must include an interface device between the PLC and the load. **See Figure 5-23.** For example, solenoids can be rated for low-power while energized, but the coil of a solenoid draws high current when energized and produces high voltage spikes (transients) when de-energized.

A few solid-state relays are available with multiple output (multicontact) switching devices. Solid-state relays with three output contacts are typically used to control medium-power loads. The low-voltage input (coil) side of a solid-state relay is connected to the output module or section of the PLC. The output side of a solid-state relay can be used to control high-voltage medium-power loads such as 3ϕ heating elements. **See Figure 5-24.**

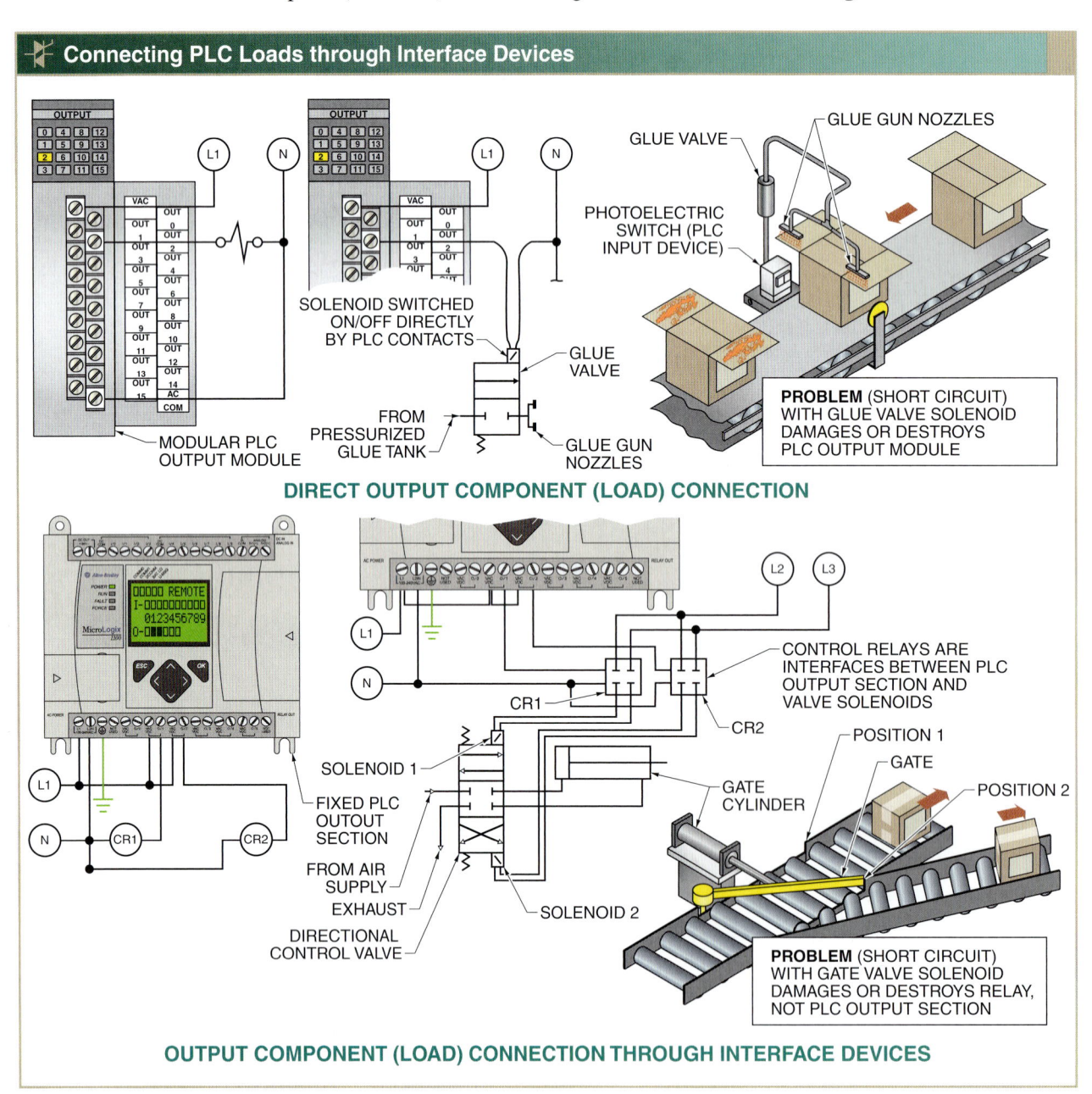

Connecting PLC Loads through Interface Devices

DIRECT OUTPUT COMPONENT (LOAD) CONNECTION

OUTPUT COMPONENT (LOAD) CONNECTION THROUGH INTERFACE DEVICES

Figure 5-23. Any load that can cause a problem to the output section or module of a PLC must include an interface device between the PLC and the load.

Solid-State Relay Output Interface Devices

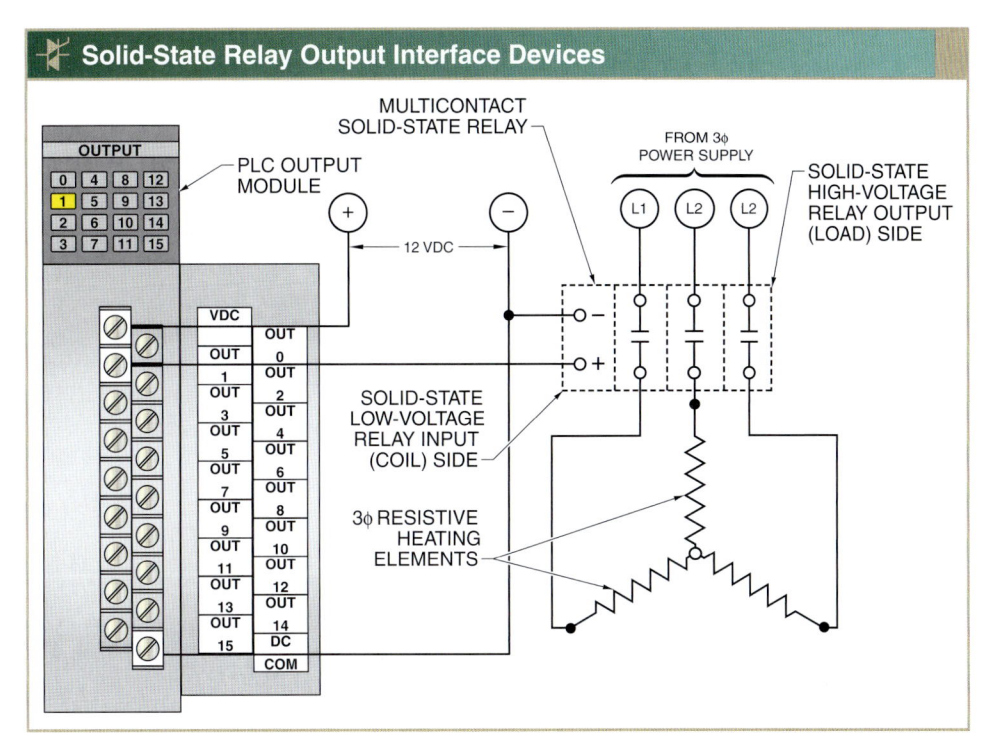

Figure 5-24. The output side of a solid-state relay can be used to control high-voltage medium-power loads such as 3ϕ heating elements.

Contactors

A *contactor* is an electrical control device that is designed to control high-power, non-motor loads. Contactors are relays that have high-current contacts. Contactors control high-power loads and/or many loads on an individual circuit (several loads connected in parallel). The most common use for contactors is in high-power lighting circuits. **See Figure 5-25.**

In lighting circuits, contactor contacts carry the high current required by the loads, and the PLC output (internal circuitry) carries the low current required by the contactor coil (typically less than 1 A). The use of a contactor allows the PLC output circuitry to be rated much lower than the loads that the PLC is controlling. Contactor ratings range from a few amperes to several thousand amperes. The advantage of using a PLC to control lamps is that every possible combination of lighting requirements can be programmed without ever having to rewire the lighting circuit.

Contactors can also be used as heating contactors. Heating elements draw high currents in order to produce heat. A 1200 W heating element will draw 10.4 A on a 115 V circuit, or 5.2 A on a 230 V circuit. The high current cannot be passed through PLC output contacts. **See Figure 5-26.**

> **⚠ WARNING**
>
> PLC modules can maintain an electrical charge after power has been removed. Always use a voltmeter to ensure there is no power at the module terminals before adding or removing devices or components.

Motor Starters

To allow a PLC to control high-power (horsepower) motors, magnetic motor starters (mechanical starters) or electric motor drives (solid-state starters) are traditionally used. A *magnetic motor starter* is a starter with an electrically operated mechanical switch (contactor) that includes motor overload protection. Magnetic motor starters consist of electrical contacts, a coil to magnetically open and close the contacts, and overload protection. Magnetic motor starters use a small control current to energize a coil to send high power to a motor. **See Figure 5-27.**

210 PROGRAMMABLE LOGIC CONTROLLERS: PRINCIPLES AND APPLICATIONS

Contactor Interfaces

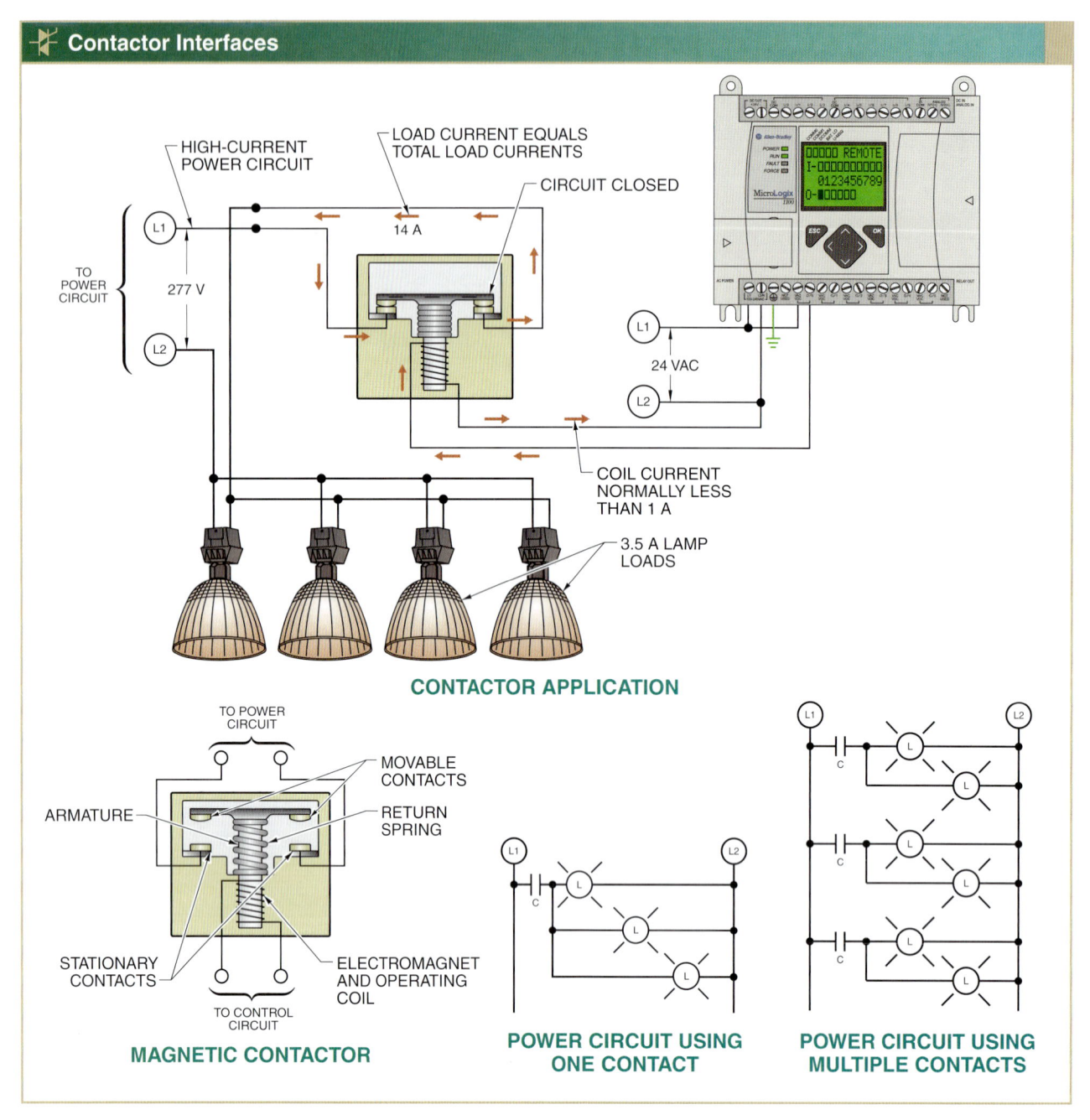

Figure 5-25. Contactors are designed to control high-power, nonmotor loads on an individual circuit, such as a lighting circuit.

The coil closes the contacts to switch a motor ON using a small control current or PLC output signal. Overload protection is provided by thermal, magnetic, or solid-state overload elements that operate a resetable overload contact that automatically removes power from the motor when an overload occurs. Magnetic motor starters include high-power contacts for switching a motor ON and OFF and typically have low-power auxiliary contacts that are used for control circuitry.

Typically, NO auxiliary contacts are wired to a PLC digital input for fault status. A motor starter fault is generated when a command is sent to a motor starter coil and the status is not received in a timely manner. Also, if the status is received as true and the command is not being sent in the same timely manner, a motor starter fault is generated.

Heating Contactor Interfaces

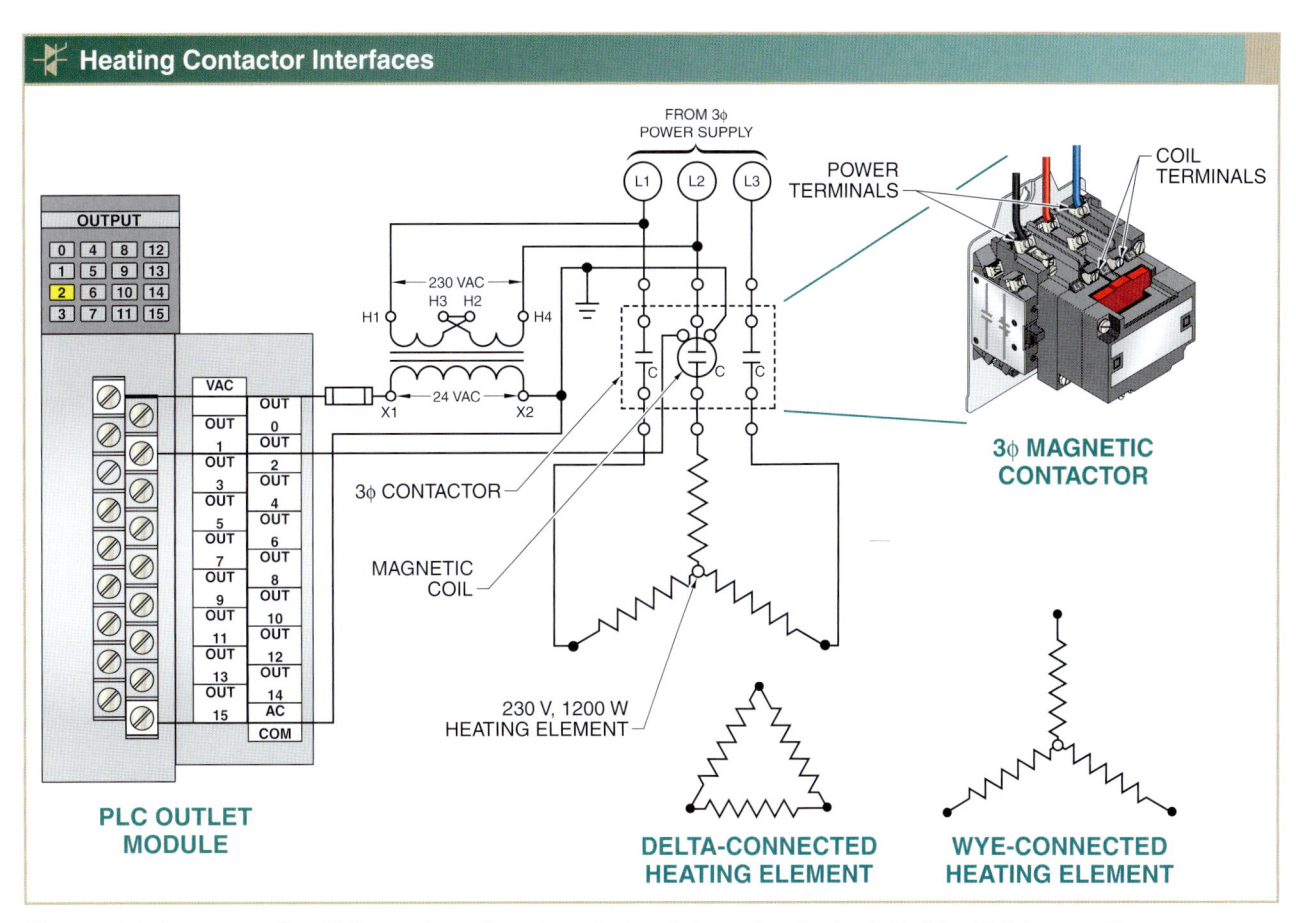

Figure 5-26. Contactors allow PLC output circuitry to be rated much lower than the loads that the PLC is controlling.

Magnetic Motor Starter Interfaces

⚠ DANGER

When a PLC is de-energized or an output contact is open, interface devices can still have dangerous voltages connected.

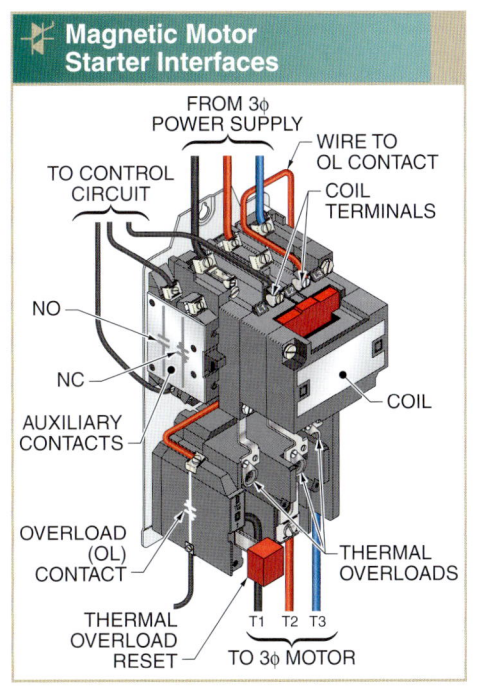

Figure 5-27. Magnetic motor starters use a small control current to energize a coil to send high power to a motor.

The voltage of a magnetic motor starter coil is typically lower than the voltage used by motors. When the coil and control circuit require lower voltage, a step-down transformer is used to reduce the high voltage to a lower voltage. **See Figure 5-28.**

In an HVAC circuit, a transformer is used to step-down the 230 V to 24 V. The stepped-down 24 V is used to power the PLC and relay connected to the PLC output terminal. When the PLC output terminal energizes the relay coil, the relay contact energizes the magnetic motor starters.

By using a transformer, relay, and motor starters as interfaces, a PLC can control high-power motors.

Step-Down Voltage Control Circuits

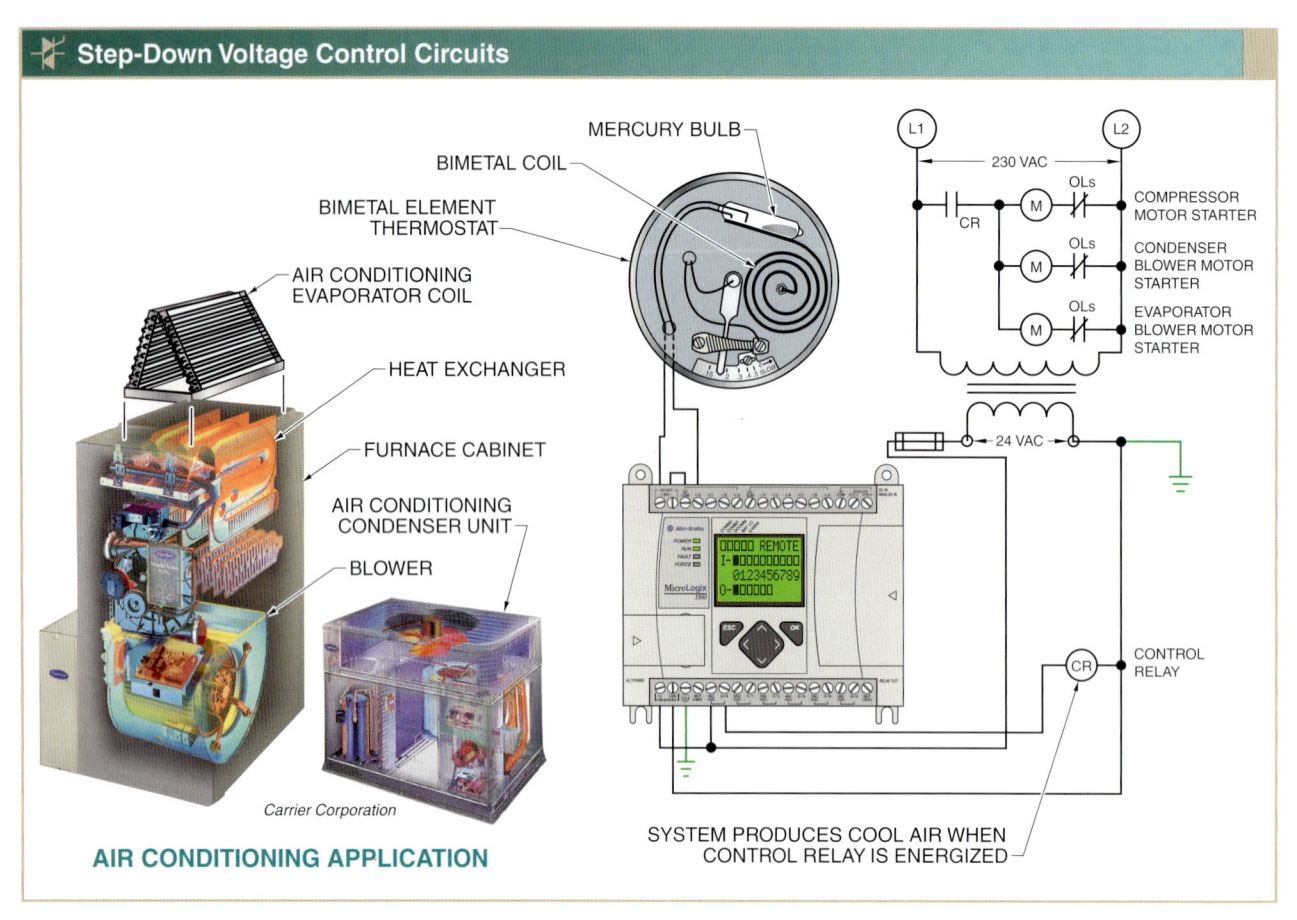

Figure 5-28. When the coil of a motor starter and the control circuit require lower voltage, a step-down transformer is used to reduce the high voltage to a low voltage.

Electric Motor Drives

Electric motor drives perform the same functions as motor starters and can also vary motor speed, reverse a motor, provide additional protection, display operating information, and interface with other electrical equipment. Electric motor drive designs are used to control any size motor, from fractional horsepower to thousands of horsepower.

AC drives control and monitor motor speed by converting incoming AC voltage to DC voltage and then converting the DC voltage back to a variable-frequency AC voltage. To change the speed of a motor, electric motor drives vary the frequency (Hz) of the electricity to the motor. A standard 60 Hz AC-rated motor operates at full speed when connected to 60 Hz, at half speed when connected to 30 Hz, and at one-quarter speed when connected to 15 Hz. **See Figure 5-29.**

⚠ WARNING

Technicians must always wear PPE around fluid power systems to protect themselves against oils from accidental discharges.

DC drives control speed by controlling the DC voltage to a motor and varying the amount of voltage and current on the armature and/or field of the motor. AC power can also be connected to DC drives because the incoming AC voltage can be converted to DC.

Electric motor drives are replacing magnetic motor starters and control circuits. Similar to magnetic motor starters and PLCs, electric motor drives typically have switches remotely located as input devices. However, unlike magnetic motor starters that require an external power supply to energize the magnetic coil, electric motor drives supply voltage to control switches for input signals. **See Figure 5-30.**

Electric Motor Drive Interfacing

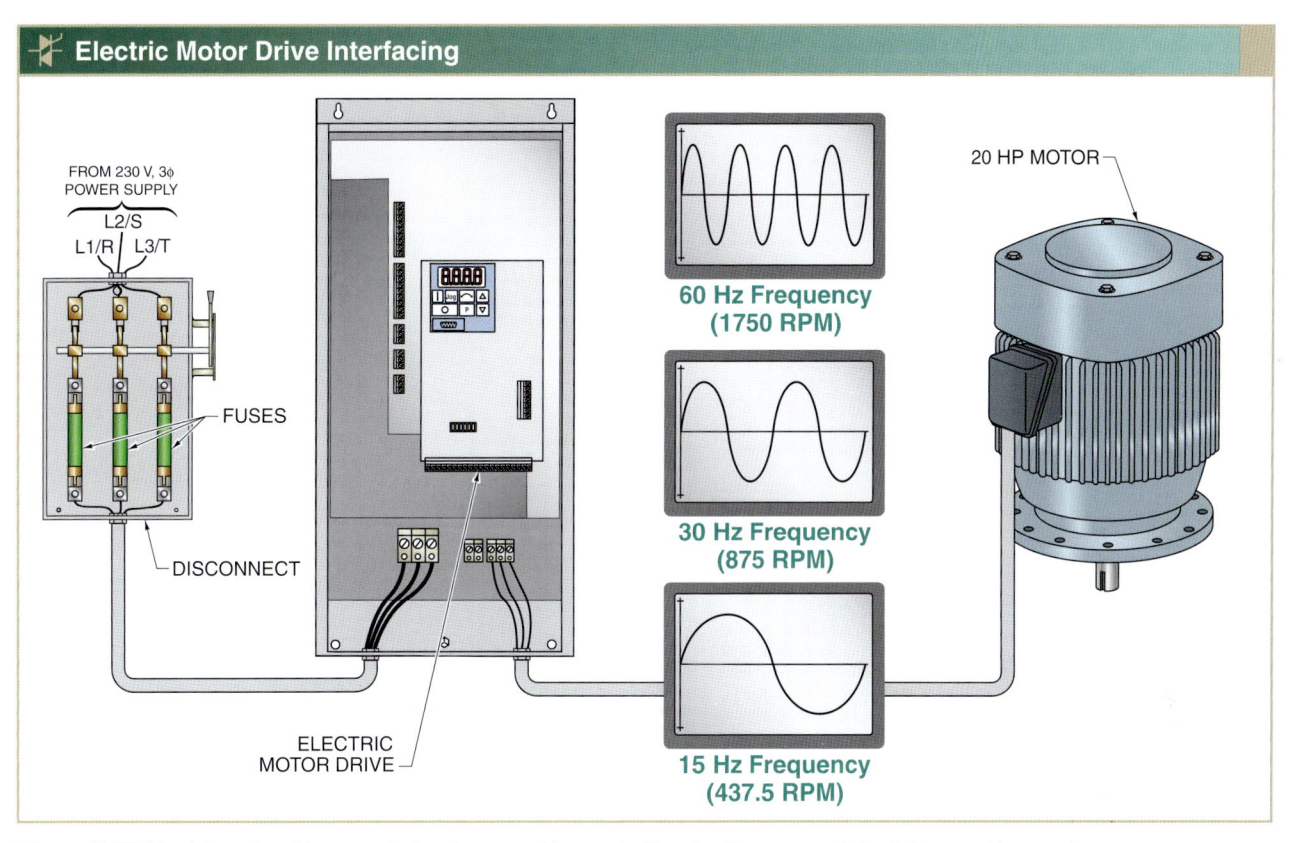

Figure 5-29. Electric motor drives control motor speed by controlling the frequency of the AC current to a motor.

Electric motor drive manufacturers typically include suggested wiring diagrams for using external switches (SW1, SW2, and SW3). Any type of switch contacts (pushbutton, pressure switch, PLC output contacts) can be used to control an electric motor drive. In an exhaust fan application, each of the three switches can be programmed using the electric motor drive keypad to set the switch parameters. For example, each switch can be used to tell the electric motor drive to control the motor at one of three different speeds.

DIGITAL AND ANALOG CIRCUITS

Basic electrical circuits include switches that turn loads ON or OFF. Most applications only require a load, such as lamps or motors, to be fully ON or fully OFF. These loads can be controlled by switches that are either open or closed. In the open position, the load is OFF, and in the closed position, the load is ON. However, some applications require an analog input signal to provide a variable range for temperature, pressure, flow rate, or level measurements.

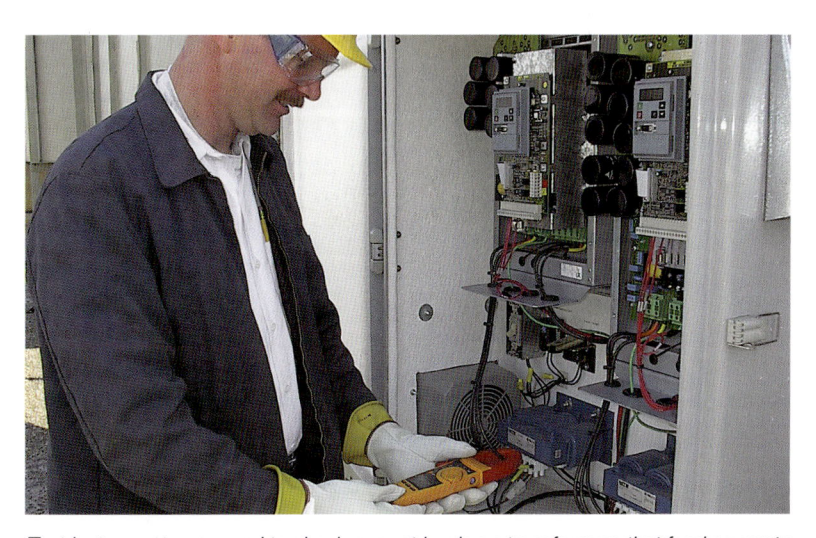

Test instruments are used to check current levels on transformers that feed power to electric motor drives.

PLCs Controlling Electric Motor Drives

CODE CONNECT

Section 430.124 of the NEC® addresses overload protection for motors powered by variable-frequency drives (VFDs). If a VFD is listed as providing motor overload protection, additional motor overload protection, such as an overload relay, is not required. However, if a VFD has a bypass circuit to allow motor operation at full voltage/full speed or powers multiple motors, then additional motor overload protection is required.

EXHAUST FAN APPLICATION

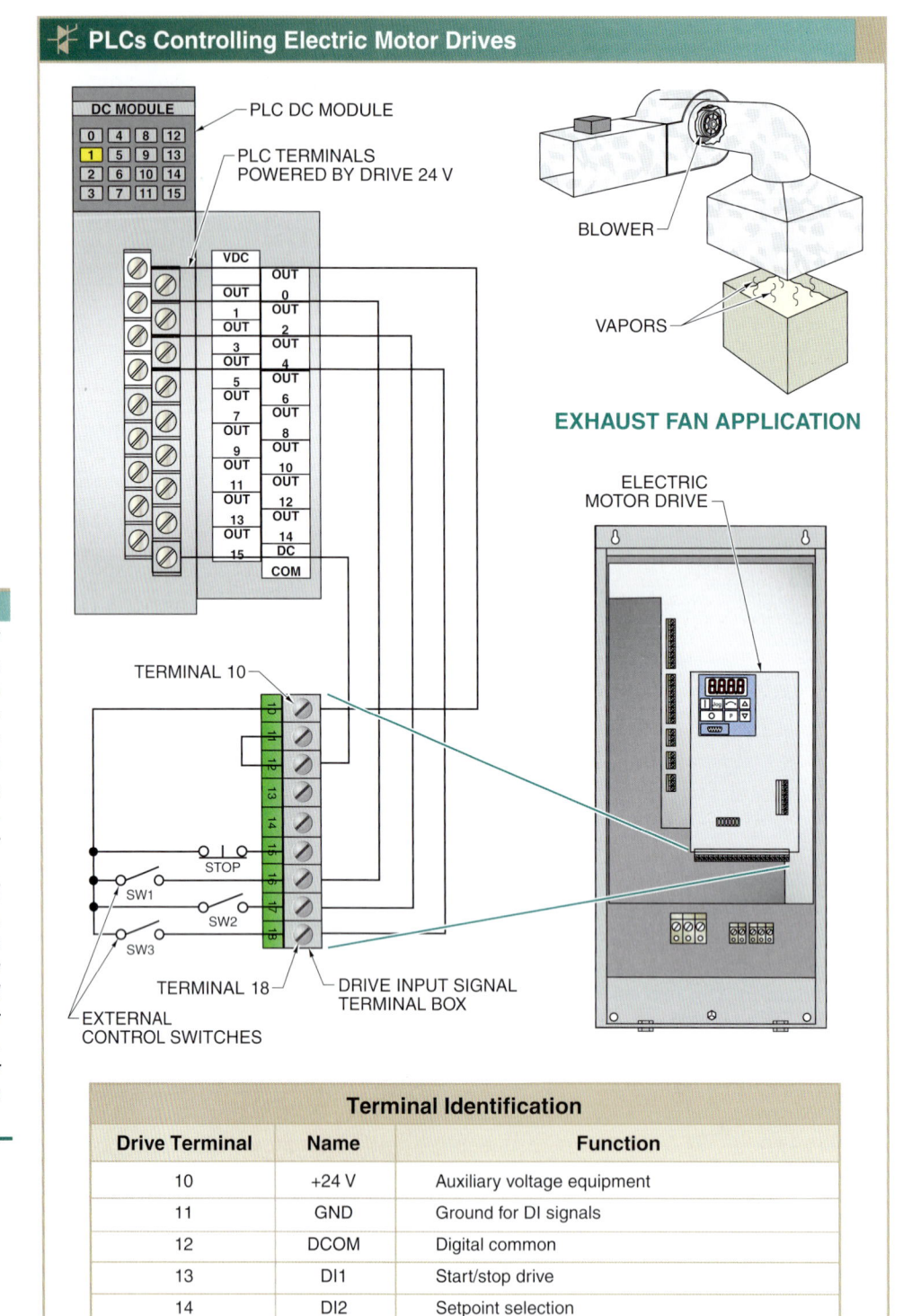

Terminal Identification		
Drive Terminal	**Name**	**Function**
10	+24 V	Auxiliary voltage equipment
11	GND	Ground for DI signals
12	DCOM	Digital common
13	DI1	Start/stop drive
14	DI2	Setpoint selection
15	DI3	Deactivation stops drive
16	DI4	Speed #1 – Low exhaust speed (Switch 1)
17	DI5	Speed #2 – Normal exhaust speed (Switch 2)
18	DI6	Speed #3 – High exhaust speed (Switch 3)

Figure 5-30. Any type of switch contacts (pushbutton, pressure switch, PLC output contacts) can be used to control an electric motor drive.

Digital Circuits

Digital circuits operate using binary signals that have only two operating states: 1 or 0, high or low, and voltage or no voltage. Digital electrical circuits have switches that have only two operating states: open or closed. A *digital electrical circuit* is a circuit in which the load is either fully ON or fully OFF. **See Figure 5-31.**

For example, a two-way switch that is used to control a lamp can be placed in only one of two positions: ON or OFF. The light, or load, can only be in one of two operating states: fully ON or fully OFF.

Analog Circuits

Digital circuits work well for most applications. However, in applications that require operating conditions between fully ON or fully OFF, to save energy, improve system performance, or to add precise control to the system, digital circuits do not work. Such applications require an analog circuit. An *analog electrical circuit* is a circuit in which the load can be in any operating condition between and including fully ON and fully OFF.

For example, a dimmer switch is used to control a lamp. The dimmer switch can be set to any position that provides full voltage or current to the load, to a position that provides no voltage or current to the load, or to a position that provides any amount of voltage and current between full and none. The advantage of the analog circuit is that the amount of light (lumens) available can be adjusted to provide better illumination and save energy. In addition, when the light is operated below its full power (voltage and current) rating, the bulb will last much longer. **See Figure 5-32.**

The digital and analog lamp control circuits are basic examples of the differences between digital and analog circuits. These circuits would not be controlled by a PLC. However, there are many commercial and industrial applications that are best controlled by an analog circuit. For example, motors use more total energy (about 62%) than all other electrical loads combined. Any motor that is turned fully ON will use a greater amount of energy, even when all the motor's power is not required for the application. The use of a PLC to control the analog circuit that controls the motor will save energy.

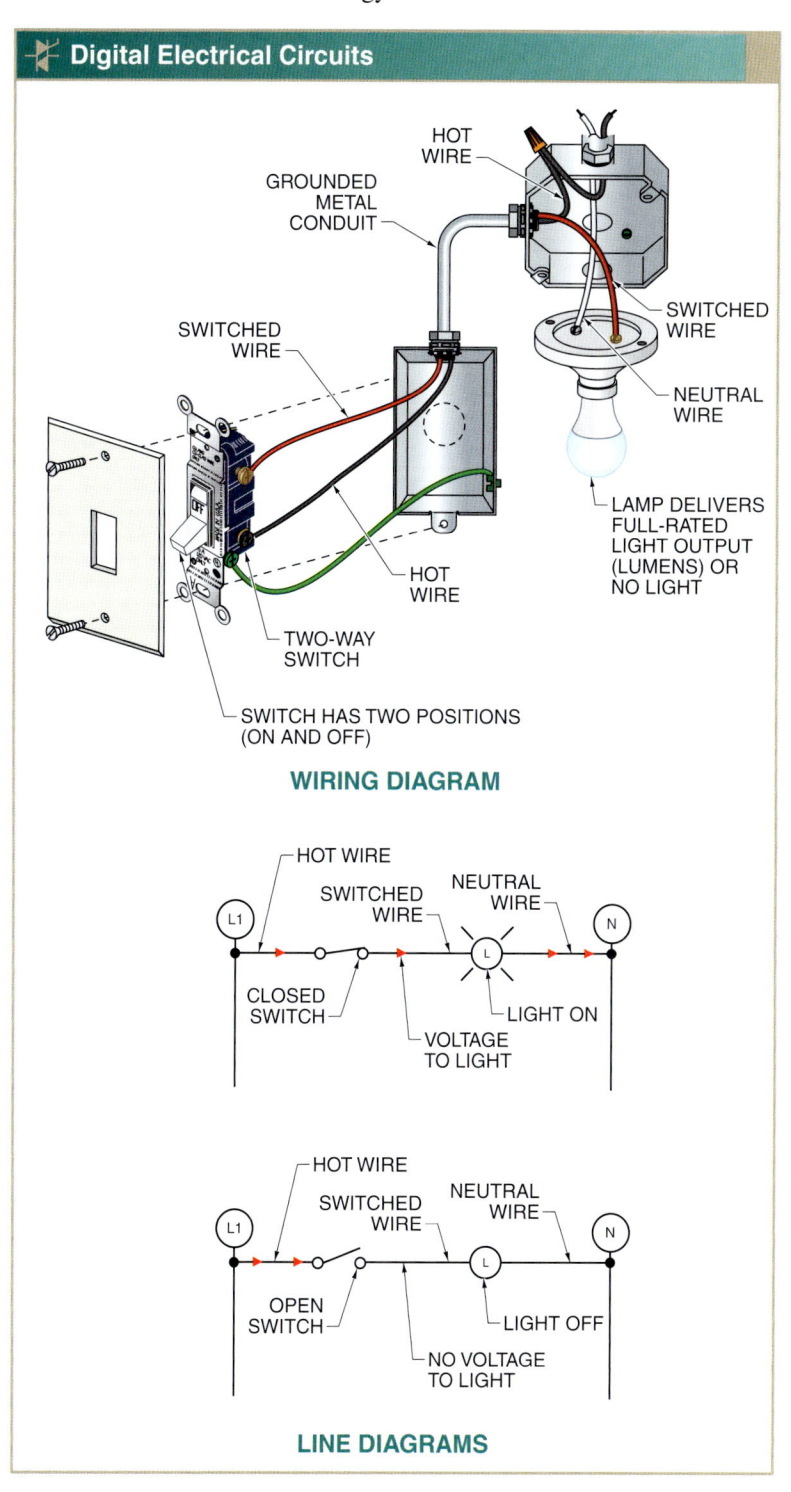

Figure 5-31. The switches in digital circuits have only two operating states, open or closed, which provide binary signals 1 or 0, high or low, and voltage or no voltage. Therefore, digital circuit loads are either fully ON or fully OFF.

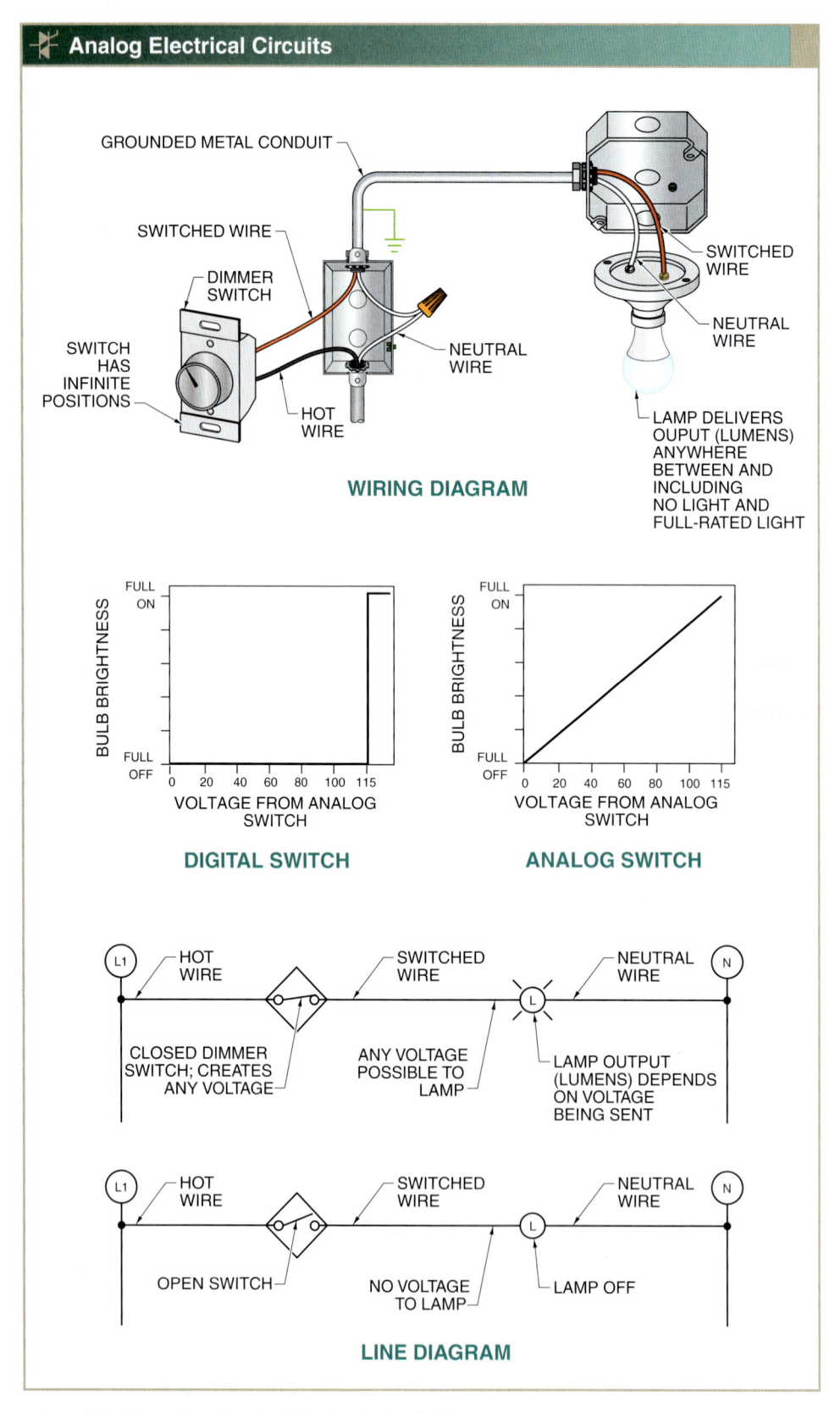

Figure 5-32. In analog electrical circuits, the load (lights, motors, or valve actuators) can be in any operating condition: fully ON, fully OFF, or anywhere in between.

Electric Motor Drive Circuits

Electric motor (variable frequency) drives are used to control the speed of electric motors and reduce the amount of power they consume. For example, when an electric motor drive is being used to control a mixer motor, the drive can be used to control motor speed anywhere between very slow to maximum. The electric motor drive outputs variable power to the motor that changes the speed of the motor. This is accomplished by the drive changing the frequency and voltage applied to the motor throughout its operating range. The motor can be turned ON and OFF and placed in forward or reverse by any digital switch connected to the drive's control terminals. **See Figure 5-33.**

An electric motor drive will accept an analog signal from a 1 kΩ to 10 kΩ variable resistor (potentiometer), a 0 VDC to 10 VDC voltage source, or a 4 mA to 20 mA current source as a control signal. The speed of the motor will be directly proportional to the analog signal received. For example, when the 0 VDC to 10 VDC analog input is used, the motor will run at 50% rated speed at 5 VDC, 75% rated speed at 7.5 VDC, and so on.

The internal components of an electric motor drive consist of integrated circuits to control all parameters of a motor application.

Electric Motor Drive with PLC Control

Control can also be added to an application by using a PLC to control a drive. A digital output (open or closed switch) from the digital output section of the PLC can be used to control the digital functions on a drive, such as turning the motor on in forward or reverse and turning the motor off. An analog output signal from the analog output section of the PLC can be used to control the analog functions (speed control of the drive). For example, a 0 VDC to 10 VDC output signal from the PLC can be connected to the electric motor drive to control the speed of the motor based on the amount of voltage the PLC outputs. **See Figure 5-34.**

The input signals to a PLC input section or module can be digital or analog. Digital input signals are typically used to turn motors and actuators ON and OFF. Analog input signals are used to send signals that represent a changing pressure, changing temperature, or a changing fluid level to the PLC. The advantage of using a PLC with both digital and analog input terminals and/or output terminals is that greater circuit control is provided, and as a result, better control and efficiency of the application occurs.

Analog Input Devices

An *analog input device* is any of the various types of sensors that measure a variable change in an environmental or operating condition and convert that change into a proportional voltage or current signal. These proportional signals are then sent to an analog processing device, such as an analog PLC input section or module. Analog input devices include temperature sensors, pressure sensors, flow sensors, position sensors, and weight sensors.

Analog Input Device Signals. There are various types of analog signals that can be sent to analog PLCs. Analog signals include 0 VDC to 10 VDC, 4 mA to 20 mA, 0 VDC to 5 VDC, −10 VDC to 10 VDC, 0 mA to 20 mA, and 1 VDC to 5 VDC. **See Figure 5-35.**

CODE CONNECT

Part X of Article 430 covers adjustable-speed drive systems. The most common type of adjustable-speed drive is a variable-frequency drive (VFD), also known as an AC drive, an inverter, or simply a drive. A VFD controls the speed of an AC motor by simultaneously adjusting the voltage and frequency applied to the motor. The use of VFDs with pumps and fans results in significant energy savings. Parts I through IX of Article 430 apply to VFD installations, unless amended by Part X.

218 PROGRAMMABLE LOGIC CONTROLLERS: PRINCIPLES AND APPLICATIONS

Analog Signals to Motor Speed Drive Control

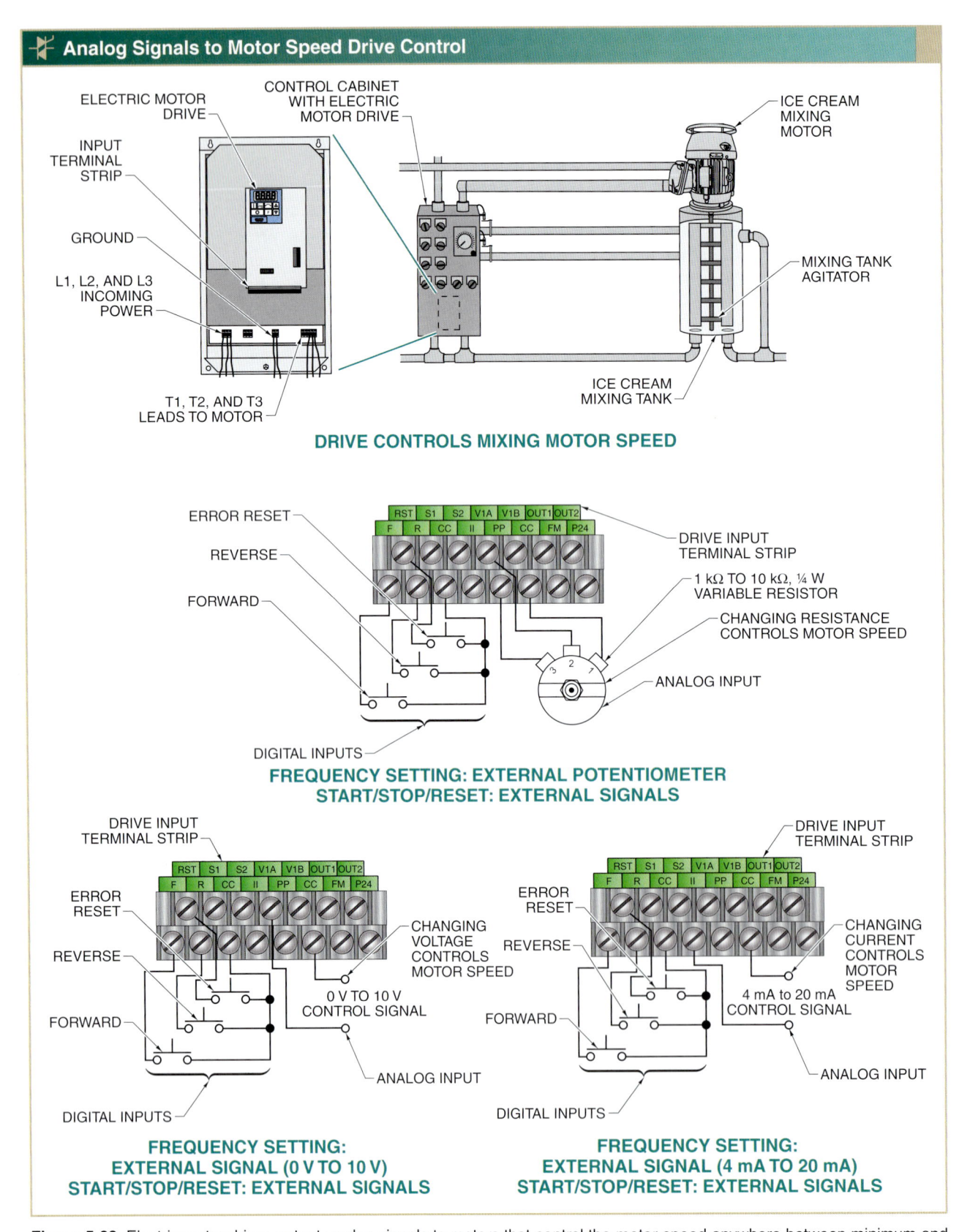

DRIVE CONTROLS MIXING MOTOR SPEED

FREQUENCY SETTING: EXTERNAL POTENTIOMETER
START/STOP/RESET: EXTERNAL SIGNALS

FREQUENCY SETTING:
EXTERNAL SIGNAL (0 V TO 10 V)
START/STOP/RESET: EXTERNAL SIGNALS

FREQUENCY SETTING:
EXTERNAL SIGNAL (4 mA TO 20 mA)
START/STOP/RESET: EXTERNAL SIGNALS

Figure 5-33. Electric motor drives output analog signals to motors that control the motor speed anywhere between minimum and maximum current.

Electric Motor Drive with PLC Control

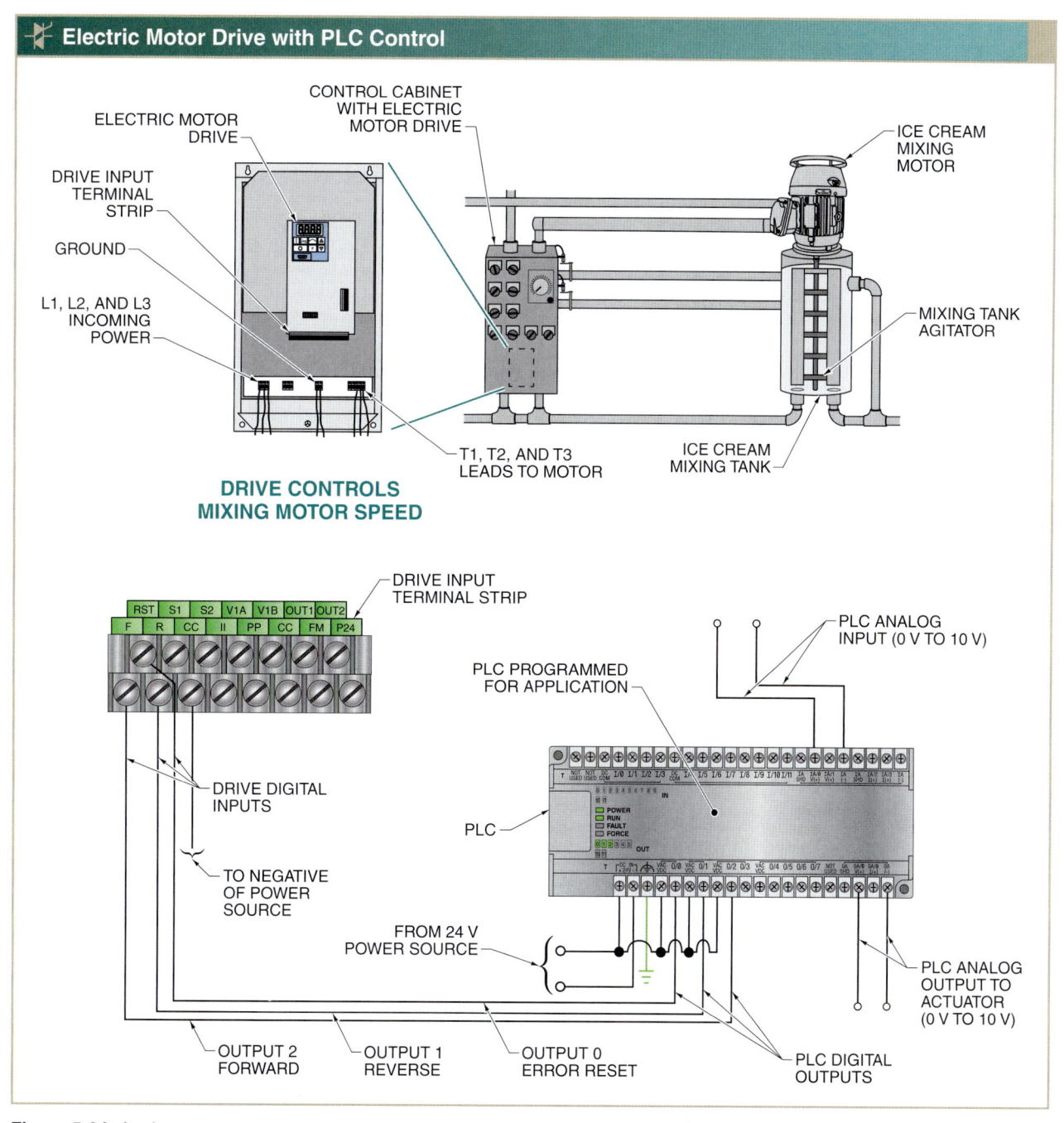

Figure 5-34. Analog sensors are used to send signals that represent a changing pressure, temperature, or fluid level to a PLC. Digital output signals from PLCs are used to control the digital functions of electric motor drives. Analog output signals are used to control the analog functions such as speed control of a motor by drives.

Some analog signals can be directly connected to a PLC analog input section or module. Other PLCs require the analog signal to first be converted into a voltage or current signal that is within the specified range of the input section or module. For example, when an analog input terminal is rated to receive a 0 VDC to 10 VDC signal and the voltage from the circuit or power source, such as a photocell, is not more than 10 VDC, the analog signal can be connected directly to a PLC analog input terminal.

Analog Input Signal Types

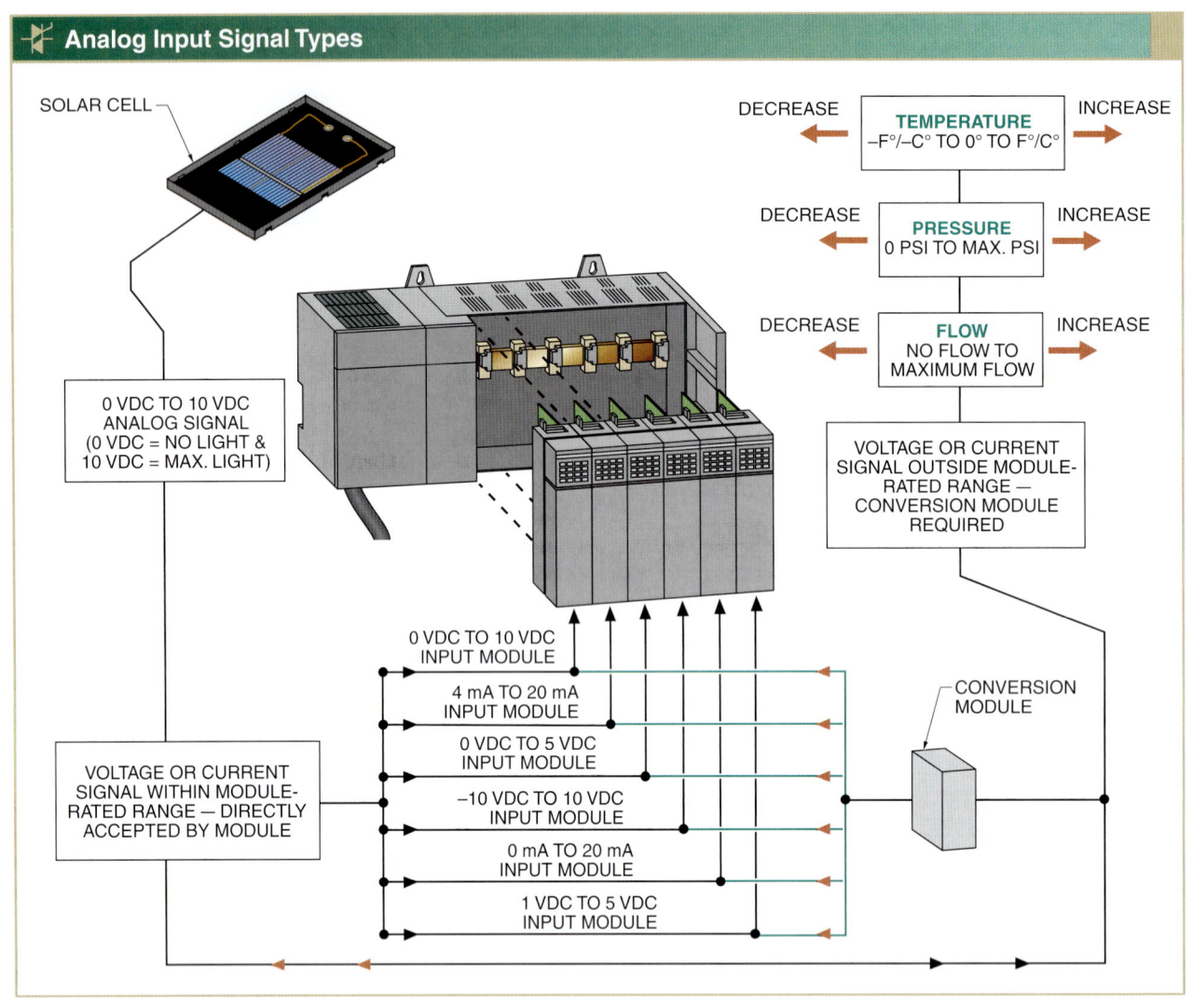

Figure 5-35. There are various types of analog signals that are sent to analog PLCs such as 0 VDC to 10 VDC, 4 mA to 20 mA, 0 VDC to 5 VDC, −10 VDC to 10 VDC, 0 mA to 20 mA, and 1 VDC to 5 VDC.

Any analog input device, such as a temperature sensor, pressure sensor, flow sensor, or solar cell, that outputs a voltage or current signal that is within the range of the input section or module of the PLC can also be connected directly to one of the analog input terminals of the PLC. For example, a temperature sensor that is rated for a 0°F to 212°F temperature range with a 4 mA to 20 mA output signal range can be connected to a 4 mA to 20 mA PLC input section or module. The temperature sensor would output 1 mA for every 13.25°F (20 mA − 4 mA = 16 mA range; 212°F ÷ 16 mA = 13.25°F). **See Figure 5-36.**

Input Device Signal Converters. When the analog output signal of a sensor is outside the range of the input section or module of the PLC, the signal will need to be converted to a voltage or current signal that is within the specified range before being sent. For this, a conversion module is used. For example, a conversion module that converts a milliamp temperature measurement signal into a voltage signal would be required to input temperature measurements to a PLC input terminal rated for voltage input. In this case, the converter may be specified to output 1 VDC for every 30°F starting at 0°F. A voltage output of the converter (the input signal to the PLC input module) of 2.38 VDC would represent 71.40°F at the sensor performing the measurement. **See Figure 5-37.**

PLC input sections or modules are designed for a specific type of analog input signal. The most common analog input module ratings are 0 VDC to 10 VDC and 4 mA to 20 mA. However, PLC input modules can be found with a variety of ratings such as 0 VDC to 5 VDC, 1 VDC to 5 VDC, –5 VDC to 5 VDC, –10 VDC to 10 VDC, 0 mA to 20 mA, and –20 mA to 20 mA.

When PLC input sections or modules are designed for a specific type of analog signal but the sensor has digital output signals, a signal converter must be used. Various types of signal converters can be used that change digital signals into analog signals, analog voltage signals into analog mA signals, or analog mA signals into analog voltage signals. Some PLCs have the converter built directly into the PLC input section or module. When a PLC input module is not designed with an internal converter, an external converter will be required.

Temperature Sensor Data

4 mA TO 20 mA		0 V TO 10 V	
Temperature*	Sensor Output†	Temperature*	Sensor Output†
–26.50	4	–42.40	0.0
–13.25	4	–21.20	0.0
0.00	4	0.00	0.0
13.25	5	21.20	1.0
26.50	6	31.80	1.5
39.75	7	42.40	2.0
53.00	8	63.60	3.0
66.25	9	74.20	3.5
79.50	10	84.80	4.0
92.75	11	106.00	5.0
106.00	12	116.60	5.5
119.25	13	127.20	6.0
132.50	14	148.40	7.0
145.75	15	159.00	7.5
159.00	16	169.60	8.0
172.25	17	190.80	9.0
185.50	18	201.40	9.5
198.75	19	212.00	10.0
212.00	20	233.20	10.0
225.25	20	254.40	10.0
238.50	20		

* in °F
† in mA

* in °F
† in V

Figure 5-36. The output voltage and current signals of analog input devices (temperature sensors, pressure sensors, and flow sensors) must be within the range of the PLC's input section or module terminals. If the signals are not within range, a converter is required.

Converter Output Spans		
Sensor Temperature*	Sensor Output†	Converter Output‡
0.00	4	0.00
13.25	5	0.44
26.50	6	0.88
39.75	7	1.32
53.00	8	1.76
66.25	9	2.20
71.40	9.7	2.38
79.50	10	2.64
92.75	11	3.08
106.00	12	3.52
119.25	13	3.96
132.50	14	4.40
145.75	15	4.84
159.00	16	5.28
172.25	17	5.72
185.50	18	6.16
198.75	19	6.60
212.00	20	7.04

* in °F
† in mA
‡ in VDC

Figure 5-37. Conversion modules are used to take an analog input signal that is outside the terminal range of the PLC and convert it into a voltage or current signal that is within the range of the input section or module.

Analog Output Devices

An *analog output device* is a load or actuator that delivers a variable position or speed according to the analog signal (voltage or current) being sent from the PLC analog output terminal. A variable output voltage or current signal can be sent to analog output devices, such as variable position valves, solenoids, or analog controlled solid-state electrical switches such as triacs, which are used in analog solid-state relays (SSRs).

Analog-controlled output devices such as variable position valves allow the valve's fluid flow to be controlled anywhere between no flow (totally closed) and maximum flow (fully open). Likewise, analog-controlled SSRs allow the output voltage applied to the load to be anywhere between no voltage and full-rated voltage.

PLC TIPS

Analog proximity sensors have noise immunity in industrial systems, which can be high in electrical noise.

Analog Output Device Signals. Proportional analog valves do not use the control signal from a PLC to power the positioning of the valve spool that controls the amount of fluid flow. For this, the valve must be connected to an external power supply that provides the power to move the valve based on the control signal provided. Analog valves are typically rated to receive only one type of analog input signal at a time, such as 0 VDC to 10 VDC, 0 VDC to 5 VDC, 4 mA to 20 mA, or 0 mA to 20 mA. However, analog valves that offer several analog input signal wiring choices are also available. **See Figure 5-38.**

Some analog output devices can be connected directly to the analog output section or module of a PLC. Other analog output devices must be connected through an interface device such as an analog SSR or relay. For example, a proportional analog fluid valve is available that can be connected directly to an analog PLC output terminal. Valve models with 0 VDC to 10 VDC or 4 mA to 20 mA are typical, with the valve's fluid output proportional to the analog input signal from the PLC. **See Figure 5-39.**

Variable position valves can be connected to a 0 VDC to 10 VDC or 4 mA to 20 mA analog signal, or an external potentiometer can be used to control the position of the valve. Regardless of what control signal is used, the valve must still receive power at all times from an external power supply.

Interface Devices. When the analog output device to be controlled requires a higher voltage, current, or power rating than the output terminal of the PLC is rated for, an interface device must be used. Typical interface devices include magnetic motor starters, motor drives, heating contactors, lighting contactors, mechanical relays (CRs), or SSRs. However, magnetic motor starters, heating contactors, lighting contactors, and mechanical relays are not used with analog signals from a PLC. Motor drives and SSRs are used with analog signals because a drive uses the varying analog signal to control the speed of a motor, while an SSR uses an analog signal to control currents well over 100 A.

Valve Power and Various Analog Signal Wiring

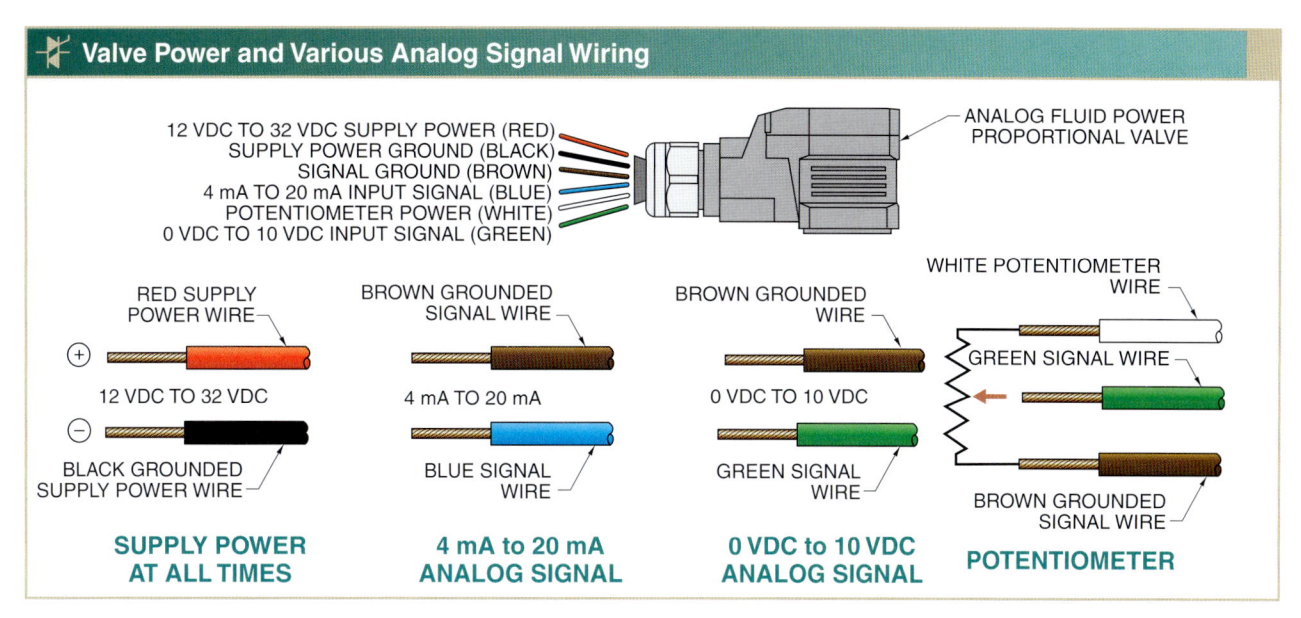

Figure 5-38. Proportional analog valves do not use the control signal from a PLC for power. An external power supply is needed to provide the power to move the valve based on the control signal provided.

Analog Proportional Valve Operation

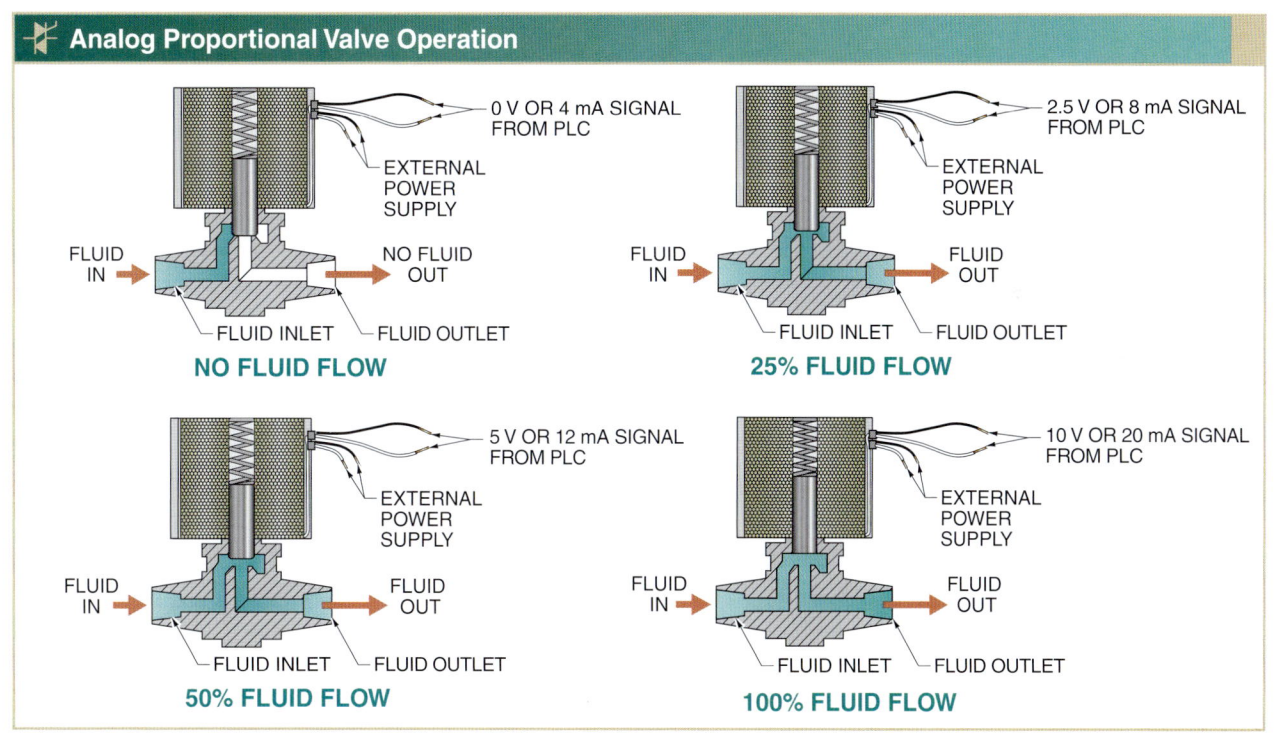

Figure 5-39. Proportional analog fluid valves can be connected directly to the analog PLC output terminals that supply 0 VDC to 10 VDC or 4 mA to 20 mA signals, with the valve's fluid output proportional to the analog input signal.

Output Device Signal Converters. Magnetic motor starters, heating contactors, lighting contactors, and CRs are not used as analog output devices. Instead, electric motor drives and SSRs are used as analog output devices. Electric motor drives are designed to accept any type of analog signal to control the speed of a motor. SSRs are available in analog models that accept a 0 VDC to 10 VDC or 4 mA to 20 mA signal that can control currents of over 100 A and at voltages of 115 V, 230 V, or 460 V. **See Figure 5-40.**

224 PROGRAMMABLE LOGIC CONTROLLERS: PRINCIPLES AND APPLICATIONS

Signal Conversion Devices

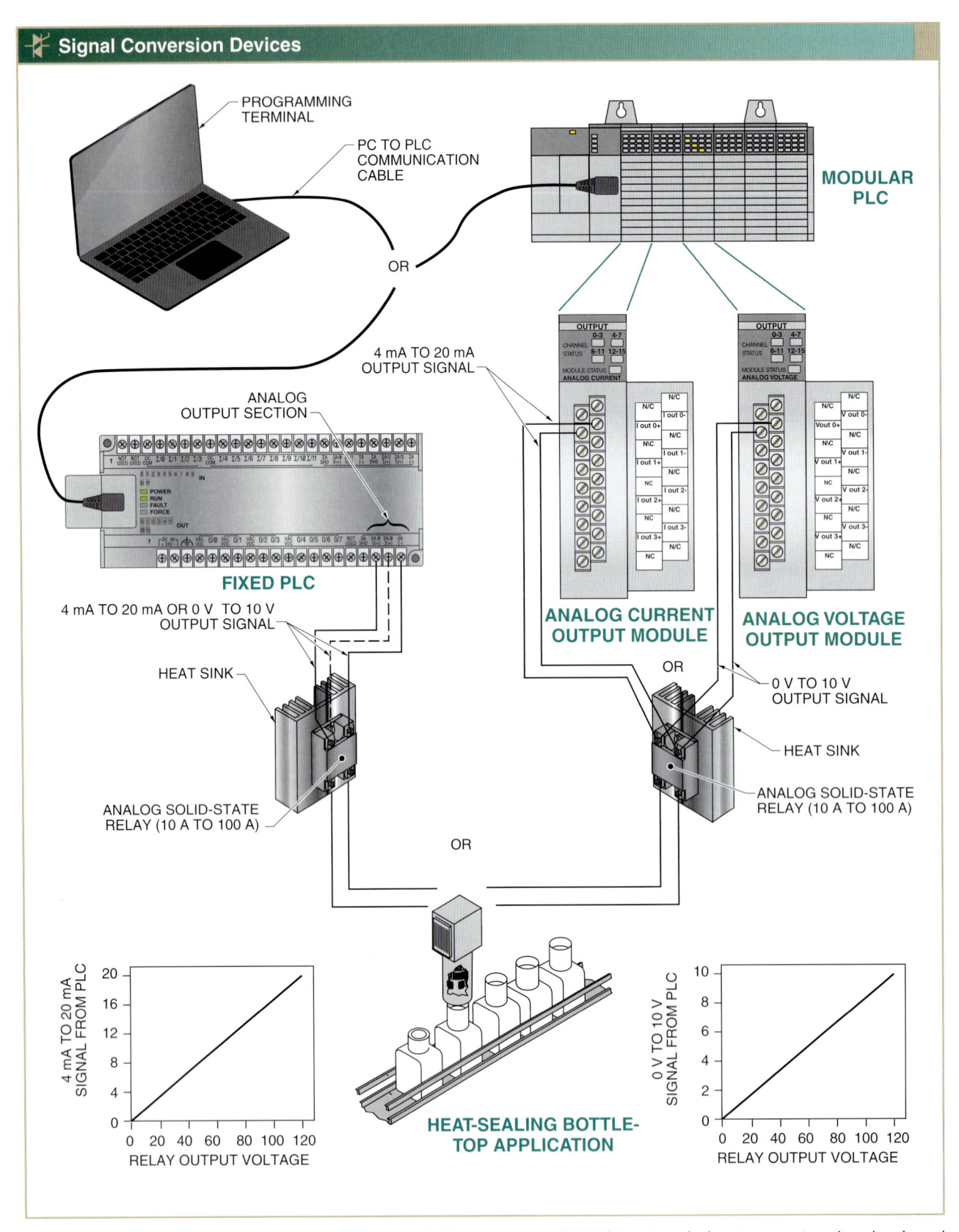

Figure 5-40. Electric motor drives and solid-state relays (SSRs) are used with analog output devices to accept analog signals and control the speed of a motor or control high amperages.

An analog SSR connected to a PLC analog output module can be used to control a heating element that seals bottle tops in an automated production line. The SSR's voltage output to the heating element is directly proportional to the analog input signal from the PLC analog output section or module. Either a 0 VDC to 10 VDC SSR or a 4 mA to 20 mA SSR can be used.

Analog SSRs control the output voltage to the load. Since heating elements have a fixed resistance (R), the higher the applied voltage (E), the higher the current (I) flow through the heating element (Ohm's law). Since power (P) is equal to the voltage (E) multiplied by the current (I), the amount of power at the heating element is based on the applied voltage and current (the power formula). **See Figure 5-41.**

Current and Power in Resistive Circuits

ENCLOSED-COIL HEATING ELEMENT TYPES
- FINNED TUBULAR
- TUBULAR
- STRIP
- RING
- BAND

OPEN-COIL HEATING ELEMENT

$I = \dfrac{E}{R}$
(VARIES PROPORTIONALLY WITH ANALOG INPUT TO SSR)
(FIXED BY RESISTANCE OF HEATING ELEMENT)

CURRENT THROUGH HEATING ELEMENT

$P = E \times I$
- P = WATTAGE OF HEATER
- E = APPLIED VOLTAGE FROM SSR
- I = CURRENT BASED ON RESISTANCE OF HEATING ELEMENT AND APPLIED VOLTAGE

HEATER OUTPUT POWER

- 0 V TO 10 V OR 4 mA TO 20 mA SIGNAL
- CONTROL CIRCUIT FROM PLC ANALOG OUTPUT SECTION OR MODULE
- ANALOG SOLID-STATE RELAY
- ENCLOSED-COIL HEATING ELEMENT
- HEAT TRANSFER FINS
- REFRACTORY INSULATION
- L1 POWER CIRCUIT VOLTAGE (E) L2
- COILED RESISTOR WIRE (R)
- TERMINAL CONNECTIONS
- CURRENT THROUGH HEATING ELEMENT VARIES PROPORTIONALLY WITH ANALOG SIGNAL FROM PLC

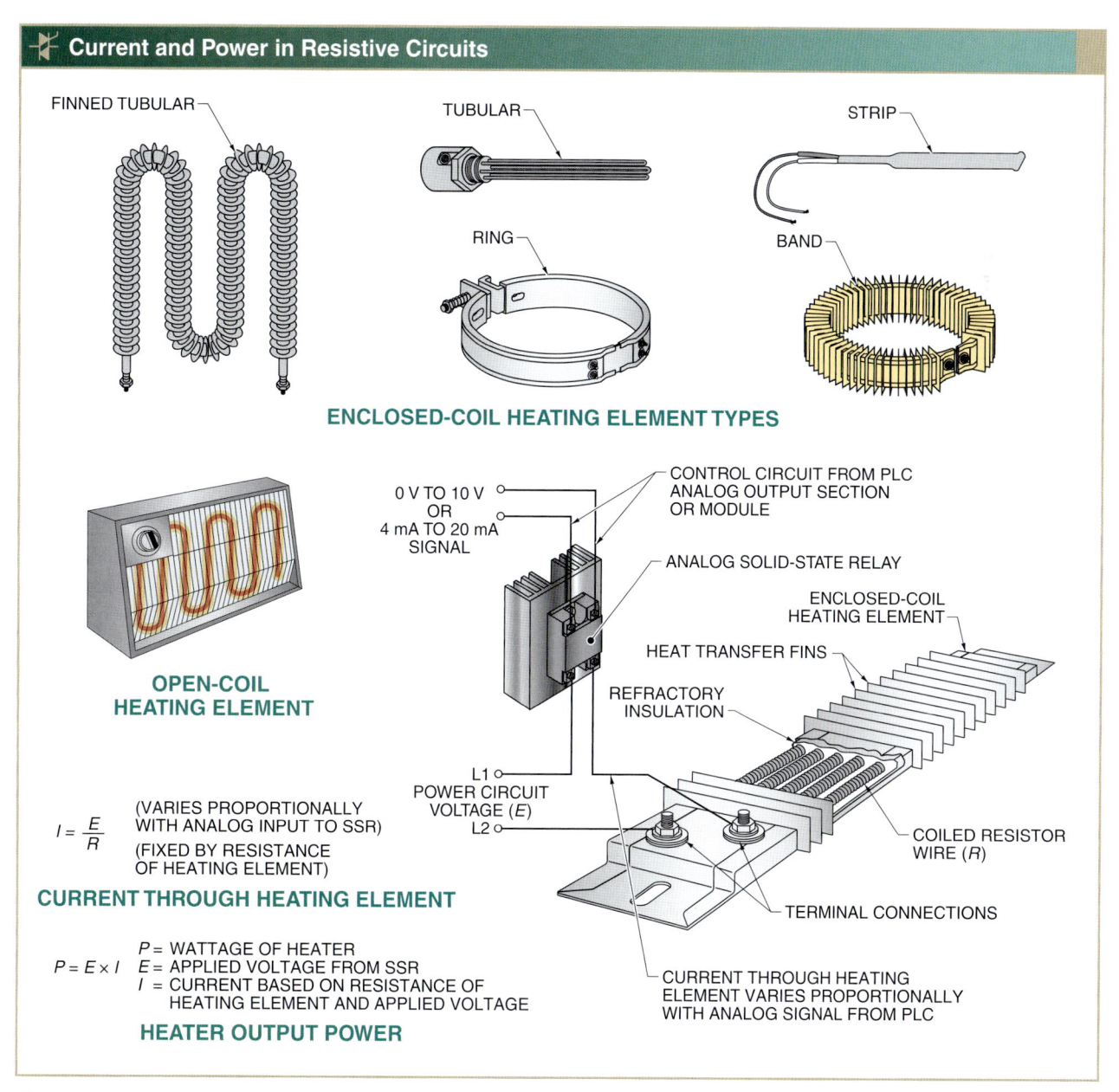

Figure 5-41. The power formula ($P = E \times I$) can be used to determine the amount of power at the heating element based on the applied voltage and current.

Chapter Review 5

Name _____ Date _____

True-False

T F 1. When a PLC is used to control a system, system operation can be monitored and easily changed by reprogramming the process control circuit.

T F 2. Circuit conditioning in an electrical system can include changing AC to DC.

T F 3. The smaller the number of basic circuits in an electrical system, the more likely a PLC will be used to reduce cost, simplify wiring, and monitor circuit operation.

T F 4. Input modules that are connected directly to a 115 VAC supply will be damaged.

T F 5. When a heating element circuit is hardwired into a PLC circuit, the heating contactor coil and heater ON lamp are typically wired in parallel.

T F 6. Solid-state relays can be rated for billions of operations.

T F 7. An analog signal is an electronic signal that has two specific quantities that change in discrete steps (ON or OFF).

T F 8. Interfacing systems allow primary systems that cannot work together to be compatible.

T F 9. Circuit conditioning is not a requirement even when the existing electrical power is not at the proper phase, voltage, or current level for the application.

T F 10. A break is the number of contact positions per pole.

T F 11. Hardwired and PLC-controlled heating element circuits respond differently.

T F 12. Contactors are required between a PLC output section and high-power load.

T F 13. In a hardwired heating element circuit, the circuit can easily be changed or modified without rewiring the control circuit.

T F 14. A basic circuit must include a protection device to ensure that the circuit operates safely and within designed electrical limits.

T F 15. Electrical systems are made up of basic and complex circuits.

T F 16. Basic electrical circuits include switches that turn loads ON or OFF.

T F 17. Digital circuits turn loads fully ON or fully OFF.

T F 18. An electric motor drive cannot accept an analog signal from a 1 kΩ to 10 kΩ variable resistor (potentiometer), a 0 VDC to 10 VDC voltage source, or a 4 mA to 20 mA current source as a control signal.

228 PROGRAMMABLE LOGIC CONTROLLERS: PRINCIPLES AND APPLICATIONS

T F 19. Analog sensors are used to send signals that represent a changing pressure, temperature, or fluid level to a PLC.

T F 20. Analog input signals are typically used to turn motors and actuators ON and OFF.

T F 21. PLC input sections or modules are designed for any type of analog input signal.

T F 22. Analog-controlled output devices such as variable position valves allow the valve's fluid flow to be controlled anywhere between no flow (totally closed) and maximum flow (fully open).

T F 23. All analog output devices must be connected through an interface device such as an analog SSR or relay.

T F 24. Magnetic motor starters, heating contactors, lighting contactors, and mechanical relays are not used with analog signals from a PLC.

T F 25. Electric motor drives are designed to accept any type of analog signal to control the speed of a motor.

Completion

_____ 1. A(n) ___ system is a primary system in which electricity is monitored and controlled to send and/or receive information, produce sound or vision, store data, control circuits, or perform other work.

_____ 2. A(n) ___ is a completely isolated circuit (contact) that a relay can switch.

_____ 3. A(n) ___ is a load whose voltage and current rating is greater than the voltage and current rating of the PLC output section.

_____ 4. PLCs and PCs are interconnected through ___ to form various types of networks.

_____ 5. A(n) ___ is an item that allows variously rated (voltage and current) components to be used together in the same circuit.

_____ 6. A(n) ___ motor starter is a motor starter that has an electrically operated mechanical switch with overload protection.

_____ 7. A(n) ___ is a system of computers, terminals, and PLCs connected by communication lines.

_____ 8. ___ is the movement of a load (in pounds) over a distance (in feet).

_____ 9. The ___ of interface devices can be mechanical or solid-state.

_____ 10. A(n) ___ is a typical interface device that allows two parts of an electrical circuit that normally would not work together to work together.

_____ 11. ___ operate using binary signals that have only two operating states.

_____ 12. A(n) ___ is a circuit in which the load can be in any operating condition between and including fully ON and fully OFF.

_____ 13. The use of a(n) ___ to control an analog circuit that controls a motor will save energy.

_____ 14. A(n) ___ output (open or closed switch) form the output section of a PLC can be used to control the digital functions on a drive, such as turning the motor in forward or reverse.

_____ 15. The advantage of using a PLC with both digital and analog input terminals and/or output terminals is that ___ is provided, and as a result, better control of the application occurs.

_____ 16. Some ___ can be directly connected to a PLC analog input section or module, while other PLCs require the analog signal to first be converted into a voltage or current signal that is within the specified range of the PLC's input section or module.

_____ 17. When PLC input sections or modules are designed for a specific type of analog signal but the sensor has digital output signals, a(n) ___ must be used.

_____ 18. A(n) ___ is a load or actuator that delivers a variable position or speed according to the analog signal (voltage or current) being sent from the PLC analog output terminal.

_____ 19. Magnetic motor starters, heating contactors, lighting contactors, and control relays are not used as ___ devices.

_____ 20. ___ SSRs control the output voltage to a load.

Multiple Choice

_____ 1. A ___ system is a fluid power system that transmits power using a gas (air).
  A. pneumatic
  B. hydraulic
  C. mechanical
  D. combination

_____ 2. A ___ is a control device that is designed to control high-power, non-motor loads.
  A. connector
  B. contactor
  C. relay
  D. transmitter

_____ 3. A(n) ___ is used to produce work, light, heat, or sound or to display information.
  A. input device
  B. output component
  C. sensor
  D. switch

_____ 4. A(n) ___ is a device that controls an electrical circuit by opening and closing contacts in another circuit.
  A. optic isolator
  B. contact regulator
  C. control relay
  D. circuit manager

_____ 5. ___ relays draw high current when energized and produce high voltage spikes when de-energized.
  A. Electromechanical
  B. Solid-state
  C. Drive
  D. Billion operation rated

_____ 6. Some applications require an analog input signal to provide a variable range for ___ or level measurements.
   A. proximity
   B. photoelectric
   C. limit
   D. pressure

_____ 7. ___ drives are used to control the speed of electric motors and reduce the amount of power they consume.
   A. Digital
   B. Analog
   C. Electric Motor
   D. Motor

_____ 8. The ___ of a motor will be directly proportional to the variable power received from a drive.
   A. rotation
   B. speed
   C. temperature
   D. energy consumption

_____ 9. Analog input devices include ___ sensors, position sensors, and weight sensors.
   A. pushbutton
   B. limit
   C. emergency
   D. temperature

_____ 10. A(n) ___ connected to a PLC analog output module can be used to control a heating element in industrial applications.
   A. interface device
   B. analog SSR
   C. PLC input
   D. solid-state relay

Proximity Sensor Detection

PLC Exercise 5.1

BILL OF MATERIALS
- TECO PLR or Micrologix™ 1100 PLC
- Selector switch
- Proximity switch
- Green light
- Red light
- Horn/alarm

Typically, different types of industries utilize proximity sensors to detect the presence or absence of objects or materials. For this lab, a proximity sensor/switch is hardwired normally open to indicate that a large format camera is level. If the magnet suspended above the sensor remains directly over the sensor, the camera is level, and a green indicator light turns ON. If the magnet is not directly over the sensor, the camera is not level, a red indicator light turns ON, and the green indicator light turns OFF. Additionally, if the camera is not level for 5 seconds, an alarm horn sounds. A switch allows the alarm horn to be turned OFF.

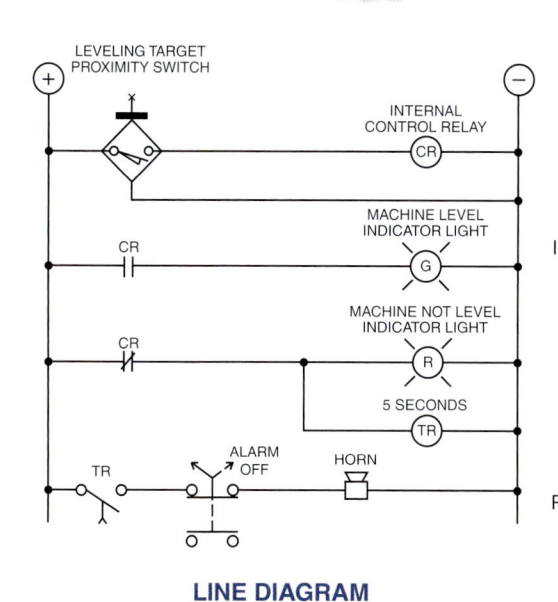

LINE DIAGRAM

T TECO Procedures

1. Create a new ladder logic or FBD program.

2. Use the TECO programming diagram or FBD to add the proper contacts and description in the Programming Grid.

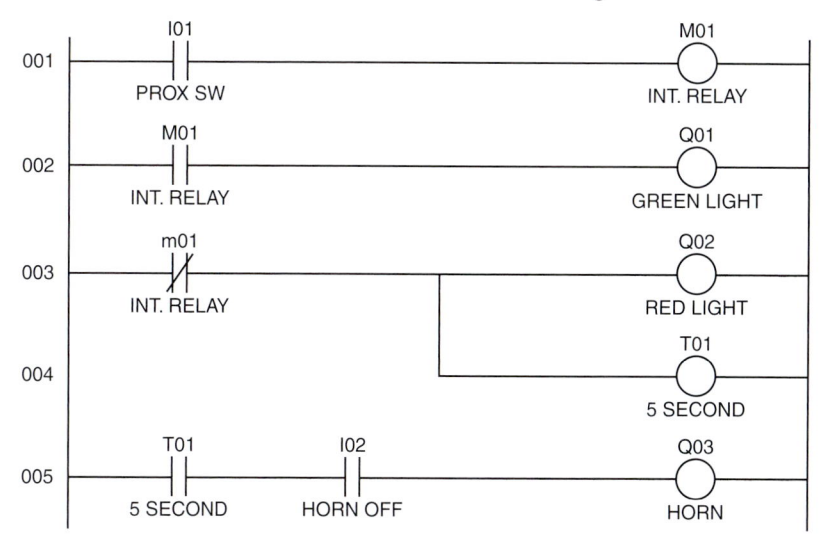

TECO PROGRAMMING DIAGRAM

231

3. Use the Edit Contact window to denote an input or output number and the Symbol button to add a description after placing the input or output in the programming grid. For an FBD, use the Comment icon to add the description.

4. To add timers to the ladder logic, select the T Contact and place in the Programming Grid. An Edit Contact/Coil window will appear. Use this window to select the timer number, add a description, and construct the time base, preset value, and preset type in this window. To add timers to an FBD, select the Tx Function Block icon and place in the Programming Grid. A Timer Function Block window will appear. Use this window to select a mode, timer number, time base, preset value, and description.

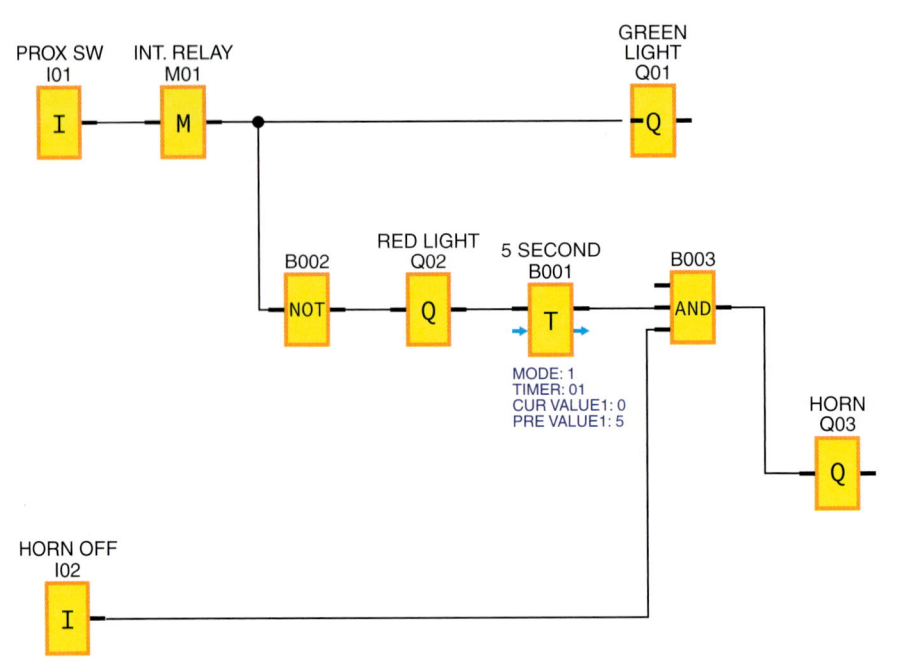

FUNCTION BLOCK DIAGRAM

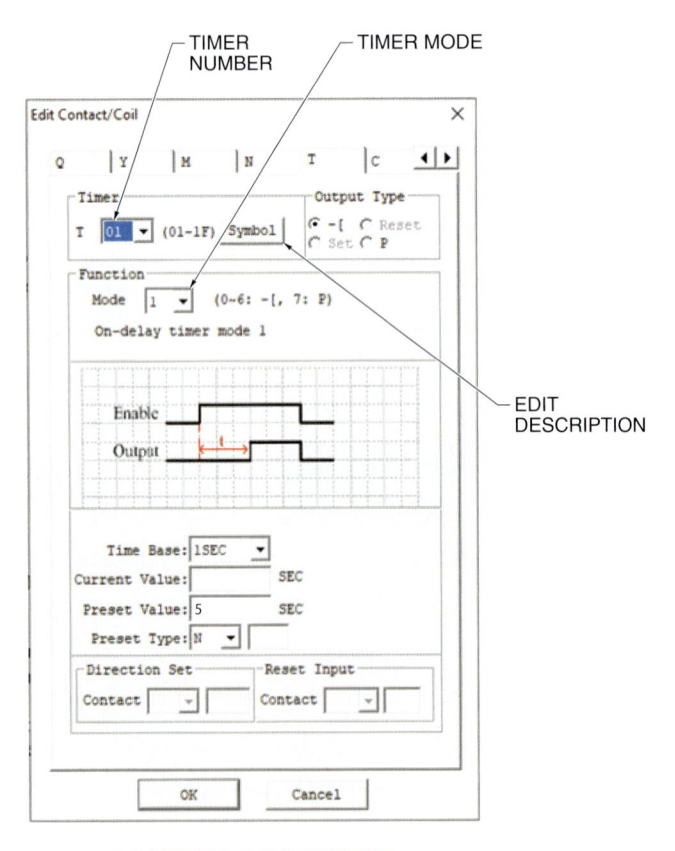

LADDER LOGIC TIMERS

5. Run the program in Simulation Mode to verify proper operation. Use the Input Status tool window to activate the inputs and control the output to simulate circuit operation.

6. Use the wiring diagram to wire the input devices and output components to the PLC.

7. Turn power to the PLC ON and link the com port.

8. Write the program to the SG2-12HR-D to download the program logic to the PLC.

9. Place the SG2-12HR-D in Run Mode. Activate the start pushbutton and monitor the I/Os on the LCD display.

10. Stop the program and use the Save As function to save and place the lab application in the required folder location.

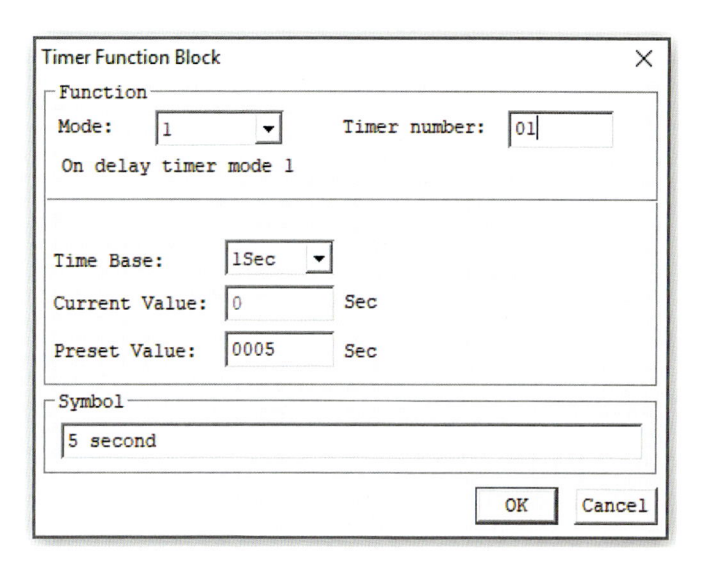

FUNCTION BLOCK TIMERS

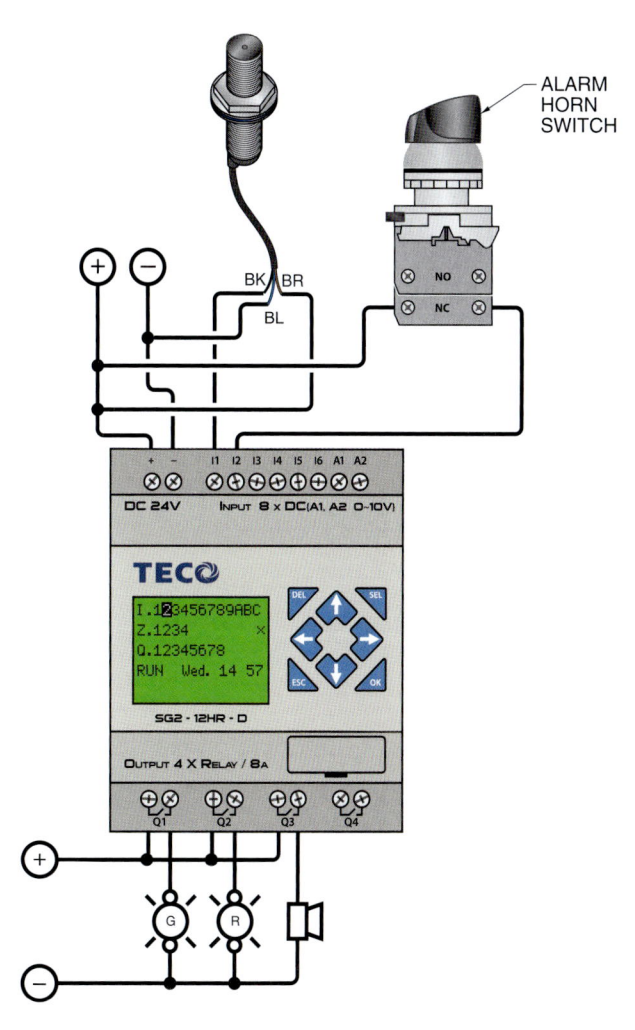

TECO WIRING DIAGRAM

234 PROGRAMMABLE LOGIC CONTROLLERS: PRINCIPLES AND APPLICATIONS

Allen-Bradley® Procedures

1. Open the RSLogix™ Micro Starter Lite program and create a new programming diagram.

2. Use the AB programming diagram to add the proper address and description in the LAD 2 programming window. To add a timer instruction, select the TON contact from the Timer/Counter tab in the Instruction toolbar. Add the TON instruction to the programming diagram and type in the timer address, time base, and preset value.

3. Verify that the programming diagram has no errors and save the project.

4. Use the wiring diagram to wire the input devices and output components to the PLC. Once all connections have been made, turn power to the PLC on.

5. Use RSLinx™ to establish communication between the PC and the Micrologix™ 1100.

6. Download the program logic to the PLC.

7. Place the PLC in Run Mode.

8. Activate input(s) and monitor them to determine if the output(s) are operating correctly.

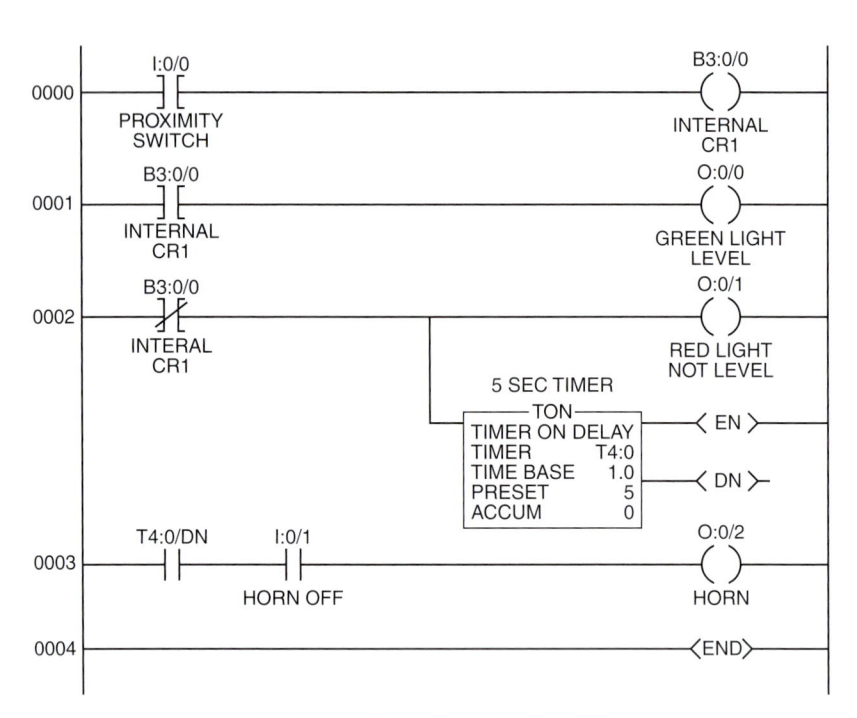

AB PROGRAMMING DIAGRAM

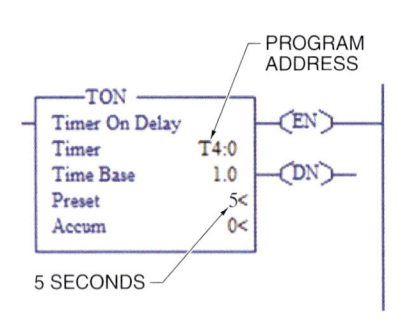

TON INSTRUCTIONS

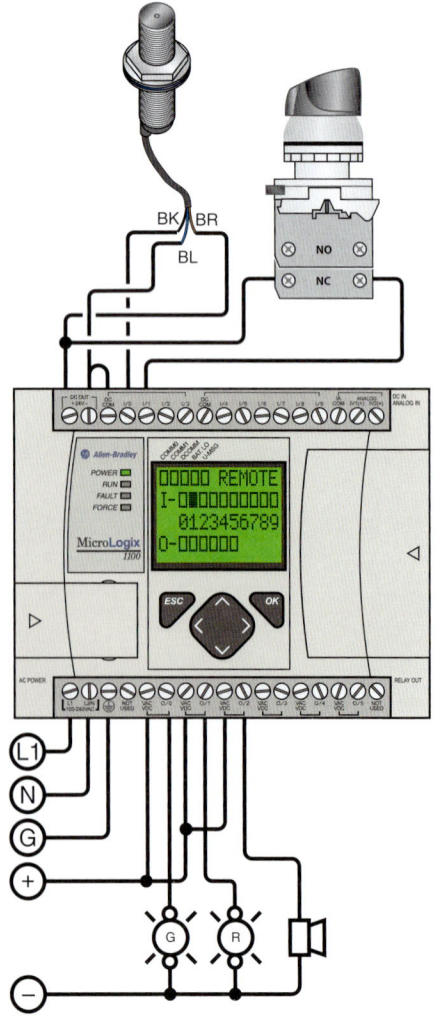

AB WIRING DIAGRAM

Chapter 5—PLC Systems, Circuits, and Interface Devices **235**

Photoelectric Switch Activation

PLC Exercise 5.2

BILL OF MATERIALS

- TECO PLR or Micrologix™ 1100 PLC
- Photoelectric switch
- Solenoid

Photoelectric switches are used in industry to detect objects in a process and activate certain components to meet process parameters. For this lab, a photoelectric switch activates a solenoid-actuated cylinder to open a gate when an object is detected. The cylinder is closed 10 seconds after an object passes through it. The photoelectric switch is hardwired normally open, and an off-delay timer is programmed.

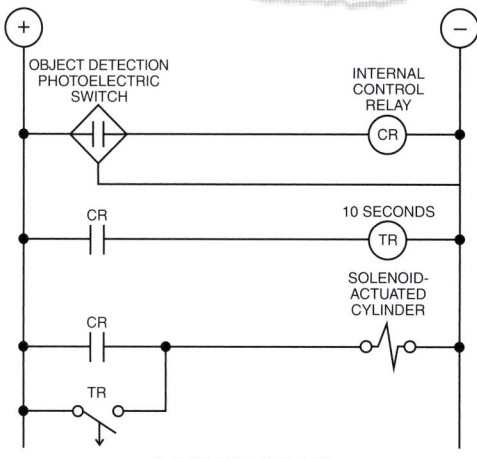

LINE DIAGRAM

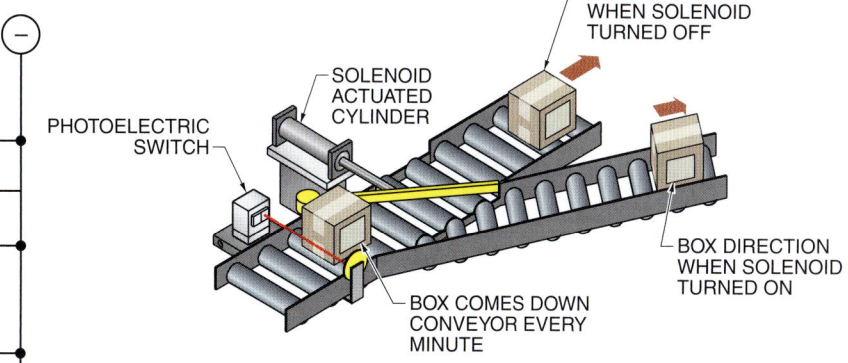

T TECO Procedures

1. Create a new ladder logic or FBD program.

2. Use the TECO programming diagram or FBD to add the proper contacts and description in the Programming Grid.

3. Use the Edit Contact window to denote an input or output number and the Symbol button to add a description after placing the input or output in the Programming Grid. For an FBD, use the Comment icon to add the description. Add a timer with the proper mode and description to the Ladder Logic Programming Grid or FBD programming area.

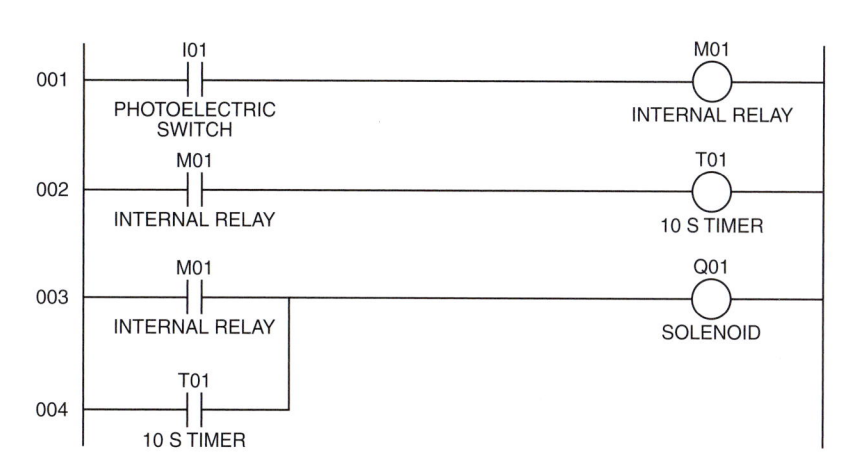

TECO PROGRAMMING DIAGRAM

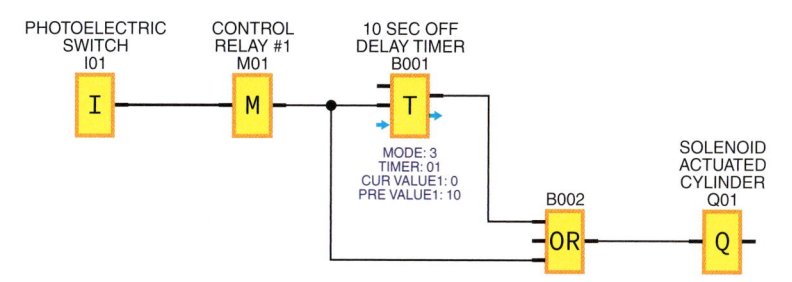

FUNCTION BLOCK DIAGRAM

236 PROGRAMMABLE LOGIC CONTROLLERS: PRINCIPLES AND APPLICATIONS

4. Run the program in Simulation Mode to verify proper operation. Use the Input Status tool window to activate the inputs and control the output to simulate circuit operation.

5. Use the wiring diagram to wire the input devices and output components to the PLC.

6. Turn power to the PLC ON and link the com port.

7. Write the program to the SG2-12HR-D to download the program logic to the PLC.

8. Place the SG2-12HR-D in Run Mode. Activate the start pushbutton and monitor the I/Os on the LCD display.

9. Stop the program and use the Save As function to save and place the lab application in the required folder location.

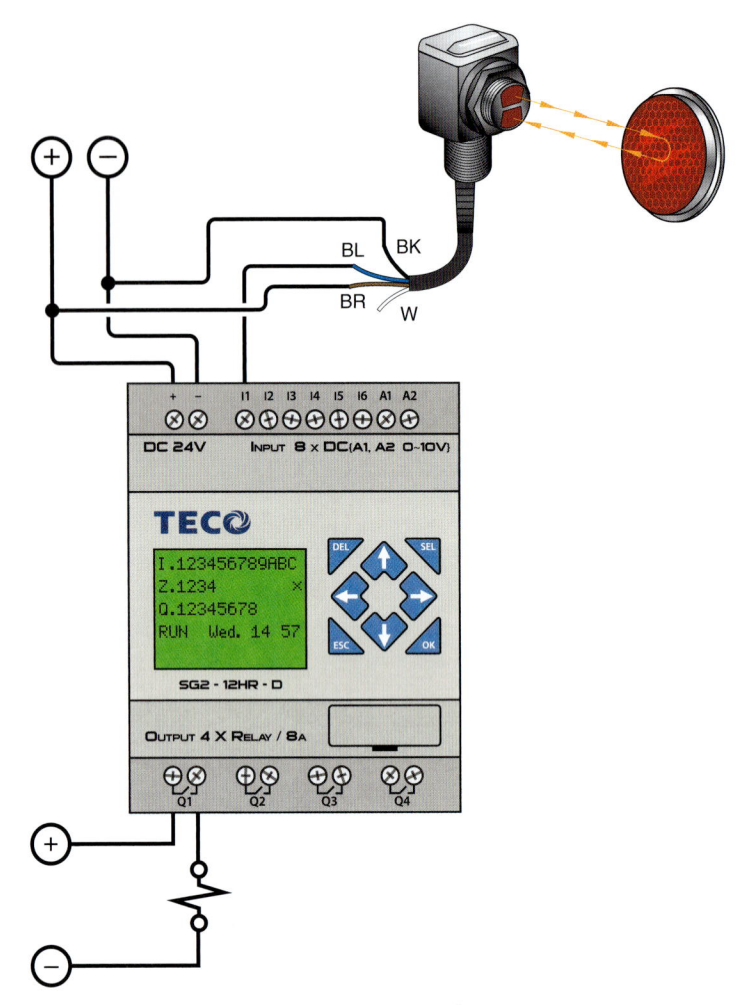

TECO WIRING DIAGRAM

AB Allen-Bradley® Procedures

1. Open the RSLogix™ Micro Starter Lite program and create a new programming diagram.

2. Use the AB programming diagram to add the proper address and description in the LAD 2 programming window. To add a timer instruction, select the TOF contact from the Timer/Counter tab in the Instruction toolbar. Add the TOF instruction to the programming diagram and type in the timer address, time base, and preset value.

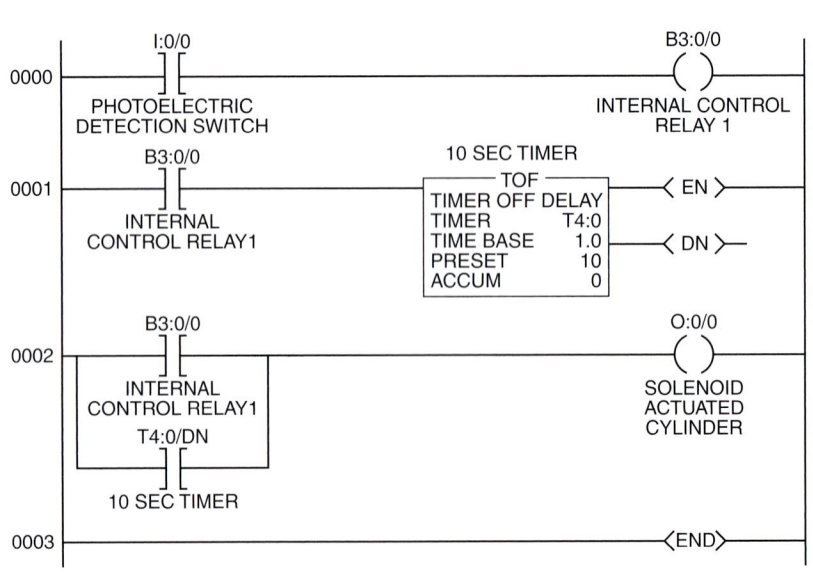

AB PROGRAMMING DIAGRAM

3. Verify that the programming diagram has no errors and save the project.

4. Use the wiring diagram to wire the input devices and output components to the PLC. Once all connections have been made, turn power to the PLC on.

5. Use RSLinx™ to establish communication between the PC and the Micrologix™ 1100.

6. Download the program logic to the PLC.

7. Place the PLC in Run Mode.

8. Activate input(s) and monitor them to determine if the output(s) are operating correctly.

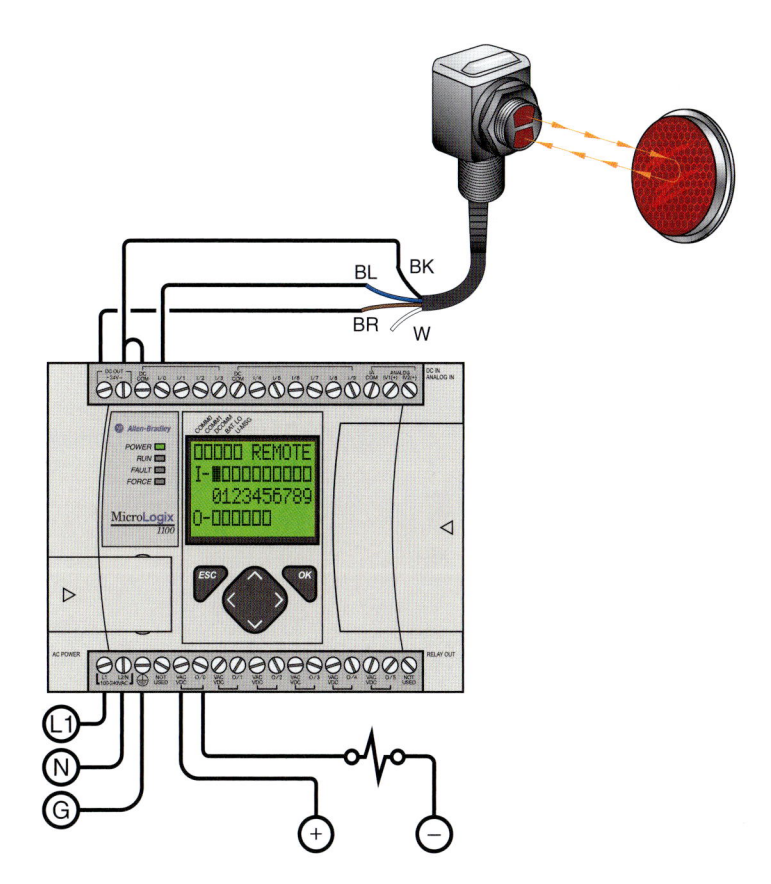

AB WIRING DIAGRAM

SIEMENS Programming and Wiring

Use the S7-1200 Easy Book to program and wire the equivalent TECO and Allen-Bradley exercises. Reference page 70 for Siemens STEP 7 programming symbols.

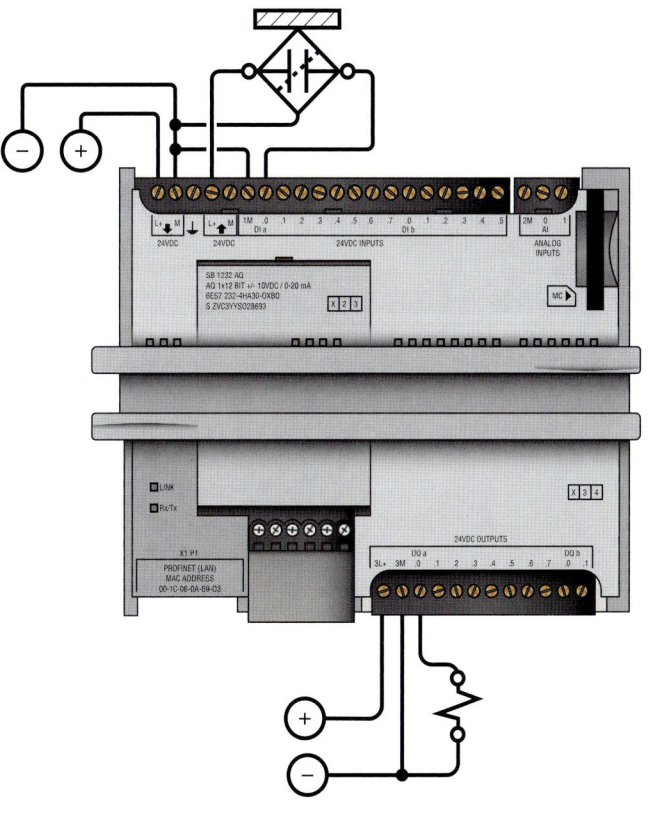

SIEMENS WIRING DIAGRAM

PLC Control of Electric Motor Drives

PLC Exercise 5.3

BILL OF MATERIALS

- TECO PLR or Micrologix™ 1100 PLC
- Electric motor drive
- NC stop pushbutton
- Two NO pushbuttons
- Motor (optional)

In many industrial processes, electric motor drives (EMDs) are used to increase motor efficiency and effectiveness for any motor application. Typically, automation allows a PLC to control an EMD. In a simple packaging application, a PLC can control an EMD's inputs for a stop, forward, and reverse motor circuit. EMD control wiring may be designed as either a sink or source operating condition, which can be controlled by a selector switch. For this lab, the PLC inputs control the stop, forward, and reverse functions for an EMD. The EMD provides input power to the input switches.

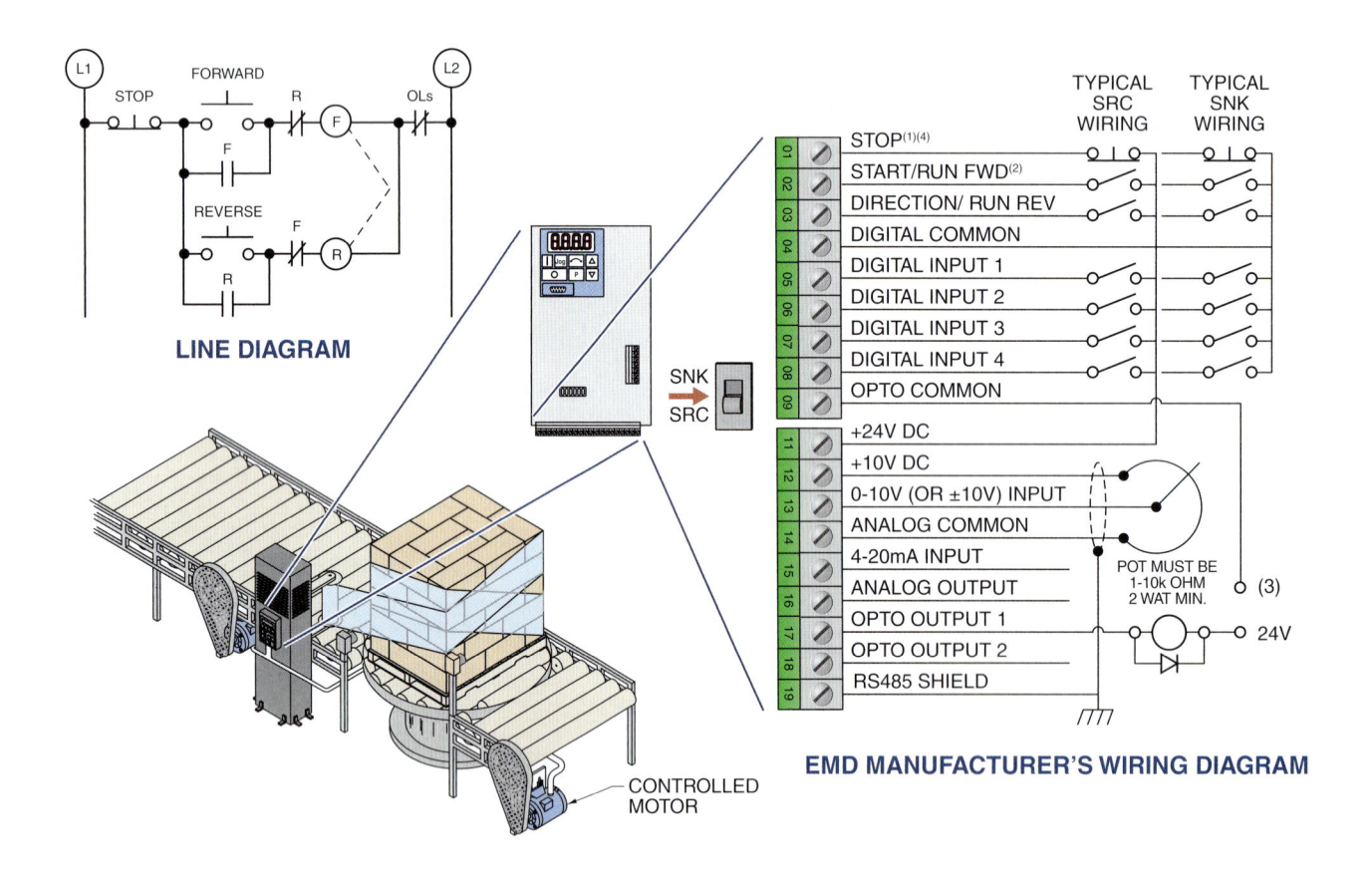

LINE DIAGRAM

EMD MANUFACTURER'S WIRING DIAGRAM

CONTROLLED MOTOR

TECO Procedures

1. Create a new ladder logic program.

2. Use the TECO programming diagram or add the proper contacts and description in the Programming Grid.

3. Use the Edit Contact window to denote an input or output number and the Symbol button to add a description after placing the input or output in the Programming Grid.

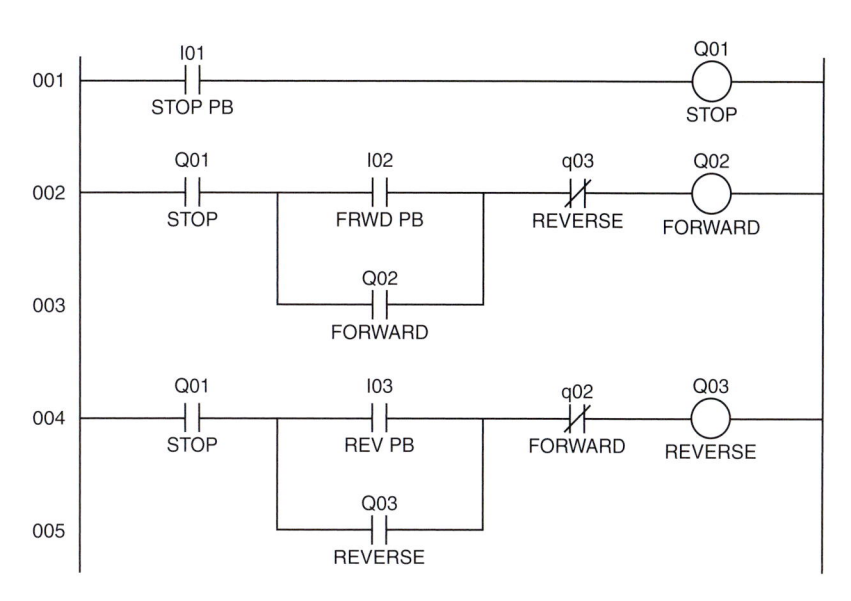

```
        I01                                              Q01
001    ─┤ ├──────────────────────────────────────────────( )──
       STOP PB                                           STOP

        Q01          I02           q03              Q02
002    ─┤ ├──────────┤ ├──────────┤/├───────────────( )──
        STOP        FRWD PB       REVERSE           FORWARD
                     Q02
003    ──────────────┤ ├──────────
                    FORWARD

        Q01          I03           q02              Q03
004    ─┤ ├──────────┤ ├──────────┤/├───────────────( )──
        STOP        REV PB        FORWARD           REVERSE
                     Q03
005    ──────────────┤ ├──────────
                    REVERSE
```

TECO PROGRAMMING DIAGRAM

4. Run the program in Simulation Mode to verify proper operation. Use the Input Status tool window to activate the inputs and control the output to simulate circuit operation.

5. Use the wiring diagram to wire the input devices to the PLC and the EMD control wiring block diagram to wire the PLC outputs to the appropriate EMD inputs. *Note:* If EMD is set in sink (SNK) position, pin 4 (negative) is used as power suppy for inputs 1 to 3 and 6 to 8. If EMD is set in source (SRC) position, pin 11 is used as power supply for inputs 1 to 3 and 6 to 8.

6. Turn power to the PLC ON and link the com port.

7. Write the program to the SG2-12HR-D to download the program logic to the PLC.

8. Place the SG2-12HR-D in Run Mode. Activate the start pushbutton and monitor the I/Os on the LCD display.

9. Stop the program and use the Save As function to save and place the lab application in the required folder location.

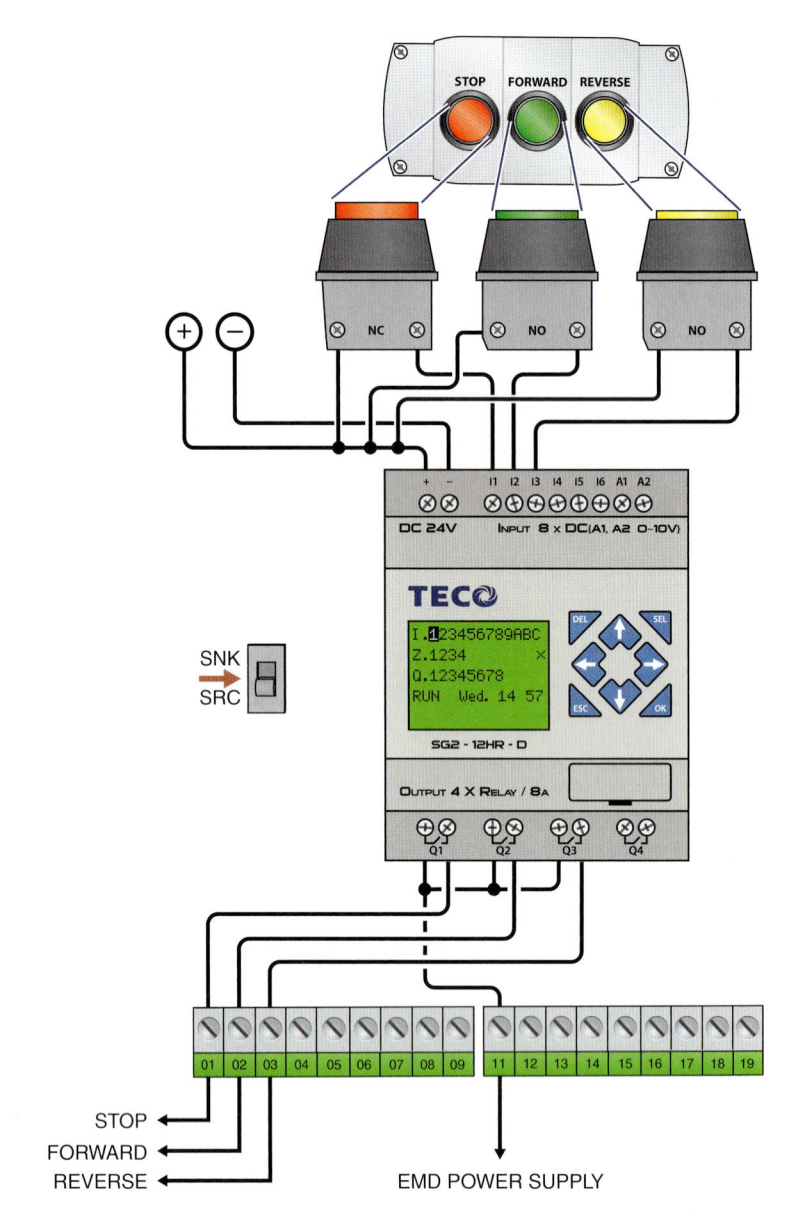

TECO WIRING DIAGRAM

Allen-Bradley® Procedures

1. Open the RSLogix™ Micro Starter Lite program and create a new programming diagram.

2. Use the AB programming diagram to add the proper address and description in the LAD 2 programming window.

3. Verify that the programming diagram has no errors and save the project.

4. Use the wiring diagram to wire the input devices to the PLC and the EMD control wiring block diagram to wire the PLC outputs to the appropriate EMD inputs. Once all connections have been made, turn power to the PLC on.

5. Use RSLinx™ to establish communication between the PC and the Micrologix™ 1100.

6. Download the program logic to the PLC.

7. Place the PLC in Run Mode.

8. Activate input(s) and monitor them to determine if the output(s) are operating correctly.

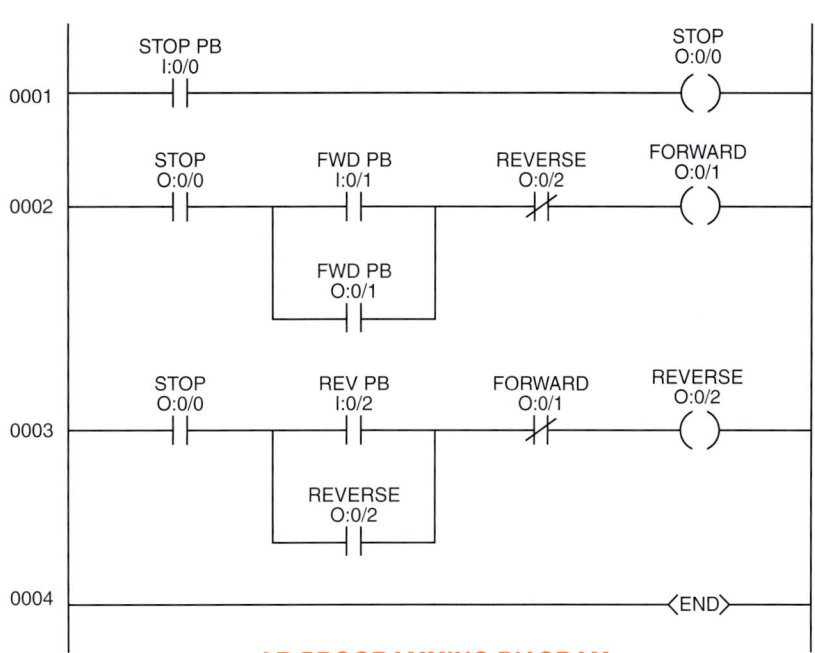

AB PROGRAMMING DIAGRAM

Rung 0001: STOP PB I:0/0 ──┤├── STOP O:0/0 ──( )──

Rung 0002: STOP O:0/0 ──┤├── FWD PB I:0/1 ──┤├── REVERSE O:0/2 ──┤/├── FORWARD O:0/1 ──( )──
   branch: FWD PB O:0/1 ──┤├──

Rung 0003: STOP O:0/0 ──┤├── REV PB I:0/2 ──┤├── FORWARD O:0/1 ──┤/├── REVERSE O:0/2 ──( )──
   branch: REVERSE O:0/2 ──┤├──

Rung 0004: ⟨END⟩

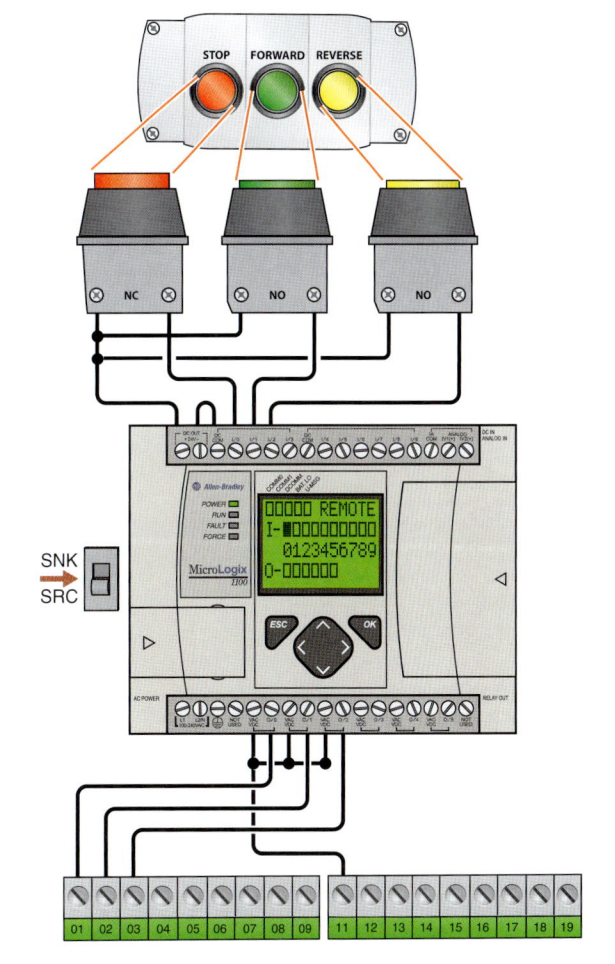

AB WIRING DIAGRAM

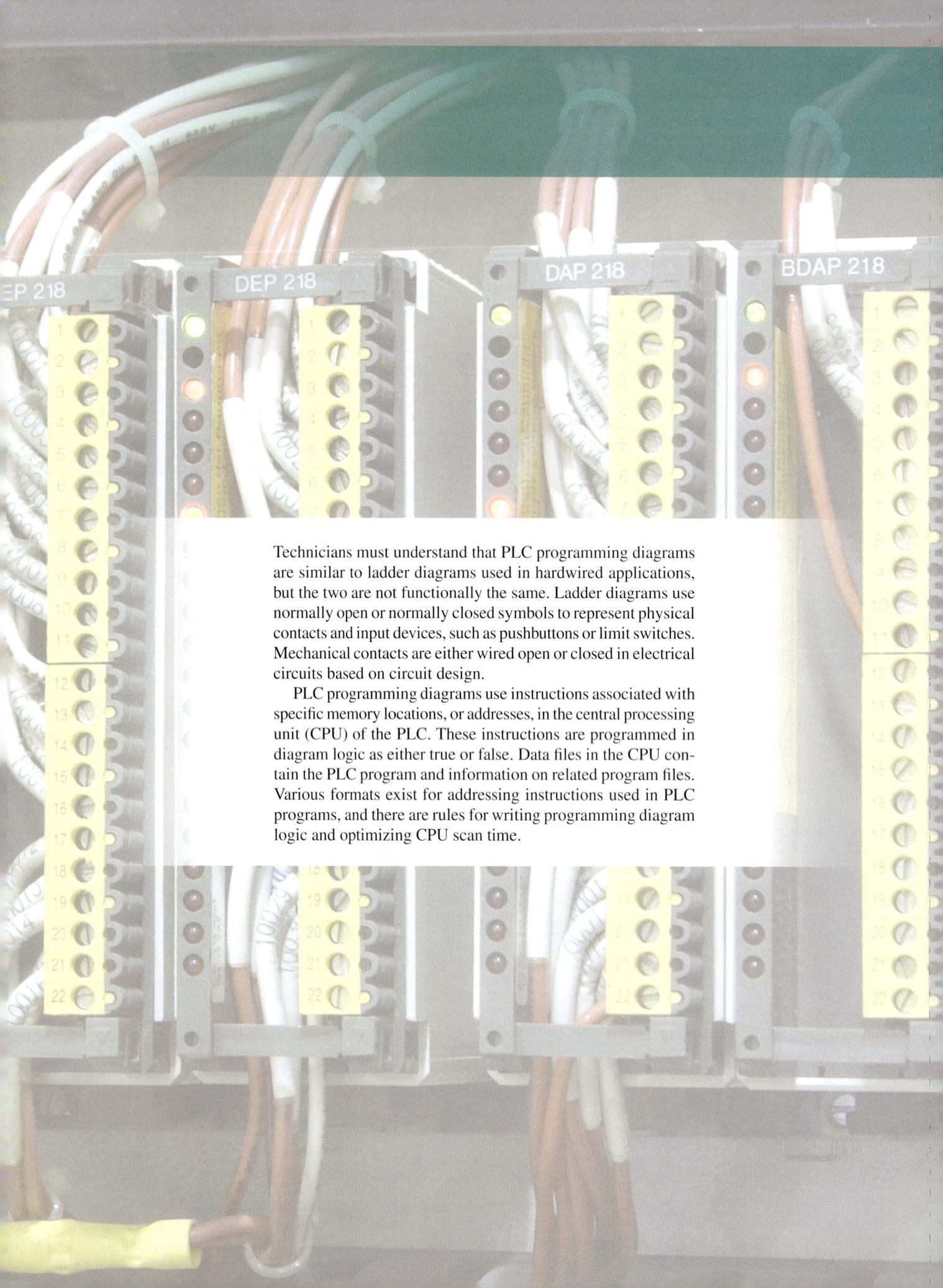

Technicians must understand that PLC programming diagrams are similar to ladder diagrams used in hardwired applications, but the two are not functionally the same. Ladder diagrams use normally open or normally closed symbols to represent physical contacts and input devices, such as pushbuttons or limit switches. Mechanical contacts are either wired open or closed in electrical circuits based on circuit design.

PLC programming diagrams use instructions associated with specific memory locations, or addresses, in the central processing unit (CPU) of the PLC. These instructions are programmed in diagram logic as either true or false. Data files in the CPU contain the PLC program and information on related program files. Various formats exist for addressing instructions used in PLC programs, and there are rules for writing programming diagram logic and optimizing CPU scan time.

PLC Programming Diagrams, Addresses, and Bit Instructions

6

Objectives:

- Describe the structure and organization of PLC program files and data table files.
- Locate a data table file and describe the structure, organization, and addressing for input, output, status, and bit files using the data tables.
- Describe the rules for writing a PLC programming diagram.
- Explain how bit instructions (XIC, XIO, OTE, OTL, and OTU) relate to real-world devices and to their respective data table addresses.
- Explain how a PLC scans a program and how the scan relates to input and output data table files.

Learner Resources
atplearningresources.com/quicklinks
Access Code: 475203

PLC PROGRAMMING DIAGRAMS

In a hardwired (non-PLC) application, the contacts of pushbuttons, selector switches, limit switches, timers, and counters, and the coils of relays, pilot lights, solenoids, and motor starters are connected together with wire. Hardwired devices and components are typically depicted using line/ladder diagrams. Line/ladder diagrams contain the logic for applications and are typically used to test and troubleshoot hardwired applications. **See Figure 6-1.**

All input devices in PLC applications (pushbuttons, selector switches, and limit switches) are wired to the input section of the PLC, and all output components (pilot lights, solenoids, and motor starter coils) are wired to the output section. All input devices and output components are represented by instruction symbols assembled into a PLC programming diagram that is written in ladder logic form.

A PLC programming diagram contains the logic for the PLC application and is used to test and troubleshoot PLC applications. PLC programming and instructions take the place of control wiring and control devices such as relays, timers, and counters. **See Figure 6-2.**

PLC TIPS

Although the outermost vertical lines of a programming diagram are referred to as power rails, power does not actually flow through these rails. The vertical rails are highlighted when the PC is online with a PLC and in the RUN mode.

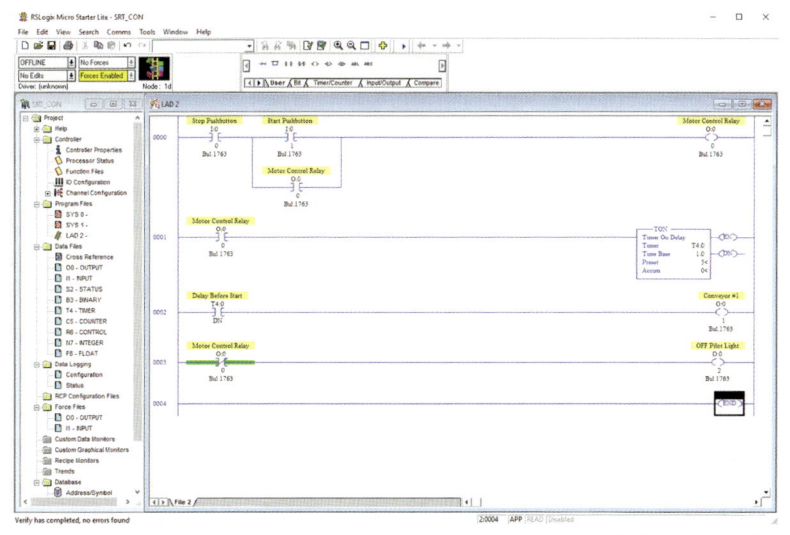

Programming diagrams are typically created electronically using PLC programming software.

243

244 PROGRAMMABLE LOGIC CONTROLLERS: PRINCIPLES AND APPLICATIONS

Hardwired Motor Controls

LINE/LADDER DIAGRAM

HARDWIRED LADDER APPLICATION

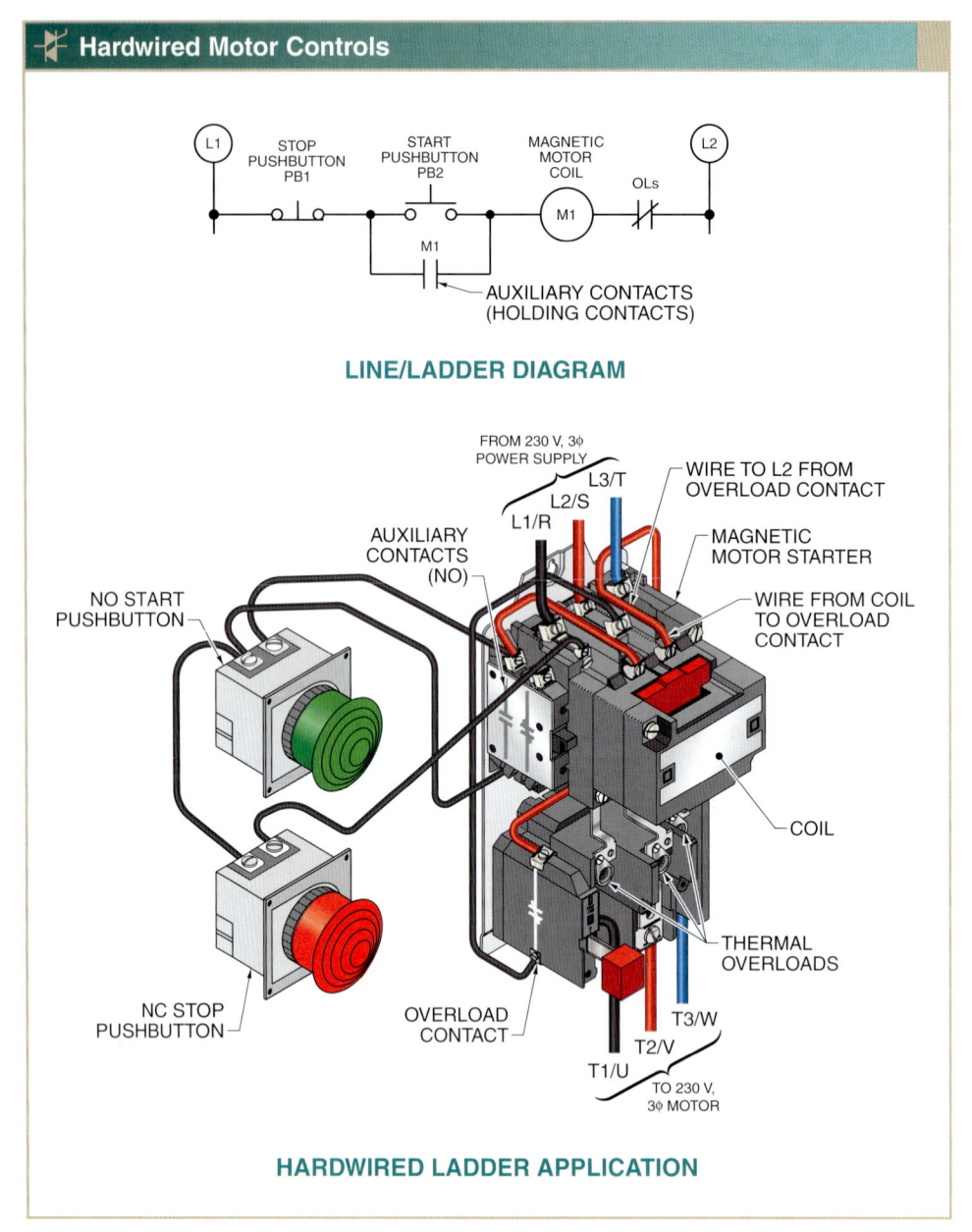

Figure 6-1. A hardwired motor control application is depicted by symbols in a ladder diagram.

A PLC programming diagram resides in the memory of a PLC. Various files within the memory of the PLC relate to the PLC program. The PLC program, related files, and instructions are addressed, assembled, and organized in a specific manner. PLC programming is similar regardless of manufacturer. However, specific programming rules and procedures must be followed for each manufacturer and model of PLC.

PROCESSOR FILES

A PLC contains a processor file in the memory of the central processing unit (CPU). A *processor file* is the section of memory in the PLC that is filled with CPU data about programming. The processor file contains the program files and data table files for specific applications. The processor file has a unique technician-assigned name and is often referred to as the "project."

PLC-Wired Motor Controls

PLC PROGRAMMING DIAGRAM

PLC-WIRED PROGRAMMING APPLICATION

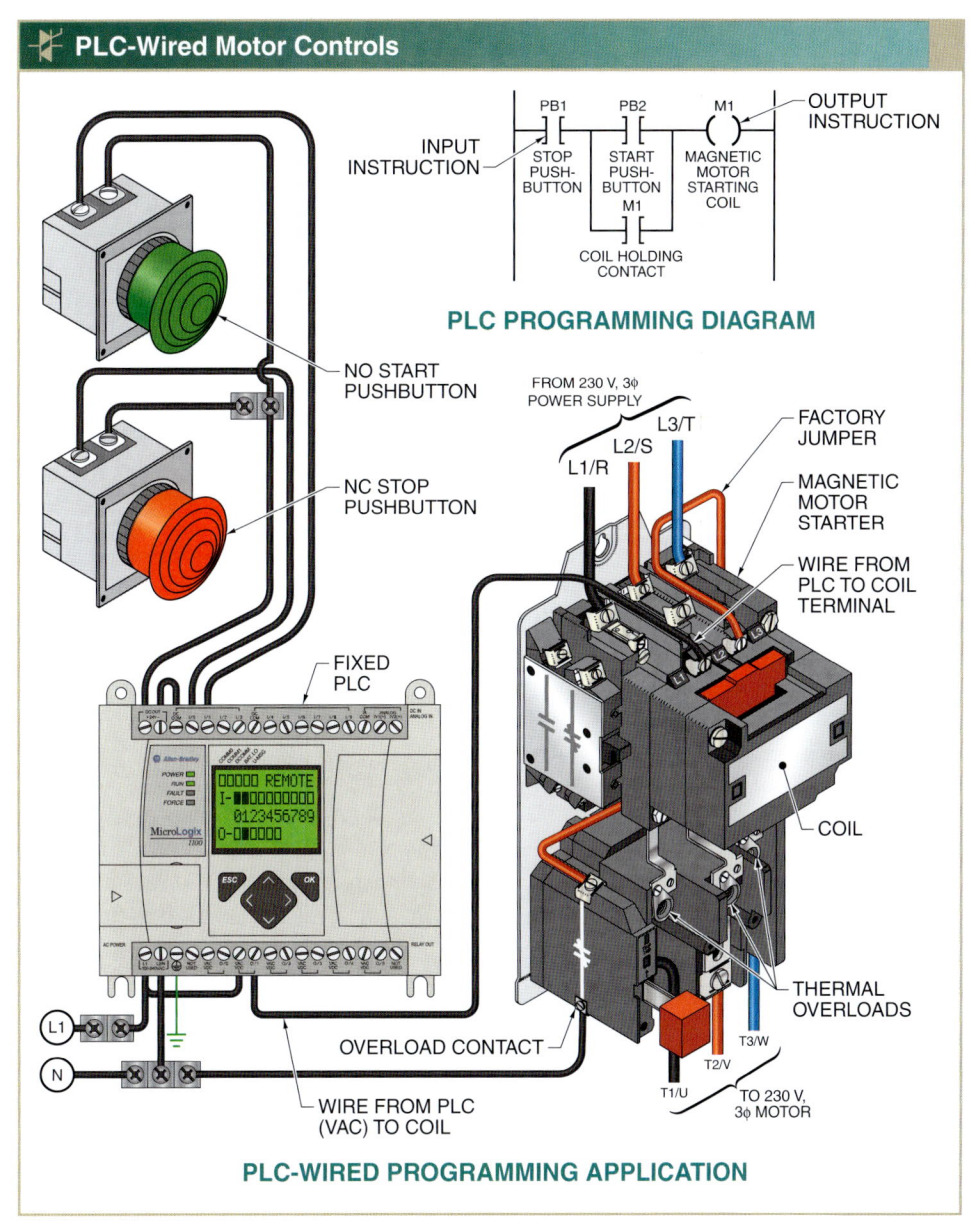

Figure 6-2. A PLC motor control application is depicted by instructions in a programming diagram.

CODE CONNECT

Nonlinear loads such as LEDs, personal computers, and CFLs generate harmonics, which can cause extra current to flow on the neutral of a three-phase, four-wire, wye system. Section 310.15(E)(3) of the NEC® requires that the neutral of a three-phase, four-wire, wye system be considered a current-carrying conductor relative to section 310.15(C)(1), when the majority of the loads are nonlinear.

Processor files are downloaded (copied) from programming devices to PLCs. Typically, processor files are thought of as file cabinets, and the program files and data table files as drawers within the cabinet. The various program files and data table files are folders within the drawers. **See Figure 6-3.**

Note: The format and addressing of PLC processor files varies from manufacturer to manufacturer and between PLC families of the same manufacturer.

Program Files

A *program file* is data in the memory of a PLC that contains technician-developed PLC programs and related information. Two hundred fifty-six program files are possible in many PLCs, and the files are numbered 0 through 255. **See Figure 6-4.** Three of the program files are always used: the system file (file 0), the reserved file (file 1), and the main program file (file 2). Files 3 through 255 are subroutine files and may or may not be used depending upon the application.

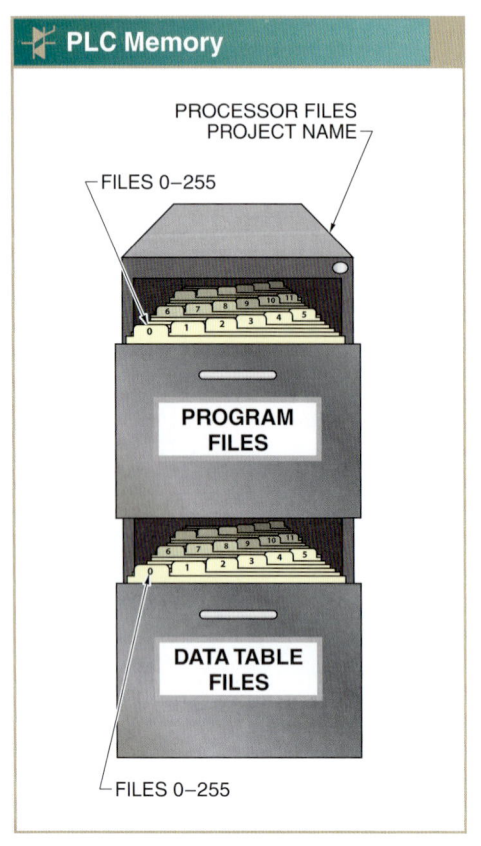

246 PROGRAMMABLE LOGIC CONTROLLERS: PRINCIPLES AND APPLICATIONS

Figure 6-3. The memory of a PLC contains the processor files, which contains the program files and data table files.

- System program file (file 0)—File 0 contains system-related information: CPU type, I/O configuration, and passwords. File 0 cannot be accessed by a technician.
- Reserved file (file 1)—File 1 is reserved by the CPU for processing data. File 1 cannot be accessed by a technician.
- Main program file (file 2)—File 2 contains the technician-developed control program for a specific application. The PLC program (programming diagram), written in ladder logic form, resides in the main program file. File 2 can be accessed by a technician.
- Subroutine program files (files 3–255)—Files 3 through 255 contain additional technician-developed control programs. Subroutine program files are activated by instructions in the main program file. Files 3 through 255 can be accessed by a technician.

Data Table Files

A *data table file* is the section of PLC memory that contains the status of the CPU, inputs and outputs, timer and counter preset and accumulated values, and other

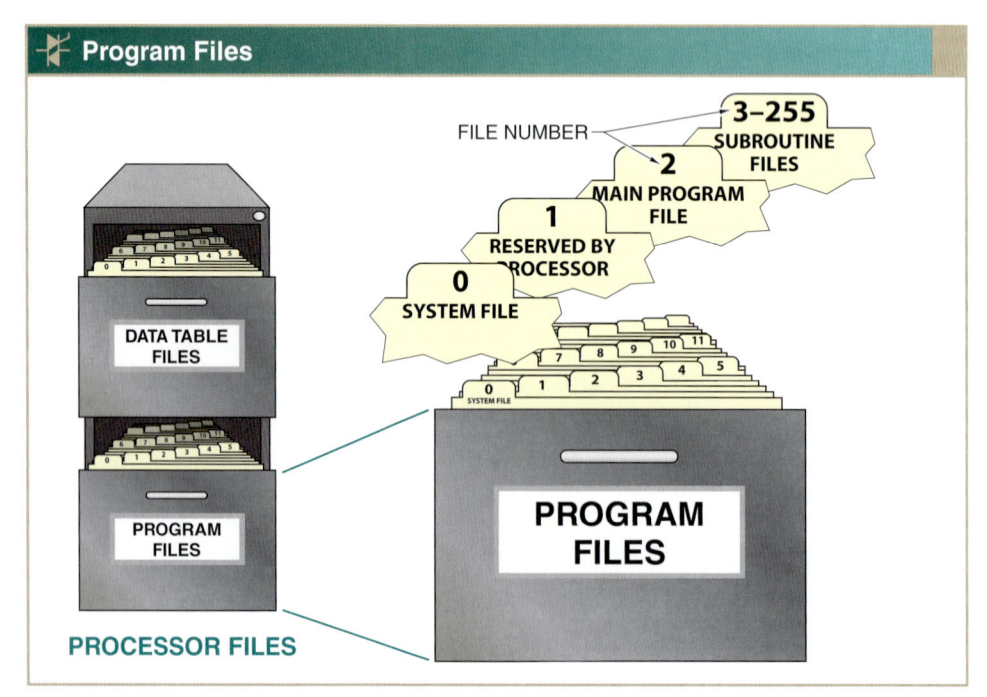

Figure 6-4. There are 256 possible program files, with the function of program files 0, 1, and 2 defaulted by the PLC.

program instruction values. There are 256 data table files, numbered 0 through 255. **See Figure 6-5.** The data table files are arranged by the type of data they contain. Nine of the data table files, numbered 0 through 8, are default files. Data table files 0 through 8 are preconfigured and the file type cannot be changed. Data table files 9 through 255 are configured by the technician.

- Output file (file 0)—File 0 contains the status of all the output terminals of the PLC. The output file is also referred to as the output image table.
- Input file (file 1)—File 1 contains the status of all the input terminals of the PLC. The input file is also referred to as the input image table.
- Status file (file 2)—File 2 contains processor (CPU) status information (powered, running, forcing, and disabling).
- Bit file (file 3)—File 3 contains relay logic that is used by the CPU.
- Timer file (file 4)—File 4 contains timer status, preset values, and accumulated values.
- Counter file (file 5)—File 5 contains counter status, preset values, and accumulated values.
- Control file (file 6)—File 6 contains information and status related to certain instructions, such as bit shift instructions and sequencer instructions.
- Integer file (file 7)—File 7 contains integer values.
- Floating point file (file 8)—File 8 contains numbers that are not integers (real numbers).
- Technician-assigned files (files 9 through 255)—Files 9 through 255 are technician-assigned files.

Data Table Files

NOTE: FILES 9–255 CANNOT BE CONFIGURED AS OUTPUT, INPUT, OR STATUS FILES.

FILE NUMBER
- 4 TIMER FILE
- 9–255 TECHNICIAN-ASSIGNED FILES
- 3 BIT FILE
- 8 FLOATING POINT FILE
- 2 STATUS FILE
- 7 INTEGER FILE
- 1 INPUT FILE
- 6 CONTROL FILE
- 0 OUTPUT FILE
- 5 COUNTER FILE
- 0 SYSTEM FILE

DATA TABLE FILES

PROGRAM FILES

PROCESSOR FILES

DATA TABLE FILES

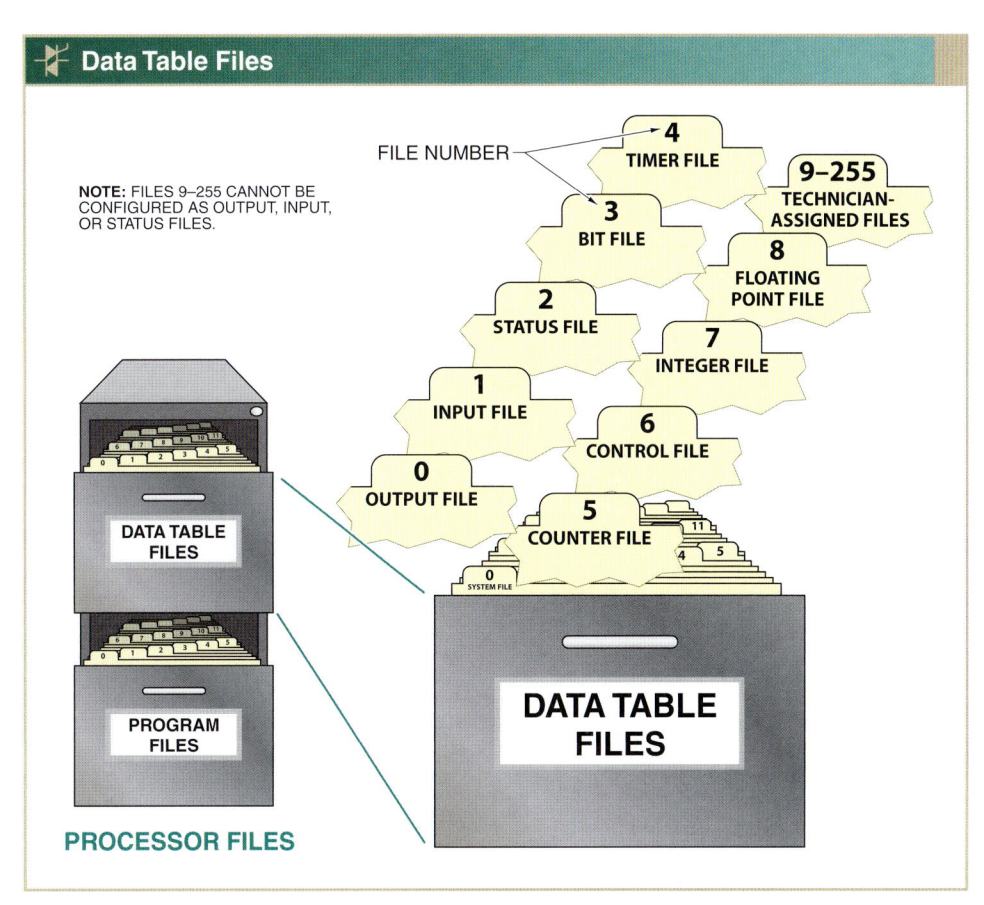

Figure 6-5. The status of the CPU, all input devices and output components, and all instructions used in the program are found in the 256 data table files.

DATA TABLE FILE ADDRESSES

Each type of data table file has a unique file identifier letter and file number. Each type of file has a variety of instructions associated with the file. The instructions within the file each have a unique address based on the file identifier and file number. **See Figure 6-6.** Different instructions use different amounts of memory. Instructions can use a bit, word, or a group of words. The address of an instruction includes the file identifier, file number, and memory location of the instruction.

DATA TABLE FILE ADDRESSING

Default Files

File Type	Identifier	File Number
Output	O	0
Input	I	1
Status	S	2
Bit	B	3
Timer	T	4
Counter	C	5
Control	R	6
Integer	N	7
Float	F	8

Technician-Assigned Files

File Type	Identifier	File Number
Bit	B	9–255
Timer	T	9–255
Counter	C	9–255
Control	R	9–255
Integer	N	9–255
Float	F	9–255

Figure 6-6. The instructions within each data table file have a unique address based on the file identifier and file number.

Input and Output File Addresses

The addressing of input devices and output components varies slightly from other instructions because the addresses represent the terminal where real-world devices are connected to a PLC. The variations are as follows:

- The output (O) and input (I) files are defaulted to file 0 and file 1 respectively and cannot be used in any other data table file location.
- The size of input and output files depends upon the number of PLC inputs and outputs.
- File numbers are typically omitted from addresses.
- Addresses contain the slot number of the module where input devices or output components are connected. Slot 0 is typically reserved for the CPU of the PLC. The maximum number of slots, using expansion racks, is considered to be 30.
- The inputs and outputs of fixed PLCs are considered slot 0 (zero) for addressing purposes.
- Addresses are formatted as slot.word/bit or as slot/bit. **See Figure 6-7.**

Status File Addresses

Status file addresses are different from other instructions because the CPU creates status files. **See Figure 6-8.** The differences are as follows:

- Status files, or "S" files, are defaulted to file 2 and cannot be used in any other data table file location.
- S files cannot be added to, deleted, or changed in any way.
- File numbers are omitted from status file addresses.
- Addresses are formatted as word/bits or as words.

PLC TIPS

When viewing certain data table files (output, input, and integer files), the radix, or base of the numbering system, can be changed. Changing the radix of a data table file changes the way information is displayed. Radix options include binary, octal, HEX/BCD, ASCII, and structured. Radix options are not available for timer and counter data files.

Input and Output File Addresses

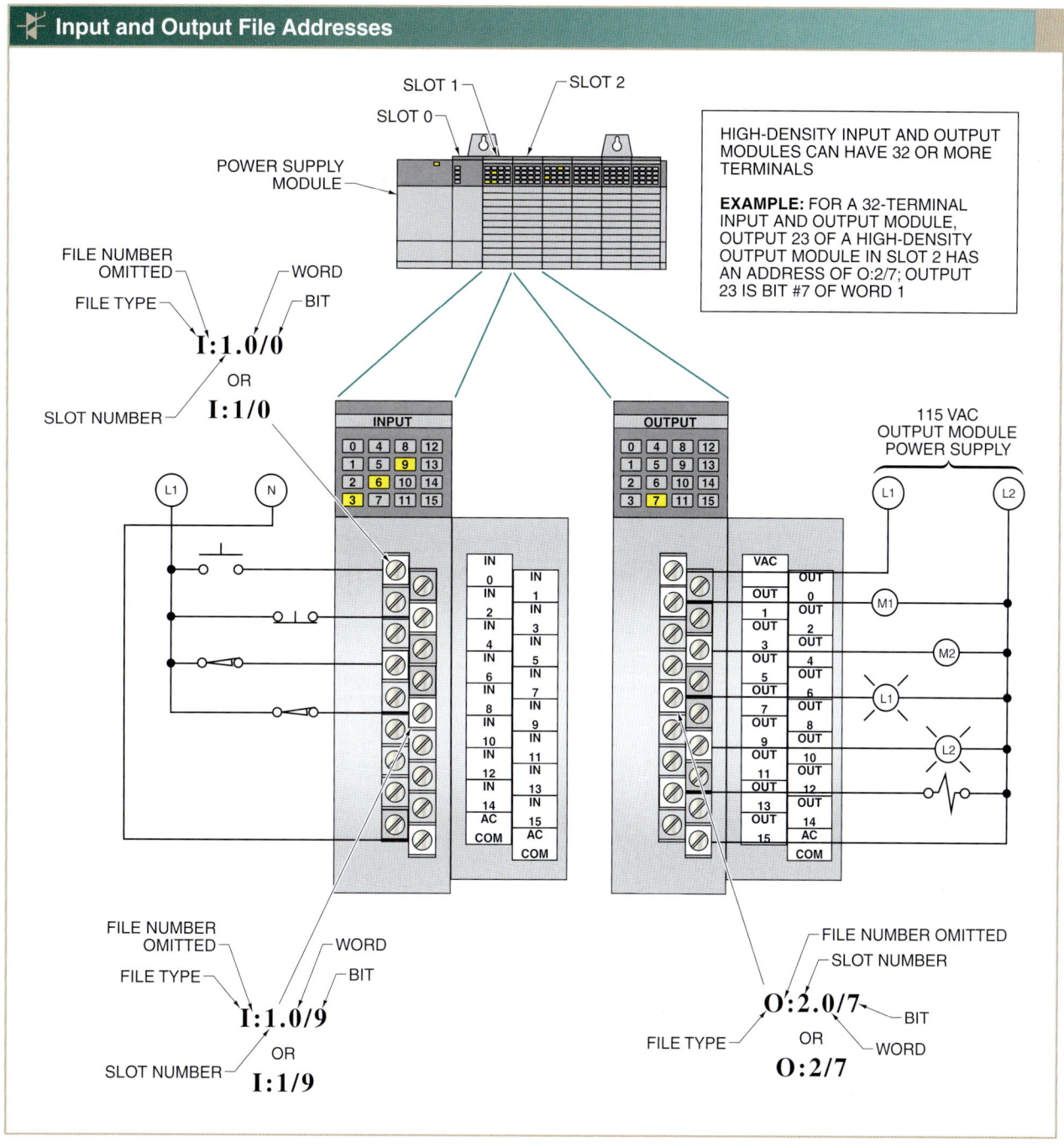

Figure 6-7. Input and output terminals are addressed as slot.word/bit or slot/bit. Standard I/O modules have 16 points (terminals) and require 1 word of memory. High-density modules have 32 points and require 2 words of memory.

HIGH-DENSITY INPUT AND OUTPUT MODULES CAN HAVE 32 OR MORE TERMINALS

EXAMPLE: FOR A 32-TERMINAL INPUT AND OUTPUT MODULE, OUTPUT 23 OF A HIGH-DENSITY OUTPUT MODULE IN SLOT 2 HAS AN ADDRESS OF O:2/7; OUTPUT 23 IS BIT #7 OF WORD 1

Bit File Addresses

Bit file addresses use words and bits. The maximum size of a bit file is 256 sixteen-bit words, or 4096 bits. **See Figure 6-9.** The specifics of bit file addressing are as follows:

- Bit files, or "B" files, are defaulted to file 3, and files 9-255 can also be configured as B files.
- File numbers are included in addresses.
- Addresses are formatted as word/bits, words, or bits.

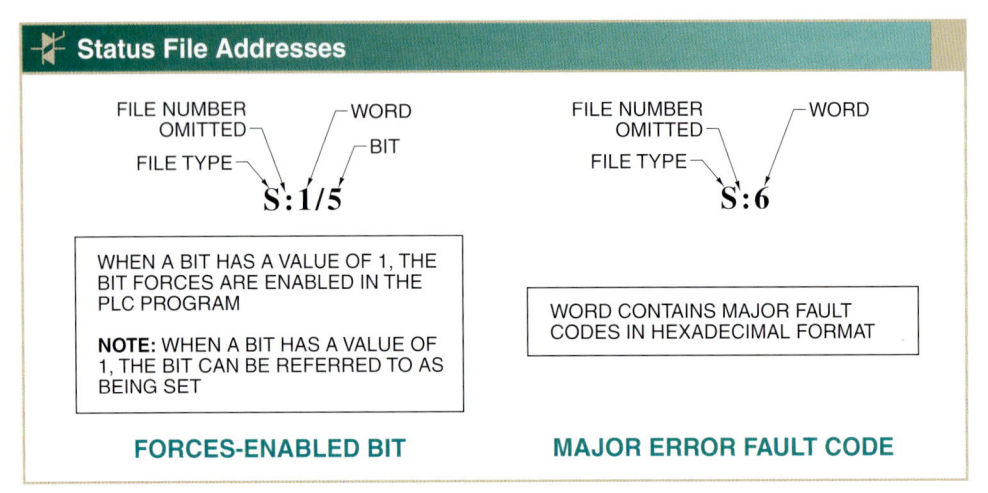

250 PROGRAMMABLE LOGIC CONTROLLERS: PRINCIPLES AND APPLICATIONS

Figure 6-8. Information in status files is formatted as word/bit or as a word. The addressing format depends on the contents of the status file.

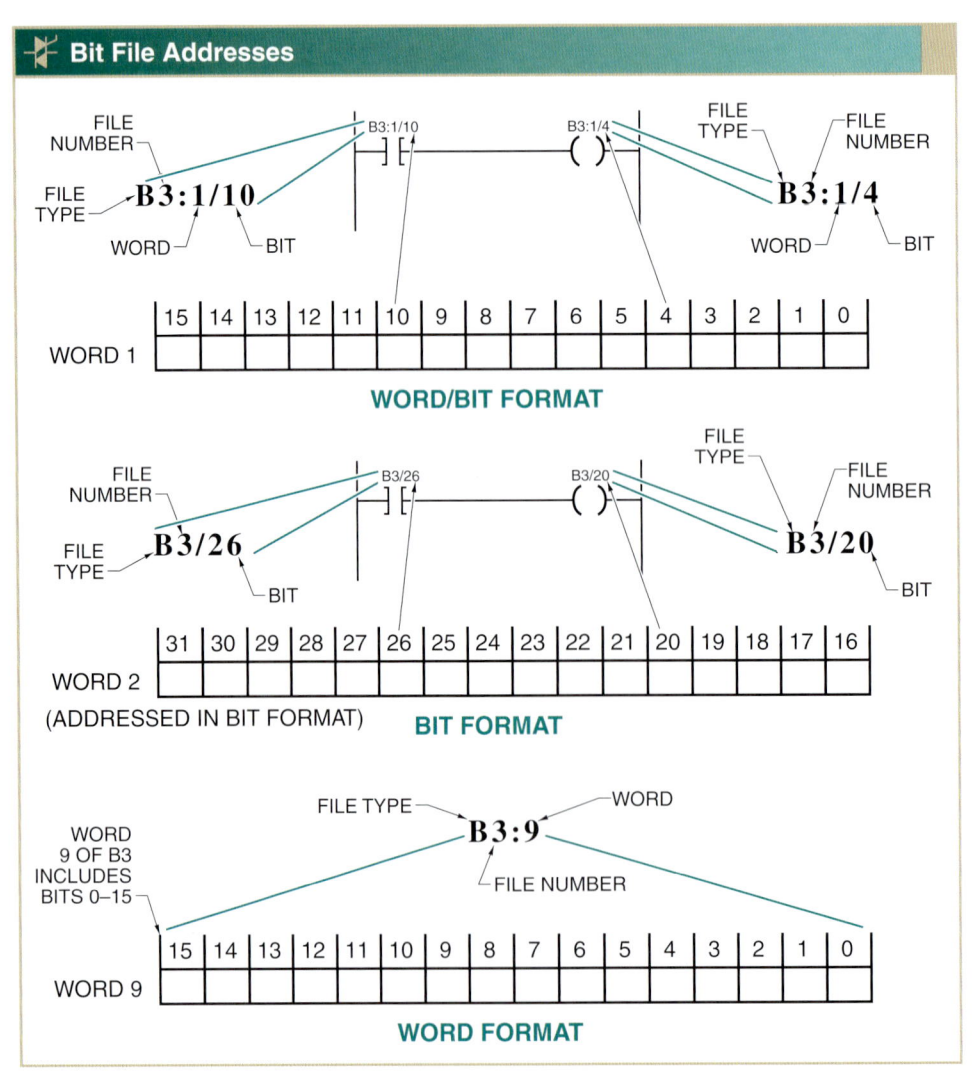

Figure 6-9. The default bit file is B3. Bit file addresses are formatted as a word/bit, bit, or word.

PROGRAMMING DIAGRAM LOGIC

PLCs are programmed in ladder logic form. Although programming diagrams resemble ladder diagrams, the two are different. A ladder diagram uses symbols to represent control devices that are hardwired together. Programming diagrams consists of instructions that represent real world PLC inputs and outputs that take the place of hardwired timers and counters. Specific rules and practices govern how instructions are assembled into programming diagrams.

Programming Diagram Rules

When writing a programming diagram for a PLC, certain rules must be followed. A PC with programming software will not allow a program to run if it does not follow the standard programming rules. **See Figure 6-10.**

Programming Rule 1. Each line of a programming diagram is referred to as a rung. Condition instructions for input devices are placed to the left on a rung. The control instructions for output components are placed to the right on a rung. Each rung must contain at least one output component control instruction. Most rungs contain both condition and control instructions.

Programming Rule 2. Only one output component can be placed on a rung. Output components can be placed in parallel but never in series.

Programming Rule 3. Input devices can be programmed in series, parallel, or in series/parallel combinations.

Programming Rule 4. Input devices and output components can be programmed in nested branches. Nested branches are branches that have varying start and/or end points.

Programming Rule 5. Nested branches cannot overlap.

Programming Rule 6. Input devices can be programmed at multiple locations. The same input device can be programmed as an XIC (examine if closed) and/or XIO (examine if open) at multiple locations.

Programming Rule 7. Output components can have various types of instructions with the same address (possibly programmed at multiple locations).

Programming Rule 8. Output components are not typically programmed at multiple locations. A special instruction is required to place an output component at multiple locations.

Bit Instructions

Bit instructions for programming diagrams look like relay contacts, relay coils, and the L1 and L2 of a ladder diagram, but do not function in the same manner. Although the two outermost vertical lines of a ladder diagram are often referred to as power rails, there is no real power present. The programming diagram and bit instructions only exist in the memory of the PLC.

The instructions in the programming diagram are either TRUE or FALSE. In order for an output instruction to be true, there must be a true line of logic preceding the output instruction. When the rung or rungs of logic preceding an output instruction is false, the output instruction is false. Programming instructions are true or false based on the type of instruction and whether the program logics make the CPU place a 1 or 0 at the data table file instruction address.

Note: Power does not flow through the input devices or output components that are represented in a programming diagram.

Control devices in a ladder diagram are either ON (closed) or OFF (open). There must be continuity through hardwired inputs in order to energize the hardwired output components. While ladder diagrams depict the flow of current, PLC programming diagrams depict logical continuity. **See Figure 6-11.**

Bit instructions are the most common type of instruction. Bit instructions are found in all data table files except floating point files. The most common type of bit instructions are XIC, XIO, and OTE.

> **⚠ CAUTION**
> The instructions for the input bits AND and OR are different from the instructions for the output words AND and OR.

252 PROGRAMMABLE LOGIC CONTROLLERS: PRINCIPLES AND APPLICATIONS

Programming Diagram Rules

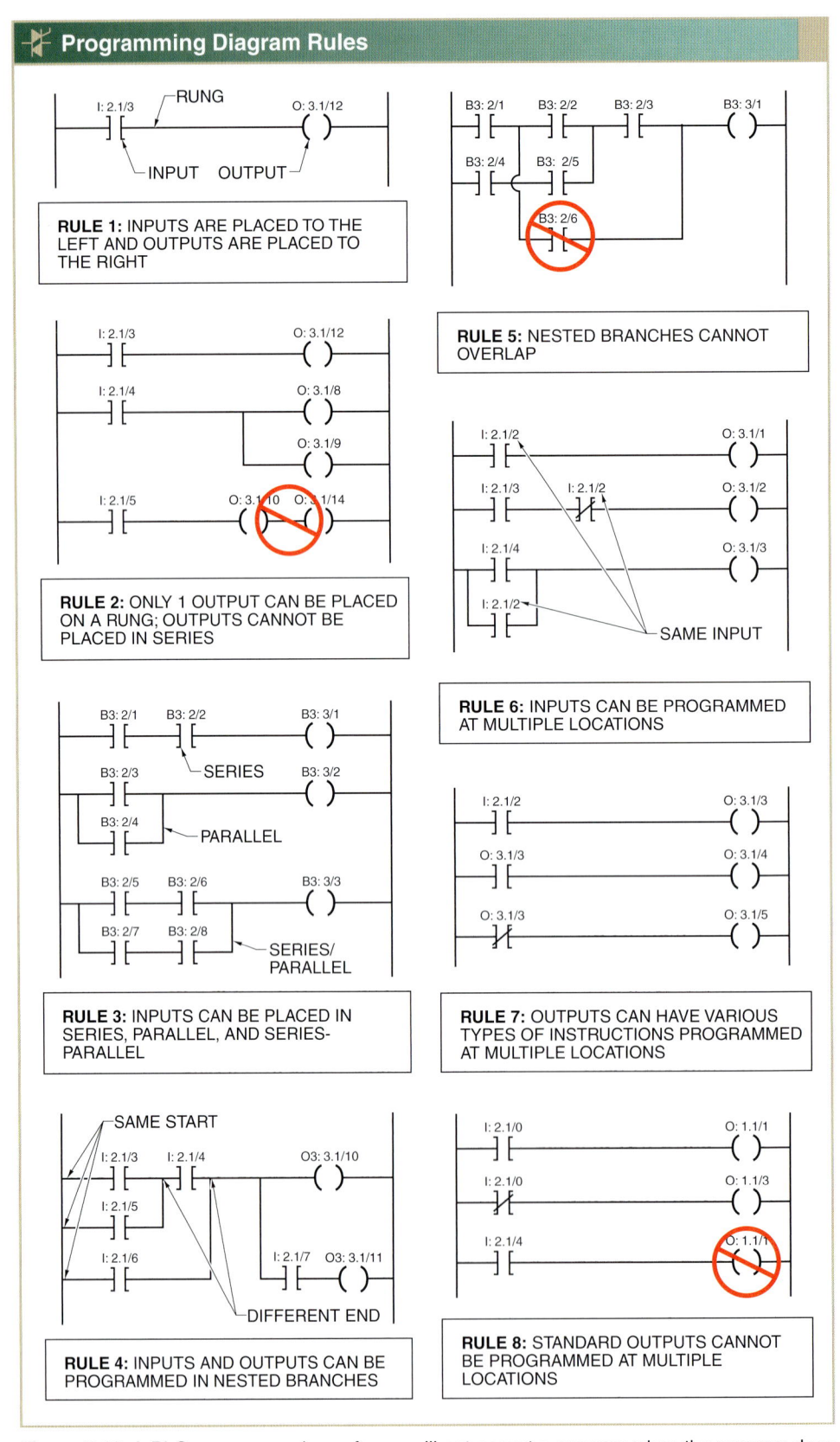

RULE 1: INPUTS ARE PLACED TO THE LEFT AND OUTPUTS ARE PLACED TO THE RIGHT

RULE 2: ONLY 1 OUTPUT CAN BE PLACED ON A RUNG; OUTPUTS CANNOT BE PLACED IN SERIES

RULE 3: INPUTS CAN BE PLACED IN SERIES, PARALLEL, AND SERIES-PARALLEL

RULE 4: INPUTS AND OUTPUTS CAN BE PROGRAMMED IN NESTED BRANCHES

RULE 5: NESTED BRANCHES CANNOT OVERLAP

RULE 6: INPUTS CAN BE PROGRAMMED AT MULTIPLE LOCATIONS

RULE 7: OUTPUTS CAN HAVE VARIOUS TYPES OF INSTRUCTIONS PROGRAMMED AT MULTIPLE LOCATIONS

RULE 8: STANDARD OUTPUTS CANNOT BE PROGRAMMED AT MULTIPLE LOCATIONS

Figure 6-10. A PLC or programming software will not accept a program when the program does not follow standard rules.

Ladder and Programming Diagrams

LADDER DIAGRAMS

OUTPUT OFF BECAUSE NO CURRENT FLOW (CONTINUITY) IN CIRCUIT

OUTPUT ON BECAUSE CURRENT FLOW (CONTINUITY) IN CIRCUIT

Data Table Address Value	XIC ⊣├─	XIO ⊣/├─	OTE ─( )─
0	False	True	False
1	True	False	True

Bit Instructions

CPU COMPARES CONDITION INSTRUCTION STATUS TO PLC PROGRAM DATA AND PLACES AN APPROPRIATE BIT IN BIT POSITION #4

Data Table B3:2 — bits: ... 0 0 1 (positions 4,3,2,1,0 area showing 0 at 4)

B3:2/2 TRUE —┤├— B3:2/3 FALSE —┤├— B3:2/4 FALSE —( )—
OUTPUT FALSE (OFF) BECAUSE NO LOGICAL CONTINUITY EXISTS

Data Table B3:2 — bits: 1 1 1

B3:2/2 TRUE —┤├— B3:2/3 TRUE —┤├— B3:2/4 TRUE —( )—
OUTPUT TRUE (ON) BECAUSE OF LOGICAL CONTINUITY

Data Table B3:2 — bits: 0 1 1

B3:2/2 TRUE —┤├— B3:2/3 FALSE —┤/├— B3:2/4 FALSE —( )—
OUTPUT FALSE (OFF) BECAUSE NO LOGICAL CONTINUITY EXISTS

Data Table B3:2 — bits: 1 0 1

B3:2/2 TRUE —┤├— B3:2/3 TRUE —┤/├— B3:2/4 TRUE —( )—
OUTPUT TRUE (ON) BECAUSE OF LOGICAL CONTINUITY

DATA TABLE ADDRESSES   **PROGRAMMING DIAGRAMS**

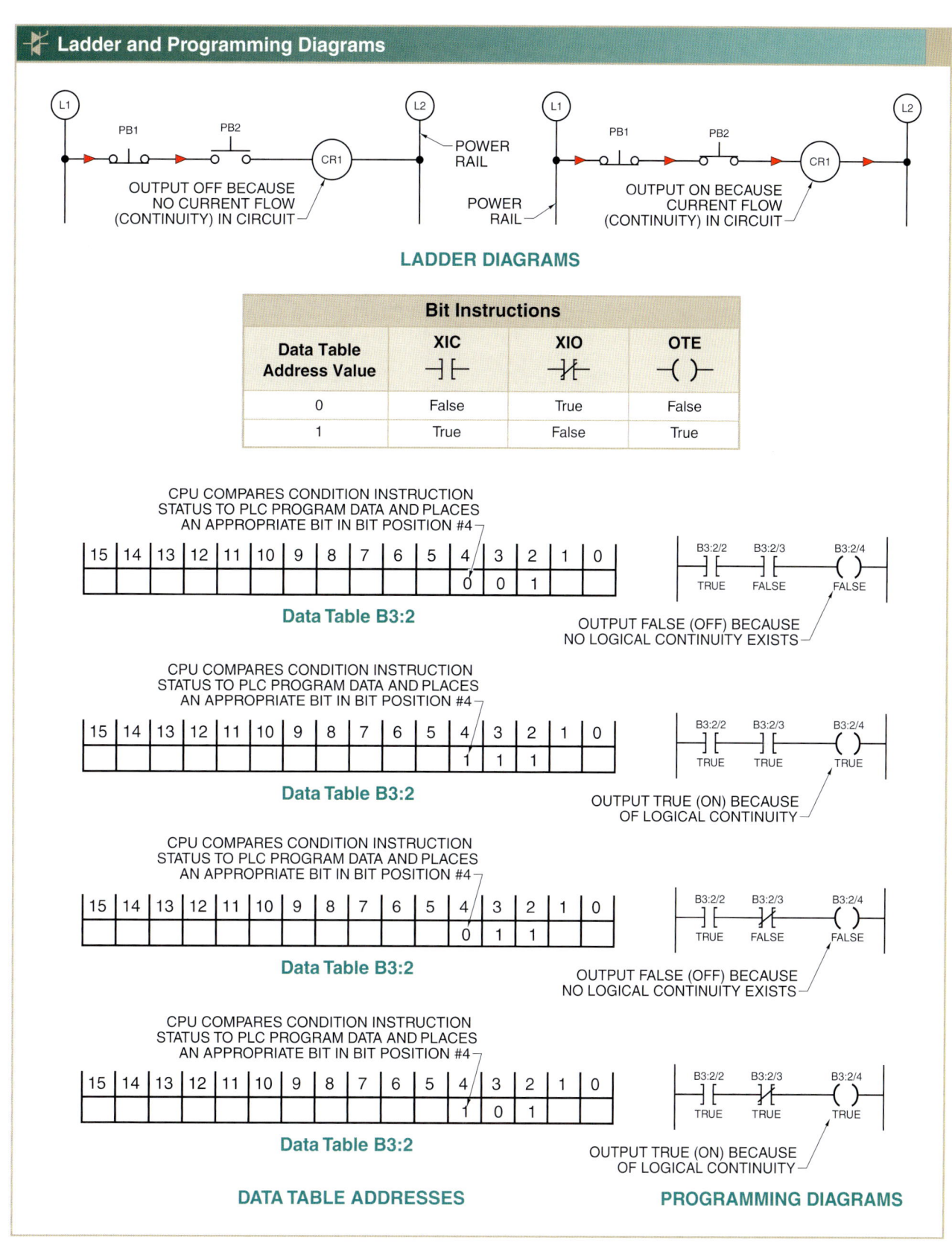

Figure 6-11. Ladder diagrams depict the flow of current through hardwired devices and components, while PLC program diagrams depict the flow of logic through instructions.

XIC (Examine if Closed). XIC instructions are input instructions. XIC instructions are true when the address of the instruction contains a "1". XIC instructions are false when the address of the instruction contains a "0".

XIO (Examine if Open). XIO instructions are also input instructions. XIO instructions are true when the address of the instruction contains a "0". XIO instructions are false when the address of the instruction contains a "1".

OTE (Output Energize). OTE instructions are output instructions. OTE instructions are true when the address of the instruction contains a "1". OTE instructions are false when the address of the instruction contains a "0".

Bit instructions can be confusing when using real-world input devices and output components with input (I) and output (O) files. The confusion stems from the fact that real-world input devices (pushbuttons, selector switches, and limit switches) can be normally closed (NC) or normally open (NO). **See Figure 6-12.** Because there are two normal conditions, two types of input bit instructions (XIC and XIO) are used.

A "0" at the instructions data table file address corresponds to no voltage being present at the terminal of the PLC for the input device or output component. A "1" at

Normally Open Input Device Instructions

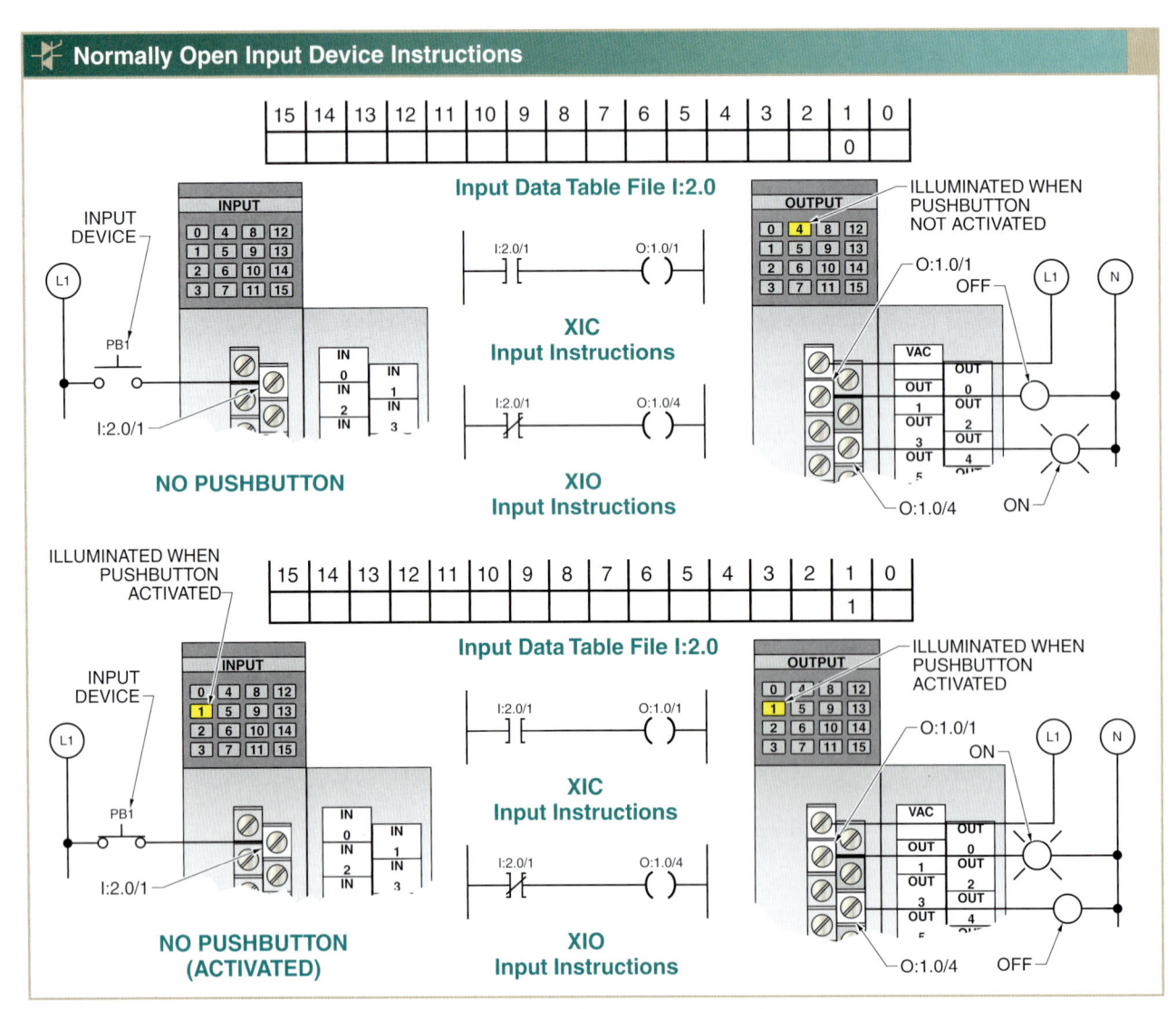

Figure 6-12. The type of input instruction, XIC or XIO, determines how the rung of a program will operate with a normally open (NO) input device.

the instructions data table file address corresponds to voltage being present at the terminal of the PLC for the input device or output component. The condition of input devices, NC or NO, and the type of input instruction, XIC or XIO, determine how a PLC program will function. **See Figure 6-13.**

One of the main benefits of a PLC is that an address can be used more than once and for more than one type of instruction.

A programming diagram that represents a hardwired 3-wire motor start/stop circuit is a good example. In 3-wire motor start/stop circuits, the address for an OTE instruction is also the address for the XIC instruction that serves as a holding contact. The XIC holding contact only exists in the memory of the PLC and is not a hardwired contact. **See Figure 6-14.**

Normally Closed Input Device Instructions

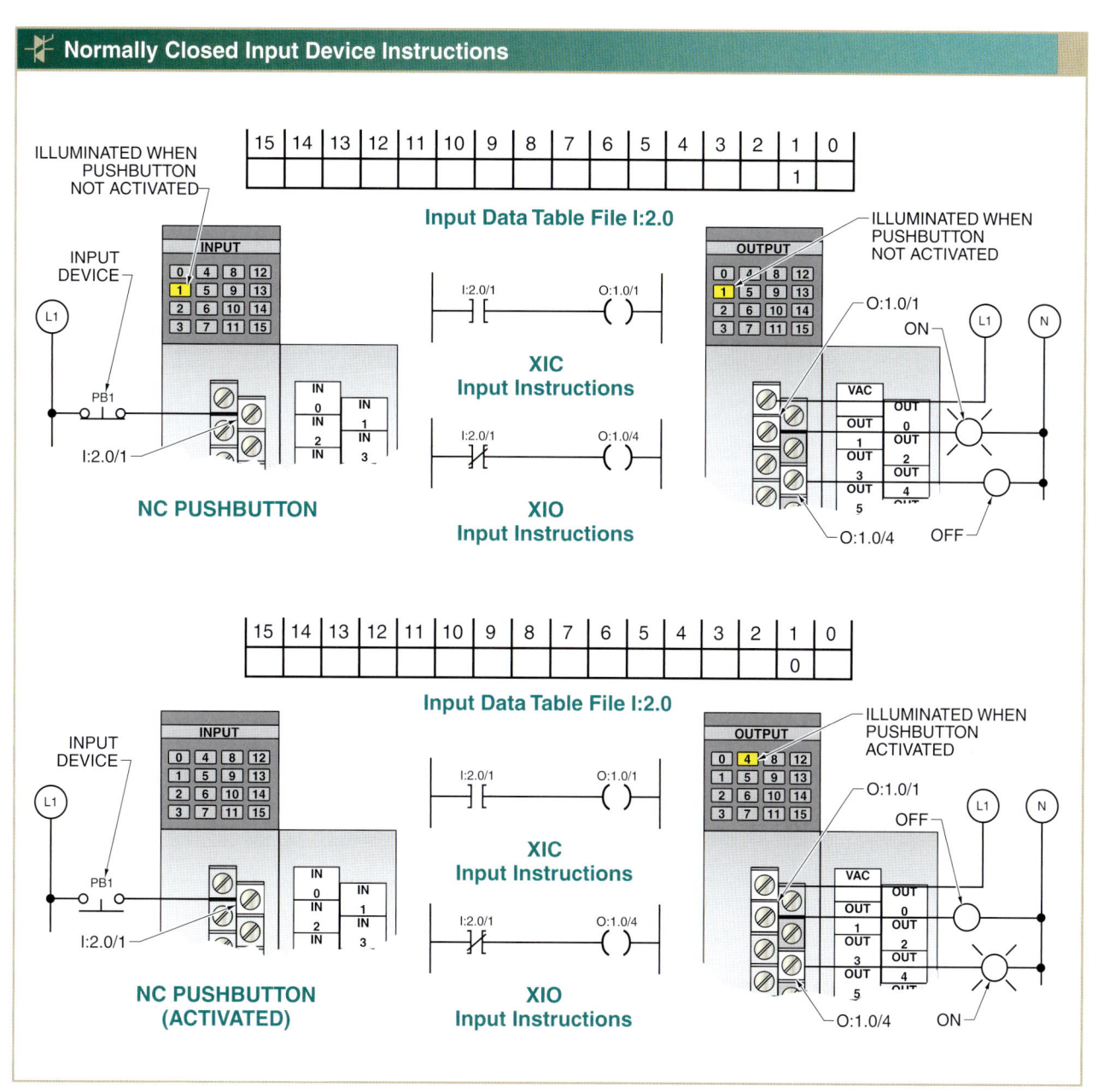

Figure 6-13. The type of input instruction, XIC or XIO, determines how the rung of a program will operate with a normally closed (NC) input device.

256 PROGRAMMABLE LOGIC CONTROLLERS: PRINCIPLES AND APPLICATIONS

XIC Holding Contacts

LADDER DIAGRAM

THREE-WIRE MOTOR START/STOP CIRCUIT

PROGRAM DIAGRAM

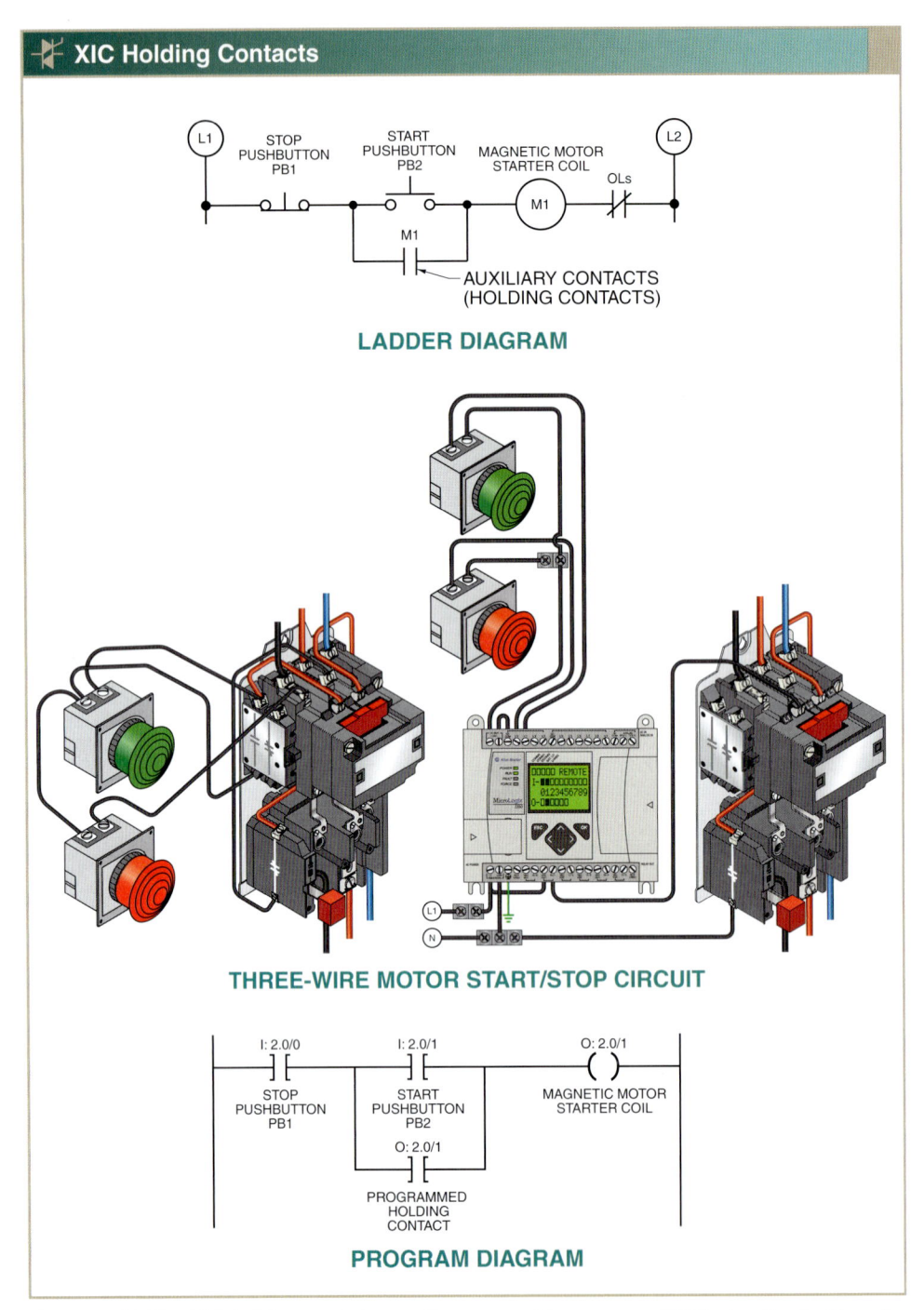

Figure 6-14. The XIC holding contact uses the same address as an OTE instruction but is not a hardwired contact and only exists in the memory of a PLC.

Output Latch and Output Unlatch. Output latch (OTL) and output unlatch (OTU) are types of bit instructions. OTL and OTU instructions differ from OTE instructions in that OTL and OTU instructions must be used together. OTL can only turn a bit ON (1 at the instruction's address), and OTU can only turn a bit OFF (0 at the instruction's address). OTL and OTU instructions share the same address. OTL and OTU can be thought of as latching relays in software. **See Figure 6-15.**

Latch (OTL) and Unlatch (OTU) Instructions

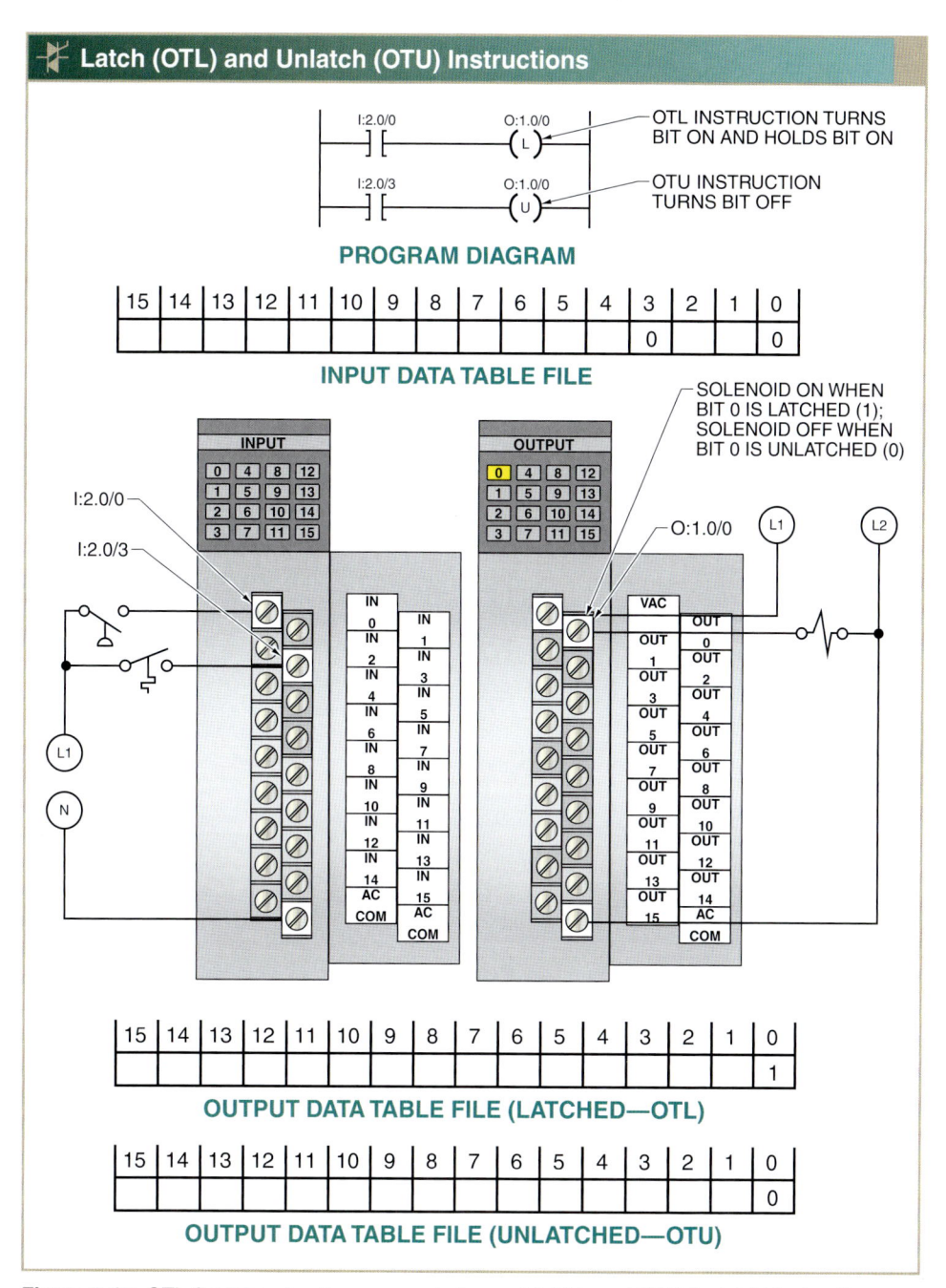

Figure 6-15. OTL (latch) instructions can only turn a bit ON, and OTU (unlatch) instructions can only turn a bit OFF.

Scan Execution

A *scan* is a method by which the CPU of a PLC looks at a program. PLCs scan the programming diagram from left to right and top to bottom. The scan begins at the top rung. When a CPU finds an instruction that is true, the CPU continues to scan the rung. When a CPU finds an instruction that is false, the CPU stops scanning the rung and goes to the next rung. For an output to affect another instruction during a specific scan, the instruction must be in a rung below the output. When an output instruction is in a rung above the output, the instruction will change state on the next scan. **See Figure 6-16.**

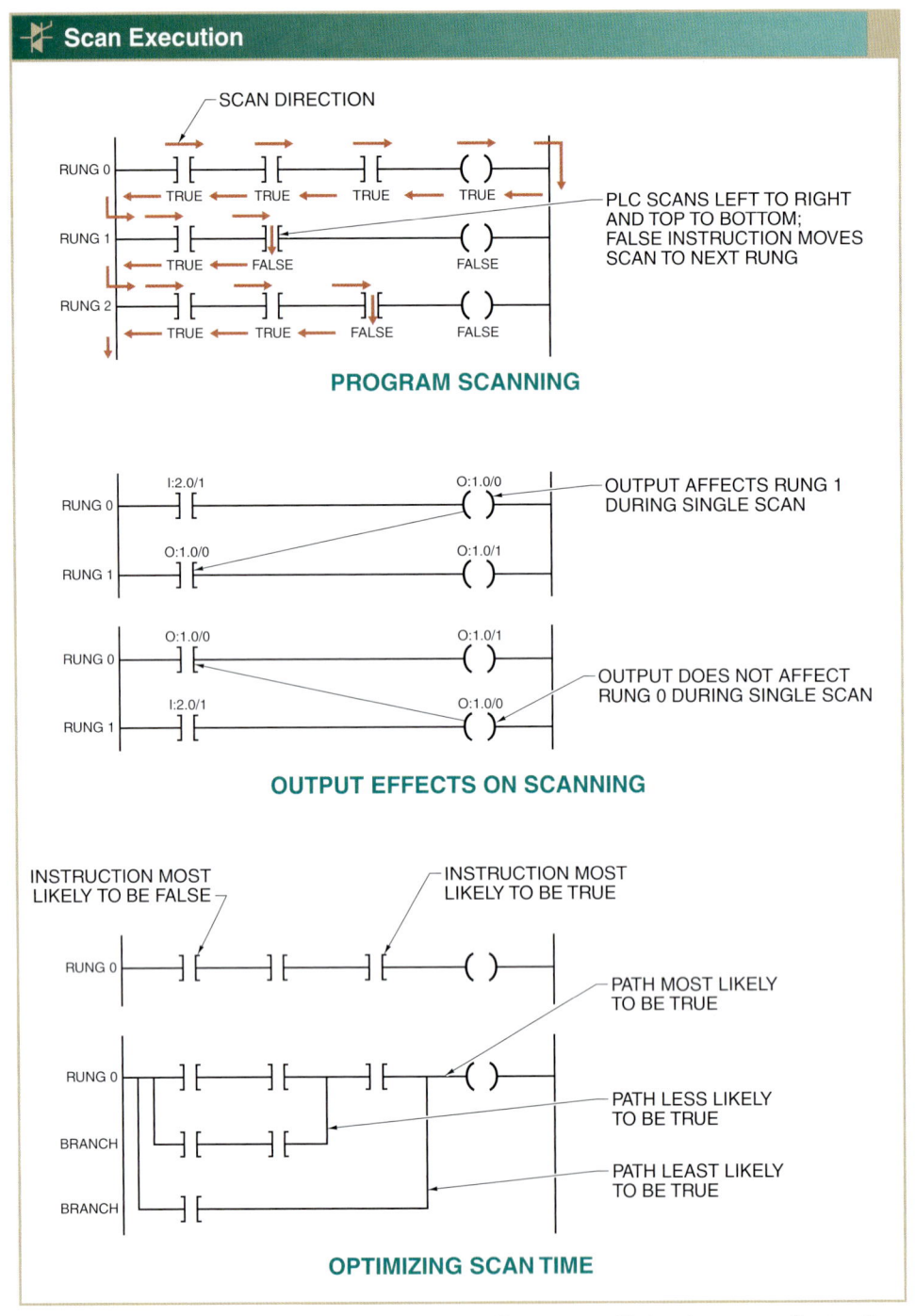

Figure 6-16. PLCs scan program diagrams in a specific manner, with the placement of instructions impacting scan times.

> ⚠ **CAUTION**
>
> *Any instruction that takes longer than 15 ms to process (whether true or false) causes a CPU poll for user interrupts to be performed.*

The placement of instructions can be used to minimize CPU scan time. When input instructions are in series, the instructions that are the most likely to be false should be on the left of the series, and the instructions that are the most likely to be true on the right of the series. When input instructions are in parallel branches, the path that is true most frequently should be on the top. The processor will not scan the lower branches unless the top is false. Minimizing scan time is critical in high-speed applications, such as with counting and sorting, in order to ensure a control application operates correctly.

PLC manufacturers provide PLC scan time worksheets that allow users to calculate the scan time for a specific program. **See Figure 6-17.** A PLC scan time worksheet is used in conjunction with a table that lists instruction execution times. The table may also specify instruction memory usage. PLC scan time worksheets are provided by manufacturers specifically for their models.

PLC Scan Time Worksheets

PLC Scan Time Worksheet

Procedure	Maximum Scan Time
1. Input scan time, output scan time, housekeeping time, and forcing.	__210__ μs (discrete) __330__ μs with forcing (analog) __250__ μs without forcing (analog)

2. Estimate program scan time:
  A. Count the number of program rungs in logic program and multiply by 6.
  B. Add up program execution times when all instructions are true. Include interrupt routines in this calculation.
	_____
	_____ μs

3. Estimate controller scan time:
  A. Without communications, add sections 1 and 2.
  B. With communications, add sections 1 and 2 and multiply by 1.05.
	_____ μs
	_____ μs

4. To determine maximum scan time in ms, divide controller scan time by 1000.	_____ ms

* If a subroutine executes more than once per scan, include each subroutine execution scan time.

Instruction Execution Times and Memory Usage

Mnemonic	False Execution Time*	True Execution Time*	Memory Usage†	Name	Instruction Type
ADD	6.78	33.09	1.50	Add	Math
CTD	27.22	32.19	1.00	Count Down	Basic
CTU	26.67	29.84	1.00	Count Up	Basic
OR	6.78	33.68	1.50	Or	Data Handling
OSR	11.48	13.02	1.00	One-Shot Rising	Basic
OTE	4.43	4.43	0.75	Output Energize	Basic
TON	30.38	38.34	1.00	Timer On-Delay	Basic
XIC	1.72	1.54	0.75	Examine If Closed	Basic
XIO	1.72	1.54	0.75	Examine If Open	Basic

* approx. μseconds
† user words

Figure 6-17. Technicians use PLC scan time worksheets to calculate the scan time for a specific PLC program.

> **PLC TIPS**
>
> *Bit instructions are used with input and output data files (external inputs and outputs), status files (values used to configure and provide controller status), bit data files (internal programmed coils), timer, counter, and control data files (various control bits), and integer files (addresses used as program requires). However, the use of the same address with multiple output instructions is not recommended.*

To calculate the execution scan time of a specific PLC program, the technician must first determine if discrete input devices are being used: 210 μs (microseconds) per discrete device, 330 μs per analog device with forcing, or 250 μs per analog device without forcing. Second, the technician estimates the program scan time. Third, the technician estimates the CPU scan time. Finally, the technician calculates the scan time in milliseconds (ms).

IEC 61131-3 STANDARD

The *International Electrotechnical Commission (IEC)* is an organization that develops international standards for electrical and electronic equipment as well as related technology. These standards cover a wide scope of electrical and electronic equipment, ranging from home appliances to fiber-optic equipment to PLCs.

Specifically, IEC 61131-3 is a standard that covers programming languages for industrial automation devices, such as process automation controllers (PACs) and discrete PLCs. IEC 61131-3 is vendor neutral and seeks to provide standardized programming languages. The standardization reduces both the time and the cost required for training programming engineers. Another benefit of standardization is the ease of transferring a program from one PLC model to another. Not all PLCs can be programmed in every language.

IEC 61131-3 identifies five common programming languages: ladder diagram (LD), function block diagram (FBD), instruction list (IL), sequential function chart (SFC), and structured text (ST). **See Figure 6-18.** Each programming language, according to IEC 61131-3, is described as follows:

- Ladder diagram (LD) is based on hardwired relay ladder logic used before the advent of PLCs. Similar to line diagrams, LDs contain simple functions between digital inputs and outputs, such as relays and magnetic motor starter circuits, that would be familiar to an electrician or maintenance technician. LD is the most common programming language and was developed in the US.

- Function block diagram (FBD) uses blocks for logic operations, such as AND and OR. The blocks are connected to provide control functionality for an application. FBD can be easier to follow for those who are not familiar with relay ladder logic. Some surveys indicate that FBD is the second most common programming language after LD.

- Instruction list (IL) uses many lines of code to address a single operation. The lines of code provide the step-by-step control functionality for an application. IL is less visual than LD and is not commonly used by maintenance technicians. It is popular in Europe, however, because a program written in IL is relatively easy to transfer between PLC platforms.

- Sequential function chart (SFC) consists of a series of rectangular boxes connected by vertical lines. Each box contains code for a specific function/step of a control application. The vertical lines represent the transition between functions. The code in a box runs until it is complete and then transitions to activate the box below it.

- Structured text (ST) looks like a high-level programming language, such as C. It has lines of code beginning with terms, such as IF, IF..THEN, and ELSE. ST is not graphical like IL but is commonly used by engineers who have experience with advanced computer programming.

IEC 61131-3 Programming Languages

LADDER DIAGRAMS (LDs)

FUNCTION BLOCK DIAGRAMS (FBDs)

```
0001   LD    IO1
0002   AND   IO2
0003   ST    QO1
```

NOTE: LD, FBD, and IL are same program, but in respective formats.

INSTRUCTION LISTS (ILs)

SEQUENTIAL FUNCTION CHARTS (SFCs)

STRUCTURED TEXTS (STs)

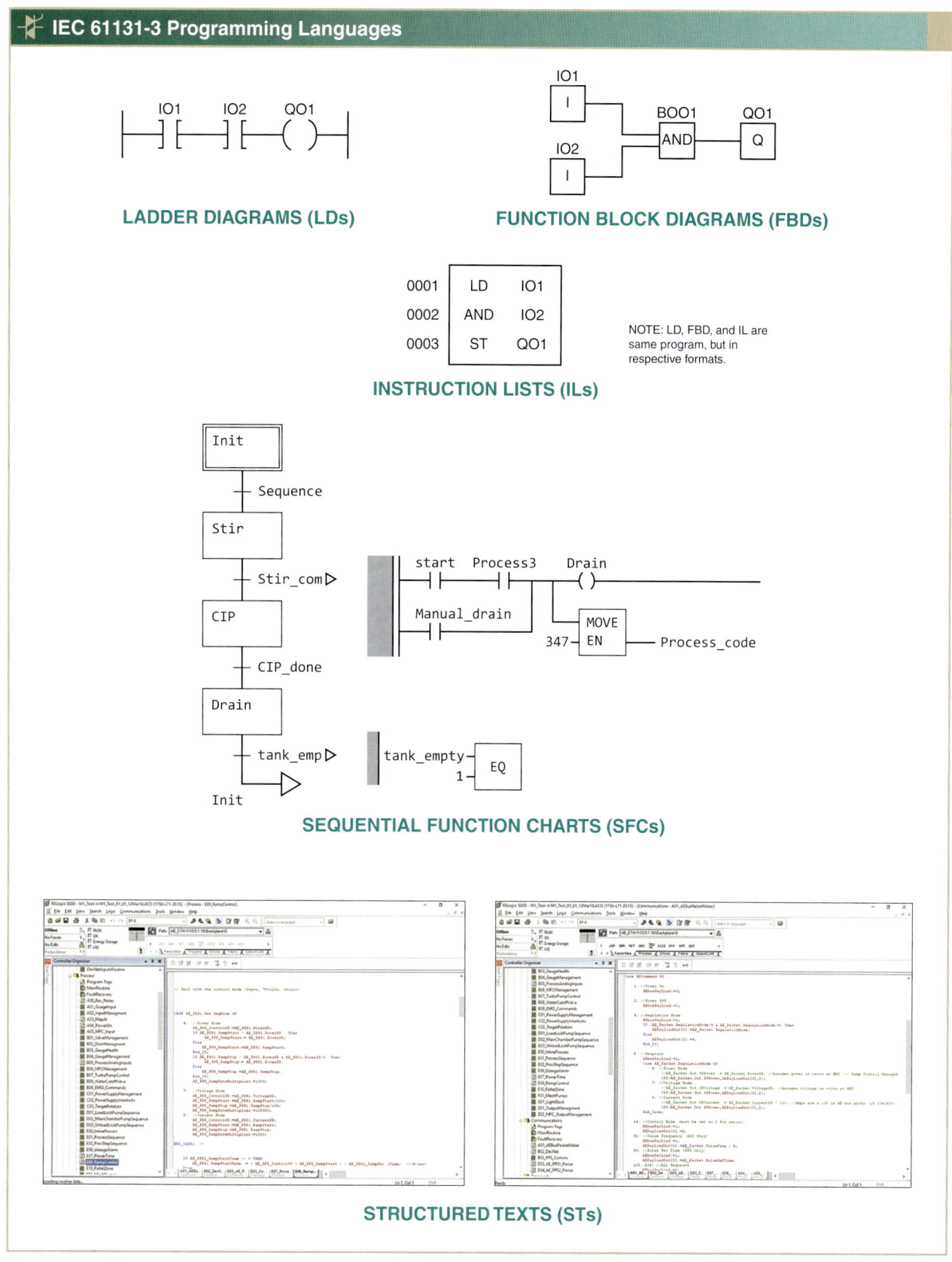

Figure 6-18. IEC standard 61131-3 identifies five common programming languages: LD, FBD, IL, SFC, and ST.

Chapter Review 6

Name _____ Date _____

True-False

T   F   1. PLC programs are typically depicted using ladder diagrams.

T   F   2. Two hundred and fifty-six program files are possible in most PLCs, and the files are numbered 0 through 255.

T   F   3. There are 256 data table files, numbered 0 through 255.

T   F   4. Output components can be placed in series but never in parallel.

T   F   5. Input devices can be programmed in series, parallel, or in series/parallel combinations.

T   F   6. Power flows through input devices and output components represented in a programming diagram.

T   F   7. Output components can have various types of instructions with the same address.

T   F   8. Data table files 3 through 255 are configured by the technician.

T   F   9. Status (S) files cannot be added to, deleted, or changed in any way.

T   F   10. XIC instructions are input instructions.

T   F   11. XIO instructions are output instructions.

T   F   12. The maximum size of a bit file is one sixteen-bit word.

T   F   13. Program file 1 is reserved by the technician for processing data.

T   F   14. The placement of programming instructions can be used to minimize CPU scan time.

T   F   15. Various PLC programming instructions can take the place of hardwired timers and counters.

Completion

_____ 1. A(n) ___ is the section of memory in the PLC that contains the status of all input devices and output components, CPU, and all instructions used in a program.

_____ 2. A(n) ___ is a type of drawing that contains the logic for an application and is typically used to test and troubleshoot hardwired applications.

_____ 3. A(n) ___ is the method by which the CPU of a PLC looks at a program.

4. A(n) ___ is the section of memory in a PLC that is filled with CPU data about programming.

5. A(n) ___ is data in the memory of a PLC that contains technician-developed PLC programs and related information.

6. Each line of a programming diagram is referred to as a(n) ___.

7. ___ branches are branches that have varying start and/or end points.

8. The ___ file contains numbers that are not integers (whole numbers).

9. Program files numbered 3 and higher are ___ program files.

10. The ___ file contains the technician-developed control program for a specific application.

Multiple Choice

1. Which of the following is not included in a PLC instruction file address?
    A. file identifier
    B. file number
    C. file size
    D. memory location

2. The ___ file contains the following system-related information: CPU type, I/O configuration, and passwords.
    A. main program
    B. reserved
    C. subroutine program
    D. system program

3. With regard to input and output file addresses, the input and output files are defaulted to file ___ and file ___ respectively and cannot be used in any other data table file location.
    A. 0; 1
    B. 1; 0
    C. 1; 2
    D. 2; 1

4. Input instructions in a programming diagram are either ___ or ___.
    A. powered; not powered
    B. ON; OFF
    C. open looped; close looped
    D. TRUE; FALSE

5. In a ___ application, the contacts of pushbuttons, selector switches, limit switches, timers, and counters, and the coils of relays, pilot lights, solenoids, and motor starters are directly connected with wire.
    A. CPU
    B. hardwired
    C. PLC
    D. programming device

Mid-Level and High-Level Alarm Systems

PLC Exercise 6.1

BILL OF MATERIALS
- TECO PLR or Micrologix™ 1100 PLC
- Two lights
- Two float/limit switches
- Two NO pushbuttons
- Solenoid
- Alarm/horn

Many processes in the food-and-beverage, petroleum, and chemical industries use sensors to actuate hydraulic or pneumatic valves to release material from tanks. For this lab, a pneumatic slide opens a loading tank to empty material automatically when a high level is reached or by manual pushbutton. All input devices are hardwired normally open.

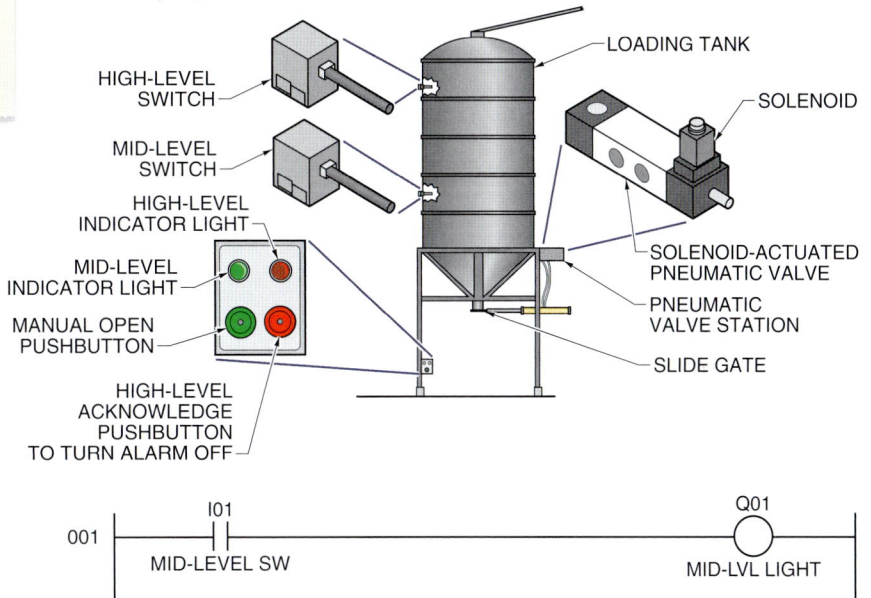

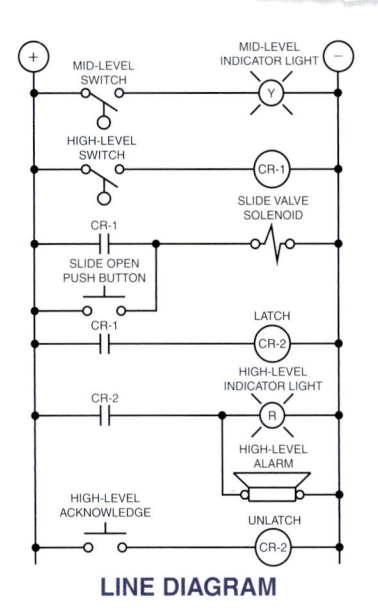

LINE DIAGRAM

TECO PROGRAMMING DIAGRAM

TECO Procedures

1. Create a new ladder logic or FBD program.
2. Use the TECO programming diagram or FBD to add the proper contacts and description in the Programming Grid.

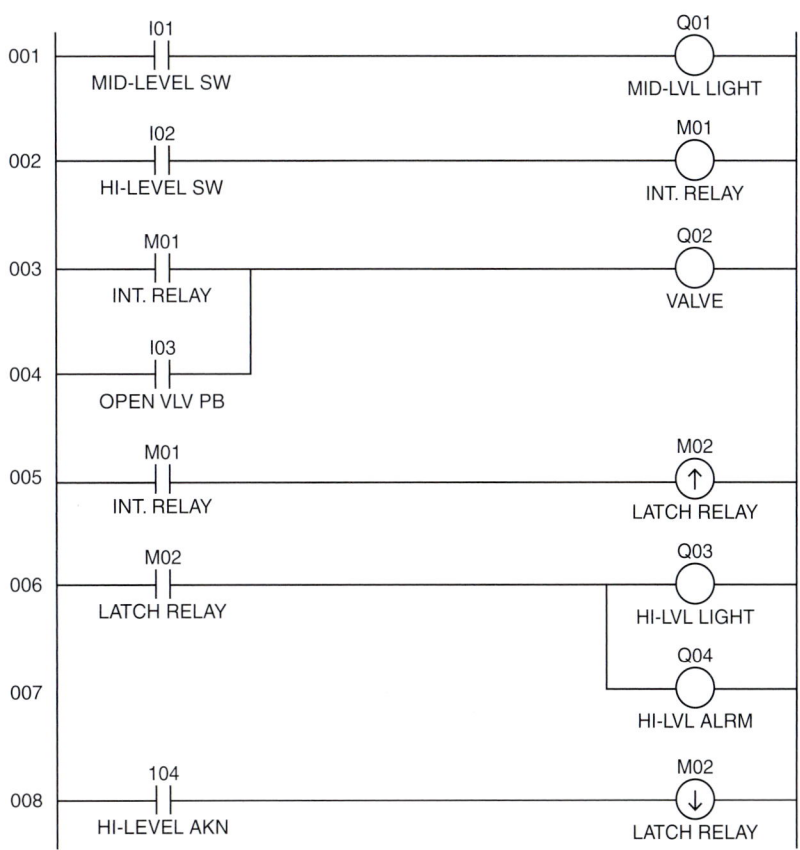

265

3. Use the Edit Contact window to denote an input or output number and the Symbol button to add a description after placing the input or output in the Programming Grid. For an FBD, use the Comment icon to add the description.

4. Run the program in Simulation Mode to verify proper operation. Use the Input Status tool window to activate the inputs and control the output to simulate circuit operation.

5. Use the wiring diagram to wire the input devices and output components to the PLC.

6. Turn power to the PLC ON and link the com port.

7. Write the program to the SG2-12HR-D to download the program logic to the PLC.

8. Place the SG2-12HR-D in Run Mode. Activate the start pushbutton and monitor the I/Os on the LCD display.

9. Stop the program and use the Save As function to save and place the lab application in the required folder location.

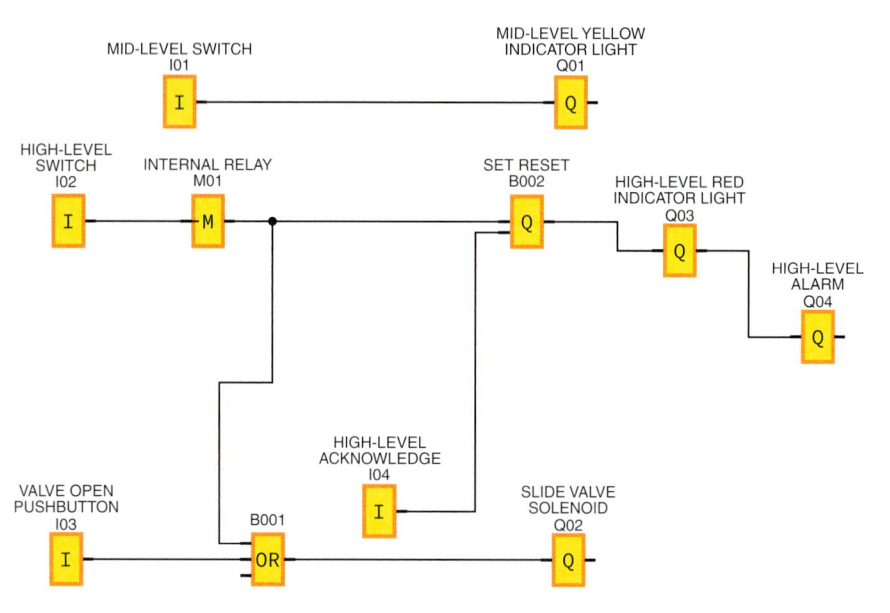

FUNCTION BLOCK DIAGRAM

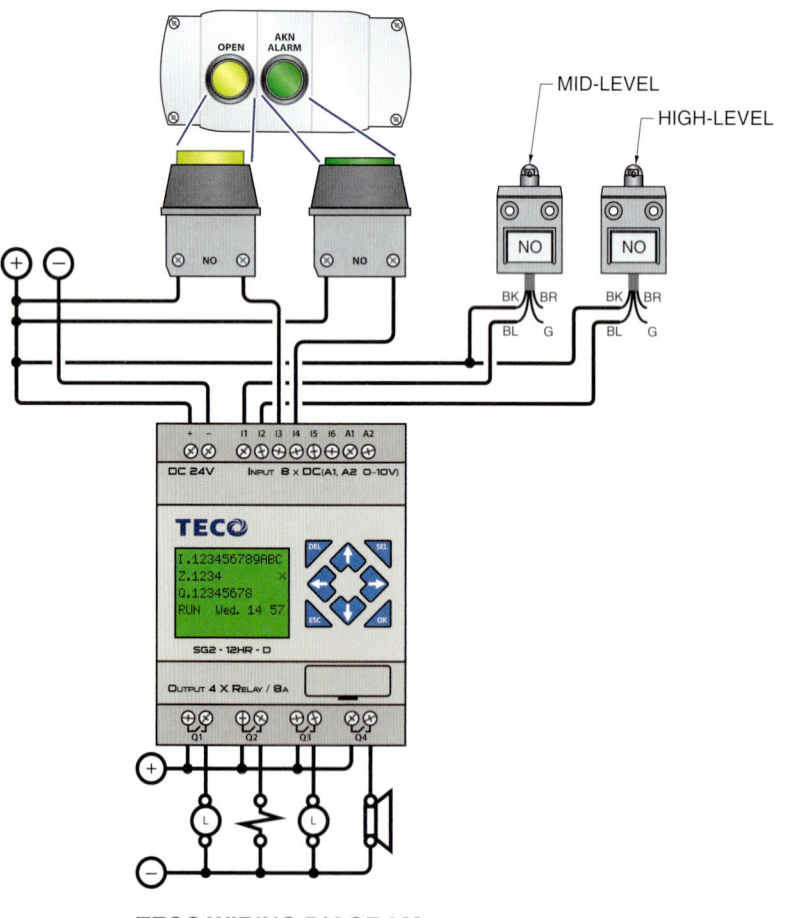

TECO WIRING DIAGRAM

Chapter 6—PLC Programming Diagrams, Addresses, and Bit Instructions

Allen-Bradley® Procedures

1. Open the RSLogix™ Micro Starter Lite program and create a new programming diagram.

2. Use the AB programming diagram to add the proper address and description in the LAD 2 programming window.

3. Verify that the programming diagram has no errors and save the project.

4. Use the wiring diagram to wire the input devices and output components to the PLC. Once all connections have been made, turn power to the PLC on.

5. Use RSLinx™ to establish communication between the PC and the Micrologix™ 1100.

6. Download the program logic to the PLC.

7. Place the PLC in Run Mode.

8. Activate input(s) and monitor them to determine if the output(s) are operating correctly.

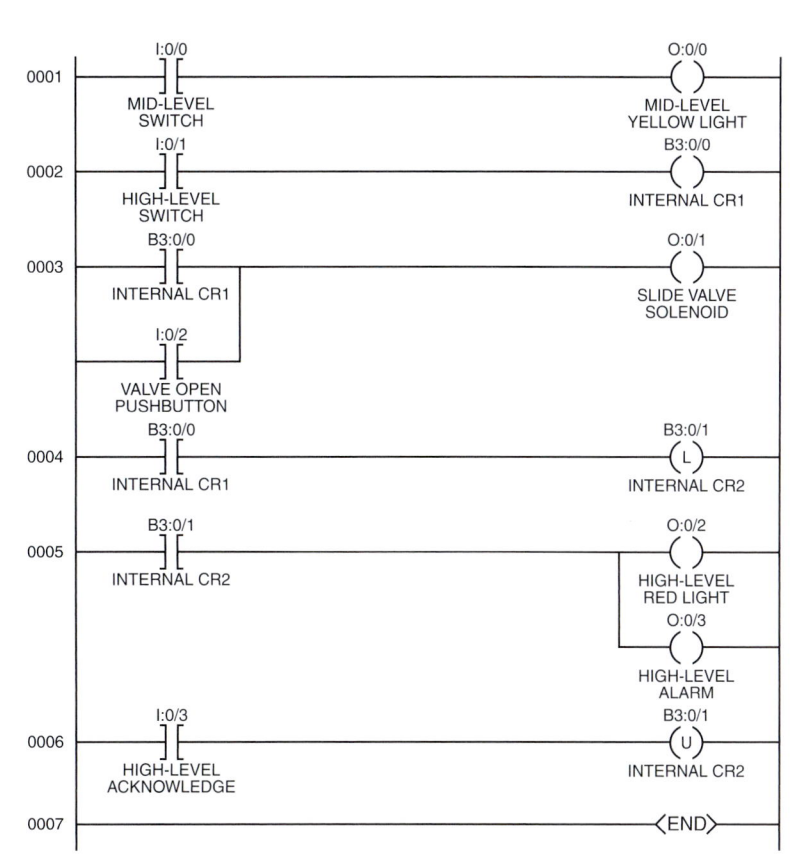

AB PROGRAMMING DIAGRAM

Rung	Input	Output
0001	I:0/0 MID-LEVEL SWITCH	O:0/0 MID-LEVEL YELLOW LIGHT
0002	I:0/1 HIGH-LEVEL SWITCH	B3:0/0 INTERNAL CR1
0003	B3:0/0 INTERNAL CR1 / I:0/2 VALVE OPEN PUSHBUTTON	O:0/1 SLIDE VALVE SOLENOID
0004	B3:0/0 INTERNAL CR1	B3:0/1 (L) INTERNAL CR2
0005	B3:0/1 INTERNAL CR2	O:0/2 HIGH-LEVEL RED LIGHT / O:0/3 HIGH-LEVEL ALARM
0006	I:0/3 HIGH-LEVEL ACKNOWLEDGE	B3:0/1 (U) INTERNAL CR2
0007		⟨END⟩

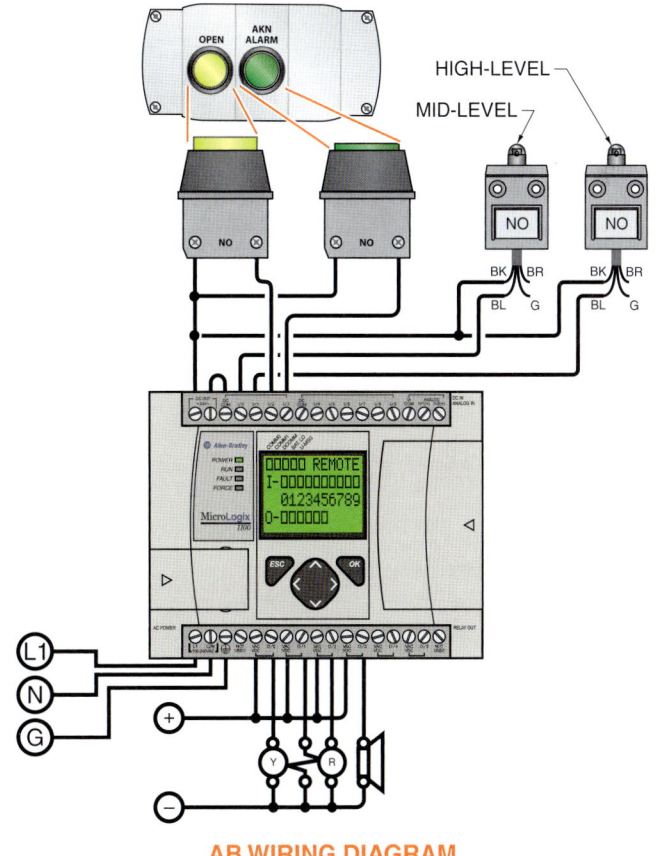

AB WIRING DIAGRAM

Conveyor Start Time Delay

PLC Exercise 6.2

BILL OF MATERIALS

- TECO PLR or Micrologix™ 1100 PLC
- Motor starter/contactor
- Light
- NC pushbutton
- NO pushbutton

In certain conveyor applications, a delayed start programmed to the motor circuit with a warning light is necessary to provide personnel with a visual warning before parts start moving. For this lab, a conveyor motor circuit is started with a 4-second delay using a 1-second time base. Before the conveyor line starts to move, a warning light turns on. The stop pushbutton is hardwired normally closed, and the start pushbutton is hardwired normally open. Do not connect output components in parallel.

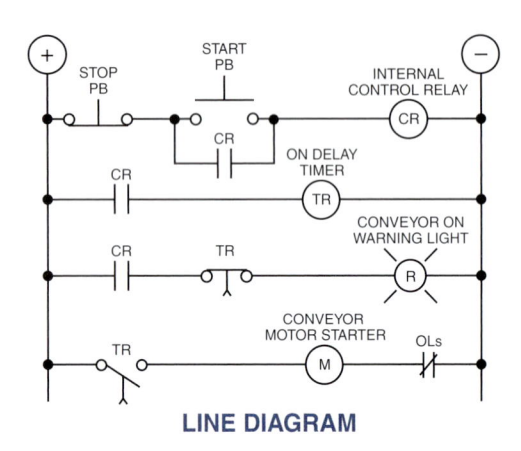

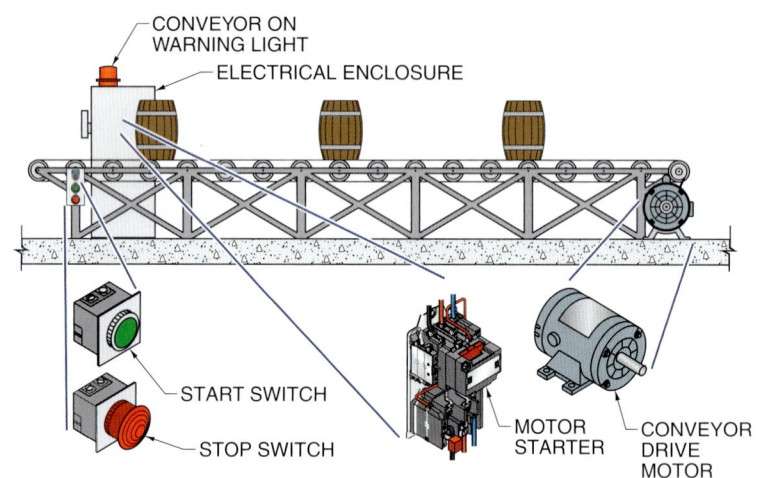

LINE DIAGRAM

TECO PROGRAMMING DIAGRAM

TECO Procedures

1. Create a new ladder logic or FBD program.

2. Use the TECO programming diagram or FBD to add the proper contacts and description in the Programming Grid.

3. Use the Edit Contact window to denote an input or output number and the Symbol button to add a description after placing the input or output in the Programming Grid. For an FBD, use the Comment icon to add the description.

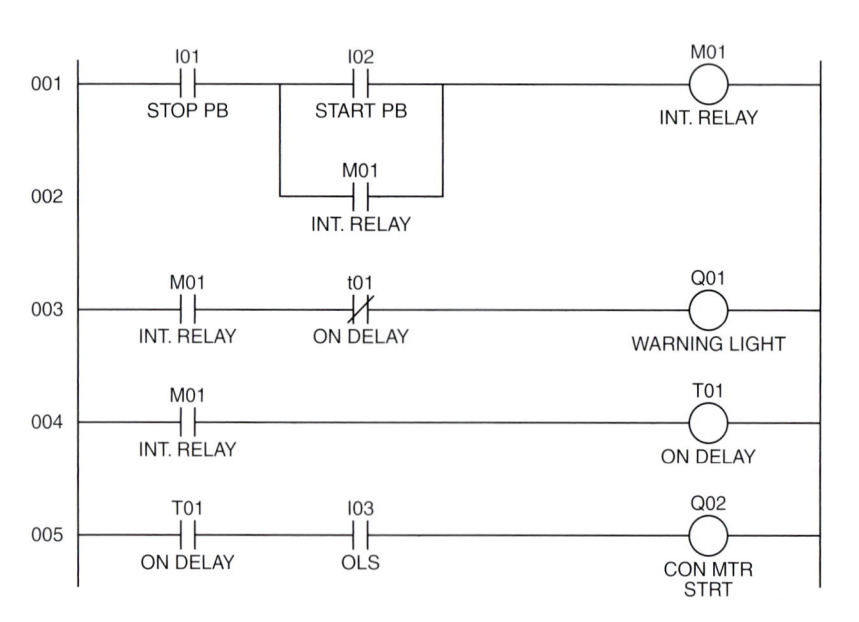

4. Select the T contact and place in the Programming Grid. Program the proper timer number and description. Construct the time base, preset value, and preset type. Using an FBD, select the Tx Function Block icon and place in the Programming Grid. Select the proper mode, timer number, time base, preset value, and description.

5. Run the program in Simulation Mode to verify proper operation. Use the Input Status tool window to activate the inputs and control the output to simulate circuit operation.

6. Use the wiring diagram to wire the input devices and output components to the PLC.

7. Turn power to the PLC ON and link the com port.

8. Write the program to the SG2-12HR-D to download the program logic to the PLC.

9. Place the SG2-12HR-D in Run Mode. Activate the start pushbutton and monitor the I/Os on the LCD display.

10. Stop the program and use the Save As function to save and place the lab application in the required folder location.

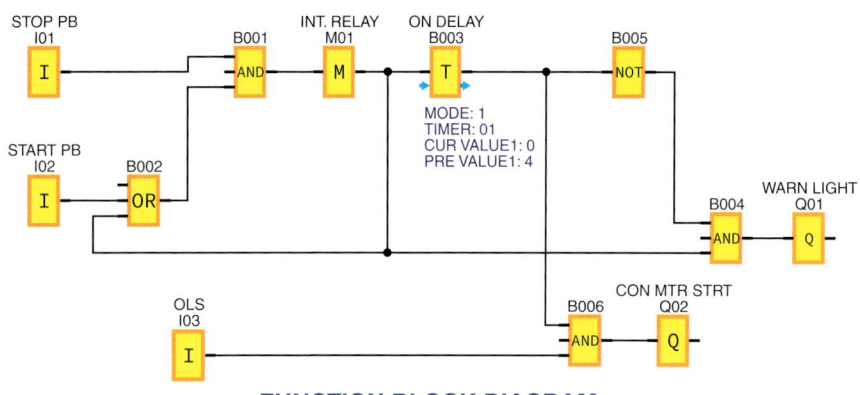

FUNCTION BLOCK DIAGRAM

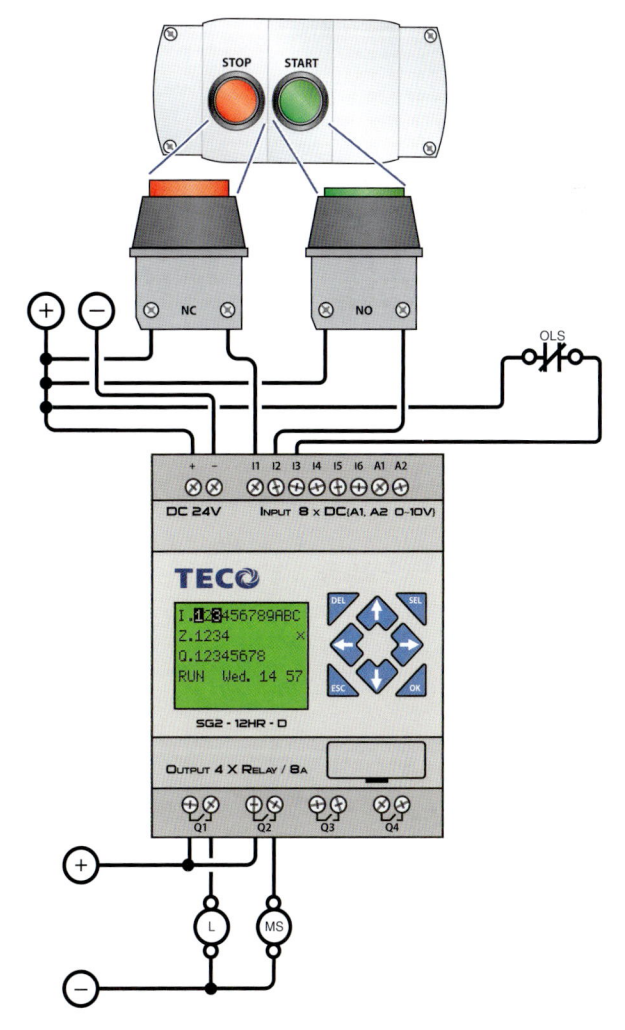

TECO WIRING DIAGRAM

Allen-Bradley® Procedures

1. Open the RSLogix™ Micro Starter Lite program and create a new programming diagram.

2. Use the AB programming diagram to add the proper address and description in the LAD 2 programming window. Add the TON instruction to the programming diagram and type in the proper timer address, time base, and preset value.

3. Verify that the programming diagram has no errors and save the project.

4. Use the wiring diagram to wire the input devices and output components to the PLC. Once all connections have been made, turn power to the PLC on.

5. Use RSLinx™ to establish communication between the PC and the Micrologix™ 1100.

6. Download the program logic to the PLC.

7. Place the PLC in Run Mode.

8. Activate input(s) and monitor them to determine if the output(s) are operating correctly.

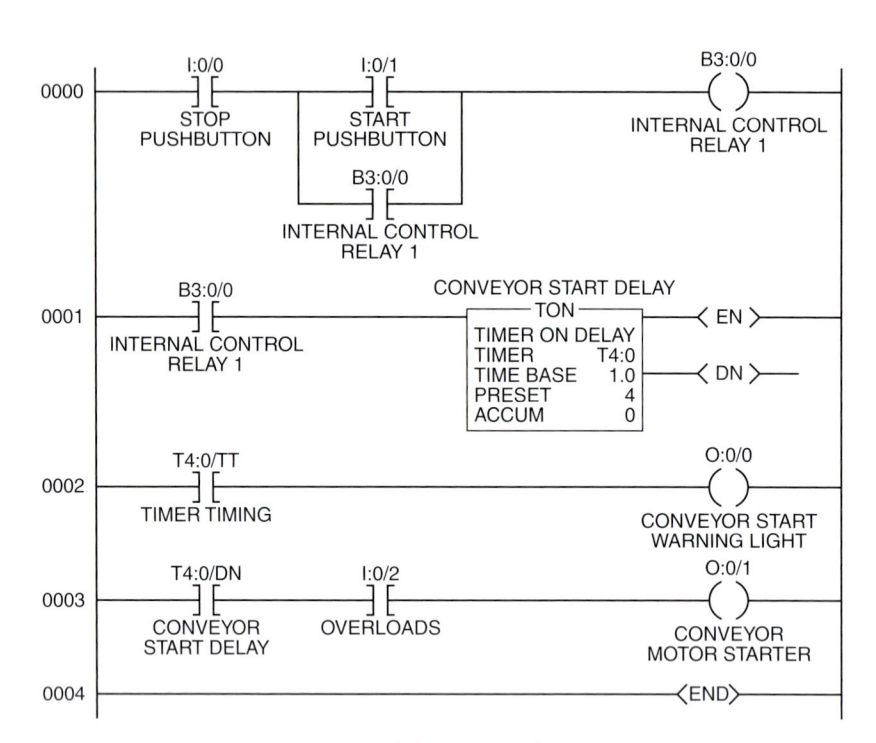

AB PROGRAMMING AREA

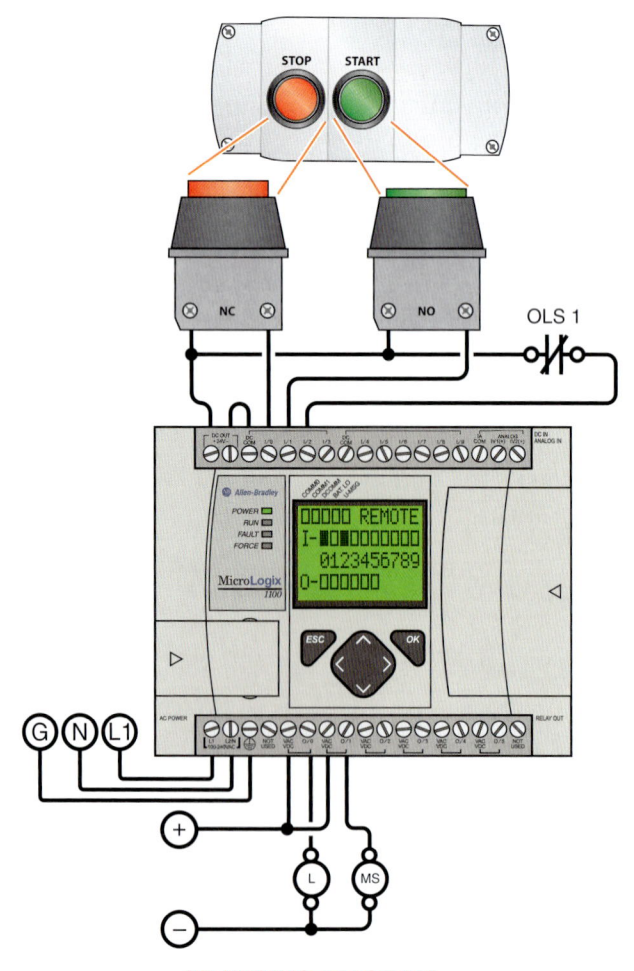

```
0000  I:0/0              I:0/1                              B3:0/0
      STOP               START                              INTERNAL CONTROL
      PUSHBUTTON         PUSHBUTTON                         RELAY 1
                         B3:0/0
                         INTERNAL CONTROL
                         RELAY 1

0001  B3:0/0                      CONVEYOR START DELAY
      INTERNAL CONTROL            TON
      RELAY 1                     TIMER ON DELAY          ⟨ EN ⟩
                                  TIMER        T4:0
                                  TIME BASE    1.0        ⟨ DN ⟩
                                  PRESET       4
                                  ACCUM        0

0002  T4:0/TT                                              O:0/0
      TIMER TIMING                                         CONVEYOR START
                                                           WARNING LIGHT

0003  T4:0/DN            I:0/2                             O:0/1
      CONVEYOR           OVERLOADS                         CONVEYOR
      START DELAY                                          MOTOR STARTER

0004                              ⟨END⟩
```

AB WIRING DIAGRAM

Off-Delay Timer for Forced Ventilation of Motors

PLC Exercise 6.3

BILL OF MATERIALS
- TECO PLR or Micrologix™ 1100 PLC
- Two motor starters/contactors
- NC pushbutton
- NO pushbutton
- Motor/load (optional)

In many industrial processes, PLCs control equipment in harsh or dusty working environments. For this lab, a motor runs a packaging line but needs to be ventilated every time it is shut off. A ventilation blower motor runs for 2 minutes after each time the motor is shut off. A start pushbutton is hardwired normally open, and a stop pushbutton is hardwired normally closed. The off-delay timer is normally open, timed to open, and is set for 2 minutes. Do not connect output components in parallel.

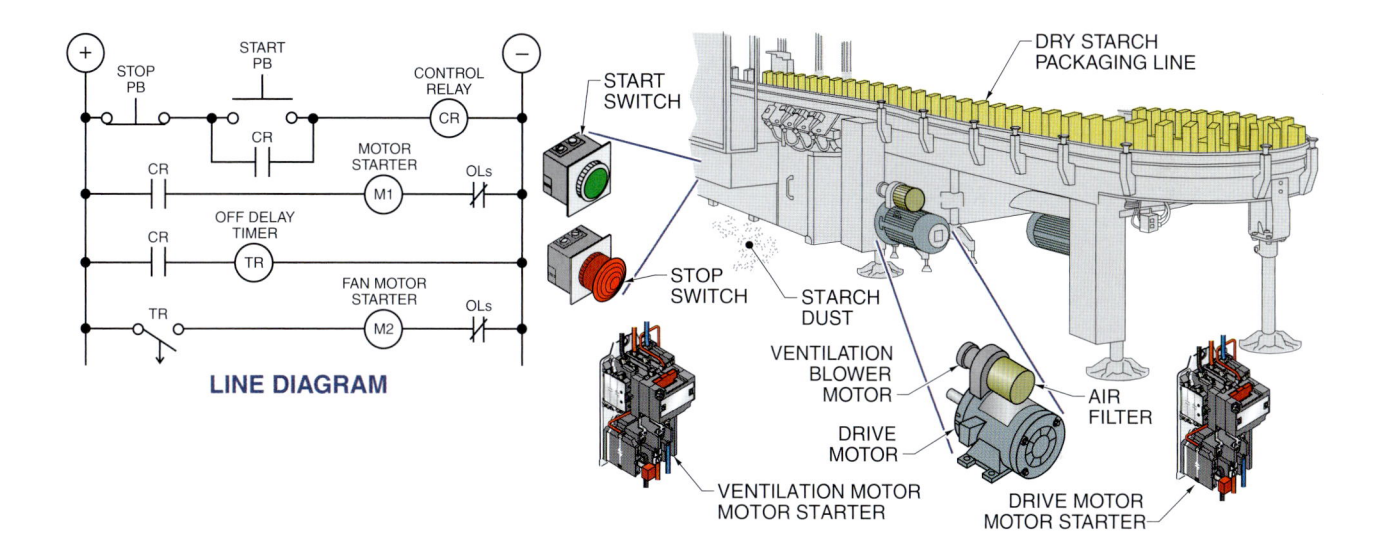

LINE DIAGRAM

T TECO Procedures

1. Create a new ladder logic or FBD program.

2. Use the TECO programming diagram or FBD to add the proper contacts and description in the Programming Grid.

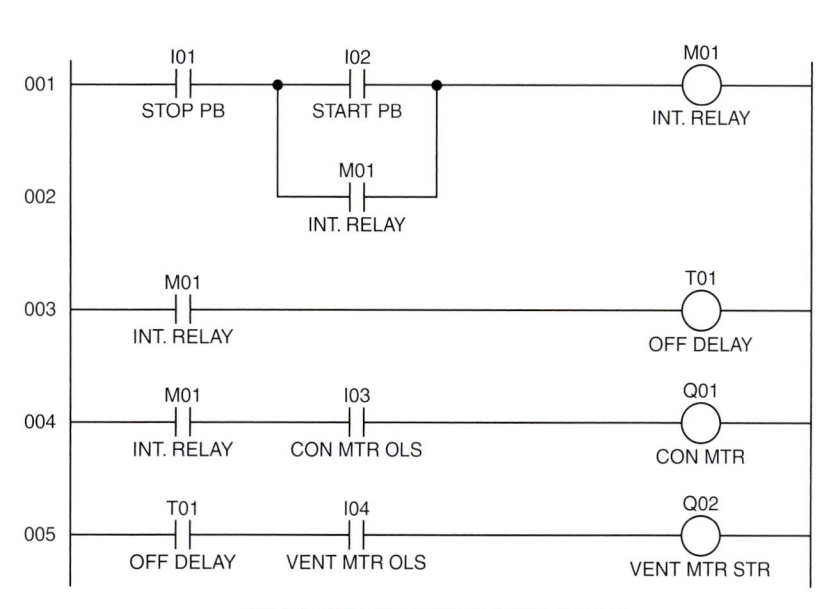

TECO PROGRAMMING DIAGRAM

3. Use the Edit Contact window to denote an input or output number and the Symbol button to add a description after placing the input or output in the Programming Grid. For an FBD, use the Comment icon to add the description.

4. Select the T contact and place in the Programming Grid. Program the proper timer number and description. Construct the time base, preset value, and preset type. Using an FBD, select the Tx Function Block icon and place in the Programming Grid. Select the proper mode, timer number, time base, preset value, and description.

5. Run the program in Simulation Mode to verify proper operation. Use the Input Status tool window to activate the inputs and control the output to simulate circuit operation.

6. Use the wiring diagram to wire the input devices and output components to the PLC.

7. Turn power to the PLC ON and link the com port.

8. Write the program to the SG2-12HR-D to download the program logic to the PLC.

9. Place the SG2-12HR-D in Run Mode. Activate the start pushbutton and monitor the I/Os on the LCD display.

10. Stop the program and use the Save As function to save and place the lab application in the required folder location.

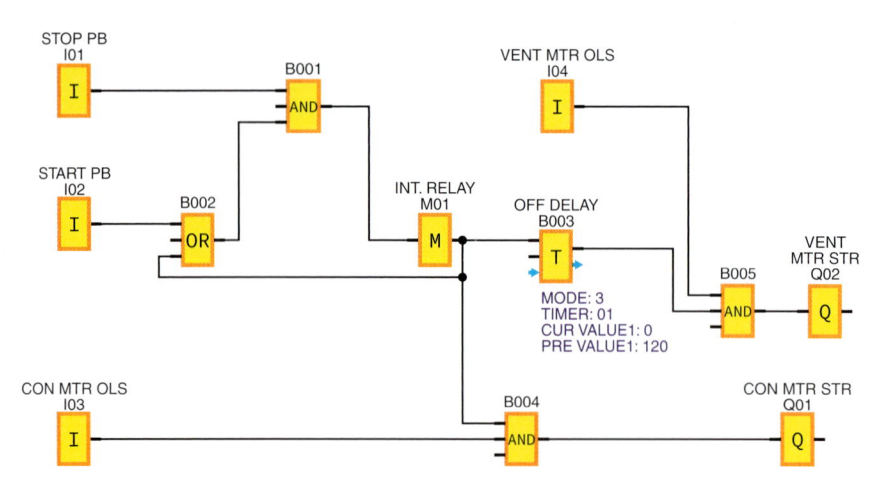

FUNCTION BLOCK DIAGRAM

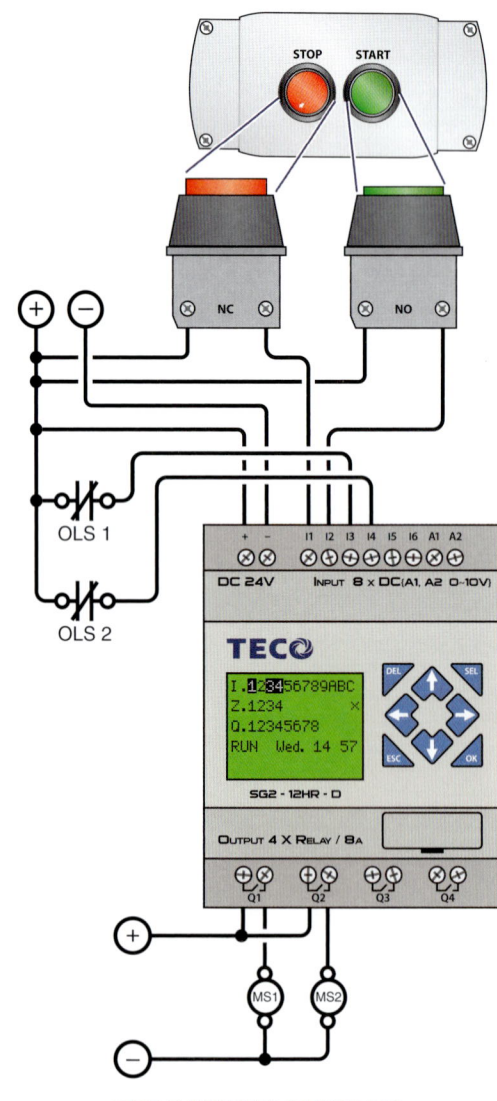

TECO WIRING DIAGRAM

Allen-Bradley® Procedures

1. Open the RSLogix™ Micro Starter Lite program and create a new programming diagram.

2. Use the AB programming diagram to add the proper address and description in the LAD 2 programming window. Add the TOF instruction to the programming diagram and type in the proper timer address, time base, and preset value.

3. Verify that the programming diagram has no errors and save the project.

4. Use the wiring diagram to wire the input devices and output components to the PLC. Once all connections have been made, turn power to the PLC on.

5. Use RSLinx™ to establish communication between the PC and the Micrologix™ 1100.

6. Download the program logic to the PLC.

7. Place the PLC in Run Mode.

8. Activate input(s) and monitor them to determine if the output(s) are operating correctly.

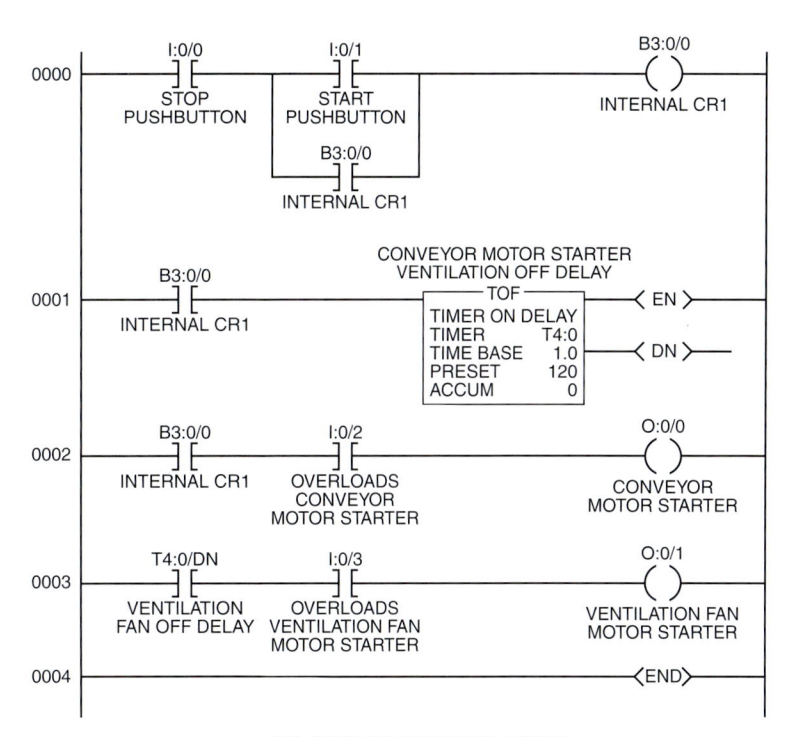

AB PROGRAMMING AREA

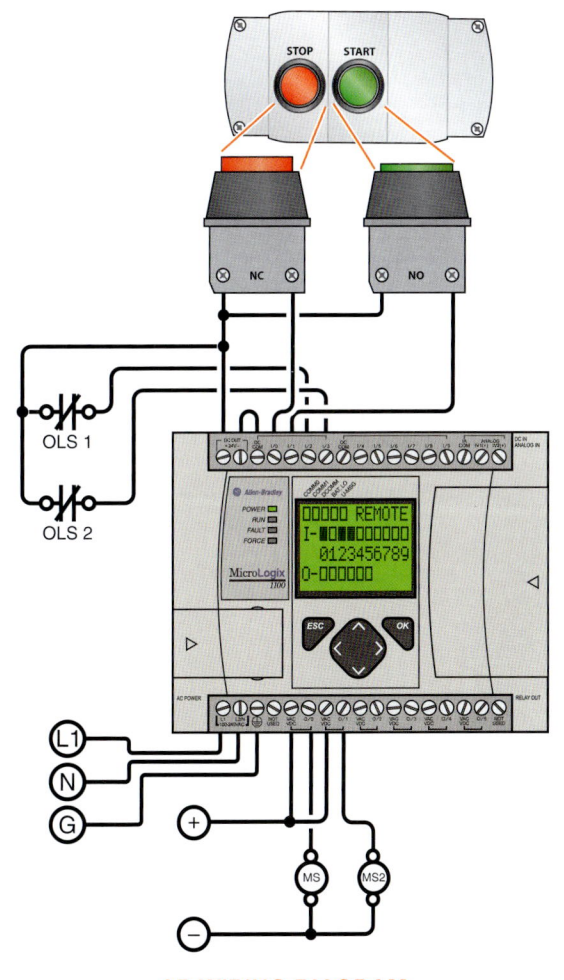

Rung 0000: I:0/0 STOP PUSHBUTTON — I:0/1 START PUSHBUTTON (parallel with B3:0/0 INTERNAL CR1) — B3:0/0 INTERNAL CR1 (output)

Rung 0001: B3:0/0 INTERNAL CR1 — TOF CONVEYOR MOTOR STARTER VENTILATION OFF DELAY, TIMER ON DELAY, TIMER T4:0, TIME BASE 1.0, PRESET 120, ACCUM 0, EN, DN

Rung 0002: B3:0/0 INTERNAL CR1 — I:0/2 OVERLOADS CONVEYOR MOTOR STARTER — O:0/0 CONVEYOR MOTOR STARTER

Rung 0003: T4:0/DN VENTILATION FAN OFF DELAY — I:0/3 OVERLOADS VENTILATION FAN MOTOR STARTER — O:0/1 VENTILATION FAN MOTOR STARTER

Rung 0004: END

AB WIRING DIAGRAM

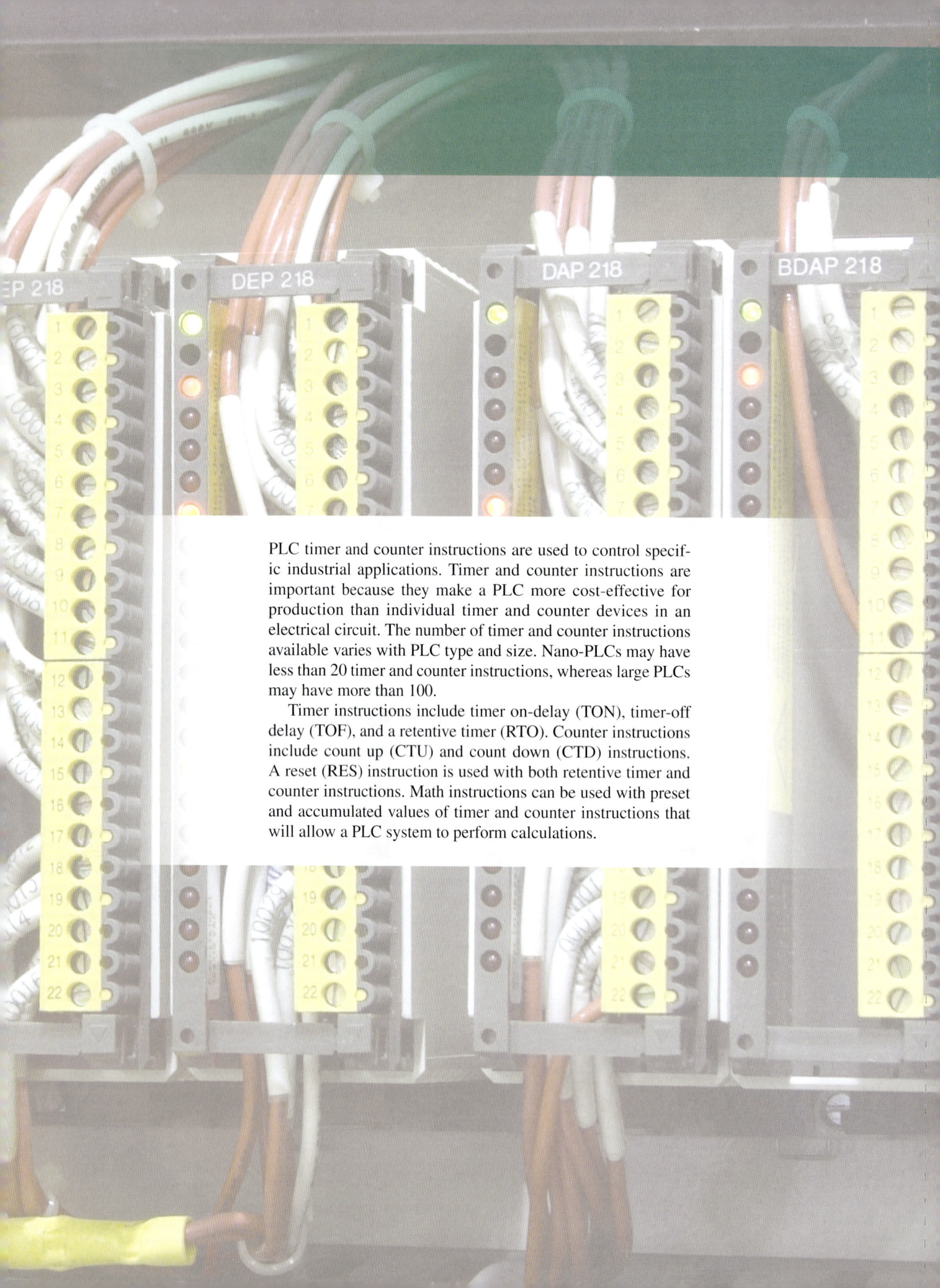

PLC timer and counter instructions are used to control specific industrial applications. Timer and counter instructions are important because they make a PLC more cost-effective for production than individual timer and counter devices in an electrical circuit. The number of timer and counter instructions available varies with PLC type and size. Nano-PLCs may have less than 20 timer and counter instructions, whereas large PLCs may have more than 100.

Timer instructions include timer on-delay (TON), timer-off delay (TOF), and a retentive timer (RTO). Counter instructions include count up (CTU) and count down (CTD) instructions. A reset (RES) instruction is used with both retentive timer and counter instructions. Math instructions can be used with preset and accumulated values of timer and counter instructions that will allow a PLC system to perform calculations.

PLC Programming Timer and Counter Instructions

7

Objectives:
- Describe the structure and addressing of timer instructions.
- Practice how to use TON, TOF, and RTO timer instructions.
- Explain the use and application of the RES instruction.
- Demonstrate how free running timers and cascaded timers are used in applications.
- Describe the structure and addressing of counter instructions.
- Practice how to use CTU and CTD counter instructions.
- Practice how to use HSC and math instructions.

Learner Resources
atplearningresources.com/quicklinks
Access Code: **475203**

TIMER AND COUNTER INSTRUCTIONS

PLC programming uses timer instructions and counter instructions to control specific functions. Nano- or micro-PLCs contain 10 to 20 programmable timers and counters, while medium and large PLCs contain hundreds of programmable timers and counters.

A *timer instruction* is a PLC programming instruction used to provide timing functions similar to mechanical or solid-state timers. The three standard timer instructions typically used during PLC programming are timer on-delay (TON), timer off-delay (TOF), and retentive timer (RTO). **See Figure 7-1.** A *counter instruction* is a PLC programming instruction used to provide counting functions similar to solid-state counters. There are two standard counter instructions that are also typically used during PLC programming. The two instructions are count up (CTU) and count down (CTD).

One additional programming instruction (reset) is typically used during timer and counter programming. Reset (RES) is a programming instruction used with RTO, CTU, and CTD instructions that zeros the accumulated value of a timer or counter.

TIMER INSTRUCTIONS

Timer instructions are frequently used to delay the start of a conveyor, control the cycle time of machinery such as packaging machines, or control the length of time a valve is open. The type of timer instruction used, TON, TOF, or RTO, depends on the specific application. Timer instructions are located in data table file 4 and are identified by the designation "T4." The designation of the first available timer is T4:0. The number of timers available in data table file 4 depends on the model of PLC. Some PLC models have additional data table files that can be configured for timer use.

Timer Instruction Words

All timer instructions consist of three words: word 0, word 1, and word 2. Each word contains 16 bits. **See Figure 7-2.** Three addressable bits of word 0 are available for use in PLC programs:
- bit 15—enable (EN)
- bit 14—timer timing (TT)
- bit 13—done (DN)

⚠ WARNING
A reset (RES) instruction cannot be used with a timer off-delay (TOF) instruction because RES always clears the status bits and accumulated value.

276 PROGRAMMABLE LOGIC CONTROLLERS: PRINCIPLES AND APPLICATIONS

Timers and Counters

HARDWIRED TIMERS AND COUNTERS
- USED FOR COUNTER OPERATIONS
- PHOTOELECTRIC SENSOR
- *Carlo Gavazzi Inc.*
- SOLID-STATE TIMER
- *Rockwell Automation, Allen-Bradley Company, Inc.*

PLC PROGRAMMED TIMERS AND COUNTERS
- *GE Fanuc Automation* — CONTAINS HUNDREDS OF PROGRAMMABLE TIMERS AND COUNTERS

TON Timer On Delay	
Timer	0
Time Base	1.0
Preset	120
Accum	0

— TIME ON-DELAY TIMER

TOF Timer Off Delay	
Timer	1
Time Base	0.01
Preset	215
Accum	0

— TIME OFF-DELAY TIMER

CTU	
Count Up	
Counter	0
Preset	173
Accum	0

— COUNT UP COUNTER

RTO Retentive Timer On	
Timer	2
Time Base	0.001
Preset	58
Accum	0

— RETENTIVE TIMER

CTD	
Count Down	
Counter	1
Preset	250
Accum	0

— COUNT DOWN COUNTER

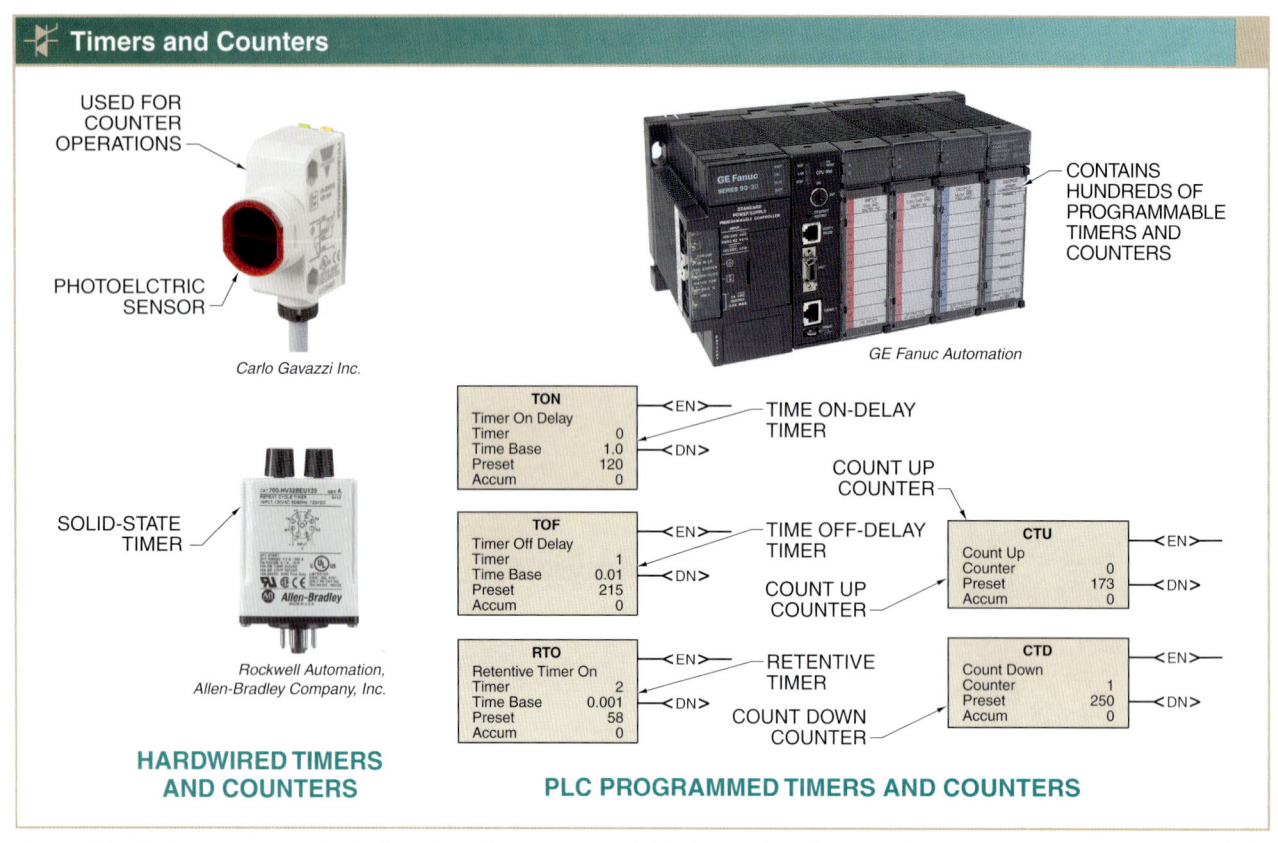

Figure 7-1. PLCs provide a cost-effective alternative to using individual hardwired timers and counters because the smallest PLCs (nanos and micros) contain 10 or more timers and counters.

Timer Instructions

Timer T4:0

	15	14	13	12	11	10	9	8	7	6	5	4	3	2	1	0
WORD 0	EN	TT	DN						← Internal PLC Use →							
WORD 1	← Preset Value →															
WORD 2	← Accumulated value →															

Addressable Bits

Description	Address
Done Bit = Bit 13	T4: 0/DN or T4: 0/13
Timer Timing Bit = Bit 14	T4: 0/TT or T4: 0/14
Enable Bit = Bit 15	T4: 0/EN or T4: 0/15

Addressable Words

Description	Address
Preset Value of Timer = Word 1	T4: 0.PRE or T4: 0.1
Accumulated Value of Timer = Word 2	T4: 0.ACC or T4: 0.2

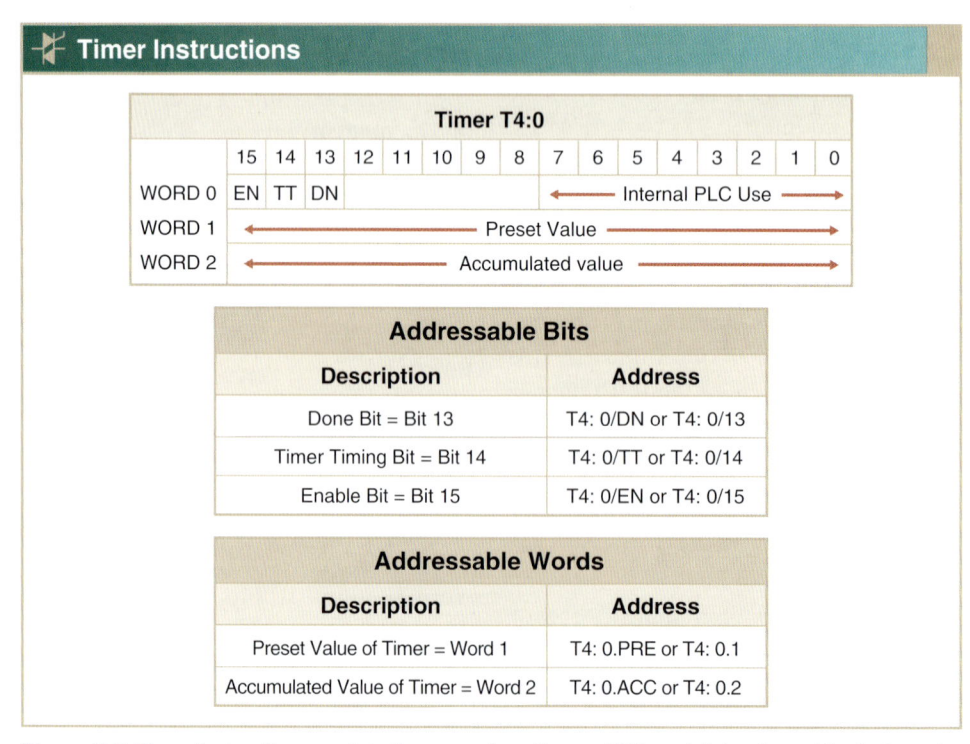

Figure 7-2. Timer instructions contain three words, with word "0" containing three bits that can be used in the PLC program.

Word 1 contains the preset value of the timer. Word 2 contains the accumulated value of the timer. The bits and words of a timer can be addressed in more than one format. When entering timer instructions, data such as addresses, time bases, presets, and accumulated values must be entered.

Timer Address. Each timer instruction has a unique address, such as T4:2. "T4" identifies the timer data table file and "2" identifies the specific timer. Timer addresses can be used for any type of timer instruction, TON, TOF, or RTO. However, a specific timer address cannot be used for more than one type of timer instruction. For example, the T4:3 address cannot be used with a TON instruction and an RTO instruction in the same program.

Time Base. A *time base* is an instruction label used to indicate the unit of time a timer uses. The available time bases vary between PLC manufacturers and models. Typical time bases that can be selected are 0.01 sec and 1.0 sec. The application and the required timing precision determine which time base is used.

Timer Preset. A *timer preset value* is the number of time base intervals a timer instruction is programmed to accomplish. The preset is multiplied by the time base to determine a timing duration. For example, a timer with a preset value of 200 and a time base of 0.01 works with a time duration of 2 sec (200 × 0.01 = 2).

Timer Accumulated Value. A *timer accumulated value* is the number of time base intervals (the length of time) that a timer has been timing. The accumulated value is typically set at 0.

Preset and accumulated values for all timers are from 0 to +32,767. A negative number for a preset or accumulated value causes a PLC processor fault.

PLC TIPS

The accumulated value of a timer cannot exceed the preset value.

Timer On-Delay Instructions

A *timer on-delay (TON) instruction* is a programming instruction used to delay the start of a machine or process for a set period of time. A TON instruction starts timing when the logic preceding the TON instruction on a rung changes from false to true. **See Figure 7-3.** When a TON instruction starts timing, the accumulated value begins to count time base intervals until the accumulated value equals the preset value. When the logic preceding a TON instruction changes from true to false, the TON instruction stops timing and the accumulated value is reset to zero.

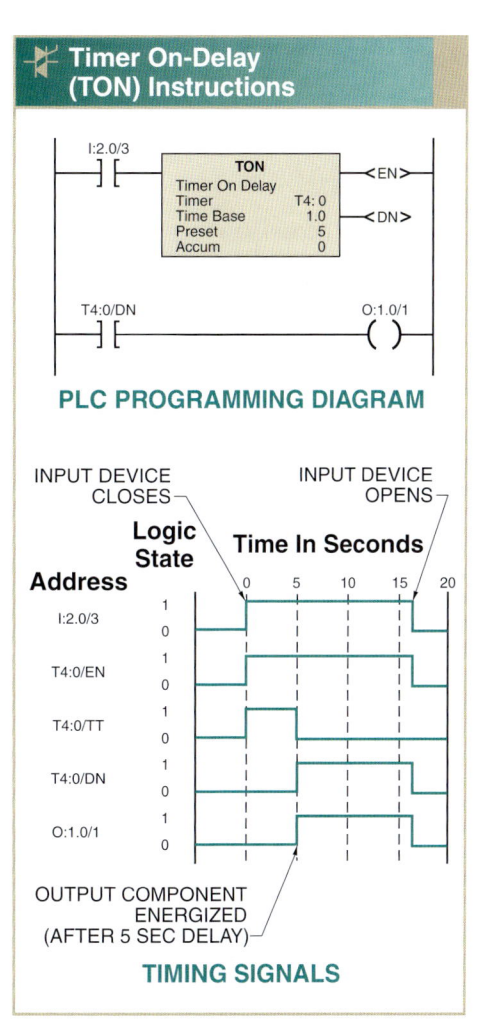

Figure 7-3. Timers following timer on-delay (TON) instructions begin timing when the logic preceding the instruction on a rung changes from false to true.

The three status bits typically used with TON instructions are the enable bit, the timer timing bit, and the done bit. TON status bits are linked to the TON instruction and change state during the timing sequence.

TON Enable (EN) Bits. An EN bit is true (1) when the logic preceding a TON instruction on a rung is true. An EN bit is false (0) when the logic preceding a TON instruction on a rung is false.

CODE CONNECT

The NEC® requires the ampacities of conductors to be derated/adjusted for a variety of conditions, including ambient temperature and the number of conductors in a raceway or cable. Section 310.15(C)(1)(b) of the NEC® exempts conductors in a raceway 24″ (600 mm) or less in length from derating.

⚠ CAUTION

Preset and accumulated values for timers range from 0 to +32,767. When a timer preset or accumulated value is a negative number, a runtime error occurs.

TON Timer Timing (TT) Bits. A TT bit is true (1) when the logic preceding a TON instruction on a rung is true and the accumulated value is less than the preset value. A TT bit is false (0) when the logic preceding a TON instruction on a rung is false or the accumulated value equals the preset value.

TON Done (DN) Bits. A DN bit is true (1) when the logic preceding a TON instruction on a rung is true and the accumulated value is equal to the preset value. A DN bit is false (0) when the logic preceding a TON instruction on a rung is false.

A PLC with a programmed timer using TON instructions can be used to replace hardwired on-delay timers. On-delay timers are used in applications such as monitoring the breathing of a patient with breathing difficulties. **See Figure 7-4.**

For example, a timer is used to sound an alarm when a patient does not take a breath within a 10 sec period of time. Hardwired on-delay timers and control relays used in patient monitoring devices have been replaced with PLCs. Original equipment manufacturers (OEMs) prefer PLCs because PLCs are compact, can perform multiple functions, contain multiple control relays and timers, and allow circuit changes without rewiring.

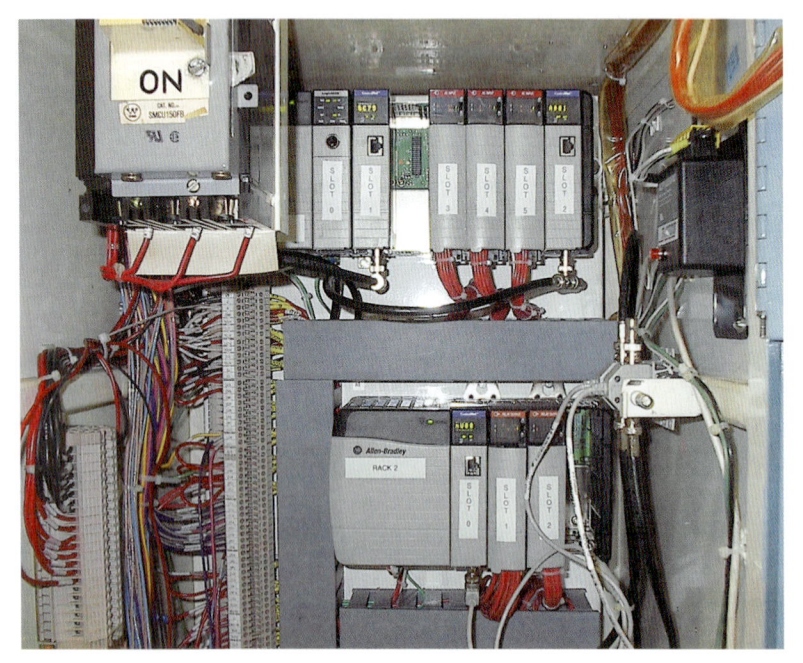

Modular PLCs can be installed in the same cabinet where one power supply controls all AC inputs and another controls all AC outputs.

PLC TIPS

When a PLC program contains timers, special care must be taken when the timer is timing (not yet reached the preset value) and power is lost, or the mode of the PLC is changed from REM run or REM test to REM program. Status bits and accumulated values are affected. When power returns or the mode of the PLC is returned to REM run or REM test, status bits and accumulated values are also affected. Timer timing (TT) can be inaccurate when any jump (JMP) instruction skips over a rung having a timer instruction. When the skip has a duration less than 2.5 sec, no timing error occurs, but when the skip is greater than 2.5 sec, timing errors occur. Programs with subroutines must use a timer instruction every 2.5 sec to prevent timing errors.

Hardwired Applications vs. PLC TON Applications

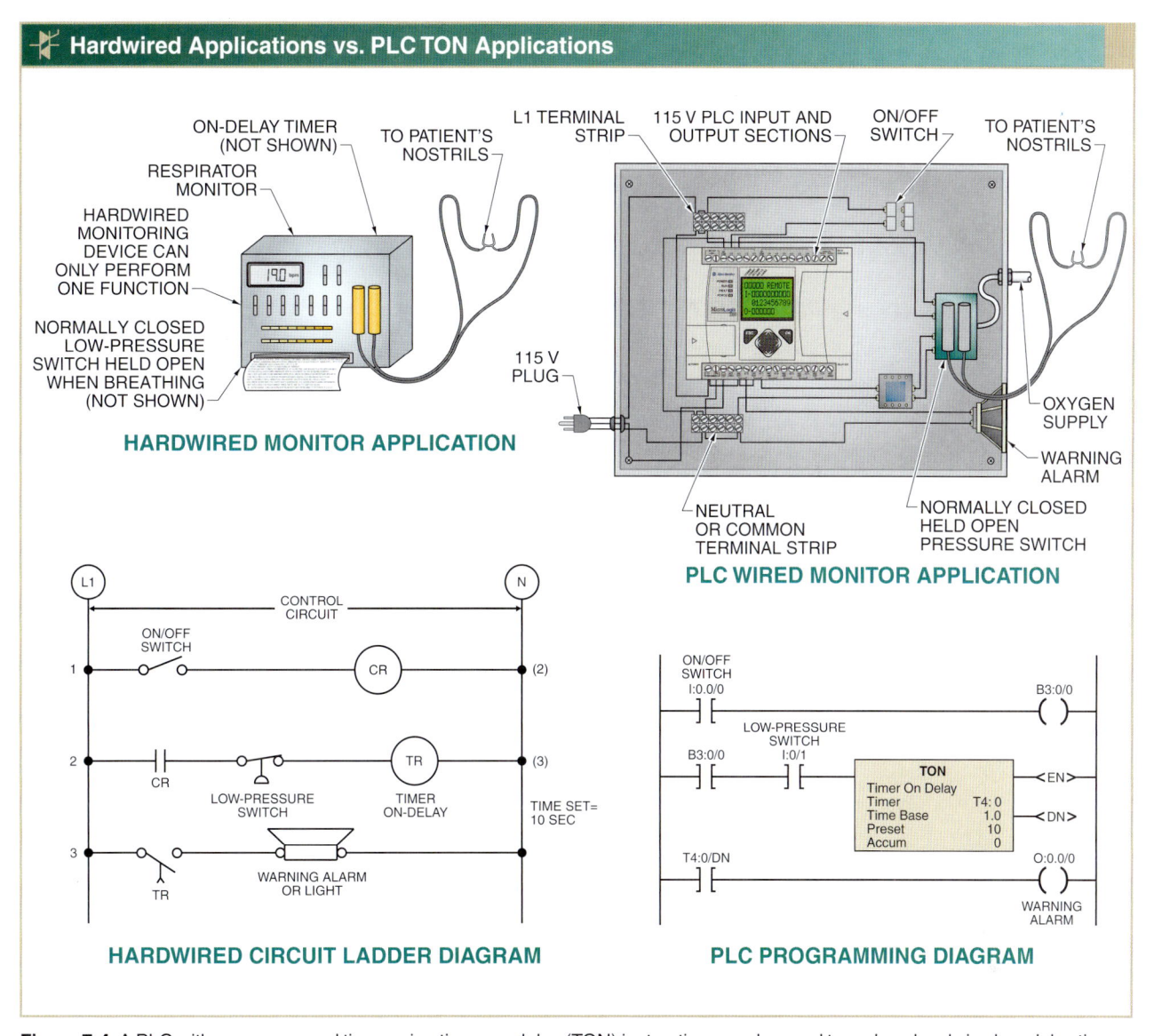

Figure 7-4. A PLC with a programmed timer using timer on-delay (TON) instructions can be used to replace hardwired on-delay timers.

Timer Off-Delay Instructions

A *timer off-delay (TOF) instruction* is a PLC programming instruction used to delay the shut down of machinery, such as an external cooling fan when a motor has been stopped. A TOF instruction starts timing when the logic preceding the instruction on the rung changes from true to false. **See Figure 7-5.** When a TOF instruction starts timing, the accumulated value begins to count time base intervals until the accumulated value equals the preset value. When the logic preceding a TOF instruction on a rung changes from false to true, the TOF instruction stops timing and the accumulated value is reset to zero.

The three status bits typically used with TOF instructions are the enable bit, the timer timing bit, and the done bit. The TOF status bits are linked to the TOF instruction and change state during the timing sequence.

TOF Enable (EN) Bits. An EN bit is true (1) when the logic preceding a TOF instruction on a rung is true. An EN bit is false (0) when the logic preceding a TOF instruction on a rung is false.

Timer Off-Delay (TOF) Instructions

PLC PROGRAMMING DIAGRAM

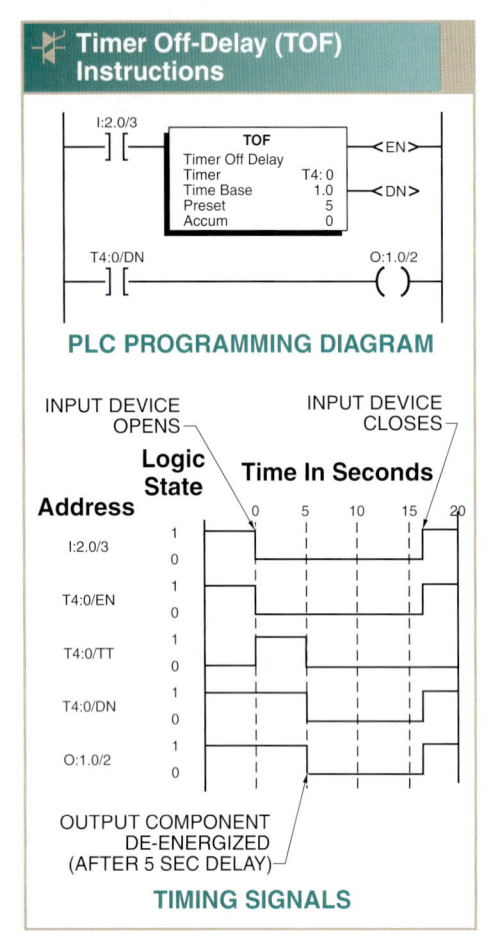

```
I:2.0/3                TOF
──┤ ├──────┬─────────────────────────┬──<EN>
           │ Timer Off Delay         │
           │ Timer          T4:0     │
           │ Time Base       1.0     │──<DN>
           │ Preset            5     │
           │ Accum             0     │
           └─────────────────────────┘
T4:0/DN                                   O:1.0/2
──┤ ├─────────────────────────────────────(  )
```

TIMING SIGNALS

- INPUT DEVICE OPENS
- INPUT DEVICE CLOSES
- OUTPUT COMPONENT DE-ENERGIZED (AFTER 5 SEC DELAY)

Address / Logic State / Time In Seconds (0, 5, 10, 15, 20)
- I:2.0/3
- T4:0/EN
- T4:0/TT
- T4:0/DN
- O:1.0/2

Figure 7-5. Timers following timer off-delay (TOF) instructions start timing when the logic preceding the instruction on a rung changes from true to false.

PLCs used with boilers have retentive timers to monitor the overall time a specific boiler has been operating and how long equipment such as induction fans have been running.

Cleaver-Brooks

PLC TIPS

Timer accuracy deals with the length of time between the moment a PLC CPU sets the timer-enable bit and the moment the CPU sets the timer-done bit. Timer accuracy depends on clock tolerance, time base, and program scan time. Typically the tolerance of a PLC clock is ±0.01%, allowing a timer to time out 0.5 sec early or late when set to a 1.0 sec time base (0.1 sec when set to 0.1 and 0.01 sec when set to 0.01 sec). A timer remains accurate when examined every 1 sec (or less) over two program scans.

TOF Timer Timing (TT) Bits. A TT bit is true (1) when the logic preceding a TOF instruction on a rung is false and the accumulated value is less than the preset value. A TT bit is false (0) when the logic preceding a TOF instruction on a rung is true or the accumulated value equals the preset value.

TOF Done (DN) Bits. A DN bit is true (1) when the logic preceding a TOF instruction on a rung is true. A DN bit is false (0) when the logic preceding a TOF instruction on a rung is false and the accumulated value equals the preset value.

A PLC with a programmed timer using TOF instructions is used to replace hard-wired off-delay timers. Off-delay timers are used in applications that require a load to remain energized after the input has been removed, such as an emergency shower. **See Figure 7-6.** An off-delay timer keeps the water flowing for 1 min after the emergency shower pushbutton is pressed. An off-delay timer used in the emergency shower circuit can be replaced with a PLC. OEMs prefer PLCs because PLCs are compact, contain multiple timers, and allow circuit changes without rewiring.

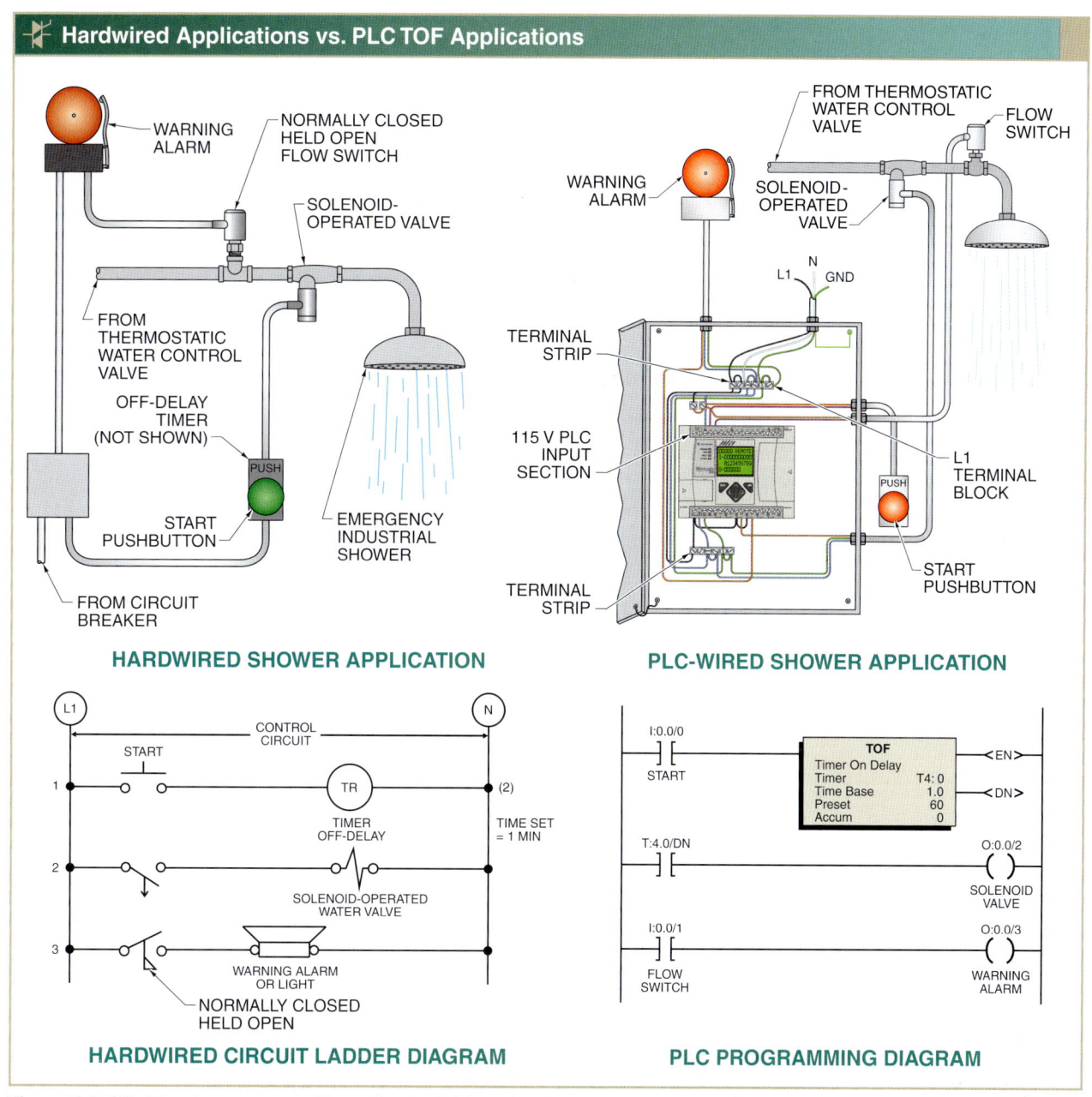

Figure 7-6. Off-delay timers are used in applications that require a load to remain energized after the input has been removed.

Retentive Timer Instructions

A *retentive timer (RTO) instruction* is a PLC programming instruction used to track the length of time a machine has been operating or to shut down a process after an accumulative time period of recurring faults. A timer starts timing when the logic preceding an RTO instruction on a rung changes from false to true. **See Figure 7-7.**

When an RTO instruction starts timing, the accumulated value begins to count time base intervals until the accumulated value equals the preset value. When the logic preceding an RTO instruction on a rung changes from true to false before the accumulated value equals the preset value, an RTO instruction stops timing and the accumulated value is retained.

Retentive Timer and Reset Instructions

PLC PROGRAMMING DIAGRAM

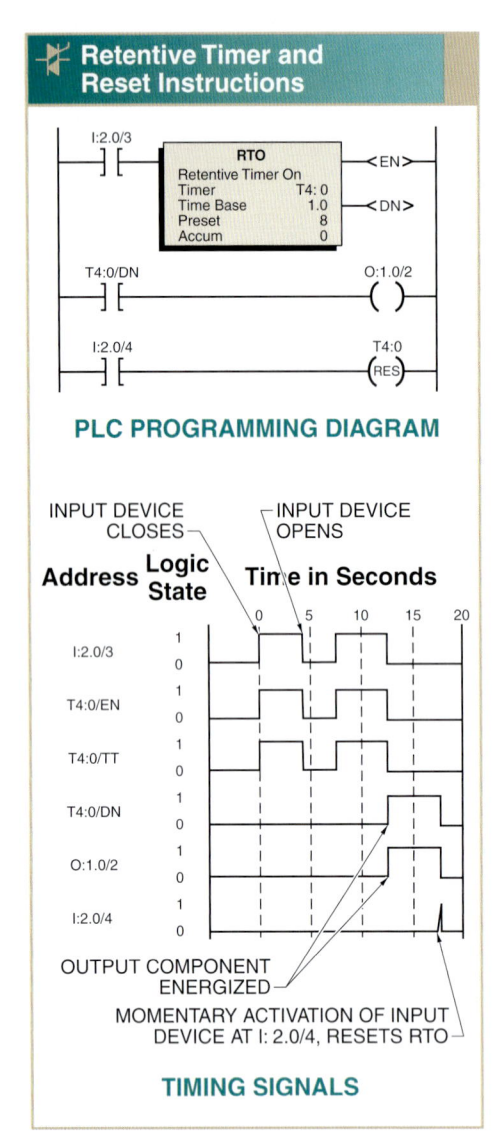

TIMING SIGNALS

Figure 7-7. Retentive timers begin timing when the logic preceding the instruction on a rung changes from false to true. Retentive timers retain their accumulated value until reset.

Unlike a TON instruction, the accumulated value is not reset on a true to false transition of logic preceding a timer instruction. An RTO instruction retains the accumulated value regardless of the number of true to false transitions. A separate reset (RES) instruction is required to reset an RTO instruction.

As with TON and TOF instructions, the three status bits typically used with RTO instructions are the enable bit, the timer timing bit, and the done bit. RTO status bits are linked to an RTO instruction and all corresponding RES instructions. RTO status bits change state during a timing sequence.

RTO Enable (EN) Bits. An EN bit is true (1) when the logic preceding an RTO instruction on a rung is true. An EN bit is false (0) when the logic preceding an RTO instruction on a rung is false.

RTO Timer Timing (TT) Bits. A TT bit is true (1) when the logic preceding an RTO instruction on a rung is true and the accumulated value is less than the preset value. A TT bit is false (0) when the logic preceding an RTO instruction on a rung is false or the accumulated value equals the preset value.

RTO Done (DN) Bits. A DN bit is true (1) when the accumulated value is equal to the preset value. A DN bit is false (0) when the accumulated value is less than the preset value.

Reset Instructions

A *reset (RES) instruction* is a PLC programming instruction used to reset timer and counter accumulated values. Typically, RES instructions are used only with RTO, CTU, or CTD instructions.

RES instructions have the same address as the timer or counter the instruction is programmed to reset. An RES instruction is true when the logic preceding the instruction on a rung is true. When used with a timer instruction, an RES instruction resets the accumulated value, DN bit, TT bit, and EN bit to zero. When an RES instruction is used with a counter instruction, an RES instruction resets the accumulated value, overflow (OV) bit, underflow (UN) bit, DN bit, count up (CU) enable bit, and count down (CD) enable bit to zero.

Special Applications

Special applications can require timers to be used in a nonstandard manner. Free running (repetitive) timers and cascaded timers are two types of timers used in special applications.

A *free running (repetitive) timer* is a type of programmed timer that is continuously timing because the XIO instruction preceding the timer is true. **See Figure 7-8.** When the accumulated value of a free running timer reaches the preset value, the instruction preceding the timer goes false, the timer is reset, and the timer starts timing again on the next scan (when the XIO instruction is true again).

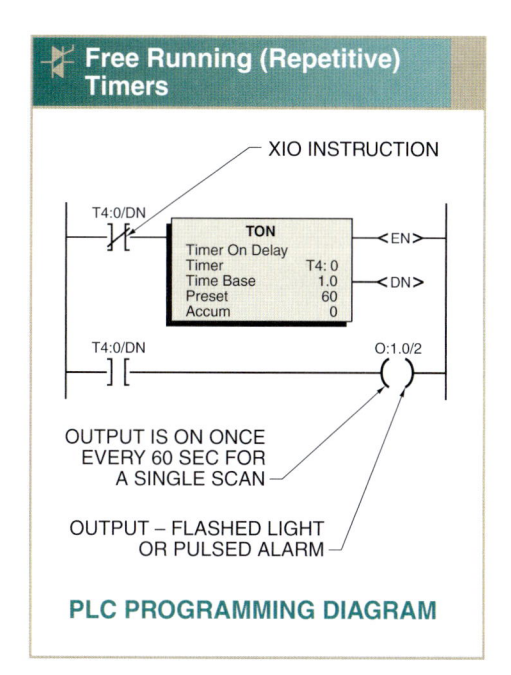

Figure 7-8. Free running timers are continuously timing because the XIO instruction preceding the timer is always true.

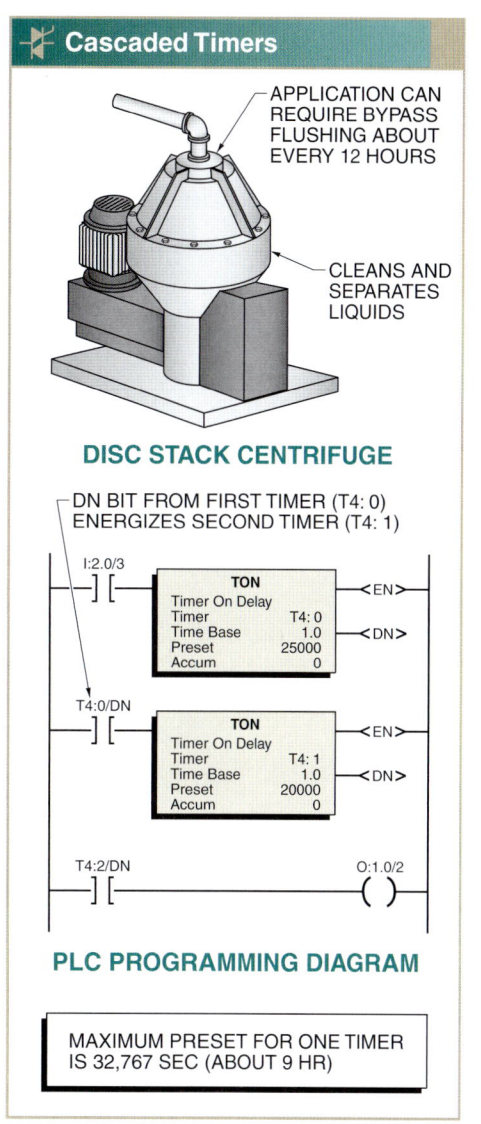

Figure 7-9. Timers are cascaded by having the DN bit of timer No. 0 activate timer No. 1. The only limit to the number of cascaded timers is the size of the memory in the PLC.

Free running timers are typically used to flash a light, or pulse an output ON at a fixed interval. The time interval is determined by the values programmed for the timer.

Timers are cascaded by having the DN bit of timer No. 0 activate timer No. 1. The only limit to the number of cascaded timers is the size of the memory in the PLC. **See Figure 7-9.** Cascaded timers are used when individual timers do not have enough time base intervals for an application. For example, an individual cascaded timer cannot be programmed for an operating time period of 45,000 sec because the maximum preset and accumulated value is 32,767 sec. However, two cascaded timers can be used to program a time period of 45,000 sec for the next bypass flushing.

Cascaded timers can also be used to start equipment sequentially, such as when starting three conveyors with a 10 sec delay between the start of each conveyor. **See Figure 7-10.** Staggering the starting of the conveyors limits the amount of inrush current that must be carried by the electrical distribution system of a building. Staggering conveyor starting also ensures that a conveyor is running before the conveyor receives product from another conveyor.

Cascading Timers for Conveyor Control

CONVEYOR APPLICATION

- CONVEYOR MOTOR 3
- CONVEYOR MOTOR 2
- CONVEYOR MOTOR 1

LADDER DIAGRAM

NORMALLY CLOSED OVERLOAD CONTACTS FOR EACH MOTOR STARTER ARE PROGRAMMED AS XIC INSTRUCTION INPUTS. (OVERLOAD CONTACTS MAY OR MAY NOT BE BROUGHT BACK TO A PLC AS INPUTS)

PLC PROGRAMMING DIAGRAM

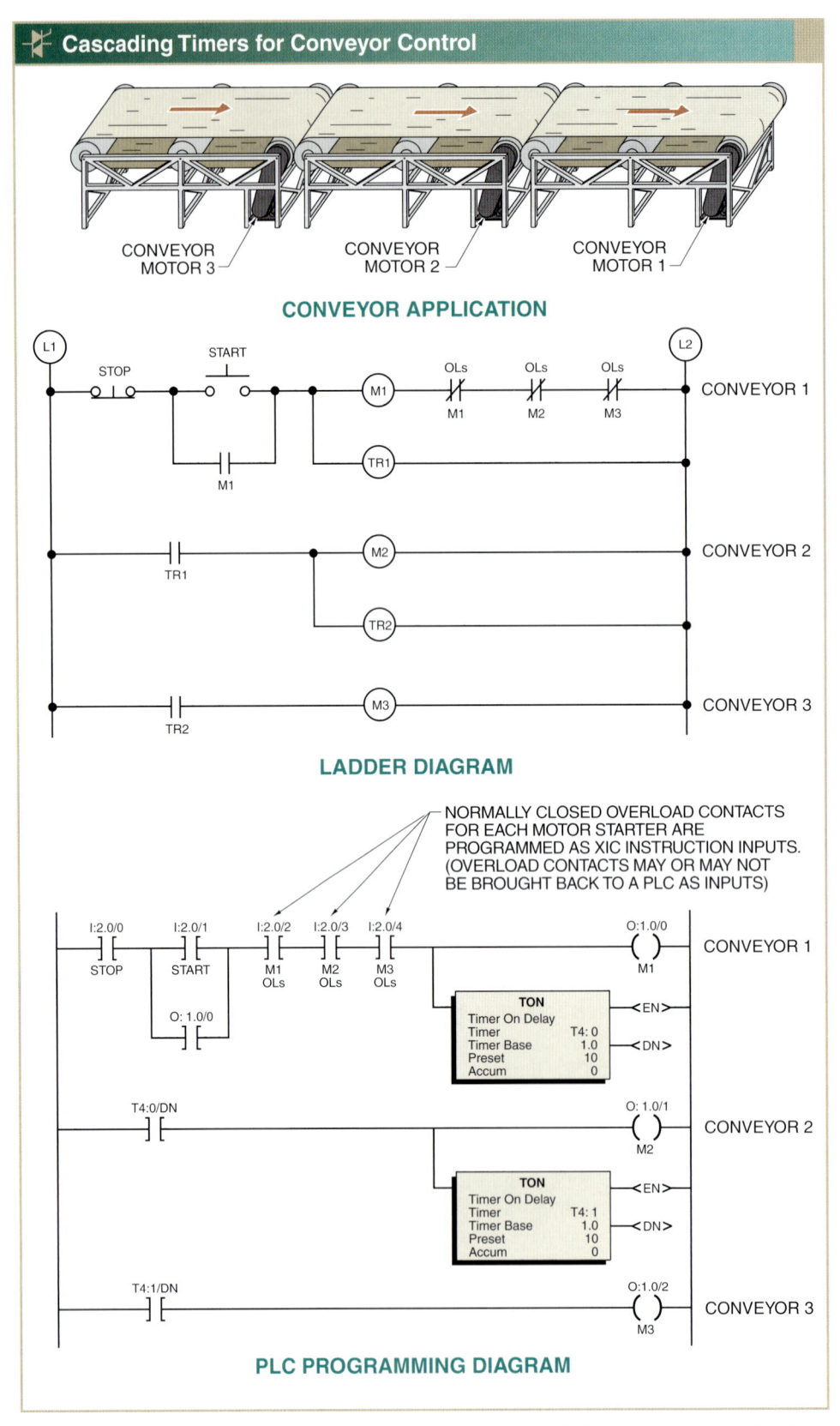

Figure 7-10. Cascaded timers are used to start equipment sequentially, such as when starting three conveyors with a 10 sec delay between the start of each conveyor.

When a PLC program contains timers, special attention must be used when a timer is timing (has not yet reached the preset value) and power is lost, or when the mode of the PLC is changed from REM run or REM test to REM program. Accumulated values and status bits, such as timer EN bit, TT bit, or DN bit, are affected. When power returns or the mode of the PLC is returned to REM run or REM test, the accumulated values and status bits default to their original values.

COUNTER INSTRUCTIONS

Although counter instructions are not used as often as timer instructions, counter instructions are typically found in PLC programs. Counter instructions can be used to count product, track inventory, or monitor the number of cars in a parking lot. Count up (CTU) instructions are used alone or in conjunction with count down (CTD) instructions having the same address. Additional counter instructions include high-speed counter instructions.

Counter instructions are located in data table file 5 and are identified by the designation C5. The designation of the first available counter is C5:0. The number of counters available in data table file 5 depends on the manufacturer and model of the PLC. Some PLCs have additional data table files that can be configured for counter use.

Counter Instruction Words

All counter instructions consist of three words: word 0, word 1, and word 2. Each word contains 16 bits. **See Figure 7-11.** Word 0 contains five bits that are available for use in a PLC program:

- bit 15—count up (CU) enable
- bit 14—count down (CD) enable
- bit 13—done (DN)
- bit 12—overflow (OV)
- bit 11—underflow (UN)

Word 1 contains the preset value of the counter. Word 2 contains the accumulated value of the counter. The bits and words of a counter can be addressed in more than one format. When entering counter instructions, items such as addresses, presets, and accumulated values must be entered.

Counter Address. Each counter instruction has a unique address, such as C5:3. "C5" identifies the counter data table file and "3" identifies the specific counter. Counter addresses are used with CTU and CTD counter instructions. Unlike timers, a specific counter address can be used for both a CTU and CTD instruction. When CTU and CTD instructions are used together, the instructions share the same address.

Counter Preset. A *counter preset value* is the number of counts a counter is programmed to reach.

Counter Accumulated Value. A *counter accumulated value* is the number of counts a counter has recorded. The accumulated value is typically set to zero. Preset and accumulated values for all counters range from −32,768 to +32,767.

PLC TIPS

Unlike a timer, the accumulated (ACC) value of a counter can exceed the preset value. ACC value is always retained until cleared by a reset (RES) instruction that has the same address as the counter.

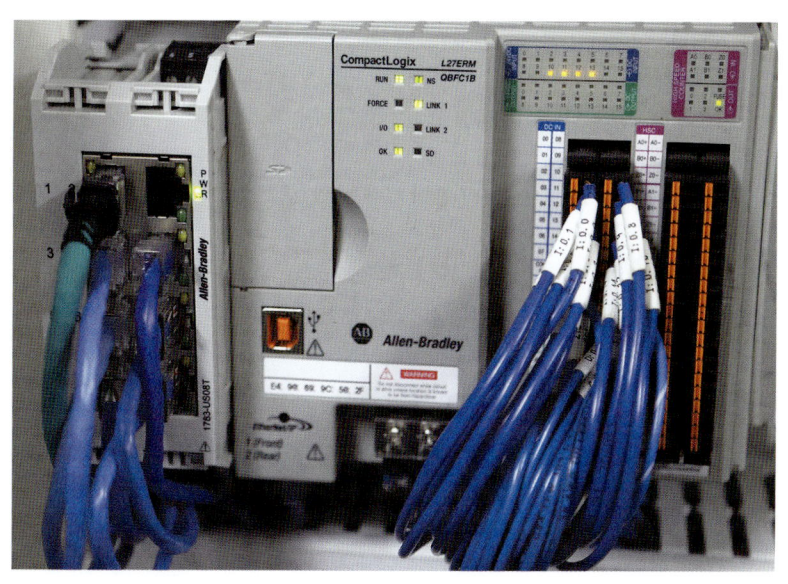

Timer and counter instructions are programmed into PLCs to operate loads at a given time during operation.

Counter Instructions

Counter C5:0

	15	14	13	12	11	10	9	8	7	6	5	4	3	2	1	0
WORD 0	CU	CD	DN	OV	UN				← Internal PLC Use →							
WORD 1	← Preset Value →															
WORD 2	← Accumulated value →															

NOTE: AVAILABLE FOR USE IN PLC PROGRAMS

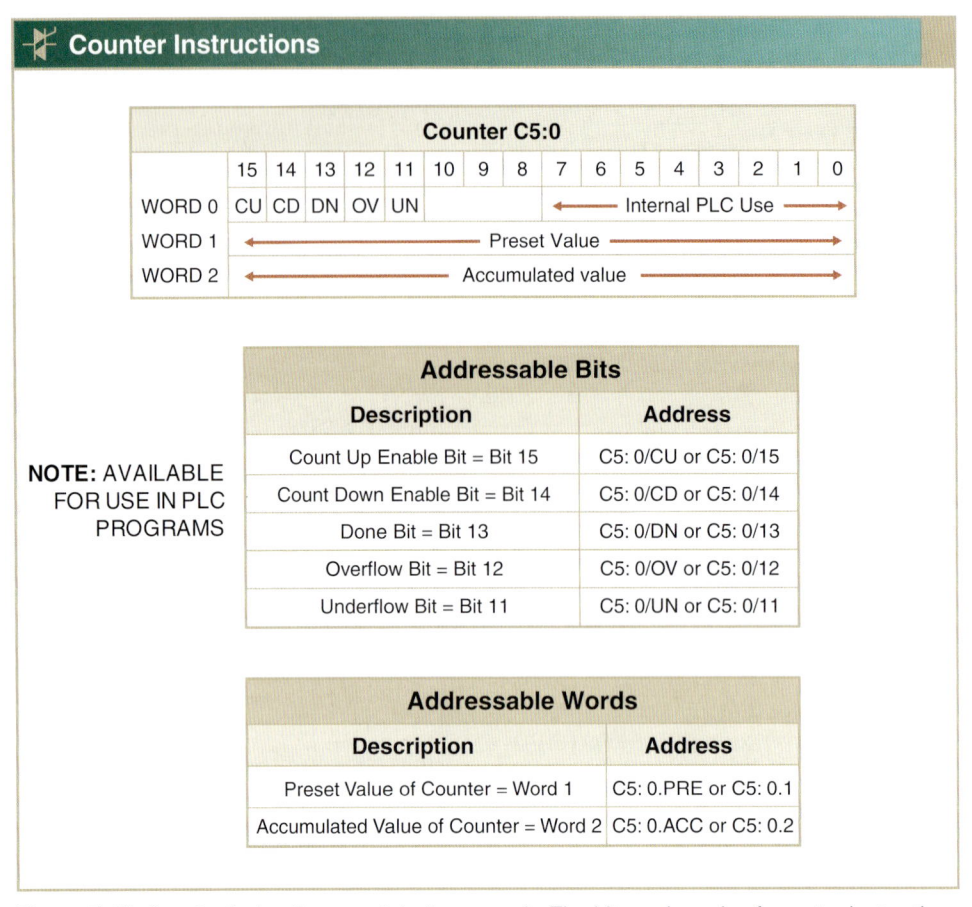

Addressable Bits

Description	Address
Count Up Enable Bit = Bit 15	C5: 0/CU or C5: 0/15
Count Down Enable Bit = Bit 14	C5: 0/CD or C5: 0/14
Done Bit = Bit 13	C5: 0/DN or C5: 0/13
Overflow Bit = Bit 12	C5: 0/OV or C5: 0/12
Underflow Bit = Bit 11	C5: 0/UN or C5: 0/11

Addressable Words

Description	Address
Preset Value of Counter = Word 1	C5: 0.PRE or C5: 0.1
Accumulated Value of Counter = Word 2	C5: 0.ACC or C5: 0.2

Figure 7-11. Counter instructions contain three words. The bits and words of counter instructions have addresses that can be used in the program of a PLC.

Baldor Electric Company

PLC counters are often used to keep track of the number of products produced by a manufacturing line for inventory and sales purposes.

Count Up Instructions

A *count up (CTU) instruction* is a PLC programming instruction used to count the number of operations or products produced by a system. A CTU instruction counts up on every false to true transition of the logic preceding the instruction on a rung. **See Figure 7-12.** The accumulated value of a CTU instruction increases by one for every false to true transition. A CTU instruction will allow a counter to continue to count past the preset value. An identically addressed RES instruction is required to reset a CTU instruction.

The three status bits typically used with CTU instructions are the count up enable bit, the done bit, and the overflow bit. The CTU status bits are linked to the CTU instruction and change state during the counting sequence.

Counter Count Up (CTU) Instructions

PLC PROGRAMMING DIAGRAM

TIMING SIGNALS

ACCUMULATED VALUE

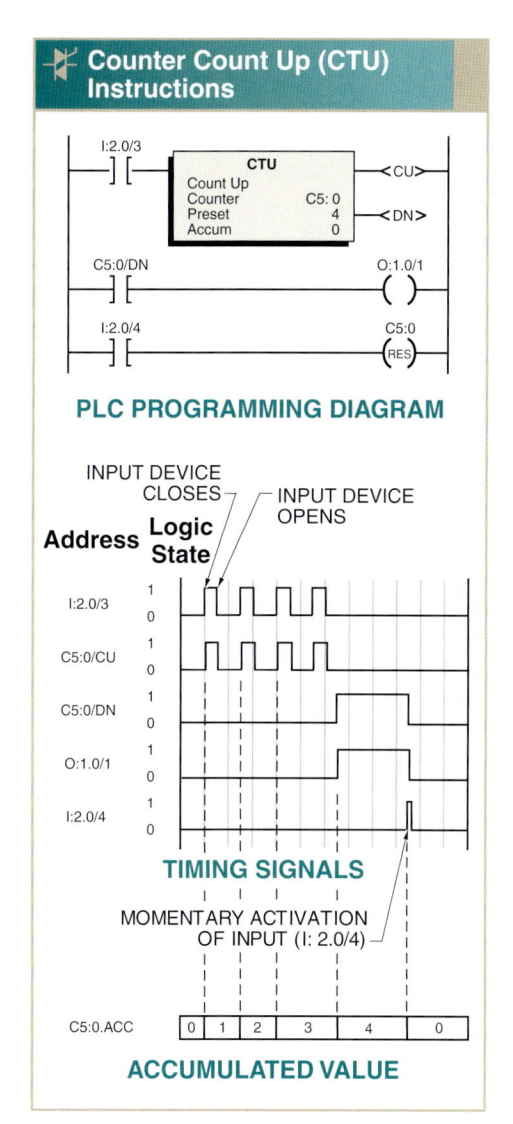

Figure 7-12. Count up (CTU) instructions increase the accumulated value incrementally on each false to true transition of the logic preceding the instruction.

CTU Count Up (CU) Enable Bits. A CU bit is true (1) when the logic preceding a CTU instruction on a rung is true. A CU bit is false (0) when the logic preceding a CTU instruction on a rung is false.

CTU Done (DN) Bits. A DN bit is true (1) when the accumulated value is equal to or greater than the preset value. A DN bit is false (0) when the accumulated value is less than the preset value.

CTU Overflow (OV) Bits. An OV bit is true (1) when the accumulated value exceeds +32,767 and wraps around to –32,768. An OV bit is false (0) when the accumulated value is less than or equal to +32,767 and has not wrapped around.

A PLC with CTU and CTD counter instructions is used to replace hardwired counters. Counters are used in applications that require items or products to be counted, such as with assembly or production lines. **See Figure 7-13.** For example, a counter can count the number of shrink-wrapped products that pass by a sensor or switch. Most hardwired counters have contacts that change state when the accumulated value equals the preset value. Hardwired counters used on production lines are being replaced with PLCs.

OEMs prefer PLCs because PLCs are compact, contain multiple counters, do not have a fixed number of contacts (the number of contacts is only limited by the memory of a PLC), have math instructions that can be used with counters, and do not require rewiring when changing circuits.

Count Down Instructions

A *count down (CTD) instruction* is a PLC programming instruction used to keep track of the number of items involved in a subprocess. Typically, CTD instructions are used with CTU instructions. For example, CTD and CTU instructions are used together to keep track of the number of automobiles on an assembly line that are in a paint spray booth.

When used together, CTD and CTU instructions have the same address, similar to OTL and OTU instructions. CTU and CTD instructions share the same status bits, preset values, and accumulated values. **See Figure 7-14.** However, each counting instruction has its own logic. A CTD instruction counts down on every false to true transition of the logic preceding the instruction on a rung. The accumulated value of a CTD instruction decreases by one for every false to true transition of the logic preceding a CTD instruction on a rung. Conversely, a CTU instruction causes the accumulated value of a CTD instruction to increase. A single identically addressed RES instruction is required to reset CTD and CTU instructions.

288 PROGRAMMABLE LOGIC CONTROLLERS: PRINCIPLES AND APPLICATIONS

Hardwired Applications vs. PLC Counter Applications

HARDWIRED INVENTORY APPLICATION

PLC-WIRED INVENTORY APPLICATION

HARDWIRED CIRCUIT LADDER DIAGRAM

PLC PROGRAMMING DIAGRAM

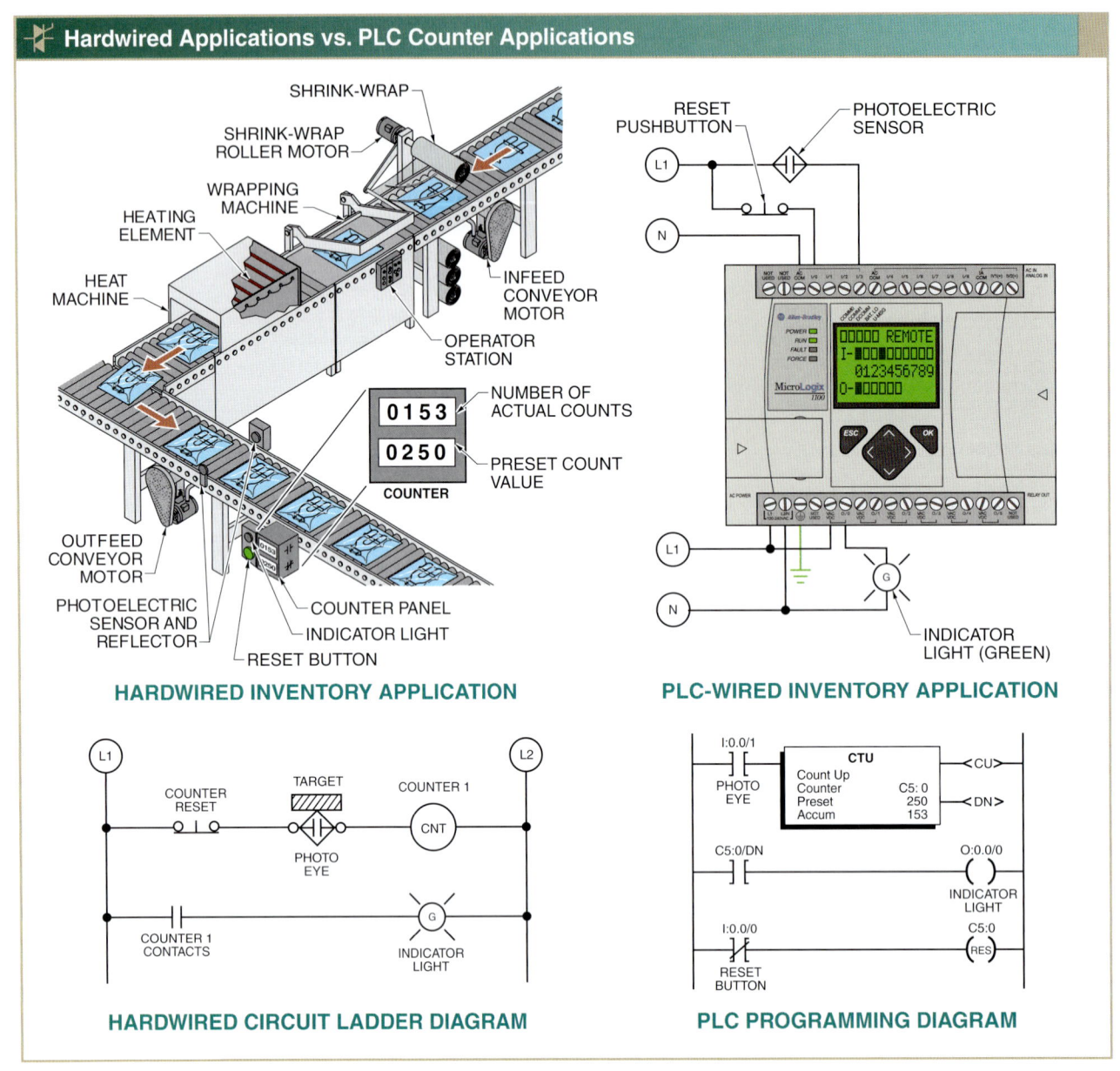

Figure 7-13. Counters are used in applications that require items or products to be counted.

Three status bits used with CTD instructions are the count down enable bit, the done bit, and the underflow bit. CTD status bits are linked to CTD instructions and change state during the counting sequence.

CTD Count Down (CD) Enable Bits. A CD bit is true (1) when the logic preceding a CTD instruction on a rung is true. A CD bit is false (0) when the logic preceding a CTD instruction on a rung is false.

CTD Done (DN) Bits. A DN bit is true (1) when the accumulated value is equal to or greater than the preset value. A DN bit is false when the accumulated value is less than the preset value.

CTD Underflow (UN) Bits. A UN bit is true (1) when the accumulated value counts down past −32,768 and wraps around to +32,767. A UN bit is false (0) when the accumulated value is greater than or equal to −32,768 and has not wrapped around.

Counter Count Down (CTD) Instructions

PLC PROGRAMMING DIAGRAM

TIMING SIGNALS

ACCUMULATED VALUE

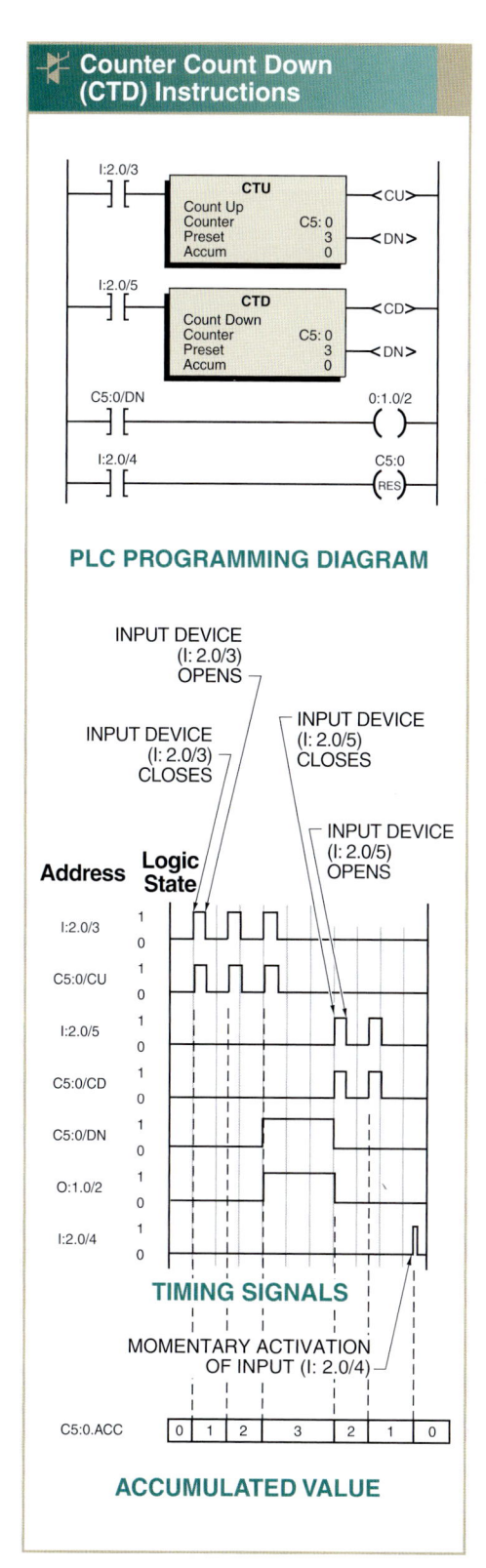

Figure 7-14. A CTD instruction counts down on every false to true transition of the logic preceding the instruction and shares the same status bits, preset values, and accumulated values as the corresponding CTU instruction.

For example, CTU and CTD instructions can be used to keep track of the cars entering and exiting a parking lot. **See Figure 7-15.** Vehicle detection sensors located at the entrance and exit act as input devices for a PLC. The preset value for the counters is the maximum number of cars the parking lot can hold.

The CTU and CTD instructions work together to track the number of cars in the parking lot at any given time. When the maximum number of cars are in the lot, the PLC turns ON a "LOT FULL" sign. The sign will turn OFF when the number of cars in the lot is less than the preset or the reset switch has been activated.

High-Speed Counter Instructions

A *high-speed counter (HSC) instruction* is a PLC programming counter instruction that can detect high-speed counts/events. Only one HSC instruction can be used per program and is fixed at the address C0. HSC instruction parameters that must be programmed are type (up or bidirectional), counter address (C5:0), and high preset (accumulated value). The accumulated value tracks the number of counts. An HSC instruction operates independently of the PLC processor scan. Consequently, an HSC instruction can respond before the processor scan is complete. For example, an HSC instruction can be used with an encoder to position a motor shaft.

A *high-speed counter load (HSL) instruction* is a PLC programming counter instruction used in conjunction with a high-speed counter instruction. An HSL instruction allows the HSC instruction to instantly update external outputs when a preset has been reached. It also allows output terminals to be instantly updated when a high or low preset is reached. HSL instruction parameters that must be programmed are source (first of data words) and length (number of elements starting from the source), which is always 5.

CTU and CTD Instruction Applications

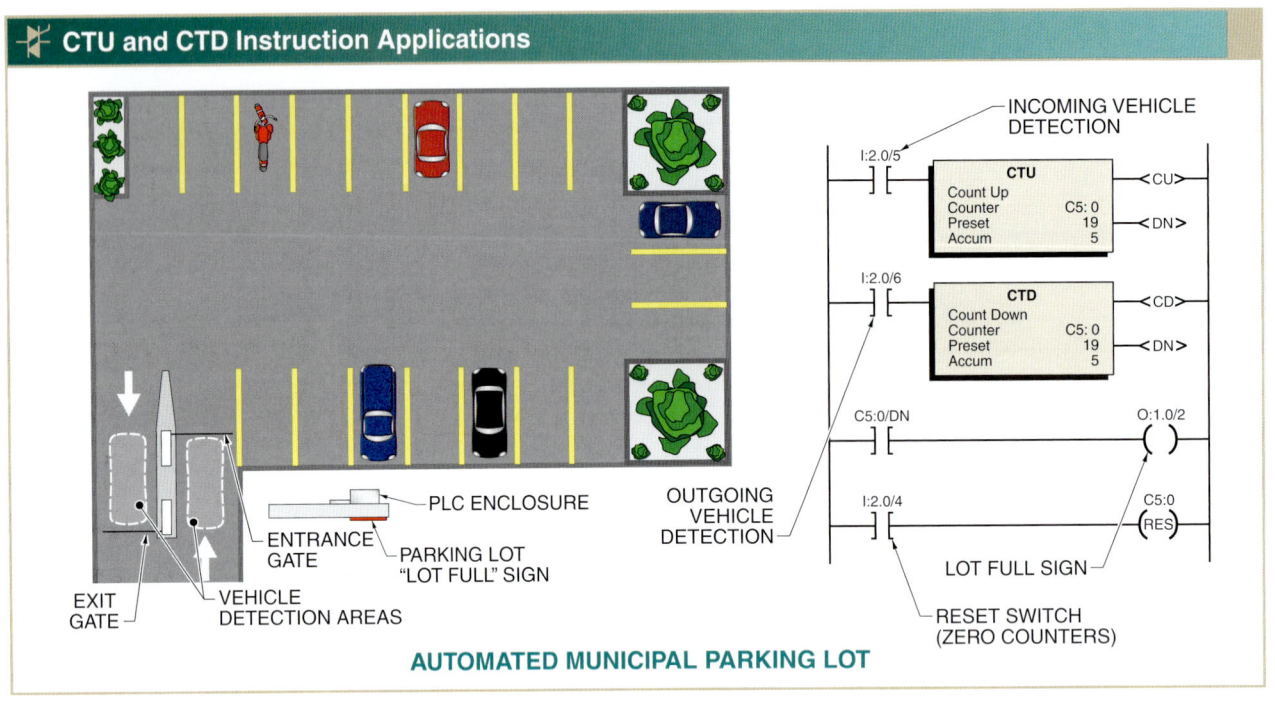

Figure 7-15. CTU and CTD instructions can be used to keep track of the cars entering and exiting a parking lot.

⚠ CAUTION
When designated for a high-speed counter instruction, C0 cannot be used as an address for any other counter instructions.

Typically, PLC OEMs require specific input terminals and specific counters in the PLC's data table to be used with high-speed counters. The number of HSC instructions available varies among PLC models and manufacturers. An OEM's instruction manual should always be consulted regarding the use and operation of HSC instructions. **See Figure 7-16.**

MATH INSTRUCTIONS

A *math instruction* is a PLC programming instruction that is used for a mathematical calculation, such as addition, subtraction, multiplication, and division. Most PLC models have basic math instructions, which are all output instructions. **See Figure 7-17.** Math instructions have two sources and a destination for the result of the calculation. The sources can be a word address or a constant. However, both sources cannot be constants. The word address can be a bit file, integer file, timer preset value, timer accumulated value, a counter preset value, or a counter accumulated value. Advanced math instructions, such as square root, sine, or cosine, can be found on specific PLCs.

High-Speed Counter (HSC) Instructions

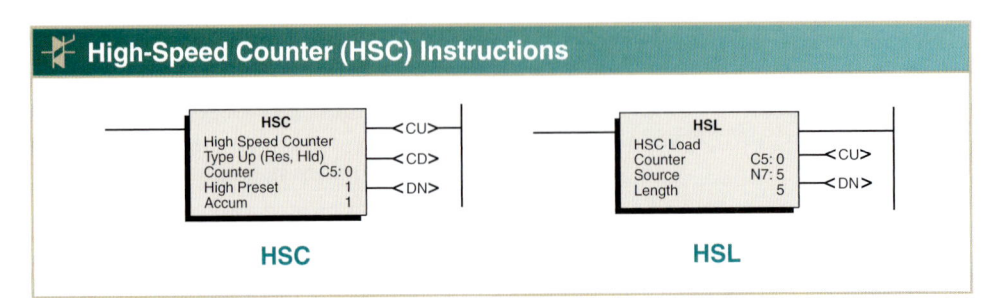

Figure 7-16. HSC instructions are used in conjunction with HSL instructions to detect high-speed counts/events but have specific requirements as shown in data tables provided by OEMs.

Math Instructions

```
     ADD
ADD
Source A    N7:10
Source B    N7:11
Dest        N7:12
```

ADD SOURCE A TO SOURCE B AND
PLACE RESULT IN THE DESTINATION

ADDITION

```
     SUB
SUBTRACT
Source A    C5:1.ACC
Source B    100
Dest        N7:10
```

SUBTRACT SOURCE B FROM SOURCE A
AND PLACE RESULT IN THE DESTINATION

SUBTRACTION

```
     MUL
MULTIPLY
Source A    T4:2.ACC
Source B    2
Dest        N7:12
```

MULTIPLY SOURCE A BY SOURCE B AND
PLACE RESULT IN THE DESTINATION

MULTIPLICATION

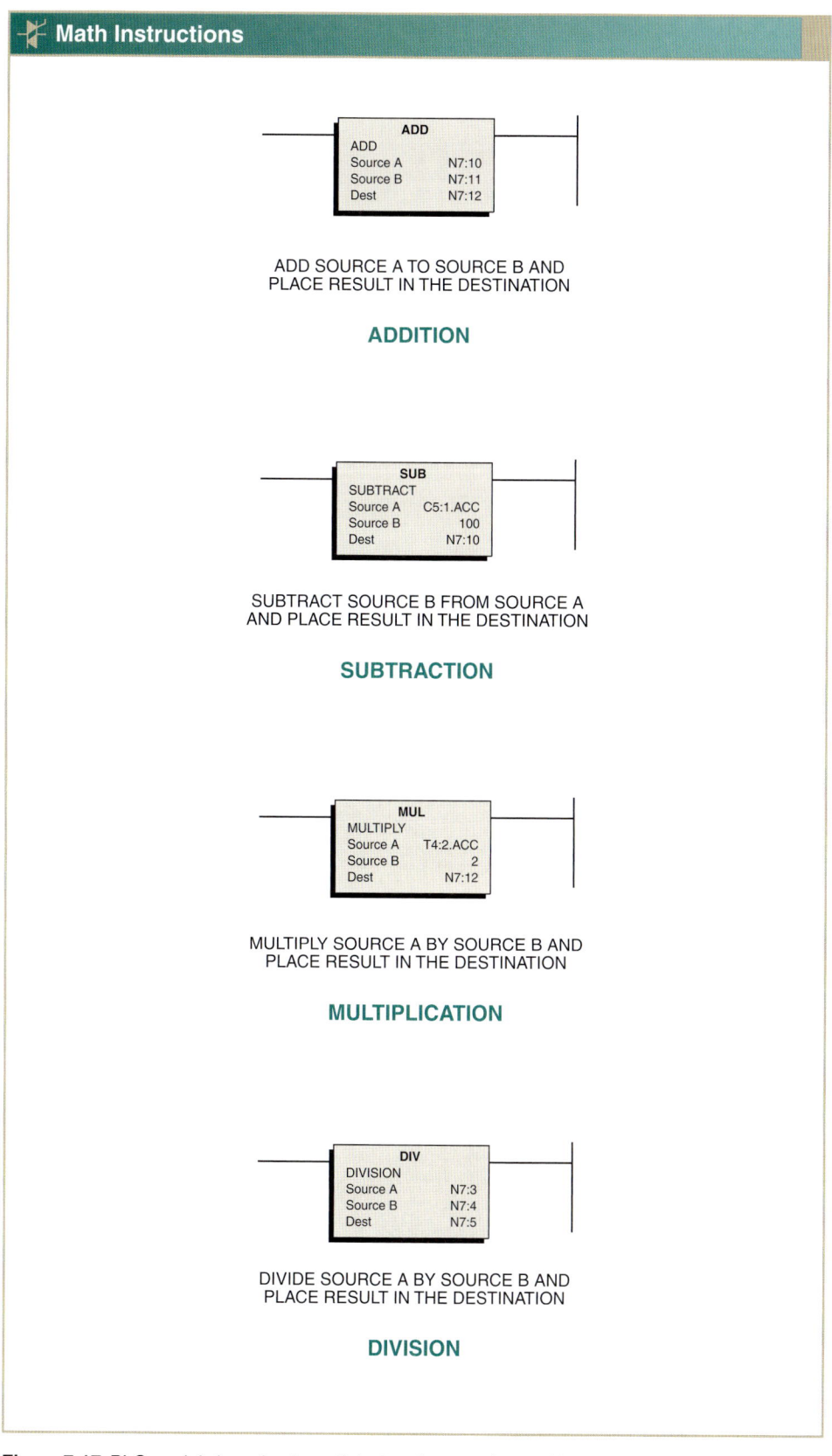

```
     DIV
DIVISION
Source A    N7:3
Source B    N7:4
Dest        N7:5
```

DIVIDE SOURCE A BY SOURCE B AND
PLACE RESULT IN THE DESTINATION

DIVISION

Figure 7-17. PLC models have basic math instructions, such as addition, subtraction, multiplication, and division, which are all output instructions that have two sources and one destination.

Chapter Review 7

Name _____ Date _____

True-False

T F **1.** All timer instructions consist of four words.

T F **2.** RES instructions are used with TOF instructions.

T F **3.** A reset instruction is required to reset an RTO instruction.

T F **4.** Counter instructions are located in data table file 5.

T F **5.** A specific timer address cannot be used for more than one type of timer instruction.

T F **6.** CTU and CTD instructions can be used together but cannot share the same address.

T F **7.** Typically, CTD instructions are always used with CTU instructions.

T F **8.** Bit instructions are used more often than timer and counter instructions.

T F **9.** Timer instructions are located in data table file 6.

T F **10.** A TOF instruction starts timing when the logic preceding the instruction on the rung changes from false to true.

T F **11.** RTO status bits change state during a timing sequence.

T F **12.** Typically, RES instructions are used only with RTO, CTU, or CTD instructions.

T F **13.** A UN bit is true when the accumulated value counts down past −32,768 and wraps around back to 0.

T F **14.** Some PLCs have more than one data table file that can be configured for timer use.

T F **15.** The timer preset value is multiplied by the timer accumulated value to determine a timing duration.

Completion

_____ **1.** A(n) ___ is the number of time base intervals (the length of time) that a timer has been timing.

_____ **2.** A(n) ___ is a PLC programming instruction used to delay the shut down of machinery, such as an external cooling fan when a motor has been stopped.

_____ **3.** A(n) ___ is a type of programmed timer that is continuously timing because the XIO instruction preceding the timer is always true.

293

_____ 4. A(n) ___ is the number of counts a counter is programmed to reach.

_____ 5. A(n) ___ is an instruction label used to indicate the unit of time a timer is using.

_____ 6. A(n) ___ is a PLC programming instruction used to track the length of time a machine has been operating or to shut down a process after an accumulative time period of recurring faults.

_____ 7. A(n) ___ is a PLC programming instruction used to provide timing functions similar to mechanical or solid-state timers.

_____ 8. A(n) ___ is a PLC programming instruction used to count the number of operations or products produced by a system.

_____ 9. A(n) ___ is a PLC programming instruction used to provide counting functions similar to solid-state counters.

_____ 10. A(n) ___ is a programming instruction used to delay the start of a machine or process for a set period of time.

Multiple Choice

_____ 1. Which of the following bits is not used with RTO instructions?
A. DN
B. EN
C. TT
D. UN

_____ 2. Which of the following bits is typically used with TON instructions?
A. CU
B. OV
C. TT
D. UN

_____ 3. Which of the following is not an application that might typically require counter instructions?
A. counting products
B. monitoring the length of time between a person's breaths
C. monitoring the number of cars in a parking lot
D. tracking inventory

_____ 4. Preset and accumulated values for all counters range from ___ to ___.
A. −32,768; 0
B. −32,768; +32,767
C. −32,767; 1
D. 0; +32,767

_____ 5. Which three instructions typically employ the same status bits?
A. CTU, CTD, RES
B. CTU, CTD, RTO
C. TON, TOF, RTO
D. TON, TOF, RES

Sequenced Start of Conveyor Motors

PLC Exercise 7.1

BILL OF MATERIALS
- TECO PLR or Micrologix™ 1100 PLC
- Three motor starters/contactors
- NO pushbutton
- NC pushbutton
- Three motors/loads (optional)

Many processes in the pulp-and-paper, plastics-and-packaging, and food-and-beverage industries have timed sequences when starting conveyor systems. For this application, three motor starters are timed to start 10 seconds apart. Conveyor 2 must start 10 seconds after conveyor 1, and conveyor 3 must start 10 seconds after conveyor 2. Timers are set using a 0.01-second time base. Overloads and the stop switch are hardwired normally closed, and the start switch is hardwired normally open. Do not connect output components in parallel.

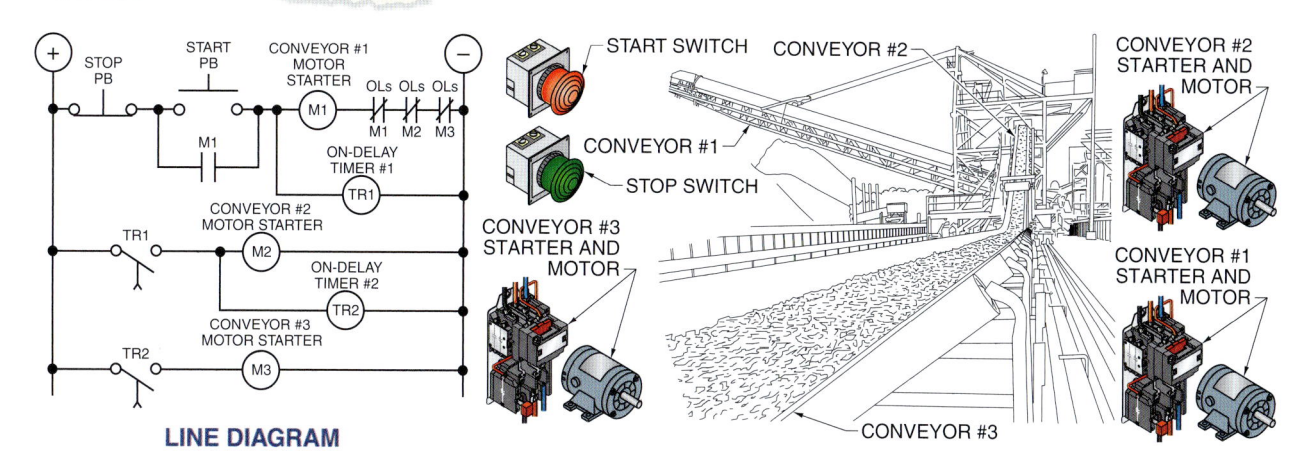

LINE DIAGRAM

T TECO Procedures

1. Create a new ladder logic or FBD program.

2. Use the TECO programming diagram or FBD to add the proper contacts and description in the Programming Grid.

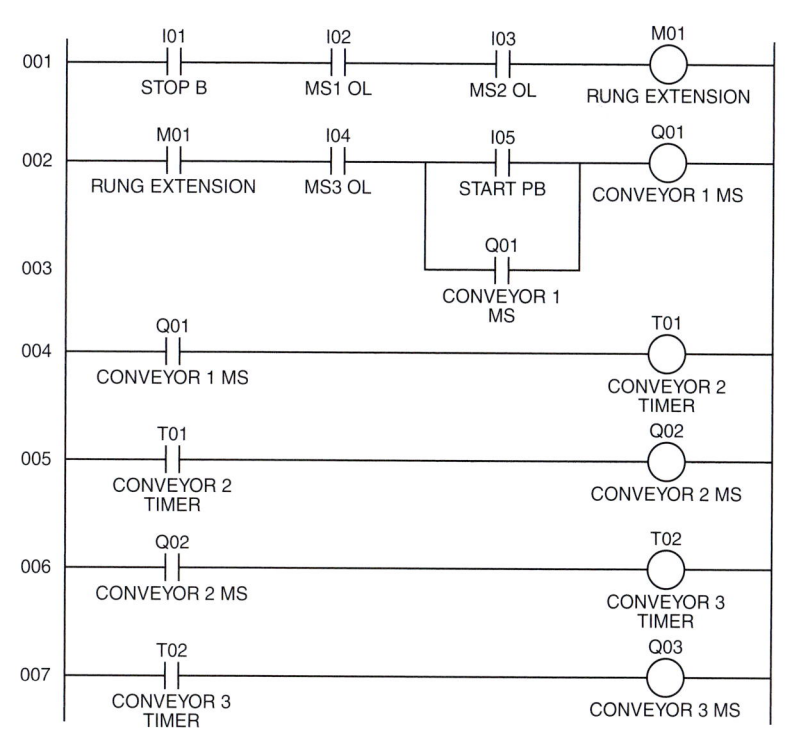

TECO PROGRAMMING DIAGRAM

295

3. Use the Edit Contact window to denote an input or output number and the Symbol button to add a description after placing the input or output in the Programming Grid. For an FBD, use the Comment icon to add the description. Add timers with the proper mode and description to the Ladder Logic Programming Grid or FBD programming area.

4. Run the program in Simulation Mode to verify proper operation. Use the Input Status tool window to activate the inputs and control the output to simulate circuit operation.

5. Use the wiring diagram to wire the input devices and output components to the PLC.

6. Turn power to the PLC ON and link the com port.

7. Write the program to the SG2-12HR-D to download the program logic to the PLC.

8. Place the SG2-12HR-D in Run Mode. Activate the start pushbutton and monitor the I/Os on the LCD display.

9. Stop the program and use the Save As function to save and place the lab application in the required folder location.

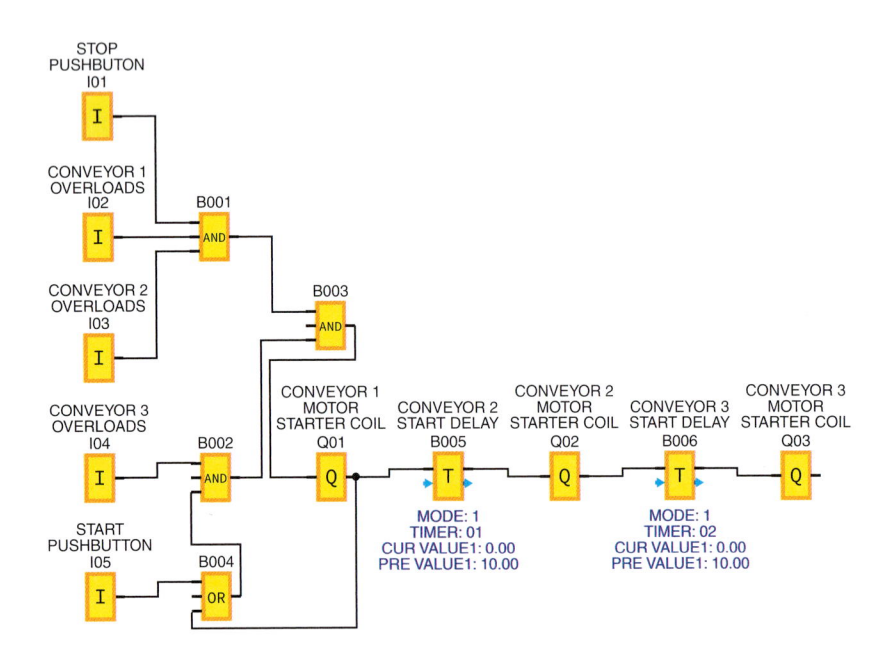

FUNCTION BLOCK DIAGRAM

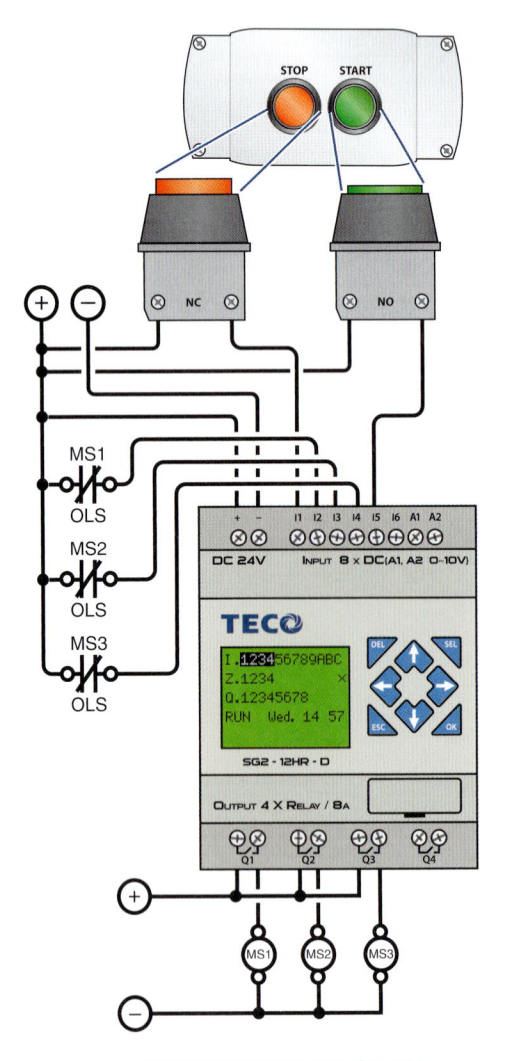

TECO WIRING DIAGRAM

Allen-Bradley® Procedures

1. Open the RSLogix™ Micro Starter Lite program and create a new programming diagram.

2. Use the AB programming diagram to add the proper address and description in the LAD 2 programming window. Add the TON instructions to the programming diagram and type in the timer address, time base, and preset value for each.

3. Verify that the programming diagram has no errors and save the project.

4. Use the wiring diagram to wire the input devices and output components to the PLC. Once all connections have been made, turn power to the PLC on.

5. Use RSLinx™ to establish communication between the PC and the Micrologix™ 1100.

6. Download the program logic to the PLC.

7. Place the PLC in Run Mode.

8. Activate input(s) and monitor them to determine if the output(s) are operating correctly.

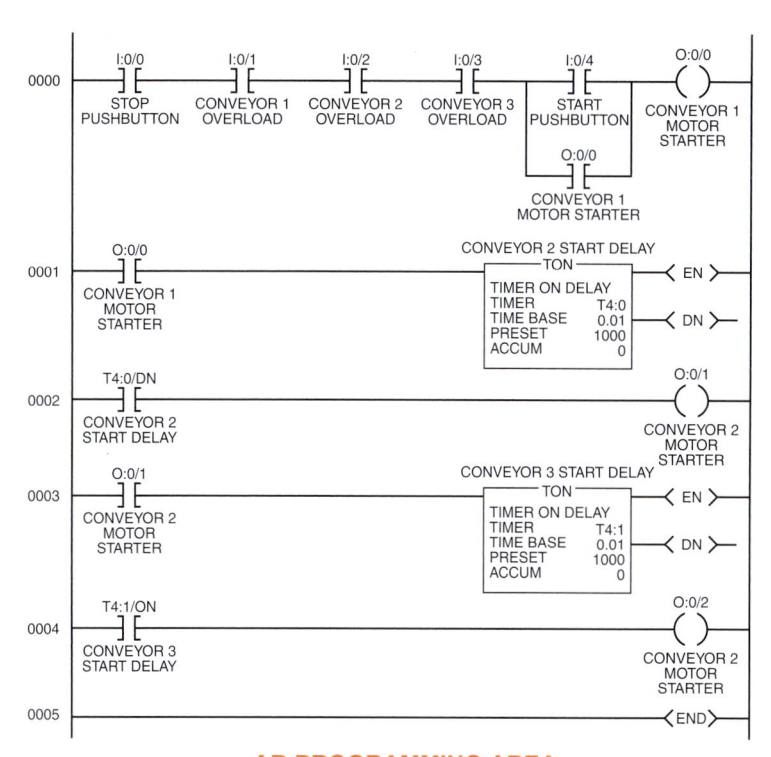

AB PROGRAMMING AREA

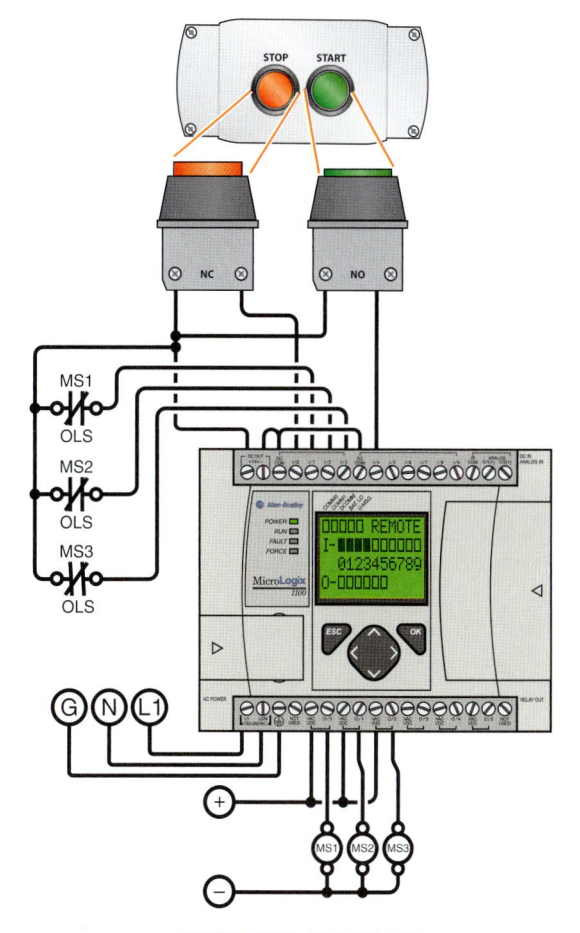

AB WIRING DIAGRAM

Sequenced Stop of Conveyor Motors

PLC Exercise 7.2

BILL OF MATERIALS
- TECO PLR or Micrologix™ 1100 PLC
- Three motor starters/contactors
- NC pushbutton
- NO pushbutton
- Three loads (optional)

In conveyor systems, a timed sequence to stop is common. For this application, three motor starters are timed to stop 5 seconds apart. Conveyor 3 must stop 5 seconds after the stop pushbutton is activated, conveyor 2 must stop 5 seconds after conveyor 3, and conveyor 1 must stop 5 seconds after conveyor 2. Timers are set to using a 0.01-second time base. Overloads and the stop switch are hardwired normally closed, and the start switch is hardwired normally open. Do not connect output components in parallel.

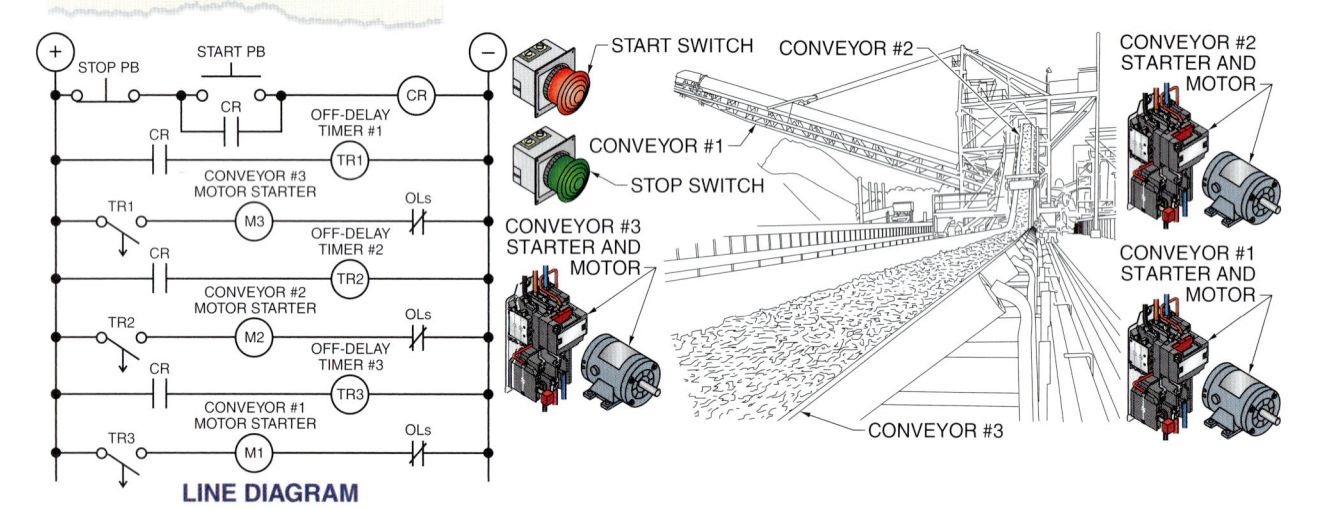

LINE DIAGRAM

T TECO Procedures

1. Create a new ladder logic or FBD program.

2. Use the TECO programming diagram or FBD to add the proper contacts and description in the Programming Grid.

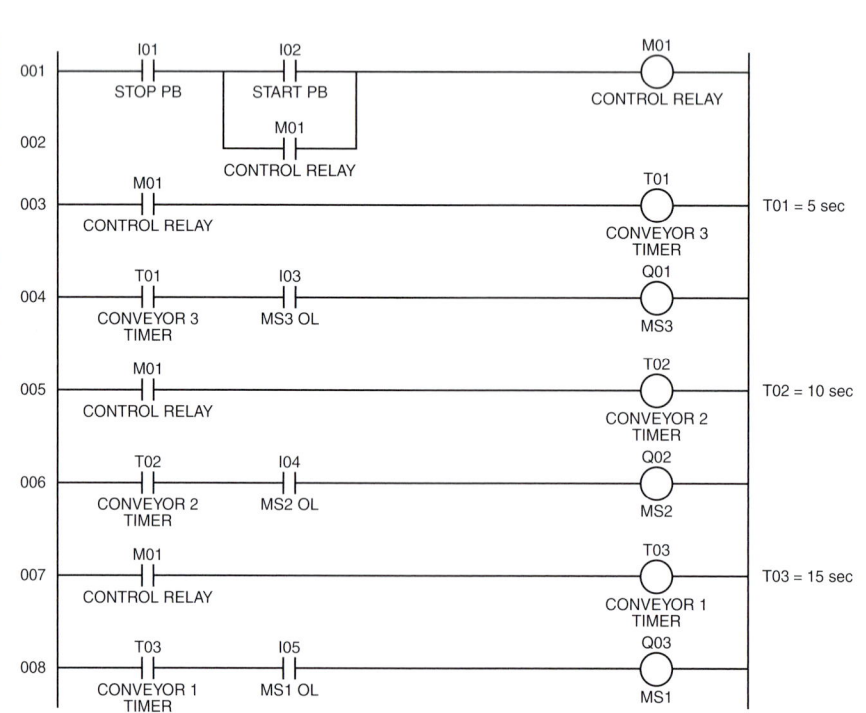

T01 = 5 sec
T02 = 10 sec
T03 = 15 sec

TECO PROGRAMMING DIAGRAM

3. Use the Edit Contact window to denote an input or output number and the Symbol button to add a description after placing the input or output in the Programming Grid. For an FBD, use the Comment icon to add the description. Add timers with the proper mode and description to the Ladder Logic Programming Grid or FBD programming area.

4. Run the program in Simulation Mode to verify proper operation. Use the Input Status tool window to activate the inputs and control the output to simulate circuit operation.

5. Use the wiring diagram to wire the input devices and output components to the PLC.

6. Turn power to the PLC ON and link the com port.

7. Write the program to the SG2-12HR-D to download the program logic to the PLC.

8. Place the SG2-12HR-D in Run Mode. Activate the start pushbutton and monitor the I/Os on the LCD display.

9. Stop the program and use the Save As function to save and place the lab application in the required folder location.

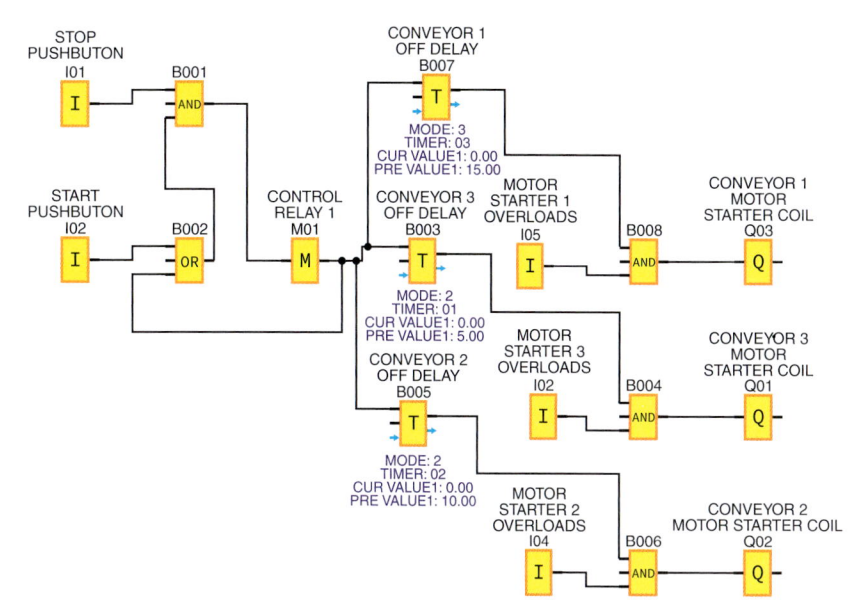

FUNCTION BLOCK DIAGRAM

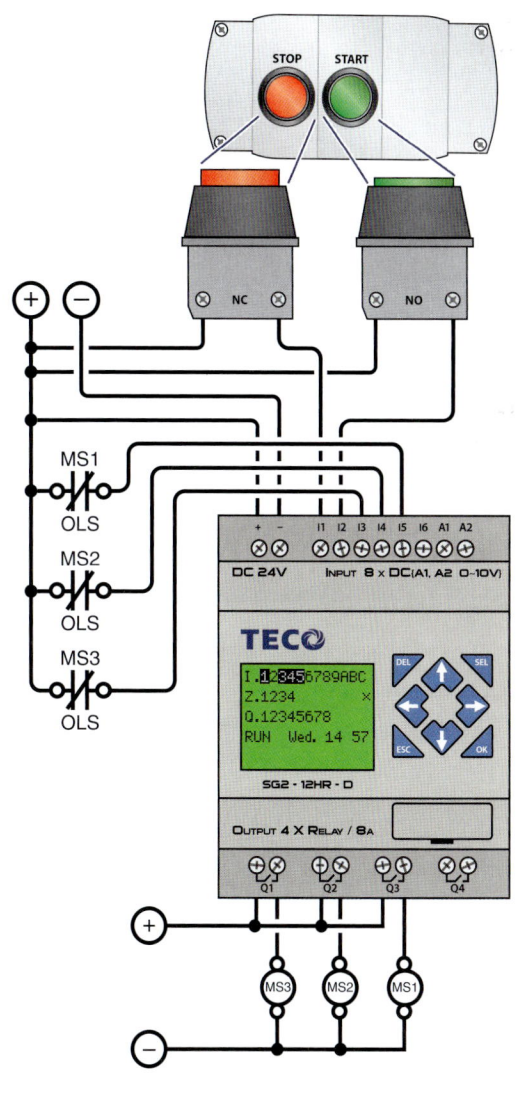

TECO WIRING DIAGRAM

Allen-Bradley® Procedures

1. Open the RSLogix™ Micro Starter Lite program and create a new programming diagram.

2. Use the AB programming diagram to add the proper address and description in the LAD 2 programming window. Add the TOF instructions to the programming diagram and type in the proper timer address, time base, and preset value for each.

3. Verify that the programming diagram has no errors and save the project.

4. Use the wiring diagram to wire the input devices and output components to the PLC. Once all connections have been made, turn power to the PLC on.

5. Use RSLinx™ to establish communication between the PC and the Micrologix™ 1100.

6. Download the program logic to the PLC.

7. Place the PLC in Run Mode.

8. Activate input(s) and monitor them to determine if the output(s) are operating correctly.

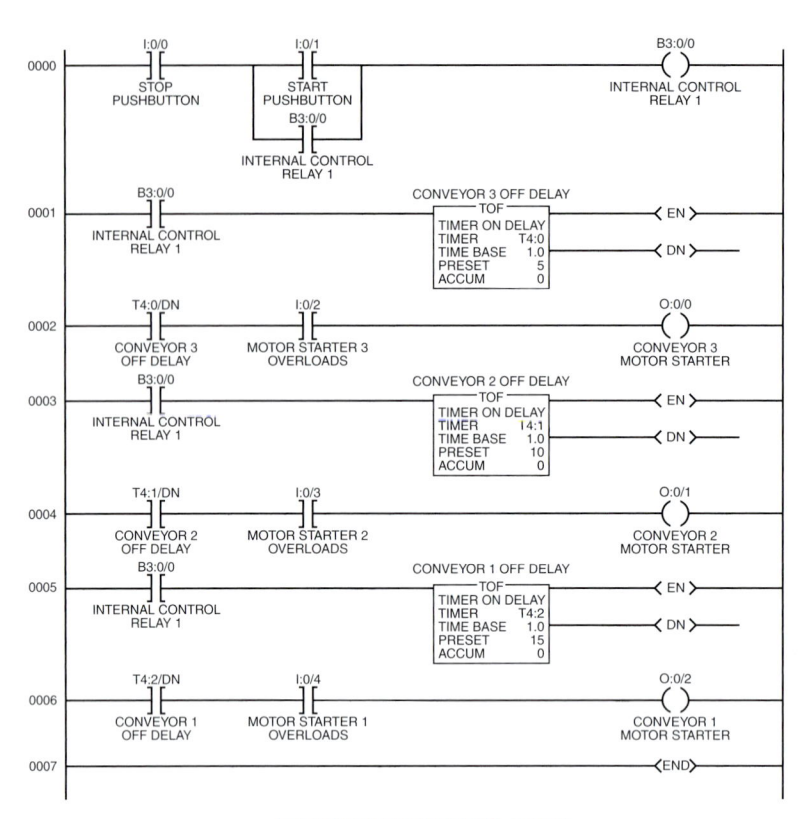

AB PROGRAMMING AREA

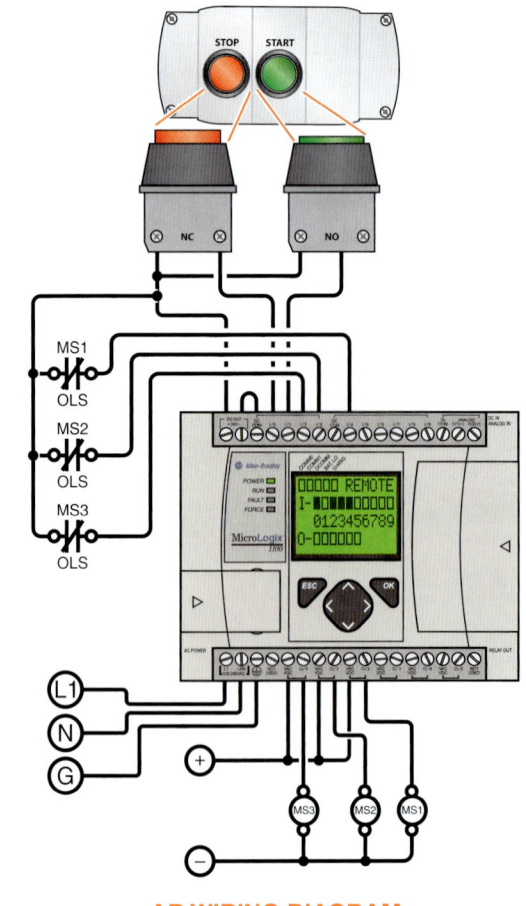

Rung 0000: I:0/0 STOP PUSHBUTTON — I:0/1 START PUSHBUTTON (parallel with B3:0/0 INTERNAL CONTROL RELAY 1) — B3:0/0 INTERNAL CONTROL RELAY 1

Rung 0001: B3:0/0 INTERNAL CONTROL RELAY 1 — CONVEYOR 3 OFF DELAY, TOF TIMER ON DELAY, TIMER T4:0, TIME BASE 1.0, PRESET 5, ACCUM 0 (EN, DN)

Rung 0002: T4:0/DN CONVEYOR 3 OFF DELAY — I:0/2 MOTOR STARTER 3 OVERLOADS — O:0/0 CONVEYOR 3 MOTOR STARTER

Rung 0003: B3:0/0 INTERNAL CONTROL RELAY 1 — CONVEYOR 2 OFF DELAY, TOF TIMER ON DELAY, TIMER T4:1, TIME BASE 1.0, PRESET 10, ACCUM 0 (EN, DN)

Rung 0004: T4:1/DN CONVEYOR 2 OFF DELAY — I:0/3 MOTOR STARTER 2 OVERLOADS — O:0/1 CONVEYOR 2 MOTOR STARTER

Rung 0005: B3:0/0 INTERNAL CONTROL RELAY 1 — CONVEYOR 1 OFF DELAY, TOF TIMER ON DELAY, TIMER T4:2, TIME BASE 1.0, PRESET 15, ACCUM 0 (EN, DN)

Rung 0006: T4:2/DN CONVEYOR 1 OFF DELAY — I:0/4 MOTOR STARTER 1 OVERLOADS — O:0/2 CONVEYOR 1 MOTOR STARTER

Rung 0007: END

AB WIRING DIAGRAM

Counting Items in a Process

PLC Exercise 7.3

BILL OF MATERIALS
- TECO PLR or Micrologix™ 1100 PLC
- Two limit/photoelectric switches
- Light/load

In the food-and-beverage, pharmaceutical, and textile industries, product or material counts are used as parameters in their overall production process. For this application, a limit or photoelectric switch is used to count a product when packaging a case. A counter will activate a control relay circuit when 12 bottles are racked into a carton. Once a carton has reached its limit, it is sent to a palletizer on a conveyor. Then a new carton is placed, and the counter is reset. A light indicator can be used to signal when the count is achieved and then reset.

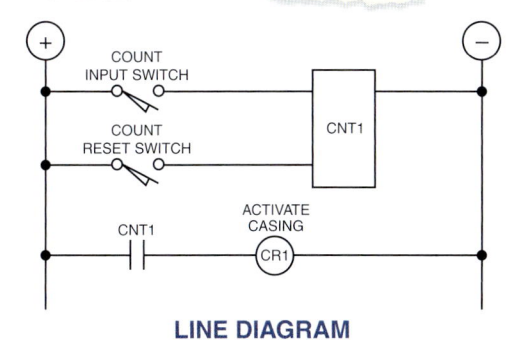

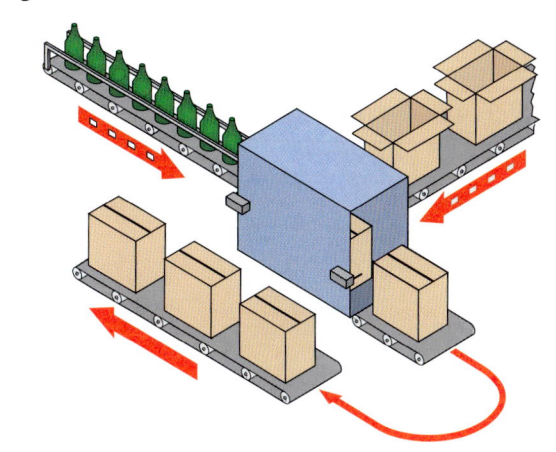

LINE DIAGRAM

TECO PROGRAMMING DIAGRAM

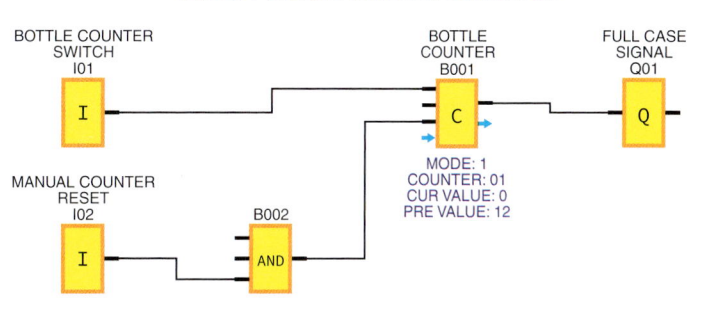

FUNCTION BLOCK DIAGRAM

T TECO Procedures

1. Create a new ladder logic or FBD program.

2. Use the TECO programming diagram or FBD to add the proper contacts and description in the Programming Grid.

3. Use the Edit Contact window to denote an input or output number and the Symbol button to add a description after placing the input or output in the Programming Grid. For an FBD, use the Comment icon to add the description.

4. To add counters to the ladder logic, select the C contact and place in the Programming Grid. An Edit Contact/ Coil window will appear. Use this window to select the

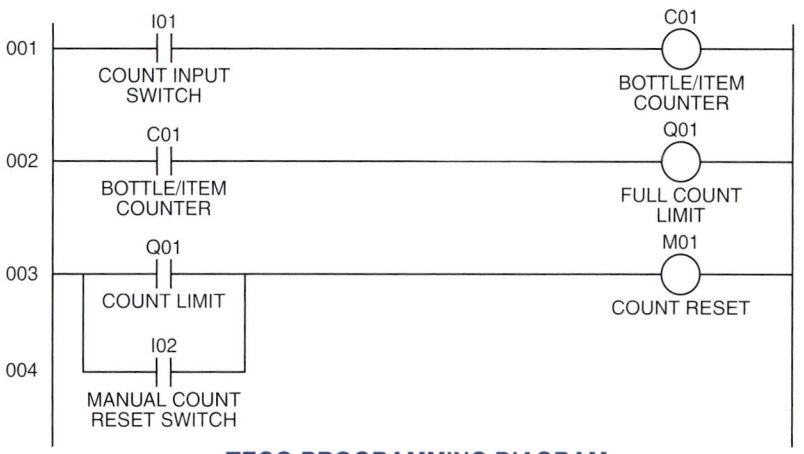

counter number, add a description, and construct the preset value and preset type in this window. To add counters to an FBD, select the Cx Function Block icon and place in the Programming Grid. A Counter Function Block window will appear. Use this window to select a mode, counter number, preset value, and description.

5. Run the program in Simulation Mode to verify proper operation. Use the Input Status tool window to activate the inputs and control the output to simulate circuit operation.

6. Use the wiring diagram to wire the input devices and output components to the PLC.

7. Turn power to the PLC ON and link the com port.

8. Write the program to the SG2-12HR-D to download the program logic to the PLC.

9. Place the SG2-12HR-D in Run Mode. Activate the start pushbutton and monitor the I/Os on the LCD display.

10. Stop the program and use the Save As function to save and place the lab application in the required folder location.

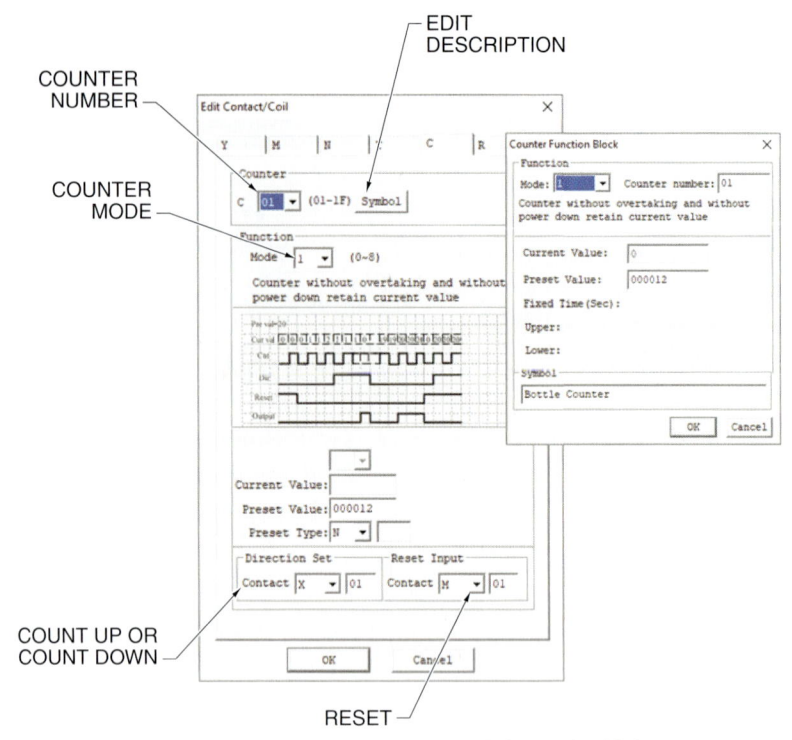

EDIT CONTACT/COIL

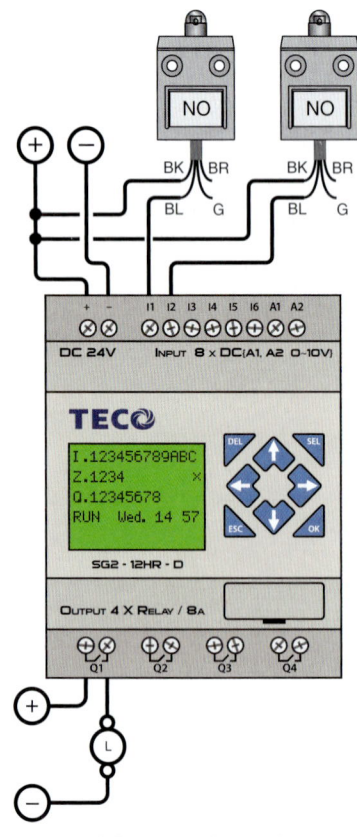

TECO WIRING DIAGRAM

Allen-Bradley® Procedures

1. Open the RSLogix™ Micro Starter Lite program and create a new programming diagram.

2. Use the AB programming diagram to add the proper address and description in the LAD 2 programming window. To add a counter instruction, select the CTU (count up) contact from the Timer/Counter tab in the Instruction toolbar. Add the CTU instruction to the programming diagram and type in the counter address and preset value.

3. Verify that the programming diagram has no errors and save the project.

4. Use the wiring diagram to wire the input devices and output components to the PLC. Once all connections have been made, turn power to the PLC on.

5. Use RSLinx™ to establish communication between the PC and the Micrologix™ 1100.

6. Download the program logic to the PLC.

7. Place the PLC in Run Mode.

8. Activate input(s) and monitor them to determine if the output(s) are operating correctly.

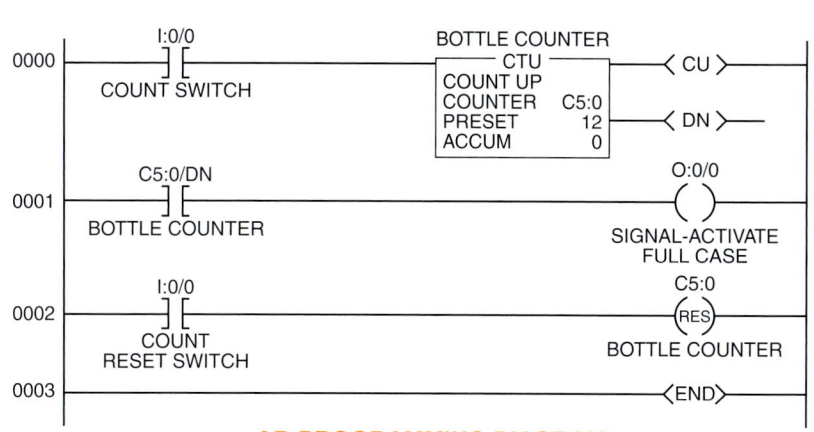

```
0000   I:0/0                          BOTTOM COUNTER
       ─┤ ├─                      ┌─────CTU─────┐      ─⟨ CU ⟩─
       COUNT SWITCH               │ COUNT UP     │
                                  │ COUNTER  C5:0│      ─⟨ DN ⟩─
                                  │ PRESET     12│
                                  │ ACCUM       0│
                                  └──────────────┘

0001   C5:0/DN                                           O:0/0
       ─┤ ├─                                            ─( )─
       BOTTLE COUNTER                                SIGNAL-ACTIVATE
                                                      FULL CASE

0002   I:0/0                                             C5:0
       ─┤ ├─                                            ─(RES)─
       COUNT                                         BOTTLE COUNTER
       RESET SWITCH

0003                                                    ⟨END⟩
```

AB PROGRAMMING DIAGRAM

```
┌─────CTU─────┐      ─⟨ CU ⟩─
│ COUNT UP     │
│ COUNTER  C5:0│      ─⟨ DN ⟩─
│ PRESET    12<│
│ ACCUM      0<│
└──────────────┘
```

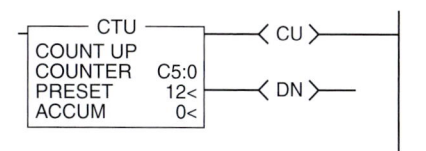

COUNTER INSTRUCTION

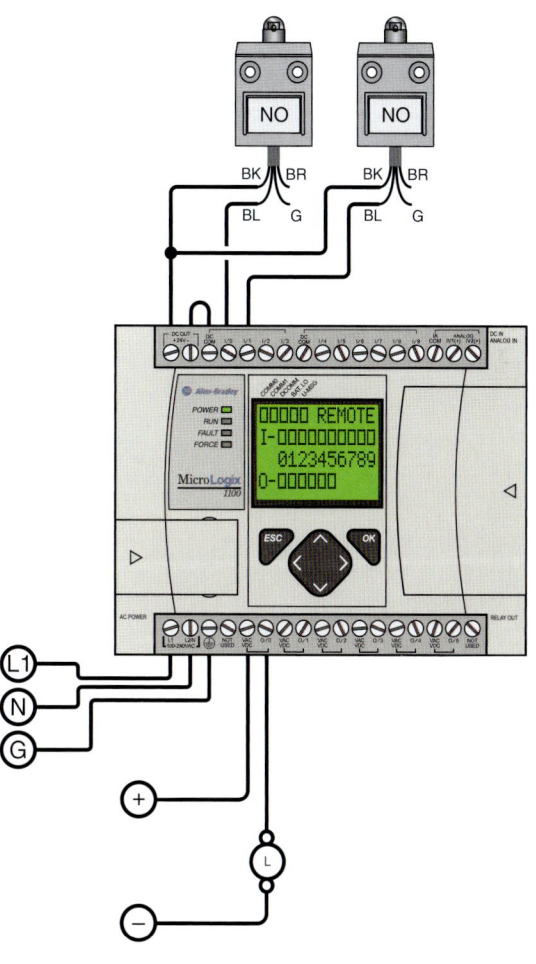

AB WIRING DIAGRAM

Maintaining Count of Material in a Cooler

PLC Exercise 7.4

BILL OF MATERIALS
- TECO PLR or Micrologix™ 1100 PLC
- Two limit switches
- NO pushbutton
- Solenoid/load

In the food-and-beverage industry, some products are counted up and down to control certain devices during a process. For this lab, racks of materials are kept and locked in a cooler. A maximum of 10 racks can be in the cooler at one time. A door-lock solenoid is energized when 10 racks are in the cooler. An in-limit switch counts up as racks enter the cooler, and an out-limit switch counts down as racks exit the cooler. A reset switch is hardwired normally open to manually reset the counter.

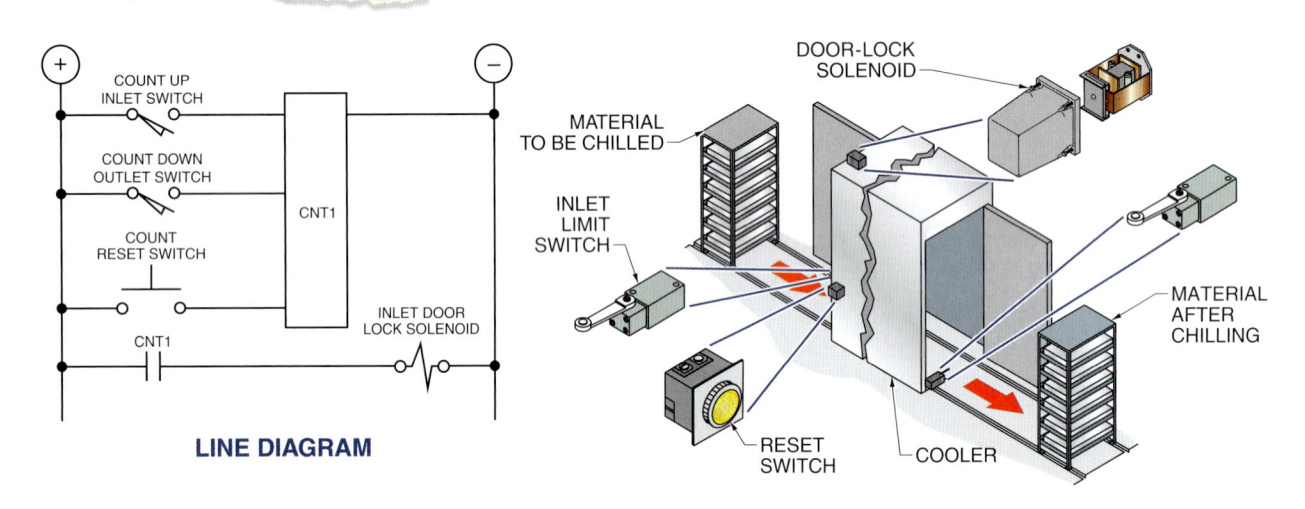

LINE DIAGRAM

TECO Procedures

1. Create a new FBD program.

2. Use the TECO programming diagram or FBD to add the proper contacts and description in the Programming Grid.

3. Use the Edit Contact window to denote an input or output number and the Symbol button to add a description after placing the input or output in the Programming Grid. For an FBD, use the Comment icon to add the description. Add the proper counter instructions.

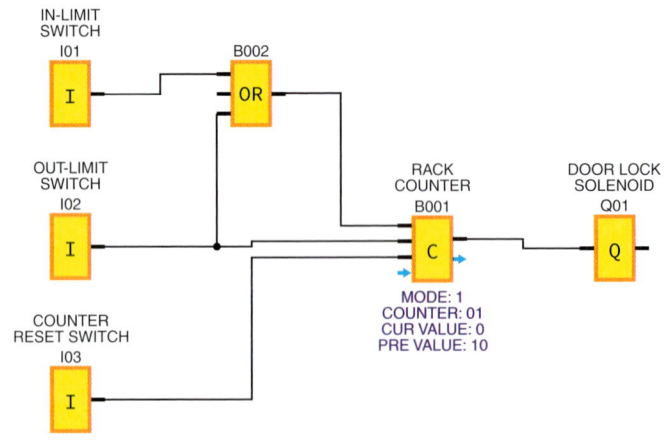

FUNCTION BLOCK DIAGRAM

MODE: 1
COUNTER: 01
CUR VALUE: 0
PRE VALUE: 10

4. Add the proper counters by selecting the C contact and place in the Programming Grid. Add a description and construct the preset value and preset type in this window. Add the proper counter to an FBD by selecting the Cx Function Block icon and place in the Programming Grid. Select a mode, counter number, preset value, and description.

5. Run the program in Simulation Mode to verify proper operation. Use the Input Status tool window to activate the inputs and control the output to simulate circuit operation.

6. Use the wiring diagram to wire the input devices and output components to the PLC.

7. Turn power to the PLC ON and link the com port.

8. Write the program to the SG2-12HR-D to download the program logic to the PLC.

9. Place the SG2-12HR-D in Run Mode. Activate the start pushbutton and monitor the I/Os on the LCD display.

10. Stop the program and use the Save As function to save and place the lab application in the required folder location.

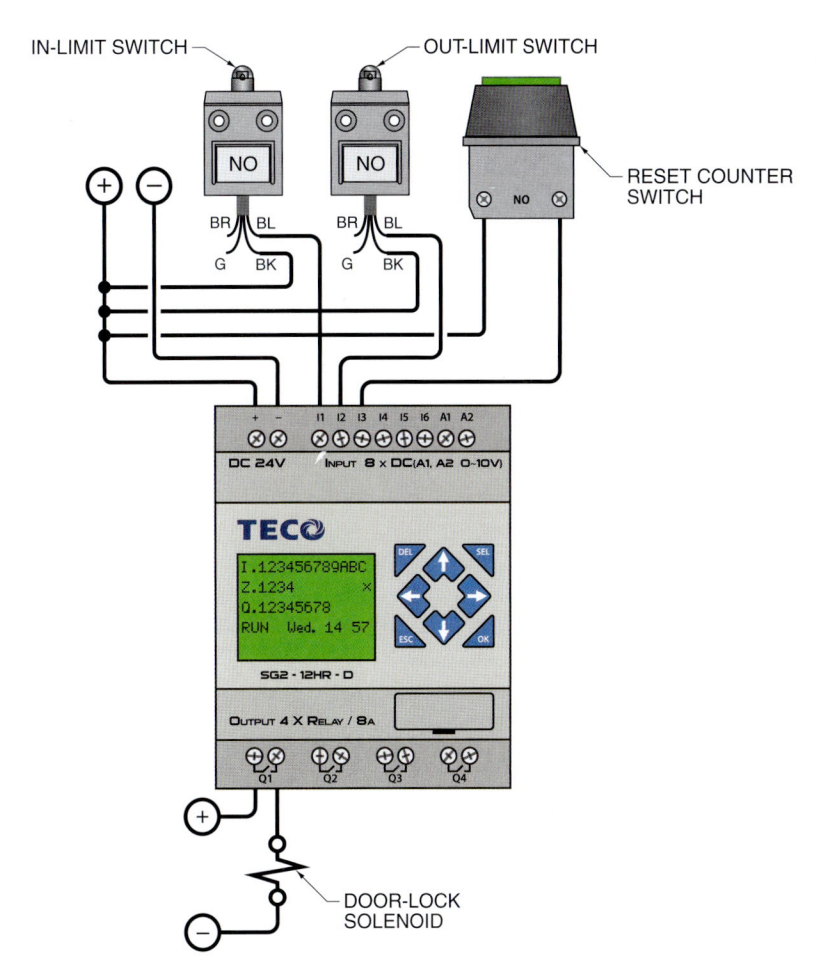

TECO WIRING DIAGRAM

Allen-Bradley® Procedures

1. Open the RSLogix™ Micro Starter Lite program and create a new programming diagram.

2. Use the AB programming diagram to add the proper address and description in the LAD 2 programming window. Add two counter instructions, select the CTU (count up) contact and CTD (count down) contact from the Timer/Counter tab in the Instruction toolbar. Add both instructions to the programming diagram and type in the counter address and preset value for each.

3. Verify that the programming diagram has no errors and save the project.

4. Use the wiring diagram to wire the input devices and output components to the PLC. Once all connections have been made, turn power to the PLC on.

5. Use RSLinx™ to establish communication between the PC and the Micrologix™ 1100.

6. Download the program logic to the PLC.

7. Place the PLC in Run Mode.

8. Activate input(s) and monitor them to determine if the output(s) are operating correctly.

AB PROGRAMMING DIAGRAM

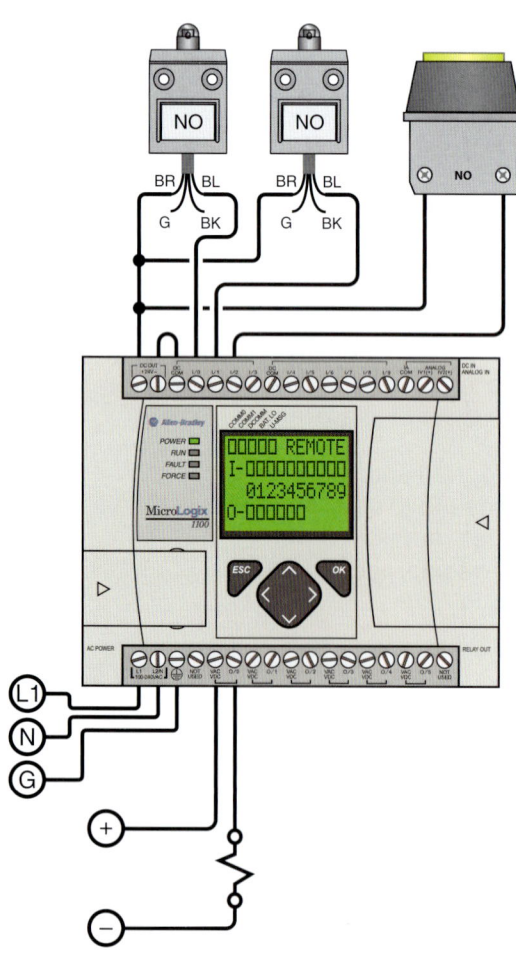

AB WIRING DIAGRAM

Using Counter and Timer Instructions in Pharmaceutical Applications

PLC Exercise 7.5

BILL OF MATERIALS

- TECO PLR or Micrologix™ 1100 PLC
- Selector switch
- Photoelectric switch
- Solenoids/lights

In the pharmaceutical industry, product fill applications are used to fill empty containers with liquid or solid medicines. For this lab, empty containers are filled with liquid medicine in a pharmaceutical plant. The empty containers come down a conveyor from a sterilizing process. Each container is filled and sent to the capping, labeling, and inspection stations. Along the conveyor, an infeed gate solenoid holds empty containers before moving them into the filling station. An outfeed gate solenoid holds four empty bottles in the filling station. A fill cylinder solenoid controls a set of hollow core cylinders that are lowered into and out of the containers. And a product dispense solenoid is then used to control the amount of liquid medicine filled into each container. Note: *Use the timing sequence table for the timing of energizing each solenoid.*

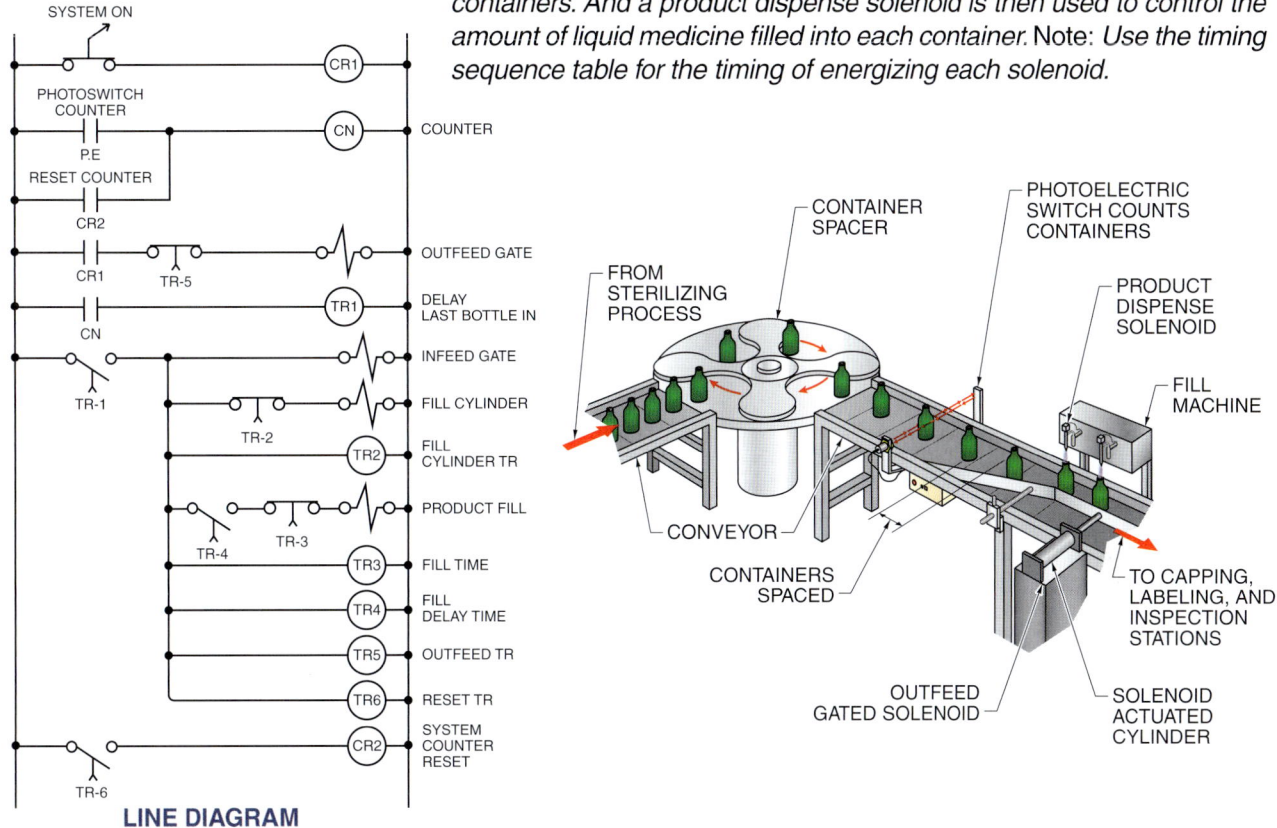

LINE DIAGRAM

Timing Sequence Table

Time in Seconds	1	2	3	4	5	6	7	8	9	10	11	12	13	14	15	16	17	18	19	20	21	22	23	24
Output gate						F															T			
Infeed gate	F						T																	
Fill cylinders	F						T									F								
Product fill	F							T							F									
Reset relay						F																		

*F = False; T = True

TIMING STARTS WHEN COUNTER IS TRUE

RESET RELAY TRUE FOR A SCAN; RESETS COUNTER AND TIMERS; ALL OUTPUTS NOW FALSE

TECO Procedures

1. Create a new ladder logic program.

2. Use the TECO programming diagram to add the proper contacts and description in the Programming Grid.

3. Use the Edit Contact window to denote an input or output number and the Symbol button to add a description after placing the input or output in the Programming Grid.

4. Add a counter with the proper mode, counter number, preset value, and description. Add timers with the proper mode and description to the Programming Grid.

5. Run the program in Simulation Mode to verify proper operation. Use the Input Status tool window to activate the inputs and control the outputs to simulate circuit operation.

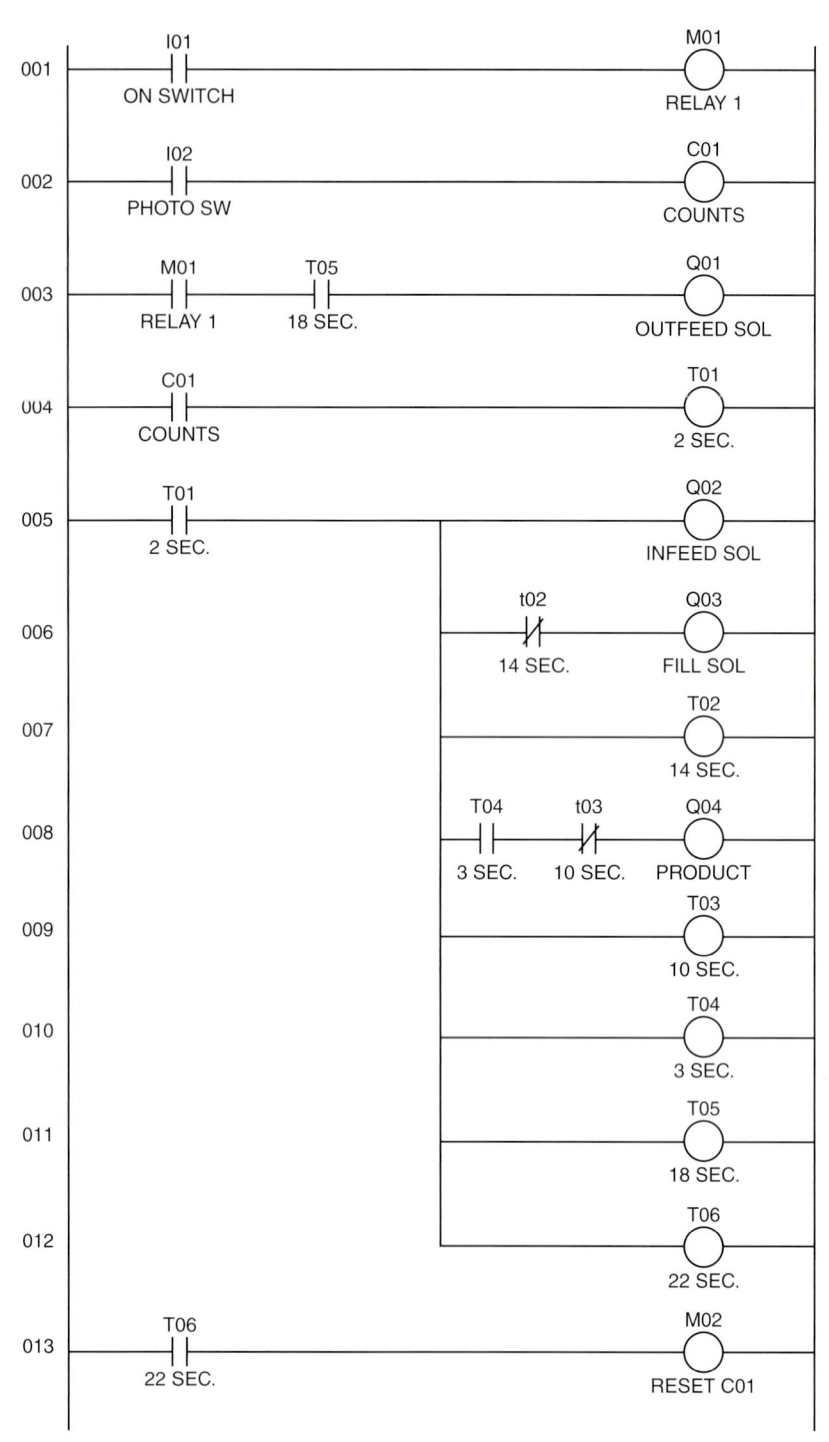

Rung	Ladder
001	I01 ON SWITCH —— M01 RELAY 1
002	I02 PHOTO SW —— C01 COUNTS
003	M01 RELAY 1 — T05 18 SEC. —— Q01 OUTFEED SOL
004	C01 COUNTS —— T01 2 SEC.
005	T01 2 SEC. —— Q02 INFEED SOL
006	t02 14 SEC. (NC) —— Q03 FILL SOL
007	—— T02 14 SEC.
008	T04 3 SEC. — t03 10 SEC. (NC) —— Q04 PRODUCT
009	—— T03 10 SEC.
010	—— T04 3 SEC.
011	—— T05 18 SEC.
012	—— T06 22 SEC.
013	T06 22 SEC. —— M02 RESET C01

TECO PROGRAMMING DIAGRAM

Chapter 7—PLC Programming Timer and Counter Instructions **309**

6. Use the wiring diagram to wire the input devices and output components to the PLC.

7. Turn power to the PLC ON and link the com port.

8. Write the program to the SG2-12HR-D to download the program logic to the PLC.

9. Place the SG2-12HR-D in Run Mode. Activate the selector switch and monitor the I/Os on the LCD display.

10. Stop the program and use the Save As function to save and place the lab application in the required folder location.

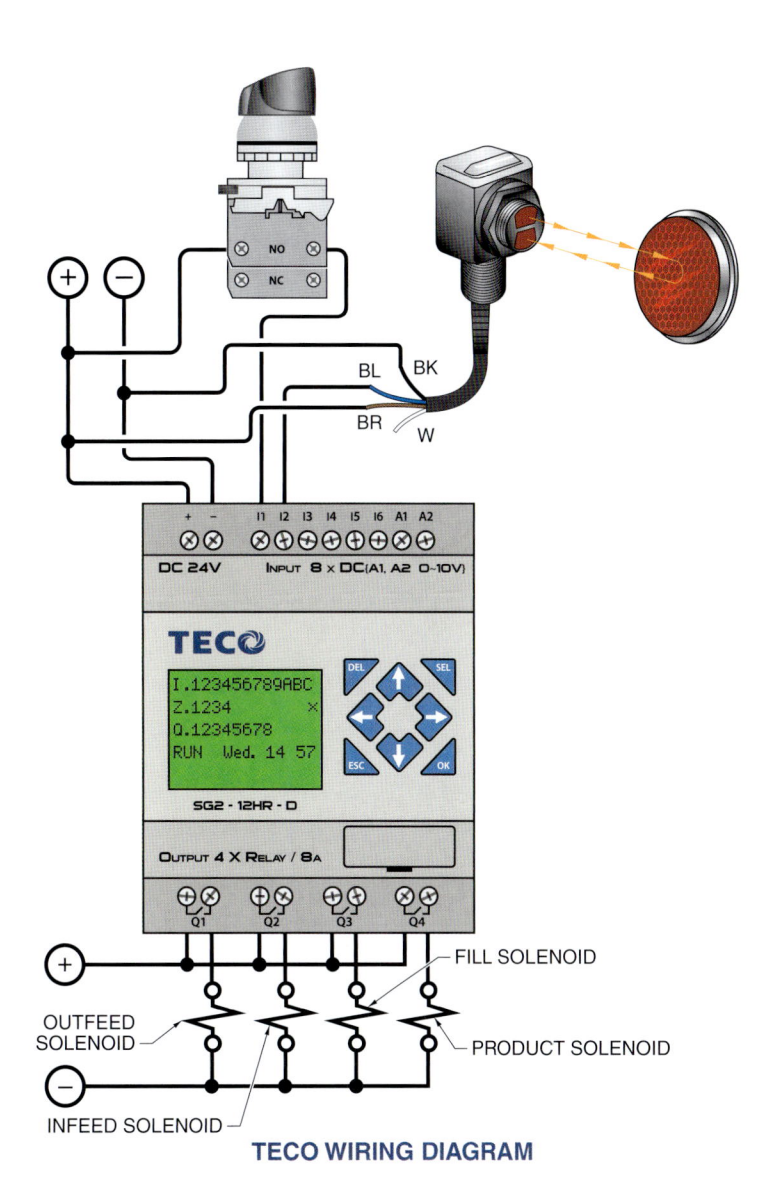

TECO WIRING DIAGRAM

Allen-Bradley® Procedures

1. Open the RSLogix™ Micro Starter Lite program and create a new programming diagram.

2. Use the AB programming diagram to add the proper addresses and descriptions in the LAD 2 programming window. Add the CTU (count up) instruction with the counter address and preset value. Add the TON instructions to the programming diagram and type in the proper timer address, time base, and preset value for each.

```
0000   ON SWITCH          RELAY 1
       I:0/0              B3:0/0
       ─] [─              ─( )─

0001   PHOTO SW                    PRODUCT COUNT
       I:0/1                       ──── CTU ────        ─⟨ CU ⟩─
       ─] [─                       COUNT UP
                                   COUNTER      C5:0
                                   PRESET         5<    ─⟨ DN ⟩─
                                   ACCUM          1<

0002   RELAY 1    OUTFEED TIMER                         OUTFEED SOLENOID
       B3:0/0     T4:4/DN                               O:0/0
       ─] [─      ─] [─                                 ─( )─

0003   COUNT DONE                  DELAY LAST PRODUCT IN
       C5:0/DN                     ──── TON ────        ─⟨ EN ⟩─
       ─] [─                       TIMER ON DELAY
                                   TIMER        T4:0
                                   TIME BASE     1.0    ─⟨ DN ⟩─
                                   PRESET         2<
                                   ACCUM          0<

0004   DELAY LAST                                       INFEED SOLENOID
       PRODUCT IN                                       O:0/1
       T4:0/DN                                          ─( )─
       ─] [─
                   FILL CYLINDER TIMER    FILL SOLENOID
                   T4:1/DN                O:0/2
                   ─]/[─                  ─( )─

                              FILL CYLINDER TIMER
                              ──── TON ────        ─⟨ EN ⟩─
                              TIMER ON DELAY
                              TIMER        T4:1
                              TIME BASE     1.0    ─⟨ DN ⟩─
                              PRESET        14<
                              ACCUM          0<

                   FILL TIME
                   DELAY       FILL TIME           PRODUCT FILL
                   T4:3/DN     T4:2/DN             O:0/3
                   ─] [─       ─]/[─               ─( )─

                              FILL TIME
                              ──── TON ────        ─⟨ EN ⟩─
                              TIMER ON DELAY
                              TIMER        T4:2
                              TIME BASE     1.0    ─⟨ DN ⟩─
                              PRESET        10<
                              ACCUM          0<

                              FILL TIME DELAY
                              ──── TON ────        ─⟨ EN ⟩─
                              TIMER ON DELAY
                              TIMER        T4:3
                              TIME BASE     1.0    ─⟨ DN ⟩─
                              PRESET         3<
                              ACCUM          0<

                              OUTFEED TIMER
                              ──── TON ────        ─⟨ EN ⟩─
                              TIMER ON DELAY
                              TIMER        T4:4
                              TIME BASE     1.0    ─⟨ DN ⟩─
                              PRESET        18<
                              ACCUM          0<

                              RESET TIMER
                              ──── TON ────        ─⟨ EN ⟩─
                              TIMER ON DELAY
                              TIMER        T4:5
                              TIME BASE     1.0    ─⟨ DN ⟩─
                              PRESET        22<
                              ACCUM          0<

0005   RESET TIMER DN                              PRODUCT COUNT
       T4:5/DN                                     C5:0
       ─] [─                                       ─(RES)─

0006                                               ─⟨END⟩─
```

AB PROGRAMMING DIAGRAM

3. Verify that the programming diagram has no errors and save the project.

4. Use the wiring diagram to wire the input devices and output components to the PLC. Once all connections have been made, turn power to the PLC on.

5. Use RSLinx™ to establish communication between the PC and the Micrologix™ 1100.

6. Download the program logic to the PLC.

7. Place the PLC in Run Mode.

8. Activate input(s) and monitor them to determine if the output(s) are operating correctly.

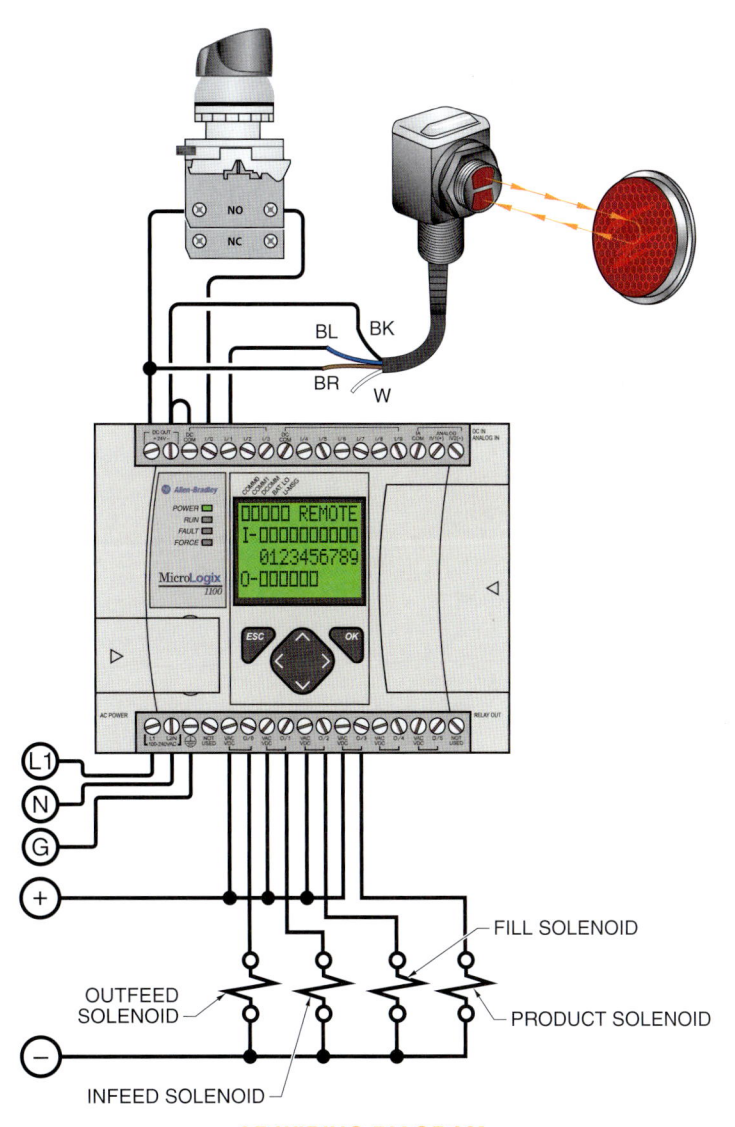

AB WIRING DIAGRAM

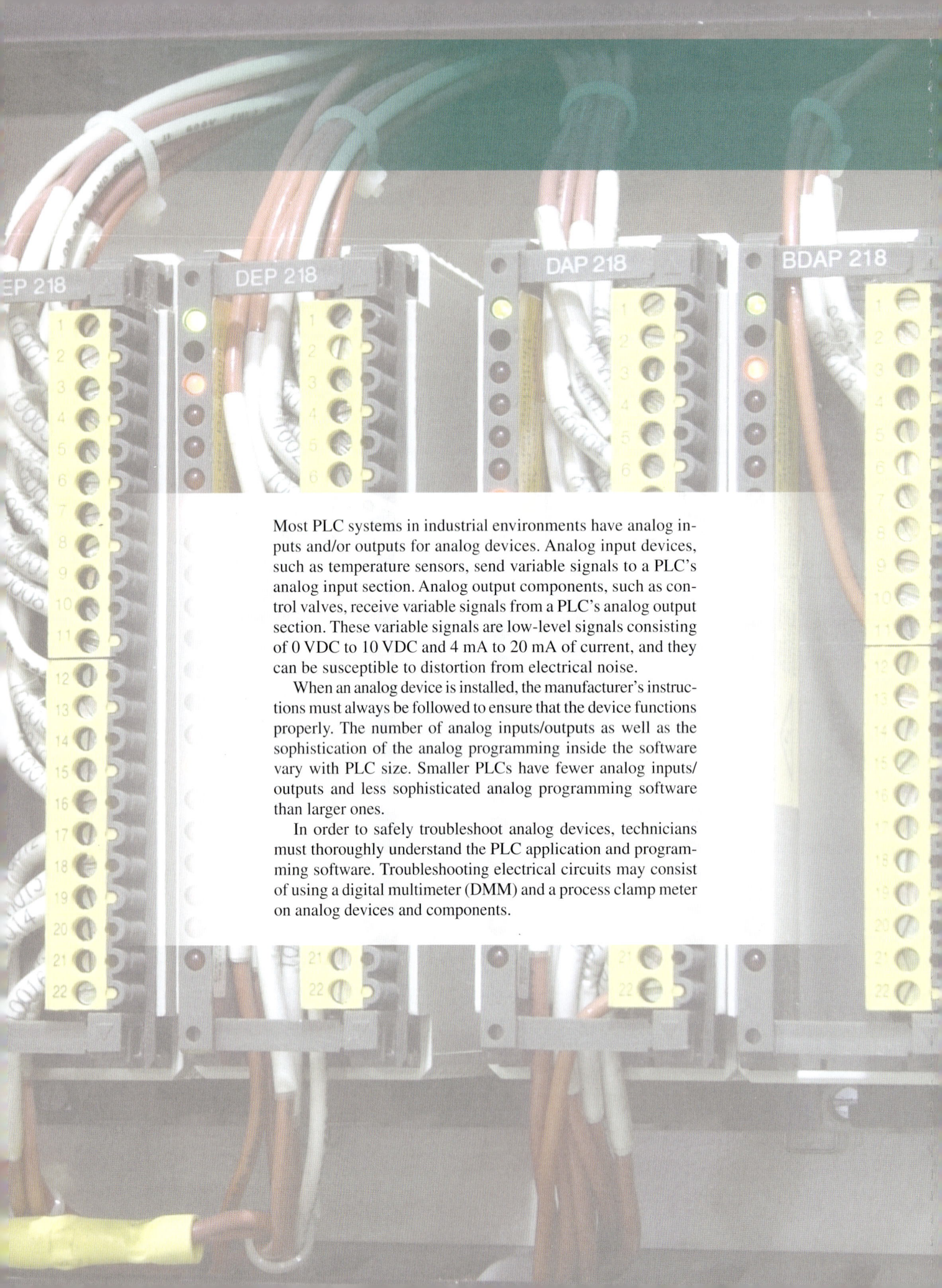

Most PLC systems in industrial environments have analog inputs and/or outputs for analog devices. Analog input devices, such as temperature sensors, send variable signals to a PLC's analog input section. Analog output components, such as control valves, receive variable signals from a PLC's analog output section. These variable signals are low-level signals consisting of 0 VDC to 10 VDC and 4 mA to 20 mA of current, and they can be susceptible to distortion from electrical noise.

When an analog device is installed, the manufacturer's instructions must always be followed to ensure that the device functions properly. The number of analog inputs/outputs as well as the sophistication of the analog programming inside the software vary with PLC size. Smaller PLCs have fewer analog inputs/outputs and less sophisticated analog programming software than larger ones.

In order to safely troubleshoot analog devices, technicians must thoroughly understand the PLC application and programming software. Troubleshooting electrical circuits may consist of using a digital multimeter (DMM) and a process clamp meter on analog devices and components.

PLC Analog Device Installation, Programming, and Troubleshooting

8

Objectives:
- Describe the applications and installation considerations of analog input devices (sensors).
- Describe the applications and installation considerations of analog output devices (actuators).
- Program a PLC for analog inputs and outputs using associated safety protocols.
- List the troubleshooting procedures required for PLC analog input devices, PLC and sensor wiring, and PLC input terminals.
- List the troubleshooting procedures required for PLC analog output devices, PLC and actuator wiring, and PLC output terminals.

Learner Resources
atplearningresources.com/quicklinks
Access Code: 475203

PLC ANALOG INPUT DEVICE AND OUTPUT DEVICE CIRCUIT WIRING

PLC installations require that analog input devices (sensors) be connected to the input terminals of a PLC. The sensors are used to measure temperature, pressure, flow, level, speed, conductivity, resistance, and other conditions of a process or environment. The sensors provide information to the PLC on how an application or process is operating. For example, a PLC controls the level of liquid in a tank by varying the position of the valve based on information from a level sensor. **See Figure 8-1.**

Analog Input Devices

Analog input devices are used in a wide variety of applications. The applications include manufacturing, commercial businesses, office buildings, wastewater treatment, irrigation, and clean/green energy sources. In manufacturing, sensors are used to measure things such as the flow of a product through a pipe or the level of a liquid in a tank. In commercial businesses and office buildings, sensors are used to measure the flow of natural gas to boilers or furnaces of HVAC systems and to control lighting. In a wastewater treatment plant, sensors measure the conductivity or the pH level of treated water. In irrigation systems such as those used for farming or on golf courses, sensors are used to measure the moisture content of the soil. When using green energy sources such as wind turbines, sensors measure the speed and direction of the wind. **See Figure 8-2.**

The analog sensor sends a variable signal that is proportional to the physical property being measured. Analog sensors can only be connected to analog input terminals on a PLC. An analog-to-digital (A-to-D) converter inside the PLC converts the current or voltage value to a number. This number represents the physical property being measured and varies as the physical property varies. This number is then manipulated or processed by the CPU of the PLC according to the program being run.

PLC TIPS

Internal or external analog-to-digital converters, also known as ADCs, typically output a complement binary number of two. Some ADCs might output a number in Gray code. Most ADCs are of the linear type, which means that the input values have a direct proportional relationship to the output number. A few ADCs are of the nonlinear type, which means that the input signal is a curve rather than a straight line.

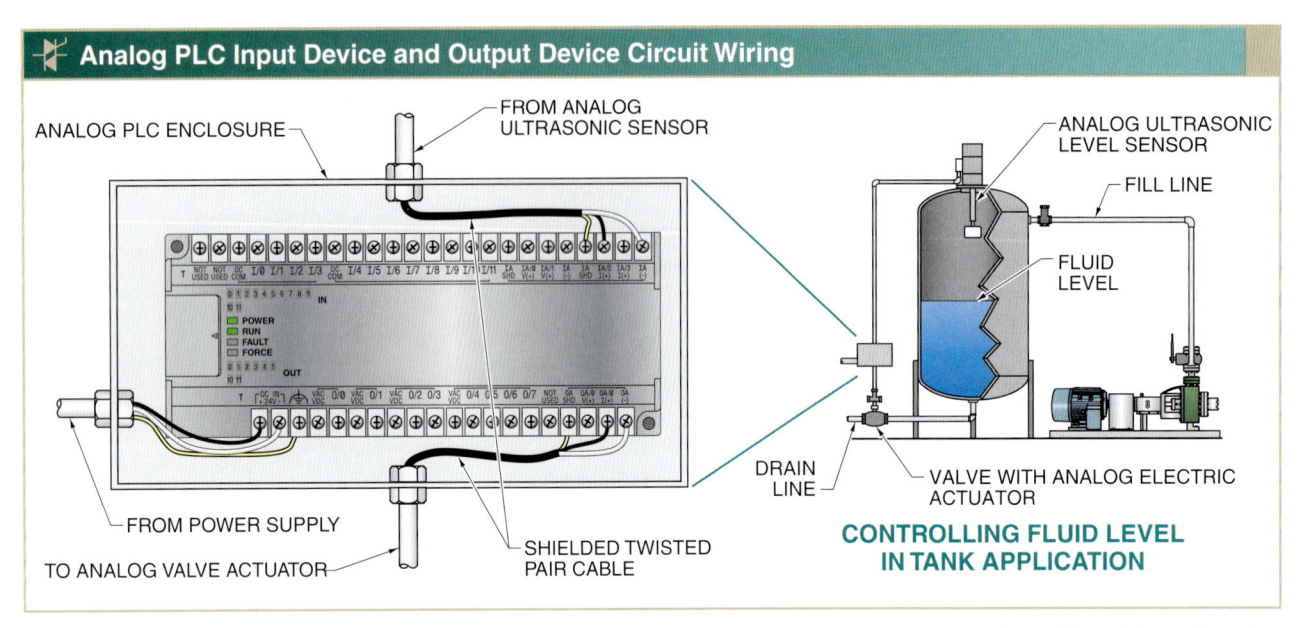

314 PROGRAMMABLE LOGIC CONTROLLERS: PRINCIPLES AND APPLICATIONS

Figure 8-1. When controlled by a PLC, a level sensor can maintain a specific level in a tank by varying the position of the outlet valve.

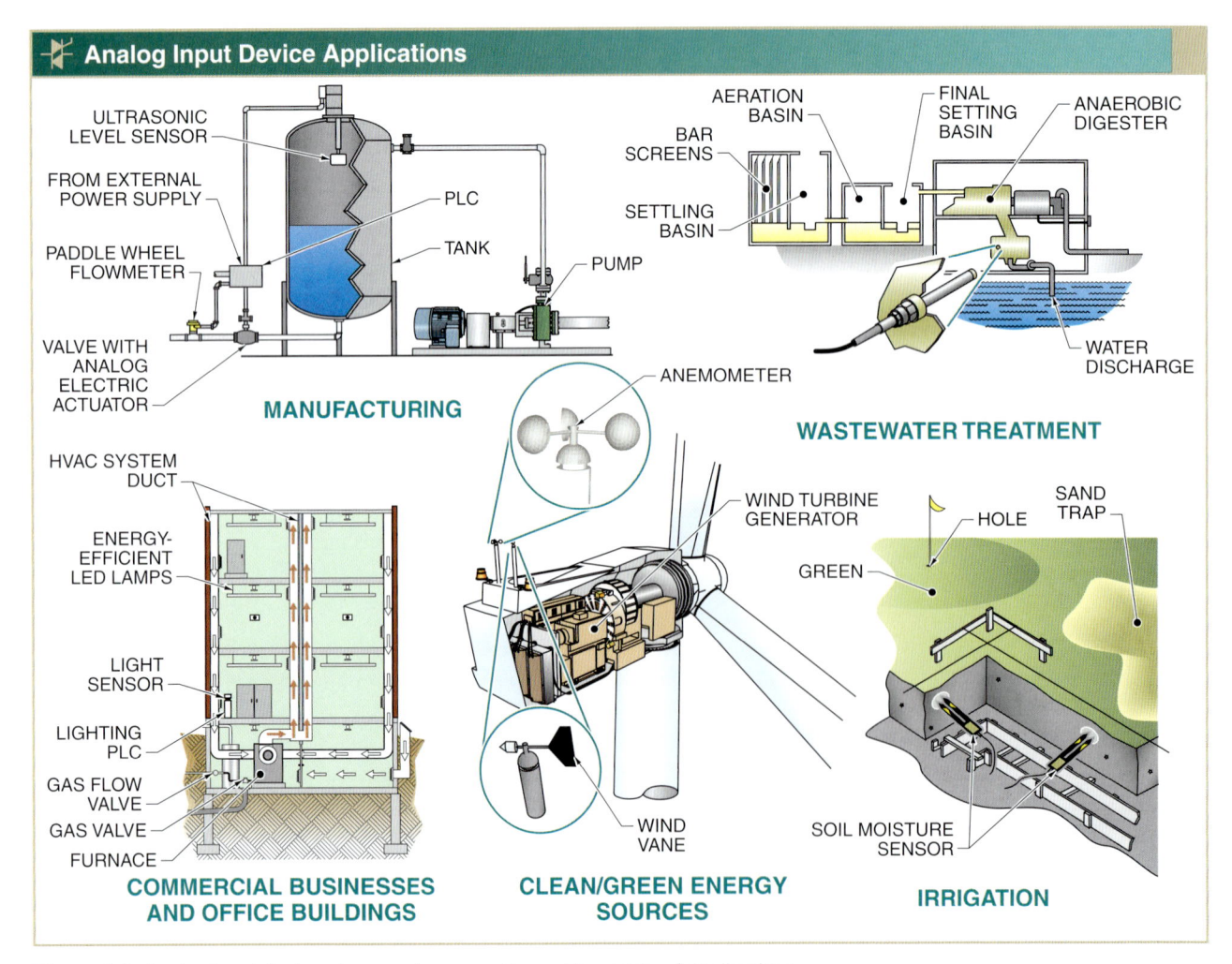

Figure 8-2. Analog input devices (sensors) are used in a wide variety of applications.

Analog Input Device Signals

PLCs with analog inputs and outputs are found in many different applications. A unique combination of applications is found when a PLC is used in a clean room to control an electric motor drive (EMD) that powers the room's fan. Clean rooms are special ultra-sterile rooms used in the manufacture of semiconductors and to support other activities. Clean room occupants must wear special clothing to prevent the introduction of contaminants. Also, clean rooms are under positive air pressure to keep contaminants out. The EMD operates the room's fan by receiving a varying amount of voltage to control the speed of the fan's motor.

For example, a sensor or series of sensors are located inside a clean room to monitor the air pressure in the room. The sensors provide an analog signal that is proportional to the air pressure. The output of the pressure sensor is connected to an analog input terminal of the PLC. The output signal is either 0 VDC to 10 VDC or 4 mA to 20 mA. Based on the PLC program, this input signal results in a specific output control signal being sent from the PLC analog output section to the EMD. The PLC analog output signal is typically 0 VDC to 10 VDC or 4 mA to 20 mA. The signal allows the EMD to operate the fan at the speed required to maintain positive air pressure in the room. **See Figure 8-3.**

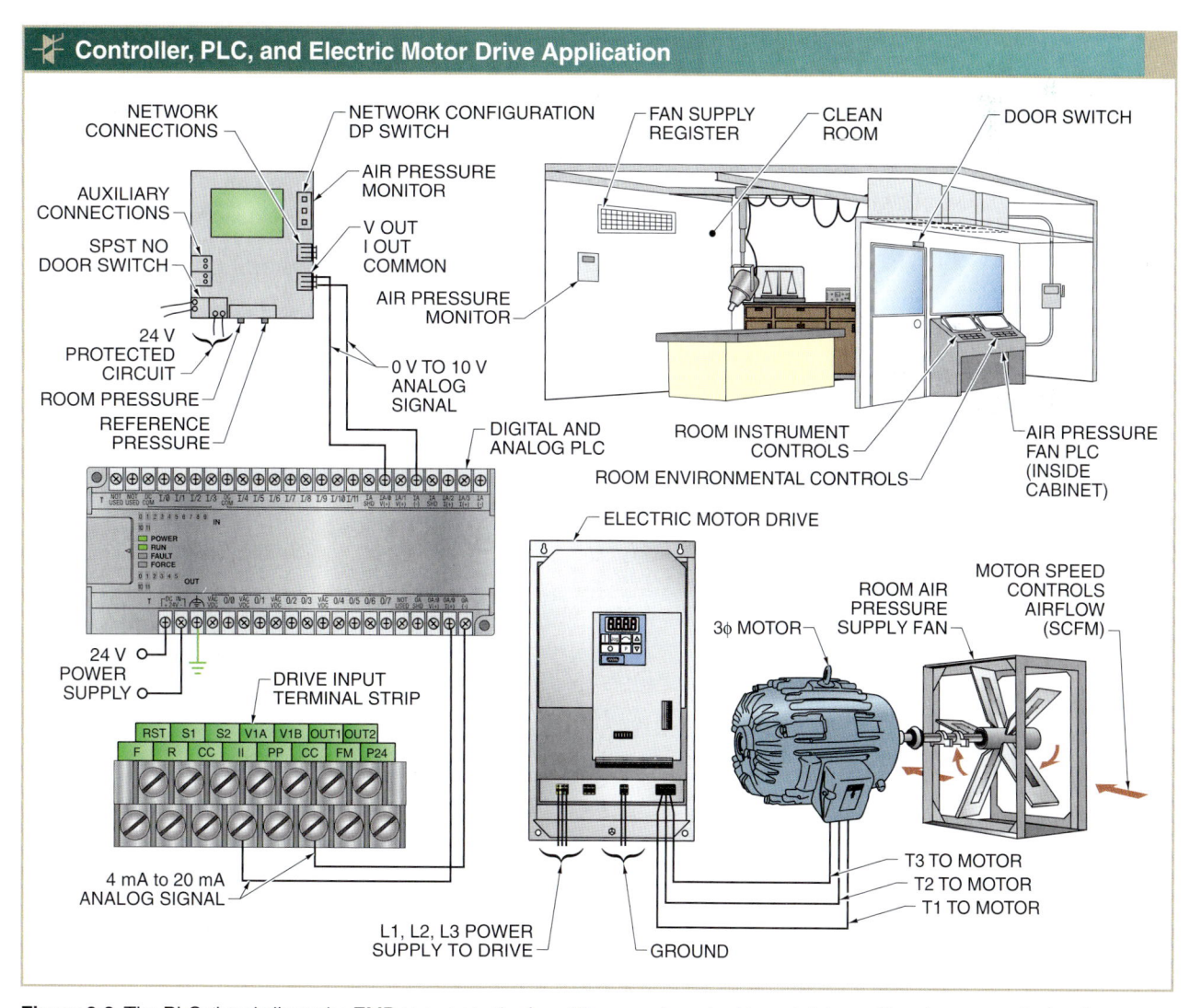

Figure 8-3. The PLC signal allows the EMD to operate the fan at the speed required to maintain positive air pressure in the clean room.

PLC Sensor Installation

Analog input devices that are connected to PLCs are always located in close proximity to the physical material or properties they are measuring. In many applications, a portion of the sensor is in direct contact with the material being measured. For example, the paddle wheel of a flowmeter is inserted into a pipe to measure the flow through the pipe.

In some applications however, the sensor is not in direct contact with the physical property being measured due to the corrosive or detrimental effect of the material. For example, the water in a cooling tower that contains biocides, which are corrosive, must be kept away from a temperature sensor. To accomplish this, a sealed well made of corrosion-resistant material is inserted into the sump of the cooling tower. The probe of the temperature sensor is then inserted into the well. The well prevents the water from coming in direct contact with the probe of the temperature sensor but conducts the temperature of the water to the probe. **See Figure 8-4.**

Sensor Installation Types

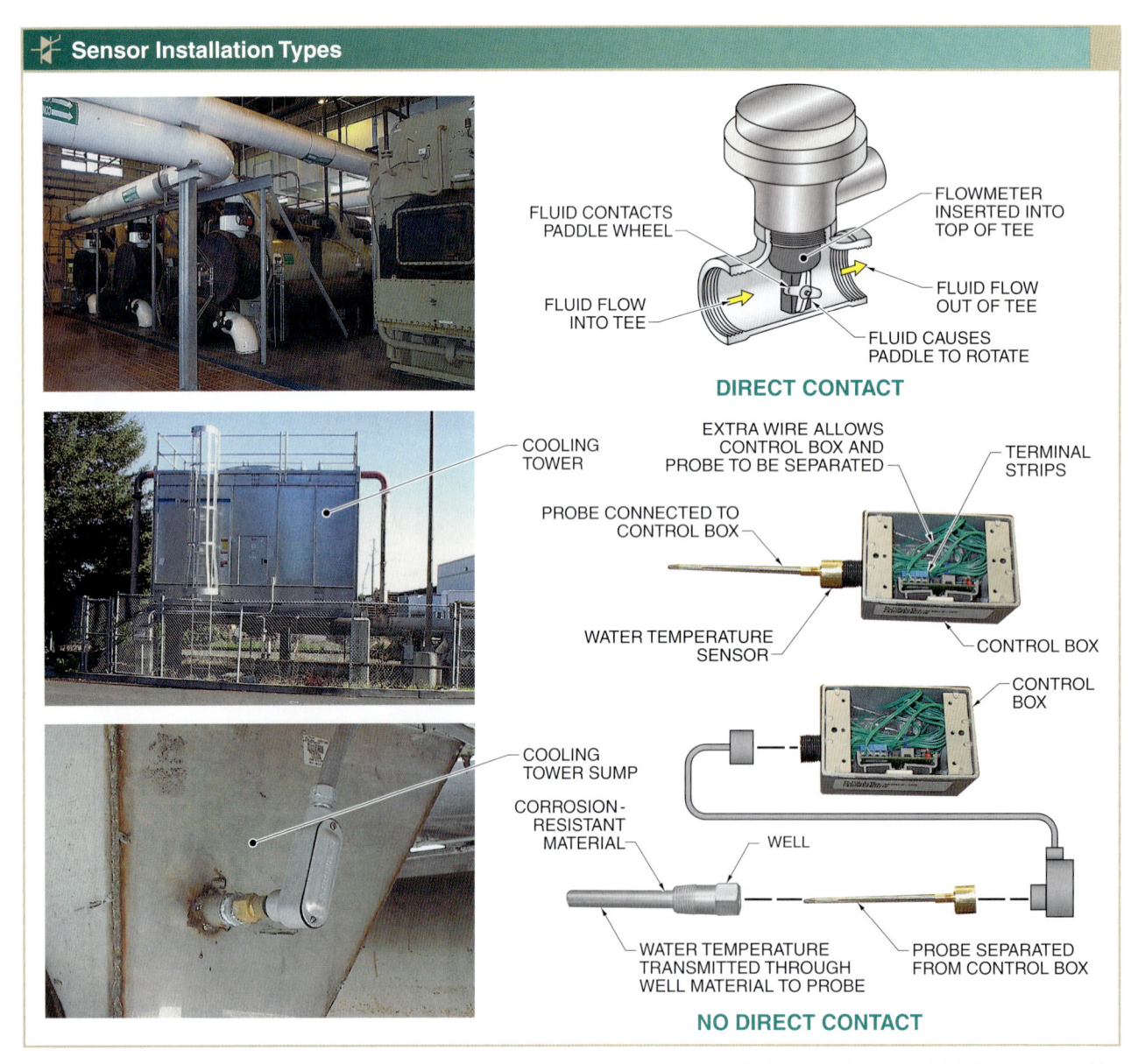

DIRECT CONTACT
- FLUID CONTACTS PADDLE WHEEL
- FLOWMETER INSERTED INTO TOP OF TEE
- FLUID FLOW INTO TEE
- FLUID FLOW OUT OF TEE
- FLUID CAUSES PADDLE TO ROTATE

NO DIRECT CONTACT
- COOLING TOWER
- EXTRA WIRE ALLOWS CONTROL BOX AND PROBE TO BE SEPARATED
- TERMINAL STRIPS
- PROBE CONNECTED TO CONTROL BOX
- WATER TEMPERATURE SENSOR
- CONTROL BOX
- COOLING TOWER SUMP
- CORROSION-RESISTANT MATERIAL
- WELL
- WATER TEMPERATURE TRANSMITTED THROUGH WELL MATERIAL TO PROBE
- PROBE SEPARATED FROM CONTROL BOX

Figure 8-4. In some applications, sensors must be protected from direct contact with the corrosive material being measured by devices such as wells.

Sensor Mounting Concerns. The methods used for mounting analog signal sensors vary. In many applications, sensor-specific installation and/or mounting components are required. For example, many paddle-wheel-type flow sensors require a special pipe fitting (plastic tee with threads) to allow the paddle wheel to be installed in-line. **See Figure 8-5.** The sensor must be oriented correctly for accurate measurement. The sensor instructions will provide specific guidelines for acceptable orientation and locations.

In some sensor installations, multiple trades will be involved. A pipefitter might install the pipe fitting for a paddle wheel flow sensor, while an electrician installs the sensor and related wiring. All sensor installation instructions must be followed to ensure that the sensor functions properly and provides accurate measurements.

Sensor Power Supplies. Analog sensors require a power supply. Some PLCs have an internal power source capable of supplying power to input sensors. Other PLCs do not have internal power sources and require that all sensors receive their power from an external power source. Still other PLCs can have either an internal power source or an external power source. When using an external power source, the type of sensor will determine the wiring configuration between the power source, sensor, and PLC. **See Figure 8-6.**

CODE CONNECT

Article 242 of the NEC® covers the requirements for installing surge-protective devices (SPDs) for systems of 1000 V or less. SPDs protect sensitive electronic equipment, such as PLCs, by limiting voltage transients. Voltage transients can be caused by lightning, utility switching, or large loads turning ON or OFF. Article 285 addresses three categories of SPDs: Type 1 SPDs, Type 2 SPDs, and Type 3 SPDs. A Type 1 SPD is connected at the line side of a service main disconnect overcurrent protective device (OCPD). A Type 2 SPD is connected anywhere on the load side of a service main disconnect OCPD. A Type 3 SPD is connected on the load side of a branch-circuit OCPD anywhere up to the equipment served.

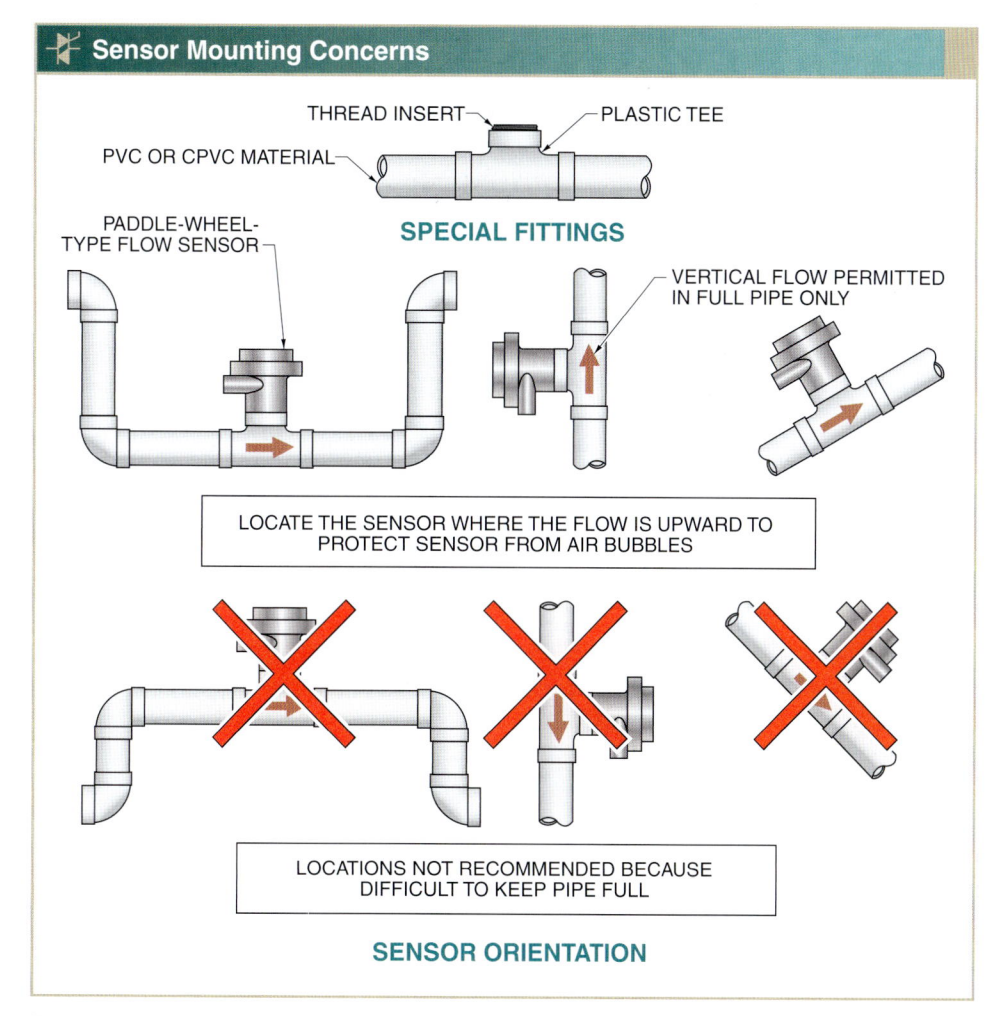

Figure 8-5. Sensor installation instructions provide information on special installation procedures, hardware, and mounting.

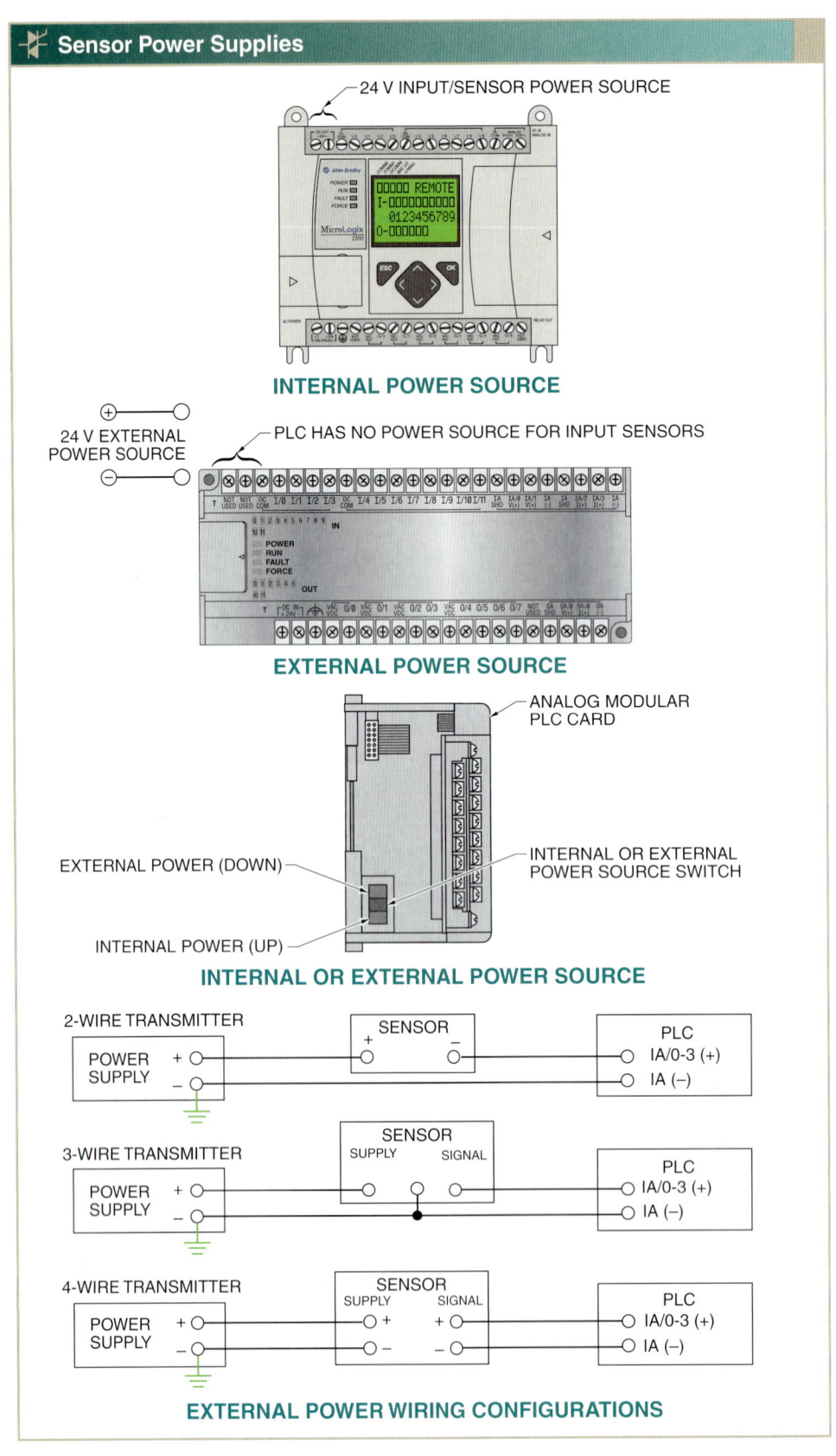

Figure 8-6. The power supply for a PLC sensor may be internal (from the PLC) or external.

Electrical Noise. The low-level analog signals from the sensor to the PLC are susceptible to electrical noise. Electrical noise can distort the analog signals and result in inaccurate measurements. Typically, shielded twisted pair (STP) cable is used to connect the sensor to the PLC. The twisting of the conductors, the foil or braided shield, and the drain wire reduce the harmful effects of electrical noise. The shield and drain wire can only be terminated at the PLC. It cannot be terminated at the sensor. The shield and drain wire can be trimmed at the sensor end and insulated with electrical tape. **See Figure 8-7.**

Depending on the application, the STP cable used for analog outputs may be installed in conduit for additional protection from environmental factors such as heat, moisture, and corrosive vapors. Metal conduit, which provides additional shielding from electrical noise, is commonly used. The STP cable must be run in a separate conduit from other energized conductors. The STP cable can pick up electrical noise when it is run in a conduit with other conductors that have 115 VAC, 230 VAC, or 460 VAC. Within the PLC enclosure, the STP cable must be separated as much as possible from other conductors. When the STP cable crosses energized higher-power conductors, it should cross them at 90° to minimize any possible electrical noise interference.

Analog Output Devices

Analog output devices are used in a wide variety of applications. The majority of PLC installations that have analog sensors also have analog output devices, usually actuators. The most common use of an analog actuator is to control the position of a valve or damper. A valve or damper may control the temperature, pressure, flow, or level of a specific application. For example, a PLC can control the flow of liquid in a process by directly varying the position of a valve stem in the discharge line of a tank. *Note:* A valve or damper is sometimes referred to as a "final element".

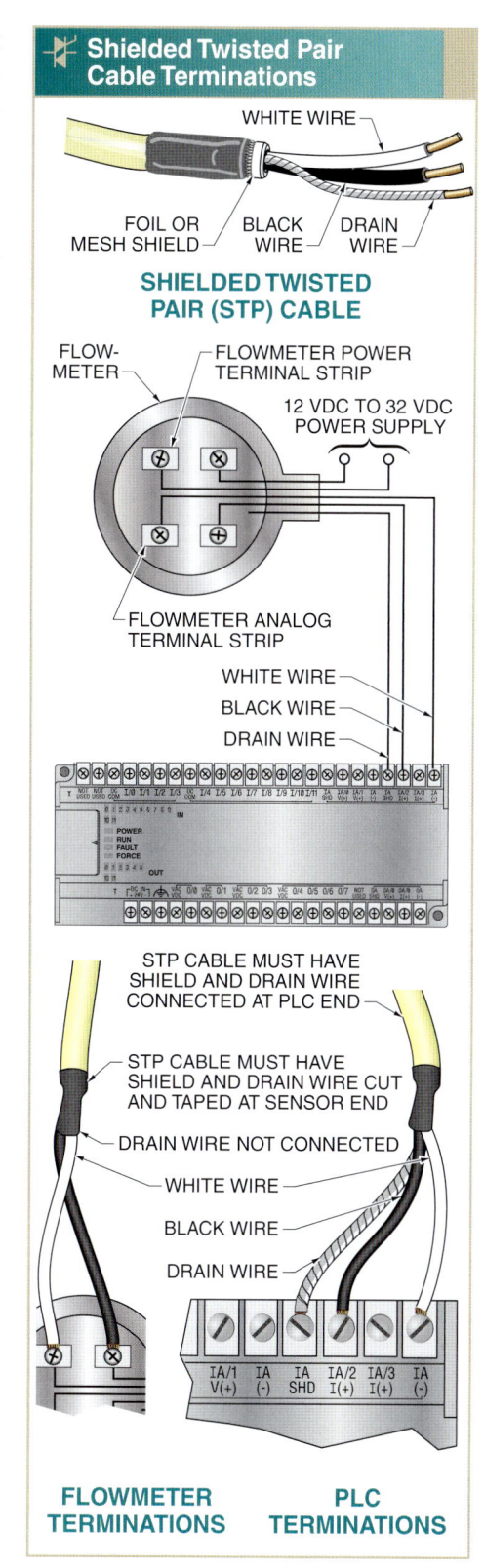

Figure 8-7. Shielded twisted pair cable reduces the harmful effects of electrical noise, but the shield and drain wires can only be terminated at the PLC end, not the sensor end.

CODE CONNECT

Section 250.8(A) of the NEC® addresses methods permitted to connect equipment grounding conductors, grounding electrode conductors, and bonding jumpers in an enclosure. Two of the methods are machine screw-type fasteners that engage two threads or are fastened with a nut, and thread-forming machine screws that engage two threads. Specifying machine screw-type fasteners and thread-forming machine screws eliminates sheet-metal screws and self-drilling "tek-type" screws as acceptable methods.

In some applications, the PLC outputs an analog signal to an intermediate piece of equipment instead of directly to an actuator. For example, a PLC can output an analog signal to a EMD that controls a printing press tensioner motor. The analog signal from the PLC allows the EMD to control the speed of the printing press 3ϕ tensioner motor. **See Figure 8-8.**

Analog Output Device Signals

Analog actuators are connected to the analog output terminals of a PLC and receive analog signals from the output section. In manufacturing, an actuator can be used to regulate the temperature of an industrial oven by controlling the position of a modulating gas valve that supplies natural

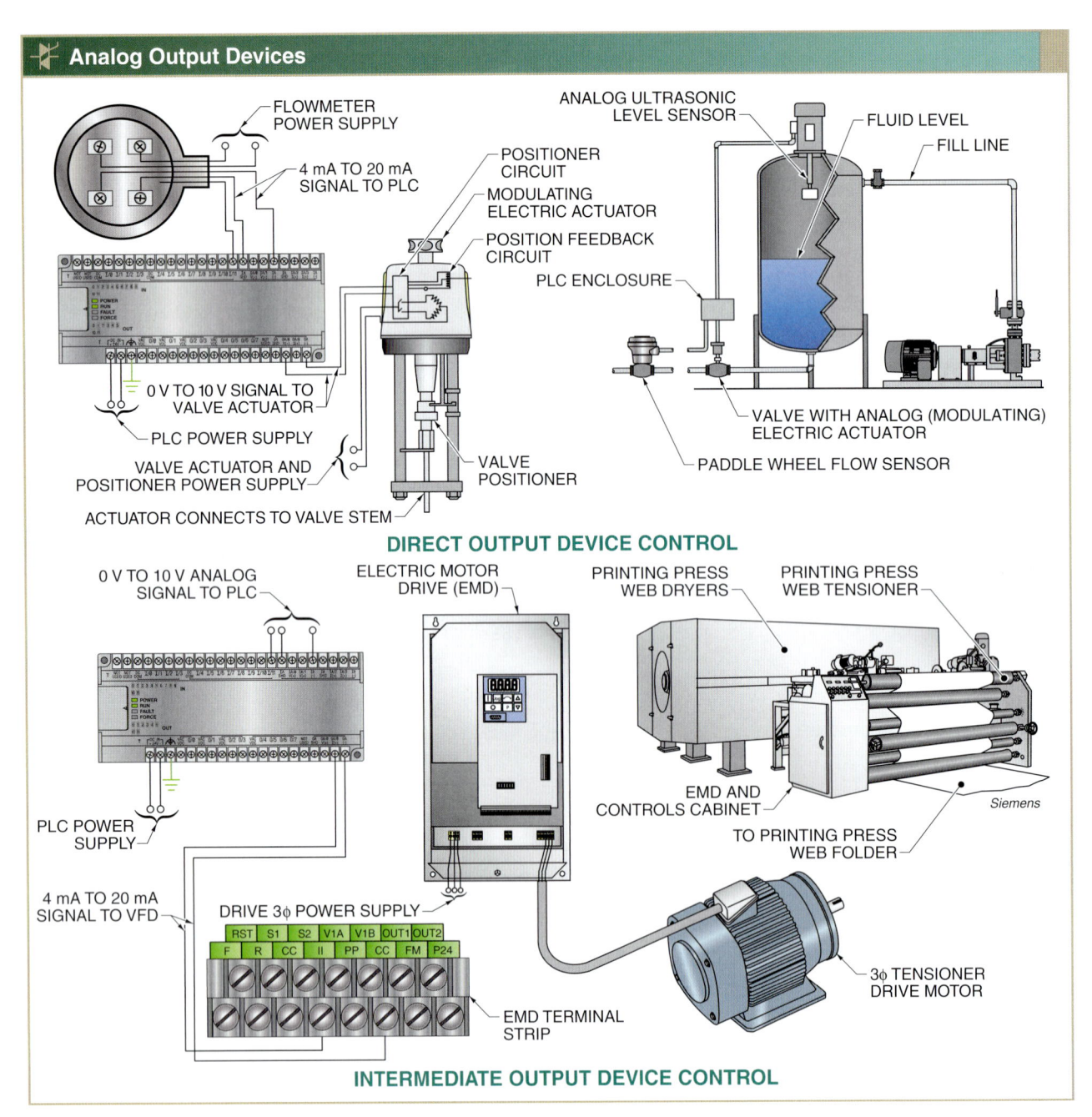

Figure 8-8. A flow sensor can measure flow rate, sending a signal to the PLC to adjust the valve position to maintain the flow rate specified by the program. An EMD can be used as an intermediate device to control the speed of a motor.

gas to the oven burners. In a commercial office building, an actuator can be used to control the flow of air in an HVAC duct by controlling the position of a damper located inside the duct. As part of an energy-saving initiative, analog output signals from a PLC can be sent to a EMD to control the speed of pump motors. **See Figure 8-9.**

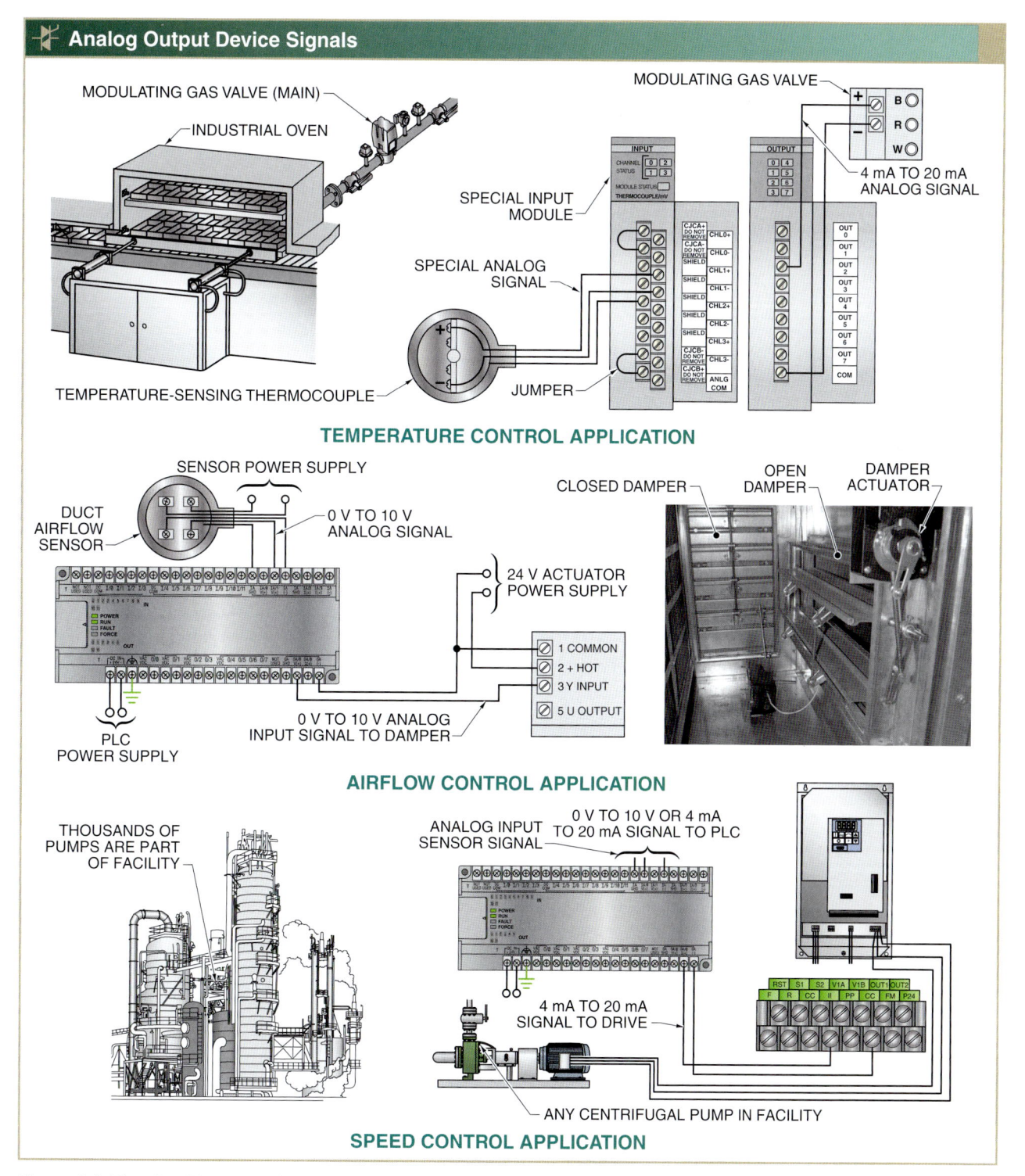

Figure 8-9. The signal from a PLC analog output terminal can be directly sent to an actuator or can be sent to another piece of equipment such as an EMD.

CODE CONNECT

Section 300.19(A) of the NEC® covers spacing intervals for supporting conductors in vertical raceways. Section 300.19(A) and associated Table 300.19(A) list the maximum distance a conductor can be run vertically in a raceway without support. Support is required at the top of a vertical raceway or as close as possible. Intermediate support may be required to avoid exceeding the maximum unsupported conductor length. Typically, multisegment wedges made of wood or other nonconductive material are inserted into the throat of a connector for support. These wedges are available from various conduit-fitting manufacturers.

Based on the PLC program, the input signal results in a specific control signal from the PLC analog output section to the actuator. The PLC's CPU executes an application-specific program designed to control the actuator. The CPU's results are a number. This number represents the level of analog output signal to be sent to the actuator. The number varies as the results of the program change. A digital-to-analog (D-to-A) converter inside the PLC converts the number into varying voltage or current values. The PLC analog output signal is typically 0 VDC to 10 VDC or 4 mA to 20 mA.

PLC Actuator Installation

PLC actuators are mounted adjacent to or directly on the valve or damper they control. Typically in the case of a valve, the actuator is mounted directly to the valve. In the case of a damper, the actuator may be mounted directly to the damper shaft, or adjacent to the damper shaft and connected by a chain or linkage. **See Figure 8-10.**

It is very important to choose the correct actuator for the type of valve or damper that the PLC is controlling. The actuator must have sufficient torque to turn, raise, or lower the valve or damper shaft. The actuator mounting method must be compatible with the valve or damper design. Adapter plates, couplings, crank arms, or linkages are sometimes required to mount the actuator to the valve or damper. Actuator installation instructions typically provide mounting information.

In some installations, multiple trades will be involved in the actuator installation. For example, a pipefitter might install the actual valve and actuator, but an electrician typically installs the conduit and wiring that carries the analog output signal from the PLC to the actuator. All actuator installation instructions must be followed to ensure the actuator functions properly. The actuator specifications will list items such as the torque output. **See Figure 8-11.**

PLC actuators require a separate power source to power the actuator motor, which varies the position of the valve or damper. The power source may be electrical or pneumatic. The electrical source can be 12 V to 480 V, AC or DC. The pneumatic source is typically 60 psi to 100 psi of air pressure. The analog output signal from a PLC provides a reference to the actuator electronics. The actuator electronics translate this signal into the position needed for the valve or damper to control temperature, pressure, flow, or other properties. **See Figure 8-12.**

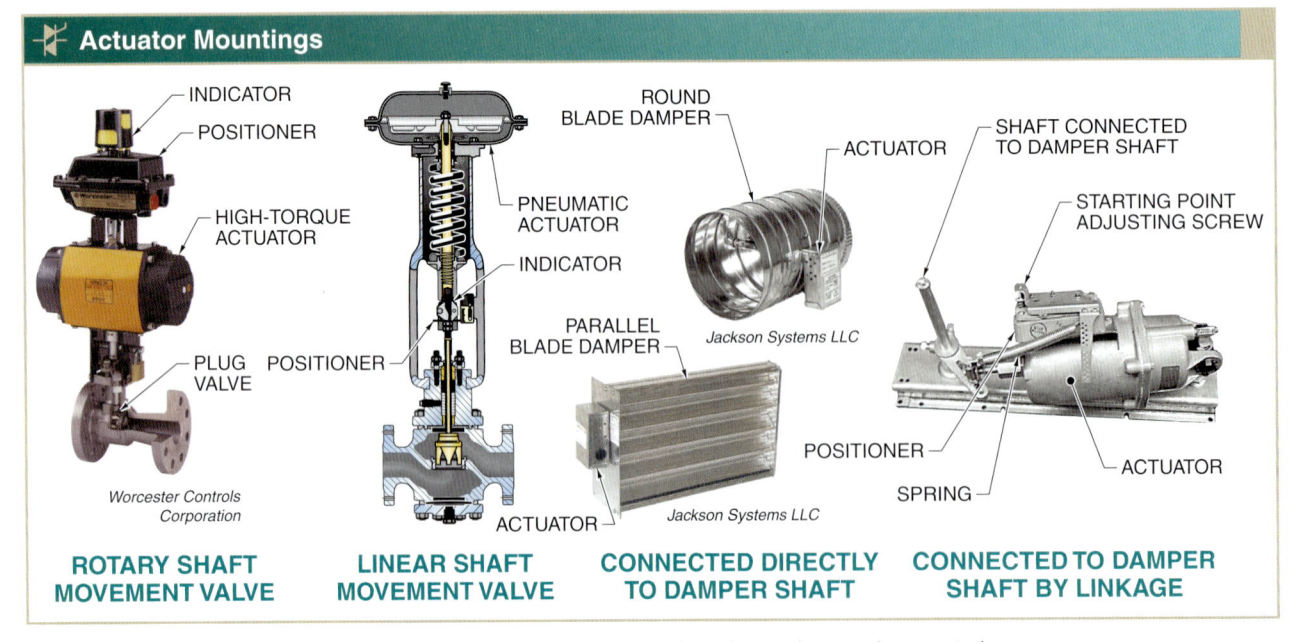

Figure 8-10. PLC actuators are mounted adjacent to or directly on the valve or damper they control.

Valve and Damper Actuator Specifications

Brand	Torque*	Holding Torque	Locked Rotor Amp Draw†	Power Consumption‡	90° Rotation Speed§
Bettis — 24 VDC	300	390	115 = 0.60, 230 = 0.30	2.50	4
Crane — 24 VDC	434	868	115 = 0.51, 230 = 0.27	2.25	13
Assured — 24 VDC	300	—	115 = 0.87, 230 = 1.00	1.70	9
ETI — 24 VDC	177	221	115 = 0.21, 230 = 0.11	0.90	11

* in inch pounds
† in AC amperage
‡ 24 VDC amperage
§ in seconds to rotate 90°

Brand	Torque*	Holding Torque	Dimensions†	Power Consumption‡	Angle of Rotation
Siemens — 24 VDC	350	175	11¾ × 4 × 2¹¹⁄₁₆	18.0 — 5.0	90°
Honeywell — 24 VDC	175	175	9¾ × 3¹⁵⁄₁₆ × 2¹⁵⁄₁₆	16.0 — 5.0	85°
DEI — 24 VDC	177	—	11¾ × 3⅞ × 3¹⁄₁₆	20.0 — 8.0	95°
InTEC — 24 VDC	45	55	4½ × 2⁷⁄₁₆ × 2⁹⁄₁₆	5.5 — 4.5	95°
VCP — 24 VDC	45	75	4⅝ × 2⁹⁄₁₆ × 2¹³⁄₁₆	2.0 — 3.0	95°

* in inch pounds
† in inches
‡ in volt amps for torque-holding torque

Figure 8-11. Valve and damper specifications provide information such as the torque required by the actuator for proper operation.

Actuator Power Sources

PNEUMATIC

- 4 mA TO 20 mA ANALOG ELECTRICAL SIGNAL
- ELECTROPNEUMATIC TRANSDUCER CHANGES ANALOG ELECTRICAL SIGNAL TO MODULATING PNEUMATIC SIGNAL
- 10 PSI TO 60 PSI ANALOG AIR SIGNAL TO VALVE ACTUATOR
- PNEUMATIC ACTUATOR
- 0 PSI TO 100 PSI SUPPLY AIR
- VALVE

ELECTRICAL

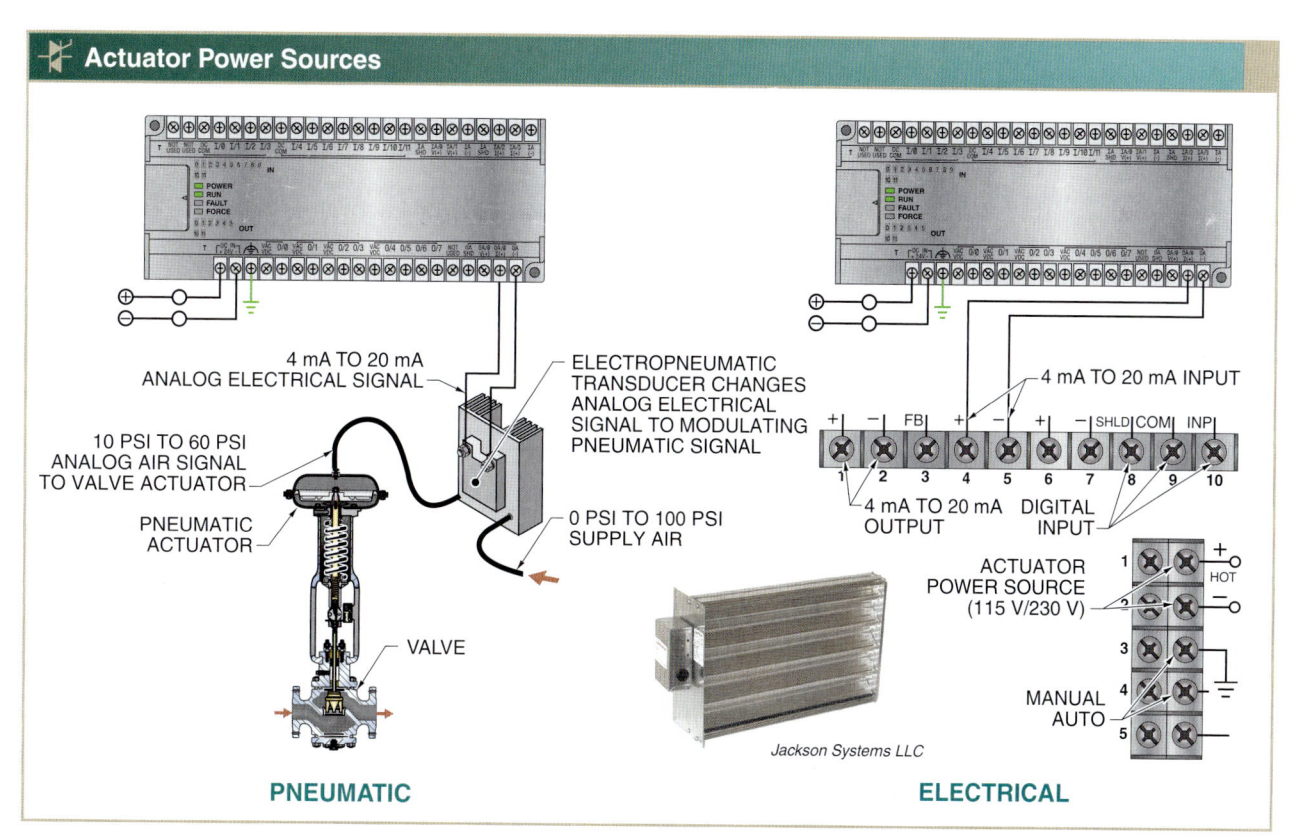

- 4 mA TO 20 mA INPUT
- 4 mA TO 20 mA OUTPUT
- DIGITAL INPUT
- ACTUATOR POWER SOURCE (115 V/230 V)
- MANUAL AUTO

Jackson Systems LLC

Figure 8-12. PLC actuators require a separate power source (electrical or pneumatic) to power the actuator motor or piston, which is used to vary the position of the valve or damper.

PROGRAMMING FOR ANALOG INPUT AND OUTPUT DEVICES

Analog input and output terminals are available on nano-, micro-, medium-, and large PLCs. The number of analog input and output terminals corresponds with the size of the PLC. Nano- and micro-PLCs have few analog inputs and outputs. Small and large PLCs (which are typically modular) are able to support hundreds of analog input and output terminals.

The sophistication and complexity of programming a PLC for analog input and output devices also corresponds with the size of the PLC. Nano- and micro-PLCs have relatively simple programming instructions with few analog options. Small and large PLCs have more complex programming instructions, including instructions specific to analog control, such as PID (proportional, integral, derivative) instructions. PID instructions have several parameters that can be adjusted to optimize analog control.

There are two methods of entering analog input and output addresses into a PID instruction. In the first method, the address of the analog input terminal and the address of the analog output terminal are entered directly into the PID instruction as Process Variable and Control Variable, respectively. In the second method, the data is transferred from the address of the analog input and output terminals to separate integer files, and then the integer files are entered as Process Variable and Control Variable. This method allows the data from the analog input or output devices to be manipulated in an integer file.

The PID instruction has a selection that allows the technician to access the PID Setup screen. The Setup screen allows the technician to adjust numerous parameters related to the PID instruction to optimize analog control. **See Figure 8-13.**

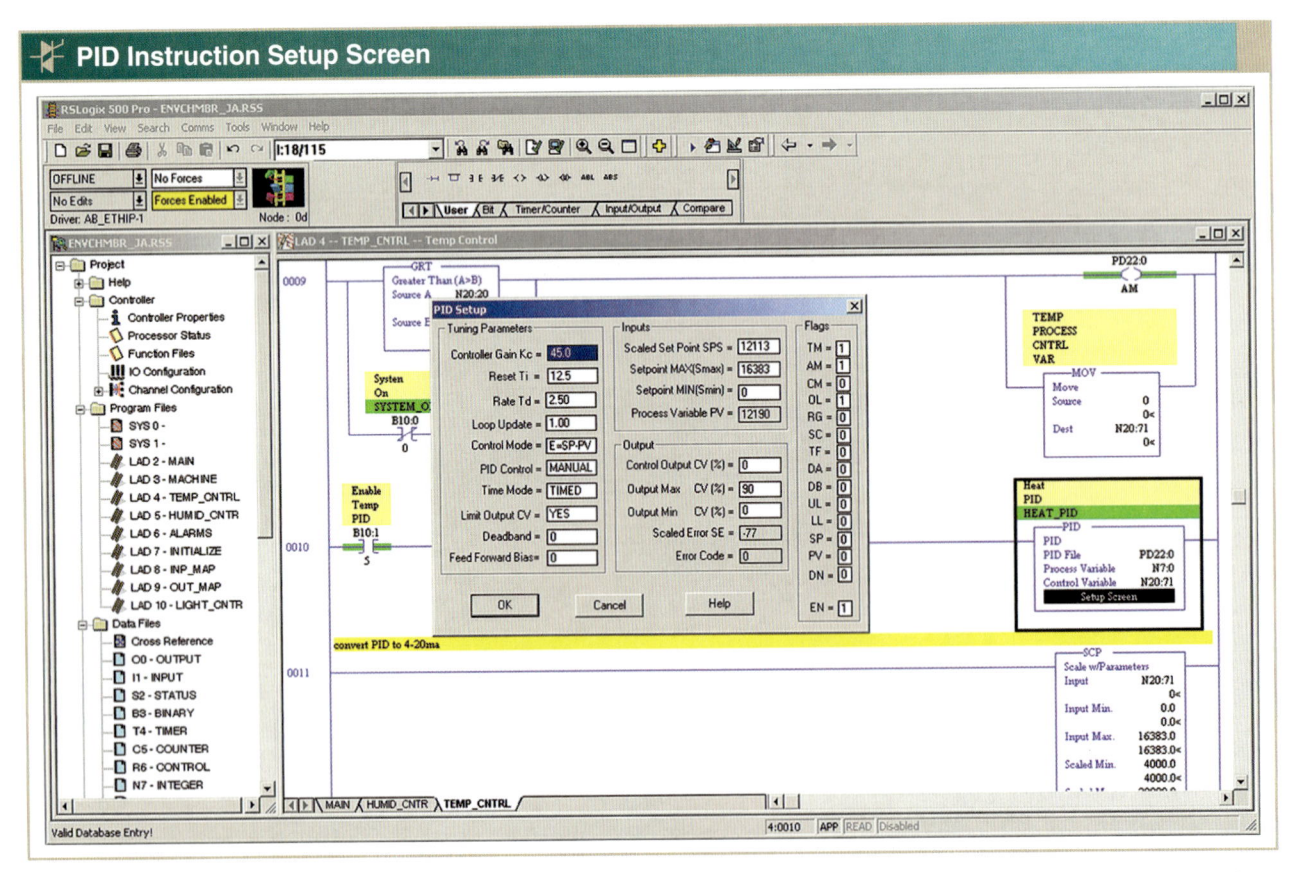

Figure 8-13. A PID instruction allows a technician to access the PID Setup screen. By using this PID Setup screen, the technician can optimize analog control by adjusting the numerous parameters related to the instruction.

It is important to understand the process being controlled when programming a PLC for analog input and output devices. Factory processes are typically dynamic. A small change in an analog-related parameter can result in a large unexpected increase in the temperature, pressure, flow, or level of a product. Unexpected changes can pose a serious safety risk to equipment and/or personnel. When a technician is unfamiliar with a process, the technician should seek the assistance of someone who thoroughly understands the process in question.

Often, nano- and micro-PLCs do not have special instructions for programming analog input and output devices. Rather, specific I/O terminals and corresponding locations in memory are designated for analog instructions. Rockwell Automation Allen-Bradley PLCs are an example of these kinds of PLCs. XIC (examine if closed) instructions are used for analog input devices. OTE (output energize) instructions are used for analog output devices. The results of the analog input A-to-D converter are processed by the CPU of the PLC, and the resulting number is sent to the D-to-A converter of the appropriate analog output terminal. A 16-bit word is reserved for each analog input and output terminal. The input and output data table files can be used to monitor analog input and output operations. **See Figure 8-14.**

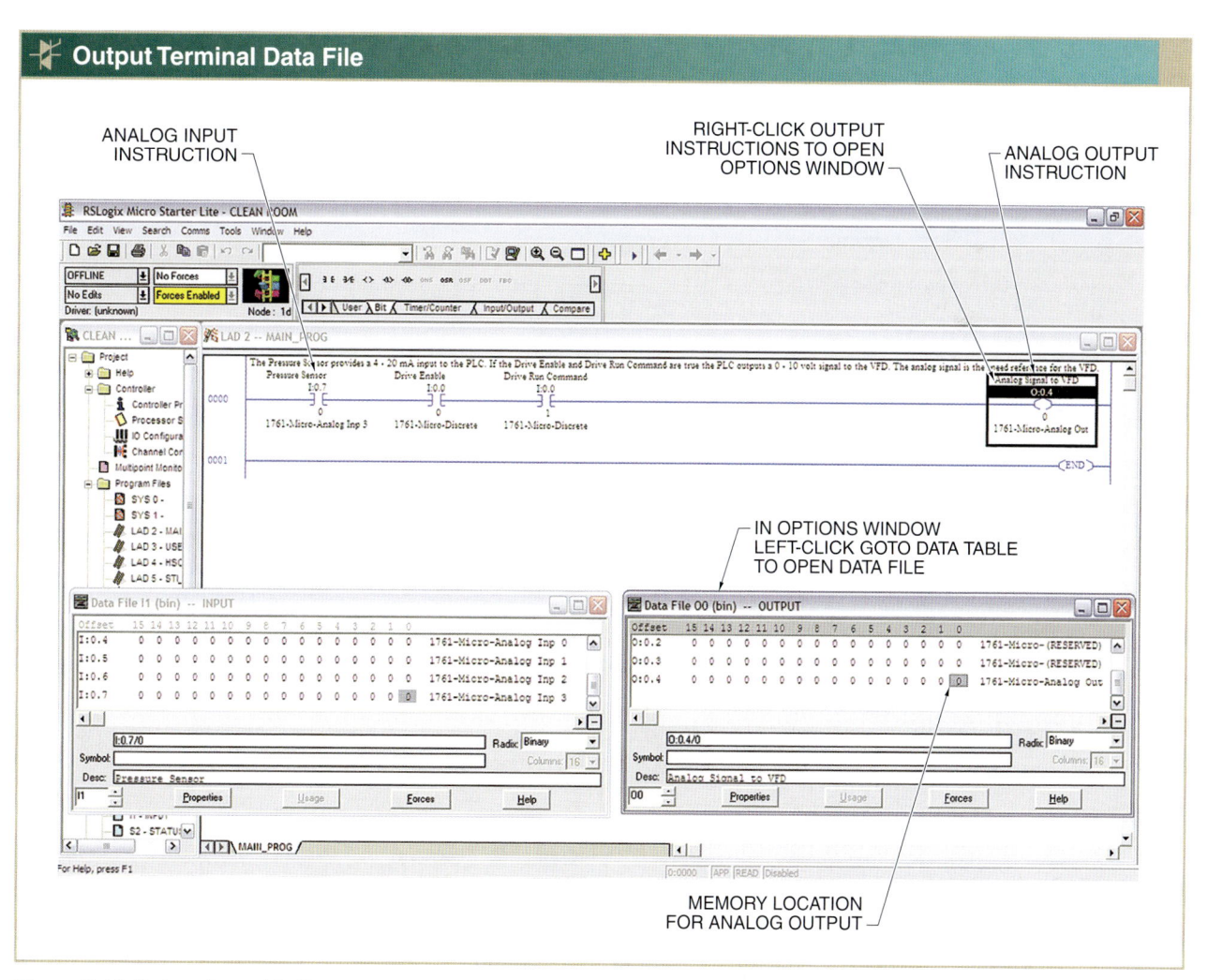

Figure 8-14. Output data table files are used to monitor PLC analog output operations.

ANALOG DEVICE PROGRAMMING SETUP

The setup dialog box for Rockwell Automation Allen-Bradley PLCs analog I/O terminals is accessed through the I/O configuration file found in the software project tree. Double-clicking on the I/O configuration file opens the I/O configuration dialog box. The I/O setup button opens the analog microcontroller setup dialog box.

PLC TIPS

Input filter response times can only be adjusted at the time the program is developed. Default for each of the input groups is typically 8 ms.

The analog microcontroller setup dialog box has a section for discrete or digital I/Os and for analog I/Os. The analog setup portion of this dialog box allows the technician to choose the output mode of the analog output signal (voltage or current), enable the analog inputs, and adjust the analog input filter settings to minimize the impact of electrical noise. **See Figure 8-15.** The analog input filter is designed to reject AC line noise that can couple or bleed into an analog input signal from high-voltage wiring. Frequencies at or above the software setting are rejected. Noise from 60 Hz or 50 Hz power sources is the primary concern.

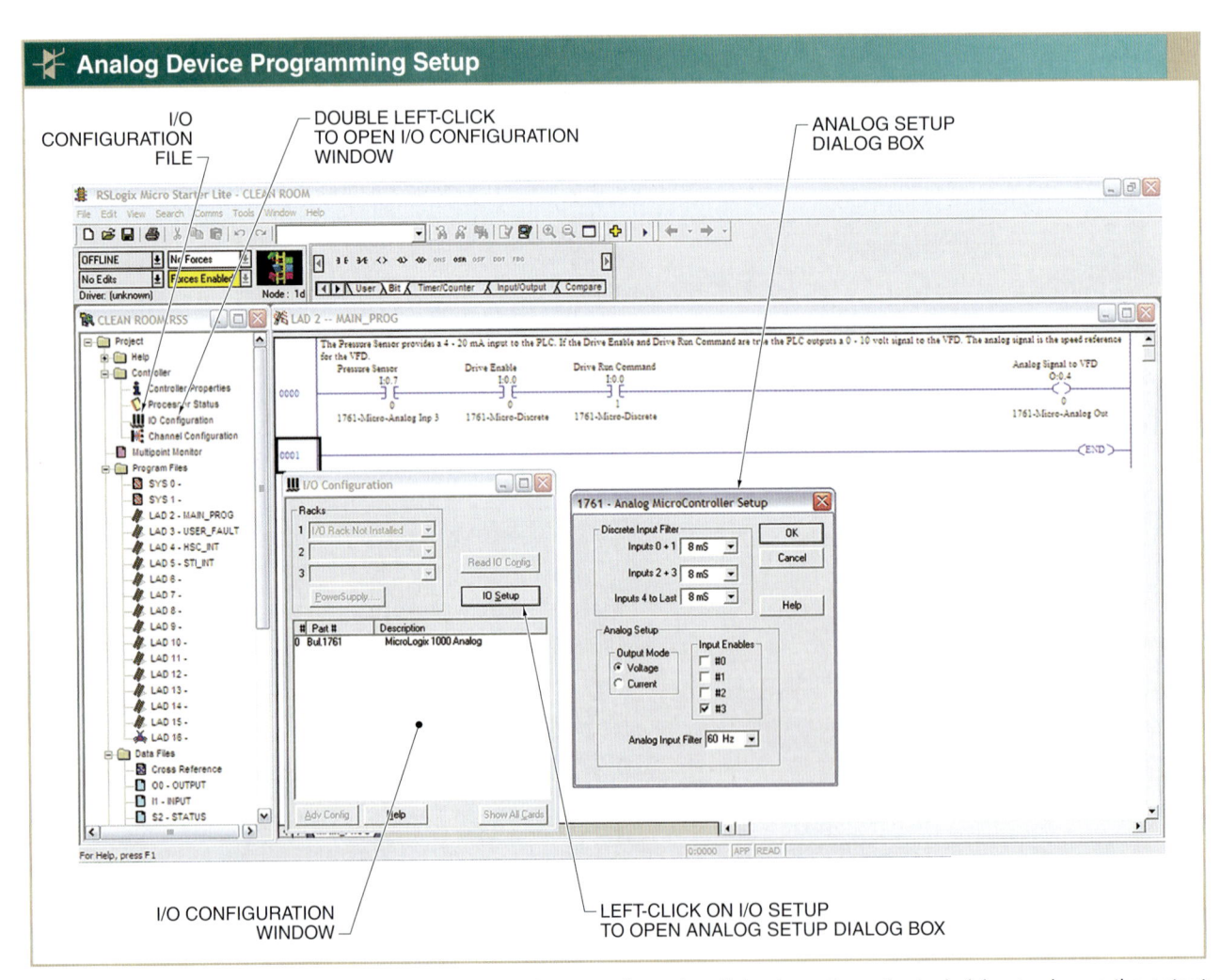

Figure 8-15. The analog setup dialog window of the microcontroller setup dialog box allows the technician to choose the output mode of the analog output signal, enable the analog inputs, and adjust the analog input filter settings to minimize the impact of electrical noise.

TROUBLESHOOTING PLC ANALOG INPUT CIRCUITS

It is not possible to cover every input-circuit troubleshooting scenario. However, the sensor, the wiring between the sensor and the PLC, and the PLC are the most common problem areas. In order to troubleshoot a PLC application with analog input signals, the technician will need a PC with the appropriate programming software, installation manuals for the PLC and sensors, a digital multimeter (DMM) capable of measuring DC voltage, a process clamp meter, and basic hand tools.

Note: It is imperative that the technician thoroughly understand the PLC application and the consequences of changing the state of a sensor or changing the state of a data source for an analog input or output device. Changes can pose serious safety risks to equipment, operational personnel, and the technician performing the troubleshooting.

Troubleshooting PLC Sensors and Sensor Wiring

The first check that a technician should perform is to verify that the PLC and sensor have proper power supply voltages. Then the technician should verify that the PLC is receiving a signal from the sensor. The two types of analog signals used in most applications are 4 mA to 20 mA or 0 VDC to 10 VDC.

Using a DMM to measure voltage and a milliamp process clamp meter set to measure current, a technician can verify that the PLC is receiving a signal from the sensor by taking a measurement at the analog input terminals of the PLC. The technician should actuate the sensor to cause it to change state, which will cause a change in current or voltage to the PLC. When a change in current or voltage is measured, the sensor and wiring from the sensor to the PLC can be eliminated as the source of the problem. See Figure 8-16.

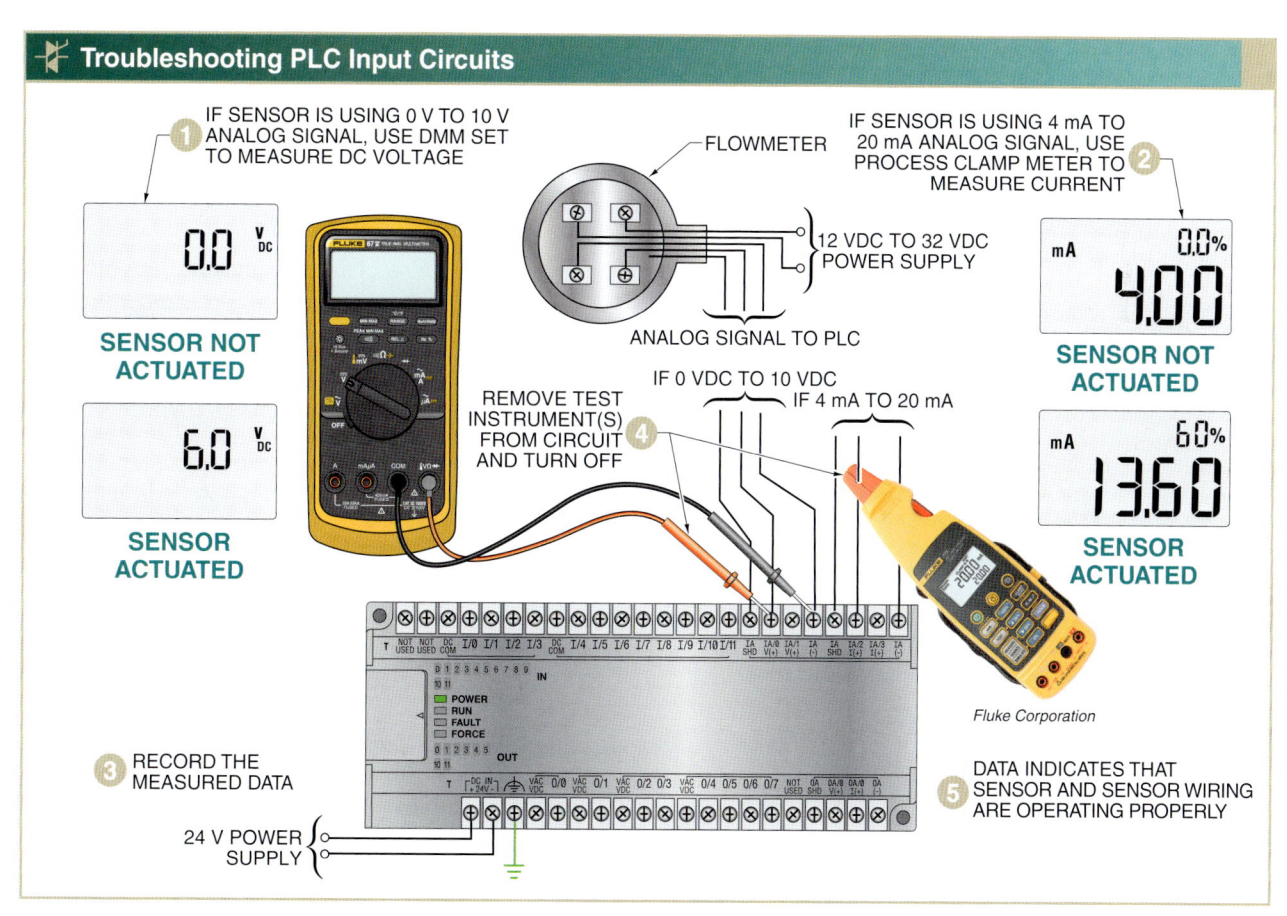

Figure 8-16. The voltage or input current from a sensor can be checked at the analog input terminals of the PLC.

To troubleshoot the input circuit of a PLC, apply the following procedures:

1. When the sensor is using 0 V to 10 V analog signals, use a DMM to measure voltage at the input terminals of the PLC.
2. When the sensor is using 4 mA to 20 mA analog signals, use a process clamp meter to measure current before the positive input terminal of the PLC.
3. Record the measured data.
4. Remove the test instrument from the circuit and turn it off.
5. Analyze the data.

When a change in current or voltage is not measured, the sensor or wiring from the sensor to the PLC is probably the source of the problem. The technician must then go to the sensor location and verify that the sensor is installed correctly. Next, the technician should open the sensor to access the terminal strip or strips. Using a DMM or a milliamp process clamp meter set to measure the appropriate voltage or current signal, the technician can verify that the sensor is outputting a signal. The technician should actuate the sensor to cause it to change state and thus output a current or voltage. If there is no change in current or voltage, the sensor is the source of the problem.

When a change in current or voltage is measured, the sensor can be eliminated as the source of the problem. In this case, the wiring between the sensor and the PLC is the likely source of the problem. The technician must check the wiring between the sensor and PLC and make repairs as needed. **See Figure 8-17.** To troubleshoot the sensor and input wiring to a PLC, apply the following procedures:

1. Verify that the correct supply voltage (24 V) is present at the power supply terminal strip of the sensor.
2. When the sensor is using 0 V to 10 V analog signals, use a DMM to measure the analog output voltage at the sensor's output terminal strip.
3. When the sensor is using 4 mA to 20 mA analog signals, use a process clamp meter to measure the positive analog output current as close as possible to the sensor's output terminal strip.
4. Record the measured data.
5. Remove the test instrument from the sensor and turn it off.
6. Analyze the data.

Troubleshooting PLC Analog Input Terminals

When a change in current or voltage is measured at the analog input terminals and the problem still exists, the technician should use the programming software and look at the analog input data table file. The technician should actuate the sensor to cause it to change state. This should cause the numbers in the analog input data table file to change. **See Figure 8-18.** When the numbers change, the PLC analog input can be eliminated as the source of the problem.

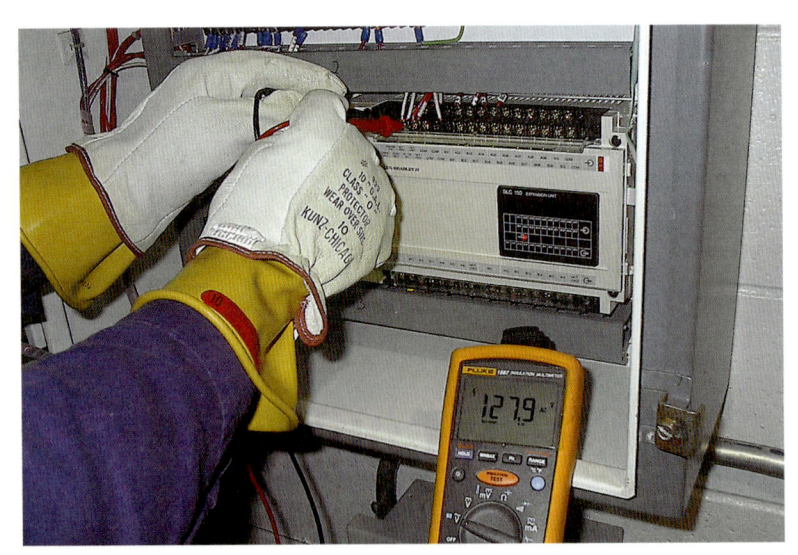

DC-powered PLCs should be checked for AC ripple using a DMM set on a low AC range. Excess ripple can negatively affect the operation of the microprocessor and memory devices typically found in a PLC.

PLC TIPS

Right-clicking the mouse on a data table file allows values in the data table to be copied and pasted and to be viewed in other numbering systems. Also, data table properties can be modified by right-clicking. A Find All function can be used to find every occurrence of a data table element.

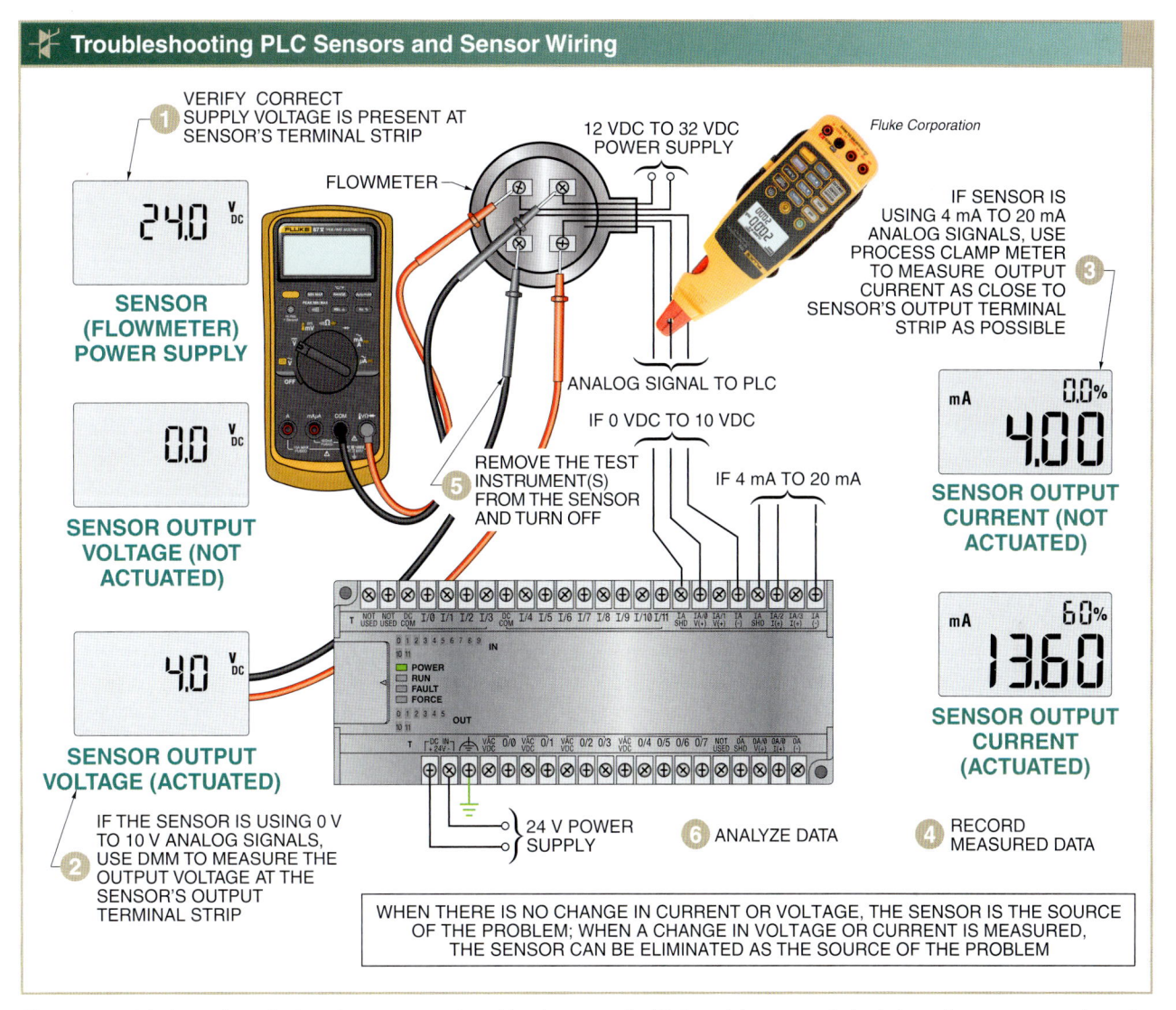

Figure 8-17. To determine whether the sensor or the wiring between the PLC and the sensor is bad, the voltage or current from the sensor must be checked at the sensor.

When the numbers in the analog input data table file do not change, there may be a problem with the software configuration for the analog I/Os, the hardware configuration for the analog I/Os, or the PLC itself. The technician should verify that the software and hardware are correctly configured. After verifying the configurations, the sensor should be actuated and the analog input data table checked. When there is still no change, the PLC analog input is the source of the problem. Depending on the PLC model, an analog input module or the entire PLC may need to be replaced.

TROUBLESHOOTING PLC ANALOG OUTPUT CIRCUITS

It is not possible to cover every analog output-circuit troubleshooting scenario. However, the actuator, the wiring between the PLC and actuator, the PLC, and the upstream data source for the analog output signal are the most common problem areas. The upstream data source for the analog output signal may be an analog input device or a file that contains the results of data manipulation created by the program.

330 PROGRAMMABLE LOGIC CONTROLLERS: PRINCIPLES AND APPLICATIONS

Troubleshooting PLC Analog Input Terminals

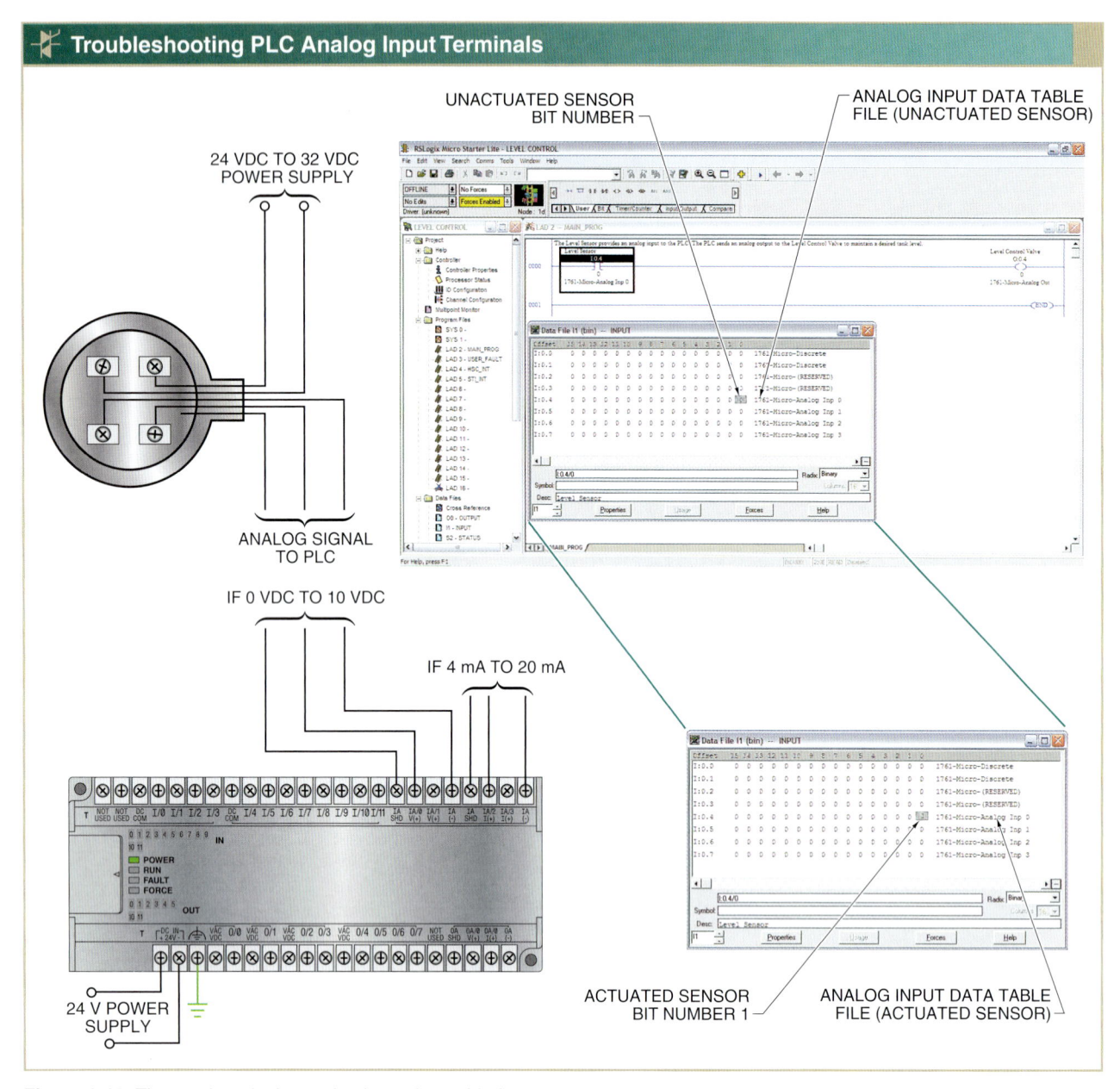

Figure 8-18. The numbers in the analog input data table file must change as the sensor is actuated.

In order to troubleshoot a PLC application with analog output devices, a technician will require a PC with the appropriate programming software, installation manuals for the PLC and actuator, a DMM capable of measuring DC voltage, a process clamp meter, and basic hand tools.

Note: It is imperative that the technician thoroughly understand the PLC application and the consequences of changing the state of a sensor or changing the state of a data source for an analog input or output device. Changes can pose a serious safety risk to equipment, operational personnel, or the technician performing the trouble-shooting procedures.

Troubleshooting PLC Actuators and Actuator Wiring

The first check that a technician should perform is to verify that the PLC is sending an analog signal to the actuator. The two most common analog output signals are 0 VDC to 10 VDC and 4 mA to 20 mA.

Using a DMM to measure voltage and a process clamp meter set to measure current, the technician can verify that the PLC is sending a signal to the actuator by taking a measurement at the analog output terminals. The technician must actuate the input sensor or other data source associated with the actuator to cause a change of state, which will cause a change in the output current or voltage levels. When a change in current or voltage is measured, the PLC can be eliminated as the source of the problem. **See Figure 8-19.**

To troubleshoot the output circuit of a PLC, apply the following procedures:

1. When the actuator is using 0 V to 10 V analog signals, use a DMM to measure voltage at the output terminals of the PLC.
2. When the actuator is using 4 mA to 20 mA analog signals, use a process clamp meter to measure current after the positive output terminal of the PLC.
3. Record the measured data.
4. Remove the test instrument from the circuit and turn it off.
5. Analyze the data.

When current or voltage is measured, the actuator or the wiring from the actuator to the PLC may be the source of the problem. The technician must go to the actuator location and verify that the actuator is installed correctly. Next, the technician should open the actuator to access the terminal strip.

> **⚠ WARNING**
>
> *Changing the state of a data source or the use of Force functions can result in sudden machine movement, causing severe injury to personnel or damage to equipment. When an input is forced, the value of the input is written to the data table, and not to the real I/O. When an output is forced, the value is written to the real I/O, not the data table. Use extreme care when using forces.*

Using a DMM to measure voltage or a process clamp meter set to measure current, the technician can verify that the actuator is receiving an output signal from the PLC. The technician must actuate the sensor or other data source associated with the actuator to cause a change of state and thus a change in the current or voltage of the output signal. When a change in current or voltage is measured, the actuator is the source of the problem.

When no change in current or voltage is measured, the wiring between the PLC and actuator is the likely source of the problem. The technician must check the wiring between the actuator and PLC and make repairs as needed. **See Figure 8-20.**

To troubleshoot the actuator and output wiring from a PLC, apply the following procedure:

1. Verify that the correct supply voltage (24 V) is present at the power supply terminal strip of the actuator (control valve).
2. When the actuator is using 0 V to 10 V analog signals, use a DMM to measure the analog input voltage at the actuator's input terminal strip.
3. When the actuator is using 4 mA to 20 mA analog signals, use a process clamp meter to measure the positive analog input current as close as possible to the actuator's input terminal strip.
4. Record the measured data.
5. Remove the test instrument from the actuator and turn it off.
6. Analyze the data.

Troubleshooting PLC Analog Output Terminals

When a change in current or voltage is not measured at the analog output terminals, the technician should use the programming software and look at the analog output data table file of the PLC. The technician should actuate the sensor or other data source associated with the actuator to cause a change of state, and thus a change in the analog output data table file. The numbers in the analog output data table file should change. When the numbers change, the software configuration for the analog I/O, the hardware configuration for the analog I/O, or the PLC output may be the source of the problem.

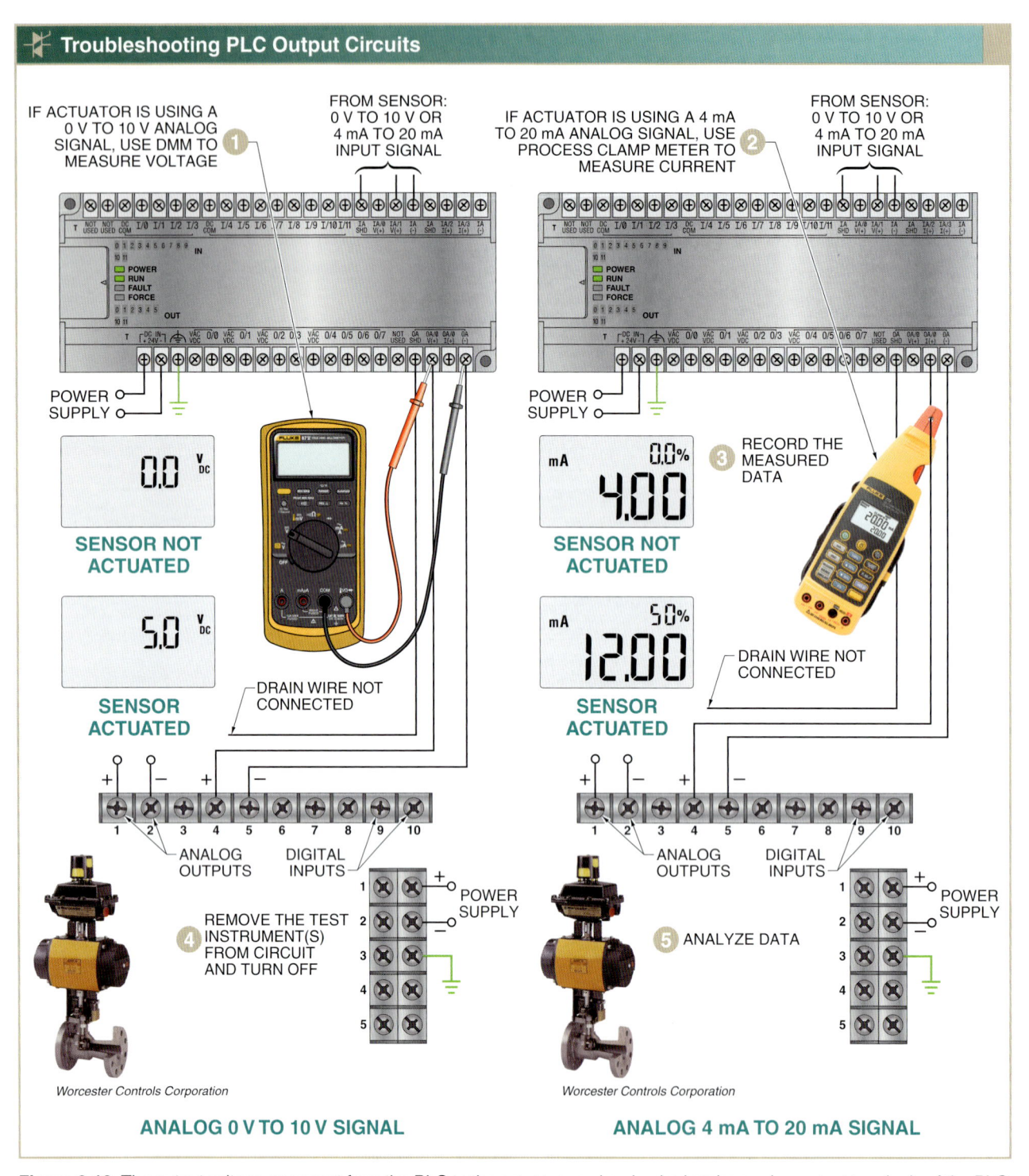

Figure 8-19. The output voltage or current from the PLC to the actuator can be checked at the analog output terminals of the PLC.

After verifying the configurations, the technician should actuate the sensor or other data source associated with the actuator again to cause a change in the analog output data table file and check for a change in voltage or current at the analog output terminals. When there is no change in voltage or current, the PLC output terminal is the source of the problem. Depending on the PLC model, an analog output module or the entire PLC may need to be replaced. **See Figure 8-21.**

Chapter 8—PLC Analog Device Installation, Programming, and Troubleshooting 333

Troubleshooting PLC Actuators and Actuator Wiring

ANALOG 0 V TO 10 V SIGNAL

① IF ACTUATOR IS USING A 0 V TO 10 V ANALOG SIGNAL, USE DMM TO MEASURE VOLTAGE

② FROM SENSOR: 0 V TO 10 V OR 4 mA TO 20 mA INPUT SIGNAL

- SENSOR NOT ACTUATED: 0.0 V DC
- SENSOR ACTUATED: 5.0 V DC

DRAIN WIRE NOT CONNECTED

ANALOG OUTPUTS — DIGITAL INPUTS

⑤ REMOVE THE TEST INSTRUMENT(S) FROM ACTUATOR AND TURN OFF

CONTROL VALVE
Worcester Controls Corporation

ANALOG 4 mA TO 20 mA SIGNAL

③ IF ACTUATOR IS USING A 4 mA TO 20 mA ANALOG SIGNAL, USE PROCESS CLAMP METER TO MEASURE CURRENT

FROM SENSOR: 0 V TO 10 V OR 4 mA TO 20 mA INPUT SIGNAL

④ RECORD THE MEASURED DATA

- SENSOR NOT ACTUATED: mA 0.0% 4.00
- SENSOR ACTUATED: mA 50% 12.00

DRAIN WIRE NOT CONNECTED

ANALOG OUTPUTS — DIGITAL INPUTS

⑥ ANALYZE THE DATA REQUIRED

CONTROL VALVE
Worcester Controls Corporation

① VERIFY THAT THE CORRECT SUPPLY VOLTAGE IS PRESENT AT THE ACTUATOR'S TERMINAL STRIP

WHEN THERE IS A CHANGE IN THE MEASURED CURRENT OR VOLTAGE, THE ACTUATOR IS THE SOURCE OF THE PROBLEM. WHEN NO CHANGE IN VOLTAGE OR CURRENT IS MEASURED, THE WIRING BETWEEN THE PLC AND ACTUATOR IS THE LIKELY SOURCE OF THE PROBLEM

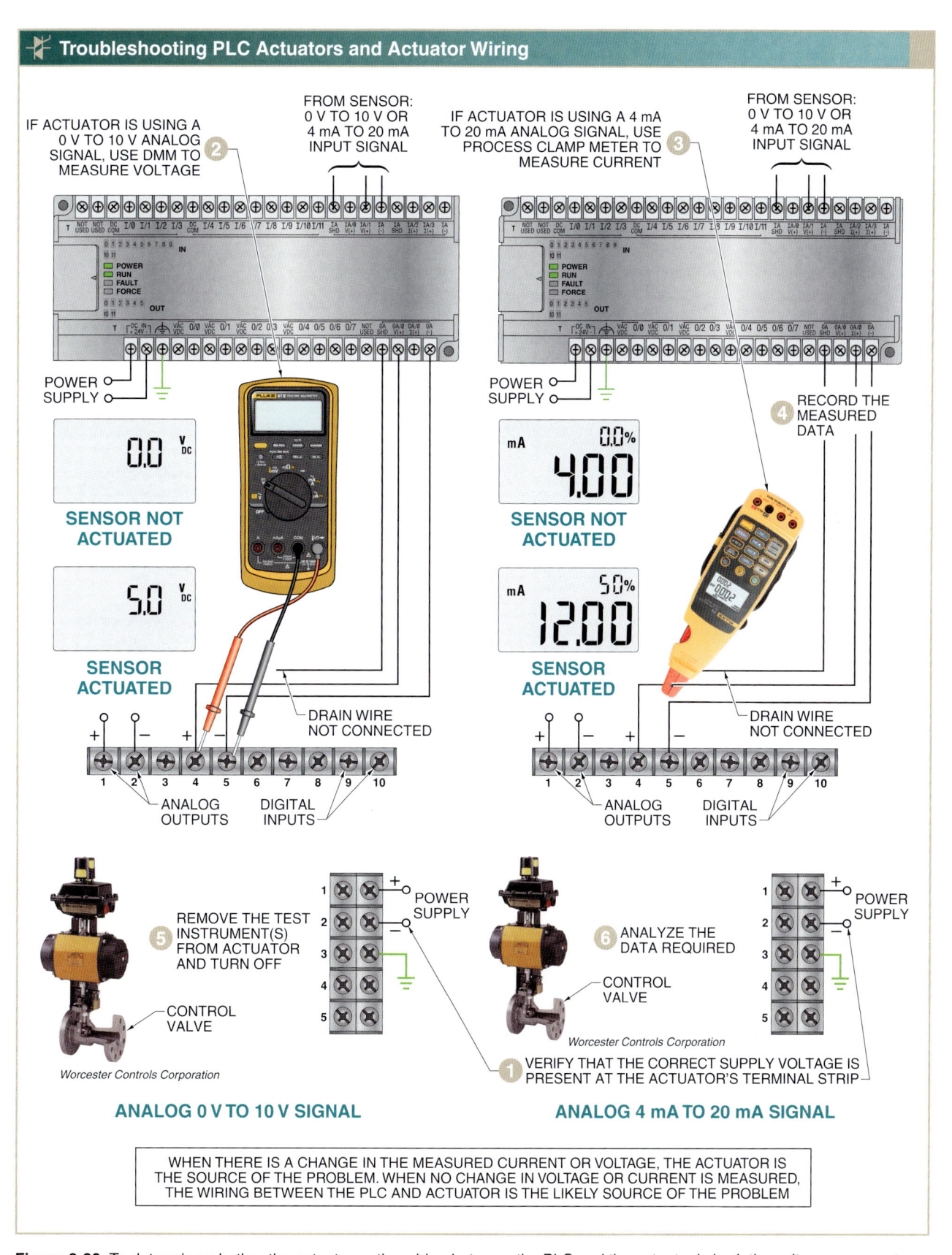

Figure 8-20. To determine whether the actuator or the wiring between the PLC and the actuator is bad, the voltage or current can be checked at the actuator.

Troubleshooting PLC Analog Output Terminals

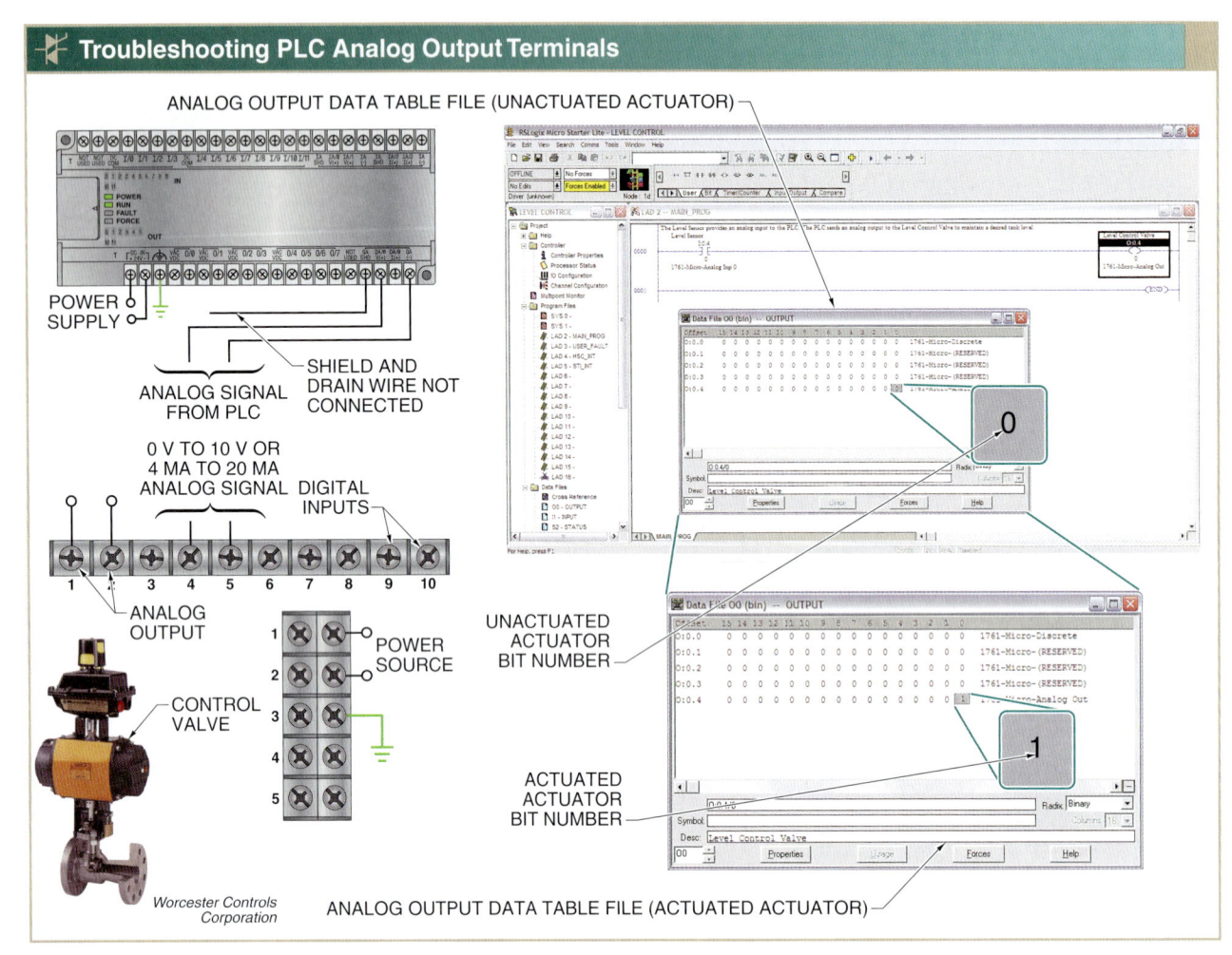

Figure 8-21. The numbers in the analog output data table file should change as the sensor associated with the analog output is actuated.

When taking voltage measurements of analog output terminals, the negative test lead should be connected to a good PLC or module ground. Other enclosure or cabinet grounds are not usable.

When the numbers in the analog output data table file do not change, there may be a problem with the upstream data source for the analog output. The technician should verify that the data source is supplying data to the analog output. The data source for the analog output may be an analog input or a file that contains the results of data manipulation instructions. When data is not being supplied correctly, the technician must troubleshoot the analog input circuit or review the data manipulation within the programming for discrepancies.

Chapter Review 8

Name _____ Date _____

True-False

T F 1. Analog sensors can only be connected to analog input terminals on a PLC.

T F 2. Analog actuators are connected to the analog output terminals of a PLC and receive digital signals from the output section.

T F 3. The most common analog input signal from an input device to a PLC is 0 VDC to 5 VDC.

T F 4. Analog input devices (sensors) can be mounted any distance from the material or properties being measured.

T F 5. Temperature sensors with probes sometimes do not have direct contact with the product due to the use of wells.

T F 6. STP cable must be run in separate conduit from other energized conductors.

T F 7. Sensors and actuators always use PLCs to power the motors and circuits of a sensor or actuator.

T F 8. The Startup screen allows a technician to adjust numerous parameters related to the PID instruction to optimize analog control.

T F 9. XIC instructions are used for analog input devices.

T F 10. When troubleshooting, a change in the current or voltage measured at a PLC's input terminal means the sensor is the source of the problem.

Completion

_____ 1. ___ can distort analog signals and result in inaccurate measurements.

_____ 2. The ___ of a dialog box allows the output mode of an analog signal (voltage or current) to be chosen, the analog inputs to be enabled, and the input filters to be adjusted.

_____ 3. A(n) ___-bit word is reserved for each analog input and output terminal in the input and output data table files.

_____ 4. The ___ or current from a PLC to an actuator can be checked at the analog output terminals of the PLC.

_____ 5. The numbers in an analog output data table file should change as the sensor associated with the ___ is actuated.

Multiple Choice

1. Fluid product sensors must be installed where the product ___.
   A. keeps the piping full
   B. empties from the piping
   C. is always on the move
   D. never moves

2. The two types of analog signals used in most applications are ___ mA to ___ mA or 0 VDC to 10 VDC.
   A. 0; 10
   B. 4; 20
   C. 10; 20
   D. 20; 4

3. The shield and drain wire can only be terminated at the ___ end.
   A. sensor
   B. actuator
   C. PLC
   D. ground

4. The two methods of entering input and output addresses into a PID instruction are either ___.
   A. the digital method or the analog method
   B. to enter them directly into the instruction line or to use separate integer files
   C. with a PC or with a PLC display (PLC buttons)
   D. the 16-bit method or the 32-bit method

5. To cause a change in state in the current or voltage to an actuator, a force must occur or ___.
   A. the PLC program must override the input device signal
   B. the data table must be opened
   C. the proper input terminal of the PLC must be jumped
   D. a change in state of the input device (sensor) must occur

Analog Sensor Control

PLC Exercise 8.1

BILL OF MATERIALS
- TECO PLR or Micrologix™ 1100 PLC
- Motor starter/contactor
- Selector switch
- Analog simulator 0–10 VDC or solar cell (6 VDC output)

The automotive industry utilizes air-compression tools along assembly lines. Analog sensors are used in these processes to control the parameters of air-compression tanks. For this lab, an OFF/ON selector switch enables an air compressor, and an analog pressure sensor maintains the pressure of the air compressor. When the pressure is less than or equal to a reference of 4.00, the compressor is started and latched ON (SET). The compressor continues to run until the analog pressure sensor is greater than or equal to a reference of 5.00. At a reference of 5.00, the compressor is latched OFF (RESET).

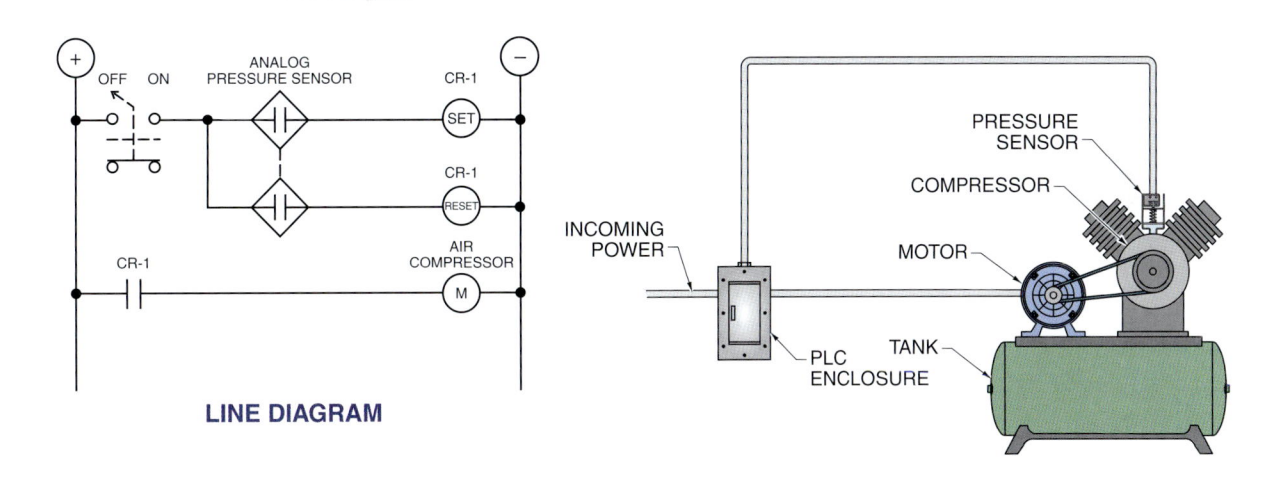

LINE DIAGRAM

TECO Procedures

1. Create a new FBD program.

2. Use the FBD to add the proper input(s), output(s), logic, and function blocks with descriptions in the Programming Area.

3. Insert the device input(s), analog input(s), output(s), logic function(s), and function block(s). The set/reset logic block (RS Trigger) is denoted SET/RST. For analog function blocks, add the Gx Function Block with the proper description, number, mode, current value, and reference value to the FBD Programming Area.

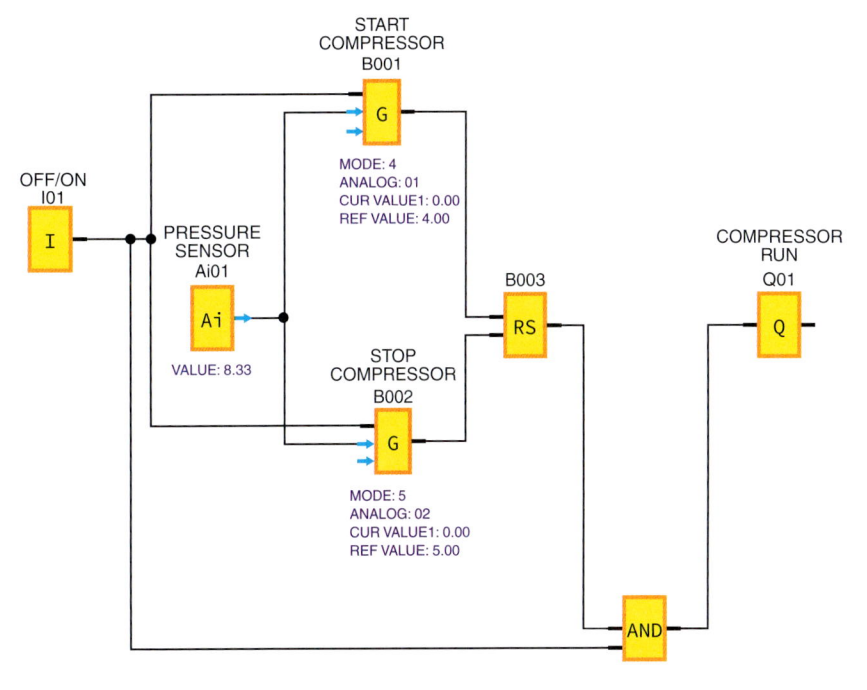

FBD PROGRAMMING AREA

4. Run the program in Simulation Mode to verify proper operation. Use the Input Status tool and Simulation Analog tool window to activate the input(s) and control the output(s) to simulate circuit operation.

5. Use the wiring diagram to wire the input devices and output components to the PLC.

6. Turn power to the PLC ON and link the com port.

7. Write the program to the SG2-12HR-D to download the program logic to the PLC.

8. Place the SG2-12HR-D in Run Mode. Activate the input(s) and monitor the I/Os on the LCD display.

9. Stop the program and use the Save As function to save and place the lab application in the required folder location.

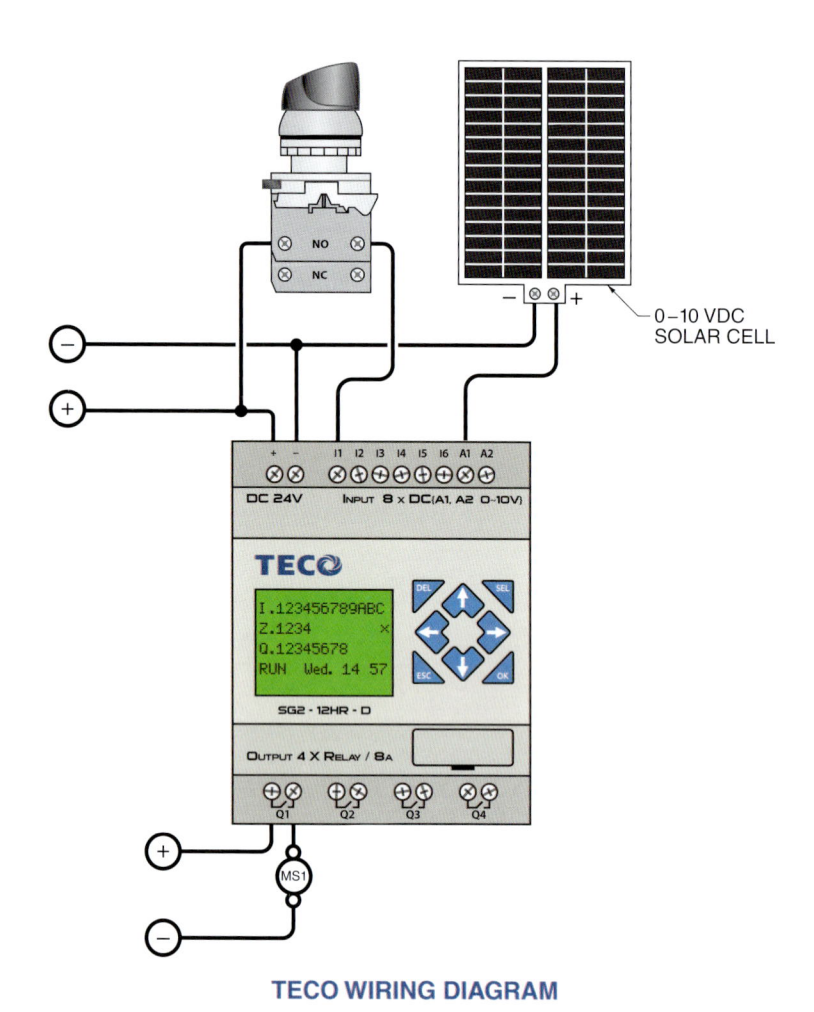

TECO WIRING DIAGRAM

Allen-Bradley® Procedures

1. Open the RSLogix™ Micro Starter Lite program and create a new programming diagram.

2. Use the AB programming diagram to add the proper address and description for each input and output instruction in the LAD 2 programming window. For analog instructions, use the LES (less than) and GRT (greater than) instruction with a word address of I:0.4 in the programming area. *Note:* I:0.4 is the word address for the IV1(+) contact on the PLC. Adjust the values of source A and source B dependent on the input value of the analog sensor/simulator.

3. Verify that the programming diagram has no errors and save the project.

4. Use the wiring diagram to wire the input devices and output components to the PLC. Connect the analog sensor positive (+) to the IV1(+) contact and the negative (−) to the IA COM contact. Once all connections have been made, turn power to the PLC on.

5. Use RSLinx™ to establish communication between the PC and the Micrologix™ 1100.

6. Download the program logic to the PLC.

7. Place the PLC in Run Mode.

8. Activate input(s) and monitor them to determine if the output(s) are operating correctly.

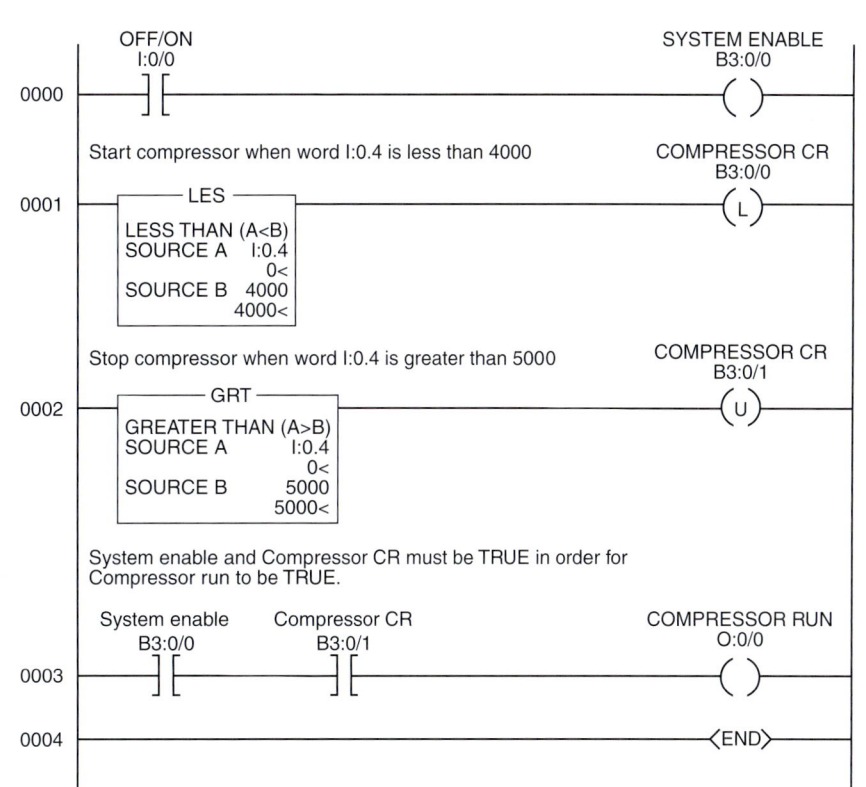

```
                OFF/ON                                          SYSTEM ENABLE
                I:0/0                                           B3:0/0
0000            ─┤ ├─                                           ─( )─

        Start compressor when word I:0.4 is less than 4000      COMPRESSOR CR
                                                                B3:0/0
0001    ┌─ LES ──────────────┐                                  ─(L)─
        │ LESS THAN (A<B)    │
        │ SOURCE A    I:0.4  │
        │               0<   │
        │ SOURCE B    4000   │
        │            4000<   │
        └────────────────────┘

        Stop compressor when word I:0.4 is greater than 5000    COMPRESSOR CR
                                                                B3:0/1
0002    ┌─ GRT ──────────────┐                                  ─(U)─
        │ GREATER THAN (A>B) │
        │ SOURCE A    I:0.4  │
        │               0<   │
        │ SOURCE B    5000   │
        │            5000<   │
        └────────────────────┘

        System enable and Compressor CR must be TRUE in order for
        Compressor run to be TRUE.

        System enable    Compressor CR                          COMPRESSOR RUN
        B3:0/0           B3:0/1                                 O:0/0
0003    ─┤ ├─────────────┤ ├─                                   ─( )─

0004                                                            ─<END>─
```

AB PROGRAMMING DIAGRAM

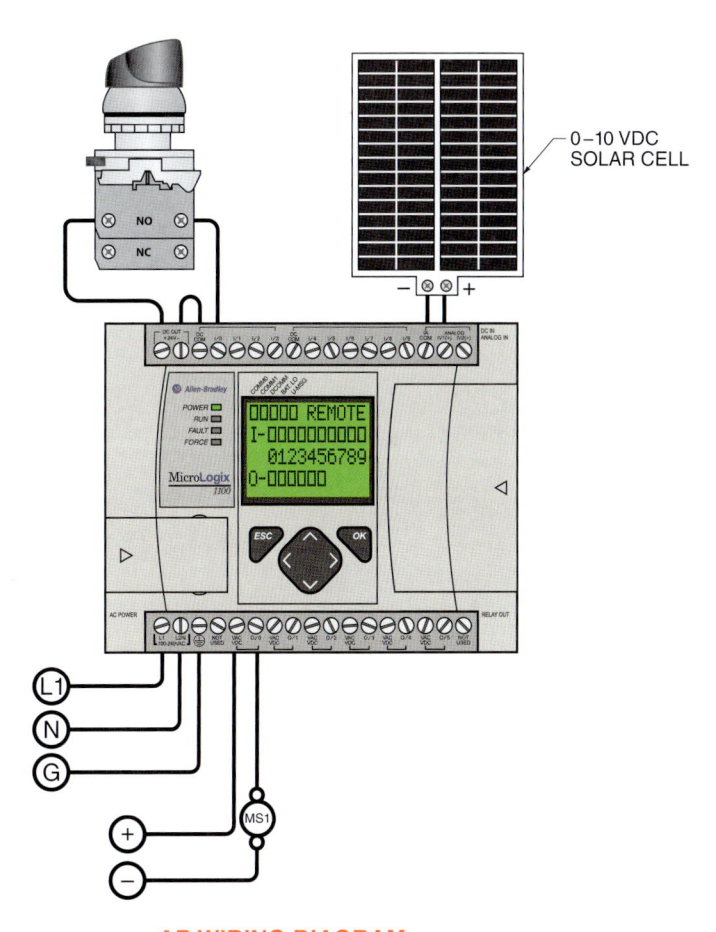

AB WIRING DIAGRAM

SIEMENS Programming and Wiring

Use the S7-1200 Easy Book to program and wire the equivalent TECO and Allen-Bradley exercises. Reference page 70 for Siemens STEP 7 programming symbols.

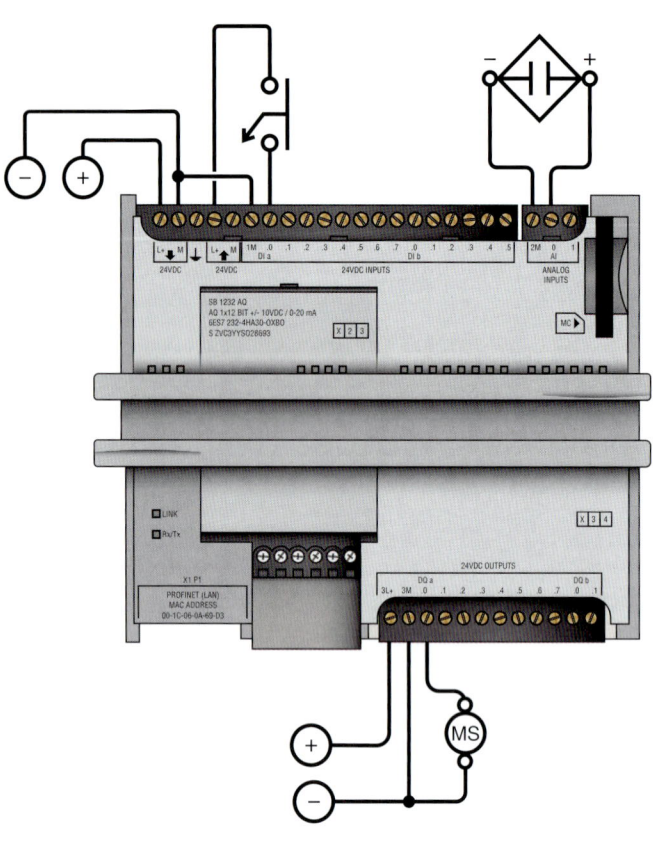

SIEMENS WIRING DIAGRAM

Chapter 8—PLC Analog Device Installation, Programming, and Troubleshooting   341

Analog Control in a Flow System

PLC Lab 8.2

BILL OF MATERIALS
- TECO PLR or Micrologix™ 1100 PLC
- Two motor starters/contactors
- NO selector switch
- Two analog simulators 0–10 VDC or solar cells (6 VDC output)

In the food-and-beverage, petroleum, and chemical industries, analog sensors are used to control flow sensors in pump systems. For this lab, a selector switch enables a supply pump that feeds liquid into a tank. The tank then feeds three filling stations. An analog flow-in sensor measures the flow into the tank. A second analog flow-out sensor measures the flow out of the tank. When the flow out of the tank is equal to or greater than the flow into the tank, an auxiliary pump is started. When the supply pump is enabled, there is a 10-second delay before the flow sensors can activate the auxiliary pump. The 10-second delay allows the tank to be partially filled.

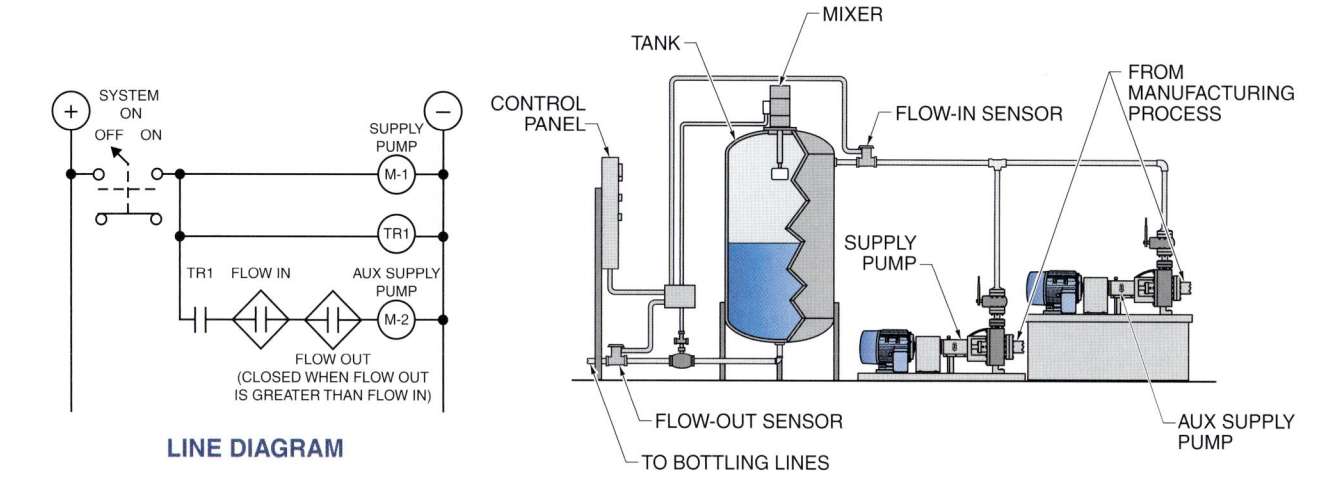

LINE DIAGRAM

T TECO Procedures

1. Based on the PLC system description and line diagram, draw the FBD logic with the proper input(s), output(s), logic, function blocks with descriptions in the Programming Area.

2. Create a new FBD program electronically.

3. Insert the device input(s), analog input(s), output(s), logic function(s), and function block(s). For analog function blocks, add the Gx Function Block with the proper

FBD PROGRAMMING AREA

description, number, mode, current value, and reference value to the FBD Programming Area. Add a timer with the proper mode, preset value, and description to the FBD Programming Area.

4. Run the program in Simulation Mode to verify proper operation. Use the Input Status tool and Simulation Analog tool window to activate the input(s) and control the output(s) to simulate circuit operation.

5. Draw the wire connections of the input devices and output components to the PLC in the wiring diagram. Then use the wiring diagram to wire the input devices and output components to the PLC.

6. Turn power to the PLC ON and link the com port.

7. Write the program to the SG2-12HR-D to download the program logic to the PLC.

8. Place the SG2-12HR-D in Run Mode. Activate the input(s) and monitor the I/Os on the LCD display.

9. Stop the program and use the Save As function to save and place the lab application in the required folder location.

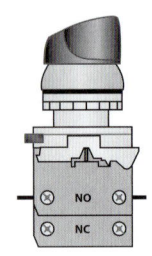

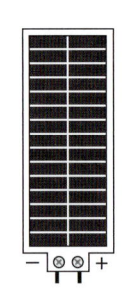

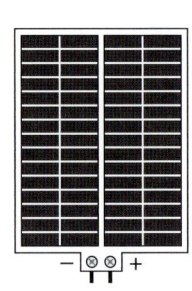

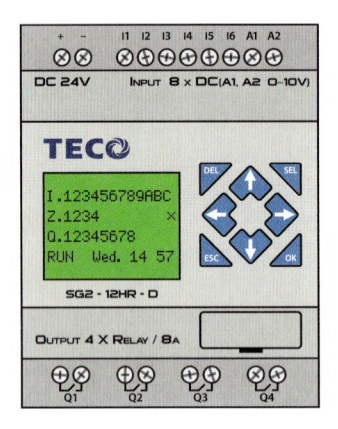

TECO WIRING DIAGRAM

Allen-Bradley® Procedures

1. Based on the PLC system description and line diagram, draw the ladder logic for RSLogix™ in the AB programming area.

2. Open the RSLogix™ Micro Starter Lite program and create a new programming diagram.

3. Use the AB programming diagram to add the proper address and description for each input and output instruction in the LAD 2 programming window. For the analog instruction, use the GEQ (greater than or equal to) instruction with a word address of I:0.4 and I:0.5 in the programming area. *Note:* Adjust the values of source A and source B dependent on the input value of the analog sensor/simulator. Add the proper timer instructions to the programming diagram and type in the timer address, time base, and preset value for each.

4. Verify that the programming diagram has no errors and save the project.

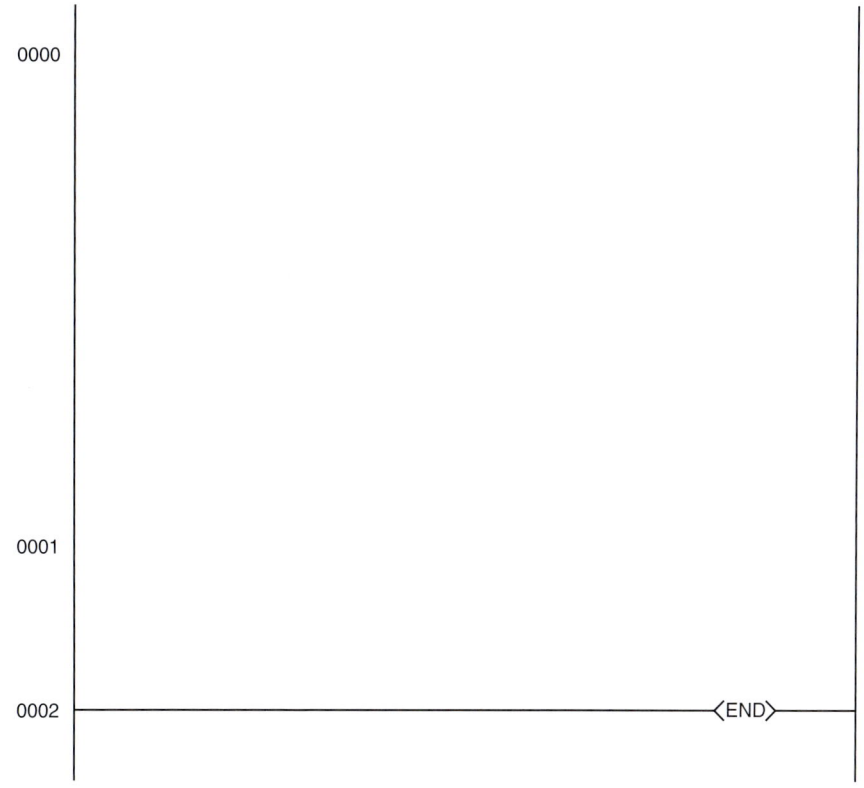

```
0000 |                                                              |
     |                                                              |
0001 |                                                              |
     |                                                              |
0002 |─────────────────────────────────────────────────────⟨END⟩────|
```

AB PROGRAMMING AREA

344 PROGRAMMABLE LOGIC CONTROLLERS: PRINCIPLES AND APPLICATIONS

5. Draw the wire connections of the input devices and output components to the PLC in the wiring diagram. Then use the wiring diagram to wire the input devices and output components to the PLC. Once all connections have been made, turn power to the PLC on.

6. Use RSLinx™ to establish communication between the PC and the Micrologix™ 1100.

7. Download the program logic to the PLC.

8. Place the PLC in Run Mode.

9. Activate input(s) and monitor them to determine if the output(s) are operating correctly.

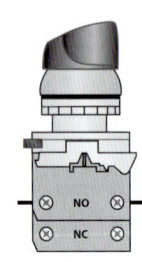

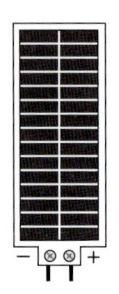

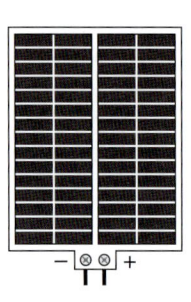

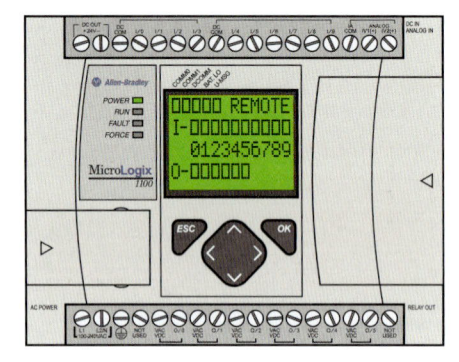

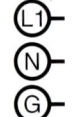

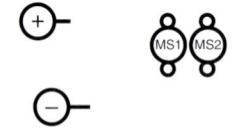

AB WIRING DIAGRAM

Analog Level Sensor Control of a Mixing Motor

PLC Lab 8.3

BILL OF MATERIALS
- TECO PLR or Micrologix™ 1100 PLC
- Motor starter/contactor
- NO selector switch
- Analog simulator (0–10 VDC) or solar cell (minimum 6 VDC output)
- Indicator light/alarm

For this lab, an OFF/ON selector switch enables a mixing motor for an ink tank. An ultrasonic level sensor turns the motor ON when the level is 3.00 or greater, which ensures that the agitator is below the surface of the ink. The mixing motor continues to run until the system is either shut OFF, the level drops below 3.00, or the level is 5.00 or greater. When the level is 5.00 or greater, the mixing motor shuts OFF, and a high-level alarm turns ON.

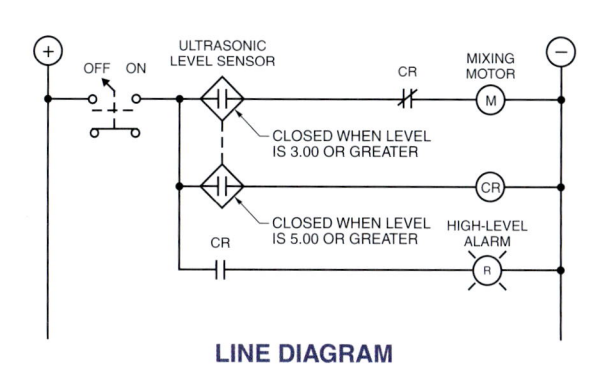

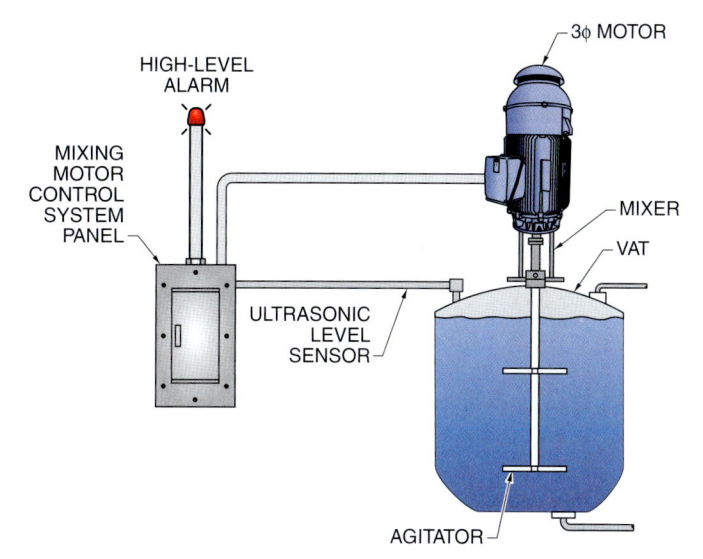

LINE DIAGRAM

T TECO Procedures

1. Based on the PLC system description and line diagram, draw the FBD logic with the proper input(s), output(s), logic, and function blocks with descriptions in the Programming Area.

2. Create a new FBD program electronically.

FBD PROGRAMMING AREA

3. Insert the device input(s), analog input(s), output(s), logic function(s), and function block(s). For analog function blocks, add the Gx Function Block with the proper description, number, mode, current value, and reference value to the FBD Programming Area.

4. Run the program in Simulation Mode to verify proper operation. Use the Input Status tool and Simulation Analog tool window to activate the input(s) and control the output(s) to simulate circuit operation.

5. Draw the wire connections of the input devices and output components to the PLC in the wiring diagram. Then use the wiring diagram to wire the input devices and output components to the PLC.

6. Turn power to the PLC ON and link the com port.

7. Write the program to the SG2-12HR-D to download the program logic to the PLC.

8. Place the SG2-12HR-D in Run Mode. Activate the input(s) and monitor the I/Os on the LCD display.

9. Stop the program and use the Save As function to save and place the lab application in the required folder location.

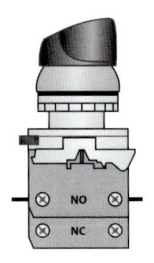

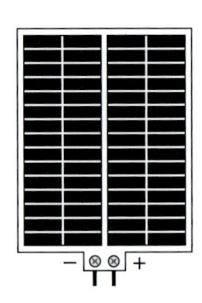

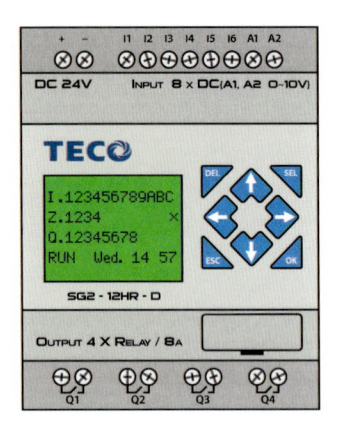

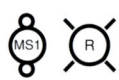

TECO WIRING DIAGRAM

Allen-Bradley® Procedures

1. Based on the PLC system description and line diagram, draw the ladder logic for RSLogix™ in the AB programming area.

2. Open the RSLogix™ Micro Starter Lite program and create a new programming diagram.

3. Use the AB programming diagram to add the proper address and description for each input and output instruction in the LAD 2 programming window. For the analog instructions, use the LES (less than) and GRT (greater than) instructions with a word address of I:0.4 in the programming area. *Note:* Adjust the values of source B dependent on the input value of the analog sensor/simulator.

4. Verify that the programming diagram has no errors and save the project.

5. Draw the wire connections of the input devices and output components to the PLC in the wiring diagram. Then use the wiring diagram to wire the input devices and output components to the PLC. Once all connections have been made, turn power to the PLC on.

6. Use RSLinx™ to establish communication between the PC and the Micrologix™ 1100.

7. Download the program logic to the PLC.

8. Place the PLC in Run Mode.

9. Activate input(s) and monitor them to determine if the output(s) are operating correctly.

AB PROGRAMMING AREA

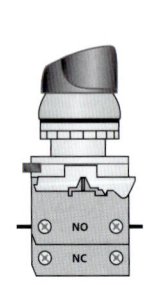

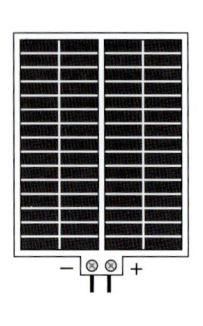

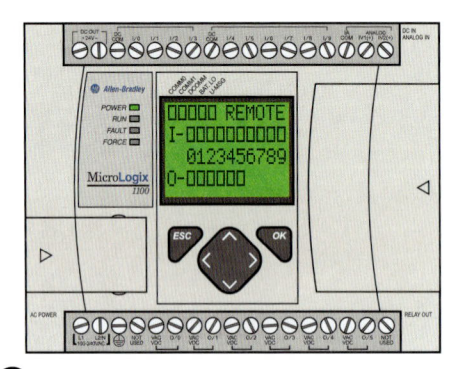

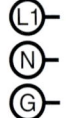

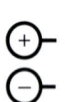

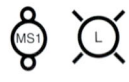

348 PROGRAMMABLE LOGIC CONTROLLERS: PRINCIPLES AND APPLICATIONS

AB WIRING DIAGRAM

Analog Sensor Monitoring of a Large Critical Motor

PLC Lab 8.4

BILL OF MATERIALS

- TECO PLR or Micrologix™ 1100 PLC
- Two motor starters/contactors
- NO pushbutton
- NC pushbutton
- Analog simulator (0–10 VDC) or solar cell (6 VDC output)
- Indicator light/alarm

Large motors are used in industry to drive machinery, actuation systems, blowers, and large conveyors. For this lab, a large critical motor has an attached cooling fan. The motor is controlled by a PLC with a 3-wire stop/start input. An integrated thermal sensor monitors the motor temperature during operation. At a level of 3 or higher, the thermal sensor turns the attached cooling fan and a high temperature warning light ON. At a level of 5 or higher, the thermal sensor shuts the motor and cooling fan OFF.

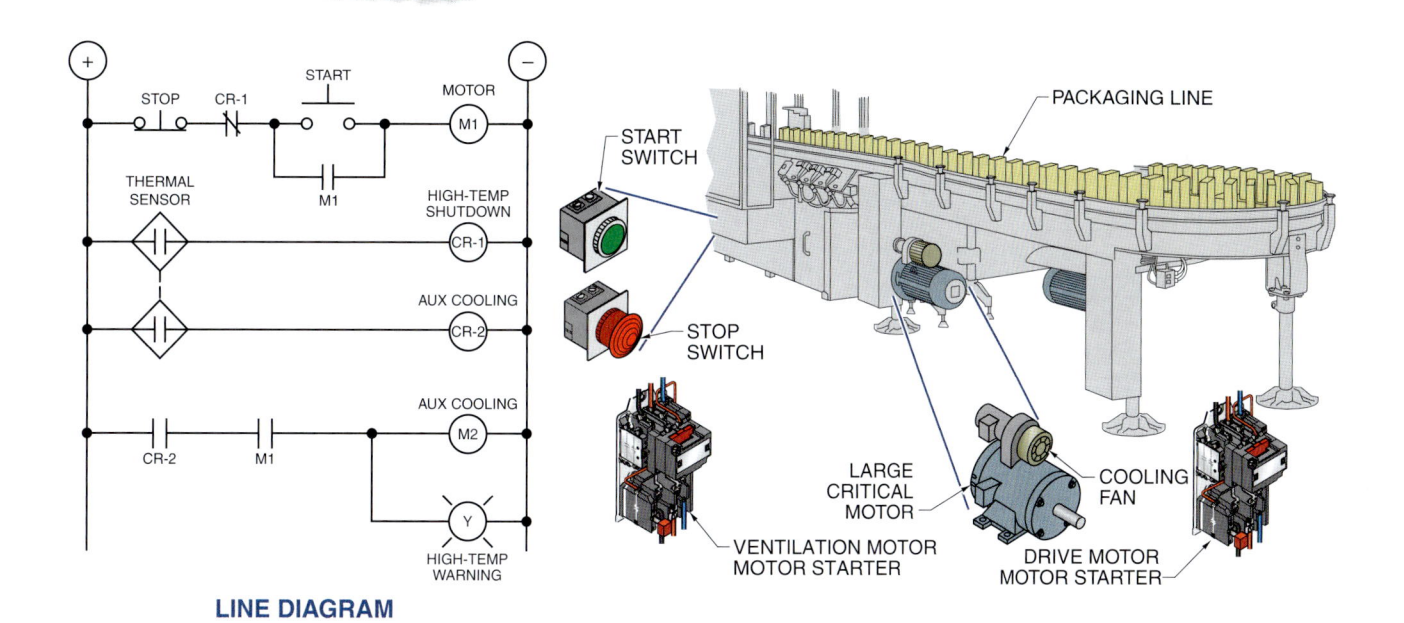

LINE DIAGRAM

TECO Procedures

1. Based on the PLC system description and line diagram, draw the FBD logic with the proper input(s), output(s), logic, and function blocks with descriptions in the Programming Area.

2. Create a new FBD program electronically.

3. Insert the device input(s), analog input(s), output(s), logic function(s), and function block(s). For analog function blocks, add the Gx Function Block with the proper description, number, mode, current value, and reference value to the FBD Programming Area.

4. Run the program in Simulation Mode to verify proper operation. Use the Input Status tool and Simulation Analog tool window to activate the input(s) and control the output(s) to simulate circuit operation.

5. Draw the wire connections of the input devices and output components to the PLC in the wiring diagram. Then use the wiring diagram to wire the input devices and output components to the PLC.

6. Turn power to the PLC ON and link the com port.

7. Write the program to the SG2-12HR-D to download the program logic to the PLC.

8. Place the SG2-12HR-D in Run Mode. Activate the input(s) and monitor the I/Os on the LCD display.

9. Stop the program and use the Save As function to save and place the lab application in the required folder location.

FBD PROGRAMMING AREA

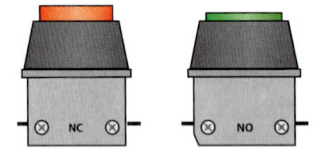

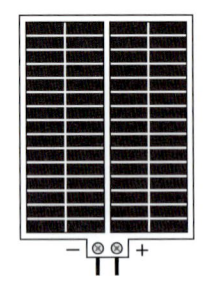

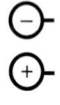

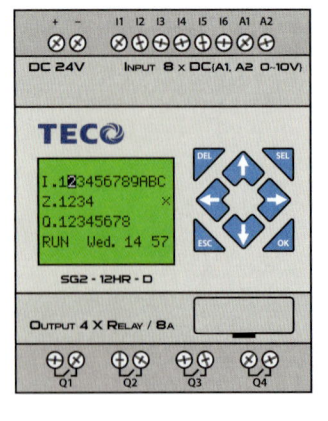

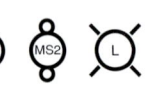

TECO WIRING DIAGRAM

Allen-Bradley® Procedures

1. Based on the PLC system description and line diagram, draw the ladder logic for RSLogix™ in the AB programming area.

2. Open the RSLogix™ Micro Starter Lite program and create a new programming diagram.

3. Use the AB programming diagram to add the proper address and description for each input and output instruction in the LAD 2 programming window. For the analog instructions, use the GRT (greater than) instruction with a word address of I:0.4 in the programming area. *Note:* Adjust the values of source B dependent on the input value of the analog sensor/simulator.

4. Verify that the programming diagram has no errors and save the project.

5. Draw the wire connections of the input devices and output components to the PLC in the wiring diagram. Then use the wiring diagram to wire the input devices and output components to the PLC. Once all connections have been made, turn power to the PLC on.

6. Use RSLinx™ to establish communication between the PC and the Micrologix™ 1100.

7. Download the program logic to the PLC.

8. Place the PLC in Run Mode.

9. Activate input(s) and monitor them to determine if the output(s) are operating correctly.

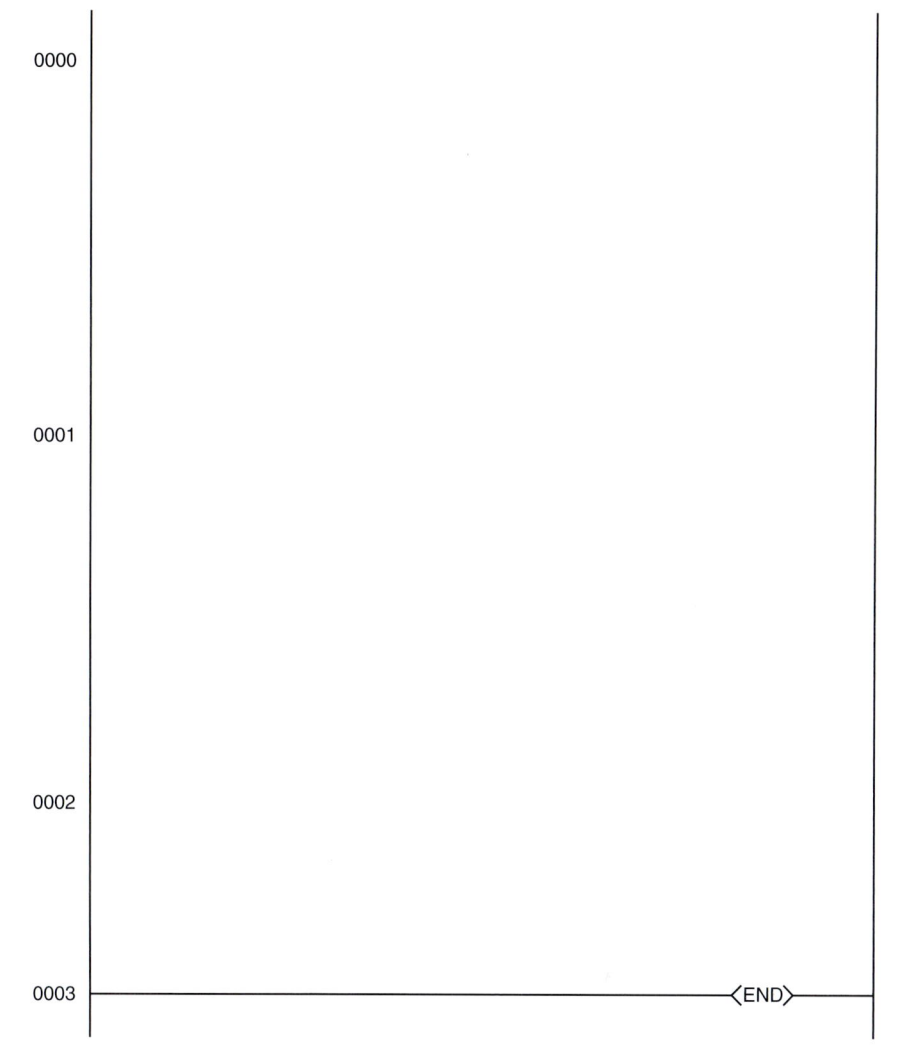

AB PROGRAMMING AREA

0000
0001
0002
0003 ─────────────────────⟨END⟩

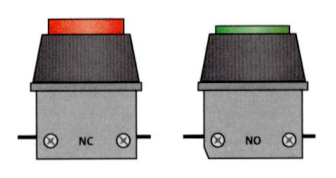

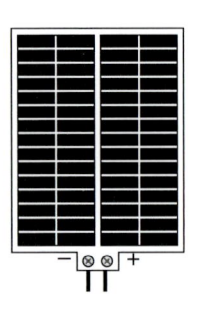

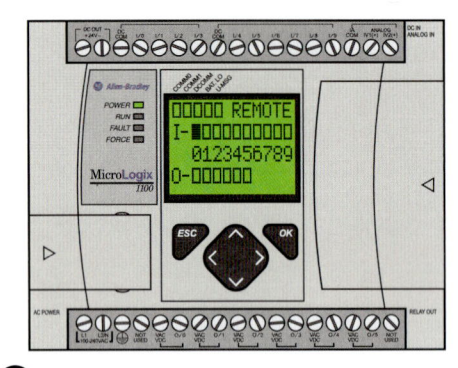

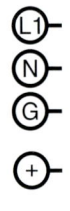

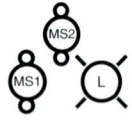

352 PROGRAMMABLE LOGIC CONTROLLERS: PRINCIPLES AND APPLICATIONS

AB WIRING DIAGRAM

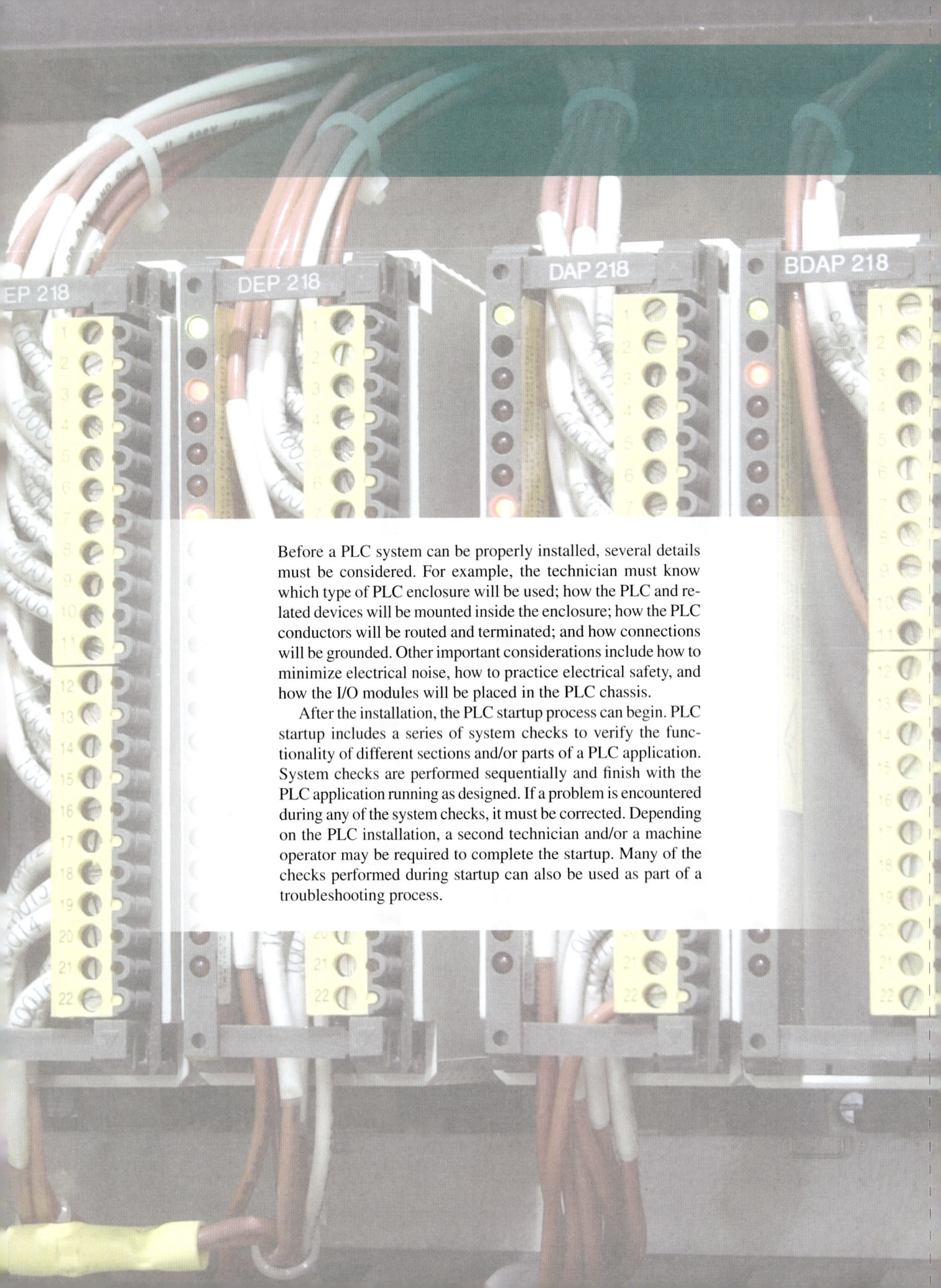

Before a PLC system can be properly installed, several details must be considered. For example, the technician must know which type of PLC enclosure will be used; how the PLC and related devices will be mounted inside the enclosure; how the PLC conductors will be routed and terminated; and how connections will be grounded. Other important considerations include how to minimize electrical noise, how to practice electrical safety, and how the I/O modules will be placed in the PLC chassis.

After the installation, the PLC startup process can begin. PLC startup includes a series of system checks to verify the functionality of different sections and/or parts of a PLC application. System checks are performed sequentially and finish with the PLC application running as designed. If a problem is encountered during any of the system checks, it must be corrected. Depending on the PLC installation, a second technician and/or a machine operator may be required to complete the startup. Many of the checks performed during startup can also be used as part of a troubleshooting process.

PLC Installations and Startup

9

Objectives:
- List the proper procedures for receiving a PLC and related components.
- Describe installation considerations for PLC enclosures and power supplies.
- Describe installation considerations regarding electrical noise and PLCs.
- Demonstrate PLC wiring methods, wiring termination methods, and I/O wiring related to PLC installations.
- Explain the checks involved in a PLC startup and the safety considerations involved in each of the checks.

Learner Resources
atplearningresources.com/quicklinks
Access Code: **475203**

PLC INSTALLATIONS

A number of issues must be considered when preparing to install a PLC. These issues include receiving a PLC, PLC enclosures, electrical noise, power supplies, safety, and wiring. All these issues must be addressed in order for the PLC to function properly. In many instances, the issues are interrelated, such as when improper wiring causes electrical noise problems. **See Figure 9-1.**

The startup of a PLC consists of a series of systematic checks. Systematic checks verify that a PLC is functioning correctly. Checks are performed in sequence to locate problems quickly, verify functionality, and fine-tune the PLC program. Many startup checks are also used to troubleshoot a PLC application. Following a set of startup procedures will result in years of trouble-free service.

⚠ **CAUTION**

Always refer to the installation instructions of the PLC manufacturer and all applicable federal, state, and local regulations to ensure a safe installation.

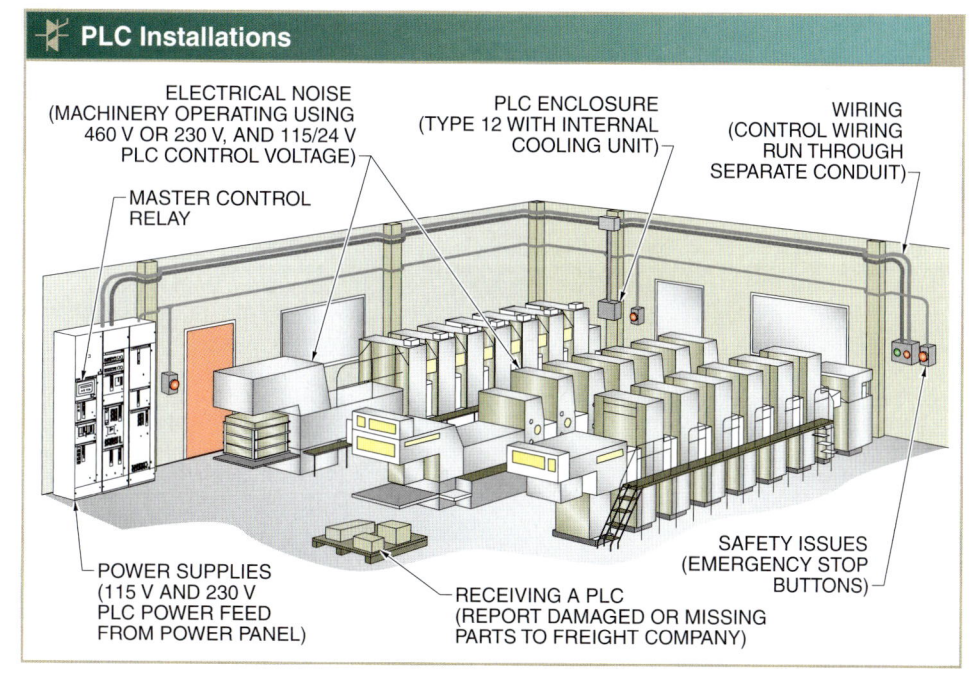

Figure 9-1. Installation issues that a technician must address include receiving a PLC, PLC enclosures, electrical noise, power supplies, safety, and wiring.

355

Receiving PLCs

When receiving a PLC and any associated hardware, there are several procedures that must be followed. These procedures help ensure the proper and safe function of the PLC.

- All items should be thoroughly inspected for any damaged or missing parts. Any problems should be reported to the freight company immediately.
- The model, rating, and configuration of the PLC and any accessories should be checked against the purchase order.
- The instructions, any other documentation, and software should be removed and any paperwork stored in a safe location for future reference.
- The PLC and related hardware should be stored in a clean, dry, secure location that conforms to the recommendations of the manufacturer. **See Figure 9-2.**

PLC Enclosures

PLCs must be mounted in an enclosure. An enclosure protects a PLC from the surrounding environment and protects technicians from accidental contact with energized parts. Typically, metal enclosures are used because metal enclosures provide shielding from electrical noise. Special nonmetallic enclosures are also available that provide shielding from electrical noise.

Enclosures conforming to NEMA Standard 250—*Enclosures for Electrical Equipment (1000 Volts Maximum)* are rated based on use and service conditions (environment). **See Figure 9-3.** The technician installing a PLC must determine the correct type of enclosure for the location and environment.

PLC TIPS

NEMA Standards publication 250 and IEC publication 60529 provide explanations of the protection provided by enclosures.

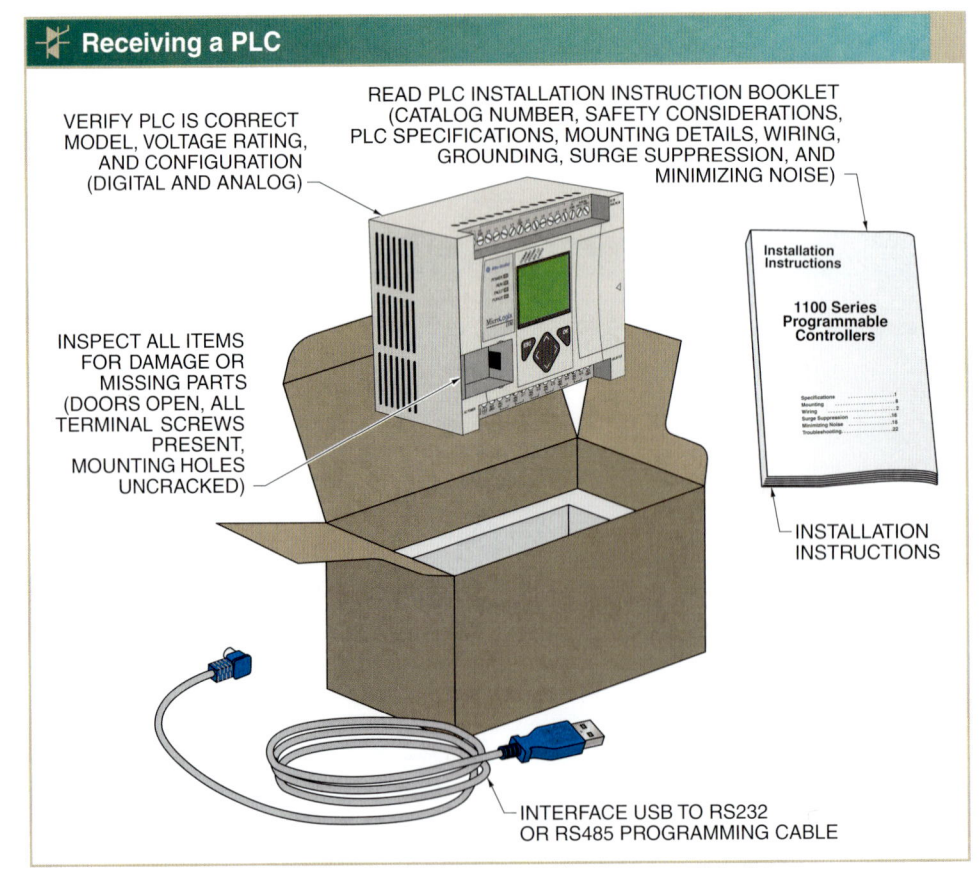

Figure 9-2. When receiving a PLC and any associated hardware, there are several procedures that must be followed.

Chapter 9—PLC Installations and Startup 357

PLC Enclosures

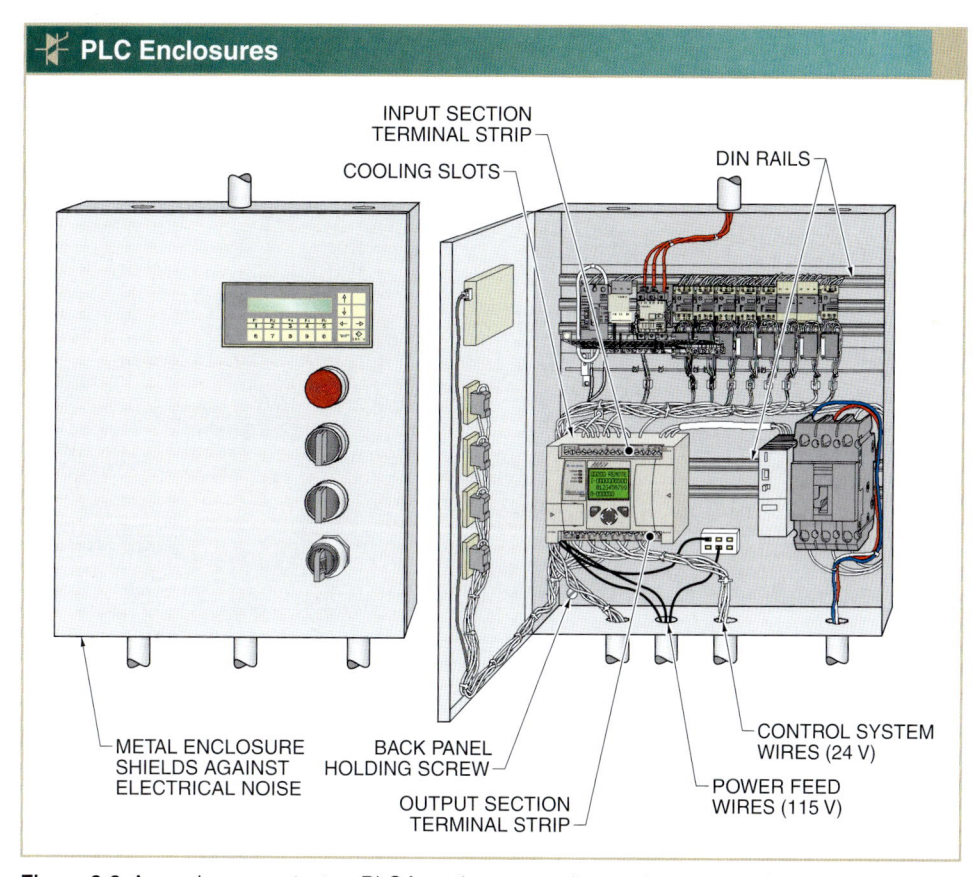

Figure 9-3. An enclosure protects a PLC from the surrounding environment and protects technicians from contact with energized parts.

Labels in figure:
- INPUT SECTION TERMINAL STRIP
- COOLING SLOTS
- DIN RAILS
- METAL ENCLOSURE SHIELDS AGAINST ELECTRICAL NOISE
- BACK PANEL HOLDING SCREW
- OUTPUT SECTION TERMINAL STRIP
- CONTROL SYSTEM WIRES (24 V)
- POWER FEED WIRES (115 V)

> **CODE CONNECT**
>
> Section 314.16 of the NEC® establishes the number of conductors permitted in outlet boxes, device boxes, junction boxes, and conduit bodies. Each has a specific volume. The total volume required by conductors, devices, and other internal fittings must not exceed the volume of the box or conduit body. Table 314.16(A) and Table 314.16(B) of the NEC® provide useful information for calculating the number of conductors allowed inside each.

Mounting. PLCs are typically mounted horizontally with the name of the manufacturer facing out and the top of the PLC facing up on the door or inside wall of the PLC enclosure. **See Figure 9-4.** PLCs are not mounted in any other way, such as at the top or at the base of an enclosure, because of the potential for heat buildup.

Typically, most manufacturers ship a debris shield with the PLC from the factory. A *debris shield* is a piece of heavy paper that covers the ventilation slots on a PLC. It prevents debris, such as metal filings or strands of wire, from entering the PLC during installation. A debris shield should be kept in place until installation is complete. Once the PLC is installed, it must be removed before the PLC is turned on in order to prevent overheating.

Frequently, PLC installation instructions provide minimum spacing requirements for all sides of a PLC and a mounting template that displays the dimensions of the PLC, the expansion modules, and the recommended mounting hardware. Expansion modules are mounted next to a PLC and are attached via a connector. Additional expansion units are connected via a ribbon cable or a connector integral to the PLC and expansion module.

Typically, the enclosure that houses a PLC has a removable back panel. The PLC can be mounted on a DIN rail attached to the back panel or mounted directly to it. If the PLC is mounted on a DIN rail, end anchors are recommended to prevent the PLC from sliding on the rail. If the PLC is mounted directly to the back panel, a mounting template can be used to lay out the mounting holes. Mounting holes may be drilled and tapped to accept threaded machine screws, or self-tapping fine-thread screws may be used. The template must be removed before mounting the PLC. **See Figure 9-5.**

> **⚠ CAUTION**
>
> PLC manufacturers typically recommend a minimum 2″ clearance between any enclosure items, PLCs, wiring duct, and terminal strips.

PLC Mounting Orientations

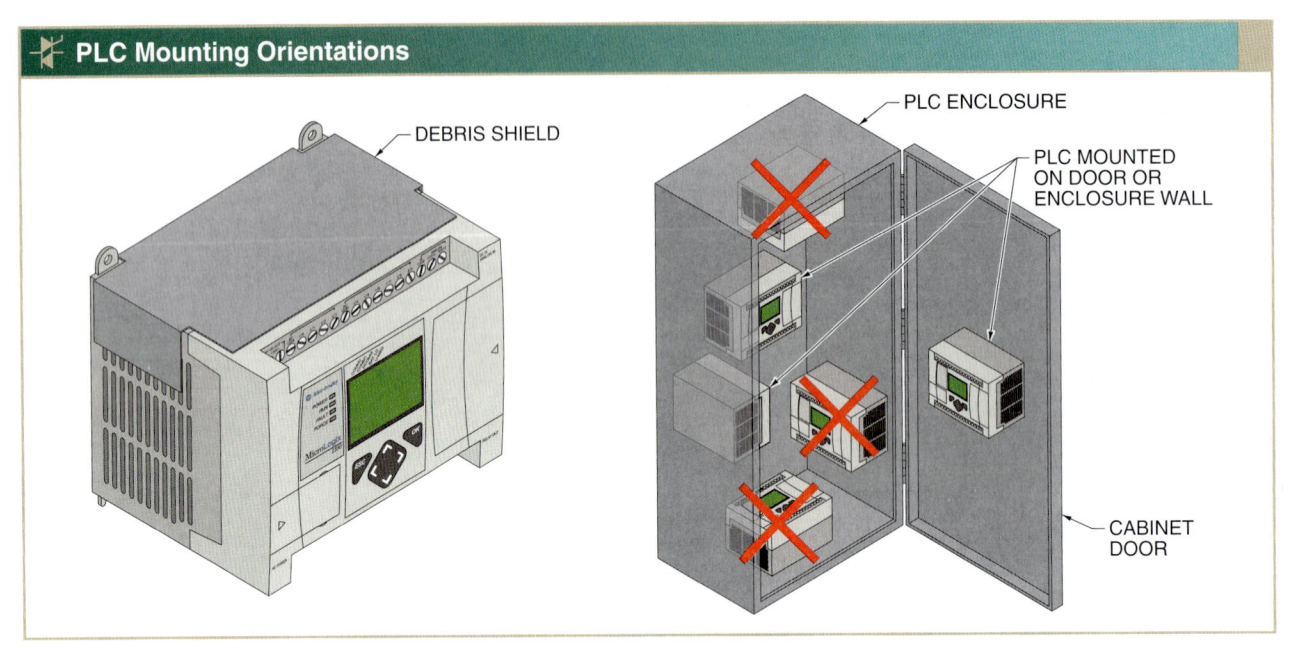

Figure 9-4. PLCs are mounted horizontally right-side up on the door or inside wall of a PLC enclosure.

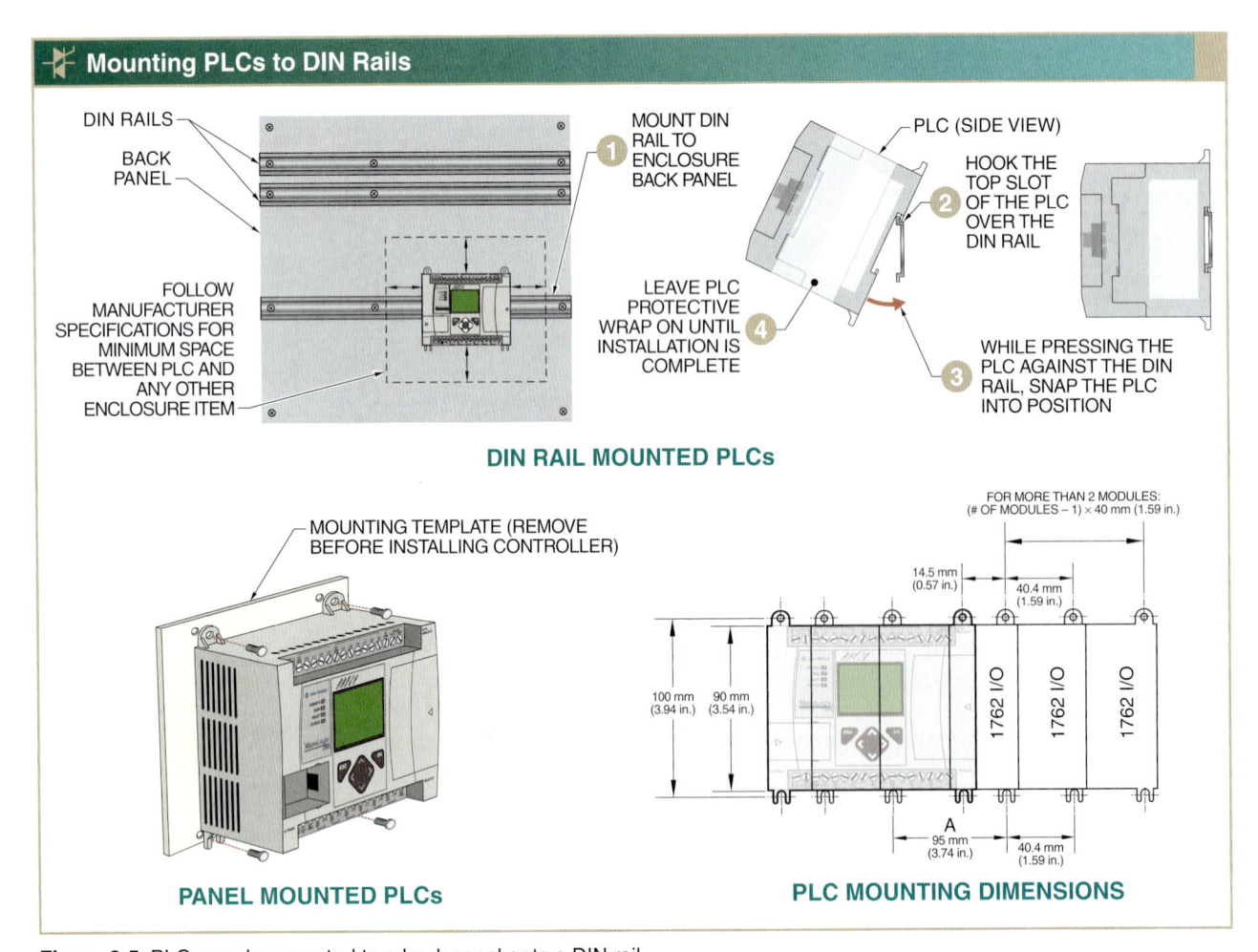

Mounting PLCs to DIN Rails

DIN RAIL MOUNTED PLCs

PANEL MOUNTED PLCs

PLC MOUNTING DIMENSIONS

Figure 9-5. PLCs can be mounted to a back panel or to a DIN rail.

In addition to the PLC, other components may be mounted in the enclosure. Fuses, power supplies, terminal strips, transformers, and wiring ducts are some common components. Technicians must follow the manufacturer's instructions when mounting and spacing these components. *Note:* PLC manufacturers typically recommend a minimum space of 2″ between any of the following components: PLCs, wiring ducts, and terminal strips.

Temperature. PLCs are designed to operate within certain temperature ranges, such as from 32°F to 140°F (0°C to 60°C). It is important that the temperature inside the enclosure remain within the temperature range rating of the PLC. A common problem with PLC enclosures is excessive heat. Electrical devices in an enclosure can cause excessive heat that may result in erratic operation of the PLC or PLC failure.

Following manufacturer recommendations for the spacing of electrical items in an enclosure typically provides adequate room for dissipating heat. When excessive heat is caused by items within an enclosure, a fan or cooling unit is added to the enclosure. When excessive heat is caused by the high ambient temperature surrounding the enclosure, fans, cooling units, and/or solar shields are added to the enclosure. **See Figure 9-6.** Some applications require a combination of cooling methods to prevent excessive heat buildup in an enclosure.

> **⚠ CAUTION**
> *High-heat-generating electrical equipment is typically mounted above a PLC in an enclosure. Devices that can produce high heat include power transistors and power diodes.*

Electrical Noise

All electrical and electronic equipment exhibits some level of immunity to electrical noise, or electromagnetic interference (EMI). Manufacturers take steps to minimize noise emissions and maximize EMI immunity when designing equipment. *Electromagnetic compatibility (EMC)* is the ability of various pieces of electrical equipment to work together with varying levels of noise emission immunity.

Regulations regarding EMI and EMC are very complicated and vary from country to country. Proper installation of a PLC will help prevent EMI problems. **See Figure 9-7.** Installation techniques can be applied and supplemental devices installed to minimize emissions and maximize immunity. Electrical noise is reduced or prevented by following the manufacturer installation instructions and standards in the NEC®.

Enclosure Temperature-Protection Methods

FANS   COOLING UNITS   SOLAR SHIELDS

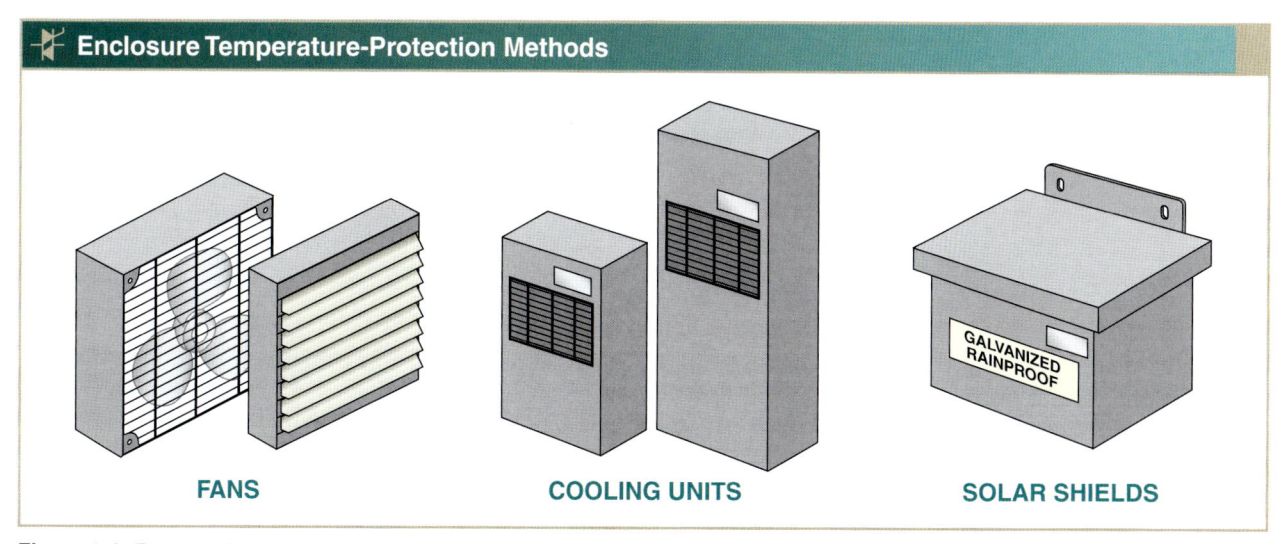

Figure 9-6. Fans, cooling units, and/or solar shields are used to prevent excessive heat buildup inside PLC enclosures.

Electromagnetic Interference (EMI) Prevention

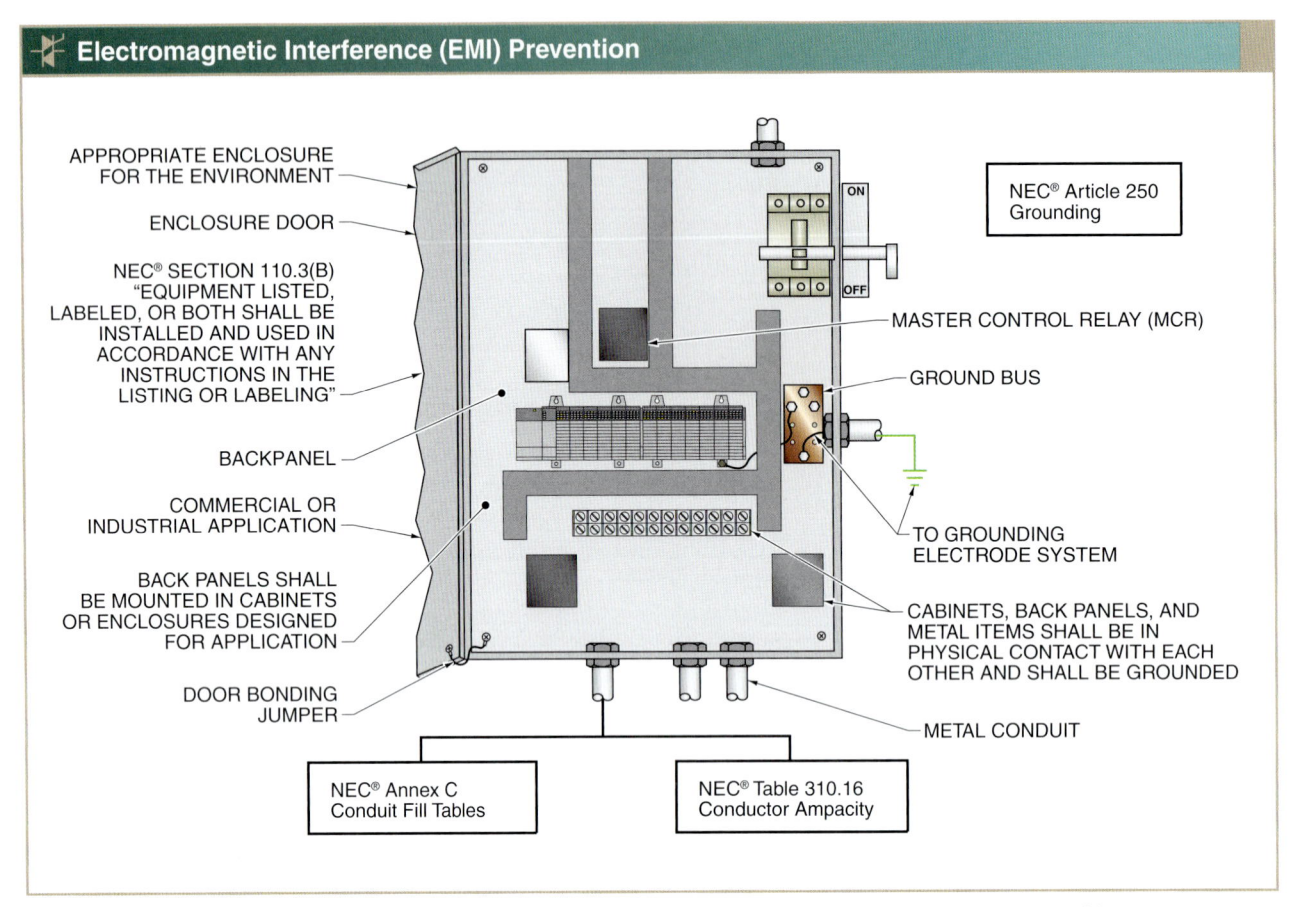

Figure 9-7. Proper installation of a PLC in an enclosure should prevent most electromagnetic interference problems.

PLC Enclosure Setup. The devices located in a PLC enclosure and the input devices and output components located in the field typically require various signal types and voltage levels. **See Figure 9-8.** Typical signal types include digital signals, analog signals, and serial communication signals.

A digital signal operates as an electrical signal that has only two states, ON or OFF. An example would be a momentary pushbutton. An analog signal operates as an electrical signal that varies over a range of values, such as between 0 VDC and 10 VDC. A typical analog signal is the speed-reference signal to an electric motor drive. A *serial communication signal* is a digital data signal from an external source. Digital data signals are generated by electric motor drives, other PLCs, and PCs.

Typical voltage levels for PLC electrically controlled equipment (not directly powered by the PLC) in the field include the following:
- 460 VAC for motors and transformers
- 230 VAC, 115 VAC, or 24 VAC for the power supply to a PLC, PLC input devices, PLC output components, and related control devices
- 5 VDC to 24 VDC for PLC input devices, PLC output components, and related control devices
- 0 VDC to 10 VDC for analog devices or serial communication

Conductors that are run from the PLC enclosure to another location should be in a metal raceway (conduit) regardless of the signal type or voltage level. Metal conduit serves as a shield, preventing conductors from radiating EMI and protecting conductors from any radiated EMI.

PLC Enclosure Setup

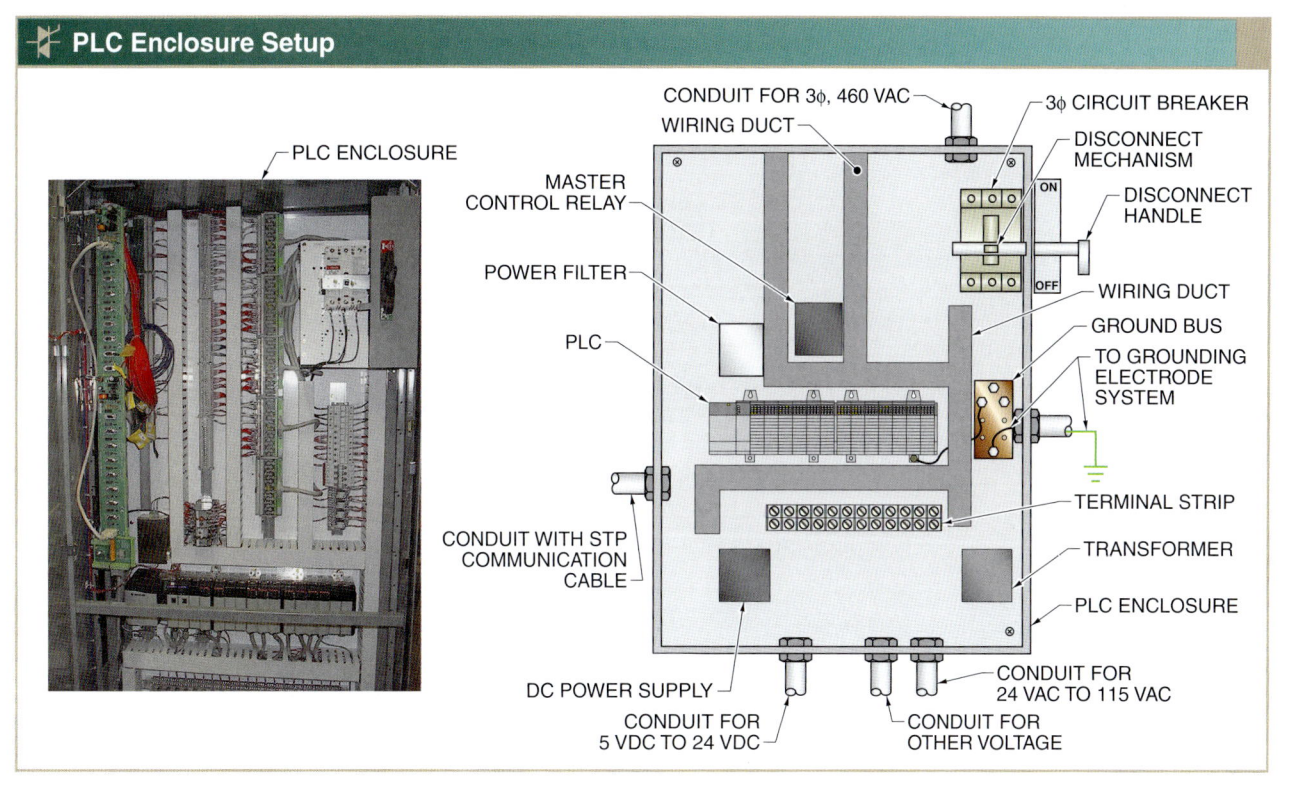

Figure 9-8. PLCs have separate metal raceways (conduits) for voltage supply wires and signal wires.

Specific items must be considered when installing conductors from a PLC to input devices and output components and within the enclosure of the PLC to minimize the harmful effects of EMI:

- Conductors for 460 VAC and 230 VAC must be installed in separate conduits.
- Conductors for 115 VAC and 24 VAC can be installed in the same conduit.
- Conductors for low level DC (5 VDC to 24 VDC) must be installed in separate conduits.
- Conductors for digital signals, analog signals, and serial communication signals must be installed in separate metal raceways based on signal type.
- Analog signals and serial communication signals must be run as shielded twisted pair cable.

Inside a PLC enclosure, analog low-level DC wiring and serial communication wiring must be separated as much as possible from other conductors and from each other. When low-level analog DC wiring or serial communication wiring crosses other conductors, it should cross at 90°. All AC conductors must be routed away from cables that connect PLC modular chassis together.

Grounding. Proper grounding is essential for the safe and reliable operation of a PLC and related equipment. Two types of grounding in PLC enclosures include, equipment grounding and electronic equipment grounding.

Equipment grounding enhances safety by providing an equal potential between all metal components of an installation and a low impedance path for fault currents. Electronic equipment grounding provides a shield to contain EMI. All electronic equipment grounds must be brought back to a single point in a PLC enclosure. The ground lug connections must be tight and mechanically sound to ensure a good electrical connection. When a grounding connection must be made to the interior of a metal enclosure, any paint coatings must be scraped away to ensure good metal-to-metal contact. **See Figure 9-9.**

⚠ CAUTION

Ensure that metal chips from drilling holes do not fall into electrical components. Remove metal filings from the enclosure.

CODE CONNECT

An equipment grounding conductor (EGC) is a conductor or enclosure that connects all noncurrent-carrying metal parts of equipment. If a ground fault, lightning strike, or line surge occurs, the EGC provides a path back to the grounding electrode system for fault currents and activates overcurrent protective devices (OCPDs). Section 250.118 of the NEC® states that an EGC shall be one or more or a combination of the following: intermediate metal conduit/Type IMC (342.60), rigid metal conduit/Type RMC (344.60), electrical metallic tubing/Type EMT (358.60), or copper, aluminum, or copper-clad aluminum conductors.

Ground Lug Connections

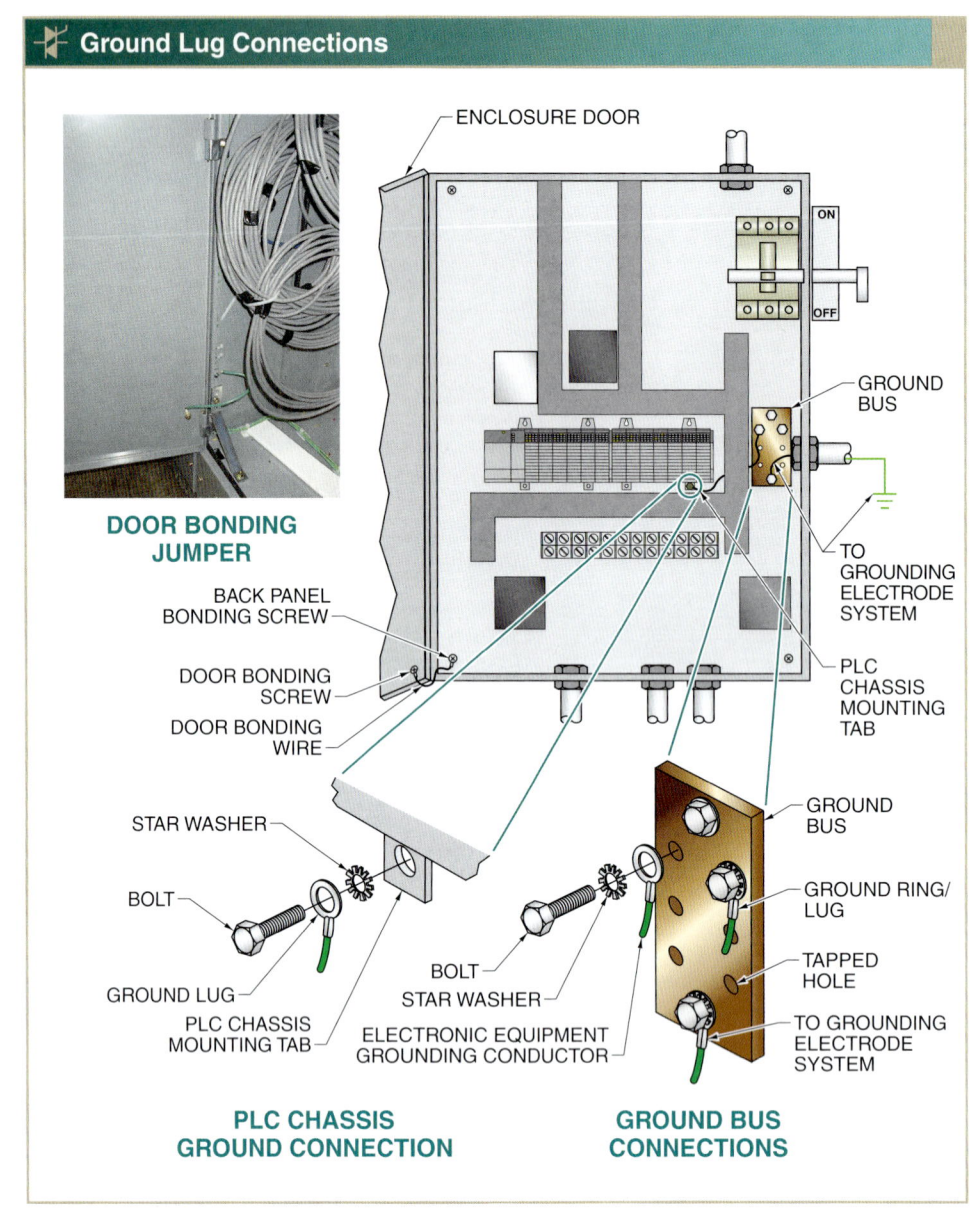

Figure 9-9. Ground lug connections must be tight and mechanically sound to ensure a good electrical connection.

Signal wires carry analog signals and serial communication signals using shielded twisted pair cable. A shield constructed of metal foil mesh or bare wire wraps around the twisted conductors. The bare wire is also referred to as a drain wire. The shield/drain wire and the twisting of the conductors provide enhanced noise reduction for wires carrying signals.

Typically the shield/drain wire is only grounded at one end (the PLC end) when connected to analog input devices. A shield/drain wire is not connected at the input device end of a signal wire. When a shield/drain wire is grounded at both ends, EMI can be introduced into the signal circuit because of the difference of potential between the two points of grounding. When a shield/drain wire is connected at both ends, current will flow through the shield/drain wire. **See Figure 9-10.**

Shielded Twisted Pair Cable Shield/Drain Wire

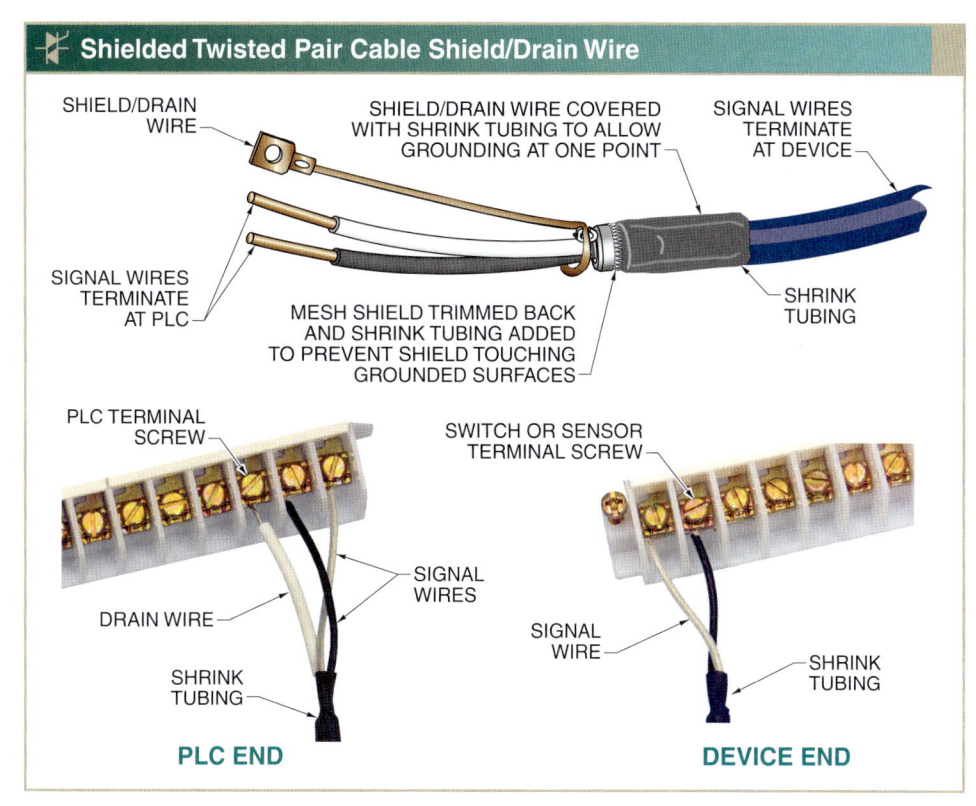

Figure 9-10. The twisting of conductors and the shield/drain wire provide enhanced noise reduction. The shield/drain wire is only grounded at one end (the PLC end) when connected to analog input devices.

PLC Power Supplies

Typically, three-phase power is brought to a PLC enclosure to feed all motor starters, devices, and components in the enclosure. Circuit breakers or fused disconnects provide a disconnecting means and overcurrent/short circuit protection for an enclosure. The primary side of a step-down transformer is connected to two phases, and the secondary side feeds the power supply of the PLC. Typically, the secondary side of a transformer is 115 VAC, with X1 being the hot-fused line and X2 being the grounded line.

PLC power supplies, input devices, output components, and related items must share a common power source. Sharing a common power source avoids ground loops and mismatched hot and neutral phases and ensures all items are powered up and powered down together. **See Figure 9-11.**

A *foreign control voltage* is a voltage that originates outside a PLC enclosure, such as voltage from an electric motor drive. PLC enclosures must be labeled to alert technicians to the presence of the foreign control voltage. Foreign control voltages are also known as "remote sources of power."

PLCs are designed to operate within a certain voltage and frequency range, such as 85 VAC to 132 VAC and 47 Hz to 63 Hz. For a PLC to operate properly, the power source must be within the voltage and frequency range and be free of power disturbances such as noise, sags, and swells. When power source voltage varies, a constant voltage transformer must be used. Within a specified input voltage range, a constant voltage transformer provides a constant voltage output regardless of supply voltage variations.

A single-phase isolation step-down transformer provides a degree of isolation from electrical noise or power disturbances that can be present on the three-phase power supply lines. When a step-down transformer does not provide enough isolation, a filter is installed between the secondary side of the transformer and the PLC.

CODE CONNECT

Section 110.14(C) of the NEC® covers temperature limitations for electrical installations. Conductors, devices, electrical equipment, and terminations all have temperature ratings. Section 110.14(C) requires that the lowest temperature rating among all of the connected conductors, devices, electrical equipment, and/or terminations be the determining temperature rating for the circuit. For example, if a conductor rated at 90°C is connected to a 60°C device, the 60°C ampacity rating of the conductor must be used.

PLC TIPS

Electrical noise filters (power filters) are typically installed immediately before a PLC power supply for protection from electrical noise. Also, some electrical noise filters protect a PLC from voltage transients (temporary voltage spikes) that can damage it.

PLC Power Supplies

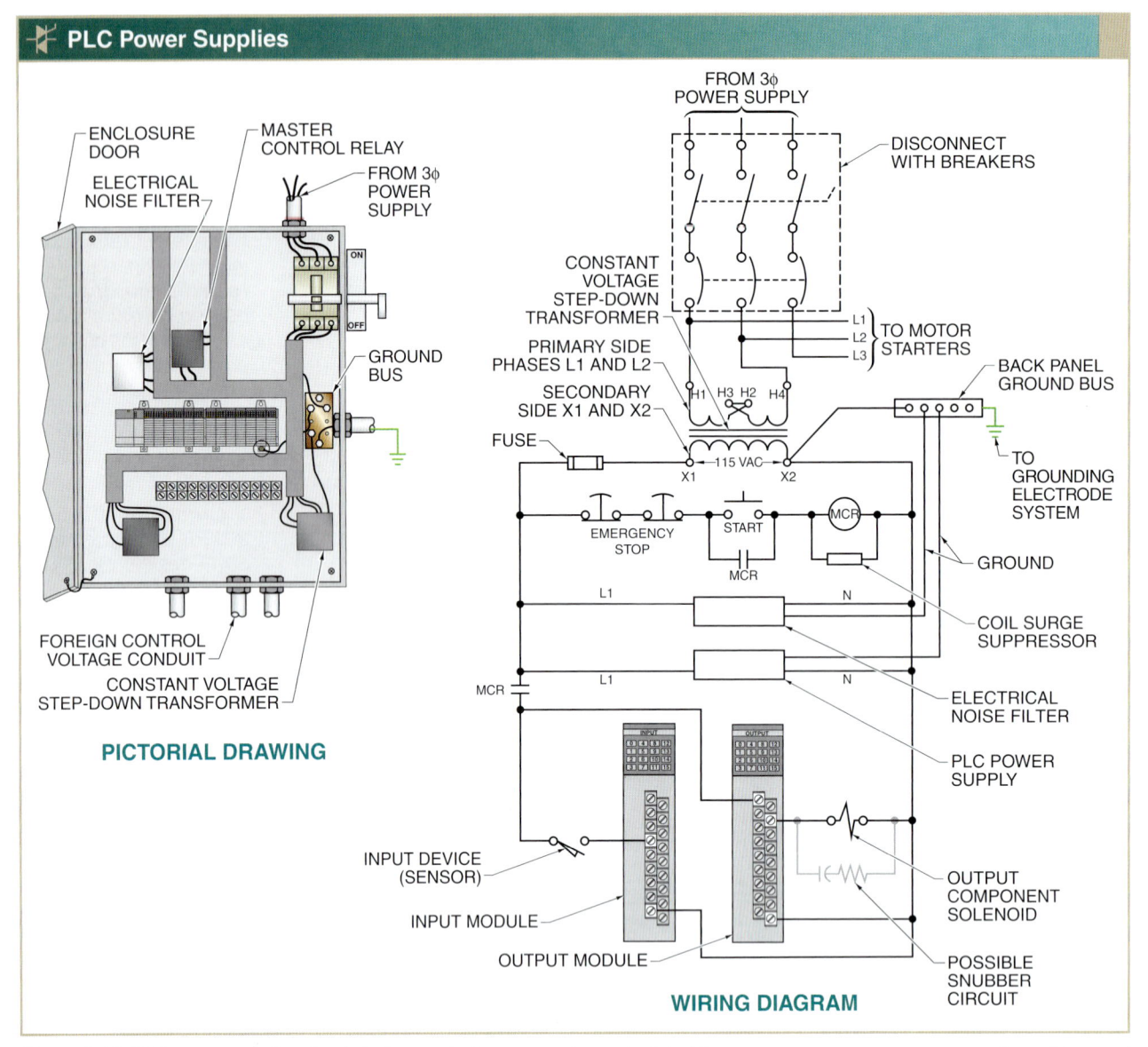

PICTORIAL DRAWING

WIRING DIAGRAM

Figure 9-11. PLC power supplies, input devices, output components, and related items use a common power source to avoid ground loops and mismatched hot and neutral phases and to ensure all items are powered up and powered down together.

PLC Installation Safety

PLCs contain solid-state components. Solid-state components operate differently than electromechanical components when power is lost or a component fails. The difference in operation presents specific safety issues when using emergency stop buttons and during power loss.

Emergency Stop Buttons. An *emergency stop button* is a special red-colored palm switch or limit switch used to stop a machine or process operation immediately to avoid physical injury or property damage. Red mushroom-head pushbuttons and overtravel limit switches are examples of emergency stop devices. It is possible for an input device or output component to fail and remain ON. Any item failing ON is a serious safety issue. Consequently, emergency stop buttons are never wired or programmed directly to a PLC. Emergency stop buttons are typically wired to master control relays (MCRs). **See Figure 9-12.**

Chapter 9—PLC Installations and Startup 365

Emergency Stop Buttons

PICTORIAL DRAWING

APPLICATION

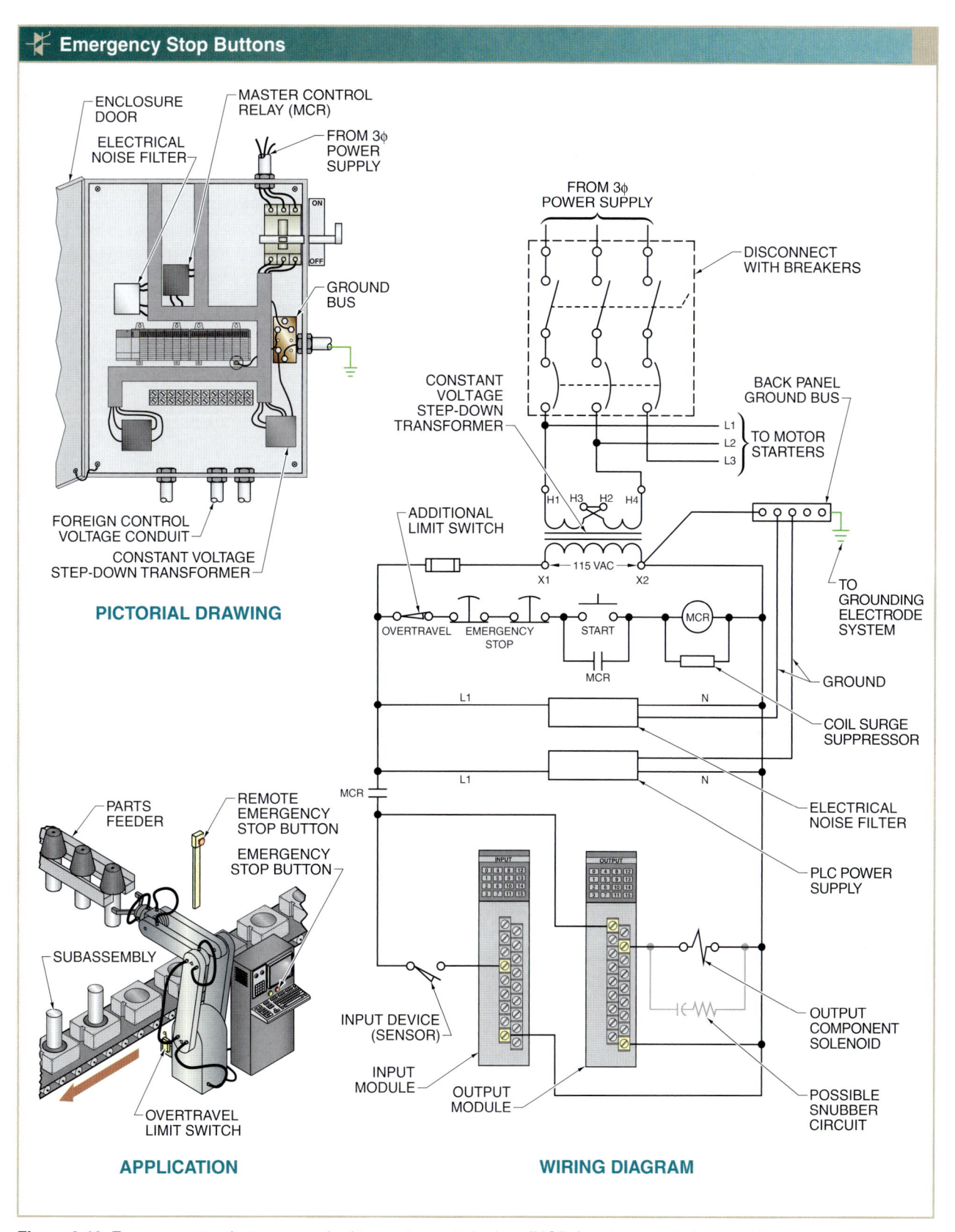

WIRING DIAGRAM

Figure 9-12. Emergency stop buttons are wired to master control relays (MCRs) and are located on machinery and process equipment to protect equipment and personnel.

366—PROGRAMMABLE LOGIC CONTROLLERS: PRINCIPLES AND APPLICATIONS

Emergency stop buttons using NC contacts are wired in series to the coil of an MCR. Whenever an emergency stop button is activated, power is removed from all input devices and output components, and the process or application controlled by the PLC stops. Power to the CPU of the PLC remains ON so status LEDs can provide up-to-date information. When a DC power supply is used for input devices and output components, the MCR contacts must be placed on the DC side to avoid turn-OFF and turn-ON delays. Emergency stop buttons use NC contacts wired in series for a fail-safe operation. In the event a wire is broken or comes off a terminal, the MCR is de-energized and power is removed from the PLC and system.

Power Loss. PLC power supplies are designed to withstand a momentary loss of power without affecting the operation of the PLC. *Hold-up time* is the length of time a PLC can tolerate a power loss without affecting operation. Hold-up time varies from 10 ms to 3 sec. **See Figure 9-13.** The operating cycle of a PLC continues during the hold-up time by scanning inputs for a change of state, scanning and updating the program, and scanning and updating outputs. The hold-up time is typically longer than the time for an input to change to its normal condition (open or closed).

When hold-up time is exceeded, the power supply of a PLC shuts down, stopping the operating cycle and the application. Long hold-up times present problems, such as when an input changes state before the power supply of a PLC shuts down. A solution to this problem is to monitor the supplied power with the PLC program and stop scanning the program upon loss of power.

PLC Wiring

The wiring from all input devices and output components must be properly terminated at the PLC in order for the PLC application to function correctly. Special attention must be paid to the wiring methods, termination methods, and the specific type of PLC input and output section designs.

PLC Wiring Methods. Various voltage levels and signal types terminate at a PLC. In order to prevent noise-related problems, conductors of different voltage levels and signal types must be separated as much as possible and must not be bundled together.

Typically, stranded conductors are used for PLC applications. Stranded wire is easy to form and resists damage from vibration.

> **⚠ WARNING**
> *Emergency stop buttons must not be programmed or connected to a PLC. Emergency stop buttons must turn OFF all machine power by turning OFF the master control relay and be labeled and placed in an easily accessible location.*

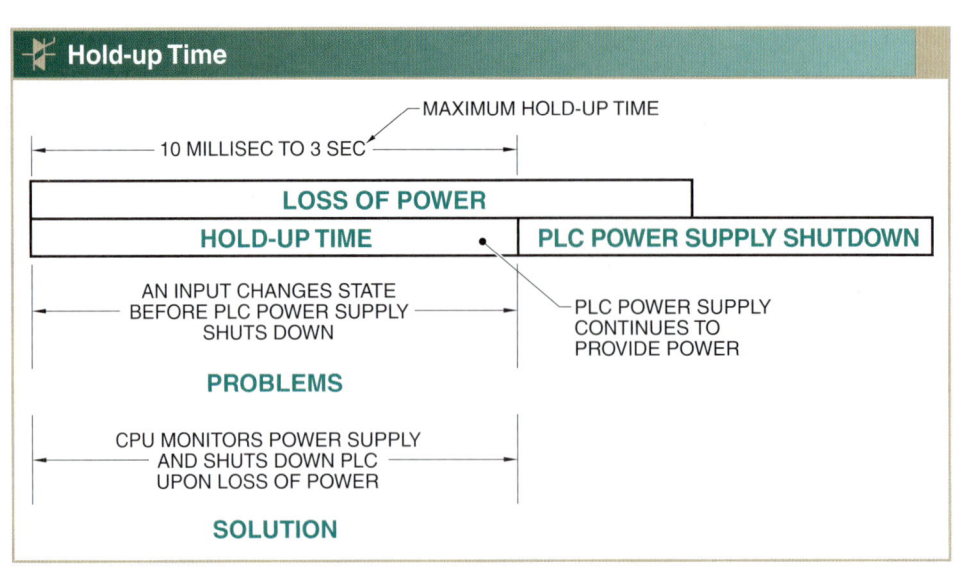

Figure 9-13. Hold-up time is the length of time a PLC can tolerate a power loss without affecting operation.

Stranded conductors can be individual conductors or part of a shielded cable. For individual conductors, the insulation color is often used to identify the type of input and output section wiring. Typically, the color blue is used for DC wiring and red is used for AC wiring. **See Figure 9-14.**

Regardless of the wiring method, a number of items must be considered when wiring a PLC. First, the technician must follow the recommendations of the manufacturer regarding wire size (AWG) and the number of wires allowed to be connected per PLC terminal. Second, all wires must be numbered and identified with a permanent smudge-proof label.

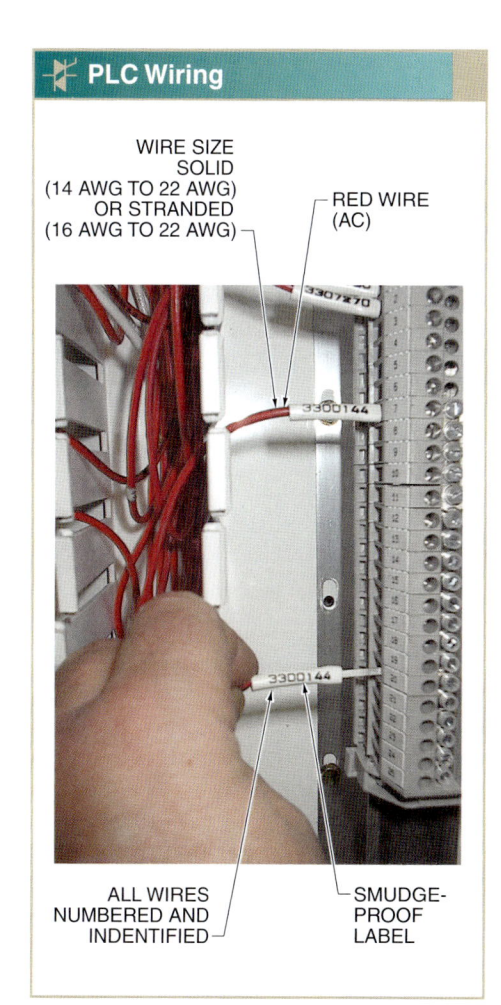

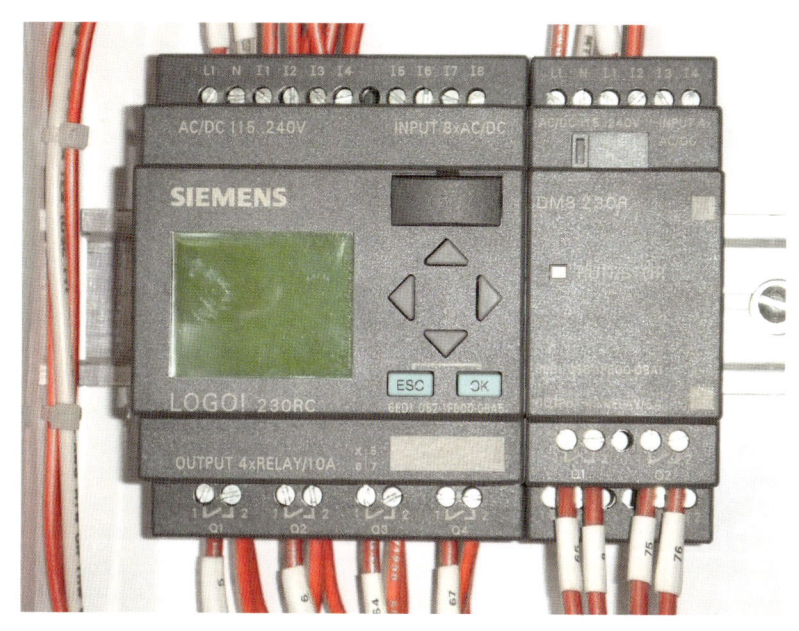

When wiring a PLC, technicians must follow manufacturer recommendations for wire size, number of wire connections per terminal, and identifying wires with permanent wire markers.

PLC Wiring

WIRE SIZE
SOLID
(14 AWG TO 22 AWG)
OR STRANDED
(16 AWG TO 22 AWG)

RED WIRE (AC)

ALL WIRES NUMBERED AND INDENTIFIED

SMUDGE-PROOF LABEL

Figure 9-14. To prevent noise-related problems, conductors of different voltage levels and signal types must be separated as much as possible and must not be cable tied together.

⚠ WARNING

A technician programming PLCs must understand the consequences that a loss of power has on the application the PLC controls. The time a system operates during power loss is called "program scan hold-up time after loss of power." Failure to fully understand the consequences can result in serious personal injury and/or property damage.

PLC Wiring Termination Methods. There are three different methods for terminating device and component wires at a PLC. **See Figure 9-15.** Device and component wires can be connected directly at a PLC, can be terminated at terminal strips, or can be connected through an interface module and prefabricated cable. With a terminal strip, a PLC is prewired to one side of a terminal strip with all device and component wires connected to the other side of the terminal strip. With an interface module, a printed circuit card with a terminal strip on one side and a modular connector and matching prefabricated cable on the other side is used.

CODE CONNECT

Section 430.102(B)(1) of the NEC® requires a disconnecting means in sight for each motor location. Section 430.102(B)(2) allows the controller (motor controller) disconnecting means to serve as the motor disconnect provided the controller disconnecting means is in sight from the motor location. Article 100 the NEC® defines the term in sight as within 50 feet and visible without any obstructions.

368—PROGRAMMABLE LOGIC CONTROLLERS: PRINCIPLES AND APPLICATIONS

PLC Termination Methods

DIRECT — SCREWS TIGHTENED TO 5 IN-LB TO 7 IN-LB, PLC, CABLE TIES

TERMINAL STRIP — OUTPUT MODULE, PREFABRICATED CABLE, TERMINAL STRIP, TERMINALS ARE NUMBERED

INTERFACE MODULE — INPUT MODULE, OUTPUT INTERFACE MODULE, INPUT INTERFACE MODULE

PIN TERMINALS

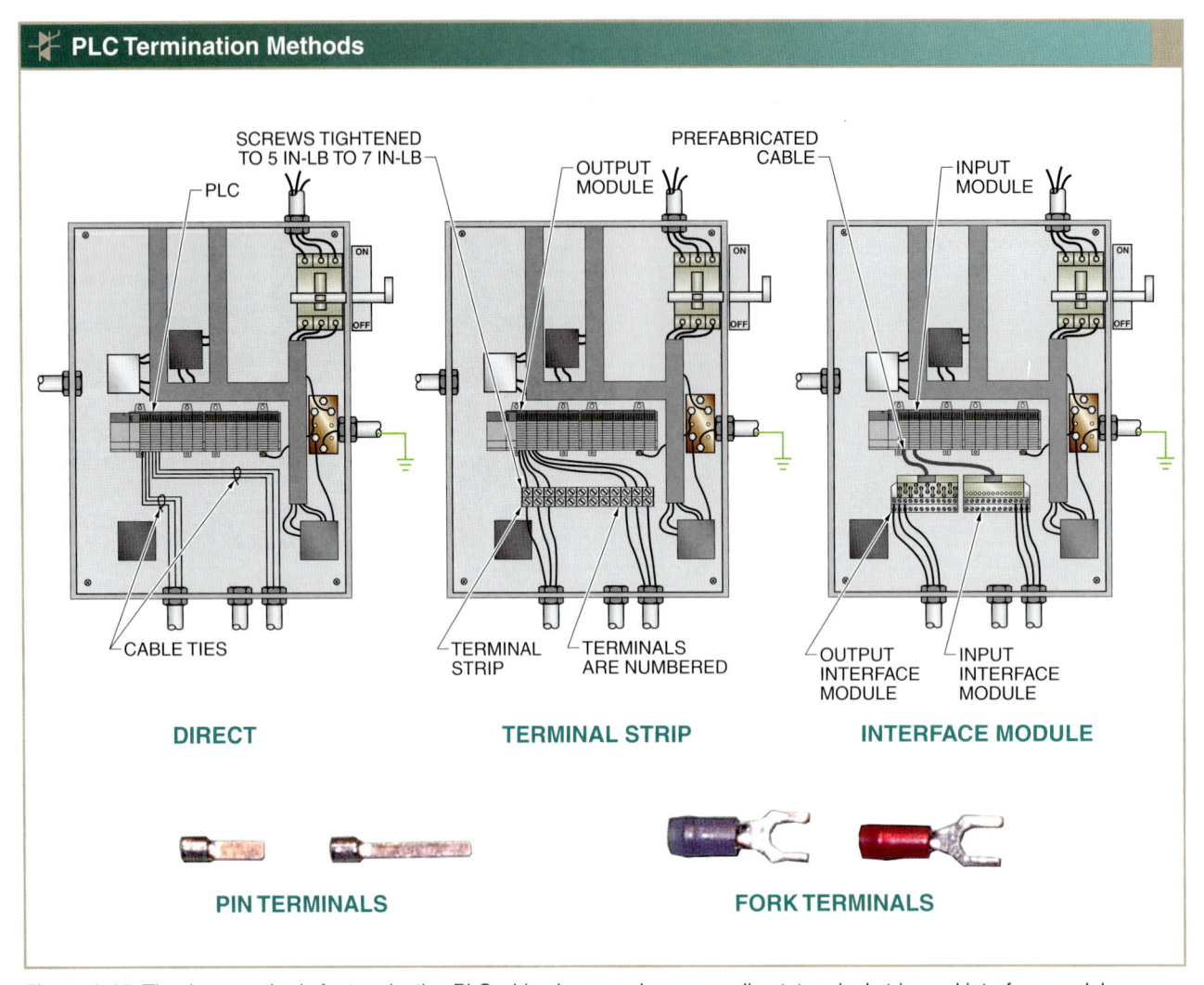

FORK TERMINALS

Figure 9-15. The three methods for terminating PLC wiring in an enclosure are direct, terminal strip, and interface module.

PLC TIPS

Some small to large PLCs can have removable terminal blocks while nano- and micro-PLCs typically have fixed points of termination.

Regardless of the termination method, a number of items must be considered when terminating wires at a PLC:

- All connections (terminal screws) must be properly tightened.
- Fork or pin terminals should be used to terminate conductors. Fork and pin terminals prevent strands of wire from slipping out from under a terminal screw and possibly grounding or shorting out.
- Wires and terminal strips must be numbered and identified with a permanent smudge-proof label.
- Cable ties should be used to secure wires and avoid strain on terminals.

When input devices and output components are connected to an interface module, a prefabricated cable must be used to connect the interface module to the PLC. Various PLC input devices and output components require different types of interface modules. **See Figure 9-16.**

PLC Input and Output Sections and Modules. Fixed PLCs use input and output sections. Modular PLCs use input and output modules. There are many types of PLC input and output sections and modules. The two main categories are digital and analog. PLC digital input and output sections and modules are used for devices and components that only have two states,

ON or OFF. For example, pushbuttons and limit switches are types of digital input devices, while relays and pilot lights are digital output components.

Interface Modules

INPUT INTERFACE MODULE
LED STATUS LIGHTS
PREFABRICATED CABLE

INPUT INTERFACE MODULE

OUTPUT INTERFACE MODULE
PREFABRICATED CABLE

OUTPUT INTERFACE MODULE

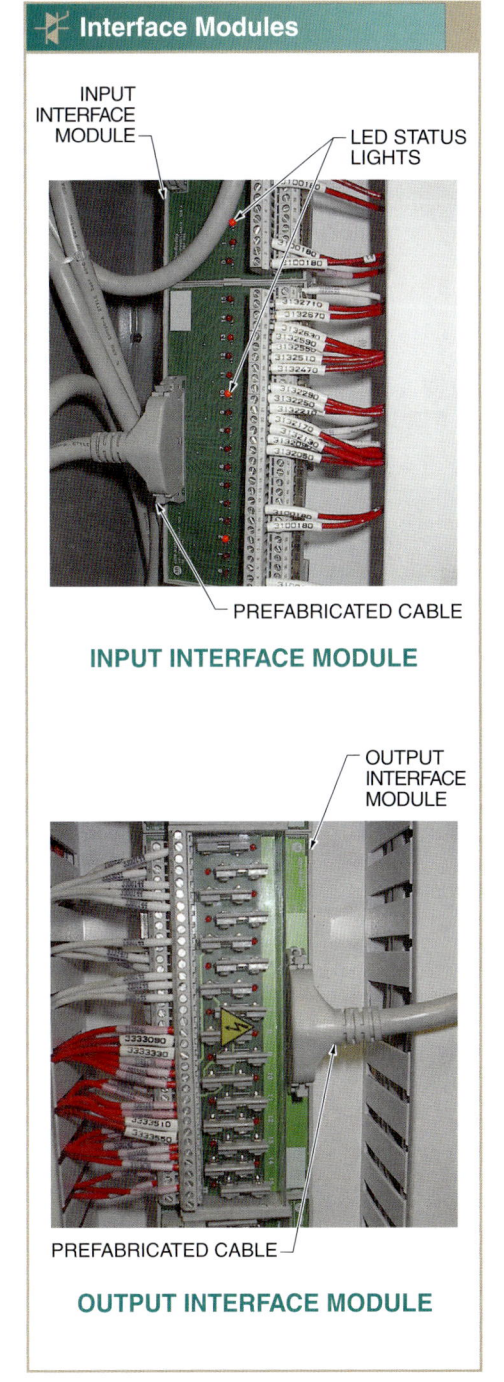

Figure 9-16. When input devices and output components are connected to an interface module, a prefabricated cable must be used to connect the interface module to the PLC.

Analog PLC input and output sections and modules are used with devices and components that vary current or voltage over a range of values. For example, temperature sensors typically send a 4 mA to 20 mA signal to a PLC to control the position of a valve. PLC input and output sections and modules are also available in a variety of voltage ratings, such as 5 VDC, 24 VDC, 24 VAC, 115 VAC, and 230 VAC. The PLC application determines the type and voltage level of the input and output sections and modules used. **See Figure 9-17.**

With a fixed PLC, input terminals and output sections are stationary. With modular PLCs, placement of input and output modules is flexible. **See Figure 9-18.** There are certain issues to consider when placing input and output modules in a PLC chassis to ensure optimum performance and prevent noise-related problems:

- Analog modules should be positioned adjacent to the PLC processor. Analog modules communicate large amounts of data to the processor and their proximity enhances communication.
- Similar types of modules should be grouped adjacent to one another. For example, all analog input and output modules should be placed together, all VAC input and output modules together, and all VDC input and output modules together.
- When there are multiple voltage levels for digital input and output modules, similar voltage levels should be grouped together. For example, all 115 VAC modules should be grouped together, all 24 VAC modules together, all 24 VDC modules together, and all analog modules should be grouped together.

When DC input devices and output components are used, the polarity of the DC voltage to the devices and components must be considered. A sinking PLC input or output has the positive (+) polarity connected to the field device. A sourcing PLC input or output has the negative (−) polarity connected to the field device. When wiring a PLC, the correct polarity must be maintained based on the type of PLC input or output module. Fixed and modular PLCs are available with sinking and sourcing input and output terminals. **See Figure 9-19.**

> **CODE CONNECT**
> Section 250.122(F) of the NEC® covers equipment grounding conductors (EGCs) for conductors in parallel. If paralleled conductors are installed in a single raceway, a single EGC is acceptable. The required size of the EGC can be found in Table 250.122 and is based on the rating of the overcurrent protective device supplying the parallel conductors. If paralleled conductors are installed in multiple raceways, an EGC must be installed in each raceway. The reason a full-size equipment grounding conductor is required for each raceway is because an individual EGC may have to carry the total fault current.

370—PROGRAMMABLE LOGIC CONTROLLERS: PRINCIPLES AND APPLICATIONS

PLC Input and Output Modules

24 V or 115 V AC Input Module

Thermocouple Input Module

Millivolt Analog Input Module

RTD Input Module

Analog Current Input Module

8-Channel Analog Input Module

Analog Current/Voltage Input/Output Module

Analog Voltage Input/Output Module

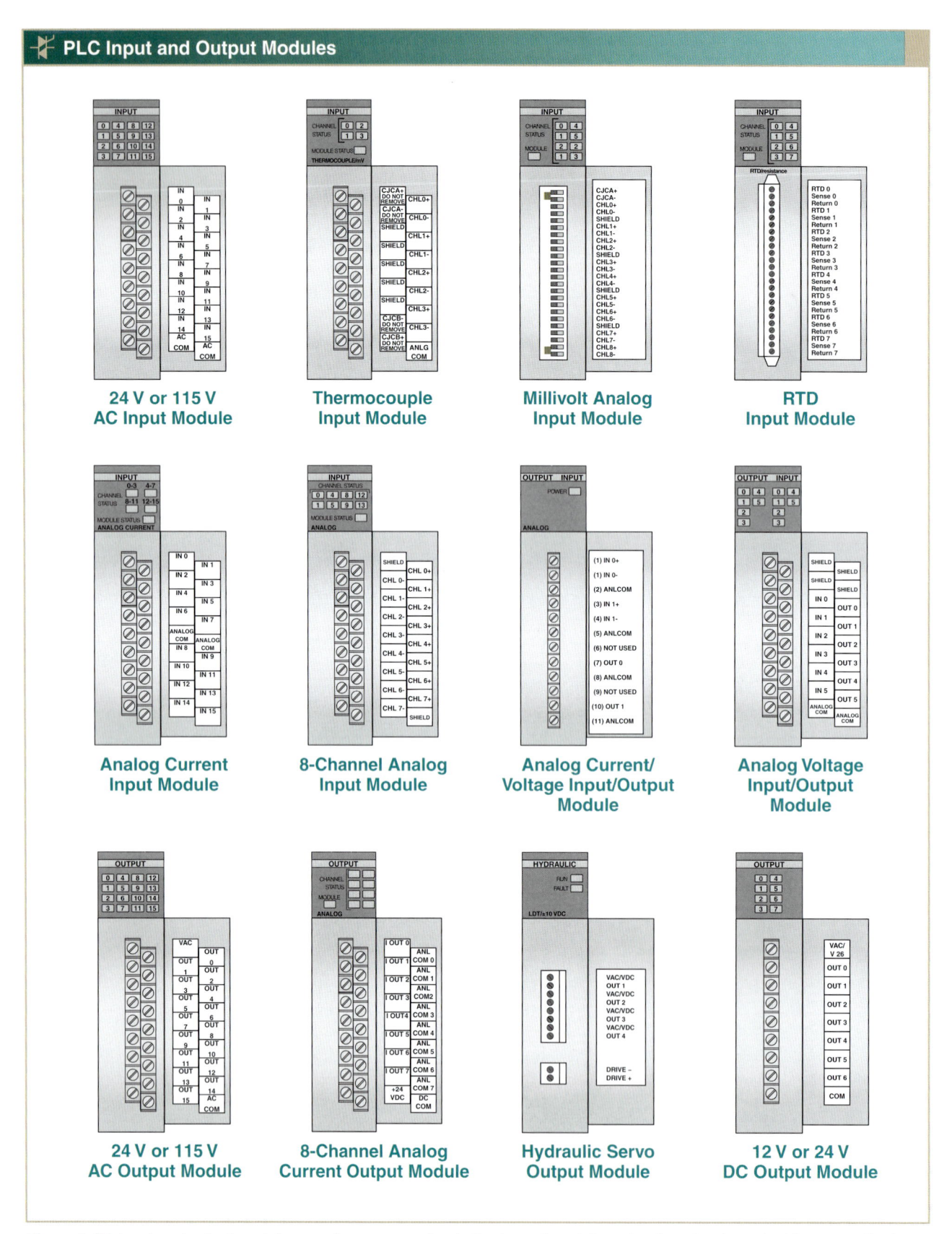

24 V or 115 V AC Output Module

8-Channel Analog Current Output Module

Hydraulic Servo Output Module

12 V or 24 V DC Output Module

Figure 9-17. Input and output modules are chosen according to the type of module and voltage level required for an application.

Input and Output Module Placement

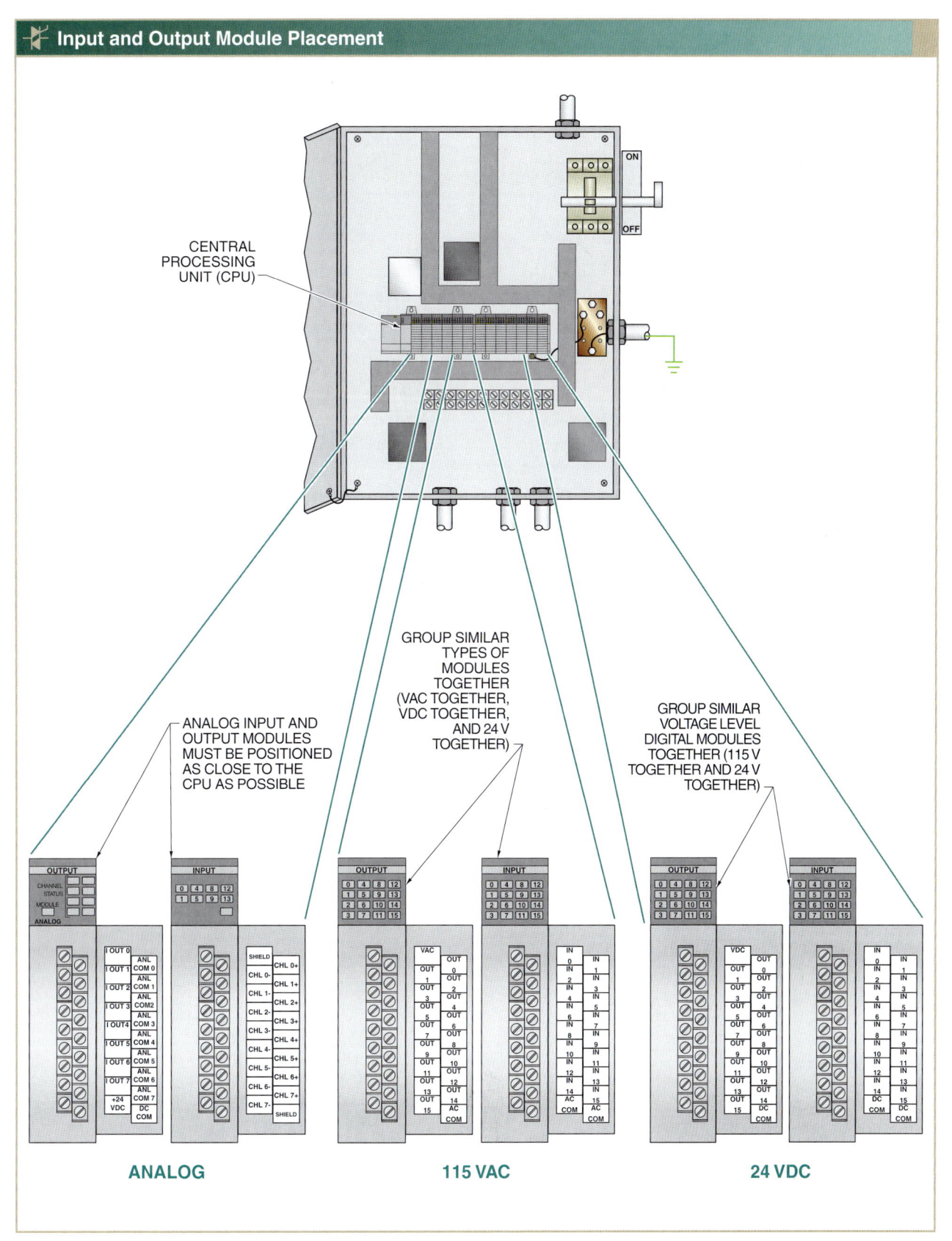

Figure 9-18. PLC input and output modules are placed in a PLC chassis to ensure optimum performance and prevent noise-related problems.

372—PROGRAMMABLE LOGIC CONTROLLERS: PRINCIPLES AND APPLICATIONS

Module Polarity (Sourcing or Sinking)

24 VDC SOURCING INPUT MODULE

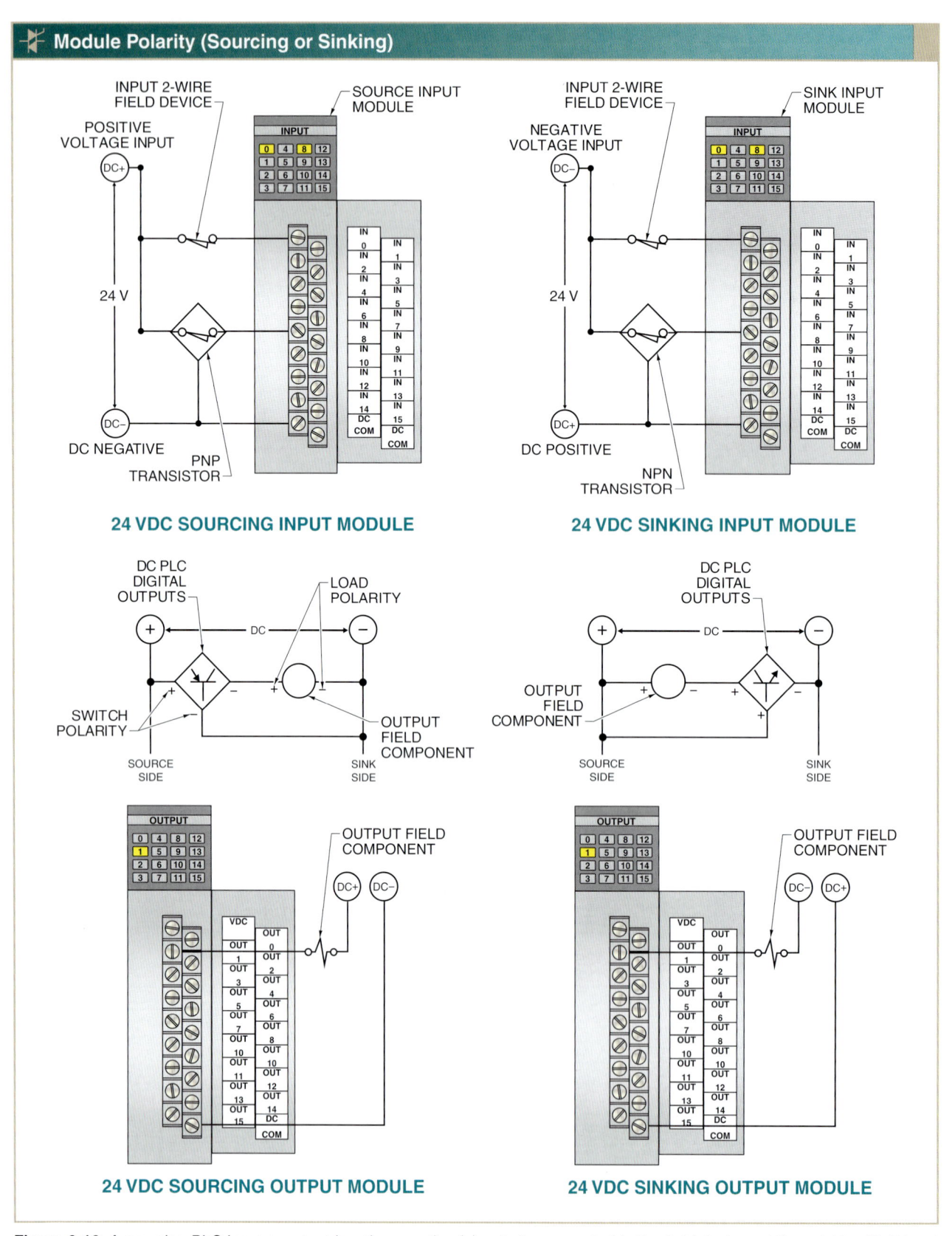

24 VDC SINKING INPUT MODULE

24 VDC SOURCING OUTPUT MODULE

24 VDC SINKING OUTPUT MODULE

Figure 9-19. A sourcing PLC input or output has the negative (−) polarity connected to the field device, while a sinking PLC input or output has the positive (+) polarity connected to the field device.

PLC STARTUP

PLC startup is a set of procedures consisting of systematic checks. Startup procedures for a PLC begin with simple checks and finish with a fully operational PLC. The procedures are designed to verify that the PLC and the machine and process equipment are installed correctly and functioning properly. Additionally, a PLC program is modified as required during the startup procedure. Consistently following startup procedures ensures that a newly installed PLC will be fully operational in the least amount of time with a minimum number of problems.

Only a technician familiar with a PLC, its related components, and the machine or process can start up a PLC. Also, for maximum safety, the technician should always refer to the recommendations and instructions of the manufacturer and any applicable federal, state, and local regulations. Failure to do so can result in serious personal injury or property damage.

A technician performing a PLC startup requires specific items: manuals for the PLC and components controlled by the PLC, a programming device (typically a notebook PC), drawings of the PLC installation, a hard copy of the PLC program, personal protective equipment (PPE), hand tools, and a digital multimeter (DMM). **See Figure 9-20.** The complexity of a PLC application can require more than one person to complete the startup procedures. A technician familiar with the application that a PLC is controlling should assist the startup technician. The assisting technician must be able to verify that the machine or process the PLC controls is working correctly and should be able to make suggestions for optimizing the process or application.

PLC TIPS

Qualified startup technicians are trained and have knowledge of the construction and operation of PLCs and PLC-controlled systems and are able to recognize and avoid hazards associated with PLC troubleshooting.

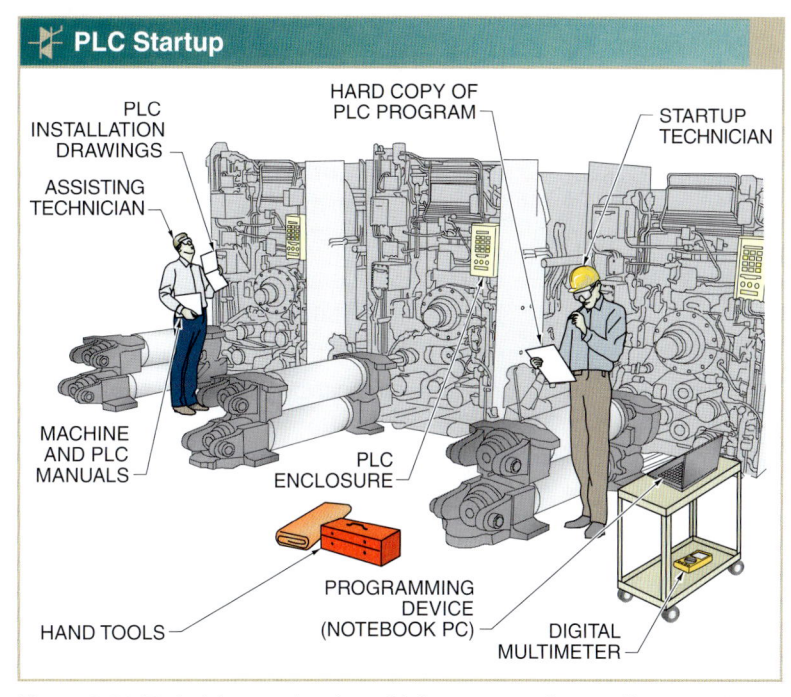

Figure 9-20. Technicians performing a PLC startup require specific startup items.

Initial PLC Checks

An *initial check* is a visual check performed by a technician before any power is applied to a PLC. The technician who performs the startup may or may not have installed the PLC. Initial checks allow a technician to become familiar with the PLC installation and minimize the hazards to personnel and equipment during the startup. **See Figure 9-21.** To perform the initial checks, apply the following procedures:

1. Lock out and tag out the incoming power to the PLC enclosure. Verify that no voltage is present at the load side of the disconnect using a digital multimeter (DMM).

2. Lock out and tag out the power to components that cause motion or cause fluid or gas to flow. Motors can be locked out at a local disconnect. While electrically operated valves must be locked out at the disconnect, the fluid or gas controlled by the valve must be locked and tagged out separately.

3. Verify that no debris or tools have been left in the PLC enclosure.

4. Verify that the PLC and related devices are mounted correctly.

5. Verify that wires are terminated securely by gently tugging to find loose wires.
6. Use drawings and wire numbers to verify that all wires are terminated at the correct location. Pay special attention that all grounds are correctly terminated.
7. Use a DMM to verify that input terminals and output terminals of the PLC are not unintentionally grounded.
8. Open control circuit breaker or remove control fuse F1 from secondary side of transformer. Remove the lockout/tagout from the incoming power disconnect. Turn the incoming power disconnect ON.
9. Using a DMM, verify that the incoming line voltage, phase-to-phase and phase-to-ground, is within the allowable voltage range.
10. Using a DMM, verify the secondary side voltage of the control transformer, phase-to-neutral and phase-to-ground, is within the allowable voltage range.

> **⚠ CAUTION**
> A PLC can have more than one mode that disables all output components. The name of the disabling mode varies among PLC manufacturers.

Automated systems being controlled by more than one PLC complicate performing input section and output section checks.

Input Section Checks

Input section (or module) checks are the second layer of checks performed on a system and PLC after the initial checks are successfully completed. An *input section check* is a check that verifies that input devices function properly and are wired to the correct PLC input terminal. **See Figure 9-22.** Typically, input section checks require two technicians and can require two-way communication if the input devices are located at a distance from the PLC. To perform input section checks, apply the following procedures:

1. Depress the emergency stop button(s) wired to the master control relay (MCR).
2. Reinstall the fuse or close the circuit breaker on the secondary side of the control transformer. Verify that the MCR does not energize. The CPU can have power, but no input device or output component should have power.
3. Load the PLC program into the PLC. Put the PLC into a mode that disables all output components such as program mode, test-continuous mode, or disable mode.
4. Test the MCR by using the start buttons and emergency stop buttons.
5. Verify that the MCR controls power to all PLC input devices and output components. Leave the MCR energized.
6. Manually actuate the input devices. Verify that when the input is actuated the corresponding LED on the PLC turns ON or changes state. If the LED does not turn ON or change state, a problem exists with the wiring, the input device, or the PLC input.

> **⚠ WARNING**
> It may not be possible to manually actuate certain input devices due to location or system configuration. When an input device is hardwired to another device, do not actuate unless the consequences are fully understood.

Chapter 9—PLC Installations and Startup **375**

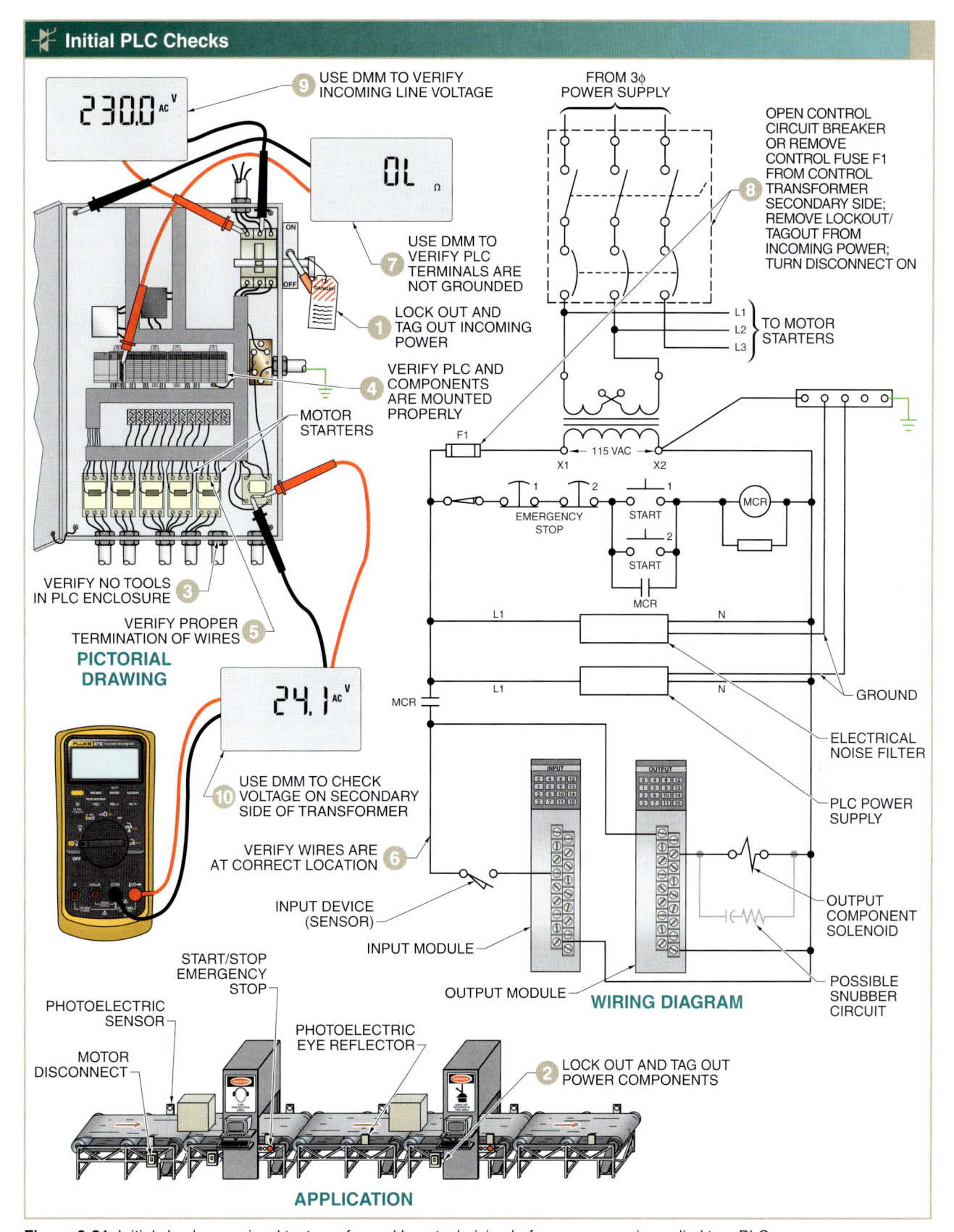

Figure 9-21. Initial checks are visual tests performed by a technician before any power is applied to a PLC.

Input Section Checks

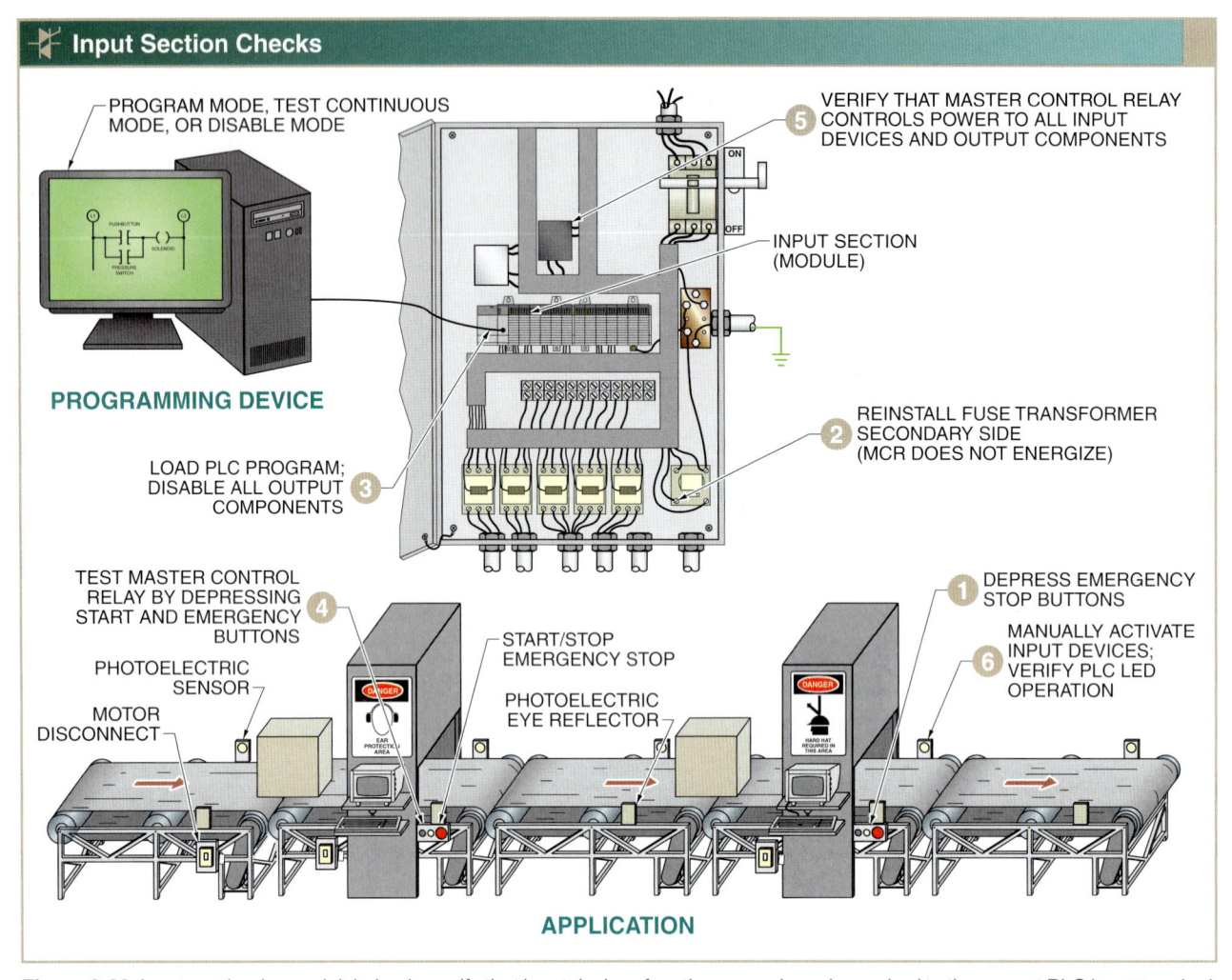

Figure 9-22. Input section (or module) checks verify that input devices function properly and are wired to the correct PLC input terminal.

⚠ **CAUTION**

A common mistake is to force a device OFF rather than removing the ON force. When a component is forced OFF, the component will not turn ON again until the OFF force is removed.

⚠ **WARNING**

When a PLC output controls more than one component through a terminal, do not force the output ON unless the consequences of the force are fully understood.

Output Section Checks

Output section (or module) checks are performed after the input checks are successfully completed. An *output section check* is a check that verifies that all output components function properly and are wired to the correct PLC output terminal. Typically, output section checks require two technicians and can require two-way communication when output components are located at a distance from a PLC.

In order to test the output components, the PLC must be in RUN mode. However, the machine or process must not be started. A startup technician is responsible for ensuring that the PLC is in RUN mode but that the machine or process does not operate. **See Figure 9-23.** To perform output section checks, apply the following procedures:

1. Depress the emergency stop button. Ensure that the machine or process will not start.
2. Place the PLC in RUN mode.
3. Release the emergency stop button(s) and press the start button so the MCR is ON.
4. Use a programming device to force each output component ON. Verify that the output component actuates and the corresponding LED on the PLC illuminates. If the output device does not turn ON, a problem exists with the wiring, output component, or the PLC output terminal.
5. After testing each output component, be sure to remove any forces.

Chapter 9—PLC Installations and Startup 377

Output Section Checks

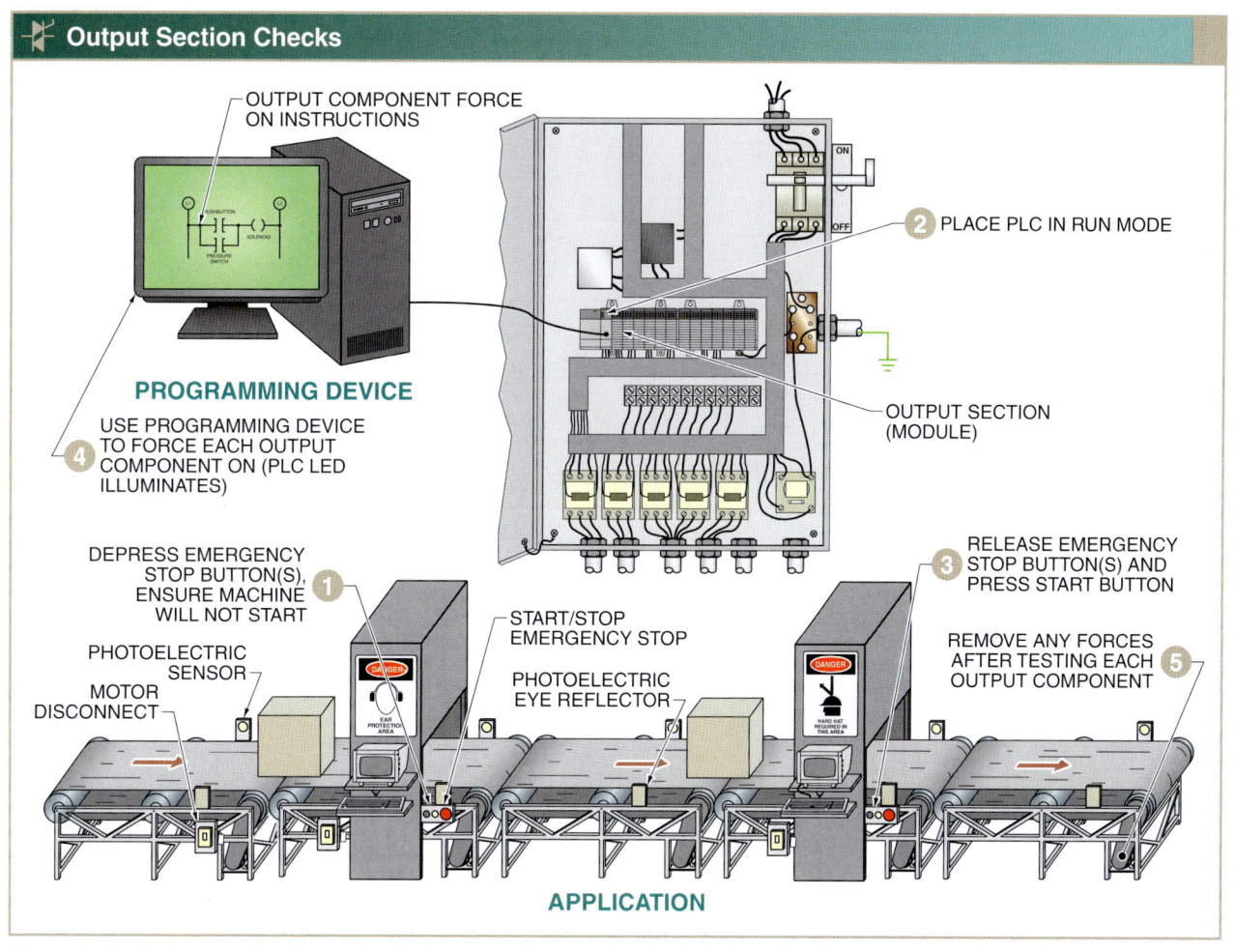

Figure 9-23. Output section (or module) checks verify that all output devices function properly and are wired to the correct PLC output terminal.

Program Checks

A *program check* is a check that verifies that a PLC program functions properly without the application or process the PLC controls actually being run. **See Figure 9-24.** To perform program checks apply the following procedures:

1. Put the PLC into a mode that allows the program to run but disables the output components, such as test-continuous mode or disable mode. Test-continuous and disable modes allow a technician to monitor the machine or process with a computer and verify the logic of the program.
2. Actuate the input device(s) that start the machine or process.
3. Use a PC to monitor the PLC program. Verify that the program works properly.
4. Make any changes necessary to the program.
5. After verifying that the program works properly, be sure to remove any forces and save any program changes.

Final Checks

Final checks are performed after all program checks are successfully completed. A *final check* is a check that verifies that an application or process that a PLC controls functions properly in RUN mode under actual conditions. There are several strategies that can be used when performing final checks that depend on the size and type of the application.

> ⚠ **CAUTION**
> To verify a PLC program, it may be necessary to force specific PLC input devices ON or OFF to simulate a system condition.

Program Checks

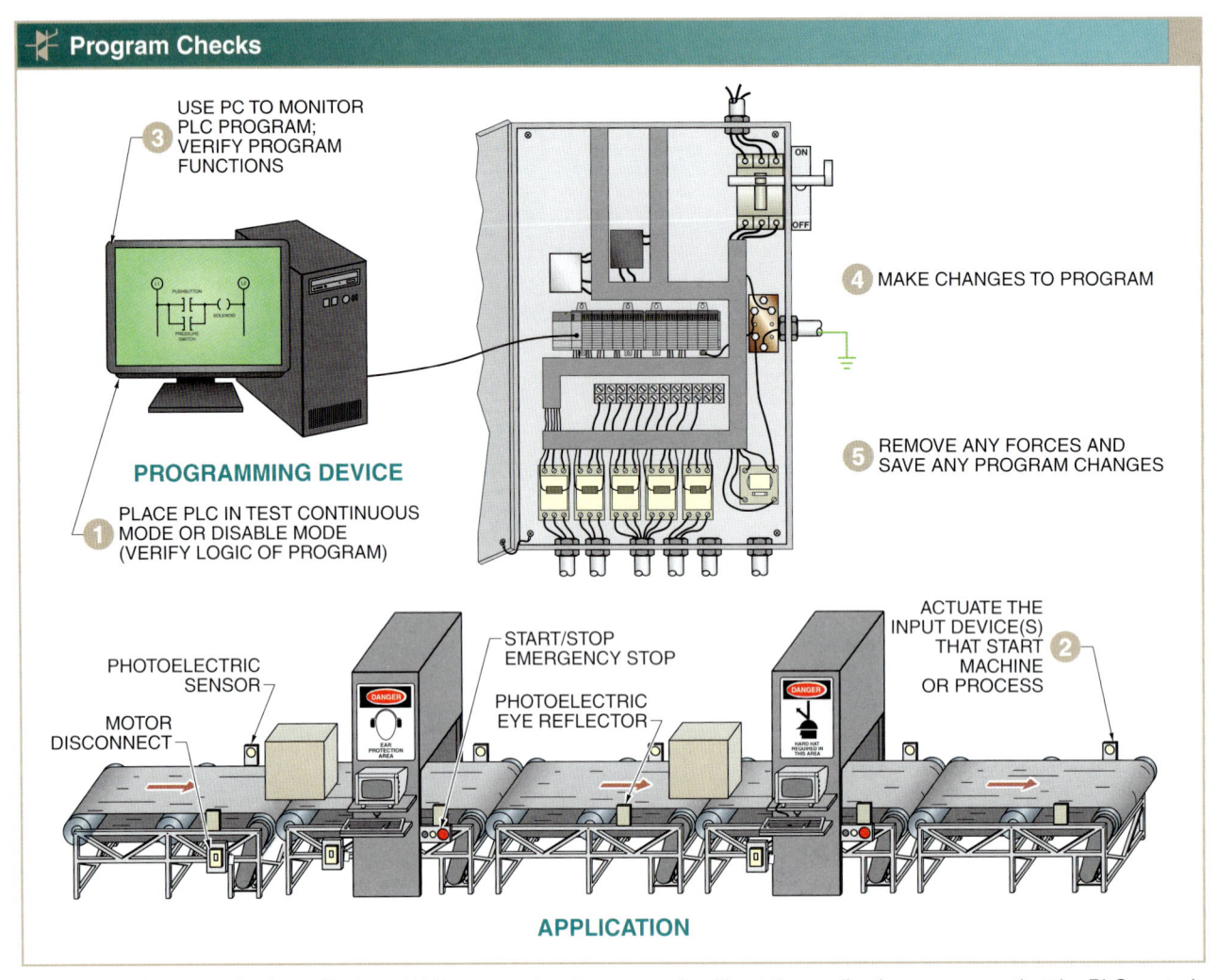

Figure 9-24. Program checks verify that a PLC program functions properly without the application or process that the PLC controls actually being run.

Typically, a small machine or process can be put in RUN mode and monitored, while large machines or processes require that the program be divided into sections, sections checked individually in RUN mode, and then checked together. Some PLC applications cannot be run with actual product because a program error could cause severe equipment damage. In these cases, a substitute product may be used. For example, a metal forming machine may use pieces of wax to simulate the metal. **See Figure 9-25.** To perform final checks apply the following procedures:

1. Depress the emergency stop button(s).
2. Remove all lockout/tagout devices.
3. Turn the electrical disconnect and fluid or gas valves ON.
4. Place the PLC in RUN mode. Release the emergency stop button(s) and press the start button so the MCR is ON.
5. Start the machine or process controlled by the PLC.
6. Monitor the PLC application and verify that the system functions correctly. When necessary, a machine operator or technician familiar with the application should assist the startup technician to verify that the system is operating correctly.
7. When necessary, make changes to the PLC program and save the changes.

8. Save the program to the EEPROM memory.
9. Save the program to the hard drive of the PC or network.
10. Print out a hard copy of the final program and store the copy in a safe location.

> **⚠ CAUTION**
> *After changes are made to a program, technicians must monitor the application to verify the results of the changes. Changes made to one part of a program impact other areas of the program.*

Final Checks

- ⑦ MAKE CHANGES TO PLC PROGRAM SAVE ANY CHANGES MADE
- ③ TURN DISCONNECT AND ALL FLUID OR GAS VALVES ON
- ② REMOVE ALL LOCKOUT/TAGOUT DEVICES
- **PROGRAMMING DEVICE**
- ⑧ SAVE PLC PROGRAM TO PLC EEPROM MEMORY
- ⑨ SAVE PROGRAM TO HARD DRIVE OF PC
- ⑩ PRINT OUT HARD COPY OF PLC PROGRAM AND STORE IN SAFE LOCATION
- ④ PLACE PLC IN RUN MODE; RELEASE EMERGENCY STOP BUTTON(S) AND PRESS START BUTTON(S) TO ENERGIZE MCR
- ⑥ MONITOR THE PLC APPLICATION AND VERIFY THAT THE SYSTEM OPERATES CORRECTLY
- ⑤ START THE MACHINE OR PROCESS CONTROLLED BY THE PLC
- ① DEPRESS EMERGENCY STOP BUTTON
- PHOTOELECTRIC SENSOR
- MOTOR DISCONNECT
- START/STOP EMERGENCY STOP
- PHOTOELECTRIC EYE REFLECTOR

APPLICATION

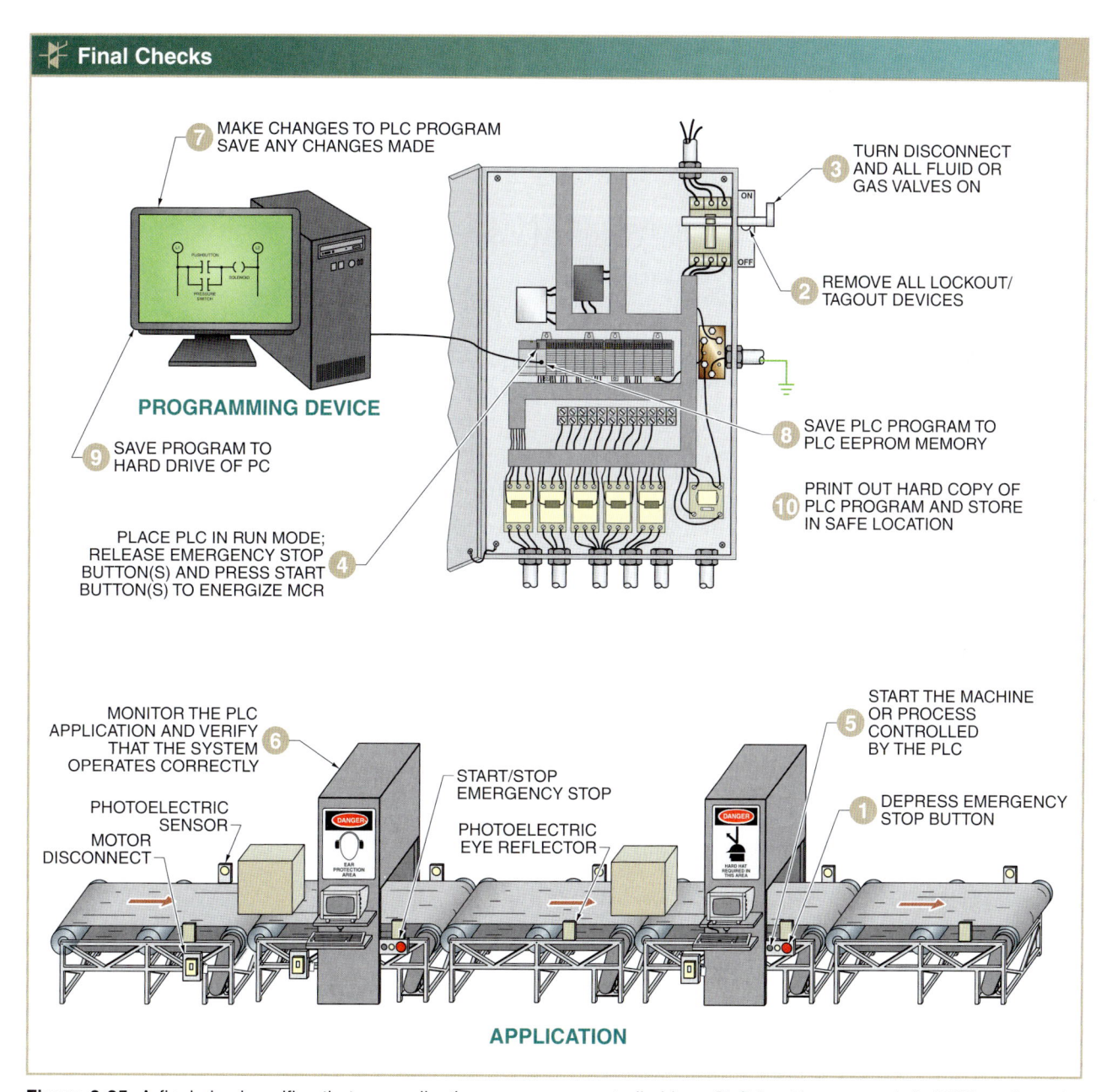

Figure 9-25. A final check verifies that an application or process controlled by a PLC functions properly in RUN mode under actual conditions.

Chapter Review 9

Name _____ Date _____

True-False

T F   1. When low-level analog DC wiring crosses AC conductors, it must cross at 90°.

T F   2. Conductors that are run from the PLC enclosure to another location should be in a metal raceway regardless of the signal type or voltage level.

T F   3. Emergency stop buttons must be wired directly to a PLC.

T F   4. Program checks are performed on a PLC and system after all section checks are successfully completed.

T F   5. When a grounding connection must be made to the interior of a metal enclosure, any paint coatings must remain intact.

T F   6. A shield/drain wire is typically connected at both ends.

T F   7. Analog modules should be positioned adjacent to the PLC processor.

T F   8. Typically, the secondary side of a control transformer is 115 VAC or 24 VAC, with X1 being the hot-fused line and X2 being the grounded line.

T F   9. For individual conductors, the color blue is used to identify AC wiring and the color red is used for DC wiring.

T F   10. A sinking PLC input has the negative polarity (–) connected to the field device.

T F   11. A PLC is typically housed in a metal enclosure.

T F   12. Items other than a PLC are sometimes housed in a PLC enclosure.

T F   13. PLCs do not contain solid-state components.

T F   14. Typically, three-phase power is brought to a PLC enclosure.

T F   15. In order to test the input devices, a PLC must be in RUN mode.

Completion

_____ 1. A(n) ___ is an electrical signal that varies over a range of values, such as between 0 VDC and 10 VDC.

_____ 2. A(n) ___ is a voltage that originates outside a PLC enclosure, such as voltage from an electric motor drive.

3. ___ is the ability of various pieces of equipment to work together with varying levels of noise emission immunity.

_____ 4. A(n) ___ is a visual check performed before any power is applied to a PLC.

_____ 5. ___ is a set of procedures consisting of systematic checks.

_____ 6. ___ is the length of time a PLC can tolerate a power loss without affecting operation.

_____ 7. A(n) ___ is a special red-colored palm switch or limit switch used to stop a machine or process operation immediately to avoid physical injury or property damage.

_____ 8. A(n) ___ is an electrical signal that has only two states, ON or OFF.

_____ 9. A(n) ___ is a check that verifies that an application or process that a PLC controls functions properly in RUN mode under actual conditions.

_____ 10. A(n) ___ is a digital signal from an external source.

Multiple Choice

_____ 1. Which of the following is not a method used for terminating device or component wires at a PLC?
  A. connection through an interface module and prefabricated cable
  B. a wireless connection
  C. a direct connection to the PLC
  D. termination at a terminal strip

_____ 2. During a final check, some PLC applications cannot be run with the actual product because ___.
  A. a program error could cause severe equipment damage
  B. technicians are not authorized to operate a PLC
  C. the number of required personnel is too high
  D. the product is too expensive to produce without an order

_____ 3. A(n) ___ check is a check that verifies that a PLC program functions properly without the application or process the PLC controls actually being run.
  A. initial
  B. input section
  C. output section
  D. program

_____ 4. A(n) ___ check is a check that verifies that input devices function properly and are wired to the correct PLC input terminal.
  A. final
  B. input
  C. output
  D. program

_____ 5. Hold-up time varies from ___ to ___.
  A. 3 ms; 10 ms
  B. 10 ms; 3 sec
  C. 3 sec; 10 sec
  D. 10 sec; 30 sec

Level Control and Mixing

PLC Lab 9.1

BILL OF MATERIALS

- TECO PLR or Micrologix™ 1100 PLC
- Two motor starters/contactors
- Solenoid
- Three float/limit switches
- Three loads (optional)

In a beverage manufacturing plant, a PLC controls a pump system mixing tank. The pump is turned ON through the normally closed, held open, low-level float switch until the high-level float switch is opened. When a mid-level float switch is reached, a mixer is turned ON, which actuates an agitator which must be below fluid level to operate. When filling stops and 45 seconds have elapsed, a drain valve is opened until the low level is reached. Low-level and mid-level float switches are hardwired normally open, and the high-level float switch is hardwired normally closed. Motor overloads are hardwired normally closed and are wired to the PLC.

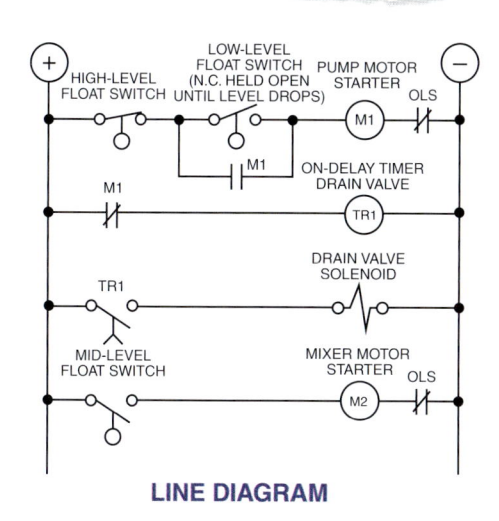

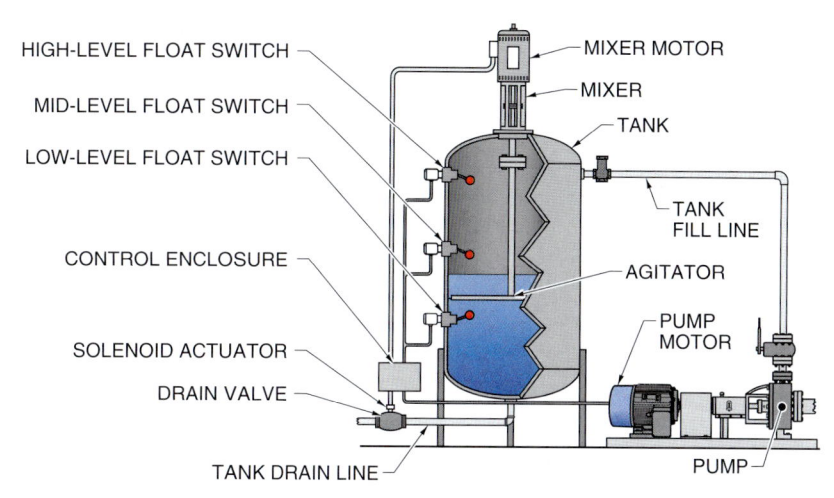

LINE DIAGRAM

TECO Procedures

1. Based on the PLC system description and line diagram, draw the TECO programming diagram and FBD logic with the proper contacts and descriptions in the Programming Grid and Programming Area.

2. Create a new ladder logic or FBD program electronically.

T01 = 45 SEC

TECO PROGRAMMING GRID

384—PROGRAMMABLE LOGIC CONTROLLERS: PRINCIPLES AND APPLICATIONS

FBD PROGRAMMING AREA

3. Use the Edit Contact window to denote an input or output number and the Symbol button to add a description after placing the input or output in the Programming Grid. For an FBD, use the Comment icon to add the description. Add timers with the proper mode and description to the Ladder Logic Programming Grid or FBD Programming Area.

4. Run the program in Simulation Mode to verify proper operation. Use the Input Status tool window to activate the inputs and control the output to simulate circuit operation.

5. Draw the wire connections in the wiring diagram between the PLC and input devices and output components. Then use the wiring diagram to wire the input devices and output components to the PLC.

6. Turn power to the PLC ON and link the com port.

7. Write the program to the SG2-12HR-D to download the program logic to the PLC.

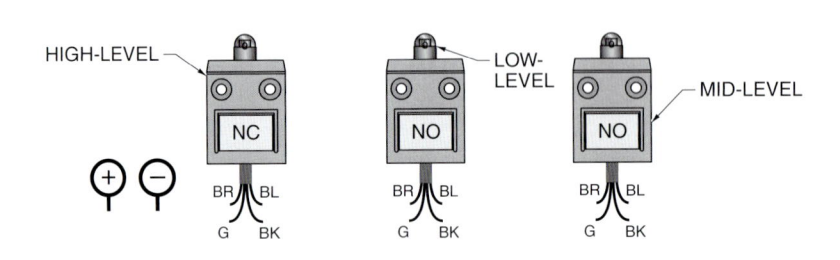

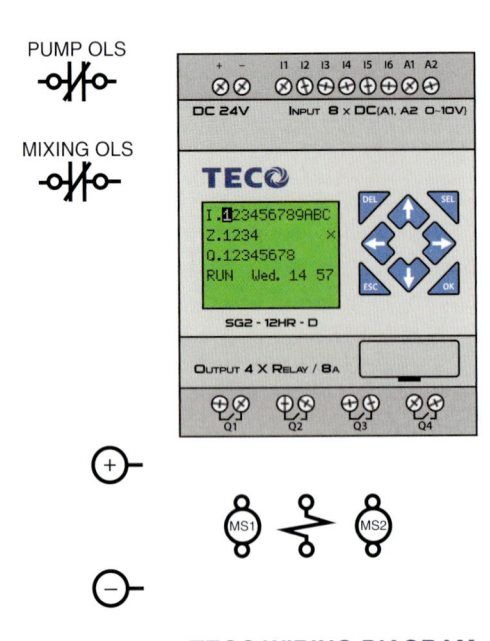

TECO WIRING DIAGRAM

Chapter 9—PLC Installations and Startup   **385**

8. Place the SG2-12HR-D in Run Mode. Activate the start pushbutton and monitor the I/Os on the LCD display.

9. Stop the program and use the Save As function to save and place the lab application in the required folder location.

AB Allen-Bradley® Procedures

1. Based on the PLC system description and line diagram, draw the ladder logic for RSLogix™ in the AB programming area.

2. Open the RSLogix™ Micro Starter Lite program and create a new programming diagram.

3. Use the AB programming diagram to add the proper address and description in the LAD 2 programming window. Add the proper timer instructions to the programming diagram and type in the timer address, time base, and preset value for each.

4. Verify that the programming diagram has no errors and save the project.

5. Draw the connections between the PLC and input devices and output components. Then use the wiring diagram to wire the input devices and output components to the PLC. Once all connections have been made, turn power to the PLC on.

6. Use RSLinx™ to establish communication between the PC and the Micrologix™ 1100.

7. Download the program logic to the PLC.

8. Place the PLC in Run Mode.

9. Activate input(s) and monitor them to determine if the output(s) are operating correctly.

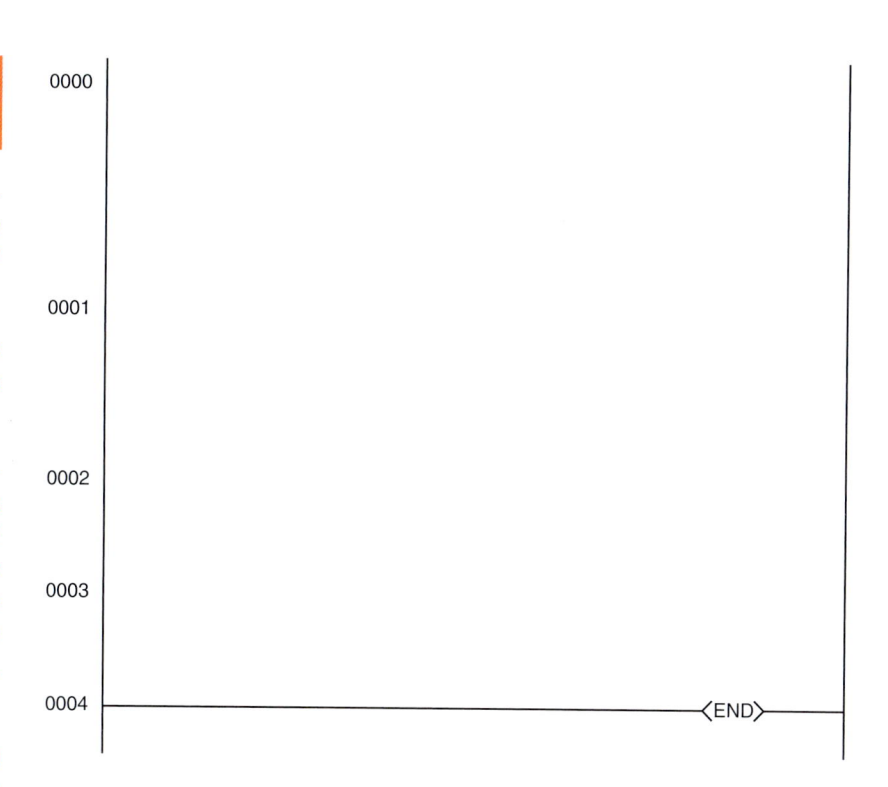

```
0000

0001

0002

0003

0004                                             ⟨END⟩
```

AB PROGRAMMING AREA

AB WIRING DIAGRAM

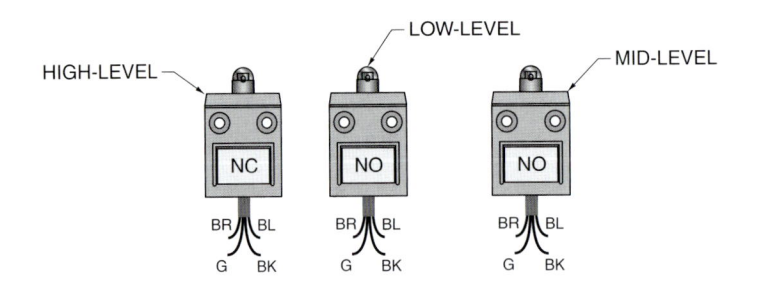

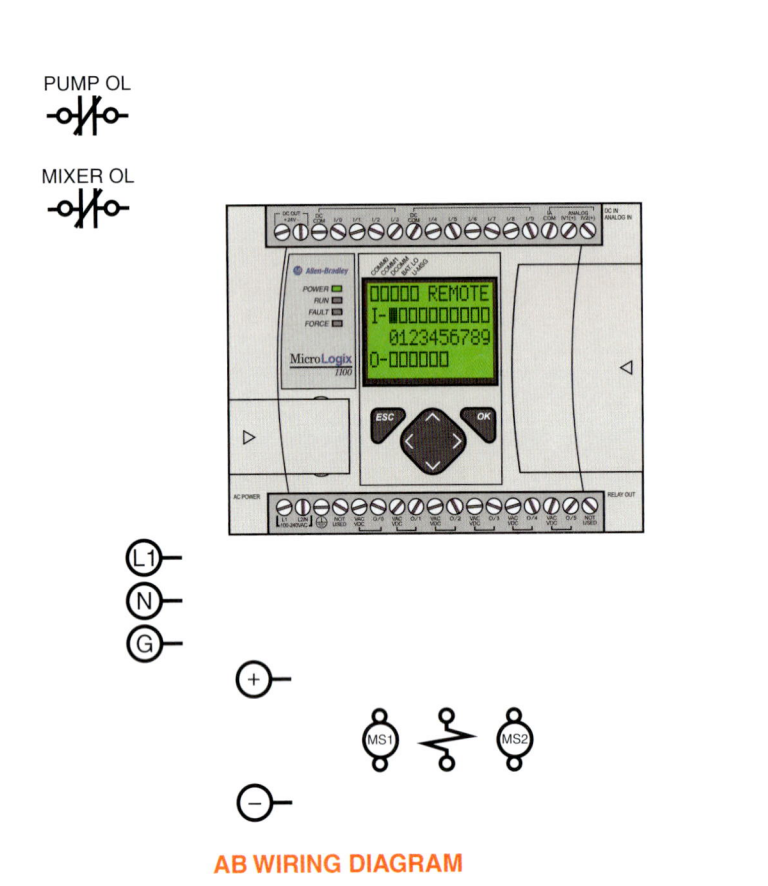

Level Control, Mixing, and Debounce

PLC Lab 9.2

BILL OF MATERIALS
- TECO PLR or Micrologix™ 1100 PLC
- Two motor starters/contactors
- Solenoid
- Selector switch
- Two float/limit switches
- Three loads (optional)

In the PLC-controlled pump system mixing tank previously described, a condition can occur where fluid entering the tank can splash up and activate the mid-level switch. For this lab, correct this condition by adding a debounce timer to the program logic. The debounce timer should be set for 3 seconds. Motor overloads are hard wired normally-closed and are wired to the PLC.

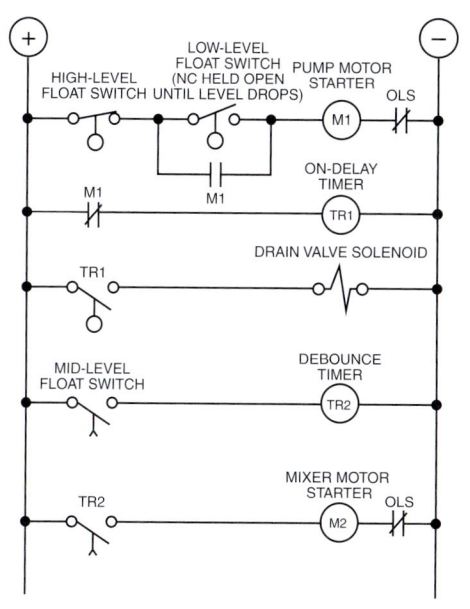

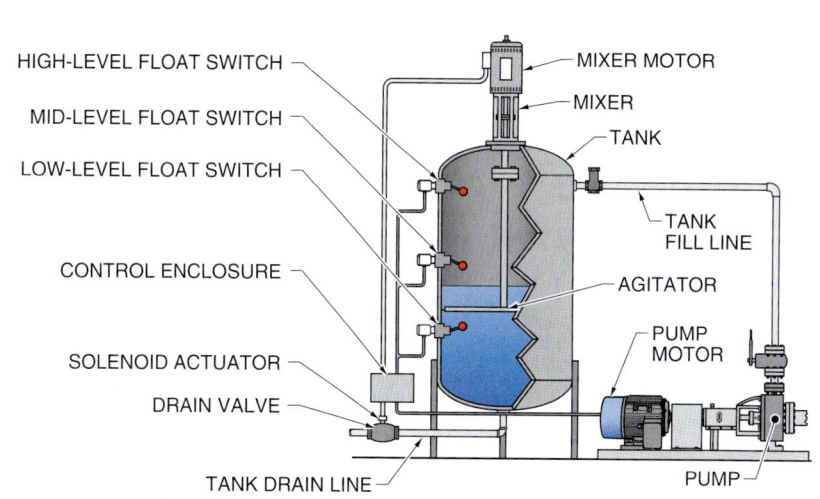

LINE DIAGRAM

TECO Procedures

1. Based on the PLC system description and line diagram, draw the TECO programming diagram and FBD logic with the proper contacts and descriptions in the Programming Grid and Programming Area.

2. Create a new ladder logic or FBD program electronically.

3. Use the Edit Contact window to denote an input or output number and the Symbol button to add a description after placing the input or output in the Programming Grid. For an FBD, use the Comment icon to add the description. Add timers with the proper mode and description to the Ladder Logic Programming Grid or FBD Programming Area.

```
001
002
003
004
005
006
```

TECO PROGRAMMING GRID

FBD PROGRAMMING AREA

Chapter 9—*PLC Installations and Startup* **389**

4. Run the program in Simulation Mode to verify proper operation. Use the Input Status tool window to activate the inputs and control the output to simulate circuit operation.

5. Draw the wire connections in the wiring diagram between the PLC and input devices and output components. Then use the wiring diagram to wire the input devices and output components to the PLC.

6. Turn power to the PLC ON and link the com port.

7. Write the program to the SG2-12HR-D to download the program logic to the PLC.

8. Place the SG2-12HR-D in Run Mode. Activate the start pushbutton and monitor the I/Os on the LCD display.

9. Stop the program and use the Save As function to save and place the lab application in the required folder location.

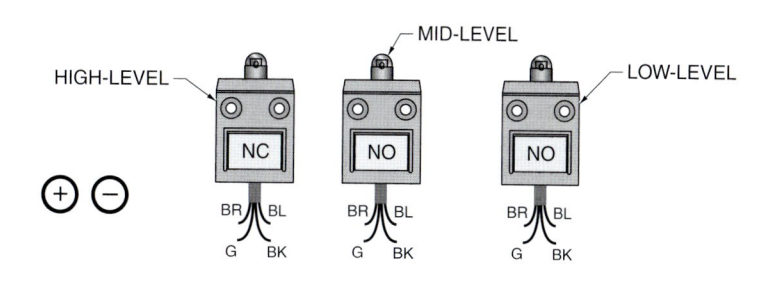

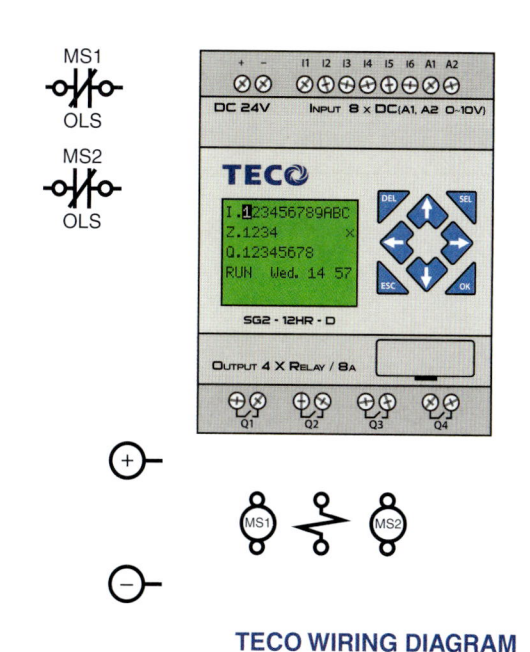

TECO WIRING DIAGRAM

Allen-Bradley® Procedures

1. Based on the PLC system description and line diagram, draw the ladder logic for RSLogix™ in the AB programming area.

2. Open the RSLogix™ Micro Starter Lite program and create a new programming diagram.

3. Use the AB programming diagram to add the proper address and description in the LAD 2 programming window. Add the proper timer instructions to the programming diagram and type in the timer address, time base, and preset value for each.

4. Verify that the programming diagram has no errors and save the project.

5. Draw the connections between the PLC and input devices and output components. Then use the wiring diagram to wire the input devices and output components to the PLC. Once all connections have been made, turn power to the PLC on.

6. Use RSLinx™ to establish communication between the PC and the Micrologix™ 1100.

7. Download the program logic to the PLC.

8. Place the PLC in Run Mode.

9. Activate input(s) and monitor them to determine if the output(s) are operating correctly.

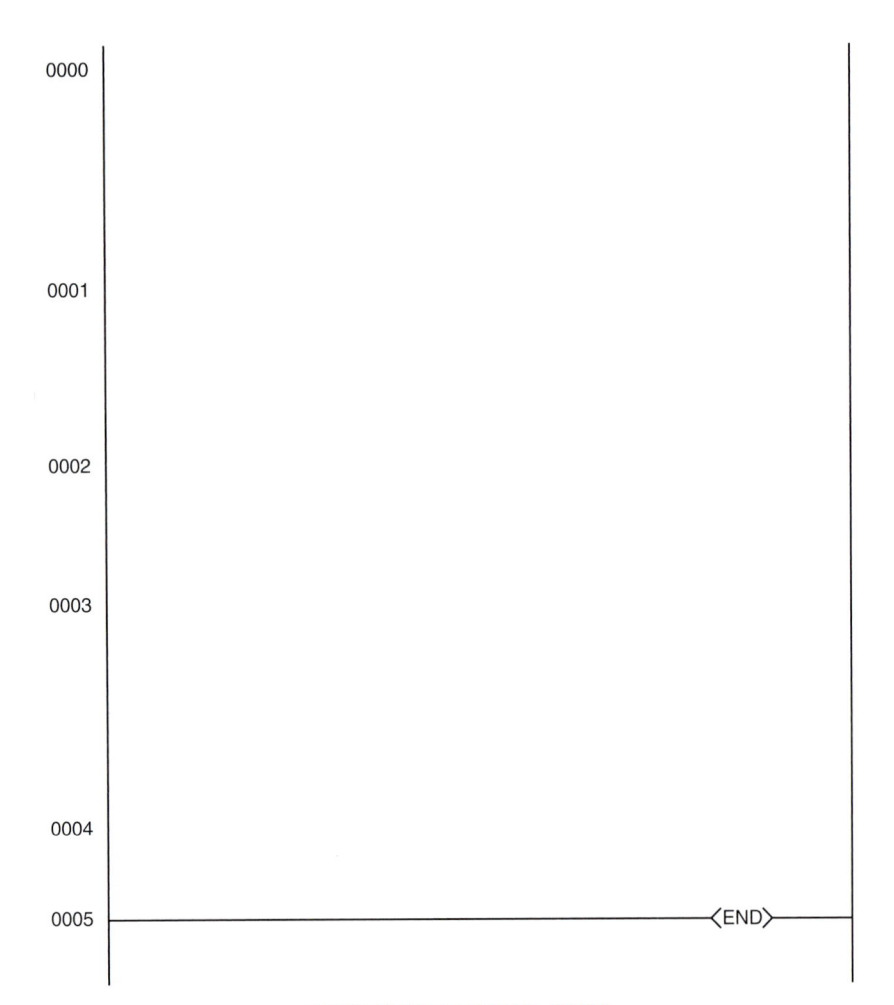

```
0000
0001
0002
0003
0004
0005 ─────────────────────⟨END⟩
```

AB PROGRAMMING AREA

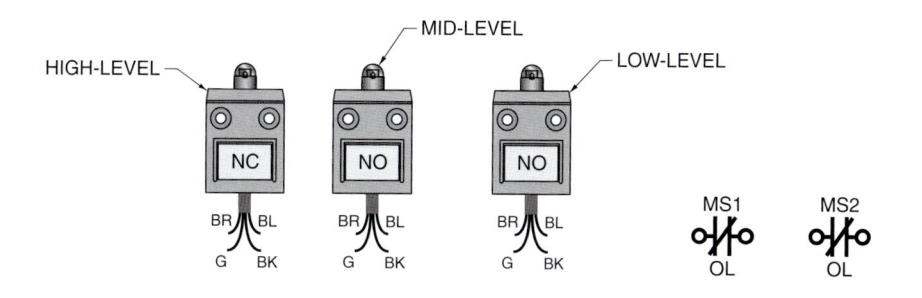

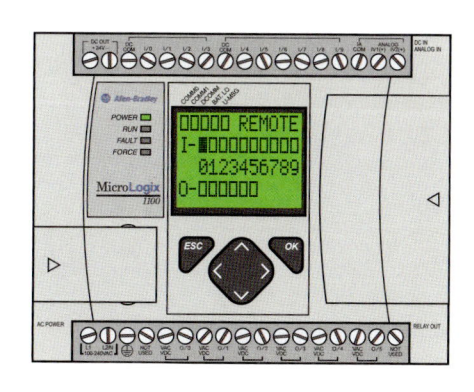

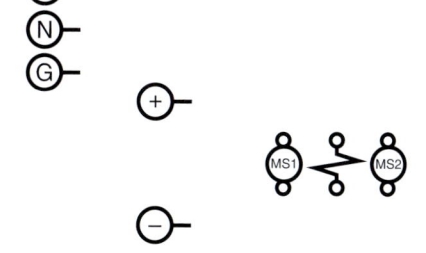

AB WIRING DIAGRAM

Packaging Products Using Two Counters

PLC Lab 9.3

BILL OF MATERIALS

- TECO PLR or Micrologix™ 1100 PLC
- Solenoid
- Selector switch
- NO pushbutton
- Two limit switches

Industries use different processes based on the requirements of the application. The food-and-beverage, pharmaceutical, and plastics-and-packaging industries utilize processes such as two product conveyor lines sent to a caser for packaging products. For this lab, two product conveyor lines are sent to a caser, which diverts 6 containers of each product for loading using a pneumatic diverter.

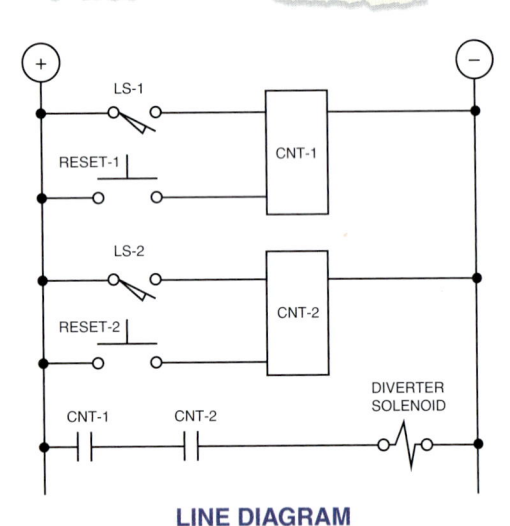

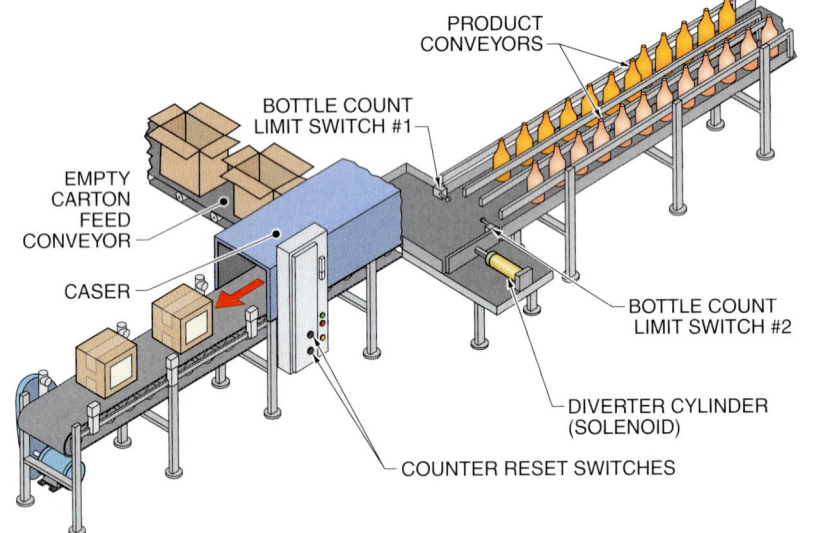

LINE DIAGRAM

T TECO Procedures

1. Based on the PLC system description and line diagram, draw the TECO programming diagram and FBD logic with the proper contacts and descriptions in the Programming Grid and Programming Area.

2. Create a new ladder logic or FBD program electronically.

001
002
003
004
005

TECO PROGRAMMING GRID

3. Use the Edit Contact window to denote an input or output number and the Symbol button to add a description after placing the input or output in the Programming Grid. For an FBD, use the Comment icon to add the description. Add counters with the proper mode and description to the Ladder Logic Programming Grid or FBD Programming Area.

4. Run the program in Simulation Mode to verify proper operation. Use the Input Status tool window to activate the inputs and control the output to simulate circuit operation.

5. Draw the wire connections in the wiring diagram between the PLC and input devices and output components. Then use the wiring diagram to wire the input devices and output components to the PLC.

6. Turn power to the PLC ON and link the com port.

7. Write the program to the SG2-12HR-D to download the program logic to the PLC.

8. Place the SG2-12HR-D in Run Mode. Activate the start pushbutton and monitor the I/Os on the LCD display.

9. Stop the program and use the Save As function to save and place the lab application in the required folder location.

FBD PROGRAMMING AREA

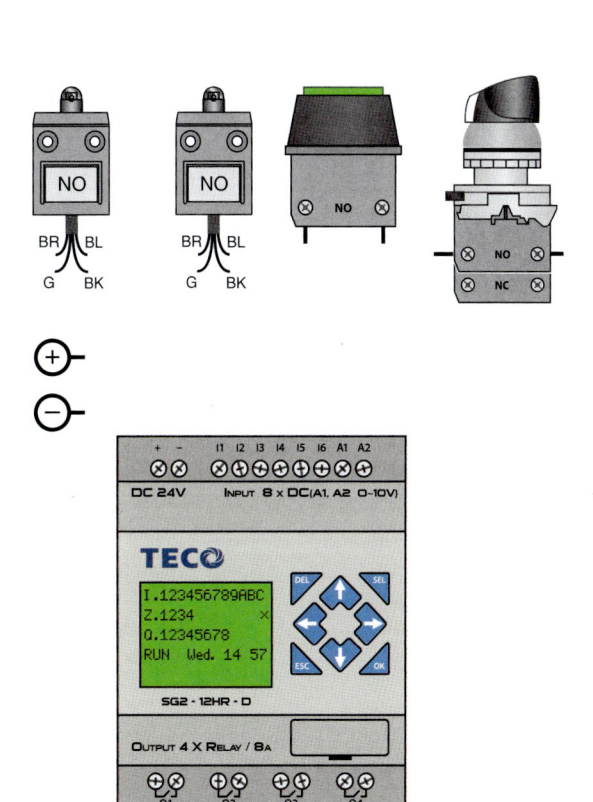

TECO WIRING DIAGRAM

Allen-Bradley® Procedures

1. Based on the PLC system description and line diagram, draw the ladder logic for RSLogix™ in the AB programming area.

2. Open the RSLogix™ Micro Starter Lite program and create a new programming diagram.

3. Use the AB programming diagram to add the proper address and description in the LAD 2 programming window. Add the proper counter instructions to the programming diagram and type in the counter address and preset value for each.

4. Verify that the programming diagram has no errors and save the project.

5. Draw the connections between the PLC and input devices and output components. Then use the wiring diagram to wire the input devices and output components to the PLC. Once all connections have been made, turn power to the PLC on.

6. Use RSLinx™ to establish communication between the PC and the Micrologix™ 1100.

7. Download the program logic to the PLC.

8. Place the PLC in Run Mode.

9. Activate input(s) and monitor them to determine if the output(s) are operating correctly.

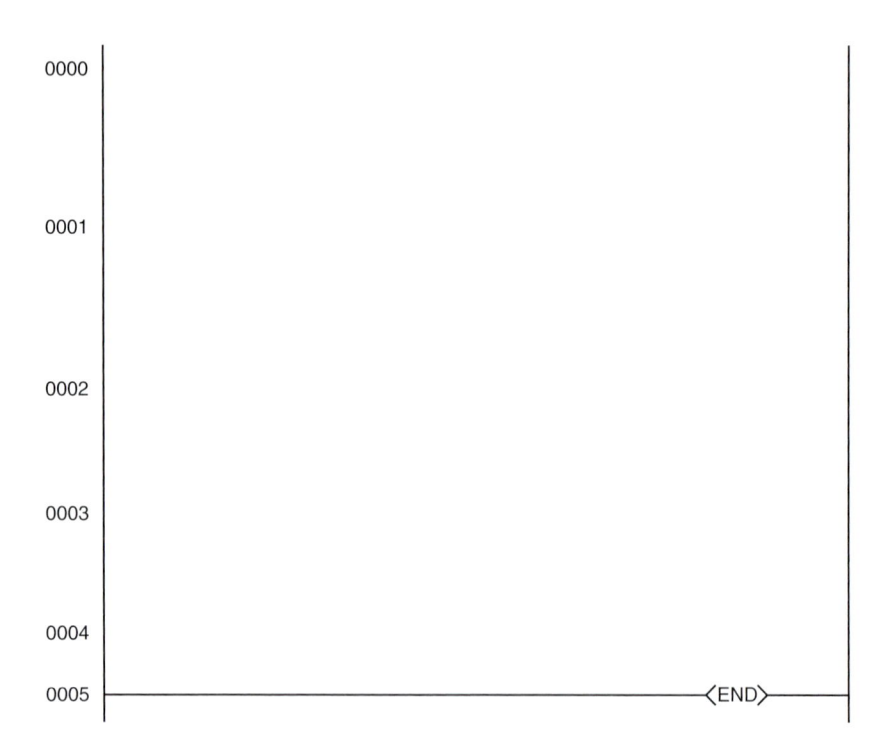

AB PROGRAMMING AREA

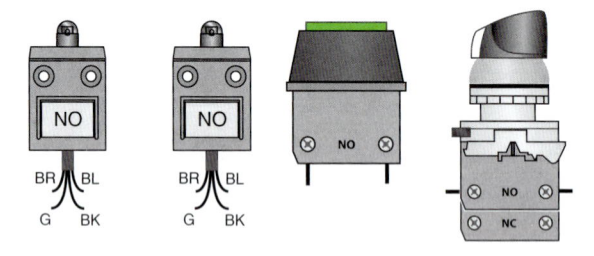

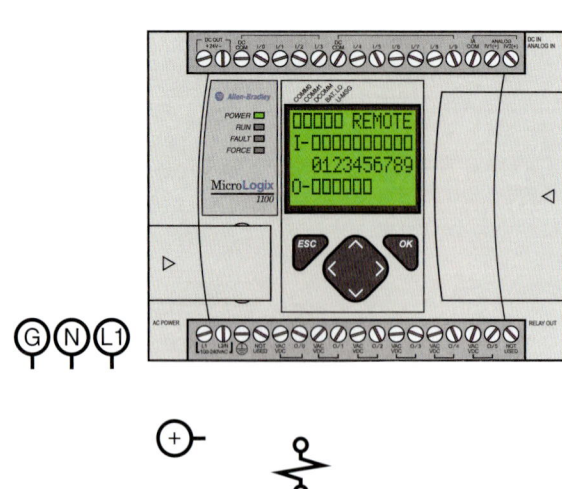

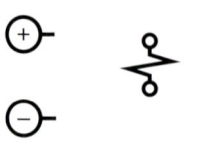

AB WIRING DIAGRAM

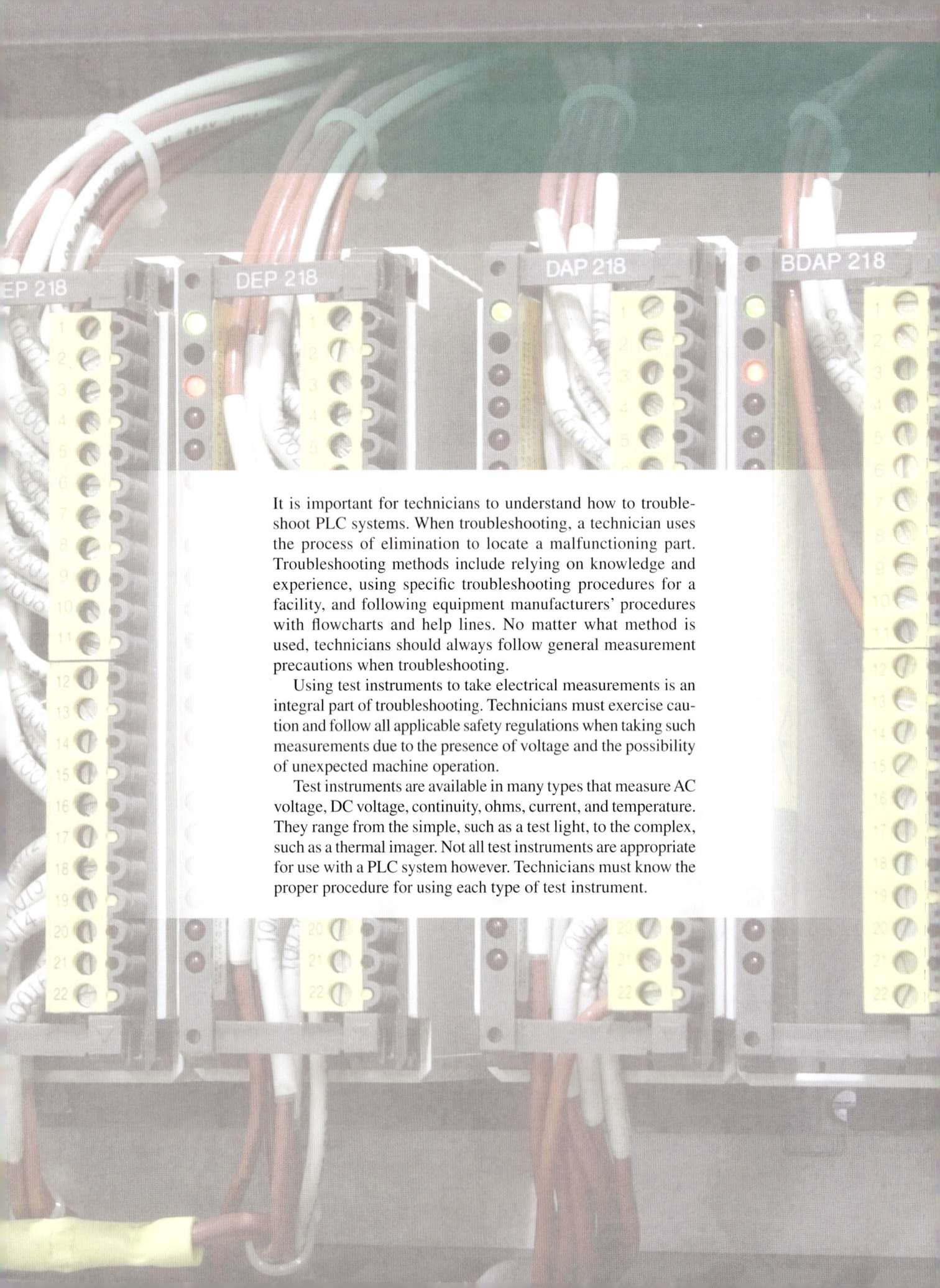

It is important for technicians to understand how to troubleshoot PLC systems. When troubleshooting, a technician uses the process of elimination to locate a malfunctioning part. Troubleshooting methods include relying on knowledge and experience, using specific troubleshooting procedures for a facility, and following equipment manufacturers' procedures with flowcharts and help lines. No matter what method is used, technicians should always follow general measurement precautions when troubleshooting.

Using test instruments to take electrical measurements is an integral part of troubleshooting. Technicians must exercise caution and follow all applicable safety regulations when taking such measurements due to the presence of voltage and the possibility of unexpected machine operation.

Test instruments are available in many types that measure AC voltage, DC voltage, continuity, ohms, current, and temperature. They range from the simple, such as a test light, to the complex, such as a thermal imager. Not all test instruments are appropriate for use with a PLC system however. Technicians must know the proper procedure for using each type of test instrument.

Troubleshooting Methods and Test Instrument Operation

10

Objectives:
- Explain what each status light on a PLC or CPU module indicates.
- Discuss the advantages of using PLC error codes and how they are displayed on a PLC.
- Identify the use of symbols and abbreviations on test instruments and their meanings.
- Explain the meaning of each CAT rating on a test instrument.
- Demonstrate how to set a test instrument to measure voltage and properly connect it to take voltage measurements on a PLC system.
- Demonstrate how to set a test instrument to measure resistance and properly connect it to take resistance measurements on a PLC system.
- Demonstrate how to set a test instrument to measure current and properly connect it to take current measurements on a PLC system.

Learner Resources
atplearningresources.com/quicklinks
Access Code: **475203**

TROUBLESHOOTING

Troubleshooting is the systematic elimination of various parts of a system or process to locate a malfunction. A *system* is a combination of components, units, or modules that are connected to perform work or meet a specific need. A *process* is a sequence of operations that accomplish desired results. A *malfunction* is the failure of a system, equipment, or part to operate properly.

Since PLCs are a major part of an electrical system and control the work performed by the electrical system of a machine or process, PLCs are the logical place to start troubleshooting. **See Figure 10-1.** The advantages of starting at the PLC include the following:

- The status lights (LEDs) of a PLC input module or a fixed LCD display provide a visual indication of which input devices are sending signals to the PLC (input status light ON) and which switches are not sending signals (input status light OFF).
- The status lights of the PLC output module or fixed LCD display indicate which output components are energized (output status light ON) and which output components are de-energized (output status light OFF).
- The operational status lights of a PLC provide a visual indication of when a PLC has power, when it is in run mode, when inputs and outputs are being forced, when a CPU fault occurs, and when the back-up battery is low.
- The PLC input section or module provides a central location for checking all input device signals for proper electrical levels using test instruments.
- The PLC output section or module provides a central location for checking all output components (lamps and solenoids) for proper signals, or system output interfaces (heating/lighting contactors or magnetic motor starters) for proper signal levels using test instruments.
- The system circuits can be monitored by connecting the PLC to a laptop. In addition to monitoring the status of circuit input devices and output components, the timers, counters, and sequencers of a PLC can also be monitored.
- Once connected to a computer, input terminals can be forced ON or OFF and output terminals can be forced ON or OFF. The ability to force input devices and output components helps to isolate faulty equipment.

⚠ **CAUTION**

Electrical gloves worn when taking measurements on energized circuits require electrical testing every six months either by an approved testing agency or in-house.

398—PROGRAMMABLE LOGIC CONTROLLERS: PRINCIPLES AND APPLICATIONS

Troubleshooting at PLC Location

INPUT SECTION STATUS LIGHTS PROVIDE VISUAL INDICATION OF INPUT DEVICE STATE (OPEN OR CLOSED)

INPUT SECTION IS CENTRAL LOCATION FOR CHECKING INPUT DEVICE SIGNALS

INPUT AND OUTPUT TERMINALS ARE FORCED DURING TROUBLESHOOTING

TO HOPPER FEED CONVEYOR MOTOR

N
L1
GND

LCD DISPLAY FOR INPUT/OUTPUT STATUS

CPU STATUS LIGHTS

HOPPER

HOPPER FEED CONVEYOR MOTOR

PHOTOELECTRIC SENSOR

SCALE CONVEYOR

SCALE

APPLICATION

PLC (INPUTS, OUTPUTS, TIMERS, COUNTERS, AND SEQUENCERS) MONITORED BY LAPTOP PC

TO SCALE CONVEYOR MOTOR

TO PHOTOELECTRIC SENSOR

TO SCALE

OUTPUT SECTION STATUS LIGHTS PROVIDE VISUAL INDICATION OF OUTPUT COMPONENT STATE (ON OR OFF)

OPERATIONAL STATUS LIGHTS PROVIDE VISUAL INDICATION OF THE STATE OF THE PLC CPU

OUTPUT SECTION IS CENTRAL LOCATION FOR CHECKING SIGNALS TO OUTPUT COMPONENTS

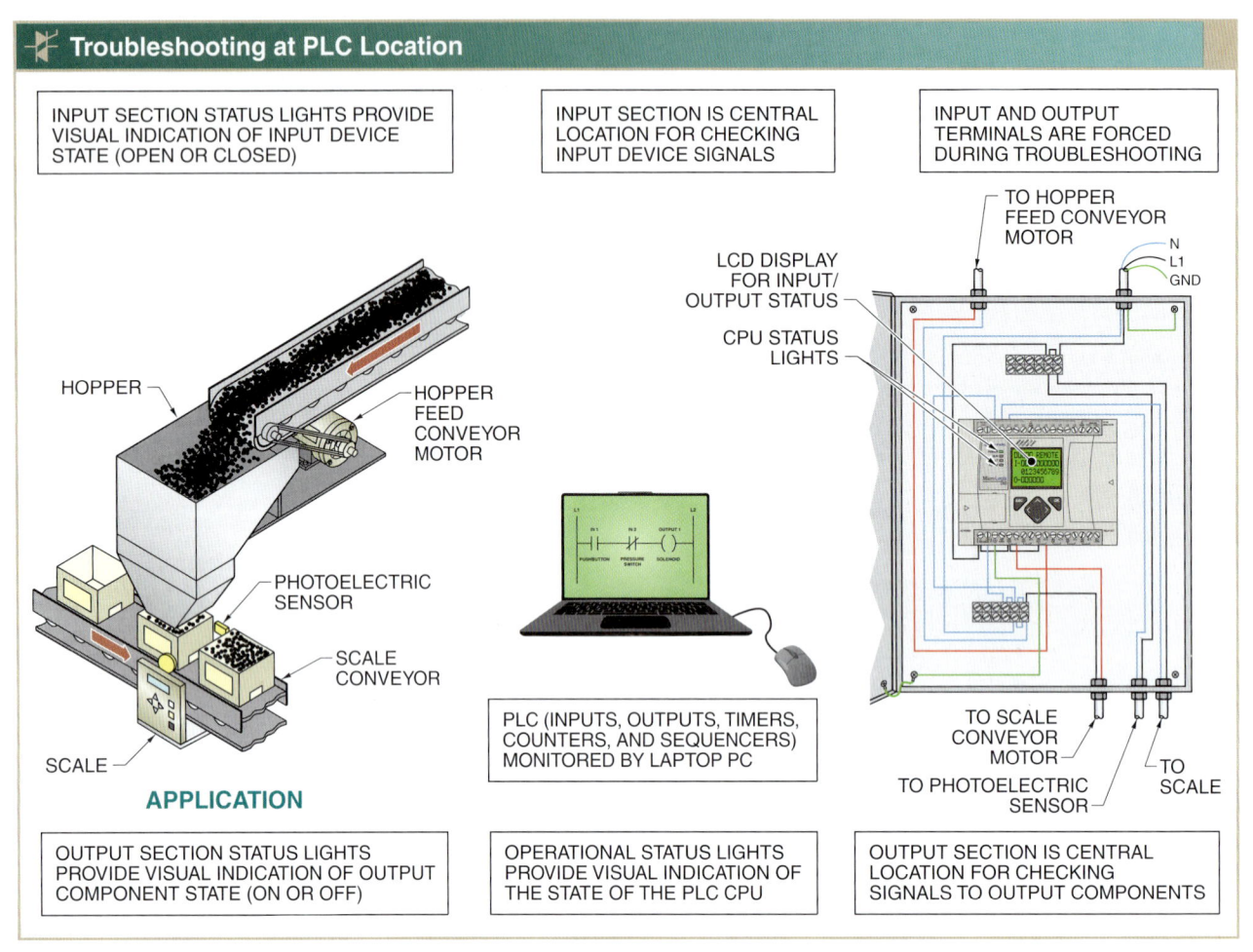

Figure 10-1. PLCs provide a centralized location for input device and output component wiring. PLCs are the logical place to start the troubleshooting process.

Troubleshooting at the PLC requires an understanding of how a PLC operates and how the machine or process should be operating. Test instruments can best be used to locate faults when the technician knows which test instrument to use for the desired information (voltage level, current draw, or resistance), where to connect the test instrument to obtain the best information (test points), and what the measured value on the test instrument means (open circuit, poor ground, or bad power supply).

TROUBLESHOOTING METHODS

Any electrical system that includes a PLC (or PLCs) is probably a large electrical system that includes multiple circuit input devices (switches) and output components (loads). In order to troubleshoot any electrical device, component, or circuit, a technician skilled in troubleshooting must follow a troubleshooting plan to find the malfunction quickly and efficiently. The larger the circuit or system, the more important an organized method of troubleshooting becomes.

When a malfunctioning component is found, the component is replaced or repaired. Preventive maintenance is then performed to prevent future problems. *Preventive maintenance* is the work performed to keep machines, assembly lines, production operations, and plant operations running with little or no downtime. Preventive maintenance programs allow equipment to be maintained in good operating condition with little downtime or troubleshooting required.

Test instruments are used to gather circuit information, but the knowledge and experience of the technician determine the usefulness of the information.

Fluke Corporation

Methods used to troubleshoot PLC circuits and systems include troubleshooting by knowledge and experience, plant procedures, manufacturer procedures, or a combination of all three methods.

Troubleshooting Using Knowledge and Experience

Troubleshooting using knowledge and experience is a troubleshooting method used for finding a malfunction in a machine or process by applying information acquired from past malfunctions. In certain circumstances, troubleshooting using knowledge and experience is only partially effective because the root cause is not corrected. For example, a fuse can blow or a circuit breaker can de-energize a part or all of a PLC-controlled system. Records may indicate that changing the fuse or resetting the breaker allows the system to continue operation, but the reason for the blown fuse or tripped circuit breaker may not be known. **See Figure 10-2.**

Troubleshooting using knowledge and experience is improved when the following conditions exist:

- Information about system components is gathered and the operation of the various primary systems is understood.
- The technician understands how input devices and output components are monitored and/or forced ON and OFF, how the program can be displayed and printed, and how system changes can be programmed into a PLC.
- All service calls are documented for future reference. Documentation includes listing all troubleshooting findings and repairs, all components checked and found to be good when a malfunction is not found or corrected, and all suggestions that may help prevent the malfunction from recurring.
- Test instruments are used to take measurements, and the measurements are documented for future use. For example, voltage measurements taken over time can indicate a power quality problem, such as transients (high-voltage spikes) present on power lines.
- Communication with other individuals (supervisors or operators) familiar with the system takes place. Supervisors or operators familiar with the system may not know the electrical, fluid power, or mechanical reasons why a machine or process works but may know about strange noises or any other unusual behavior that occurred before the malfunction.

CODE CONNECT

Article 450 of the NEC® covers the installation of transformers, both dry-type and liquid-filled. The following information must be listed on a transformer nameplate per section 450.11: manufacturer's name, kVA rating, frequency, primary and secondary voltages, impedance of transformers 25 kVA and larger, clearances for transformers with ventilation openings, amount and type of insulating liquid if used, and insulation temperature class for dry-type transformers.

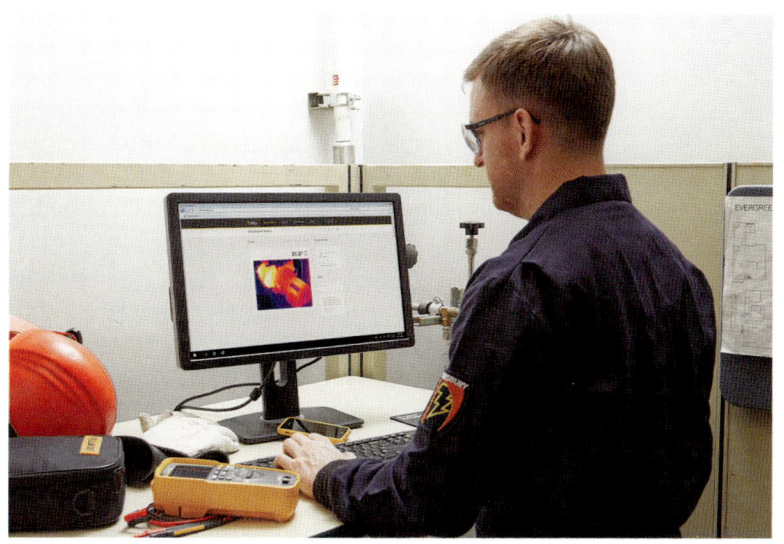

Fluke Corporation

Test instruments gather information about system components such as the temperature of a motor or shaft, which can be documented for future use for preventive maintenance and monitored during operation.

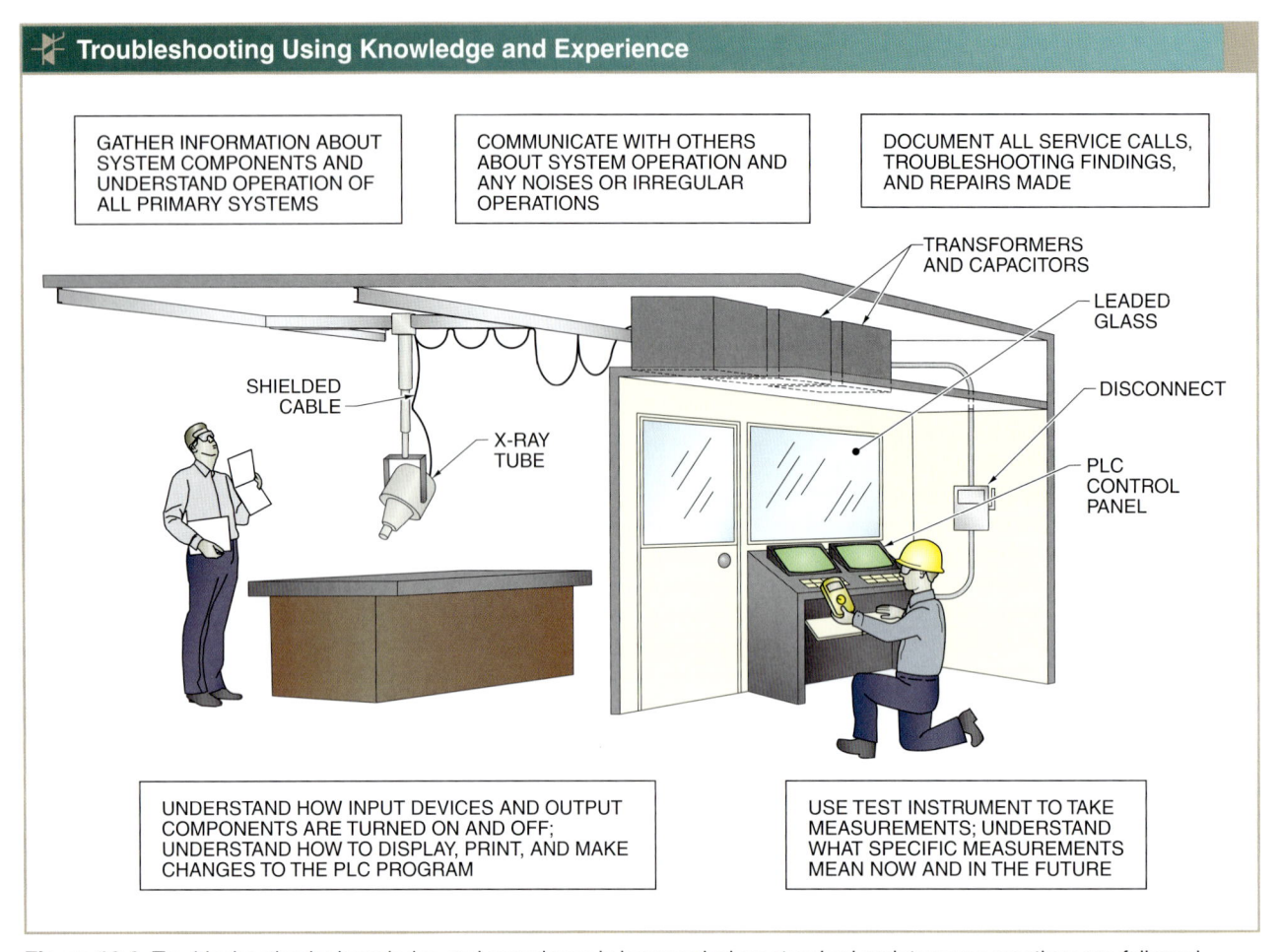

Figure 10-2. Troubleshooting by knowledge and experience is improved when standard maintenance practices are followed.

Facility Troubleshooting Procedures

Troubleshooting using facility procedures is a method of finding malfunctioning equipment using the procedures recommended by company personnel. Most facilities have procedures for troubleshooting machinery or processes. Facility procedures are typically developed by engineers, supervisors, or operators and are used to ensure safe and efficient troubleshooting of equipment by plant maintenance personnel. System troubleshooting procedures are specific to the machine or process in use by the company. **See Figure 10-3.**

The advantage of having and using facility procedures (even when the technician knows what is wrong and how to fix the problem) is that a procedure list provides an ordered checklist of things to do. For example, commercial airplane pilots know everything that must be done before takeoff and may have performed these procedures hundreds of times, but the pilots still follow preflight checklist procedures. PLC systems (like airplane systems) are often complex, with some PLCs controlling multimillion-dollar machines or processes. Using a checklist ensures that all steps are followed.

PLC TIPS

Between the time power is turned ON to a PLC and the time it takes the PLC to establish communication with connected programming devices being used for troubleshooting, the only form of communication to indicate faults to a technician is through LEDs.

Troubleshooting Using Facility Procedures

> **⚠ WARNING**
>
> **Defective Module Replacement**
>
> 1. To ensure orderly machine shutdown, inform the machine operator that the power will be turned OFF.
> 2. Turn OFF, lock out, and tag the disconnect switch feeding power to the machine.
> 3. Using a DMM set to measure voltage, test to ensure that no voltage is present at the power terminals of the programmable controller.
> 4. Remove the conductor connected to each terminal screw. Mark each conductor with the same number as the terminal screw. If removable strips are present, removing individual conductors is not necessary.
> 5. Pull the module locking lever out and down until it is perpendicular to the face of the module.
> 6. Slide the module out and away from the receptacle.
> 7. Insert the replacement module into the receptacle and lock it in place using the locking lever.
> 8. Reconnect the conductors to the terminal screws.
> 9. Clear the area of tools and any debris resulting from the maintenance call.
> 10. Inform the machine operator that the power will be turned ON.
> 11. Place all machine selector switches in the manual position.
> 12. Remove the lock and tag from the disconnect switch. Make sure all employees are clear of the machine.
> 13. Turn power ON.
> 14. Restart the machine.
> 15. Cycle the machine through one operation using the manual switches.
> 16. If the machine operates properly, place the selector switches in the automatic position.
> 17. Cycle the machine automatically 10 times.
> 18. Inform supervisor on duty if machine is not operating properly.
> 19. Fill out a company repair report if the machine is not operating properly. Place one copy of the report in the maintenance folder inside the machine cabinet and return one copy to supervisor on duty.
> 20. Remain with the operator until the machine is back in operation.

Figure 10-3. Facility (plant) procedures are specific to the system or process used by a company.

Manufacturer Troubleshooting Procedures

Troubleshooting using manufacturer procedures is a method of finding malfunctioning equipment by using the procedures recommended by the machine or process manufacturer. Manufacturer procedures differ from facility procedures in that manufacturer procedures are typically specific to an individual piece of equipment or section of equipment. **See Figure 10-4.**

For example, PLCs have status indicator lights that provide a visual display of the operating conditions of the PLC. The troubleshooting section of a PLC's operations manual lists the meanings of the various status light conditions along with possible problems and suggested corrective actions. **See Figure 10-5.**

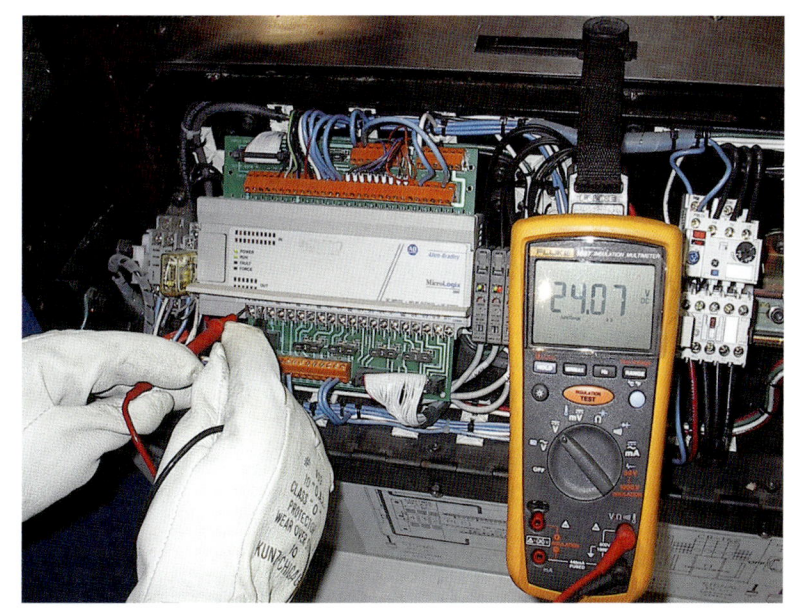

Taking voltage measurements of a PLC is a required procedure in all manufacturer troubleshooting procedures.

402—PROGRAMMABLE LOGIC CONTROLLERS: PRINCIPLES AND APPLICATIONS

Troubleshooting Using Manufacturer Procedures

⚠️ **WARNING**

Programmable Controller Troubleshooting and Replacement

1. If there is no indication of power (status lights OFF) on the programmable controller, measure the voltage at the incoming power terminal on the power supply module.
2. If correct voltage is present, replace power supply on programmable controller.
3. Remove power from programmable controller.
4. Disconnect the power lines from the power supply terminals.
5. Disconnect the processor power cable from the power supply output terminal.
6. Remove the four mounting screws on the power supply from the main panel.
7. Grasp the power supply firmly and pull out.
8. Press the replacement power supply into the main panel.
9. Replace and tighten the four mounting screws.
10. Connect the processor power cable.
11. Connect the power lines.
12. Turn power ON.

Figure 10-4. Manufacturer procedures vary from facility procedures in that manufacturer procedures are shorter and generally refer to a specific piece of equipment or part.

PLC CPU STATUS INDICATOR LIGHT CONDITIONS

Status Indicator Light	Problem	Possible Cause	Corrective Action
POWER / PC RUN / CPU FAULT / FORCED I/O / BATTERY LOW	No power or low system power	Blown fuse, tripped CB, or open circuit	Test line voltage at power supply. Line voltage must be within ±10% of the controller's rated voltage. Check for proper power supply jumper connections when voltage is correct. Replace the power supply module when the module has power coming into it but is not delivering the correct power.
POWER / PC RUN / CPU FAULT / FORCED I/O / BATTERY LOW	Programmable controller not in run mode	Improper mode selected for system operation	Place in run mode. Ensure that all personnel are clear before placing the system in run mode.
POWER / PC RUN / CPU FAULT / FORCED I/O / BATTERY LOW	Fault in controller	Fault memory module, memory loss or memory error, normally caused by a high-voltage surge, short circuit, or improper grounding	Turn power OFF and restart system. Remove power and replace the memory module when fault indicator is still ON. Load backup program on new memory module and reboot system.
POWER / PC RUN / CPU FAULT / FORCED I/O / BATTERY LOW	Fault in controller due to inadequate or no power	Loss of memory when power was OFF and battery charge was inadequate to maintain memory	Replace battery and reload program.
POWER / PC RUN / CPU FAULT / FORCED I/O / BATTERY LOW	System does not operate as programmed	Input device(s) or output component(s) in forced condition	Monitor program and device(s) and determine forced input and output component(s). Disable forced input device(s) or output component(s) and test system.
POWER / PC RUN / CPU FAULT / FORCED I/O / BATTERY LOW	System does not operate	Defective input device, input module, output component, output module, or program	Monitor program and check condition of status lights on the input and output modules. Reload program when there is a program error.

Figure 10-5. PLC status indicator lights (LEDs) provide a visual display of operating conditions.

All PLCs and most electrical equipment have written documentation and a troubleshooting section in the back of the installation and operations manual. The manufacturer information needs to be saved in an orderly way. Having a sign-out sheet available so any information removed from maintenance or engineering files is tracked and recorded is helpful. Verifying that files are kept up-to-date is part of all preventive maintenance programs. When manufacturer information is not kept as hard copies (or as available network information), technicians can look to manufacturer web sites for available product service manuals.

Troubleshooting Using Manufacturer Flowcharts. Some PLC manufacturers include flowcharts with the PLCs to aid in troubleshooting. A *flowchart* is a diagram that shows a logical sequence of steps for a given set of conditions. Flowcharts help a troubleshooter follow a logical path when trying to solve a problem. Flowcharts use symbols and interconnecting lines to provide analytical direction to the troubleshooting process. **See Figure 10-6.**

Symbol shapes used in flowcharts include ellipses, rectangles, diamonds, and arrows. An *ellipse symbol* is a symbol in a flowchart that indicates the beginning and end of a section of a chart. A *rectangle symbol* is a symbol in a flowchart that contains a set of instructions. A *diamond symbol* is a symbol in a flowchart that contains a question, worded so that the answer can be a "yes" or a "no." An *arrow symbol* is a symbol in a flowchart that indicates the direction to follow through the rest of the chart based on the answers to the questions.

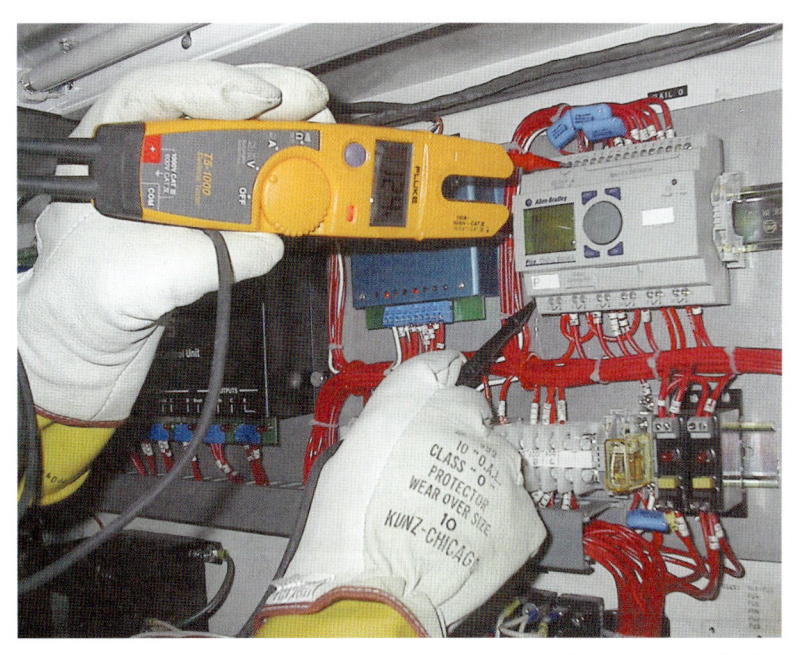

A technician must always wear the properly rated insulating gloves with leather protectors when troubleshooting a PLC while the power is on to ensure that no part of the technician's body makes contact with the circuit or leads.

Troubleshooting Using Manufacturer Help Lines. In addition to manufacturer written procedures, most PLC manufacturers have help lines for technicians to call when an existing problem cannot be solved. When possible, the technician should make the call from the location of the PLC. Before calling a manufacturer help line, the technician should have specific manufacturer information ready:

- PLC processor type and model/serial numbers—The model number is required so the manufacturer customer service person understands what product is being used. The serial number is also important because there can be specific problems with groupings of units that were manufactured at the same time and the manufacturer will be familiar with any problems that may have occurred.

- PLC software and program being used, including the software version—Most manufacturers offer software downloads to upgrade software already in operation. The technician should check to see if any software upgrades have been downloaded before calling.

PLC TIPS

Technicians may need the assistance of an operator when troubleshooting a process controlled by a PLC. Technicians need to understand how a process is supposed to operate and may need an operator's knowledge in order to troubleshoot it.

Troubleshooting Using Manufacturer Flowcharts

OUTPUT TROUBLESHOOTING FLOWCHART

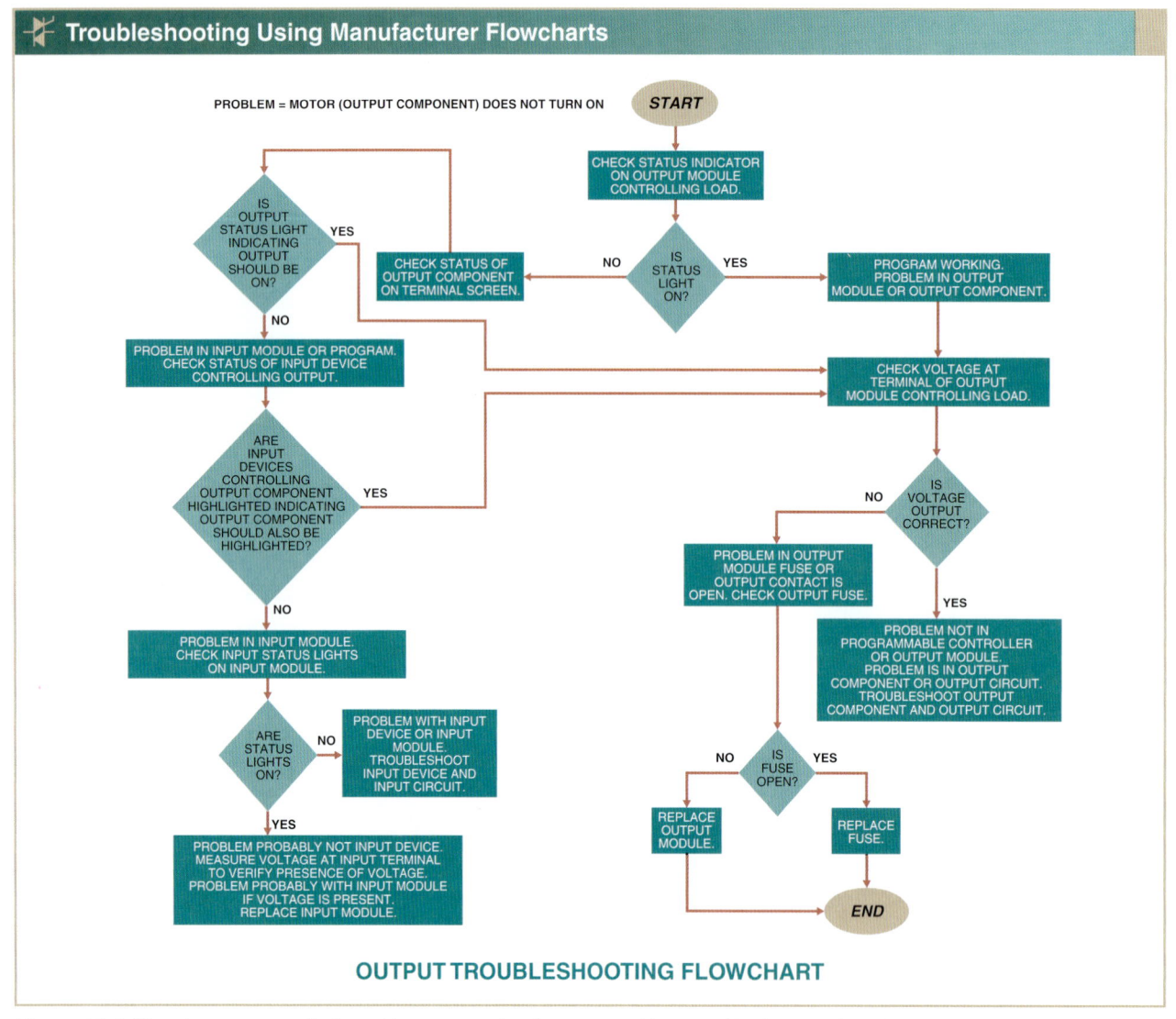

Figure 10-6. Flowcharts use symbols and interconnecting lines to provide a troubleshooter with a logical path for problem solving.

- Condition of LED status lights (ON or OFF)—For example, are any of the POWER, RUN, CPU FAULT, and BATTERY LOW lights on, solid, or flashing? Checking all the fault lights can help locate problems that might typically be overlooked. Also, checking the status of input and output LEDs is helpful. Knowing which output terminals are on (or not on) helps isolate the problem to specific circuits. When PLC processor error codes are indicated, the operations manual should be consulted for the meaning of the code. **See Figure 10-7.**
- PLC control circuit electrical measurements—Test instrument measurements should be recorded and ready to be conveyed to the service person on the help line.

MEASUREMENT PRECAUTIONS

Some electrical measurements (voltage and current) are taken with the power ON while others (continuity and resistance) are taken with the power OFF. There is always a chance of an electrical shock and/or spark when taking measurements with the power ON. All test instruments are designed for specific applications and have specific features and limits.

TECO PLC ERROR CODES

Error Code	Explanation	Error Action
ROM ERROR	System ROM/Flash memory check error	STOP SG2
Vpd ERROR	Power down circuits check error	STOP SG2
PROG ERROR	Ladder/FBD code invalid in EEPROM	STOP SG2
LOGIC ERROR	FBD code logic check error	STOP SG2
EXT ERROR	Expansion I/O error (When I/O alarm is disabled in "SET" of the main function, the alarm cannot appear.)	STOP SG2
COMM ERROR	RS485-type communication error	Warning only
RTC ERROR	RTC check or work error	Warning only
EMPTY PACK	Memory pack is empty when reading from the memory pack	Warning only
MEM ERROR	Memory packs check error when writing to the memory pack	Warning only

Figure 10-7. TECO PLC error codes indicate problems, such as memory errors, program errors, logic errors, and communication errors.

The user manual of a test instrument details specifications and features, proper operating procedures, safety precautions, warnings, and allowed applications. **See Figure 10-8.** The user manual should always be consulted before using any test instrument and the test instrument used only after the information provided is completely understood and can be properly applied. **See Appendix.**

Because PLCs are the central control for both machines and processes, troubleshooting PLCs requires measuring different voltage and current types (AC and/or DC), various voltage levels (5 V, 12 V, 24 V, 115 V), and different current levels (from a few mA up). In general, any voltage 36 V or higher and any current 6 mA or higher is dangerous. Thirty-six volts at 6 mA can cause a spark or electrical shock, and any voltage and/or current high enough to produce a spark can cause a fire or an explosion. Only qualified individuals, with the proper training and work experience, should measure and test PLC circuits.

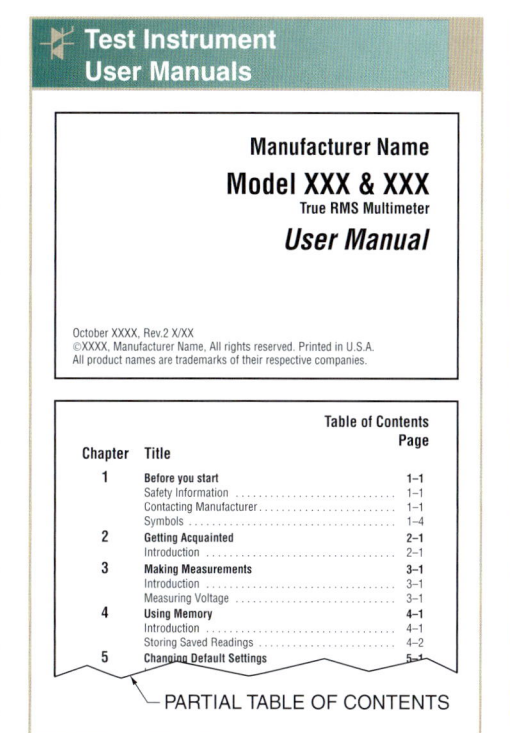

Figure 10-8. The user manual for a test instrument details specifications and features, proper operating procedures, safety precautions, warnings, and allowed applications.

⚠ WARNING

Technicians must follow proper electrical measurement practices when working on energized circuits for the sake of both accuracy and safety. This consists of wearing and using the appropriate PPE, inspecting test instruments and leads for damage, and tagging and removing damaged items from use.

Before taking any voltage measurement, technicians must use the three-point test procedure. The procedure is as follows:

1. *Test a known energized circuit.*
2. *Test the target circuit.*
3. *Test the known energized circuit again.*

This verifies that the test instrument is functioning properly both before and after the target energized circuit is tested.

Conditions can change quickly as voltage and current levels change in individual PLC-controlled circuits. **See Figure 10-9.** General safety precautions required when using test instruments on a PLC-controlled system include the following:

- A PLC-controlled system must include a sufficient number of emergency stop buttons that totally stop the machine or process operation in case of an emergency.

- Technicians must know where emergency stop buttons are and be prepared to use them at all times. Emergency stop buttons are important when troubleshooting because troubleshooting is performed on machines and processes that are not properly operating and the troubleshooting process itself can produce unexpected results, such as loads suddenly turning ON.

- When a system (circuit) does not have to be energized, such as when taking resistance measurements or changing fuses, all equipment and circuits to be tested or serviced should be locked out and tagged out.

- A technician should never assume a test instrument is operating correctly. For example, a test instrument that will be measuring voltage should be checked on a known energized voltage source before a measurement is taken on an unknown voltage source. After taking a measurement on the unknown voltage source, the test instrument should be retested on a known source to verify that the instrument is still operating properly.

- The technician should ensure that the test leads of a test instrument are

Test Instrument Safety Precautions

ENSURE TEST INSTRUMENT TEST LEADS ARE CONNECTED TO METER PROPERLY FOR MEASUREMENT TAKEN

LOCKOUT AND TAGOUT SYSTEM WHEN WORK DOES NOT REQUIRE AN ENERGIZED SYSTEM

TECHNICIANS MUST KNOW THE LOCATION OF EMERGENCY STOP BUTTONS

- TEST INSTRUMENT (DMM)
- TEST LEADS
- PHOTOELECTRIC S5
- RINSE AREA
- DRY AREA
- PHOTOELECTRIC S4
- WASH AREA
- PHOTOELECTRIC S3
- SOAP AREA
- PHOTOELECTRIC S2
- WET-DOWN AREA
- MAIN DISCONNECT (NOT SHOWN)
- PHOTOELECTRIC S1
- PLC CONTROL PANEL
- WHEEL PULLER
- EMERGENCY STOP 5
- EMERGENCY STOP 4
- EMERGENCY STOP 3
- EMERGENCY STOP 2
- SYSTEM START/SYSTEM STOP/CAR STOP/ CAR JOG/EMERGENCY STOP 1

NEVER ASSUME TEST INSTRUMENT IS OPERATING CORRECTLY (TEST BEFORE AND AFTER MEASUREMENT)

PLC SYSTEMS MUST HAVE SUFFICIENT NUMBER OF EMERGENCY STOP BUTTONS

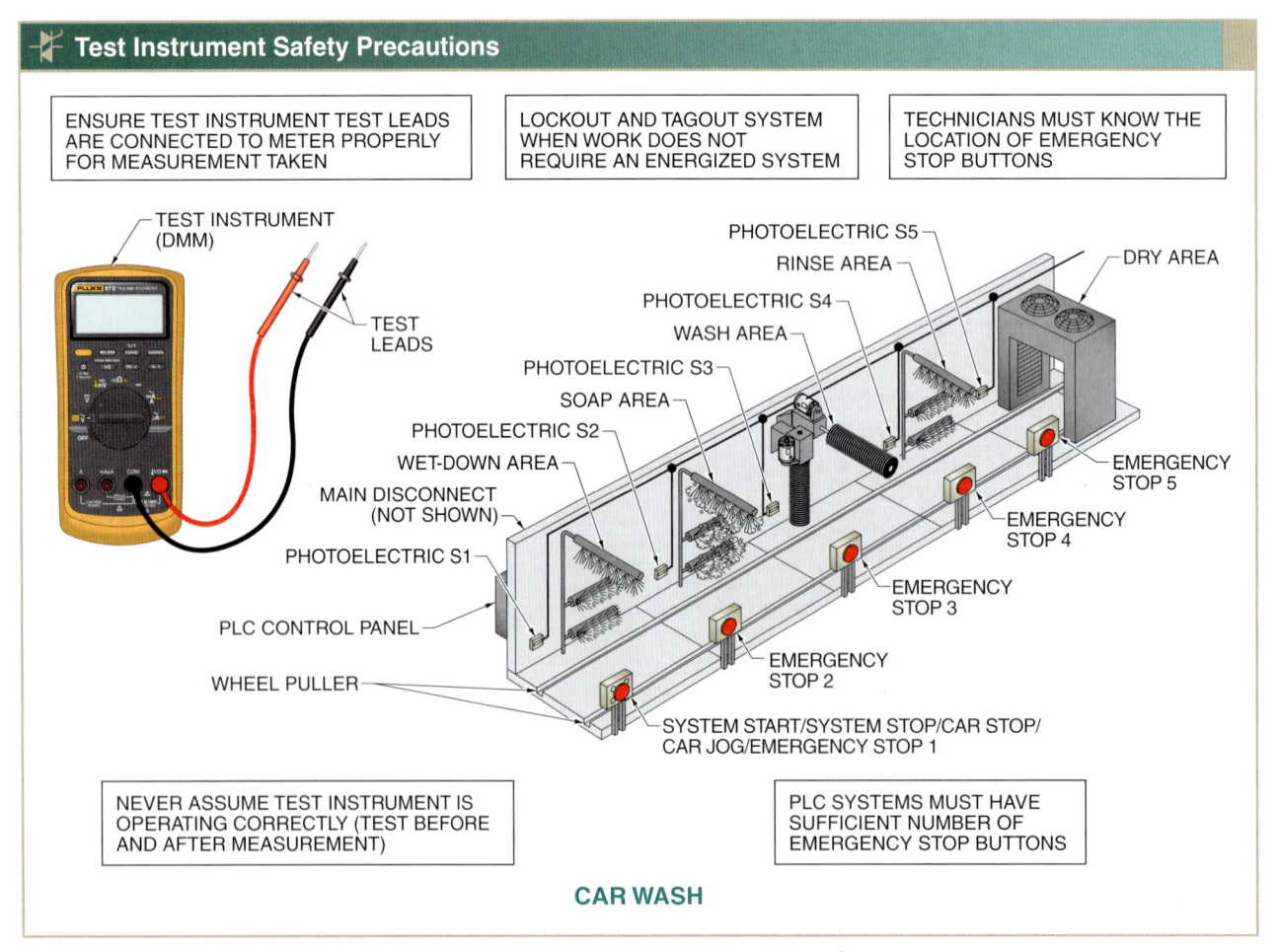

CAR WASH

Figure 10-9. Several precautions must be taken when using test instruments on a PLC-controlled system.

correctly connected. Test leads that are not connected to the correct jacks are dangerous. For example, attempting to measure voltage while the test leads are in the current jacks produces a short circuit because current jacks have very little resistance.

IMPROPER WIRE TERMINATIONS

When troubleshooting, technicians need to be aware of the possibility of improper wire terminations made on PLCs, input devices, and output components. Various problems can arise as the result of improper wire terminations such as intermittent signals and the danger of electrical shock due to exposed wiring.

Types of improper wire terminations include loose terminations and excessive insulation removal from a conductor. Loose terminations may cause intermittent signals from devices to the PLC or to components from the PLC. Since these signals happen at irregular intervals, they can be difficult to pinpoint during troubleshooting. Excessive insulation removal from a conductor leaves the conductor exposed to technicians, which presents an electrical shock safety hazard, or to other conductors, which may cause a short circuit. If an exposed conductor encounters another conductor, a spark or arc can damage the PLC. **See Figure 10-10.**

Improper Wire Terminations

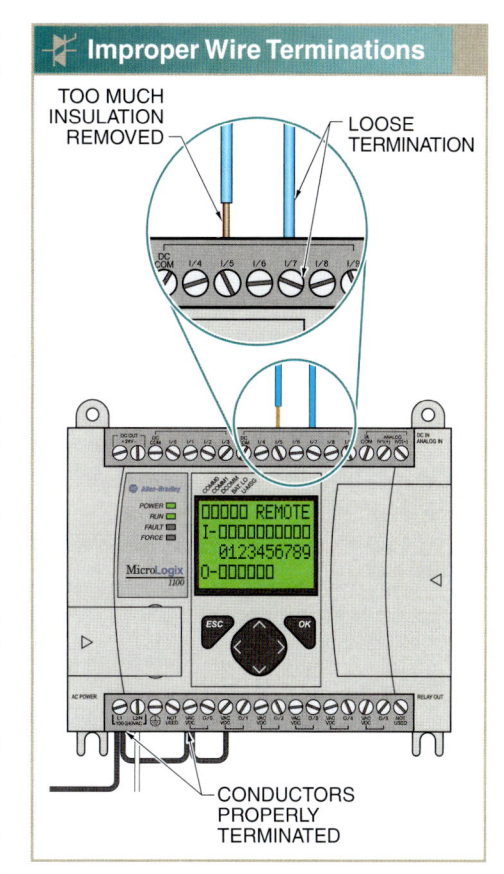

Figure 10-10. Technicians must be aware of the possibility of improper wire terminations, such as loose terminations and excessive insulation removal, when troubleshooting PLCs, devices, and output components.

METER ABBREVIATIONS, SYMBOLS, AND RATINGS

Test instruments typically use standard abbreviations and symbols to represent quantities, units, and electrical properties. In order to properly use any test instrument, the technician must understand the abbreviations and symbols on the instrument. As test instrument markets become more global, simpler abbreviations and symbols are becoming more commonplace.

In addition to understanding the abbreviations, symbols, and proper usage of test instruments, a technician should know where a test instrument can and cannot be safely used. The CAT (category) rating of a test instrument classifies the environment in which a test instrument can safely take measurements.

PLC TIPS

On advanced DMMs, special functions can be selected with the function switch, such as capacitance, inductance, and temperature functions. Advantages to using advanced DMMs consist of recording measurements and easy-to-read display values. Standardized colors are used for test leads to efficiently denote electrical functions, such as the red lead equals hot and the black lead equals common.

CODE CONNECT

Section 250.122 of the NEC® contains the requirements for sizing equipment grounding conductors. Typically, the size of an equipment grounding conductor is based on the ampere rating of the overcurrent protective device ahead of the equipment. This information is found in Table 250.122. However, if the ungrounded conductors are increased in size from their required minimums, Section 250.122(B) requires that the equipment grounding conductor be increased proportionally. Ungrounded conductors may be increased in size from their required minimums to compensate for voltage drop.

Electrical Test Instrument Abbreviations

Electrical test instruments use abbreviations to indicate what can be measured, such as VAC, Hz, mA, or °F, and to indicate measurement features, such as MIN MAX. An *abbreviation* is a letter or combination of letters that represent a word. Abbreviations can be used individually, such as V for volts and A for amps, or in combination with prefixes, such as mV for millivolt or kV for kilovolt. **See Figure 10-11.**

Electrical Test Instrument Symbols

Electrical test instruments use symbols to indicate information. A *symbol* is a graphic element that represents a quantity, unit, device, or component. Symbols provide quick recognition and are independent of language because a symbol can be interpreted regardless of the language a person speaks. Symbols can be used individually, such as the sine wave used to represent AC and the two straight lines (one dashed) used to represent DC, or in combination, such as the sine wave with two straight lines to represent AC or DC. **See Figure 10-12.**

Test instruments are used to measure electrical quantities, such as the voltage in a circuit. When voltage is a variable in electrical formulas, such as Ohm's law ($E = I \times R$) or the power formula ($P = E \times I$), voltage is represented by the capital letter "E," for electromotive force. However, when using a test instrument to measure voltage, the unit symbol "V" is used on the meter and meter display.

For each electrical quantity being measured by a test instrument, it is important to know the unit of measurement and the abbreviations used to represent the electrical quantity on both the test instrument and in the electrical formula. **See Figure 10-13.**

CAT Ratings

The International Electrotechnical Commission (IEC) 61010 standard classifies the applications in which test instruments can be used into four overvoltage installation categories (CAT I–CAT IV). The IEC categorizes the magnitude of transient voltages a test instrument must be able to withstand when used on energized systems such as distribution systems. Test instruments are designed and marked for the maximum voltage and category in which the instrument can safely be used.

Selected Test Instrument Abbreviations

AC	Alternating current or voltage
DC	Direct current or voltage
V	Volts
mV	Millivolts
kV	Kilovolts
A	Amperes
mA	Milliamperes
µA	Microamperes
W	Watts
kΩ	Kilohms
MΩ	Megohms
Hz	Hertz
kHz	Kilohertz
µF	Microfarads
nF	Nanofarads
°F	Degrees Fahrenheit
°C	Degrees Celsius

LOG	Readings are being recorded
LO	Low
AUTO-V	Automatic volts
LoZ	Low input impedance
nS	Nanosiemens (1×10^{-9}) or 0.000000001 siemens
MEM	Memory
MS	Time display in minutes:seconds
HM	Time display in hours:minutes

RPM	Revolutions per minute
COM	Common
OL	Overload
T	Time
LSD	Least significant digit
MAX	Maximum
MIN	Minimum
AVG	Average
TRIG	Trigger
V_{avg}	Average voltage
V_p	Peak voltage
V_{p-p}	Peak-to-peak voltage
V_{rms}	Root-mean-square (rms) voltage
HiZ	High input impedance
dB	Decibel
dBV	Decibel volts
dBW	Decibel watts

Figure 10-11. Abbreviations are used individually or in combination with prefixes.

Selected Test Instrument Symbols

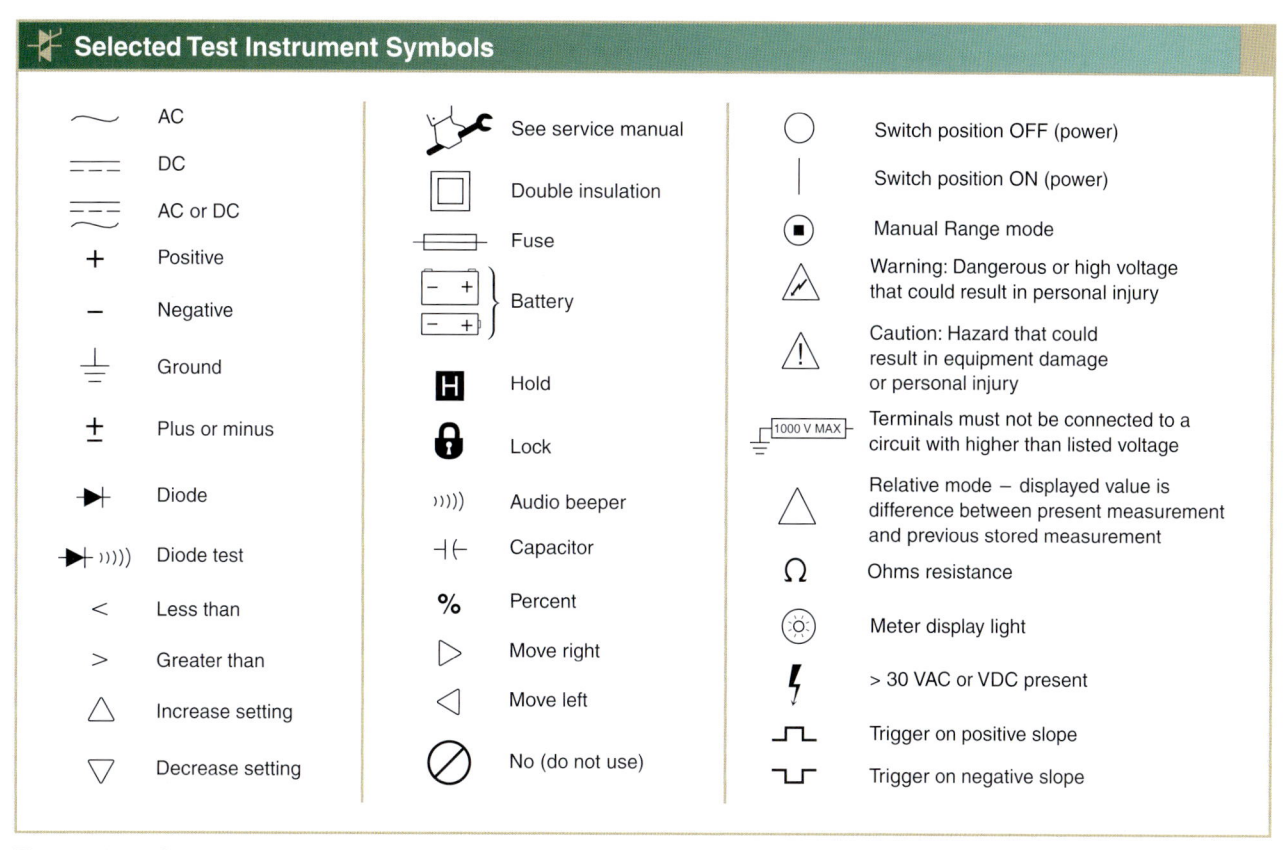

Symbol	Meaning	Symbol	Meaning	Symbol	Meaning
∼	AC	(wrench)	See service manual	○	Switch position OFF (power)
=	DC	□	Double insulation	|	Switch position ON (power)
≅	AC or DC	—[ ]—	Fuse	⊙	Manual Range mode
+	Positive	(battery)	Battery	⚠	Warning: Dangerous or high voltage that could result in personal injury
−	Negative	H	Hold	⚠	Caution: Hazard that could result in equipment damage or personal injury
⏚	Ground	🔒	Lock	1000 V MAX	Terminals must not be connected to a circuit with higher than listed voltage
±	Plus or minus	)))))	Audio beeper	△	Relative mode − displayed value is difference between present measurement and previous stored measurement
▶│	Diode	—|(—	Capacitor		
▶│)))))	Diode test	%	Percent	Ω	Ohms resistance
<	Less than	▷	Move right	(light)	Meter display light
>	Greater than	◁	Move left	⚡	> 30 VAC or VDC present
△	Increase setting	⊘	No (do not use)	⎍	Trigger on positive slope
▽	Decrease setting			⎌	Trigger on negative slope

Figure 10-12. Symbols provide quick recognition and meaning regardless of the language spoken.

Electrical Quantities

VOLTAGE (MEASUREMENT ABBREVIATION)

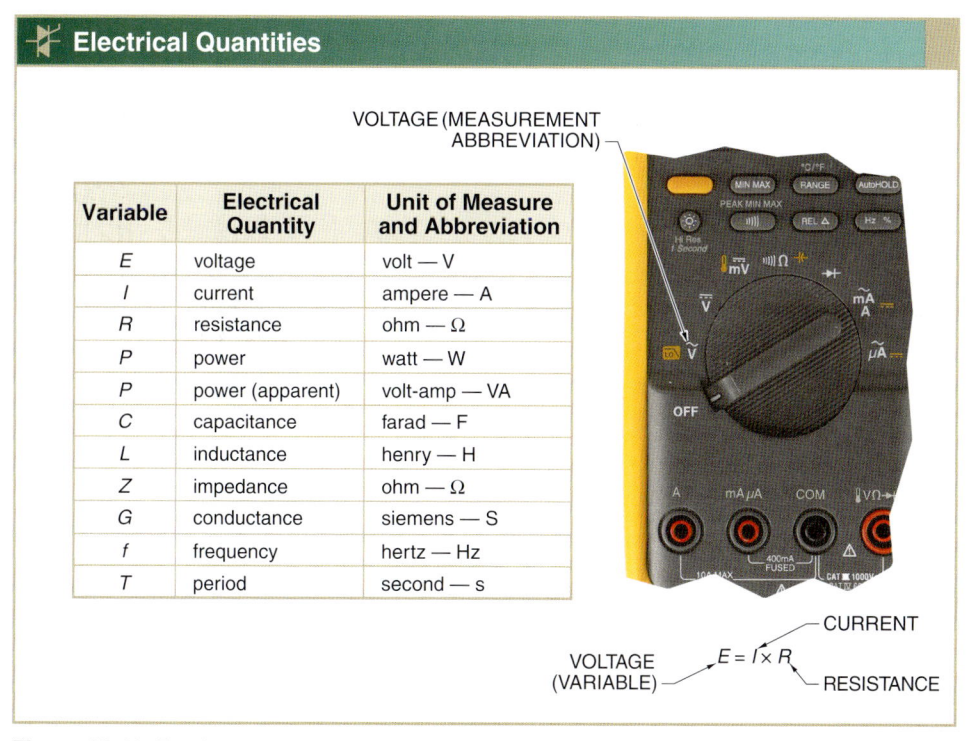

Variable	Electrical Quantity	Unit of Measure and Abbreviation
E	voltage	volt — V
I	current	ampere — A
R	resistance	ohm — Ω
P	power	watt — W
P	power (apparent)	volt-amp — VA
C	capacitance	farad — F
L	inductance	henry — H
Z	impedance	ohm — Ω
G	conductance	siemens — S
f	frequency	hertz — Hz
T	period	second — s

$$E = I \times R$$

VOLTAGE (VARIABLE) — CURRENT — RESISTANCE

Figure 10-13. Test instruments are used to measure electrical quantities. A technician should be able to recognize both the unit of measurement and the abbreviation used to represent the quantity.

Applications require the CAT rating of a test instrument to be the same or higher than the application category. A test instrument with a rating of CAT III or higher must be used when taking measurements on electrical systems that include a PLC so that the test instrument can be used anywhere on the PLC-controlled machine or process. When a test instrument is to be used outdoors for taking measurements on PLC-controlled input devices and output components, a test instrument with a CAT IV rating must be used. **See Figure 10-14.**

TEST INSTRUMENTS AND MEASUREMENT PROCEDURES

Test instruments are used to measure electrical properties such as voltage, resistance, current, frequency, and power. Test instruments are also used to check electrical components such as diodes and capacitors. Attachments are added to test instruments to measure nonelectrical quantities such as temperature, pressure, and speed.

Basic usage test instruments like voltage testers can be used to quickly verify that power is present at a location, that fuses

> **⚠ DANGER**
>
> *Before touching any part of a circuit after using a test light, measure for voltage in the circuit using another test instrument to verify that voltage is not present.*

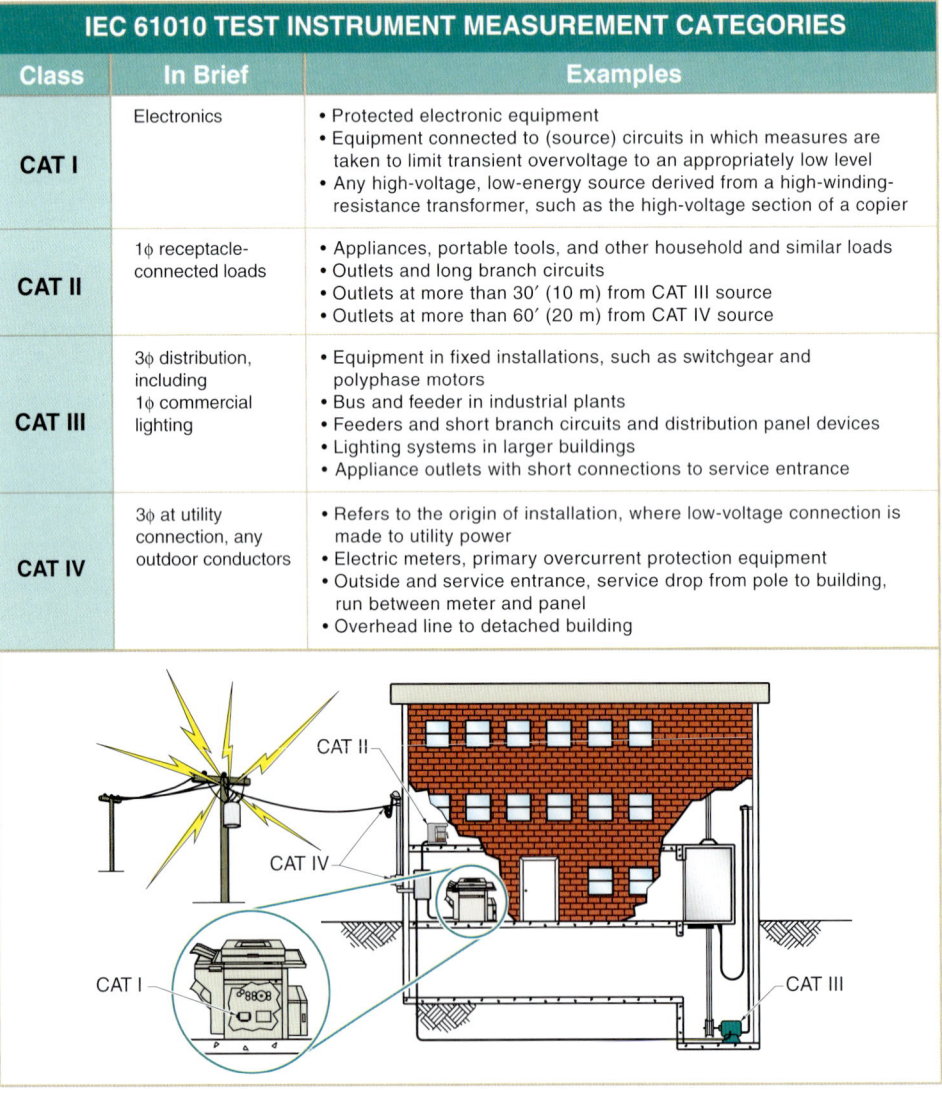

IEC 61010 TEST INSTRUMENT MEASUREMENT CATEGORIES

Class	In Brief	Examples
CAT I	Electronics	• Protected electronic equipment • Equipment connected to (source) circuits in which measures are taken to limit transient overvoltage to an appropriately low level • Any high-voltage, low-energy source derived from a high-winding-resistance transformer, such as the high-voltage section of a copier
CAT II	1φ receptacle-connected loads	• Appliances, portable tools, and other household and similar loads • Outlets and long branch circuits • Outlets at more than 30′ (10 m) from CAT III source • Outlets at more than 60′ (20 m) from CAT IV source
CAT III	3φ distribution, including 1φ commercial lighting	• Equipment in fixed installations, such as switchgear and polyphase motors • Bus and feeder in industrial plants • Feeders and short branch circuits and distribution panel devices • Lighting systems in larger buildings • Appliance outlets with short connections to service entrance
CAT IV	3φ at utility connection, any outdoor conductors	• Refers to the origin of installation, where low-voltage connection is made to utility power • Electric meters, primary overcurrent protection equipment • Outside and service entrance, service drop from pole to building, run between meter and panel • Overhead line to detached building

Figure 10-14. The IEC 61010 standard classifies the applications in which test instruments and meters can be used into four overvoltage installation categories.

and circuit breakers are operational, and that a system is grounded. However, when the exact voltage level must be known, the minimum and/or maximum voltage over time must be recorded, or the peak/rms voltage must be compared, a voltmeter with additional operational modes (MIN MAX, PEAK, and RELATIVE) must be used.

A voltage tester and other electrical meters are test instruments used to verify that power is present (or not present) at a specific location. A continuity tester is a test instrument used to identify open or closed connections when power is OFF. An ammeter is a test instrument used to determine how much a circuit or component is loaded (current draw). Temperature measurements are taken to locate loose or high-resistance connections.

Test Lights

A *test light* is a test instrument with a bulb, typically neon, that is connected to two test leads to provide a visual indication of when voltage is present. **See Figure 10-15.** The test light bulb lights up when voltage is present in the circuit being tested. Test lights can also include several different bulbs used to indicate approximate voltage level (115 VAC or 230 VAC).

The advantage of using test lights is that test lights are inexpensive, small enough to carry in a pocket, and easy to use. The disadvantage of test lights is that test lights have a limited voltage-indicating range and can only determine that voltage is present in a circuit, not the actual voltage. Test lights have a very limited usage when testing PLC circuits.

Voltage Testers

A *voltage tester* is a test instrument that indicates when voltage is present at a test point. Voltage testers indicate the approximate voltage amount and type of voltage (AC or DC) in a circuit by the movement of a pointer (and by vibration on some models). When a voltage tester includes a solenoid, the solenoid vibrates when the tester is connected to AC voltage. Some voltage testers include a colored plunger or other indicator, such as a light that indicates the polarity of the test leads as positive or negative when measuring a DC circuit. **See Figure 10-16.**

> **PLC TIPS**
>
> *Ninety-nine percent of test lights are for AC use only. Test lights can produce a false positive due to resistance values built into the test light by the manufacturer.*

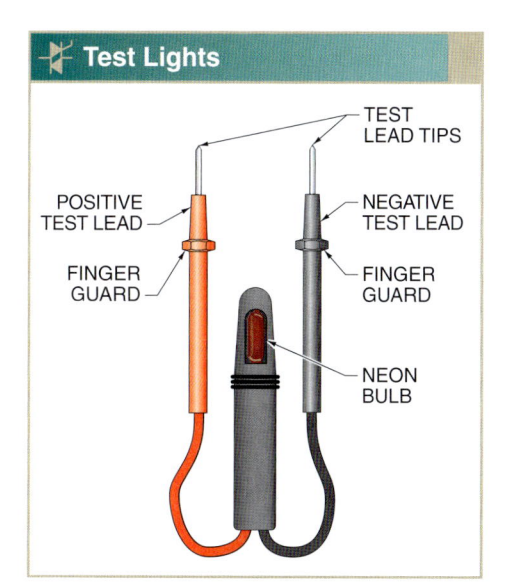

Figure 10-15. Test lights provide a visual indication when voltage is present in non PLC circuits but do not indicate the amount of voltage.

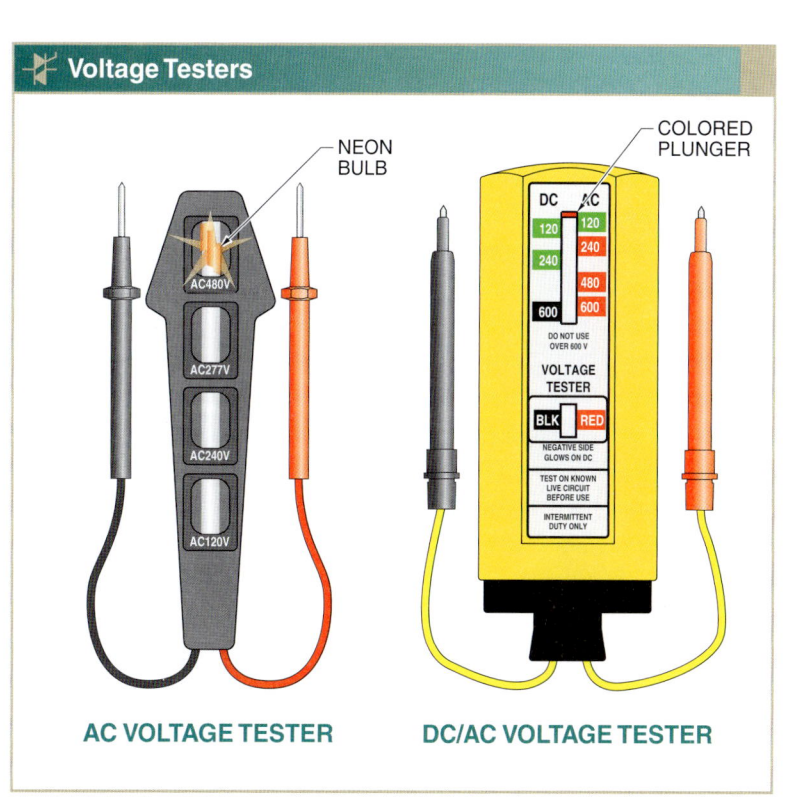

Figure 10-16. Voltage testers indicate the approximate voltage amount and type of voltage (AC or DC) in a circuit.

⚠ WARNING

Low-impedance solenoid voltage testers and some DMMs can activate PLC circuits when used on input sections or modules because of a low-impedance characteristic. This creates a potential safety hazard to personnel and/or equipment.

The advantage of using a voltage tester with a vibrating solenoid is that technicians can concentrate on the placement of the test leads instead of reading the tester. Voltage testers with solenoids have a low enough impedance that solenoid voltage testers can be used to test ground-fault circuit interrupters (GFCIs) for proper operation when connected between the hot and ground slot.

The disadvantage of using solenoid voltage testers is that their low impedance affects electronic equipment and circuit operation. Because of low impedance, a solenoid voltage tester should only be used to test AC voltages, branch circuit fuses, and circuit breakers. However, solenoid voltage testers should not be used to test input or output sections or modules of a PLC. A voltage tester can be used to test the PLC power supply. After using a voltage tester to take a measurement, verify voltage tester operation by taking a measurement on a known power source.

Voltage Tester Measurement Procedures

Before taking any measurements using a voltage tester, verify that the voltage tester is designed to take measurements on the circuit being tested. **See Figure 10-17.** Refer to the operating manual of the test instrument for all measuring precautions, limitations, and procedures. To take a voltage measurement using a voltage tester, apply the following procedures:

1. Verify that the voltage rating of the voltage tester is higher than the highest potential voltage.
2. Connect the common test lead (black) to the point of testing (neutral or ground).
3. Connect the voltage test lead (red) to the point of testing (ungrounded conductor). The pointer of the voltage tester indicates a voltage reading and vibrates when the current in the circuit is AC. The indicator shows a voltage reading and does not vibrate when the current in the circuit is DC.
4. Record the voltage measurement displayed.
5. Remove the voltage tester and test leads from the circuit.

Voltmeters

A *voltmeter* is a test instrument that measures voltage. Voltage is either direct current (DC) or alternating current (AC). Unlike voltage testers, which are not typically used around PLCs, voltmeters are used to test all parts of PLC circuits. A multimeter that includes both AC and DC voltage measurement functions is the best voltmeter for most basic PLC troubleshooting. A *multimeter* is a portable test instrument that is capable of measuring two or more electrical properties. Multimeters are either analog or digital.

An *analog multimeter* is a portable test instrument that uses electromechanical components to display measured values. A *digital multimeter (DMM)* is a portable test instrument that uses electrical components to display measured values. DMMs are the most common multimeter used when troubleshooting PLC systems. **See Figure 10-18.**

Determining Voltage Type

A meter can be set to measure AC voltage and be connected to a DC voltage without damaging the meter, circuit, or circuit components; however, the measured results will not be correct. The opposite is also true. A meter can be set to measure DC voltage and be connected to an AC voltage without damaging the meter, circuit, or circuit components; again however, the measured results will not be correct.

When the type of voltage (AC or DC) is unknown, a measurement should be taken with the meter set to measure DC voltage. The meter test leads should be reversed and the measurements taken again. When the voltage is DC at the test point, the two measured values will be the same, but one will have a negative (−) reading and the other a positive (+) reading. **See Figure 10-19.**

⚠ WARNING

Solenoid testers are known for inductive kickback when removed from the circuit being tested. The inductive kickback can damage a PLC.

Voltage Tester — Measurement Procedures

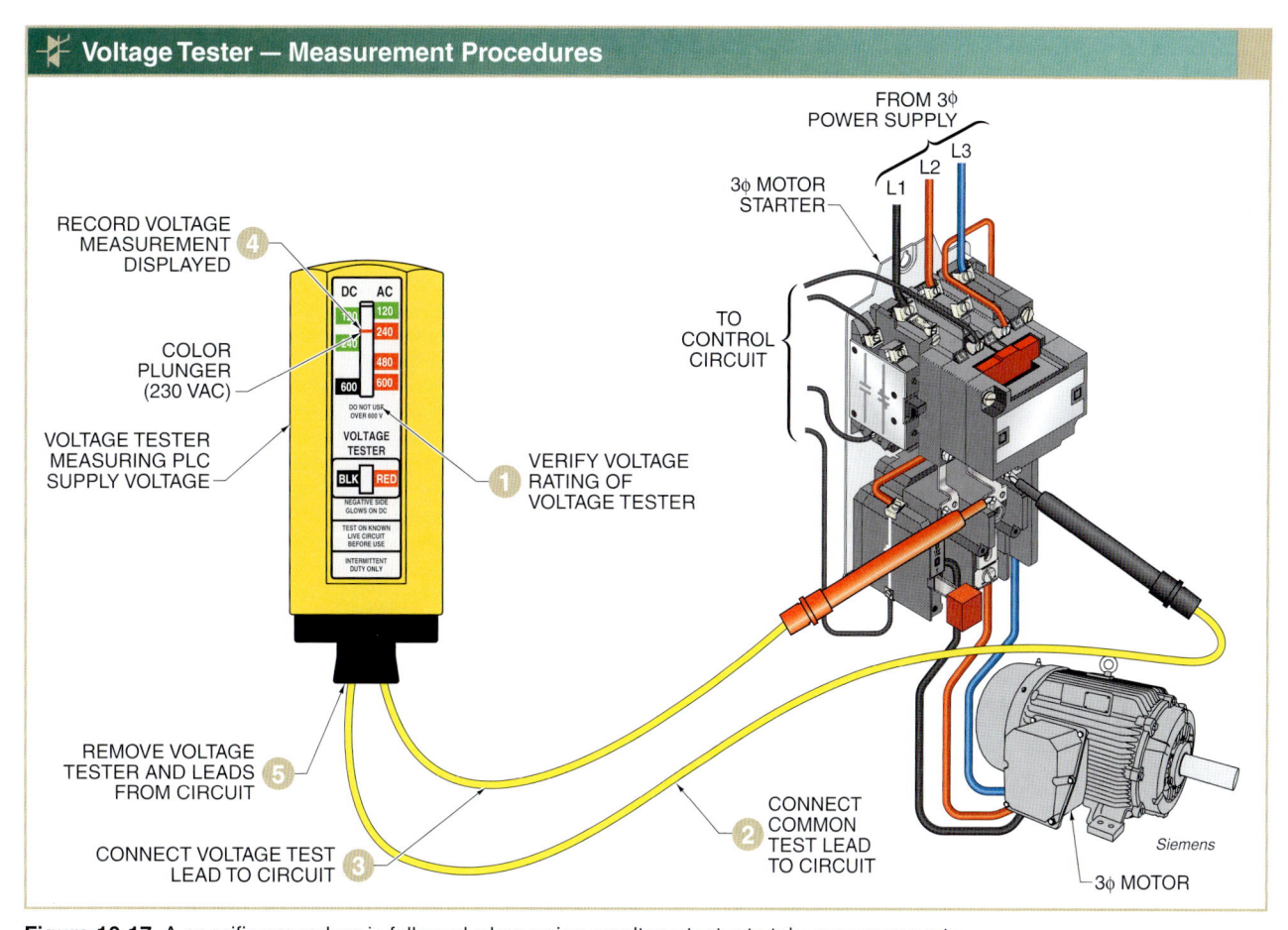

- RECORD VOLTAGE MEASUREMENT DISPLAYED ④
- COLOR PLUNGER (230 VAC)
- VOLTAGE TESTER MEASURING PLC SUPPLY VOLTAGE
- REMOVE VOLTAGE TESTER AND LEADS FROM CIRCUIT ⑤
- CONNECT VOLTAGE TEST LEAD TO CIRCUIT ③
- ① VERIFY VOLTAGE RATING OF VOLTAGE TESTER
- ② CONNECT COMMON TEST LEAD TO CIRCUIT
- FROM 3φ POWER SUPPLY — L1, L2, L3
- 3φ MOTOR STARTER
- TO CONTROL CIRCUIT
- 3φ MOTOR
- Siemens

Figure 10-17. A specific procedure is followed when using a voltage tester to take measurements.

Digital Multimeters (DMMs)

- ACCESS ADDITIONAL FUNCTIONS
- MEASURE RESISTANCE
- MEASURE mVDC VOLTAGE
- MEASURE DC VOLTAGE
- MEASURE AC VOLTAGE
- FUNCTION SWITCH
- AMPERAGE TERMINAL
- MILLI/MICRO AMPERAGE TERMINAL
- DIGITAL DISPLAY
- MEASURE CAPACITANCE
- DIODE TEST
- MEASURE AC AND DC CURRENT
- COMMON TERMINAL
- VOLTAGE TERMINAL (JACK)

> **⚠ DANGER**
>
> *Always connect test leads perpendicular to the terminals. PLC power supply terminals, input terminals, and output terminals are spaced close together and it is important to ensure that the test leads of a meter do not short the two adjacent terminals when taking measurements.*

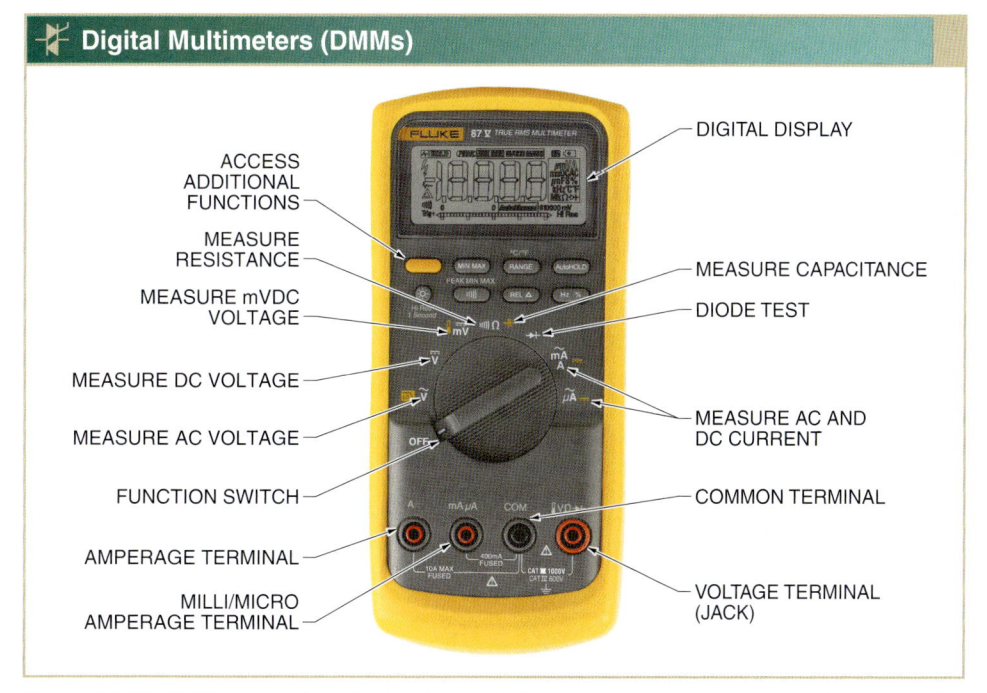

Figure 10-18. DMMs are portable test instruments that measure two or more electrical properties and display the measured properties as numerical values.

Measuring DC Voltage

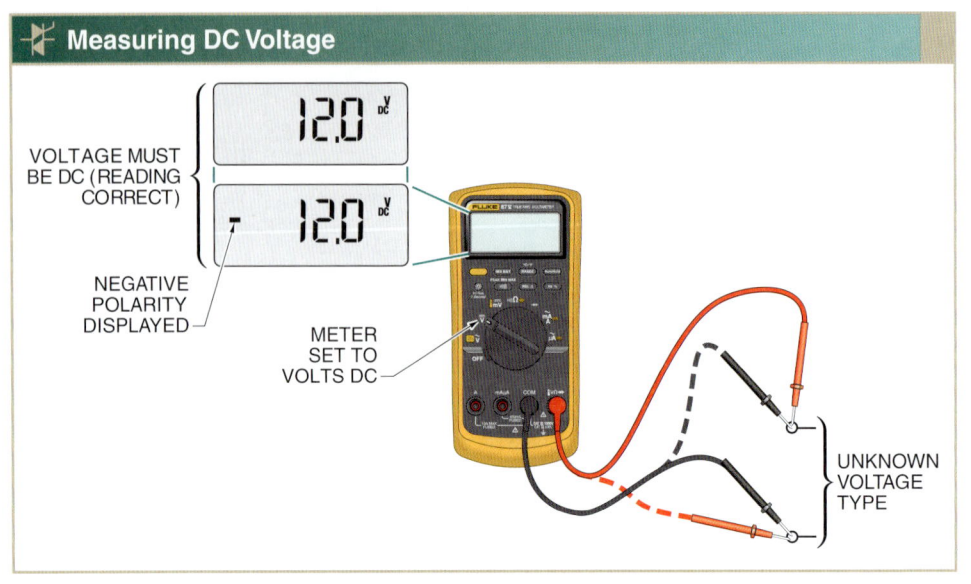

Figure 10-19. When the voltage is DC at the test point, the two measured values will be the same, but one will have a negative (−) reading and the other a positive (+) reading.

When the two measured voltage values do not indicate a DC voltage, the meter should be set to measure AC and the measurements should be retaken. When the voltage at the test point is AC, both readings will be the same (for example, 117.5 VAC and 117.5 VAC and neither one will be negative). **See Figure 10-20.**

When voltage measurements do not clearly indicate whether a voltage is DC or AC, a graphic display voltmeter (scope type) must be used to observe the voltage waveform. The viewed waveform may show that the voltage includes both AC and DC elements.

AC Voltage Measurement Procedures

When checking or troubleshooting a PLC system, loads, circuits, or individual components, voltage measurements must be taken. **See Figure 10-21.** A DMM is typically used to take voltage measurements. Many PLCs include AC voltages for powering internal circuitry and for operating AC output components such as motor starters.

A true-rms DMM should be used when taking AC voltage measurements in a PLC circuit or system. True-rms meters are designed to take accurate measurements in circuits that contain electronic devices (PLCs, electric motor drives, and computers). The technician should review the operating manual of the DMM before taking a voltage measurement because AC voltage measurement procedures may vary slightly with different DMMs.

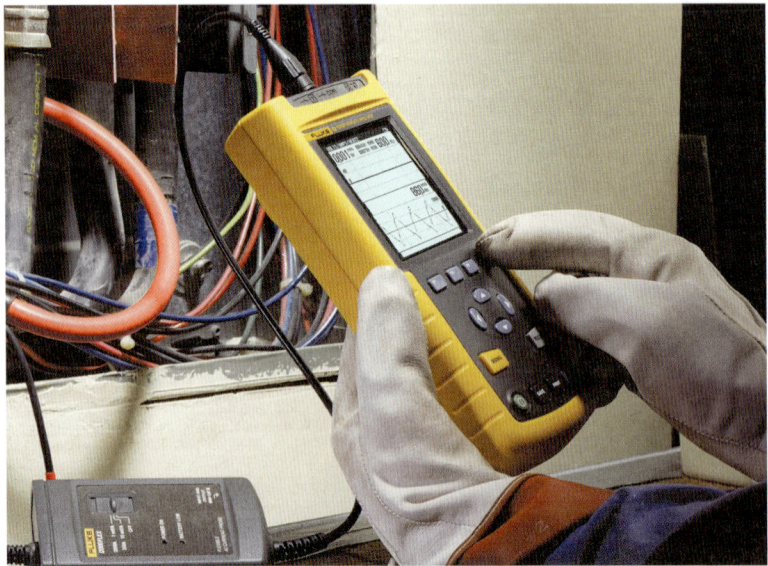

Fluke Corporation

Graphic display test instruments provide a visual display of the waveform in a circuit.

Measuring AC Voltage

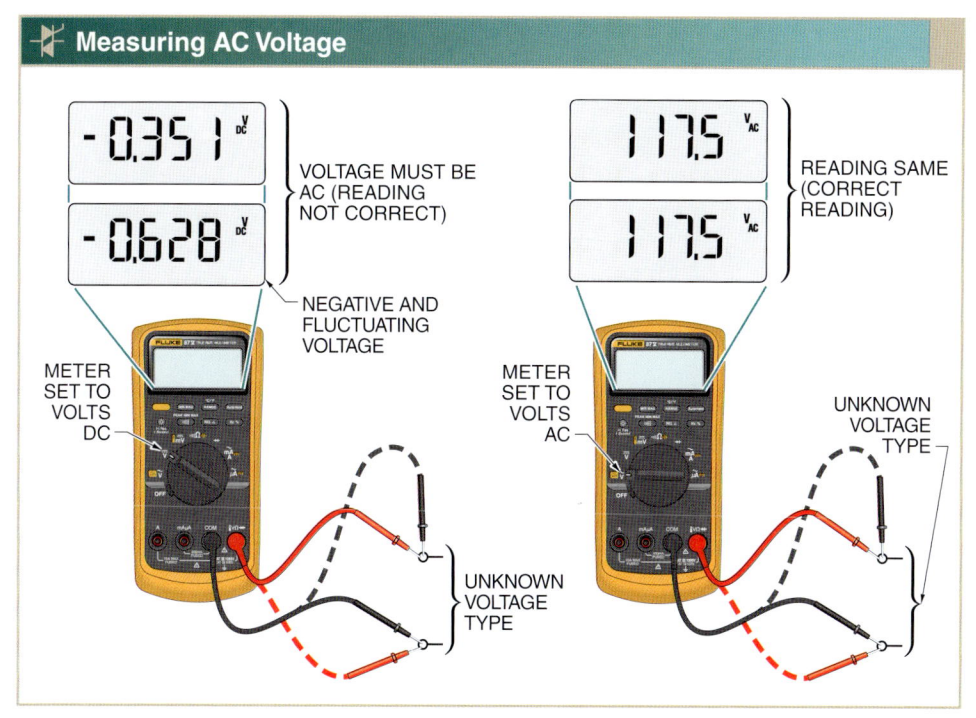

Figure 10-20. When the two measured voltage values do not indicate a DC voltage, the meter should be set to measure AC and the measurements should be retaken.

Before taking any measurements using a DMM, technicians must verify that the DMM has a voltage rating higher than the highest potential voltage in the circuit being tested. Technicians must refer to the operating manual of the test instrument for all measuring precautions, limitations, and procedures. To take an AC voltage measurement using a DMM, apply the following procedures:

1. Set the function switch of the DMM to AC voltage. Set the range to the highest voltage setting if voltage in the circuit is unknown.

   Note: Most DMMs power up in auto-range mode, which automatically selects a measurement range based on voltage present.

2. Plug the black test lead into the common jack.

3. Plug the red test lead into the voltage jack.

4. Connect the test leads to the circuit by connecting the common test lead (black) to the neutral, ground, or negative terminal and then connecting the voltage test lead (red) to the hot or positive terminal.

5. Record the voltage measurement displayed.

In addition to procedural steps 1–5, the following advanced features on some DMMs provide helpful steps:

6. Press the RANGE button to select a specified fixed measurement range.

7. Press the HOLD button to capture a stable measurement. The recorded measurement can be viewed later.

8. Press the MIN MAX button to capture the lowest or highest rms measurement. The DMM beeps each time a new reading is recorded.

9. Press the REL button to set the DMM to a specific reference value. The set reference value becomes the new zero point reference. Subsequent measurements are displayed as the difference between the reference value and the measurement as a positive or negative value.

10. Remove the DMM and test leads from the circuit.

11. Turn DMM off.

CODE CONNECT

A 120/240 V, three-phase, four-wire delta system has one transformer winding grounded to create a neutral point. This creates a higher voltage in one phase that is opposite from the grounded winding. This phase is referred to as the high leg in the NEC® and is approximately 208 V. Section 408.3(E)(1) requires that the B phase be the high leg. Section 110.15 requires that the B phase be identified with an orange outer finish.

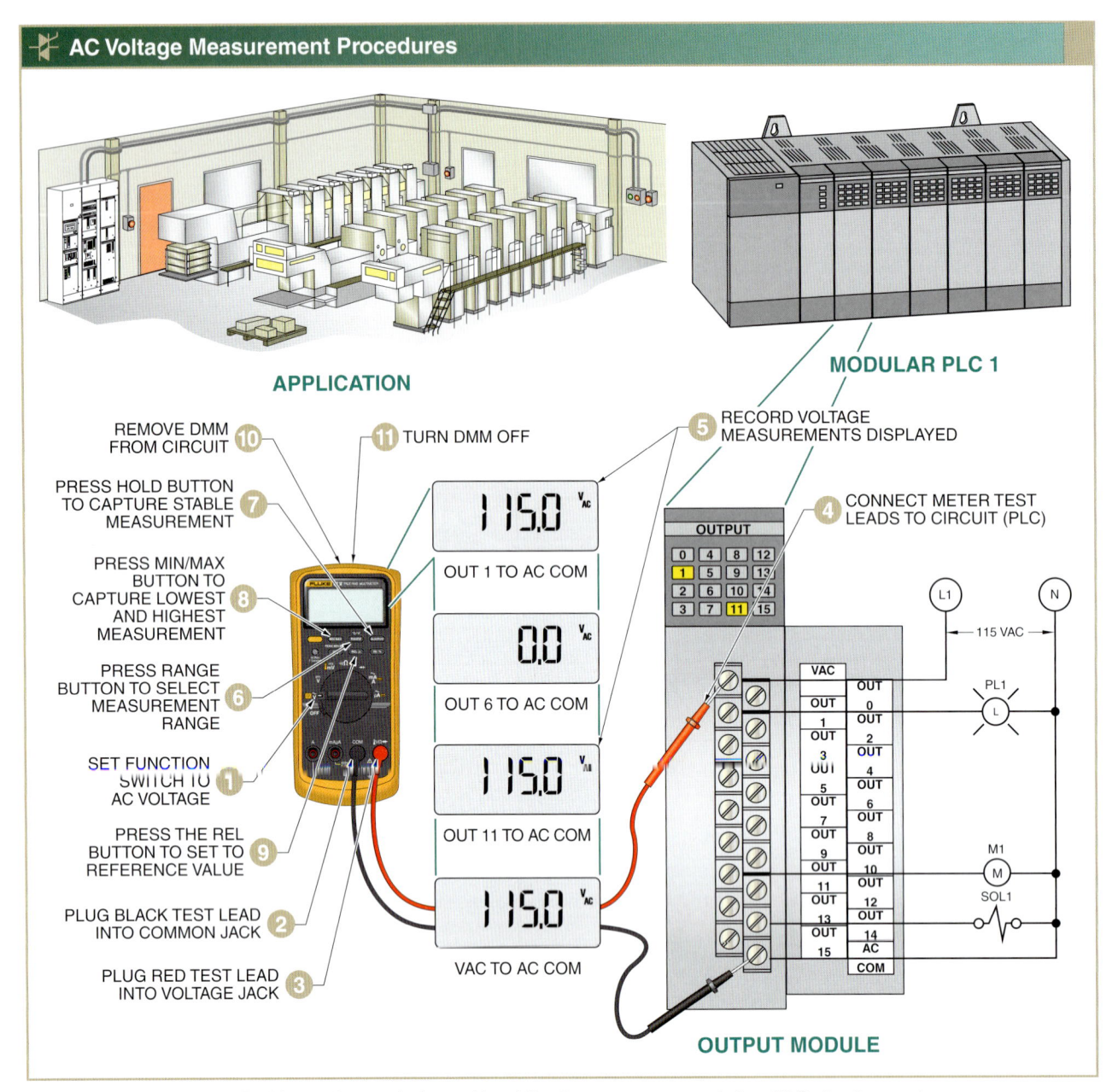

416—PROGRAMMABLE LOGIC CONTROLLERS: PRINCIPLES AND APPLICATIONS

Figure 10-21. A true-rms DMM must be used when taking AC voltage measurements in a PLC circuit or system.

DC Voltage Measurement Procedures

Many PLC systems include DC voltages somewhere. Typically, PLC input devices are DC-powered (24 VDC) and PLC output components that require a DC voltage (DC motors or solenoids) can be supplied with DC voltage. **See Figure 10-22.** As when measuring AC voltages, caution must be exercised when taking any circuit measurement that has low DC voltage (12 VDC, 24 VDC). Measurements of DC voltages exceeding 50 V and DC measurements near any battery require extra caution to prevent an electrical shock or spark, which can cause a fire or an explosion.

DC Voltage Measurement Procedures

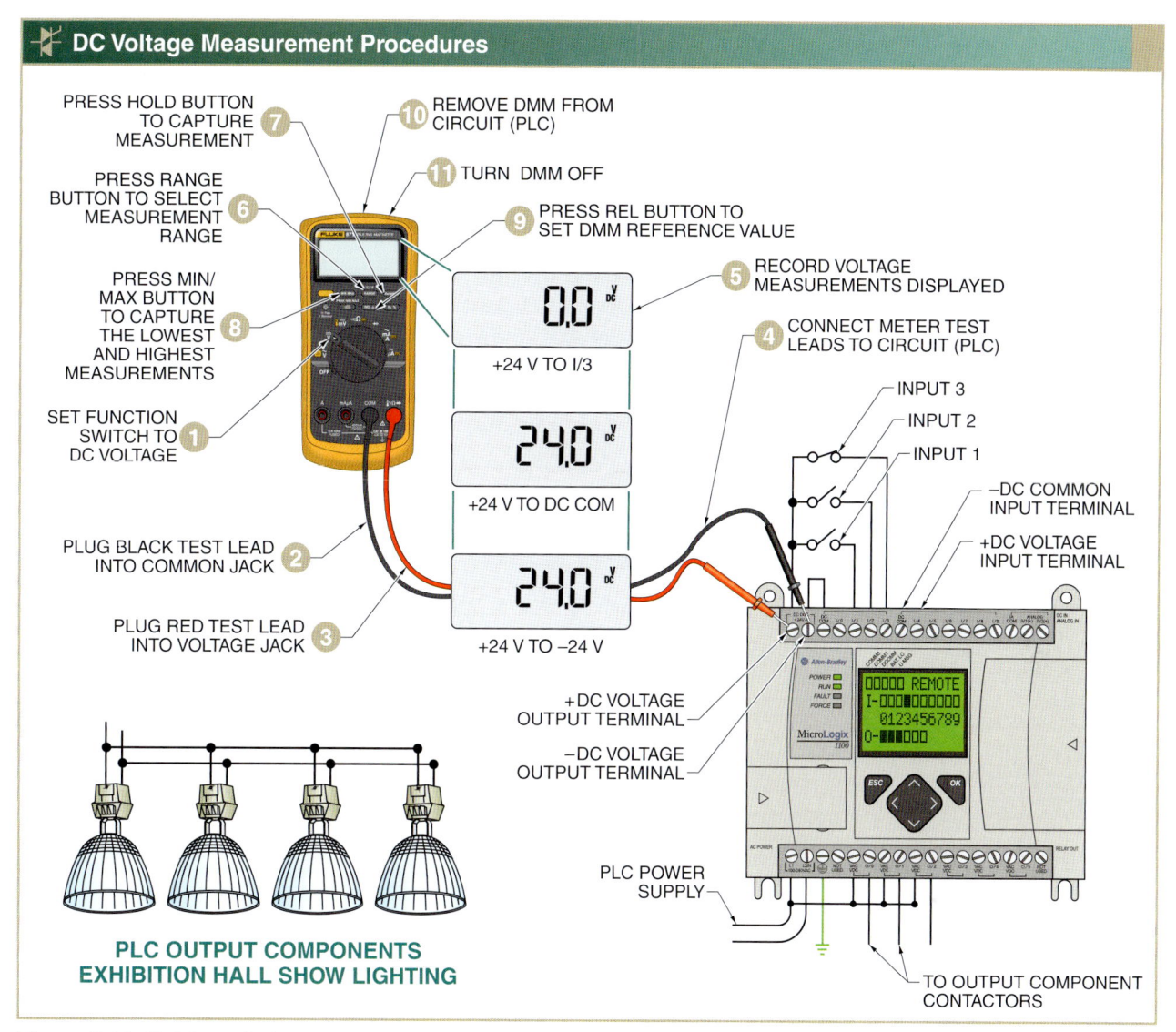

Figure 10-22. PLC input devices are typically powered by DC voltages such as 24 VDC.

Before taking any DC voltage measurements using a DMM, technicians must verify that the DMM has a voltage rating higher than the highest potential voltage in the circuit being tested. Technicians must refer to the operating manual of the test instrument for all measuring precautions, limitations, and procedures. To take a DC voltage measurement using a DMM, apply the following procedures:

1. Set the function switch to DC voltage. If the DMM includes more than one DC setting, select the highest setting. For example, some DMMs include a VDC setting and an mV DC setting. The VDC setting should be used.

2. Plug the black test lead into the common jack.

3. Plug the red test lead into the voltage jack.

4. Connect the test leads to the circuit. The black test lead is connected to the negative polarity test point (circuit ground) and the red test lead is connected to the positive polarity test point. Reverse the test leads when a negative sign (–) appears to the left of the displayed measurement.

5. Record voltage measurement displayed.

418—PROGRAMMABLE LOGIC CONTROLLERS: PRINCIPLES AND APPLICATIONS

CODE CONNECT

Table 250.66 of the NEC® provides information on the sizing of alternating current grounding electrode conductors. A grounding electrode conductor connects a grounding electrode to the grounded conductor of an electrical system. Grounding electrode conductors are sized based on the size of the largest ungrounded service conductor or equivalent area for parallel service conductors. Table 250.66 provides information for copper, aluminum, and copper-clad aluminum conductors.

In addition to procedural steps 1–5, the following advanced features on some DMMs provide helpful steps:

6. Press the RANGE button to select a specific fixed measurement range. If the voltage measurement is within the range of a lower VDC setting, such as the mV DC, a more accurate measurement can be obtained by changing the setting. Disconnect the positive (red) test lead before disconnecting the negative (black) test lead from the circuit. Change the DMM setting, reconnect the DMM to the circuit at the same test points, and record the measurement displayed.

7. Press the HOLD button to capture a stable measurement. The recorded measurement can be viewed at any time.

8. Press the MIN MAX button to capture the lowest and highest measurement. The DMM beeps each time a new reading is recorded.

9. Press the REL (relative) button to set the DMM to a specific reference value. Measurements above and below the reference value are displayed.

10. Remove the DMM and test leads from the circuit.

11. Turn DC voltage meter (DMM) OFF.

Continuity Testers

A *continuity tester* is a test instrument that tests for a complete path for current to flow. For example, a closed switch that is operating properly has continuity; however, an open switch does not have continuity. **See Figure 10-23.** Continuity tests can be performed on any nonpowered mechanical switch, such as the dry contact terminals or alarm contacts on a PLC.

Some test instruments test for continuity using a continuity test mode. The continuity test mode is commonly used to test electrical input devices such as switches, fuses, connections, and individual conductors. A test instrument set to continuity test mode emits an audible response (beeps) when there is a complete path for current to flow. Indication of a complete path can be used to determine the condition of a component as open or closed. For example, a good fuse should have continuity, whereas a bad fuse will not.

A continuity test is used to check input switches to ensure their proper operation before they are connected to the input section or module of a PLC. However, a continuity test cannot be done to check the operation of a photoelectric switch.

Some photoelectric switches have normally open and normally closed contacts that can be set for light operation or dark operation. When a photoelectric switch is set to light-operated mode, the contacts switch position when the target is missing (removed from the beam). When a photoelectric switch is set to dark-operated mode, the contacts switch position when the target is present (breaking the beam).

Continuity Tester Measurement Procedures

Continuity is tested with a test instrument set on the "continuity test" mode. Before taking any continuity measurements using a continuity tester, ensure the meter is designed to take measurements on the circuit being tested. **See Figure 10-24.** To take continuity measurements with a continuity tester, apply the following procedures:

1. Set the DMM function switch to continuity test mode as required. Most test instruments have the continuity test mode and resistance mode sharing the same function switch position.

2. With the circuit de-energized, connect the test leads across the component being tested. The position of the test leads is arbitrary.

3. When there is a complete path (continuity), listen for the beep of the tester. When there is no continuity (open circuit), the continuity tester will not beep.

4. After completing all continuity tests, remove the continuity tester and test leads from the circuit or component being tested.

5. Turn the continuity tester OFF to prevent battery drain.

Chapter 10—Troubleshooting Methods and Test Instrument Operation   419

Continuity Testers

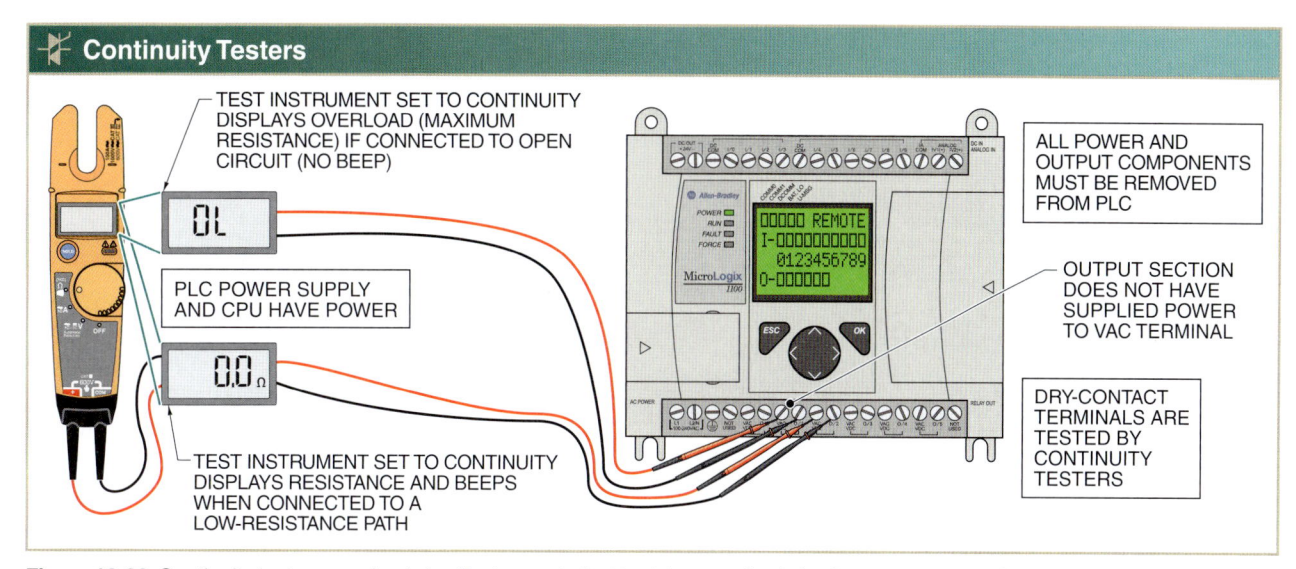

- TEST INSTRUMENT SET TO CONTINUITY DISPLAYS OVERLOAD (MAXIMUM RESISTANCE) IF CONNECTED TO OPEN CIRCUIT (NO BEEP)
- PLC POWER SUPPLY AND CPU HAVE POWER
- TEST INSTRUMENT SET TO CONTINUITY DISPLAYS RESISTANCE AND BEEPS WHEN CONNECTED TO A LOW-RESISTANCE PATH
- ALL POWER AND OUTPUT COMPONENTS MUST BE REMOVED FROM PLC
- OUTPUT SECTION DOES NOT HAVE SUPPLIED POWER TO VAC TERMINAL
- DRY-CONTACT TERMINALS ARE TESTED BY CONTINUITY TESTERS

Figure 10-23. Continuity testers are simple test instruments that test de-energized circuits or components for a complete path for current.

Continuity Tester Measurement Procedures

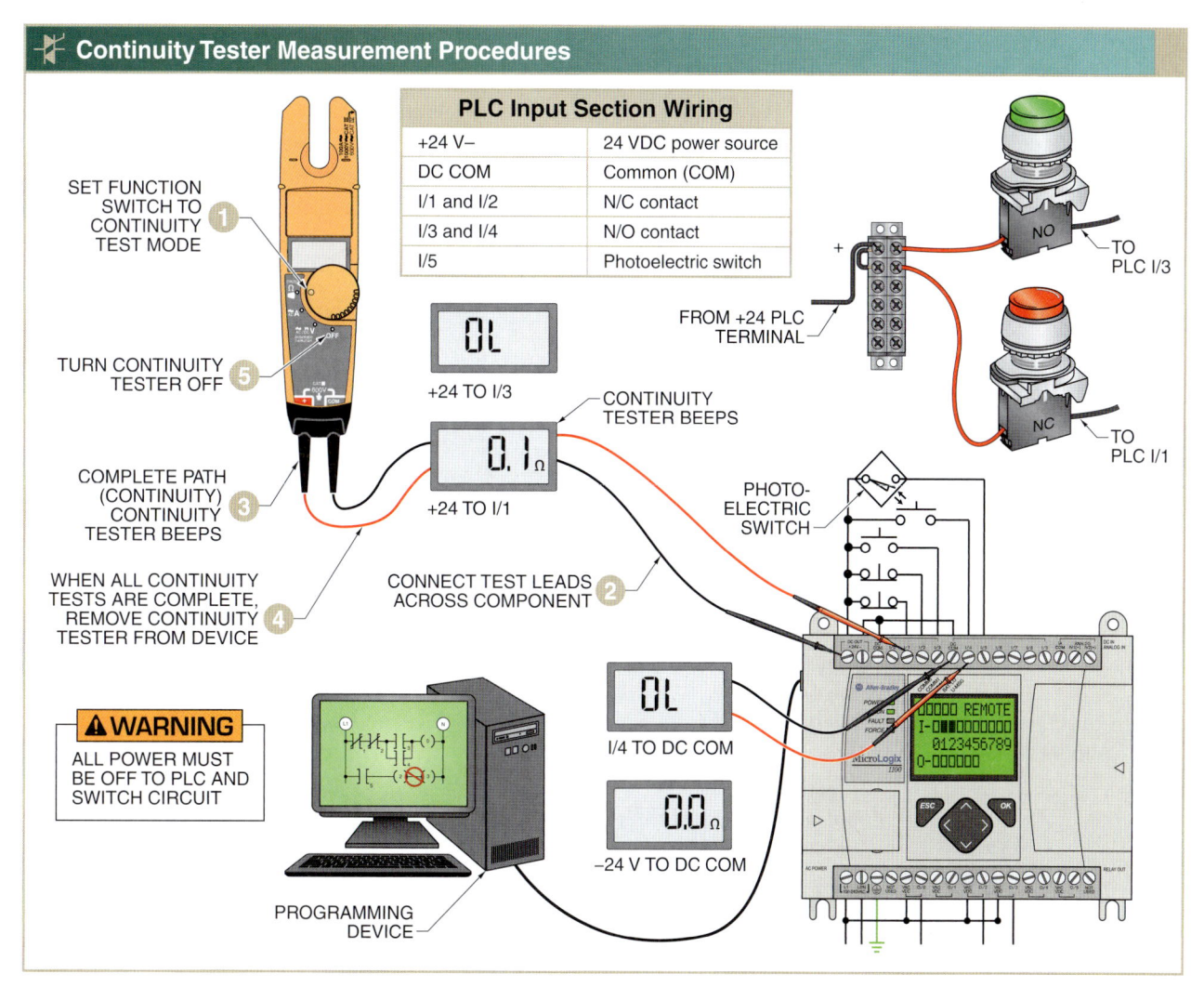

PLC Input Section Wiring

+24 V–	24 VDC power source
DC COM	Common (COM)
I/1 and I/2	N/C contact
I/3 and I/4	N/O contact
I/5	Photoelectric switch

- ① SET FUNCTION SWITCH TO CONTINUITY TEST MODE
- ② CONNECT TEST LEADS ACROSS COMPONENT
- ③ COMPLETE PATH (CONTINUITY) CONTINUITY TESTER BEEPS
- ④ WHEN ALL CONTINUITY TESTS ARE COMPLETE, REMOVE CONTINUITY TESTER FROM DEVICE
- ⑤ TURN CONTINUITY TESTER OFF

- +24 TO I/3
- +24 TO I/1 — CONTINUITY TESTER BEEPS
- I/4 TO DC COM
- –24 V TO DC COM
- FROM +24 PLC TERMINAL
- PHOTOELECTRIC SWITCH
- TO PLC I/3
- TO PLC I/1
- PROGRAMMING DEVICE

⚠ WARNING ALL POWER MUST BE OFF TO PLC AND SWITCH CIRCUIT

Figure 10-24. A continuity test can be used to check the operation of start and stop pushbuttons.

⚠ CAUTION

A continuity tester must only be used on de-energized circuits or equipment. Any voltage applied to a continuity tester causes damage to the test instrument and/or harm to the technician.

Ohmmeters

An *ohmmeter* is a test instrument that measures the resistance of a device or circuit. Ohmmeter resistance measurements are taken to determine the resistance of de-energized devices, components, or circuits. **See Figure 10-25.** The significance of a resistance measurement depends on the component being tested.

Ohmmeters

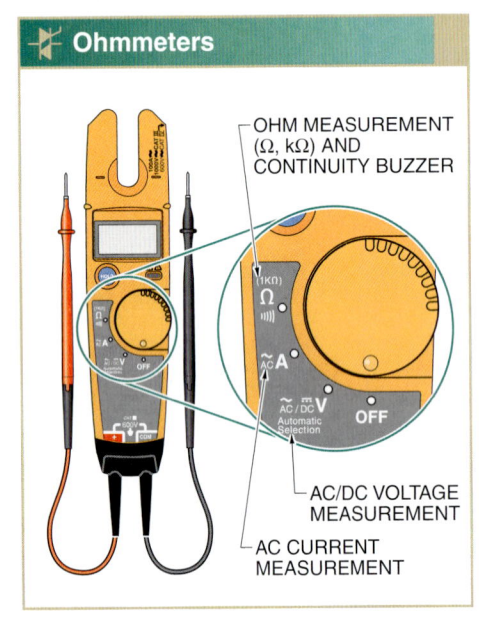

- OHM MEASUREMENT (Ω, kΩ) AND CONTINUITY BUZZER
- AC/DC VOLTAGE MEASUREMENT
- AC CURRENT MEASUREMENT

Figure 10-25. Ohmmeters measure the amount of resistance (in ohms) in de-energized circuits, devices, or components.

In general, resistance of any one component varies over time and from component to component. Slight resistance changes are not typically critical but may indicate a pattern. For example, as resistance increases, the current passing through a component decreases and the power produced also decreases. Likewise, as resistance decreases, current increases and the power produced also increases. Increased power means increased heat and insulation breakdown.

Ohmmeter Measurement Procedures

Before taking any resistance measurements using an ohmmeter, technicians must verify that the ohmmeter has an output voltage rating higher than the highest potential voltage output required for the circuit being tested. Technicians must refer to the operating manual of the test instrument for all measuring precautions, limitations, and procedures. **See Figure 10-26.** To take a resistance measurement using an ohmmeter, apply the following procedures:

1. Verify that all power is OFF to the circuit and remove the component being tested from the circuit.

2. Set the eletrical tester function switch to the resistance mode (ohmmeter) as required. Test instruments display OL and the ohm symbol (Ω) when set to the resistance mode.

3. Plug the black test lead into the common jack when required.

4. Plug the red test lead into the resistance jack when required.

5. Ensure that the batteries of the test instrument are in good condition. Most digital meters display the battery symbol when the batteries are low.

6. Zero ohmmeter when required.

7. Connect the test leads across the circuit or component being tested. Ensure that the test leads are correctly connected.

8. Record the resistance measurement displayed on the ohmmeter.

9. After completing all resistance measurements, remove the ohmmeter and test leads from the circuit or component.

10. Turn the ohmmeter OFF to prevent battery drain.

Ohmmeter Measurement Procedures

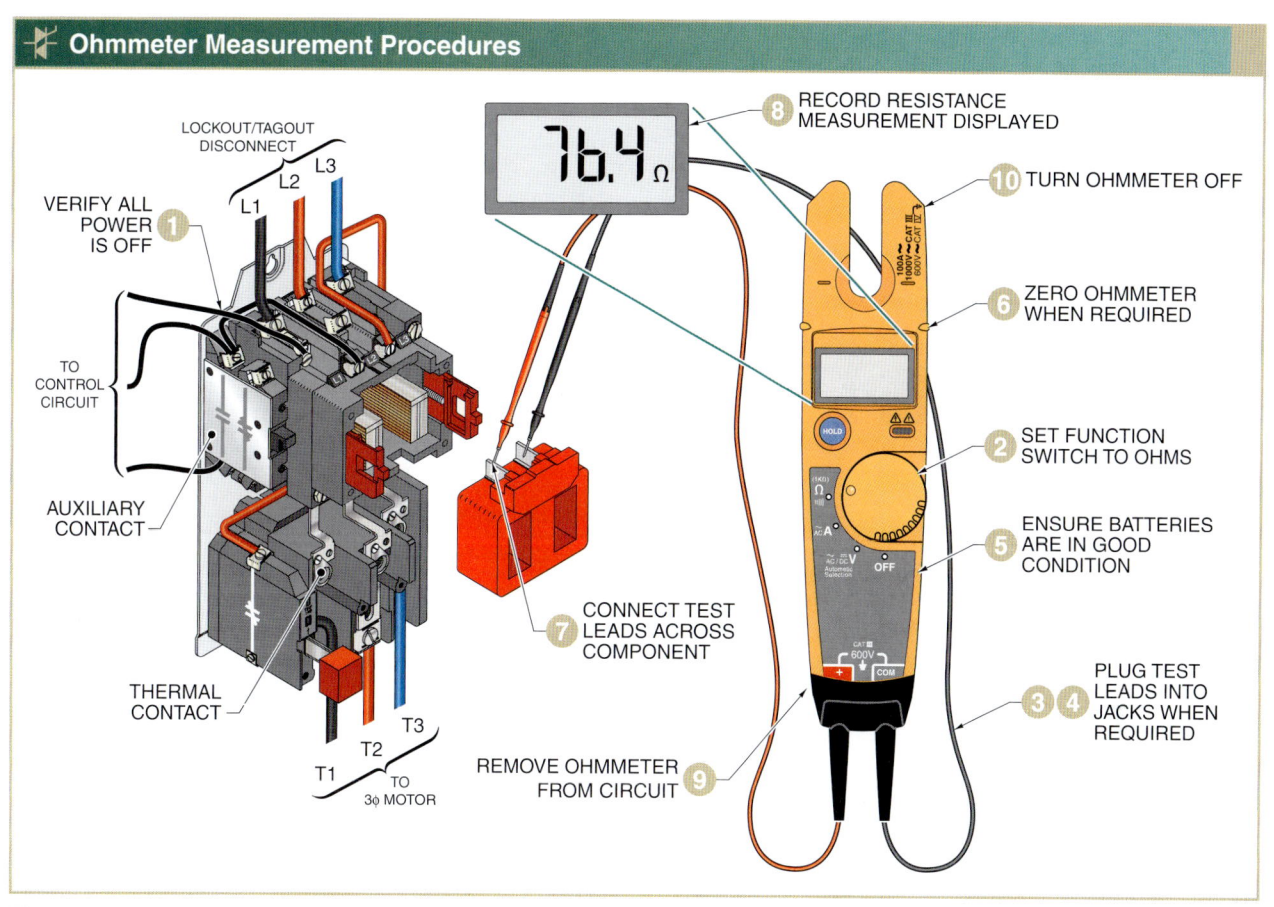

Figure 10-26. Technicians must always verify that circuits, devices, or components do not have voltage before taking any resistance measurements.

Ammeters

An *ammeter* is a test instrument that measures the amount of current in an electrical circuit. Amperage measurements are used to determine the amount of circuit loading or the condition of an electrical component (load). The more electrical energy required, the higher the current usage.

Current measurements provide more information about the condition of a load than voltage measurements. For example, 230 V can be measured when a new motor has been placed in service, when the motor has been in service for a long time with worn parts, when lightly loaded, or when heavily loaded. However, when current measurements are taken, the current will vary with each of the motor conditions.

Current is typically measured using clamp-on ammeters or multimeters with clamp-on current probe accessories. Small amounts of current can be measured using a multimeter connected as an in-line ammeter. **See Figure 10-27.**

When a clamp-on current probe accessory is used with a multimeter, the multimeter functions as a clamp-on ammeter. Clamp-on current probe accessories are available for multimeters that allow for the measurement of AC and DC current. A multimeter connected as an in-line ammeter should only be used to measure small amounts of current in a circuit by inserting the test leads of the ammeter in series with the component or components being tested.

> **⚠ DANGER**
>
> In-line current measurements are the most dangerous measurements to take of any electrical measurement. Always ensure that the function switch position matches the jack connections of the test leads. In-line ammeters can be damaged if the test leads are connected to measure current and a voltage measurement is taken.

Ammeters

Current Measurement Devices

	Clamp-On Ammeter	DMM with Clamp-On Current Probe Accessory	In-Line Ammeter
Range	Up to 3000 A	Up to 1000 A	Less than 10 A
Procedure	Measurement does not require opening circuit	Measurement does not require opening circuit	Measurement requires opening circuit
Functions	Some voltage and resistance	MIN MAX, relative, etc.	MIN MAX, relative, etc.

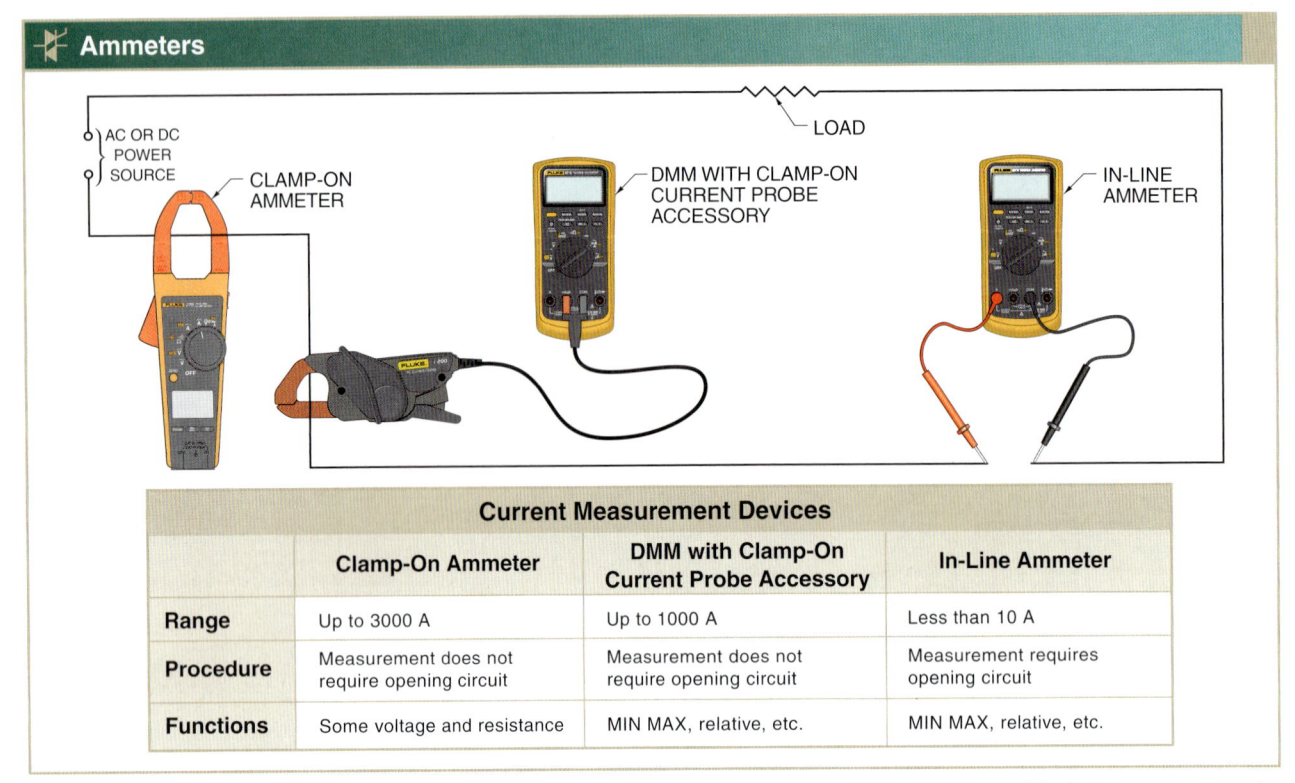

Figure 10-27. Current measurements are typically measured using clamp-on ammeters or multimeters with clamp-on current probe accessories. In low-current applications, in-line ammeters can be used.

PLC TIPS

Milliamp process clamp meters can be used to measure milliamp (mA) signals from analog inputs and to analog outputs. Milliamp process clamp meters allow measurements to be taken without having to open an electrical circuit for in-line current measurements.

Clamp-On Ammeter Measurement Procedures

Clamp-on ammeter current measurements can be taken on loads that are connected to PLC output sections or modules to ensure the loads do not exceed the current rating of the PLC output section or module. Before taking any current measurements using a clamp-on ammeter, technicians must verify that the clamp-on ammeter has a current rating higher than the highest potential current in the circuit being tested. Technicians must refer to the operating manual of the test instrument for all measuring precautions, limitations, and procedures. **See Figure 10-28.**

To take a current measurement using a clamp-on ammeter, apply the following procedures:

1. Determine if the current in the circuit to be measured is AC or DC.
2. Select the type of clamp-on ammeter needed to measure the circuit current. When both AC and DC measurements are required, select an ammeter that can measure both AC and DC.
3. Ensure that the current range of the ammeter is high enough to measure the maximum current that exists in the circuit being tested.
4. Set the ammeter function switch to the proper current range (400 mA, 10 A, 200 A, or 600 A). Select a setting as high or higher than the highest possible circuit current when there is more than one test position. Or if the circuit current is unknown, select the highest setting.
5. When required, plug a clamp-on current probe accessory into a multimeter. The black test lead of the clamp-on current probe accessory is plugged into the common jack. The red test lead is plugged into the mA jack for current measurement accessories that produce a current output. The red test lead is plugged into the voltage (V) jack for current measurement accessories that produce a voltage output. The current measurement accessories that produce current output are designed to measure

AC only and generally deliver 1 mA to the meter for every 1 A of measured current (1 mA/A). Current accessories that produce a voltage output are designed to measure AC or DC current and deliver 1 mV to the meter for every 1 A of measured current (1 mV/A).

6. Open the clamp-on meter or probe accessory jaws by pressing against the trigger.

7. Enclose one conductor in the center of the jaws. Ensure that the jaws are completely closed before taking any current measurements.

8. Record the current measurement displayed on the clamp-on ammeter or the multimeter with the clamp-on probe accessory.

9. Remove the clamp-on ammeter or clamp-on accessory from the circuit.

10. Turn the ammeter OFF.

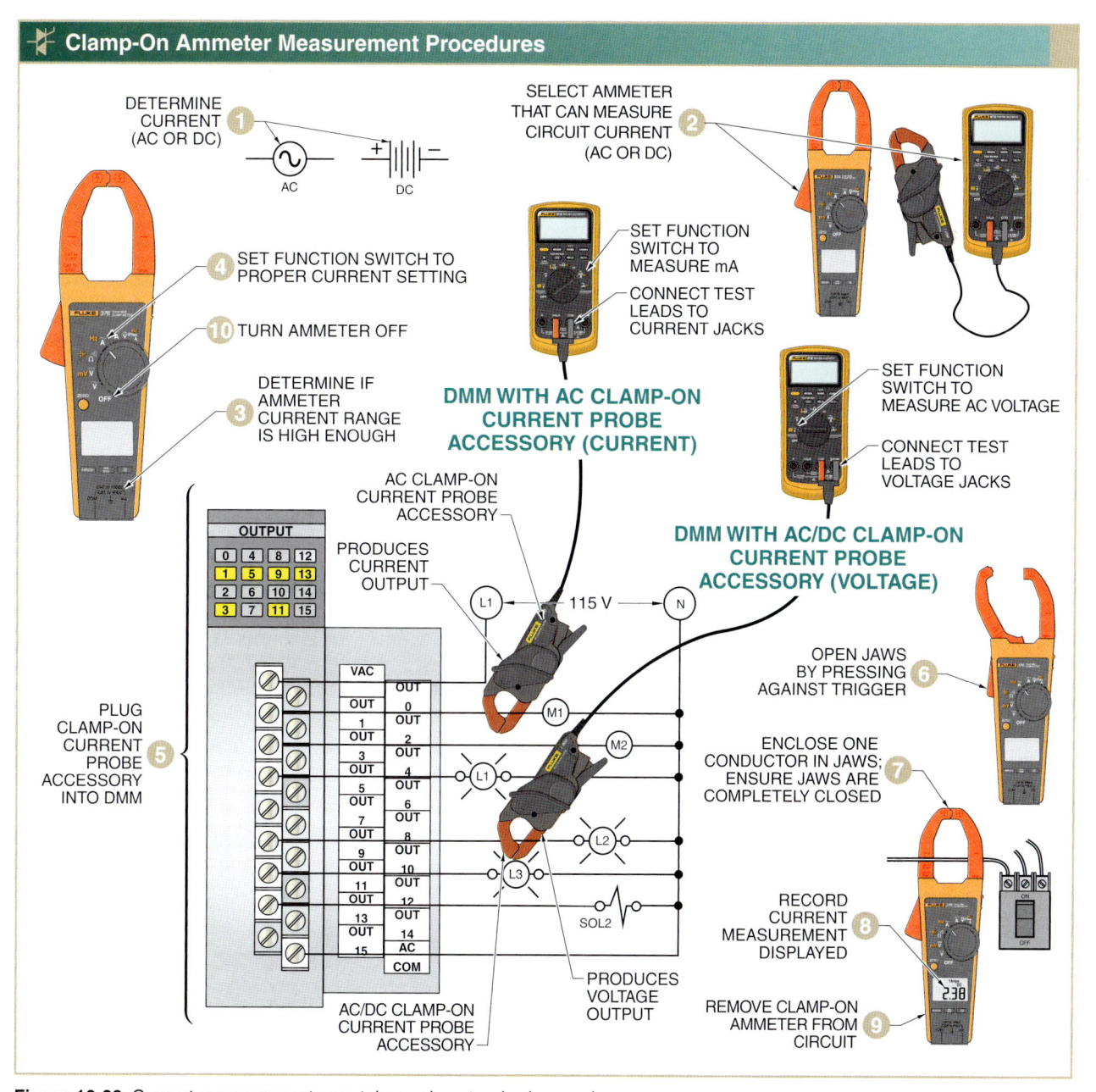

Figure 10-28. Current measurements are taken using standard procedures.

Clamp-on ammeters are used to measure the current (magnetic field) flowing through one isolated conductor.

In-Line Ammeter Measurement Procedures

In-line ammeter readings are taken when current is typically less than 1 A and an exact current measurement is required. For example, individual PLC input switches do not use much current, but total current increases as more input devices are added. For PLCs that supply a voltage output to be used to power the input switches, there is a limit to the current that the PLC power supply can deliver.

To measure the amount of current through an individual input switch or the total of all input switches, an in-line current measurement is taken. For individual switches, the ammeter is placed in-line with only one switch. For total input switch current, the ammeter is placed in-line with the power supply delivering power to all the switches.

In-line current measurements can be taken on DC or AC circuits. The only difference is the setting of the meter to DC amps or AC amps. Before taking any current measurements using an in-line ammeter, technicians must verify that the in-line ammeter has a current rating higher than the highest potential current in the circuit being tested. **See Figure 10-29.** Technicians must refer to the operating manual of the test instrument for all measuring precautions, limitations, and procedures. To take a current measurement using an in-line ammeter, apply the following procedures:

1. Set the function switch of the in-line ammeter to the proper position for measuring DC current (A or mA/μA). Select a setting with a high enough rating to measure the highest possible circuit current when the ammeter has more than one position.

2. Plug the black test lead into the common jack.

3. Plug the red test lead into the current jack. The current jack may be marked A or mA/μA.

4. Turn the power to the circuit or component being tested OFF and discharge all capacitors if possible.

5. Open the circuit at the test point and connect the test leads to each side of the opening. For DC current, the black (negative) test lead is connected to the negative side of the opening, and the red (positive) test lead is connected to the positive side of the opening. Reverse the black and red test leads when a negative sign appears to the left of the measurement displayed.

6. Turn the power to the circuit being tested ON and the in-line ammeter ON.

7. Record the current measurement displayed.

8. Turn circuit power OFF, remove the ammeter and test leads from the circuit, and reconnect wire to PLC.

9. Turn in-line ammeter OFF.

In-Line Ammeter Measurement Procedures

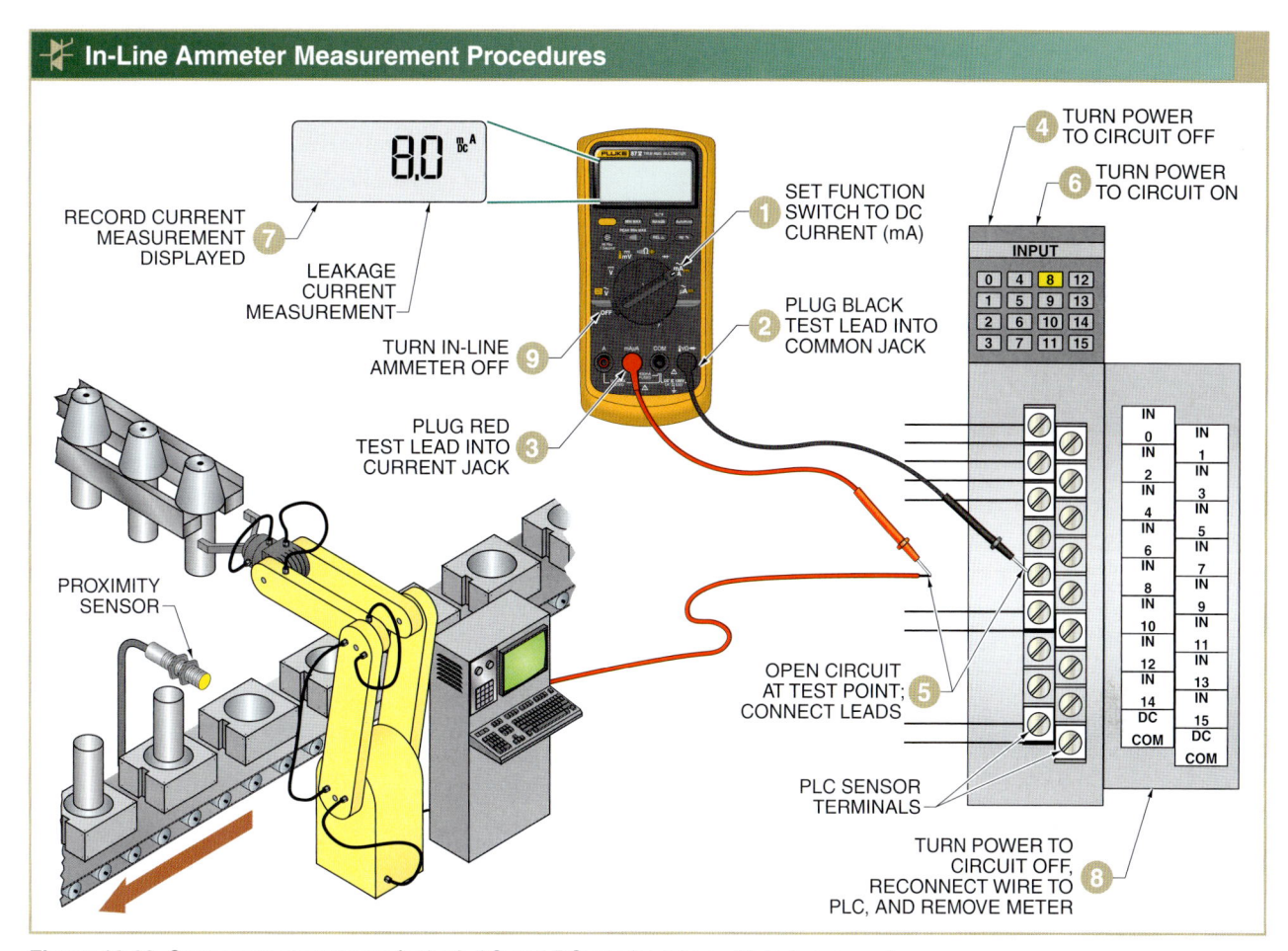

Figure 10-29. Current measurements for both AC and DC can be taken with in-line ammeters.

PLC TIPS

Use the MIN MAX recording function of a multimeter to record, over time, the current of an output component connected to PLC output terminals to ensure the current rating of the output section or module is not being exceeded.

Temperature Test Instruments

Electrical circuits are used to transfer, switch, control, or convert electrical energy into other forms of energy, such as light, sound, or mechanical motion. Converting electrical energy into another energy form produces heat. Electrical heat lamps and electric heating elements produce heat intentionally.

Unwanted heat is produced in an electrical circuit any time current flowing through the circuit encounters resistance, such as when current flows through conductors and encounters a bad connection or a faulty switch. The higher the resistance of the conductor, bad connection, or faulty switch, the greater the amount of heat produced at that point.

All PLCs produce heat and all PLCs can be damaged by excessive heat. For most applications, proper spacing of the PLC and components will allow normal convection cooling to keep the PLC within the specified operating range. Typical PLC temperature operating specifications call for the air in the enclosure around the PLC (ambient air) to be kept within a range of 32°F to 140°F (0°C to 60°C).

Excessive heat produced by loose connections and undersized conductors causes problems within an electrical system but does not cause problems with the PLC unless the heat is produced near the PLC. However, excessive heat produced by poor spacing of PLCs and heat-producing components such as control transformers around the PLC can damage it. Manufacturers include spacing guidelines in installation and operation manuals to prevent this problem.

When heat is a problem because of tight space limits or there is a high ambient temperature, a fan-cooled enclosure should be used (or a cooling fan added). However, when bringing outside air into an enclosure, an air filter should always be added to the air openings.

Test instruments can be used to measure the amount of heat at different locations within an electrical system. By taking temperature measurements throughout an electrical circuit, existing problems can be found and future problems can be avoided.

Temperature measurements are taken using contact and noncontact temperature test instruments. Noncontact temperature instruments are typically infrared or thermal imaging. An *infrared temperature meter* is a meter that measures heat energy by measuring the infrared energy emitted by a material, and displays the temperature as a numerical value. All materials emit infrared energy in proportion to the temperature of the material.

A *thermal imager* is a meter that measures heat energy by measuring the infrared energy emitted by a material and displays the temperature as a color-coded thermal picture. **See Figure 10-30.** Because heat can be a cause of PLC problems and failure, temperature measurements must be taken to ensure that the temperature in a PLC enclosure is within set limits.

⚠ DANGER

In-line ammeters not marked as fused cannot be used for taking current measurements.

Noncontact Temperature Instruments

INFRARED METER

THERMAL IMAGER

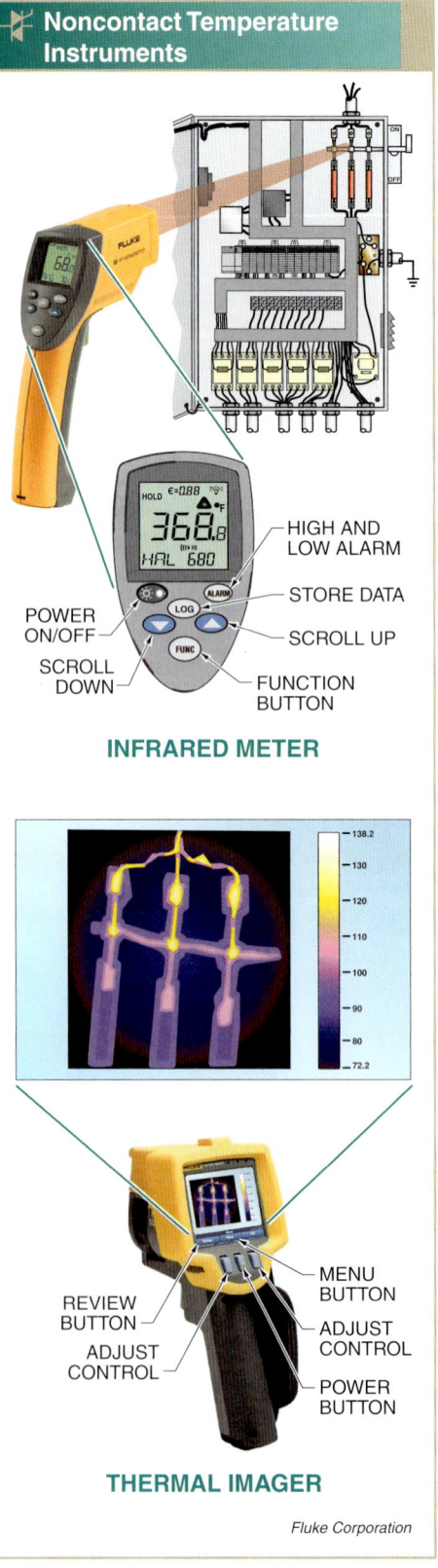

Fluke Corporation

Figure 10-30. Noncontact temperature instruments measure heat by measuring the infrared energy emitted by a material.

Infrared Temperature Meter Measurement Procedures

Before taking any temperature measurements using an infrared temperature meter, verify that the meter has a temperature rating higher than the highest potential temperature in the enclosure being tested. **See Figure 10-31.** Refer to the operating manual of the test instrument for all measuring precautions, limitations, and procedures. To take a temperature measurement using an infrared temperature meter, apply the following procedures:

1. Set the infrared temperature meter to measure degrees Fahrenheit (°F) or degrees Celsius (°C).
2. Aim the infrared temperature meter at the area being measured. The meter is focused based on the distance between the object and the meter.
3. Take an ambient temperature reading (area around suspect spot) for reference.
4. Take temperature readings of any area suspected to have temperatures above ambient temperature. Temperature rises of 50°F (10°C) or more must be investigated. Temperature rises of 100°F (38°C) or more require immediate shutdown of the system and repair of the fault.
5. Repeat procedures for all additional measurement areas that are required.
6. Turn the infrared temperature meter OFF.

Infrared Temperature Meter Measurement Procedures

① SET INFRARED TEMPERATURE METER TO MEASURE DEGREES FAHRENHEIT

AMBIENT TEMPERATURE: 87.4 °F

② AIM METER AT AREA TO BE MEASURED

③ TAKE AMBIENT TEMPERATURE READING FOR REFERENCE

④ TAKE TEMPERATURE READING OF SUSPECT AREAS

SUSPECT AREA TEMPERATURE: 192.8 °F

⑤ REPEAT PROCEDURES FOR ALL ADDITIONAL MEASUREMENT AREAS

⑥ TURN INFRARED TEMPERATURE METER OFF

Finding Temperature Reading

What is the temperature reading if the temperaure of an area is 145°F and the ambient temperature is 60°F?

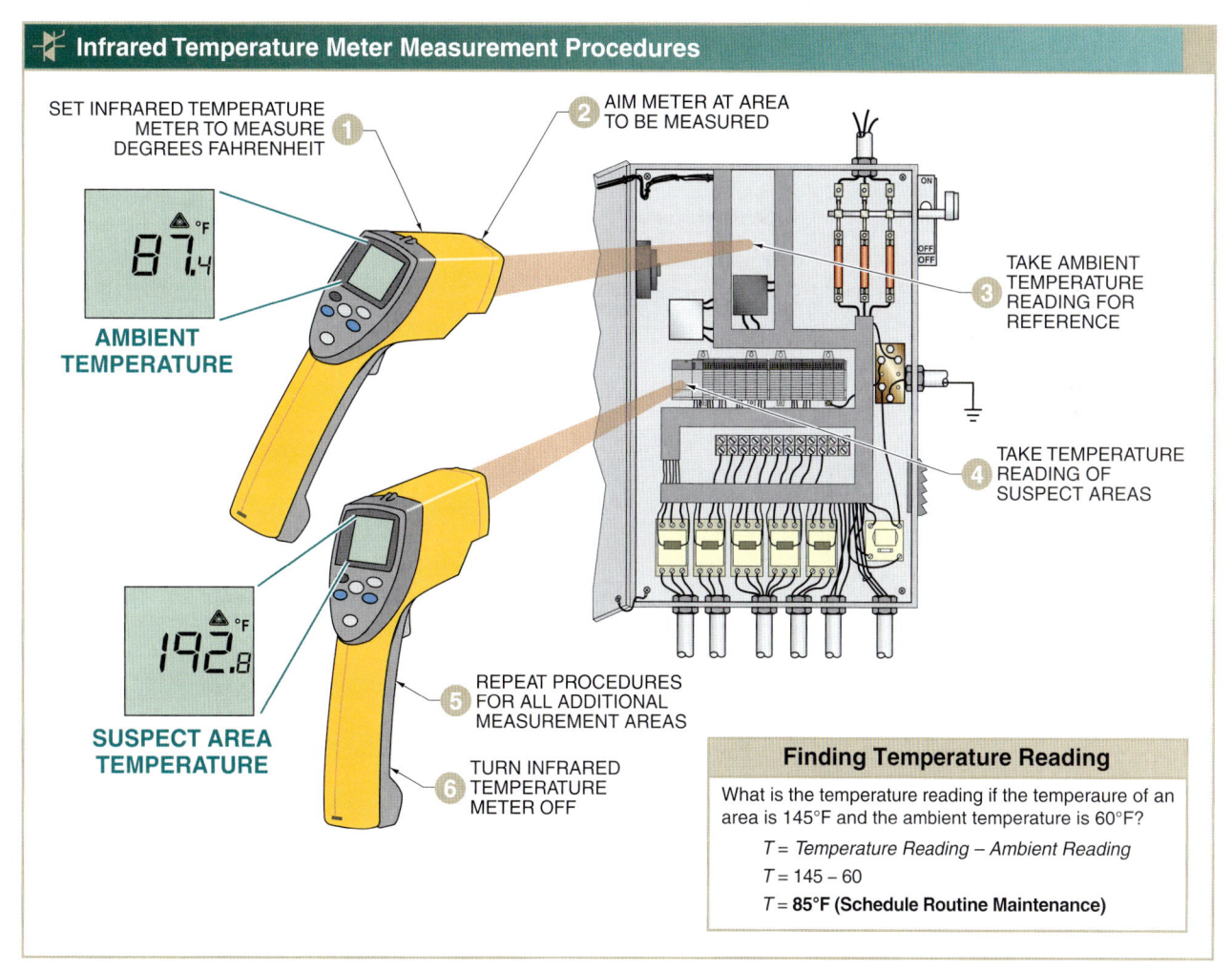

T = Temperature Reading − Ambient Reading
T = 145 − 60
T = 85°F (Schedule Routine Maintenance)

Figure 10-31. Infrared temperature measurements prevent problems by locating unwanted heat in electrical equipment enclosures before the heat can cause PLC or equipment failure.

Chapter Review 10

Name _____ Date _____

True-False

T　F　　1. All electrical measurements should be taken with the power ON.

T　F　　2. A malfunction is the failure of a system, equipment, or part to operate properly.

T　F　　3. PLCs are a logical place to begin troubleshooting an electrical system.

T　F　　4. The International Electrotechnical Commission (IEC) 61010 standard classifies the applications in which test instruments can be used into five overvoltage installation categories (CAT I–CAT V).

T　F　　5. A meter can be set to measure AC voltage and be connected to a DC voltage without damaging the meter, circuit, or circuit components.

T　F　　6. A good fuse should have continuity, whereas a bad fuse will not.

T　F　　7. All PLCs produce heat and all PLCs can be damaged by excessive heat.

T　F　　8. A continuity test is used to check input switches to ensure their proper operation before they are connected to the input section or module of a PLC.

T　F　　9. PLC systems very rarely include any DC voltages.

T　F　　10. The disadvantage of using solenoid voltage testers is that their high impedance affects electronic equipment and circuit operation.

T　F　　11. Voltage measurements provide more information about the condition of a load than current measurements.

T　F　　12. When the type of voltage (AC or DC) is unknown, a measurement should be taken with the meter set to measure DC voltage.

T　F　　13. In general, any voltage 50 V or higher and any current 6 mA or higher is dangerous.

T　F　　14. An ellipse symbol is a symbol in a flowchart that contains a question, worded so that the answer can be a "yes" or a "no."

T　F　　15. An infrared temperature meter is a meter that measures heat energy by measuring the infrared energy emitted by a material, and displays the temperature as a numerical value.

Completion

_____ 1. A(n) ___ is a diagram that shows a logical sequence of steps for a given set of conditions.

_____ 2. A(n) ___ is a test instrument that tests for a complete path for current to flow.

_____ 3. A(n) ___ is a test instrument with a bulb that is connected to two test leads to provide a visual indication of when voltage is present.

_____ 4. A(n) ___ is a graphic element that represents a quantity, unit, or component.

_____ 5. A(n) ___ is a combination of components, units, or modules that are connected to perform work or meet a specific need.

_____ 6. A(n) ___ is a letter or combination of letters that represent a word.

_____ 7. A(n) ___ is a test instrument that measures the resistance of a device or circuit.

_____ 8. A(n) ___ is a meter that measures heat energy by measuring the infrared energy emitted by a material and displays the temperature as a color-coded thermal picture.

_____ 9. A(n) ___ is a portable test instrument that uses electrical components to display measured values.

_____ 10. A(n) ___ is a sequence of operations that accomplish desired results.

Multiple Choice

_____ 1. ___ is/are the work performed to keep machines, assembly lines, production operations, and plant operations running with little or no downtime.
   A. Facility procedures
   B. Manufacturer procedures
   C. Preventive maintenance
   D. Troubleshooting

_____ 2. A(n) ___ symbol is a symbol in a flowchart that contains a set of instructions.
   A. arrow
   B. diamond
   C. ellipse
   D. rectangle

_____ 3. Troubleshooting using ___ is a method of finding malfunctioning equipment using the procedures recommended by company personnel.
   A. facility procedures
   B. knowledge and experience
   C. manufacturer flowcharts
   D. manufacturer help lines

_____ 4. A(n) ___ is a test instrument that measures the amount of current in an electrical circuit.
   A. ammeter
   B. continuity tester
   C. ohmmeter
   D. voltmeter

_____ 5. Troubleshooting using ___ is a method of finding a malfunction in a machine or process by applying information acquired from past malfunctions.
   A. facility procedures
   B. knowledge and experience
   C. manufacturer flowcharts
   D. manufacturer help lines

Testing for Continuity and Measuring Resistance

PLC Lab 10.1

BILL OF MATERIALS

- TECO PLR or Micrologix™ 1100 PLC
- NO pushbutton
- NC pushbutton
- Limit switch
- Indicator light

Continuity tests are performed when testing mechanical switches (contacts) and electrical connections. These tests are performed with all power OFF. Continuity tests are used to determine which contacts are normally open (NO) and which contacts are normally closed (NC) on all types of switches. A test instrument, such as a DMM, beeps when continuity exists. Resistance tests are used to determine the resistance of a wire, connection, or mechanical switch. Resistance tests are also performed with all power OFF. For this lab, continuity tests and resistance tests will be performed on mechanical switches.

T Testing Continuity Procedures

1. Ensure power is OFF to the PLC, devices, and components. Connect the DMM test leads to the proper jacks and set the function switch of the DMM to continuity test mode.

2. Verify that the DMM continuity mode is operating correctly by touching the two test leads together. The DMM should beep when the two test leads touch.

3. Use the continuity tester to check the mechanical contacts of the NO and NC pushbuttons. Which pushbutton has continuity when the button is not pushed? Which pushbutton has continuity when the button is pushed?

_____

4. Use the continuity tester to check the mechanical contacts of the limit switch. Which lead of the limit switch is common? Which lead of the limit switch is closed when the switch is actuated? Which lead of the limit switch is open when the switch is actuated?

Variable	Electrical Quantity	Unit of Measure and Abbreviation
E	voltage	volt — V
I	current	ampere — A
R	resistance	ohm — Ω
P	power	watt — W
P	power (apparent)	volt-amp — VA
C	capacitance	farad — F
L	inductance	henry — H
Z	impedance	ohm — Ω
G	conductance	siemens — S
f	frequency	hertz — Hz
T	period	second — s

_____

431

5. With the DMM set to resistance (Ω) test mode, touch the two test leads of the DMM together and then press the relative mode (REL) button to zero the digital display. The REL button subtracts the resistance of the test leads from the resistance reading, which allows only the resistance of the mechanical contacts to be displayed.

6. Use the resistance test mode to check the mechanical contacts of the pushbuttons and the limit switch. What are the resistance measurements of the mechanical contacts in order?

_____

7. Create a single-rung programming diagram using a TECO or Micrologix™ 1100 with one input device (connected) and one output component (not connected). The output contacts should close when the input is closed. Turn power ON, download the program to the PLC, and place the PLC in RUN mode.

8. Set the DMM to continuity test mode and connect the test leads across the designated output terminal. Does the continuity test indicate continuity when the input is actuated or when the input is not actuated?

_____

9. What is the resistance of the output contacts when the contacts are closed? What is the reading of the digital display when the contacts are open?

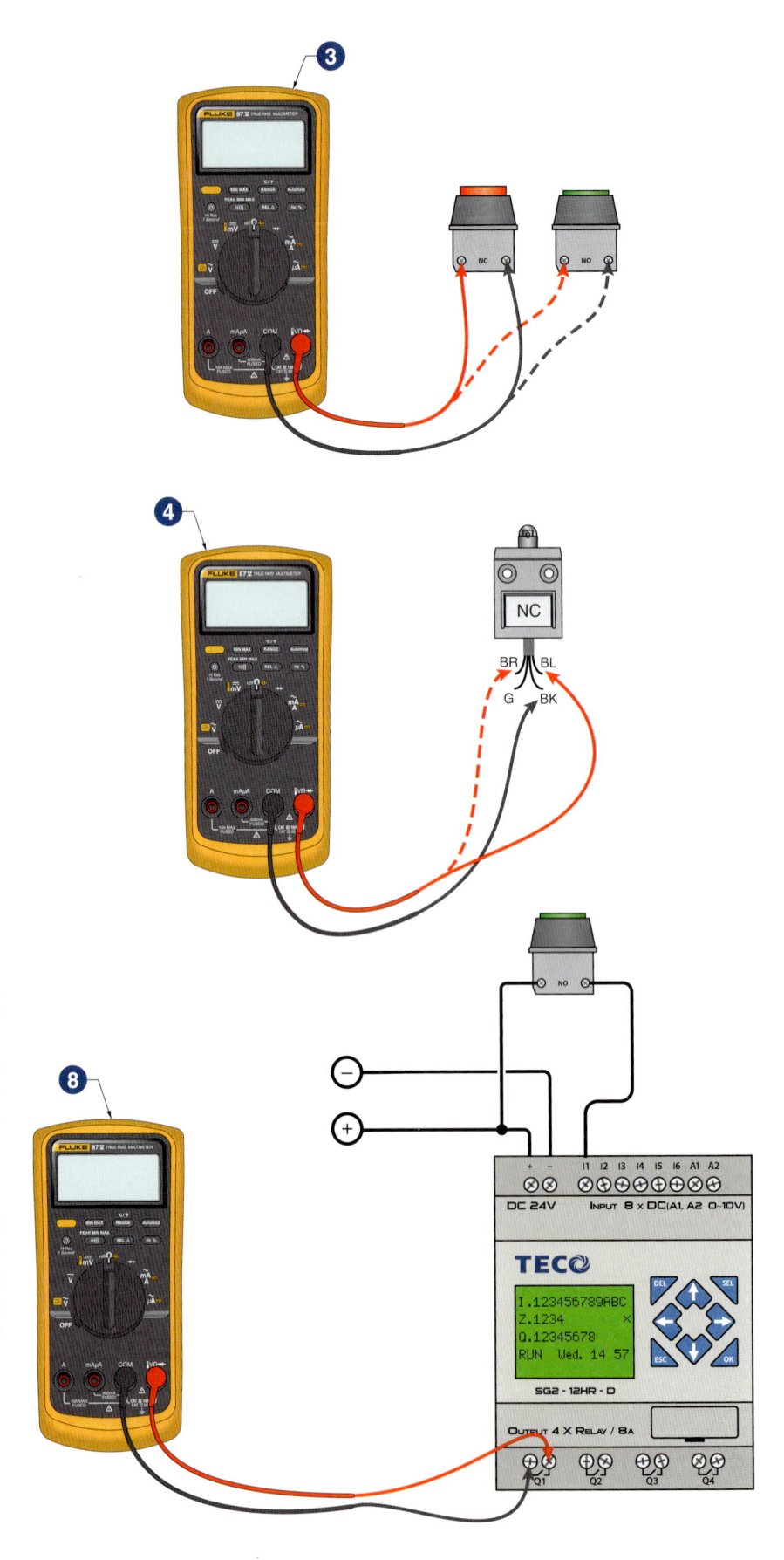

_____

Measuring Voltage in PLC Circuits

PLC Lab 10.2

BILL OF MATERIALS
- TECO PLR or Micrologix™ 1100 PLC
- NO pushbutton
- Indicator light

Voltage measurements are performed at a specific location to verify that voltage is present, at the correct level, and of the right type (AC or DC). With PLC circuits, PPE is required when voltage levels are 50 V or higher at or around the location of testing. For this lab, a single-rung diagram will be programmed where one input controls one output, and a DMM will be used to measure voltage.

Note: Amount of voltage will vary based on PLC type and model.

T Measuring Voltage Procedures

1. Based on the PLC system description and line diagram, use the ladder logic for either a TECO PLR or Micrologix™ 1100 PLC.

2. Verify that the programming diagram has no errors and save the project.

3. Wire and then operate the input to control the output circuit to determine correct operation. Test the circuit for all possible conditions.

4. Connect the DMM test leads to the proper jacks for DC voltage measurement. Set the function switch of the DMM to DC voltage.

5. Use the wiring diagram to connect the DMM test leads to the proper locations to measure DC voltage.

LINE DIAGRAM

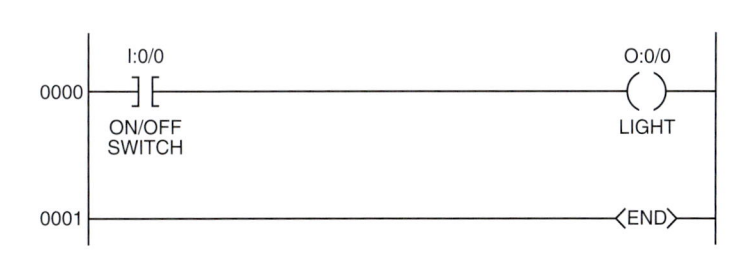

TECO PROGRAMMING DIAGRAM

AB PROGRAMMING DIAGRAM

6. Connect the DMM test leads across the supplied power terminals to the PLC. What does the DMM display when the supply voltage to the PLC is measured?

_____

_____

_____

7. Reverse the DMM test leads across the supplied power terminals of the PLC. What does the DMM display when the test leads are reversed across the supply voltage to the PLC?

_____

_____

_____

8. Connect the DMM test leads to measure the voltage sent to the PLC from the input switch. What is the voltage when the input switch is open? What is the voltage when the switch is closed?

_____

_____

_____

9. Connect the DMM test leads to measure the voltage sent to the output component from the PLC. What is the voltage when output 1 is ON? What is the voltage when output 1 is OFF?

_____

_____

_____

AB WIRING DIAGRAM

Measuring Current in PLC Circuits

PLC Lab 10.3

BILL OF MATERIALS

- TECO PLR or Micrologix™ 1100 PLC
- Selector switch
- Indicator light

A current measurement is performed at a specific location to determine the amount of current flowing in a circuit or a circuit branch. Current measurements are always performed in-line with any device or component with power ON. PPE is required when voltage levels are higher than 50 V at or around the location of testing. For this lab, a single-rung diagram is programmed where one input controls one output, and a DMM is used to measure current.

T Measuring Voltage Procedures

1. Based on the PLC system description and line diagram, use the ladder logic for either a TECO PLR or Micrologix™ 1100 PLC.

2. Verify that the programming diagram has no errors and save the project.

3. Wire and then operate the input to control the output circuit to determine correct operation. Test the circuit for all possible conditions.

4. Connect the DMM test leads to the proper jacks for in-line DC current measurement. Set the function switch of the DMM to DC milliamp current.

5. Use the wiring diagram to connect the DMM test leads to the proper locations to measure DC current.

LADDER DIAGRAM

TECO PROGRAMMING DIAGRAM

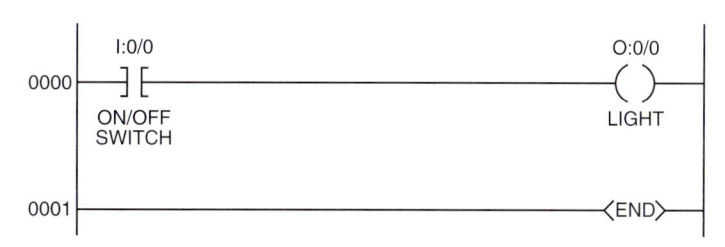

AB PROGRAMMING DIAGRAM

6. Connect the DMM test leads to measure current sent to the PLC from the input switch. What is the current when the input switch is open? What is the current when the switch is closed?

_____

_____

_____

7. Connect the DMM test leads between the light output terminal and the light. What does the DMM display in milliamps when the current flow to the light is measured?

_____

_____

_____

Note: The current measured when the switch is closed or when the load is ON is called the operating current. The current measured when the switch is open or when the load is OFF is called the leakage current. Leakage current may exist when solid-state devices are used.

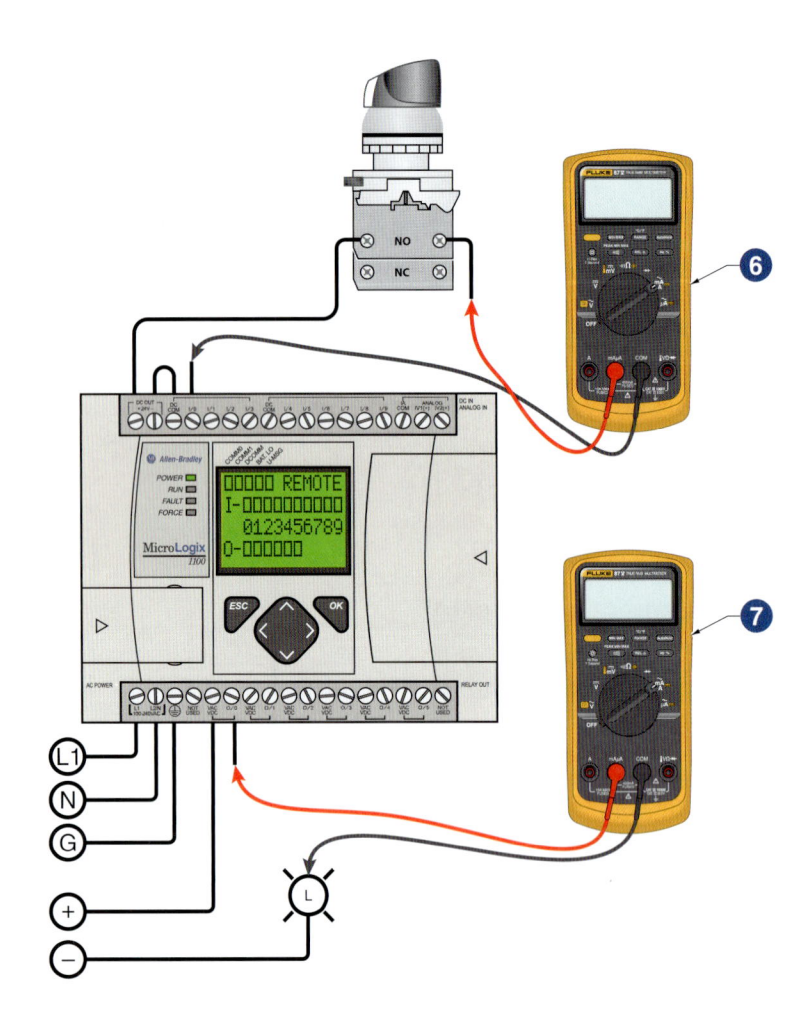

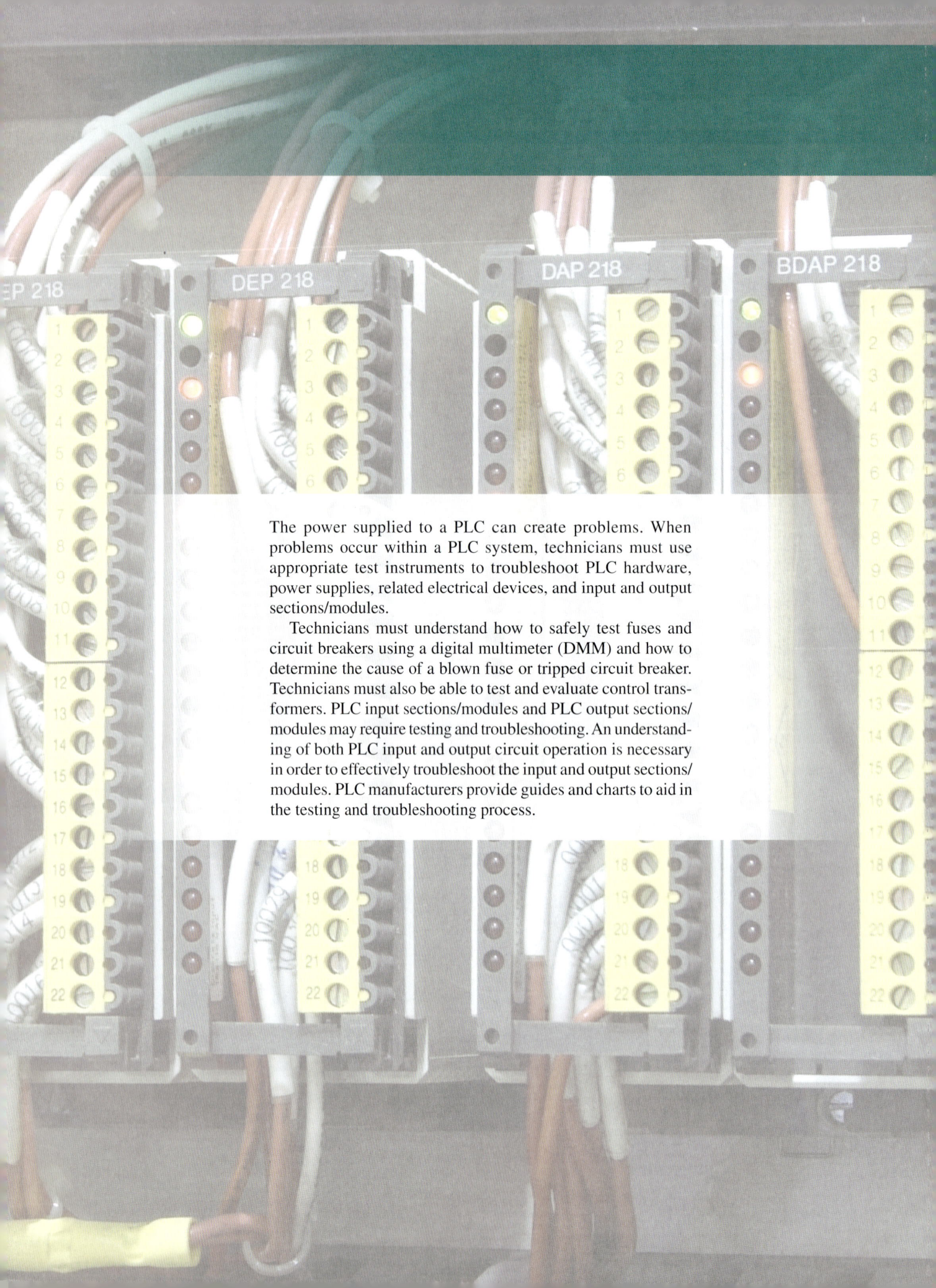

The power supplied to a PLC can create problems. When problems occur within a PLC system, technicians must use appropriate test instruments to troubleshoot PLC hardware, power supplies, related electrical devices, and input and output sections/modules.

Technicians must understand how to safely test fuses and circuit breakers using a digital multimeter (DMM) and how to determine the cause of a blown fuse or tripped circuit breaker. Technicians must also be able to test and evaluate control transformers. PLC input sections/modules and PLC output sections/modules may require testing and troubleshooting. An understanding of both PLC input and output circuit operation is necessary in order to effectively troubleshoot the input and output sections/modules. PLC manufacturers provide guides and charts to aid in the testing and troubleshooting process.

Testing and Troubleshooting Electrical Devices and PLC Hardware

11

Objectives:

- Identify the differences between sag, swell, undervoltages, overvoltages, and transients.
- Demonstrate how to test a fuse/circuit breaker and how to connect a test instrument to determine if it is good or bad.
- Demonstrate how to test a control transformer and how to connect a test instrument to determine if it is good or bad.
- Demonstrate how to test a PLC power supply and how to connect a test instrument to determine if it is good or bad.
- Demonstrate how to test a PLC input/output section or module and how to connect a test instrument to determine if it is good or bad.

Learner Resources
atplearningresources.com/quicklinks
Access Code: 475203

PLC HARDWARE PROBLEMS

Unlike computers and most electronic equipment, PLCs are designed to operate in harsh commercial or industrial environments and control high-power loads. When properly installed, programmed, and maintained, PLCs operate with little or no problems for many years. However, hardware problems do eventually develop in most electrical systems and understanding how to detect hardware problems using logical testing and troubleshooting processes is important.

Detecting PLC hardware problems requires an understanding of the PLC-controlled systems, including the incoming power, PLC input sections or modules, input devices, PLC output sections or modules, and output components. In order to test and troubleshoot effectively, the technician must be able to choose and operate the correct test instruments, know where to connect them, and understand what each measured value means.

PLC TIPS

Some PLC applications do not require that input device circuits have MCR protection; however, an MCR must be in series with the supplied power when power is removed from all inputs.

POWER SUPPLY PROBLEMS

All electrical devices and components must be powered by a power supply that can deliver the correct voltage level with enough current capacity. When a PLC is used in a system, several individual power supplies must be considered. First, a PLC must be powered for internal circuitry operation to deliver output power for operating PLC input devices and/or output components. Often, the power supplied to a PLC is different than the power supplied to PLC input devices. For example, PLCs that have 115 VAC supply power can deliver 24 VDC output power for powering PLC input devices, such as proximity switches, photoelectric switches, and pushbuttons. External DC power can be used for output components. **See Figure 11-1.**

In addition to the power supplied to a PLC, a PLC can often control electrical systems that include a combination of different voltage types (1ϕ AC, 3ϕ AC and/or DC) at different voltage levels (12 V, 24 V, 115 V, 208 V, 230 V, or 460 V). A problem in any one of the system power supplies affects the operation of the total system.

PLC power supply problems typically fall into one of two categories: loss of power and low voltage. Loss of power is an easy problem to troubleshoot because loss of power involves blown fuses, tripped circuit

⚠ CAUTION

When using an external DC output power supply, interrupt the DC output side rather than the AC line side of the supply to avoid the additional delay of power supply turn-off.

breakers, or bad connections. Low voltage (voltage sags/undervoltage) or power interruptions (momentary, temporary, or sustained) are more difficult to troubleshoot.

Blown fuses and open circuit breakers are the most common power supply problems because overload devices are designed to be the weakest link in an electrical system in order to protect the circuit and circuit components. Fuses and circuit breakers have fixed ratings (amperes), and once the rating is exceeded, fuses blow and circuit breakers trip, eliminating voltage from the system.

Typically, undervoltage problems are a result of the continued addition of loads to a circuit or PLC-controlled system over time. *Undervoltage* is a drop in voltage of more than 10% (but not to 0 V) below the normal rated line voltage for a period of time longer than 1 minute.

The second and more difficult power supply problem to troubleshoot involves power quality problems, such as transients (high-voltage spikes), overvoltages, and voltage fluctuations. An *overvoltage* is an increase in voltage of more than 10% above the normal rated line voltage for a period of time longer than 1 minute. The reason power quality problems are harder to detect is that power quality problems typically occur randomly and often require specialized test instruments (power quality meters) to detect.

Fluke Corporation
An energized power supply circuit in an industrial environment can be controlled by a PLC and tested for the presence of voltage. Voltages can range from 3φ 230 VAC to 460 VAC.

Power Supply Voltages

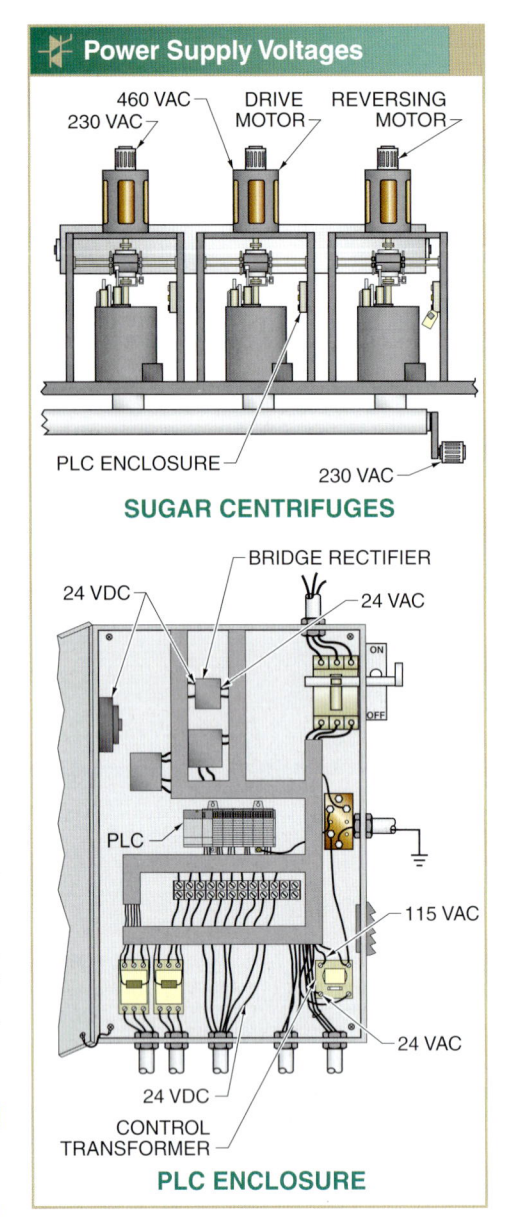

SUGAR CENTRIFUGES

PLC ENCLOSURE

Figure 11-1. PLCs are designed for use with all common supply voltages, such as 230 V, 208 V, 115 V, and 24 V.

For any reoccurring problems, such as damaged PLC power supplies, input and/or output sections or modules, or other normally reliable system components, a check of the supplied power is required. A check of the supplied power should be performed over time (8 hr, 12 hr, or 24 hr) using a test instrument with a recording function, such as the MIN MAX recording function on digital multimeters (DMMs) or the record function on

Chapter 11—Testing and Troubleshooting Electrical Devices and PLC Hardware **441**

power quality meters. DMMs can record voltage or current over time and power quality analyzers can record voltage, current, power (VA, watts, power factor), transients, and harmonics over time. **See Figure 11-2.**

⚠ CAUTION
To prevent unwanted potentials across the logic ground of a PLC, the DC neutral of an external power source must be isolated from the PLC ground.

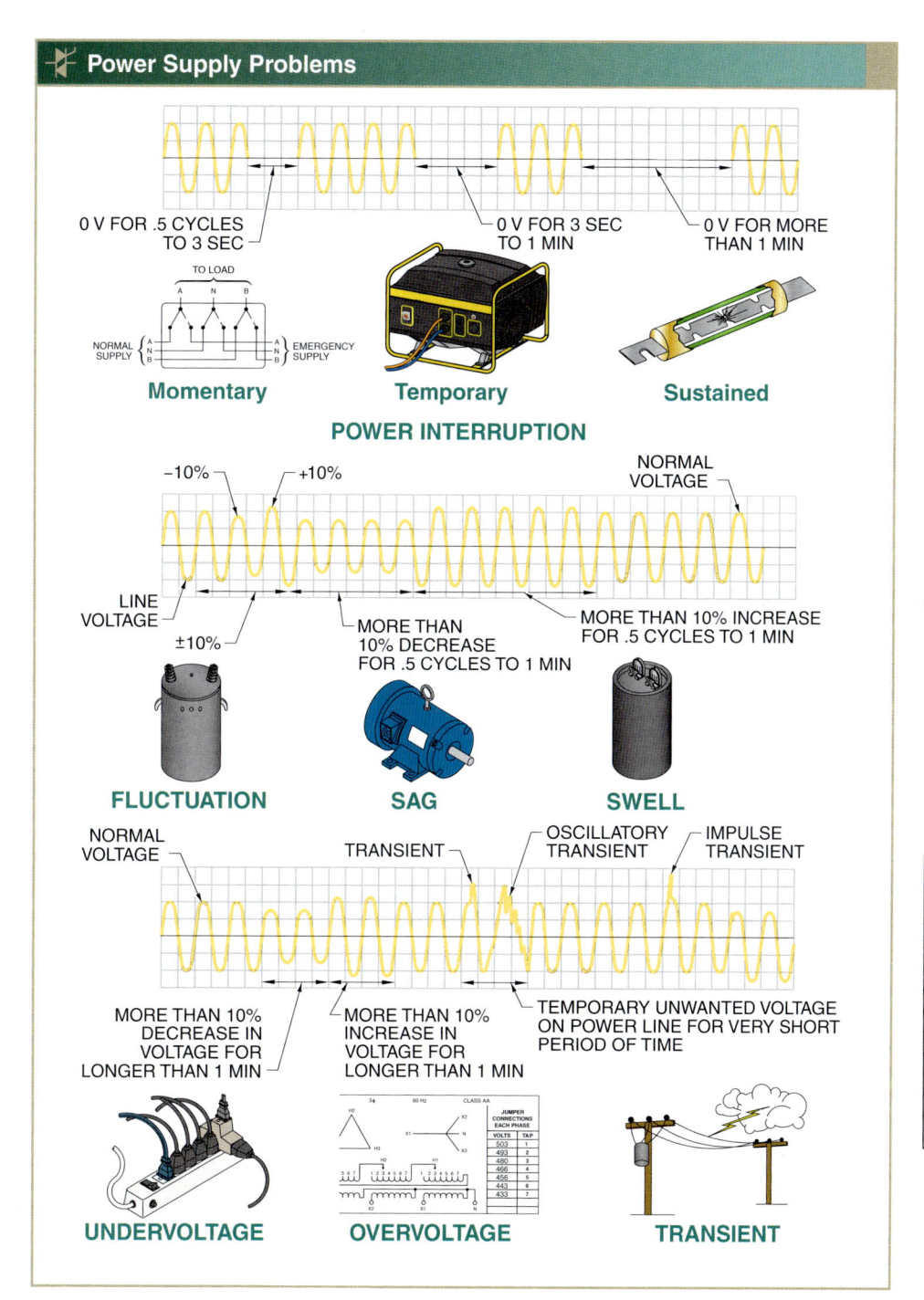

Power Supply Problems

POWER INTERRUPTION
- Momentary — 0 V FOR .5 CYCLES TO 3 SEC
- Temporary — 0 V FOR 3 SEC TO 1 MIN
- Sustained — 0 V FOR MORE THAN 1 MIN

FLUCTUATION — ±10%
SAG — MORE THAN 10% DECREASE FOR .5 CYCLES TO 1 MIN
SWELL — MORE THAN 10% INCREASE FOR .5 CYCLES TO 1 MIN

UNDERVOLTAGE — MORE THAN 10% DECREASE IN VOLTAGE FOR LONGER THAN 1 MIN
OVERVOLTAGE — MORE THAN 10% INCREASE IN VOLTAGE FOR LONGER THAN 1 MIN
TRANSIENT — TEMPORARY UNWANTED VOLTAGE ON POWER LINE FOR VERY SHORT PERIOD OF TIME (OSCILLATORY TRANSIENT, IMPULSE TRANSIENT)

CODE CONNECT
Article 100 of the NEC® provides two definitions related to fault conditions. The NEC® defines the term interrupting rating as the maximum current (fault current) at a specific voltage that a device, such as a fuse or circuit breaker, can safely interrupt. The NEC® defines the term short-circuit current rating as the fault current at a specific voltage that electrical equipment, such as a switchboard, can withstand without exceeding acceptable damage levels.

⚠ DANGER
If the power supply to a 24 VDC PLC falls below 18 VDC for a period exceeding the CPU hold-up time, the PLC will turn OFF and will not turn back on until the voltage is increased to 20 VDC.

Figure 11-2. When problems occur with PLC power supplies, input and/or output modules, or any system components, a check of the supplied power should be performed over time using a test instrument with a recording function.

Testing Fuses and Circuit Breakers

Fuses and circuit breakers are used in electrical circuits to protect the circuits from short circuits and overloads. **See Figure 11-3.** A *short circuit* is an overcurrent condition in which the current of a circuit leaves the normal current-carrying path by going around the load back to the power source via ground or another uncontrolled conductor. Short circuits typically occur when conductor insulation is damaged, which allows a current-carrying conductor to come in contact with a neutral conductor, ground conductor, or any grounded noncurrent-carrying metal parts.

PLC Short Circuits and Overloads

SHORT CIRCUIT — HOT WIRE TO GROUND SHORT

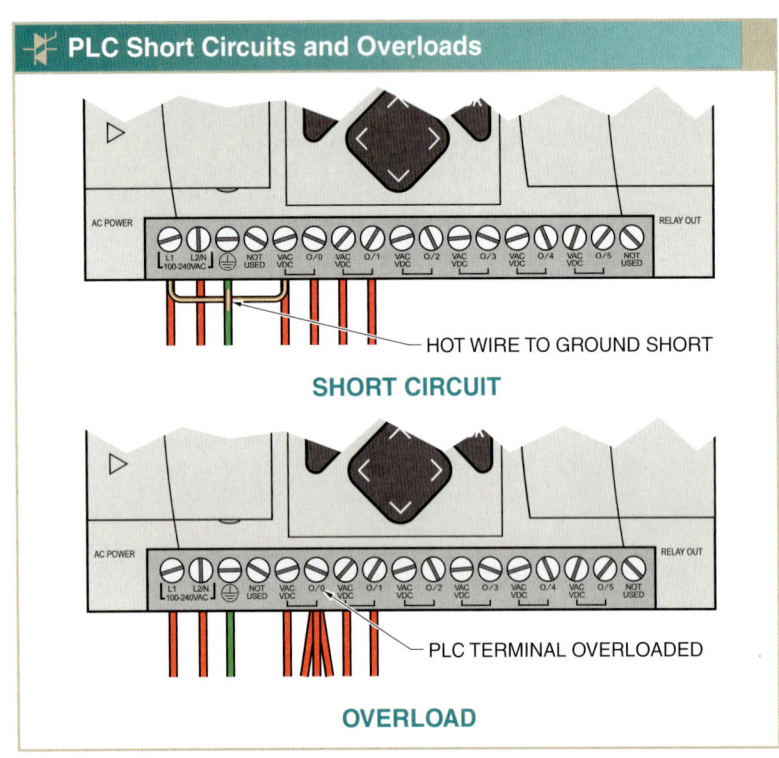

OVERLOAD — PLC TERMINAL OVERLOADED

Figure 11-3. Short circuits typically occur when conductor insulation is damaged, which allows a current-carrying conductor to come in contact with any grounded noncurrent-carrying metal parts. An overload is an overcurrent condition that occurs when circuit current exceeds normal PLC operating current and/or designed circuit current.

An *overload* is an overcurrent condition that occurs when circuit current rises above the normal current level at which the PLC and/or circuit is designed to operate. An overcurrent condition typically occurs when too many loads are added to a circuit or when several high-starting current devices, such as motors, are turned ON at the same time.

The technician should understand why a fuse opens or a circuit breaker trips. After replacing a fuse or resetting a circuit breaker, the circuit current should always be checked. Measuring the current is the best way of determining how much load is on a circuit. The higher the measured current, the more a circuit is loaded. It is best to turn all loads OFF before replacing a fuse or resetting a circuit breaker. After any fuse is replaced or circuit breaker is reset, individual circuit loads are turned back ON and the current of the circuit is monitored.

In general, circuit current should not exceed 80% of the fuses or circuit breaker rating per the NEC®. The 80% rule is more important in PLC circuits because the closer the percentage is to the maximum rating of the conductors, the higher the temperature of the conductors. The higher conductor temperatures, the higher the temperature will be around a PLC inside an enclosure.

Testing Fuses Using Voltage Measurements

A *fuse* is an overcurrent protection device that includes a fusible link that melts and opens a circuit when an overcurrent condition occurs. Fuses are connected in series with the circuit being protected. Fuses protect a circuit from overloads and short circuits. Electrical circuits include fuses to protect the incoming power supply circuit, control circuits, and individual components such as PLC power supplies, PLC sections or modules, input devices, and output components. Fuses used to protect AC or DC circuits are tested using test instruments that can measure voltage. **See Figure 11-4.**

Once a fuse is replaced, the circuit is returned to normal operation, and current measurements are taken to compare against the rating of the fuse or circuit breaker.

⚠ CAUTION

Use only replacement fuses of the type and rating specified for the PLC by the manufacturer. Improper fuse selection can result in equipment damage. PLC memory will not be lost because PLCs typically have EEPROM or a back-up battery.

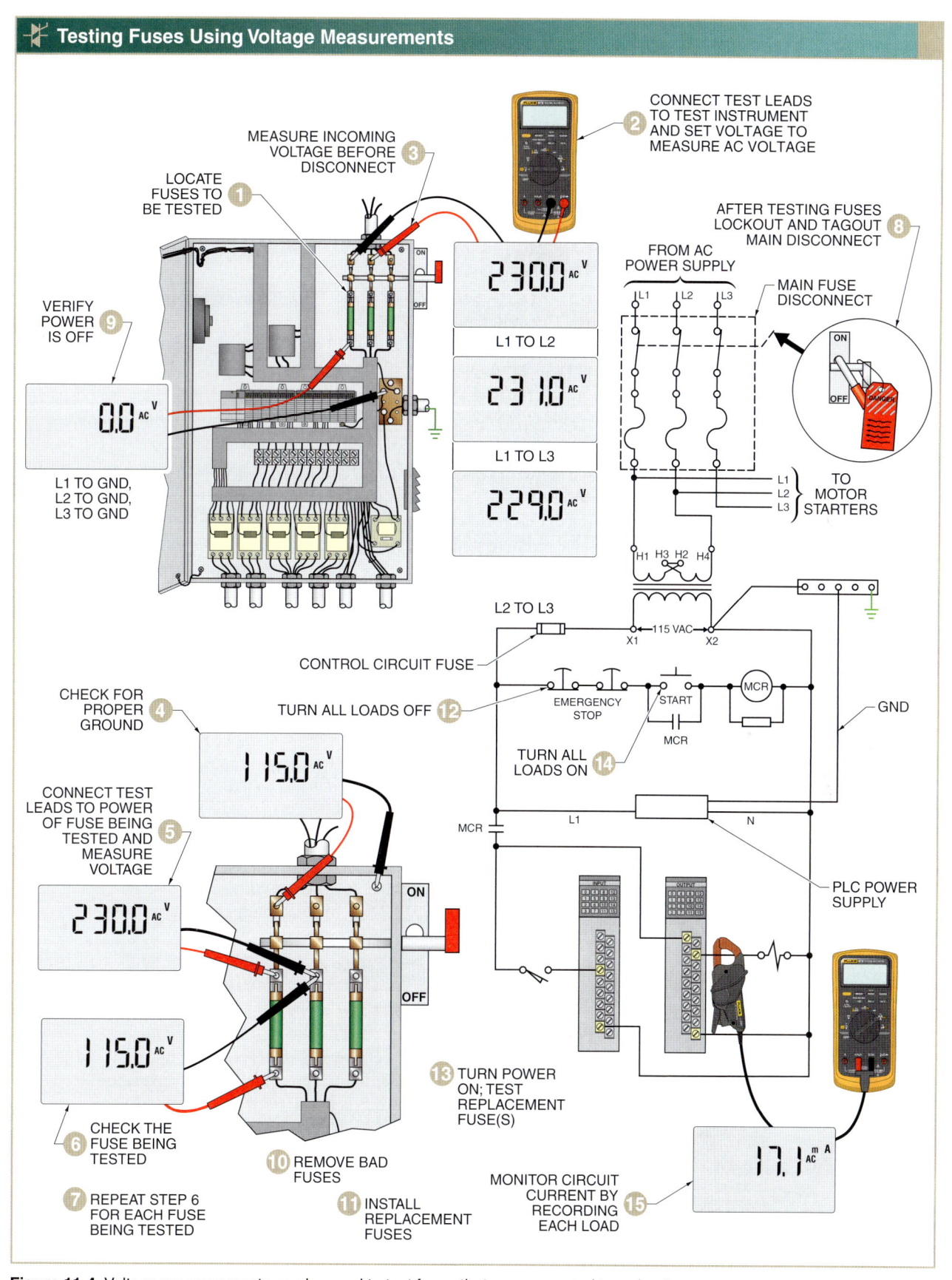

Figure 11-4. Voltage measurements can be used to test fuses that are connected to a circuit.

To test for a blown fuse in an energized enclosure using a voltage tester, apply the following procedure:

1. Locate the fuses to be tested. The main power fuses are typically located near the main disconnect (the lockout/tagout location). There will be one fuse for each hot (ungrounded) current-carrying conductor (L1, L2, or L3). The control circuit fuses are typically located on (or at) the step-down control transformer. Control transformer fuses are located on the output side (lower-voltage secondary side), but fuses can also be found on the input side of the transformer. Individual equipment fuses are located with the fused piece of equipment.

2. Connect test leads to the test instrument (DMM) and set the DMM to measure voltage type (VAC or VDC) and for the highest voltage possible in the circuit.

3. Measure the incoming supply voltage before the disconnect. The voltage of the circuit should be within +5% to −10% of the rating of the components operating on the circuit. When voltage is not present, or the voltage is not within specifications, the problem is in the circuit before the disconnect and must be corrected.

4. Check for proper ground to ensure the safety of personnel working around the electrical system and to reduce system noise and other electrical problems. All noncurrent-carrying metal parts must be grounded. Verify that the system is grounded by measuring the voltage before the fuse of any current-carrying conductor (L1, L2, or L3) and ground (any noncurrent-carrying metal). If the system is grounded, there is measured voltage between a hot conductor and ground. If there is no voltage measured, the system is not grounded and the problem must be corrected to ensure the system will operate safely.

5. Connect the test leads of the DMM to the incoming power of the fuse being tested and measure the voltage just before the fuse. For any circuit containing two or more fuses (high-voltage 1ϕ and all 3ϕ circuits), the measurement will be taken on the input side (side that is energized at all times) of the fuses. For a circuit containing only one fuse (low-voltage 1ϕ and DC circuits) the measurement will be taken on the input to the fuse and the neutral (grounded) conductor. There should be a voltage measurement regardless of whether the fuse is good or bad (melted open). If there is no voltage measured, that part of the system is not powered and the problem must be corrected.

6. Check the fuse being tested by moving the red test lead to the output side (side going to protected circuit/equipment) of the fuse being tested and the black (common) test lead on another phase. If the fuse is good, the voltage measured will be equal to the supply voltage.

Fluke Corporation

Voltage measurements are typically used to determine if a fuse is blown. Proper PPE must be worn when taking voltage measurements.

If the fuse is bad, there is no voltage measured. On some circuits, if the fuse is bad, it is possible to measure a voltage less than the supply voltage because of voltage feedback in the system. In this case, if the voltage measured across the line and load of the same fuse equals the supply voltage or anything less than the supply voltage, the fuse is considered bad.

7. Repeat step 6 for each fuse to be tested.
8. After testing each fuse, turn power OFF and lock and tag out the system.
9. Verify power is OFF at a fuse by measuring the voltage between the load side of the fuse and ground. There must be no voltage measured or the circuit is not OFF.
10. Using a fuse puller, remove any fuse that tested bad. All suspect fuses must be checked with an ohmmeter to verify fuse condition.
11. Replace bad fuses with good fuses of the same voltage, amperage, and interrupting ratings.
12. Before turning power back ON, turn all loads in the circuit OFF.
13. Turn power ON and test the replaced fuses using step 6. If the power to a circuit or component has not been restored, the "power on" light of the PLC will not turn ON.
14. Turn the circuit loads back ON, one at a time. The system must be reenergized in stages to isolate any potential problems.
15. Monitor circuit current so the current draw of each load is known.

PLC TIPS

Cable and wire management requires routing cables and wires from one point to another for greatest efficiency. Technicians physically manage wires with horizontal and vertical wireways in enclosures, racks, trays, and conduit outside enclosures.

Fluke Corporation

Technicians use test instruments with multiple attachments to measure every aspect of control circuits to find the source of a malfunction.

When a test instrument measuring voltage identifies a blown fuse, the blown fuse should be removed using a fuse puller and the condition of the fuse verified by using a test instrument set to measure resistance.

Testing Fuses Using Resistance Measurements

Fuses removed from a circuit can be tested using a test instrument (DMM) set to measure resistance. **See Figure 11-5.** To test fuses with an ohmmeter or DMM, apply the following procedure:

1. Turn power to the circuit OFF and lock out and tag out disconnect.
2. Verify power is OFF at the fuses by measuring the voltage at the fuses to be removed.
3. Using a fuse puller, remove all fuses to be tested and place them on a nonconductive surface.
4. Set the ohmmeter to measure resistance (low setting) or continuity.
5. Connect the test leads to each end of the fuse under test. The resistance of a good fuse should be 0 Ω (or close to 0 Ω). If the continuity function is used, the fuse should show continuity (complete path). A bad (open) fuse will show an open loop (OL) on resistance or no continuity.
6. Repeat step 5 for each fuse being tested.
7. Turn the ohmmeter OFF to prevent battery drain.

⚠ WARNING

Ensure that no voltage is present in a circuit when resistance measurements are taken.

CODE CONNECT

Section 240.60(B) of the NEC® requires that fuse holders for current-limiting fuses be designed to prevent the insertion of noncurrent-limiting fuses. Class R (rejection) fuses are current limiting and have a high current interrupting rating. Class R fuses have a ring on one of the ferrule ends or a notch in one of the fuse blades. Only fuses with this rejection feature can be inserted into fuse holders intended for current-limiting fuses.

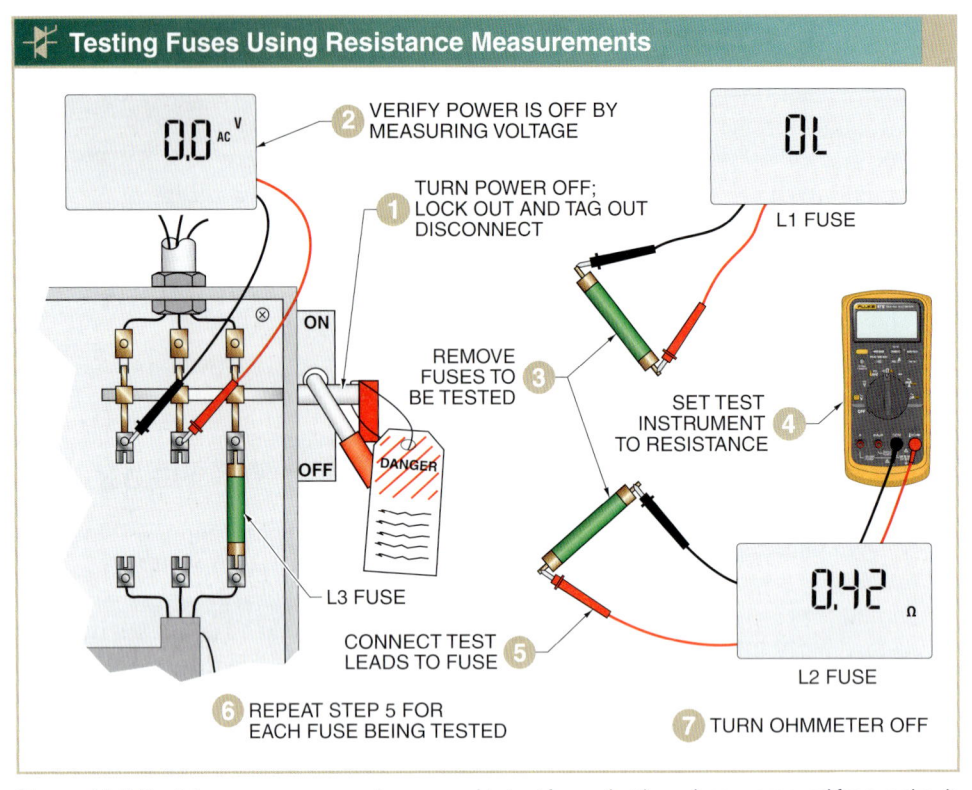

Figure 11-5. Resistance measurements are used to test fuses that have been removed from a circuit.

Troubleshooting Circuit Breakers

A *circuit breaker (CB)* is a reusable overcurrent protection device that opens a circuit automatically at a predetermined overcurrent. CBs are connected in series with the circuit being protected. CBs (like fuses) protect a circuit from overloads and short circuits. CBs are magnetically and/or thermally operated and must be reset after an overcurrent. Circuit breakers perform the same function as fuses and are basically tested the same way. **See Figure 11-6.** To test circuit breakers using a voltage tester, apply the following procedure:

⚠ CAUTION
- Follow all electrical safety practices and procedures per NFPA 70E®.
- Check and wear personal protective equipment (PPE) for the procedure being performed.
- Perform only authorized procedures.
- Follow all manufacturer recommendations and procedures.

1. Locate the circuit breaker to be tested. The main power circuit breaker is typically located at the main disconnect (the lockout/tagout location). There is one circuit breaker pole for each hot (ungrounded) current-carrying conductor. However, unlike fuses where there can be more than one fuse, a 3φ circuit breaker is one assembled device with one ON/OFF/TRIP lever. Additional circuit breakers are typically used in the rest of the circuit where needed.

2. Connect the test leads to the DMM and set DMM to measure the highest possible voltage and voltage type (VAC or VDC) in the circuit.

3. Connect the test leads of the DMM to the incoming power before the circuit breaker and measure the voltage. The voltage of the circuit should be within +5% to −10% of the rating of the components operating on the circuit. When voltage is not present, or the voltage is not within

specifications, the problem is in the circuit before the circuit breaker and must be corrected.

4. For the safety of personnel working around the electrical system and to reduce system noise and other electrical problems, all noncurrent-carrying metal parts must be grounded. Verify that the system is grounded by measuring the voltage before the circuit breaker of a current-carrying conductor (ungrounded) and ground (any noncurrent-carrying metal). If a system is grounded, there is measured voltage between a hot conductor and ground. If there is no voltage measured, the system is not grounded and the problem must be corrected to ensure the system will operate safely.

5. Before taking any voltage measurements, check the position of the circuit breaker lever. If the circuit breaker is in the ON position, proceed to step 7. If the circuit breaker lever is in the tripped position, do not immediately move the circuit breaker lever to the ON position. The circuit breaker must be reset first.

6. When the circuit breaker lever is in the tripped position and all loads are OFF, reset the circuit breaker (pull the CB handle into the OFF position and then into the ON position).

7. Connect the test leads of the test instrument across the circuit breaker being tested by placing one test lead (common) on the output side (side going to protected circuit/equipment) of the phase to be tested and one test lead (voltage) on the input side (side that is energized at all times) of another phase to verify the breaker has been reset. If the circuit breaker is good, voltage will be measured across the circuit breaker. If the circuit breaker is bad, there will be no voltage or a voltage less than the supply voltage because of voltage feedback though the system. *Note:* The phase of the circuit breaker being tested is the phase with the common test lead.

8. Repeat step 7 for each phase of circuit breaker or circuit breakers to be tested.

9. After a circuit breaker is reset, turn the loads back ON, one at a time, to isolate any potential problems.

10. Monitor circuit current so the current draw of each load is known.

⚠ WARNING

When a circuit breaker has tripped, turn all loads in the circuit to the OFF condition and check for short circuits before resetting.

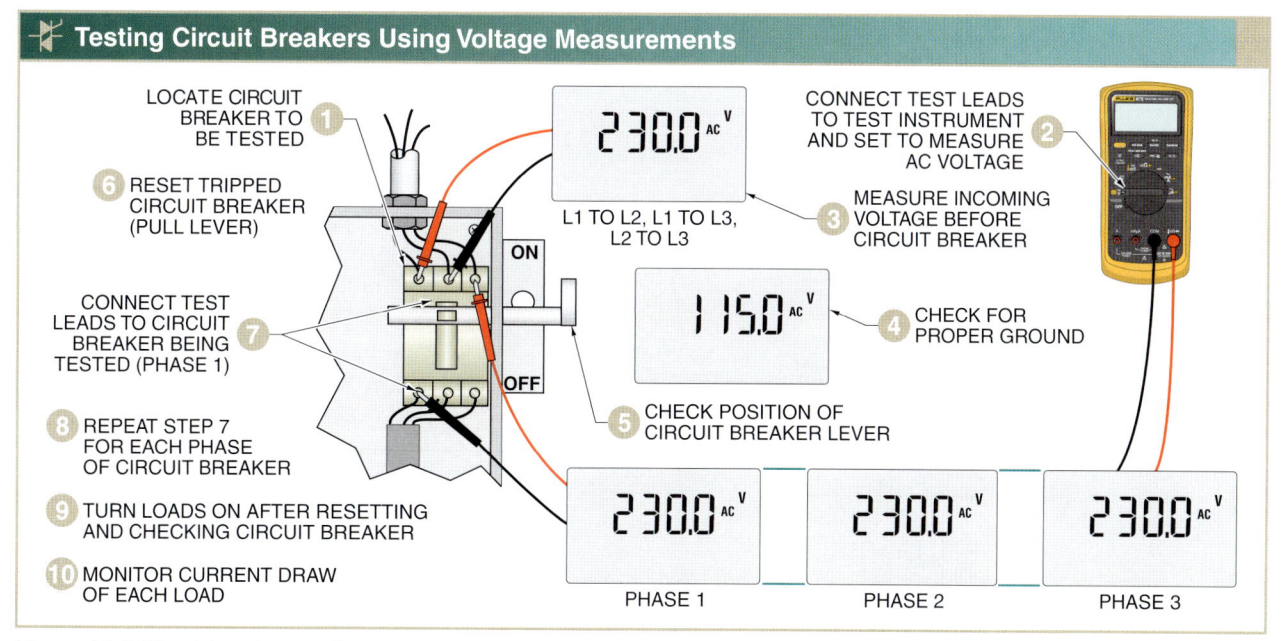

Figure 11-6. Circuit breakers perform the same function as fuses and are basically tested the same way.

Troubleshooting Control Transformers

In most circuits in which the load is rated higher than 115 VAC (208 V, 230 V, or 460 V) the control circuit is operated at a lower voltage level than the loads. A step-down control transformer is used to step down the voltage to a level required in the control circuit. Typically, the secondary side of the control transformer is rated at 120 VAC for most industrial control panels, but today 12 VAC and 24 VAC are becoming more common. **See Figure 11-7.**

PLC TIPS

Control transformers are self-air-cooled constant-potential transformers used to step down the voltage to signal and control circuits in PLC enclosures.

Most control transformers have a dual-voltage primary side. A dual-voltage primary side allows a transformer to be configured for either one of two voltages, such as 230 VAC or 460 VAC. The primary coils of the transformer are connected in series for higher voltage and parallel for lower voltage. The terminals on the primary side of a control transformer are marked with an "H" (H1, H2, H3, and H4). The terminals on the secondary side of the control transformer are marked with an "X" (X1 and X2).

All control transformers have a fixed power output rating in volt-amps (VA) or kilovolt-amps (kVA). When the power rating of a transformer is exceeded by placing too great a load on the transformer, the voltage will start to decrease and circuit problems will develop. **See Figure 11-8.** To test a control transformer using voltage measurements, apply the following procedure:

1. Locate the control transformer to be tested. The control transformer receives power on the primary side from the main fuse or circuit breaker disconnect.

2. Connect the test leads to the test instrument and set the multimeter to measure AC voltage at the highest possible voltage in the circuit.

3. Measure the incoming supply voltage into the primary side of the transformer (H1 and H4). The measurement should be within +5% to –10% of the rating of the transformer's primary side. If there is no voltage, or the voltage is not within specifications, the problem is before the transformer and must be corrected. The voltage at the primary side of the transformer affects the voltage on the secondary side of the transformer. Control transformers (like all transformers) have a primary side to secondary side voltage ratio, such as 2:1, or 4:1. For example, a transformer with a 2:1 voltage ratio that has 240 V on the primary side will deliver 120 V to the secondary side, a transformer with 230 V on the primary side will deliver 115 V to the secondary side, and a transformer with 220 V on the primary side will deliver 110 V to the secondary side.

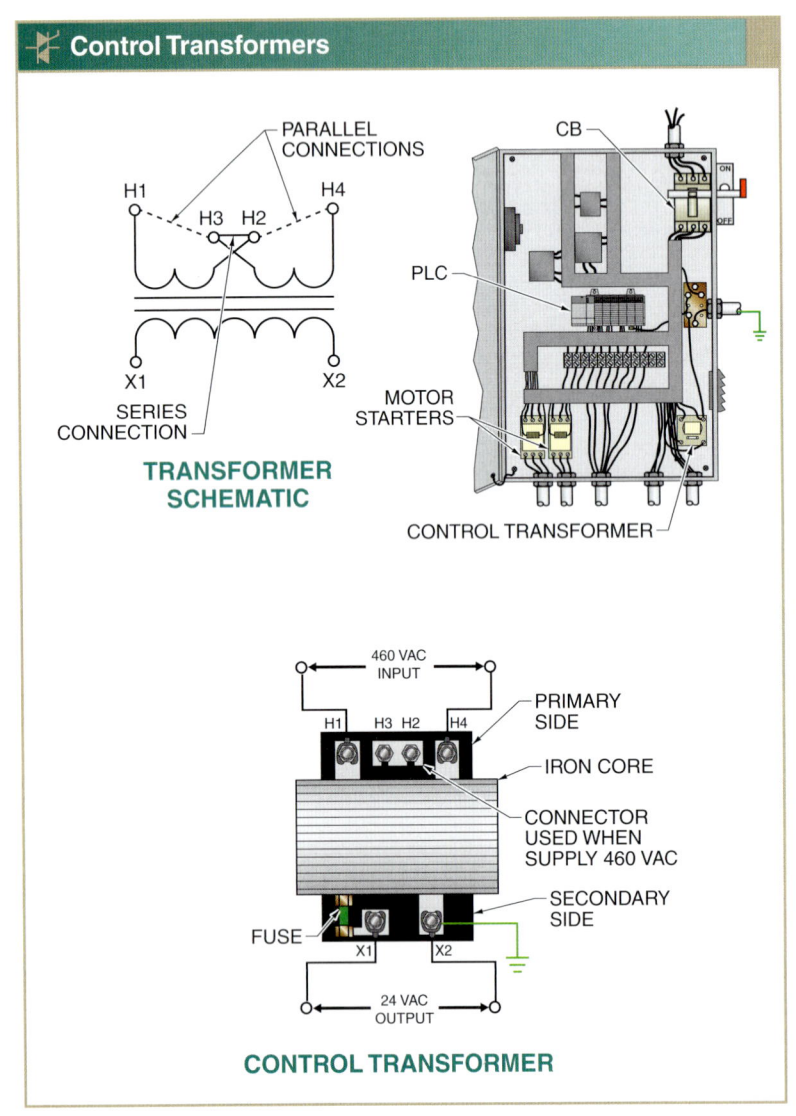

Figure 11-7. Control transformers are typically used to step down 115 VAC, 208 V, 230 V, or 460 V to 24 VAC for control circuit use in a PLC enclosure.

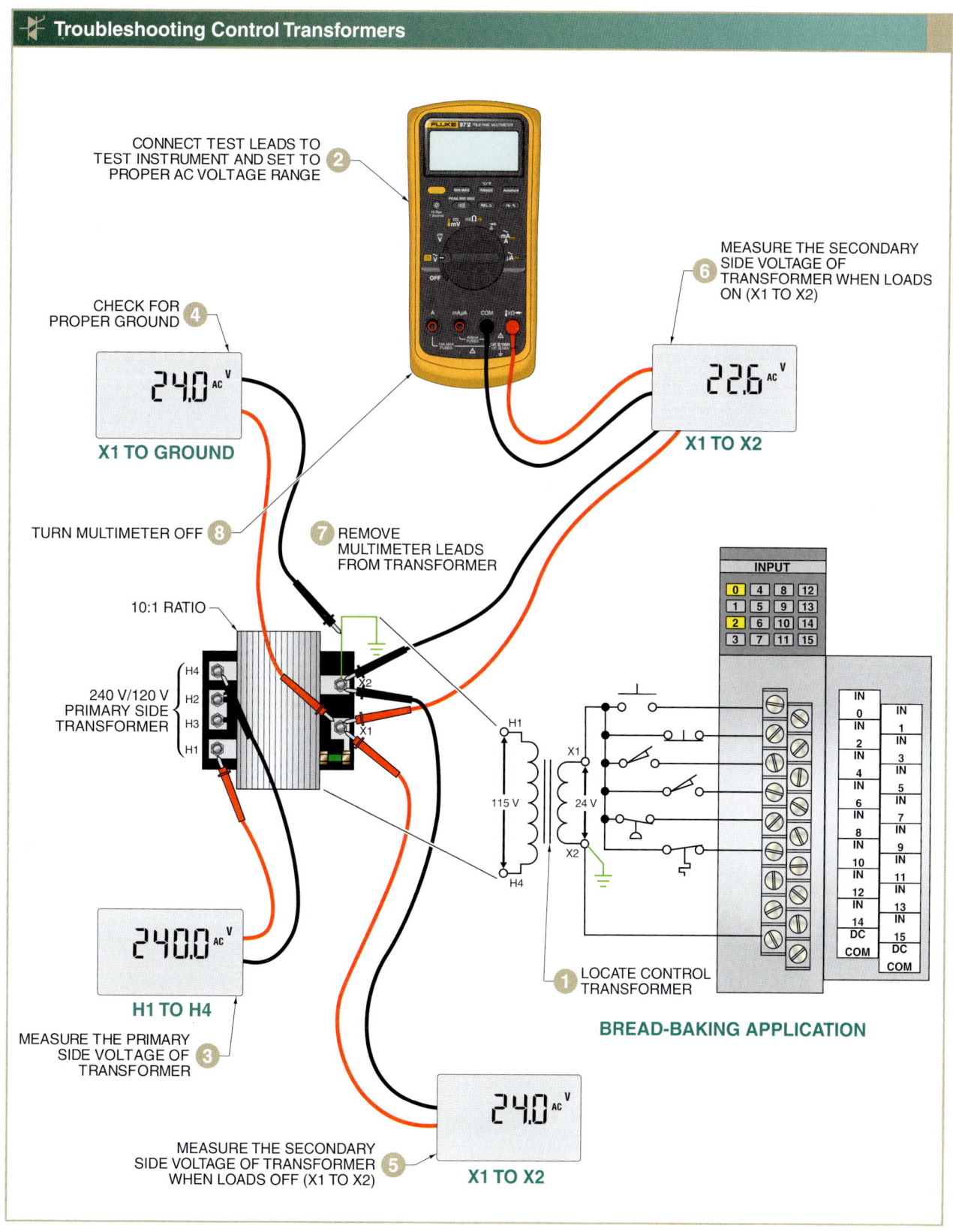

Figure 11-8. When the power rating of a transformer is exceeded by placing too great a load on the transformer, the voltage on the secondary side will start to decrease and circuit problems will develop.

4. For the safety of personnel working around electrical systems and to reduce system noise and other electrical problems, verify that all noncurrent-carrying metal parts are grounded. Once power travels through a transformer, any prior ground is lost. To reestablish ground, one side (X2) on the output of a transformer is grounded to establish a ground for a new circuit. To verify that the secondary side of a transformer is grounded, measure the voltage between a hot (ungrounded) conductor on the secondary side of the transformer (X1) and ground (any noncurrent-carrying metal). If the system is properly grounded, the resulting step-down voltage (12 V, 24 V, or 120 V) will be measured between a hot conductor (X1) and ground. If there is no voltage measured, the system is not grounded and the problem must be corrected to ensure the system operates safely.

5. Move the test leads of the multimeter directly to the X1 and X2 terminals on the secondary side of the transformer. Measure the secondary side control circuit voltage. If there is no voltage measured on the secondary side, but the proper voltage was measured on the primary side, the control transformer is bad. However, before changing the transformer, ensure the transformer secondary side (X1) fuse (often mounted on the transformer) is not blown.

6. When the measured voltage from the secondary side (X1 and X2 terminals) of a transformer is correct, measure the voltage when the circuit is fully loaded. An overloaded transformer problem only occurs when a transformer is fully loaded. The output voltage from a transformer should not drop more than 10% from a no-load condition to a fully loaded condition. A multimeter with a MIN MAX recording mode is typically used to take load measurements from the secondary side of a transformer.

7. Remove the multimeter leads from the transformer.

8. Turn the multimeter OFF.

Testing Power Supplied to PLC

All PLCs have a specified input voltage rating. The voltage rating may be a single fixed rating (12 VDC, 24 VDC, or 115 VAC), a dual voltage rating (12/24 VDC, 115/230 VAC), or a voltage range rating such as 10 VDC to 30 VDC. When PLC specifications are not available from installation manuals, operation manuals, manufacturer catalogs, or web sites, then the voltage rating listed on the PLC and the following guidelines should be used:

- When a PLC has a single fixed voltage rating, then the voltage supplied to the PLC must remain within a +5% to –10% voltage range at all times.

- When a PLC has a dual voltage rating, a selector switch or movable links on the PLC are used to set the PLC to one of the possible voltage ratings. When a 115/230 VAC rated PLC is set to the 115 VAC position, the PLC cannot be connected to 230 VAC without first resetting the voltage selector switch to the 230 VAC position. Due to safety concerns, a PLC that has a dual voltage rating is shipped with the PLC set to the higher voltage rating.

- When a PLC has a voltage range, the voltage supplied to the PLC can be any voltage within the range, but voltages at either end of the rating should be avoided.

The power being supplied to a PLC should be the first item checked when troubleshooting a PLC hardware problem.

When PLC specifications are available, the manufacturer will list the exact voltage range in which a PLC properly operates. For example, a 115 VAC rated PLC may have an 85 VAC to 132 VAC listed range. Although the PLC should operate satisfactorily within the specified range, any voltage at either end of the rating requires the power supply feeding the PLC to be checked and corrected if possible. **See Figure 11-9.** To test the input voltage to a PLC using voltage measurements, apply the following procedure:

1. Locate the PLC input (sensor) power supply terminals to be tested.
2. Connect the test leads to the DMM and set the DMM to measure voltage type, VAC or VDC at the highest possible voltage setting.
3. Measure the incoming supply voltage into the PLC at the PLC power supply input terminals. The voltage rated must be within +5% to –10% of the rating of the PLC power supply. If there is no voltage, or the voltage is not within specifications, the problem is upstream (the circuit feeding the PLC) and must be corrected.
4. Remove the DMM test leads from the PLC input terminals and turn the DMM OFF.

⚠ WARNING

Do not expose the terminals or CPU of a PLC to electrostatic charges. Always discharge electrostatic charges before working with a PLC.

CODE CONNECT

Section 250.118(5) of the NEC® allows flexible metal conduit (FMC) to be used as the equipment grounding conductor for circuit conductors under the following conditions:

- Listed fittings are used to terminate the FMC.
- The overcurrent protective device protecting the conductors in the FMC is 20 A or less.
- The maximum size of the FMC is 1¼".
- The total length of the FMC or combination FMC, FMT, and LFMC does not create a ground fault path longer than 6'.

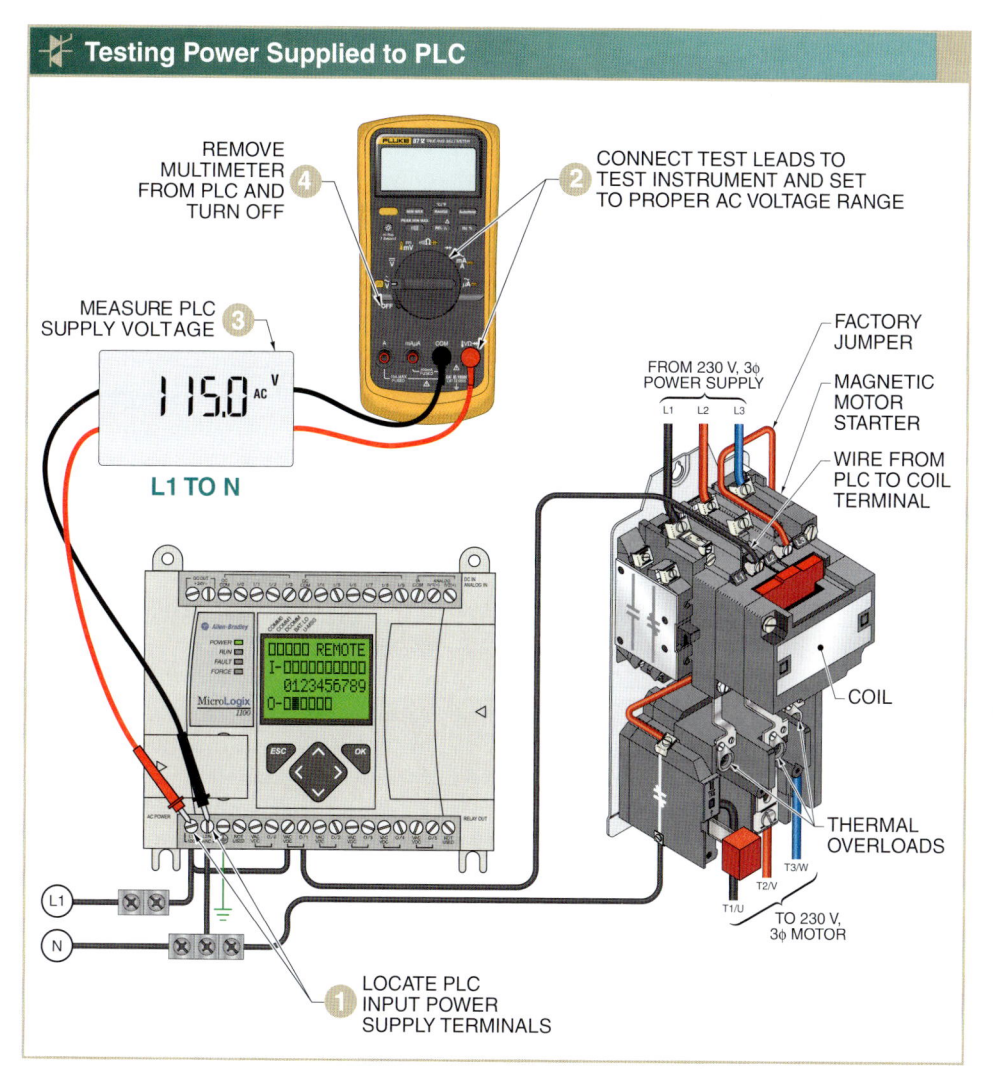

Figure 11-9. PLCs can have a single fixed voltage rating or a dual voltage rating. Dual voltage rated PLCs (115/230 VAC) have a selector switch (or movable links) to set the PLC to one of the possible voltage ratings.

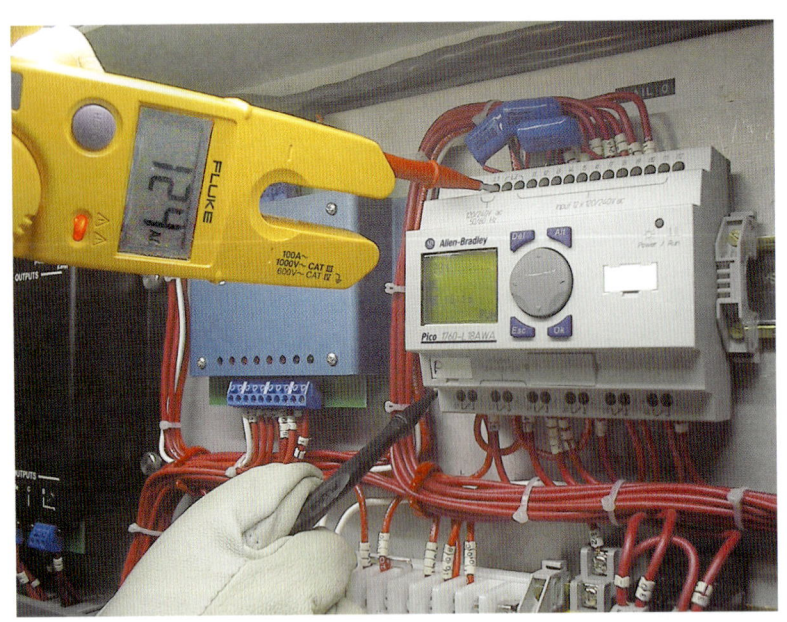

The output power supplied to PLC input devices must be within +5% to −10% of the rating of the PLC in order for the PLC and input devices to operate properly.

Testing PLC Input Power Supplies

Some PLCs supply voltage (24 VDC sensor power) that can be used to supply power to input devices. **See Figure 11-10.** When a PLC supplies voltage to input devices, the input (sensor) power supply will have both a voltage rating and a current rating. For example, typical PLC input power supply ratings are 24 VDC/250 mA. The input power supply voltage rating is what the voltage output from the PLC will actually be, but the current rating is a range. For example, a 250 mA rating means that any amount of current from 1 mA to 250 mA can safely be drawn from the power supply.

If loads that draw more than 250 mA are connected to a PLC input (sensor) power supply, the power supply will be overloaded. The listed current rating of an input power supply must be accepted as the maximum amount of current that can safely be provided from a PLC power supply. When loads are connected to a PLC that exceed the current rating of the PLC input power supply, the voltage output of a PLC will decrease. The greater the overload, the greater the voltage drop. To test the output voltage of a PLC input power supply using voltage measurements, apply the following procedure:

1. Locate the PLC input (sensor) power supply terminals to be tested.
2. Turn power OFF to the PLC supplied power section. Verify power is OFF by taking an AC voltage measurement at the PLC supplied power section L1 and L2.
3. Remove all input device wires from the PLC input (sensor) power supply.
4. Set the test instrument to the highest possible voltage setting (AC or DC) based on the circuit voltage.
5. Turn the PLC power ON.
6. Measure the supply voltage out of the PLC at the PLC input (sensor) power supply terminals. The voltage should be within +5% to −10% of the rating of the PLC input power supply. If there is no voltage, or the voltage is not within specifications, the problem is in the PLC internal power supply. The power supply and/or the PLC needs to be replaced.
7. Turn the PLC power OFF. Verify that the power is OFF by taking a voltage measurement at the PLC power supplied section terminals.
8. Reconnect the input devices to the PLC input (sensor) power supply. Input devices are connected one at a time to determine if an overload problem exists.
9. Turn the PLC power ON.
10. Measure the voltage at the PLC input (sensor) power supply terminals. A voltage drop of more than 5% indicates that the output power is overloaded. If the voltage does drop, a current measurement must be taken to determine whether there is still current available for the input circuits.

PLC TIPS

The general specifications for a micro-sized PLC indicate the 24 VDC sensor power available for input devices is typically about 250 mA. The nominal current per input terminal allowed is 8.5 mA.

Testing PLC Input (Sensor) Power Supplies

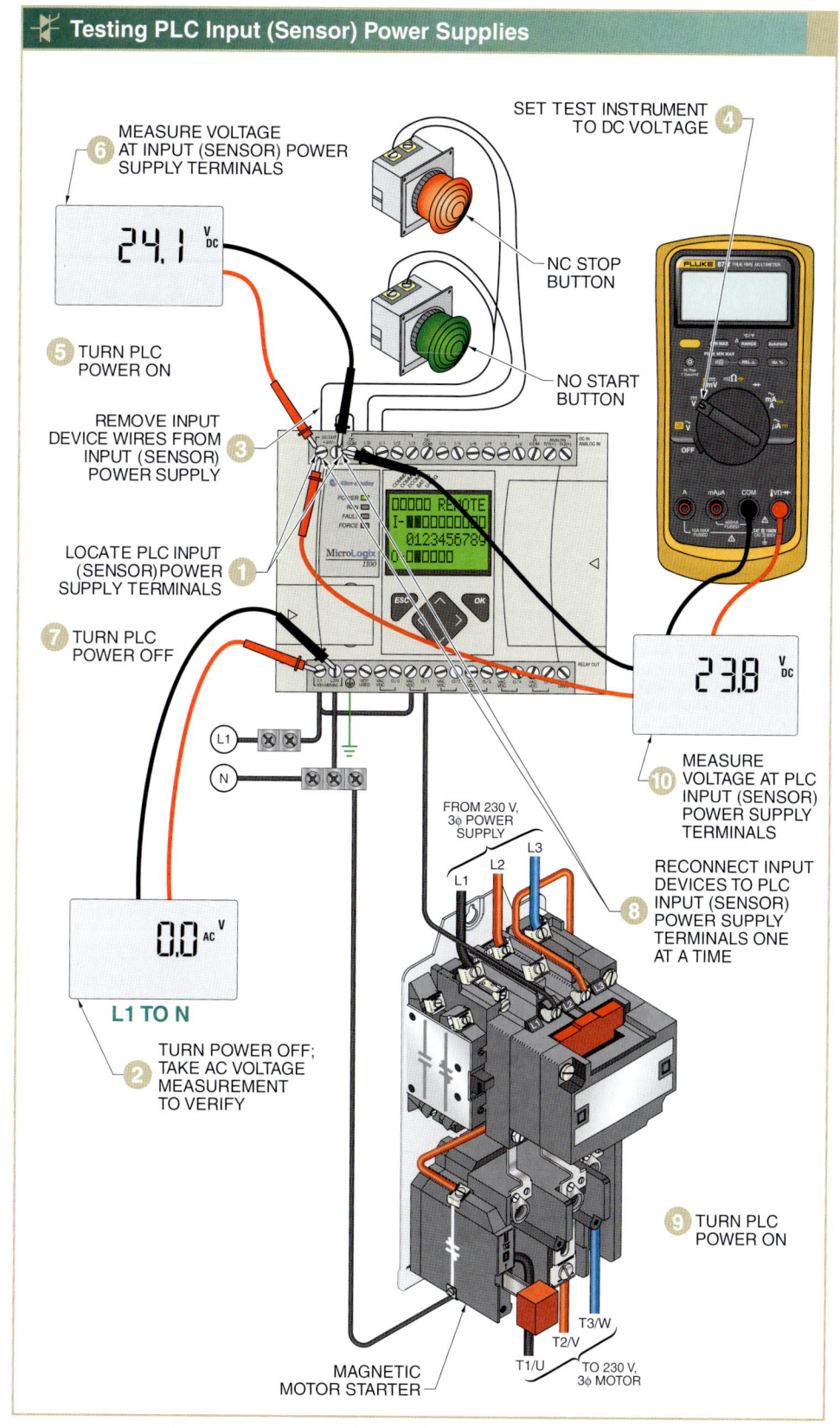

Figure 11-10. Input (sensor) power supplies of PLCs are typically used to supply voltage to input devices connected to the PLCs.

CODE CONNECT

Section 240.83(D) of the NEC® covers the use of circuit breakers as switches for lighting loads. In certain installations, such as warehouses, the circuit breaker(s) in a panel may be used to turn the lights ON and OFF. Special circuit breakers are used in these applications because of the additional wear due to the switching. Circuit breakers listed, labeled, and marked HID can be used as switches for high-intensity discharge lighting, 120 V fluorescent lighting, or 277 V fluorescent lighting. Circuit breakers listed, labeled, and marked SWD can only be used as switches for 120 V fluorescent lighting or 277 V fluorescent lighting.

An in-line ammeter is typically used to test for input (sensor) power supply current. Some applications will require a clamp-on ammeter to be used instead of performing an in-line current measurement. **See Figure 11-11.** To test the input (sensor) power supply system of a PLC using an in-line ammeter, apply the following procedure:

1. Turn power supplied to the PLC OFF. Verify power is OFF by taking a voltage measurement at the PLC power supplied section terminals.

2. Set the ammeter to take an in-line current measurement. Set the ammeter to the highest possible current setting. Use only an ammeter that has fused current jacks.

Measuring PLC Input (Sensor) Power Supply Current

- ⑤ MEASURE THE INPUT (SENSOR) POWER SUPPLY CURRENT
- ③ CONNECT IN-LINE AMMETER TEST LEADS IN SERIES WITH INPUT DEVICES
- ⑦ REMOVE IN-LINE AMMETER FROM CIRCUIT; RECONNECT INPUT DEVICES TO PLC INPUT (SENSOR) POWER SUPPLY
- NC STOP BUTTON
- NO START BUTTON
- L1 TO N
- ① TURN POWER OFF; TAKE VOLTAGE MEASUREMENT TO VERIFY
- ④ TURN PLC POWER ON
- ② CONNECT TEST LEADS TO IN-LINE AMMETER; SET TO MEASURE DC CURRENT
- ⑥ TURN POWER SUPPLIED TO PLC OFF
- ⑧ TURN PLC POWER ON
- FROM 230 V, 3φ POWER SUPPLY
- MAGNETIC MOTOR STARTER
- TO 230 V, 3φ MOTOR

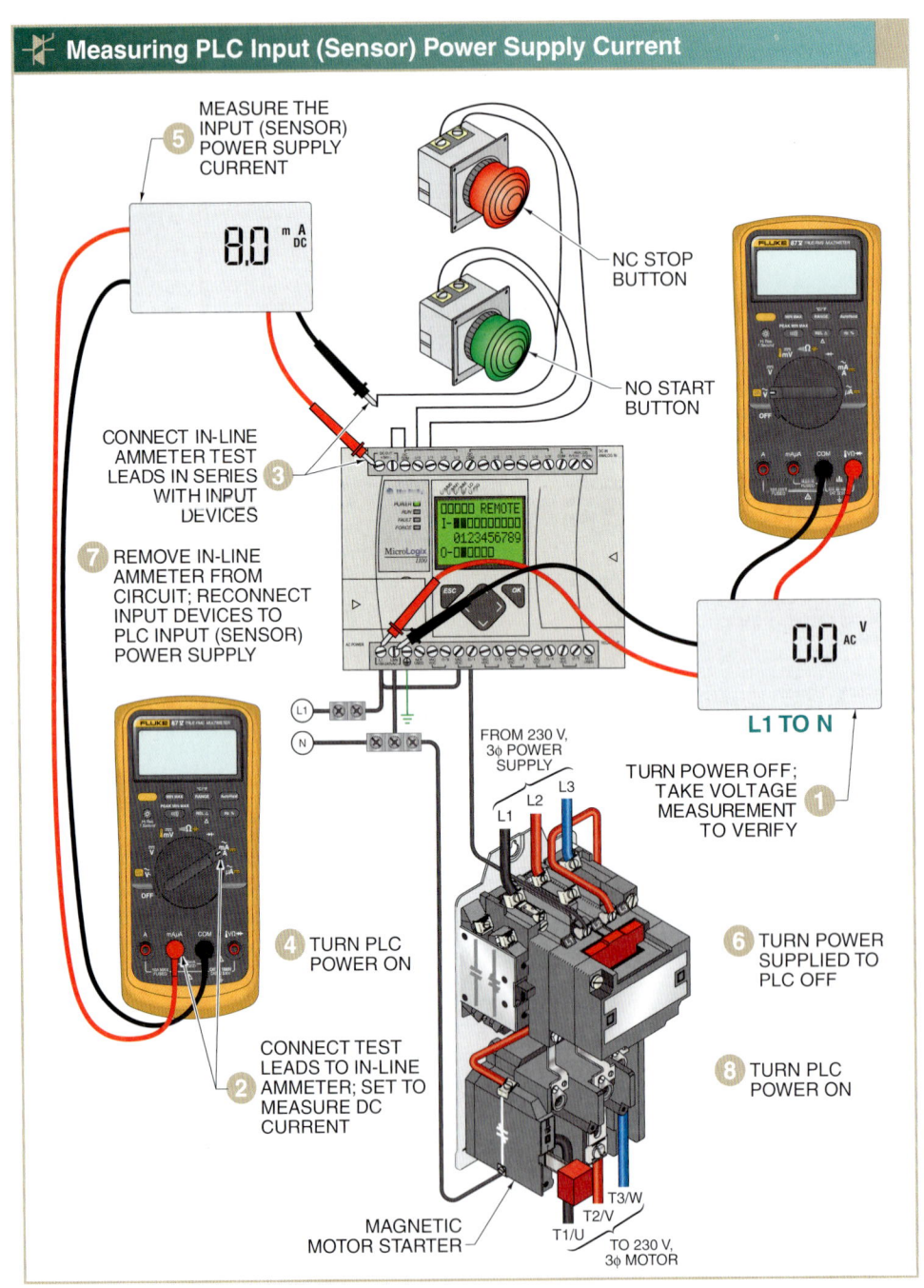

Figure 11-11. The listed current rating of a power supply must be accepted as the maximum amount of current that can safely be provided from a PLC.

3. Connect the in-line ammeter in series with the input devices connected to the PLC input (sensor) power supply.
4. Turn PLC power ON.
5. Measure the current from the PLC to the input devices. At no time should the measured current exceed the rated current of the PLC input (sensor) power supply. A meter with a MIN MAX recording mode can be used to record current measurements over time.
6. Turn power supplied to the PLC OFF. Verify power is OFF by taking a voltage measurement at the PLC supplied power section terminals.
7. Remove the in-line ammeter from the circuit and reconnect the input devices to the PLC input (sensor) power supply.
8. Turn PLC power ON.

TROUBLESHOOTING PLC INPUT SECTIONS OR MODULES

All PLCs have an input section or module that input devices (switches and sensors) are connected to. Input devices can be of a digital type, such as open/closed, or analog type, such as 0 to 10 VDC or 4 mA to 20 mA. Digital input devices are the most common type of input device because digital devices represent the largest group of mechanical and solid-state switches, such as pushbuttons, pressure switches, selector switches, photoelectric and proximity switches, and limit switches. Analog input devices are input devices that send a varying input signal to a PLC that can be any changing variable, such as voltage, current, or resistance.

When a PLC-controlled system is not operating properly, the PLC input devices must be checked. Without the proper input signals, the output components (loads) of the PLC will not operate as designed. Troubleshooting a PLC input section or module and correcting a problem requires knowing the following:

- how the parts of an input section or module work
- what to look for when troubleshooting an input section or module—this includes understanding the CPU status lights, the input section or module status lights, how open and closed input devices are displayed on a computer monitor, and how to use the PLC program to monitor and/or force input devices to change operating state
- how to use test instruments to take measurements at an input section or module and what the measurements mean
- how to monitor circuit input devices using the PLC displayed program and data files
- how to correct input circuit problems, such as leakage current problems

PLC Input Circuit Operation

Input devices, such as limit switches and temperature switches are connected to the input section or module of a PLC. When an input signal is sent to a PLC, the PLC detects the current flow and voltage and processes the signal. **See Figure 11-12.**

The input devices and the PLC input section or module must be checked any time the output components of a system are not operating.

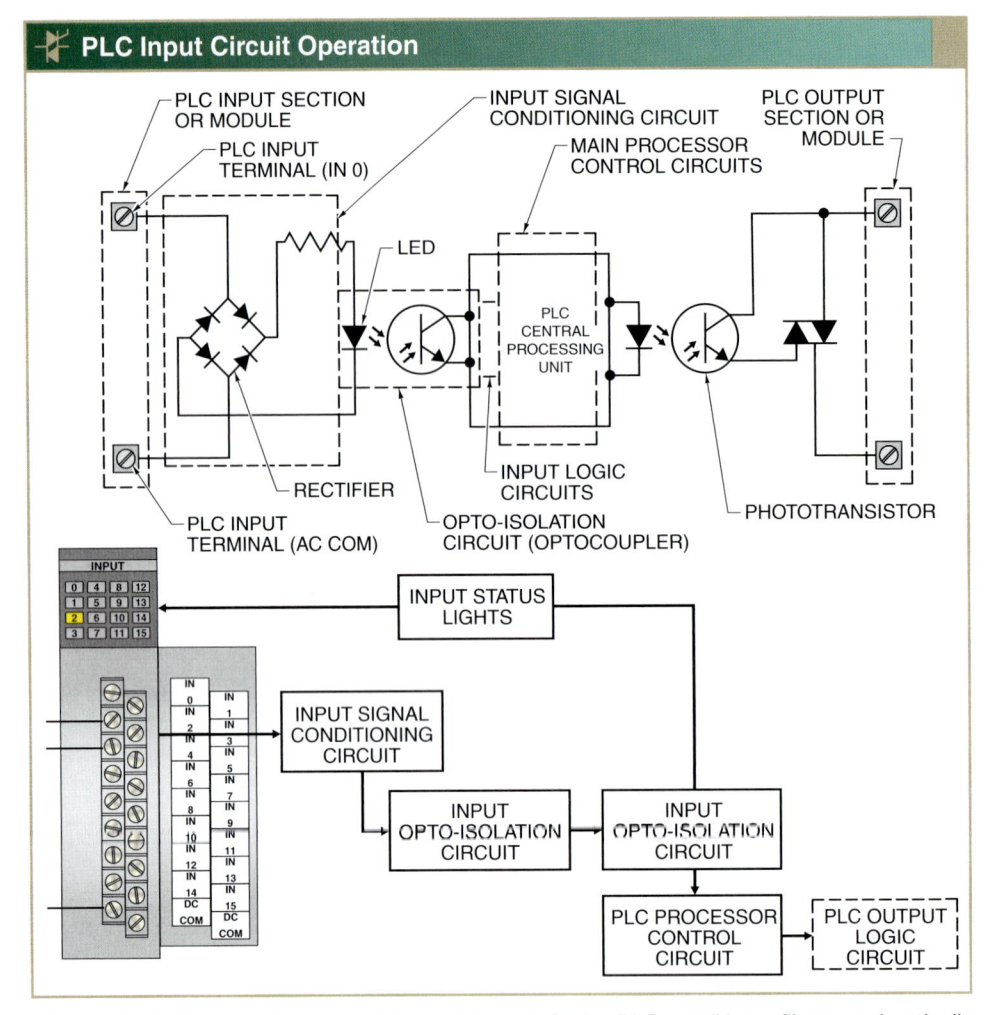

Figure 11-12. When an input signal is sent to a PLC, the PLC conditions, filters, and optically isolates the signal.

The input signal goes into an input signal conditioning circuit that filters the signal. Filtering includes removing contact bounce signals, rectifying AC to DC, and/or eliminating electromagnetic interference. *Contact bounce* is the rapid making and breaking of mechanical switch contacts when switch contacts are closed. Contact bounce occurs so quickly (in milliseconds) that it is typically not a problem in mechanical circuits (except for contributing to switch wear), but can be a problem in high-speed electronic circuits that can detect the contact bounce as several switch closings. Typically, contact bounce is only a problem in counting applications. *Electromagnetic interference (EMI)* is unwanted electrical noise generated by electrical and electronic equipment, such as lighting ballasts and transformers.

From the input signal conditioning circuit, the input signal is sent to an opto-isolation (optocoupler) circuit. The opto-isolation circuit receives the incoming signal and converts the signal into a transmitted light signal that is sent to a light-sensitive receiver circuit for processing. Opto-isolation circuits help protect PLC circuits from incoming transients and other unwanted electrical signals that can travel on metal conductors but not light beams. Opto-isolating circuits only allow signals to flow in one direction.

From the opto-isolation circuits the input signal is sent to the logic circuits of the PLC. The logic circuits process the signal so it can be sent to the PLC main processor control circuits and the input status light (LED) circuits, which light up on the PLC input section or module when the signal is received.

From the logic circuit, the input signal is sent to the PLC processor control circuits (CPU). The processor control circuits execute the preprogrammed logic and send the appropriate signals as required to the PLC internal timers, counters, and output sections or modules.

Digital Input Filtering

When PLC digital inputs receive signals from a variety of input devices, such as pushbuttons, selector switches, relay contacts, and photoelectric switches, unwanted pulses (contact bounce signals) occur and are detected by the PLC digital input. Contact bounce signals are caused by input devices rapidly changing state from OFF to ON and ON to OFF. This can be misinterpreted by the PLC and may lead to counting errors or operational malfunctions.

To counteract contact bounce, PLCs have digital input filters. These are software functions that allow a technician to configure the length of time an input signal must be ON or OFF before the PLC digital input recognizes the signal. **See Figure 11-13.** Typically, several input devices are opened or closed simultaneously, so groups of digital inputs are adjusted together. The longer the length of time set by the technician, the greater the signal filtering effect. However, the length of time can also be decreased in order to detect high-speed input signals. *Note:* EMI can also generate unwanted pulses that affect PLC digital inputs.

AC modules are indicated by red wire and use LEDs for signal status.

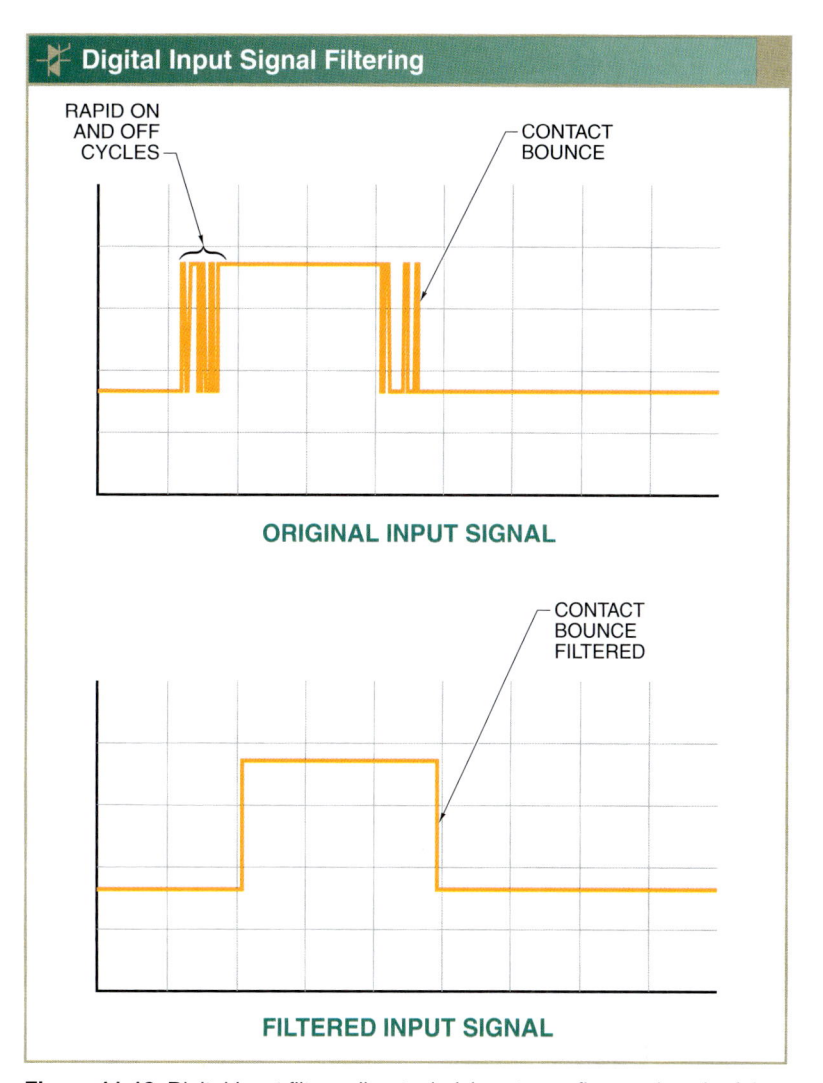

Figure 11-13. Digital input filters allow technicians to configure a length of time input signals must be ON or OFF before the signal is recognized by the PLC.

Troubleshooting Input Sections or Modules

Troubleshooting the input section or module of a PLC requires knowledge of how the programmed circuit should be operating and how the PLC input section or module works. The PLC program can be viewed on the monitor or printed out. Using a programming diagram when troubleshooting a PLC input section or module aids in understanding how the input devices are connected and which loads are being controlled. Manufacturers also provide troubleshooting charts to help isolate input section or module problems. **See Figure 11-14.**

Troubleshooting Input Modules

When an Input Circuit LED Is	And the Input Device Is	And	Probable Cause	Recommended Action
On	On/Closed/Activated	Input device will not turn off.	Device is shorted or damaged.	Verify device operation. Replace device.
		Program operates as though it is off.	Input circuit is damaged.	Verify proper wiring. Try other input circuit. Replace module.
			Input is forced off in program.	Check the FORCED I/O or FORCE LED on processor and remove forces.
		Program operates as though it is off and/or the input circuit will not turn on.	Input device off-state leakage current exceeds input circuit specification.	Check device and input circuit specifications. Use load resistor to bleed-off current.
			Input device is shorted or damaged.	Verify device operation. Replace device.
			Input circuit is damaged.	Verify proper wiring. Try other input circuit. Replace module.
Off	Off/Open/Deactivated	Program operates as though it is on.	Input is forced on in program.	Check processor FORCED I/O or FORCE LED and remove forces. Verify proper wiring. Try other input circuit. Replace module.
			Input circuit is damaged.	Verify proper wiring. Try other input circuit. Replace module.

Figure 11-14. When troubleshooting PLC input sections, programming diagrams and manufacturer troubleshooting charts indicate how input devices are connected and which output components are being controlled.

Manufacturer troubleshooting charts are very helpful in isolating a problem in a circuit. When using a manufacturer troubleshooting chart, the manufacturer assumes the troubleshooter/technician knows the following:

- how to take voltage measurements in the system as suggested
- how to take in-line current measurements in order to determine if there is a leakage current problem with an input device
- how to wire inputs into the PLC (the difference between wiring a current sink and current source input device)
- what a load (bleeder) resistor is and how to connect it into the circuit to solve a leakage current problem
- what the PLC status LEDs mean

Testing Input Sections or Modules

Signals and information are sent to a programmable controller using input devices such as pushbuttons, limit switches, level switches, and pressure switches. The input devices are connected to the input section or module of the programmable controller. When a PLC does not receive the proper information from input devices (opening and closing contacts as designed) or the input section or module is not operating correctly, the controlled system cannot operate properly. **See Figure 11-15.** To troubleshoot an input section or module, apply the following procedure:

1. Locate the PLC input section or module to be tested. Set the test instrument to the highest possible voltage setting and voltage type (VAC or VDC) in the circuit.

2. Measure the input supply voltage or external power supply to ensure that there is power supplied to the input devices. Test the main power supply of the PLC when there is no sensor power present.

3. Measure the voltage from the control switch. Connect the meter directly to the same terminal screw to which the input device is connected. The voltmeter should read the supply voltage when the

control switch is closed. The voltmeter should read the full supply voltage when the control device uses mechanical contacts. The voltmeter should read nearly the full supply voltage when the control device is solid-state. Full supply voltage is not read because 0.3 V to 1.5 V is dropped across the solid-state control device. The voltmeter should read zero or little voltage when the control switch is open.

4. Monitor the status indicators on the input section or module. The status indicators should illuminate when the voltmeter indicates the presence of supply voltage.

5. Monitor the input device symbol on the programming terminal monitor. The symbol should be highlighted when the instruction is true even when an XIO instruction is used and no voltage is present.

Testing Input Modules

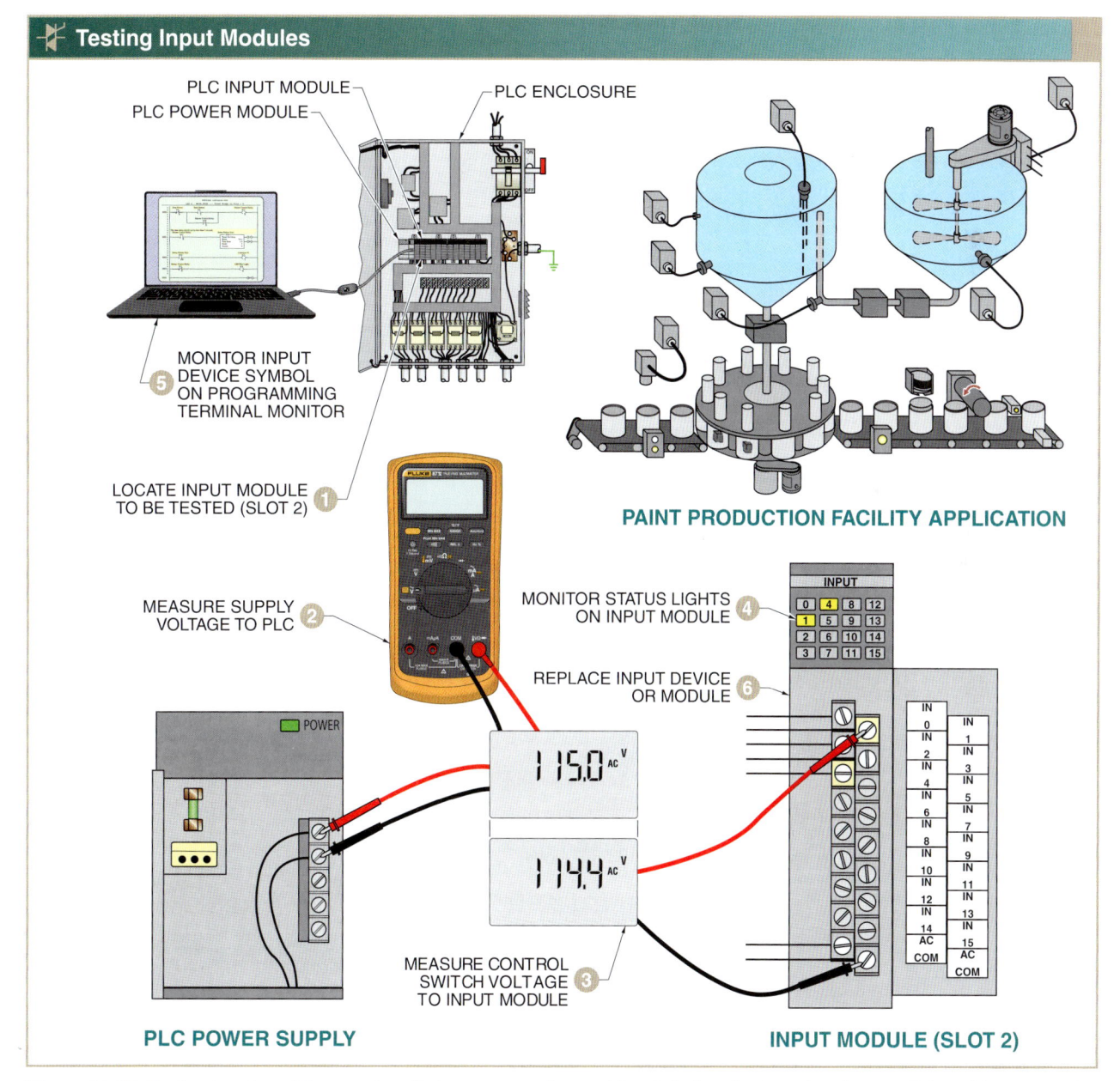

Figure 11-15. Testing input modules requires that power supplies and input devices be tested and status lights of input modules and symbols on computer monitors be checked.

6. Replace the control device if the control device does not deliver the proper voltage. Replace the input section or module if the control device delivers the correct voltage, but the status indicator does not illuminate.

Monitoring Input Devices

Input devices such as pushbuttons, limit switches, pressure switches, and temperature switches are connected to the input section or module of a PLC. Input devices send information and data concerning circuit and process conditions to the PLC. The PLC processor receives the information from the input devices and executes the program. All input devices and the PLC program must operate correctly for the circuit to operate properly. A major advantage to using a PLC to control a circuit/system is that all the PLC inputs and outputs can be monitored. The ability to monitor the inputs of a circuit is an advantage when troubleshooting a problem. **See Figure 11-16.** To troubleshoot an input device, apply the following procedure:

1. Place the PLC in TEST mode. Having a PLC in TEST mode prevents the output components from turning ON. Output components are only turned ON when a controller is placed in RUN mode.

2. Monitor the input devices using the input status indicators (located on each input section or module), the programming terminal monitor, or the data file. A *data file* is a group of data values (inputs, timers, counters, and outputs) that is displayed as a group and whose status may be monitored.

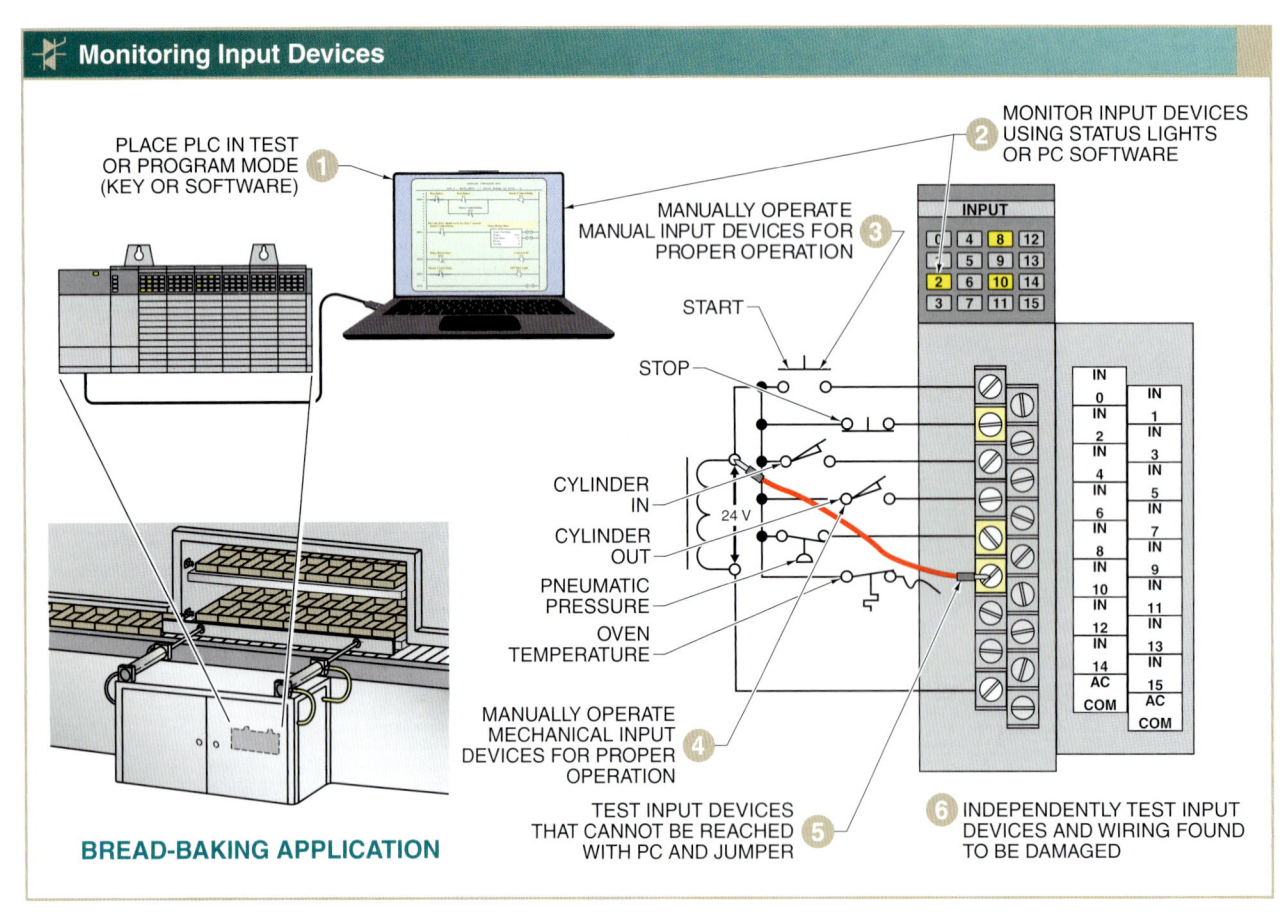

Figure 11-16. All input devices and the PLC program must operate correctly in order for an automated circuit to operate properly.

3. Manually operate input devices (such as pushbuttons or selector switches) to test for proper operation. When a normally open input device is closed, the input status indicator lights located on an input section or module light up and the input symbol is highlighted in the control circuit on the monitor screen. The bit status on the programming terminal monitor screen should be set to 1 (a high), which indicates a presence of voltage. When a normally closed input device is open, the input status indicator light located on the input section or module turns OFF and the input symbol is no longer highlighted in the control circuit on the monitor screen. The bit status on the programming terminal monitor screen should be set to 0 (a low), which indicates an absence of voltage. Test each manual input device for proper operation.

4. Manually operate mechanical input devices (such as limit switches) if they are located in a safe and convenient place. However, never reach into a machine when manually operating an input device. Always use a nonconductive device. Automatic input devices such as pressure and temperature switches cannot be manually operated.

5. Test input devices that cannot be reached by using the PLC display screen and data files or by disconnecting the input device from the PLC and using a jumper to simulate the input device opening and closing. Test each mechanical and automatic input device for proper operation with the PLC program.

6. Test input devices and wiring found to be damaged. When a manual, mechanical, or automatic input device terminal shows proper operation in the PLC program when a jumper is used to simulate an energized input but does not show proper operation when the actual input device is reconnected to the assigned input terminal, the input device or wiring must be considered damaged and independently tested.

Input Leakage Current Problems

There are two basic types of switches, mechanical and solid-state. When mechanical switches open, the two switch contacts physically move and separate. The resistance between the contacts is so high that no current flows through the open contacts.

In a solid-state switch, there are no moving contacts. Instead, a solid material is used that can be switched from a very high resistance to a very low resistance. There are numerous advantages to solid-state switching, such as an extremely long operation life, much faster operating speed (ON/OFF), and no contact bouncing. However, there are a few disadvantages to solid-state switching. Because solid-state "contacts" never go to 0 Ω of resistance when closed, there is always a voltage drop, which produces heat at the contacts. Solid-state devices that control loads can be cooled by heat sinks and/or forced air currents. Proper heat sinking and cooling is required with solid-state devices that control loads to eliminate any potential heat problems. **See Figure 11-17.**

CODE CONNECT

As the number of current-carrying conductors increases in a raceway or cable, the amount of heat increases. If the number of current-carrying conductors exceeds three, the NEC® requires that the ampacity of the conductors be derated/adjusted to counter the effects of increased heating. Table 310.15(C)(1) provides the necessary derating percentages based on the number of conductors in a raceway or cable.

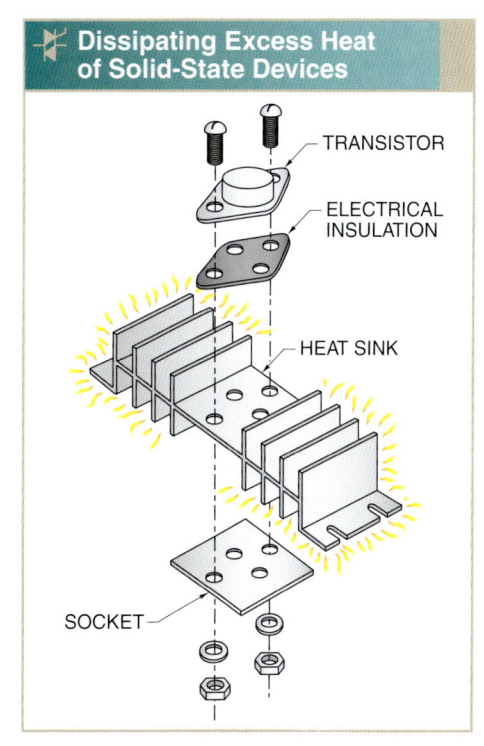

Figure 11-17. Proper heat sinking and cooling is required with solid-state devices to eliminate any potential heat problems.

A second problem is that when contacts are "open" there is always some current that will still flow through the high-resistance solid-state material. The current that flows through an open solid-state switch is called leakage current. For most loads, leakage current is not a problem because the current is not high enough to energize the load. Loads such as solenoids, incandescent lamps, motor starter coils, contactor coils, and motors will never be energized by the leakage current from a solid-state switch.

However, other loads, such as PLC input circuitry, require a very small amount of current to turn ON and can be affected by solid-state device leakage current. PLC input terminals only require a small amount of current to signal the presence of an input signal from an input device. Thus, leakage current from a solid-state switch such as a proximity switch can send a false signal to a PLC input section or module. **See Figure 11-18.**

To determine if there is a leakage current problem, the OFF-state leakage current can be measured and that value can be compared to the specified minimum operating current of the PLC input section or module. When there is a leakage current problem, the problem can be solved by adding a load or bleeder resistor. A bleeder resistor is connected in parallel with a PLC input.

The bleeder resistor acts as an additional lower resistance load, which allows the leakage current to flow through the lower resistance path. Typically, a 5 kΩ to 20 kΩ resistor is used to solve a leakage current problem. **See Figure 11-19.** To test the amount of leakage current with an in-line ammeter, apply the following procedure:

1. Turn the PLC power OFF. Verify that the power is OFF by taking a voltage measurement at the PLC power supplied section terminals.

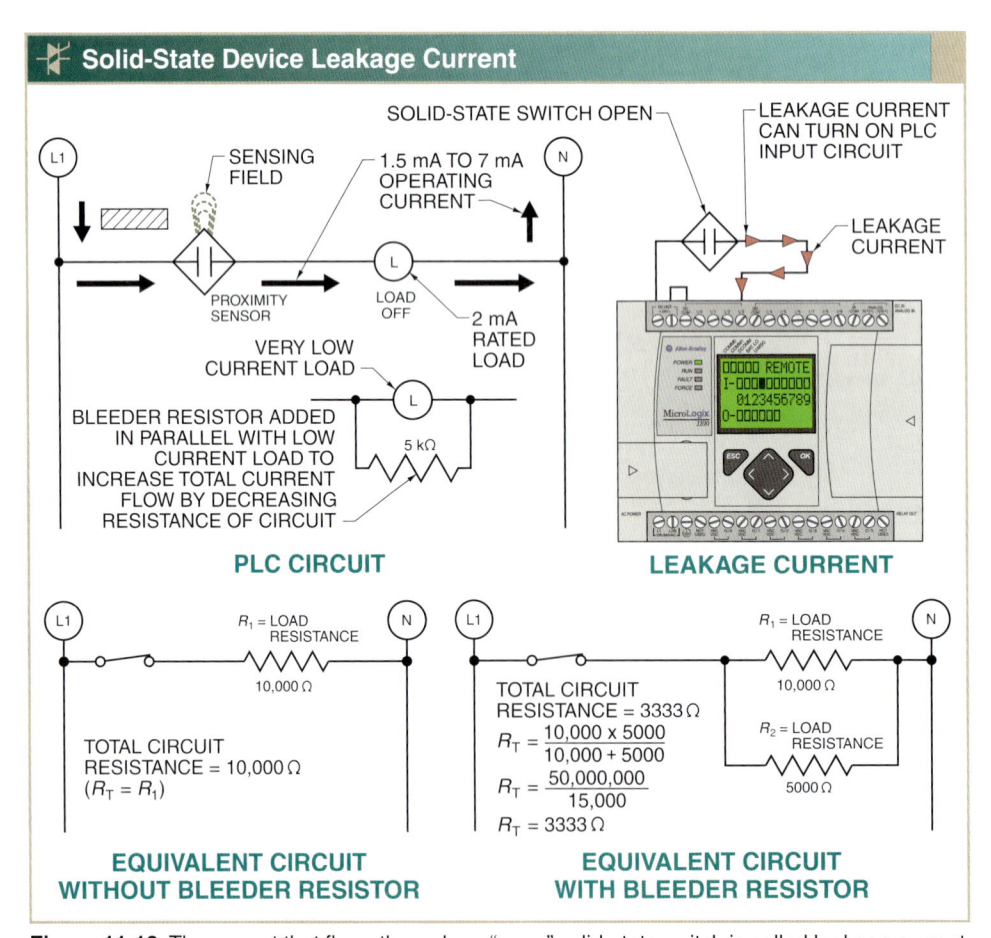

Figure 11-18. The current that flows through an "open" solid-state switch is called leakage current. Leakage current can turn on the input circuitry of a PLC, affecting system operation.

2. Set the ammeter to take an in-line current measurement. Although leakage currents will be very low (a few mA or µA), set the meter to the highest possible current rating to start with so the meter is not damaged when the switch is ON and full operating current flows. Use only an ammeter that has fused current jacks.

3. Connect the in-line ammeter in series with the solid-state input switch connected to the PLC input section or module.

4. Turn the PLC power ON.

5. Measure the current when the switch is in the ON and OFF operating states. The amount of current measured when the switch is "open" is the leakage current. The amount of current measured when the switch is "closed" is the operating current.

6. Turn the PLC power OFF. Verify that the power is OFF by taking a voltage measurement at the PLC input (sensor) power supply terminals.

7. Remove the in-line ammeter from the circuit and reconnect the solid-state input device to the PLC input section or module.

8. Turn PLC power ON.

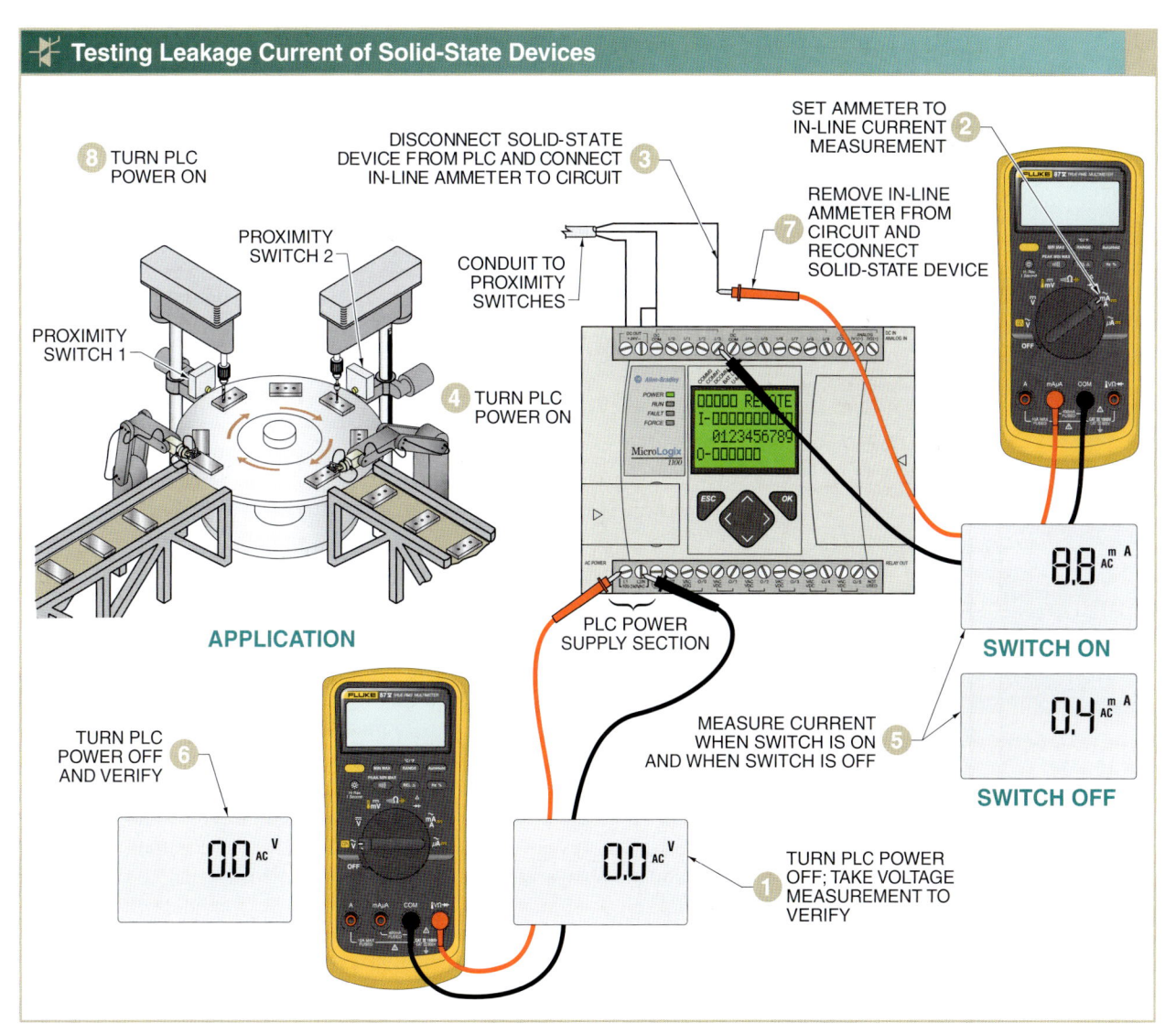

Figure 11-19. Testing the leakage current of solid-state devices requires that the OFF-state leakage current measurement be compared to the specified minimum operating current of the PLC input module.

TROUBLESHOOTING PLC OUTPUT SECTIONS OR MODULES

All PLCs have an output section or module that circuit output components (loads) are connected to. The output components perform all work required, such as producing light, heat, sound, mechanical linear force (solenoids), and mechanical rotating force (motors).

Lower-powered loads, such as small wattage lamps and solenoids can be connected directly to the PLC output section or module. Higher-powered loads, such as higher-wattage lamps (or multiple lamps on the same output), heating elements, and motors are connected through an interface (such as a relay, lighting contactor, heating contactor, motor starter, or motor drive).

When a PLC-controlled system is not operating properly and the input circuits have been checked, the PLC output section or modules and output components must be checked.

When a PLC-controlled system is not operating properly and the input section or module has been checked, the PLC output section or module and output components need to be checked. Although it is more likely that an output will fail before an input will, the outputs cannot properly operate unless the inputs controlling them are operating. A quick check of the PLC displayed program and input LED status lights will indicate if the inputs controlling the outputs are correct.

Troubleshooting a PLC output section or module and correcting a problem requires the troubleshooter/technician to know the following:

- how the parts of the output section or module work
- what to look for when troubleshooting an output section or module—this includes an understanding of the PLC status LEDs, the output status LEDs, how energized and de-energized outputs are displayed on the computer monitor, and how to use the PLC program to monitor and/or force outputs to change their operating condition
- how to use test instruments to take measurements at the output section or module and what the measurements mean
- how to monitor circuit outputs using the PLC displayed program and data files
- how to correct output problems, like using interfaces (relays, starters, etc.) to help protect the PLC output contacts (mechanical or solid-state)

Output Circuit Operation

Output components are connected to the output section or module of the PLC. The output signal flows from the PLC processor control circuits out to the output drivers in the opposite direction of the input signal. **See Figure 11-20.**

From the PLC processor control circuits the output signal is sent to the PLC logic circuit. The logic circuit processes the signal so that it can be sent to the PLC opto-isolation circuit and the output status light (LED) circuit, which lights up on the PLC output section or module when the output signal is received. Any time the PLC output LED status light is ON, the output device connected to that output terminal should also be ON.

PLC Output Circuit Operation

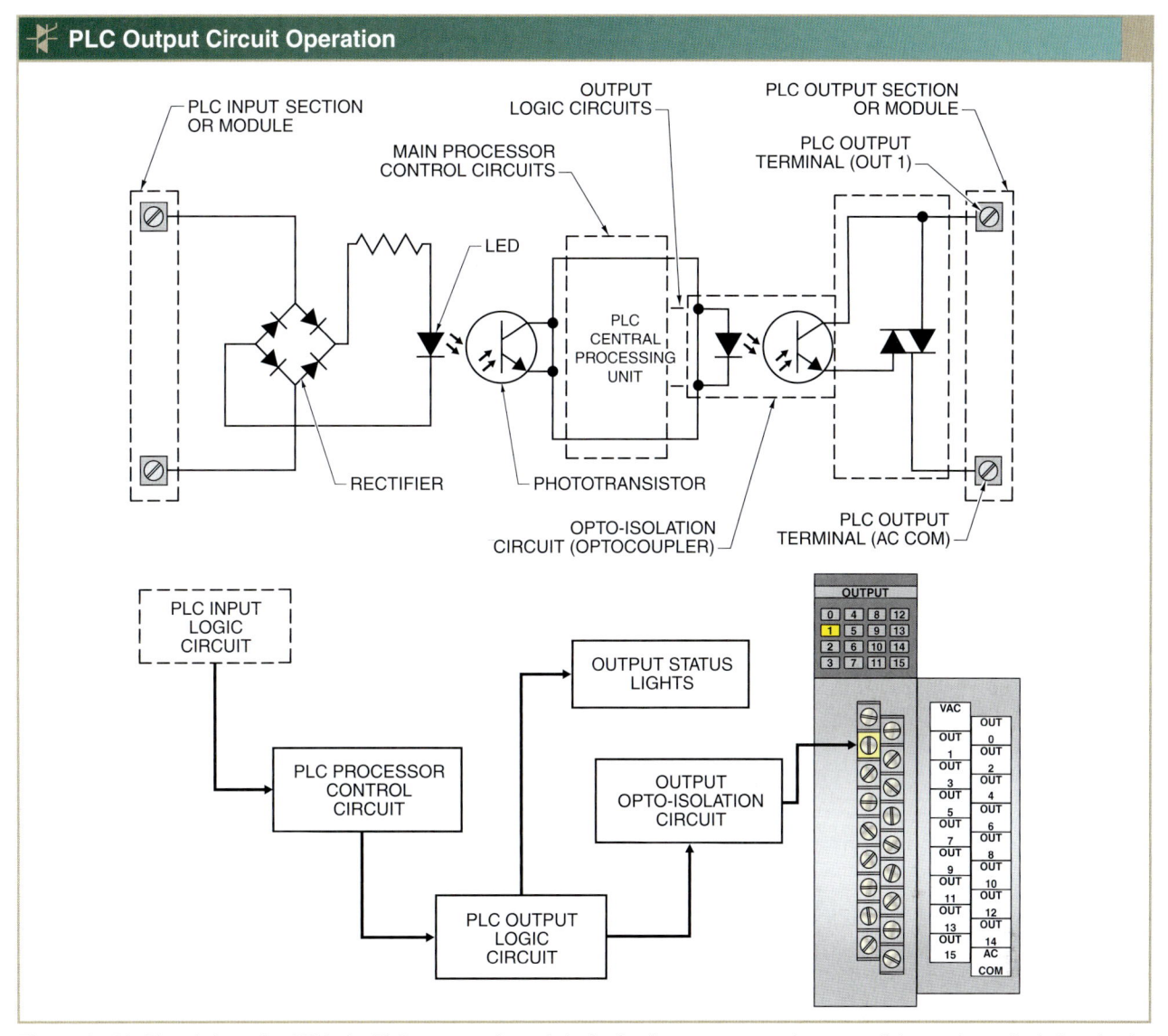

Figure 11-20. Signals from the CPU of a PLC are sent through logic circuits to output section status lights and opto-isolation circuits before being sent to output terminals.

From the logic circuit the output signal is sent to the opto-isolation circuit. The opto-isolation circuit takes the outgoing signal and converts the signal into a transmitted light signal that is sent to a light-sensitive receiver circuit for processing. The opto-isolation circuit sends a light signal that helps protect the PLC circuits from incoming transients or other unwanted electrical signals that can be produced when outputs short or have high inrush currents.

From the opto-isolation circuit the output signal is sent to the output drivers. The output drivers turn the corresponding output terminals ON and OFF. The outputs may be any of the following:
- mechanical contacts (relays) that can switch either AC or DC
- triacs that can switch low-current AC loads
- transistors that can switch low-current DC loads
- other types of outputs (analog, etc.)

⚠ WARNING

If a PLC processor is returned to the factory, all programs in the PLC memory will be lost and communication configurations will be returned to default settings.

Troubleshooting Output Sections or Modules

Troubleshooting the output section or module of a PLC requires knowledge of how the programmed circuit should be operating and how the PLC output section or module works. The programmed circuit can be viewed on the monitor and/or printed out. Using the programmed circuit when troubleshooting the PLC output section or module is necessary to understand when the output components are to be energized based on input device conditions. Manufacturers also provide troubleshooting charts to help isolate output section or module problems. **See Figure 11-21. See Appendix.**

Manufacturer troubleshooting charts are very helpful in isolating a problem in a circuit. When using a manufacturer troubleshooting chart, the manufacturer assumes the troubleshooter/technician knows the following:

- how to take voltage measurements in the system as suggested
- how to take in-line current measurements in order to determine if there is a leakage current problem with an output component
- how to wire outputs out of the PLC and the difference between wiring a current sink and current source output component
- what a load (bleeder) resistor is and how to connect the resistor into a circuit to solve a leakage current problem
- what the PLC status LEDs mean

PLC TIPS

PLCs and PLC input devices can operate at low voltages (12 VDC or 24 VDC). PLC output components typically operate at higher voltages (115 VAC, 230 VAC, or 460 VAC) that can cause an electrical shock. Always wear required PPE when working around any electrical circuit (relays, contactors, or motor starters) that has high voltages.

Troubleshooting Output Modules

When an Output Circuit LED Is	And the Output Component Is	And	Probable Cause	Recommended Action
On	On/Energized	Program indicates that the output circuit is off or the output circuit will not turn off.	Programming problem	Check for duplicate outputs and addresses using the search function. If using subroutines, outputs are left in their last state when not executing subroutines. Use the force function to force output off. If this does not force the output off, output circuit is...
Off	Off/De-energized	Program indicates that the output circuit is on or the output circuit will not turn on.	Output is forced off in program	Check processor FORCED I/O or FORCED LED and remove forces.
			Output circuit is damaged	Use the force function to force the output on. If this forces the output on, then there is a logic/programming problem. If this does not force the output on, the output circuit is damaged. Try other output circuit. Replace the module.

Figure 11-21. When troubleshooting PLC output sections, programming diagrams and manufacturer troubleshooting charts help the technician understand how the input devices are connected and which output components are being controlled.

Testing Output Sections or Modules

A PLC turns the output components (loads) in a circuit ON and OFF according to the program. The output components are connected to the output section or module of the programmable controller. When an automated system is not producing quality work, the problem may lie in the output section or module, output component, or controller. **See Figure 11-22.** To troubleshoot an output section or module, apply the following procedure:

1. Locate the PLC output section or module to be tested.
2. Measure the supply voltage at the output section or module to ensure that there is power supplied to the output components. Test the main power supply of the PLC when there is no power.
3. Measure the voltage to the output component. Connect the meter directly to the same terminal screw to which the output component is connected. The voltmeter should read the supply voltage when the output terminal is energized. The voltmeter should read the full supply voltage when the output terminal uses mechanical contacts. The voltmeter should read nearly the full supply voltage when the output terminal is solid-state. Full supply voltage is not read because 0.3 V to 1.5 V is dropped across the solid-state control device. The voltmeter should read zero or little voltage when the output is not energized.
4. Monitor the status indicator lights on the output section or module. The status indicator lights illuminate when the voltmeter indicates the presence of supply voltage at a terminal.
5. Monitor the output component symbol on the programming terminal monitor. The symbol is highlighted when voltage is present at the output terminal.
6. If the PLC output status lamp indicates the output is ON and the voltmeter does not indicate the correct voltage, replace the output module.

PLC TIPS
In PLC systems high in electrical noise and surges, technicians can install panel mount filters to provide noise and surge protection. Panel mount filters monitor high-frequency electrical disturbances and diminish frequencies that may disrupt a PLC system, which prolongs system operation.

Monitoring Output Components

Output components such as motor starters, solenoids, contactors, and lights are connected to the output section or modules of a programmable controller. An output component performs the work required for an application. The processor energizes and de-energizes the output components according to the program. All output components must operate correctly for a circuit to operate properly. A major advantage to using a PLC to control a circuit or system is that all PLC inputs and outputs can be monitored. The ability to monitor circuit input devices and output components aids in troubleshooting a problem. **See Figure 11-23.** To troubleshoot an output component, apply the following procedure:

1. Monitor the output components using the output status indicators (located on each output section or module), the programming terminal monitor, or the data file.
2. Activate the input device that controls the first output component. Check the program displayed on the monitor screen to determine which input device activates which output device.
3. Select the next output component and test the component when the status indicator light and associated bit status match. Continue testing each output component until all components have been tested.
4. Troubleshoot the input device and output component when the status indicator and output component status do not match.

⚠ DANGER
Output components can turn ON when a controller is in RUN mode.

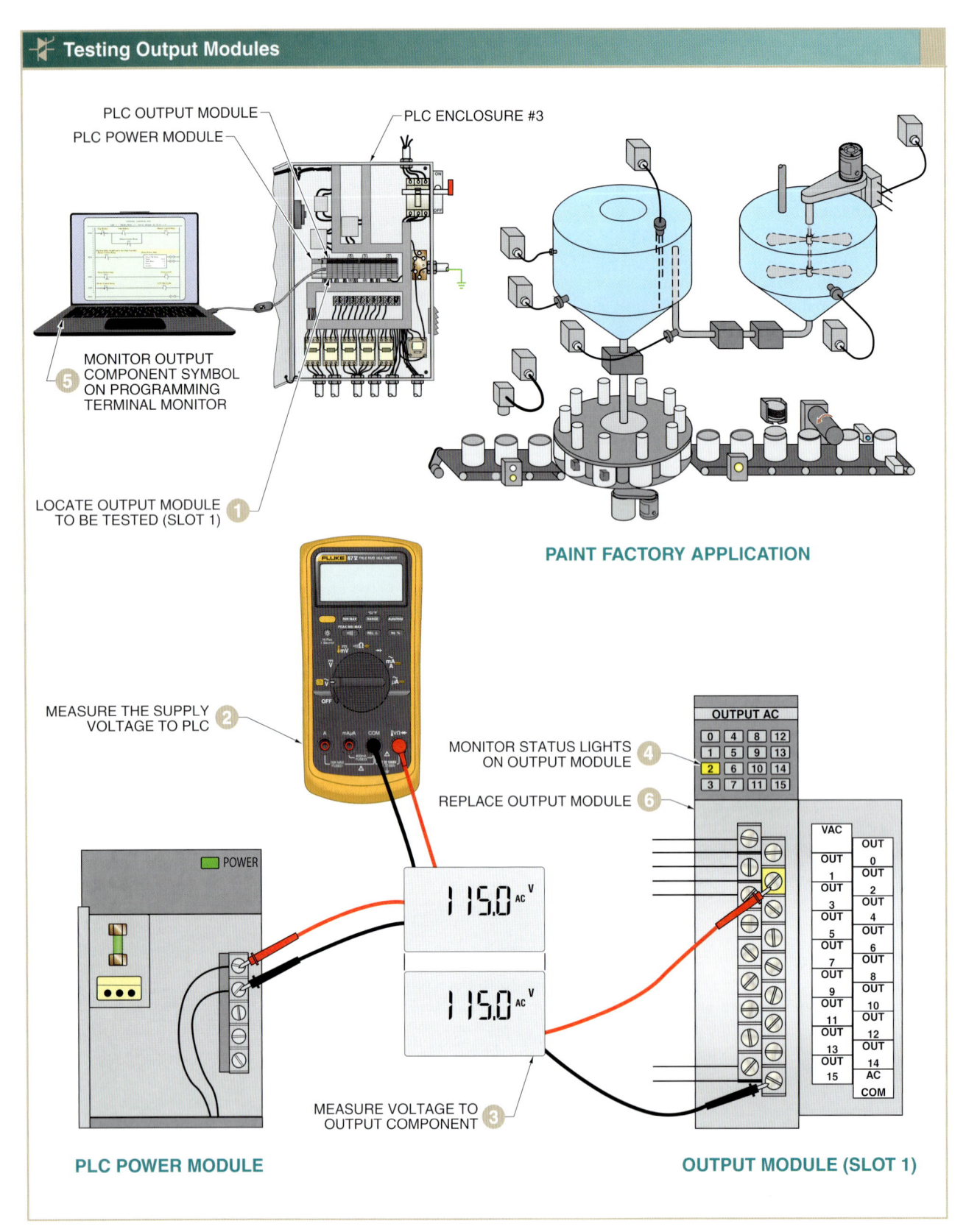

Figure 11-22. Testing output modules requires that power supplies and output components be tested and status lights of output modules and symbols on computer monitors be checked.

Monitoring Output Components

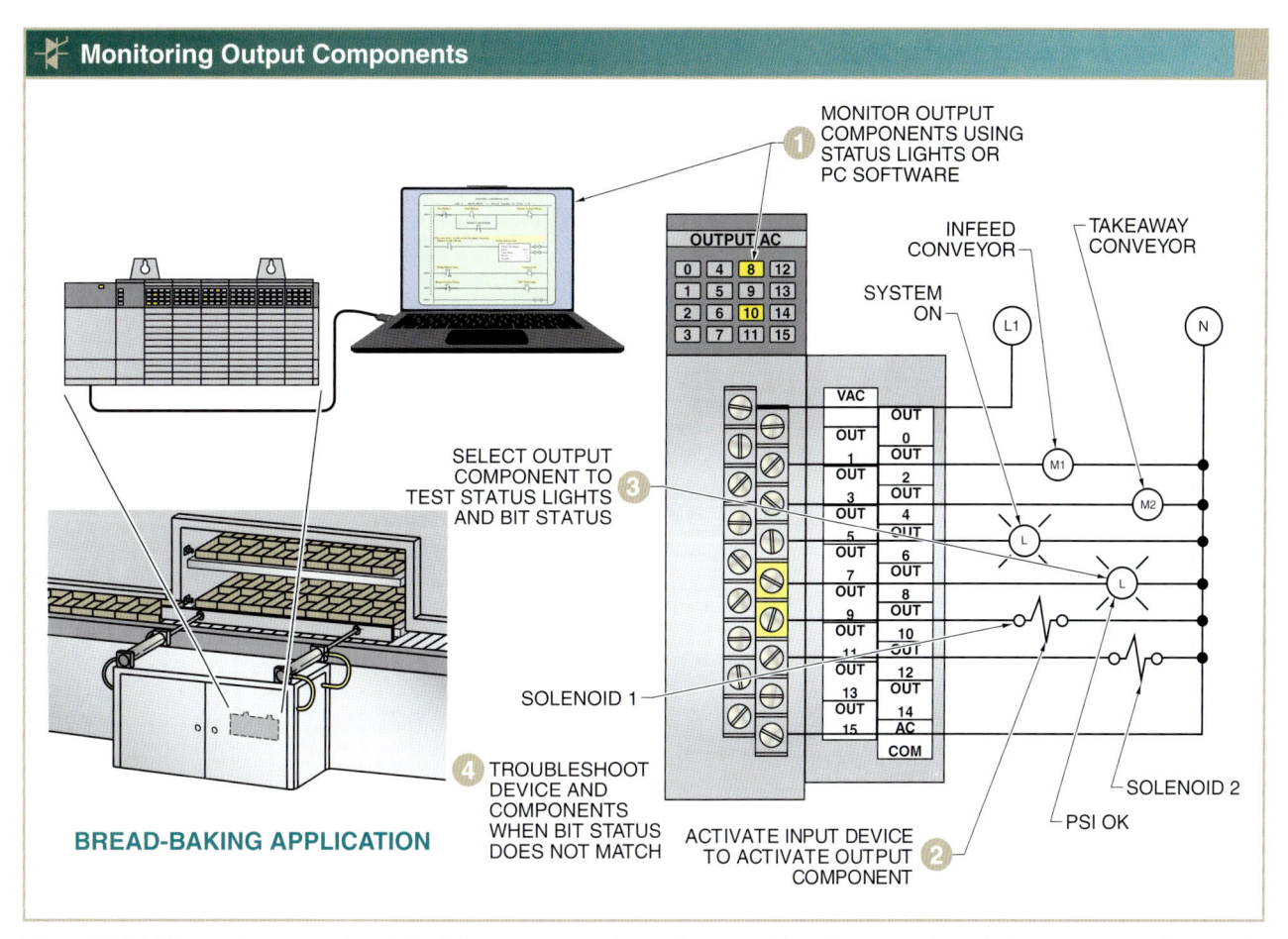

Figure 11-23. All output components and the PLC program must operate correctly in order for an automated system to operate properly.

Chapter Review 11

Name _____ Date _____

True-False

T F **1.** PLC hardware problems require knowledge about troubleshooting incoming power, PLC input sections or modules, input devices, PLC output sections or modules, and output components.

T F **2.** Measuring the circuit resistance is the best way of determining the load on a circuit.

T F **3.** In general, circuit current should not exceed 25% of the fuses or circuit breaker rating for the best circuit performance.

T F **4.** The first step used to test fuses with an ohmmeter is to turn power to the circuit OFF and lock and tag out the disconnect.

T F **5.** Circuit breakers perform the same function as fuses and are basically tested the same way.

T F **6.** When loads are connected to a PLC that exceed the current rating of the PLC output power supply, the voltage output of a PLC will increase.

T F **7.** Digital input devices are the most common type of input device because they represent the largest group of mechanical and solid-state switches.

T F **8.** Opto-isolating circuits only allow signals to flow in one direction.

T F **9.** To determine if there is a leakage current problem, measure the OFF-state leakage current and compare that value to the specified minimum operating current of the PLC input section or module.

T F **10.** Timer instruction requires knowledge of which test instruments to use, how to use test instruments, where to connect test instruments, and what each measured value means.

T F **11.** Most control transformers have a dual-voltage primary side which allows it to be configured for either one of two voltages.

T F **12.** Grounding the input section or module of a PLC requires knowledge of how the programmed circuit should be operating and how the PLC input section or module works.

T F **13.** In a solid-state switch, there are no moving contacts.

T F **14.** All PLCs have output terminals that circuit components (loads) are connected to.

T F **15.** A PLC only turns output components (loads) ON and OFF according to analog signals.

Completion

_____ 1. A(n) ___ is an overcurrent condition in which the current of a circuit leaves the normal current-carrying path by going around the load back to the power source via ground.

_____ 2. A(n) ___ is an overcurrent condition that occurs when circuit current rises above the normal current level at which the PLC and/or circuit is designed to operate.

_____ 3. A(n) ___ is an overcurrent protection device that includes a fusible link that melts.

_____ 4. A(n) ___ is a reusable overcurrent protection device that opens a circuit automatically.

_____ 5. ___ is the rapid making and breaking of mechanical switch contacts.

_____ 6. ___ is the unwanted electrical noise generated by electrical and electronic equipment, such as lighting ballasts and transformers.

_____ 7. When voltage measurements identify a blown fuse, the fuse is removed using a(n) ___.

_____ 8. When a PLC supplies a sensor voltage, the input (sensor) power supply will have both a voltage rating and a(n) ___ rating.

_____ 9. A(n) ___ circuit sends a light signal that helps protect the PLC circuits from incoming transients or other unwanted electrical signals.

_____ 10. All electrical devices and components must be powered by a(n) ___ that can deliver the correct voltage level with enough current capacity.

Multiple Choice

_____ 1. Two categories of power supply problems are loss of power and ___.
   A. high resistance
   B. low voltage
   C. default current
   D. overvoltage

_____ 2. Blown fuses and ___ are the most common power supply problems.
   A. open circuit breakers
   B. analog signals
   C. disable commands
   D. closed jumper links

_____ 3. Two basic types of switches are mechanical switches and ___ switches.
   A. source
   B. transistor
   C. time base
   D. solid-state

_____ 4. When the power rating of a transformer is exceeded by placing too great a load on the transformer, the voltage will start to ___ and circuit problems will develop.
   A. improve
   B. increase
   C. decrease
   D. none of the above

_____ 5. Fuses and ___ are used in electrical circuits to protect the circuits from short circuits and overloads.
   A. circuit breakers
   B. parallel-connected devices
   C. jumper links
   D. solenoids

Troubleshooting Solid-State Sensors

PLC Lab 11.1

BILL OF MATERIALS
- TECO PLR or Micrologix™ 1100 PLC
- Proximity sensor/photoelectric switch
- Motor starter/contactor
- Load (optional)

Proximity and photoelectric sensors are commonly used in PLC circuits. Both types of sensors are typically 3-wire designs. Two of the wires must always be connected to a power supply to develop a "sensing" field, regardless of whether the switch is open or closed or the load is ON or OFF. All solid-state sensors leak some current, but it is generally measured in microamps and is not significant. The current that flows through the sensor when the load is ON is called the operating current. The total current is the sum of the operating current and the leakage current. Proximity and photoelectric switches involve several types of current that must be understood when installing and troubleshooting the switches. For this lab, a proximity switch is used to control a motor starter, and the operating current, the leakage current, and the total current need to be measured for correct operation.

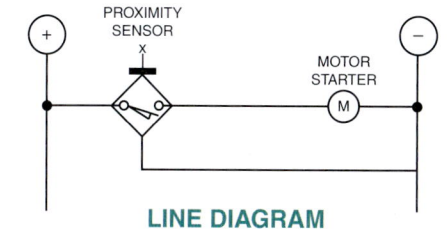

LINE DIAGRAM

T TECO Procedures

1. Based on the PLC system description and line diagram, draw the TECO programming diagram and FBD logic with the proper contacts and descriptions in the Programming Grid and Programming Area.

2. Create a new ladder logic or FBD program electronically.

3. Use the Edit Contact window to denote an input or output number and the Symbol button to add a description after placing the input or output in the Programming Grid. For an FBD, use the Comment icon to add the description.

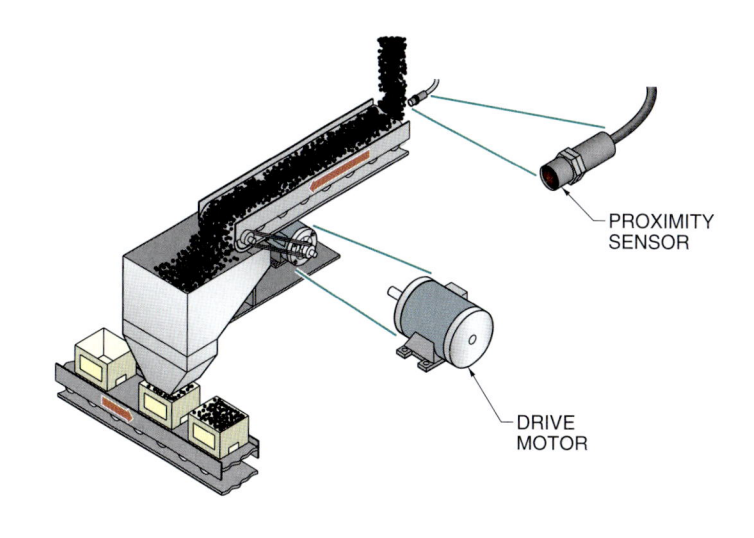

TECO PROGRAMMING DIAGRAM

FBD PROGRAMMING AREA

473

4. Run the program in Simulation Mode to verify proper operation. Use the Input Status tool window to activate the input and control the output to simulate circuit operation.

5. Use the wiring diagram to wire the input devices and output components to the PLC.

6. Turn power to the PLC ON and link the com port.

7. Write the program to the SG2-12HR-D to download the program logic to the PLC.

8. Place the SG2-12HR-D in Run Mode. Activate the proximity switch and monitor the I/Os on the LCD display.

9. Stop the program and use the Save As function to save and place the lab application in the required folder location.

10. Turn power OFF and set a DMM to measure milliamperage. Turn power ON, run the program, and measure the operating current of the proximity switch when the switch is actuated and the load is ON. What is the operating current in milliamps?

_____

11. Measure the leakage current of the proximity switch when the switch is not actuated. What is the leakage current in milliamps?

_____

12. Turn power OFF and connect the DMM leads to measure the total current of the proximity switch. Turn power ON and run the program. What is the total current in milliamps?

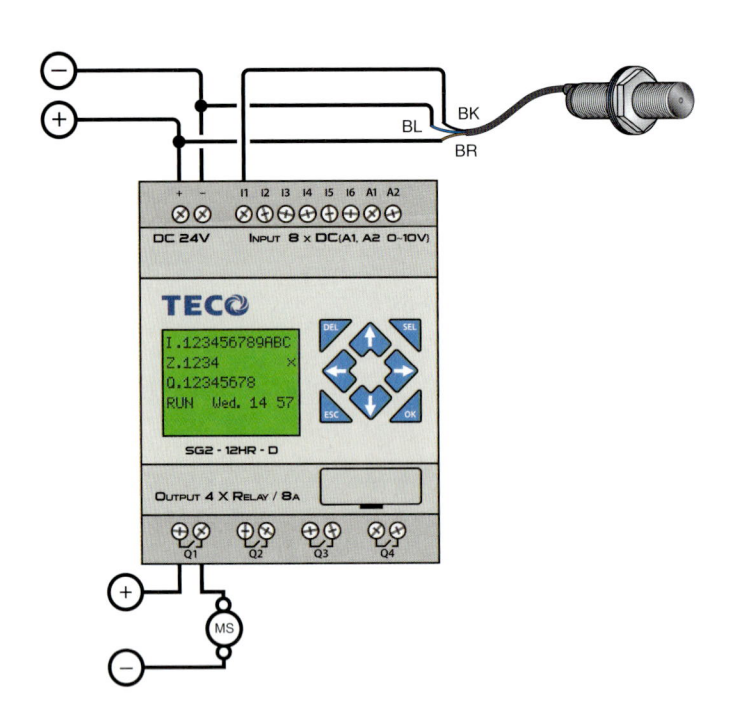

_____

TECO WIRING DIAGRAM

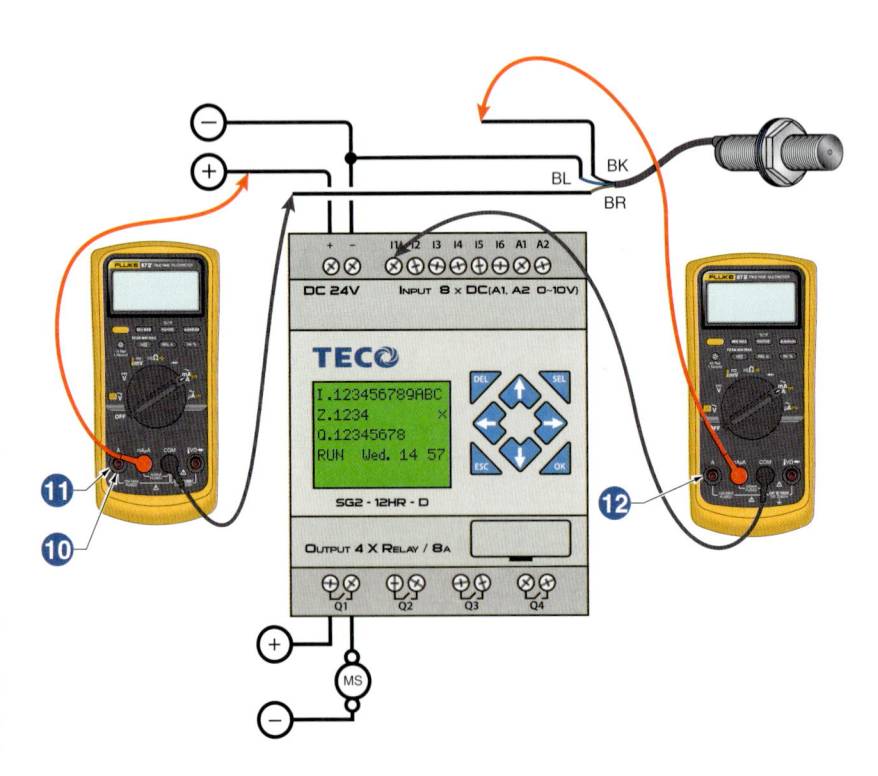

Allen-Bradley® Procedures

1. Based on the PLC system description and line diagram, draw the ladder logic for RSLogix™ in the AB programming area.

2. Open the RSLogix™ Micro Starter Lite program and create a new programming diagram.

3. Use the AB programming diagram to add the proper address and description in the LAD 2 programming window.

4. Verify that the programming diagram has no errors and save the project.

5. Use the wiring diagram to wire the input devices and output components to the PLC. Once all connections have been made, turn power to the PLC on.

6. Use RSLinx™ to establish communication between the PC and the Micrologix™ 1100.

7. Download the program logic to the PLC.

8. Place the PLC in Run Mode.

9. Activate input(s) and monitor them to determine if the output(s) are operating correctly.

0000

0001 ─────────────────────────⟨END⟩

AB PROGRAMMING DIAGRAM

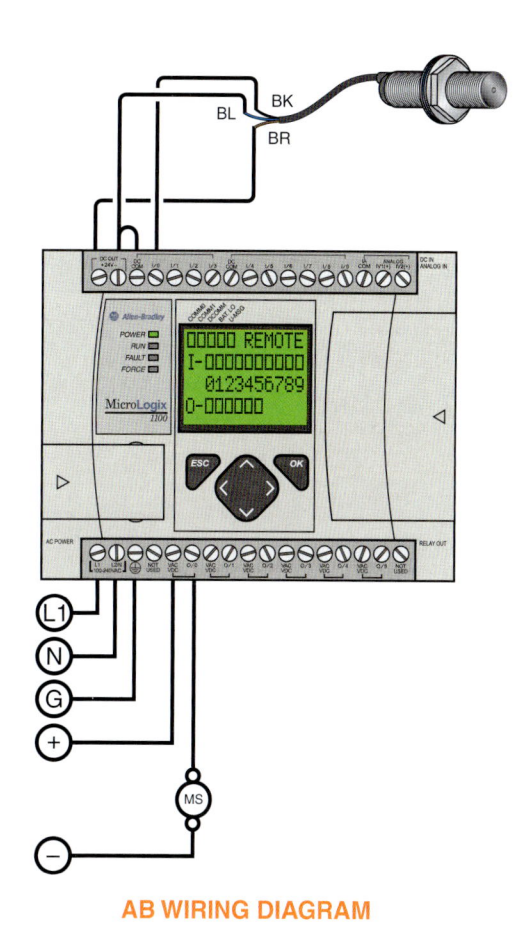

AB WIRING DIAGRAM

10. Turn power OFF and set a DMM to measure milliamperage. Turn power ON, run the program, and measure the operating current of the proximity switch when the switch is actuated and the load is ON. What is the operating current in milliamps?

_____

11. Measure the leakage current of the proximity switch when the switch is not actuated. What is the leakage current in milliamps?

_____

12. Turn power OFF and connect the DMM leads to measure the total current of the proximity switch. Turn power ON and run the program. What is the total current in milliamps?

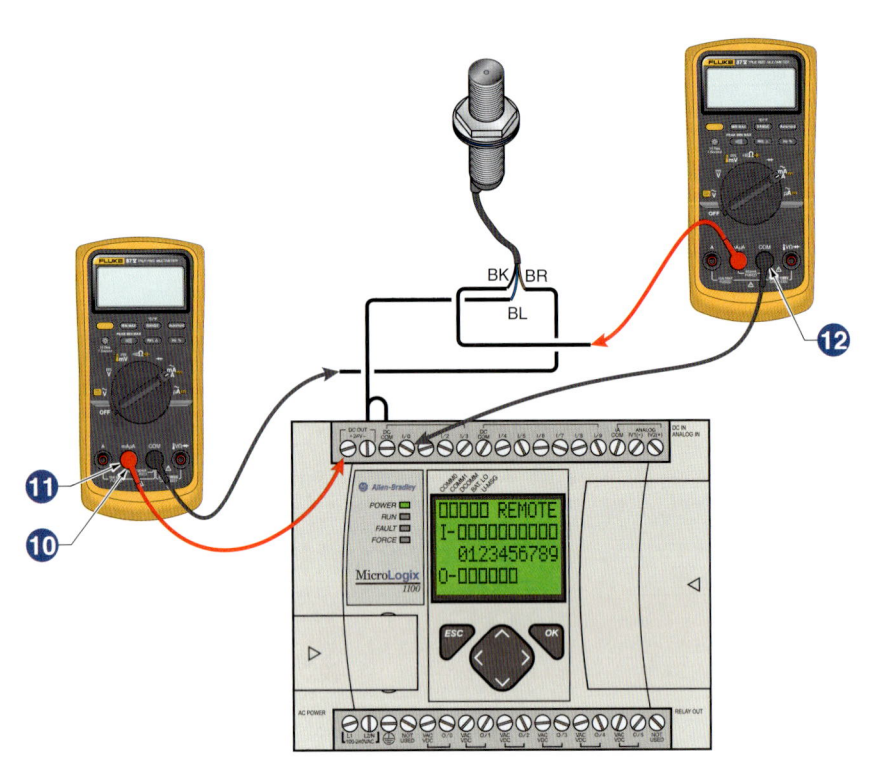

_____

Troubleshooting Current Sink Solid-State Switches

PLC Lab 11.2

BILL OF MATERIALS
- Micrologix™ 1100 PLC
- Photoelectric Sensor
- Two indicator lights

Some proximity switches and photoelectric switches have both current source and current sink outputs. Some PLCs have input sections that can be connected to both. The Micrologix™ 1100 PLC can have input terminals 0, 1, 2, and 3 connected through one DC common for either sink or source, and terminals 4, 5, 6, 7, 8, and 9 with another DC common connected for the opposite (sink or source). For this lab, a photoelectric switch is connected to a Micrologix™ 1100 PLC. The switch is connected as both a sink input to control an indicator light and as a source input to control another indicator light. The operating current, the leakage current, and the total current need to be measured for correct operation.

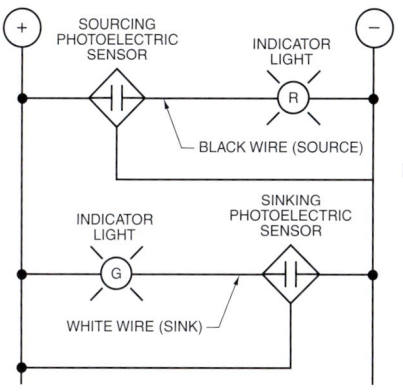

LINE DIAGRAM

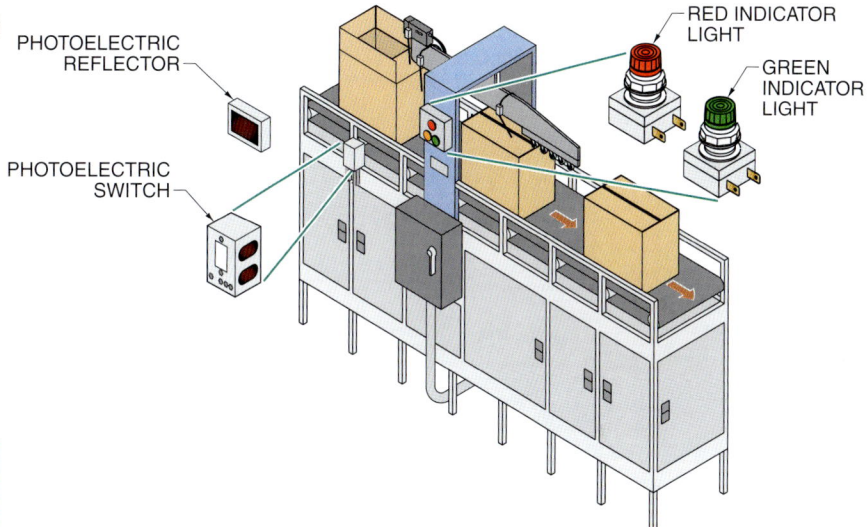

Allen-Bradley® Procedures

1. Based on the PLC system description and line diagram, draw the ladder logic for RSLogix™ in the AB programming area.

2. Open the RSLogix™ Micro Starter Lite program and create a new programming diagram.

3. Use the AB programming diagram to add the proper address and description in the LAD 2 programming window.

4. Verify that the programming diagram has no errors and save the project.

AB PROGRAMMING DIAGRAM

478—PROGRAMMABLE LOGIC CONTROLLERS: PRINCIPLES AND APPLICATIONS

5. Use the wiring diagram to wire the input devices and output components to the PLC. Once all connections have been made, turn power to the PLC on.

6. Use RSLinx™ to establish communication between the PC and the Micrologix™ 1100.

7. Download the program logic to the PLC.

8. Place the PLC in Run Mode.

9. Activate input(s) and monitor them to determine if the output(s) are operating correctly.

10. Turn power OFF and set a DMM to measure milliamperage. Turn power ON, run the program, and measure the operating current of the photoelectric sensor when the switch is actuated and the load is ON. What is the operating current in milliamps?

   _____

11. Measure the leakage current of the photoelectric sensor in contact I/0 when the sensor is not actuated. What is the leakage current in milliamps?

   _____

12. Measure the leakage current of the photoelectric sensor in contact I/4 when the sensor is not actuated. What is the leakage current in milliamps?

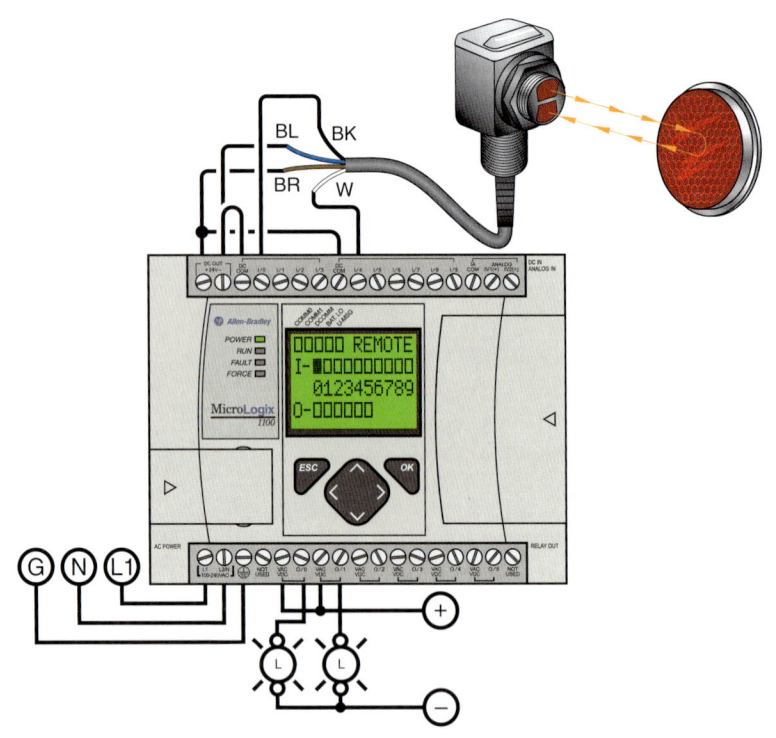

AB WIRING DIAGRAM

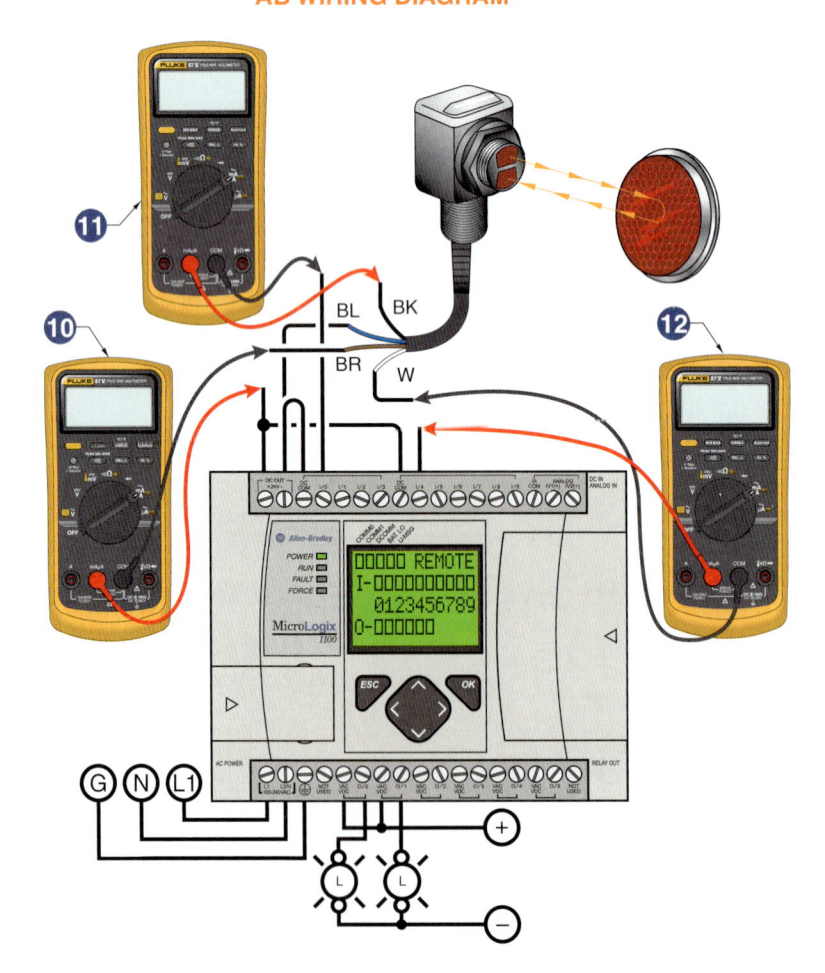

Chapter 11—Testing and Troubleshooting Electrical Devices and PLC Hardware   479

Troubleshooting PLC System Problems

PLC Lab 11.3

Troubleshooting a problem in a PLC system requires an understanding of how the circuit is interconnected and how the measurements are taken. It also requires an understanding of the technical data on the circuit devices and components. For this lab, a maintenance report states that a motor has become unreliable. It continually turns OFF and can be restarted only after the overload circuit is reset. The drive motor is single-phase 230 V.

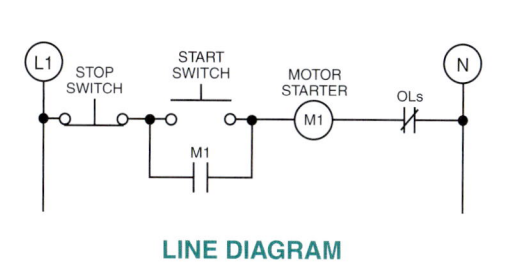

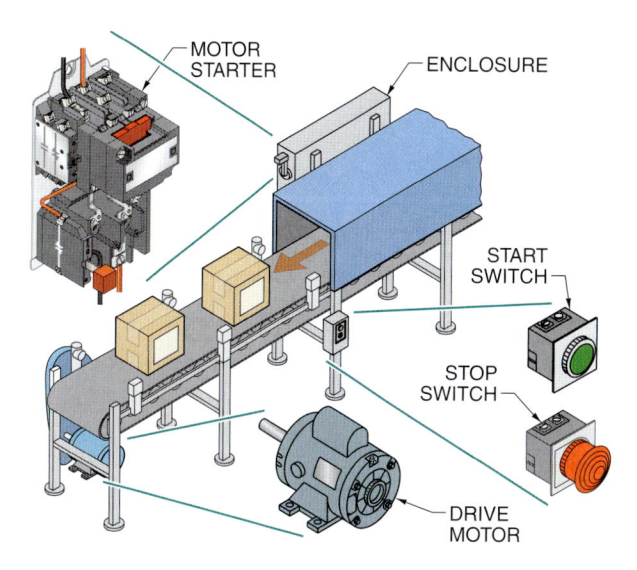

LINE DIAGRAM

STEP 1—Gathering the Required Information

_____ 1. What is the current draw of the motor when it is fully loaded at 230 V but is not operating at service factor amperage (SFA)?

_____ 2. What is the number of the thermal heater unit that should be installed in the starter to protect the motor from the maximum motor current draw not at SFA?

_____ 3. What is the number of the thermal heater unit that should be installed in the starter to protect the motor from the maximum motor current draw up to the SFA?

_____ 4. What is the PLC output number assigned to the motor starter? See STEP 2 detailed drawing.

_____ 5. What is the magnetic motor starter coil voltage rating? See STEP 2 detailed drawing.

AC MOTOR THERMALLY PROTECTED

MOD 38DKLXX24			
HP 1/3	HZ 60		
V 115/230	PH 1		
RPM 1725	CODE L		

LEAD	LOW VOLT	HIGH VOLT
BROWN	5	4
WHITE	2	5
RED	5	5
BLACK	A	A

A 6.6/3.3	SF 1.35
SFA 7.2/3.6	FR 56
AMB 40C	INSUL CLASS A	NEMA DESIGN
TIME RATING CONT.	
SER. NO. MWT	

CCW ROTATION AS SHOWN. TO REVERSE ROTATION INTERCHANGE BLACK & RED

GROUND IN ACCORDANCE WITH LOCAL AND NATIONAL ELECTRICAL CODES. KEEP FINGERS AND FOREIGN OBJECTS AWAY FROM OPENINGS AND ROTATING PARTS.

LUBRICATION: AFTER 3 YRS NORMAL OR 1 YR HEAVY DUTY SERVICE ADD OIL ANNUALLY. USE ELECTRIC MOTOR OR SAE 10 OIL.

MADE IN U.S.A.

Thermal Unit Current Ratings

Motor Full-Load Current (Amps)			Thermal Unit (Number)
1 Unit (Heater)	2 Unit (Heater)	3 Unit (Heater)	
2.15 – 2.40	2.15 – 2.40	1.98 – 2.32	B3.30
2.41 – 2.72	2.41 – 2.72	2.33 – 2.51	B3.70
2.73 – 3.15	2.73 – 3.15	2.52 – 2.99	B4.15
3.16 – 3.55	3.16 – 3.55	3.00 – 3.42	B4.85
3.56 – 4.00	3.56 – 4.00	3.43 – 3.75	B5.50
4.01 – 4.40	4.01 – 4.40	3.76 – 3.98	B6.25
4.41 – 4.88	4.41 – 4.88	3.99 – 4.48	B6.90

STEP 2—Taking Measurements Using a Test Instrument

What are the normal voltage measurements at meters 2 and 5 and the maximum current measurements at meters 1, 3, and 4 when the motor is operating at its highest service factor rating?

_____ 6. Meter 1 reading

_____ 7. Meter 2 reading

_____ 8. Meter 3 reading

_____ 9. Meter 4 reading

_____ 10. Meter 5 reading

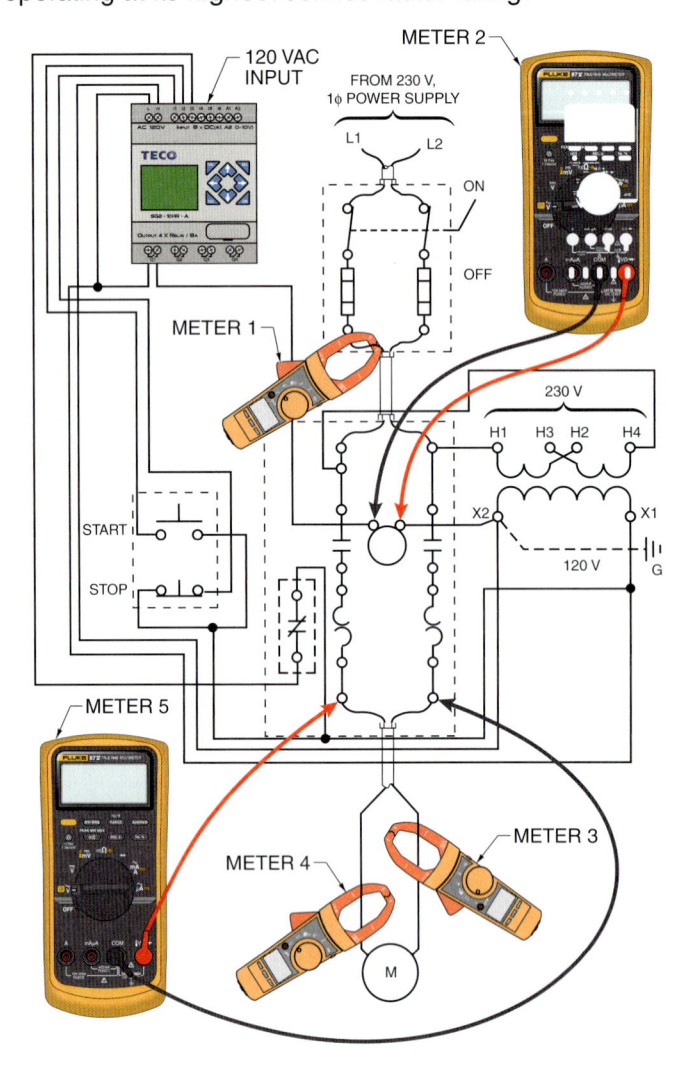

STEP 3—Understanding Meter Readings

Motor readings that are not what they should be indicate a problem exists.

_____ 11. If the current measurements taken are within the nameplate rating, what meter function is used to measure current over time to determine if an infrequent problem exists?

_____ 12. Would a low-voltage measurement on meter 2 cause the overloads to trip?

_____ 13. Would a low-voltage measurement on meter 2 cause the motor to stop and require a restart using the start pushbutton?

_____ 14. Installing a thermal heater unit numbered B3.70 would cause the overloads to trip. Would they trip at a current value below or above the motor nameplate rated current?

_____ 15. Would a problem, such as an open circuit or a loose connection, with the wiring in the overload contact circuit going to the PLC input terminal cause the motor overloads to trip as if an overload occurred?

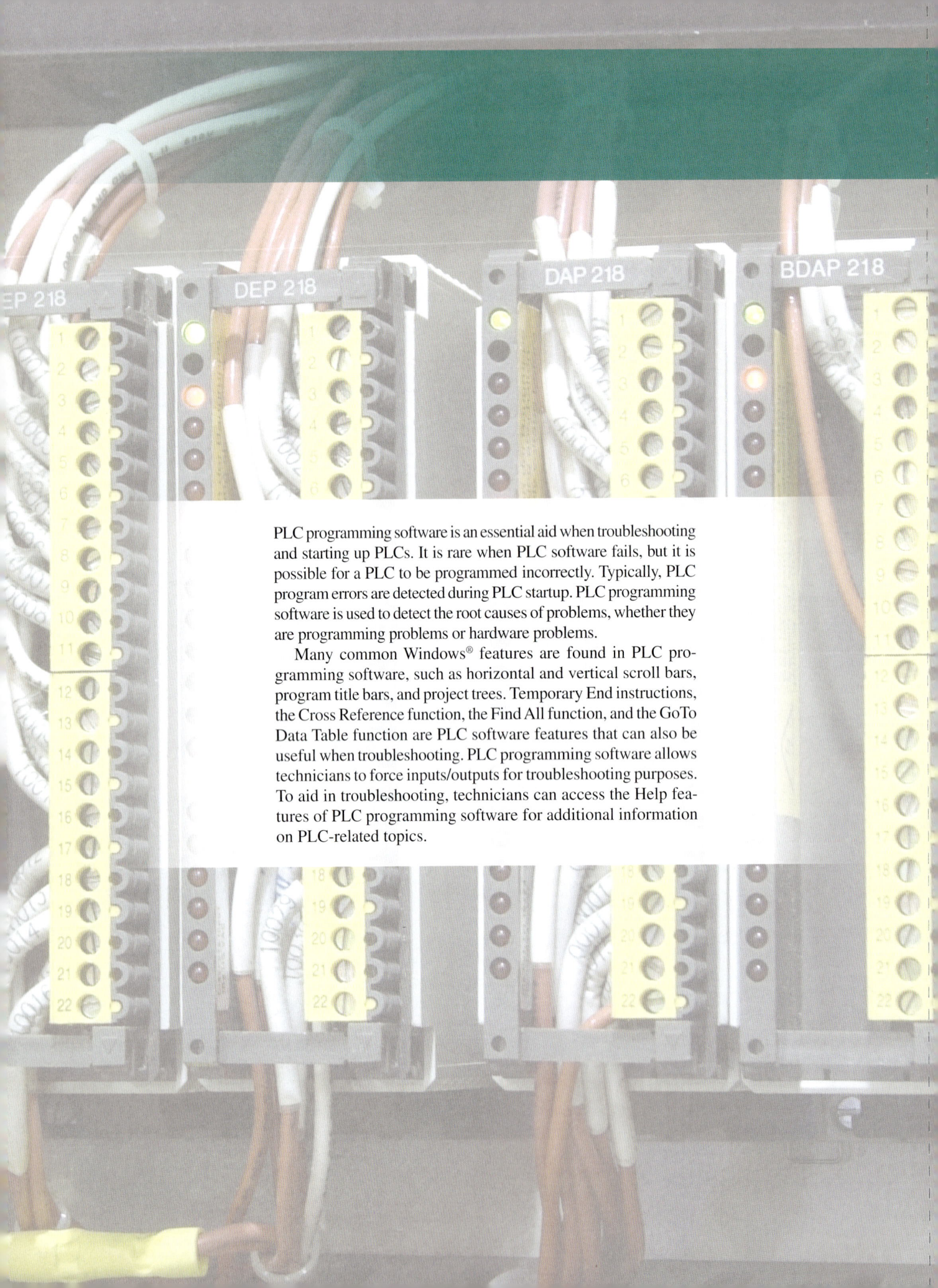

PLC programming software is an essential aid when troubleshooting and starting up PLCs. It is rare when PLC software fails, but it is possible for a PLC to be programmed incorrectly. Typically, PLC program errors are detected during PLC startup. PLC programming software is used to detect the root causes of problems, whether they are programming problems or hardware problems.

Many common Windows® features are found in PLC programming software, such as horizontal and vertical scroll bars, program title bars, and project trees. Temporary End instructions, the Cross Reference function, the Find All function, and the GoTo Data Table function are PLC software features that can also be useful when troubleshooting. PLC programming software allows technicians to force inputs/outputs for troubleshooting purposes. To aid in troubleshooting, technicians can access the Help features of PLC programming software for additional information on PLC-related topics.

Troubleshooting with PLC Software

12

Objectives:
- Demonstrate how to use Windows® features to view a PLC program.
- Demonstrate how to use programming software features to view a PLC program.
- Demonstrate how to safely and systematically debug a PLC program using the TND instruction, the Cross Reference function, the Find All function, and the Goto Data Table function.
- Explain how to safely use the Force function when troubleshooting.
- Explain the possible consequences of forcing an input or forcing an output.
- Describe how to use the Help features in PLC programming software.

Learner Resources
atplearningresources.com/quicklinks
Access Code: 475203

PLC PROGRAMMING SOFTWARE

Programming software is used to write PLC programs for a variety of machine applications and to control processes. PLC programming software is thoroughly tested and debugged by the software manufacturer before the software is released to the end users (PLC programmers, system integrators, and maintenance personnel). During the installation and startup of a specific application, the PLC program is debugged and changes are made to the program to enable proper machine or process operation.

Typically PLC software does not fail or break, but a PLC program can be programmed incorrectly. Occasionally, over an extended period of time, a programming flaw can become evident because the flaw occurs only under a unique set of circumstances. In order to use PLC software for troubleshooting, a technician must understand how to view PLC programs, use specific features of PLC programming software to debug a program, use the Force function of the software, and use the software Help features.

VIEWING PLC PROGRAMS

A PLC program is commonly referred to as a project. The PLC program or project contains all of the files—program files, data table files, and documentation—for a specific PLC. The first step in troubleshooting a PLC application is to view the program. **See Figure 12-1.**

In order to effectively troubleshoot a PLC application, the technician must be thoroughly familiar with the various viewing features of the PLC programming software. Most programming software is Windows®-based and uses several of the functions found in other Windows® applications such as Word®, PowerPoint®, and Excel®. In addition, the PLC programming software has its own unique features for viewing a PLC program or project. Although the features vary between PLC manufacturers, there are several features common to all.

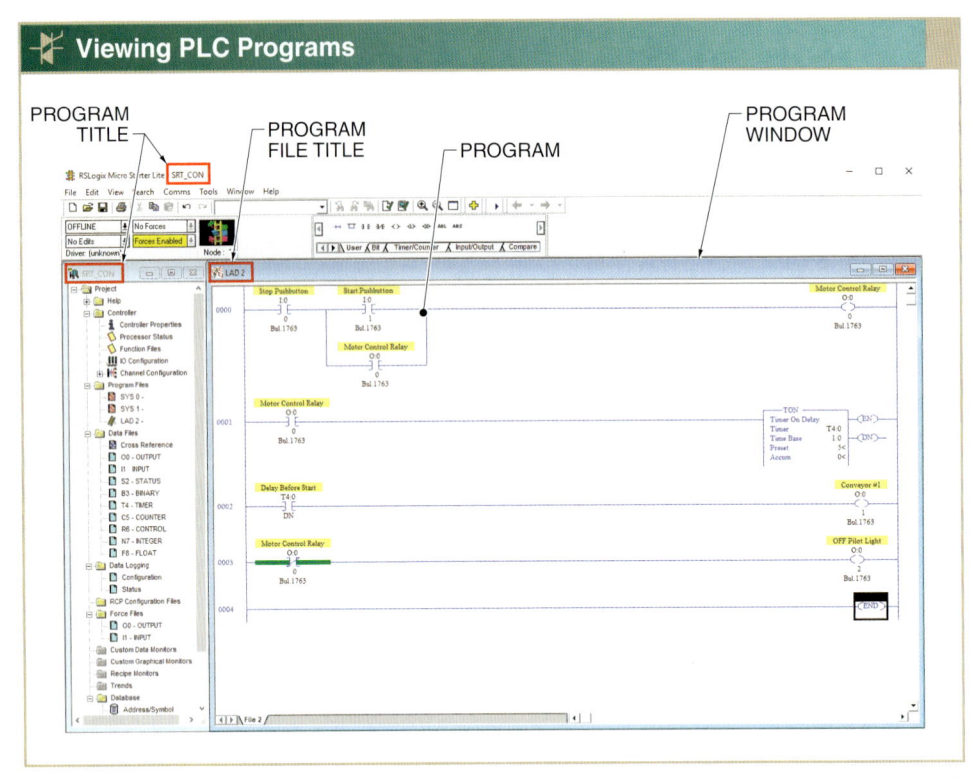

Figure 12-1. The viewing screen contains multiple elements that allow a program to be viewed in a window.

Windows® Features

Because many of the viewing features found in Windows®-based programs are used in PLC programming software, technicians are generally familiar with most of the PLC software viewing features; for example, multiple windows can be open at the same time. The manipulation of the PLC programming windows is something that all technicians must thoroughly understand. **See Figure 12-2.** The common Windows® screen elements used in PLC programming software also include horizontal and vertical scroll bars, title bars, and project trees.

Horizontal and Vertical Scroll Bars. A *scroll bar* is a screen element at the bottom of a window or to the right of a window that allows a technician to move the contents of a window horizontally or vertically on the screen by dragging the bars. Scroll buttons can also be used to move the contents of a window by clicking on the buttons for right/left or up/down movement.

Title Bars. A *title bar* is a shaded horizontal bar at the top of a window that contains the title of the window. A title bar is typically blue when the window is active and gray when the window is inactive. Title bars can be used to move the placement of a window by positioning the mouse pointer on the title bar, left-clicking, and dragging the mouse (window) to the new position.

Project Trees. A *project tree (project view)* is a visual representation of all the files and folders that make a PLC program. The project tree consists of folders and files that contain all of the information about the PLC program named in the title bar. In front of each folder, there is an icon containing a plus sign (+) or a minus sign (−). A plus sign indicates that a folder contains files that are not visible yet (not opened). A minus sign indicates that the contents of a folder are visible (opened).

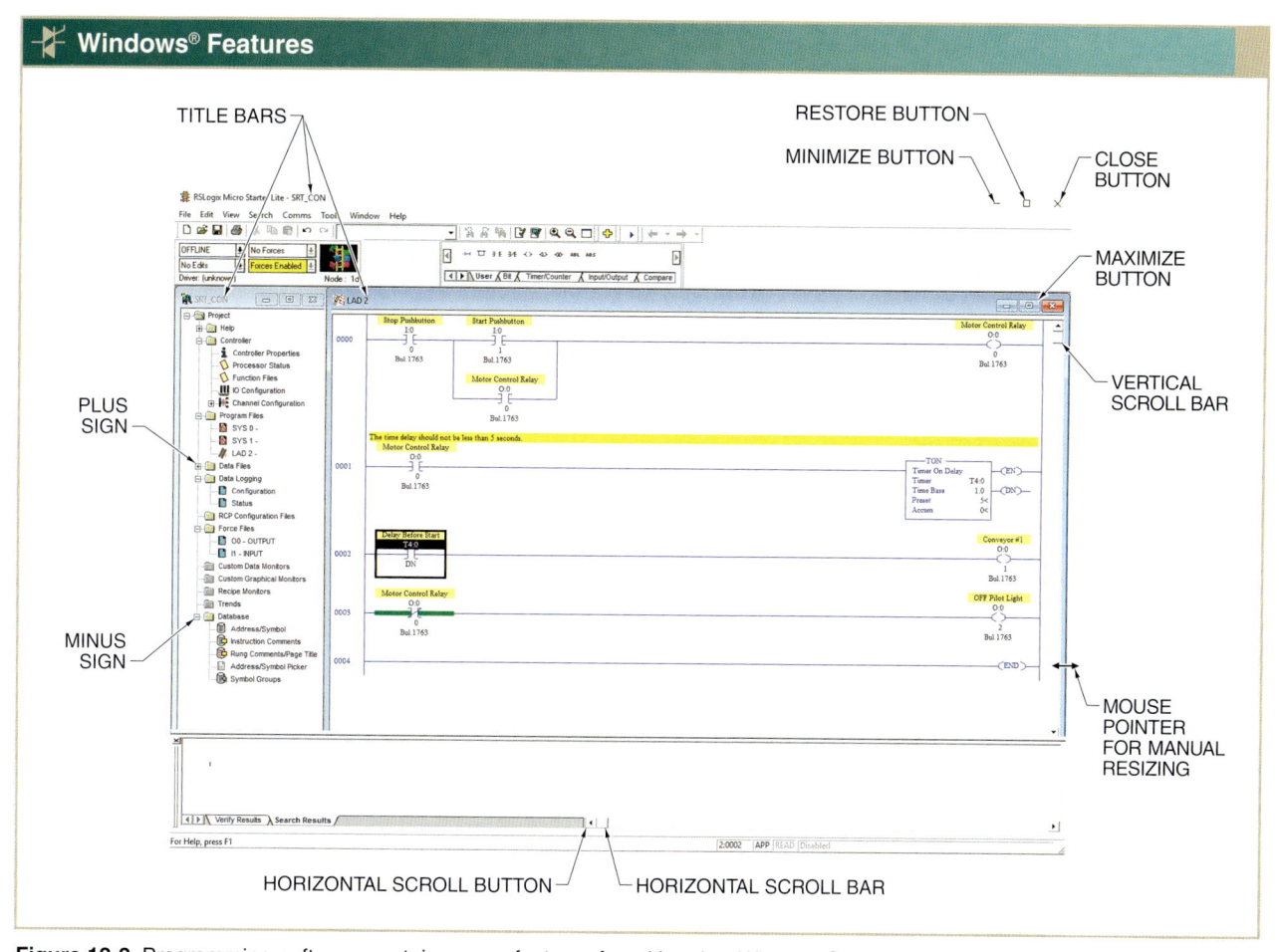

Figure 12-2. Programming software contains many features found in other Windows®-based software products.

Windows on a monitor can be maximized and minimized. A minimize button collapses the window down to the taskbar. The taskbar is part of the Windows® operating software and is horizontal along the bottom of the screen. Positioning the mouse pointer on the minimized window of the taskbar and left-clicking expands the window to its preminimized size.

The maximize button is a two-state button. A maximize button enlarges a window to full size (the window utilizes the entire monitor screen). A restore button appears after a window has been maximized. The restore button restores the window to the premaximized size. A close button closes its window. When multiple windows are open, a technician can select which windows to work with or close.

Windows may be manually resized when the mouse pointer is placed on the edge of a window and the pointer changes into a double-headed arrow. Manual resizing is done by left-clicking and holding the mouse button while moving the mouse to resize the window.

PLC Programming Software Features

Programming software allows a technician to view all of the elements of a program, program files, data table files, and documentation. Various menus, toolbars, and windows are used to program, install, and troubleshoot a PLC using a PC. Technicians must understand the PC screen elements of programming software in order to perform tasks such as programming, installing, and troubleshooting PLCs. **See Figure 12-3.**

PLC TIPS

PLC programming software has project trees that include wizards, online tools, and direct access to system and data windows.

The common viewing features of programming software include the menu bar, icon bar, online bar, instruction toolbar, project tree, ladder view, results window, and status bar. The *menu bar* is the section of application software containing headings for pull-down menus that is placed across the top of a screen. The *icon bar* is the section of software containing buttons for functions that are frequently used.

The *online bar* is the section of software that contains information on the processor mode, edit status, force status, and communication drivers of a PLC. The *online icon* is a button that is animated when a PC is online with a PLC, such as when changing the mode of a PLC to remote programming, remote run, test continuous, or test single scan. The online icon is inactive when the PC is offline.

The *instruction toolbar* is the section of the software that contains various instruction buttons. Instructions are displayed as abbreviations (also known as mnemonic symbols). Tabs are used to group instructions according to type. The project tree is the section of the software that contains all of the files for a project. Specific files can be opened or closed by left-clicking on the + or − sign in front of the name of the file. Right-clicking on a file opens a menu for that file.

The *ladder view,* or programming diagram view, is the section of the programming window that contains the programming diagram. The programming diagram is created and edited in the ladder view section of the programming window. The *results window* is a small window section of the software that

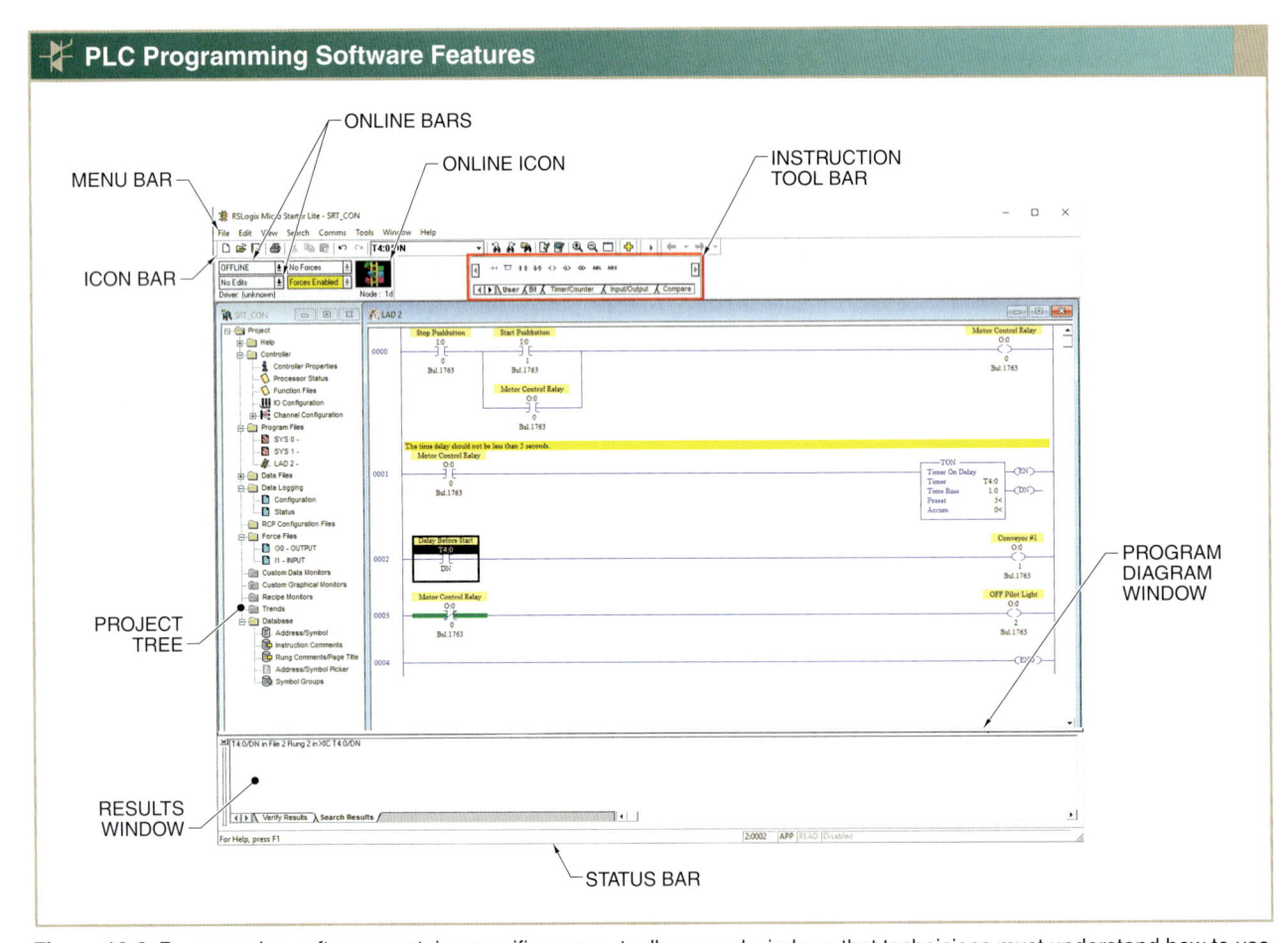

Figure 12-3. Programming software contains specific menus, toolbars, and windows that technicians must understand how to use.

Chapter 12—Troubleshooting with PLC Software   **487**

opens up with the results of a Find All or a Verification (either file or project). The *status bar* is the section that contains current status information and provides information on how to use the software.

Another Windows® feature found in programming software is the dialog box. Dialog boxes are small windows that open to allow users to make choices and enter information. Dialog boxes are typically accessed from the menu bar. Dialog boxes contain check boxes, radio buttons, and text boxes for entering information. Some dialog boxes contain multiple pages, which are identified by tabs. Programming software dialog boxes are available for system communications, viewing properties, and system options. **See Figure 12-4.**

DEBUGGING PLC PROGRAMS

PLC programs are written to control machine applications and processes. The programmer writes a program based on equipment specifications, data on the sequence of operations, and system requirements. Quite often the information is initially incomplete or is changed during the programming process. During the startup of a PLC the program is debugged. *Debugging* is a task that consists of eliminating errors or malfunctions and verifying that a PLC program operates correctly and meets the specifications and requirements of the system.

During debugging, changes are made to the program to optimize machine or process performance. **See Figure 12-5.**

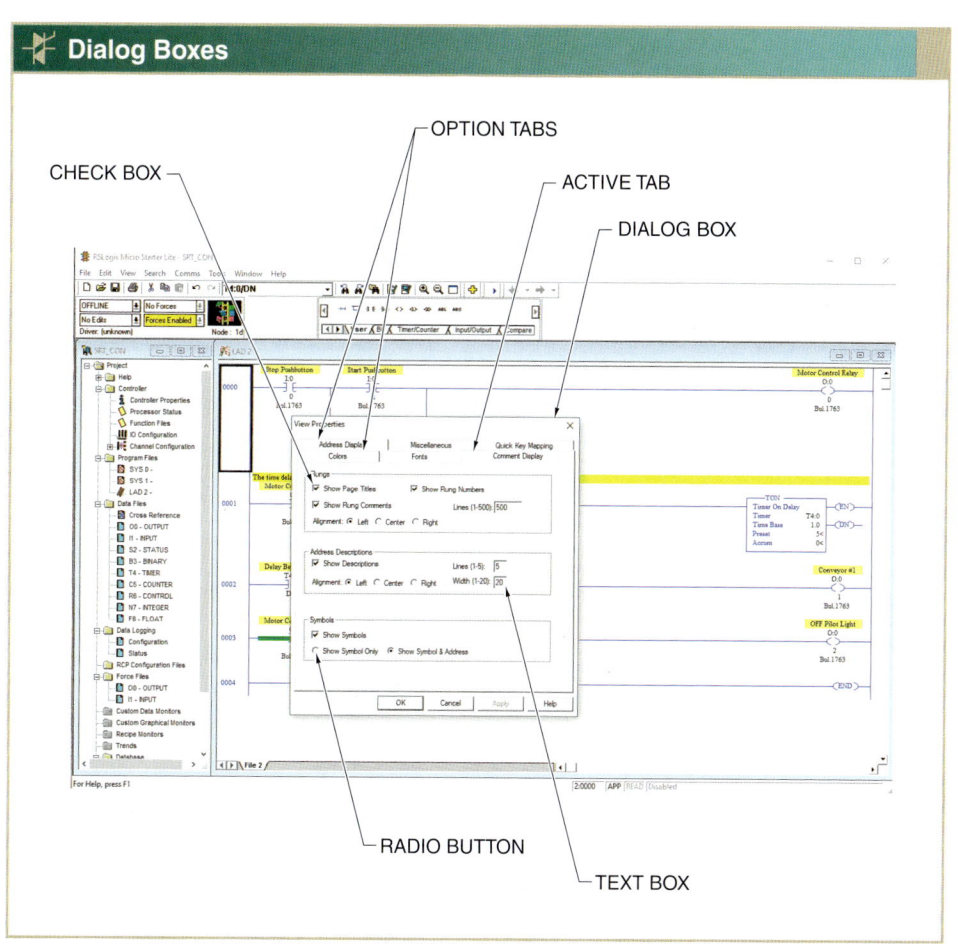

Figure 12-4. Programming software contains dialog boxes that allow the user to make choices and enter information. Some dialog boxes contain multiple pages.

PLC programming software features include temporary end (TND) instructions and Cross Reference, Find All, and Goto Data Table functions. The Cross Reference, Find All, and Goto Data Table functions can be accessed as follows:

1. Position the mouse pointer on an instruction in the ladder view window.
2. Left-click the mouse on the instruction.
3. Right-click the mouse on the black section of the highlighted instruction to access a menu of various functions.

Temporary End Instructions

A *temporary end (TND) instruction* is an output instruction that is used to debug a PLC program. A rung containing a TND instruction is added to a program to stop the program from scanning a PLC. When the rung containing the TND instruction is true, the PLC processor stops scanning the rest of the program, updates the I/O with information that has been scanned, and continues scanning back at the beginning of the program (rung 0). **See Figure 12-6.**

When a rung containing a TND instruction is false, the PLC processor continues scanning the program until the scan reaches a true TND instruction or reaches an end of program (END) instruction. When a TND instruction is programmed without an input instruction, the instruction is always true.

Typically the debugging process starts at the beginning of a program. By moving the TND instruction from top to bottom in the program, a technician can test parts of the program and sequentially change the position of a TND instruction or add TND instructions until the entire program has been tested. When the debugging process has been completed, any remaining TND instructions are removed.

⚠ CAUTION

When a temporary end of program instruction is inside a nested subroutine, execution of all nested subroutines is terminated.

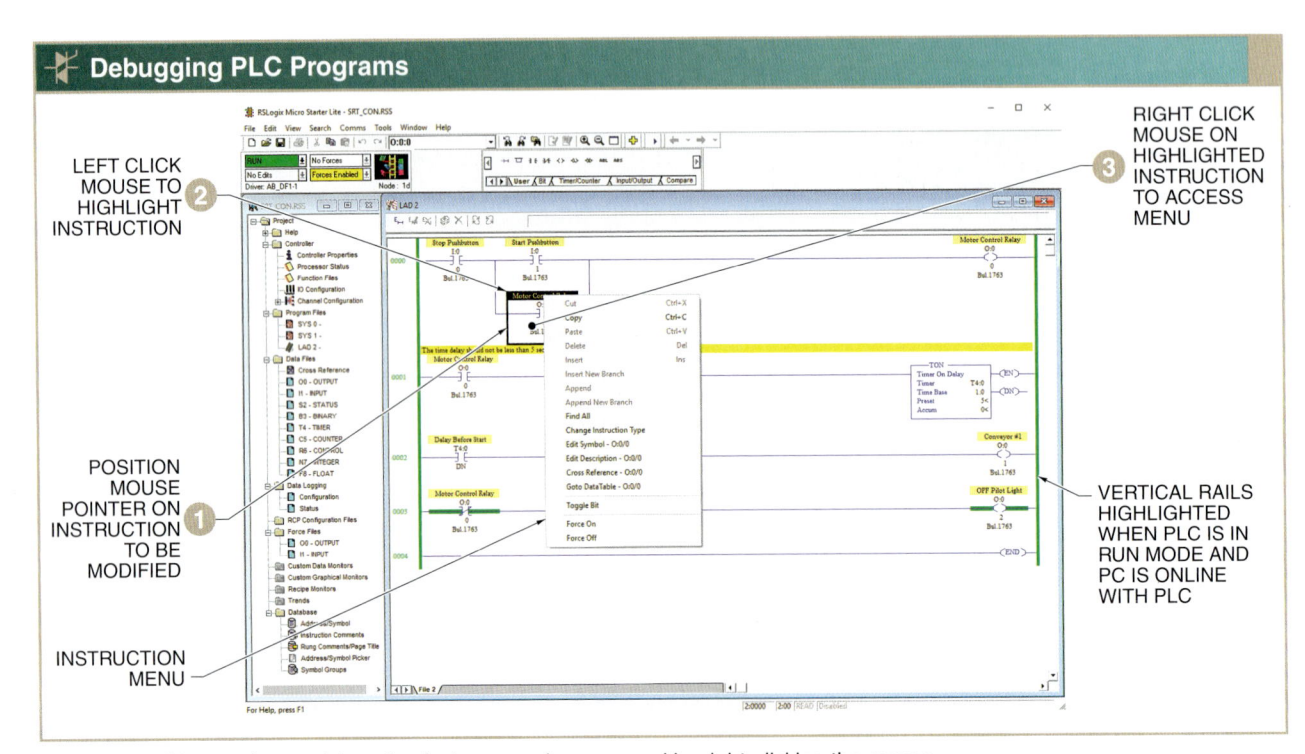

Figure 12-5. Many software debugging features can be accessed by right-clicking the mouse.

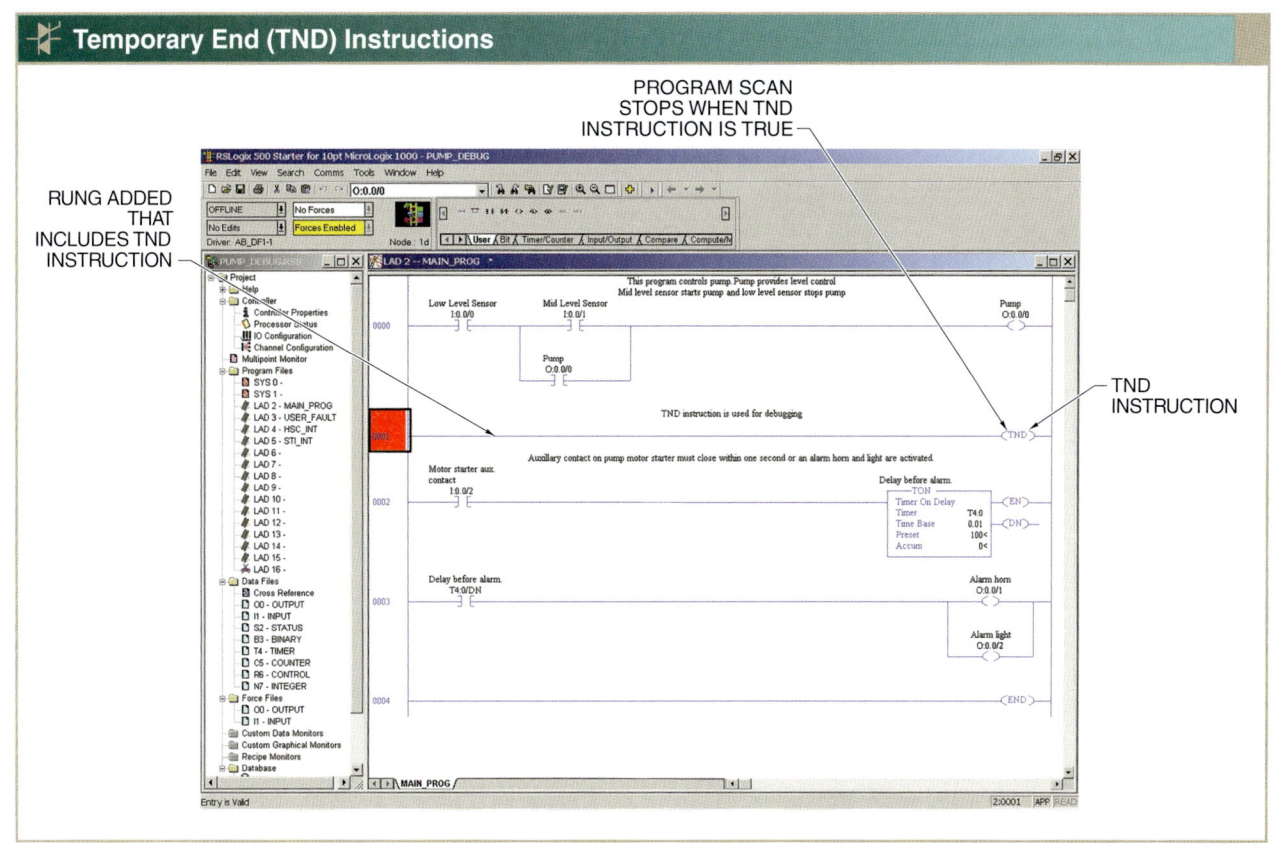

Figure 12-6. Temporary end (TND) instructions are useful when debugging programs, especially large programs that contain several different machines or processes.

Accessing the Cross Reference Function

When debugging a PLC, it is often necessary to locate each instance of a particular address. The Cross Reference function searches the entire program (all program files) to locate each instance of a selected address, regardless of the instruction type. **See Figure 12-7.** The Cross Reference function is accessed as follows:

1. Position the mouse pointer on an instruction in the ladder view window.

2. Left-click the mouse on the instruction to highlight.

3. Double-left-click the mouse on the Cross Reference title in the project tree to open the Cross Reference Report window. *Note:* The Cross Reference function can be used while a PLC is online or offline.

The results of the Cross Reference function are displayed in the Cross Reference Report. The Cross Reference Report contains all of the addresses used in an entire program. The addresses are displayed in the same order as the data table files O, I, S, and B. When an address has a description, the description is displayed. The instruction type, program file, and rung number for each instance where the address is found are also listed. Each occurrence of the address is displayed sequentially, starting with program file 2 and rung 0.

PLC TIPS

Passwords provide protected access to a program and prevent changes from being made. Each program can contain two passwords. A user password allows access to the PLC program. A master password overrides a user password and applies to all PLCs in a facility.

490 — PROGRAMMABLE LOGIC CONTROLLERS: PRINCIPLES AND APPLICATIONS

Accessing the Cross Reference Function

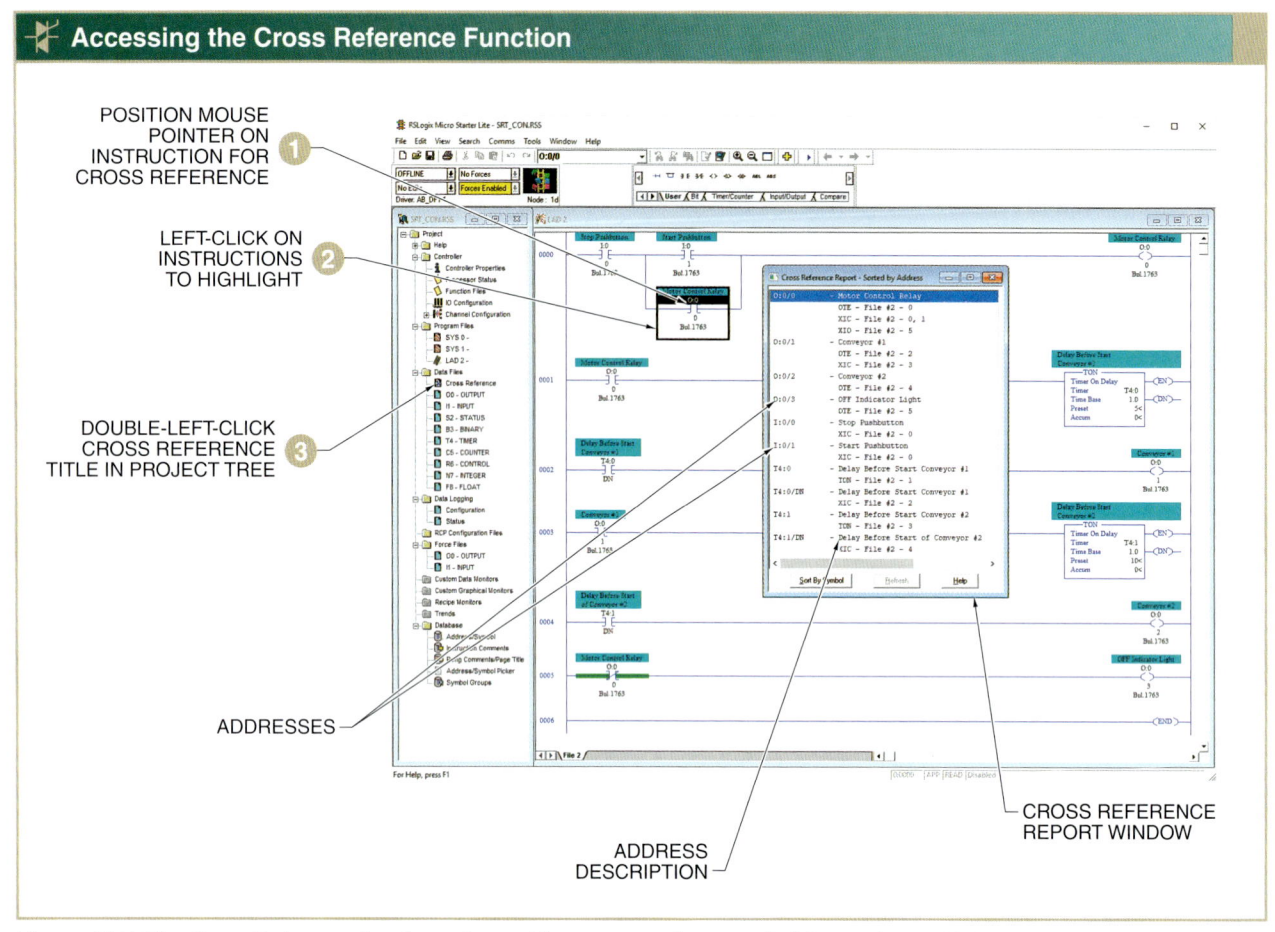

Figure 12-7. The Cross Reference function only provides a cross reference of all instructions with the same address. The results are displayed in the Cross Reference Report window.

Accessing the Find All Function

The Find All function is similar to the Cross Reference function. The Find All function searches an entire program (all program files) and locates each instance of a selected address, regardless of the instruction type. **See Figure 12-8.** The Find All function can be accessed as follows:

1. Position the mouse pointer on an instruction in the ladder view window.
2. Left-click the mouse on the instruction to highlight the instruction's address.
3. Right-click the mouse on the highlighted instruction to access the menu.
4. Position the mouse pointer on the Find All bar in the menu.
5. Left-click the mouse on the Find All bar to access the results window.

Note: A Find All function can be used online or offline.

Unlike the Cross Reference function, the Find All function only displays the instructions with the same address. The results of a Find All function are displayed in the results window. The address, program file number, rung number, and instruction type are listed in the results window.

The instructions appear in descending order in the results window, starting with the instruction immediately after the initially selected instruction and moving left to right and top to bottom through the program. The initially selected instruction appears last in the list. Moving the mouse pointer over an instruction in the results window and left-clicking highlights the instruction. When working with a large program, the Find All function is particularly helpful.

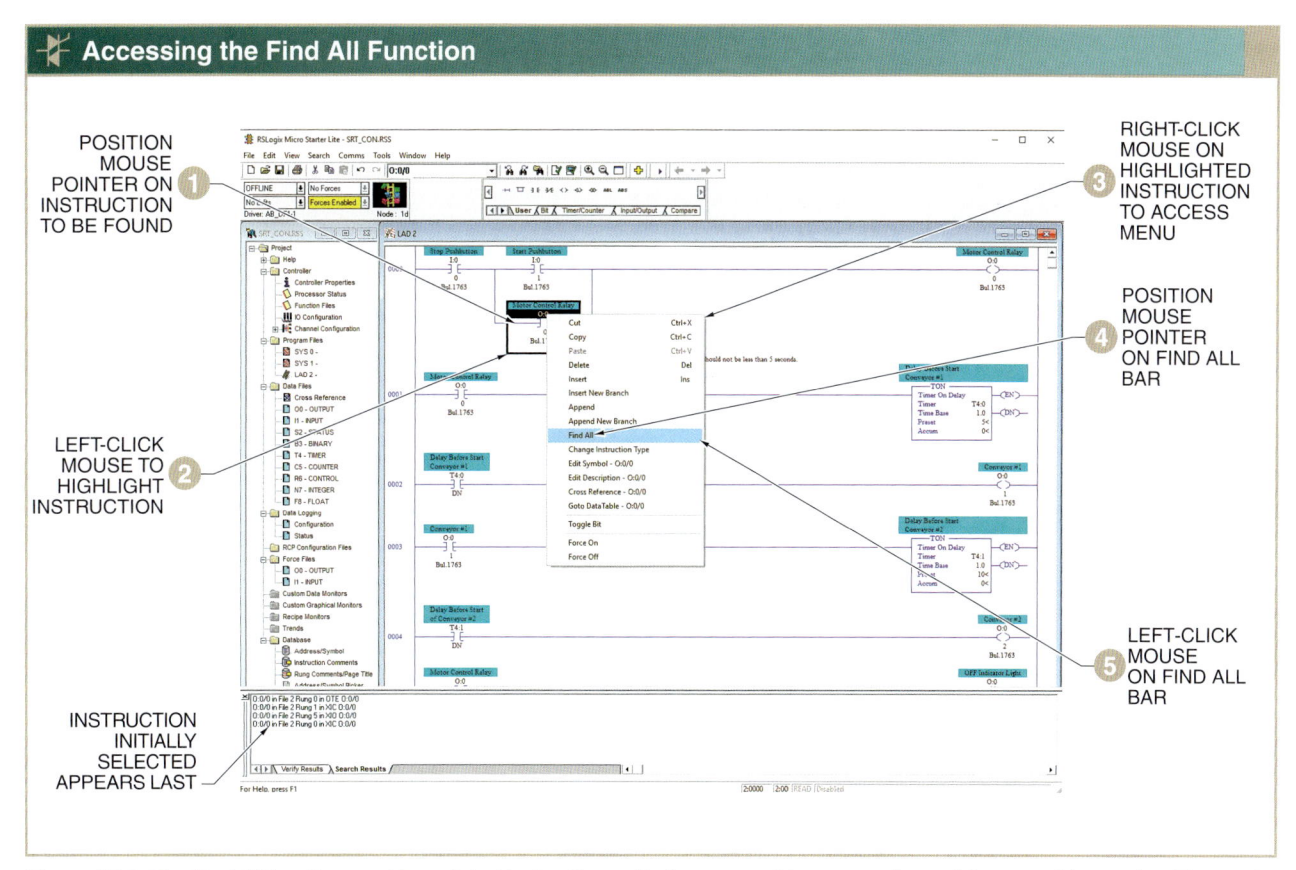

Figure 12-8. The Find All function provides a list of instructions with the same address regardless of the type of instruction. The results are displayed in the results window.

Accessing the Goto Data Table Function

As part of the debugging process of a PLC and system, it is sometimes necessary to monitor the bit status of certain instructions. The Goto Data Table function displays the data table for the selected instruction. **See Figure 12-9.** The status of individual bits can be monitored, such as the presets and accumulated values of timers and counters. The Goto Data Table function is accessed as follows:

1. Position the mouse pointer on an instruction in the ladder view window.
2. Left-click the mouse on the instruction to highlight.
3. Right-click the mouse on the highlighted instruction to access the instruction menu.
4. Position the mouse pointer on the Goto Data Table bar in the menu.
5. Left-click the mouse on the Goto Data Table bar to access the Data Table window. *Note:* The Goto Data Table function can be used online or offline.

The complete data table for the type of instruction (Output or Input) is displayed as a window with Status and Radix boxes included. The address of the instruction selected and any descriptions are shown in the boxes and highlighted. A menu in the data table can access other data table files. Also a drop-down menu allows the radix to be changed. Changing the radix of a data table file changes the method (typically a numbering system) used to display the file status.

PLC TIPS

To remove a password from a PLC program, double-left-click on "Controller Properties" in the project tree. Single-left-click on the "Passwords" tab. Single-left-click on the "Remove" button. Type the current password for authorization, and do not enter a new password. Press the Enter key instead.

Accessing the Goto Data Table Function

[Screenshot of RSLogix Micro Starter Lite programming software showing the Goto Data Table function access process with the following annotations:]

- ① POSITION MOUSE POINTER ON SELECTED INSTRUCTION
- ② LEFT-CLICK MOUSE TO HIGHLIGHT INSTRUCTION
- SELECTED INSTRUCTION ADDRESS
- DATA TABLE
- RADIX CAN BE CHANGED
- ADDRESS DESCRIPTION
- ARROWS ACCESS OTHER DATA TABLES
- ③ RIGHT-CLICK HIGHLIGHTED INSTRUCTION TO ACCESS INSTRUCTION MENU
- ④ POSITION MOUSE POINTER ON GOTO DATA TABLE BAR
- ⑤ LEFT-CLICK MOUSE ON GOTO DATA TABLE BAR TO ACCESS DATA TABLE

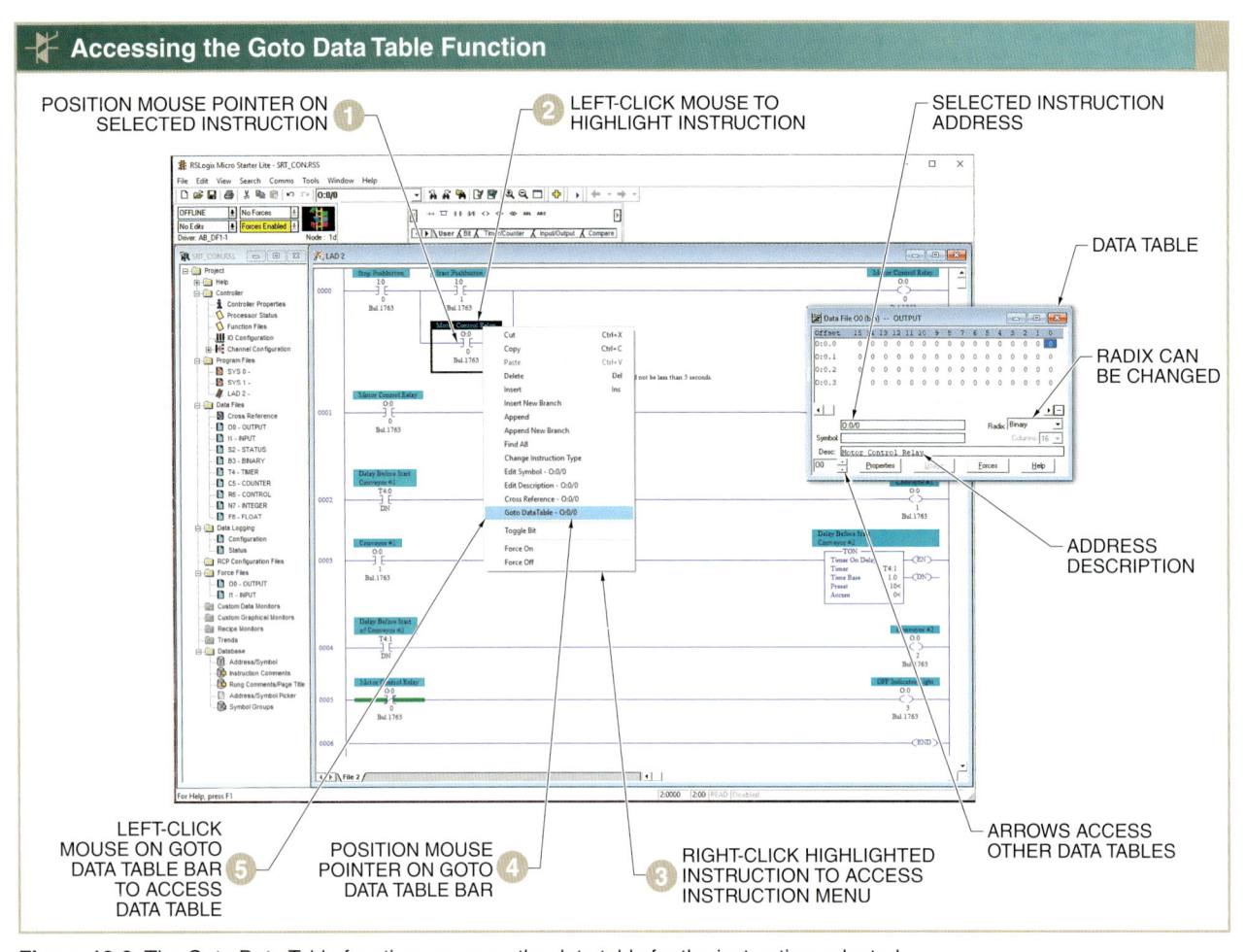

Figure 12-9. The Goto Data Table function accesses the data table for the instruction selected.

⚠ **WARNING**

Always use extreme caution when using the Force function. Use of the Force function can result in unexpected machine operation, possibly injuring personnel or damaging equipment.

FORCE FUNCTIONS

The Force function is a feature of PLC programming software that allows a technician to simulate the state of an input device or output component as ON or OFF. The software can force an input device or output component ON or OFF regardless of the actual state of the device or PLC logic controlling the component. The Force function only operates with actual input devices and output components.

There are several instances where technicians typically use the Force function. Technicians must fully understand the consequences of the Force function before using it. Technicians must also follow company policy regarding the Force function. The Force function is used during maintenance, troubleshooting, and startup debugging.

The Force function is used to gain access to a section of a machine or process for servicing. When troubleshooting, the Force function can simulate an energized input device to determine if the program is running properly and the output components are working. Forcing helps determine whether the actual input device is damaged or broken. A force can be left in place until the input device is repaired or replaced when it is safe to do so. Forcing used during startup debugging verifies that the wiring to input devices and output components is in working order. **See Figure 12-10.** The Force function is accessed as follows:

1. Position the mouse pointer on an instruction in the programming diagram window.

2. Left-click the mouse on the instruction to highlight.
3. Double-left-click the mouse on the output or input title under the Force files in the project tree to open the corresponding Force Data File.
4. Left-click the mouse on the instruction address in the Force Data File that will be forced.
5. Right-click the mouse on the highlighted instruction address in the Force Data File to access the instruction menu.
6. Position the mouse on the desired force function and left-click.

Force On or Force Off instructions apply forces to a specific input or output instruction. Remove Force removes forces from an instruction. When a technician is finished with the Force function, all forces must be removed. A common mistake is to choose the Force Off option instead of the Remove Force option. When a program contains a force and the program is saved, uploaded, or downloaded, the force is also saved, uploaded, or downloaded.

Forcing Inputs

The input terminals of a PLC can be forced ON or OFF using programming software. When a force is applied to an input instruction, the programming software writes the force to the input data table. The force affects all instructions with the same address and program logic functions. An ON or OFF appears adjacent to all instances of the forced instructions with the same address. Also, any instruction that is true as a result of the force is highlighted. When a force is applied, the POWER, RUN, and FORCE processor status LEDs of the CPU are illuminated. The LCD display rectangle for the forced input terminal appears empty. *Note:* PLCs with LCD displays use both LEDs and the display to provide status information. **See Figure 12-11.**

> **⚠ CAUTION**
> When an output terminal is turned ON as a result of a forced input, the status LED for the output terminal lights up.

> **PLC TIPS**
> When no forces are in place for the instruction, the menu options are Force On and Force Off. When a force is in place for the instruction, the menu options are Force Off, Remove Force, and Remove All Forces.

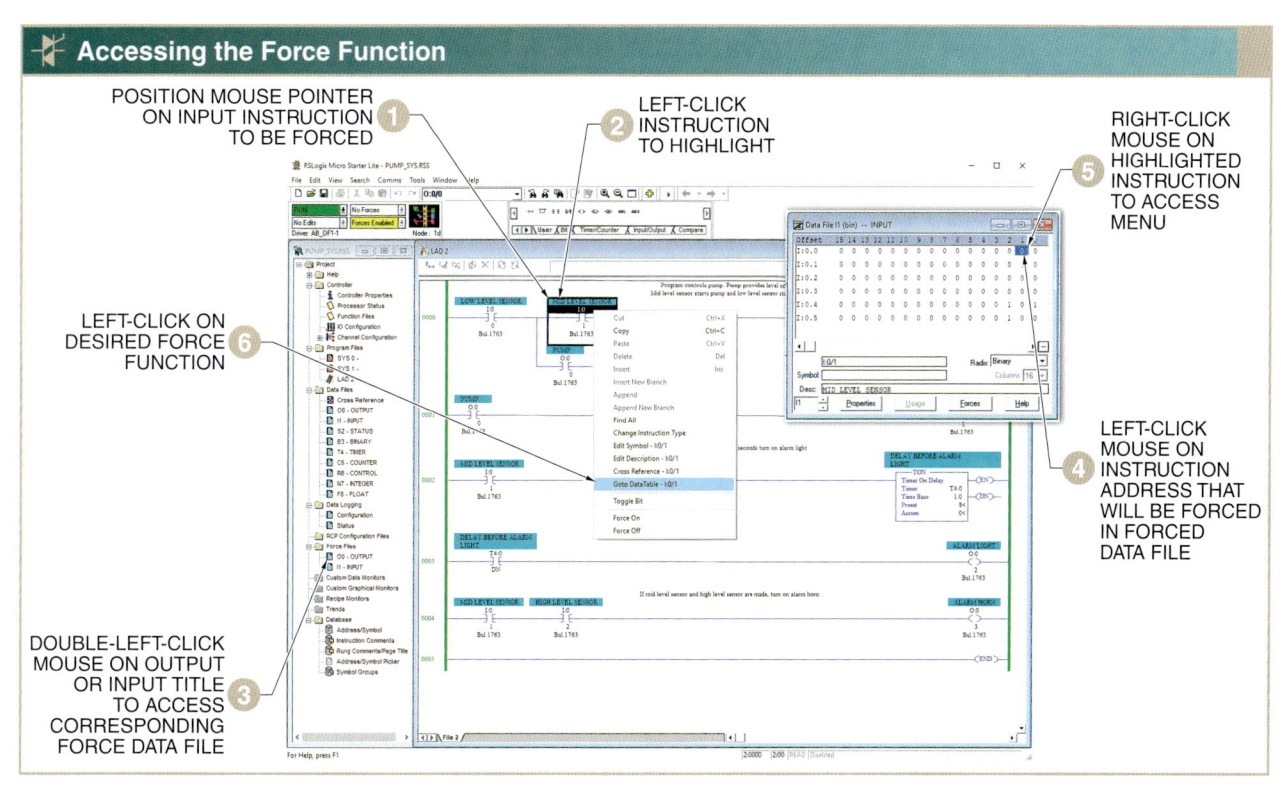

Figure 12-10. The menu options for the Force function depend on whether or not a force is already being applied.

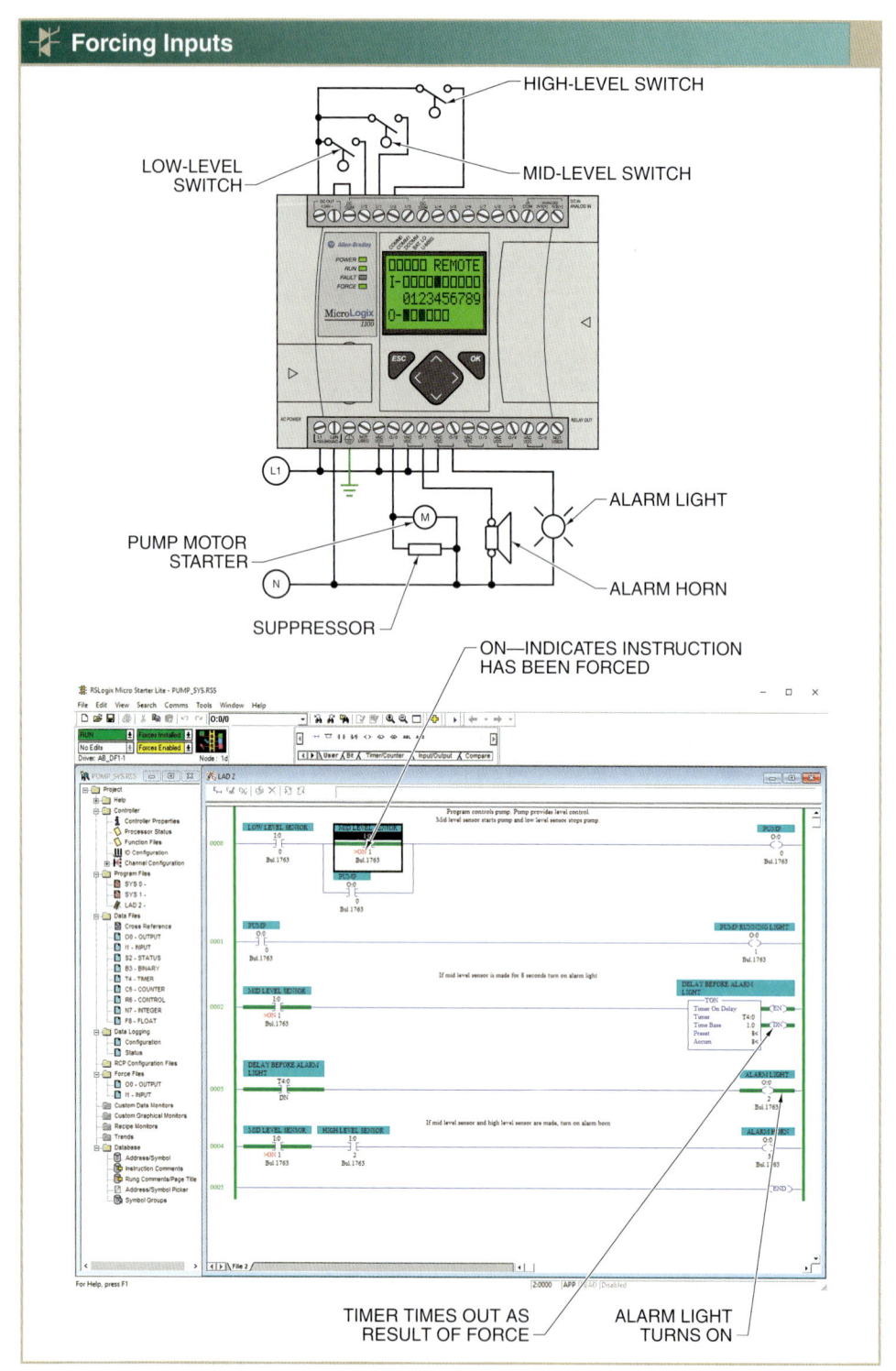

Figure 12-11. A forced input affects all input instructions with the same address. A single forced input can affect many instructions throughout a program.

Forcing Outputs

The output terminals of a PLC can also be forced ON or OFF using programming software. When a force is applied to an output instruction, the programming software writes directly to the output terminal, not to the output data table. Program logic does not react to an output Force function.

Although an ON or OFF appears adjacent to all addresses of the forced instruction, only the output component is affected. Also an output instruction is not highlighted when it is forced. When a force is applied, the POWER, RUN, and FORCE processor status LEDs of the CPU light up. The LCD display rectangle for the forced output terminal is filled in. **See Figure 12-12.**

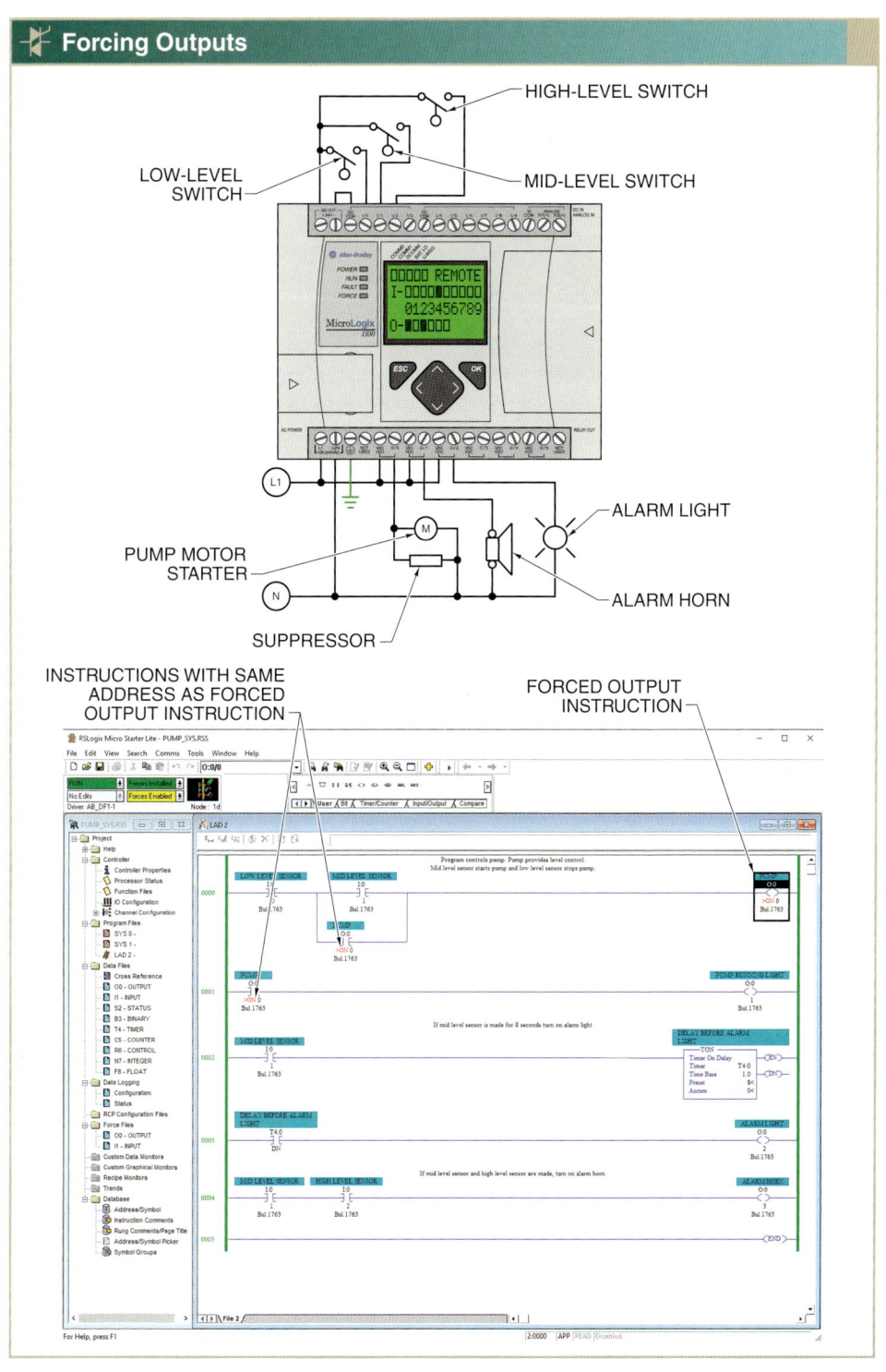

Figure 12-12. A forced output only affects the output instruction. Instructions with the same address as the forced output have an ON or OFF adjacent to them, but they are not affected.

Large modular PLC systems can have extensive program files exceeding thousands of inputs and outputs. Technicians must not only understand a PLC system but also efficiently know the programming software.

CODE CONNECT

Section 350.60 of the NEC® covers the grounding and bonding requirements for liquidtight flexible metal conduit, or Type LFMC. Type LFMC is commonly referred to by its product name Sealtite®. If Type LFMC is used to make a connection to a piece of equipment or machinery where flexibility is necessary to diminish the transmission of vibration, then an equipment grounding conductor must be installed. Likewise, if Type LFMC is used to make a connection to a piece of equipment or machinery that allows the equipment to be moved after installation, the same requirement applies.

SOFTWARE "HELP" FEATURES

Programming software contains several Help features. The Help features provide assistance to end users, PLC programmers, system integrators, and technicians. Help topics include addressing, communication configuration, instruction comments, and power supply loading. The Help features consist of a searchable index of topics, links to web sites maintained by the software manufacturer, and menus accessed by the F1 key. **See Figure 12-13.** All Help features can be used while a PLC is online or offline. The searchable index of topics is accessed as follows:

1. Position the mouse pointer on Help in the menu bar.
2. Left-click the mouse.
3. Position the mouse pointer on the Contents bar.
4. Left-click the mouse. A dialog box titled Help Topics will appear in the program diagram window.

A list of topics can be found under the Contents tab. Topics can be entered and searched for under the Index tab. The Contents tab and the Index tab generate windows that display help information on selected topics. The windows with the help information sometimes contain links to related topics found in the searchable index.

The manufacturers of PLC programming software maintain web sites to provide assistance to technicians. The manufacturer websites provide product updates, technical support, downloadable manuals and documentation, and access to software user groups. Typically the websites are accessible through the Help menu. When a PC with PLC programming software has Internet access, the menu takes the user directly to the support website.

The F1 key is another Help feature found in programming software. The F1 key generates a window that displays help information similar to the searchable index of topics. When the F1 key is pressed, a window with help information is displayed for the dialog box, instruction, view, or window that has been selected. **See Figure 12-14.**

PLC TIPS

When a PLC project or program is verified, the programming software checks that the program complies with basic programming rules. Errors are displayed in the results window. When creating a program, the program should be frequently verified to locate any possible errors.

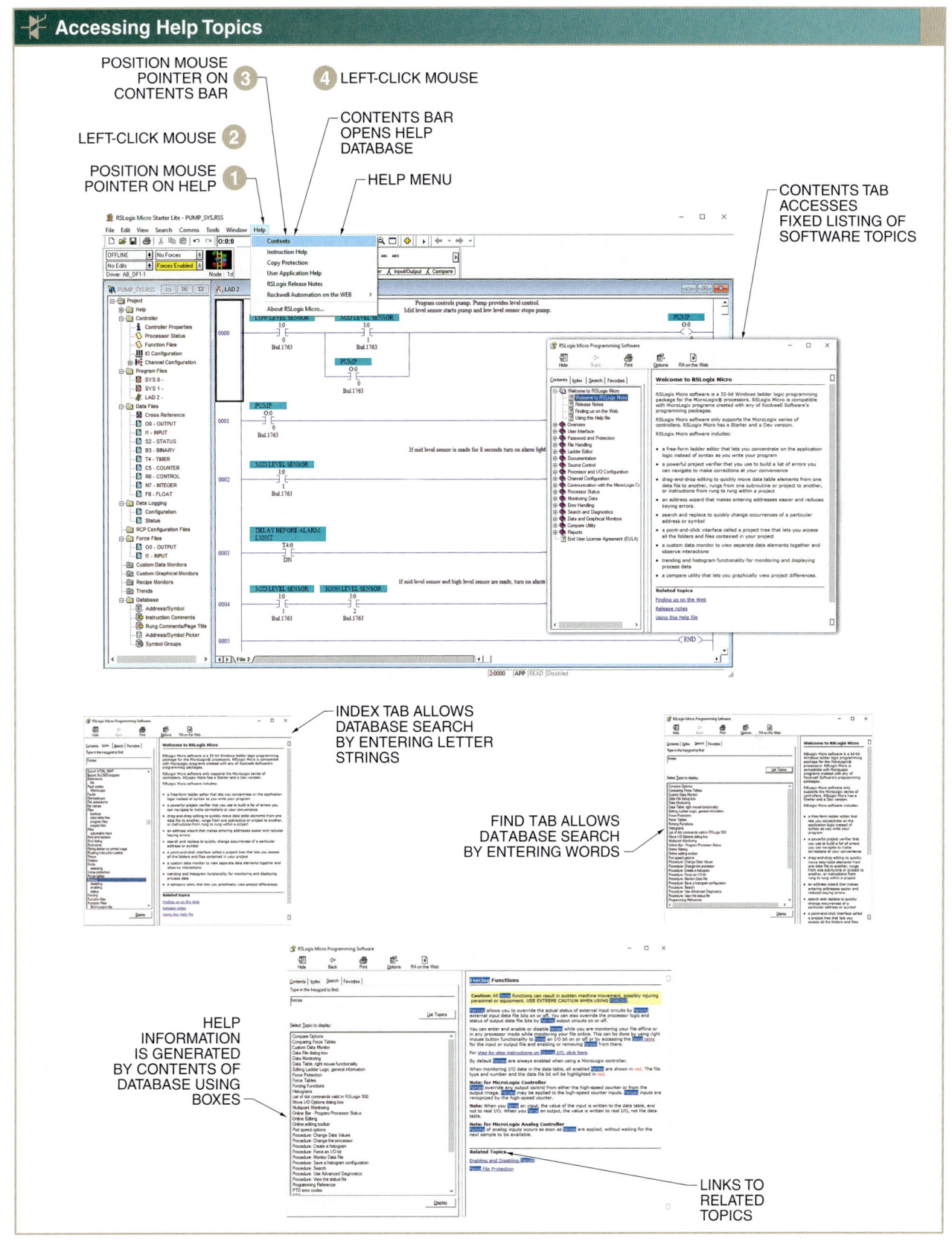

Figure 12-13. The Help Topics dialog box uses more than one method to locate information.

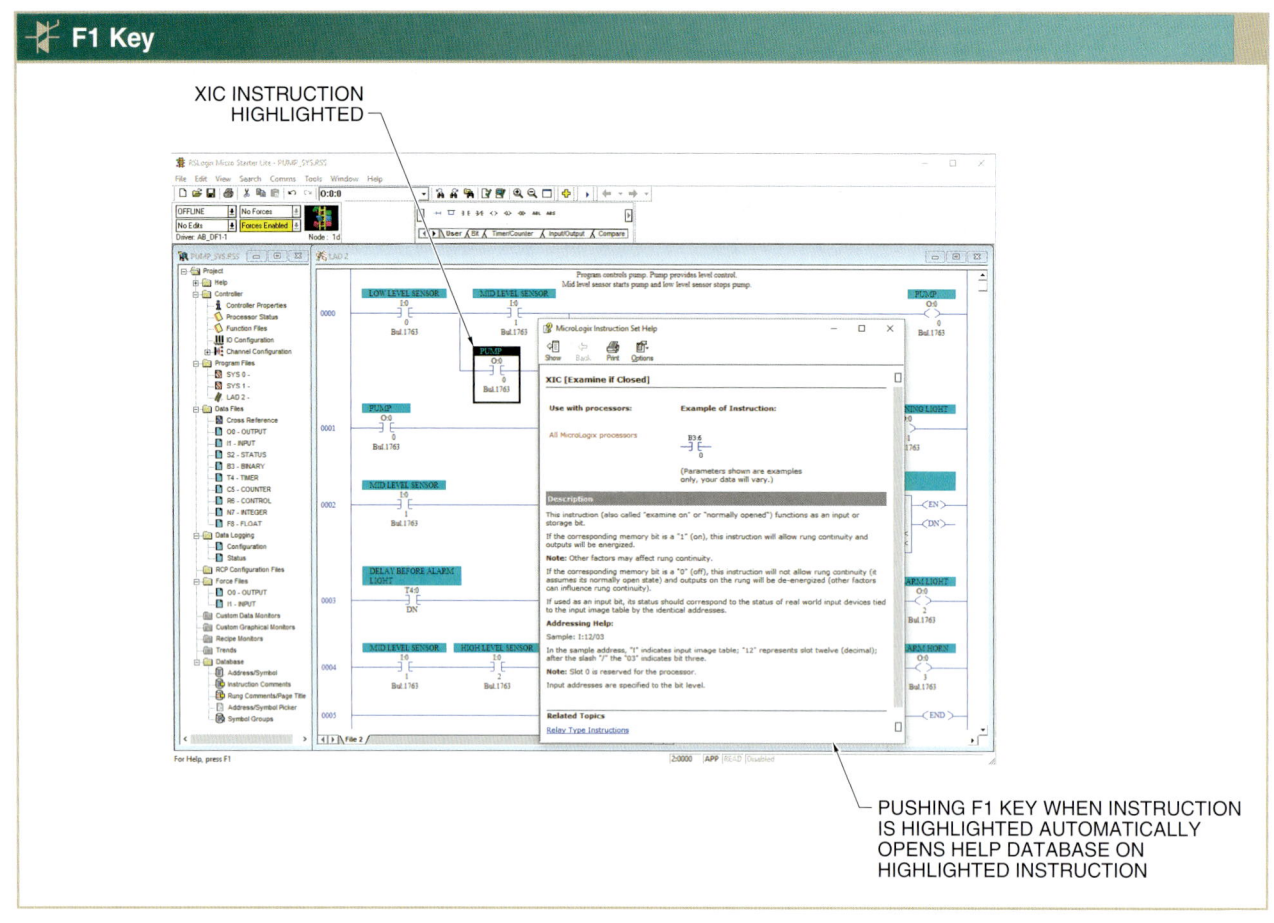

498—PROGRAMMABLE LOGIC CONTROLLERS: PRINCIPLES AND APPLICATIONS

Figure 12-14. The F1 key generates a window containing help information based on what dialog box, instruction, view, or window is selected.

Chapter Review 12

Name _____ Date _____

True-False

T F **1.** PLC software does not typically fail or break.

T F **2.** A PLC program is commonly referred to as a project.

T F **3.** A scroll bar is a shaded horizontal bar at the top of a window that contains the title of a window.

T F **4.** The ladder view is the section of the programming window that contains the programming diagram.

T F **5.** The project tree is the section of application software containing headings for pull-down menus that is placed across the top of a screen.

T F **6.** The first step in troubleshooting a PLC application is to view the program.

T F **7.** Unlike the Find All function, the Cross Reference function only displays the instructions with the same address.

T F **8.** When a technician is finished with the Force function, all forces must be removed.

T F **9.** The debugging process typically starts at the beginning of a program.

T F **10.** The status bar is the section that contains current status information and provides information on how to use the software.

T F **11.** When an input terminal is forced, the LCD display rectangle for the forced input terminal is filled in.

T F **12.** When a temporary end (TND) instruction is programmed without an input instruction, the instruction is always true.

T F **13.** Both the input terminals and output terminals can be forced ON or OFF using programming software.

T F **14.** When the debugging process is completed, any remaining temporary end (TND) instructions are removed.

T F **15.** When using PLC programming software, pressing the F8 key generates a window that displays Help information.

Completion

_____ **1.** The ___ function displays the data table for the selected instruction.

_____ **2.** The ___ features consist of a searchable index of topics and links to web sites maintained by software manufacturers.

_____ 3. The ___ function is a feature of PLC programming software that allows a technician to simulate the state of an input device or output component as ON or OFF.

_____ 4. The PLC ___ contains all of the files, program files, data table files, and documentation for a specific PLC.

_____ 5. A(n) ___ instruction is an output instruction that is used to debug a PLC program.

_____ 6. The ___ is the section of software containing buttons for functions that are frequently used.

_____ 7. ___ is a task that consists of eliminating errors or malfunctions and verifying that a PLC program operates correctly and meets the specifications and requirements of the system.

_____ 8. The ___ function searches the entire program (all program files) to locate each instance of a selected address regardless of the instruction type.

_____ 9. The ___ is a small window section of PLC software that opens up with the results of a Find All or a Verification.

_____ 10. The ___ is a button that is animated when a PC is online with a PLC.

Multiple Choice

_____ 1. The ___ is the section of software that contains information on the processor mode, edit status, force status, and communication drivers of a PLC.
      A. instruction toolbar
      B. ladder view
      C. online bar
      D. project tree

_____ 2. The ___ function searches an entire program (all program files) and locates each instance of a selected address, regardless of the instruction type.
      A. Cross Reference
      B. Find All
      C. Force
      D. Help

_____ 3. ___ are small windows that open to allow users to make choices and enter information.
      A. Check boxes
      B. Dialog boxes
      C. Instruction toolbars
      D. Menu bars

_____ 4. The ___ is the section of programming software that contains various instruction buttons.
      A. instruction toolbar
      B. ladder view
      C. menu bar
      D. online bar

_____ 5. Using the ___ function, a technician can monitor the status of individual bits.
      A. Cross Reference
      B. Find All
      C. Force
      D. Goto Data Table

Wind Speed Analog Sensors

PLC Lab 12.1

BILL OF MATERIALS
- TECO PLR or Micrologix™ 1100 PLC
- Selector switch
- Motor starter/contactor
- Indicator light
- Analog simulator 0–10 VDC or solar cell (10 VDC or less)

In the renewable energy industries, rotating equipment, such as motors and generators on wind turbines, have a maximum safe speed. Exceeding the maximum safe speed can result in damage to the equipment due to excessive centrifugal force. For this lab, a wind speed sensor on a wind turbine is used to slow the turbine when the wind speed exceeds or is equal to a certain level. The sensor activates a brake to slow the turbine and cause an alarm light to come ON.

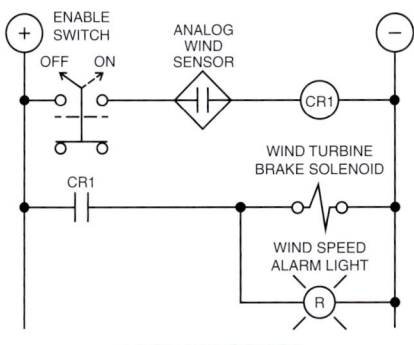

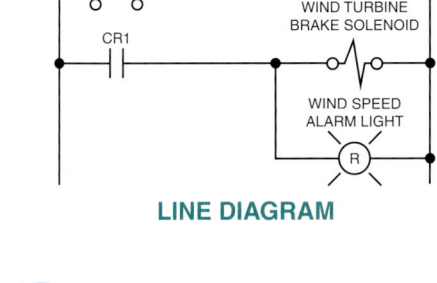

LINE DIAGRAM

T TECO Procedures

1. Based on the PLC system description and line diagram, draw the FBD logic with the proper input(s), output(s), and logic function block(s) with descriptions in the Programming Area.

2. Create a new FBD program electronically.

3. Insert the device input(s), analog input(s), output(s), and analog function block(s). For analog function blocks, add the Gx Function Block with the proper description, number, mode, current value, and reference value to the FBD Programming Area.

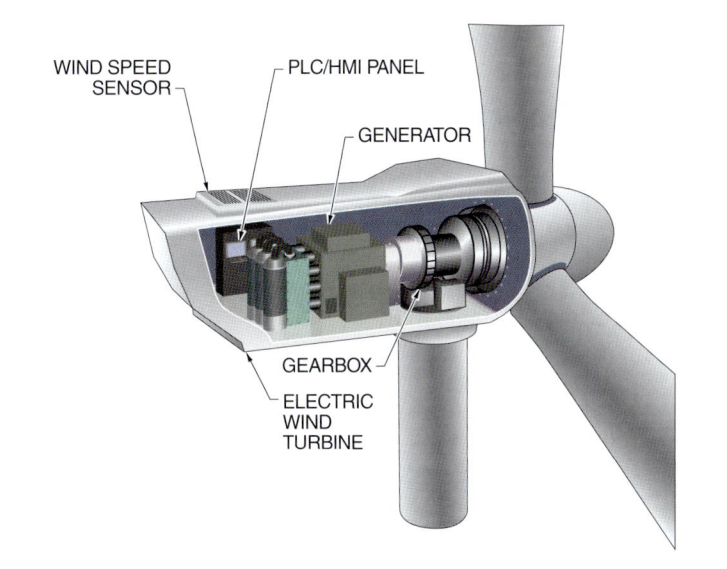

FBD PROGRAMMING AREA

4. Run the program in Simulation Mode to verify proper operation. Use the Input Status tool and Simulation Analog tool window to activate the input(s) and control the output(s) to simulate circuit operation.

5. Draw the wire connections of the input devices and output components to the PLC in the wiring diagram. Then use the wiring diagram to wire the input devices and output components to the PLC.

6. Turn power to the PLC ON and link the com port.

7. Write the program to the SG2-12HR-D to download the program logic to the PLC.

8. Place the SG2-12HR-D in Run Mode. Activate the input(s) and monitor the I/Os on the LCD display.

9. Stop the program and use the Save As function to save and place the lab application in the required folder location.

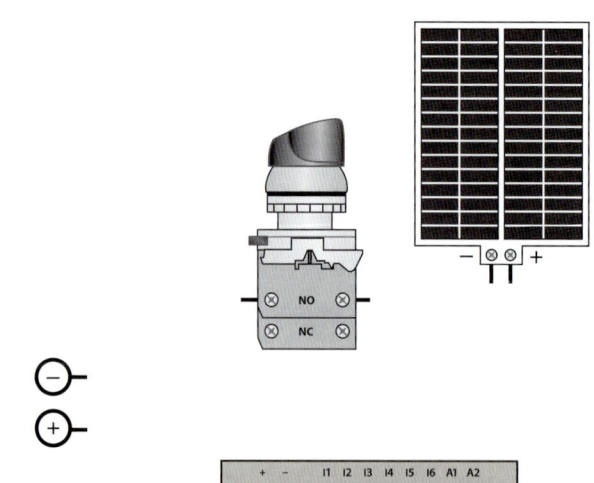

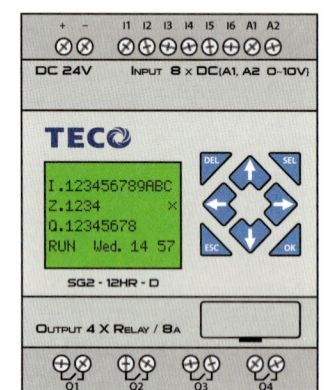

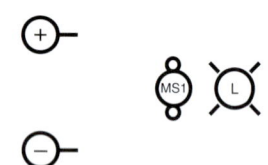

Note: When using solar cells rated less than 10 VDC, a series connection may be made to increase voltage up to 10 VDC.

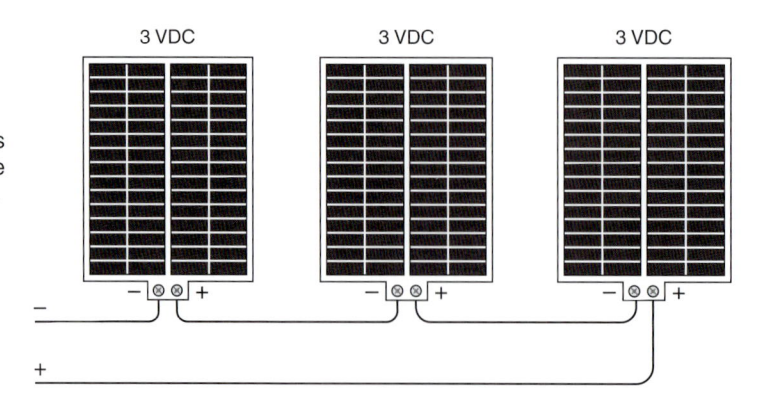

TECO WIRING DIAGRAM

AB Allen-Bradley Procedures

1. Based on the PLC system description and line diagram, draw the ladder logic for RSLogix™ in the AB programming area.

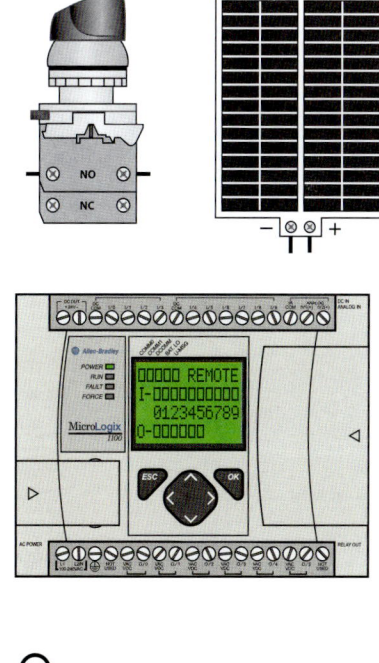

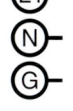

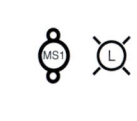

2. Open the RSLogix™ Micro Starter Lite program and create a new programming diagram.

3. Use the AB programming diagram to add the proper address and description for each input and output instruction in the LAD 2 programming window. For the analog instruction, use the GEQ (greater than or equal to) instruction with a word address of I:0.4 in the AB programming area. *Note:* Adjust the values of source A dependent on the input value of the analog sensor/simulator.

4. Verify that the programming diagram has no errors and save the project.

5. Draw the wire connections of the input devices and output components to the PLC in the wiring diagram. Then use the wiring diagram to wire the input devices and output components to the PLC. Once all connections have been made, turn power to the PLC on.

6. Use RSLinx™ to establish communication between the PC and the Micrologix™ 1100.

7. Download the program logic to the PLC.

8. Place the PLC in Run Mode.

9. Activate input(s) and monitor them to determine if the output(s) are operating correctly.

AB PROGRAMMING AREA

AB WIRING DIAGRAM

Ultrasonic Level Sensors Used in Filling Applications

PLC Lab 12.2

BILL OF MATERIALS

- TECO PLR or Micrologix™ 1100 PLC
- Selector switch
- Motor starter/contactor
- Indicator light
- Analog simulator 0–10 VDC or solar cell (10 VDC or less)

In an industrial filling application, a holding tank contains a fluid that is used in a continuous process. For this lab, an ultrasonic level sensor is used to maintain the liquid level in a tank between a reference of 4.00 and 5.00. When the fluid level is less than or equal to 4.00, a fill valve is opened to add fluid to the tank to maintain a minimum level. When the level is greater than or equal to 5.00, a drain valve is opened to prevent the tank from overflowing.

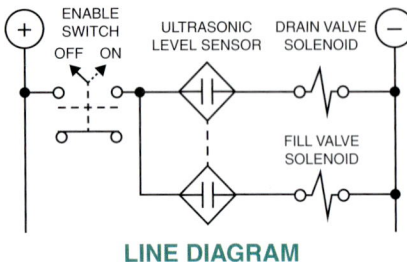

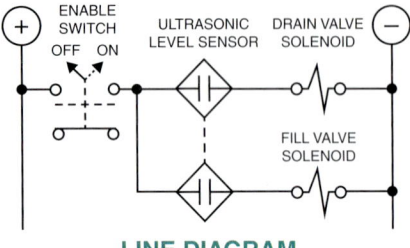

LINE DIAGRAM

T TECO Procedures

1. Based on the PLC system description and line diagram, draw the FBD logic with the proper input(s), output(s), and logic function blocks with descriptions in the Programming Area.

2. Create a new FBD program electronically.

3. Insert the device input(s), analog input(s), output(s), and analog function block(s). For analog function blocks, add the Gx Function Block with the proper description, number, mode, current value, and reference value to the FBD Programming Area.

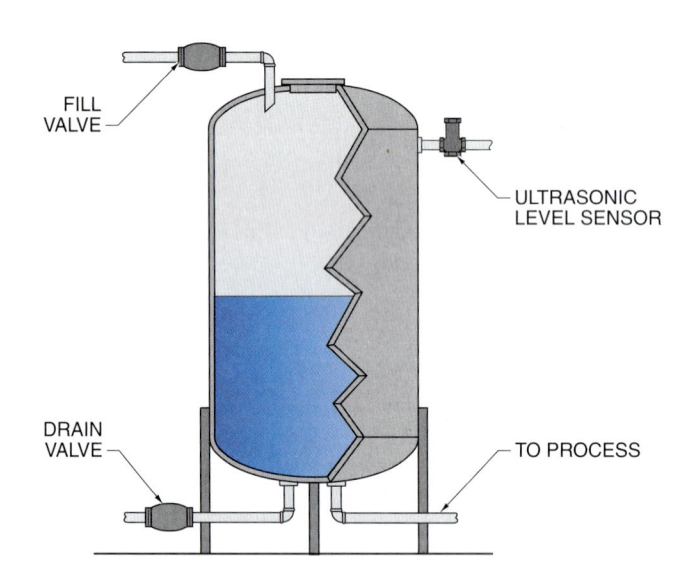

FBD PROGRAMMING AREA

4. Run the program in Simulation Mode to verify proper operation. Use the Input Status tool and Simulation Analog tool window to activate the input(s) and control the output(s) to simulate circuit operation.

5. Draw the wire connections of the input devices and output components to the PLC in the wiring diagram. Then use the wiring diagram to wire the input devices and output components to the PLC.

6. Turn power to the PLC ON and link the com port.

7. Write the program to the SG2-12HR-D to download the program logic to the PLC.

8. Place the SG2-12HR-D in Run Mode. Activate the input(s) and monitor the I/Os on the LCD display.

9. Stop the program and use the Save As function to save and place the lab application in the required folder location.

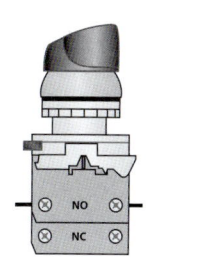

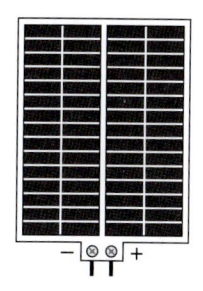

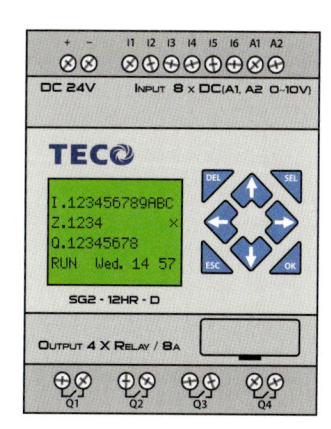

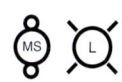

TECO WIRING DIAGRAM

Allen-Bradley Procedures

1. Based on the PLC system description and line diagram, draw the ladder logic for RSLogix™ in the AB programming area.

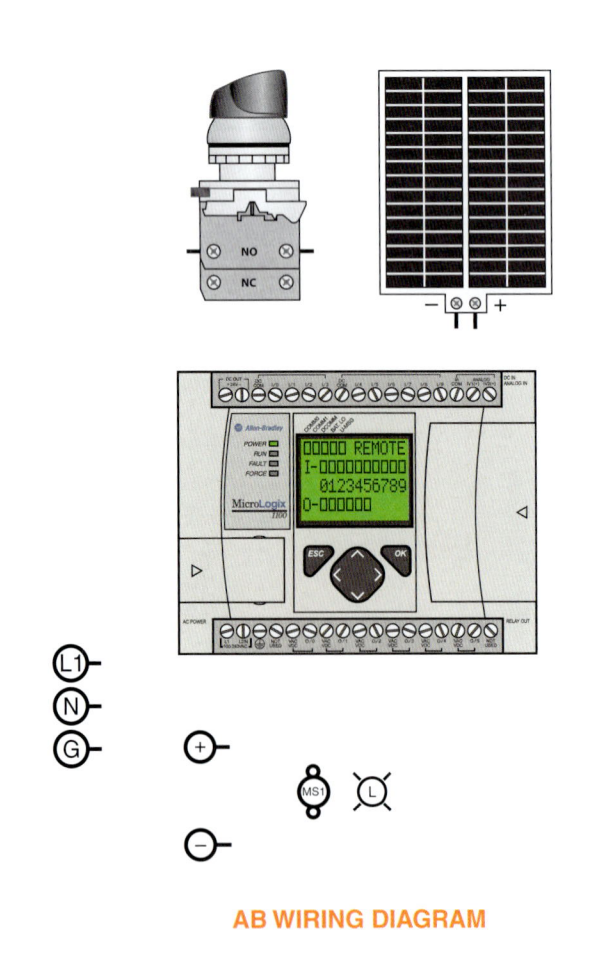

2. Open the RSLogix™ Micro Starter Lite program and create a new programming diagram.

3. Use the AB programming diagram to add the proper address and description for each input and output instruction in the LAD 2 programming window. For the analog instructions, use the GEQ (greater than or equal to) and LEQ (less than or equal to) instructions with a word address of I:0.4 in the AB programming area. *Note:* Adjust the values of source A dependent on the input value of the analog sensor/simulator.

4. Verify that the programming diagram has no errors and save the project.

5. Draw the wire connections of the input devices and output components to the PLC in the wiring diagram. Then use the wiring diagram to wire the input devices and output components to the PLC. Once all connections have been made, turn power to the PLC on.

6. Use RSLinx™ to establish communication between the PC and the Micrologix™ 1100.

7. Download the program logic to the PLC.

8. Place the PLC in Run Mode.

9. Activate input(s) and monitor them to determine if the output(s) are operating correctly.

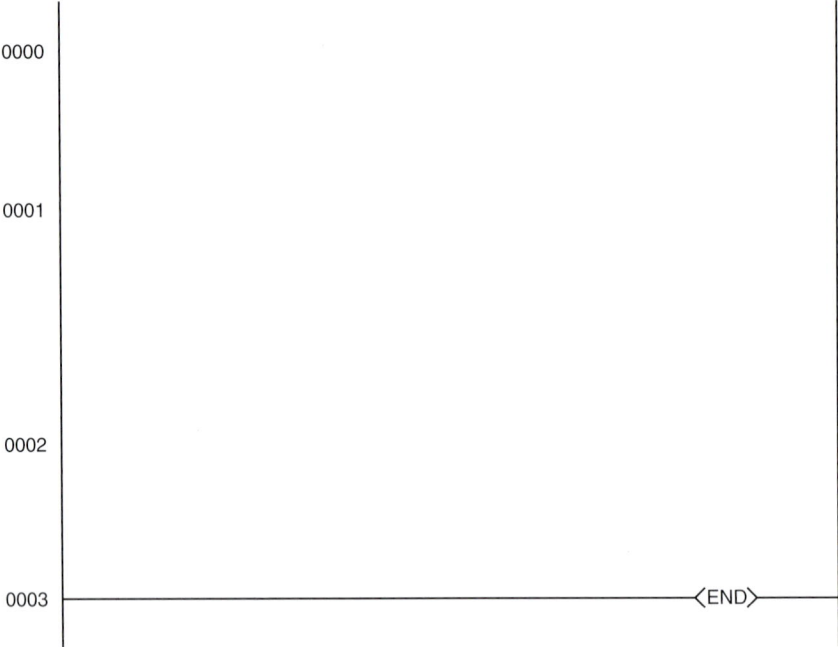

AB PROGRAMMING AREA

AB WIRING DIAGRAM

Chapter 12—Troubleshooting with PLC Software 507

Using Sensors to Optimize a Water Utility

PLC Lab 12.3

BILL OF MATERIALS
- TECO PLR or Micrologix™ 1100 PLC
- Selector switch
- Motor starter/contactor
- Indicator light
- Two analog simulators 0–10 VDC or solar cells (10 VDC or less)

Watering a golf course can be a large expense. Not only do owners spend money on the cost of the water but also on the cost of the electricity to pump the water around the golf course. To conserve money and natural resources, sensors are used to measure the moisture content of the soil and the wind speed. The wind speed and moisture content are measured to determine the optimal time to turn on the sprinklers. For this lab, a moisture sensor is set to turn ON when the moisture content is less than or equal to a reference. A wind sensor is set to turn ON when the wind speed is less than or equal to a reference. The moisture sensor and the wind sensor must be ON in order for the sprinkler solenoid to turn ON.

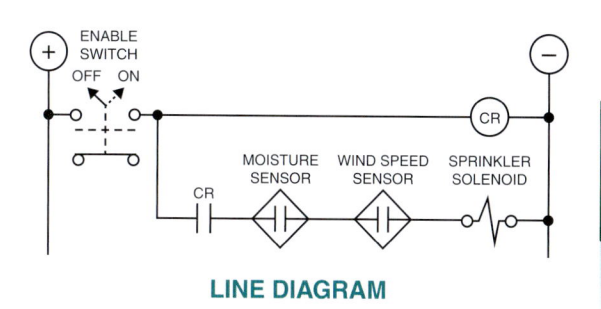

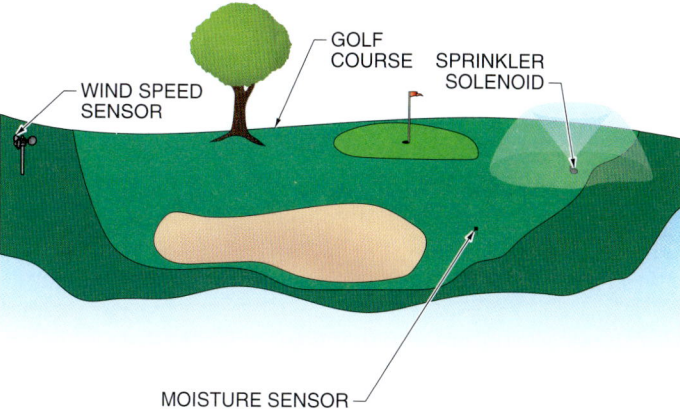

LINE DIAGRAM

T TECO Procedures

1. Based on the PLC system description and line diagram, draw the FBD logic with the proper input(s), output(s), and logic function blocks with descriptions in the Programming Area.

2. Create a new FBD program electronically.

3. Insert the device input(s), analog input(s), output(s), and analog function block(s). For analog function blocks, add the Gx Function Block with the proper description, number, mode, current value, and reference value to the FBD Programming Area.

FBD PROGRAMMING AREA

4. Run the program in Simulation Mode to verify proper operation. Use the Input Status tool and Simulation Analog tool window to activate the input(s) and control the output(s) to simulate circuit operation.

5. Draw the wire connections of the input devices and output components to the PLC in the wiring diagram. Then use the wiring diagram to wire the input devices and output components to the PLC.

6. Turn power to the PLC ON and link the com port.

7. Write the program to the SG2-12HR-D to download the program logic to the PLC.

8. Place the SG2-12HR-D in Run Mode. Activate the input(s) and monitor the I/Os on the LCD display.

9. Stop the program and use the Save As function to save and place the lab application in the required folder location.

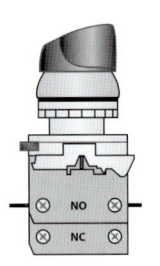

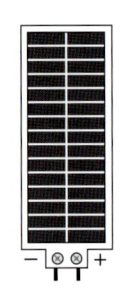

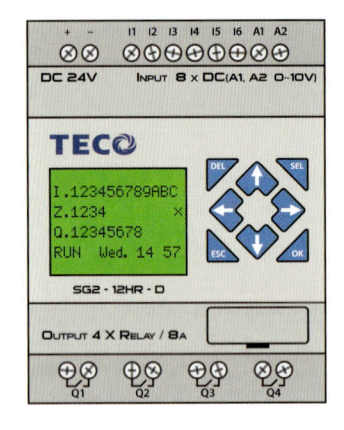

TECO WIRING DIAGRAM

Chapter 12—Troubleshooting with PLC Software 509

Allen-Bradley Procedures

1. Based on the PLC system description and line diagram, draw the ladder logic for RSLogix™ in the AB programming area.

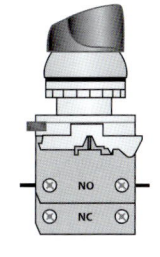

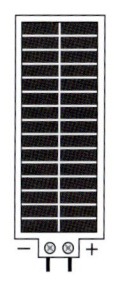

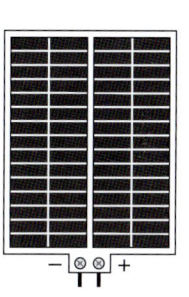

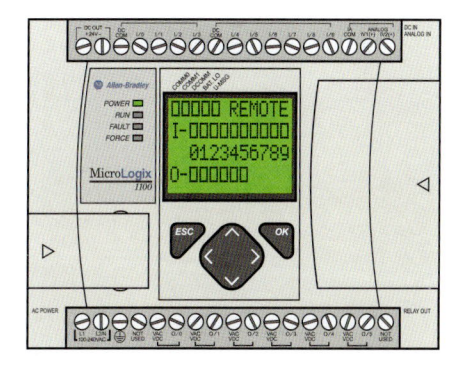

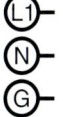

2. Open the RSLogix™ Micro Starter Lite program and create a new programming diagram.

3. Use the AB programming diagram to add the proper address and description for each input and output instruction in the LAD 2 programming window. For the analog instruction, use the LEQ (less than or equal to) instructions with a word address of I:0.4 and I:0.5 in the AB programming area. *Note:* Adjust the values of source A dependent on the input value of the analog sensor/simulator.

4. Verify that the programming diagram has no errors and save the project.

5. Draw the wire connections of the input devices and output components to the PLC in the wiring diagram. Then use the wiring diagram to wire the input devices and output components to the PLC. Once all connections have been made, turn power to the PLC on.

6. Use RSLinx™ to establish communication between the PC and the Micrologix™ 1100.

7. Download the program logic to the PLC.

8. Place the PLC in Run Mode.

9. Activate input(s) and monitor them to determine if the output(s) are operating correctly.

0000

0001

0002 ─────────────────────────⟨END⟩

AB PROGRAMMING AREA

AB WIRING DIAGRAM

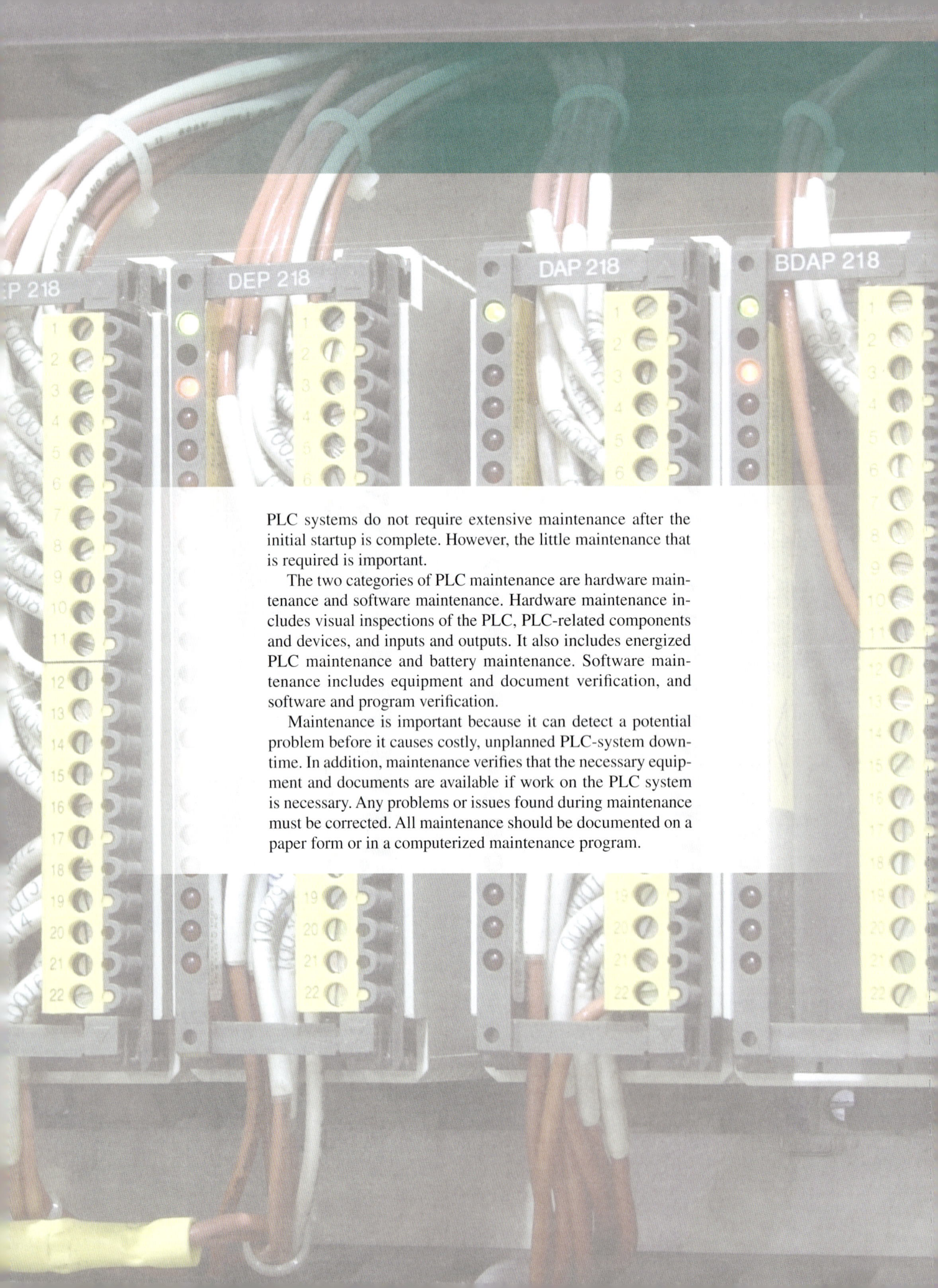

PLC systems do not require extensive maintenance after the initial startup is complete. However, the little maintenance that is required is important.

The two categories of PLC maintenance are hardware maintenance and software maintenance. Hardware maintenance includes visual inspections of the PLC, PLC-related components and devices, and inputs and outputs. It also includes energized PLC maintenance and battery maintenance. Software maintenance includes equipment and document verification, and software and program verification.

Maintenance is important because it can detect a potential problem before it causes costly, unplanned PLC-system downtime. In addition, maintenance verifies that the necessary equipment and documents are available if work on the PLC system is necessary. Any problems or issues found during maintenance must be corrected. All maintenance should be documented on a paper form or in a computerized maintenance program.

PLC System Maintenance

13

Objectives:

- Explain how a PLC and PLC system are maintained.
- List the steps involved in safely performing a visual inspection of a PLC as part of hardware maintenance.
- List the steps involved in safely performing energized maintenance of a PLC as part of hardware maintenance.
- List the steps involved in safely performing battery maintenance of a PLC as part of hardware maintenance.
- Explain the process of equipment and documentation verification as part of PLC software maintenance.
- Explain the process of software and program verification as part of PLC software maintenance.

Learner Resources
atplearningresources.com/quicklinks
Access Code: **475203**

PLC SYSTEM MAINTENANCE

After installation and startup, PLCs do not require a large amount of maintenance. PLC system maintenance consists of two main areas: hardware and software. Hardware maintenance covers the PLC, input devices and output components, PLC enclosure, and PLC-related components. Software maintenance covers the programming device and related hardware, the PLC program, documentation for the PLC application, and a method to back up the PLC program.

A problem found during maintenance must be corrected before the problem leads to PLC failure and equipment downtime or hampers future troubleshooting efforts. System maintenance must be scheduled so maintenance does not interfere with the machine or process the PLC controls. As part of PLC system maintenance, power to a PLC is turned OFF. PLC system maintenance is typically performed annually and always recorded using maintenance logs. **See Figure 13-1. See Appendix.**

The following guidelines should be observed when performing PLC system maintenance:

- Only qualified technicians familiar with the PLC, related components, and the machine or process being controlled can perform PLC maintenance.
- For maximum safety, the technician should follow the recommendations and instructions of the machine and PLC manufacturer and any applicable federal, state, and local regulations.

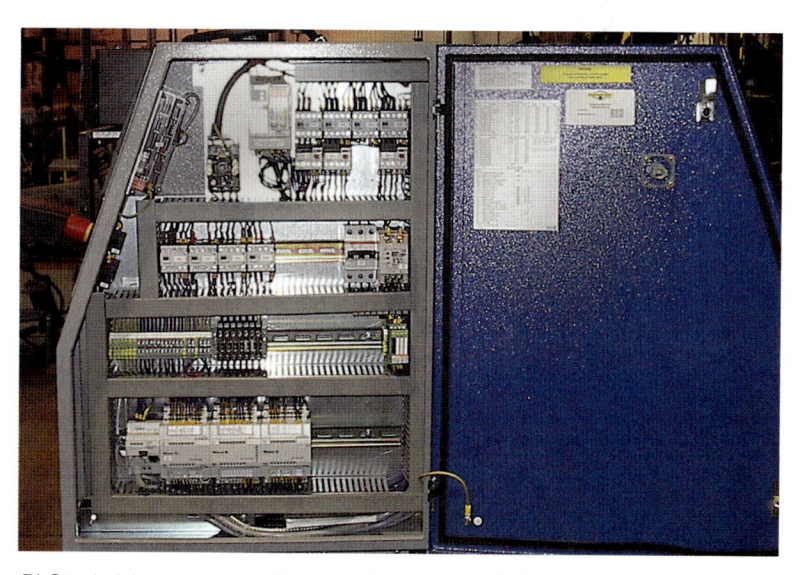

PLC technicians are required to understand the controlled machine or process (primary system), understand PLC programming and hardware, and maintain all electronic and paper records.

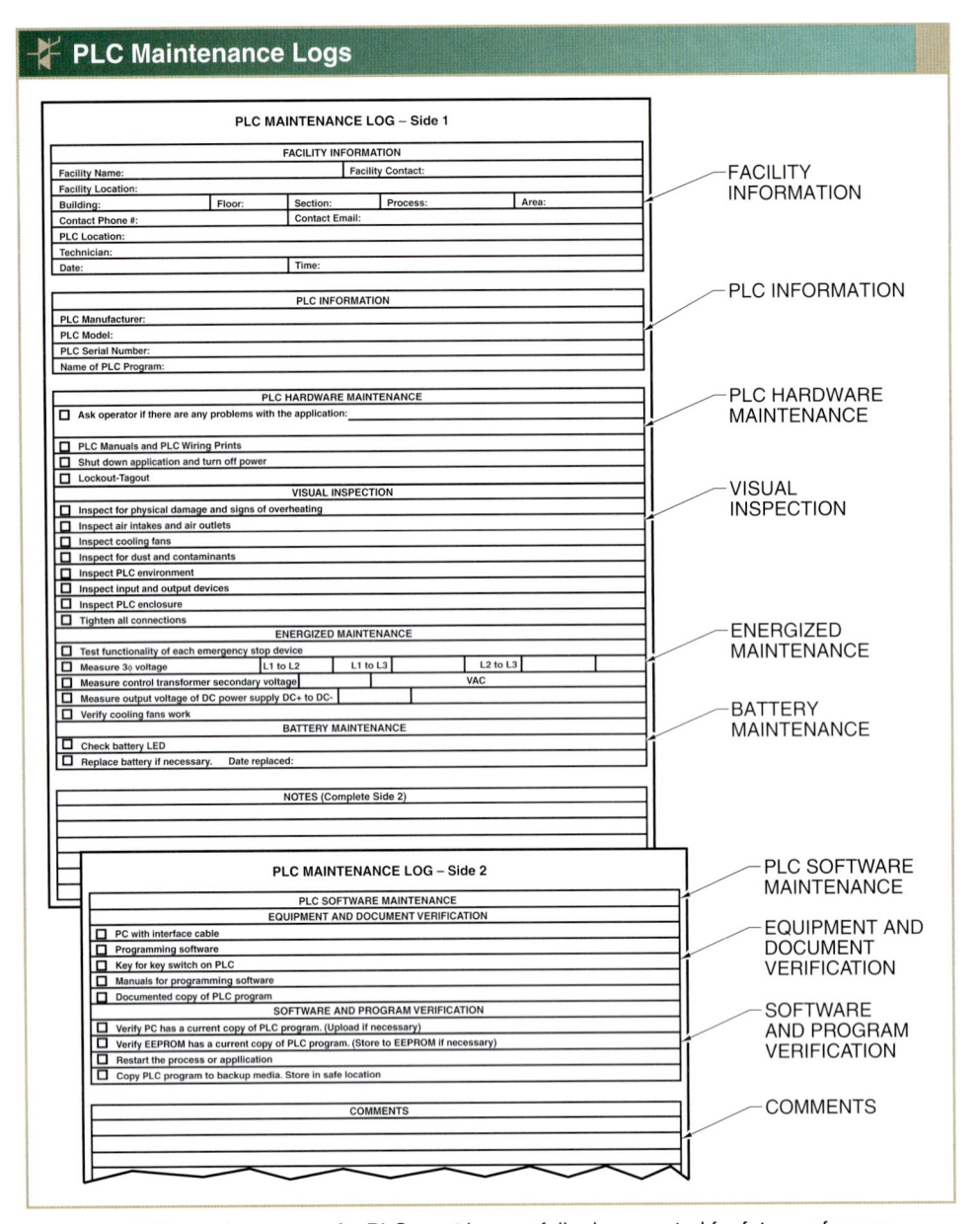

Figure 13-1. The maintenance of a PLC must be carefully documented for future reference.

PLC TIPS

Every electrical control panel in a facility must be opened when being audited, and all PLCs must have their brand, model, and other important information written down. Important PLC information typically includes machine or area name; PLC program name; network name; network node address; PLC brand and model number; copy of program with descriptions; availability of spare; availability of EEPROM; last program modification date and back-up date; last EEPROM storage date; and last battery change date.

- Before performing PLC system maintenance, the technician must talk to the machine or process operators who run the PLC-controlled system about any problems. When an operator reports a problem, the technicians must locate, verify, and correct the problem.

- Technicians performing PLC system maintenance must understand how to safely stop the machine or process, turn the PLC OFF, and lockout and tagout all energy sources. **See Figure 13-2.**

PLC Maintenance Guidelines

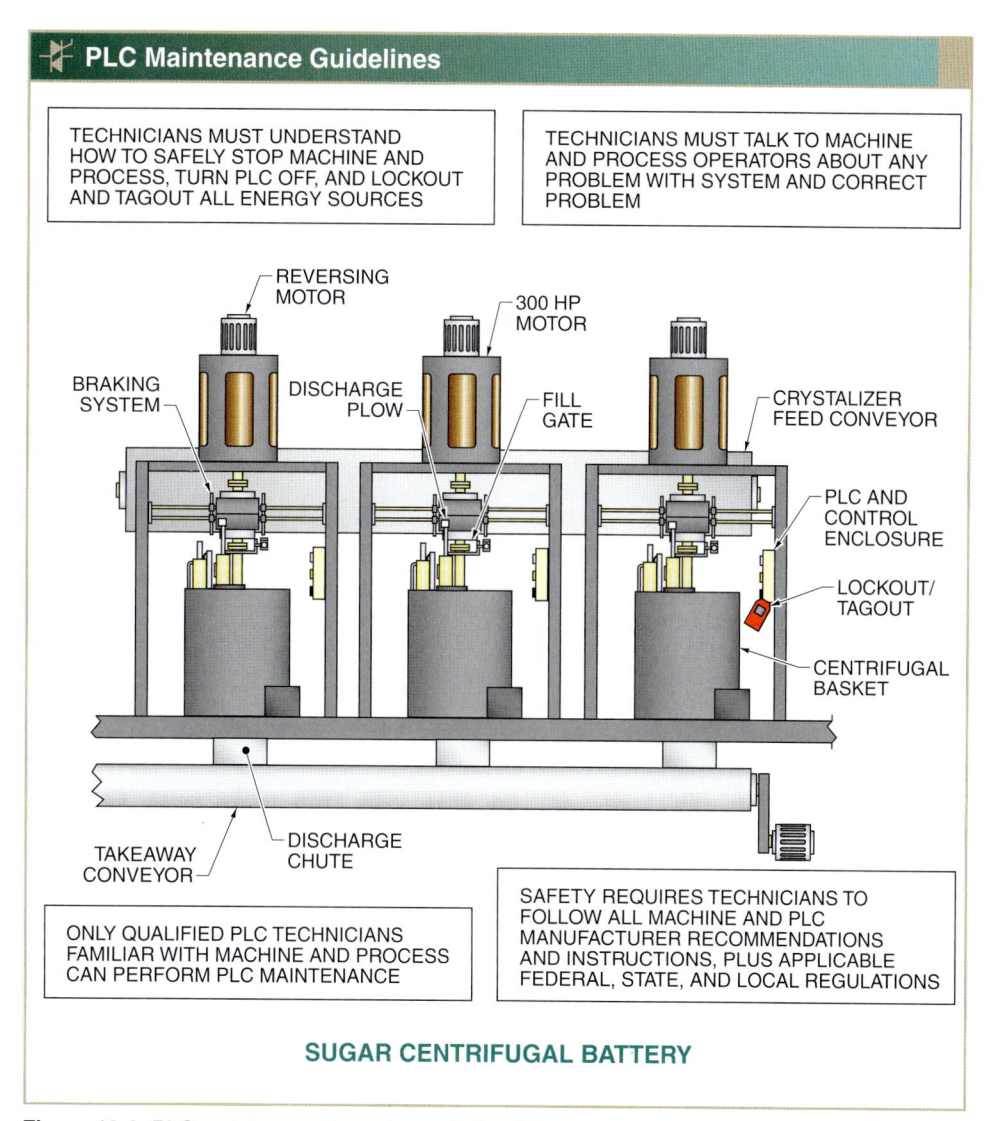

Figure 13-2. PLC maintenance is performed after the application is stopped and a lockout/tagout is applied.

CODE CONNECT

Article 440 of the NEC® covers air conditioning and refrigeration equipment, commonly referred to as heating, ventilation, air conditioning, and refrigeration (HVACR). Included in Article 440 are the requirements for sizing conductors and overcurrent protective devices. Manufacturers of commercial HVACR equipment list the minimum circuit ampacity (MCA) and the maximum overcurrent protection (MOP) on equipment nameplates. MCA is used to size the supply conductors that feed the HVACR equipment. MOP is the maximum size overcurrent device (fuse or circuit breaker) allowed. These values are calculated by the HVACR equipment manufacturers per the NEC® and applicable UL standards.

PLC HARDWARE MAINTENANCE

Hardware maintenance is maintenance that consists of a visual inspection of the PLC, energized maintenance procedures, and battery maintenance. To perform the PLC hardware maintenance, a technician must have the PLC manuals and the PLC wiring prints. Before any visual inspection is performed, power to the PLC must be turned OFF. Any foreign control voltages present in the PLC enclosure must also be turned OFF. A technician performing PLC hardware maintenance may require the assistance of a second technician or an operator because some PLC applications must be stopped in a specific sequence to prevent equipment damage. After all power is turned OFF, a lockout/tagout must be applied to the disconnect. **See Figure 13-3.**

PLC TIPS

Some PLCs utilize an HTML-based human-machine interface (HMI) for monitoring machine or process status using a web browser. The main page menu of the PLC appears when the address of the PLC is entered into the browser.

Chapter 13—PLC System Maintenance   513

PLC Hardware Maintenance

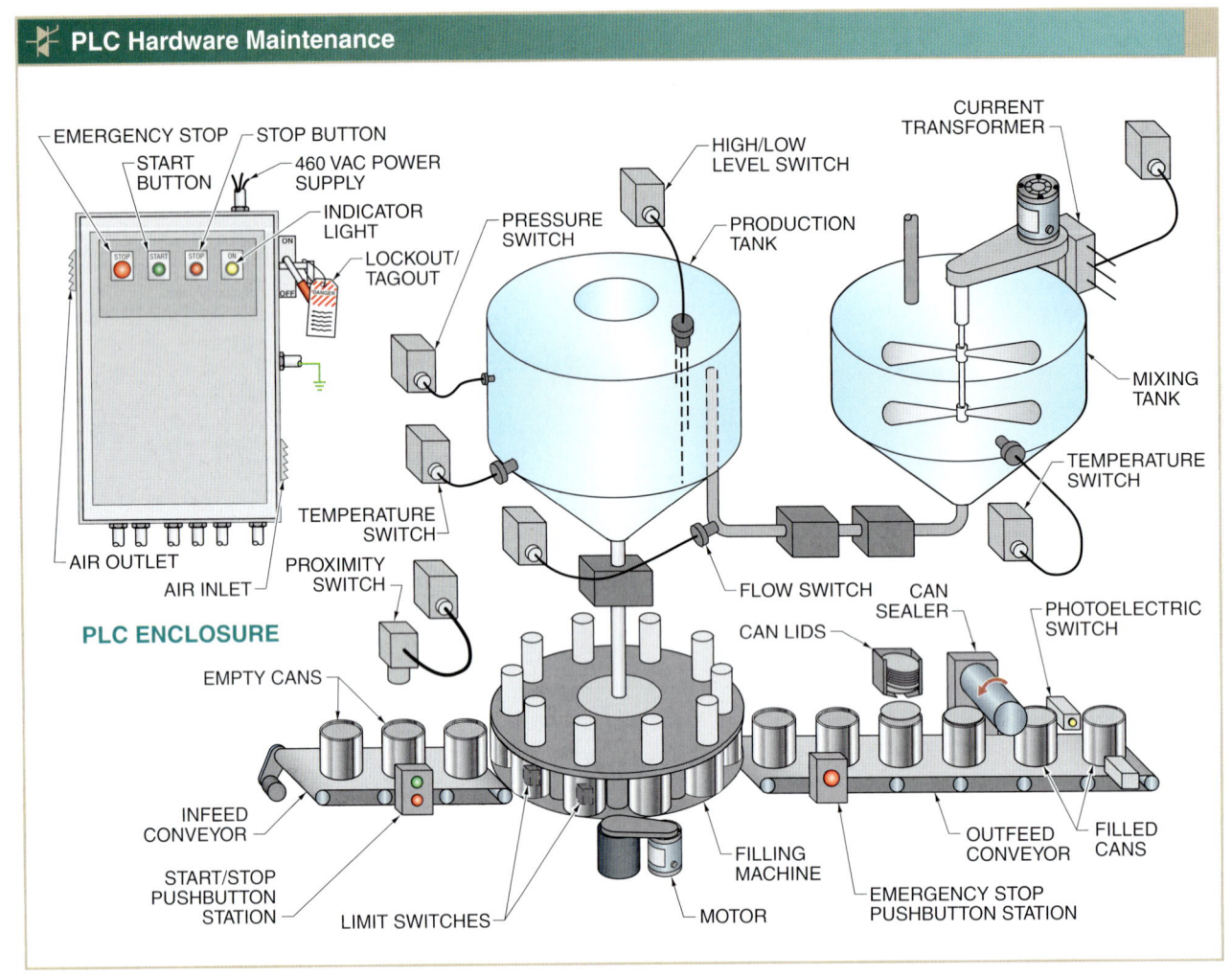

Figure 13-3. PLC hardware maintenance consists of visual inspection of a PLC, energized maintenance procedures, and battery maintenance.

Visual Inspections

A thorough visual inspection of the PLC, PLC-related components, and input devices and output components must be done periodically. **See Figure 13-4.** Before performing a visual inspection, a DMM is used to verify that no voltage is present in a PLC enclosure. To perform a visual inspection, the following procedure is applied:

1. Inspect the PLC and PLC-related components (terminal strips, circuit breakers, relays, and motor starters) for physical damage and signs of overheating.

2. Inspect air intakes and outlets if present. Air intakes and outlets must be free of dirt, dust, and obstructions. Intakes and outlets must have filters to prevent contaminants from entering the enclosure. Replace the air intake and air outlet filters per the manufacturer's schedule or when dirty.

3. Inspect cooling fans if present. Fan housings and fan blades must be free of dirt, dust, and obstructions. When not powered, cooling fans rotate freely by hand and make no noise. A fan that does not rotate freely or is noisy requires replacement. Typically fans are replaced every 3-to-5 years.

4. Inspect PLC and PLC-related components for accumulation of dirt or contaminants. Dust and contaminants

can interfere with the cooling of devices and components and cause heat-related problems. Dust and contaminants can also be conductive, causing shorts to ground. Use an electrostatic discharge (ESD) safe vacuum to remove dust and contaminants from PLCs and enclosures. Do not use compressed air to remove dust or contaminants. Compressed air blows dust or contaminants into inaccessible areas and creates additional problems.

Visual Inspections

- ⑦ INSPECT PLC ENCLOSURE FOR DAMAGE OR WEAR
- ③ INSPECT COOLING FAN
- ② INSPECT AIR OUTLET AND FILTER (NOT SHOWN)
- ④ INSPECT PLC FOR DIRT AND CONTAMINANTS
- ⑧ TIGHTEN ALL CONNECTIONS
- ① INSPECT TERMINAL STRIPS, CIRCUIT BREAKERS, RELAYS, AND MOTOR STARTERS FOR OVERHEATING
- ④ INSPECT TERMINAL STRIPS, CIRCUIT BREAKERS, RELAYS, AND MOTOR STARTERS FOR DIRT AND CONTAMINANTS
- ① INSPECT PLC FOR OVERHEATING
- ② INSPECT AIR INTAKE AND FILTER
- ⑥ INSPECT INPUT CONDUITS FOR DAMAGE
- ⑤ VERIFY ENVIRONMENT AROUND PLC ENCLOSURE IS STABLE
- ⑥ INSPECT INPUT DEVICES FOR DAMAGE
- ⑥ INSPECT OUTPUT COMPONENTS FOR DAMAGE

PLC ENCLOSURE

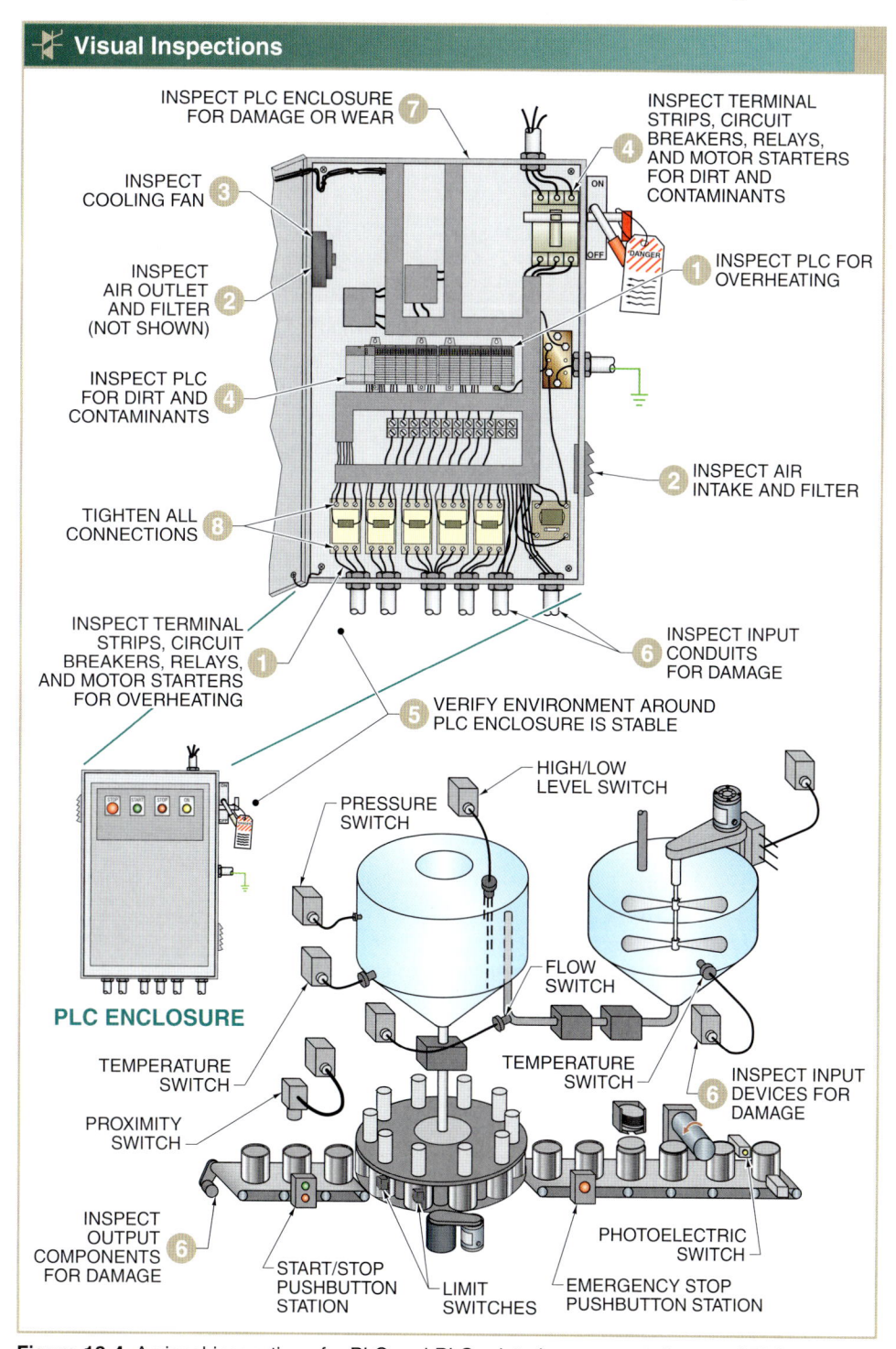

PRESSURE SWITCH, HIGH/LOW LEVEL SWITCH, FLOW SWITCH, TEMPERATURE SWITCH, TEMPERATURE SWITCH, PROXIMITY SWITCH, START/STOP PUSHBUTTON STATION, LIMIT SWITCHES, PHOTOELECTRIC SWITCH, EMERGENCY STOP PUSHBUTTON STATION

Figure 13-4. A visual inspection of a PLC and PLC-related components is part of PLC hardware maintenance.

CODE CONNECT

Section 300.21 of the NEC® requires fire-stopping where electrical installations penetrate fire-rated ceilings, floors, or walls. Approved fire-stopping materials must be used to return the ceiling, floor, or wall to its original rating before penetration. A variety of approved fire-stopping means are available, including foam, caulk, and putty.

5. Verify that the environment surrounding the PLC enclosure has not changed. For example, a source of heat or EMI such as a welding machine may have been added near the enclosure. A change in the environment may require a change in the type of enclosure, cooling method, or method of dealing with EMI.

6. Inspect input devices and output components for any physical damage or excessive wear. Also, inspect conduits and cables that feed devices and components for any physical damage or abuse.

7. Inspect the PLC enclosure for damage or uncapped holes. When the cover of an enclosure has a gasket, verify that the gasket is not cracked or damaged.

8. Tighten all connections inside the PLC enclosure. Connections can be loose from the original installation or can become loose over time due to vibration or due to the heating and cooling of conductors as loads are turned ON and OFF. Loose connections are one of the most common sources of intermittent PLC problems.

Energized PLC Maintenance

Energized PLC maintenance is performed after the visual inspection is successfully completed. *Energized maintenance* is maintenance that verifies that the master control relay (MCR) is functioning correctly, voltages are within specified range, and no abnormal conditions exist. **See Figure 13-5.** To perform energized maintenance, apply the following procedure:

1. Close and secure the PLC enclosure door. Remove the lockout/tagout.

2. Turn the disconnect ON. Stand to the side of a PLC enclosure when energizing in case of a major electrical fault or failure.

3. Open the door of the PLC enclosure. Most disconnects have an interlock connected to the enclosure door to prevent the door from opening when the disconnect is ON. Typically there is a screw near the disconnect handle that allows a technician to defeat the interlock and open the door when the disconnect is ON.

4. Before testing the emergency stops and MCR, verify that the machine or process will not be started by energizing the MCR. Typically a separate start button is used for the machine or process.

5. Test the MCR using the MCR start button and the emergency stop pushbuttons. Verify that the emergency stop pushbuttons de-energize the MCR, and that when the MCR is de-energized, there is no power to the PLC input devices and output components. Test each emergency stop pushbutton and any devices used as emergency stops, such as safety curtain limit switches. When finished, leave the MCR de-energized.

6. Measure and record the incoming three-phase voltage, the control transformer secondary voltage, and the output of the DC power supply (if present). Verify that the voltages are within the specified range.

7. Verify that the enclosure cooling units and cooling fans, when present, are operating. Replace cooling units or fans that do not work. Also replace fans that are noisy.

8. Check the PLC power supply and processor LEDs. The number of LEDs and what the LEDs represent vary between PLC models and PLC manufacturers. Common LEDs include POWER, RUN, FAULT (FLT), BATTERY (BATT), and FORCE. POWER and RUN must be ON. FAULT, BATTERY, and FORCE should be OFF. When the state of these LEDs on the CPU is different, a technician must determine the cause.

Battery Maintenance

Battery maintenance is performed after energized maintenance is successfully completed. *Battery maintenance* is maintenance that verifies that a PLC battery has not exceeded its rated life. Many PLCs use lithium batteries to provide a back-up power source for RAM memory in the event of a power outage. Lithium batteries last between 2 yr and 5 yr. Lithium batteries are replaced as required.

Rockwell Automation Inc.
Modular PLC processor cards have back-up battery power supplies and back-up capacitor power for retaining RAM memory.

Energized PLC Maintenance

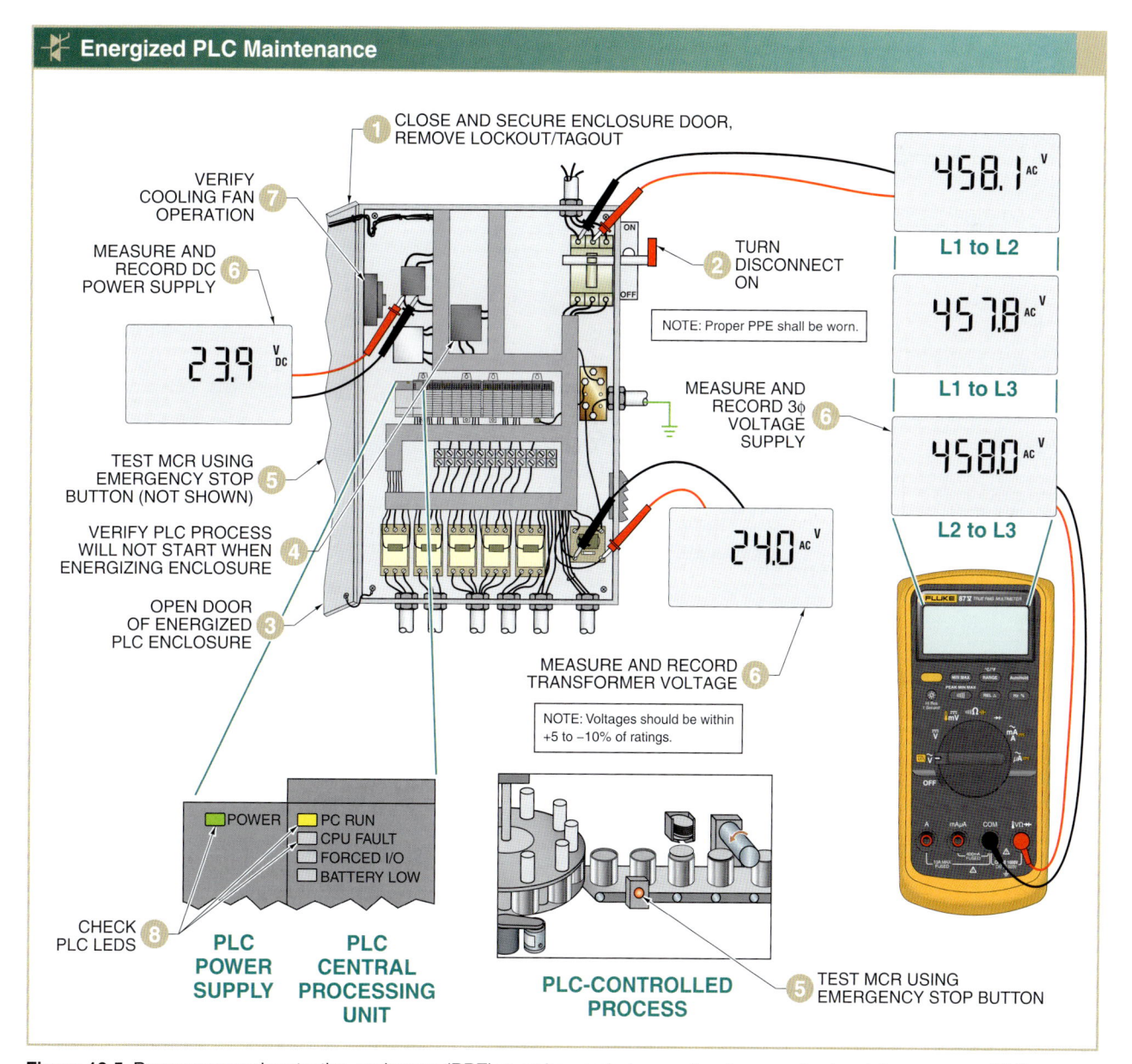

1. CLOSE AND SECURE ENCLOSURE DOOR, REMOVE LOCKOUT/TAGOUT
2. TURN DISCONNECT ON — NOTE: Proper PPE shall be worn.
3. OPEN DOOR OF ENERGIZED PLC ENCLOSURE
4. VERIFY PLC PROCESS WILL NOT START WHEN ENERGIZING ENCLOSURE
5. TEST MCR USING EMERGENCY STOP BUTTON (NOT SHOWN) / TEST MCR USING EMERGENCY STOP BUTTON
6. MEASURE AND RECORD DC POWER SUPPLY — 23.9 V DC
   MEASURE AND RECORD 3φ VOLTAGE SUPPLY — 458.1 AC V (L1 to L2), 457.8 AC V (L1 to L3), 458.0 AC V (L2 to L3)
   MEASURE AND RECORD TRANSFORMER VOLTAGE — 24.0 AC V
   NOTE: Voltages should be within +5 to –10% of ratings.
7. VERIFY COOLING FAN OPERATION
8. CHECK PLC LEDS — PLC POWER SUPPLY (POWER); PLC CENTRAL PROCESSING UNIT (PC RUN, CPU FAULT, FORCED I/O, BATTERY LOW)

PLC-CONTROLLED PROCESS

Figure 13-5. Proper personal protective equipment (PPE) must be used when performing energized maintenance on a PLC.

518—PROGRAMMABLE LOGIC CONTROLLERS: PRINCIPLES AND APPLICATIONS

The processor of a PLC monitors battery voltage level. When battery voltage falls below a threshold, the "LOW" LED is turned ON and the battery must be replaced. There are two different procedures for replacing PLC batteries, depending on the location of the battery. One procedure is used when the battery is mounted behind a hinged door or cover and can be easily removed. Another procedure is used when the battery is mounted internally on the card of the CPU. **See Figure 13-6.**

Battery behind Hinged Door. To replace a lithium battery located behind a hinged door or cover, apply the following procedure:

1. Ensure power to the PLC is ON and will remain ON during the battery replacement. When power is OFF during battery replacement, RAM memory is lost.
2. Open the hinged door or cover that provides access to the battery.
3. Disconnect the old battery and remove from the holder.

PLC Battery Replacement

BATTERY BEHIND HINGED DOOR (FRONT VIEW)
1. ENSURE POWER TO PLC REMAINS ON
2. OPEN DOOR OF CPU
3. DISCONNECT AND REMOVE BATTERY
4. INSTALL NEW BATTERY AND CONNECT
5. APPLY LABEL WITH NEW BATTERY DATE
6. PROPERLY DISPOSE OF OLD BATTERY

PROCESSOR CARD MOUNTED BATTERY (SIDE VIEW)
1. TURN PLC POWER OFF; LOCKOUT AND TAGOUT DISCONNECT
2. REMOVE PROCESSOR CARD FROM CHASSIS
3. DISCONNECT AND REMOVE BATTERY
4. INSTALL NEW BATTERY AND CONNECT
5. INSTALL PROCESSOR CARD; REMOVE LOCKOUT/TAGOUT
6. APPLY LABEL WITH NEW BATTERY DATE
7. PROPERLY DISPOSE OF OLD BATTERY

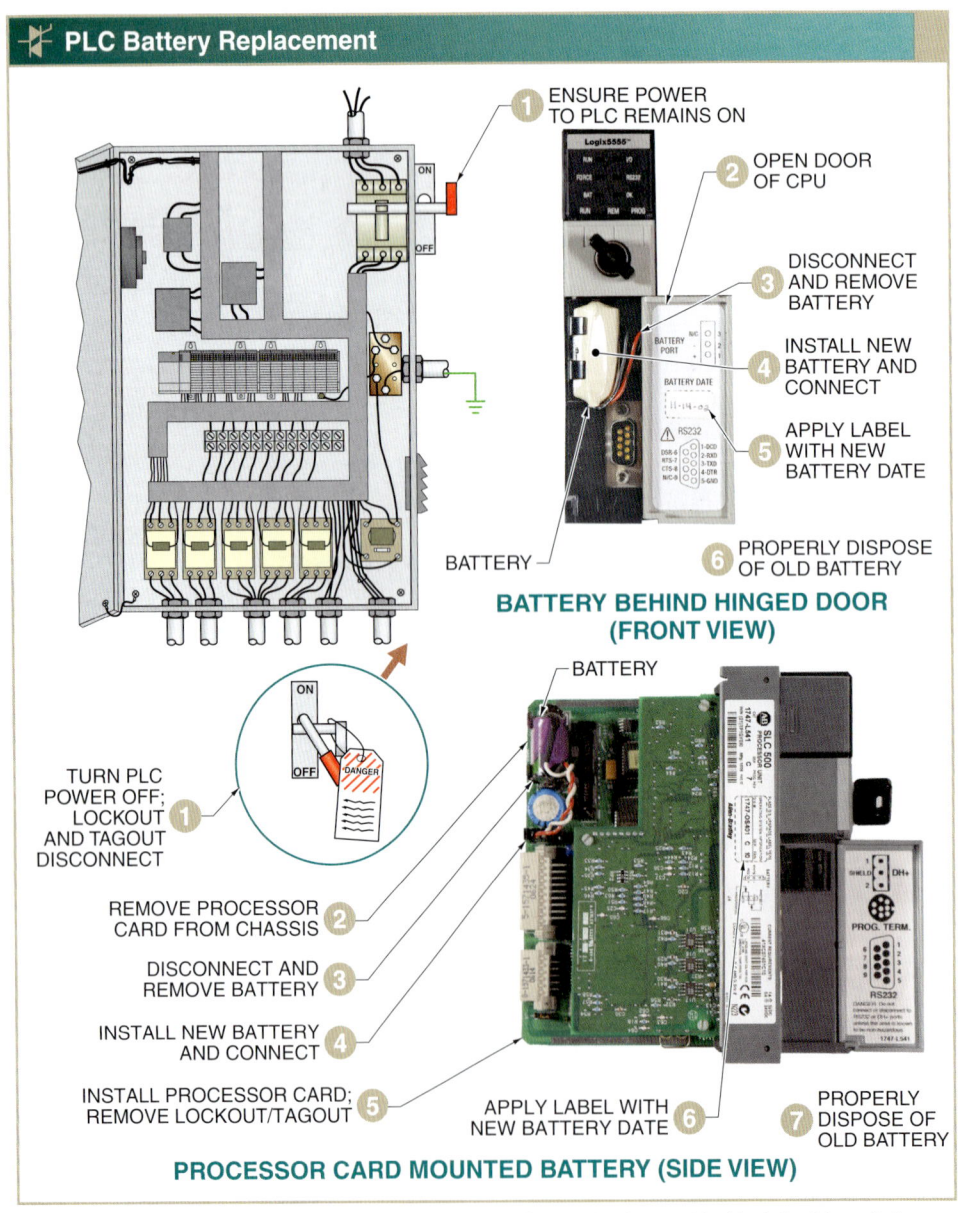

Figure 13-6. Back-up batteries for the CPU of a PLC may be located behind the hinged door or cover or mounted on the processor card.

4. Install the new battery in the holder and connect it.

5. Apply a label to the PLC noting the installation date of the new battery. Also record the date in the maintenance log.

6. Dispose of the old battery per manufacturer instruction, local codes, or EPA regulations.

Battery on Processor Card. To replace a lithium battery that is mounted internally on a modular processor card of a PLC, apply the following procedure:

1. Back up the PLC program and turn the power to the PLC OFF. After the power is OFF, lockout and tagout the disconnect. Use a DMM to verify that voltage is not present.

2. Remove the processor card from the PLC chassis.

3. Disconnect the old battery and remove the battery from the holder.

4. Install the new battery in the battery holder and connect the battery.

5. Reinstall the processor card. Remove the lockout/tagout and turn the power ON.

6. Apply a label to the PLC noting the installation date of the new battery. Also record the date in the maintenance log.

7. Dispose of the old battery per manufacturer instructions, local codes, or EPA regulations.

> **PLC TIPS**
>
> *Processor cards typically have a capacitor that provides power to RAM memory while the battery is changed out. The capacitors of a processor card provide 30 min of power. As long as the battery is replaced within the time limit of the capacitor, RAM memory will not be lost, and the machine or process will not be disabled.*

PLC SOFTWARE MAINTENANCE

After PLC startup is completed along with any needed changes, PLC software remains virtually maintenance free. *Software maintenance* is maintenance performed to ensure that a technician has the correct equipment and software to troubleshoot future PLC problems. Because PLCs are very dependable, months or years can pass between problems. However, when PLC software is not maintained between problems, items such as equipment, documents, and/or software can be misplaced. PLC software maintenance consists of equipment and documentation verification, along with software and program verification. PLC software maintenance is performed after all PLC hardware maintenance is completed.

Equipment and Documentation Verification

In order to use PLC programming software, specific equipment and documents are required. *Equipment and document verification* is a process that ensures that a technician has the correct equipment and documents for the selected PLC programming software. If even one item is missing, a technician may not be able to use the programming software. The type of equipment, the amount of equipment, and the number of documents required vary. Typically, the larger a PLC is, the more equipment and documents the technician requires. Equipment and document verification is an essential part of PLC software maintenance.

Regardless of the size or type of PLC installation, a PC with an interface cable is required for programming. **See Figure 13-7.** Laptops are very popular for programming PLCs because laptops can be taken to the location of a PLC installation. Other equipment that can be required during PLC programming are PCMCIA interface cards, CDs with programming software, programming software downloaded from manufacturer's websites, and keys for changing modes.

> **⚠ DANGER**
>
> *Follow manufacturer recommendations for storage, handling, and disposal of lithium batteries. Improper handling of lithium batteries can pose a safety hazard. Lithium batteries cannot be disposed of in the general trash because lithium batteries contain pressurized sulfur dioxide gas and lithium thionyl chloride, which vaporize upon exposure to air and are highly toxic.*

> **⚠ CAUTION**
>
> *Handle processor cards by the edges and avoid touching any solid-state components on the card. Processor cards are damaged by electrostatic charges (static electricity).*

PCMCIA interface cards fit into auxiliary slots on a laptop. Interface cards typically include an interface cable that connects the PCMCIA card to the PLC. CDs contain programming software. Some programming software will not run unless the original CD is inserted into the PC. Such security features are designed into programming software by some manufacturers to prevent violation of software copyrights. Keys allow a technician to change the mode of a processor to RUN, REMOTE, or PROGRAM. *Note:* Not all PLC processors have a key switch.

In addition to PLC manuals and wiring diagrams (electrical prints used for hardware maintenance), technicians require manuals for the programming software, manuals for the interface cards and cables, and a documented copy of the PLC program (electronic and paper) in order to perform PLC software maintenance. The manuals serve as references and provide answers in the event of a software problem or a problem when connecting the PC to the PLC.

Copies of the PLC program typically contain various descriptions and comments that explain how the machine or process operates and identify the specific input and output devices. Although similarities exist, program documentation varies among programming software manufacturers. PLC programming documentation includes address descriptions, instruction comments, rung comments, and program titles. **See Figure 13-8.**

The various types of descriptions and comments serve different purposes:

- Address descriptions provide information about a specific address. All instructions with the same address, regardless of the type of instruction, will have the same description throughout the program.

- Instruction comments provide information about a specific instruction associated with a specific address. Only the same type of instruction with the same address will have the same comment throughout the program.

Figure 13-7. A PC, an interface cable, and sometimes an interface card are required to connect a PC to a PLC and use the PLC programming software.

Program Documentation

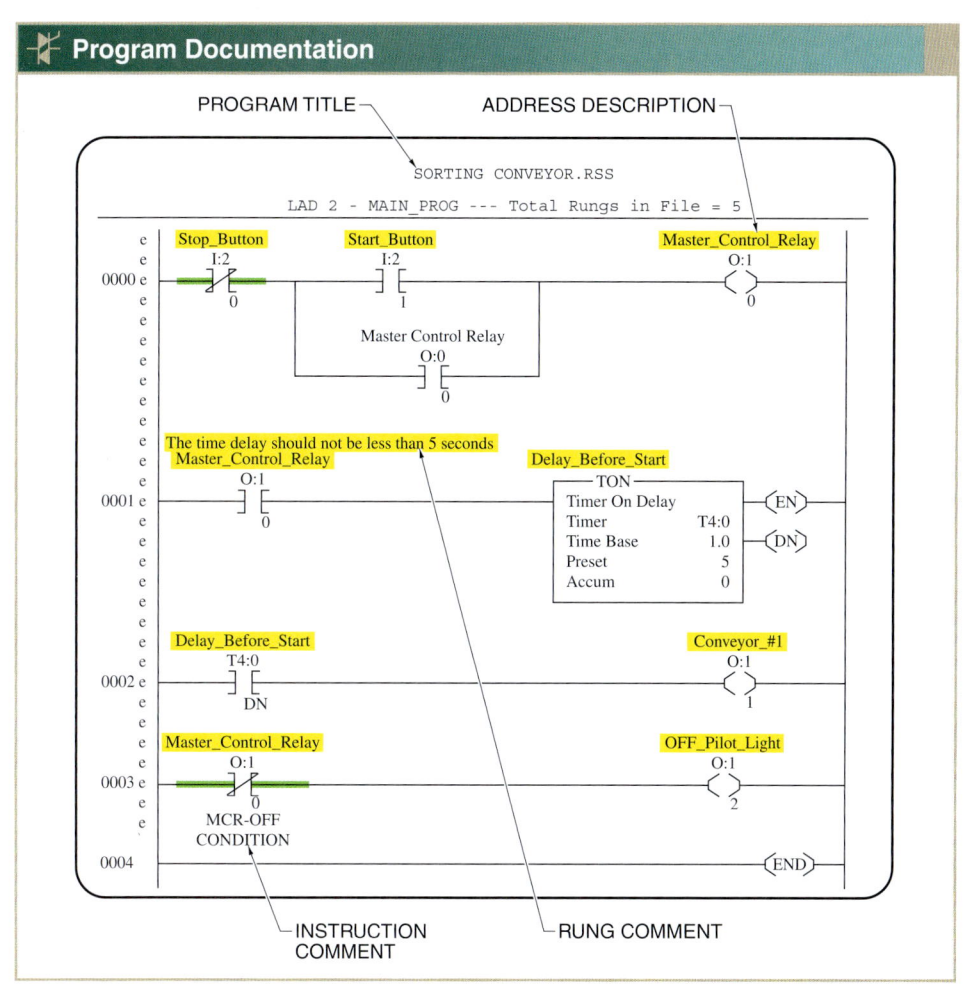

Figure 13-8. A documented program contains program titles, address descriptions, rung comments, and instruction comments.

- Rung comments provide information about a specific rung in a program. Rung comments can be linked to a specific rung or to a specific output.
- Program titles provide information about a group of related rungs. Typically, program titles are used at the top of a page. Program titles can also be linked to a specific rung or to a specific output terminal.

Software and Program Verification

Software and program verification is a process that ensures that a PLC can be accessed by a PC and that the EEPROM chip (when present) has the correct PLC program (back-up program). Software and program verification is performed after the equipment and documentation verification is successfully completed. As part of software and program verification, the technician must connect the PC to the PLC and switch the PLC between RUN mode and PROGRAM mode.

Technicians can perform a variety of tasks using PC and PLC programming software. Although the specific procedures and terms for connecting a PC to a PLC vary from PLC manufacturer to manufacturer, the vast majority of PLCs have the following function capabilities:

- Upload—*Upload* is a computer operation where a copy of a PLC program (ladder logic only, no documentation) is transferred from a PLC to a PC.

- Download—*Download* is a computer operation where a copy of a PLC program (ladder logic only, no documentation) is transferred from a PC to a PLC.
- Compare—*Compare* is a PLC software operation where the PLC program in the memory of the PLC is compared to the PLC program in the PC. The differences between the versions of the program are detailed in a report.
- Run mode—*Run mode* is a PLC state of operation where the operating cycle of the PLC is running continuously to control a machine or process.
- Program mode—*Program mode* is a PLC state of operation where the operating cycle of the PLC is not running, but the CPU is ready to accept information.
- On-line monitoring—*On-line monitoring* is a PLC software operation where a PC is monitoring the PLC program and graphically displaying the state of all input and output terminals (TRUE or FALSE).

Connecting a PC to a PLC. The first step in software and program verification is connecting the PC to the PLC and verifying that the PC has the current copy of the program that is in the PLC. Certain issues must be considered before connecting a PC to a PLC. **See Figure 13-9.**

- Although laptops can run on batteries, a laptop should always be plugged into a 115 VAC receptacle. Laptop batteries can discharge rapidly and a laptop PC could turn OFF unexpectedly, causing a problem.
- The technician must verify that a compatible USB or serial port is available for the PLC programming software. Typically, a USB port or serial port (COM 1) is used to connect the PC to the PLC. There can be compatibility conflicts between the PLC programming software and other devices that also use the same USB port.
- Some PLC programming software requires technicians to set the baud rate (transmission speed) for the serial port of a PC to interface with the PLC. The technician must verify that the baud rate is set correctly.
- Some interface cables have an in-line interface box with DIP switches. The DIP switches must be set correctly.

PLC TIPS

Always use AC power from a PLC enclosure for laptop computers when the enclosure provides power for external use. Using AC power from the PLC enclosure avoids ground potential differences that can occur between facility electrical systems and enclosure grounding. Ground potential differences often times cause hardware failures with either the PLC or PC.

To connect a PC to a PLC, apply the following procedure:

1. Connect the interface cable from the PC USB or serial port to the PLC.
2. Turn the PC ON and open the PLC programming software.
3. Open the program for the PLC that the PC is connected to and select ON-LINE MONITORING. *Note:* When a PLC program is not on the PC proceed to step 5.
4. Use ON-LINE MONITORING to verify that the PC has the most current copy of the PLC program. ON-LINE MONITORING will be active if the program is identical but will not be active if the PC does not have a current copy of the program.
5. When required, use the PC to upload the PLC program from the PLC. When an outdated copy of the program is on the PC, name the current program (program in the PLC) as a revision, such as: outdated = palletizer, current = palletizer_rev1. When no recent program for the PLC exists on the PC, name the program to reflect the machine or process the PLC controls, such as "box conveyor" or "bottler."

Connecting a PC to a PLC

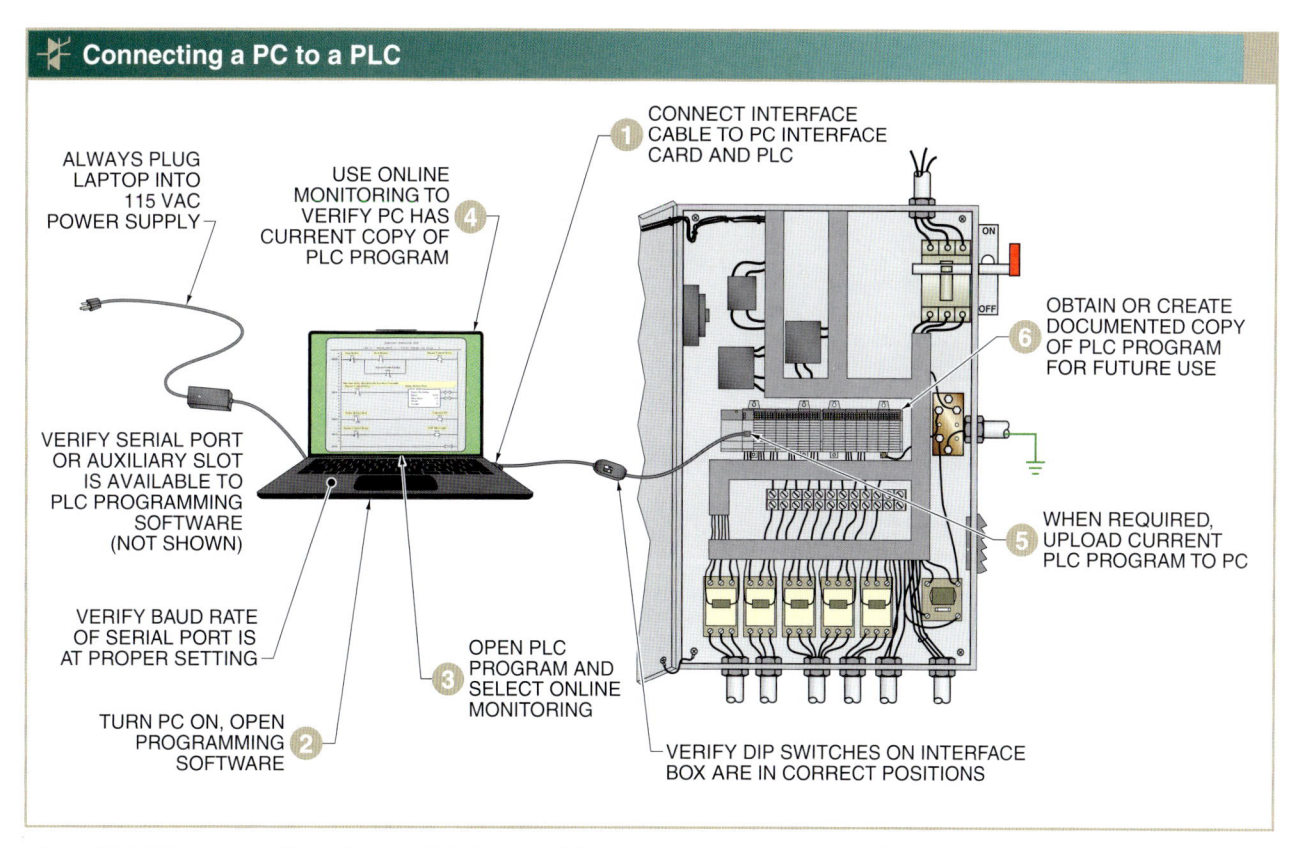

Figure 13-9. When connecting a PC to a PLC, the technician must be aware of certain technical issues and follow a series of steps.

6. Programs uploaded from a PLC are considered undocumented. Technicians must obtain a documented copy of the PLC program for future use. Obtain a documented copy of the program for future use from the original equipment manufacturer (OEM) or the system integrator who designed the machine or process. PLC programs must be transferred to a PC and paper copies printed.

EEPROM Verification. The second step in software and program verification is EEPROM verification. Many PLCs have electrically erasable programmable read-only memory (EEPROM) chips to provide a backup to the RAM memory. A copy of the PLC program (ladder logic only, no documentation) resides on the EEPROM. A PLC can be configured to load a program from the EEPROM to the CPU upon power-up or upon detection of a memory error. EEPROM chips can be an integral part of a PLC or a separate unit that plugs into a socket. **See Figure 13-10.**

Because EEPROM provides a back-up copy of the processor program, technicians must verify that the EEPROM contains a copy of the same program the PLC CPU is running. For PLCs that have integral EEPROMs, this is not necessary. These PLCs are designed so the integral EEPROM always has a copy of the processor program and the program is automatically copied between the EEPROM and the processor.

PLCs with separate EEPROMs must be configured to copy the program between the EEPROM and the PLC processor. It is possible for a PLC with a separate EEPROM to have a different program in the processor than in the EEPROM. Different programs present problems when a processor program is lost or corrupted and must be reloaded from the EEPROM.

EEPROM Verification

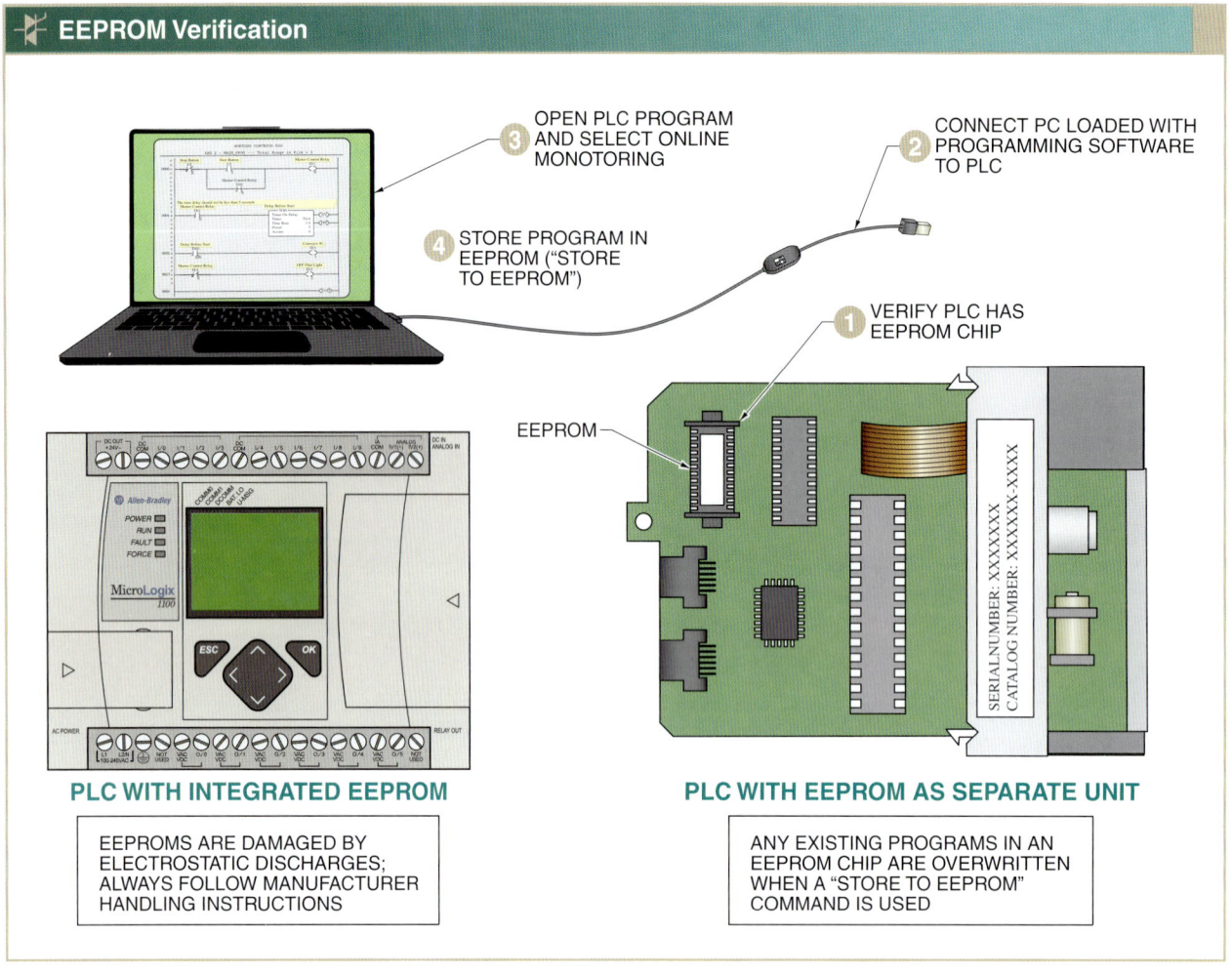

Figure 13-10. PLCs have electrically erasable programmable read-only memory (EEPROM) chips to provide a backup to RAM memory.

EEPROM verification confirms that the program in the EEPROM is the same as the program in the processor. To perform EEPROM verification, apply the following procedure:

1. Verify that a PLC has an EEPROM chip installed.
2. Connect the PC with the PLC programming software to the PLC.
3. Open the PLC program for the PLC that the PC is connected to and select ON-LINE MONITORING.
4. Choose STORE TO EEPROM (COPY TO EEPROM or TRANSFER TO EEPROM) from the programming software menu. A "STORE TO EEPROM" command copies the program in the CPU of the PLC to the EEPROM chip. Any existing programs in an EEPROM chip are overwritten.

Program Backup. The third step in software and program verification is to back up the PLC program that resides in the PC. The PLC program must be backed up because PCs can malfunction and lose information. Program backup is also the final step in PLC maintenance. **See Figure 13-11.** To perform a PC program backup, apply the following procedure:

1. When a PLC is not in the RUN mode, put the PLC in the RUN mode. Disconnect the PC from the PLC.
2. Remove any emergency stop conditions and press the start button so the MCR is ON.

Chapter 13—*PLC System Maintenance* **525**

3. Start the machine or process that the PLC controls. Monitor the machine or process and verify correct operation.
4. Copy the PLC program to a CD-ROM or USB portable drive (memory stick). Store the copy of the PLC program in a safe location that is separate from the PC.

⚠ WARNING

All programs that have been edited must be verified for errors (controller RAM memory) before being stored to a memory module. Programs with errors must never be saved to the EEPROM of a controller or to a memory module.

Program Backup

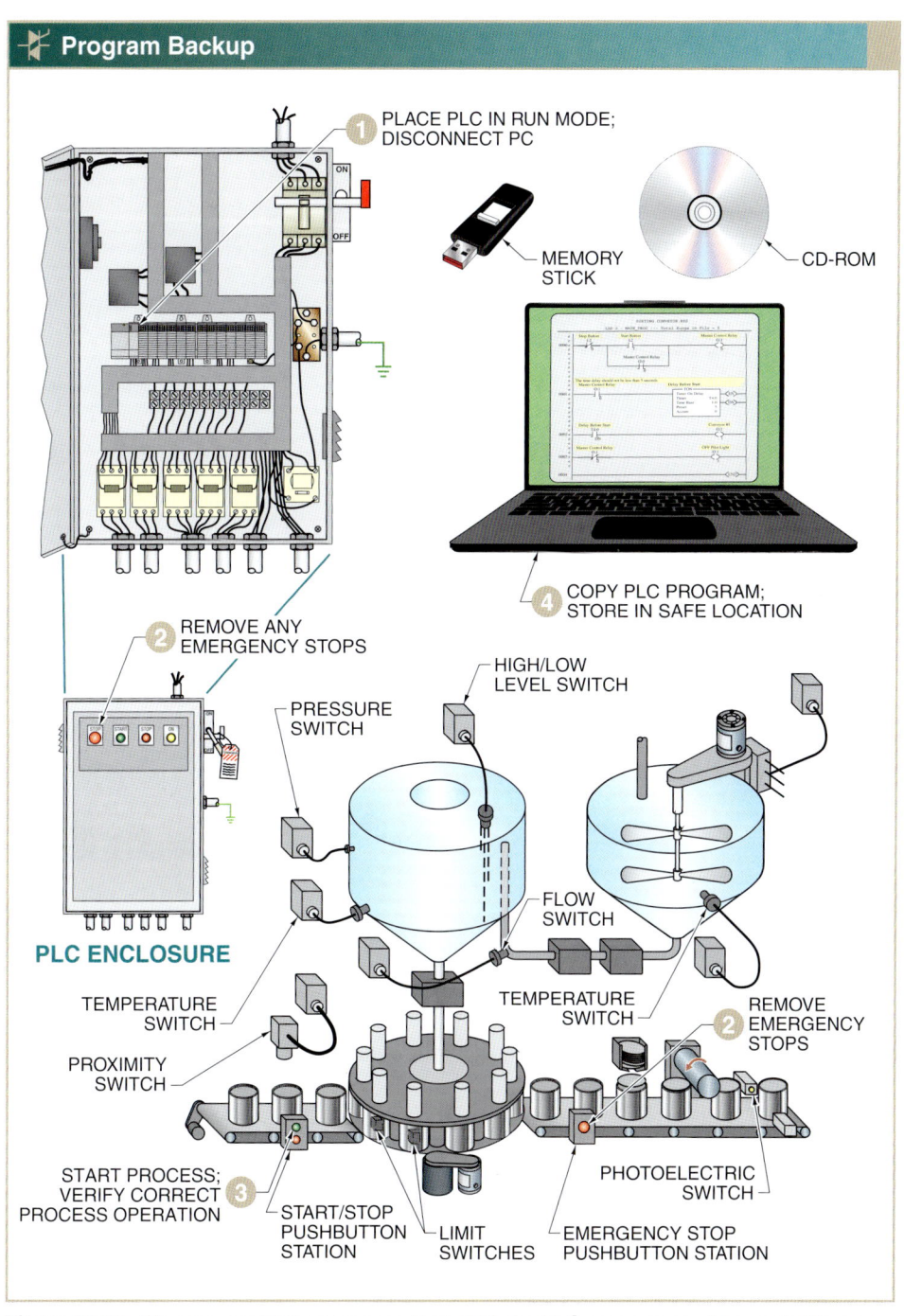

1. PLACE PLC IN RUN MODE; DISCONNECT PC
2. REMOVE ANY EMERGENCY STOPS
3. START PROCESS; VERIFY CORRECT PROCESS OPERATION
4. COPY PLC PROGRAM; STORE IN SAFE LOCATION

- MEMORY STICK
- CD-ROM
- PLC ENCLOSURE
- PRESSURE SWITCH
- HIGH/LOW LEVEL SWITCH
- FLOW SWITCH
- TEMPERATURE SWITCH
- TEMPERATURE SWITCH
- PROXIMITY SWITCH
- START/STOP PUSHBUTTON STATION
- LIMIT SWITCHES
- PHOTOELECTRIC SWITCH
- REMOVE EMERGENCY STOPS
- EMERGENCY STOP PUSHBUTTON STATION

Figure 13-11. After the machine or process is started, the PLC program must be copied to a separate medium and stored in a safe location.

Chapter Review 13

Name _____ Date _____

True-False

T F 1. When performing a visual inspection, a technician should use an electrostatic discharge safe vacuum to remove dust and contaminants from the PLC and enclosure.

T F 2. After performing a visual inspection, a technician should use a DMM to verify that no voltage is present in the PLC enclosure.

T F 3. Although laptops can run on batteries, a laptop connected to a PLC should always be plugged into a 115 VAC receptacle.

T F 4. All PLC processors have a key switch.

T F 5. It is possible for a PLC to have a different program in the processor than in the EEPROM chip.

T F 6. PLC system maintenance is typically performed annually.

T F 7. Compressed air can be used to remove dust or debris from a PLC.

T F 8. The processor of a PLC monitors battery voltage level.

T F 9. Only the same type of instruction with the same address will have the same comment throughout a PLC program.

T F 10. A PLC battery may be mounted behind a door or to a modular processor card.

T F 11. Rung comments provide information about a group of related rungs.

T F 12. Enclosure cooling fans are typically replaced every five-to-seven years.

T F 13. Battery maintenance must be performed before energized maintenance.

T F 14. When programming PLCs, a technician typically uses a PC and interface cable.

T F 15. Program backup is the final step in PLC maintenance.

Completion

_____ 1. ___ is maintenance that verifies that the master control relay (MCR) is functioning correctly, voltages are within specified range, and no abnormal conditions exist.

_____ 2. ___ is a process that ensures that a PLC can be accessed by a PC and that the EEPROM chip (when present) has the correct PLC program (back-up program).

_____ 3. Maintenance performed to ensure that a technician has the correct equipment and software to troubleshoot future PLC programs is referred to as ___.

528—PROGRAMMABLE LOGIC CONTROLLERS: PRINCIPLES AND APPLICATIONS

_____ 4. ___ is a PLC software operation where the PLC program in the memory of the PLC is compared to the PLC program in the PC.

_____ 5. ___ is a process that ensures that a technician has the correct equipment and documents for the selected PLC programming software.

_____ 6. Maintenance that consists of a visual inspection of the PLC, energized maintenance procedures, and battery maintenance is called ___.

_____ 7. ___ is a PLC state of operation where the operating cycle of the PLC is not running, but the CPU is ready to accept information.

_____ 8. ___ is maintenance that verifies that a PLC battery has not exceeded its rated life.

_____ 9. ___ is a computer operation where a PLC program is copied from a PLC to a PC.

_____ 10. Many PLCs have ___ chips to provide a backup to RAM memory.

Multiple Choice

_____ 1. A technician performing PLC hardware maintenance may require the assistance of a second technician or operator because ___.
   A. some PLC applications must be stopped in a specific sequence
   B. the equipment is too heavy to be handled by one person
   C. the technician does not know how to perform a lockout/tagout
   D. the technician is unfamiliar with maintenance procedures

_____ 2. Which of the following sequences demonstrates the appropriate order for maintenance operations?
   A. battery maintenance, visual inspection, energized maintenance, equipment and document verification, software and program verification
   B. energized maintenance, visual inspection, battery maintenance, software and program verification, equipment and document verification
   C. equipment and document verification, software and program verification, visual inspection, battery maintenance, energized maintenance
   D. visual inspection, energized maintenance, battery maintenance, equipment and document verification, software and program verification

_____ 3. ___ is a PLC state where the operating cycle of the PLC is running continuously to control a machine or process.
   A. Download monitoring
   B. On-line monitoring
   C. Program mode
   D. Run mode

_____ 4. Lithium batteries typically last ___.
   A. three-to-nine months
   B. six months to one year
   C. one-to-three years
   D. two-to-five years

_____ 5. The first step in software and program verification is connecting the PC to the PLC and ___.
   A. backing up the PLC program that resides in the PLC
   B. performing EEPROM verification
   C. switching to RUN mode
   D. verifying that the PC has the current copy of the program that is in the PLC

PLC Sequencing of Industry Equipment

PLC Lab 13.1

BILL OF MATERIALS

- TECO PLR or nano-/micro-PLC
- Rotary selector switch
- Three motor starters/contactors
- Three NC limit switches

In a medium-sized industrial plant, three air compressors supply the compressed air for process functions. The compressor outputs connect to a common manifold and run to the plant floor. For this lab, a sequencer control cabinet holds a PLC and motor starters. The cabinet varies the order in which the compressors start (first, second, or third) to equalize the number of hours each one runs. There are three pressure switches in the compressed air piping running to the plant floor. The pressure switches are NC and open on a rise in pressure. The pressure switch (PS) settings are staggered so the compressors do not all start at once. As the demand for compressed air increases (pressure drops), the additional compressors start to meet the demand.

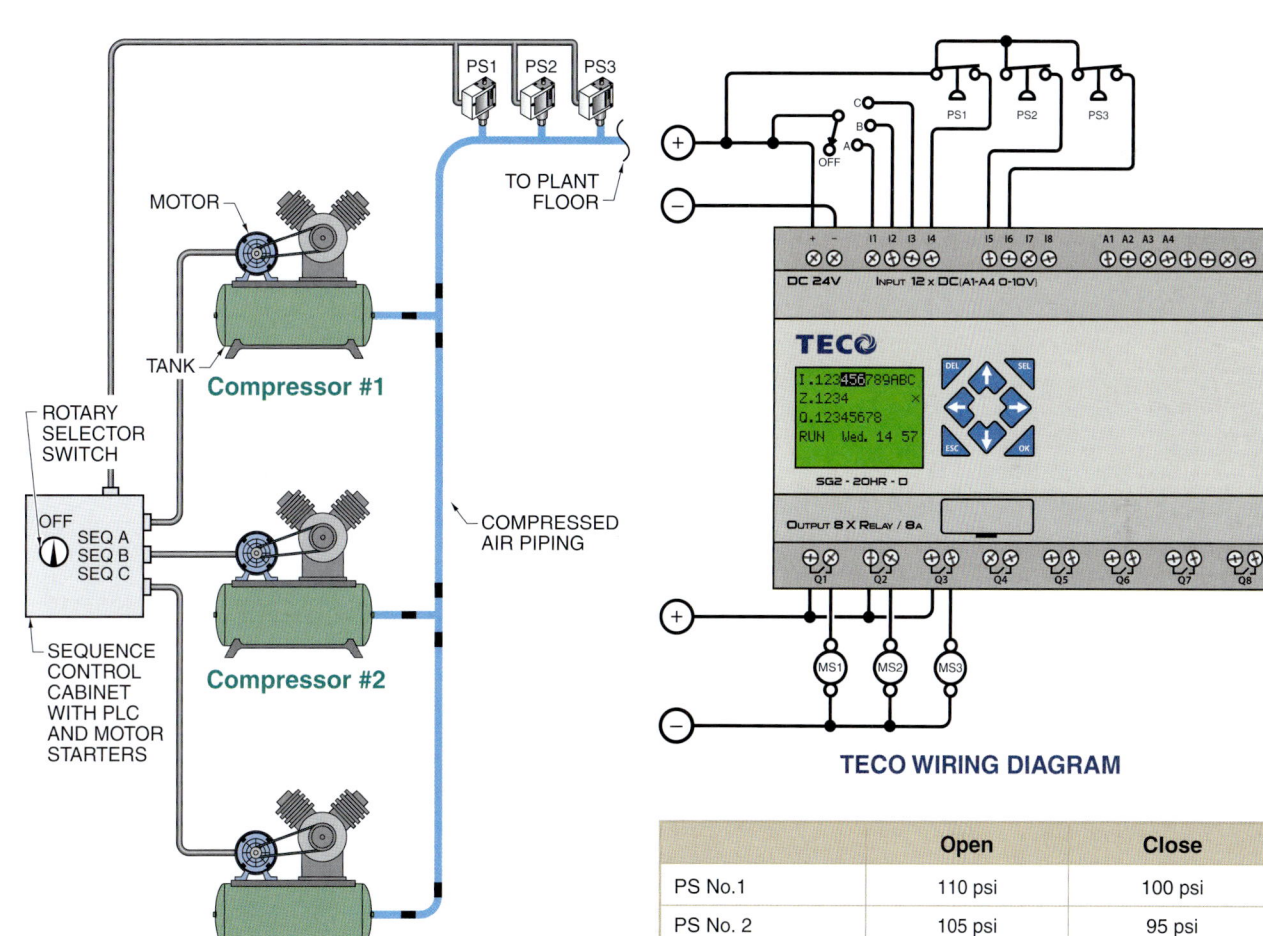

TECO WIRING DIAGRAM

	Open	Close
PS No.1	110 psi	100 psi
PS No. 2	105 psi	95 psi
PS No. 3	100 psi	90 psi

SEQ A	SEQ B	SEQ C
Comp No. 1	Comp #2	Comp #3
Comp No. 2	Comp #3	Comp #1
Comp No. 3	Comp #1	Comp #2

TECO Procedures

1. Based on the PLC sequence description, pressure switch settings, and detailed drawing, draw a line diagram with the proper inputs, control relays, and outputs for the PLC circuit. *Note*: Based on the type or model of PLC used, line diagrams and programming logic answers may vary.

2. Create a new ladder logic program electronically.

3. Use the Edit Contact window to denote an input or output number and the Symbol button to add a description after placing the input or output in the Programming Grid.

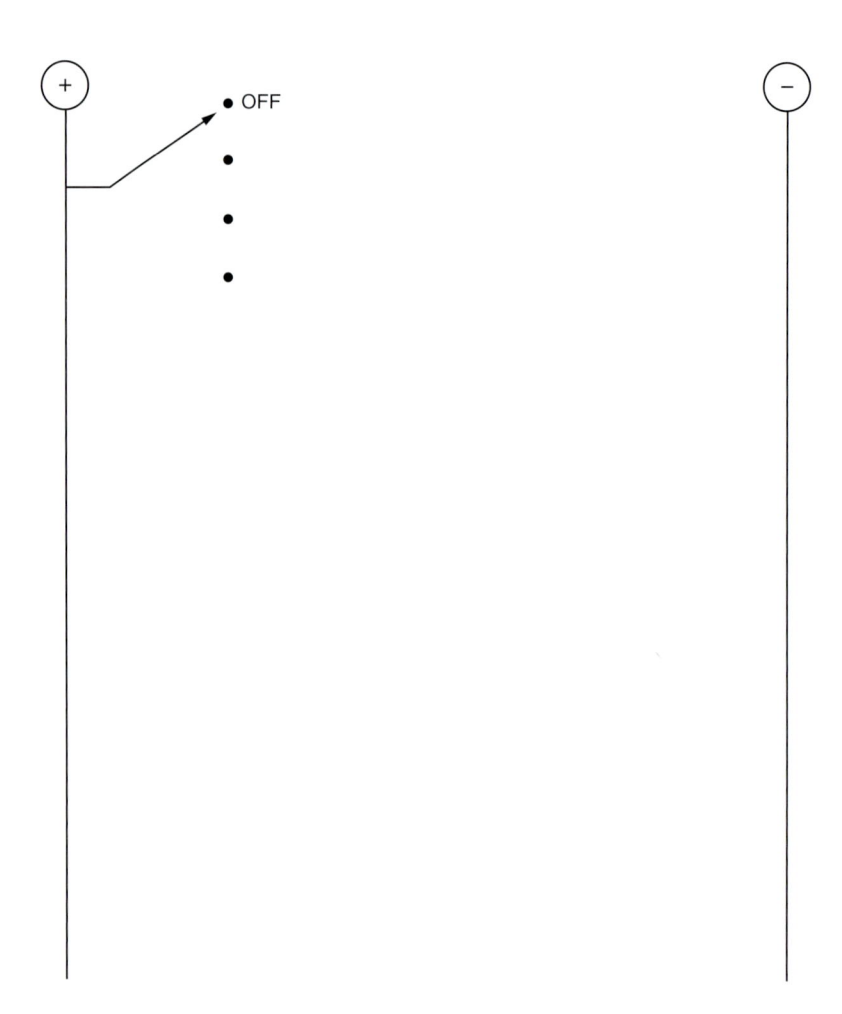

LINE DIAGRAM

4. Run the program in Simulation Mode to verify proper operation. Use the Input Status tool to activate the input(s) and control the output(s) to simulate circuit operation.

5. Turn power to the PLC ON and link the com port.

6. Write the program to the SG2-20HR-D to download the program logic to the PLC.

7. Place the SG2-20HR-D in Run Mode. Activate the input(s) and monitor the I/Os on the LCD display.

8. List the PLC model used:

   _____

   _____

001
002
003
004
005
006
007
008
009
010
011
012
013
014
015

TECO PROGRAMMING DIAGRAM

PLC Sequencing of Industry Equipment, Part 2

PLC Lab 13.2

BILL OF MATERIALS

- TECO PLR or nano-/micro-PLC
- Rotary selector switch
- Three indicator lights
- NO pushbutton
- Three motor starters/contactors
- Four NC limit switches

The three air compressors are critical to how the plant process functions. For this lab, it is necessary to have a means to identify when a pressure switch has a fault. A pressure switch fault will trigger when a switch with a lower closing pressure closes before a switch with a higher closing pressure. To identify these faults, a fourth pressure switch (low-low pressure) is added in-line to the compressed air piping. When a pressure switch has faulted, a corresponding light must turn ON and remain so until a clear fault switch is activated. Pressure switches are normally closed and open on the rise in pressure.

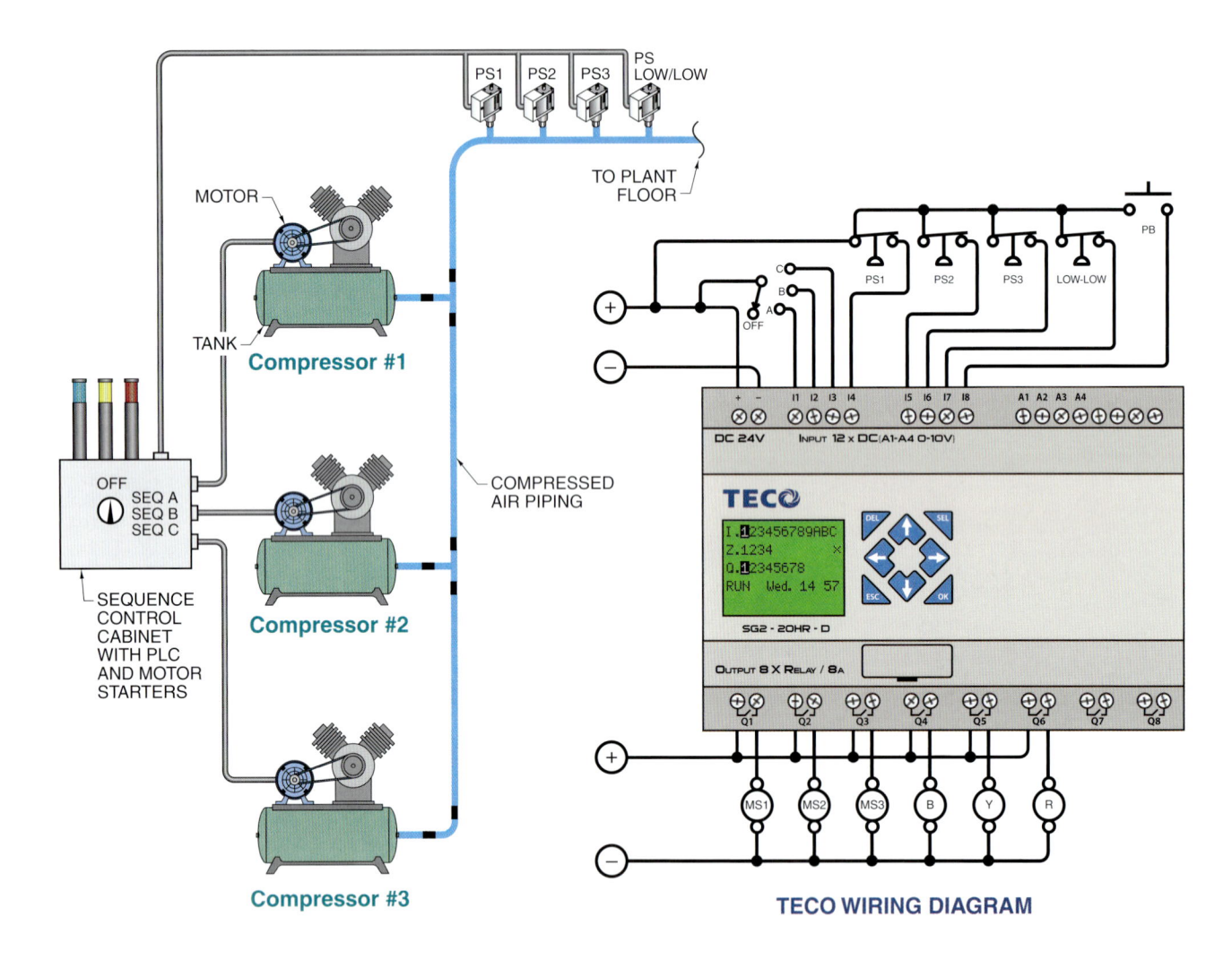

TECO WIRING DIAGRAM

	Open	Close
LOW/LOW	100 psi	85 psi

T TECO Procedures

1. Use the line diagram from PLC Lab 13.1 and the PLC circuit description to add the required input devices, control relays, and output components to identify faulted pressure switches for the PLC circuit. *Note:* Based on the type or model of PLC used, line diagrams and programming logic answers may vary.

2. Create a new ladder logic program electronically.

3. Use the Edit Contact window to denote an input or output number and the Symbol button to add a description after placing the input or output in the Programming Grid.

4. Run the program in Simulation Mode to verify proper operation. Use the Input Status tool to activate the input(s) and control the output(s) to simulate circuit operation.

LINE DIAGRAM

001
002
003
004
005
006
007

TECO PROGRAMMING DIAGRAM
(continuted on next page)

5. Turn power to the PLC ON and link the com port.

6. Write the program to the SG2-20HR-D to download the program logic to the PLC.

7. Place the SG2-20HR-D in Run Mode. Activate the input(s) and monitor the I/Os on the LCD display.

8. List the PLC model used: _____

_____

TECO PROGRAMMING DIAGRAM

PLC Sequencing of Industry Equipment, Part 3

PLC Lab 13.3

For this lab, compare the cost of general-purpose relays, timers, latching relays, and miscellaneous relays versus the cost of the TECO SG2-20HR-D PLR used.

TECO Procedures

1. Create an inventory of the number of general-purpose relays, timers, latching relays, miscellaneous relays, and bases used in the line diagram from PLC labs 13.1 and 13.2.

2. Fill in the spreadsheet below with the inventory.

3. Use the link for the FACTORYMATION website by clicking on the Learner Resources tab in the Internet Resources. Find the make, model number, and pricing of the inventory and TECO PLR used. Guidelines for pricing are as follows:

   - Relay coils are 24 VDC.
   - General-purpose relays should have an indicating LED and diode for coil suppression.
   - Multiple relays may need to be used if relays do not have enough poles.
   - PLC inputs are DC, and outputs are relay.

Item	Make	Model #	QTY	$ Each	Total
General purpose ice cube relay, 3PD					
Relay base with screw terminals					
General purpose ice cube relay, 4PDT					
Relay base with screw terminals					
Multi-function time relay, DPDT					
Relay base with screw terminals					
Latching relay, DPDT					
Relay base with screw terminals					
Misc relay					
Relay base with screw terminals					
Grand Total					

Item	Make	Model #	QTY	$ Each	Total
PLC	TECO	SG2-20HR-D	1		
PLC Expansion Unit					
Grand Total					

Math Instructions Used for Programming PLCs

PLC Lab 13.4

BILL OF MATERIALS

- Micrologix™ 1100 PLC
- Four sensors (limit/photoelectric/proximity)
- NO pushbutton
- Light/load

Math instructions, such as add (ADD), subtract (SUB), multiply (MUL), and divide (DIV), are used in various industries to meet specific parameters during production. They are also used by many municipalities for traffic operations. In this lab, a parking lot has 24 parking spaces, 2 entrances, and 2 exits. All have sensors, but the lot does not have gates. Each entrance counts each car that enters, while each exit counts each car that leaves. When the total number of cars exceeds 24, a light indicator or "LOT FULL" sign is illuminated. All input devices are hardwired normally open, and a reset pushbutton is wired to reset all counters.

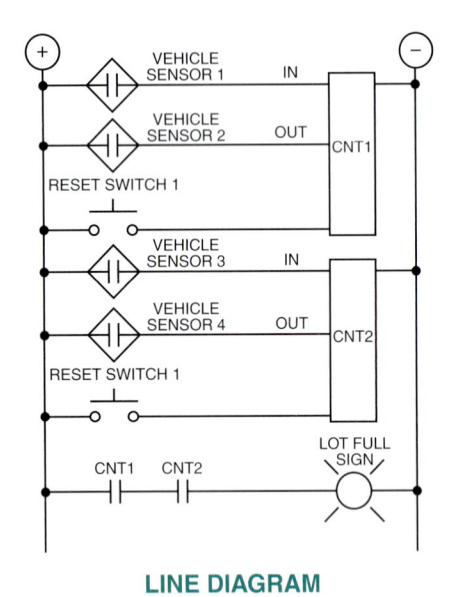

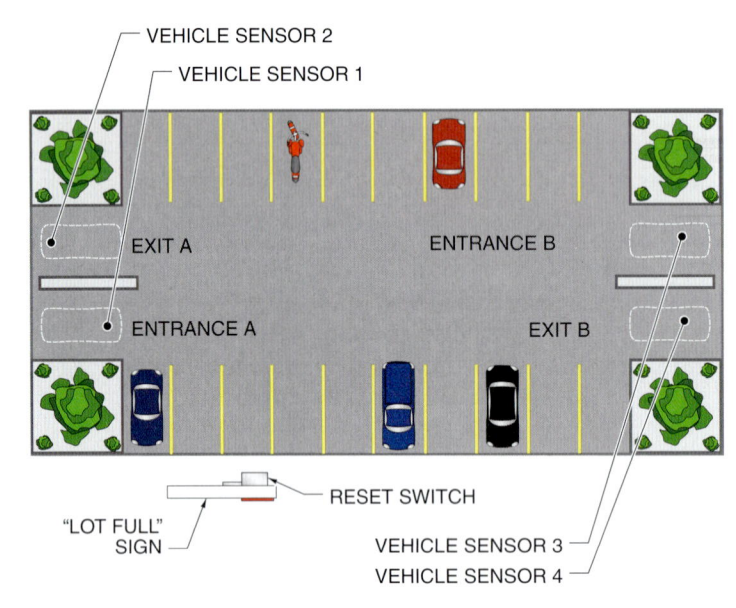

LINE DIAGRAM

Allen-Bradley Procedures

1. Based on the PLC system description and line diagram, draw the ladder logic for RSLogix™ in the AB programming area.

2. Use the AB programming diagram to add the proper address and description in the LAD 2 programming window. Add four counter instructions by selecting the CTU (count up) instruction or CTD (count down) instruction from the Timer/Counter tab in the Instruction toolbar for each. Add each instruction to the programming diagram and type in the counter address and preset value. The ADD instruction takes the accumulate (accum) of each CTU instruction and provides an integer destination of N7:0. Use a GRT (greater than) instruction with a source A of N7:0 and a source B of 24 to control the output.

3. Verify that the programming diagram has no errors and save the project.

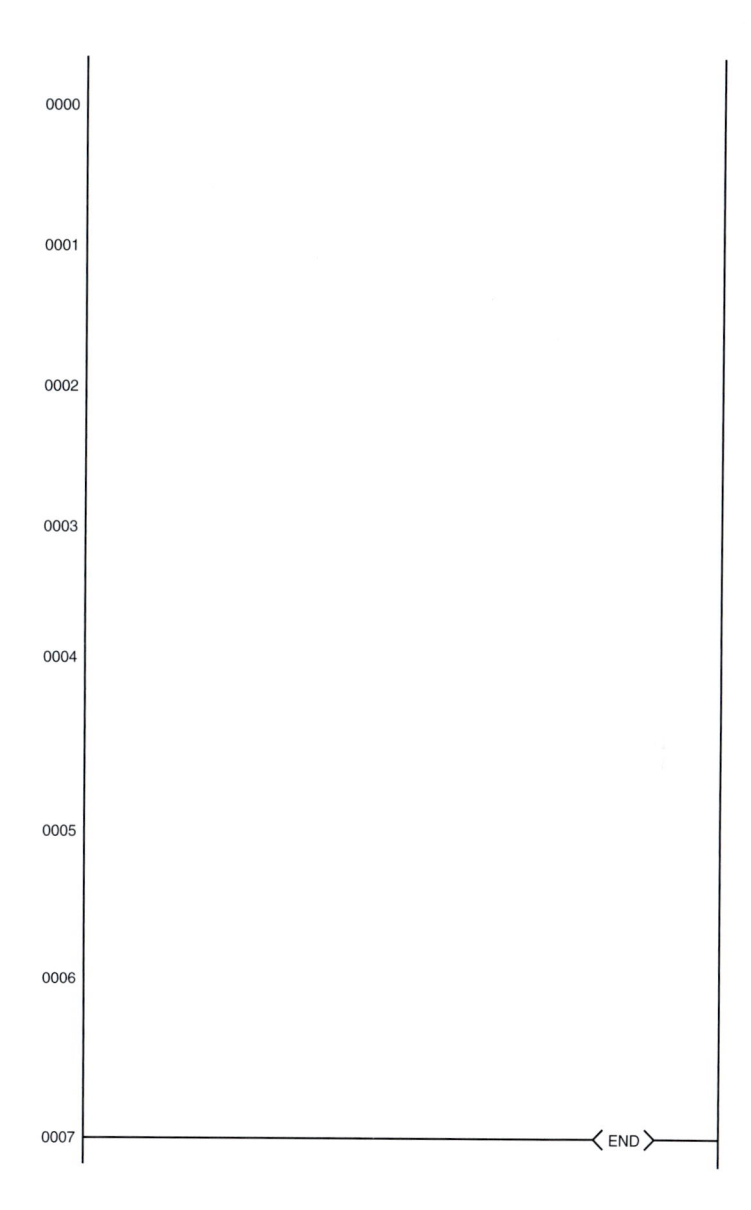

```
0000
0001
0002
0003
0004
0005
0006
0007                                           ⟨END⟩
```

AB PROGRAMMING DIAGRAM

4. Use the wiring diagram to wire the input devices and output components to the PLC. Once all connections have been made, turn power to the PLC on.

5. Use RSLinx™ to establish communication between the PC and the Micrologix™ 1100.

6. Download the program logic to the PLC.

7. Place the PLC in Run Mode.

8. Activate input(s) and monitor them to determine if the output(s) are operating correctly.

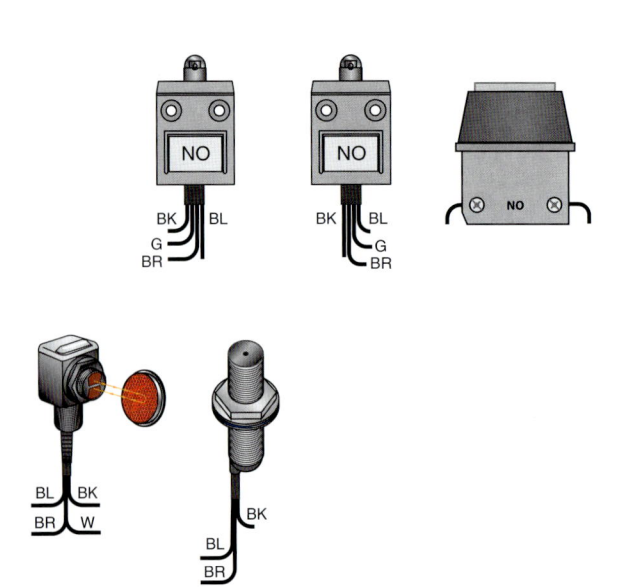

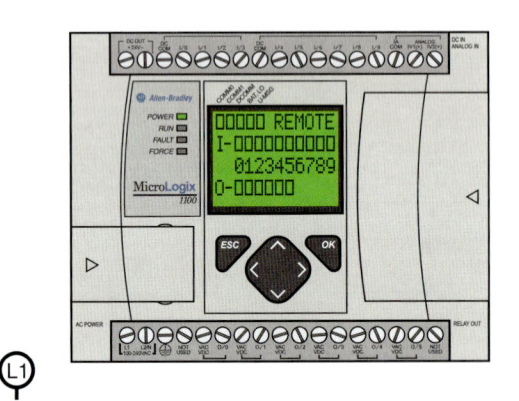

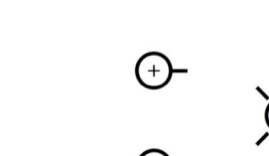

AB WIRING DIAGRAM

Sensor Actuation and Counting Instructions

PLC Lab 13.5

BILL OF MATERIALS
- Micrologix™ 1100 PLC
- Three sensors (limit/photoelectric/proximity)
- NO pushbutton
- Light
- Motor starter/contactor

For this lab, a PLC circuit controls a gate to a parking lot using vehicle detection sensors. When a vehicle is detected at the entrance, a control relay actuates a motor starter to raise the gate, and a second sensor closes the gate once the vehicle has entered. The vehicle sensor at the entrance also counts each vehicle that enters up to 19, and a vehicle sensor at the exit counts down as each vehicle leaves. Once the count reaches 19, a "LOT FULL" sign is enabled. An NO pushbutton is wired to reset the counter. All input devices are hard-wired normally open.

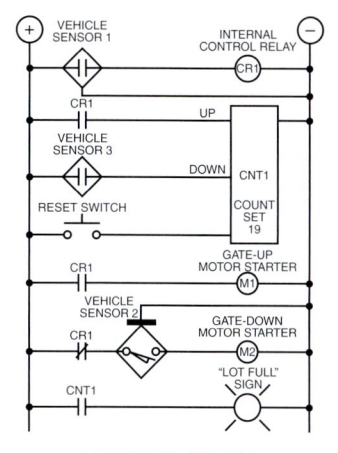

LINE DIAGRAM

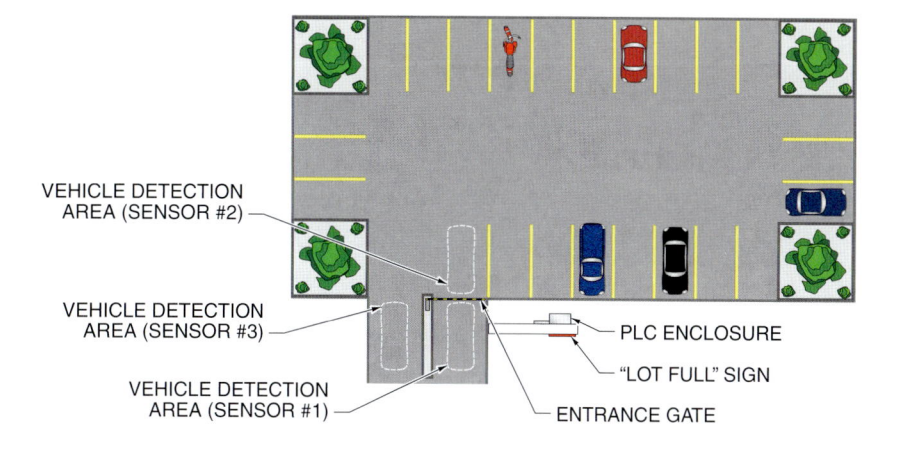

Allen-Bradley Procedures

1. Based on the PLC system description and line diagram, draw the ladder logic for RSLogix™ in the AB programming area.

2. Open the RSLogix™ Micro Starter Lite program and create a new programming diagram.

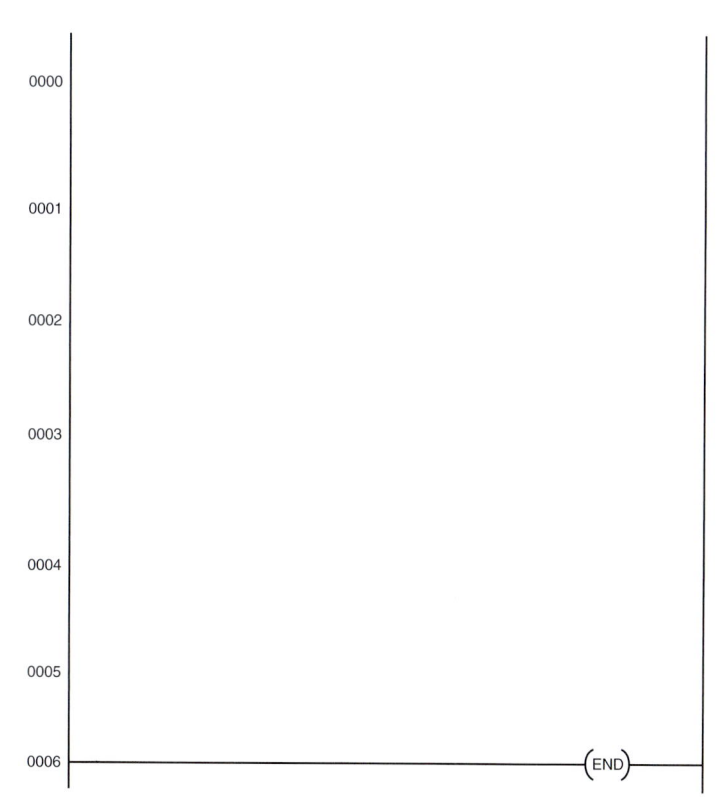

AB PROGRAMMING DIAGRAM

3. Use the AB programming diagram to add the proper address and description for each input and output instruction in the LAD 2 programming window. Add the proper counter instructions from the Timer/Counter tab in the Instruction toolbar. Add each instruction to the programming diagram and type in the counter address and preset value for each.

4. Verify that the programming diagram has no errors and save the project.

5. Draw the wire connections of the input devices and output components to the PLC in the wiring diagram. Then use the wiring diagram to wire the input devices and output components to the PLC. Once all connections have been made, turn power to the PLC on.

6. Use RSLinx™ to establish communication between the PC and the Micrologix™ 1100.

7. Download the program logic to the PLC.

8. Place the PLC in Run Mode.

9. Activate input(s) and monitor them to determine if the output(s) are operating correctly.

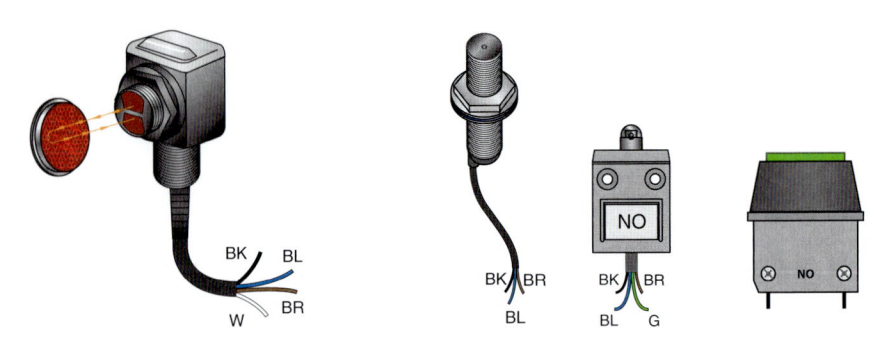

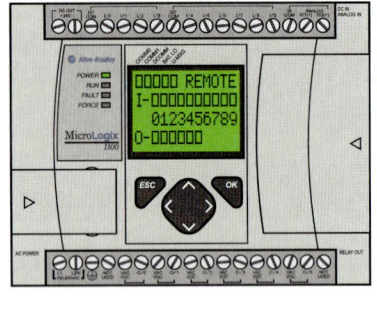

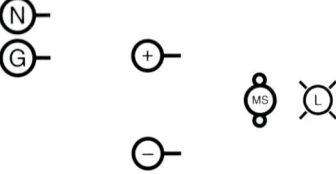

AB WIRING DIAGRAM

Sensor Actuation and Output Flashing

PLC Lab 13.6

BILL OF MATERIALS
- Micrologix™ 1100 PLC
- Three sensors (limit/photoelectric/proximity)
- NO pushbutton
- Light
- Motor starter/contactor

For this lab, a PLC circuit controls a gate to a parking lot using vehicle detection sensors. When a vehicle is detected at the entrance, a control relay actuates a motor starter to raise the gate, and then a second sensor closes the gate once the vehicle has entered. The vehicle sensor at the entrance also counts each vehicle that enters up to 19, and a vehicle sensor at the exit counts each vehicle that leaves. Once the count reaches 19, a "LOT FULL" sign is enabled and flashes ON and OFF. An NO pushbutton is wired to reset the counter. All input devices are hardwired normally open.

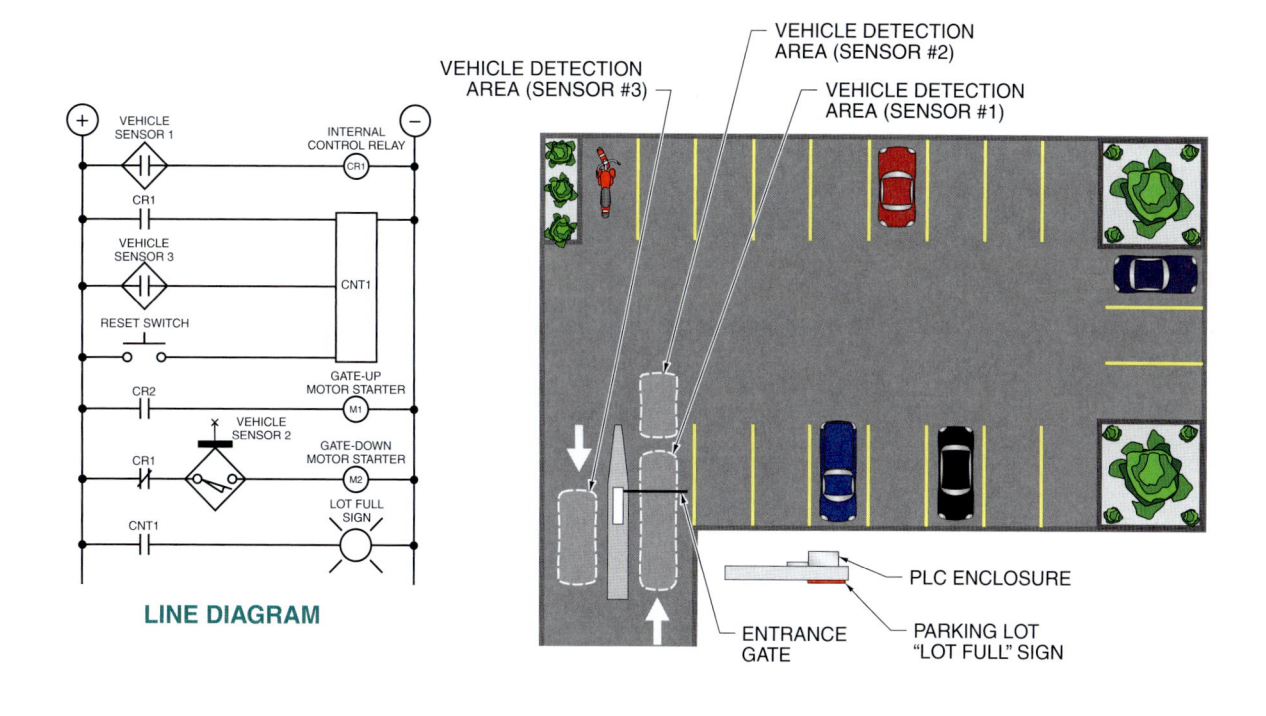

LINE DIAGRAM

Allen-Bradley Procedures

1. Based on the PLC system description and line diagram, draw the ladder logic for RSLogix™ in the AB programming area.

2. Open the RSLogix™ Micro Starter Lite program and create a new programming diagram.

3. Use the AB programming diagram to add the proper address and description for each input and output instruction in the LAD 2 programming window. Add the proper counter instructions from the Timer/Counter tab in the Instruction toolbar. Add each instruction to the programming diagram and type in the counter address and preset value for each. The address for the free-running clock is S:4. Each bit has a different clock rate. Bits S:4/0 through S:4/15 are 20, 40, 80, 160, 320, 640, 1280, 2560, 5120, 10240, 20480, 40960, 81920, 163840, 327680, and 655360 ms respectively. The "Lot Full" sign flashes at a clock rate of 5120 ms.

4. Verify that the programming diagram has no errors and save the project.

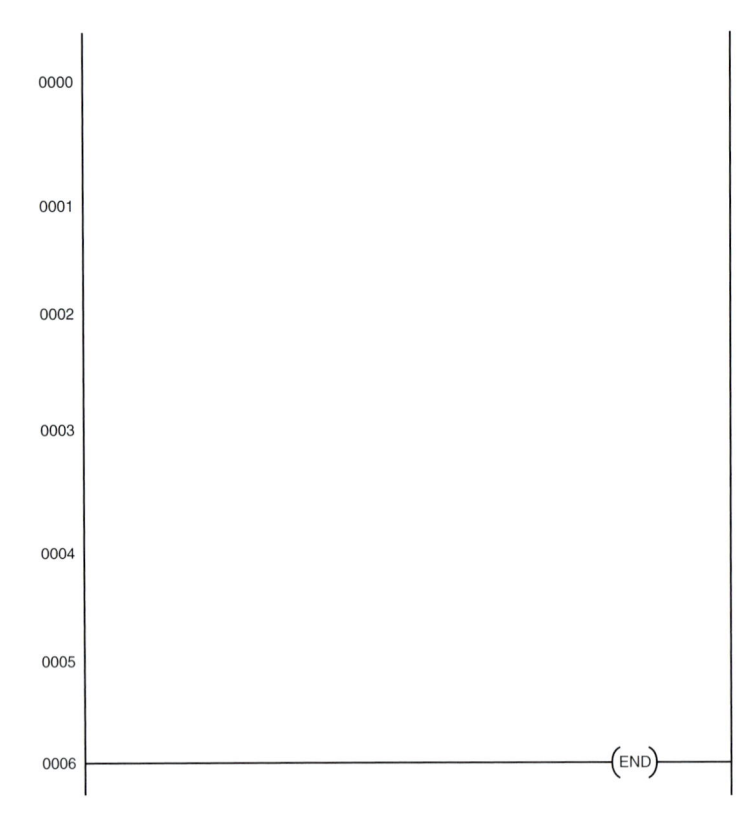

AB PROGRAMMING DIAGRAM

Chapter 13—*PLC System Maintenance* **543**

5. Draw the wire connections of the input devices and output components to the PLC in the wiring diagram. Then use the wiring diagram to wire the input devices and output components to the PLC. Once all connections have been made, turn power to the PLC on.

6. Use RSLinx™ to establish communication between the PC and the Micrologix™ 1100.

7. Download the program logic to the PLC.

8. Place the PLC in Run Mode.

9. Activate input(s) and monitor them to determine if the output(s) are operating correctly.

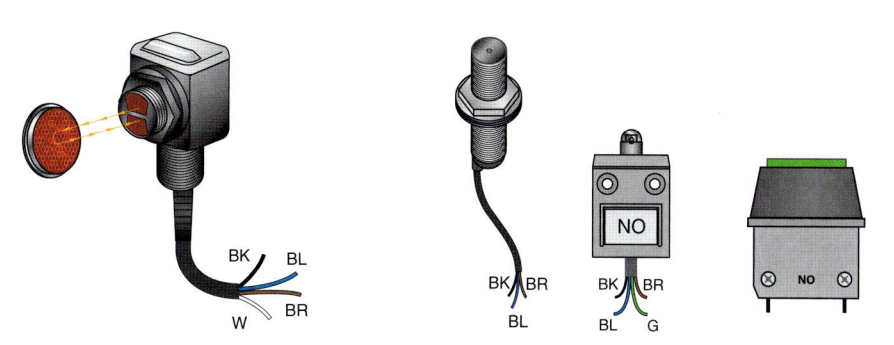

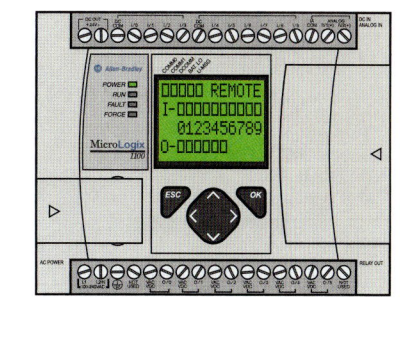

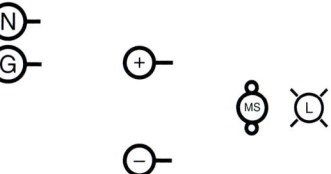

AB WIRING DIAGRAM

Appendix

Electrical Shock Effects 546
Ohm's Law 546
Conductor Ratings 546
Power Formula 547
Impedance 547
Switch Symbols 548
Solid-State Switch Symbols 548
Normally Open (NO) and Normally Closed (NC) Switches 549
Switch Operator Conditions 549
Expanded (Special) Programming Symbols 550
Bit, Timer, and Counter Instructions 550
File, Sequencer, and Move Instructions 550
Math, Compare, Conversion, and Logical Instructions 551
Program Control, Interrupt, and I/O Message Instructions 551
"PID" and Communication Instructions 551
Logic Functions 552
Truth Tables 552
Units of Memory 553
Electromechanical Relays 553
PLCs Controlling Electric Motor Drives 554
Temperature Sensor Data 555
Converter Output Spans 556
Data Table File Addressing 556
PLC Scan Time Worksheets 557
Timer Instructions 558
Timer On-Delay (TON) Instructions 558
Timer Off-Delay (TOF) Instructions 558
Retentive Timer and Reset Instructions 559
Cascaded Timers 559
Free Running (Repetitive) Timers 559
Counter Instructions 560
Counter Count Up (CTU) Instructions 561
Counter Count Down (CTD) Instructions 561
High-Speed Counter (HSC) Instructions 562
Math Instructions 562
Selected Test Instrument Abbreviations 563
Selected Test Instrument Symbols 563
Electrical Quantities 564
IEC 61010 Test Instrument Measurement Categories 564

Electrical Shock Effects

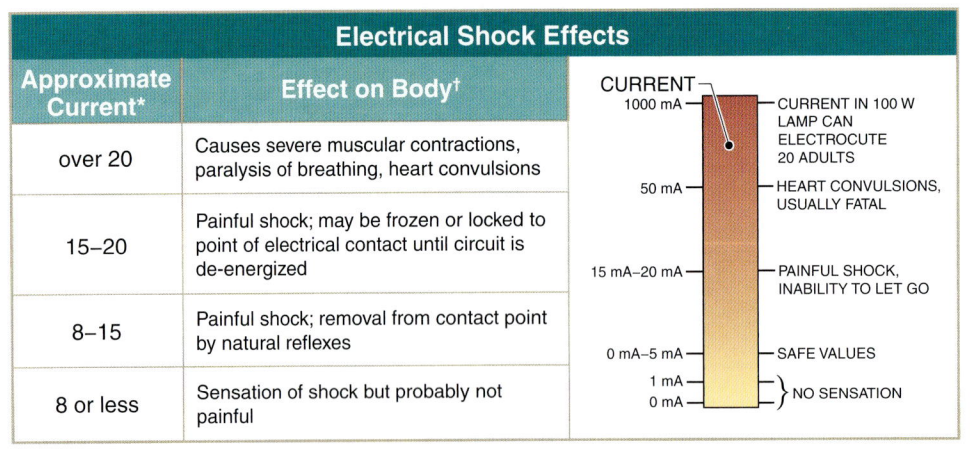

Approximate Current*	Effect on Body†
over 20	Causes severe muscular contractions, paralysis of breathing, heart convulsions
15–20	Painful shock; may be frozen or locked to point of electrical contact until circuit is de-energized
8–15	Painful shock; removal from contact point by natural reflexes
8 or less	Sensation of shock but probably not painful

* in mA
† effects vary depending on time, path, amount of exposure, and condition of body

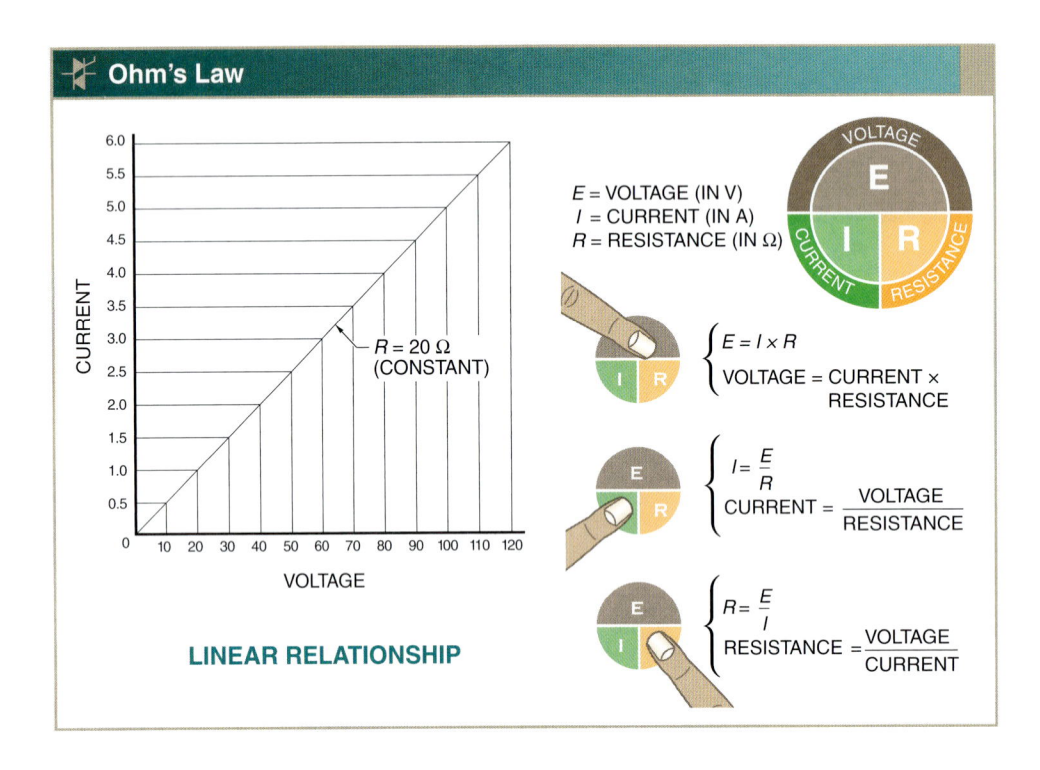

CURRENT
- 1000 mA — CURRENT IN 100 W LAMP CAN ELECTROCUTE 20 ADULTS
- 50 mA — HEART CONVULSIONS, USUALLY FATAL
- 15 mA–20 mA — PAINFUL SHOCK, INABILITY TO LET GO
- 0 mA–5 mA — SAFE VALUES
- 1 mA
- 0 mA — NO SENSATION

Ohm's Law

E = VOLTAGE (IN V)
I = CURRENT (IN A)
R = RESISTANCE (IN Ω)

$E = I \times R$
VOLTAGE = CURRENT × RESISTANCE

$I = \dfrac{E}{R}$
CURRENT = $\dfrac{\text{VOLTAGE}}{\text{RESISTANCE}}$

$R = \dfrac{E}{I}$
RESISTANCE = $\dfrac{\text{VOLTAGE}}{\text{CURRENT}}$

$R = 20\ \Omega$ (CONSTANT)

LINEAR RELATIONSHIP

Conductor Ratings

AWG	Copper Ampacity	Aluminum Ampacity	Dia*
22	—	—	25.0
20	—	—	32.0
18	—	—	40.0
16	—	—	51.0
14	15	—	64.0
12	20	15	81.0
10	30	25	102.0
8	40	30	128.0
6	55	40	162.0

* in mils (Rounded to nearest whole number.)

Power Formula

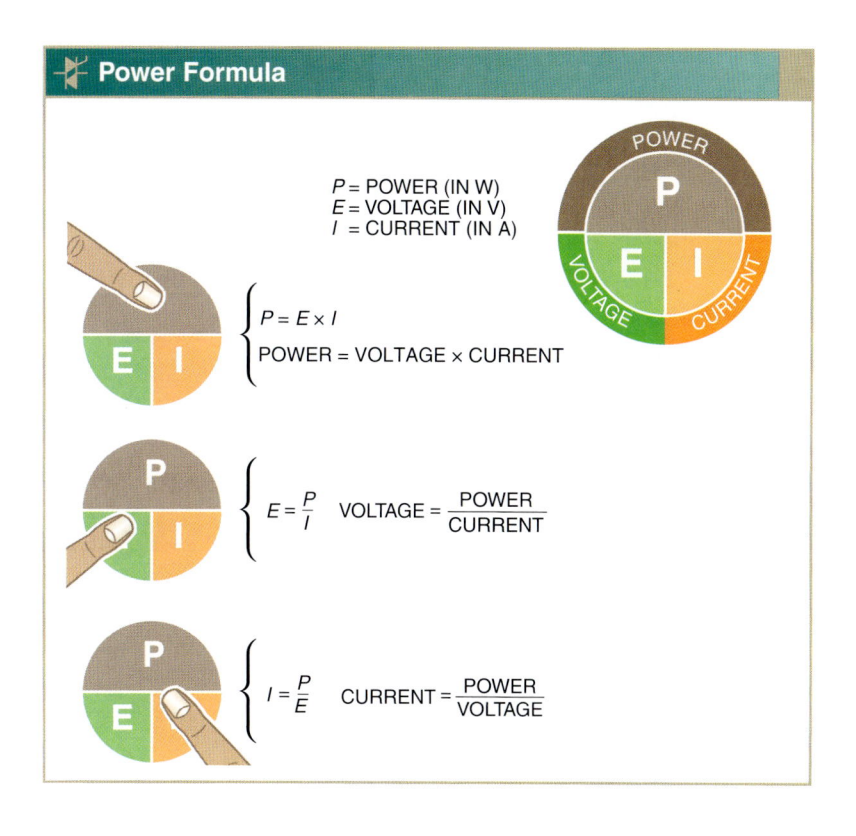

P = POWER (IN W)
E = VOLTAGE (IN V)
I = CURRENT (IN A)

$P = E \times I$
POWER = VOLTAGE × CURRENT

$E = \dfrac{P}{I}$ VOLTAGE = $\dfrac{\text{POWER}}{\text{CURRENT}}$

$I = \dfrac{P}{E}$ CURRENT = $\dfrac{\text{POWER}}{\text{VOLTAGE}}$

Impedance

I = CURRENT (IN A)
E = VOLTAGE (IN V)
Z = IMPEDANCE (IN Ω)

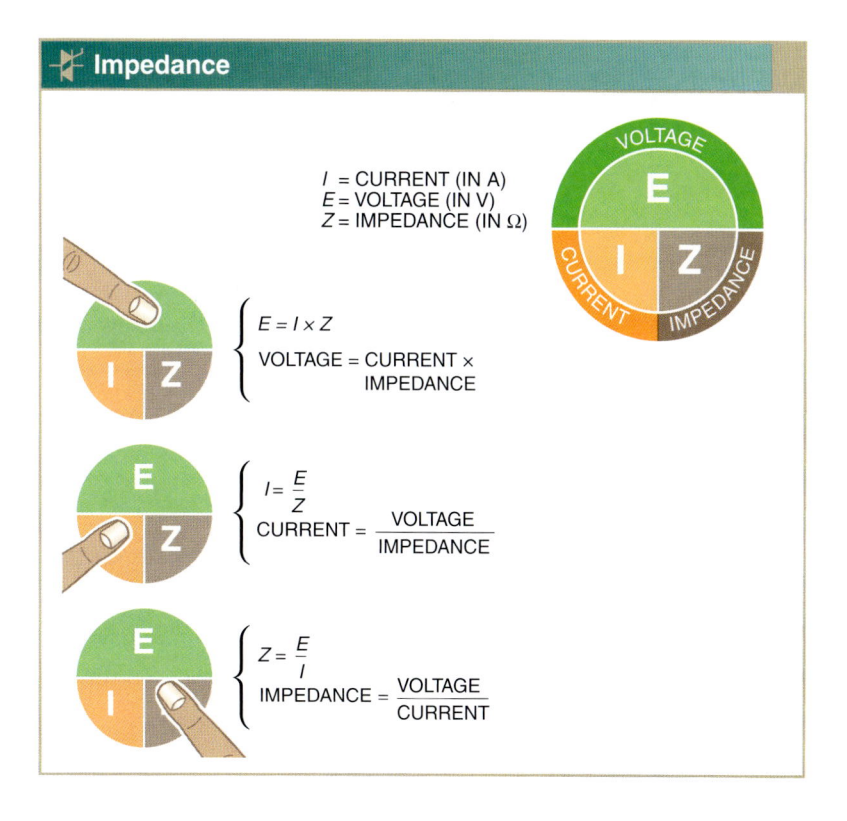

$E = I \times Z$
VOLTAGE = CURRENT × IMPEDANCE

$I = \dfrac{E}{Z}$
CURRENT = $\dfrac{\text{VOLTAGE}}{\text{IMPEDANCE}}$

$Z = \dfrac{E}{I}$
IMPEDANCE = $\dfrac{\text{VOLTAGE}}{\text{CURRENT}}$

Switch Symbols

Device	Part	Abbr.	Symbol	Function/Notes
PUSHBUTTON	Single-circuit, normally open (NO)	PB or PB-NO	MANUAL OPERATOR / TERMINALS	To make (NO) or break (NC) a circuit when manually depressed; one of the simplest and most common forms of control; typical pushbutton consists of one or more contact blocks, an operator, and a legend plate
LIMIT SWITCH (MECHANICAL)	Normally open (NO)	LS	MECHANICAL OPERATOR	To convert mechanical motion into an electrical signal; limit switches accomplish this conversion by using some type of lever to open or close contacts within the limit switch enclosure
TEMPERATURE SWITCH	Normally open (NO)	TEMP SW or TS	TEMPERATURE OPERATOR	To respond to temperature changes; temperature switch may be used to maintain a specified temperature within a process, or to protect against overtemperature conditions
PRESSURE SWITCH	Normally open (NO)	PS	PRESSURE OPERATOR	To open or close contacts in response to pressure changes in media such as air, water, or oil; electrical contacts may be used to start or stop motors of fans, open or close dampers or louvers, or signal a warning light or alarm
FLOW SWITCH	Normally open (NO)	FLS	FLOW OPERATOR	To sense the movement of a fluid; fluid may be air, water, oil, or other gases or liquids; flow switch is a control switch that is usually inserted into a pipe or duct; element will move and activate contacts whenever the fluid flow is sufficient to overcome a spring tension
LEVEL SWITCH	Normally open (NO)	LEVEL SW or LS	LEVEL OPERATOR	To measure and respond to the level of material; material may be water, oil, paint, granules, or other solids; level switches are control devices
FOOT SWITCH	Normally open (NO)	FTS	FOOT OPERATOR	To allow free use of hands while providing for manual control of a machine; many foot switches include a guard to prevent accidental operation

Solid-State Switch Symbols

Device	Part	Abbr.	Symbol	Function/Notes
LIMIT SWITCH (SOLID-STATE)	Normally open (NO)	PROX	SOLID-STATE OR	To detect the presence or absence of an object; solid-state limit switches are known as proximity switches; proximity switches eliminate the need to touch the object; switch uses solid-state components to start or stop the flow of current; three basic types of proximity switches are inductive, capacitive, and magnetic (Hall effect)
	Normally closed (NC)	PROX	OR	

Appendix **549**

Normally Open (NO) and Normally Closed (NC) Switches

Device	Part	Abbr.	Symbol/Condition — Normal	Symbol/Condition — Activated	Function/Notes
SWITCH (MANUAL)	Normally open (NO)	S	MANUAL OPERATOR / TERMINALS	MANUAL OPERATOR	Control contacts are contacts that switch low currents; power contacts are contacts that switch high currents
	Normally closed (NC)	S	TERMINALS / MANUAL OPERATOR	MANUAL OPERATOR	
LIMIT SWITCH (MECHANICAL)	Normally open (NO)	LS	TERMINALS / MECHANICAL OPERATOR	MECHANICAL OPERATOR	To convert mechanical motion into an electrical signal; limit switches accomplish this conversion by using some type of lever to open or close contacts within the limit switch enclosure
	Normally closed (NC)	LS	MECHANICAL OPERATOR / TERMINALS	MECHANICAL OPERATOR	

Switch Operator Conditions

Device	Part	Abbr.	Symbol	Notes
LIMIT SWITCH (MECHANICAL)	Normally open (NO)	LS	MECHANICAL OPERATOR	Shown on diagram in normal condition with nothing touching switch operator
	Normally open, held closed (NO)	LS	OPERATOR IN CLOSED CONDITION	Shown on diagram in activated condition in which something (guard in place) is touching switch operator
	Normally closed (NC)	LS	MECHANICAL OPERATOR	Shown on diagram in normal condition with nothing touching switch operator
	Normally closed, held open (NC)	LS	OPERATOR IN CLOSED CONDITION	Shown on diagram in activated condition in which something (guard in place) is touching switch operator
TEMPERATURE SWITCH	Normally open (NO)	TEMP SW	100°F TEMPERATURE OPERATOR / 80°F AIR TEMPERATURE	Shown on diagram in a condition in which the temperature around the switch is less than the actual setting of the temperature switch
	Normally open, held closed (NO)	TEMP SW	100°F OPERATOR IN CLOSED CONDITION / 110°F AIR TEMPERATURE	Shown on diagram in a condition in which the temperature around the switch is greater than the actual setting of the temperature switch
	Normally closed (NC)	TEMP SW	100°F TEMPERATURE OPERATOR / 80°F AIR TEMPERATURE	Shown on diagram in a condition in which the temperature around the switch is less than the actual setting of the temperature switch
	Normally closed, held open (NC)	TEMP SW	100°F OPERATOR IN OPEN CONDITION / 110°F AIR TEMPERATURE	Shown on diagram in a condition in which the temperature around the switch is greater than the actual setting of the temperature switch

Expanded (Special) Programming Symbols

Component	Symbol
Equal-to contact	─]=[─
Not-equal-to contact	─]≠[─
Greater-than or equal to contact	─]≥[─
Less-than contact	─]<[─
Addition	─(+)─
Subtraction	─(−)─
Multiplication	─(×)─
Division	─(÷)─
End	─(END)─
Set or latch	─(SET)─ ─(L)─
Reset or unlatch	─(RSET)─ ─(U)─
Read high speed clock	─(RHC)─ or ─[RHC]─
Timer	─(TMR)─ or ─[TMR]─
Counter	─(CNT)─ or ─[CNT]─

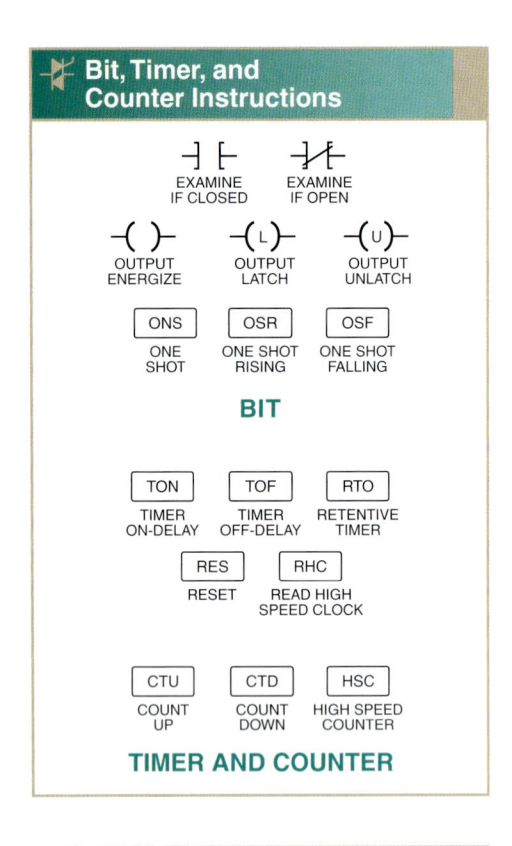

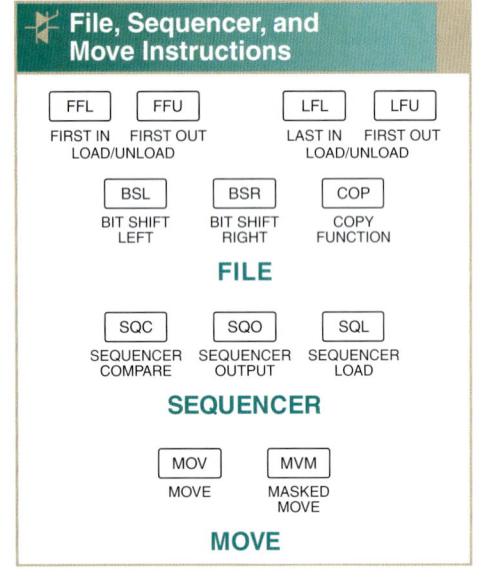

Bit, Timer, and Counter Instructions

BIT
- ─] [─ EXAMINE IF CLOSED
- ─]/[─ EXAMINE IF OPEN
- ─( )─ OUTPUT ENERGIZE
- ─(L)─ OUTPUT LATCH
- ─(U)─ OUTPUT UNLATCH
- ONS — ONE SHOT
- OSR — ONE SHOT RISING
- OSF — ONE SHOT FALLING

TIMER AND COUNTER
- TON — TIMER ON-DELAY
- TOF — TIMER OFF-DELAY
- RTO — RETENTIVE TIMER
- RES — RESET
- RHC — READ HIGH SPEED CLOCK
- CTU — COUNT UP
- CTD — COUNT DOWN
- HSC — HIGH SPEED COUNTER

File, Sequencer, and Move Instructions

FILE
- FFL — FIRST IN FIRST OUT LOAD/UNLOAD
- FFU — FIRST IN FIRST OUT LOAD/UNLOAD
- LFL — LAST IN FIRST OUT LOAD/UNLOAD
- LFU — LAST IN FIRST OUT LOAD/UNLOAD
- BSL — BIT SHIFT LEFT
- BSR — BIT SHIFT RIGHT
- COP — COPY FUNCTION

SEQUENCER
- SQC — SEQUENCER COMPARE
- SQO — SEQUENCER OUTPUT
- SQL — SEQUENCER LOAD

MOVE
- MOV — MOVE
- MVM — MASKED MOVE

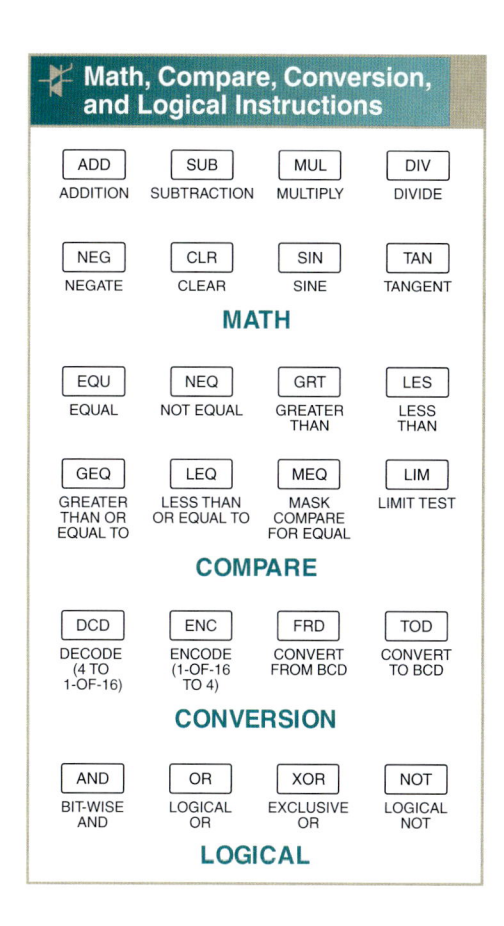

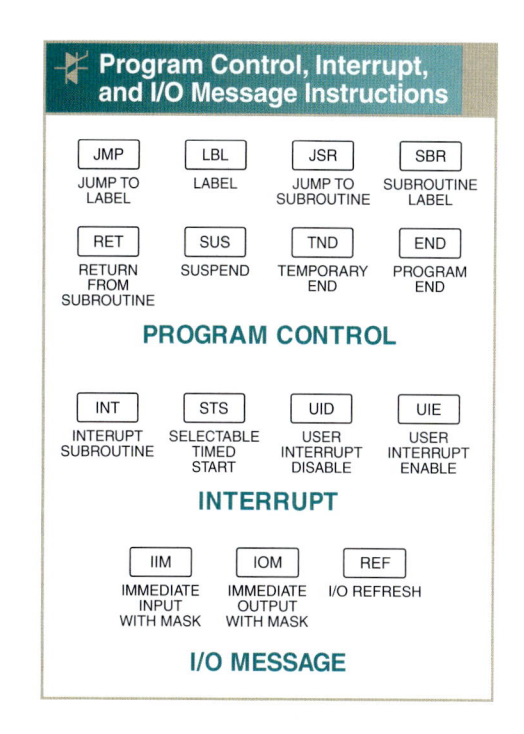

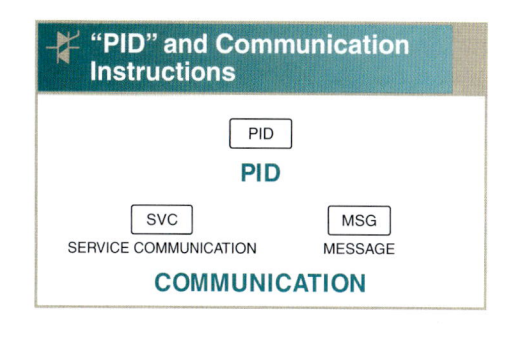

Math, Compare, Conversion, and Logical Instructions

ADD	SUB	MUL	DIV
ADDITION	SUBTRACTION	MULTIPLY	DIVIDE

NEG	CLR	SIN	TAN
NEGATE	CLEAR	SINE	TANGENT

MATH

EQU	NEQ	GRT	LES
EQUAL	NOT EQUAL	GREATER THAN	LESS THAN

GEQ	LEQ	MEQ	LIM
GREATER THAN OR EQUAL TO	LESS THAN OR EQUAL TO	MASK COMPARE FOR EQUAL	LIMIT TEST

COMPARE

DCD	ENC	FRD	TOD
DECODE (4 TO 1-OF-16)	ENCODE (1-OF-16 TO 4)	CONVERT FROM BCD	CONVERT TO BCD

CONVERSION

AND	OR	XOR	NOT
BIT-WISE AND	LOGICAL OR	EXCLUSIVE OR	LOGICAL NOT

LOGICAL

Program Control, Interrupt, and I/O Message Instructions

JMP	LBL	JSR	SBR
JUMP TO LABEL	LABEL	JUMP TO SUBROUTINE	SUBROUTINE LABEL

RET	SUS	TND	END
RETURN FROM SUBROUTINE	SUSPEND	TEMPORARY END	PROGRAM END

PROGRAM CONTROL

INT	STS	UID	UIE
INTERUPT SUBROUTINE	SELECTABLE TIMED START	USER INTERRUPT DISABLE	USER INTERRUPT ENABLE

INTERRUPT

IIM	IOM	REF
IMMEDIATE INPUT WITH MASK	IMMEDIATE OUTPUT WITH MASK	I/O REFRESH

I/O MESSAGE

"PID" and Communication Instructions

PID

PID

SVC	MSG
SERVICE COMMUNICATION	MESSAGE

COMMUNICATION

Logic Functions

Logic Function	Circuit Description	Ladder Diagrams	PLC Programming Diagrams	Digital Logic/PLC Logic Block Programming Diagrams
AND	**ENERGIZED** Output energized if all inputs are activated. **DE-ENERGIZED** Output de-energized if any one input is deactivated.			
OR	**ENERGIZED** Output energized if one or more inputs are activated. **DE-ENERGIZED** Output de-energized if all inputs are deactivated.			
NOT	**ENERGIZED** Output energized if input is not activated. **DE-ENERGIZED** Output de-energized if input is activated.			
NOR	**ENERGIZED** Output energized if no inputs are activated. **DE-ENERGIZED** Output de-energized if one or more inputs are activated.			
NAND	**ENERGIZED** Output energized unless all inputs are activated. **DE-ENERGIZED** Output de-energized if all inputs are activated.			

Truth Tables

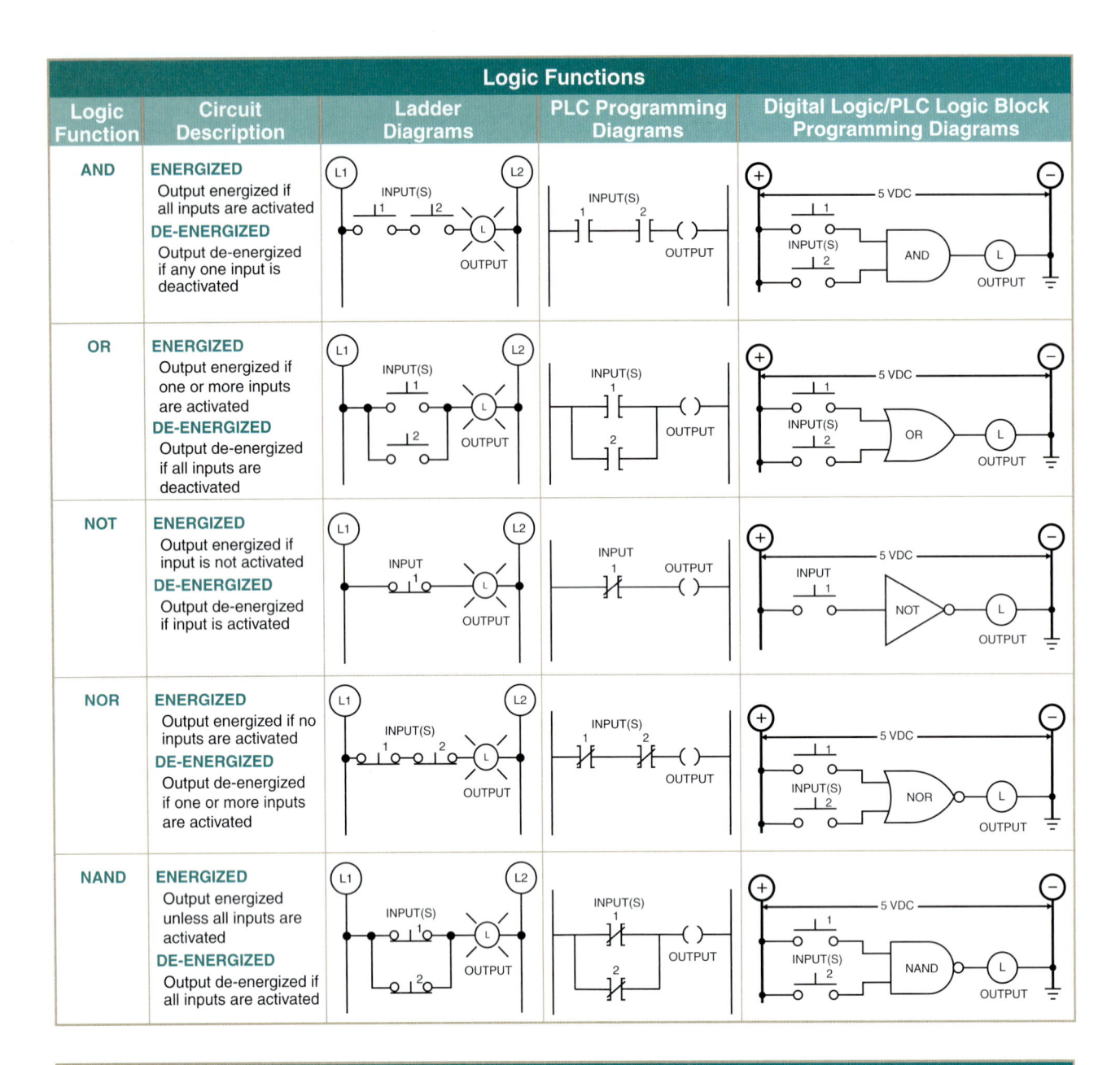

AND

Inputs		Output
A	B	Y
0	0	0
0	1	0
1	0	0
1	1	1

OR

Inputs		Output
A	B	Y
0	0	0
0	1	1
1	0	1
1	1	1

NOT

Inputs	Output
A	Y
0	1
1	0

NOR

Inputs		Output
A	B	Y
0	0	1
0	1	0
1	0	0
1	1	0

NAND

Inputs		Output
A	B	Y
0	0	1
0	1	1
1	0	1
1	1	0

XOR

Inputs		Output
A	B	Y
0	0	0
0	1	1
1	0	1
1	1	0

Units of Memory

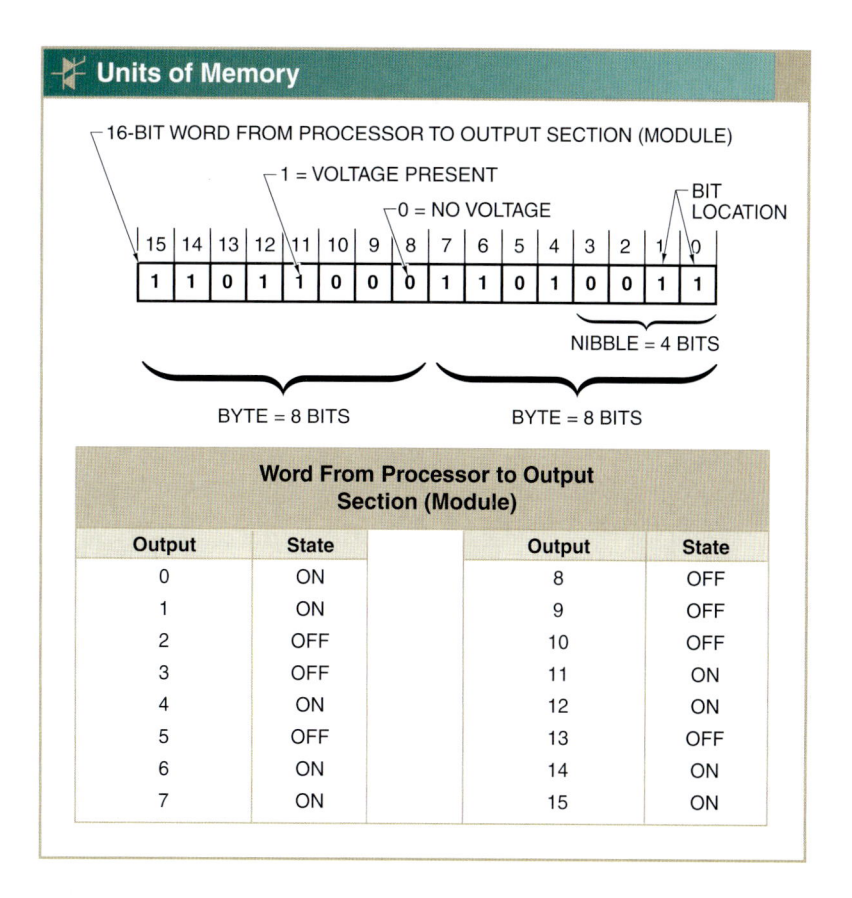

16-BIT WORD FROM PROCESSOR TO OUTPUT SECTION (MODULE)
1 = VOLTAGE PRESENT
0 = NO VOLTAGE
BIT LOCATION

15	14	13	12	11	10	9	8	7	6	5	4	3	2	1	0
1	1	0	1	1	0	0	0	1	1	0	1	0	0	1	1

NIBBLE = 4 BITS
BYTE = 8 BITS
BYTE = 8 BITS

Word From Processor to Output Section (Module)

Output	State	Output	State
0	ON	8	OFF
1	ON	9	OFF
2	OFF	10	OFF
3	OFF	11	ON
4	ON	12	ON
5	OFF	13	OFF
6	ON	14	ON
7	ON	15	ON

Electromechanical Relays

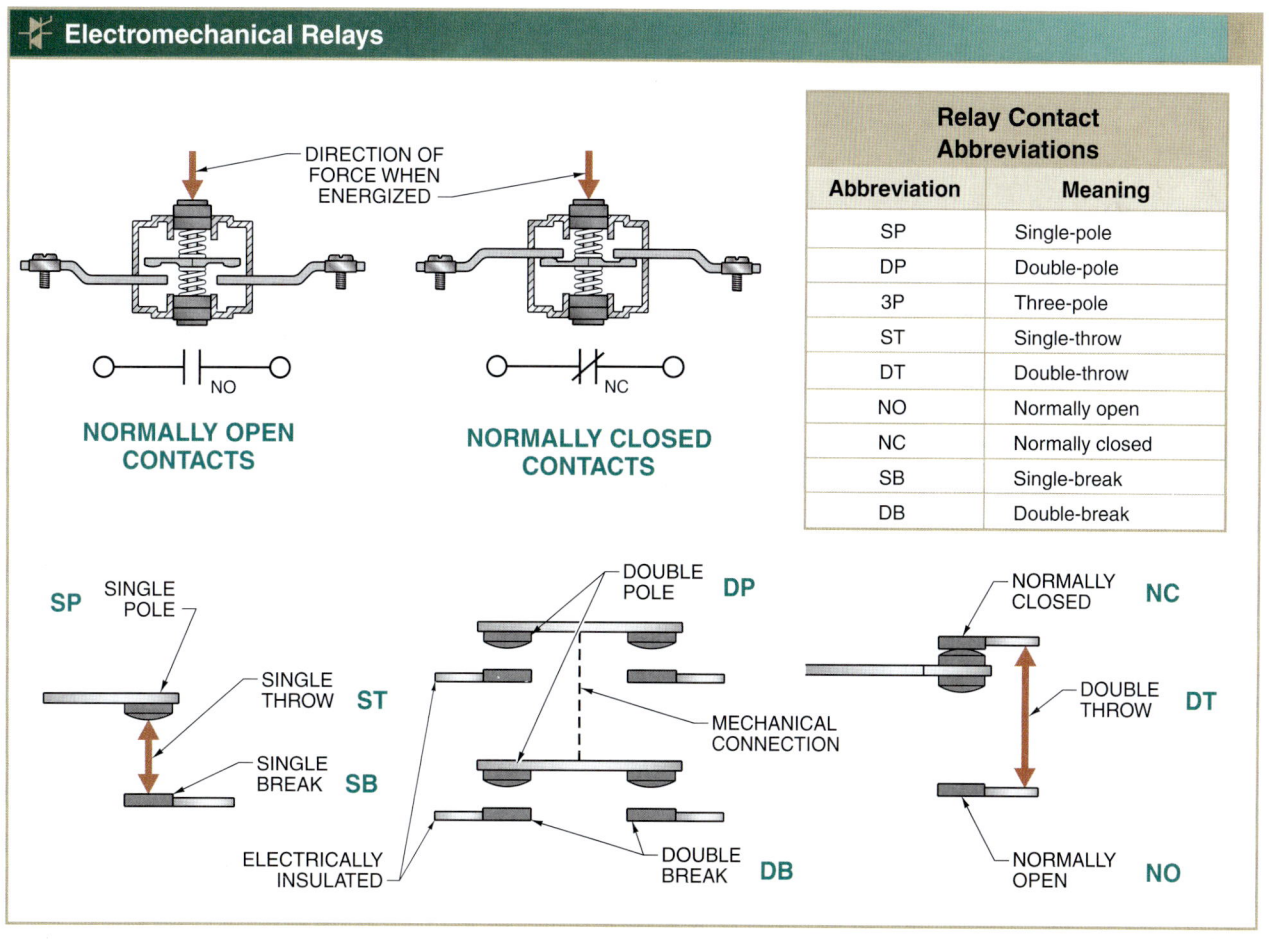

DIRECTION OF FORCE WHEN ENERGIZED

NORMALLY OPEN CONTACTS

NORMALLY CLOSED CONTACTS

Relay Contact Abbreviations

Abbreviation	Meaning
SP	Single-pole
DP	Double-pole
3P	Three-pole
ST	Single-throw
DT	Double-throw
NO	Normally open
NC	Normally closed
SB	Single-break
DB	Double-break

SP SINGLE POLE
ST SINGLE THROW
SB SINGLE BREAK
ELECTRICALLY INSULATED

DP DOUBLE POLE
MECHANICAL CONNECTION
DB DOUBLE BREAK

NC NORMALLY CLOSED
DT DOUBLE THROW
NO NORMALLY OPEN

PLCs Controlling Electric Motor Drives

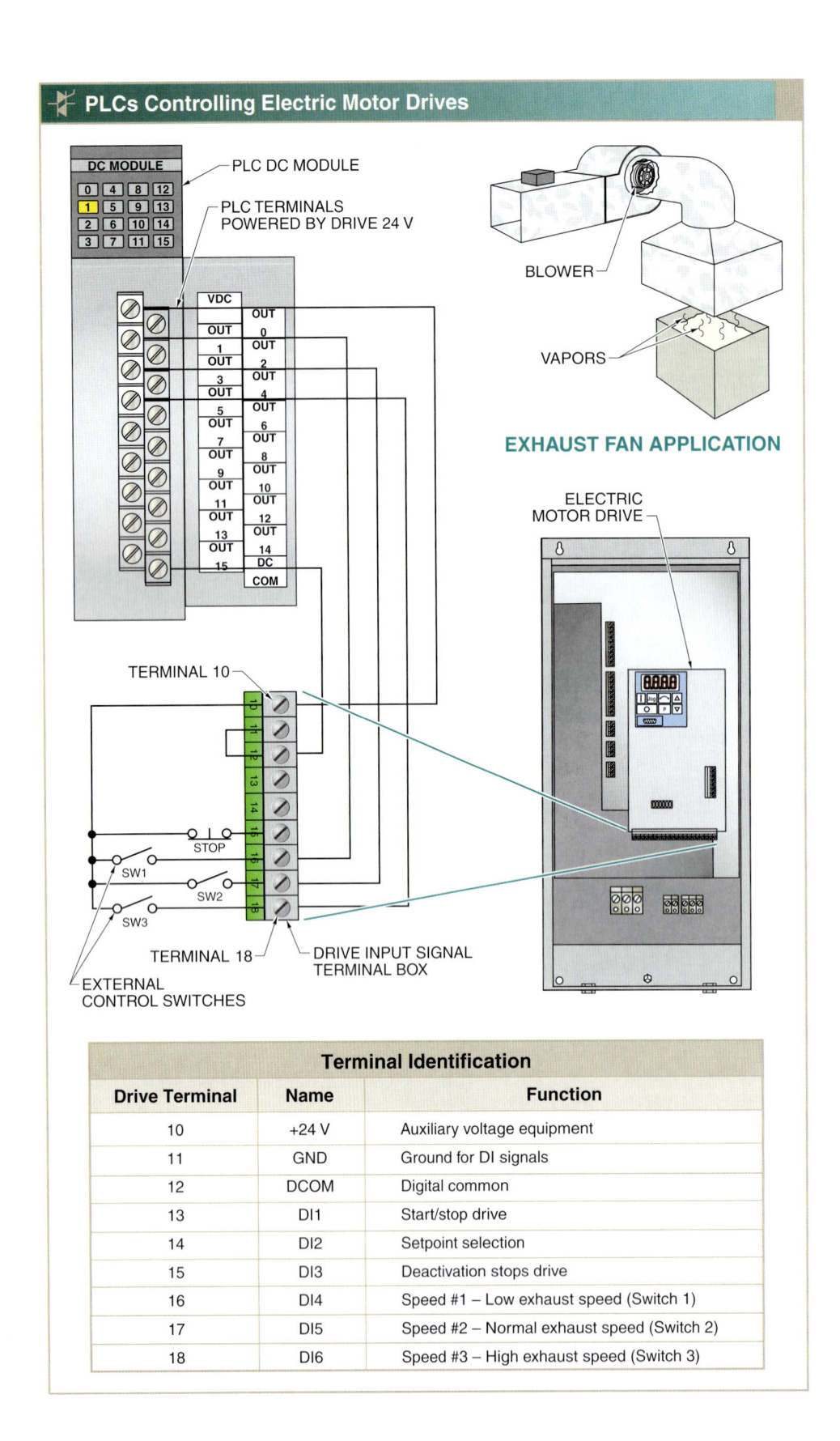

EXHAUST FAN APPLICATION

Terminal Identification		
Drive Terminal	**Name**	**Function**
10	+24 V	Auxiliary voltage equipment
11	GND	Ground for DI signals
12	DCOM	Digital common
13	DI1	Start/stop drive
14	DI2	Setpoint selection
15	DI3	Deactivation stops drive
16	DI4	Speed #1 – Low exhaust speed (Switch 1)
17	DI5	Speed #2 – Normal exhaust speed (Switch 2)
18	DI6	Speed #3 – High exhaust speed (Switch 3)

Temperature Sensor Data

4 mA TO 20 mA	
Temperature*	Sensor Output[†]
−26.50	4
−13.25	4
0.00	4
13.25	5
26.50	6
39.75	7
53.00	8
66.25	9
79.50	10
92.75	11
106.00	12
119.25	13
132.50	14
145.75	15
159.00	16
172.25	17
185.50	18
198.75	19
212.00	20
225.25	20
238.50	20

* in °F
[†] in mA

0 V TO 10 V	
Temperature*	Sensor Output[†]
−42.40	0.0
−21.20	0.0
0.00	0.0
21.20	1.0
31.80	1.5
42.40	2.0
63.60	3.0
74.20	3.5
84.80	4.0
106.00	5.0
116.60	5.5
127.20	6.0
148.40	7.0
159.00	7.5
169.60	8.0
190.80	9.0
201.40	9.5
212.00	10.0
233.20	10.0
254.40	10.0

* in °F
[†] in V

Converter Output Spans

Sensor Temperature*	Sensor Output†	Converter Output‡
0.00	4	0.00
13.25	5	0.44
26.50	6	0.88
39.75	7	1.32
53.00	8	1.76
66.25	9	2.20
71.40	9.7	2.38
79.50	10	2.64
92.75	11	3.08
106.00	12	3.52
119.25	13	3.96
132.50	14	4.40
145.75	15	4.84
159.00	16	5.28
172.25	17	5.72
185.50	18	6.16
198.75	19	6.60
212.00	20	7.04

* in °F
† in mA
‡ in VDC

Data Table File Addressing

Default Files

File Type	Identifier	File Number
Output	O	0
Input	I	1
Status	S	2
Bit	B	3
Timer	T	4
Counter	C	5
Control	R	6
Integer	N	7
Float	F	8

Technician-Assigned Files

File Type	Identifier	File Number
Bit	B	9–255
Timer	T	9–255
Counter	C	9–255
Control	R	9–255
Integer	N	9–255
Float	F	9–255

PLC Scan Time Worksheets

PLC Scan Time Worksheet

Procedure	Maximum Scan Time
1. Input scan time, output scan time, housekeeping time, and forcing.	__210__ µs (discrete) __330__ µs with forcing (analog) __250__ µs without forcing (analog)

2. Estimate your program scan time:
   A. Count the number of program rungs in your logic program and multiply by 6.
   B. Add up your program execution times when all instructions are true. Include interrupt routines in this calculation.
	_____
	
	_____ µs

3. Estimate your controller scan time:
   A. Without communications, add sections 1 and 2.
   B. With communications, add sections 1 and 2 and multiply by 1.05.
	_____ µs
	_____ µs

4. To determine your maximum scan time in ms, divide your controller scan time by 1000.	_____ ms

* If a subroutine executes more than once per scan, include each subroutine execution scan time.

Instruction Execution Times and Memory Usage

Mnemonic	False Execution Time*	True Execution Time*	Memory Usage†	Name	Instruction Type
ADD	6.78	33.09	1.50	Add	Math
CTD	27.22	32.19	1.00	Count Down	Basic
CTU	26.67	29.84	1.00	Count Up	Basic
OR	6.78	33.68	1.50	Or	Data Handling
OSR	11.48	13.02	1.00	One-Shot Rising	Basic
OTE	4.43	4.43	0.75	Output Energize	Basic
TON	30.38	38.34	1.00	Timer On-Delay	Basic
XIC	1.72	1.54	0.75	Examine If Closed	Basic
XIO	1.72	1.54	0.75	Examine If Open	Basic

* approx. µseconds
† user words

Timer Instructions

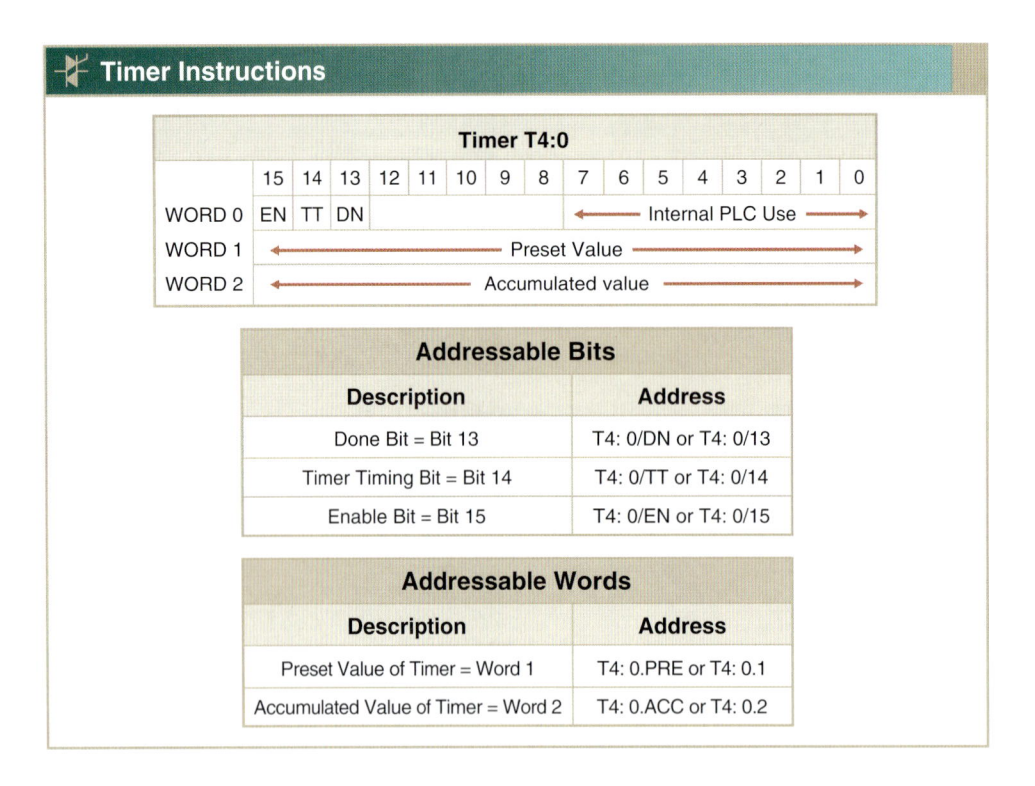

Timer T4:0

	15	14	13	12	11	10	9	8	7	6	5	4	3	2	1	0
WORD 0	EN	TT	DN						← Internal PLC Use →
WORD 1	← Preset Value →
WORD 2	← Accumulated value →

Addressable Bits

Description	Address
Done Bit = Bit 13	T4: 0/DN or T4: 0/13
Timer Timing Bit = Bit 14	T4: 0/TT or T4: 0/14
Enable Bit = Bit 15	T4: 0/EN or T4: 0/15

Addressable Words

Description	Address
Preset Value of Timer = Word 1	T4: 0.PRE or T4: 0.1
Accumulated Value of Timer = Word 2	T4: 0.ACC or T4: 0.2

Timer On-Delay (TON) Instructions

PLC PROGRAMMING DIAGRAM

I:2.0/3 — TON
- Timer On Delay
- Timer: T4:0
- Time Base: 1.0
- Preset: 5
- Accum: 0
- <EN>
- <DN>

T4:0/DN — O:1.0/1

TIMING SIGNALS

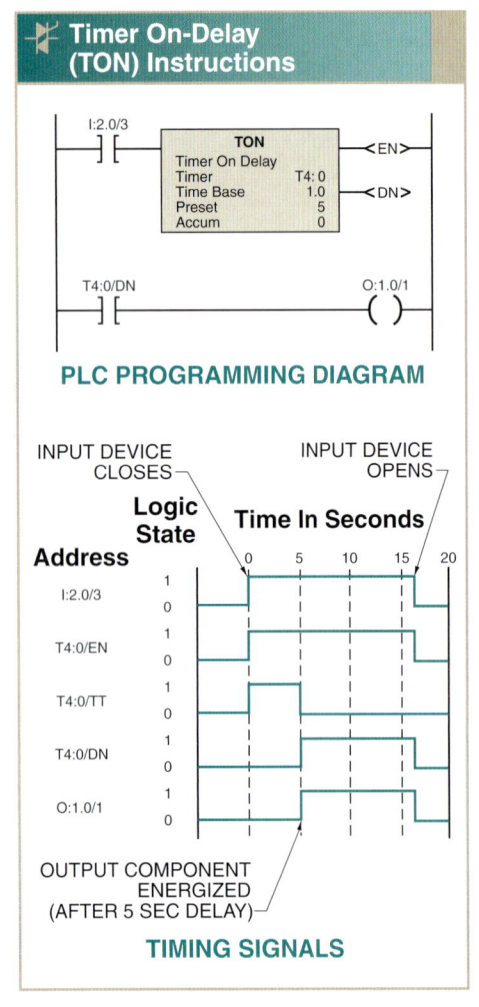

INPUT DEVICE CLOSES / INPUT DEVICE OPENS

Addresses: I:2.0/3, T4:0/EN, T4:0/TT, T4:0/DN, O:1.0/1

OUTPUT COMPONENT ENERGIZED (AFTER 5 SEC DELAY)

Timer Off-Delay (TOF) Instructions

PLC PROGRAMMING DIAGRAM

I:2.0/3 — TOF
- Timer Off Delay
- Timer: T4:0
- Time Base: 1.0
- Preset: 5
- Accum: 0
- <EN>
- <DN>

T4:0/DN — O:1.0/2

TIMING SIGNALS

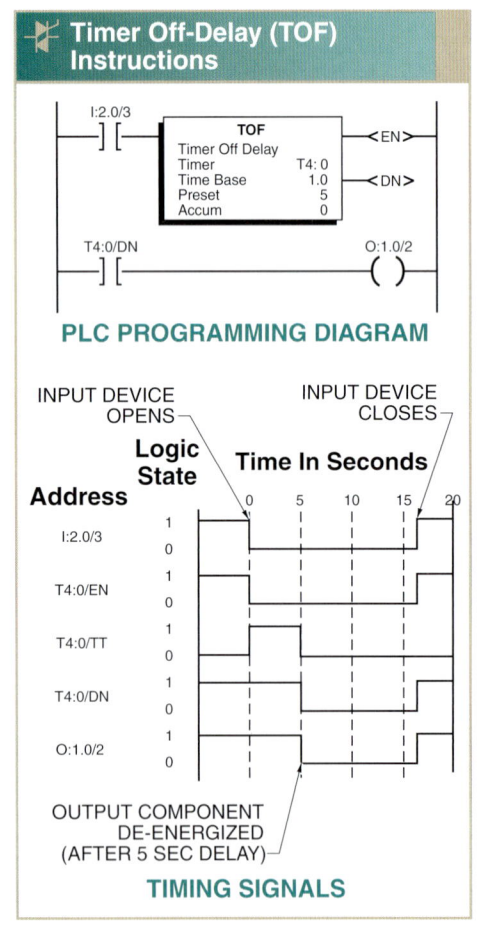

INPUT DEVICE OPENS / INPUT DEVICE CLOSES

Addresses: I:2.0/3, T4:0/EN, T4:0/TT, T4:0/DN, O:1.0/2

OUTPUT COMPONENT DE-ENERGIZED (AFTER 5 SEC DELAY)

Appendix **559**

Retentive Timer and Reset Instructions

PLC PROGRAMMING DIAGRAM

```
I:2.0/3                RTO
──┤ ├──────────    Retentive Timer On    ⟨EN⟩
                   Timer         T4:0
                   Time Base      1.0    ⟨DN⟩
                   Preset           8
                   Accum            0

T4:0/DN                                  O:1.0/2
──┤ ├───────────────────────────────────( )

I:2.0/4                                  T4:0
──┤ ├───────────────────────────────────(RES)
```

TIMING SIGNALS

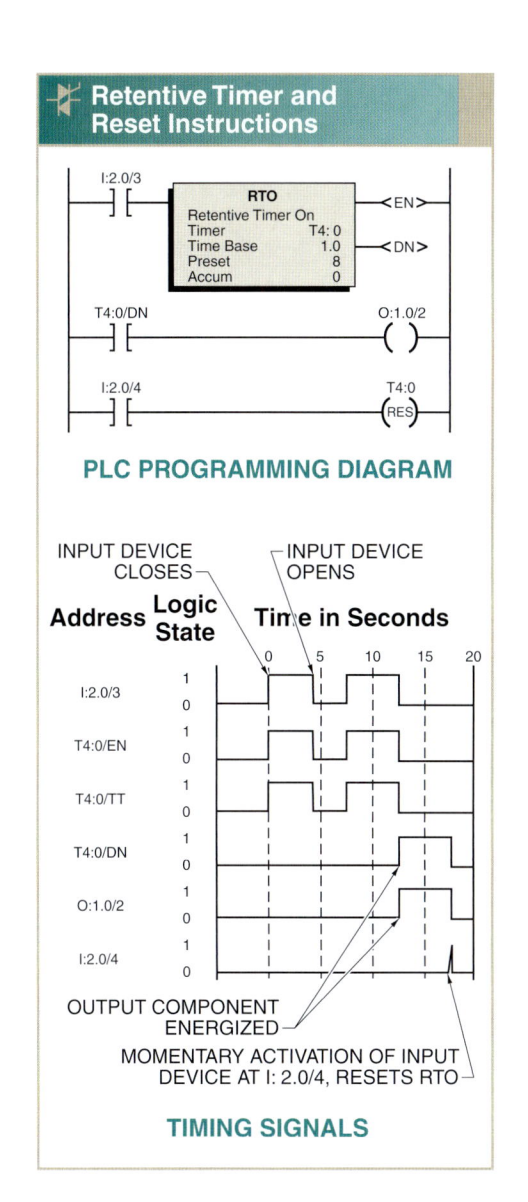

- INPUT DEVICE CLOSES
- INPUT DEVICE OPENS

Address — Logic State — Time in Seconds (0, 5, 10, 15, 20)

- I:2.0/3
- T4:0/EN
- T4:0/TT
- T4:0/DN
- O:1.0/2
- I:2.0/4

OUTPUT COMPONENT ENERGIZED

MOMENTARY ACTIVATION OF INPUT DEVICE AT I:2.0/4, RESETS RTO

Cascaded Timers

DISC STACK CENTRIFUGE

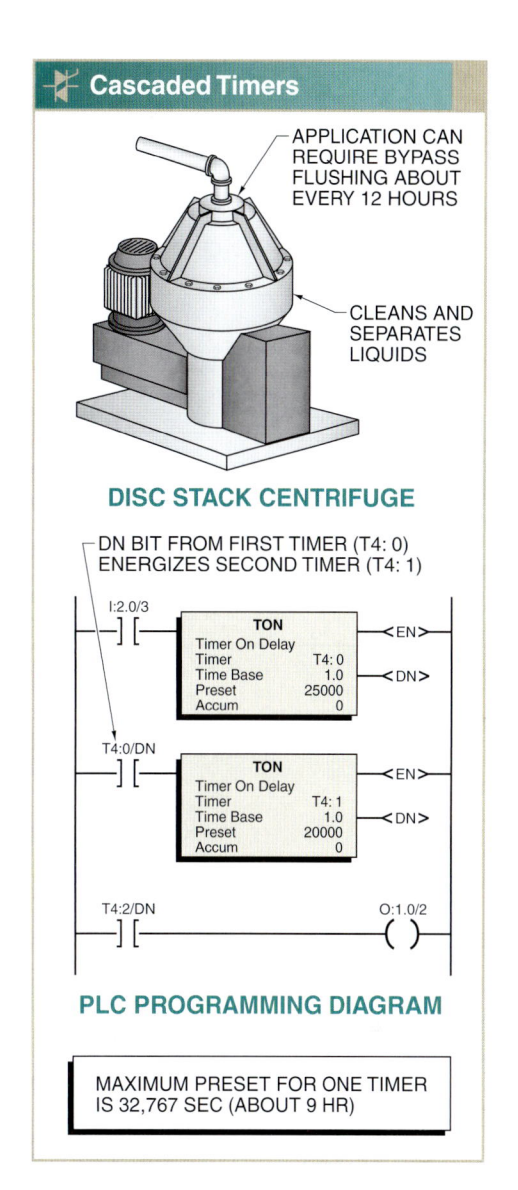

- APPLICATION CAN REQUIRE BYPASS FLUSHING ABOUT EVERY 12 HOURS
- CLEANS AND SEPARATES LIQUIDS

DN BIT FROM FIRST TIMER (T4:0) ENERGIZES SECOND TIMER (T4:1)

```
I:2.0/3                TON
──┤ ├──────────    Timer On Delay        ⟨EN⟩
                   Timer         T4:0
                   Time Base      1.0    ⟨DN⟩
                   Preset       25000
                   Accum            0

T4:0/DN                TON
──┤ ├──────────    Timer On Delay        ⟨EN⟩
                   Timer         T4:1
                   Time Base      1.0    ⟨DN⟩
                   Preset       20000
                   Accum            0

T4:2/DN                                  O:1.0/2
──┤ ├───────────────────────────────────( )
```

PLC PROGRAMMING DIAGRAM

MAXIMUM PRESET FOR ONE TIMER IS 32,767 SEC (ABOUT 9 HR)

Free Running (Repetitive) Timers

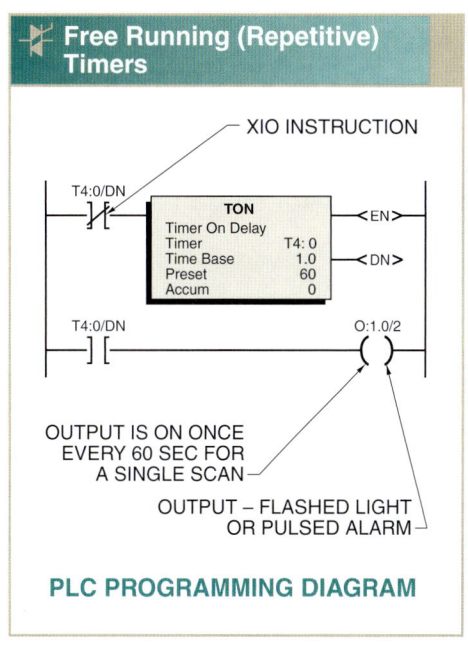

XIO INSTRUCTION

```
T4:0/DN                TON
──┤/├──────────    Timer On Delay        ⟨EN⟩
                   Timer         T4:0
                   Time Base      1.0    ⟨DN⟩
                   Preset          60
                   Accum            0

T4:0/DN                                  O:1.0/2
──┤ ├───────────────────────────────────( )
```

OUTPUT IS ON ONCE EVERY 60 SEC FOR A SINGLE SCAN

OUTPUT – FLASHED LIGHT OR PULSED ALARM

PLC PROGRAMMING DIAGRAM

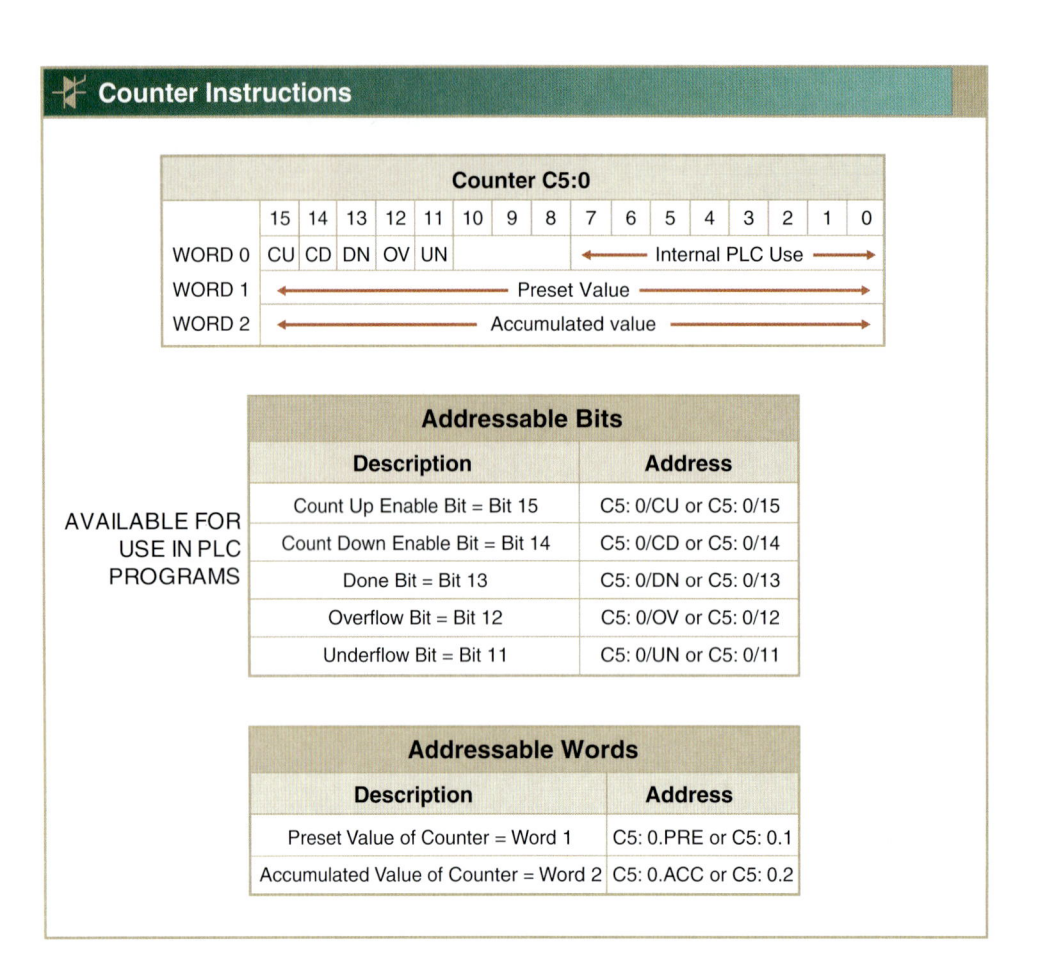

Counter Instructions

Counter C5:0

	15	14	13	12	11	10	9	8	7	6	5	4	3	2	1	0
WORD 0	CU	CD	DN	OV	UN				←——— Internal PLC Use ———→
WORD 1	←————————————— Preset Value —————————————→
WORD 2	←——————————— Accumulated value ———————————→

AVAILABLE FOR USE IN PLC PROGRAMS

Addressable Bits

Description	Address
Count Up Enable Bit = Bit 15	C5: 0/CU or C5: 0/15
Count Down Enable Bit = Bit 14	C5: 0/CD or C5: 0/14
Done Bit = Bit 13	C5: 0/DN or C5: 0/13
Overflow Bit = Bit 12	C5: 0/OV or C5: 0/12
Underflow Bit = Bit 11	C5: 0/UN or C5: 0/11

Addressable Words

Description	Address
Preset Value of Counter = Word 1	C5: 0.PRE or C5: 0.1
Accumulated Value of Counter = Word 2	C5: 0.ACC or C5: 0.2

Appendix **561**

Counter Count Up (CTU) Instructions

PLC PROGRAMMING DIAGRAM

TIMING SIGNALS

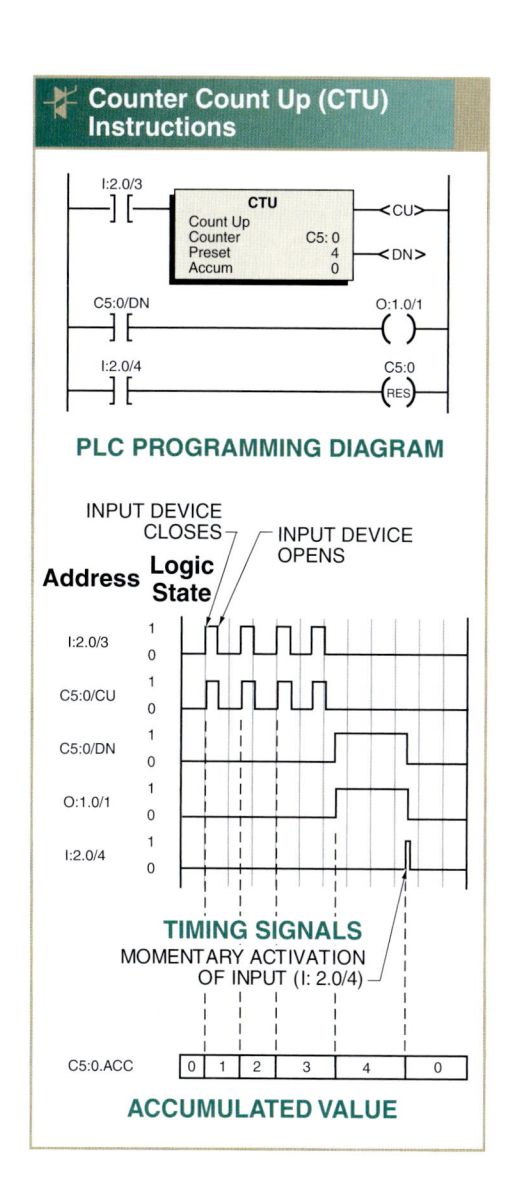

ACCUMULATED VALUE

Counter Count Down (CTD) Instructions

PLC PROGRAMMING DIAGRAM

TIMING SIGNALS

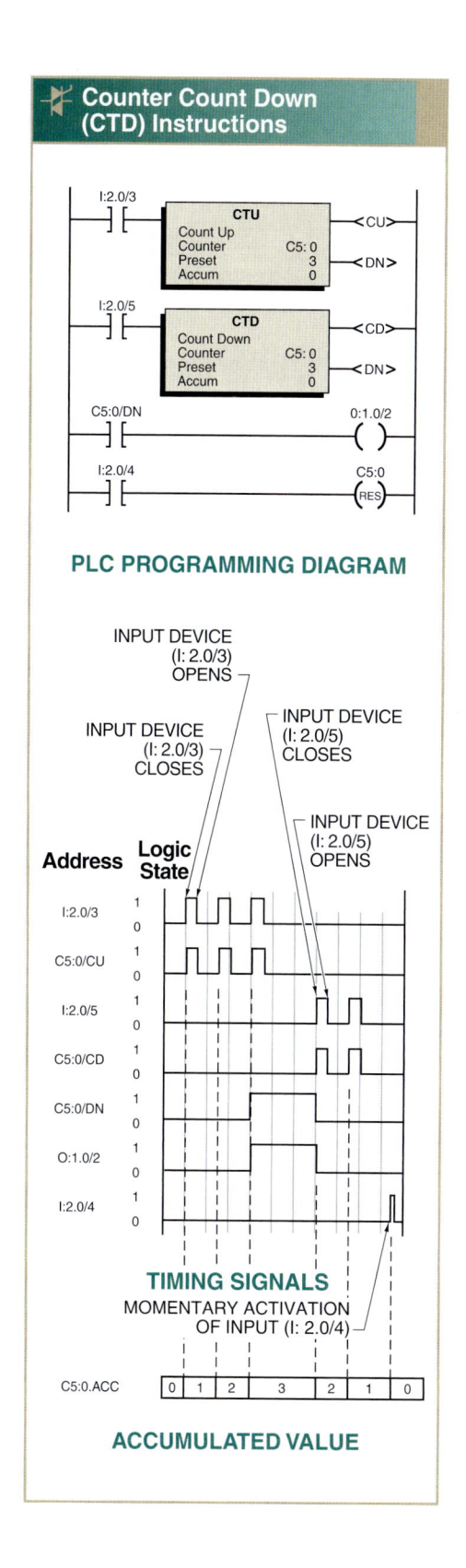

ACCUMULATED VALUE

High-Speed Counter (HSC) Instructions

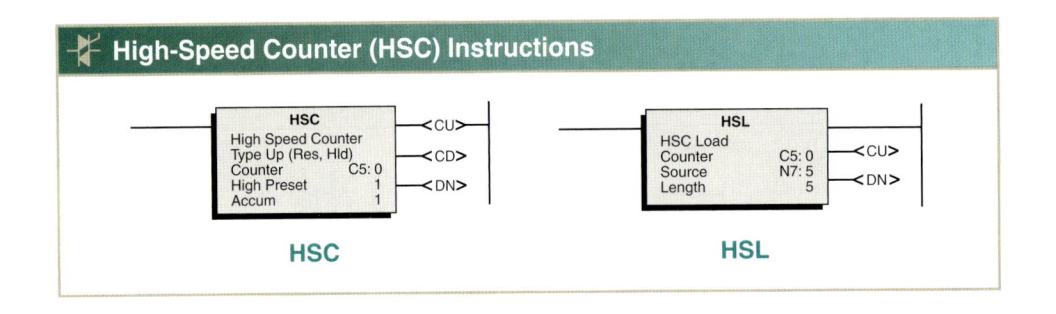

```
       HSC
High Speed Counter       <CU>
Type Up (Res, Hld)
Counter         C5:0     <CD>
High Preset        1
Accum              1     <DN>
```
HSC

```
       HSL
HSC Load
Counter         C5:0     <CU>
Source          N7:5
Length             5     <DN>
```
HSL

Math Instructions

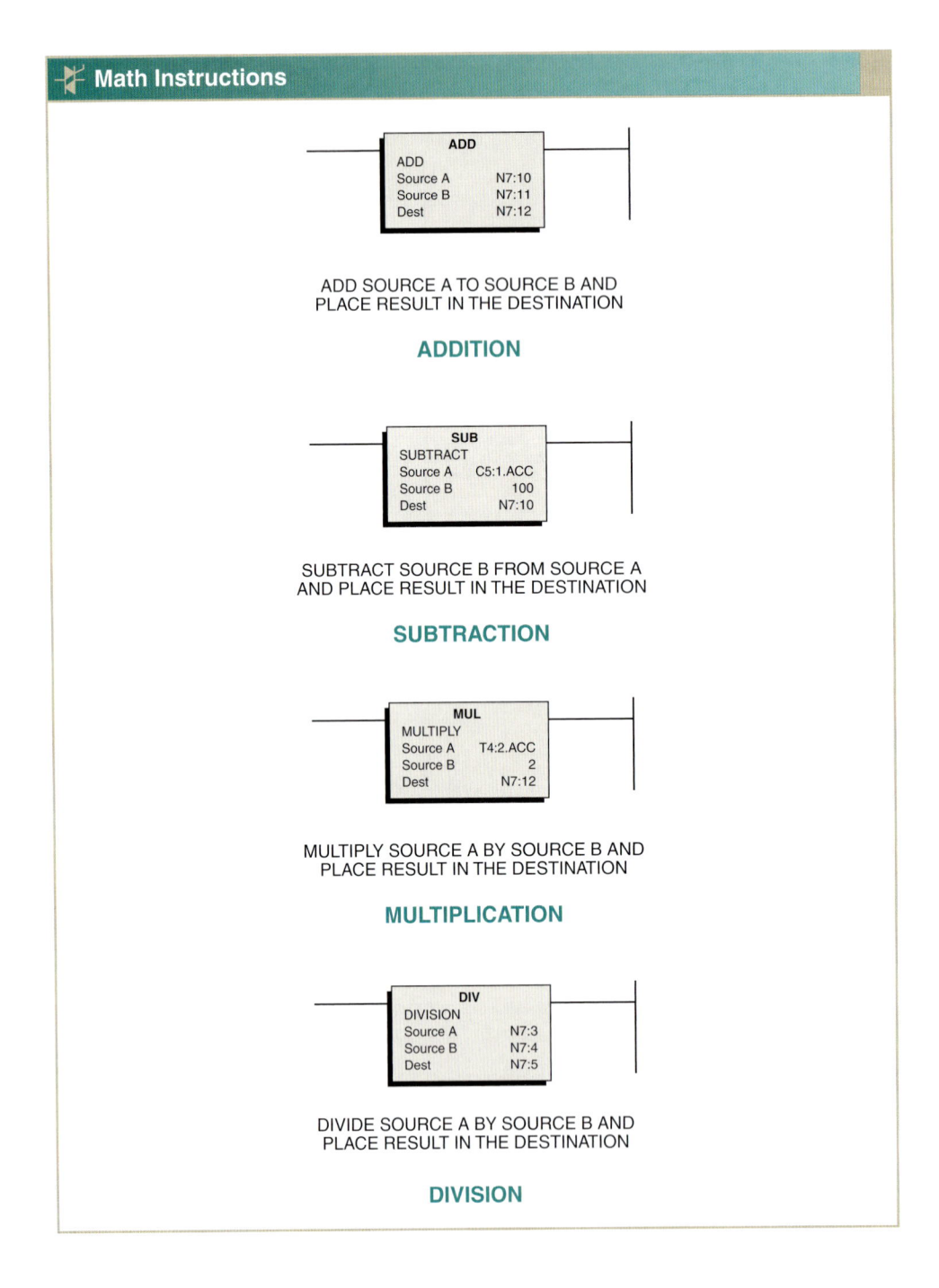

```
       ADD
ADD
Source A    N7:10
Source B    N7:11
Dest        N7:12
```

ADD SOURCE A TO SOURCE B AND
PLACE RESULT IN THE DESTINATION

ADDITION

```
       SUB
SUBTRACT
Source A    C5:1.ACC
Source B         100
Dest            N7:10
```

SUBTRACT SOURCE B FROM SOURCE A
AND PLACE RESULT IN THE DESTINATION

SUBTRACTION

```
       MUL
MULTIPLY
Source A    T4:2.ACC
Source B           2
Dest            N7:12
```

MULTIPLY SOURCE A BY SOURCE B AND
PLACE RESULT IN THE DESTINATION

MULTIPLICATION

```
       DIV
DIVISION
Source A    N7:3
Source B    N7:4
Dest        N7:5
```

DIVIDE SOURCE A BY SOURCE B AND
PLACE RESULT IN THE DESTINATION

DIVISION

Selected Test Instrument Abbreviations

AC	Alternating current or voltage	RPM	Revolutions per minute
DC	Direct current or voltage	COM	Common
V	Volts	OL	Overload
mV	Millivolts	T	Time
kV	Kilovolts	LSD	Least significant digit
A	Amperes	MAX	Maximum
mA	Milliamperes	MIN	Minimum
µA	Microamperes	AVG	Average
W	Watts	TRIG	Trigger
kΩ	Kilohms	V_{avg}	Average voltage
MΩ	Megohms	V_p	Peak voltage
Hz	Hertz	V_{p-p}	Peak-to-peak voltage
kHz	Kilohertz	V_{rms}	Root-mean-square (rms) voltage
µF	Microfarads	HiZ	High input impedance
nF	Nanofarads	dB	Decibel
°F	Degrees Fahrenheit	dBV	Decibel volts
°C	Degrees Celsius	dBW	Decibel watts

LOG	Readings are being recorded	
LO	Low	
AUTO-V	Automatic volts	
LoZ	Low input impedance	
nS	Nanosiemens (1×10^{-9}) or 0.000000001 siemens	
MEM	Memory	
MS	Time display in minutes:seconds	
HM	Time display in hours:minutes	

Selected Test Instrument Symbols

∼	AC
═	DC
≡	AC or DC
+	Positive
−	Negative
⏚	Ground
±	Plus or minus
▶│	Diode
▶│)))	Diode test
<	Less than
>	Greater than
△	Increase setting
▽	Decrease setting
🔧	See service manual
▫	Double insulation
⎓	Fuse
⊟	Battery
H	Hold
🔒	Lock
)))))	Audio beeper
⊣⊢	Capacitor
%	Percent
▷	Move right
◁	Move left
⊘	No (do not use)
○	Switch position OFF (power)
│	Switch position ON (power)
⦿	Manual Range mode
⚡	Warning: Dangerous or high voltage that could result in personal injury
⚠	Caution: Hazard that could result in equipment damage or personal injury
1000 V MAX	Terminals must not be connected to a circuit with higher than listed voltage
△	Relative mode − displayed value is difference between present measurement and previous stored measurement
Ω	Ohms resistance
☼	Meter display light
⚡	> 30 VAC or VDC present
⎍	Trigger on positive slope
⎎	Trigger on negative slope

Electrical Quantities

Variable	Name	Unit of Measure and Abbreviation
E	voltage	volt — V
I	current	ampere — A
R	resistance	ohm — Ω
P	power	watt — W
P	power (apparent)	volt-amp — VA
C	capacitance	farad — F
L	inductance	henry — H
Z	impedance	ohm — Ω
G	conductance	siemens — S
f	frequency	hertz — Hz
T	period	second — s

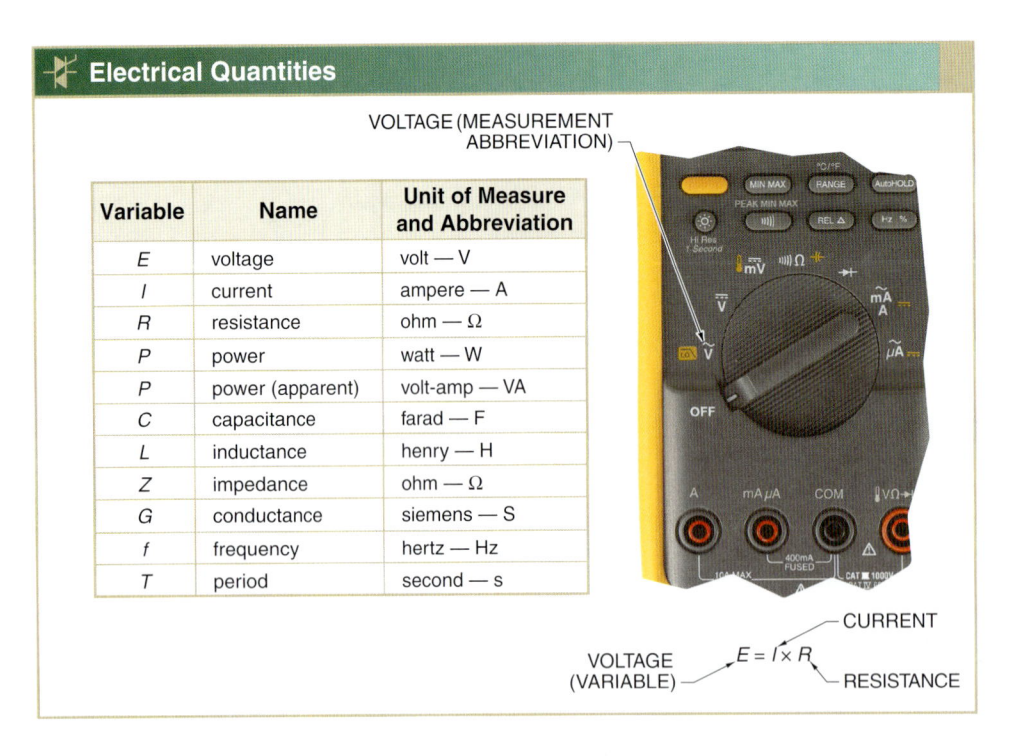

VOLTAGE (MEASUREMENT ABBREVIATION)

$E = I \times R$

VOLTAGE (VARIABLE) — CURRENT — RESISTANCE

IEC 61010 Test Instrument Measurement Categories

Class	In Brief	Examples
CAT I	Electronics	• Protected electronic equipment • Equipment connected to (source) circuits in which measures are taken to limit transient overvoltage to an appropriately low level • Any high-voltage, low-energy source derived from a high-winding-resistance transformer, such as the high-voltage section of a copier
CAT II	1φ receptacle-connected loads	• Appliances, portable tools, and other household and similar loads • Outlets and long branch circuits • Outlets at more than 30′ (10 m) from CAT III source • Outlets at more than 60′ (20 m) from CAT IV source
CAT III	3φ distribution, including 1φ commercial lighting	• Equipment in fixed installations, such as switchgear and polyphase motors • Bus and feeder in industrial plants • Feeders and short branch circuits and distribution panel devices • Lighting systems in larger buildings • Appliance outlets with short connections to service entrance
CAT IV	3φ at utility connection, any outdoor conductors	• Refers to the origin of installation, where low-voltage connection is made to utility power • Electric meters, primary overcurrent protection equipment • Outside and service entrance, service drop from pole to building, run between meter and panel • Overhead line to detached building

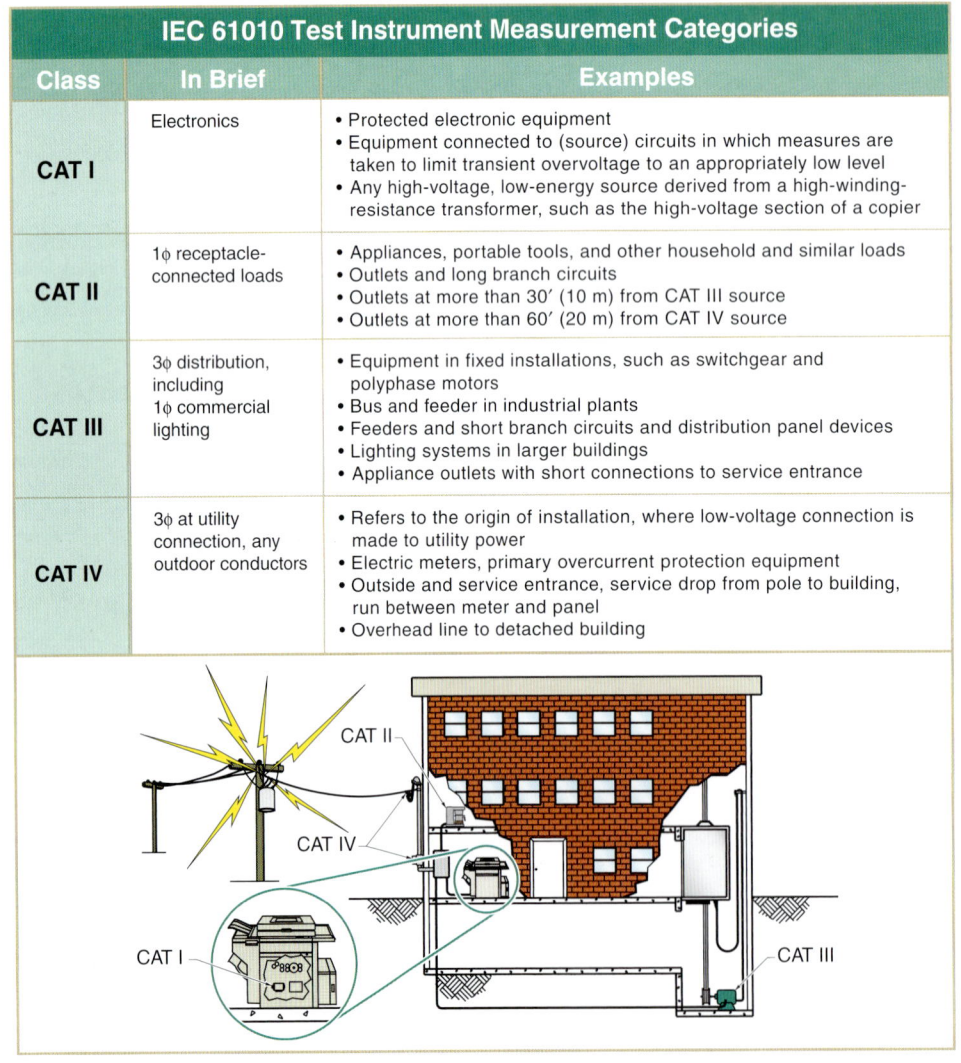

Glossary

A

abbreviation: A letter or combination of letters that represent a word.

AC sine wave: A symmetrical waveform that contains 360 electrical degrees and has one positive alternation and one negative alternation per cycle.

AC voltage: Voltage in a circuit that has current that reverses its direction of flow at regular intervals.

alternating current (AC): Current that reverses direction of flow at regular intervals.

alternation: Half of a cycle.

ammeter: A test instrument that measures the amount of current in an electrical circuit.

ampere: One coulomb passing a given point in an electrical circuit in one second.

analog electrical circuit: A circuit in which the load can be in any operating condition between and including fully ON and fully OFF.

analog input device: Any of the various types of sensors that measure a variable change in an environmental or operating condition and convert that change into a proportional voltage or current signal.

analog multimeter: A portable test instrument that uses electromechanical components to display measured values.

analog output device: A load or actuator that delivers a variable position or speed according to the analog signal (voltage or current) being sent from the PLC analog output terminal.

analog signal: An electronic signal that has continuously changing quantities (values) between defined limits.

AND circuit logic: The control logic developed when two or more normally open switches or contacts are connected in series to control a load.

arc flash hood: An eye and face protection device that covers the entire head with plastic and material.

arrow symbol: A symbol in a flowchart that indicates the direction to follow through the rest of the chart based on the answers to the questions.

automatically operated switch: A switch that is activated independent of a person or object.

B

battery maintenance: Maintenance that verifies that a PLC battery has not exceeded its rated life.

bit: The smallest unit of memory.

break: A place on a contact that opens or closes an electrical circuit.

building grounding: The connection of an electrical system to earth ground through a GEC to grounding electrodes, the metal frame of a building, concrete-encased electrodes, or underground metal water pipes.

byte: A group of 8 bits.

C

central processing unit (CPU): A section of a PLC that houses the processor (brain) of the PLC.

circuit breaker (CB): A reusable overcurrent protection device that opens a circuit automatically at a predetermined overcurrent.

combination circuit: A combination of series- and parallel-connected devices and components.

combination system: A system that interconnects two or more primary systems to combine the individual advantages of each system to meet the requirements of a given application.

compare: A PLC software operation where the PLC program in the memory of the PLC is compared to the PLC program in the PC.

contact bounce: The rapid making and breaking of mechanical switch contacts when switch contacts are closed.

contactor: An electrical control device that is designed to control high-power, non-motor loads.

continuity tester: A test instrument that tests for a complete path for current to flow.

control logic: A plan of operation as designed with a PLC program for an electrical circuit or system.

control relay: A device that controls an electrical circuit by opening and closing contacts in another circuit.

coulomb: The practical unit of measurement for a specific quantity of electrons.

count down (CTD) instruction: A PLC programming instruction used to keep track of the number of items involved in a subprocess.

counter accumulated value: The number of counts a counter has recorded.

counter instruction: A PLC programming instruction used to provide counting functions similar to solid-state counters.

counter preset value: The number of counts a counter is programmed to reach.

count up (CTU) instruction: A PLC programming instruction used to count the number of operations or products produced by a system.

current (I): The rate at which electrons flow through a conductor.

cycle: One complete positive alternation and one complete negative alternation of a waveform.

D

data: The information that is stored in the memory of a CPU.

data file: A group of data values (inputs, timers, counters, and outputs) that is displayed as a group and whose status may be monitored.

data table file: The section of PLC memory that contains the status of the CPU, inputs and outputs, timer and counter preset and accumulated values, and other program instruction values.

DC voltage: Voltage in a circuit that has current that flows in one direction only.

debris shield: A piece of heavy paper that covers the ventilation slots on a PLC.

debugging: A task that consists of eliminating errors or malfunctions and verifying that a PLC program operates correctly and meets the specifications and requirements of the system.

diamond symbol: A symbol in a flowchart that contains a question, worded so that the answer can be a "yes" or a "no."

digital electrical circuit: A circuit in which the load is either fully ON or fully OFF.

digital multimeter (DMM): A portable test instrument that uses electrical components to display measured values.

digital signal: An electronic signal that has two specific quantities that change in discrete steps (ON or OFF).

direct current (DC): Current that flows in only one direction.

direct hardwiring: A wiring method where the power circuit and the control circuit are wired point-to-point.

disable command: A special software override that prevents one or all input or output devices from operating.

double-break (DB) contact: A contact that breaks (opens) an electrical circuit in two places.

download: A computer operation where a copy of a PLC program (ladder logic only, no documentation) is transferred from a PC to a PLC.

E

electrical enclosure: A housing that protects wires and equipment and prevents personal injury by accidental contact with energized circuits.

electrical energy: Energy created with the flow of an electric charge.

electrically erasable programmable read-only memory (EEPROM): A type of nonvolatile memory that is retained when power is lost.

electrical noise: Any unwanted signal present on power lines.

electrical system: A primary system that produces work by transmitting and controlling the flow of electricity through conductors (wires).

electromagnetic compatibility (EMC): The ability of various pieces of electrical equipment to work together with varying levels of noise emission immunity.

electromagnetic interference (EMI): Unwanted electrical noise generated by electrical and electronic equipment, such as lighting ballasts and transformers.

electromechanical relay: A switching device that has sets of contacts that are closed by magnetic force.

electronic equipment grounding: The connection of electronic equipment, such as PLCs, to earth ground to reduce the chance of electrical shock through grounding the equipment and all non-current-carrying exposed metal parts.

electronic system: A primary system in which electricity is monitored and controlled to send and/or receive information, produce sound or vision, store data, control circuits, or perform other work.

ellipse symbol: A symbol in a flowchart that indicates the beginning and end of a section of a chart.

emergency stop button: A special red-colored palm switch or limit switch used to stop a machine or process operation immediately to avoid physical injury or property damage.

energized maintenance: Maintenance that verifies that the master control relay (MCR) is functioning correctly, voltages are within specified range, and no abnormal conditions exist.

equipment and document verification: A process that ensures that a technician has the correct equipment and documents for the selected PLC programming software.

equipment grounding: The connection of machinery electrical systems to earth ground to reduce the chance of electrical shock by grounding all non-current-carrying exposed metal.

F

face shield: An eye and face protection device that covers the entire face with a plastic shield and is used for protection from flying objects.

final check: A check that verifies that an application or process that a PLC controls functions properly in RUN mode under actual conditions.

fixed PLC: A PLC that has a set number and type of input and output (I/O) terminals.

flowchart: A diagram that shows a logical sequence of steps for a given set of conditions.

fluid power system: A primary system that produces work by transmitting fluid (gas or liquid) under pressure through pipework.

force command: A special software override that simulates opening or closing an input device or turns an output component ON or OFF.

foreign control voltage: A voltage that originates outside a PLC enclosure, such as voltage from an electric motor drive.

form factor: The physical configuration used to connect the components of a PLC into a housing.

free running timer: A type of programmed timer that is continuously timing because the XIO instruction preceding the timer is true. Also known as a repetitive timer.

fuse: An overcurrent protection device that includes a fusible link that melts and opens a circuit when an overcurrent condition occurs.

G

goggles: An eye protection device with a flexible frame that is secured to the face with an elastic headband.

grounding: The connection of all exposed non-current-carrying metal parts to earth.

H

handheld programming device: A separate keypad device that is not an integral part of a PLC.

hardware maintenance: Maintenance that consists of a visual inspection of the PLC, energized maintenance procedures, and battery maintenance.

high-power load: A load whose voltage and current rating is greater than the voltage and current rating of the PLC output.

high-speed counter (HSC) instruction: A PLC programming counter instruction that can detect high-speed counts/events.

high-speed counter load (HSL) instruction: A PLC programming counter instruction used in conjunction with a high-speed counter instruction.

high-voltage spike: A type of electrical noise that is produced when inductive loads such as motors, solenoids, and coils are turned OFF.

hold-up time: The length of time a PLC can tolerate a power loss without affecting operation.

housekeeping and overhead: The section of the operating cycle during which the PLC performs memory management and updates timers and counters.

human machine interface (HMI): A color or monochrome display panel with a keypad and buttons that shows the status of a process or application in real time.

hydraulic system: A fluid power system that transmits power using a liquid (typically oil).

I

icon bar: The section of software containing buttons for functions that are frequently used.

impedance (Z): The total opposition to current flow in an AC series circuit.

informational system: A primary system that monitors operations, displays quantities (values), and indicates the status of machines and processes.

infrared temperature meter: A meter that measures heat energy by measuring the infrared energy emitted by a material and displays the temperature as a numerical value.

initial check: A visual check performed by a technician before any power is applied to a PLC.

input scan: The section of the operating cycle during which the PLC examines the input devices for the absence or presence of voltage and records a 0 or 1 at the corresponding input data table location.

input section: A grouping of terminal screws that receives signals from input devices and sends the signals to the CPU.

input section check: A check that verifies that input devices function properly and are wired to the correct PLC input terminal.

instruction toolbar: The section of the software that contains various instruction buttons.

integrated programming device: A device that consists of a small liquid crystal display (LCD) and a small keypad or set of buttons that are part of a PLC.

interface device: An item that allows variously rated components to be used together in the same circuit.

interfacing system: A system or device that allows primary systems to work and/or communicate as one.

International Electrotechnical Commission (IEC): An organization that develops international standards for electrical and electronic equipment as well as related technology.

J

jumper link: A device used to select the power rating of a dual-voltage PLC.

L

ladder diagram: *See* line diagram.

ladder view: The section of the programming window that contains the programming diagram. Also known as a programming diagram view.

line diagram: A diagram that has a series of single lines (rungs) that indicate the logic of a control circuit and how the control devices are interconnected. Also known as a ladder diagram.

lockout: The process of removing the source of electrical power and installing a lock, which prevents the power from being turned ON.

low impedance ground: A grounding path that contains very little resistance to the flow of fault current to ground.

M

magnetic motor starter: A starter with an electrically operated mechanical switch (contactor) that includes motor overload protection.

malfunction: The failure of a system, equipment, or part to operate properly.

manually operated switch: Any switch that requires a person to physically change the state of the switch.

math instruction: A PLC programming instruction that is used for a mathematical calculation, such as addition, subtraction, multiplication, or division.

mechanically operated switch: A switch that detects the physical presence of an object.

mechanical system: A primary system in which power is transmitted through gears, belts, chains, shafts, couplings, and linkages.

memory: A part of the CPU where program files are loaded for execution and data files are stored for fast access.

menu bar: The section of application software containing headings for pull-down menus that is placed across the top of a screen.

modular PLC: A PLC that has a variable number of input and output (I/O) terminals based on the number of cards or modules placed into a chassis or rack.

multimeter: A portable test instrument that is capable of measuring two or more electrical properties.

N

NAND circuit logic: The control logic developed when two or more normally closed switching contacts are connected in parallel to control a load.

National Electrical Code® (NEC®): A code book of electrical standards that indicates how electrical systems must be installed and how work must be performed.

National Fire Protection Association® (NFPA®): A national organization that provides guidance in assessing the hazards of products of combustion.

network control system: A system of computers, terminals, databases, and PLCs connected by communication lines.

NFPA 70E®: *Standard for Electrical Safety in the Workplace:* The electrical safety requirements for employee workplaces that are necessary for the safeguarding of employees in pursuit of gainful employment.

nibble: A group of 4 bits.

NOR circuit logic: The control logic developed when two or more normally closed switches or contacts are connected in series to control a load.

NOT circuit logic: The control logic developed when a normally closed switch or contact is connected to control a load.

O

ohmmeter: A test instrument that measures the resistance of a device or circuit.

Ohm's law: The direct relationship between voltage (E), current (I), and resistance (R) in an electrical circuit.

online bar: The section of software that contains information on the processor mode, edit status, force status, and communication drivers of a PLC.

online icon: A button that is animated when a PC is online with a PLC, such as when changing the mode of a PLC to remote programming, remote run, test continuous, or test single scan.

on-line monitoring: A PLC software operation where a PC is monitoring the PLC program and graphically displaying the state of all input and output terminals (TRUE or FALSE).

OR circuit logic: The control logic developed when two or more normally open switches or contacts are connected in parallel to control a load.

output scan: The section of the operating cycle during which the PLC energizes or de-energizes output components based on the information in the output data table.

output section: The section that sends signals from the CPU to components that perform work.

output section check: A check that verifies that all output components function properly and are wired to the correct PLC output terminal.

overload: An overcurrent condition that occurs when circuit current rises above the normal current level at which the PLC and/or circuit is designed to operate.

overvoltage: An increase in voltage of more than 10% above the normal rated line voltage for a period of time longer than 1 minute.

P

parallel-connected devices and components: Two or more devices or components that are connected so that there is more than one flow path for current to take.

personal protective equipment (PPE): Clothing and/or equipment worn by technicians to reduce the possibility of injury in the work area.

pictorial drawing: A drawing that resembles a photograph or three-dimensional picture.

PLC hardware problem: A condition where a PLC program is correct for the application, but the system is not working properly because there is a problem with inputs, outputs, wiring (loose connections, cross talking, incorrect wiring), a short or open circuit exists somewhere in the system, or another hardware problem exists.

PLC software problem: A condition where a PLC is properly installed and all inputs and outputs are working, but there is a problem with the PLC program.

PLC startup: A set of procedures consisting of systematic checks.

pneumatic system: A fluid power system that transmits power using a gas (air).

point-to-point wiring: A wiring method where each component in a circuit is connected (wired) directly to the next component as specified by wiring or line diagrams.

polarity: The positive (+) or negative (−) state of an object.

pole (contact): A completely isolated circuit that a relay can switch.

power (P): The rate of energy consumption or conversion in an electrical circuit or system in a given amount of time.

power supply: A section of a PLC that converts the voltage of a power source to the low-level voltage (typically 5 VDC) required by the CPU and the related electronics inside the PLC.

power supply module: A self-contained, regulated or nonregulated voltage unit that supplies DC or AC voltage to a PLC.

preventive maintenance: The work performed to keep machines, assembly lines, production operations, and plant operations running with little or no downtime.

primary system: A system that transmits and controls the movement of some form of energy.

process: A sequence of operations that accomplish desired results.

processor file: The section of memory in the PLC that is filled with CPU data about programming.

program check: A check that verifies that a PLC program functions properly without the application or process the PLC controls actually being run.

program file: Data in the memory of a PLC that contains technician-developed PLC programs and related information.

programmable logic controller (PLC): An industrial computer control system that is preprogrammed to carry out automatic operations.

program memory: The maximum number of programming instructions a CPU can hold.

programming device: A device with a keypad that is used to write and enter a user-developed control program into a PLC.

programming diagram view: *See* ladder view.

program mode: A PLC state of operation where the operating cycle of the PLC is not running, but the CPU is ready to accept information.

program scan: The section of the operating cycle during which the PLC examines the PLC program, compares the program to the status of the inputs, uses the comparison to determine what output components will be energized or de-energized, and records a 0 or 1 at the corresponding output data table location.

project tree: A visual representation of all the files that make a PLC program. Also known as a project view.

project view: *See* project tree.

protective clothing: Clothing that provides protection from contact with sharp objects, hot equipment, and harmful materials.

Q

qualified person: An individual with the necessary education and training who is familiar with the construction and operation of electrical equipment and devices.

R

random access memory (RAM): A type of memory that permits accessing the storage medium to store and retrieve program files and data table files.

rectangle symbol: A symbol in a flowchart that contains a set of instructions.

repetitive timer: *See* free running timer.

reset (RES) instruction: A PLC programming instruction used to reset timer and counter accumulated values.

resistance (R): The opposition to current flow.

results window: A small window section of software that opens up with the results of a Find All or a Verification (either file or project).

retentive timer (RTO) instruction: A PLC programming instruction used to track the length of time a machine has been operating or to shut down a process after an accumulative time period of recurring faults.

run mode: A PLC state of operation where the operating cycle of the PLC is running continuously to control a machine or process.

S

safety glasses: An eye protection device with special impact-resistant glass or plastic lenses, reinforced frames, and side shields.

scan: A method by which the CPU of a PLC looks at a program.

scan time: The length of time a processor takes to execute a program once.

scroll bar: A screen element at the bottom of a window or to the right of a window that allows a technician to move the contents of a window horizontally or vertically on the screen by dragging the bars.

serial communication signal: A digital data signal from an external source.

series-connected devices and components: Two or more devices or components that are connected so that there is only one flow path for current to take.

service communications: The section of the operating cycle during which the PLC communicates with other devices, such as handheld programmers or PCs.

shielded cable: A type of cable that uses an outer conductive jacket (shield) to surround the inner conductors that carry the signals.

short circuit: An overcurrent condition in which the current of a circuit leaves the normal current-carrying path by going around the load back to the power source via ground or another conductor uncontrolled.

single-break (SB) contact: A contact that breaks (opens) an electrical circuit in one place.

single-phase AC voltage: A type of electrical supply that contains only one alternating voltage waveform (sine wave).

sink: The negative power supply terminal of a DC-powered PLC.

snubber circuit: An electrical circuit designed to suppress voltage spikes.

software and program verification: A process that ensures that a PLC can be accessed by a PC and that the EEPROM chip (when present) has the correct PLC program (back-up program).

software maintenance: Maintenance performed to ensure that a technician has the correct equipment and software to troubleshoot future PLC problems.

solid-state relay: A relay that uses electronic switching devices in place of mechanical contacts.

source: The positive terminal of a DC-powered PLC.

static electricity: An electrical charge at rest.

status bar: The section that contains current status information and provides information on how to use the software.

symbol: A graphic element that represents a quantity, unit, device, or component.

system: A combination of components, units, or modules that are connected to perform work or meet a specific need.

T

tagout: The process of placing a danger tag on the source of electrical power, which indicates that the equipment cannot be operated until the danger tag is removed.

temporary end (TND) instruction: A PLC programming, output instruction used to debug a PLC program.

terminal block: A device used to interconnect wires from electrical devices.

terminal strip: A strip of adjacent terminal screws to which electrical devices are connected.

test light: A test instrument with a bulb, typically neon, that is connected to two test leads to provide a visual indication of when voltage is present.

thermal imager: A meter that measures heat energy by measuring the infrared energy emitted by a material and displays the temperature as a color-coded thermal picture.

thermocouple: A temperature sensor made of two dissimilar metals that are joined at the end where heat is to be measured.

three-phase AC voltage: A type of electrical supply that combines three alternating waveforms (sine waves) each displaced 120 electrical degrees (one-third of a cycle) apart.

throw: The number of closed contact positions per pole.

time base: An instruction label used to indicate the unit of time a timer is using.

timer accumulated value: The number of time base intervals (the length of time) that a timer has been timing.

timer instruction: A PLC programming instruction used to provide timing functions similar to mechanical or solid-state timers.

timer off-delay (TOF) instruction: A PLC programming instruction used to delay the shutdown of machinery, such as an external cooling fan when a motor has been stopped.

timer on-delay (TON) instruction: A programming instruction used to delay the start of a machine or process for a set period of time.

timer preset value: The number of time base intervals a timer instruction is programmed to accomplish.

title bar: A shaded horizontal bar at the top of a window that contains the title of a window.

troubleshooting: The systematic elimination of various parts of a system or process to locate a malfunction.

U

undervoltage: A drop in voltage of more than 10% (but not to 0 V) below the normal rated line voltage for a period of time longer than 1 minute.

upload: A computer operation where a copy of a PLC program (ladder logic only, no documentation) is transferred from a PLC to a PC.

V

voltage (E): The electromotive force or pressure in a conductor that allows electrons to flow.

voltage drop: The amount of voltage consumed by a device or component as current passes through the object.

voltage tester: A test instrument that indicates when voltage is present at a test point.

voltmeter: A test instrument that measures voltage.

W

wiring diagram: A technical diagram that displays the connection of all devices and components to a PLC.

word: A unit of memory that consists of 16 bits.

word address: A designated name or number in a PLC program for a particular input or output.

work: The movement of a load (in pounds) over a distance (in feet).

Index

Page numbers in italic refer to figures.

A

abbreviations, *408*, 408
AC (alternating current), 28
accumulated (ACC) values, 277, 278, 285
AC drives. *See* variable-frequency drives (VFDs)
AC power, 522
AC ripple, *328*
AC sine waves, 77, *78*
actuators
  installation of, *322*, 322, *323*
  troubleshooting, 330–331, *332*, *333*
  uses of, 319
AC (alternating current) voltage, 76, 77–80, *78*, *79*, *80*
AC voltage measurements, *411*, 414–415, *415*, *416*
ADCs (analog-to-digital converters), 313
address descriptions, 520, *521*
addresses
  counters and, 285
  file and, *248*, 248–249, *249*, *250*
  timer and, 277
adjustable-speed drive systems, 217
air conditioning and refrigeration equipment, 513
air intakes and outlets, 514, *515*
alarms, 22, *23*
Allen-Bradley PLCs, *10*, 10
alternating current (AC), 28
alternating current (AC) voltage, 76, 77–80, *78*, *79*, *80*
alternations, 77, *78*
aluminum conductors, *88*, 88, 92
ammeters
  clamp-on, 421–423, *423*, *424*
  defined, 411, 421–424
  in-line, 421, *422*, 424, *425*
amperes, 28, *29*
analog and digital circuits. *See* electrical circuits
analog electrical circuits, 213, 215, *216*
analog input devices
  applications of, 312, 313–314, *314*
  circuits and, 217, 219–221, *220*, *221*, *222*
  overview of, 8, *9*, 21, *22*
  programming, 324–326, *325*, *326*
  signals, *315*, 315
  troubleshooting, *327*, 327–329, *328*, *329*, *330*, 455
analog modules, 369
analog multimeters, 412
analog output devices
  applications of, 312, 319–320, *320*
  circuits and, 222–225, *223*, *224*, 225
  programming, *324*, 324–326, *325*, *326*
  signals, 320–322, *321*
  troubleshooting, 329–334, *332*, *333*, *334*
analog output signals, *219*
analog sections, 369
analog signals, 76, 98, 188, *219*, 360
analog solid-state relays (SSRs), *225*, 225
analog-to-digital (A-to-D) converters, 313
AND circuit logic, *134*, 134, *135*
approach boundaries, *28*, 28
arc-flash boundaries, *28*, 28
arc flash hoods, *42*, 42
arc-rated protective clothing, *41*, 41, *42*
arc suppression blankets, 42
arrow symbols on flowcharts, 403
A-to-D (analog-to-digital) converters, 313
automatically operated switches, 119–123, *120*

B

backplanes, *14*, 14
backups, program, 524–525, *525*
basic electrical circuits, 197–198, *198*, *199*, *200*, *201*
batch processes, 5
battery maintenance, 517–519, *518*
baud rates, 522
bit file addresses, 249, *250*
bit files (file 3), *247*, 247
bit instructions
  programming diagram logic and, 251, 254–256, *255*, *256*, *257*, 260
  programming symbols and, 125–126, *126*
bits, 172, *173*
bleeder resistors, *462*, 462
boiler operation, 6, *7*
bonding, 100
Boolean algebra, 130
break current ratings, 87
breaks, *206*, 206
building grounding, *31*, *32*, 32
bytes, 172, *173*

C

capacitors, 158
car wash facilities, 3
cascaded timers, *283*, 283, *284*
CAT (category) ratings, 407, 408–410, *410*

CD (count down) enable bits, 285, *286*, *288*
cement production, 6
central processing units (CPUs), *159*, 163–164, *164*
chassis, 33
chassis grounds, 76
circuit breakers
   defined, 446
   lockout/tagout, *43*, 43
   operation of, 440
   switches for lighting, 453
   testing of, 442, 446–447, *447*
circuit conditioning, *193*, 193
circuit operation, *455*, 455–457, *456*, 464–465, *465*
circuits. *See* electrical circuits
clamp-on ammeters, 421–423, *422*, *423*, *424*
class R (rejection) fuses, 445
clean rooms, *315*, 315
close buttons, *485*, 485
coils, 87–88, *88*
combination circuits, 100–102
combination systems, 190–191, *191*
common (polarity), 77, *78*
communication during troubleshooting, 399
communication instructions, *128*, 128
communications cables, 92, *93*
compare function, 522
compare instructions, 126, *127*
complex electrical circuits, 201–202, *203*, *204*
concentric knockouts, 100
conductor ampacities, 278
conductors, *88*, 88–89
conductor supports, 322
conduit, 360
conduit roof support systems, 131
conduit sealing, 161
connection inspections, *515*, 516
connections
   input/output, 14–17, *15*, *16*, *17*
   sinking and sourcing, 18–20, *19*, *20*
   switching devices, *18*, 18
contact bounce, 456, 457
contactor interfaces, 204, *210*, *211*
contactors
   input/output connections, 14, *15*, 22, *23*
   interface devices, 202, *205*, 209
contacts (poles), 205, *206*
contaminants, 514–515
Contents tab, 496
continuity testers, 411, 418, *419*, 420
continuous current ratings, 87
control circuits, 202, *205*
control files (file 6), *247*, 247
control logic, 126. *See also* logic functions

control relays (CRs), 89, 205, *206*
control transformers, 199, *448*, 448–450, *449*
conversion instructions, *127*, 127
conversion modules, 220
converters, 220–221, *222*, 223–225, *224*
copper conductors, *88*, 88, 92
COPSs (critical operations power systems), 32
coulombs, 28
count down (CD) enable bits, 285, *286*, *288*
count down (CTD) instructions, 275, *276*, 287–289, *289*, *290*
counter accumulated values, 285
counter addresses, 285
counter files (file 5), *247*, 247
counter instructions
   count down (CTD), 275, *276*, 287–289, *289*, *290*
   count up (CTU), 275, *276*, 286–287, *287*, *288*
   defined, 274–275, *276*, 285, *285*, *286*
   high-speed counter (HSC), 289–290, *290*
   symbols and, 125–126, *126*
   words, 285, *286*
counter preset values, 285
count up (CU) enable bits, 285, *286*, *287*
count up (CTU) instructions, 275, *276*, 286–287, *287*, *288*
CPUs (central processing units), *159*, 163–164, *164*
critical operations power systems (COPSs), 32
Cross Reference function, *488*, 488, 489, *490*
Cross Reference Reports, 489
CRs (control relays), 89, 205, *206*
CTD (count down) instructions, 275, *276*, 287–289, *289*, *290*
CTU (count up) instructions, 275, *276*, 286–287, *287*, *288*
CU (count up) enable bits, 285, *286*, *287*
current
   combination circuits and, *101*, 101, *102*
   DC input switching and, 84–86, *85*, *86*
   defined, 82–83
   electrical property, 28–29, *29*
   inductive kickback, 87–88, *88*
   input ratings, 83, *84*
   Ohm's law calculation and, *90*, 90
   output ratings, *87*, 87
   parallel circuits and, 97–98, *98*
   series circuits and, *94*, 94
current-limiting fuses, 445
current sinking, *85*, 85–86, *86*
current sourcing, *85*, *86*, 86
cycles, 77, *78*

D

daisy-chained circuits. *See* series circuits
dampers, 319

danger tags, *43*, 43
data, 171
data files, 460
data table files, 171, 246–249, *247*, *248*, *249*, *250*
DB (double-break) contacts, *206*, 206
DC (direct current), 28
DC input switching, 84–86, *85*, *86*
DC (direct current) voltage, 76, 77, *78,* 80
DC voltage measurements, *411*, 412, *414*, 416–418, *417*
debris shields, 357
debugging PLC programs, 487–491, *488*, *489*, *490*, *491*, *492*
dedicated space requirements, 171
derating conductors, 461
diagrams, 129–133, *130*, *131*, *132*, *133*
dialog boxes, *487*, 487
diamond symbols on flowcharts, 403
digital electrical circuits, 213, *215*, 215
digital input devices, *21*, 21, 455
digital input filtering, *457*, 457
digital modules, 369
digital multimeters (DMMs). *See* DMMs (digital multimeters)
digital output signals, *219*
digital signals, 188, 360
DIN rails, 357, *358*
direct current (DC), 28
direct current (DC) voltage, 76, 77, *78* 80
direct hardwiring, *129*, 136–140, *137*, *138*, *139*
disable commands, 26, 27
disable safety considerations, 26, 26–27
disconnecting means, 43, 367
disconnects, 191
DMMs (digital multimeters)
  AC measurements and, 414–415, *415*, *416*
  DC measurements and, *417*, 417–418
  defined, 412, *413*
documentation, 399, 519–521, *521*
done (DN) bits
  counters and, 285, *286*, 287, 288
  retentive timer instructions and, 280, 282
  timer off-delay (TOF) instructions and, *281*
  timer on-delay (TON) instructions and, 278, *279*
  timers and, 275, *276*
double-break (DB) contacts, *206*, 206
double-throw contacts, *206*, 206
download function, 522
drain wire, 362, *363*
drives. *See* variable-frequency drives (VFDs)
dual-voltage primary sides, 448
dual voltage ratings, 80, *81*
dust, 514–515

E

eccentric knockouts, 100
EEPROM (electrically erasable programmable ready-only memory), 171–172, *173*, 523–524, *524*
EGCs (equipment grounding conductors), 362, 407
electrical circuits
  analog, 215, *216*
  basic, 197–198, *198*, *199*, *200*, *201*
  complex, 201–202, *203*, *204*
  digital, *215*, 215
  digital vs analog, 213
  electric motor drive, *217*, 217, *218*, *219*
  interfacing, 202, 204, *205*
electrical enclosures, 38. *See also* enclosures
electrical energy, 194–196, *197*
electrically erasable programmable read-only memory (EEPROM), 171–172, *173*, 523–524, *524*
electrical measurement precautions, 404–407, *405*, *406*
electrical noise
  PLC installation and, 359–362, *360*, *361*, *362*, *363*
  sensor installation and, *319*, 319
  suppression of, 35, *35*, *36,*
electrical noise filters, 363
electrical noise suppression, 35, *35*, *36*
electrical power consumption, 194–196, *197*
electrical power sources, 322, *323*
electrical properties, 28–29, *29*, *30*
electrical shock, 27, *27*, *28*
electrical symbols, 119–123, *120*, *121*, *122*, *123*, *124*
electrical systems, *187*, 187, *188*
electrical test instruments. *See* test instruments
electric motor drive circuits, *217*, 217, *218*, *219*
electric motor drives, 212–213, *213*, *214*
electromagnetic compatibility (EMC), 359
electromagnetic interference (EMI), 456
electromechanical relays, 205–206, *206*, *207*
electronic equipment grounding, *31*, *33*, 33–34, *34*
electronic systems, *187*, 188, 193
electrostatic discharge, 45
elevators, 5
ellipse symbols, 403
EMC (electromagnetic compatibility), 359
emergency stop buttons, 364–366, *365*, 406
EMI (electromagnetic interference), 456
employment opportunities, 6–7, *7*
enable (EN) bits
  counters, 285, *286*
  timers, 275, *276*, *278*, *279*, 282
enclosures
  contents of, *26*, 83
  defined, *38*, 38
  inspection of, 515, *516*

enclosures (*continued*)
  PLC installation and, 356–359, *357*, *358*, *359*
  and safety considerations, 23
  setup of, 360–361, *361*
energized PLC maintenance, 516, *517*
energy-and-power industry, 5
energy management systems, 188
equipment and document verification, 519–521, *520*, *521*
equipment grounding, *31*, 32–33, *33*, 361
equipment grounding conductors (EGCs), 362, 407
error codes, *405*
examine if closed (XIC) instructions, *254*, 254, *256*, 325
examine if open (XIO) instructions, *254*, 254
expanded input devices, 124, *126*
expanded output components, 124
external power sources, 317, *318*
eye protection, *40*, 41–42, *42*

F

face shields, *42*, 42
fans, 514, *515*
fault conditions, 441
FBD (function block diagram) programming language, 260, *261*
file instructions, *127*, 127
files 0 through 8, 245–247, *246*, *247*
files 3 through 255 (subroutine program files), *246*, 246
files 9 through 255 (technician-assigned files), *247*, 247
final elements, 319
final PLC checks, 377–379, *379*
Find All function, *488*, 488, *490*, *491*
fire alarm systems, 25
fire-stopping, 515
fixed form factors, 158, *159*, *170*, 170
fixed voltage ratings, 80, *81*
flexible metal conduit (FMC), 451
floating point files (file 8), *247*, 247
flowcharts for troubleshooting, 403, *404*
flow switches, *120*
fluid power systems, *187*, 188, *189*, 193
FMC (flexible metal conduit), 451
F1 keys, 496, *498*
food-and-beverage industry, *5*, 5
foot switches, *120*
force commands, *26*, 26–27
Force functions, 492–495, *493*, *494*, *495*
force safety considerations, *26*, 26–27
foreign control voltages, 363
fork terminals, *368*, 368
form factors, 170
free running (repetitive) timers, 282–283

functional block language, 130
function block diagram (FBD) programming language, 260, *261*
function switches, 81
fuse markings, *80*, 80
fuses, 440, 442–445, 443, *444*, *446*
fuses, selecting, 73

G

GFCIs (ground-fault circuit interrupters), 75
GFPE (ground-fault protection of equipment), 75
glass production, 6
goggles, 42
Goto Data Table function, *488*, 488, 491, *492*
graphic display test instruments, *414*, 414
ground-fault circuit interrupters (GFCIs), 75
ground-fault protection of equipment (GFPE), 75
grounding
  electrical noise and, 361–362, *362*, *363*
  prepunched knockouts and, 100
  safety practices and, *30*, 30–34, *31*, *32*, *33*, *34*
grounding conductors, 34
grounding electrode conductors, 418
ground potential differences, 522

H

handheld programming device memory modules, 173
handheld programming devices, 165–166, *166*
hardware maintenance
  batteries, 517–519, *518*
  defined, 510, 513, *514*
  energized PLCs, 516, *517*
  visual inspections, 514–516, *515*
hardware problems, 75, *76*, 439. *See also specific hardware*
hardwired circuits, 83
hardwired devices, 243, *244*
hardwiring, direct. *See* direct hardwiring
hardwiring using terminal strips, 140, *141*, *142*
hasps, *43*
hazardous (classified) locations, 48
heaters, 132
heating contactors, 209, *211*
heating, ventilation, air conditioning, and refrigeration (HVACR), 513
Help features, 496, *497*, *498*
help lines for troubleshooting, 403–404, *405*
high legs, 415
high-power loads, 202
high-pressure switches, *123*

high-speed counter (HSC) instructions, 289–290, *290*
high-speed counter load (HSL) instructions, 289
high-voltage spikes, 37
HMIs (human-machine interfaces), 166–167, *167*, 513
HOLD buttons, 415, 418
hold-up times, 162, *366*, 366
horizontal scroll bars, 484, *485*
horns, 22, *23*
housekeeping and overhead, 173, *174*
HSC (high-speed counter) instructions, 289–290, *290*
HSL (high-speed counter load) instructions, 289
human-machine interfaces (HMIs), 166–167, *167*, 513
HVACR (heating, ventilation, air conditioning, and refrigeration), 513
hydraulic systems, 188

I

icon bars, *486*, 486
IEC (International Electrotechnical Commission), 260
IEC publication 60529 for enclosures, 356
IEC 61010 standard, 408, *410*
IEC 61131-3 standard, 260
IL (instruction list) programming language, 260, *261*
impedance, *91*, 91
Index tab, 496
inductive kickback, 87–88, *88*
industrial environments and PLC use, *5*, 5–6, *6*
informational systems, 189, *190*
infrared temperature meters, *426*, 426–427, *427*
initial PLC checks, 373–374, *375*
in-line ammeters
  input power supply current testing, 454
  leakage current testing, 462–463, *463*
  measurement procedure, 424, *425*
  not fused, 426
  use of, 421, *422*
in-line current measurements, 421, 424, *425*
input circuit operation, *455*, 455–457, *456*
input circuits, 74
input connections, 14–17, *15*, *16*, *17*
input current ratings, 83, *84*
input devices
  analog. *See* analog input devices
  complex electrical circuits and, 202, *204*
  digital, *21*, 21
  programming diagram rules and, 251, *252*
  programming symbols, 124, *125*
  troubleshooting, *460*, 460–461
input file addresses, 248
input files (file 1), *247*, 247
input forcing, 493, *494*

input leakage current, *461*, 461–463, *462*, *463*
input modules, 3, 368–369, *369*, *370*, *371*, *372*
input/output (I/O) message instructions, 127–128, *128*
input/output (I/O) modules, 4, 13–14, *14*
input/output (I/O) switching devices, *18*, 18
input power supply testing, 452–455, *453*, *454*
input scans, 173, *174*
input section checks, 374, *376*
input sections
  defined, 158, *159*
  internal circuitry, 158, *160*
  micro-PLCs and, 12, *13*
  nano-PLCs and, 10, 11, *12*
  section wiring, 160, *161*
  troubleshooting
    circuit operation, 455–457, *456*
    digital input filtering, *457*, 457
    leakage current, *461*, 461–463, *462*, *463*
    overview, *455*, 455, 457–461, *458*, *459*, *460*
    starting at PLC, 397, *398*
  wiring, 368–369, *369*, *370*, *371*, *372*
input signals, 217, 219–220, *220*, *221*
input switching, 84–86, *85*, *86*
input terminals, 77, 328–329, *330*, 397, *398*
input voltage ratings, 80–81, *82*
inspecting PLCs, 45–48, *46*, *47*
installation. *See* PLC installation
instruction comments, 520, *521*
instruction list (IL) programming language, 260, *261*
instruction toolbars, *486*, 486
instruction words, 275–277, *276*, 285, *286*
insulation removal, *407*, 407
insulators, 88
integer files (file 7), *247*, 247
integrated programming devices, *165*, 165
interface cables, *520*, 520
interface cards, *520*, 520
interface devices
  analog output, 222
  contactors and, 209, *210*, *211*
  electric motor drives and, 212–213, *213*, *214*
  electromechanical relays, 205–206, *206*, *207*
  motor starter interfaces, *211*, 212
  motor starters and, 209–211
  overview, 204
  solid-state relays, 206–208, *207*, *208*, *209*
  system interfacing and, 191–194, *192*, 193, *195*, *196*
interface modules, 367, *368*
interfacing electrical circuits, 202, 204, 205
interfacing electric motor drives, 212–213, *213*, *214*
interfacing systems, 190, *191*
internal circuitry, 158, *160*, 161

internal power sources, 317, *318*
International Electrotechnical Commission (IEC), 260
IEC 61010 standard, 408, *410*
IEC 61131-3 standard, 260
interrupting rating, 441
interrupt instructions, 127–128, *128*
inverters. *See* variable-frequency drives (VFDs)
I/O (input/output) message instructions, 127–128, *128*
I/O (input/output) modules, *4*, 13–14, *14*
I/O (input/output) switching devices, *18*, 18
isolation transformers, 74

J

jumper links, 45

K

kilohms, 88
knockouts, 100

L

labeled equipment, 169
ladder diagram (LD) programming language, 260, *261*
ladder (line) diagrams, 129–133, *130*, *131*, *132*, *133*, 251
ladder views, *486*, 486
laptops, 519, *520*
large PLCs, *169*, 169. *See also* medium/large PLCs
LD (ladder diagram) programming language, 260, *261*
leakage current, *461*, 461–463, *462*, *463*
level switches, *120*
LFMC, 496
light energy consumption, 196, *197*
limited approach boundaries, *28*
limit switches
  as input devices, *21*, 21
  symbols for, *120*, *121*, 121, *122*, *124*
line (ladder) diagrams, 129–133, *130*, *131*, *132*, *133*, 251
liquidtight flexible metal conduit, 496
listed equipment, 169
lithium batteries, 519
lockout, 42–44, *43*, *44*
logical instructions, *127*, 127
logic circuits, 464, *465*
logic functions, 133–136, *134*, *135*, *136*
LOGO! 8 nano-PLCs, 10–11, *12*
loose terminations, *407*, 407
loss of power problems, 439
low impedance grounds, 32
low-impedance solenoid voltage testers, 412
low voltage problems, 440

M

magnetic motor starters, 14–15, *15*, 209–210, *211*
main program files (file 2), 245–246, *246*
maintenance
  hardware. *See* hardware maintenance
  overview of, 510–512, *511*, *512*, *513*
  software. *See* software: maintenance
make current ratings, 87
malfunctions defined, 397
manually operated switches, 119–123, *120*, *121*
manuals. *See* documentation
manufacturer troubleshooting procedures, *401*, 401–404, *402*, *404*, *405*
master passwords, 489
math instructions, 126, *127*, 274, 290, *291*
maximize buttons, *485*, 485
maximum current ratings, 87
maximum demand, 86
maximum overcurrent protection (MOP), 513
MCA (minimum circuit ampacity), 513
mechanical contacts, *18*, 18
mechanical contact switches, *82*, 82
mechanical limit switches, *120*
mechanically operated switches, 119–123, *120*, *121*
mechanical switches, 461
mechanical systems, *187*, 189, *190*
medium/large PLCs
  brands of, 10
  connections and, *17*, 17–18, *20*, 20
  power supplies and, 8, 13–14, *14*
  rack-mounted, 13–14, *14*
  system setup, 8, *9*
medium PLCs, *169*, 169. *See also* medium/large PLCs
megohms, 88
memory, 171–172, *172*, *173*
memory modules, 173
menu bars, *486*, 486
metal conduit, 360
Micro810™ nano-PLC, 10–11, *12*
MicroLogix 1100, 11–12, *13*
micro-PLCs
  brands of, *10*
  classification, *169*, 169
  connections and, 15, *16*, *19*
  power supply modules and, 11–12, *13*
  system setup, 8, *9*
micro-sized PLCs, 452
milliamp process clamp meters, 422
minimize buttons, *485*, 485
minimum circuit ampacity (MCA), 513
MIN MAX buttons, 415, 418, 425
minus signs, 484, *485*

mnemonic symbols, 486
modular cards, 3
modular form factors, 158, *159*, 170, *171*
modular PLCs, *278*, *496*, *517*
MOP (maximum overcurrent protection), 513
motor overloads, 199
motors, 195, *197*
motor starter interface devices, 204
motor starter overloads, 132
motor starters, 204, *205*, 209–211, *211*, *212*
mounting enclosures, 357–359, *358*
mounting sensors, *317*, 317
move instructions, *127*, 127
multimeters, 412
multiple lockout/hasps, *43*

N

NAND circuit logic, *134*, *136*, 136
nano-PLCs
   classification of, *169*, 169
   PLC system setup and, 8, *9*
   power supply modules and, 10–11, *12*
   sourcing connections, *19*
National Electrical Code® (NEC®), 23, 24, *25*
National Electrical Manufacturers Association (NEMA)
   250 standard for enclosures, 356
National Fire Protection Association (NFPA), 24
NC (normally closed) input device instructions, *255*
NC (normally closed) switches, 120, *121*, *124*
NEC® (National Electrical Code®), 23, 24, *25*
NEC® standards for EMI, 359, *360*
NEC® table values, 81
negative I/Os, 82
negative polarity, 77
negative switching. *See* current sinking
NEMA (National Electrical Manufacturers Association)
   250 standard for enclosures, 356
nested branches, 251, *252*
network control systems, 201
neutral-to-ground connections, 31
NFPA (National Fire Protection Association), 24
NFPA 70E: Standard for Electrical Safety in the Workplace, 24, 39
NFPA 70: National Electrical Code® (NEC®), 23, 24, *25*
nibbles, 172, *173*
NO (normally open) input device instructions, *254*
noncontact temperature instruments, *426*, 426
nonlinear loads, 165
nonvolatile memory, 171
NOR circuit logic, *134*, *135*, *136*
normally closed (NC) input device instructions, *255*

normally closed (NC) switches, 120, *121*, *124*
normally open (NO) input device instructions, *254*
normally open (NO) switches, 119–120, *121*, *124*
NO (normally open) switches, 119–120, *121*, *124*
NOT circuit logic, *134*, *135*, 135
NPN transistor switching. *See* current sinking

O

ohmmeters, *420*, 420, *421*
Ohm's law, 72, *89*, 89–90, *90*, *91*
online bars, *486*, 486
online icons, *486*, 486
on-line monitoring, 522
operating cycles, 172–173, *173*, *174*
opto-isolation circuits, *465*, 465
OR circuit logic, *134*, 134, *135*
OTE (output energize) instructions, 254, 325
OTL (output latch), 256, *257*
OTU (output unlatch), 256, *257*
output circuit operation, 464–465, *465*
output circuits, 74
output components
   complex circuits and, 202
   programming diagram rules and, 251, *252*
   programming symbols and, 124, *125*
   troubleshooting, 467, *469*
   types of, 22, *23*
output connections, 14–17, *15*, *16*, *17*
output current ratings, *87*, 87
output devices. *See* analog output devices
output drivers, 465
output energize (OTE) instructions, 254, 325
output file addresses, 248
output files (file 0), *247*, 247
output forcing, 494–495, *495*
output image table. *See* output files (file 0)
output latch (OTL), 256, *257*
output modules
   rack-mounted, *14*, 14
   troubleshooting, *464*, 464–467, *465*, *466*, *468*, *469*
   wiring, 368–369, *369*, *370*, *371*, *372*
output scans, 173, *174*
output sections
   checking, 376, *377*
   defined, *159*, 161, *162*
   micro-PLCs and, 12, *13*
   nano-PLCs and, 10, 11, *12*
   troubleshooting, *464*, 464–467, *465*, *466*, *468*, *469*
   wiring, 368–369, *369*, *370*, *371*, *372*
output signals, 222, *223*
output terminals, 77, 331, 332, *334*, 334

output unlatch (OTU), 256, *257*
output voltage ratings, 81–82, *82*
overcurrent protection, 73
overflow (OV) bits, 285, *286*, 287
overhead (and housekeeping), 174
overload protection, 214
overloads, 199, *442*, 442
overvoltages, 440, *441*

P

padlocks, *43*
panel mount filters, 467
parallel circuits, *96*, 96–100, *97*, *98*, *99*
parallel conductors, 123
passwords, 489, 491
PCMCIA interface cards, *520*, 520
PC to PLC connections, 522–523, *523*
personal computers (PCs), *167*, 167–168, *168*
personal protective equipment (PPE), *22*, 39–42, *40*, *41*, *42*
petroleum-and-chemical industry, 5
photoelectric switches, *21*, 21
photovoltaic systems, 22
pico PLCs. *See* nano-PLCs
pictorial drawings, 128, *129*
"PID" instructions, *128*, 128
pin terminals, *368*, 368
plastics-and-packaging industry, 5
PLC enclosures. *See* enclosures
PLC inspections, 45–48, *46*, *47*
PLC installation
  electrical noise and, 359–362, *360*, *361*, *362*, *363*
  enclosures and, 356–359, *357*, *358*, *359*, 360–361, *361*
  general guidelines, 39, *40*
  overview, 354–355, *355*
  power supplies and, 363, *364*
  receiving PLCs, *356–357*, 356–357
  safety and, 364–366, *365*, *366*
  wiring and. *See* wiring methods
PLC maintenance logs, *512*
PLC memory, 171–172, *172*, *173*
PLC programming diagrams, *132*, *133*, *134*
PLC programming symbols, 124–127, *125*, *126*, *127*, *128*
PLCs (programmable logic controllers)
  actuator installation, *322*, 322, *323*
  analog input devices. *See separate entry*
  analog output devices. *See separate entry*
  classifications of, *169*, 169–170, *170*, *171*
  counter instructions. *See separate entry*
  defined, 2–3
  development of, 157, *158*
  diagrams, 129–133, *130*, *131*, *132*, *133*

electrical principles and, 73, *74*
how they work, 3–7, *4*, *5*, *6*, *7*
installation of. *See* PLC installation
logic functions, 133–136, *134*, *135*, *136*
maintenance. *See separate entry*
operating cycles, 172–173, *173*, *174*
power supply modules and, 10
problems, *74*, 74–75, *76*
programming diagrams and, 242, *243*, 243–244, *245*
programming software, 482–487, *484*, *485*, *486*, *487*. *See also* troubleshooting: software
safety. *See separate entry*
sensor installation, *316*, 316–319, *317*, *318*, *319*
startup of. *See* PLC startup
symbols. *See separate entry*
system setup, 8, *9*
testing power to, 450–455, *451*, *452*, *453*, *454*
timer instructions. *See separate entry*
PLC scans. *See* operating cycles
PLC sections
  central processing units (CPUs), *159*, 163–164, *164*
  input sections. *See separate entry*
  output sections. *See separate entry*
  power supplies, *159*, 162–163, *163*
  programming devices, 164–168, *165*, *166*, *167*, *168*
PLC startup
  final checks, 377–379, *379*
  initial checks, 373–374, *374*
  input checks, *375*
  input section checks, 374, *376*
  output section checks, 376, *377*
  overview, 354–355, 373, *373*
  program checks, 377, *378*
PLC systems
  electrical power consumption, 194–196, *197*
  interfacing, 191–194, *192*, *193*, *195*, *196*
  primary systems, *187*, 187–191, *188*, *189*, *190*, *191*
PLC technicians, 6–7, *7*
PLC-wired circuits, *138*, *140*, *143*
plus signs, 484, *485*
pneumatic power sources, 322, *323*
pneumatic systems, 188
PNP devices, *86*, 86
PNP transistor switching, *85*, *86*, 86
point-to-point wiring, 136, *137*
polarity, 77, *78*
poles (contacts), 205, *206*
portable PLCs, 76, *77*
positive I/Os, 82
positive polarity, 77
positive switching, *85*, *86*, 86
power consumption, 194–196

power defined, 91
power filters, 363
power formula, *91*, 91
power loss to PLCs, *366*, 366
power rails, 243
power sources, actuator, 322, *323*
power supplies
  defined, *159*, 162–163, *163*
  PLC installation, 363, *364*
  problems with
    control transformers, *448*, 448–450, *449*
    fuses and circuit breakers, *442*, 442–447, *443*, *444*, *446*, *447*
    overview, 439–441, *440*, *441*
    power to PLC, 450–455, *451*, *452*, *453*, *454*
  sensors and, 317, *318*
power supply modules
  defined, 8, *10*
  medium/large PLCs and rack-mounted I/O modules, 13–14, *14*
  micro-PLCs, 11–12, *13*
  nano-PLCs, 10–11, *12*
  PLCs, 10
  terminal blocks, 10, *11*
  terminal strips, 8, *11*
power supply voltage ratings, 80, *81*
PPE (personal protective equipment), *22*, 39–42, *40*, *41*, *42*
preset values, 277, 278, 285
pressure switches, *120*, *123*
preventive maintenance, 398
primary systems, *187*, 187–191, *188*, *189*, *190*, *191*
processes defined, 397
processing plants, 6
processor cards, 519
processor control circuits, 457, 464, *465*
processor files, 244–247, *246*, *247*
program backups, 524–525, *525*
program checks, 377, *378*
program control instructions, 127–128, *128*
program files, 245–246, *246*
programmable logic controllers (PLCs). *See* PLCs (programmable logic controllers)
program memory, 164
programming device memory modules, 173
programming devices, 164–168, *165*, *166*, *167*, *168*
programming diagram logic
  bit instructions, 251, 254–256, *255*, *256*, *257*, 260
  rules, 251, *252*
  scan execution, 257–260, *258*, *259*
programming diagrams, 242, *243*, 243–244, *245*. *See also* programming diagram logic
programming diagram views, *486*, 486

programming software, 482–487, *484*, *485*, *486*, *487*. *See also* troubleshooting: software
program modes, 522
program scan hold-up time after loss of power, 367
program scans, 173, *174*
programs/projects, 483–487, *484*, *485*, *486*, *487*
program titles, *521*, 521
program verification, 521–525, *523*, *524*, *525*
projects. *See* processor files *and* programs/projects
project trees (views), 484–485, *486*, 486
protective clothing, *40*, *41*, 41
proximity sensors, 222
proximity switches, *21*, 21
pulp-and-paper industry, 5
pushbuttons, *21*, 21, *120*

Q

qualified persons, 24

R

rack-mounted input/output (I/O) modules, 13–14, *14*, *17*, 17–18, *20*
radices, 248
random access memory (RAM), 171
RANGE buttons, 415, 418
ratings, meter, 407, 408–410, *410*
rectangle symbols on flowcharts, 403
rectified voltage, 8, *10*
refrigeration equipment, 513
relative humidity sensors, 21, *22*
relays, 201, 202, *205*
REL (relative) buttons, 415, 418
remote sources of power, 363
repetitive (free running) timers, 282–283
reserved files (file 1), 245–246, *246*
reset (RES) instructions, 275, 282
resistance
  combination circuits and, 100–101, *102*
  conductors and, *88*, 88–89
  defined, 29, *30*, 88
  impedance and, *91*, 91
  Ohm's law and, *89*, 89–90, *90*, *91*
  parallel circuits and, 96–97, *97*
  power formula and, *91*, 91
  series circuits and, 92, *93*
resistance measurements, 80, 445, *446*
resistors, 94
restricted approach boundaries, *28*
results windows, *486*, 486–487

retentive timer (RTO) instructions, 275, 281–282, *282*
retentive timers, *280*
reusable danger tags, *43*
Rockwell Automation Allen-Bradley PLCs, 325, 326
rung comments, *521*, 521
rungs, 251, *252*
run mode, 522

S

safety
   general PLC considerations, 23–28, *24*, *25*, *26*, *27*, *28*
   PLC installations and, 364–366, *365*, *366*
   rules for, *39*, 39
safety circuits, 39
safety glasses, *40*, 41, *42*
safety labels, *27*, 27
SB (single-break) contacts, *206*, 206
SCADA (supervisory control and data acquisition) systems, 5–6, *6*, *7*
scan execution, 257–260, *258*, *259*
scans. *See* operating cycles
scan time, 164
scan time worksheets, *259*, 259
scroll bars, 484, *485*
SCRs, *82*
Sealtite, 496
selector switches, *21*, 21
sensor installation, *316*, 316–319, *317*, *318*, *319*
sensors. *See* analog input devices
sensor troubleshooting, *327*, 327–328, *329*
sequencer instructions, *127*, 127
sequential function chart (SFC) programming language, 260, *261*
serial communication signals, 360
serial ports, 522
series circuits, *92*, 92–95, *93*, *94*, *95*
service communications, 173, *174*
setup dialogs, *326*, 326
SFC (sequential function chart) programming language, 260, *261*
SG2 nano-PLCs, 10–11, *12*
shield/drain wire, 362, *363*
shielded cables, *35*, 35, *36*
shielded twisted pair (STP) cable, *35*, 35, *319*, 319
short-circuit current ratings, 441
short circuits, *442*, 442
Siemens PLCs, 10, *12*, 163
signaling towers, 22, *23*
signals, analog, *315*, 315, 320–322, *321*
sine waves, 77, *78*
single-break (SB) contacts, *206*, 206

single-phase AC voltage, 77–78, *78*, *79*
single-phase isolation step-down transformers, 363
single-throw contacts, *206*, 206
sink (polarity), 77
sinking connections, 18–20, *20*
sinking I/Os, 82
sinks, 86
smart production processes, 6
snubber circuits, *37*, 37
software
   maintenance
     defined, *510*, 519
     equipment and documentation verification, 519–521, *520*, *521*
     software and program verification, 521–525, *523*, *524*, *525*
   PLC programming, 482–487, *484*, *485*, *486*, *487*
   problems, 74–75, *75*
   program verification, 521–525, *523*, *524*, *525*
   troubleshooting. *See* troubleshooting: software
software updates, 403
solar cells, 21, *22*
solenoid voltage testers, 412
solid-state relays (SSRs)
   analog, *225*, 225
   interface devices and, 206–208, *207*, *208*, *209*
   load switching and, 89
solid-state switches
   leakage current and, *461*, 461–462, *462*
   output voltage ratings and, 82, *82*, 95
   symbols for, 123, *124*
   voltage drop and, 18, *18*
source (polarity), 77
sources, 86
sourcing connections, 18–20, *19*, *20*
sourcing I/Os, 82
spacing intervals, 322
SPDs (surge-protective devices), 317
Standard for Electrical Safety in the Workplace. *See NFPA 70E: Standard for Electrical Safety in the Workplace*
start pushbuttons, 15, *15*, 17, *17*
startup. *See* PLC startup
startup technicians, 373
static electricity, 33, *37*, 37
status bars, *486*, 487
status bits, 286–287, 288
status file addresses, *248*, 250
status files (file 2), *247*, 247
status lights, 397, *398*, 401, *402*, 405
stop pushbuttons, 15, *15*, 17, *17*
STP (shielded twisted pair) cable, *35*, 35, *319*, 319

stranded conductors, 366
structured text, 130, 260, *261*
subroutine program files (files 3 through 225), *246*, 246
supervisory control and data acquisition (SCADA) systems, 5–6, *6*, *7*
supplied power sections, 10, 12
supply (polarity), 77
surge-protective devices (SPDs), 317
surge suppressors, 74
switching devices, *18*, 18, 84–86, *85*, *86*
symbols
    defined, 408
    electrical, 119–123, *120*, *121*, *122*, *123*, *124*
    electrical test instruments and, 408, *409*
    flowcharts and, 403, *404*
    PLC programming, 124–127, *125*, *126*, *127*, *128*
system checks, 354
system files (file 0), 245–246, *246*
system interfacing, 191–194, *192*, *193*, *195*
systems defined, 397

T

table values, 81
tagout, 42–44, *43*, *44*
tag ties, *43*
task bars, 485
technician-assigned files (files 9 through 255), *247*, 247
TECO PLCs, 10, *12*
temperature ratings per NEC®, 363
temperature requirements for enclosures, *359*, 359
temperature sensors, 220, *221*
temperature switches, *120*, *121*, *123*
temperature test instruments, 425–427, *426*, *427*
temporary end (TND) instructions, 488, *489*
terminal blocks, 10, *11*
terminal strips, 8, *11*, 140, *141*, *142*, 367, *368*
termination methods, 367–368, *368*, *369*
testing. *See* troubleshooting
test instruments
    abbreviations, symbols, and ratings, 407–410, *408*, *409*, *410*
    AC voltage measurements, 414–415, *415*, *416*
    ammeters. *See separate entry*
    continuity testers, 411, 418, *419*
    current measurements, *213*
    DC voltage measurements, 412, *414*, 416–418, *417*
    function switches, 81
    measurement precautions, 404–407, *405*, *406*
    ohmmeters, *420*, 420, *421*
    overview, 396
    temperature and, 425–427, *426*, *427*
    test lights, 411
    troubleshooting using experience, *399*, 399
    troubleshooting using manufacturer procedures, *401*
    voltage testers, 410, *411*, 411–412, *413*
    voltage type determination, 412, *414*, 414, *415*
    voltmeters, 412
test leads, 407, 413
test lights, 411
textile industry, *5*, 5
thermal imagers, *426*, 426
thermocouples, 129, *131*
three-phase AC voltage, 77–78, *78*, *79*
three-phase motor starters, *79*
three-point test procedures, 405
three-wire input switch circuits, *85*, 85
throws, *206*, 206
time bases, 277
timer accumulated values, 277, 278
timer accuracy, 280
timer addresses, 277
timer files (file 4), *247*, 247
timer instructions
    defined, 274–275, *276*, 285
    reset (RES), 282
    retentive timer (RTO), 275, 281–282, *282*
    special applications, 282–285, *283*, *284*
    symbols for, 125–126, *126*
    timer off-delay (TOF), 275, *276*, 279–280, *280*, *281*
    timer on-delay (TON), 275, *276*, 277, 277–278, *279*
    words, 275–277, *276*
timer off-delay (TOF) instructions, 275, *276*, 279–280, *280*, *281*
timer on-delay (TON) instructions, 275, *276*, 277, 277–278, *279*
timer preset values, 277, 278
timer timing (TT) bits, 275, *276*, 278, 280, 282
tinted, arc-rated face shields, 42
title bars, 484, *485*
TND (temporary end) instructions, 488, *489*
TOF (timer off-delay) instructions, 275, *276*, 279–280, *280*, *281*
TON (timer on-delay) instructions, 275, *276*, 277, 277–278, *279*
traffic lights, 3
transformer installation, 399
transistors, *82*
triacs, *82*
troubleshooting
    analog input devices, *327*, 327–329, *328*, *329*, *330*
    analog output devices, 329–334, *332*, *333*, *334*
    facility procedures, 400, *401*
    flowcharts and, 403, *404*

troubleshooting (*continued*)
  input sections/modules
    circuit operation, 455–457, *456*
    digital input filtering, *457*, 457
    leakage current, *461*, 461–463, *462*, *463*
    overview, *455*, 455, 457–461, *458*, *459*, *460*
  knowledge and experience, *399*, 399, *400*
  manufacturer procedures, *401*, 401–404, *402*, *404*, *405*
  output sections/modules, *464*, 464–467, *465*, *466*, *468*, *469*
  overview of, 396–399, *398*, *399*
  software
    debugging programs, 487–491, *488*, *489*, *490*, *491*, *492*
    Help features, 496, *497*, *498*
troubleshooting charts, 457–458, *458*, *466*, 466
true-rms DMMs (digital multimeter), 414, *416*
TT (timer timing) bit. *See* timer timing (TT) bits
two-wire input switch circuits, *85*, 85

U

underflow (UN) bits, 285, *286*, 288
undervoltage problems, 440, *441*
upload function, 521
USB ports, 522
user passwords, 489
user's manuals, *405*, 405

V

valves, 319
variable-frequency drives (VFDs), 214, 217
vertical scroll bars, 484, *485*
visual inspections, 514–516, *515*
volatile memory, 171
voltage
  AC voltage, 76, 77–80, *78*, *79*, *80*
  combination circuits and, *101*, 101
  DC voltage, 76, 77, *78*, 80
  defined, *29*, 29, 76, 77
  Ohm's law calculation, *90*, 90
  parallel circuits and, *99*, 99–100
  PLC input ratings, 80–81, *82*
  PLC output ratings, 81–82, *82*
  PLC power supply ratings, 80, *81*
  series circuits and, 94, *95*
voltage drops, 94–95, *95*, *99*, 99–100, *101*, 101
voltage measurements
  analog output terminals and, *334*, 334
  control transformers and, 448–450, *449*
  fuses and, 442, 442–443, *443*, *444*
  input power supplies and, 452, *453*
voltage range ratings, 80, *81*
voltage testers, 410, *411*, 411–412, *413*
voltage types, 412, *414*, 414, *415*
voltmeters, 411, 412

W

watts, 91
Windows, 484–485, *485*
wire terminations, *407*, 407
wiring input sections, 160, *161*
wiring diagrams, 129, *130*, *132*
wiring methods
  direct hardwiring, 129, 136–140, *137*, *138*, *139*
  hardwiring using terminal strips, 140, *141*, *142*
  input and output sections and modules, 368–369, *370*, *371*, *372*
  methods of, 366–367, *367*
  PLC-wired circuits, 140, *143*
  terminations, 367–368, *368*, *369*
word addresses, 127
words, 172, *173*
work, 187

X

XIC holding contacts, 255, *256*
XIC (examine if closed) instructions, *254*, 254, *256*, 256, 325
XIO (examine if open) instructions, *254*, 254